Lecture Notes in Artificial Intelligence 929

Subseries of Lecture Notes in Computer Science
Edited by J. G. Carbonell and J. Siekmann

Lecture Notes in Computer Science
Edited by G. Goos, J. Hartmanis and J. van Leeuwen

Springer

Berlin
Heidelberg
New York
Barcelona
Budapest
Hong Kong
London
Milan
Paris
Tokyo

F. Morán A. Moreno
J.J. Merelo P. Chacón (Eds.)

Advances in Artificial Life

Third European Conference on Artificial Life
Granada, Spain, June 4-6, 1995
Proceedings

Springer

Series Editors

Jaime G. Carbonell
School of Computer Science
Carnegie Mellon University
Pittsburgh, PA 15213-3891, USA

Jörg Siekmann
University of Saarland
German Research Center for Artificial Intelligence (DFKI)
Stuhlsatzenhausweg 3, D-66123 Saarbrücken, Germany

Volume Editors

Federico Morán
Departamento de Bioquímica, Universidad Complutense de Madrid
E-28040 Madrid, Spain

Alvaro Moreno
Departamento de Lógica y Filosofia de la Ciencia, Universidad del País Vasco
Apdo. 1249, E-20080 San Sebastián, Spain

Juan Julián Merelo
Departamento de Electrónica, Universidad de Granada
Campus Fuentenueva, E-18071 Granada, Spain

Pablo Chacón
Centro de Investigaciones Biológicas, CSIC
Velázquez 144, E-28006 Madrid, Spain

CR Subject Classification (1991): I.2, J.3, F.1.1–2, G.2, H.5.1, I.5.1, J.4, J.6

ISBN 3-540-59496-5 Springer-Verlag Berlin Heidelberg New York

CIP data applied for

Printed in Germany

Typesetting: Camera ready by author
SPIN: 10486232 06/3142 – 543210 – Printed on acid-free paper

¿Qué es la vida? Un frensí.
¿Qué es la vida? Una ilusión,
una sombra, una ficción,
el mayor bien es pequeño;
que toda vida es sueño,
y los sueños, sueños son.
(Calderón de la Barca. *La Vida es Sueño.* 1635)

... La vida es siempre un jugarse la vida al naipe de unos ciertos valores,
y, por eso, toda vida tiene estilo – bueno o malo, personal o vulgar,
creado originalmente o recibido del contorno –.
(José Ortega y Gasset. *Introducción a Velázquez.* San Sebastián, 1947)

Preface

Despite its short history, Artificial Life (ALife) is already becoming a mature scientific field. By trying to discover the rules of life and extract its essence so that it can be implemented and investigated in artificial media in different media, ALife research is leading us to a better understanding of a large set of interesting biology-related problems, such as self-organization, emergence, the origins of life, self-reproduction, computer viruses, learning, growth and development, animal behavior, ecosystems, autonomous agents, adaptive robotics, etc.

Understanding biological phenomena is a complex task and, like all disciplines of the *sciences of complexity*, it needs a multidisciplinary approach. From its origins, ALife has been oriented in two main directions: applying of *biologically inspired* solutions to the development of new techniques and methods; and the use of artificial media (particularly computation) to model, replicate, and investigate life processes.

ALife is now turning into a consolidated field. The number of online (WWW, USENET newsgroup, ftp sites, and forums in services like CompuServe or America Online) and offline (books and journals) is already huge and, besides, is increasing every day. ALife is slowly slipping into the mainstream of technology. More and more companies use biologically inspired techniques to solve problems, optimize solutions, simulate or model real-life situations, etc. Autonomous robots with biologically-based or evolved behavior are becoming more widely studied and used.

On the other hand, theoretical biology and ALife are coming closer and closer. Although, at this point, only a level of *awareness* has been reached between them, soon these fields will interact to a greater extent and probably profit from this interaction. It is our opinion, and that of many others, that the future *survival* of ALife is highly related to the presence of *people from biology* in the field. As stated above, the big success of the biologically inspired development of new techniques and methods could strongly influence and bias the future direction of ALife. the big success of biologically-inspired applications in technology could bias ALife excessively in this direction. We are convinced that the approaches and methods of ALife are important in understanding *life*, and we should work to maintain the interest and involvement of biologists, biochemists, geneticist, ecologists, and many other *naturalscientists* in the exciting and important field of ALife.

This book contains the 71 papers, including oral presentations and invited lectures, presented at the 3rd European Conference on Artificial Life, ECAL'95, held in Granada (Spain) from 4 to 6 June, 1995. ECAL'95 was organized by the Universidad Complutense de Madrid, Universidad del País Vasco, and Universidad de Granada. More than 160 papers were received for oral presentation at the Conference. All submitted papers were rigorously reviewd by members of the Program Committee, and only those with a minimum of two positive referee reports were accepted for oral presentation and included in this publication. This process is not perfect, but it is some guarantee of the scientific quality of the papers included in this volume, and that it therefore represents some of the best work currently being done in the field of ALife.

The book is organised according to the eight scientific sessions of ECAL'95, preceeded by the opening invited lecturer given by Prof Peter Schuster. In addition to the *classical* topics of ALife we have paid special attention to issues related to biology. Topics like the origins of life, the evolution of metabolism, protein design, or ecosystem behavior, are good examples of this.

We would like to thank very much all the authors for their contributions, and for their patience and efforts in re-formating their manuscripts in time to produce this book version of the proceedings for the Conference. We also wish to express our gratitude to all members of the Organization, Local, and Program Committee. Most of them had between 8 and 12 submitted papers to review and all of whom worked very hard to do a careful and thorough job.

We would also like to thank all the sponsors of ECAL'95: the International Society for the Study of the Origin of Life (ISSOL), the Spanish RIG IEEE Neural Networks Council, Silicon Graphics Spain, IBM Spain, the Parque de las Ciencias de Granada, and in particular the organisations which have contributed towards the costs of the Conference, the EC-DG-XII Biotechnology, the DGICYT (MEC, Spain), the CICYT (MEC, Spain), the Junta de Andalucia, and the Universidad de Granada. Our special thanks to all the unamed people who have worked these Organisations to help make ECAL'95 a success. Finally, but not least, we are very grateful for the all support and hard work of the Technical Secretariat, GESTAC, during the organisation and running of the Conference.

Madrid, April 1995

Federico Morán
Alvaro Moreno
Juan Julián Merelo
Pablo Chacón

Organization Committee

Federico Morán	U. Complutense Madrid (E)	Chair
Alvaro Moreno	U. País Vasco, San Sebastián (E)	Chair
Juan J. Merelo	U. Granada (E)	Secretary
Pablo Chacón	U. Complutense, Madrid (E)	
Arantza Etxeberria	U. País Vasco, San Sebastián (E)	
Julio Fernández	U. País Vasco, San Sebastián (E)	
Tim Smithers	U. País Vasco, San Sebastián (E)	
Carme Torras	U. Politec. Catalunya, Barcelona (E)	

Program Committee

Francisco Varela	CNRS/CREA, Paris (F)	Chair
Riccardo Antonini	U. Rome (I)	
Michael Arbib	USC, Los Angeles, CA (USA)	
Randall D. Beer	CWRU, Cleveland, OH (USA)	
Wolfgang Banzhaf	Dortmund University (D)	
George Bekey	USC, Los Angeles, CA (USA)	
Hugues Bersini	ULB, Brussels (B)	
Paul Bourgine	CEMAGREF, Antony (F)	
Rodney Brooks	MIT, Cambridge, MA (USA)	
Scott Camazine	Wissenschaftskolleg, Berlin (D)	
Peter Cariani	MEEI, Boston, MA (USA)	
Michael Conrad	Wayne State U., Detroit, MI (USA)	
Jaques Demongeot	U. J. Fourier, La Tronche (F)	
Jean-L. Deneubourg	ULB, Brussels (B)	
Michael Dyer	UCLA, Los Angeles, CA (USA)	
Claus Emmeche	U. of Rosekilde, (DK)	
Walter Fontana	U. of Vienna, (A)	
Brian C. Goodwin	Open U., Milton Keynes (UK)	
Ricard Guerrero	U, Barcelona (E)	
Reinhart Heinrich	Humbolt Univ. Berlin (D).	
Pauline Hogeweg	U. of Utrecht, (NL)	
Philip Husbands	U. of Sussex, Brighton (UK)	
George Kampis	ELTE Univ. Budapest (H)	
Kunhiko Kaneko	University of Tokyo (JP)	
Hiroaki Kitano	Sony Comp. Sci. Lab., Tokyo (JP)	
John Koza	Stanford U., CA (USA)	
Chris Langton	Santa Fe Institute, NM (USA)	
Antonio Lazcano	UNAM, México (MX)	
Pier L. Luisi	ETHZ, Zurich (CH)	
Pattie Maes	MIT, Cambridge, MA (USA)	
Pedro C. Marijuán	U. Zaragoza, (E)	
Maja J. Mataric	MIT, Cambridge, MA (USA)	

Barry MacMullin	Dublin City U., Dublin (IE)
Robert M. May	Oxford U. (UK).
Eric Minch	Stanford U., CA (USA)
Melanie Mitchell	Santa Fe Institute, NM (USA)
Andrés Moya	U. Valencia, (E)
Francisco Montero	U. Complutense, Madrid (E)
Jim D. Murray	U. of Washington, Seattle, WA (USA)
Juan C. Nuño	U. Politecnica de Madrid, (E)
Julio Ortega	U. Granada (E)
Domenico Parisi	CNR, Roma (I)
Mukesh Patel	Politecnico di Milano, Milan (I)
Howard Pattee	SUNY, Binghamton, NY (USA)
Francisco J. Pelayo	U. Granada (E)
Juli Peretó	U. Valencia, (E)
Rolf Pfeifer	U. Zurich-Irchel, Zurich (CH)
Alberto Prieto	U. Granada (E)
Steen Rasmussen	LANL, Los Alamos, NM (USA)
Tom Ray	ETL (JP)
Robert Rosen	Dalhousie U. Halifax (CA)
Chris Sander	EMBL, Heidelberg (D)
Peter Schuster	IMB, Jena (D)
Hans Paul Schwefel	Dortmud U. (D)
Mosher Sipper	Tel Aviv U., Tel Aviv (IL)
Luc Steels	VUB, Brussels (B)
John Stewart	Institut Pasteur, Paris (F)
Jon Umérez	SUNY, Binghamton, NY (USA)
Alfonso Valencia	CNB/CSIC, Madrid (E).
Günter Wagner	Yale U., CT (USA).
Hans V. Westerhoff	U. Amsterdam (NL).
William C. Winsatt	U. of Chicago, (USA)
René Zapata	LIRM, Montpellier (F)

Contents

3. Adaptive and Cognitive Systems

4. Artificial Worlds

5. Robotics and Emulation of Animal Behavior

6. Societies and Collective Behavior

7. Biocomputing

8. Applications and Common Tools

Opening Lecture

Artificial Life and Molecular Evolutionary Biology

Peter Schuster

Institut für Molekulare Biotechnologie e.V., PF 100813
D-07708 Jena, Germany

Abstract. Artificial life is extending the scope of molecular evolutionary biology as it tries to complement natural life on earth by searching for systems with properties that are sufficient to allow for evolution. Evolution is characterized by specific forms of dynamics that are based on the capability of replication. RNA molecules form a toy universe calles the *RNA world* that shares many features with current scenarios of procaryotic life. Since RNA is able to unite the properties of genotypes (sequences) and phenotypes (spatial structures) within the same molecule, RNA sequence to structure mappings present the key to an understanding of evolutionary dynamics. Such mappings dealing with RNA secondary structures allow straightforward investigations assisted by computer simulation and mathematical analysis. The main result of these studies is summarized in the principle of *shape space covering*: only a small fraction of sequence space has to be searched in order to find a sequence that folds into a predefined structure.

1. Evolutionary Dynamics

The pioneering works of Sol Spiegelman [1] and Manfred Eigen [2] in the late sixties initiated the novel discipline of *molecular evolutionary biology*. It contrasted and complemented conventional *molecular evolution* by adding dynamical components to the essentially phylogenetic issues of sequence data comparisons as initiated and developed, for example, by Magaret Dayhoff [3]. Meanwhile experiments with replicating molecules in the test tube have shown that evolution in the sense of Charles Darwin's principle is no priviledge of cellular life: optimization of properties related to the *fitness* of replicating molecules is observed routinely *in vitro* with *naked* RNA molecules in evolution experiments.

In the late eighties Christopher Langton and his colleagues from Los Alamos National Laboratory and the Santa Fe Institute made an attempt to create a novel discipline named *artificial life*. Its objective among others is to design and realize new systems sharing some essential properties with current living systems. In this respect artificial life has some overlap with molecular evolution.

A central issue of research in both areas deals with the dynamics of evolutionary processes.

Darwinian dynamics consisting of optimization through mutation and selection was found to be just one feature of evolutionary systems. Others being, for example, suppression of selection by (mutualistic) interaction through RNA catalysis. Evolutionary dynamics may also give rise to complex dynamical phenomena like oscillations or deterministic chaos.

To model or to describe the dynamics underlying the evolutionary process is a complex problem that can be visualized on several different conceptual levels. Figure 1 presents a sketch illustrating the three most relevant genetic aspects of molecular evolution:

- the mapping of genotypes into phenotypes,
- the dynamics of populations, and
- the dynamics of the support of populations.

The three processes are viewed best in three different abstract metric spaces: the *shape space*, the *concentration space*, and the *sequence space*. The three spaces are useful concepts in (molecular) evolutionary biology. They refer to a given class of biopolymers (e.g. proteins, RNA or DNA molecules of given chain length n, monomer composition, etc.): the sequence space is the space of all possible sequences or genotypes, the shape space is the space of all possible structures (commonly at some coarse grained resolution), and the concentration space being identical with the metric spaces of (stochastic or deterministic) dynamic variables used in chemical kinetics or populations genetics. The notion of distance in sequence space is of crucial importance in evolution: it is given by the (smallest) number of individual mutations required for the interconversion of two genotypes. If point mutations (single base exchanges) dominate, the relevant distance is the Hamming metric. The support of the population is a subset of points that contains the genotypes instantaneously present in the population. New genotypes form by mutation, existing ones may die out and thus, the population support travels through sequence space in the course of evolution.

Each of the three processes listed above is visualized best in one of the three abstract spaces which are so to say used for projections of the complex evolutionary process that highlight a given aspect:

- Genotype phenotype mapping assigns a particular phenotype to every genotype. In molecular evolution this is tantamount to folding biopolymer sequences into structures. Accordingly the *shape space* is predestined to be used for the description. Precisely, it is the mapping from sequences into structures and further into function that is relevant for evolution: genotype phenotype mapping provides the (kinetic) parameters which enter population dynamics.
- Population dynamics describes temporal evolution of the population variables (particle numbers, genotype frequencies, concentrations, etc.). It is

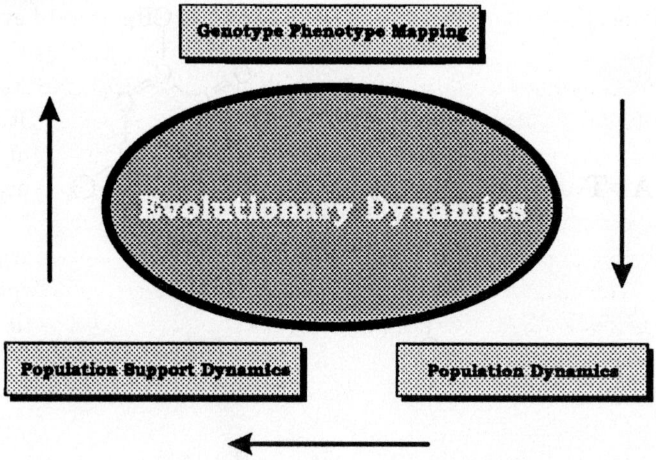

Fig.1: Evolutionary dynamics in sequence, shape and concentration space.

properly described in the conventional concentration space of chemical kinetics. The number of possible genotypes is huge and thus the majority of them will neither be materialized in an evolution experiment nor in nature. Only a small subset of all possible genotypes will be present at a given instant in the population. Whenever a new variant is formed by mutation or some genotype dies out the concentration space changes, a new dimension housing the variable describing the frequency of the new variant is added or the obsolete dimension is removed, respectively.

- Population support dynamics is a process to be visualized in sequence space since it deals with migrating sets of genotypes. Moreover, it provides the input for genotype phenotype mapping as it defines the areas in sequence space where new genotypes appear.

The three projections of evolutionary dynamics onto the three abstract spaces form a conceptual cycle in the sense that each of the three processes provides the input for the next one: genotype phenotype mapping provides the parameters (fitness, for example, being one of them) for population dynamics. Population dynamics deals with temporal alterations in concentrations and thus hands the information on arriving new and disappearing old genotypes over to the support dynamics. Support dynamics in turn transfers changes in the population support to the genotype phenotype mapping in order to make the new parameters available for population dynamics.

A=T

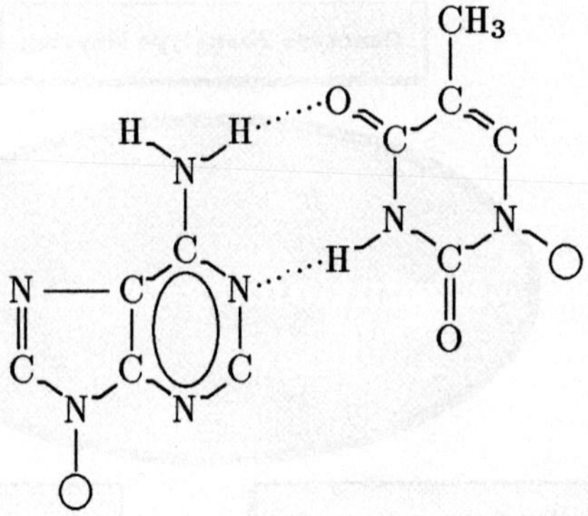

G≡C

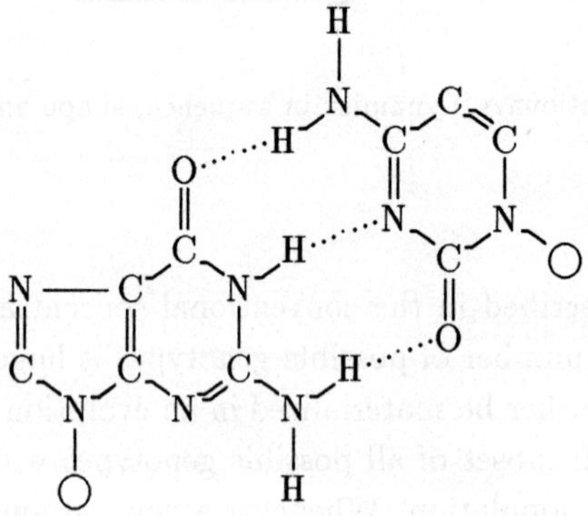

X≡K

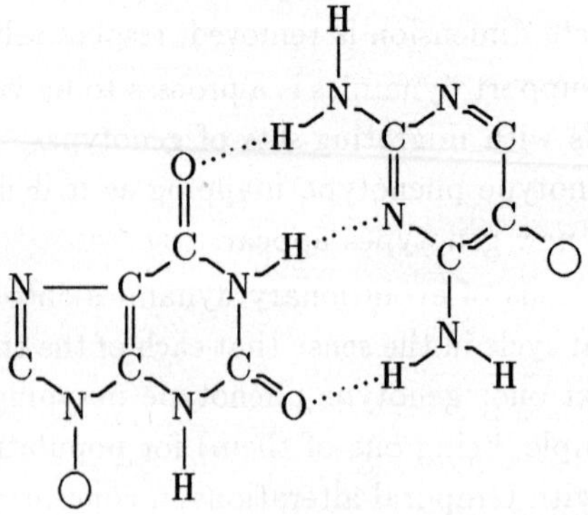

Fig.2: The two natural Watson Crick base pairs (**AT** and **GC**) and one artificial analogue (**XK**) which can be incorporated enzymatically into DNA or RNA and thus extends the conventional base pairing logic to a six letter alphabet.

2. Template Chemistry and Replication of Molecules

Molecular recognition makes use of specific binding. Discrimination between correct and incorrect partner molecules occurs through the strength of the associations. Purine and pyrimidine bases form highly specific patterns of hydrogen bonds in the double helical structures of nucleic acids (see for example figure 2). The bases in double helices are arranged in rather fixed geometrical positions and only very few patterns of hydrogen bonds fit into the predefined spatial arrangement. These configurations define the complementarity rules in DNA double helices: **A** has to oppose **T** and **G** has to oppose **C**. The two natural base pairs, however, do not exhaust the physical possibilities: two additional base pairs would be possible without seriously challenging the uniqueness of complementary assignments. Steven Benner and coworkers [4] synthesized nucleotides derived from the bases xanthine and 2,6-diaminopyridine. The artificial nucleotides were incorporated into RNA and DNA by enzymatic reactions and formed a third base pair (**X**≡**K** in figure 2) which is not in conflict with the two natural base pairs.

As Steven Benner and coworkers have demonstrated the genetic alphabet can be extended to a new logic based on six letters. Why did evolution never make a successful attempt to use a third (or perhaps a forth) base pair? A speculative explanation given by Eörs Szathmáry [5] sees in the choice of two base pairs a compromise between replication fidelity, decreasing with the number of base pairs, and catalytic capacity of RNA which is supposed to increase with the number of base pairs. We mention an alternative less obvious explanation derived from sequence structure statistics (section 3): structures in a two letter alphabet are too sensitive to mutation whereas a six or more letter alphabet cannot provide sufficient mutational variability. In addition these structures are thermodynamically less stable on the average since the base pairing probabilities are smaller in six or eight letter alphabets.

The principle of template induced polynucleotide replication is double helix formation from single strands by sequential addition of mononucleotides to a growing complementary strand. There seems to be no *a priori* reason why replication should need a protein catalyst. Extensive studies by Leslie Orgel and coworkers (see ref. [6] and references 1-5 quoted therein) have indeed shown that template induced synthesis of RNA strands complementary to the template can be achieved under suitable conditions without an enzyme. Enzyme-free template induced replication of RNA, however, is exceedingly difficult. In the early experiments with homopolymer templates (poly-**U**, poly-**C**) the conditions which allow for efficient synthesis of one strand are usually very poor conditions for the synthesis of the complementary strand. The synthesis leads to formation of stable plus-minus double strands which do not dissociate and therefore represent dead ends for enzyme-free replication. Günter von Kiedrowski [7] did the first successful enzyme-free replication experiment involving two trinucleotides as building blocks and a self-complementary hexamer (plus and minus strands are identical) as template. The bond between the two trinucleotides is formed by means of a water soluble carbodiimide as condensating agent:

$$\mathrm{CCG+CGG+GGCGCC} \longrightarrow\longrightarrow \begin{matrix}\mathrm{CCGCGG}\\ \vdots\vdots\vdots\vdots\vdots\vdots\\ \mathrm{GGCGCC}\end{matrix} \rightleftharpoons 2\,\mathrm{GGCGCC}$$

At first the hexanucleotide forms a double-helical ternary complex with the two trinucleotides. Bond formation between the trinucleotides is facilitated by the geometry of the double helix, and a binary complex of two identical, self-complementary hexanucleotides is obtained. The reaction equation shown above indicates already the intrinsic problem of template induced replication: the double-helical binary complex formed in the copying process has to dissociate in order to provide templates for further replication. High stability implies very low dissociation constants. Thus we recognize the crucial role of the enzyme in molecular evolution experiments which it plays by readily separating the double-helical replication complex into template and growing strand.

Recent studies by Julius Rebek and coworkers have shown that template induced replication is not a priviledge of nucleic acids [8-10]. They used other classes of rather complex templates carrying the complementary units and obtained replication under suitable conditions. Complementarity is based on essentially the same principle as in nucleic acids: specific hydrogen bonding patterns allow to recognize the complementary digit and to discriminate all other letters of the alphabet. Template chemistry and molecular replication are rapidly growing fields at present [11]. Seen in the light of molecular evolution organic chemistry provides an ample and rich source for the extension of principles from nature to new classes of molecules in the spirit of artificial life.

In vitro replication of RNA molecules by means of specific enzymes, so-called replicases, proved to be sufficiently robust in order to allow for detailed kinetic studies on its molecular mechanism [12]. These investigations revealed the essential features of template induced RNA synthesis by virus specific replicases which like the conventional photographic process follows a complementarity rule:

$$\textit{activated monomers} + \textit{plus strand} \xrightarrow{\textit{replicase}} \textit{minus strand} + \textit{plus strand}$$

$$\textit{activated monomers} + \textit{minus strand} \xrightarrow{\textit{replicase}} \textit{plus strand} + \textit{minus strand}$$

Both steps together yield the basis of multiplication at the molecular level. The replicating entity thus is the plus-minus-ensemble. In addition to correct complementary replication several fundamental side reactions were detected, for example

- double strand formation from plus and minus strand [13] ,
- *de novo* RNA synthesis [14-16], and

- mutation [17, 18] comprising in essence three classes of processes, point mutations or single base exchanges, deletions and insertions.

Mutation and selection yield populations which depending on the rate of mutation may either settle in optimal regions of sequence space (and form *molecular quasispecies*) or migrate by a diffusion like mechanism [17,18]. The two regimes of localized quasispecies and radomly drifting populations are separated by sharp *error thresholds*.

It seems necessary to stress a fact which is often overlooked or even ignored by theorists and epistemologists. Molecular replication is anything but a trivially occuring function of molecules. As we have seen above, even in cases were the (already hard to meet) molecular requirements for template action are fulfilled replication still depends on low stability of the complex formed by association of the complementary, plus and minus, molecular species. There is no replication when the complex does not readily dissociate. This a kind of *Scylla-and-Charybdis* problem since weak complex formation implies weak mutual recognition between complements tantamount to ineffectiveness in template reactions. There are three solutions to the problem at the present state of our knowledge:

(1) the strength of complementary interaction weakens as a consequence of bond formation between the two units associated with the template (usually monomer and growing chain, or two oligoners),
(2) the separation of plus and minus species is achieved by a third reaction partner (like the enzyme in present day replication of RNA viruses), or
(3) the plus-minus-duplex is replicated as an entity as it happens with natural DNA replication (requiring, however, a highly elaborate machinery involving thirty enzymes or more).

Only the first solution seems to be achievable in prebiotic chemistry and early evolution.

The copying instruction in computing is a very simple function. Chemistry and early biological evolution are radically different from computer science in this respect: replication has to find a simultaneous solution to all requirements which is generally in conflict with common physical chemistry. Working compromises between contradicting demands are rare, and hence only highly elaborate structures might be able to replicate efficiently or even to replicate at all.

3. Molecular Phenotypes

Phenotypes are commonly understood as organisms which are formed in a highly complex unfolding process by instructions from the genotype. In case of higher multi-cellular organisms the phenotype is formed through the process of embryonic development. At present molecular developmental biology reveals the

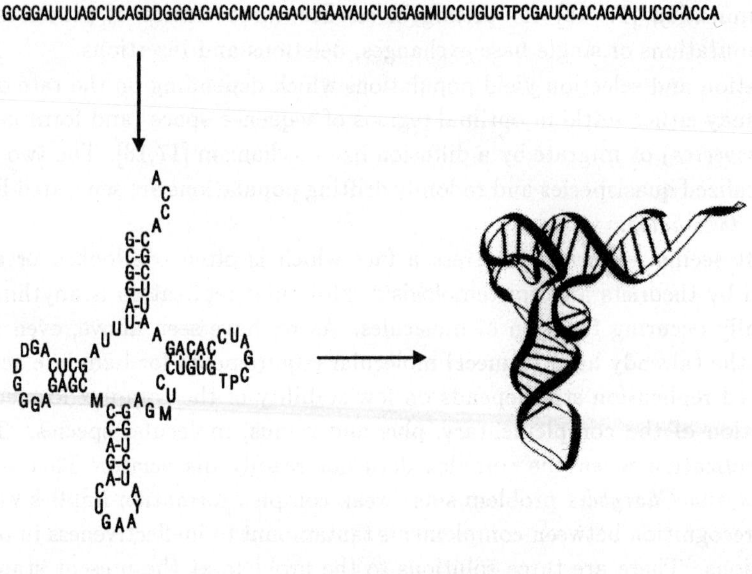

Fig.3: Folding of RNA sequences into secondary structures and spatial structures.

genetic details of embryonic morphogenesis but we are still far away from an understanding of entire process.

Molecular evolution reduces unfolding of the phenotype to the ultimate core: genotypes and phenotypes of RNA molecules in evolution *in vitro* are two features of the same molecule; the sequence is the genotype, the spatial structure is the phenotype. Unfolding of the phenotype is then tantamount to folding the RNA sequence into a three-dimensional structure. The present state of knowledge allows detailed analysis and theoretical approaches to genotype phenotype mappings only in this highly reduced case. On the other hand molecular evolution experiments provide a solid experimental basis for the theoretical approach. In addition, RNA molecules in test tube evolution experiments show all the essential features of a Darwinian scenario and interpretation of data yielded and yields conclusive answers to open questions concerning the mechanism of evolution.

3.1 Genotype Phenotype Mapping

The approach to evolutionary dynamics followed here requires global information on the relation between genotypes and phenotypes. We understand this relation as a mapping from the space of genotypes into the space of phenotypes. In constant environment every genotype unfolds to yield a defined phenotype. The mapping, however, does not need to be invertible: commonly we shall have more genotypes than evolutionarily distinguishable phenotypes and thus many genotypes may form the same phenotype.

Sol Spiegelman [1] suggested to consider spatial structures of RNA molecules as their phenotypes in the test tube experiments. Genotype phenotype mapping then is tantamount to the mapping of RNA sequences into spatial structures. Only few full three-dimensional structures of RNA molecules are presently known with sufficient precision. As shown in figure 3 the conversion of sequences can be formulated in two steps with an intermediate coarse grained *secondary structure*. The secondary structure is a listing of all Watson Crick type and **GU** base pairs which are commonly retained on folding the two-dimensional pattern into a spatial structure. RNA secondary structures are also meaningful for predicting RNA reactivities since RNA function has been discussed successfully in biochemistry over more than 30 years. It has to be said, however, that many features of RNA reactivity and catalysis cannot be understood without a detailed knowledge of the full three-dimensional structures.

In order to study the mapping of RNA sequences into secondary structures one needs the concept of sequence space: every sequence is represented as a point in sequence space and the distance between sequences is the Hamming distance, the (minimal) number of point mutations converting one sequence into an other sequence. RNA sequence to structure mapping has been investigated in great detail [19 - 23].

The development of a fast prediction algorithms for RNA secondary structures [23] made large scale statistical studies possible which revealed several global features of RNA sequence to structure mappings:

- The number of sequences is much larger than the number of RNA secondary structures.
- Individual structures differ largely with respect to their frequencies.
- Sequences folding into the same secondary structure are distributed randomly in sequence space.
- Structural and evolutionary properties of RNA molecules depend strongly on the base composition (**A**:**U**:**G**:**C** ratio) of the sequences.

According to structure statistics [19] the four letter alphabet appears to be an optimal compromise between too high sensitivity against mutation (as found in the two letter alphabet) and too low thermodynamical stability (six letter alphabet).

RNA secondary structures may be considered as a kind of toy universe where we have computational access to genotype phenotype mappings. Are the results obtained in this model system representative for more complex systems or not? At the present state of our knowledge it seems very likely that extensive studies on three-dimensional structures of nucleic acids or proteins will provide similar results. Extensions of the approach presented here to entire organisms seem to be possible in the case of viroids and simple RNA viruses.

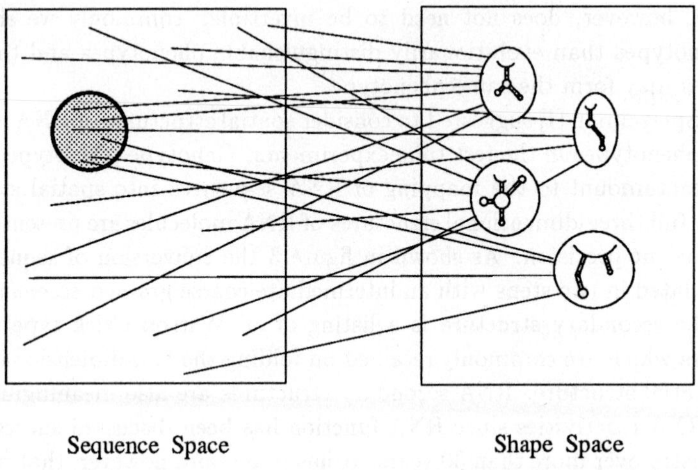

Fig.4: Covering of RNA shape space by a small fraction of sequence space.

In the following two subsections we consider two important features of the RNA world discovered in the studies on RNA secondary structures: *shape space covering* and *neutral networks* which have direct implications for evolutionary dynamics.

3.2 Shape Space Covering

The results obtained for the mapping of RNA sequences into secondary structures were combined into the general principle of *shape space covering* describing a regularity in sequence space which governs evolutionary dynamics [24]. Only a small fraction of all possible sequences which is found in the environment of any arbitrarily chosen reference sequence has to be searched in order to find a sequence that folds into a given structure (Figure 4). Shape space covering applies to all common structures. To define *common* we divide the number of sequences by the number of distinct structures to obtain the average number of sequences folding into the same structure. Every structure is common when it is formed by more sequences than the average. The radius of the shape space covering sphere can be computed by simple statistics.

The shape space covering conjecture was verified by means of a concrete example, **GC** only sequences of chain length $n = 30$. All approximately 10^9 sequences were folded and analyzed [25]. The results are shown in figure 5. Simple statistics yields values for the shape space covering radius which agree well with the numerical data.

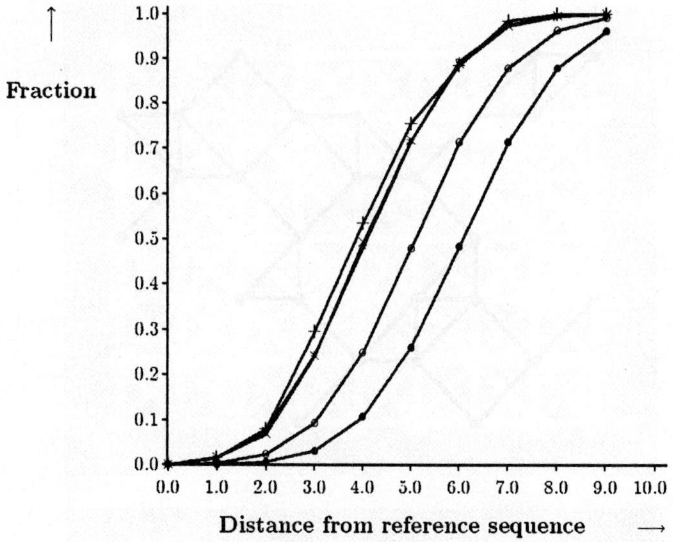

Fig.5: Verification of shape space covering in the sequence space of **GC** only sequences of chain length $n = 30$. The five curves refer to the most common structure ($\times$), the top five ($*$), the top ten ($+$), the top 2625 most common structures ($\circ$) and all 22 718 common structures ($\bullet$).

3.3 Neutral Networks

All sequences folding into the same secondary structure are considered as a *neutral network*. In order to visualize neutral networks neighbouring sequences are connected. Two classes of neighbours are defined which refer to single point mutations in single stranded parts of the structure and to base pair exchanges in the double helical stacks. In figure 6 we show two characteristic examples of neutral networks above and below a percolation threshold [26] which can be evaluated by application of random graph theory to the problem of sequence to structure mappings.

One important result of the random graph approach deals with the structure of neutral networks below the percolation threshold. The generic decomposition of the network into components yields a *giant* component that is much larger than all other components. This largest component spans a fraction of sequence space which comes close to the maximum possible area. The sequence of components in neutral networks of RNA secondary structures can be interpreted in terms of structural regularities: whenever we found deviations form the predicitions of random graph theory this was caused by specific features of the RNA

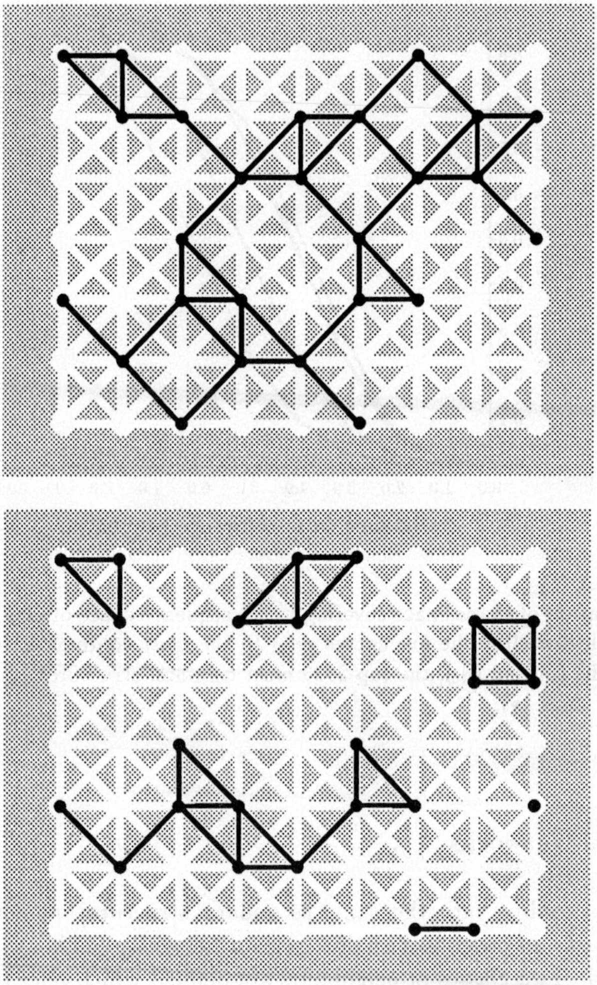

Fig.6: Neutral networks and percolation threshold. Graphs representing neutral networks shown in both figures connect sequences which are nearest neighbours in sequence space and which fold into identical (secondary) structures. The network in the upper part of the figure is above the percolation threshold and consists of a single component. The lower part shows an example of a disintegrated network below threshold that is commonly split into a largest (giant) component and many islands in sequence space.

base pairing logic [25,26]. For example, two largest components of approximately equal size are found instead of a single giant component in case the structures contain either large loops (with five or more members) or two dangling ends. In both cases structures are easily altered by single base substitutions which allow to form an additional base pair. These specific structural features are idosyncrasies of RNA molecules and would not be found with other biopolymers.

4. Optimization on Landscapes

Sequence to structure mappings represent the first part of the unfolding of genotypes in order to yield the kinetic parameters of population dynamics. The second part being the evaluation of phenotypes. In molecular evolutionary biology this is tantamount to a mapping from structures to function:

$$\textit{sequence} \Longrightarrow \textit{structure} \Longrightarrow \textit{function}$$

The whole mapping thus assigns a function that can be expressed in kinetic parameters, for example in fitness values. We are thus dealing with a mapping from sequence space into the real numbers which is commonly denoted as a *(fitness) landscape* [27]. Evolutionary dynamics in simple systems following the Darwinian principle can be understood as an optimization problem or an adaptive walk on a fitness landscape [28].

A model system based on RNA sequence to secondary structure mappings was conceived about ten years ago and computer simulations were performed in order to study the basic mechanisms of optimization on realistic landscapes [29, 30]. These studies have shown that the error threshold phenomenon of replication-mutation dynamics is observed also on realistic landscapes and indicated great relative importance of neutral evolution in population dynamics.

Based on the recent knowledge on the internal regularities of sequence to structure mappings as expressed in the properties of neutral networks the computer simulations were repeated in order to learn more about the mechanism of neutral evolution [31]. These studies confirmed that populations migrate on neutral networks by means of a diffusion like process. Transitions between two neutral networks of equal fitness [32] were shown to occur via sequences that are compatible with both structures. A proof was derived which states that the set of compatible sequences for two arbitrarily chosen structures, called the *overlap*, is never empty. The size of the overlap plays the essential role in transitions between structures of equal fitness.

5. Breakdown of Darwinian Scenarios

Evolutionary dynamics visualized as an adaptive walk of a population on a complex multipeaked landscape optimizing its fitness thereby is equivalent to the conventional Darwinian scenario. This concept is based on several implicit assumptions two of them are crucial for the validity of the model:

(1) constant environment and
(2) independent reproduction.
The condition of constant environment may be relaxed to the requirement that environmental changes have to occur on a slower time scale than the adaptation of populations. In ecosystems part of the environmental change is caused by coevolving species. These changes are caused by the same adaptive mechanism and there is no *a priori* reason that they should occur more slowly than adaptation in the reference population. Landscape models of coevolution could easily fail therefore and should be considered with special care. Independent reproduction means that the rate constant of replication does not depend on the concentration of the reproducing entities. Host parasite systems and symbiotic interactions are examples of such dependencies that lead to a breakdown of the landscape scenario.

Current molecular evolution experiments are almost always dealing with constant environments and the replication rate constants are commonly independent of the concentrations of RNA molecules. In nature, however, none of these conditions is usually fulfilled.

6. Artificial Life Extending Biology

Molecular evolution hase become an established field and produces already spin-offs with interesting technological aspects [33]. What has it contributed to the theory of evolution? Firstly, it has shown that replication and evolutionary adaptation are not exclusive priviledges of cellular life. Polynucleotides were found to replicate with suitable enzymes in cell-free assays. Small oligonucleotides can be replicated even without protein assistence. Other replicators which are not based on nucleotide chemistry were found as well. Template chemistry starts to become a fascinating field in its own right.

Secondly, the old tautology debate on biological fitness has come to an end. Fitness in molecular systems can be measured independently of the survival in an evolution experiment. Evolutionary processes may be described and analysed in the language of physics.

Thridly, the probability argument against a Darwinian mechanisms of evolution is invalidated by the experimental proof of target oriented adaptation found for evolution experiments with RNA molecules. Evolutionary search is substantially facilitated by the properties of sequence space and shape space, in particular by the fact that the evolutionarily relevant part of shape space is covered by a tiny fraction of sequence space.

The knowledge acquired in molecular evolution allows ample extensions. In the near future many more non-natural systems will be designed and synthesized which fulfil the principles of molecular genetics and thus are capable of evolution but, apart from that, have little in common with their living counterparts.

Molecular evolution, in essence, has established the basic kinetic mechanisms of genetics. In the case of RNA replication and mutation the reaction mechanisms were resolved to about the same level of details as with other polymerization reactions in physical chemistry. Several aspects of a comprehensive theory of evolution, however, are still missing. For example, the integration of cellular metabolism and genetic control into a comprehensive theory of molecular genetics has not yet been achieved. Morphogenesis and development of multicellular organisms need to be incorporated into the theory of evolution. No satisfactory explanations can be given yet for the mechanisms leading to the origin of real novelties often addressed as the great jumps in evolution [34] .

Nevertheless the extension of molecular biology into organic and biophysical chemistry as described here is predestined to be part of the still heterogeneous discipline of artificial life. Principles are still taken from the living world, but the material carriers of the essential properties are new and unknown in the biosphere. At the same time this very active area of research spans a bridge from chemistry to biology and also sheds some light on burning questions concerning the origin of life.

Acknowledgements

Financial support by the Austrian *Fonds zur Förderung der wissenschaftlichen Forschung* (Projects S 5305-PHY, P 8526-MOB, and P 9942-PHY) and by the Commission of the European Communities (Contract PSS*0396) is gratefully acknowledged.

References

1. Spiegelman, S.: An approach to the experimental analysis of precellular evolution. Quart.Rev.Biophys. **4** (1971) 213-253

2. Eigen, M.: Selforganization of matter and the evolution of biological macromolecules. Naturwissenschaften **58** (1971) 465-523

3. Dayhoff, M.O., Park, C.M.: Cytochrome-c: Building a phylogenetic tree. In: Dayhoff, M.O., ed. Atlas of protein sequence and structure: Vol.4, pp. 7-16. National Biomedical Research Foundation: Silver Springs (Md.) 1969

4. Piccirilli, J.A., Krauch, T., Moroney, S.E., Benner, A.S.: Enzymatic incorporation of a new base pair into DNA and RNA extends the genetic alphabet. Nature **343** (1990) 33-37

5. Szathmáry, E.: Four letters in the genetic alphabet: a frozen evolutionary optimum ? Proc.R.Soc.Lond.B **245** (1991) 91-99

6. Wu, T., and Orgel, L.E.: Nonenzymatic template-directed synthesis on oligodeoxycytidylate sequences in hairpin oligonucleotides. J.Am.Chem.Soc. **114** (1992) 317-322

7. von Kiedrowski, G.: A self-replicating hexadeoxynucleotide. Angew.Chem.Int.Ed.Engl. **25** (1986) 932-935

8. Tjivikua, T., Ballester, P., Rebek, J., Jr.: A self-replicating system. J.Am.Chem.Soc. **112** (1990) 1249-1250

9. Nowick, J.S., Feng, Q., Tjivikua, T., Ballester, P., Rebek, J., Jr.: Kinetic studies and modeling of a self-replicating system. J.Am.Chem.Soc. **113** (1991) 8831-8839

10. Hong, J.-I., Feng Q., Rotello, V., Rebek, J., Jr.: Competition, cooperation, and mutation: improving a synthetic replicator by light irradiation. Science **255** (1992) 848-850

11. Orgel, L.E.: Molecular replication. Nature **358** (1992) 203-209

12. Biebricher, C.K. and Eigen, M.: Kinetics of RNA replication by Qβ replicase. In: Domingo, E., Holland, J.J. and Ahlquist, P., eds. RNA Genetics. Vol.I: RNA directed virus replication, pp.1-21. CRC Press: Boca Raton (Fl.), 1988

13. Biebricher, C.K., Eigen, M., Gardiner Jr., W.C.: Kinetics of RNA replication: plus-minus asymmetry and double-strand formation. Biochemistry **23** (1984) 3186-3194

14. Sumper, M., Luce, R.: Evidence for *de novo* production of self-replicating and environmentally adapted RNA structures by bacteriophage Qβ replicase. Proc.Natl.Acad.Sci.USA **72** (1975) 162-166

15. Biebricher, C.K., Eigen, M., Luce, R.: Product analysis of RNA generated *de novo* by Qβ replicase. J.Mol.Biol. **148** (1981) 369-390

16. Biebricher, C.K., Eigen, M., Luce, R.: Template-free RNA synthesis by Qβ replicase Nature **321** (1986) 89-91

17. Eigen, M., McCaskill, J., Schuster, P.: The molecular quasispecies – An abridged account. J.Phys.Chem. **92** (1988) 6881-6891

18. Eigen, M., McCaskill, J., Schuster, P.: The molecular quasispecies. Adv.Chem.Phys. **75** (1989) 149-263

19. Fontana, W., Konings, D.A.M., Stadler, P.F., Schuster, P.: Statistics of RNA secondary structures. Biopolymers **33** (1993) 1389-1404

20. Fontana, W., Stadler, P.F., Bornberg-Bauer, E.G., Griesmacher, T., Hofacker, I.L., Tacker, M., Tarazona, P., Weinberger, E.D., Schuster, P.: RNA foling and combinatory landscapes. Phys.Rev.E. **47** (1993) 2083-2099

21. Bonhoeffer, S., McCaskill, J.S., Stadler, P.F., Schuster, P.: RNA multi-structure landscapes. Eur.Biophys.J. **22** (1993) 13-24

22. Tacker, M., Fontana, W., Stadler, P.F., Schuster, P.: Statistics of RNA melting kinetics. Eur.Biophys.J. **23** (1994) 29-38

23. Hofacker, I.L., Fontana, W., Stadler, P.F., Bonhoeffer, L.S., Tacker, M., Schuster, P.: Fast folding and comparison of RNA secondary structures. Mh.Chem. **125** (1994) 167-188

24. Schuster, P., Fontana, W., Stadler, P.F., Hofacker, I.L.: From seqeunces to shapes and back: a case study in RNA secondary structures. Proc.Roy.Soc.Lond.B **255** (1994) 279-284

25. Grüner, W.: Doctoral Thesis. Universität Wien, 1994

26. Reidys, C.: Dotoral Thesis. Friedrich-Schiller-Universität Jena, 1995

27. Schuster, P., Stadler, P.F.: Landscapes: complex optimization problems and biopolymer structures. Computers Chem. **18** (1994) 295-324

28. Kauffman, S.A., Levin, S.: Towards a general theory of adaptive walks on rugged landscapes. J.Theor.Biol. **128** (1987) 11-45

29. Fontana, W., Schuster, P.: A computer model of evolutionary optimization. Biophys.Chem. **26** (1987) 123-147

30. Fontana, W., Schnabl, W., Schuster, P.: Physical aspects of evolutionary optimization and adaptation. Phys.Rev.A **40** (1989) 3301-3321

31. Huynen, M., Fontana, W., Stadler, P.F.: Unpublished results. Santa Fe (N.M.), 1994

32. Weber, J., Reidys, C., Forst, C., Schuster, P.: Unpublished results. Jena (Germany), 1995

33. Schuster, P.: How to search for RNA structures. Theoretical concepts in evolutionary biotechnology. J.Biotechnology, in press 1995

34. Eigen, M., Schuster, P.: Stages of emerging life - Five principles of early organization. J.Mol.Evol. **19** (1985) 47-61

1. Foundations and Epistemology

Artificial Life Needs a Real Epistemology

H. H. Pattee

Systems Science and Industrial Engineering Department
State University of New York at Binghamton
Binghamton, New York 13902-6000

Abstract Foundational controversies in artificial life and artificial intelligence arise from lack of decidable criteria for defining the epistemic cuts that separate knowledge of reality from reality itself, e.g., description from construction, simulation from realization, mind from brain. Selective evolution began with a description-construction cut, i.e., the genetically coded synthesis of proteins. The highly evolved cognitive epistemology of physics requires an epistemic cut between reversible dynamic laws and the irreversible process of measuring initial conditions. This is also known as the measurement problem. Good physics can be done without addressing this epistemic problem, but not good biology and artificial life, because open-ended evolution requires the physical implementation of genetic descriptions. The course of evolution depends on the speed and reliability of this implementation, or how efficiently the real or artificial physical dynamics can be harnessed by nondynamic genetic symbols.

Life is peculiar, said Jeremy. As compared with what? asked the spider. [15]

1. What Can Artificial Life Tell Us About Reality?

When a problem persists, unresolved, for centuries in spite of enormous increases in our knowledge, it is a good bet that the problem entails the nature of knowledge itself. The nature of life is one of these problems. Life depends on matter, but life is not an inherent property of matter. Life is peculiar, obviously, because it is so different from nonliving matter. It is different, not so obviously, because it realizes an intrinsic epistemic cut between the genotype and phenotype. Our knowledge of physics, chemistry, molecular biology, genetics, development, and evolution is enormous, but the question persists: Do we really understand how meaning arises from matter? Is it clear why nonliving matter following inexorable universal laws should acquire symbolic genes that construct, control, and evolve new functions and meanings without apparent limit? In spite of all this knowledge, most of us still agree with Jeremy.

Where we find disagreement is on the answer to the spider's question. Artificial life must ask this question: With what do we compare artificial life? The founding characterizations of artificial life comparing "*life-as-it-could-be*" with "*life-as-we-know-*

it" [29,30], or "implementation-independent" computer life with space-time-energy-dependent material life [20] was a creative beginning, but this highly formal view of life was immediately questioned. Such abstract characterizations do not clearly separate science fiction and computer games from physical reality [35]. On the other hand, the idea of life in a computer does stimulate the philosophical imagination.

An alternative view of artificial life uses computation to control robots in a real physical world. Although in this approach the more fundamental philosophical issues are not as apparent, it has the enormous advantage in a practical sense of using the physical world at face value. As Brooks [6] understates the point: "It is very hard to simulate the actual dynamics of the real world."

My first answer to the spider's question is that we can only compare life to nonlife, that is, to the nonliving world from which life arises and evolves. Artificial life must be compared with a real or an artificial nonliving world. Life in an artificial world requires exploring what we mean by an alternative physical or mathematical reality. I want to follow Dennett's [17] suggestion that we use artificial life as a "prosthetically controlled thought experiment" that may provide some insights into such foundational questions. Metaphysical questions, like whether reality is material, formal, or mental, are empirically undecidable, but nevertheless, discussion of these concepts are an important part of scientific discovery. Historically we have seen concepts of reality shifting ground and new horizons discovered, especially with the advent of quantum theory, computation theory, and cosmology. The question is: Can artificial life add some new ideas to the problem of knowledge and the epistemic cut or will it only increase the confusion?

2. Life Requires an Epistemic Cut

The first problem for life in a computer is to recognize it. How peculiar does artificial life have to be? That is, how will we distinguish the living parts of the computation from the nonliving parts? And what are the parts? I have argued for many years that life is peculiar, fundamentally, because it separates itself from nonliving matter by incorporating, within itself, autonomous epistemic cuts [32,33,34,38,39]. Metaphorically, life is matter with meaning. Less metaphorically, organisms are material structures with memory by virtue of which they construct, control and adapt to their environment. Evolution entails semantic information [19], and open-ended evolution requires an epistemic cut between the genotype and phenotype, i.e., between description and construction. The logical necessity of this epistemic cut is the fundamental point of von Neumann's [48] self-replicating automaton. It is this type of *logical* argument that gives some validity to the concept of formal life, implementation-independent life, or life in a computer.

It is not clear how far such logical arguments can take us. As von Neumann warned, if one studies only formal life, " . . . one has thrown half the problem out the window, and it may be the more important half." In spite of all our knowledge of the chemical properties of the components of the genotype and phenotype, no one knows the answer to von Neumann's "most intriguing, exciting and important question of why the molecules or aggregates which in nature really occur . . . are the sorts of things they are." In fact, this question is the best reason I know for studying artificial life where we can invent different "sorts of things" and see how they behave.

2.1 The Epistemic Cut Requires Implementation

What does implementing a description mean? Descriptions are nondynamic, stored structures that do nothing until they are interpreted and implemented. In life-as-we-know-it this means *translating and constructing what is described.* We know that this is a very complex process in real life involving DNA, messenger RNA, transfer RNA, coding enzymes, ribosomes, and a metabolism to drive the entire synthesis process. It is therefore not clear what total implementation-independence or formalization of artificial life can tell us. It is precisely the effectiveness of implementation of genetic descriptions that evolution by natural selection is all about. Complete formalization would indeed throw half the problem out the window, as von Neumann says.

The central problem of artificial life, as theoretical biology, is to separate the essential aspects of this implementation from the frozen accidents or the incidental chemistry and physics of the natural world that might have been otherwise. Of course all these levels of detail are useful for the problems they address, but to answer the question of why these molecules are the "sorts of things" they are requires abstracting just the right amount.

It is not generally appreciated in artificial life studies why formal self-replication is only half the problem. All of evolution, emergence, adaptation, and extinction, depends on how quickly and efficiently the variations in the genotype can be implemented in phenotypic functions. How does a symbolic-sequence space map into a physical-function space? In spite of all the physical and chemical knowledge we have, it still appears unreasonably fortuitous that only linear sequences of nucleotides are sufficient to instruct the synthesis of all the structural proteins including their self-folding and self-assembling properties, and all the coordinated, highly specific and powerful enzymes that control the dynamics of all forms of life. It is significant that even at the simplest level the implementation entails a computational intractable problem - the polymer folding problem.

The advantage of the autonomous robotics approach to artificial life is that it avoids the most intractable computational problems in the same way that real life does - it harnesses the real physics. However, robotics does not face the more fundamental construction and self-assembly problems. The question is how much can we learn from computational models alone about such efficient implementations of genetic information? Such questions depend largely on our epistemology of computation, that is, how we think of measurements and symbols constraining a dynamics. The same problem exists for all physical systems, as I will discuss in Sec. 3. I will survey some current concepts of computation after outlining what I mean by an epistemology and summarizing the standard epistemic principles of physical theory.

In traditional philosophy epistemic cuts are viewed as problems only at the cognitive level. They are called problems of reference or how symbols come to "stand for" or to "be about" material structures and events [7,22,52]. I have always found the complementary problem much more fundamental: How do material structures ever come to be symbolic? I think if we fully understood how molecules become messages in cells we would have some understanding of how messages have meaning. That is why the origin of life problem is important for philosophy.

3. What Is an Epistemology?

An epistemology is a theory or practice that establishes the conditions that make knowledge possible. There are many epistemologies. Religious mystics, and even some physicists [55], believe that higher knowledge is achieved by a state of ineffable oneness with a transcendent reality. Mystics do not make epistemic cuts. While this may work for the individual, it does not work for populations that require *heritable* information or common knowledge that must be communicable [5]. Knowledge is potentially useful information *about* something. Information is commonly represented by *symbols.* Symbols *stand for* or are *about* what is represented. Knowledge may be about what we call reality, or it may be about other knowledge. It is the *implementation* of "standing for" and "about" - the process of executing the epistemic cut - that artificial life needs to explore.

Heritable, communicable, or objective knowledge requires an epistemic cut to distinguishes the knowledge from what the knowledge is about. By *useful* information or knowledge I mean information in the evolutionary sense of information for construction and control, measured or selected information, or information ultimately necessary for survival. This is contrasted with ungrounded, unmeasured, unselected, hence, purely formal or syntactic information. My usage does not necessarily imply higher-level cognitive concepts like understanding and explanation, neither does it exclude them. I am not troubled by the apparent paradox that primitive concepts may be useful without being precisely understood. I agree with C. F. von Weizsäcker [49], "Thus we will have to understand that it is the very nature of basic concepts to be practically useful without, or at least before, being analytically clarified."

3.1 The Epistemology of Physical Theory

The requirement for heritable or objective knowledge is the separation of the subject from the object, the description from the construction, the knower from the known. Hereditary information originated with life with the separation of description and construction, and after 3.6 billion years of evolution this separation has developed into a highly specialized and explicit form at the cognitive level. Von Neumann [47] states this epistemology of physical theory clearly: " . . . we *must* always divide the world into two parts, the one being the observed system, the other the observer. The boundary between the two is arbitrary to a very large extent . . . but this does not change the fact that the boundary *must* be put somewhere, if the method is not to proceed vacuously . . ." In physical theory, the observer is *formally* related to the observed system only by the *results* of measurements of the observables defined by the theory, but the formulation of the theory, the choice of observables, the construction of measuring devices, and the measurement process itself cannot be formalized.

No matter where we divide the world into observed and observer, the fundamental condition for physical laws is that they are invariant to different observers or to the frames of reference or states of observers. Laws therefore hold everywhere - they are universal and inexorable. In addition to the invariance or symmetry principles, the laws must be separated from the initial conditions that are determined only by measurement. The distinction between laws and initial conditions can also be expressed in terms of information and algorithmic complexity theory [9]. Algorithmic complexity of information is measured by the shortest program on some Turing-like machine that can compute this information. Laws then represent information about the world that can be enormously shortened by algorithmic compression. Initial conditions represent information that cannot be so compressed.

Mystical and heritable epistemologies are not necessarily incompatible. They simply refer to different forms of knowledge [18]. For example, Penrose [40] agrees that this separation of laws is "historically of vital importance" but then expresses more mystically the "very personal view" that "when we come ultimately to comprehend the laws . . . this distinction between laws and boundary conditions will dissolve away."

3.2 Incomplete Knowledge - the Necessity of Statistical laws

The epistemology of physics would be relatively simple if this were all there were to it, but laws and initial conditions alone are not enough to make a complete physical theory that must include measurement. Measurement and control require a third category of knowledge called boundary conditions or constraints. These are initial conditions that can be compressed *locally* but that are neither invariant nor universal like laws. When such a constraint is viewed abstractly it is often called a rule; when it is viewed concretely it is often called a machine or hardware.

Both experience and logic teach us that initial conditions cannot be *measured*, nor boundary conditions *constructed*, with the deterministic precision of the formal dynamical laws. This uncertainty requires a third category of knowledge we call *statistical laws*. Statistical laws introduce one of the great unresolved fundamental problems of epistemology. The dynamical laws of physics are all symmetric in time and therefore reversible, while statistical laws are irreversible. Formally, these two types of laws are incompatible. It is even difficult to relate them conceptually. From Bernoulli and Laplace to the present day this problem persists. As Planck [41] says, "For it is clear to everybody that there must be an unfathomable gulf between a probability, however small, and an absolute impossibility." He adds, "Thus dynamics and statistics cannot be regarded as interrelated." Von Neumann [48] agreed with Planck but cautioned, " . . . the last word about this subject has certainly not been said and it is not going to be said for a long time." Thirty years later, Jaynes [26] says about the interpretation of probability in quantum theory, " . . . we are venturing into a smoky area of science where nobody knows what the real truth is."

What types of boundary conditions or constraints can "self-organize" from deterministic dynamical laws, and what types can only "emerge" from a statistical bias on a heritable population distribution (i.e., natural selection) is a central problem in evolution theory and an active study in artificial life [30]. As with all such problems, the issue depends on the existence of an epistemic cut.

3.3 Measurement Defines an Epistemic Cut

Like it or not, the epistemic cut in physical theory falls in Planck's "unfathomable gulf" between dynamical and statistical laws. The possible trajectories of the world are described dynamically by reversible, noiseless laws, but any explicit knowledge of a trajectory requires observations or measurements described by irreversible, noisy statistical laws. This is the root of the measurement problem in physical theory. The problem arises classically, where it is often discussed using the thought experiments such as Maxwell's demon [31], and in quantum theory where the formal treatment of the measurement process only makes matters worse [50]. Von Neumann [47] described the problem in this way: An epistemic cut must separate the measuring device from what is measured. Nevertheless, the constraints of the measuring device are also part of the world. The device must therefore be describable by universal dynamical laws, but this is possible only at the cost of moving the epistemic cut to exclude the measurement. We then need a new observer and new measuring devices - a vacuous regress.

When we distinguish the Turing-von Neumann concept of programmable computation from other less well-defined concepts, we will see in Sec. 5.6 that when described physically a "step" in the computation must be a measurement. The completion of a measurement is indicated by a record or memory that is no longer a part of the dynamics except as a incoherent (nonintegrable) constraint.

It is important to understand that invariance and compressibility are not themselves laws, but are necessary epistemic conditions to establish the heritability, objectivity and utility we require of laws. As P. Curie [14] pointed out, if the entire world in all its details were really invariant there would be nothing to observe. No epistemic cut would be possible, and therefore life could not exist, except perhaps in a mystical sense. It is only because we divide or knowledge into two categories, dynamical laws and initial conditions, that invariance itself has any meaning [24,54]. How we choose this cut intellectually is largely a pragmatic empirical question, although there is also a strong aesthetic component of choice [42].

The point is that invariance and compressibility are general epistemic requirements for evolution that preceded physical theory. They are both "about" something else, and therefore they require a cut between what does not change and what does change, and between the compression and what is compressed. How life, real or artificial, spontaneously discovers an invariant, compressible, and hence evolvable, description-construction cut is the origin of life problem. However it happened, it is clear that compressibility is necessary to define dynamical laws and life. Without compressibility life could not adapt or evolve, because there is no way to adapt to endlessly random (incompressible) events.

4. Artificial Life Requires an Artificial Physics

How is this physical epistemology relevant for artificial life? The important point is that physical epistemology is a highly evolved and specialized form of the primitive description-construction process. The cognitive role of physical epistemology appears to be far removed from the constructive function of genes, but both define a fundamental epistemic cut. Great discoveries have been made in physics without understanding the mechanisms that actually implement the epistemic cut, because physics does not need to study the epistemic cut itself. Measurement can simply be treated as an irreducible primitive activity. That is why in most sciences the epistemic cut appears sharp - we tend to ignore the details of constructing the measurement devices and record only the results. The reality is that physical theory would remain in a primitive state without complex measuring devices, and in fact most of the financial resources in physics are spent on their construction.

Unlike physical theory, great discoveries in the evolution of natural and artificial life are closely related to understanding how the description-construction process can be most efficiently *implemented*. The course of evolution depends on how rapidly and efficiently an adaptive genotype-phenotype transformation can be discovered and how reliably it can be executed [11,12].

Real and artificial life must have arisen and evolved in a nonliving milieu. In real life we call this the real physical world. If artificial life exists in a computer, the computer milieu must define an artificial physics. This must be done explicitly or it will occur by default to the program and hardware. What is an artificial physics or physics-as-it-might-be? Without principled restrictions this question will not inform philosophy or physics, and will only lead to disputes over nothing more than matters of taste in computational architectures and science fiction. If an epistemology-as-we-know-it in physics has evolved from life itself, we must consider this a fundamental restriction. What we now distinguish as the three essential categories of knowledge: laws, initial conditions, and statistics, we need to represent in computational models of artificial life.

This means that artificial laws must correspond to algorithmically compressible subsets of computational events, and initial conditions must refer to incompressible events determinable only by measurements by organisms. In other words, any form of artificial life must be able to detect events and discover laws of its artificial world. Defining a measurement in a computer is a problem. I discuss this in Sec. 5.6. Also, autonomy requires what I call *semantic closure* [39]. This means the organism's measurement, memory, and control constraints must be *constructed* by the genes of the organism from parts of the artificial physical world. Of course consistency requires that all activities of the organisms follow the laws they may discover. Whether such organisms are really alive or only simulated is a matter of definition. A more objective and important question is how open-ended is such computational life. No consensus can be expected on this question unless there is consensus on what computation means.

5. What Is Computation?

There are two fundamentally different views of computation, the mathematical or formal view and the physical or hardware view. Barrow [2] sees these views arising from "two great streams of thought" about physical reality. The traditional view is based on symmetry principles, or invariance with respect to observers. The currently popular view is based on an abstract concept of computation. Roughly, the symmetry view is based on the established physical epistemology that I outlined above with statistical measurement playing an essential role. The computational view emphasizes a dynamical ontology, with logical consistency and axiomatic laws playing the essential role. The one view sees computation as a locally programmable, concrete, material process strictly limited by the laws of physics. The other view sees computation as an universal, abstract, dynamics to which even the laws of physics must conform.

5.1 Formal Computation

The ontological view of computation has some roots in the historical ideal of formal symbol manipulation by axiomatic rules. The meaning of a formal system in logic and mathematics as conceived by Hilbert is that all the procedures for manipulating symbols to prove theorems and compute functions are axiomatically specified. This means that all the procedures are defined by idealized unambiguous rules that do not depend on physical laws, space, time, matter, energy, dissipation, the observer's frame of reference, or the many possible semantic interpretations of the symbols and rules. The founders of computation theory were mostly logicians and mathematicians who, with the significant exceptions of Turing and von Neumann, were not concerned with physical laws. Ironically, formal computational procedures are now called "effective" or "mechanical" even though they have no epistemic relation to physical laws. These procedures are justified only by intuitive thought experiments. This weakness is well-known, but is usually ignored. As Turing [46] noted, "All arguments which can be given [for effective procedures] are bound to be, fundamentally, appeals to intuition and for this reason rather unsatisfactory mathematically."

This complete conceptual separation of formal symbol manipulation from its physical embodiment is a characteristic of mathematical operations as we normally do them. All symbol strings are discrete and finite, as are all rewriting steps. Steps may not be analyzed as analog physical devices. Proofs do not allow statistical fluctuation and noise. The concepts of set and function imply precise symbol recognition and complete determinism in rewriting all symbols. Formal computation is, *by definition*, totally implementation-independent [27].

This formal view of computation appears to contribute little to understanding the nature of epistemic cuts because formal systems are self-contained. Symbols and rules have no relation to measurement, control, and useful information. In fact, purely formal systems must be free of all influence other than their internal syntax, otherwise they are in error. To have meaning they must be informally interpreted, measured, grounded, or selected *from the outside*. "Outside" of course is established only by an epistemic cut. It is for this reason that formal models can be programmed to *simulate* everything, except perhaps the ineffable or mystical, since all the interpretation you need to define what the simulation means can be freely provided from outside the formal activity of the computer.

5.2 Laplacean Computation

An extension of formal computation is the Laplacean ideal which, as a thought experiment, replaces the epistemic cut with an in-principle isomorphism between the formal computational states and physical states. Such thought experiments often lead to apparent paradoxes precisely because an isomorphism is a formal concept that does not define how to execute the epistemic cut. Maxwell's demon and Schrödinger's cat are famous examples. The demon forces us to clearly state how measured information is distinguished from physical entropy, and the cat forces us to decide when a measurement occurs. These distinctions both require defining epistemic cuts between the knower and the known. It is significant that there is still no consensus on where the cut should be placed in both cases [31,50].

5.3 Computation in the Wild

A further elaboration of formal computation has become popular more recently as a kind of backward Laplacean ideal. That is, the Laplacean isomorphism is interpreted as its converse: computation does not provide a map of the universe, the universe is a map of a computation, or "IT from BIT" as Wheeler [51] states it. This ontological view is what Dietrich [16] calls "the computer in the wild." Historians might try to blame this view on Pythagoras, but its modern form probably began with the shift in the view of mathematics as a pure logical structure to more of a natural science after the failure of pure logic to justify its foundations. The ontological view also arose from the ambiguous relation of information to entropy in the contexts of cosmology, quantum theory, and algorithmic complexity theory [9,57]. Toffoli [45] describes computation this way: "In a sense, nature has been continually computing the 'next state' of the universe for billions of years; all we have to do - and actually all we can do - is 'hitch a ride' on this huge ongoing computation, and try to discover which parts of it happen to go near where we want."

This confounding of formal rules that arise from constraints, and dynamics that describe physical laws, leads to ambiguous questions like, "Is there a physical phenomenon that computes something noncomputable? Contrariwise, does Turing's thesis . . . constrain the physical universe we are in?" (Chaitin, [8]). This speculative association of formal theorems with physical laws is sometimes called the strong Church-Turing thesis. It leads to the argument that if there were a natural physical process that could not be Turing-computed, then that process could be used as a new computing element that violates the thesis [12].

The strong Turing-Church thesis is also used to try to equate formal Turing-equivalence between two symbol systems, with fitness equivalence between two physical implementations of the formal systems. The argument is that because there are many Turing-equivalent formalisms, like cellular automata and artificial neural nets, that there is no significant difference in the behavior of their different physical implementations. Of course from the biological perspective this is not the case, because it is precisely the overall *efficiency* of the physical implementation that determines survival. The significant processes in life at all levels, from enzyme catalysis to natural selection, depend on statistical biases on the rates of change of noisy population distributions, whereas formal equivalence is neither a statistical bias, rate-dependent, noisy, nor a population distribution.

The believers in strong artificial intelligence have further popularized the computer metaphor by defining brains and life as just some kind of computer that we do not yet understand. This view is labelled *computationalism*. It replaces the Laplacean isomorphism with an identity. Like Toffoli, Dietrich [16] believes that, "every physical system in the universe, from wheeling galaxies to bumping proteins, is a special purpose computer in the sense that every physical system in the universe is implementing some computation or other." According to this view, the brain is a computer *by definition*. It is our job to figure out what these physical systems are really computing. Thus, according to Churchland and Sejnowski [10], ". . . there is a lot we do not yet know about computation. Notice in particular that once we understand more about what sort of computers *nervous systems* are, and how they do whatever it is they do, we shall have an enlarged and deeper understanding of what it is to compute and represent." Hopfield [23] extends this vague, generalized view of computation to evolution: "Much of the history of evolution can be read as the evolution of systems to make environmental measurements, make predictions, and generate appropriate actions. This pattern has the essential aspects of a computational system."

This undifferentiated view of the universe, life, and brains as all computation is of no value for exploring what we mean by the epistemic cut because it simply includes, *by definition*, and without distinction, dynamic and statistical laws, description and construction, measurement and control, living and nonliving, and matter and mind as some unknown kinds of computation, and consequently misses the foundational issues of what goes on within the epistemic cuts in all these cases. All such arguments that fail to recognize the necessity of an epistemic cut are inherently mystical or metaphysical and therefore undecidable by any empirical or objective criteria [36,37,43].

5.4 The Programmable Physical Computer

The formal view of computation would be conceivable as long as Turing's [46] condition that every symbol is "immediately recognizable" (i.e., perfectly precise measurement) and Gödel's [21] condition of perfect mechanism (i.e., perfect determinism) were possible. However, even though we have no way of knowing if nature is ultimately deterministic or not, we do know that measurement must at some stage be irreversible, and the results of measurement cannot be used to violate the 2nd law of thermodynamics. Hence, useful measurements are dissipative and subject to error an violate the assumptions of Laplace, Turing, and Gödel.

The physical view of computation is little more than the engineering view that recognizes the hardware constraints as a necessity for implementing any symbol manipulation. This view holds that statistical physical laws are both the foundation and limitation of computation. Programmable hardware is inherently slow and noisy. Most of the peculiar design features of the computer are to overcome these limits. It is a wonder of technology that these limits have been extended so far. Actually, it was Turing [46] who first justified the use of bits as the highest signal-to-noise symbol vehicle, and of course von Neumann [48] believed that any rigorous theory of computation must have its roots in thermodynamics. He did not think of computers as implementation-independent: "An automaton cannot be separated from the milieu to which it responds. By that I mean that it is meaningless to say that an automaton is good or bad, fast or slow, reliable or unreliable, without telling in what milieu it operates." The same is true for natural and artificial life.

5.5 Limits of Physical Computation

The requirement that computation must satisfy physical laws, especially the 2nd law of thermodynamics, is seldom questioned, but is nevertheless largely ignored by both formalists and computationalists. On the other hand, hardware designers are acutely aware of the practical physical limits of speed, reliability and dissipation. Theories of reversible (dissipationless) computation have been proposed over the last few decades [3,28], but they are essentially thought experiments with idealized dynamical constraints. No one knows how to build a useful programmable computer along these lines.

Bennett, [4] argues that the source of irreversibility in measurement is erasure rather than the measurement itself. This interpretation is possible if the measured results remain unused on the physical side of the epistemic cut. In any case, the basic condition is that our use of measured information cannot lead to a violation of the second law of thermodynamics. Therefore the addition of any new measured information that is actually used to decreases the accessible states (i.e., entropy decrease, useful work, natural selection, control of dynamics, etc.) must be compensated by information loss (i.e., entropy increase, noise, dissipation, increased accessible states, etc.) in some aspect of the measuring process [57].

5.6 Analog Dynamics

The ontological computationalists will argue that normal programmable computation is just our *interpretation* of a constrained physical dynamical system. They claim that all dynamical systems, likewise, can be interpreted as computing because they are *implementing* some functions [16,45]. This may be the case for analog computers where the dynamics maps initial states to final states without programs, measurements, or intermediate steps, but this is too great an abstraction for describing a programmable computer.

The issue is what we mean by implementation of a function, and how we define a step. If a computer is to be an implementation of formal logic or mathematics, then it must implement *discrete symbols* and perform *discrete steps* in the rewriting of these symbols to and from memory according to a sequence of rules or a program. This is what formal logic and mathematics is about. This is what Turing/von Neumann programmable computers do. It is also the case that any implementation of such symbolic computational steps must be a law-based system with physical dynamics, so the question is: What corresponds to a symbol and a step? Physical dynamics does not describe symbols and steps. They are not in the primary categories of knowledge called laws and initial conditions. A *step* can only be defined by a measurement process, and a *symbol* as a record of a measurement. Therefore, a programmable computation can be described in physical terms only as a dynamical system that is internally constrained to regularly perform a sequence of simple measurements that are recorded in memory. The records subsequently act as further constraints. Since the time of measurement, by definition, has no coherence with the time of the dynamics, the sequence of computational steps is *rate-independent*, even though all physical laws are rate-dependent. As in all arguments about when measurement occurs, this also depends on where the epistemic cut is placed.

The ontological computer-in-the-wild is a physical system that may be interpreted as a dynamical analog device that parallels some other process. Such analog computers were common before the development of the programmed digital computer. They cannot be classified easily because all systems are indeed potential analogs. Furthermore, what aspects of the system are to be interpreted as computation are not crisply defined as are symbolic, rule-based, programmed computers. It should be clear that these are two extremes that only produce confusion by being lumped together. In rate-dependent dynamical analogs no memory is necessary, and one epistemic cut is made at the end when the final result is measured. In rate-independent programmed computation each step is a measurement recorded in memory. There are innumerable possibilities for machines with all degrees of constraints in between these extremes, but none have general utility.

6. The Epistemology of Organisms

Living systems as-we-know-them use a hybrid of both discrete symbolic and physical dynamic behavior to implement the genotype-phenotype epistemic cut. There is good reason for this. The source and function of genetic information in organisms is different from the source and function of information in physics. In physics new information is obtained only by measurement and, as a pure science, used only passively, to know *that* rather than to know *how*, in Ryle's terms. Measuring devices are designed and constructed based on theory. In contrast, organisms obtain new genetic information only by natural selection and make active use of information to know *how*, that is, to construct and control. Life is constructed, but only by trial and error, or mutation and selection, not by theory and design. Genetic information is therefore very expensive in terms of the many deaths and extinctions necessary to find new, more successful descriptions.

This high cost of genetic information suggests an obvious principle that there is no more genetic information than is necessary for survival. What affects this minimum? The minimum amount of genetic or selected information will depend largely on how effectively this information can be implemented using the parts and the dynamics of physical world. For example, some organisms require genetic instruction for synthesizing amino acids from smaller molecules, but if all amino acids are available as environmental parts, there is no need for these genes. At the next level, if the information that determines the linear sequence is sufficient constraint to determine the folding and self-assembly of proteins then no further folding information is necessary. However, in some cases, when the self-folding is not stable, additional genes for membrane or scaffolding proteins to further constrain the folding are necessary.

This minimum genetic information principle should not be confused with algorithmic compression of information. Algorithmic compression is defined only in a formal context on unselected information. Compressiblity across an epistemic cut can only be interpreted informally as the efficiency of implementation of selected information in a physical milieu. No such minimum information principle can apply to formal or programmable computation. Formal computation requires, by definition, complete informational control. That is, no self-folding or any other law-based dynamics can have any effect on formal symbol manipulation. Any such effect is regarded either as irrelevant or an error.

The success of evolution depends on how quickly and effectively organisms can adapt to their environment. This in turn depends on how efficiently the sequence space of genes can transform, gradually, the control or function space of phenotypes. Efficiency here includes the search problem, i.e., how to find good descriptions [44], and the construction problem, i.e., how to reliably assemble parts according to the description [11,13].

As I mentioned in Sec. 2.1, it is important to recognize that these implementation problems, if treated formally, are combinatorially complex. The search space is enormous and the number of degrees of freedom of an enzyme is large, so that even though polymer folding is the simplest possible natural process that implements the genotype-phenotype transformation, a purely computational mapping is impractical. Even the two-dimensional folding of RNA is NP-complete, and *ab initio* computation of detailed protein folding appears out of reach. To make matters worse, folding requires finding only a stationary state. A quantum dynamical model of enzyme catalysis has not even been formulated..

The only practical computational approach to these combinatorially complex problems is to use "reverse biological engineering" and simulate the natural dynamics with artificial neural nets [25,56], and natural selection in the form of genetic algorithms to evolve the connection weights in the nets [53]. There is no doubt that these techniques derived from life-as-we-know-it are of practical engineering value. However, I would call them virtual dynamical analogs implemented by programmed computers. Adlelman [1] has used real DNA molecules in a "massively parallel" chemical search for a solution of the Hamiltonian path problem. It is a matter of taste whether this should be called molecular computing or chemical graph theory.

7. Conclusions.

If artificial life is to inform philosophy, physics, and biology it must address the implementation of epistemic cuts. Von Neumann recognized the *logical* necessity of the description-construction cut for open-ended evolvability, but he also knew that a completely axiomatic, formal, or implementation-independent model of life is inadequate, because the course of evolution depends on the *speed*, *efficiency*, and *reliability* of implementing descriptions as constraints in a dynamical milieu.

Many nonlinear dynamical models of populations of interacting units, like cellular automata, Kauffman-type networks, and games of life have been interpreted as genetic populations with or without a genotype-phenotype mapping. These populations compete, cooperate, and coevolve in an artificial environment. However, where there is a genotype-phenotype mapping this is usually a fixed program in which the evolution of efficiency of implementation does not arise. Of course implementation-independent self-organization may play essential roles in the origin of life and in limiting the possibilities for natural selection. The significance of these roles needs to be determined.

Implementation of a description means constructing organisms with the parts and the laws of an artificial physical world. Some epistemic principles must restrict physics-as-it-could-be if it is to be any more than computer games. In order to evolve, organisms must discover by selection or measurement some "compressible" genetic descriptions of this artificial physical world. Compressibility is a formal concept that is not strictly applicable across epistemic cuts. Compressibility across an epistemic cut simply corresponds to efficiency of implementation. There is no general way to measure how far we can compress the genetic information that is necessary to implement a biological function, because this depends on the physical laws and the quality of function necessary for survival. For the same reason we cannot in general specify how much information is necessary to construct a measuring device. The amount of information for construction of the measuring device is incommensurable with the survival value of the information obtained by the measurement. This is generally the case for all biological structure-function relations.

To evolve, organisms must efficiently implement these descriptions as constructions. A fundamental limitation for computer life is that evolution can only reflect the complexity of the artificial physical world in which organisms live. An epistemic cut affords the *potential* for efficient implementation and open-ended evolution, but in a simple world, efficient implementations will be limited and life will also remain simple.

Hybrid symbolic-dynamic systems, like life-as-we-know-it and computer-controlled robots actually address the problem of efficient implementation of control instructions in the real world, but robots are still a long way from implementing efficient memory-controlled *constructions* of real life that self-assemble at all levels, from polymer folding to multicellular development. Real world dynamics will always have some advantages for efficient implementations, because there are necessary but gratuitous inefficiencies of programmed computer simulations that are missing in reality, as well as significant unpredictable efficiencies of reality missing in the simulations.

References

1. Adelman, L. M., 1994, Molecular computation of solutions to combinatorial problems, *Science*, 266,1021-1024.
2. Barrow, J. D., 1991, *Theories of Everything*, Oxford University Press, p. 203.
3. Bennett, C. H., 1982, The thermodynamics of computation - a review, *Int. J. Theor. Phys.*,21, 905-940.
4. Bennett, C. H., 1987, Demons, engines, and the second law, *Sci. Am.*, 257, 108-116.
5. Born, M., 1964, Symbol and reality, *Universitas* 7, 337-353. Reprinted in Born, M., *Physics in my Generation*, Springer-Verlag, NY, pp. 132-146.
6. Brooks, R., 1992, Artificial life and real robots. In *Toward a Practice of Autonomous Systems*, F. Varela and P. Bourgine, eds, MIT Press, Cambridge, MA, *pp.3-10.*

7. Cassirer, E., 1957, *The Philosophy of Symbolic Forms, Vol 3: The Phenomena of Knowledge*, Yale Univ. Press, New Haven, CT.
8. Chaitin, G., 1982, Gödel's theorem and information, *Int. J. of Theor. Phys.* 21, 941-953.
9. Chaitin, G., 1987, *Algorithmic Information Theory*, Cambridge University Press.
10. Churchland P. S. and Sejnowski, T. J., 1992, *The Computational Brain*, MIT Press.
11. Conrad, M., 1983, *Adaptability*, Plenum, New York.
12. Conrad, M., 1989, The brain-machine disanalogy, *BioSystems*, 22, 197-213.
13 Conrad, M., 1990, The geometry of evolution, *BioSystems*, 24, 61-81.
14. Curie, P., 1908, *Oeuvres*, Gauthier-Villars, Paris, p. 127.
15. Curtis, C. and Greenslet, F., 1945, *The Practical Cogitator*, Houghton Mifflin, Boston, p. 277.
16. Dietrich, E. 1994, ed.,*Thinking Computers and Virtual Persons*, Academic Press, NY, p. 13.
17. Dennett, D., 1994, Artificial life as philosophy, *Artificial Life* 1, 291-292.
18. Eddington, A., 1928, *The Nature of the Physical World*, Macmillan, New York, p. 260.
19. Eigen, M., 1992, *Steps Toward Life*, Oxford University Press.
20. Fontana, W., Wagner, G., and Buss, L. W., 1994, Beyond digital naturalism, *Artificial Life*, 1, 211-227.
21. Gödel, K., 1964, Russell's mathematical logic, and What is Cantor's continuum problem? In P. Benacerraf and H. Putnam, eds., *Philosophy of Mathematics*, Prentice-Hall, Englewood Cliffs, NJ, pp. 211-232, 258-273.
22. Harnad, S., 1990, The symbol grounding problem, *Physica D* 42, 335-346.
23. Hopfield, J. J. 1994, Physics, computation, and why biology looks so different, *J. Theor. Biol.* 171, 53-60.
24. Houtappel, R. M. F., Van Dam, H., and Wigner, E. P., 1965, The conceptual basis and use of the geometric invariance principles, *Rev. Mod. Physics* 37, 595-632.
25. Hunter, L., 1993, ed., *Artificial Intelligence and Molecular Biology*, AAAI/ MIT Press, Menlo Park, CA.
26. Jaynes, E., 1990, Probability in quantum theory. In W. H. Zurek, ed., *Complexity, Entropy, and the Physics of Information*, Addison-Wesley, Redwood City, CA, p. 382.
27. Kleene, S. C., 1952, *Introduction to Metamathematics*, Van Nostrand, Princeton, NJ.
28. Landauer, R., 1986, Computation and physics: Wheeler's meaning circuit, *Found. Phys.*, 16, 551-564.
29. Langton, C., 1989, Artificial life, In C. Langton, ed., *Artificial Life*, Addison-Wesley, Redwood City, CA, pp. 1-47.
30. Langton, C., Taylor, C., Farmer, J., and Rasmussen, S., eds., *Artificial Life II*, Addison-Wesley, Redwood City, CA.

31. Leff, H. S. and Rex, A. F., eds., 1990, *Maxwell's Demon, Entropy, Information, Computing*, Princeton Univ. Press, Princeton, NJ.
32. Pattee, H. H., 1969, How does a molecule become a message? *Developmental Biology Supplement* 3, 1-16.
33. Pattee, H. H., 1972, Laws, constraints, symbols, and languages. In C. H. Waddington, ed., *Towards a Theoretical Biology 4*, Edinburgh Univ. Press, pp. 248-258.
34. Pattee, H. H., 1982, Cell psychology: An evolutionary approach to the symbol-matter problem, *Cognition and Brain Theory* 5(4), 325-341.
35. Pattee, H. H., 1988, Simulations, realizations, and theories of life. In C. Langton, ed., *Artificial Life*, Addison-Wesley, Redwood City, CA, pp. 63-77.
36. Pattee, H. H., 1989,The measurement problem in artificial world models, *BioSystems*, 23, 281-290.
37. Pattee, H. H., 1990, Response to E. Dietrich's "Computationalism," *Social Epistemology* 4(2), 176-181.
38. Pattee, H. H., 1993, The limitations of formal models of measurement, control, and cognition, *Applied Math. and Computation*, 56, 111-130.
39. Pattee, H. H., 1995, Evolving self-reference: matter, symbols, and semantic closure, *Communication and Cognition* - AI 12(1-2), 9-28.
40. Penrose, R., 1989, *The Emperor's New Mind*, Oxford University Press, p. 352.
41. Planck, M., 1960, *A Survey of Physical Theory*, Dover, New York, p. 64.
42. Polanyi, M., 1964, *Personal Knowledge*, Harper & Row, NY.
43. Rosen, R., 1986, Causal structures in brains and machines, *Int. J. General Systems*, 12, 107-126.
44. Schuster, P., 1994, Extended molecular evolutionary biology: Artificial life bridging the gap between chemistry and biology, *Artificial Life*, 1, 39-60.
45. Toffoli, T. 1982, Physics and computation, *International J. of Theoretical Physics* 21, 165-175.
46. Turing, A., 1936, On computable numbers, with an application to the Entscheidungsproblem, *Proc. Lond. Math. Soc.* 42, 230-265.
47. von Neumann, J., 1955, *Mathematical Foundations of Quantum Mechanics*, Princeton Univ. Press, Princeton, NJ, Chapter VI.
48. von Neumann, J., 1966, *The Theory of Self-reproducing Automata*, A. Burks, ed., Univ. of Illinois Press, Urbana, IL.
49. von Weizsäcker, C. F., 1973, Probability and quantum mechanics, *Brit. J. Phil. Sci.* 24, 321-337.
50. Wheeler, J. and Zurek, W., 1983, *Quantum Theory and Measurement*, Princeton Univ. Press, Princeton, NJ.
51. Wheeler, J. A., 1990, Information, physics, quantum: The search for links. In W. H. Zurek, ed., Addison-Wesley, Redwood City. CA.
52. Whitehead, A. N., 1927, *Symbolism: Its meaning and Effect*, Macmillan, NY.
53. Whitley, D. and Hanson, T., 1989, Optimizing neural networks using faster, more accurate genetic search. In *Proceedings of the Third International Conference on GAs*, Morgan Kauffman, pp. 391-396.
54. Wigner, E. P., 1964, Events, laws, and invariance principles, *Science* 145, 995-999.
55. Wilber, K., 1985, *Quantum Questions*, Shambala, Boston, MA.
56. Zuker, M., 1989, On finding all suboptimal folding of large RNA sequences using thermodynamics and auxiliary information, *Nucleic Acids Res.* 9, 133.
57. Zurek, W. H., 1990, *Complexity, Entropy, and the Physics of Information*, Addison-Wesley, Redwood City, CA.

Grounding and the Entailment Structure in Robots and Artificial Life

Erich Prem

The Austrian Research Institute for Artificial Intelligence
Schottengasse 3, A-1010 Vienna, Austria
erich@ai.univie.ac.at

Abstract. This paper is concerned with foundations of ALife and its methodology. A brief look into the research program of ALife serves to clarify its goals, methods and subfields. It is argued that the field of animat research within ALife follows a program which is considerably different from the rest of ALife endeavours. The simulation – non-simulation debate in behavior-based robotics is revisited in the light of ALife criticism and Simon's characterization of *the sciences of the artificial.* It reveals severe methodological problems, or dangers at least, which can only be overcome by reconsidering naturalness in the study of ALife. A comparison of ALife with other such sciences like Artificial Intelligence and Cognitive Science shows similar problems in these areas. Grounding, embodiment, and building real models are ways to overcome the dangers of unbounded artificiality.

Key words. Artificial life, epistemology, robotics, simulation, symbol grounding

1 The Research Program of Artificial Life

1.1 Goals

A widely accepted list of criteria to be fulfilled by ALife has been established by Chris Langton at the Artificial Life workshop in 1987. He defines Artificial Life as

> ...the study of man-made systems that exhibit behaviors characteristic of natural living systems. [Langton 89, p.1]

According to Langton, ALife (among other things)

- tries to synthesize life-like behaviors within artificial systems,
- locates *life-as-we-know-it* within "the larger context of *life-as-it-could-be,*"
- studies the ongoing dynamic behavior rather than a final result,
- constructs systems which exhibit emergent phenomena
- and are not controlled by some central processing unit.

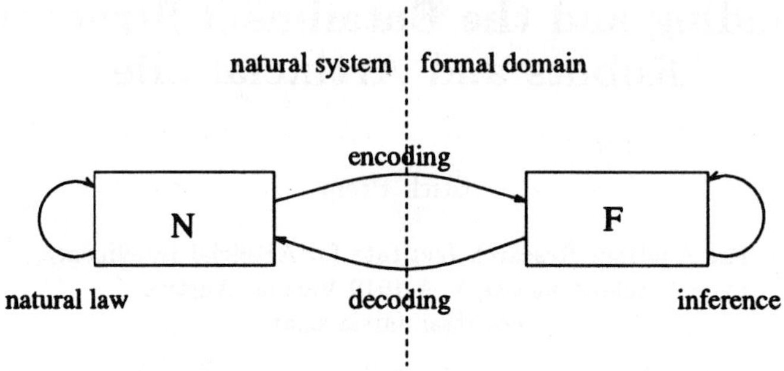

Fig. 1. Model of a natural system (after [Casti 92, p.29])

Figure 1 serves to make the idea of modeling phenomena of life as it is pursued in ALife more clear. *N* refers to some part of the natural world, e.g. to a group of animals. *F* denotes some formalism which is supposed to capture some essential characteristics of *N*, e.g. its behavior, emergent properties, or the dynamics of the system. The encoding of elements from *N* in *F* is usually done by a system designer who decides which phenomena of *N* are encoded in the formalism (e.g. which set of behaviors of an animal and what these behaviors amount to and how they are encoded ...). Langton says that the object of study in ALife is to a great extent *F*, *not* *N*! *F* is the "man-made system", which ALife studies.

1.2 Methods

The central methodology of ALife research is the construction of artifacts which possess the above characteristics, with a major emphasis on computer simulation.

According to the previously mentioned characterization of ALife, its methodology essentially conforms to behaviorism. However, lines of argumentation in ALife papers are very often distinct from a pure behaviorist program. Where behaviorism is rather clear about what is allowed in terms of entities used in explanations, ALifers are usually not so strict. They *do* recur to ideas which biology has about the generation of certain behaviors, i.e. about internal structures and processes. This methodology can be better understood by depicting it as being metaphor based, as it is done in Fig. 2.

Here we have two different natural systems: the computer and the living system. The living system is described in the world of formal systems mainly based on its I-O activity, but in ALife practice usually also with respect to internal and external structures (e.g. sensory organs, nervous systems, individuals, etc.).

The system designer selects a set of phemonea which she wishes to model. These entities, which comprise at least some kind of input and output, become encoded in the formalism. A computer program then is constructed so as to produce the output from the input. The designer also selects (maybe after the program has terminated or during its run) which entities of the formal system (the program) correspond to entities in the natural system. A characteristic

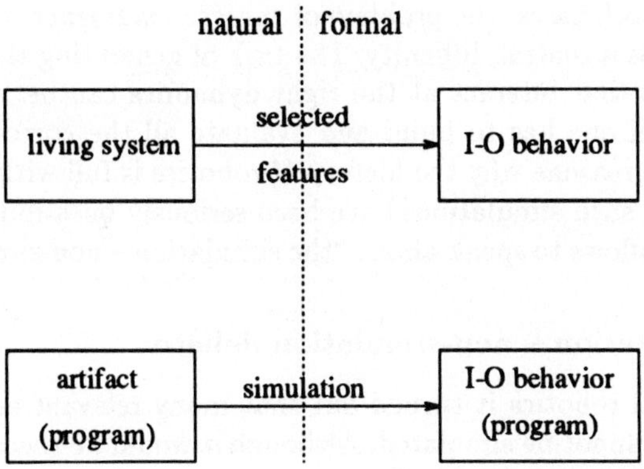

Fig. 2. Metaphor of a system (cf. [Rosen 85, p.81])

feature of these simulations is that not all entities decode into something in the real world.

Since there is no direct connection between the system and its simulation, this approach is called a metaphor. Of course, having a real model where all entities of the formal system decode into something about the natural system would be much better. However, the complexity of living system does often not allow to do so. Such models as in Fig. 1 are still the ideal being aimed at.

2 Robots in ALife

2.1 Robotics can be a part of ALife

A considerable part of ALife research is concerned with the construction and programming of robots in order to study behavioral and social phenomena (see e.g. [Resnick 89, Mataric & Marjanovic, Steels 94].)

There is a perspective from which it makes good sense to include the study of single behavior-based robots or animats in the domain of Artificial Life. Proponents of this kind of embodied AI (Artificial Intelligence) have continually pointed out that a concrete behavior of a robot is created through the interaction of many independent behavioral modules and through the robot's interaction with the physical world [Brooks 91, Steels & Brooks 93]. The great number of interacting modules and their complex way of interaction makes it difficult, if not impossible, to describe an observed robot behavior in terms of the modules which generate it. An observer of such a system is forced to change the level of description from entities which generate the behavior to rules which contain "higher" (behavioral) concepts: an emergent phenomenon has occured.

Experiments in which groups of such robots interact and exhibit global phenomena can all the more count as a subfield of ALife for working without global control.

However, in both cases, the problem of *inverse emergence* or *behavior generation* (Langton) is a central difficulty. The task of generating the correct set and kinds of modules that interact at the right dynamics can be very difficult and time-consuming if one has to build and evaluate all the corresponding robots. This is one of the reasons why the history of robotics is full with computer simulations. However, such simulations have been seriously questioned in the past to an extent which allows to speak about "the simulation – non-simulation debate".

2.2 The simulation – non-simulation debate

In behavior-based robotics it turned out that many relevant aspects of the behavior of agents cannot be simulated. Although it would be theoretically possible to develop an elaborated model of an agent and its environment this is simply not practical. Even if we would succeed in the development of such a model, it is likely that the actual behavior of the robot is influenced by some properties of the environment which have not been foreseen. Therefore, it is much more practical to actually use a test robot for the evaluation of a theory.

As an example for this statement one may consider the simulation of a simple robot which walks around in a room. A full-blown simulation of the robot with the movement of its whiskers, body, legs and its interaction with different surfaces amounts to an incredibly difficult piece of software. The physical properties of a robot's body alone are so complicated to calculate that one rather sticks to building the robot in a smaller scale.

Note that there are two different aspects of this problem here. Firstly, the system consisting of the robot and its environment is very difficult to describe. Secondly, the important characteristics of the robot (those which we actually wish to study) are drastically changed depending on only slight variations of the environment or the interaction of the robot and its environment. This second aspect is it, which is the real argument against any simulation, not the mere difficulty of developing a good model. If exactly those features of a system which are subject to non-linear or chaotic influences are those which we want to study, then any simulation increases the danger that we simply miss the salient features of the system behavior in a simulation.

The term "simulation" is moreover very often used for something that is not a simulation of something at all. The field of robotics is full with "calculation examples" which are not *about* anything in the real world. This means that many such so-called simulations are not validated in the real world, i.e. using the system which they are pretending to actually simulate. But unless this is not done *one cannot speak about a simulation.*

Instead of talking about a simulation it would be better to speak about using a (mathematical) model of some phenomenon instead of the "real" system to predict its behavior. How can you know if you have built such a model if you did not validate it with respect to the real system? The answer is that there *is no such thing like the theoretical validation of a model.* A model of a natural phenomenon cannot be justified by arguments that entirely rely on the formal aspects of it. It can only be shown to be (approximately) appropriate through comparison with what it is supposed to model. (Of course, so-called "simulations" can still serve as important metaphors for the clarification of ideas. However, this is usually not the purpose with which they are made up and a different story not considered in this paper.)

2.3 Entailment in formalisms

The deeper reason for the problems related to this argument about the virtues of simulations lies in the fact that any formalism (by definition) has only those properties which have been designed in it by some careful engineer (cf. [Rosen 91] or [Pattee 89]). Formal structures are completely arbitrary in what the formalism entails. Physically embodied systems do not share this property with formal models. Such systems are *open* to interaction (causal dependencies) that go beyond the purely syntactic entailments in any formalism. ("Physics" is not just computation, see e.g. [van Gelder 94] in the context of the question as to whether cognition is computation.)

One argument that is often referred to in this debate concerns the confusion about different levels of simulation. This argument usually takes the form "We did not simulate ***this part*** of the real world, but on a lower level of granularity, we could have done so." Consider again the case of the small robot with its whiskers. It takes a great deal of work and anticipation to model, for example, that the whisker of a robot can be bent or broken. Of course, this *can* be included in such a simulation, however, it will be done only after a test run with the real robot where the whisker has actually been bent or broken.

It is true that the hypothesis of reductionism is that there exists one formalism which models every natural system. The level at which this formalism operates is, however, too far away from any practically useful model (i.e. it operates below atoms on electrons, quarks, etc.). Let aside the arguments that exist against any reductive theory (see e.g. [Rosen 91]), it follows that appropriate models will always have to be constructed for that special case in which a model is needed.

Now the argument of behavior-based roboticists is simply that it is much better not to waste one's time and efforts in the design of complex formalisms which must be validated. Instead, one should use models which *automatically* contain at least some of the entailment structure of the systems to be modeled.

2.4 Entailment in robotics

This can be done by using physical models, for example. A smaller built vehicle is not a mere mathematical (formal) model of a vehicle. Instead, it is open to physical interactions to which a closed formal system would not be open and it can be assumed that it possesses a great deal of the original system's entailment structure (namely those physical properties which are not affected by the scale). Of course, such physical models do also have to be validated with respect to what these scaling properties are. However, in behavior-based robotics there exists evidence that such physical models are much more likely to model the relevant aspects of a system rather than formal computer simulations which exhibit a great deal of phenomena and characteristics simply not available in the real world and do not possess other features which the real world has.

It is worthwile to stress the important difference between behavior-based robotics (or collective robotics) and the rest of ALife. The mere fact of using a physically embodied robot ensures that a great deal of the characteristics of the model are in accordance with what is being modeled. This means that the metaphor from Fig. 2 changes to what is depicted in Fig. 3.

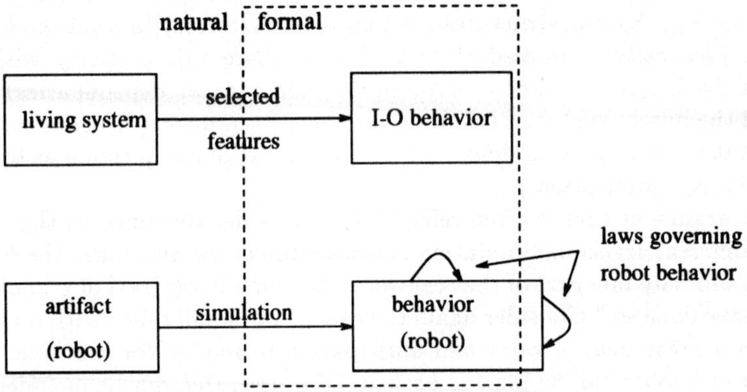

Fig. 3. Entailment in embodied systems. The model is no longer governed by formal rules alone, but also by material implication, i.e. the same "physics" as the modeled system.

In a behavior-based robot the rules governing the system behavior are not completely formal. They depend to a large extent on the physical world and on what physics does to the system. Since the natural system, the behavior of which is actually simulated, is also subject to this very same physics, it can be expected that such an embodied model will automatically contain a large amount of the entailment structure which is governing the original system. In other terms, it will automatically be a better model.

As I have argued in the beginning, animat research can be seen as a subarea of ALife. But, apart from that, what has all this concern about material implication, entailment, etc. to do with Artificial Life, which—after all—is a science of the *artificial*? Is ALife not mainly concerned with F instead of N from Fig. 1? Or, in other terms, what is the object of study in ALife?

3 ALife: A Science of the Artifical ?

3.1 Simon's sciences of the artificial

In a widely acknowledged book H. Simon gives an account of what he calls "the science of the artificial", its subject matter, and the methodology of this area of research. As opposed to the natural scientist who seeks to describe *how things are*, the scientist of the artificial

> is concerned with how things *ought* to be—how they ought to be in order to *attain goals*, and to function. [Simon 69, p.7]

An essential characteristic of a science of the artificial is its concern with construction, i.e. with synthesis. The construction is further aided by the fact that a scientist of the artificial is mainly interested in imitating appearance, which can be be characterized in terms of functions, goals, and adaptation. For Simon, every adaptive system can also be easily viewed as being artificial. The reason lies in the fact that adaptivity can be regarded as attaining the goal of being adapted, i.e. functioning to some extent. Even more, a factorization of adaptive systems into descriptions of goals and functions would allow us to greatly simplify the description of complex systems. But if *function* is of central interest to the scientist of the artificial, he needs mainly be concerned with the I-O behavior of the system:

> We might look toward a science of the artificial that would depend on the relative simplicity of the interface as its primary source of abstraction and generality. [Simon 69, p.12]

Simon's book is important to Artificial Life not only because it is quoted by Chris Langton and others in the field. Indeed, the characterization above seems to fit very well to a great part of ALife. The real value of Simon's contribution lies in his further description of how a science of the artificial can and should proceed. Simon is one of the first proponents of simulations as a source of new knowledge, not only in deriving truths about a system which may be implicit in a set of axioms. For Simon, the question as to whether

> simulation can be of any help to us when we do not know very much initially about the natural laws that govern the behavior of the inner system [Simon 69, p.20]

must be answered positively, and not only so:

> The more we are willing to abstract from the detail of a set of phenomena, the easier it becomes to simulate the phenomena. [Ibd.]

As a consequence of these considerations ALife and other similar sciences abstract from internal structures as long as one succeeds in getting the behavior right. A further consequence is that getting the behavior right can be studied by means of computer simulations.

If one refers to Simon (like Chris Langton does) as a proponent of the sciences of the artificial, one should also point out the dangerous threats of doing so. I have argued elsewhere [Prem 95a] that one of the problems of regarding AI as a science of the artificial consists in the consequence to consider intelligence as the sum of a set of specialiced domains. This view stems from the need of clearly defining what is meant by *X*. Let *X* be *intelligence, cognition, life* in

AI, Cognitive Science, and ALife, respectively. As soon as we are forced to make explicit the requirements for a technical system achieving X, are we reducing the phenomenon of X to a descriptive list of single aspects of what it means to possess X. The problem is that in a science of the artificial, which is supposed to support the construction of a functioning system, this list will automatically be taken to be a complete and sufficient specification of what it means to be *intelligent*, *cognitive*, or *alive*. Once such an approach has been initiated it cannot be escaped by simply arguing that something would be missing on the list, because each such criticism will in turn support the ongoing process of reduction. Each "turn of the screw" produces another specification of "how things ought to be", another research program in a science of the artificial.

This, of course, is a problem only, if one believes that the essence of life, intelligence, and cognition lies in the breadth of the term, i.e. in the innumerable phenomena associated with it. If one thinks, however, that an intelligence test measures intelligence, one may be satisfied with the reductive strategy outlined above.

Another of Simon's proposals to cope with complex systems consists in describing (and treating) them as so-called "nearly decomposable systems". Since this characterization is of interest in the present context, it will be briefly described in the next section.

3.2 Nearly decomposable systems

The hierarchical organization of many man-made as well as of natural systems eases, according to Simon, their description very much. In the description and prediction of such systems it suffices to deal with decomposable or "nearly-decomposable" systems. We can then distinguish between the interactions *among* subsystems and the interactions *within* subsystems (i.e. among the parts of those subsystems). If the intracomponent linkages are much stronger than intercomponent linkages, then the corresponding systems can be easily described by nearly-decomposable matrices. Very often then, these matrices will contain many instances of the same kind of matrix, i.e. it is possible to describe the system as being made up of many parts which possess the same characteristics.

This feature is so important for ALife, because in such systems the short-run behavior of subsystems will be approximately independent of other components and the long-run behavior of one component depends only in an aggregate way on other components. Let us compare this feature with what Langton says about the methodology of ALife.

> Artificial Life studies natural life by attempting to capture the behavioral essence of the constitutent components of a living system, and endowing a collection of artificial components with similar behavioral repertoires. If organized correctly, the aggregate of artificial parts should exhibit the same dynamic behavior as the natural system. [Langton 89, p.3]

This statement directly follows Simon's claim that

> [r]esemblance in behavior of systems without identity of the inner systems is particularly feasible if the aspects in which we are interested arise out of the *organization* of the parts, independently of all but a few properties of the individual components. [Simon 69, p.21]

The research program of ALife is therefore bound with a set of assumptions made about the characteristics of describing and simulating living systems. Again, all these considerations seem to suggest a high degree of abstraction, replacement of a system by its I-O behavior, and aggregating building blocks of mainly functional decompositions in order to genereate the desired behavior.

I shall now argue why there is evidence from a subarea of ALife, i.e. from behavior-based robotics, that the study of living systems is better not pursued in such a way.

3.3 What ALife can learn from robotics

One of the main goals for switching from more classical robot architectures to the new behavior-based ones was to get the interaction dynamics of the robot right for human purposes. In order to achieve this, two main steps had to be undertaken: (i) physical embodiment of robots as described previously and (ii) replacement of a functional modularization by behavior generating modules [Brooks 85].

Although the separation of functions into modules seemed to be a good idea in robotics, it turned out to be the major hindrance in getting the robot's interaction dynamics with the world right. Research in behavior-based robotics provides evidence that getting this interaction speed right is not just some luxurious feature so as to achieve commercially interesting behaviors. The rate of interaction influences to a very large extent the concrete kind of behavior which can be observed [Smithers 91].

It is true, of course, that in a system comprised of functional modules, these parts will usually only interact at a rather low rate compared to the intracomponent linkage. This is all the more true, if the modules serve to provide complex functions. The modules in behavior-based robots, however, do very often interact at high rates and in complex manners. Moreover, they are often fully connected with each other, which can make them a system which is not decomposable into its parts. It is quite obvious that there are a number of difficulties associated with viewing functions as decomposable in natural systems, because functions are often not separated in modules with clear interfaces. An example is a bird's wing which is airfoil and engine at the same time [Rosen 93].

An additional problem stems from the fact that these robots cannot be judged apart from a concrete physical environment in which they are put. Only when people started building their robots did it turn out that the architectures which seemed to be the right ones in simulations did not work in the real world. Instead of continuing this criticism I would like to make a positive statement about the possible contribution of behavior-based robotics to ALife.

This contribution can consist in recognizing the importance which the *automated* generation of the right entailment structure in a model has, and how important it can be, in general, to stick to models and not mere simulations. This reacknowledgement of material implication also happened in AI, another science of the artificial. This development is briefly oulined below so as to justify that *Alife is to behavior-based robotics as AI is to grounded connectionism.*

3.4 Other sciences of the artificial

AI is one of the fields explicitely mentioned in [Simon 69], in 1980 Simon extended his set of "sciences of the artificial" by adding Cognitive Science to it [Simon 80]. The reasons for doing so lie again in the adaptiveness of cognitive systems, in the physical symbol systems hypothesis, and in a commitment to a description of human intelligence as a compound of functional modules. It seems quite natural that human intelligence is in a sense of a simple structure, if one assumes low interaction dynamics between the modules. This simply follows from the way in which these modules have been generated. It is an epistemological side-effect, not an ontological one.

Let aside all this criticism, one of the hardest attacks on this science of the artificial is Searles "Chinese Room" argument, which will not be repeated here. Searle attacks symbolic AI on the basis of the specific character of symbols. Symbols have no other meaning than that which the programmer has given to it through connecting them to other, however again, symbolic descriptions. This diagnosis is, of course, in close accord with the lack of any entailment in formal systems, except for that, which has been explicetely designed. In 1990 Stevan Harnad propsed "symbol grounding" as a means for equipping arbitrary symbol tokens with non-arbitrary interpretations [Harnad 90]. He suggests to create some kind of correlational semantics, e.g. by means of neural networks, to enrich the arbitrary formalism with non-arbitrary meaning.

As of today, it seems quite clear that grounding cannot overcome the problems related to intentionality [Prem 95b]. However, grounding can contribute to the problem of narrowing the meaning of symbols, i.e. restricting the possible interpretations of these entities. This is because measurement devices are connected to the neural networks which then extract invariances among the measurements with respect to a given symbol. Thereby, the system's entailment structure becomes directly influenced by the real world. Symbol grounding can therefore be regarded as the automated construction of formal models of natural domains [Prem 95b, Prem 95c]. A similar argument has long been made by connectionists in discussions which center around so-called "radical connectionism" [Clark 93, p.31]: If networks are "immediately grounded" in the real world, they will be able to take advantage of the features of the real world and provide systems with better representations of the environment.

Without the danger of over-interpreting these developments in Cognitive Science and AI one can safely say that they show a tendency towards reality, not only with respect to the models which are generated. Instead, it is the automated generation of such models which is supported by means of exploiting physics.

3.5 Naturalness Revealed in the Study of ALife

From what I argued in this paper it must be concluded that the science of Artificial Life is either not a science of the artificial or that there are severe problems associated with viewing it to be so. I have given several reasons to support this claim, most of which were practical ones drawn from experiences with behavior-based robots. The practical reason follows from the problem of how to generate the right behavior. But this problem is not completely separate from the generation of the correct entailment structure in ALife models. BBR *is* concerned with reality and it has good reasons to do so: it wants to get the dynamics right. Therefore, it could not remain a pure science of the artifical.

From the fact that ALifers want to study "life as a property of the organization of matter" (Langton) it does not follow that matter need not be considered at all. It can help a great deal, as I have pointed out with the examples above, in getting this organization right. Therefore, as opposed to the claim that the "material is irrelevant" [Langton 89a, p.21], I suggest that *matter matters.*

4 Could Alife Be a "Life-as-it-could-be"?

There seems to exist an argument that can be used to refute the base line of this paper. It is contained in Langton's definition of ALife research, namely that ALife was not only studying (carbon-based) life on earth, but "life as it could be" [Langton 89a, p.1]. However, ALife has so far undertaken only very few efforts to clarify the meaning and the epistemological status of this deviation in the program of studying living systems. ("Deviation" with respect to the path of traditional biology.) To me it seems that the research program vaguely specified by life-as-it-could-be needs at least some clarification as to what "life" actually is. Langton and others in the field try to solve the problem by reference to some phenomenological aspects of living systems like dynamic, chaotic behavior, self-steering processes, directedness, goal-orientation, emergence... Consequently, life-as-it-could-be comprises a wide set of systems that exhibit these phenomena. But then, life-as-it-could-be is only the set generated by a crude extrapolation from rudimentary characterizations of living systems replacing some more serious attempt to answer Schrödinger's still burning question "What is life ?" [Schrödinger 67].

The set of systems that so becomes subject of ALife is better characterized as "it could be life" than "life as it could be". It is, however, equally justified to deny some elements of this set that they are alive. This is because the reasons, why such superficial features should be accepted as the essential characteristics of life remain vague. This definition of life-as-it-could-be is based on life-as-we-know-it, which in turn means that we cannot gain an understanding of the latter from studying the former. More specifically, reference to some phenomena living systems exhibit cannot explain why a simulation of these phenomena should be called alive. That such nice patterns of dynamic patterns are sufficient for attributing life to a system simply does not follow from the fact that systems which have been traditionally called alive exhibit such neat dynamical phenomena.

"Life" in Life-as-it-could-be has always been judged on life-as-we-know-it. But this again is something natural. So far, I can see no reason how one can learn something about real life through a methodology which studies a broader and much less clearly defined set of systems than–conventionally–living systems. The reason why artificiality is such a grand enticement to (not only) computer scientists seems to arise from mere despair about the complexities of real life. But, at least to my personal understanding of the scientific method, desparation is not a scientific argument.

5 Conclusion

In this paper I have argued along three lines. I have reminded the reader why embodiment of robots, instead of their simulations, turned out to be useful for the field of robotics. Embodiment serves to produce the right behavior-based on the correct entailment structure in models, something which can maybe achieved by grounding, too. Secondly, I have tried to convince the reader why ALife should not be considered what Simon calls a *science of the artificial.* Thirdly, I have argued that life-as-it-could-be is an ill defined term, if it is supposed to rescue the research program of ALife into a domain where naturalness is not important. This is so, because it is and must be based on life-as-we-know-it as long as there is no better (necessary and sufficient) characterization of life.

The practical upshot of all this lies in the fact that Artificial Life must indeed be more concerned with *natural* systems than it may itself consider to have to. Building robots, building models (instead of simulations), and grounding systems may be ways for ALife in the future.

6 Acknowledgements

The Austrian Research Institute for Artificial Intelligence is supported by the Austrian Federal Ministry of Science and Research.

References

[Brooks 85] Brooks R.A.: A robust layered control system for a mobile robot, AI-Laboratory, Massachusetts Institute of Technology, Cambridge, MA, AI-Memo 864, 1985

[Brooks 91] Brooks R.A.: Intelligence without reason, in Proceedings of the 12th International Conference on Artificial Intelligence, Morgan Kaufmann, San Mateo, CA, pp. 569–595, 1991

[Casti 92] Casti J.: Reality rules, John Wiley & Sons, N.Y., 1992

[Clark 93] Clark A.: Associative engines, A Bradford Book, MIT Press, Cambridge, MA, 1993

[van Gelder 94] van Gelder T.: What might cognition be if not computation?, Indiana University, Research Report 75, 1994

[Harnad 90] Harnad S.: The symbol grounding problem, Physica D, 42, pp.335–346, 1990

[Langton 89] Langton C.G.(ed.): Artificial life, Addison-Wesley, Reading, MA, 1989

[Langton 89a] Langton C.G.: Artifical life, in [Langton 89], 1–47

[Mataric & Marjanovic] Mataric M.J., Marjanovic M.J.: Synthesizing complex behaviors by composing simple primitives, in Proc. of the Europ. conf. on Artif. Life, ECAL'93, Bruxelles, 1993

[Pattee 89] Pattee H.H.: Simulations, realizations, and theories of life, in [Langton 89], 63–77

[Prem 95a] Prem E.: New AI: Naturalness revealed in the study of artificial intelligence, in Dorffner G.(ed.), Neural networks and a new AI, Chapman & Hall, London, to appear, 1995

[Prem 95b] Prem E.: Symbol grounding and transcendental logic, in Niklasson L. & Boden M.(eds.), Current trends in connectionism, Lawrence Erlbaum, Hillsdale, NJ, pp. 271–282, 1995

[Prem 95c] Prem E.: Dynamic symbol grounding, state construction and the problem of teleology, paper submitted for publication. See also Prem E.: Symbol Grounding, PhD Thesis (in German), University of Technology, Vienna, 1994

[Resnick 89] Resnick M.: LEGO, Logo, and life, in [Langton 89], 397–406

[Rosen 85] Rosen R.: Anticipatory systems, Pergamon, Oxford, UK, 1985

[Rosen 91] Rosen R.: Life itself, Columbia University Press, New York, Complexity in ecological systems series, 1991

[Rosen 93] Rosen R.: Bionics revisited, in Haken H., Karlqvist A., Svedin U.(eds.): The machine as metaphor and tool, Springer, Berlin, 87–100, 1993

[Schrödinger 67] Schrödinger E.: What is life and mind and matter, Cambridge University Press, Cambridge, UK, 1967

[Simon 69] Simon H.A.: The sciences of the artificial, MIT Press, Cambridge, MA, 1969 (expd.ed. 1980)

[Simon 80] Simon H.A.: Cognitive science: The newest science of the artificial, Cognitive Science, 4(1), 33–46, 1980

[Smithers 91] Smithers T.: Taking eliminative materialism seriously: A methodology for autonomous systems research, in: Varela F.J. et al., Towards a practice of autonomous systems, p. 31–40, 1991

[Steels & Brooks 93] Steels L., Brooks R.A.(eds.): The 'Artificial Life' route to 'Artificial Intelligence'. Building situated embodied agents, Lawrence Erlbaum Ass., New Haven, 1993

[Steels 94] Steels L.: Emergent functionality in robotic agents through on-line evolution, in Brooks R.A., Maes, P.(eds.): Artificial Life IV, Proc. of the fourth int. workshop on the synthesis and simulation of living systems, MIT Press, 1994

Mean Field Theory of the Edge of Chaos

Howard Gutowitz[1,2] and Chris Langton[2]

[1] Ecole Supérieure de Physique et Chimie Industrielles, Laboratoire d'Electronique, 10 rue Vauquelin, 75005 Paris

[2] The Santa Fe Institute, 1399 Hyde Park Road, Santa Fe, NM 87501

Abstract. Is there an Edge of Chaos, and if so, can evolution take us to it? Many issues have to be settled before any definitive answer can be given. For quantitative work, we need a good measure of complexity. We suggest that *convergence time* is an appropriate and useful measure. In the case of cellular automata, one of the advantages of the convergence-time measure is that it can be analytically approximated using a generalized mean field theory.

In this paper we demonstrate that the mean field theory for cellular automata can 1) reduce the variablity of behavior inherent in the λ-parameter approach, 2) approximate convergence time, and 3) drive an evolutionary process toward increasing complexity.

1 Introduction

Cellular automata are dynamical systems in which space, time, and the states of the system are discrete. Each cell in a regular lattice changes its state with time according to a rule which is local and deterministic. All cells on the lattice obey the same rule. This class of dynamical systems has been extensively studied as a model of natural systems in which large numbers of simple individuals interact locally so as to give rise to globally complex dynamics.

Study of cellular automata has given rise to the "edge of chaos" (EOC) hypothesis. In its basic form, this is the hypothesis that in the space of dynamical systems of a given type, there will generically exist regions in which systems with simple behavior are likely to be found, and other regions in which systems with chaotic behavior are to be found. Near the boundaries of these regions more interesting behavior, neither simple nor chaotic, may be expected.

Early evidence for the existence of an edge of chaos has been reported by Langton [Lan86, Lan90] and Packard [Pac88]. Langton based his conclusion on data gathered from a parameterized survey of cellular automaton behavior. Packard, on the other hand, used a genetic algorithm [Hol75] to evolve a particular class of complex rules. He found that as evolution proceeded the population of rules tended to cluster near the critical region identified by Langton.

The validity of some of these results have been called into question by Mitchell, Crutchfield, and coworkers ([MHC93, MCH94, DMC94]) These authors performed experiments in the spirit of Packard's. They found that while their genetic algorithm indeed produced increasingly complex rules, the cellular automata generated could not be considered to reside at a separatrix between simple and chaotic dynamical regimes.

The edge of chaos is a stimulating idea in that it promises to provide a framework in which to relate methods and results originating in biology, physics, and computer science. Yet the very generality and cross-disciplinary nature of the edge of chaos concept has lead to enormous difficulties of communication between workers from different background attempting to make these intuitions rigorous by the standards of their discipline.

We can divide the evolution toward the edge of chaos issue into several sub-issues:

- *Existence*: is there an edge of chaos? That is,
  - can the space of cellular automata be parameterized in such a way that cellular automata with similar parameter values have similar behavior?,
  - do there exist transitional values of these parameters where behavior changes rapidly, for instance from periodic and simple to chaotic? Are complex behaviors typically found at such a transition?
- *Uniqueness*: is there one edge of chaos, or many? The answer will depend, among other things, on the parameterization chosen. Here we will use mean-field parameters.
- *Characterization:* If an edge of chaos exists, what are its numerical characteristics? What is its form, how do rules *at* the edge of chaos behave? What is the relationship between computation theoretic and dynamical systems theoretic characterizations? What, indeed, is an appropriate definition of complexity for these systems?
- *Constructibility*: How can the edge of chaos be located? Can evolutionary processes be used to find the edge(s) of chaos?

In this paper we clarify a subset of the issues surrounding the edge of chaos theme. We restrict ourselves to study of the edge of chaos in the context of cellular automata. It is in this context that these issues were originally brought forward. We note, however, that an important advance was made by Suzuki and Kaneko [KS93, SK94] who used the logistic map, with a well-defined transition of chaos, as the basis for genetic evolution. They found that under a suitable dynamic, evolution indeed proceeded toward this transition.

Of the many objections that have been raised to EOC, we focus mainly on one: that the large variability of behavior for cellular automata with a given λ value makes the identification of a sharp "edge" of chaos difficult. We show that variability can be drastically reduced by moving to a more refined parameterization of cellular automata. When this new parameter space is projected into the λ parameter space, sharper transitions are found at the same locations predicted by earlier work using λ alone.

We use the mean field theory as our primary tool. It provides us with an analytic handhold on the relationship between parameterizations of the space of cellular automata, and the evolutionary construction of complex rules. The mean field theory allows us to 1) locate with some precision regions of rule space which contain particularly complex rules, and 2) to perform genetic evolution in such a way as to at once produce increasingly complex rules, and to approach a region in parameter space where the domains of simple and chaotic rules adjoin.

2 Review of and Mean Field Theory

2.1 The λ Parameter

To define λ, we consider CA with a symmetric quiescent condition (homogeneous neighborhoods do not change), and we arbitrarily pick one of the states and call it the *reference state* $\mathcal{R}$. If we let P($\mathcal{R}$) be the percentage of transitions to state $\mathcal{R}$ in the rule table, then we define the parameter λ as: $\lambda = 1.0 - P(\mathcal{R})$. We now use λ to generate random CA transition rules as follows. Pick a value for $\mathcal{R}$ and divide the remaining probability of $1.0 - P(\mathcal{R})$ equally among the remaining states. Walk through the rule table filling in the transitions using these probabilities. Thus, for $\lambda = 0.0$, *all* the transitions save the non-$\mathcal{R}$ homogeneous neighborhoods will be to state $\mathcal{R}$, for $\lambda = 0.5$, half of the transitions will be to state $\mathcal{R}$ and the remaining transitions will be equally distributed among the other states, and for $\lambda = 1.0$, *none* of the transitions will be to $\mathcal{R}$, and the other states will be represented equally in the transition table. Twisting the λ knob generating a series of transition rules and studying the behavior of CA run under the rule-tables so-generated is a simple way to search for structure and interesting regions in CA rule-space. Previous work has shown that λ serves to isolate regions of rule space in which dynamically simple or complex rules are located. λ is a crude measure of the behavior of CA; rules with identical λ may have widely different behavior. The approach taken here is to limit the variability of behavior by using a set of parameters, the mean field parameters, which refine the λ parameter.

2.2 Mean Field Theory

The mean field theory has mainly been used [GVK87] to characterize individual cellular automata. Here the mean field theory is used to characterize the full space of cellular automaton rules. We view the mean field theory as a parametric representation of the space of cellular automata[Gut90]. Each setting of parameter values corresponds to a subset of cellular automaton rules. As the parameter values are changed, the properties of the corresponding cellular automata change as well. The results below suggest that small changes in parameter values typically lead to small changes in the properties of the cellular automata described. At certain distinguished values of the parameters, however, changes in behavior can be abrupt.

The mean field theory describes how cellular automata act on probability measures of a particular type. A probability measure is an assignment of probability to blocks of cell states of all sizes. In empirical simulations block probabilities are estimated by the frequency with which these blocks occur in large configurations. By definition, the probability of a block b, at a given time t+1,${p_b}^{t+1}$ is the sum of the probabilities of the blocks B at time t which map to b under the cellular automaton τ, ${p_b}^{t+1} = \sum_{B|\tau(B)\to b} {P_B}^t$. Let τ be a one-dimensional cellular automaton with radius r with two states per cell, labeled 0 and 1. If b is the cell state 1, the probability of a 1 at time t is the sum of the probabilities at time t of blocks B of length 2r+1 which lead to 1 under the rule, ${p_1}^{t+1} = \sum_{B|\tau(B)\to 1} {P_B}^t$.

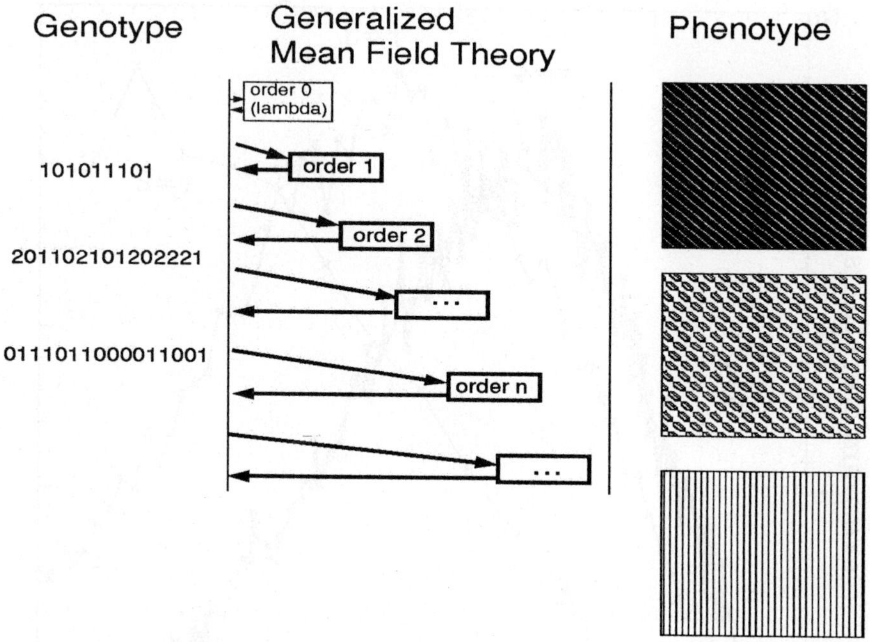

Figure 1. Relationship of λ , mean field theory, and generalized mean field theory. As order of approximation increases, so does the fidelity with which the generalized mean field theory represents the behavior of cellular automata. The generalized mean field theory can be used to approximate the behavior of rules. Conversely, given some behavior, the mean field theory can be used to find rules which have that behavior. This figure indicates that in the current context, rule tables are the genotype, and the space-time patterns the phenotype.

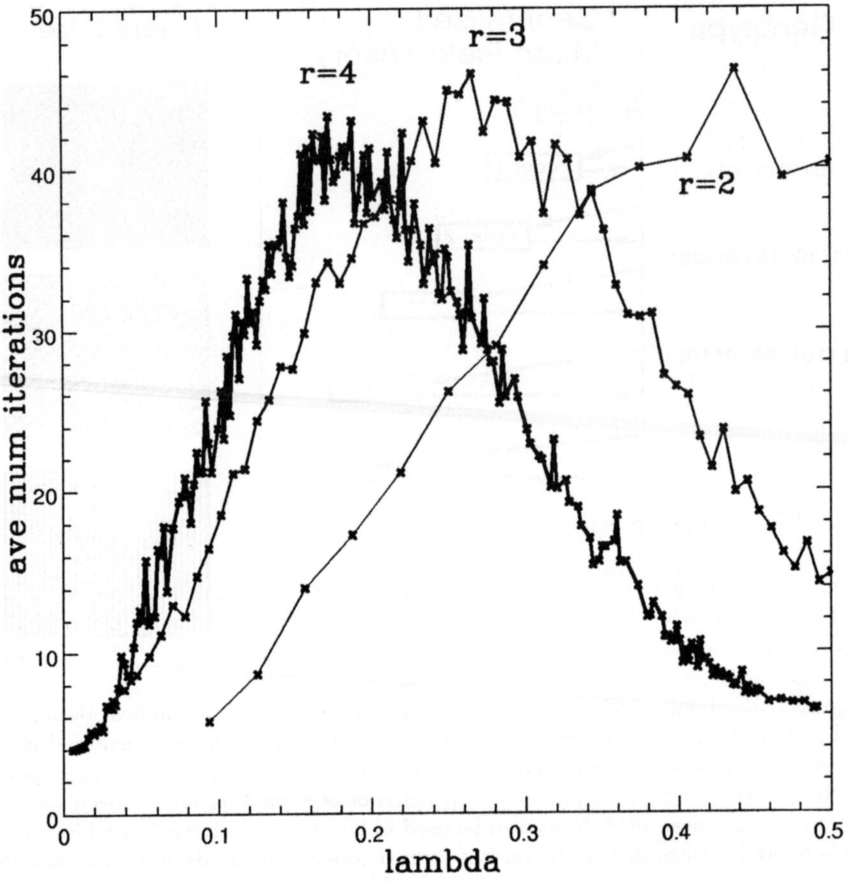

Figure 2. Convergence in mean field approximation. In this figure the convergence time in mean field approximation is plotted parametric in λ. For each radius, 2-4, approximately 0.5 x 10^6 rules were generated, and their convergence time in 10th-order mean field approximation calculated. The convergence time was averaged over all rules with a specified λ value. These curves are in good quantitative agreement with the studies of Li et al. [ea90], in which the "critical" values of λ were located using purely empirical means.

Let $\#0(B)$ and $\#1(B)$ denote the number of cells in state 0, 1 respectively in a block B. If we assume that the probability for a cell to be in a given state is uncorrelated with the states of other cells, then the probability of a block is just the product of the probabilities of the states of the cells in the block, $P_B = (p_1)^{\#1(B)}(1-p_1)^{\#0(B)}$. Substituting this into the above, we obtain the mean field equation

$$p_1{}^{t+1} = \sum_{B|\tau(B)\to 1} (p_1{}^t)^{\#1(B)}(1-p_1{}^t)^{\#0(B)}. \tag{1}$$

This mean field equation is a real-valued dynamical system which approximates the action of a cellular automaton on uncorrelated probability measures. In effect, the mean field theory represents the action of a cellular automaton combined with a noise process which removes all correlations between cell states after each application of the automaton.

3 The Mean-Field Refinement of the Parameterization

To see how the mean field theory supplies a parameterization of cellular automata, we rewrite equation 1 as follows. First note that only blocks B which lead to a 1 under the rule contribute to the sum on the r.h.s. of equation 1. Second note that two blocks B and B' such that $\#1(B) = \#1(B')$ contribute to the sum in exactly the same way. Let $\mathbf{a_i}$ be the number of blocks which contribute to the sum and contain i cells in state 1. Then equation 1 is written

$$p^{t+1} = f(\mathbf{a_0}, \cdots, \mathbf{a_{2r+1}}; p^t) = \tag{2}$$

$$\sum_{i=0}^{2r+1} \mathbf{a_i}\, (p^t)^i (1-p^t)^{2r+1-i}. \tag{3}$$

All cellular automata with the same coefficient values $\mathbf{a}$ have the same behavior in mean field approximation. A cellular automaton which gives rise to the coefficient values $\mathbf{a}$ is said to occupy the mean field class determined by $\mathbf{a}$.

Improvement of the mean field theory requires that the assumption that no correlations between cell states are generated by cellular automata be relaxed. A sequence of generalizations [GVK87]. of the mean field theory has been devised in which correlations are represented in terms of the probability of blocks of length l. In the following, we will use 2nd and 10th order mean field theory to approximate the behavior of rules. Thus, correlations included in the probability of blocks of size 2 or 10 respectively are included in the calculations.

4 The Convergence-Time Measure of Complexity

There is a growing body of evidence that the phenomenology of computation is related to phenomenology of phase transitions. In particular, if a system is to be capable of universal computation, it must be capable of exhibiting arbitrarily

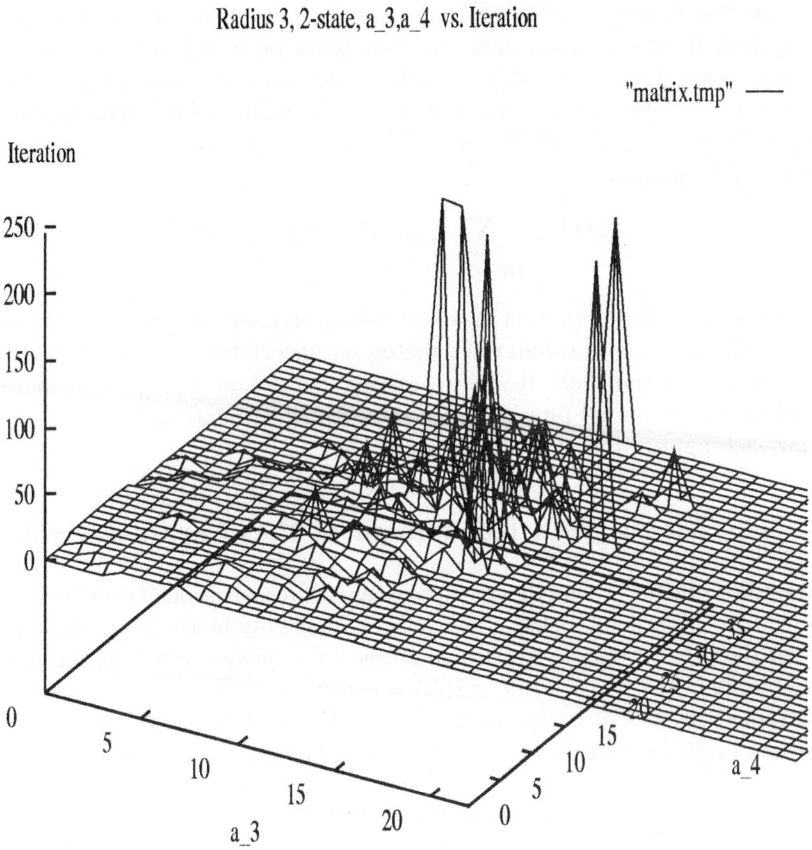

Figure 3. Convergence time parameterized by the mean field theory. Here the radius-3 rules of figure 2 are divided according to their mean-field parameter values. There are 8 mean field parameters for radius-3, 2-state rules. This figure shows only those rules for which 6 of the parameter values are 0, a_0, a_1, a_2 and a_5, a_6, a_7. The other two parameters, a_3 and a_4 are allowed to vary. The height of the surface in this figure corresponds to the convergence time averaged over all rules with the given mean-field parameter values. It is clear that for most settings of these parameters, convergence time is small. Convergence is large at a sharp peak located approximately at $a_3 = 15$ and $a_4 = 25$.

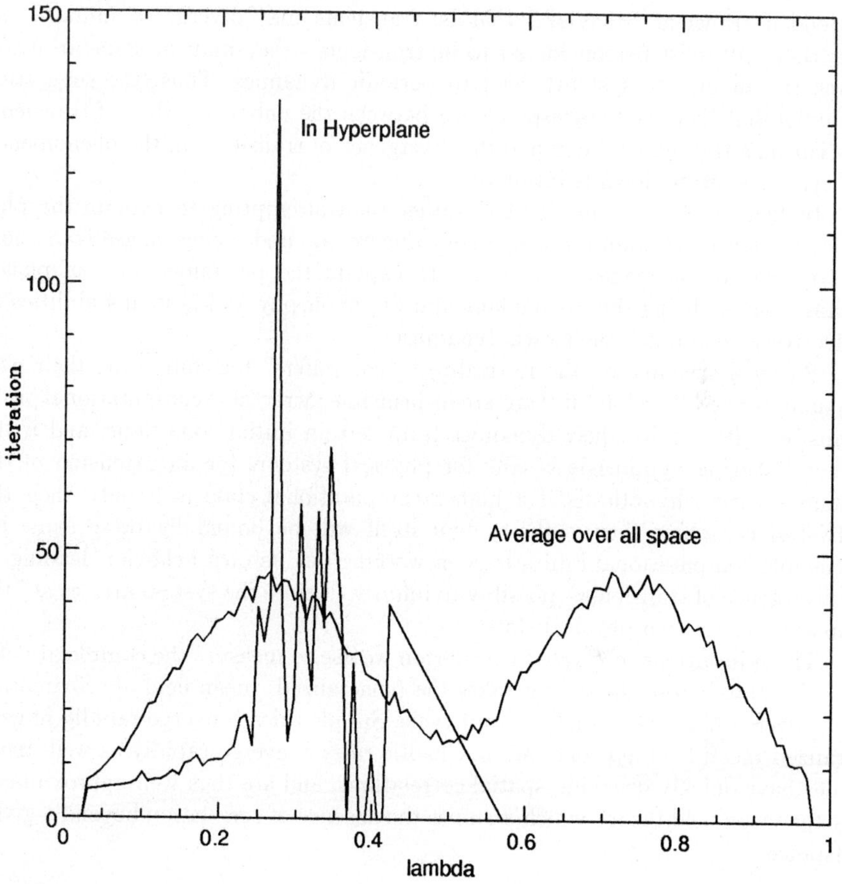

Figure 4. Projection of the mean-field parameter space onto the λ parameter space. In this figure convergence time of rules in the $\mathbf{a} = (0, 0, 0, a_3, a_4, 0, 0, 0)$ hyper-plane is compared with the convergence time of all rules in the sample, parameterized by λ. In the hyperplane, $\lambda = a_3 + a_4$. It is seen that the peak at $\lambda =$ approximately 0.27 is much sharper in the mean-field hyperplane.

extended transient behavior - that is, transients may diverge to infinity, and yet they still must be considered to be transients - they may, at some arbitrary time (in an infinite system), lead to periodic dynamics. Thus, the suggestion [Lan90] that there is a correspondence between the unboundability of transients in Turing's Halting Problem and the divergence of transients in the phenomenon of critical slowing down is intuitive.

In fact, as pointed in [Lan90], rather than attempting to explain the phenomenology of computation as being due to an underlying phase-transition, in many ways it makes more sense to explain the phenomenology of phase-transitions as being due to fundamental limits on physical system's abilities to effectively "compute" their own dynamics.

Physical systems are the most direct "computers" for computing their own dynamics - and therefore, if there are in-principle intractable computational problems associated with their dynamics from certain initial conditions, and if the Church-Turing hypothesis is true for physical systems (or an extension of the Church-Turing hypothesis if a higher computational class is found), then the physical computer that is the system itself will be bound by these same in-principle computational limitations on working out its own behavior, leading to a divergence of transients - possibly to infinity - before the system arrives at "the answer" for its own physical state.

These intuitions motivate the criterion we use to measure the complexity of a cellular automaton: the time it takes the (generalized) mean field approximation to converge to a fixed point or limit cycle. Simple rules converge rapidly in generalized mean-field approximation. Chaotic rules converge rapidly as well: these rules have quickly decaying spatial correlations, and are thus well-approximated by the mean field theory which assumes the absence of correlation beyond a given distance.

5 Results

Our results are shown in figures 2-4. In figure 2, the convergence time in 10th-order mean field theory is studied for rules of radius 2-4. Increasing line width indicates increasing radius. Many rules were generated at each possible value of λ, and the convergence-time averaged over the set of rules with a given λ value. For each radius a broad peak in convergence time is found. The location and width of these peaks is in close correspondence with previous empirical results using the λ parameter, e.g. those Li et al. [ea90]. Thus, our convergence-time criterion provides an analytic measure of complexity in quantitative agreement with empirical measures.

Figure 3 shows an example of how the mean field theory can be used to refine the λ parameter analysis. Here two mean field parameters, a_3, and a_4 for radius-3 rules are varied, while the other parameters are held constant. To generate this figure, all rules from figure 2 which happened to fall in this hyperplane in mean-field parameter space were selected. The height of the surface is this figure gives the average convergence time for rules with specified mean-field parameter values.

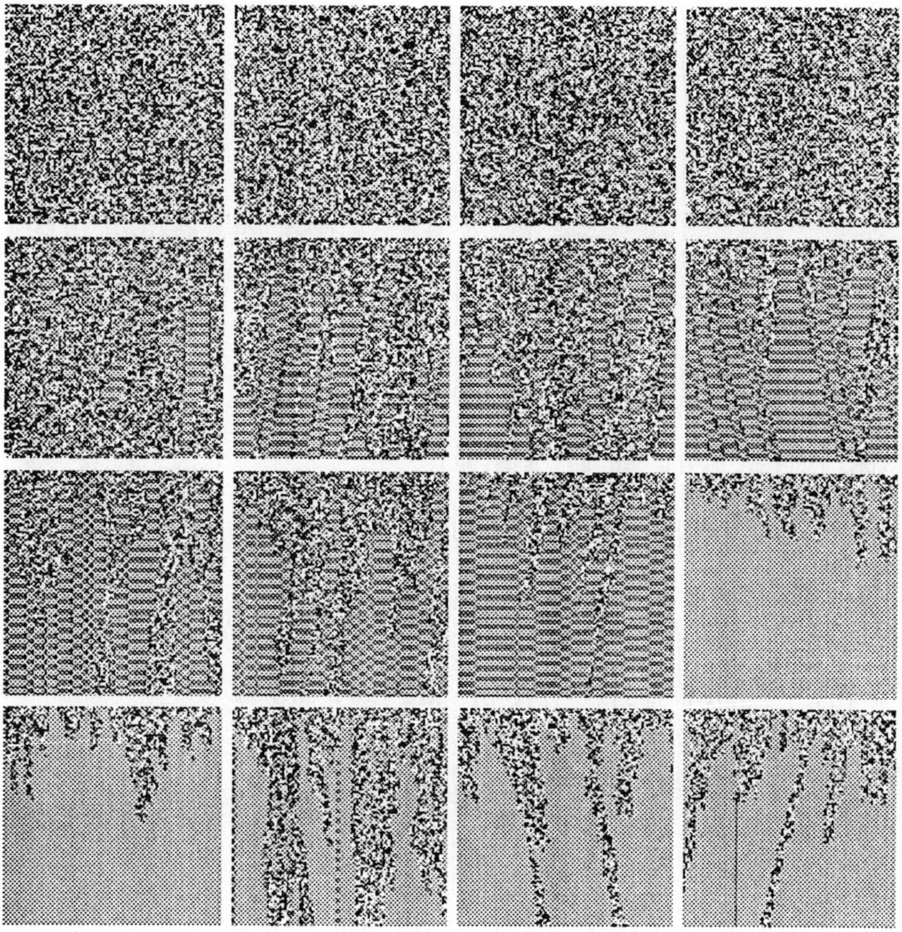

Figure 5. Results of a typical genetic algorithm experiment. In this experiment 8-state radius-1 rules were selected using a genetic algorithm. Fitness was proportional to the the number of iterations required for the order-2 mean field theory to converge. Using a simple genetic algorithm (population size =100), rules exhibiting qualitatively "complex" behavior were found. Each panel here shows the space-time pattern generated by the best-fit individual found at some generation of the genetic algorithm. (Every even-numbered generation up to 20, then generations 24,32,37,44, and 60) The initial population for this run was random. Similar results are obtained if the initial population has other statistics, e.g. consists of all-0 rule tables.

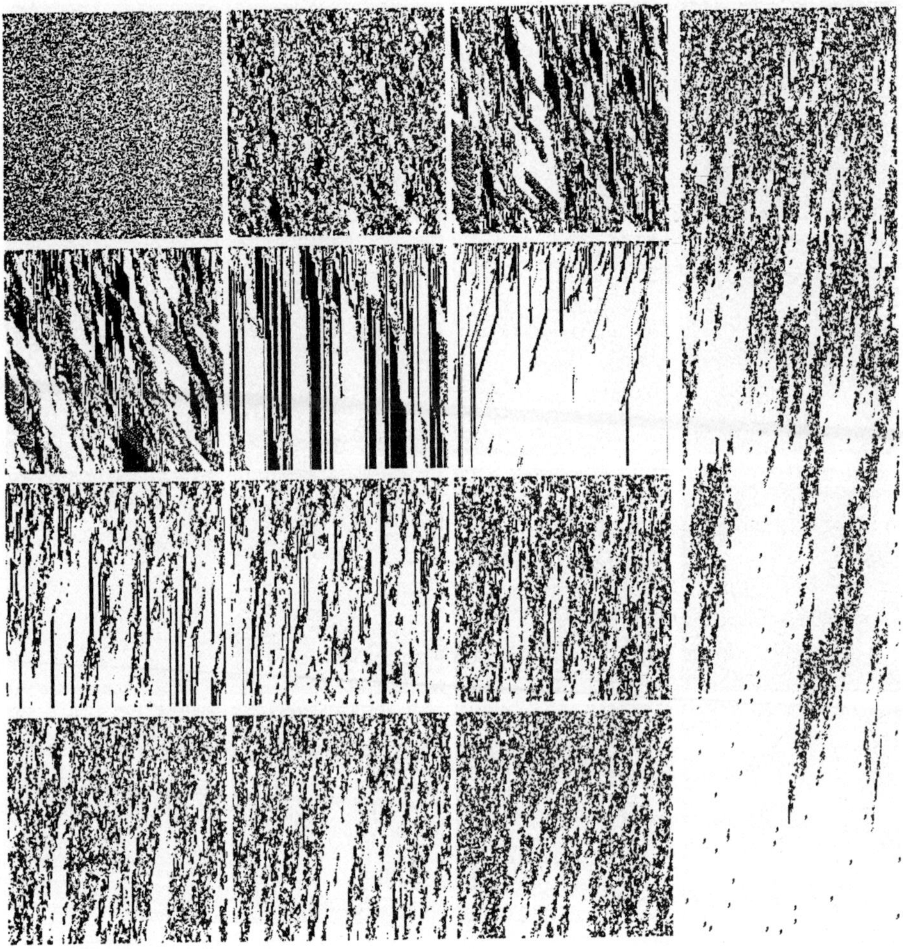

Figure 6. Evolution of Complex 2-D Hexagonal Rules. In this experiment 2-D, 2-state, hexagonal rules were evolved using the convergence-time fitness criterion. The convergence time was measured by tracking the density of 1's in 150 × 150 hexagonal grid. Population size 20. Shown on the left are space-time cuts through the pattern generated by the most-fit individual at several generations of a simple genetic algorithm. (Small panels, generations: 0, 2, 5, 17, 18, 33, 44, 64, 69, 73, 87, 91, 98, 106, 119. large panel: generation 175) The panel on the right shows the most-fit individual found in the experiment. These data were obtained by Patrice Simon (ESPCI, Paris).

Note that by restricting consideration to this hyperplane, we greatly sharpen the region in which rules of high convergence time are found.

This point is emphasized in figure 4. In this figure, the hyperplane is projected into the λ parameter space. For this hyperplane, $\lambda = a_3 + a_4$. It is seen that the sharp mean-field peak is in the same location as the λ parameterized peak. By increasing the order of the mean field theory, we should be able to locate with greater precision the region in parameter space in which high-convergence- time reside.

6 Genetic Evolution

The final two figure (figures 5 and 6) give an indication of how the convergence-time measure can be used to drive an evolutionary process. Figure 5 shows a selection of space-time patterns generated by cellular automata found during a run of the genetic algorithm. As the number of generations increases, rules with longer and longer convergence time (higher "fitness") are found. Each panel shows the most-fit individual at the corresponding generation of the genetic algorithm. These are 8-state, 1-D, nearest-neighbor CA. Convergence time is measured in 2nd-order mean field approximation.

Figure 6 shows the results of a genetic evolution similar to that of figure 5. Here, however, 2-state rules on a hexagonal lattice are used. Further, fitness is not measured using a mean-field approximation to the convergence time, but simply the empirical convergence time as measured by a Monte Carlo method. That the same kind of evolution toward increasing complex rules is found suggests that the convergence-time measure is useful for finding complex rules, independent of the number of states, number of dimensions, neighborhood size, etc. of the rules.

References

[DMC94] Rajarshi Das, Melanie Mitchell, and James P. Crutchfield. A genetic algorithm discovers particle-based computation in cellular automata. In *Proceedings of the Third Parallel Problem-Solving From Nature*, march 1994.

[ea90] W. Li et al. Transition phenomena in CA rule space. *Physica D*, 45:77, 1990.

[Gut90] Howard Gutowitz. Introduction (to cellular automata). *Physica D*, 45:vii, 1990.

[GVK87] H. A. Gutowitz, J. D. Victor, and B. W. Knight. Local structure theory for cellular automata. *Physica D*, 28:18–48, 1987.

[Hol75] J.H. Holland. *Adaptation in Natural and Artificial Systems*. University of Michigan Press, Ann Arbor, MI, 1975.

[KS93] Kunihiko Kaneko and Junji Suzuki. Evolution toward the edge of chaos in an imitation game. In Langton et al., editor, *Artificial Life III*. Adddison-Wesley, 1993.

[Lan86] Christopher G. Langton. Studying artificial life with cellular automata. *Physica D*, 22:120–149, 1986.

[Lan90] C. G. Langton. Computation at the edge of chaos. *Physica D*, 42, 1990.

[MCH94] Melanie Mitchell, James P. Crutchfield, and Peter T. Hraber. Dynamics, computation, and the edge of chaos: A re-examination. In G. Cowan, D. Pines, and D. Melzner, editors, *Complexity:Metaphors, Models, and Reality*, volume 19, Reading, MA, 1994. Santa Fe Institute Studies in the Sciences of Complexity, Proceedings, Addison-Wesley.

[MHC93] Melanie Mitchell, Peter T. Hraber, and James P. Crutchfield. Revisiting the edge of chaos: Evolving cellular automata to perform computations. *Complex Systems*, 7:89–130, 1993.

[Pac88] N. H. Packard. Adaptation toward the edge of chaos. In A. J. Mandell J. A. S. Kelso and M. F. Shlesinger, editors, *Dynamic Patterns in Complex Systems*, pages 293 – 301, Singapore, 1988. World Scientific.

[SK94] Junji Suzuki and Kunihiko Kaneko. Imitation games. *Physica D*, 75:328–342, 1994.

Escaping from the Cartesian Mind-Set: Heidegger and Artificial Life

Michael Wheeler

School of Cognitive and Computing Sciences,
University of Sussex, Brighton BN1 9QH, U.K.
Telephone:+44 1273 678524
Fax:+44 1273 671320
E-Mail:michaelw@cogs.susx.ac.uk

Abstract. In this paper, I propose a neo-Heideggerian framework for A-Life. Following an explanation of some key Heideggerian ideas, I endorse the view that persistent problems in orthodox cognitive science result from a commitment to a Cartesian subject-object divide. Heidegger rejects the primacy of the subject-object dichotomy; and I set about the task of showing how, by adopting a Heideggerian view, A-Life can avoid the problems that have plagued cognitive science. This requires that we extend the standard Heideggerian framework by introducing the notion of a *biological background*, a set of evolutionarily determined practices which structure the norms of animal worlds. I argue that optimality/ESS models in behavioural ecology provide a set of tools for identifying these norms, and, to secure this idea, I defend a form of adaptationism against enactivist worries. Finally, I show how A-Life can assist in the process of mapping out biological backgrounds, and how recent dynamical systems approaches in A-Life fit in with the neo-Heideggerian conceptual framework.

Keywords: adaptationism, Cartesian, enactivism, Heidegger, significance, subject-object dichotomy.

1 Introduction

The philosopher Martin Heidegger wouldn't have known an animat if one had bitten him. Nevertheless, in what follows, I shall argue that there are compelling reasons for researchers in Artificial Life to adopt a recognizably Heideggerian philosophical position.[1] However, it is not possible to take Heideggerian philosophy 'off the shelf', so to speak, and to plug it straight into A-Life, without first making some principled modifications to Heidegger's own ideas (details later). These modifications are significant; but they do not affect the fundamental nature of the philosophical perspective. Given this, perhaps the eventual framework on offer here should be described as 'neo-Heideggerian'. But whatever the most appropriate label is, there is no doubt that the position demands an outright rejection of what can be identified as *Cartesian* ways of thinking. Of course, one can give up on Cartesian assumptions without necessarily going on to become a fully-paid-up Heideggerian. But (as I shall argue), a recognizably Heideggerian approach might be a fruitful option to explore. And at least one explicitly anti-Cartesian philosophical framework, which is already influential in large areas of philosophical and practical work in A-Life, draws on Heideggerian ideas. Here I refer to the *enactive* approach to cognitive science, as proposed by Varela, Thompson, and Rosch [23].[2] However, there are differences between enactivism and the position developed here. I shall highlight these differences when and where they are relevant to the argument.

[1] Henceforth I shall abbreviate 'Artificial Life' to 'A-Life' and 'Artificial Intelligence' to 'AI'. For the purposes of this paper, I am concerned with A-Life as a distinctive attempt to explain evolutionary and adaptive systems, including (ultimately) the phenomena that we group together with labels such as intelligence, mind, and cognition.

[2] For explicit reference to the Heidegger-enactivism-connection, see [23, p.11 and p.149]. Of course, enactivism also builds on the prior work of Maturana and Varela [15].

2 On "Being and Time"

In this exegetical section, I shall attempt to explain some key ideas from division 1 of Heidegger's "Being and Time" [12]. It is, of course, impossible, in the space available here, to do any real justice to this complex material, and many important Heideggerian concepts (such as 'falling' and 'projection') will be ignored altogether. My goal is to do just enough to open up the Heideggerian space for readers unfamiliar with Heideggerian philosophy.

Heidegger uses the term 'Dasein' (literally 'being-there') to characterize human being (the way in which humans are). Dasein, as one kind of entity amongst others, embodies a special way of being, namely *self-interpretation.* The idea is that Dasein always *takes a stance on itself,* not through conscious decision, but through unreflective engagement in norm-laden ways of acting. The normativity here is secured by the fact that *being-with* other Daseins is fundamental to Dasein's being-in-the-world; that is, Dasein is inextricably embedded in culture and and society. For example, in 'being a father,' in a particular culture, at a particular time in history, one *should* behave in certain ways. We might sum this up by saying that Dasein exists in that it is socialized into certain ways of acting — the background practices that define cultures.

According to Heidegger, the fundamental character of the entities with which Dasein has dealings in everyday practical activity is that of *equipment.* We encounter entities as being *for*, for instance, throwing, writing, transportation, or working. Equipment possesses its own kind of being — *readiness-to-hand* — and our most 'primordial' (i.e., direct and revealing) understanding of equipment comes about not through some detached intellectual or theoretical consideration, but through our everyday manipulations of such entities.

The function of a piece of equipment is the role it plays in Dasein's normal way of life. To see the structure of this functionality, consider Heidegger's example of the hammer. The hammer is involved in an act of hammering; that hammering is involved in making something fast; the making something fast is involved in achieving protection against the weather; and finally, protection against the weather is to shelter Dasein. The totality of such *involvements* is a relational whole (a domain of significance, or, in Heidegger's terminology, a *world*), defined by reference to Dasein's skills and practices. And because these practices are the norm-laden, defining features of cultures and societies, it is with socialization into a culture that certain entities become significant as equipment. So the activity of hammering a fence post into the ground articulates one of the hammer's pre-existing socio-cultural significances.[3] Thus, in a hermeneutical process, the socio-cultural biases embodied in Dasein's practices are constitutive elements in the process through which Dasein makes sense of things. They are the 'prestructures' of understanding, a product and expression of a specific place in socio-cultural history. And because the significances at issue are conceptually prior to individual Dasein's activities, the process of bringing forth significance *presupposes* the referential whole of the totality of involvements.

Corresponding to everyday understanding as an activity-based phenomenon is Heidegger's concept of *circumspection.* This is identified as a feature of our typical everyday activity, for which divisions between subject and object have no ontological purchase. We are *absorbed* in our activities in such a way that we skilfully *cope* with developments and eventualities without being explicitly aware of events or objects. It is on the basis of this phenomenological observation that Heidegger rejects 'mental content' (understood in terms of a subject relating to objects) as a feature of our everyday situated activity. Everyday activities (e.g., hammering) display directedness; but this directedness is not the traditional directedness of the internal states (or the thoughts) of a subject towards objects. Rather, it is directedness in the sense that Dasein's activity articulates the referential links in the totality of involvements.

When equipment malfunctions or impinges on Dasein's activity in such a way that Dasein cannot immediately switch to a new mode of coping, that equipment exhibits a way of being that Heidegger calls *un-ready-to-hand.* Under these circumstances, there emerges a separation between Dasein and the piece of equipment. But still, the deliberate actions that Dasein takes to remedy

[3] It should be noted that to the extent that naturally-occurring objects (e.g., rocks) show up in a totality of involvements (i.e., when Dasein *uses* such entities as tools, missiles etc.), then they too are significant as equipment.

the situation take place in the context of Dasein's everyday concerns, so the entity has not yet been revealed as a decontextualized, value-free object. On other occasions, we deal with entities as objects of theoretical investigation by the physical sciences. Then the entities concerned have been completely decontextualized from everyday practice, and a new way of being has been revealed — the *present-at-hand* — in which entities show up as value-free objects, and a characterization of the encounters in terms of a subject-object dichotomy is entirely appropriate to the phenomena. In fact, as Heidegger points out, the investigation of entities by the physical sciences should be conceptualized as a process of *recontextualization*. Dasein's scientific practices are themselves skilled activities involving theory-construction, observation and experimentation. But these are practices of a sort different from the activities of absorbed coping or context-embedded problem-solving, and, accordingly, they reveal entities in a different mode of being. Heidegger captures this process by saying that the physical sciences have "the character of depriving the world of its worldhood in a definite way" [12, p.94].[4]

So Heidegger does not deny that there are properties of nature which explain why it is that you can make hammers out of wood or metal, but not out of steam or soup. These underlying causal properties constitute the explanatory domain of the physical sciences. The physical sciences can explain 'how it is' that the hammer does what it does, but not 'what it is' that the hammer is, what its significance is in our world; i.e., the present-at-hand cannot explain significance. Finding present-at-hand objects in nature requires precisely that we strip away the significance corresponding to our everyday practices. But the present-at-hand as a way of being for entities makes sense only with reference to Dasein's understanding of presence-at-hand as a way of being at all. Thus the realist assumptions embedded in the background practices of the physical sciences (standardly conceived) cannot be used to justify the sort of metaphysical realism that demands the philosophical primacy of the subject-object distinction in all explanatory projects.

This completes our lightning tour of the Heideggerian space. We must now turn our attention to why such ideas might be acutely relevant to A-Life.

3 Catching the Neo-Cartesians

Descartes' most famous conclusion was that mental states and processes constitute an essentially independent realm of operations, which causally interacts with the material world on an intermittent basis [8]. As a result of this relationship, which was never adequately explained by Descartes (or anyone else), the cognizer (the thing that thinks) was, in effect, severed from its body and its environment. It might be thought that a problematic divide between the mind on one side and the body/environment on the other couldn't possibly carry over to the received view in cognitive science, in which representations are ultimately physical states of physical systems, and computations are ultimately physical processes occurring in those same physical systems. But as well as enshrining substance-dualism, Descartes' account of perception and action embraces an *explanatory* divide between mind and world, according to which a subject relates to an 'external' world of objects via some interface. And orthodox cognitive science is Cartesian, precisely in the sense that it preserves this explanatory divide; i.e., it maintains that the subject-object dichotomy must be a fundamental feature of any account of perceptually guided situated activity. As we have seen, the primacy of the subject-object dichotomy is precisely what Heidegger denies.[5]

The Cartesian explanatory divide manifests itself in various ways. For instance, for Descartes, perception, thought, and action must be temporally distinct and theoretically separable. The cognizer's body is essentially a courier system, delivering sensory-input-messages to, and collecting motor-output-messages from, the thinking thing. And that picture of perception, thought and action bears an uncanny resemblance to the sense-plan-act cycle adopted in orthodox cognitive

[4] Most commentators on Heidegger interpret his comments concerning the present-at-hand as applying to the *natural* sciences, rather than the more restricted grouping of the physical sciences. For reasons which will become clear in section 6, I interpret his comments as applying to the more restricted grouping only.

[5] The conclusion that orthodox cognitive science is fundamentally Cartesian in one respect or another has also been reached in (at least) [9, 10, 11, 22, 23]. Although my own route to the conclusion follows a slightly different path, I claim only to be adding to an existing body of evidence (see also [24]).

science, in which a perception module delivers representations of the world to a central reasoning system (the thing that thinks) which, after some typically complex information processing operations, delivers representations of the desired actions to the action-mechanisms.[6]

This Cartesian decoupling of thought from sensory-motor control has some important, interlocking implications. First, some of the really hard problems in achieving real-time sensory-motor activity (e.g., those stemming from the uncertainty of sensors and motors) are conceptualized as implementation difficulties to be solved by the engineers, and not as issues which should be centre-stage in cognitive science. Situated roboticists in and around the A-Life movement have long stressed that real-time sensory-motor activity is not a peripheral issue in intelligent behaviour (e.g., [4]). Second, our theoretical models of cognitive systems can exhibit what we might call arbitrary temporality with respect to each other and to environmental events. That is, although the whole sequence of sense-plan-act events have to happen in the right order, and fast enough to get a job done, there are no constraints on how long each operation takes, or on how long the gaps between operations are. Moreover, this account of cognition includes no reference to intrinsic rhythms or rates of change, temporal phenomena which are part and parcel of biological or robot embodiment [22].

In addition to the foregoing worries, it seems that the notorious *frame problem* may well be an artefact of Cartesian assumptions. The frame problem rears its ugly head whenever orthodox cognitive science attempts to answer the following question: how, given particular sets of circumstances, goals and actions, does an agent (animal or robot) come to take account of those state-changes in its world (self-induced or otherwise) which really matter, whilst ignoring those which do not? For the Cartesian cognizer, two processes that would appear to be crucial to success here are the retrieval of just those internal representations which are appropriate to a given situation, and the updating — in appropriate ways — of just those representations which need to be updated.

In toy worlds (e.g., simulated blocks-worlds, or tightly constrained 'real' environments), AI-designers have been able to 'overpower' the frame problem, either by taking comprehensive and explicit account of the effects of every action or change, or by working on the assumption that nothing changes in a scenario unless it is explicitly said to change by some operator-definition. But, for agents embedded in uncertain, dynamic environments, the aforementioned tactics of 'make explicit' or 'ignore' are just not plausible options. And, as things stand, an appeal to specialized relevancy heuristics just won't do; the processing mechanisms would still have to solve the problem of accessing just those relevancy heuristics which are relevant in the current context. An infinite regress threatens. In the face of such problems, orthodox AI has remained firmly entrenched in abstracted sub-domains of intelligence, where the contexts defining what counts as relevant can be pre-defined, and the frame problem can be ignored.

What happens if we allow Heideggerian phenomenology to guide our scientific approach to the mechanisms underlying situated activity?[7] From a Heideggerian perspective, the mechanisms that we postulate to explain everyday situated activity cannot be bearers of representational content. This follows directly from the fact that, for Heidegger, ongoing coping is not a matter of subjects realizing intentional states about objects. That is, it is a mistake, so this story goes, to treat background capacities and skills as sharing a conceptual form with beliefs and desires. In everyday activity, I use the hammer in certain culturally-determined ways, without realizing any occurrent beliefs/representations (explicit or implicit, conscious or unconscious) that that is what I should do with hammers. But now if situated activity is not a matter of storing, retrieving and updating representations of the world in appropriate ways, then the frame problem itself just disappears.

[6] In many ways, this point can be seen as a restaging of Brooks' analysis of classical AI [4]. But it is worth stressing the fact that the account of situated activity adopted by mainstream connectionists preserves the most fundamental commitments of the classical vision [25].

[7] The thought that we might allow a phenomenological framework to constrain a scientific-explanatory framework will be intuitively unappealing to some theorists, on the grounds that ordinary experience often has no access to the nature of the mechanisms identified by science. I do not have the space to reply in depth to this complaint. But it seems that a *cognitive* science, perhaps unlike any other science, really does have a responsibility to confront the deliverances of experience (hence the clamour to resist eliminative materialism). In any case, the plausibility of the move I suggest rests, in the end, on the overall explanatory pay-off we get from adopting the neo-Heideggerian framework.

Of course, there has to be some explanation of the mechanisms that enable ongoing coping — it's not magic. Without such an explanation, the orthodox cognitive scientist is entitled to see the Heideggerian move as mere sleight of hand. I shall return to this challenge in section 9. For now the essential point is that the Heideggerian move may have the potential to liberate autonomous agent researchers from the frame problem, precisely through its denial that subject-object (representational) intentionality should be part of any explanation of situated activity.

4 Animal Worlds

To take further steps along this road, we need to confront a prima facie tension between A-Life and Heideggerian philosophy. Two hallmarks of most autonomous agent work in A-Life have been (i) an incremental approach to the synthesis and understanding of situated agents (start with simple but complete agents, and work up), and (ii) the thought that at least most of what we think of as 'higher' cognitive capacities will yield to the same explanatory principles as prove useful in the case of 'simpler' capacities. But, for Heidegger, non-human animals and infants are not socialized into cultures, so they are not Dasein. This means that non-human animals and infants (let alone animats) are not embedded in domains of significance. Thus it would seem that Heideggerian concerns are irrelevant to A-Life accounts of situated activity in non-human animals (or animats), and that A-Life's commitment to explanatory continuity will, in the end, stand in opposition to the anthropocentric tendencies in Heidegger's philosophy. Can we resolve these tensions?

Whenever Heidegger tackles the question of 'animality' — the way-of-being of non-human animals (henceforth just 'animals') — it seems hard to resist the conclusion that his characterization of the nature of worlds, and his account of what is required for world-possession, are both incomplete. The problem is that animals do not merely show up as meaningful entities in Dasein's everyday world (e.g., as pets, pests, or, for some people, dinner); animals also co-ordinate their own activities with respect to the objects, events, and situations with which they deal on an everyday basis, in ways which are difficult to dismiss as devoid of significance. As Heidegger admits, the lizard has *some sort* of relationship with the rock on which it basks in the sun, and, whatever that relationship is, it is not of the same genre as the relationship that one rock bears towards a second rock on top of which *it* 'sits' (see [7, pp.52-4]). So to get a real grip on what it takes to be embedded in a domain of significance, it seems that we need to understand animality.

But can a Heideggerian framework supply us with the right conceptual tools for the job? Since the process of revealing entities as present-at-hand requires the capacity to do something equivalent to physical science or philosophy, it would be rather bizarre to think that animals ever deal with entities as present-at-hand. So any account taking the subject-object dichotomy as primary is not appropriate for characterizing the ways in which animals are embedded in their ecological worlds. However, it seems much more plausible to suppose that at least some animal worlds have analogues of the ready-to-hand and non-intentional directedness. For example, wild primates use twigs and branches in threat displays, leaves in bodily care, and lumps of wood or stone in food preparation [5]. These seem to be obvious examples of naturally-occurring entities used as equipment, where the use of such equipment articulates significant aspects of an animal's world. But this neo-Heideggerian picture of animality can make sense only if there exists a background of significance which is not a reflection of socio-cultural embeddedness.

5 Enactivism and the Biological Background

The painfully obvious move is to appeal to evolution as a source of significance. But can we reconcile Heidegger's hermeneutical account of significance with the naturalism of an evolutionary account? In other words, is there a *biological background* (a body of evolutionarily-determined capacities and abilities), which functions in ways analogous to the socio-cultural background?

What appears to be an appeal to this sort of phenomenon is made in the enactivism of Varela et al. [23]. For example, cognition is conceptualized as *embodied action*, where the term 'embodied' means (i) that cognition depends on the experiences brought forth through particular sets of sensory-motor systems, and (ii) that individual sets of sensory-motor capacities are "themselves embedded in a more encompassing biological, psychological, and cultural context" (pp.172-3). And human scientists are held to perform acts of theoretical reflection "out of a given background (in the Heideggerian sense) of biological, social, and cultural beliefs and practices" (p.11).

Now, as argued above, the idea that there could be such a thing as a background of biological practices, *in the Heideggerian sense,* is far from obvious. Heidegger himself certainly didn't recognize the existence of one. So some work has to be done to explain how this is supposed to work. But it seems that when the term 'background' figures explicitly in certain key enactive explanations of how ongoing situated activity brings forth meaning, it is *not* the Heideggerian sense of the term which Varela et al. have in mind.

For example, consider *Bittorio* [23, pp.89-91 and pp.151-7]. *Bittorio* is a ring of eighty elementary processing units (cellular automata). The experimenter stipulates (A) the possible states into which each unit is able to move (in this case there are just two possible states, 0 and 1), (B) the rules governing the way each of those units change state as a result of its local interactions with neighbouring units (here this is determined by a pair of Boolean functions for each unit), and (C) the way in which the network is coupled to a random milieu (i.e., the way in which individual cells change state if they encounter a randomly specified 0 or 1 external to the ring). A and B in conjunction define the system's *operational closure* and C defines a form of *structural coupling.*

For certain sets of internal dynamics, the combination of closure and coupling resulted in suggestive evolutions in the spatio-temporal configuration of the network. For example, in one case, an encounter with a 0 or a 1 at a particular cell-location would perturb the spatial periodicities of the ring from one stable pattern to another. But a second perturbation at the same cell would restore the first configuration. The upshot of this was that, over time, finite, odd sequences of such perturbations resulted in changes in configurations, whereas finite, even sequences effectively left the system's global state unchanged. So *Bittorio* had become a 'finite odd sequence recognizer', despite the fact that it had not been supplied with any program enabling it to recognize such sequences. As Varela et al. put it, the system's activity performs a minimal kind of 'interpretation', meaning that it "selects or brings forth a domain of significance out of the background of its random milieu" (p.156). *Bittorio* is offered as a simple example of an operationally closed system bringing forth "significance from a background" (p.156). But although there are Heideggerian echoes here, the random mileu to which Bittorio is coupled cannot be a background in the Heideggerian sense of the term. Dasein's activities bring forth significance from a background of practice, in the sense that those activities articulate aspects of that background. A Heideggerian background is structured and normative. A random mileu is neither of these things.

Now let's move from relatively simple cellular automata to embodied, situated, sensory-motor activity, and consider Varela et al.'s interpretation of Skarda and Freeman's work on the neural dynamics underlying odor recognition and discrimination in rabbits [19]. The conclusion of Skarda and Freeman's study is that it is the presence of chaotic dynamics in the olfactory bulb which enables the system to add new odors to its repertoire of learned odors. A chaotic background state prevents convergence to previously learned neural response-states and allows the generation of new states. Varela et al. (p.175) describe the situation as follows: "emergent patterns of activity seem to be created out of a background of incoherent or chaotic activity into a coherent attractor ... smell is not a passive mapping of external features but a creative form of enacting significance on the basis of the animal's embodied history." Notice that although, in terms of spatial location, the background here is 'agent-internal', the actual form of the process is consistent with the *Bittorio* model. The agent's ongoing activity brings forth significance from an essentially meaningless background. Here we are dealing with a biological system (the rabbit's olfactory bulb), so it is conceivable that this is, for the enactivists, a case of a biological background. If so, then, as in the *Bittorio* example, this use of the term 'background' is not Heideggerian.

Let me make it clear that I am not attempting to outlaw all non-Heideggerian uses of the term 'background'. The point of these examples from enactivism is merely to show that the phenomenon that I wish to describe — the neo-Heideggerian biological background — has certain essential features which are not addressed by these specific enactive accounts of the bringing forth of significance. What we require for a biological background is a set of evolutionarily-determined skills, activities and practices. These practices structure a normative domain of holistic significance, which is then presupposed by individual meaning-events. So, taking the bit between the teeth, I propose that to map out a specific biological background, we should appeal to the biological sub-discipline of behavioural ecology.

6 Natural Selection and Normativity

Behavioural ecology is concerned with questions of function, rather than questions of causal mechanism; i.e., it is an attempt to understand *why* (as opposed to how) evolved organisms behave the way they do [14]. A common theoretical tool in behavioural ecology is the *optimality model.* To construct an optimality model, one has to assume that Darwinian natural selection is the key process in evolution. The modeller then decides on (i) a 'currency' for the model (e.g., in foraging models, possible currencies could be rate of food intake, food-finding efficiency, or risk of starvation), and (ii) the set of constraints (bodily and environmental) operating on the set of possible phenotypic behaviours. (The 'unconstrained' phenotype would be immortal, unassailable by predators, permanently reproductive and so on.) For the model to have real explanatory bite, at least some of the constraints assumed in the model should be supported by independent empirical evidence. The next step is to produce quantitative assessments of the costs and benefits, to an individual, of a specific behaviour pattern, and to combine those assessments with the prediction that, in an ideal situation, natural selection will have produced adaptively fit agents, so an individual will act in a way that tends to maximize its own net benefit. The specific behavioural predictions generated by the model are then tested against empirical data.

In multi-agent scenarios, what counts as the best strategy for an individual will depend on the strategies adopted by the other agents in the relevant eco-system. So a strategy is considered optimal when it is *evolutionarily stable*; i.e., when the ecological situation is such that if the strategy is adopted by most members of a population, then that population cannot be invaded by a mutant playing a rare alternative strategy [16]. So fitness is maximized only in the sense that individuals not adopting the ESS do worse. This can be thought of as a form of "competitive optimization" [18]. So my use of the term 'optimality models' can be taken to include ESS models.

To see how an optimality model might be used to map out the normative constraints on biological domains of significance, consider a study of nesting behaviour in digger wasps by Brockman, Grafen, and Dawkins [3]. Female digger wasps either (i) dig burrows in which to lay their eggs, or (ii) enter existing burrows and, if necessary, fight the sitting tenants for the right of ownership. The fitnesses of these strategies are frequency dependent, but it always pays to adopt the rarer strategy (i.e., dig if most individuals are entering, and vice versa). From field data, Brockman et al. calculated the proportions with which, at equilibrium, each of the two behaviours should occur.

If we focus on the static case of a population at equilibrium, the expected frequency distribution could be realized in many different ways. For example, it may be that x% of wasps always dig, whilst $100 - x$% always enter, in which case each individual in the population is adopting one or the other of two *pure* strategies. Alternatively, it may be that each and every individual in the population adopts the same *mixed strategy,* digging x% of the time, and entering $100 - x$% of the time. Here each individual in the population is adopting the ESS. Finally, individuals may be adopting different mixed strategies. In spite of these theoretical possibilities, only one of the arrangements is to be expected, because, on the way to equilibrium, natural selection will favour the ESS over other strategies; so individuals are expected to adopt the ESS. The significance of this conclusion is that the model clearly makes a prediction about what each individual *should* do, given that the ecological costs and benefits have been identified correctly.

To test the model, Brockman et al. retrieved empirical data from two populations of wasps. It was discovered that digging/entering behaviour conformed to the model in one population, but not in the other. So, if the ecological niches of the two populations were (in all relevant respects) equivalent, this result would tell us that the population diverging from the non-optimal behaviour is unstable, in the sense that that population could be invaded by mutants adopting the mixed strategy ESS. But Brockman et al. (pp.493-4) argue that the data are not consistent with the hypothesis that the population is behaving in accordance with the model, but away from the equilibrium. It is likely that the divergence occurs because the ecological situations at the two sites are different. This would mean not that the original model is entirely useless, but that the relevant ecological variations would have to be identified, and incorporated into a revised model.

It is significant that whilst behavioural ecology embodies a *scientific-theoretical* approach to animal behaviour, it does not treat animals as value free objects (i.e., as present-at-hand). Thus this area of biology (at least) cannot be treated as equivalent to the physical sciences.[8] In fact, it seems that for the discipline of behavioural ecology to make any sense, the capacity of animals to open up domains of significance has to be assumed. Behavioural ecologists treat animals as autonomous agents who adopt strategies with fitness consequences for both the 'strategy adopting' animal itself, and the other animals with whom it interacts. And although the activity of behavioural ecology can be characterized as the taking of a stance by us (by Dasein) towards animals, the fact is that once we adopt that stance, what we find is not 'up to us'. What we find depends on how the predictions of the model match the observed behaviour. So, in the case of well-constructed, detailed models, we can have confidence that the behavioural norms we postulate are actual features of animal worlds, and not arbitrary projections from our world. These norms can be thought of as the evolutionary prestructures of animal understanding. Of course, humans are just as embedded in evolutionary history as are non-human animals; so we should expect our activity to be codified not only by cultural norms, but also by evolutionary norms. Indeed, the significances articulated in Dasein's building of a shelter (Heidegger's own example, see section 2) are most naturally cashed out by an appeal to evolutionarily-determined, survival-based norms.

7 In Defence of Adaptationism

The proposal to use optimality models as theoretical tools to map out normative biological backgrounds will no doubt have raised the hackles of those who are sceptical of adaptationist accounts of evolution. In the context of this paper, it is significant that the critics of adaptationism include the enactivists. This important debate deserves much more space than I can give it here, but one line of argument, as presented by Varela et al. [23, pp.180-214], revolves around the existence of evolutionary processes such as linkage and pleiotropy (the multiple effects of interdependent genes), morphogenesis, and random genetic drift. The sophisticated adaptationist will acknowledge that these (and other) factors may all play a role in evolution, but she will argue that the central engine of evolutionary change is still Darwinian selection (natural or sexual). So the use of optimality models does not imply that there are no evolutionary processes other than selection. But it does suggest that most phenotypic traits in most populations of organisms can be explained by natural or sexual selection (i.e., non-selective evolutionary processes can, in most cases, be ignored). By contrast, the critic argues that, when taken together, these factors cast doubt on the adaptationist program, and suggests that we need to 'change the logical geography' of our theories. The first recommendation is that we move to a proscriptive rather than a prescriptive logic; i.e., we see evolution as selecting-against unfit behaviours, rather than selecting-for increasing fitness. So selection is not a process that "guides and instructs in the task of improving fitness" [23, p.195]. Second, evolution should be conceptualized not as a process of optimizing, but as a process of satisficing (finding satisfactory rather than optimal solutions).

How should the adaptationist respond? As Varela et al. point out, adaptationists are not committed to the absurd thesis that present-day animals are optimal solutions to some set of adaptive problems posed in the primeval swamp. So the notions of adaptation and fitness need to be relativized to individual ecological niches. This is uncontentious. But the real challenge still remains. The critics' point is that selection is not a goal-driven process of optimization, 'guiding' organisms towards some peak in a pre-defined fitness landscape. I think the right response to this is to say that, indeed, no adaptationist should think of the process of selection in these terms.

We can say that a trait has been *selected-for* when we have discovered that it has conferred a fitness advantage on those individuals who possess it. That is, in a given population, in a given ecological niche, those individuals who have evolved that trait (by whatever mechanisms) end up with better survival and reproductive prospects than those who have not. Given the assumption that resources are limited, individuals with the new trait will take over the population. Notice that it is the subsequent evolutionary trajectory of the population, once the new trait has appeared, that allows us to describe the process as selection-for.

[8] For a discussion of the thought that large areas of biology might be hermeneutical disciplines, see [2].

However, we can redescribe 'the same' event as follows: Through an evolutionary process (such as random genetic drift), a new trait comes into being. Given the assumption of limited resources, individuals are in competition; so, if the newly-appeared trait enables the individuals who possess it to obtain an increased share of the resources, then at least some of the traits that previously satisficed will no longer be viable. Hence individuals with the new trait will take over the population. So certain pre-existing traits have been *selected-against*. From what has been said so far, there seems to be no reason (apart from taste) to tell one of these stories rather than the other. But, if that is right, then the notions of selection-against and satisficing seem to be concepts to which a sophisticated adaptationism can happily subscribe. Selection-for implies selection-against and vice versa; and many important evolutionary changes will mean that ongoing satisficing will result in the appearance of increasingly fit organisms.

What implications does this have for our proposed deployment of optimality models? Such models do not assume, nor do they attempt to demonstrate, that animal behaviour is always optimal [18]. However, as long as the ecological factors assumed by a particular optimality model — the various constraints according to which the model was constructed — remain constant, then that model will allow us to identify the evolutionary consequences of occurrent sub-optimal behaviour; i.e., populations adopting that behaviour will be unstable against invasion by fitter mutants, *if such mutants ever evolve*. In other words, if, by whatever evolutionary mechanisms you care to mention, fitter mutants evolve, existing individuals (or lineages) will — in order to continue satisficing — be forced to develop fitter strategies, or they will tend to be selected against. To be maximally resistant to the threat of invasion by fitter strategies (i.e., to have the best chances of maintaining the integrity or stability of the individual or the group through time), animals *should* adopt the optimal or evolutionarily stable strategy. And whilst optimality models are essentially idealized systems, it is worth making the point that field biologists are widely impressed by just how adaptively fit animal behaviour actually is.

Of course, if an animal's ecological niche changes in ways which render the various constraints built into an optimality model inappropriate, then the previously identified norms no longer apply. Coevolutionary scenarios make this sort of change entirely possible, even likely. But then it seems that what has happened is that the animal's biological background has shifted in significant ways.

8 The Role of A-Life 1: Synthetic Ecologies

How can A-Life help with the philosophically potent project of mapping out biological backgrounds? One central theme in A-Life research involves the construction of synthetic ecological contexts, in which the behaviour of animats can be observed. Such simulations afford the theorist considerable advantages in experimental control and repeatability. When the field biologist is investigating behaviour in natural ecologies, her control over what she believes to be the key factors affecting that behaviour is, in the vast majority of cases, minimal or, at best, partial. By contrast, in A-Life simulations, the parameters underlying the nature of the 'physical' environment (e.g., the rate at which food is replenished) and those driving basic aspects of the animats' behaviour (e.g., the rate at which the animats' 'metabolisms' require food-intake) are under direct experimental control. And, of course, a simulation can be run over and over again from similar initial conditions, in order to test the robustness of hypotheses. Where A-Life simulations are designed to be sensitive to existing biological knowledge, they are, in effect, potentially powerful ways of doing theoretical biology (cf. [17]). And if experiments involving animats in synthetic ecologies are performed with the specific goal of investigating the functional aspects of animal behaviour (e.g., [6]), then the A-Lifers concerned are investigating features of biological backgrounds.

An exciting area of debate here involves empirical A-Life research which concludes that certain significant features of animal worlds are not to be explained by adaptive function at all. Such work would no doubt please the critics of adaptationism. For example, te Boekhorst and Hogeweg [21] use a synthetic eco-system, based on an existing natural habitat, to investigate the formation of travel parties in orang-utans. Orang-utans are relatively solitary animals. But they sometimes aggregate and travel together. Te Boekhorst and Hogeweg argue that, for an optimality approach, there is something of a dilemma here. Food trees will be emptied much sooner when visited by parties of orang-utans, meaning that the animals have to travel further to obtain enough food. These costs have been cited as the reasons why orang-utans tend to be rather solitary animals. But then why should travel parties form at all? The results from te Boekhorst and Hogeweg's simulation suggest that travel parties are to be explained as emergent properties of simple individual foraging behaviours in interaction with the structure of the environment, and not as social groups which are actively created and maintained by the animals to secure some fitness advantage.

This may all be true. But what would follow? Even if orang-utan travel parties are not *actively* created and maintained by the creatures, their formation may still have positive adaptive consequences in the creature's world as a whole (although, perhaps, not with regard to foraging). There seems no reason why emergent behaviours have to be function-less, or why a pattern of behaviour which is a side-effect of a different activity cannot have adaptive consequences in other areas of the animal's world. At the very least, the formation of travel parties cannot be seriously disadvantageous (with respect to the orang-utan's entire ecological world), or, as everyone admits, they would have been selected against. The view on offer here does commit me to saying that if a behaviour really does have no adaptive consequences at all in a creature's world, then that behaviour plays no role in structuring the creature's domain of significance. But, to me at least, that does not seem wildly unreasonable.

9 The Role of A-Life 2: the Liberation of Mechanism

I shall end this paper by suggesting that recent moves in A-Life to explain the mechanisms underlying autonomous agent behaviour using dynamical systems theory are consonant with a neo-Heideggerian conceptual framework.[9]

Husbands et al. [13] place a (simulated) visually-guided robot in a (simulated) circular arena with a black wall and a white floor and ceiling. A genetic algorithm is used to develop a control system made up of sensors (including vision), motors, and non-standard artificial neural networks. The networks feature continuous-time processing, real-valued time delays on the connections between the units, noise, and connectivity which is not only both directionally unrestricted and highly recurrent, but also not subject to symmetry constraints. From any randomly chosen position in the arena, the robot's task, as encoded in the genetic algorithm's evaluation function, is to reach the centre of the arena as soon as possible and then to stay there.

Let's focus on one successfully evolved agent. The robot has two 'eyes' (each with a single pixel), but only ever uses one eye at a time. Minimal monocular vision is no basis for building representations of the world. So storage and retrieval metaphors are unsuitable for this style of evolved solution. The robot succeeds in its task by exploiting its own movements to create variations in light inputs. Thus the ways in which the sensory-motor mechanisms are coupled to the world enable the robot to complete the task, given its ongoing activity as a whole agent.

To gain an understanding of the robot's activity, an analysis of the dynamics present in the significant visuo-motor pathways in the neural network is combined with a two-dimensional state space representing (to the observer) the robot's visual world (the visual signals which would be received at different positions in the world). The result is a phase portrait predicting the way in which the robot will tend to move through its visuo-motor space. In an environment of wall-height 15 (the environment in which the control system was evolved) the phase portrait features a single point attractor. This attractor corresponds to a very low radius circle about the centre of the world. The whole state space is, in effect, a basin of attraction for this attractor. So the model predicts that the robot will *always* succeed at its task, a prediction which was borne out by empirical demonstration. The next stage was to investigate the adaptiveness of the control system by analysing the behaviour of the robot in an arena with wall-height 5 (i.e., in an environment for which the control dynamics were not specifically evolved). The change in wall-height means a

[9] Here I concentrate on an example by Husbands, Harvey, and Cliff [13]. I have used this example before, to make related points [25]. Despite its simplicity, it is a powerful source of ideas. For other examples of dynamical systems theory at work in A-Life, see [1, 20]. I do not claim that any of these theorists necessarily share my philosophical orientation. For a more detailed discussion of the issues raised in this section, see [24].

change in the structure of the robot's visuo-motor state space. The same process of analysis now yields a phase portrait featuring two point attractors, both corresponding to successful behaviours. Once again the model was confirmed by empirical demonstration.

For our purposes, the key feature of this analysis is that the visuo-motor space does not conform to the principles of any subject-object divide. The space is not supposed to exist on the external side of some interface, waiting for internal recovery by the robot. Neither is it supposed to exist as a robot-internal structure. It is an observer's characterization of an *agent-environment system*, a pattern of situated activity. It is true that to construct the explanatory space, it was necessary for Husbands et al. to gain an understanding of the internal dynamics of the neural network (i.e., the aforementioned dynamics in the significant visuo-motor pathways). But the internal dynamics alone are not an adequate explanation of the embedded behaviour. They are merely one stage in the process of analysis. In the complete picture, the mechanisms underpinning the explanatory space are spread out over the neural network, the sensory-motor systems, *and* the environment. This is a way of thinking about autonomous agent behaviour which does not assume the Cartesian subject-object dichotomy as primary. One might gesture at the proposed connection here in the following thought: the conceptual hyphens in Heidegger's concept of 'being-in-the-world' are partners of the empirical hyphen in the notion of an agent-environment system.

10 Conclusions

I have endeavoured to develop and defend a Heideggerian approach to A-Life. Heideggerian phenomenology suggests that the subject-object dichotomy is appropriate only when an agent is 'decoupled' from ongoing coping. Extending this principle to our scientific explanations of situated activity requires that we explain, *without recourse to representational intentionality,* (i) significance in the worlds of humans and other animals, and (ii) the mechanisms underlying situated activity.

My claim has been that an extension to the standard Heideggerian framework, in the form of the biological background, can meet the first of these demands. Although Heidegger was wrong to think that 'being embedded in a culture' is the only source of significance, the fundamental process through which ongoing activity brings forth significance from a background of capacities and practices carries over from the strictly Heideggerian socio-cultural background to the neo-Heideggerian biological background. A-Life has a key role to play in the project of understanding the biological backgrounds that define different animal worlds. And A-Lifers are already beginning to investigate alternative ways of explaining mechanism. This is still a relatively unexplored space, and, without a doubt, there will be difficult problems to solve as the behavioural capacities of animats become increasingly complex. But no one ever said that escaping from the Cartesian mind-set was going to be easy.

Acknowledgements

This work was supported by British Academy award no. 92/1693. Many thanks to Maggie Boden for helpful comments on an earlier version of this paper.

References

1. R. D. Beer. Computational and dynamical languages for autonomous agents. Forthcoming in van Gelder, T. and Port, R. (Eds.), *Mind as Motion.* M.I.T. Press.
2. M.A. Boden. The case for a cognitive biology. In *Minds and Mechanisms: Philosophical Psychology and Computational Models*, chapter 4, pages 89–112. The Harvester Press, Brighton, 1981.
3. J. Brockman, A. Grafen, and R. Dawkins. Evolutionarily stable nesting strategy in a digger wasp. *Journal of Theoretical Biology*, 77:473–496, 1979.
4. R. A. Brooks. Intelligence without reason. In *Proceedings of the Twelth International Joint Conference on Artificial Intelligence*, pages 569–95, San Mateo, California, 1991. Morgan Kauffman.
5. D. L. Cheney and R. M. Seyfarth. *How Monkeys See The World — Inside the Mind of Another Species.* University of Chicago Press, Chicago and London, 1990.

6. P. de Bourcier and M. Wheeler. Signalling and territorial aggression: An investigation by means of synthetic behavioural ecology. In D. Cliff, P. Husbands, J.-A. Meyer, and S. W. Wilson, editors, *From Animals to Animats 3: Proceedings of the Third International Conference on Simulation of Adaptive Behavior*, pages 463–72, Cambridge, Massachusetts, 1994. M.I.T. Press / Bradford Books.
7. J. Derrida. *Of Spirit — Heidegger and The Question.* University of Chicago Press, Chicago and London, 1989. Translated by G. Bennington and R. Bowlby.
8. R. Descartes. Meditations on the first philosophy in which the existence of God and the real distinction between the soul and the body of man are demonstrated. In *Discourse on Method and the Meditations.* Penguin, London, 1641. Translated by F.E. Sutcliffe in 1968.
9. H. L. Dreyfus. *Being-in-the-World: A Commentary on Heidegger's Being and Time, Divison 1.* M.I.T. Press, Cambridge, Massachusetts and London, England, 1991.
10. I. Harvey. Untimed and misrepresented: Connectionism and the computer metaphor. Cognitive Science Research Paper 245, University of Sussex, 1992.
11. J. Haugeland. *Artificial Intelligence — The Very Idea.* M.I.T. Press / Bradford Books, Cambridge, Massachusetts and London, England, 1985.
12. M. Heidegger. *Being and Time.* Basil Blackwell, Oxford, England, 1926. Translated by J. Macquarrie and E. Robinson in 1962.
13. P. Husbands, I. Harvey, and D. Cliff. Circle in the round: State space attractors for evolved sighted robots. Forthcoming in *Robotics and Autonomous Systems.*
14. J. R. Krebs and N. B. Davies. *An Introduction to Behavioural Ecology.* Blackwell Scientific, Oxford, 2nd edition, 1987.
15. H. Maturana and F. J. Varela. *The Tree of Knowledge: The Biological Roots of Human Understanding.* New Science Library, Boston, 1987.
16. J. Maynard Smith. *Evolution and the Theory of Games.* Cambridge University Press, Cambridge, 1982.
17. G.F. Miller. Artificial life as theoretical biology. Forthcoming in Boden, M.A., (ed.) *The Philosophy of Artificial Life.* Oxford University Press.
18. G. A. Parker and J. Maynard Smith. Optimality theory in evolutionary biology. *Nature*, 348:27–33, 1990.
19. C. A. Skarda and W. J. Freeman. How brains make chaos in order to make sense of the world. *Behavioral and Brain Sciences*, 10:161–195, 1987.
20. T. Smithers. What the dynamics of adaptive behaviour and cognition might look like in agent-environment interaction systems. Revised version of a paper presented at the workshop 'On the Role of Dynamics and Representation in Adaptive Behaviour and Cognition', December 1994, San Sebastian, Spain.
21. I.J.A. te Boekhorst and P. Hogeweg. Effects of tree size on travelband formation in orang-utans: Data analysis suggested by a model study. In R. Brooks and P. Maes, editors, *Proceedings of Artificial Life IV*, pages 119–129, Cambridge, Massachusetts, 1994. M.I.T. Press.
22. T. J. van Gelder. What might cognition be if not computation? Technical Report 75, Indiana University, Cognitive Sciences, 1992.
23. F. J. Varela, E. Thompson, and E. Rosch. *The Embodied Mind: Cognitive Science and Human Experience.* M.I.T. Press, Cambridge, Massachusetts, and London, England, 1991.
24. M. Wheeler. From robots to Rothko: the bringing forth of worlds. Forthcoming in Boden, M.A., (ed.) *The Philosophy of Artificial Life*, Oxford University Press.
25. M. Wheeler. From activation to activity: Representation, computation, and the dynamics of neural network control systems. *A.I.S.B. Quarterly*, 87:36–42, 1994.

Semantic Closure: A Guiding Notion to Ground Artificial Life

Jon Umerez

Dept. of Systems Science & Industrial Engineering
State University of New York (SUNY) at Binghamton
P.O. Box 6.000, Binghamton, NY 13902-6000, USA
E-mail: jumerez@bingvaxa.cc.binghamton.edu

Abstract.- The lack within AL of an agreed-upon notion of life and of a set of criteria for identifying life is considered. I propound a reflection upon the codified nature of the organization of living beings. The necessity of a guiding notion based on the coding is defended. After sketching some properties of the genetic code I proceed to consider the issue of functionalism as strategy for AL. Several distinctions ranging from plain multiple realizability to total implementation independence are made, arguing that the different claims should not be confused. The consideration of the semantic and intrinsically meaningful nature of the code leads to discuss the "symbol grounding" in AL. I suggest the principle of *Semantic Closure* as a candidate for confronting both problems inasmuch as it can be considered an accurate guiding notion to semantically ground Artificial Life.

1 Introduction

I will address, in essence, two problems which may be different in character but might turn out to have closely related answers. These two problems are:

* the lack of agreement upon a useful notion of life for Artificial Life (AL), not necessarily a definition but at least a pragmatic "guiding" notion of it; and

* the absence of any real method or criterion in AL for "grounding" the interpretation of computer patterns and processes or robotic behaviors, beyond subjective and controversial ones.

A generalized and explicitly semiotic consideration of the codified nature of living organization (the relation between the dynamical and the informational aspects of actual living systems) can help to advance forward in both of these problems. Therefore, the main purpose of this contribution is to bring to AL the discussion around the relevance of the genetic code in biological life (BL) and its potential relevance for AL. Questions to begin with could be: is the genetic code necessary? Is some kind of code necessary? Is it a condition of possibility? Is it, on the contrary, contingent and particular (just a local feature of life on Earth)? Can we then imagine a sort of life with no code? Of a comparable variety and complexity? And so on.

This kind of questions are especially well suited for AL to deal with. For biology, in particular theoretical biology, could be interested on them but it really does not make a difference to BL. Knowing whether the genetic code is a condition of possibility for life can be interesting and even illuminating for biology but it really does not matter too much if it is not. After all, biology has to deal with the only one instance of life we know, which happens to rely on a codified relation. So, whether it is contingent or not does not diminish a bit the necessity to study it in order to reveal its functioning and its origin. This is probably why (among other reasons like the role of theory in empirical sciences) biology is much more interested in discovering how the genetic code could actually have arisen and which is the ultimate nature of the coding than in discussing its necessity or contingency.

However, this is not the case within AL, which is born with a very explicit [19] commitment to universality by *transcending the "one-case" study of biological life* into an abstract exploration of any kind of life. To this respect it is not irrelevant if some kind of code is necessary or contingent, i.e., whether it is a condition of possibility for life in its most general sense or not. Precisely because we would like not to limit ourselves to the only example of life we know and, therefore, because we could imagine a kind of life without any coded relation between its main components. Or we could not? Consequently, AL should be concerned not only with problems of origins (origin of life, metabolism, genetic information, and genetic code) but with other fundamental ones such as the very "logic" of life[1]. In BL we can imagine, and even describe with detail and reproduce through experiments, different scenarios for the origin of life and its development and further complexification towards the configuration known today. We have very powerful hypothesis of RNA and protein worlds or of a cell/system world. We have extremely accurate experimental and theoretical/mathematical models [7, 12, 24]. We also have in AL very interesting simulations of self-organizing dynamical networks and reproducing informational structures [1, 20, 21, 22, 51]. There are, of course, in both realms, BL and AL, very insightful guesses upon the way in which proteins and nucleic acids came to end linked (from frozen accident to stereo-chemical affinities hypothesis through the study of plausible more simple codes or relics of possibly different ones). However, what BL does not need to do, but AL certainly does, is to ask the prior question: could it be, whenever or wherever, any phenomenon worth the name of life, in any possible definition of it, without any kind of coded relation between its basic constituents?

2 A Guiding Notion of Life

2.1 Statement of the Problem

As epistemological discussions in AL state, we hardly have any more than several "shopping-lists" [8, 10, 11, 30, 31] when it comes to figure out a definition of life. This lack by itself is enough to weaken any possible claim of universality [32]. This is not so dramatic for biology, whose practitioners are not always very keen of making such claims [25, 26]. However, for AL is the contrary inasmuch as its reliance to an universal theory of life is constitutive of the research program.

The situation is even worst for AL if we take into consideration one further difference with respect BL, where we have an effective enough intuitive or pragmatic notion of life which is totally absent in AL. Biology, even if it lacks a precise and unanimously agreed upon definition of life, is able to use those lists of properties in a very coherent and almost unproblematic (unanimous) way when it is needed to

---

[1] I am well aware that claims of considering the "formal"/"logical"/"abstract" aspects of life are made and accepted from the very beginning in AL, but I am referring here to other kind of more substantial and not only formal "logic" or abstraction.

identify a system as alive or not[2] (in practice, not in theory [24], as pointed out in [29] and [32] in relation to the main paradigms in theoretical biology). This is probably so because, dealing with natural physical entities, it is not so difficult to relatively weight these properties and opportunistically assign them more or less value with respect to the others, and still being able to keep a coherent enough intuitive notion. In AL, by the contrary, we have no intuitive idea able to stop us from choosing and highlighting any of these properties of our own convenience in relative or complete detriment of any other. Being an artificial (and often just computational) realm we lack any flexibility for "playing" with the relevance of these properties of the living in an ambiguous but practically efficient and uncontroversial way. We lack a valid guiding notion which should indicate the more or less fuzzy but coherent borderlines of applicability of the notion of "lifelike". An guiding notion which, in biology, has arisen from a long term acquaintance with living systems in their fully developed forms ("life-as-we-know-it"). A familiarity which, of course, we cannot still have with their (potential) artificial counterparts.

However, we can not wait 200 years for a new (artificial)biology to emerge. We need our own intuitive notion of artificial life right now in order to use it as a "guiding" notion through the increasing wealth of computer processes and robotic behaviors eager to receive that name. I am afraid the only place we can look for insight and suitable materials for such a pragmatic notion is, still, biological life, BL.

2.2 Usefulness of the List(s) of Properties

We already have and use these lists in AL. Even if they are insufficient, as we said before, we should not forget the usefulness of relying upon reasonable lists of properties in order to characterize and evaluate the results we obtain through computer and robotic modeling. We have to realize that, while we have nothing better (and this "something better", like a proper definition of life, is more likely to come from the very development of AL than from any other source) these lists are good enough to begin with. They can constitute the basis of the guiding notion we are looking for. At least as far as they are rigorously used and critically considered. In my opinion we do have some of these rigorous lists with well defined general properties of living systems which we can use[3]. What we lack instead is a thoroughly conscious use of them when it amounts to describe the features of any particular artificial system held to be "lifelike". Therefore, even if we have very useful and appropriate lists, the problem is that we do not use them but superficially. If any of this lists or any blend of them is going to be of any use for getting a guiding notion of life applicable to the artificial realm, we will have to play fair with them. To be honest in this context does not only mean to believe what one is saying but to explicitly give the way for others to judge if there is sufficient reason for this belief. For instance, if we want to claim that we have artificially implemented some self-reproducing or adaptive phenomenon, we should be able to explicit what do we mean by these concepts (or to refer to some explicit definition of the property) and to acknowledge how fully does the implementation satisfy, say, the definition of the biological counterpart or some potentially more general definition of it. This way, instead of being discussing if any particular proposal is or not A*Life*, we could proceed to discuss whether it really

2 This does not mean that some discussion upon the relevance of some of these properties is not needed in each case, especially in the more controversial ones, like viruses.

3 [8, 10, 11, 25, 30, 31].

accomplishes its promises with respect to particular properties of BL, letting aside the wider discussion on which of these properties are essential or more important.

What I advocate for is to separate two levels of discussion:

* one should be concerned with the practical and strict evaluation of the success in duplicating a given property of living systems under given definitions of it (where the discussion could range from different definitions or approaches upon the very property to different views about the practical way of implementation);

* the other should engage the further reaching discussion about the relative relevance of each of these properties, their ultimate accurate characterization, and ways of interrelating them; altogether with the making of the choices among them and their potential fusion in a definition.

Right now it is not easy to foresee how could we resolve the issues implied in the second level. It would be good just to agree upon some kind of guiding notion as will be explained next (also because of the second problem involved here: the grounding of interpretation). However, the main lesson I would like to draw from this distinction is that *the second kind of discussion should not interfere with the first one neither the latter be propelled to the level of the former*. The more conceptual discussions about different properties should not obstruct the development of new artificial programs and devices, neither these ones should pretend to disdain these discussions by making unconstrained claims pertaining to the conceptual realm.

This is not a renounce to make AL or even to pursue a better understanding of "life-as-it-could-be". On the contrary, this is the best way to advance in our research about the potentially universal features of life and to become closer to the goal *in the doing*. Thus, by getting the more uncontroversial instances of particular facets of life, of particular items on those lists (even without claiming them to be instances of life) we will be able to advance step after step in this direction. This path is going to be rewarding as far as we keep in mind the previous distinction and do not expect that a simulation of life can become a realization just by refining it [41] or that by getting a mimesis of enough properties of life we can reach life itself [47].

2.3 A Guiding Notion

Even if it seems paradoxical, when looking for a "guiding" notion we should not search among the simple and elementary ones but among the more complex and complete of them. A "guiding" notion has to be something which indicates the way, which points to a further goal and, simultaneously, something what allow us to measure how far we are from reaching that goal. And then, once this notion is identified, we have to proceed to ask about the conditions of possibility for getting such a goal. In AL in particular, we should ask which are the conditions of possibility for life. For life as we know it to begin with but in a way which allows us to transcend it. This means not only to ask which is the abstract logic of life but which its actual conditions of possibility in some physical universe are. Not only to ask which are the conditions of possibility for its origin (chemical evolution) and development (prebiotic evolution) on Earth but, also, which are the conditions of possibility for life (as we know it) to exist in this particular *milieu*. Because the answers we could be able to give to this foundational questions are the ones which will allow us to build a general (perhaps universal) concept of life-as-it-could-be.

In order to choose any such notion one has to try to look at the higher level of generality. In this respect, Dobzhansky was extremely accurate when he wrote that "nothing in Biology makes sense except in the light of evolution" [5] as the title of a short piece of wise and lucid divulgence against creationism. Similarly, we could ask if anything in AL could make sense except in the light of evolution? We can try to be a little more specific and notice that, naturally, what Dobzhansky had in mind was Darwinian evolution by natural selection[4]. Following Sober's suggestion [49], *this view of life* could be abstracted using Lewontin's [23] characterization:

> "This three principles [(phenotypic variation), (differential fitness), (fitness is heritable)] embody the principle of evolution by natural selection. (...) The generality of the principles of natural selection means that any entities in nature that have variation, reproduction, and heritability may evolve." ([23] p. 1]).

The importance of these principles has been frequently addressed in the AL literature but in most of the cases the very interrelationship which links the three of them is not taken into account. For instance, Langton [19], in an interesting attempt to generalize these notions, asserts that "the *genotype/phenotype* distinction" is "(t)he most salient characteristic of living systems"[5] and yet he does not consider at all the intricate relation between the two. Another example may be Sober's [49] attempt to find biological properties liable to be abstracted and his *right* concern with the possible material limitations to implement "hereditary mechanisms", while he is not considering first the very *abstract* logic of this heritability. More examples could be pointed out sharing the same characteristic: significant mention of the principles governing biological evolution without linking them together. A somewhat misleading lack of realization of the importance and the nature of the code.

As to the importance, to consider biology "between history and physics" and the transition point as the "most uniquely biological area" [48] is certainly adequate:

> "This transition between physics and history is identified with the introduction of the genetic mechanism and it has long been recognized that the origin of the code is central to our understanding of this transition. (...) It is, therefore, this important transitional region between physics and history, the region most uniquely biological, about which we can make the fewest definitive statements." ([48] p. 275).

As to the nature, it shoul be enough to mention the key feature of constituting the transition point between these two different realms of description and explanation, some of whose attributes are going to be discussed next.

Therefore, I would like to argue that the code is (at least) one of such conditions of possibility for life I was mentioning before and, in a more practical side, I would like to submit to consideration the adoption of the code as a guiding ideal for the intuitive notion of life in AL. We could use a general or abstract version of the genetic code as a kind of ruler against which we could compare our results and measure how close we are to instantiate something similar. This could help us to find a "meta-criterion" able to deal with the normal situation where some properties are better instantiated than others, by offering a reference point to assign relative values to each of them. A mainly heuristic meta-criterion which we could dispose of once we had (if we ever have) any better or more general one.

---

4 I want to mention that the argument I intend to make should work the same either if we accept Dobzhansky's and many others' position on "natural selection as the main factor that brings evolution about" ([5] p.127) or if we are more sympathetic with critical claims defending a more important role of developmental processes and morphological constraints. My point about the relevance of the code qualifies for both approaches.

5 "from the behavior generation point of view" is added ([19] p. 22).

3 Some Characteristics of the Code

In order to illustrate better the nature of this proposal it would help to stress briefly some characteristics of the code and submit them, as well as any other I might have failed to identify, to discussion upon their relevance.

Before going any further, I would like to make clear that when I am defending the necessity to consider the genetic code as guiding notion to intuitively grasp life I am not trying to substitute what some people in AL call "carbon-chauvinism" for some form of "genetic-imperialism" which is a more real danger (and more feared by lots of conscious biologists). I want to make clear that what I consider important from the genetic code is more the "code". Specially because this name can be misleading of what I think is the key nature of the code (even more in its abstract form) as a semiotic relation between the genetic and the metabolic aspects of a cell, where no component alone -nucleic acid or protein- has any primacy or leading role.

Let's then begin by recalling which the nature of a code is. In order to have a code of any kind we need a situation where something refers to something by virtue of something else [33, 36, 50]. Each of these 'something' can be replaced by any physical entity or process whenever the association among the first two only arises through the mediation of the third one [28]. This mediated aspect of the connection is which confers the code its semiotic nature. It is, then, epistemologically advantageous to consider the genetic code thoroughly as a code and not just as a metaphor. Let's mention now (without going into detail) some of the well known and more relevant characteristics of the code which we should keep in mind when making AL.

3.1 Point of Transition and Linkage

One of the most fascinating characteristics of the code is its intermediate nature with respect to an important set of scientific approaches, even some largely dichotomic ones. This connecting feature implies that the code shares some properties with both of the realms it is tying up [48]. A kind of mixture whose nature and principles we should certainly study.

If we take the point of view of the origin of life, the code lies in the transition between what it is usually referred to as chemical evolution and biological (Darwinian) evolution [7, 12] but neither approach alone can account for it. We can address the issue from both sides of the gap (temporally forward and backward [12]), trying to get more sophisticated protocells on the one side or more elemental genetic strings on the other, and we certainly come close to overcome the gap. But finally we always have to add something "*ex machina*" to actually get a viable living system. So the code lies just where both the RNA and the protein "worlds" come to an end.

Taking the point of view of the theoretical paradigms somehow in competition in biology nowadays we can see that the code is not only between both of them but that it is what ultimately is connecting their respective realms of explication. This refers to what can be generally called the Self-Organizational and the Informational paradigms [29, 32]. Like in the origin of life problem, the self-organizing approach by itself does not seem to be able to account for the informational properties of life and the informational approach tends to ignore the self-organizing properties which underlie and allow the very expression of the genetic information. The code is precisely linking the heritable information stored in the genome with the network of catalytic reactions in the metabolism that make possible the adaptation to/in the environment. As M. Smith [24] found appropiate to conclude:

"I started by suggesting that there are two ways of viewing living things: as dissipative structures, and as entities capable of transmitting hereditary information. These two aspects of life are brought together by the genetic code." (p. 124]

Another interesting point of view is a more physically oriented one which considers the code as a very specific connection between the microscopic (molecular) and the macroscopic levels with interesting considerations on irreversibility [37, 40, 43] and noise dependence [48]:

"This [encoding] resulted in a radical shift in the character of the system, as a mechanism was now available for molecular events to be amplified to the macroscopic domain. Since thermal noise is a characteristic feature of the molecular level it also became possible for random thermal events to have macroscopic consequences so that the large scale system became incompletely determinate over the long term." ([48] p. 269).

If we take the point of view of the computer sciences we find in the code an almost unique "entanglement" between software and hardware [16, 17] amounting to the tight and peculiar connection between the symbolic and the dynamic processes following, respectively, rate-independent rules and rate-dependent laws [37, 38, 39] in a complementary way [39, 41, 42].

In an abstract sense, the biological code represents a genuine instantiation of the classical dichotomy between matter and symbol incorporating more or less hidden clues to unveil the origin of, transition to, and relation between symbols.

3.2 Some Advantages of Both Worlds

The intermediate nature of the code also amounts to the ability of adding together some of the advantages of the two realms it is linking. On the one hand, we find a maximal use of physical/material constraints which enables the system not having to encode everything in a symbolic way. This permits to overcome the limits that would come very early from a totally explicit and purely informational strategy [29]. On the other hand, we also find that this physico-chemical dynamics works not only loosely guided by global constraints of a self-organizing origin but also very specifically directed by particular constraints controlling specific and local components [50]. This also allows to surpass the limitations of a just dynamical self-organizing strategy. Already 30 years ago Pattee pointed out the importance of the "problem of the evolution of control in complex organizations" suggesting that:

"From this point of view, the question of the origin of life becomes the problem of understanding elementary control processes, and formulating a theory of the evolution of molecular control." ([35] pp. 405-6).

A distinction between non-specific constraints giving rise to global patterns (i.e., micro-macro emergence of order in dissipative systems) and specific constraints effecting local control should be further investigated [50].

3.3 Open-ended Evolution

The simultaneous unity and diversity of living systems which is so striking when wholly considered has also its roots in the code and is performed through the distinction and interaction between genotype and phenotype. This separation between a symbolic description and a dynamical construction has been consider, after von Neumann [34], (see [19]) the logical condition for an open-ended evolution able to transcend a generic complexity threshold . As Pattee [41] summarizes it:

"Von Neumann's kinematic description of the logical requirements for evolvable self-replication should be a paradigm for artificial life study. He chose the evolution of complexity as the essential characteristic of life that distinguishes it from non-life, and then argued that symbolic instructions and universal construction were necessary for heritable, open-ended evolution." ([41] p. 69).

The relevance of open-endedness of evolution as a significant characteristic of life, its practical meaning and ways to achieve it should then be seriously considered.

3.4 Implementation Independence as a Consequence of the Code

This is, of course, a general quality of codes and hence it is not privative of the genetic one. Nevertheless, it is necessary to reflect upon it in the frame of AL. Any code or any symbol system is, by its very definition, independent of any particular implementation of it. An independence which ranks in several degrees according to the amount of abstraction and formalization, as it is obvious. In this sense, *implementation independence* means that the particular physical (material) support or realization (structure) of a particular sign (function) is irrelevant for the meaning it conveys (the function it performs). This is related with the, more or less, arbitrary (not causal) and mediated (through interpretation) nature of the coded relation. The extreme implication is that the structure neither helps or constraints functions.

We probably are not very often aware of this circumstance due to our every day acquaintance with symbols and meanings through natural language and other ways of representation. This is a circumstance which permeates even our own perceptual skills. However we should realize that no physical process or entity is by itself independent of its implementation. It is only due to our cognitive abilities that we eventually can make abstraction of the particular instances we experience and extract some general patterns that in some cases become universal laws. This is also why we may reproduce and mimic almost every physical phenomena using enough powerful computational means. That is, we can model a planetary system in a lot of different supports and even learn something from the model. Still nobody would claim in that and similar cases that the model is really instantiating the planetary system (not even that a non-analog or purely computational model is instantiating any of its basic properties like mass or any of its operations like the gravity force). This is due to the realization that the independence from the implementation is heteronomously instilled and has to do with our ability to symbolically connect naturally diverse things.

Nevertheless this is not the case when life or mind are the phenomena involved in the artificial modeling and implementation. Here whether we can obtain real instantiations in artificial means or not is highly controversial. The difference is prompted by the essential and original coded nature of both of these phenomena. The fact that they involve, by themselves, some kind or some amount of coded relations not dependent on our modeling make the implementation independence to appear as well as independent of our modeling. This is true to some extent and this is eventually what makes any sort of functionalist approach possible. We will therefore discuss this issue in the context of the functionalist claims. Nonetheless, it is worthy to stress already that *the possibility of implementation independence is not a primitive of any system but a property of some kind of systems endowed with a code.* The questions are that we should not confuse *whose* code are we dealing with, the system's or ours, and that we should be aware that the very attribute of embodying a code might be, by itself, not implementation independent.

The lesson to extract from this survey upon the code is that these features rely on having a semantically closed relation between two kind of components. A semantical dimension which is not imported from any external agent but arises autonomously in the system only mediated by its relation with the environment.

4 Functionalist Interpretation, Implementation Independence

Already together with the very original and straightforward declaration of intentions in AL by Langton [19] - "life as form"-, some warning was made [6, 41] in order to restrain such expectations. Later on more formalist claims have been made (i.e. [45]) and more arguments have been given to be cautious (i.e. [3, 4, 8, 9]). That original statement has been mostly interpreted, both to criticize and to defend it, in computationalist terms and has been generally related to the functionalist approach in the philosophy of psychology and AI.

Although important aspects of the critical reminders have been lost in the way, some kind of general awareness of the complex nature of the problem is also being recognized, not only from the more critically warning sides but also from the moderately optimistic or "positive" critics so well (i.e. [14, 15, 49]). This is a value judgment which can be impugned or opposed. Anyway, this position will be taken as starting point because, on the one hand, it is not necessary to repeat recent argumentations against computationalism to which no much can be added. On the other, it can be more helpful trying to go a step further in the way of this awareness of the problems with a purely computationalist approach. The idea is that the most moderate critics still miss some fundamental points and that the most radical ones still leave us room to try to refine the commitment to some kind of functionalism. A form, perhaps, of semiotic functionalism as Cariani [3] puts it:

> "The result is a phenomenally grounded *semiotic functionalism* incorporating syntactic, semantic, and pragmatic relations, rather than a logically grounded computational functionalism which only considers syntactic operations as its constituent relations." ([3] p. 784).

4.1 From Multiple-Realizability to Implementation Independence

First of all, a brief reflection upon the roots of the (different) functionalist claims is due in order to clarify the implicit assumptions they harbor and make some distinctions. Thus, for instance, one thing is to say that we can challenge the identity hypothesis because we can show examples of different states of brain resulting in the same mental phenomenon or vice versa, or saying that different physical properties result in the same relative fitness. Another very different thing is to say that we can ascribe mental phenomena or fitness functions to something else than populations of neurons or populations of organisms, whichever its actual configuration is. Of course this second claim can be made if it is argued for, but it has not to be confused with the first one nor can it be defended with arguments supporting the first one.

Both, multiple realizability (MR) and implementation independence(II), are matters of degree and, to certain extent, both concepts share an intermediate region were they overlap. However, MR is a more generic concept which can be seen as including II as a more strict or narrow case which needs further conditions. MR just means that a particular function can be realized in multiple structures. It only means that the material structure has *no totally determinant causal power* over the function, without implying anything about the specific materiality of these realizations. Or, in other words, without implying necessarily that this materiality could be any or that it has no influence at all. So we can consider cases (actually the most common ones) of MR implemented in the *same* materiality but with different structural conformations, as the examples above. Or we can have cases of MR implemented in a wider but limited number of material supports. Or in the extreme case, the one overlapping with II, MR implemented in any material basis. The difference is that II (even within practical limits) implies that the materiality is irrelevant (or should be), while MR does not. MR can be conceived as taking into account different materially constrained ways of instantiating various more or less similar functions. To MR, that function is not *totally* determined by the structure does not mean that *is not determined at all*.

The transition from one to the other was almost inadvertently made very early by the philosophy of psychology and AI. In the philosophy of psychology the comparison with computers and robotic counterparts appeared to be an ideal recourse to check different hypothesis about, in general, the issues of consciousness and intentionality or, in particular, the hypothesis of the identity between several mental and brain states (i.e., pain). This contrasting procedure had the sane role of avoiding a too "self-sensitive" way of discussing about these problems. In the case of AI and its related philosophy this move was still more natural. In general, this brought about a focus on functional organization, hence, functionalism. As Putnam put it long ago:

> "The positive importance of machines was that it was in connection with machines, computing machines in particular, that the notion of functional organization first appeared. Machines forced us to distinguish between an abstract structure and its concrete realization. Not that that distinction came into the world for the first time with machines. But in the case of computing machines, we could not avoid rubbing our noses against the fact that what we had to count as to all intents and purposes the same structure could be realized in a bewildering variety of different ways; that the important properties were not physical-chemical." ([44] pp. 299-300).

It could not be more explicitly said. However, as it can be easily guessed from the presence of the modifier -positive-, there is a second part in this paragraph:

> "The negative importance of machines, however, is that they tempt us to oversimplification. The notion of functional organization became clear to us through systems with a very restricted, very specific functional organization. So the temptation is present to assume that we must have that restricted and specific kind of functional organization." ([44] p. 300).

This temptation turned out too overwhelming to be eluded. The particular course the *surrender* took was a very natural one: if the functional organization of computing machines was just a purely formal and syntactic (symbolic) one, it was tempting (and certainly simplifying) to "assume that we must have *that* restricted and specific kind of functional organization". To fall under the temptation required only this small (but very important) step: to consider mind/intelligence (or life) as having a purely computational organization or, at least, open to be fully described in those terms; just like the machines we introduced in order to check hypothesis and/or provide prosthetic devices. This approach gave rise to the symbol grounding problem [13], which (as will be argued) is a problem inherent to any purely computationalist and totally implementation independence approach and not, due to such issues as consciousness and intentionality, a particular problem for cognitive sciences and AI.

4.2 Question of Degree or Deep Conceptual Issue

That step is well described in our field by, among others, Sober [49]. Nevertheless I can hardly accept his rendition of this leap in terms of degree of "liberalism" in the use of definitions. I am afraid that something else is involved here, a deeper conceptual problem. For, let's see if I have his presentation right: from the realization that mental states are multiply realizable we come to (supposedly) overwhelm the identity hypothesis and from here we come to decide that intelligence (or mind or whatever you want) is just some representational and computational phenomenon which, by principle, can be instantiated in any material support. Then, of course we have to be cautious not to go too far or limit ourselves to endorse the "extreme" position illustrated in the idea of passing the Turing Test. As Sober puts it, there is a question of choosing "how much abstraction is permissible".

However, as has been said before, there is a step missing here. Together with the abandonment of the identity hypothesis we have to make another hypothesis about the purely "formal" nature of psychological states which allows us to infer the independence of implementation (which, as we may recall, can not be considered a primitive by itself). This intermediate and hypothetical step is certainly more than a bigger or smaller amount of *liberalism* in deciding how much abstraction do we want. This is a very specific, definite and assertive claim which has to be independently sustained *apart from* the generic justification of functionalism as such.

We do not need to enter here in the dispute about the grounds and pertinence of this substitute hypothesis in the case of psychological phenomena. But it is important to stress that in the biological counterpart we can ensue a different path. In different words, any claim of independence between biological function and structure that can be made (and a lot of very judicious ones are made) does not have to entail as a necessary logical *sequitur* the consequence of the implementation independence[6]. Not at least necessarily in the same terms (formalist, computationalist) as in psychology or AI (independently of its accuracy in these fields). If any such claim is to be made it has to proceed from scratch, i.e., it has to offer specific (biological) support to hold the (computational) hypothesis that provides the connection between the simple multiple realizability and the pure implementation independence.

4.3 Formalist Approaches in AL

From the previous point we can conclude that if we have to consider the issue of functionalism for AL we should not just import the positions and arguments from the discussions within AI and the cognitive sciences (as Keeley [18] is right to complain). We will find, of course, that we end confronting some of the same difficulties but it would be better if we had our own starting point to address them.

Nevertheless, when we turn to AL looking for this kind of specific biological argumentation we do not find much. We *do* have some general postulates [45] or foundational declarations [19] with neither a proper argumentative basis nor a biologically specific starting point. Langton's account is illustrative of how the step described before is taken the same as in cognitive sciences and AI, *via assumption*:

> "The work of Church, Kleene, Gödel. Turing, and Post formalized the notion of a logical sequence of steps, leading to the realization that the essence of a

---

6 This is neither the place to undertake the task of investigating the usage of functional ideas in biology but it is at least necessary to mention that they have a pretty specific one which can be very diverse from the most common in psychology. Both of them share an original association to holistic considerations in interpreting the relevant biological and psychological properties (or phenomena) as belonging to the system as a whole and not to particular and distinguishable structures of it. In philosophy of biology there is plenty of literature discussing the suitability of functional explanations and its teleological nature. Another similarly interesting point which I can not address here is the analysis of the fascinating and slippery route conveying from this originally antireductionist stand of functionalism in both biology and psychology to ultimately reductionist approaches in AL and AI where life and mind are nothing but syntactically describable and computationally, i.e. mechanically, implementable behavioral attributes. A route where the only unchanged landscape that remains at the end is just the appellation of functionalism.

mechanical process -the "thing" responsible for its dynamic behavior- is not a thing at all, but an abstract control structure, or "program" -a sequence of simple actions selected from a finite repertoire. Furthermore, it was recognized that the essential features of this control structure could be captured within an abstract set of rules -a formal specification- without regard to the material out of which the machine was constructed. The "logical form" of a machine was separated from its material basis of construction, and it was found that "machineness" was a property of the former, not of the latter. Of course, the principle [sic] assumption made in Artificial Life is that the "logical form" of an organism can be separated from its material basis of construction, and that "aliveness" will be found to be a property of the former not of the latter." [19] pp. 10-11).

We also have some attempts to illustrate this move with biological examples whose interpretation is open to discussion. For instance, the examples in Sober [49] and others, like the fitness one, are of a very much "humble" kind of multiple realizability than the one implied by the formalist claim: all the instances able to "realize" the fitness function share all the characteristics of "life-as-we-know-it" and, therefore, they are more an illustration of biological diversity than one of a radical implementation independence of life itself or even of any abstract idea of fitness. It is like saying that you can realize the function of being a table in a multitude of ways but with the constraint that all of them have to have four legs and be made of wood; you still have, of course, tones of different tables and you can claim multiple realizability, but you are leaving out a lot more instances that common sense would allow as realizing the function of being a table. Well, we can not forget that the differences between a cockroach and a zebra[7] [49] are more like the differences between different four-leg tables made of wood than like the differences between the whole range of possible tables. And we are still, even in the broadest of the examples with tables, in a physical and not in a computational realm. In this example Sober himself is falling prey of a very subtle form of the Shoe/Fly Fallacy that he is denouncing [49]. It seems that he has mistaken the generic idea of multiple realizability with the narrower one of implementation independence and, therefore, he is arguing in favor of the latter with arguments relevant only for the former.

There are, of course, many other kind of examples which are not taken from the biological realm but directly from the computational one and then are claimed to satisfy some abstract version of a biologically relevant property. The problem with this procedure is that it already takes for granted what still needs to be argued for. The intermediate step which should justify the implementation independence is lacking.

It is precisely the code which allows this formalist claims to be uttered and which gives the assumption-carrying step its appearance of plausibility. It is the code which makes so easy to surrender to the mechanistic parallel and so natural to jump over the intermediate and missing step of endorsing the sufficiency of a "logical form" without even being aware of it. As well as in AI it was natural to content oneself with a logical description when dealing with high level cognitive behaviors (the paradigmatic case of chess), in AL it seemed equally natural to be satisfied with a logical description of the so called "central dogma" of molecular biology. Nevertheless, the problems of AI with more elemental cognitive and not only properly cognitive behaviors (from pattern recognition to movement) should put us in guard and teach us that there is something more to life (from the proper understanding

---

7 Due to all the common biological constraints and similarities they still share despite their diversity; affinities which are not only material but organizational as well.

of evolutionary process to the mechanisms of development). Or, more important yet, we need to acknowledge the nature of the code and realize that it does not justify a purely formalist and mechanistic kind of step. The coded nature is just what in the last term makes this jump possible but this does not account for itself, this requires further conditions. Moreover, there is a long way from primitive and specific codes to completely formal and universally algorithmic ones, a similar but broader and deeper gap than the one "connecting" natural language and mathematical formal logic.

4.4 Disanalogies and Analogies Between AL and AI

As Keeley [18] has noticed, we have here an important question about the analogies and disanalogies between AI and AL. Although what Keeley labels the Sober Analogy between AI and AL can be accepted to a great extent, he is right in criticizing the Global Replacement Strategy. This is precisely what is meant in the previous point: it is not accurate nor useful to import assumptions from another field, whatever germane and inspiring it is, without offering specific grounding from our own field. It is curious, however, that the main disanalogy Keeley is pointing to is just the same one Sober himself recognizes: that a behavioral (Turing-Test-like) account of biological properties or phenomena can be appropriate for biology even if it is not so for psychology. It is precisely in this point about the validity of a more or less "behaviorally" tested functionalism for AL where I disagree with both of them. Unfortunately, Keeley's optimistic diagnose and his feeling of relief in discounting "one less obstacle to overcome" somehow naïve and premature. For realizing that in AL we do not have to deal with such difficult matters as consciousness or intentionality does not amount to say that we do not have to cope with a more elementary version of the very same problem underlying those two: the attribution of meaning. The problem of giving account of the internally coded nature of living systems and, in particular, how to design computational replicas of them keeping as faithfully as possible this mixed symbolic/dynamic characteristic; and, moreover, how to reproduce the closed quality of the relation relative to semantic considerations.

4.5 Symbol Grounding Problem and Artificial Life

This problem of attribution of meaning has been defined in AI under the designation of "symbol grounding problem" [13] as a way to acknowledge and face the necessity to give account of such issues as consciousness, intentionality or, more generally, reference. At this respect it is extremely accurate to notice that:

> "Traditional philosophy sees this relation as the problem of reference, or how symbols come to stand for material structures (...). I have always found the complementary question of how material structures ever came to be symbolic much more fundamental." ([42], p. 11).

Harnad [14, 15] has tried to suggest the application to AL of the solutions he proposed for AI. Regardless of the aforementioned problems of adapting solutions from other fields and of the question that those solutions were already earlier and more "biologically" proposed in our field (i.e. [3, 4, 37, 41]), we can select at least some possible points of discussion, under which the quoted remark will be underlying.

Harnad's suggestion of introducing sensorimotor transduction and, therefore, the consideration of analog or non-symbolic aspects together with a more exigent kind of test is highly relevant and well asserted. The problem with this solution is that the identification of what was lacking in a pure Turing-Test-like approach is not enough. And, unfortunately, it seems that both his test [14] and his heuristic criterion [15] are too unspecific and still miss the point of accounting for the crucial way in which the symbols are *actually* grounded in biological systems through the genotype/phenotype distinction and under the guidance of evolution by natural selection. Hence, they could lead AL to a dead end by the impossibility of satisfying such criteria.

With respect to the sensorimotor transduction, if we just postulate the necessity of a connection with the environment (as Sober [49] is also right to do), without any further specification, we are not going to be able to rule out a kind of behavior or device pretty close to the plainly computational Turing machine and which does not seem to bear symbols any better grounded than a bare computer does. Something like a thermostat or, in general, what Cariani [2, 3, 4] calls *fixed-robotic* devices could satisfy that condition. Even if we introduce some kind of adaptive regulation with respect to the environment between sensors and effectors, we would still have to distinguish between *syntactically* and *semantically adaptive* devices depending on whether the adaptability refers just to the connection or to the very sensors and effectors too [2, 3, 4]. The proposal of an still behavioral but complete test like the Total Turing Test (TTT), instead of the classical one which was intended to judge just the symbolic behavior, seems to require more demanding conditions:

> "The TTT accordingly requires the candidate system, which is now a robot rather than just a computer, to be able to interact robotically with (i.e., to discriminate, identify, manipulate, and describe) the objects, events, and states of affairs that its symbols are systematically interpretable as denoting, and it must be able to do it so its symbolic performance coheres systematically with its robotic performance." ([15] p.547).

The problem is that its too "cognitive taste" would make it somehow difficult to extract from here a test for life. Besides that, this would certainly be a more precise requirement (regardless the *crucial* difficulty of defining operationally the interactions requested) inasmuch as we could specify what the coherence between symbolic and robotic performances amounts to in a purely behavioral and external way. It seems we *do* have an heuristic criterion to fulfill this ultimate testing task:

> "Chances are that whatever could slip by the Blind Watchmaker across evolutionary generations undetected is alive. (..) whatever living creatures are, they are what has successfully passed through the dynamic Darwinian filter that has shaped the biosphere." ([15] pp. 293-4).

This would indeed amount to a very accurate test from, at least, a biological point of view (not in vain it brings us back to Dobzhansky's fortunate formulation). Nevertheless, it cannot be easily avoided the feeling that we have not go too far from the questions in the Introduction if we still have to rely in Darwinian evolution *as such*. In *real* and *ages-long* Darwinian evolution. For, which could be the ways to slip under natural selection (and/or any other mechanism for evolution)?, or which are the conditions for evolution we could try to identify *here* and *now* without having to wait millions of years?, or which effectively heuristic subsets of this general criterion could we distinguish and use in order to evaluate our actual devices or programs?

It seems that the only way out from this riddle is to complete this kind of external and behavioral test with some internal criteria which could provide us with a manageable shortcut in terms of evolutionary time scale. Of course, the claim is that a proper understanding and application of the intermediate nature of the code (point of transition and linkage between form and matter, essential characteristic of life, and condition of evolution) is such kind of useful criterion, especially if we can elicit from it its abstract form and generalize it to other realms.

5 *Semantic Closure*: Guiding and Grounding Concept for AL

The *semantic closure principle* [38] is precisely intended to designate and explain this autonomous and intrinsically meaningful interrelation between symbol and matter that is characteristic of the living organization. This idea has its roots in a radical approach to biological systems begun already long ago. An approach which takes thoroughly into consideration the epistemological reflection upon the main principles and concepts of physics and relates them to the ones belonging to biology and other fields. All this amounts to a deep and far reaching concern and study of the more subtle relations between matter and symbol at several levels of organization; i.e., beginning with physics where the issue of *measurement* plays a key role, continuing with biology where the aspect of *control* turns to be central, and ending with cognitive and social systems where the concepts of *communication* and *computation* become highly significant. In Physics is where those concepts are defined the best and the most objectively (despite the open problem of measurement) inasmuch as they do not belong to the system but are introduced by the scientist. In biology the intrinsic functional nature of living systems imposes the difficulty of considering such concepts as an integral part of the system. Finally in cognitively developed systems this concepts acquire an even more independent status which becomes complexely entangled with the other two realms. The development of these four pivotal ideas, altogether with some other related ones, constitutes the main skeleton of an account of living organization and evolving systems that implies an inherently hierarchical and complementary approach, taking explicitly into account the interrelation between different levels as well as their dynamical and symbolic characteristics[8]. These concepts have been further developed and completed with new insights and specific applications by researchers as Minch [27], Cariani [2] and others.

It would, hence, require more than a full paper and certainly more space than I have available here to explain in detail this concept. Besides that, we *do* have those papers because original [38, 39] and more recent [42, 43] first hand sources are easily available. Also, the main purpose of this contribution is not the semantic closure by itself but to argue in favor of *considering it* as a helpful notion for the task to create and identify artificial life. Moreover, I have already mentioned some of the relevant features of the genetic code from which this principle is primarily abstracted. Therefore, I will just remind the essence of the principle by quoting two passages.

The principle derives from the code:

> "Looking more closely at how this comes about in the cell we see that this type of symbol-matter-function dependence is an exceptional kind of interdependence which I call *semantic closure*. We can say that the molecular strings of genes only become symbolic representations if the physical symbol tokens are, at some stage of string processing, *directly* recognized by translation molecules (tRNA's and synthetases) which thereupon execute specific but arbitrary actions (protein synthesis). The semantic closure arises from the necessity that the translation molecules are themselves referents of the gene strings." ([38] p. 333).

Anyway, it can be generalized to a broader abstract concept:

> "By general S. C. I mean the relation between two primitive constraints, the generalized measurement-type constraints that map complex patterns to simple actions and the generalized linguistic-type constraints that control the sequential

---

[8] See Umerez [50] for a more complete list of references that establish the background of the concept and widen its scope.

construction of the measurement constraints. The relation is semantically closed by the necessity for the linguistic instructions to be read by a set of measuring devices to produce the specified actions or meaning." ([39] p. 272).

Consequently, in the semantic closure we have an abstract principle which condenses the main features of the living organization we could easily agreed upon at least as reference points.

6 Conclusion

This contribution was mainly intended as a double invitation to take the genetic code faithfully into consideration in the context of AL and to open a discussion upon the convenience and ways to adopt its logic as an heuristic guiding notion to identify life-as-it-could-be. To endow this discussion with some contention to begin with, the principle of *semantic closure* defined and developed by H. Pattee has been proposed as an adequate condensation of such an essential logic. It is necessary to stress that this is not (it could be, this is another discussion) a "implementation-dependence" thesis or a TTTT [14, 15], it is more an abstract condition to be fulfilled by any implementation in an approach similar to the "medium-dependence in multiple possible media" of Emmeche [8] or the semantic functionalism of Cariani [3]. Nevertheless, what is plausible is that maybe not any medium is going to be able to fulfill this condition or not in any way and, hence, we will have to ask which could be the right ones among the many available. From what we already know a computer-robotic one seems to be an open, powerful and versatile enough medium to try to accomplish the task. But we have still to work hard in order to set the right conditions.

I do not want to finish without mentioning a parallel avenue of research related to the code in AL which has not been addressed here but which is not less fascinating and promising. I mean the innumerable possibilities to "experiment" *in silico* that are becoming available, i.e., the challenge to address in a different "experimental" realm the most intriguing biological issues (as the origin and development of the code). We have all the possibilities to devise computational versions of biological experiments which could take too long or would be to intricate to develop *in vitro* or in nature. For instance, the exploration of different codes with different number of nucleotides and amino acids and with different relations among them (from more physically dependent to more arbitrary ones) and the possibility to let them compete. There is equally the possibility to very freely pursue tests with very different RNA editing processes between transcription and translation of genetic strings [46]. All we could eventually learn about *life-itself* would indeed compensate the tremendous effort this work requires in designing the right and faithful models.

Acknowledgments

The author is specially indebted to H. Pattee for his ideas and comments and to our group of discussion, in particular to L. Rocha. He also thanks very helpful comments by A. Etxeberria. He is recipient of a postdoctoral grant from the MEC (Spain) and member of the research project 230 HA168/93 funded by the DGCYT (MEC).

References

1. Brooks, R. & Maes, P. (eds.) (1994) *Artificial Life IV*. Cambridge, MA: MIT Pr.
2. Cariani, P. (1989) *On the Design of Devices with Emergent Semantic Functions*. Ph. D. Dissertation, SUNY.
3. Cariani, P. (1992a) Emergence and artificial life. In C.G. Langton et al. (eds.) *Artificial Life II*, pp. 775-97.
4. Cariani, P. (1992b) Some epistemological implications of devices which construct their own sensors and effectors. In F.J. Varela & P. Bourgine (eds.) *Toward a Practice of Autonomous Systems*, pp. 484-93.
5. Dobzhansky, Th. (1973) Nothing in biology makes sense except in the light of evolution. *American Biology Teacher* **35**: 125-129.
6. Drexler, K.E. (1989) Biological and nanomechanical systems: Contrasts in evolutionary capacity. In C.G. Langton (ed.) *Artificial Life*, pp. 501-19.
7. Dyson, F. (1985) *Origins of Life*. Cambridge, MA: Cambridge University Press.
8. Emmeche, C. (1992a) Modeling life: A note on the semiotics of emergence and computation in artificial and natural living systems. In T.A. Sebeok & J. Umiker-Sebeok (eds.) *Biosemiotics.*, Berlin: M. de Gruyter, pp. 77-99.
9. Emmeche, C. (1992b) Life as an abstract phenomenon: Is AL possible? In Varela & Bourgine (eds.)*Toward a Practice of Autonomous Systems*, pp. 466-74.
10. Emmeche, C. (1993) Is life as a multiverse phenomenon? In C.G. Langton et al. (eds.) *Artificial Life II*, pp. 553-68.
11. Farmer, J.D. & Belin, A. (1992) Artificial Life: The coming evolution. In C.G. Langton et al. (eds.) *Artificial Life II*, pp. 815-40.
12. Fox, S. (1988) *The Emergence of Life*. New York, NY: Basic Books.
13. Harnad, S. (1990) The symbol grounding problem. *Physica D* **42**: 335-346.
14. Harnad, S. (1993) Artificial Life: synthetic versus virtual. In C. Langton (ed.) *Artificial Life III*, pp. 539-552.
15. Harnad, S. (1994) Levels of functional equivalence in reverse bioengineering. *Artificial Life* **1**: 293-301.
16. Hofstadter, D. (1979) *Gödel, Escher, Bach: an Eternal Golden Braid*. NY: Basic.
17. Hofstadter, D. (1985) *Metamagical Themas*. New York, NY: Basic Books.
18. Keeley, B.L. (1993) Against the global replacement: On the application of the philosophy of AI to AL. In C. Langton (ed.) Artificial Life III, pp. 569-587.
19. Langton, C.G. (1989) Artificial Life. In Langton (ed.) *Artificial Life*, pp. 1-47.
20. Langton, C.G. (ed.) (1989) *Artificial Life*. Reading, MA: Addison-Wesley.
21. Langton, C.G. (ed.) (1993) *Artificial Life III*. Reading, MA: Addison-Wesley.
22. Langton, C.G., Farmer, J.D., Rasmussen, S. & Taylor, C.E. (eds.) (1992) *Artificial Life II*. Reading, MA: Addison-Wesley.
23. Lewontin, R.C. (1970) The units of selection. *Ann. Rev. Ecol. Syst.* **1**: 1-18.
24. Maynard Smith, J. (1986) *The Problems of Biology*. Oxford: Oxford U. P.
25. Mayr, E. (1982) *The Growth of the Biological Thought*. Cambridge: Harvard UP.
26. Mayr, E. (1988) *Towards a New Philosophy of Biology*. Cambridge: Harvard UP.
27. Minch, E. (1988) *The Representation of Hierarchical Structure in Evolving Networks*. Ph. D. Dissertation, SUNY.
28. Moreno, A., Etxeberria, A., & Umerez, J. (1993) Semiotics and interlevel causality in biology. *Rivista di Biologia - Biology Forum* **86(2)**: 197-209.
29. Moreno, A., Etxeberria, A. & Umerez, J. (1994) Universality without Matter? In R. Brooks & P. Maes (eds.) *Artificial Life IV*, pp. 406-10.

30. Moreno, A., Fernández, J. & Etxeberria, A. (1990a) The necessity of a definition for life. Poster to the *2nd Workshop on Artificial Life*, Santa Fe: February '90.
31. Moreno, A., Fernández, J. & Etxeberria, A. (1990b) Cybernetics, Autopoiesis, and definitions of life. In R. Trappl (ed.) *Cybernetics and Systems '90*, Singapore: World Scientific, pp. 357-64.
32. Moreno, A., Umerez, J. & Fernández, J. (1994) Definition of life and research program in Artificial Life. *Ludus Vitalis. J. of Life Sciences*, **II (3)**: 15-33.
33. Morris, C.W. (1938) Foundations of the Theory of Signs. O. Neurath, R. Carnap & C.W. Morris (eds.) *International Encyclopedia of Unified Science* **1(2)**.
34. Neumann, J. von (1966) *The Theory of Self-Reproducing Automata*. A. Burks (ed.), Urbana, ILL: U. of Illinois Pr.
35. Pattee, H.H. (1965) Experimental approaches to the origin of life problem. In F.F. Nord (ed.) *Advances in Enzymology* **27**: 381-415, New York: Wiley.
36. Pattee, H.H. (1969) How does a molecule become a message? *Dev. Biol. Sup.* **3**: 1-16.
37. Pattee, H.H. (1972) Laws and constraints, symbols and languages. In C.H. Waddington (ed.) *Towards a Theoretical Biology 4. Essays*, Edinburgh: Edinburgh University Press, pp. 248-58.
38. Pattee, H.H. (1982) Cell Psychology: An evolutionary approach to the symbol-matter problem. *Cognition and Brain Theory* **5(4)**: 325-341.
39. Pattee, H.H. (1985) Universal principles of measurement and language functions in evolving systems. In J.L. Casti & A. Karlqvist (eds.) *Complexity, Language, and Life*, Berlin: Springer-Verlag, pp. 268-81.
40. Pattee, H.H. (1987) Instabilities and information in biological self-organization. In F.E. Yates (ed.) *Self-Organizing Systems. The Emergence of Order*, New York, NY: Plenum, pp. 325-38.
41. Pattee, H.H. (1989) Simulations, realizations, and theories of life. In C.G. Langton (ed.) *Artificial Life*, pp. 63-77.
42. Pattee, H.H. (1995a) Evolving self-reference: matter, symbols, and semantic closure. *Communication and Cognition-Artificial Intelligence* **12(1-2)**: 9-27.
43. Pattee, H.H. (1995b) Artificial Life Needs a Real Epistemology. This volume.
44. Putnam, H. (1975) *Mind, Language, and Reality. Philosophical Papers, vol. 2.* Cambridge, MA: Cambridge U. Pr.
45. Rasmussen, S. (1992) Aspects of information, life, reality, and physics. In C.G. Langton, et al. (eds.) *Artificial Life II*, pp. 767-73.
46. Rocha, L. (1995) Contextual GA: Evolving developmental rules. This volume.
47. Rosen, R. (1994) On Psychomimesis. *J. Theor. Biol.* **171**: 87-92.
48. Smith, T.F. & Morowitz, H.J. (1982) Between History and Physics. *Journal of Molecular Evolution* **18**: 265-82.
49. Sober, E. (1992) Learning from functionalism - Prospects for strong Artificial Life. In C.G. Langton et al. (eds.) *Artificial Life II*, pp. 749-65.
50. Umerez, J. (1994) *Jerarquías Autónomas.- Un estudio sobre el origen y la naturaleza de los procesos de control y de formación de niveles en sistemas naturales complejos*. Ph.D. Dissertation, University of the Basque Country.
51. Varela, F.J. & Bourgine, P. (eds.) (1992) *Toward a Practice of Autonomous Systems. Proceedings of the 1st ECAL.* Cambridge, MA: MIT Press.

The Inside and Outside Views of Life

George Kampis

Dept. of Philosophy of Science, Eötvös University, Budapest, H-1088 Rákóczi u. 5., Hungary. Phone/FAX: (36) 1 266 4954

Abstract. Abstract. Biology is, better than anything else, about existence in time. Hence biological reality cannot be defined without reference to a temporally situated observer. The coupled or detached character of this observer (with respect to the own time variable of the system) provides a link between the observer and the observed. This connections delimits the kinds of scientific descriptions that can be given at all by an observer. In particular, two fundamentally different forms of description, corresponding to different epistemological attitudes and different philosophies of science, called endo- and exo-physics, can be distunguished. Two old puzzles, the Omniscience Problem (illustrated here on the example of Internal Chemistry) and the Chameleon Problem (originally an argument against philosophical functionalism) are reconsidered in the light of these distinctions. As application, the question, in what sense computer models of life can be suitable for studying life, is examined.

1. Introduction

Let us begin with some trivia. One of the permanent problems both in the sciences and in philosophy of science is that of "objectivity", and, in relationship to that, the status of the observer. Twentiest century philosophy, and within that, the logical positivist programme of the Vienna Circle (or more precisely, its failure) has most eminently shown that the observer cannot be excluded from a scientific theory. It must be acknowledged that the way we *look* determines what we *see*, or rather it co-determines the latter, in conjunction with what *there is*. As a result, the modern conceptions of science, beginning with Popper's model concept, put forward a constrained and conditional concept of objectivity — or, as a border case, reject objectivity and impartiality as a whole. Kuhn's much-debated notion of "paradigm", the notions of social constructivism, the "strong" programme of the sociology of scientific knowledge, naturalized epistemology, and several other enterprises can be aligned along the same axis.

2. Endophysics

Endophysics is a new approach to the old problems of observing. Endophysics literally means "inner physics" (or internal physics). It offers a study of systems that have enclosed observers in them. The originators of the idea are O.E. Rössler [1987] and D. Finkelstein [Finkelstein and Finkelstein 1983] who also mention numerous earlier theories where a similar concept was already implicit, but without a suitable name at hand. The science fiction story "Simulacron Three" by Galouye [1964] may have been the first (though not scientific) of these publications. The story depicted, among other things, the logical problems that intelligent beings simulated on a computer would face when trying to discover their Universe. Among the better known predecessors of endophysics there is another short story, one by S. Lem, about his favourite Trurl and Calapacius [Lem 1979], in which the two protagonists construct what we would today call a Virtual Reality machine. Not being able to see the situation from the outside, the user gets trapped in the machine, lost in a hyperspace of newer and newer illusory realities which are entered without a hope for returning to the initial, original reality.

Of course, that is not the whole message of endophysics, yet the story offers a good example. Endophysics is about the limits (and forms) of knowledge.

In contrast to endophysics, its counterpart *exo*physics can be defined as an external, birds-eye view of an object domain. The birds-eye, or nonparticipatory view implies a detaching from the constraints that characterize the observed, while the endo view means submersion in the pool of the same constraints. Also, it is clear that the words "internal" and "external" refer to epistemological committments rather than footholds; to give an internal description, one does not have to move in, or, to take the exo stance, one doesn't need to move out.

The endo/exo distinction and the implied interface problem (or translation problem, or equivalence problem) of descriptions obtainable by the two methods has important applications in various domains. To map the domain-specific consequences is the closer concern of endophysics research. The publications on endophysics include three books [Rössler 1992, Kampis and Weibel 1993, Atmanspacher and Dalenoort 1994], that bring contributions on cosmology [Finkelstein 1993], dynamical systems [Kampis 1993], mathematics [Löfgren 1993], and the methodology of modelling [Kampis 1994]. Also here belong some chapters of a recent book on algorithmic physics [Svozil 1993].

3. Observers of Life

The question of the observer is particularly important for biology and evolution theory. Certain aspects of this fact are widely known and their existence taken as granted. There is, just to pick a random example, the recognition that our entire knowledge of evolution is based on historical guesses of directly unobservable events. This leaves much room for rival interpretations, an important epistemological problem exemplified by the semi-recent saltational debate.

Besides that, however, there are a more subtle forms of the involvement of the observer that can be discovered in biological theory. One of them will be the focus of our paper. It is the observer's apparent *omniscience* in the models of dynamical life processes, something that stands in a striking contrast to the abundantly complex and concealing nature of the subject to which these descriptions refer. The concern here is that life's natural complexity and its essential involvement in termporal relations may challenge the "Platonic" (or, rather, Parmenidean) block universe concept of modeling that requires transparency and a *prior* explicit knowledge of all qualities, as if they were permanent. I will call this the *Omniscience Problem.*

To the genesis of this problem (sometimes dubbed as the problem of emergence vs. determinism, of open vs. closed universe, and so on) we can quote, in a wider bio-philosophical or plain philosophical context, Morgan [1923], Bergson [1944], Whitehead [1929] and many other authors. Also Popper's [1965] and the follow-up literature have a direct relevance. Nagel's [1961] brings a review of the classical conceptions; more recent works include [Cariani 1989], Emmeche [1993] and Kampis [1991] where newer developments are reviewed. Finally, we will see that a quite different literature on philosophical functionalism converges to the same point.

In the sequel I am trying to apply the "endo" approach of philosophy of science to the observer problem of the theory of life, with consequences for computer modelling and potentially for the whole enterprise of Artificial Life.

4. Endo- and Exophysics, and Biology

Needless to say, it is not necessary to restrict objects to those physicists are interested in. In this way endo-"physics" can mean endo-biology as well.

We said that endo and exo standpoints differ in the way they reflect the constraints of the observed. Two fundamental constraints particularly sensitive to the choice of the end/exo state are related to *space* and *time*. In particular, an endo observer is bound to local observations whereas the exo observer can take global perspectives, along both types of coordinate.

Central to our subsequent discussion is the status of the time parameter (although the whole story could be re-told on the spatial side just as well). With respect to time, the endo strategy amounts to subjecting the observer to the effect of time, whereas the exo strategy amounts to decoupling the observers, and hence the observations, from perceptual, flowing time.

To give a not entirely random example in order to illustrate the spirit, the first (endo) situation is typical for students of evolution, and the second (exo) for those of ecology; in the first case, the species change simultaneously to the observation process (the observer is embedded in the evolution of the system), whereas in the second, all species changes (and, together with that, the evolutionary time scale) are cancelled — in other words, evolution is about how species come about, ecology is about what those are already there do.

The time-dependent evolutionary and time-independent ecological views are not the only ones possible. As the major part of actual biological evolution is already behind us, or at least that is what most biologists assume in their daily work, past evolution lends itself to another exo study, that of historical reconstruction. The issue gets even more complicated if time scales are also taken into account. What amounts to decoupled viewing at one time scale may count as internal perspective on another.

For instance, for observers of ecology, events on the evolutionary time scale are simultaneously available (since all species are just there, and hence the observer's standpoint is decoupled with respect to their genesis and extinction), but the ecological events are not.

The basic situation is clear: chosing any given scale or time-frame, the observer keeps a certain universe fixed, with respect to which he will find himself in an exo situation (hence he can view it globally), whereas the rest, if there is anything left, is approached from what is the "time-internal" perspective of the given situation.

5. Internal Chemistry, and the Imperative of Standpoint

Sometimes observer relativity means arbitrariness of standpoint; sometimes not. In the latter sense the choice between endo- or exo-physics can be less spontaneous.

The difference between the two types of scientific construction, endo and exo, together with the case for the imperative choice of the standpoint can be best presented on the example of biochemistry, or indeed chemistry of any kind. Here I am taking an idea elaborated in [Rössler 1984] and Kampis [1991] in detail.

Chemical reactions produce new chemicals by consuming others. As every chemist knows, in Nature the reaction $A + B \longrightarrow C$ takes place in two phases; in phase one A and B are present, from which in phase two (sometimes via some intermediate complex) the end product C is formed. In this phase, both A and B are absent.

A complex network of reactions can embody a large transition system of such productions, yet the whole network can be started by just a few chemicals. That is, we don't have to buy all elements of the reaction matrix except for a few, shake them well, and wait long enough. At the time subsequent reactions take place the initial substrates are in general no more available, and so on; there is no way to meet or view the complete set as a whole.

This phenomenon is universal wherever reactions take place, but it goes unnoticed (as unimportant) in many cases, such as in small molecular chemistry, where chemical compounds are drawn from a simple set of very few types, but in exchange their amounts are exuberant (so there are many more tokens, or exemplars, than there are types of the chemicals; the concern is not with sequentiality but with mass balance). This can be contrasted to macromolecular chemistry, where the situation is quite the opposite. Of the astronomic numbers of macromolecular possibilities just a very sparse subset is realized (or can ever be realized, in the lifetime of this Universe), and those that are realized are represented with usually a few molecules or a single piece each. Consider the fact that, strictly speaking, every individual DNA strand is unique, and so on.

To fix things: the usual description that operates on the complete set of chemicals (such as a reaction kinetic equation) is clearly exophysical (time-detached, a posteriori, global) and the intuitive picture we gave above is endophysical (time-bound, synchronous, local).

Now, philosophically speaking, it is completely artificial (yet can be useful on practical grounds) to speak about the whole exo universe of chemical compounds. The corresponding entity is nowhere present in our world. Thus, in the presented example the endo viewpoint is clearly privileged as a *natural* one. (In this spirit endophysics can be used as part of a constructivist critique of state-space approaches, see [Kampis 1991].)

The exo matrix offers computational powers (such as the use of algebra and, based on that, of differential calculus), something that would be wrong not to utilize once it is there. At the same time, when applying the computational model, the complex epistemological structure of the problem must be also kept in mind. When interpreting the computation results, one has to be conscious about having computed on postulated elements, like C, that have not been there yet, when A and B came to react with each other.

This unusual *anticipatory* character of chemical models and the resulting, necessarily incomplete endo/exo transcription is reflected in some unique traits of the reaction matrix models, such as their extreme sparsity, that is, the overwhelming amount of zeroes, marking currently nonexistent molecules that must be postulated, recorded, and carried anyway. There is another unmatched property; unlike with electric or mechanical devices, here a full zero state implies the material nonexistence of the system [Rossler 1984].

6. A Complex System "Verifies" Every Model, Or, Nature is a Chameleon

Another and even more disquieting problem that exists in the endo domain is that the complexity of describing can result, under certain circumstances, in a full collapse of the model-based approach as such.

This statement must be considered against the background of philosophical functionalism. Philosophical functionalism amounts to the assumption that the "essence" of things is independent from their material realization. The idea was introduced in Wiener's cybernetics [1948], where the famous statement "information is immaterial" was forwarded, and in Putnam's [1960]. It was re-introduced or rediscovered in several later works, where the computer metaphor, that is, the separation of hardware from software, as a universal principle applicable every all material systems was suggested.

The debate on functionalism is a publication industry of its own, and I will not discuss it at all. I confine myself to but one relevant idea, directly related to the problem of endo- and exophysics. It is a provocative antithesis to functionalism, coming from Rosen [1977, 1993] and from a more recent work of Putnam himself [1988]. Both authors use essentially the same argument. They show that every natural system can behave like any other. More precisely, the argument is about the ability of arbitrary natural systems to simulate every finite automaton, in Putnam's case, and every dynamical system, in Rosen's. For the sake of simplicity, let us now disregard the finer details. Form the broadest point of view, both arguments are based on a simple mathematical trick, on a re-partitioning of the state space of a given system, achieved by using a complex and arbitrary transformation, to make it look like a different state space.

The moral, in a nutshell, is that if you can do this, then by changing the way we look, we can see anything as compared to what there is — in other words, that the activity of observing can involve a transformation applied to unknown reality, and, as a consequence of that, this transformation can change the apparent "nature" of this "reality". The argument concludes that *every* model can account for some (transformed) aspect of an underlying reality.

Although Putnam's proof involves a mixture of physical and mathematical elements and was criticised on various grounds, the thesis "Nature is a Chameleon" was a revelation for engineers, systems scientists and practical modelers, who always suspected something like that. This Chameleon Thesis, taken to the extreme, says that modeling becomes impossible as we lose selection criteria for models. We have something like an applied version of the infamous "anyting goes": reality seems to "verify" every model.

Let us accept the Chameleon Thesis for a while, and let us turn to an application of endo/exo concepts.

Using these words, it will be immediately clear that a decoupled, external observer can be free from the Chameleon problem. It only bothers a constrained observer. By using temporal (spatial, ..., etc.) independence ensured by the decoupling, the external observer can view the entire movie of the given system's interactions, and select those models that fit a posteriori. Nature may be a Chameleon but certain camouflages are more important than others. If we wait long enough, they can be identified.

That is, we conclude that the "Chameleon thesis" is implicitly based on an endo, or internal, consideration. Consequently, in a further twist, the functionalist perspective which it denies must imply an exo view. If it does so, it cannot offer direct temporal reference.

Thus, the situation is this: either we want temporal coupling, take an endo stance, and lose the direct describability of complex reality, or we drop temporal fidelity, go to the exo standpoint, impose a detached view, and this long look gives us a valid functionalist perspective. To put this even more plainly: informational detachment is only possible if accompanied with temporal (exophysical) detachment.

7. Life, Complexity and Computer Simulations

The close relationship we have seen to exist between observer standpoints, models formulated by a given observer, and the outcomes of the Omniscience (viz. Chameleon) problems permits us to analyze the role of computer simulations in the theory of life.

Let us first summarize the earlier insights of this paper. We argued that the logic of observation is compatible with two scientific strategies that are essentially different. The observer may adopt an externalist, decoupled, time-global, omniscient, functionalist — in fact somewhat "Platonistic" — view. Using a further assumption, that computers are universal functionalist tools, this line of thought opens a way for computer simulations such as those praised in the Alife enterprise. Having seen this on the case of chemistry, we understand that if efficiency of description is what we hold important, this enforces us to follow an externalist epistemological strategy.

Alternatively, we may chose the complementary vista, which stays on the endo side, and resigns from functionalistic computer models, in order to express the built-in temporal constraints (such as those of genesis and decay) lost otherwise. The imperative of the standpoint may suggest this strategy in cases where, like in Internal Chemistry or evolution, the situation is obligate.

I would guess that the temptation to understand reality as *we* know it (in the Kantian sense of *sub specie hominis*) should bias our expressions towards the latter, internal, alternative.

Acknowledgment

The paper was written with the support of the research grant "Cognitive Biology" (#T6428) of OTKA, Hungary. Partial support of the Department of Logic and Philosophy of Science, University of Basque Country, San Sebastian, is acknowledged. The author wishes to thank Prof. O.E. Rössler for numerous discussions on the subject during an earlier research stay at the Dept. of Theoretical Chemistry, University of Tübingen, Germany.

References

1. Atmanspacher, H. and Dalenoort, G.J. (ed.) 1994: End/Exo Problems in Dynamical Systems, *Springer*, Berlin.
2. Bergson, H. 1944: Creative evolution, *The Modern Library*, New York (French orig. in 1912).
3. Cariani, P. 1989: On the Design of Devices with Emergent Semantic Functions, Ph.D. dissertation, *Dept. of Systems Sci.*, SUNY at Binghamton.
4. Emmeche, C. 1993: The Garden in the Machine: the Emerging Science of Artificial Life, *Princeton UP*, Princeton NJ.
5. Finkelstein, D. and Finkelstein, S.R. 1983: Computer Interactivity Simulates Quantum Complementarity, Int. J. Theor. Phys. **22**, 753-779.

6. Galouye, D.F. 1964: Simulacron Three, *Gregg Press*, Boston.
7. Kampis, G. 1991: Self-Modifying Systems: A New Framework for Dynamics, Information, and Complexity, *Pergamon*, Oxford-New York, pp 543+xix.
8. Kampis, G. (ed) 1992: Creativity in Nature, Mind and Society, World Futures: TheJnl. of General Evolution, **32**/2-3.
9. Kampis, G 1993: From Dynamics to Information. The Endophysical Perspective of Change. in: Kampis and Weibel 1993.
10. Kampis, G. 1994: Biological Evolution as a Process Viewed Internally, in: End/Exo Problems in Dynamical Systems (H. Atmanspacher, G.J. Dalenoort ed.), *Springer*, Berlin.
11. Kampis, G. and Weibel, P. (ed.) 1993: Endophysics. A New Apporach to the Observer-Problem with Applications in Physics, Biology and Mathematics, *Aerial Press*, Santa Cruz, CA.
12. Lem, S. 1979: Non Serviam, in: A Perfect Vacuum, *M. Secker and Warburg*, London.
13. Löfgren, L. 1993: Language, Information, and Endophysics. in: Kampis and Weibel 1993.
14. Morgan, C. Lloyd 1923: Emergent evolution, *Williams and Norgate*, London.
15. Nagel, E. 1961: The Structure of Science; Problems in the Logic of Scientific Explanation. *Harcourt, Brace & World*, New York.
16. Popper, K.R. 1965: Das Elend des Historizismus. *J.C.B. Mohr*, Tübingen.
17. Putnam, H. 1960: Minds and Machines, in: (S. Hook ed.) Dimensions of Mind, *New York Univ. Press*, New York.
18. Putnam, H. 1988: Representation and Reality, *MIT Press*, Boston.
19. Rosen, R. 1977: Complexity as a System Property, Int.J. General Systems **3**, 227-232.
20. Rosen, R. 1993: Life Itself, *Columbia Univ. Press*, New York.
21. Rössler, O.E. 1987: Endophysics, in: Real Brains - Artificial Minds (ed. Casti, J. and Karlquist, A.), *North-Holland*, New York.
22. Rössler, O.E. 1992: Endophysik. Die Welt des Inneren Beobachters, *Merve*, Berlin.
23. Rössler, O.E. 1984: Deductive Prebiology, in: Molecular Evolution and Protobiology (ed. Matsuno, K., Dose, K., Harada, K, and Rohlfing, D.L.), *Plenum*, New York, pp. 375-385.
24. Svozil, K. 1993: Randomness and Undecidability in Physics, *World Scientific*, Singapore.
25. Whitehead, A.N. 1929: Process and Reality, *Free Press*, New York.
26. Wiener, N. 1948: Cybernetics, Or Control and Communication in the Man and the Machine, *MIT Press*, Boston.

2. Origins of Life and Evolution

Prebiotic Chemistry, Artificial Life, and Complexity Theory: What Do they Tell us About the Origin of Biological Systems?

Antonio Lazcano
Facultad de Ciencias, UNAM
Apdo. Postal 70-407
Cd. Universitaria, México 04510, D.F.
MEXICO

ABSTRACT

Although the origin of self-sustaining, autoreplicating systems capable of undergoing Darwinian evolution is still unknown, a research program based on the hypothesis of chemical and precellular evolution has provided a framework within which the abiotic synthesis of biochemical monomers and membrane components, the experimental study of replicative systems, and the interactions between different ribozymes and a potentially wide range of substrates including amino acids, can be incorporated into a coherent historical narrative of evolutionary events. The significance of mathematical models and computer-based simulations of autocatalytic cycles based on complexity theory to the study of the origin of life will be considerable enhanced when experimental evidence supporting them becomes available.

keywords: prebiotic chemistry, precellular evolution, RNA world

1. Introduction

How the transition from the non-living to the living took place is of course still unknown. Nonetheless, a fruitful experimental approach to this issue has developed following the ideas suggested by A. I. Oparin over seventy years ago, according to which the emergence of living systems was the result of a lengthy step-wise process of abiotic synthesis of organic molecules that eventually self-organized into colloidal systems from which the first heterotrophic, anaerobic bacteria originated (Oparin, 1924, 1938). This hypothesis, which was the outcome of the analysis of the issue within a Darwinian framework, has received strong support not only from the laboratory simulations of possible prebiotic conditions, but also from a wide range of results from fields ranging from astronomy to molecular cladistic analysis.

Although many of Oparin's original premises about the nature of the first organisms have been superseded and require a radical redefinition, the suggestion that life resulted from a process of prebiotic chemical evolution is still favoured by many biologists. In the past recent years this view, which is now considered by some as a disputable orthodoxy, has encountered healthy resistance not only because of the disputes on the environmental conditions of the primitive Earth, but also from alternative suggestions that include the possibility that (a) the organic compounds from which life arose were bought in by comets or meteorites (Chyba et al, 1990); (b) that chemically active mineral surfaces such those of crystalline clays led to the first replicating systems (Cairns-Smith, 1982); or (c) that pyrite may have played a direct role in generating sulphur-based metabolic cycles from which primordial autotrophic forms of life eventually evolved (Wächtershäuser, 1988). Another approach that has become increasingly popular in the past few years, specially among scientists working on physical sciences, artificial intelligence, parallel computing, and the dynamics of complex adaptative systems, attempts to understand the origin of life in terms of complexity theory. The purpose of this paper is to outline some of the most significant results based on the idea of chemical evolution described above, pointing out both its advantages and limitations. A brief (and very likely, biased) discussion on the significance of artificial life and complexity theory to the issue of the origins of life is also included.

2. Prebiotic synthesis of organic compounds

The possibility that the origin of life was preceded by a period of chemical abiotic synthesis of organic compounds was received with considerable skepticism when it was first suggested by Oparin (1924, 1938). However, the existence of organic molecules of potential prebiotic significance in interstellar clouds and cometary nuclei, and of a copious array of small molecules of considerable biochemical importance in carbonaceous meteorites, provides conclusive evidence that such process took place not only before the appearance of life, but even before the origin of the Earth itself. Some of these compounds, like those found in the interstellar environment, may have played a minor role in the origin of life, but their presence in places where star and planetary formation is occuring is significant. In particular, the presence of a large array of proteinic and non-proteinic amino acids, carboxylic acids, purines, pyrimidines, hydrocarbons, and other molecules that have been found in the 4.5 x 10^9 years-old carbonaceous chondrites support the idea that comparable processes took place in the primitive Earth (Oro et al, 1990).

The hypothesis of a prebiotic phase of chemical evolution has found considerable support in laboratory simulations. The first succesful synthesis of organic compounds under plausible primordial conditions was accomplished by the action of electric discharges acting for a week over a mixture of CH_4, NH_3, H_2, and H_2O, yielding racemic mixtures of several proteinic amino acids, as well as hydroxy acids, urea, and other organic molecules (Miller, 1953). A few years later, Oro (1960) showed that adenine, a purine that plays a central role in both genetic processes and biological energy utilization, was a major product of the non-enzymatic condensation of HCN. The potential role of HCN as a precursor in prebiotic chemistry was further supported when it was found that the hydrolytic products of one of its polymers included amino acids, purines, and orotic acid, an important biosynthetic intermediate in the anabolic routes of both uracil and cytosine (Ferris et al, 1978). The laboratory synthesis under possible primitive conditions of oligomers of these compounds, as well as of additional molecules including tricarboxilic acids, fatty acids, alcohols, porphyrins, and a number of coenzymes has been reviewed elsewhere (Oro et al, 1990).

The fact that a number of chemical components of contemporary forms of life can be synthesized non-enzymatically in the laboratory does not necessarily implies by itself that that they were also essential for the origin of life, or that they were available in the prebiotic environment. Moreover, the lack of agreement on the chemical constituents of the primitive atmosphere has opened major debates, since the possibility that it consisted of a mixture of much less reducing gases such CO_2, N_2, and H_2O is now favoured by many planetary scientists. However, the issue is far from solved, and other alternatives remained open and have not been explored in detail. For instance, impacts of iron-rich asteroids may have enhanced the reducing conditions (Chyba et al, 1990), or tadditional sources of hydrogen such as pyrite ma have also been available, since in the presence of ferrous iron, two sulfide ions (SH^-) would be converted into a disulfide ion (S_2^-) releasing molecular hydrogen (Drobner et al, 1990).

Critical acknowledgement of the problems faced in prebiotic chemistry should not lead to a complete disqualification of more than three decades of experimental simulations, nor of the basic premises underlying Oparin's hypothesis. Even if the ultimate source of these organic compounds turn out to be comets and meteorites (Chyba et al, 1990), recognition of their extraterrestrial origin would not imply by itself a reappraisal of panspermia, since an epoch of intense collisions is now recognized as part of the early evolution of the Earth and other solar system bodies. Moreover, the stricking correlation between the compounds which are produced in prebiotic simulations and also found in carbonaceous chondrites, and those forming part of biological macromolecules should not be underscored, since it is strongly suggestive of the possibility that such compounds were the starting material for the origins of life.

3. The origin of replication

No molecule is alive by itself, but all contemporary life forms require an intracellular molecular apparatus capable of expressing and, upon reproduction, transmitting to the progeny information capable of undergoing evolutionary change. How this ubiquitious genetic system originated is one of the major unsolved problems in contemporary biology and, perhaps not surprisingly, has become one the major questions in the study of the origin of life.

It is generally accepted that complex DNA genomes were preceded by simpler molecules which made copies of themselves by reactions simpler than extant nucleic acid replication. The possibility that RNA was used in early cells genetic material prior to the evolutionary emergence of DNA has been expressed independently by many authors. As reviewed elsewhere (Lazcano, 1994), this hypothesis is supported by several independent lines of evidence, which include (a) the central role that different RNA molecules play in protein biosynthesis, and the well-known fact that this process can take place in the absence of DNA; (b) the fact that the biosynthesis of deoxyribonucleotides always proceeds via the enzymatic reduction of ribonucleotides, i. e., in all organisms deoxyribonucleotides are formed directly or indirectly (as in the case of dTTP) from a cellular pool of ribonucleotides; and (c) last but not least, the discovery of catalytic RNA molecules, i. e., ribozymes.

The existence of a primordial replicating and catalytic apparatus devoid of both DNA and proteins and based solely on RNA molecules, a stage now refered to as the RNA world, was first suggested by Woese (1967), Crick (1968), and Orgel (1968). Replication and transcription of nucleic acids are both based on the non-enzymatic hydrogen-bonding of complementary bases, but how RNA replicated during this hypothetical primitive stage is unknown. A template-dependent ribozymic RNA polymerase has not yet been discovered, but the application of powerful molecular biology techniques has led to the synthesis of an artificial ribozyme able to catalyze repeatedly the ligation of short oligonucleotides complementary to an RNA template (Doudna and Szostak, 1989). The study of non-enzymatic template-catalyzed polymerization reactions of activated derivatives of ribonucleotides suggests that protein-free replication of genetic information is possible, but all efforts to achieve polyribonucleotide self-replication under possible prebiotic conditions have failed (Orgel, 1987). The problem is further aggravated by the fact that the abiotic availability of RNA is still an open question. It is possible that primitive chemical synthesis of RNA components may be achieved, but the rather low half-life of ribose under prebiotic conditions have led credence to the idea that RNA was preceded by a simpler organic replicating systems (Joyce et al, 1987).

The chemical nature of the first genetic polymers and the catalytic agents that may have preceded the RNA world are still unknown, and at the time being only semi-educated guesses may be made. The problems involved with the prebiotic synthesis and accumulation of D-ribose, and with the chain-elongation inhibition reaction that takes place when both the D- and L-ribose forms are present, have led to the suggestion that RNA was preceded by simpler polymers of prochiral open-ring nucleoside analogues based on acyclic, flexible compounds such as glycerol (Joyce et al, 1987). However, the study of such molecules yielded results far from encouraging. Other candidates viewed with considerable interest are the so-called polypeptide nucleic acids, or PNAs, which are linear molecules in which the sugar-phosphate skeleton is replaced by an uncharged peptide-like backbone. These polymers are formed by achiral amino acid units linked by amide bonds, to which the bases are attached by methylene carbonyl linkages (Nielsen et al, 1991). Such molecules form very stable complementary duplexes both with themselves and with nucleic acids, and because of their peptide-like nature they may well be endowed with catalytic activity. Since amino acids may have been amongst the most abundant organic compounds in the prebiotic environment, study of the prebiotic availability of PNAs and their possible evolutionay significance is worthy of serious pursuit.

Even if our current descriptions of the origins of life are strongly hindered by our lack of understanding of the nature of the first genetic material, recent evidence suggests that non-enzymatic replication may be a widespread phenomenon in different synthetic chemical systems, some of which lack the familiar nucleic acid-like structure (Orgel, 1992). One such system, which has been designed by Julius Rebek and his group at MIT, consists of a self-complementary product of the chemical reaction between an amino-adenosine and a complex aromatic ester, whose product can catalyze the coupling of more of these molecules in a non-aqueous solvent such as chloroform (Hong et al, 1992). A completely different chemical replicator was developed by Pier Luigi Luisi and collaborators at the ETH-Zürich, and is formed by synthetic micelles containing lithium hydroxyde stabilized by octanoic acid sodium salt as surfactant. These reverse micelles are contained in an organic solvent, which acts as a substrate for the formation of additional structural components. The micelle-mediated synthesis of additional octanoic acid from the octanoic acid octyl ester is hydrolyzed yielding octanoic acid salt and 1-octanol (Bachmann et al, 1992).

Some authors have argued that the formation of membranes took place after the development of a replicating system (Eigen and Schuster, 1978). Other hypothesis on the origin of life are strongly committed to the presence of phase-separated polymolecular systems in which replication was not a self-contained property, but arose because of the interaction between different components of the prebiotic soup (Lazcano, 1994). Whether a genetic machinery could developed or not unless it was embedded in a self-organizing system is a largely open question. However, the lack of compartments would severely limit the possible cooperative interaction between the different molecules forming the replicative apparatus and its possible preferential selecion and accumulation, and would also lead eventually to their dispersal. It is possible that autocatalytic micelles may have existed during primitive times and provided a microenvironment favourable for the development of nucleic acid-like replication (Orgel, 1992). However, a reappraisal of their evolutionary significance will require not only their formation using potentially prebiotic reactants, but also a coherent description of the hypothetical transition from replicative membranous vesicles, to the present stage in which cellular reproduction and biological inheritance are based not on autocatalytic membranes but on nucleic acid replication.

4. The emergence of biological order

A criticism frequently raised against Darwinian explanations is the inability of historical descriptions based solely on natural selection to account for the origin of a wide range of adaptations. How did such complex traits as ribosome-mediated protein synthesis evolve through the accumulation of small variations, each of which must represent an adaptative improvement on the previous one, until they came to serve their current purpose? The origin of gene expression remains one of the most difficult problems in studying the emergence of life, but what may be important insights on possible intermediate evolutionary are rapidly accumulating. For instance, the efficient abiotic synthesis of uracil derivatives to which amino acid side groups can be chemically linked suggests that polyribonucleotides with protein-like properties could represent a bridge between an RNA world and an RNA/protein world (Robertson and Miller, 1995). Evidence suggesting that initial steps of the genetic code became stablished during the RNA world stage include the discovery of highly specific binding sites in RNA molecules for arginine (Yarus, 1988) and valine (Majerfeld and Yarus, 1994). More recently, the *in vitro* selection from a random pool of a ribozyme that catalyzes its own aminoacylation (Illangasekare et al, 1995).

A number of authors have, however, suggested that nucleic acid-like replication and protein biosynthesis developed after the stablishement of self-perpetuating autocatalytic cycles ancetsral to the first living organisms (Wächtershäuser, 1988; Morowitz, 1992; Kauffman, 1993). It is reasonable, of course, to assume that the prebiotic broth must have been a extremely complex mixture in which a large number of as yet undescribed changes were constantly taking place, leading to the stablishment of huge networks of interrelateded catalysts and reaction products. The idea of under such conditions cycles became stablished and eventually led to the first living organisms, however, it is not without precedents: it was first suggested by Ycas (1955) and, in a way, by Oparin (1924, 1938), who argued long before the genetic role of nucleic acids was known that networks of reactions that could be considered as a form of primitive metabolism preceding the emergence of the first living systems first evolved within coacervates.

An intellectual kinship can be recognized between the above suggestions and the ideas developed within the framework of complexity theory, according to which life was the outcome not of self-replicating molecules, but resulted from the evolution of self-sustaining closed networks of catalyzed reactions that produced copies of themselves. Once a certain level of complexity was reached in a large enough collection of random catalytic polypeptides, the system underwent a sudden phase transition that lead to systems with greater complexity and enhanced catalytic properties (Kauffman, 1993). This approach is increasingly popular among physicists and computer scientists, and is part of a tradition that includes the comparison of living systems with physical and physico-chemical systems far from equilibrium such as the Bénard convection pattern and the Beloussov-Zhabotinsky reactions. As summarized recently by Christian de Duve (1995), the development of complexity theory and artificial life is related to cellular automatons, catastrophe theory, von Bertalanfy's general system theory, chaos and fractals, and so on. Although complexity theory has achieved remarkable simulations using by computer-assisted theoretical analysis, such *in silico* description of autocatalytic cycles lacks compelling experimental data. No protein is known to replicate, and from a biological standpoint it is reasonable to assume that regardless of the chemical complexity of the prebioth environment, life could have not evolved in the absence of a genetic replicating mechanism insuring the maintenance, stability, and diversification of its basic components.

5. The synthesis of life

"There still remains, however, the problem of the artificial synthesis of organisms, but for its solution a very detailed knowledge of the most intimate; internal structure of living things is essential" wrote A. I. Oparin in his 1938 book on the origins of life, "even the synthesis of comparatively simple organic combinations can be accomplished only when one possesses a more or less complete understading of the atomic arrangement of their molecules. This, of course; would apply even more so in the case of such complex systems as organisms. We are still to far removed from such a comprehensive knowledge of the living organism to even dream of attempting their chemical synthesis."

Perhaps not surprsingly, there has been considerable speculation over the years as to how long it took for life to arise. Most of the estimates have assumed a slow, step-wise process involving periods of time of billions of years, but there is convincing paleontological evidence showing that a complex, highly diversified microbiota was already in existence 3.5 x 10^9 years ago (Schopf, 1993). In fact, the chemistry of prebiotic reactions is fast and robust, and no examples of slowly synthesized molecules are know. The main bottlenecks in the origin of life and early stages of metabolic evolution appear to have been (a) the appearance of replicating systems; (b) the emergence of protein synthesis; and (c) the evolutionary development of the stater primitive types from which latter proteins evolved through gene duplication and divergence. How long it took for ribosome-mediated protein synthesis to appear is unknown, but it must have developed before the prebiotic amino acids were destroyed. Because of the rapid exhaustation of prebiotically synthesized compounds, it is doubtful that the emergence of self-sustaining cells could have been delayed for lengthy periods of time. In fact, the chemistry and bacterial genetics are such that the time require for life to appear and evolve to the cyanobacteria-like fossils found in the Warrawoona formation could have been of 10 million years of less (Lazcano and Miller, 1994).

Is it possible to compress such a relatively small geological period into a laboratory timescale? Is the artificial synthesis of new life forms likely to be achieved in the near future? The hypothesis of chemical evolution will not be disproved by our inability to achieve the synthesis of life under simulated prebiotic conditions. The goal of evolutionary biology is not to reproduce under *in vitro* conditions the entire sequence of events, but to construct a coherent historical narrative supported by experimental observations, paleontological evidence, and sequence comparisons. If the synthesis of artificial life forms does indeed take place, very likely it will be derived from directed evolution experiments with catalytic RNA molecules. As emphasized by de Duve (1995) if life "is ever to be created artificially, it will be by a chemist, not by a computer". The experimental results achieved in the past few years using *in vitro* systems based on RNA molecules subjected to cyles of selection, amplification, and mutation under directed evolution conditions (cf. Gesteland and Atkins, 1993) strongly suggest that a system based on ribozymes and exhibiting some of the essential characteristics of extant organisms may be feasible. If such goal is achieved, our description of the emergence of the biosphere would still be hindered by our lack of understanding of the origin of RNA and its predecessors.

6. Conclusions

Mainstream ideas on the origin of life have been developed mainly within the historical framework of an evolutionary analysis, which attempts to weave together a large number of miscellaneous observations and experimental findings. Herein lies the heruistic value of such approach. Because of its vey nature, evolutionary biology is a highly ecclectic discipline, and the possibility that it may be enrichened by the results achieved so far by complexity theory analysis should not be underestimated.

Indeed, the development of complexity theory and of computer-assisted mathematical models for generating patterns and forms have reached levels of considerable sophistification. Such mathematical excersises may not be totally accurate representations of reality, but they may provide stimulating models on the origin of some of the basic traits of biological systems. However, for the study of the origin of life these results are suggestive, but not compelling. At the time being these models may have considerable application in computer science, but their significance for the understanding the emergence of biological order appears to be limited to providing an analogy. Whether they can lead to the development of new approaches to major biological issues is still an open question, and the development of a strong and convincing research program based on this approach will probably depend strongly on empirical support so far unavailable.

Acknowledgements

I am indebted to the organizers of the 3rd. European Meeting on Artificial Life for their kind invitation to present my views on the emergence of life. Parts of this manuscript were completed during a leave of absence as Visiting Professor during which I enjoyed the hospitality of Patrick Forterre and his associates at the Institut de Génétique et Microbiologie at the Université de Paris-Sud, Orsay.

References

- Bachmann, P. A., Luisi, P. L., and Lang, J. (1992) *Nature* **357**: 57-59

- Cairns-Smith, A. G. (1982) *Genetic Takeover and the Mineral Origins of Life* (Cambridge University Press, Cambridge)

- Chyba, C. F., Thomas, P. J., and Sagan, C. (1990) *Science* **249**: 366-373

- Crick, F. H. C. (1968) *Jour. Mol. Biol.* **38**, 367-379

- de Duve, Ch. (1995) *Vital Dust: life as a cosmic imperative* (Harper Collins Publ. Co., New York)

- Doudna, J. A. and Szostak, J. W. (1989) *Nature* **339**: 519-522

- Drobner, E., Huber, H., Wächtershäuser, G., Rose, D., and Stetter, K. O. (1990) *Nature* **346**: 742-744

- Eigen, M. and Schuster, P. (1978) *Naturwissenschaften* **65**: 341-369

- Ferris, J. P., Joshi, P. C., Edelson, E. H., and Lawless, J. M. (1978) *J. Mol. Evol.* **11**: 293-311

- Gesteland, R. F. and Atkins, J. F. (eds.) *The RNA World* (Cold Spring Harbor Laboratory Press, Cold Spring Harbor, NY)

- Hong, J. I., Feng, Q., Rotello, V., and Rebek, J. Jr. (1992) *Science* **255**: 848-850

- Illangasekare, M., Sanchez, G., Nickles, T., and Yarus, M. (1995) *Science* **267**: 643-647

- Joyce, G. F., Schwartz, A. W., Orgel, L. E., and Miller, S. L. (1987) *Proc. Natl. Acad. Sci. USA* **84**: 4398-4402

- Kauffman, S. A. (1993) *The Origins of Order: self-organization and selection in evolution* (Oxford University Press, Oxford), 709 pp.

- Lazcano, A. (1994) In S. Bengtson (ed), *Early Life on Earth: Nobel Symposium No. 84* (Columbia University Press, New York), 60-69

- Lazcano, A. and Miller, S. L. (1994) *Jour. Mol. Evol.* **39**: 546-554

-Majerfeld, I. and Yarus, M. (1994) *Nature Struct. Biol.* **1**: 287-289

- Miller, S. L. (1953) *Science* **117**: 528-529

- Morowitz, H. J. (1992) *Beginnings of Cellular Life: metabolism recapitulates biogenesis* (Yale University Press, New Haven)

- Nielsen, P. E., Egholm, M., Berg, R. H., and Buchardt, O. (1991) *Science* **254**: 1497-1500

- Oparin, A. I. (1924) *Proiskhodenie Zhisni* (Moscovky Rabotchii, Moscow)

- Oparin, A. I. (1938) *The Origin of Life* (MacMillan, New York)

- Orgel, L. E. (1968) *Jour. Mol. Biol.* **38**, 381-393

- Orgel, L. E. (1987) *Cold Spring Harbor Symp. Quant. Biol.* **52**, 9-16

- Orgel, L. E. (1992) *Nature* **358**, 203-209

- Oro, J. (1960) *Biochem. Biophys. Res. Comm.* **2**: 407-411

- Oro, J., Miller, S. L., Lazcano, A. (1990) *Annu. Rev. Earth Planet. Sci.* **18**: 317-356

- Robertson, M. D. and Miller, S. L. (1995) *Science* (in press)

- Schopf, J. W. (1993) *Science* **260**: 640-646

- Wächtershäuser, G. (1988) *Microbiol. Rev.* **52**, 452-484

- Woese, C. R. (1967) *The Genetic Code : the molecular basis for gene expression* (Harper and Row, New York)

- Yarus, M. (1988) *Biochemistry* **28**: 980-988

- Ycas, M. (1955) *Proc. Natl. Acad. Sci. USA* **41**: 714-716

Compartimentation in replicator models

Juan C. Nuño[1,2], Pablo Chacón[2,3], Alvaro Moreno[4] and Federico Morán[2]

[1] Cátedra de Matemáticas, ETSI de Montes, U.P.M., 28040 Madrid, Spain
[2] Dpto. Bioquímica y Biología Molecular I, Facultad de Químicas. U.C.M., 28040 Madrid, Spain
[3] Centro de Investigaciones Biológicas. C.S.I.C, Velázquez 144, 28006 Madrid, Spain
[4] Dpto. Lógica y Filosofía de la Ciencia, Facultad de Filosofía, U.P.V., Apdo. 1249, 20080 San Sebastián, Spain

Abstract. Recently a model formed by self-replicative units with catalytic capabilities evolving in an extended system has been presented. It has been shown that under particular conditions this model exhibits spatial *compartimentation* without any kind of membrane. In the framework of *ALife* , we suggest that this model can allow a global growth in the complexity of those models based on the hypothesis of the so-called *RNA-world.* However, this increase has got a limit defined by the impossibility of expressing the informational potentialities into functional complexity when a unique type of entity is involved in the system.

1 Introduction

Theoretical biology faces the origin of life as a specific problem while Artificial Life (*ALife*) intends to solve the broader problem of how a living organization could become manifest (no matter where). The main theoretical biology goal is to follow the clues of the origin of a very specific Living Organization (LO), without asking about the generality of this organization or why these components and not others. Biology tries to explain, from the laws of physics and chemistry, the origin of an organization supported by complex molecular components (DNA, RNA, proteins, ...) and, as an intermediate step, the synthesis of these components from simpler forms of organization is investigated.

Most of the work done on the origin of life, in the frame of theoretical biology, is an attempt to explain what kind of process (and from what components) originated the most primitive forms of life (i.e. the terrestrial minimal life [1]) from which all the today's organisms have evolved. It is commonly accepted that in this terrestrial biogenesis there would exist contingent local facts that could differ from those (hypothetically) happening in other places. On the other hand, it is also reasonable to think that some of the processes that took place during the origin of life on the Earth are necessary and, therefore, universal. Then, which of these processes are local? and, which are universal?

From the *ALife* perspective the problem of the origin of life is addressed in a different way. It is intended to propose abstract models (that refer to unspecific components) that are able to produce more complex forms. The starting point for these models should be coherent with a set of properties that are always present in molecular systems, and usually explained by physics and/or chemistry. But apart from this constraint, the biogenesis models in *ALife* do not restrict to *natural* kinds of components, in contrast with what happens in the framework of theoretical biology. In fact, the functionalism (predominant in *ALife*) defends the thesis of the radical separation between the idea of LO and the physical components supporting this organization [2]. Biologists, on the contrary, define the logic of the LO from a set of components that implicitly determine the organization. This has been forgotten in *ALife* even when the idea of LO is borrowed from theoretical biology. In addition, its claim on the material unspecificity of the primitive components would imply an enormous extension of the definition of LO, along with a change in the scale of the definition of the primitives. Likely, this is one of the most controversial points between biology and *ALife* . As a matter of fact, biology defines the LO assuming a higher degree of organization implicit in the primitive components. Certainly the situation is a bit different in the theory of the origin of life because in this field, what matter is to get a sequence of abstract and mutually connected stages, which allows the evolution towards progressively complex organizations.

Thus, if it is intended that the two aforementioned research programs converge, two related problems must be solved. On the one hand: what kind of component can support an abstract organization? On the other hand: how must the abstract models define the component characteristics in order to endow them with the capabilities of bringing about more complex organization forms? That is why the more capabilities for complexification the prebiotic *ALife* models have, the more specifications will have the components of such models for the origin of life.

In theoretical biology, many works and models on the origin of life deal with the so-called *RNA-world* [3]. This is based on the evidence of combined template and catalytic actions in some RNA molecules [4],[5]. If we agree that evolution is the driving force in the origin of life, the main question to be answered is what are the selective and evolutive features of a population of molecular species under prebiotic conditions. The first theoretical attempt of studying this problem was carried out by M. Eigen two decades ago [6]. He translated the biological problem to a mathematical model, formulating the molecular evolution in terms of differential equations of the different species concentration. Although the model was conceived to deal with RNA-like molecules, it can, in principle, account for any kind of abstract entity holding the assumptions of the model. Eigen's model is based on three important features; two are individual characteristics: *self-replication* and *fitness*. The third one is the existence of a constraint imposed by the environment to the whole population.

Can RNA-world based models be reformulated in abstract terms? What are the fundamental assumptions under which a theory of evolution should be developed? Since the original formulation appeared, a lot of work has tried to analyze and generalize this formulation. Following the original idea, we present here a model of a population of species that evolve in a formal *world*. Components of the system are assumed to be (by simplicity) self-replicative polymers evolving in a medium where interactions occur by diffusion (e.g., an aqueous medium). We assume also that these polymers are endowed with a basic template capacity, and have got sequence-dependent catalytic capabilities (in accordance with the RNA-world hypothesis in which self-replication is not helped by external catalyzers). If, in addition, we add other considerations about physico-chemical constraints, the number of different polymeric species that can be candidates for this kind of models decreases drastically. Therefore, the model we present here can be considered as an abstract *RNA-like world*.

In the following, we will comment the main characteristics of those models of populations formed by *general* self-replicative species, and we will suggest that diffusion can be the cause of spatial compartimentation. We will discuss also the implications of this compartimentation in the origin of life. We hope that this discussion may help to bridge the gap between biology, together with its standard works on the origin of life, and those that *ALife* has begun to develop.

2 Replicator models

Self-reproduction is an essential characteristics of the *living beings*. Without this capacity there is no way to maintain indefinitely the LO and, at the same time, explore new possibilities. Whereas *reproduction* involves the creation of a new organization, *replication* is referred to the process by which a structure (e.g. a molecule) is copied. Here we are concerned with the latter. In this context, self-replication is the process from which a given entity creates a new entity *similar* to the former one. Although the limits of this similarity are not totally specified, we will say that self-replication occurs when there is a way to relate the offspring with the original entity. Notice that following this definition self-replication is intrinsically *error-prone*, allowing the emergence of new entities and, therefore, *evolution*.

Every abstract entity that can replicate itself will be referred to as a *replicator* [7]. Obviously, many different entities satisfy this generic definition. It can be proven rigorously that the evolution laws of these systems are equivalent under some general conditions [8]. For instance, the temporal evolution of an ecosystem when rabbits and foxes coexist can be similar to the evolution of a population formed by RNA-like molecules evolving in a tank reactor if an error-free self-replication is assumed. The description of such *replicator models* depends on important aspects that decide the subsequent development of the theory. What are the basic characteristics of a model conceived to explain the temporal evolution of a population of formal replicators evolving in a hypothetic prebiotic scenario? Let us enumerate and comment the most relevant:

(*i*) *The macroscopic observable must be described.* Concentration is the macroscopic observable that is studied usually in chemical kinetics. That was also the variable M. Eigen uses in his original formulation [6]. Thus, generally, replicator models describe the temporal evolution of the concentration of every replicator in the system. If, as remarked in the previous section, replicators are thought to be polymeric chains builded up from different types of digits, the number of different replicators that could be present simultaneously in the system can be enormous (e.g., if the chain length is about 100, and the different types of digits are 4, as occurs in present day nucleic acids, this number is astronomic: $4^{100} \approx 10^{60}$). This characteristic, and the fact that the dimensionality of the system changes with time (appearance or extinction of a replicator) imposes serious limitations to the treatment of these models.

(*ii*) *The physical mechanisms that take place during the process of self-replication.* There is a great controversy about this point. On the one hand, we observe that the most simple way self-replication occurs in Nature is *via* a template. Hence, we will assume here that our abstract polymers self-replicate *via* a template: a particular component of the chain reacts preferably with another component (its complement). E.g., in current RNA, guanine (G) reacts with *larger probability* with citosine (C), and uracil (U) with timine (T). Notice that the reaction among not complementary pairs is also allowed. On the other hand, when we have relatively complex structures self-replication is a template process that involves catalytic action. However, the RNA-world hypothesis discards the existence of any kind of external catalyzer working during self-replication. The matter is that replication is a catalyzed process, but must be made without contribution of any *external* catalyzer, so it must be *internal.* The only way to solve this paradoxical situation seems to assume that replicators have got catalytic activity. Then, it is necesary that replicators play both template and catalytic role. The discovery of the catalytic activity of some current RNA, the so-called *ribozymes* [4], [5], has offered a plausible answer to this problem. In summary, self-replication must be necessarily carried out through the catalytic help of other replicators, giving rise to a *catalytic network* (see an example in figure 1).

(*iii*) *The existence of some kind of restriction in the accessibility of the phase space, i.e., the macroscopic variables cannot get all the possible values.* The influence of constraints on a population was pointed out by Darwin more than hundred years ago. Whereas in the absence of a restriction the total population increases indefinitely, any limitation in the sources yields necessarily to the selection of one or several types of replicators. Thus, any model about the origin of life should include some kind of restriction. There are several ways this selective process can be implemented [9]. For example, one possibility is to have a closed system in which *only energy* can be interchanged with the surroundings [10]. In this context, the total mass must be conserved and the only way to keep the system out of equilibrium is maintaining the affinity of the reactions different from zero by controlling the external flux of energy. As it will be discussed below, this can be done through the reaction of the regeneration of *active* or *rich-energy* monomers (the building blocs from which the replicators are built up). This is a charecteristic endergonic process where *inactive* monomers are regenerated with a finite velocity using the external supply of energy, by means of any kind of primitive transduction process. For instance, Skulachev [11] has proposed a plausible mechanism for a primitive use of UV radiation as a primary source of energy for monomer activation. Conclusions about the dynamics of catalytic networks evolving under this constraint have been obtained recently [10], [12].

(*iv*) *The particular boundary conditions under which the system is forced to evolve.* Related with the previous point is the problem of defining the boundary conditions of the system: the relation with the environment and the spatial geometry. For example, if, as stated before, a closed system is assumed, then there is no flux of matter through its borders. The choice of one or other topology depends on the nature of the problem (e.g., it can be assumed one, two, or three dimensional spaces, and considering periodic or non-periodic boundary conditions). The influence of boundary conditions on the system dynamics can be critical, i.e., slight environment disturbances can cause important changes in the system behavior. Although this point is considered of secondary importance in many studies, it could contain the explanation for relevant aspects of the theory of molecular evolution.

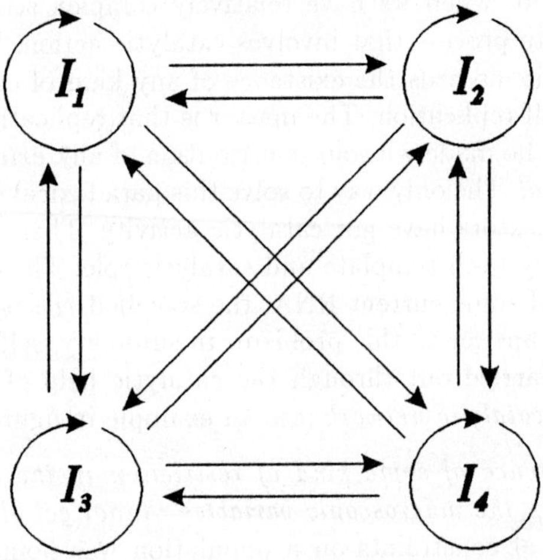

Figure 1. Schematic example of catalytic network. Replicator I_k ($k = 1,2,3,4$) can react with the catalytic help of other replicators I_j ($j \neq k$) to produce a copy similar to itself. In addition, replicators have got a lifetime. Then, they can degrade producing energy-poor monomers (process not shown in the figure).

According to these points, mathematical models have been developed within the framework of theoretical biology. From these models, the general behavior of catalytic networks has been extensively analysed during the last two decades and many relevant properties has been derived (see, for example, the recent study [13], and references therein).

In his pioneer work, Eigen and Schuster studied with special emphasis a particular network architecture, the *hypercycle*, in a continuous stirring tank reactor (i.e., a homogeneous medium) [14]. A hypercycle is a catalytic network formed by replicators that are cyclically linked. Contrary to what happens in populations of replicators without catalytic capabilities [15], hypercycles steps torward diversity, allowing the coexistence of different replicators through cooperation. But, the strong selective properties of hypercycles prevent their later evolution and growth (by increasing the number of replicators involved in the organization). Moreover, it has been proved that highly membered hypercycles become very sensitive to fluctuations and they are eventually unstable [16].

Due to unspecificity, in a prebiotic scenario, general catalytic networks are certainly more realistic than hypercycles. However, they are also unstable. It can be proven that coexistence among all the replicators is favoured when the catalytic interactions are symmetric and self-catalysis is not allowed [17]. If catalytic constants are different enough from each other, the state of coexistence of all the replicators involved in the network becomes unstable, and the system evolves towards a less membered network (maybe by losing more than one element). Therefore, any mutation that affects the catalytic efficiency of replicators can eventually destroy the network. This fact limits seriously the possibility of evolution of catalytic nets. The question then is how to solve the limitations of these models to increase the information content and therefore, explain more complex organizations?

3 The role of diffusion

There is an aspect, included in the previous assumptions, that must be more deeply taken into account. That is the diffusive forces acting on all the species present in the system. As it was previously stated, models based in the RNA-world hypothesis must consider populations of replicators evolving on an extended environment (that eventually could be inhomogeneous). Indeed, replicators are diluted in a liquid medium where, without any physical constraint, the diffusion through it is possible. Thus, in principle, diffusive forces should be considered in the study of evolving populations in prebiotic conditions. Nevertheless, many replicator models are analysed assuming that the population evolves in experimental devices with any kind of artificial mechanism of homogenization, e.g., the Continuous Stirring Tank Reactor (CSTR) models [9]. Therefore, the conclusions obtained from them (as those about catalytic networks remarked above) are only valid in the homogeneous limit.

Diffusion tends to homogenize the medium when acting on the replicators as a single force. Therefore, intuitively, one would say that systems under diffusion will approach asymptotically to a homogeneous state (and then, the diffusive terms could be removed from the theoretical description). However, in replicator models diffusion is coupled with a *local reaction field*, dependent on the replicator properties. Then, a direct question arises: is this coupling between local reactions and diffusion able to cause symmetry breaking and spatial pattern formation?

In his pioneer work, and much to the surprise of scientific community, Turing proved that this coupling can bring about the formation of spatial structures for a particular choice of the local field [18]. Since then, the analysis of many models describing physical and chemical systems has demonstrated that self-organization is not as an rare event as was thought several decades ago [19].

A similar reasoning might guide Boerlijst and Hogeweg [20] to discover that replicator models can suffer also self-organization when evolving in an extended system. Essentially, they showed that a local reaction based in self-replication coupled with diffusive forces can produce pattern formation. In particular, they studied a hypercycle formulated using cellular automata techniques, and they proved that diffusion induces the formation of spiral patterns. Perhaps, from an evolutive point of view, the main consequence is that under diffusion hypercyles are more resistant to parasites (i.e., any replicator which takes advantage of being a member of the network without favouring any of the network elements) (a recent discussion about that can be found in [21]). This fact revived the hypercycle organization as a relevant concept to evolutionary biology. However, the importance of their work on the rest of selective and evolutive features of replicator models has not been studied in its total extent.

Can other network architectures form spatial pattern? What kind of patterns? In the next section we will present a model that exhibit pattern formation. This patterns could be directly related with a compartimentation process during the origin of Life.

4 Clustering by diffusion

Chacón and Nuño have presented a model of a population of replicators evolving in a closed extended system [22]. The model is designed accordingly with a previously proposed theoretical scheme [10]. It addresses the temporal evolution of a population of replicators in a closed system. In agreement with the assumptions stated in section 2, the following reactions can take place:

$$I_i + I_j + \mu^* \xrightarrow{k_{ij}} 2I_i + I_j \tag{1}$$

$$I_i \xrightarrow{\delta_i} \mu \tag{2}$$

$$\mu \xrightarrow{\gamma} \mu^* \tag{3}$$

Each replicator I_i, in the presence of the activated monomer μ^*, can self-replicate with the catalytic help of I_j with rate k_{ij} (1). Replicators can degrade producing inactivated monomers μ with a rate δ_i, (2), and this byproduct can be recyclated in μ^*, with a rate γ (3).

Replicators and monomers are embedded in a medium, where they diffuse with diffusion coefficients D and d, respectively. In a first approach, the probability of mutation is set to be zero. Reaction (3) maintains the system far from equilibrium by means of an endergonic reaction (for example, by taking energy from the environment).

According with the above reaction scheme, the time evolution of the species concentration is governed by the following system of partial differential equations:

$$\dot{x}_i(\mathbf{r}) = x_i(\mathbf{r})[a(\mathbf{r})\sum_j^n k_{ji}x_j(\mathbf{r}) - \delta_i] + D\nabla^2 x_i(\mathbf{r})$$

$$\dot{a}(\mathbf{r}) = \gamma b(\mathbf{r}) - [a(\mathbf{r})\sum_i^n\sum_j^n k_{ji}x_i(\mathbf{r})x_j(\mathbf{r})] + d\nabla^2 a(\mathbf{r}) \quad (4)$$

$$\dot{b}(\mathbf{r}) = \sum_i^n \delta_i x_i(\mathbf{r}) - \gamma b(\mathbf{r}) + d\nabla^2 b(\mathbf{r})$$

where x_i, a and b are the concentrations of the replicators and activated and inactivated monomers, respectively. Clearly, the structure of the equations maintains the total concentration ($\sum_i^n x_i + a + b$) constant within the system.

Under particular conditions, system (4) presents inhomogenous solutions that account for pattern formation [22]. But, contrary to the hypercycle model, the patterns are radically different. Instead of spirals, consequence of the cyclic coupling among the members of the network, symmetrical or almost symmetrical stationary clusters appear for a particular choice of the parameters. These clusters are regions with high concentration of the total amount of replicators and low concentration of their constituents (monomers) separated by areas highly populated by monomers. This pattern remains stable although monomers can move over the whole space. An example of these structures is shown in figure 2.

It is worthy to mention that, as occurs in hypercycle models, a minimun number of replicators (four) is needed for the system to exhibit pattern formation. Moreover, the network architecture and the catalytic weights have a high influence on the system dynamics: even for catalytic networks formed by four replicators only special choices of the catalytic constants allow pattern formation. This fact may be a consequence of the local dynamics: the formation of clusters seems to be possible only when the dynamics for the species concentrations is chaotic [28], [27]. Similar behavior has been reported in Coupled Map Lattice models [23].

5 Discussion

In the light of what has been said in the previous section, one is tempted to establish a straightforward relation between the formation of spatial clusters and the origin of a permanent compartimentation in the system. Nevertheless, the permanence of dissipative structures needs of a constant influx of energy into the system. In the previous model an *energy* contribution is required during the conversion of poor-energy monomers into rich-energy monomers. The mechanism is similar to that described by Sagan and Ponamperuma for the formation of ATP, CTP, GTP and UTP from inactivated monomers [24]. This process could

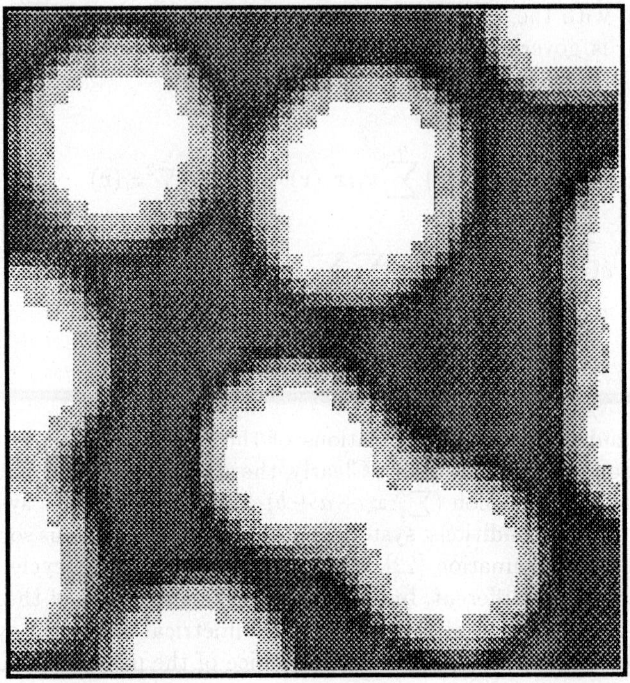

Figure 2. Snapshot of the total concentration of a four-membered catalytic network evolving in a bidimensional domain with non-flux boundary conditions. The diffusion coefficients are $D = 10^{-2}$ and $d = 1$. The matrix of catalytic constants is:

$$\begin{pmatrix} 0.5 & 1.6 & 0 & 2.2 \\ 1.5 & 1.0 & 2.0 & 0 \\ 0.5 & 0 & 0.6 & 0.4 \\ 0.1 & 0 & 0 & 0 \end{pmatrix}$$

The values of the rest of the parameters are: $\delta_i = 0.1$ for $i = 1, 2, 3, 4$ and $\gamma = 1$. White represents areas with a high amount of replicators, whereas dark grey represents areas with low concentration. Notice that the system has not reached a stationary state (some of the clusters are still splitting to get a quasisymmetric disposition)

be thought as a first step for the fixing of a metabolism dedicated to create the activated units needed for self-replication of molecules. It is well known that present day cells take advantage of energy by means of a mechanism coupled with the membrane (during the process of transduction carried out by ATPases)[25], a so complex mechanism that is hard to imagine that it could appear during the prebiotic stage. However, it turns out even more risky to formulate the question about how it could go further, from these structures, towards the origin of a biological membrane.

The interest of this compartimentation process is that it presents some of the functional advantages of cellular systems, without requiring the relatively complex structure of a membrane. The appearance of a membrane in simple molecular systems is not a rare phenomenon. There are models for which the construction of the membrane in very simple systems is a result of the *metabolic* activity of the own system [26]. But for such very simple systems, their isolation means an obstacle for a possible increase in their complexity (since there is no information content inside the system). The main problem when considering the process of compartimentation with a membrane is that as the complexity of the encapsulated systems increases, the set of functional conditions that the membrane should fulfil increases too. Already, for relatively low-complexity systems, the functional requirements for the membrane to act as an interface unit are very high (mechanisms for selective permeability and transduction of energy). Accordingly, the origin of a membrane with these functional characteristics requires that the replicators could translate the information in specifying the construction of a sufficiently complex metabolic network. Hence, when a too simple system is enclosed within a membrane, it will not have any chance to evolve towards the complexity needed in a prebiotic scenario.

That is why the possibility of compartimentation without membrane is very interesting for molecular systems of an intermediate complexity. Contrary to other models that explain the emergence of heterogeneous structures that have not functional consequences for the system, the appearance of a compartimentation can play a *casual role* in the behavior of the whole system. Although this question might be investigated with some detail, our conjecture is that the compartimentation could have allowed the replicators to increase their informational content.

Let us comment briefly some consequences that spatial compartimentation could have on the subsequent evolution of population of replicators. The searching efficiency of a replicator model depends on the diversity of the system. Thus, during the process of evolution, having many different *testing banks* will increase the possibility of appearance of a useful combination. This can be solved in two different ways: (a) simultaneous coevolution of many networks in different spatial scenarios, without intercommunication; or (b) evolution of a single network that *self-compartimentates* in different patches within the same extended system. Both solutions (a) and (b) are not mutually exclusive, but (b) certainly looks more realistic and attractive since it allows both evolution and communication, being the latter an aspect that could become very important to generate higher order organization among sufficiently differentiated patches of the same system.

On the other hand, the selective features can be also altered, as the following example shows. Assume that each replicator develops a specific way of regenerating the rich-energy monomers. Thus, in a medium without compartimentation, all the members of the population would take advantage of a larger production of activated units developed by a particular replicator. On the contrary, if compartimentation takes place, and eventually one replicator is linked to a metabolic activity greater than the average within a compartment, only those replicators belonging to this compartment will exploit this selective advantage, to the detriment of the rest of the population. Work along this line is currently in progress.

Acknowledgements

This work has been supported in part by grants PB92-0456, PB92-0908 and PB92-0007 from the DGICYT (Spain), and UPV 003.230-HA 160/94 from the University of the Basque Country (Spain).

References

1. Fleischaker, G.; Three models of a minimal cell. Colloquium on Prebiological Organization. University of Maryland. April 8-10, (1987)
2. Langton, Ch.; Artificial Life. In *Artificial Life I* pp 1-47, Ed. Ch. Langton, Addison Wesley, (1989).
3. Joyce, G.F. 1989. *Nature* **338**, 217-224
4. Doudna, J.A. & J.W Szostak. *Nature* **339** (1989) 519
5. Cech, T.R. *Nature* **339**, (1989) 507-508
6. Eigen, M. *Naturwissenschaften* **58** (1971) 465
7. Dawkins, R. *The Extended Phenotype.* Freeman, San Francisco. (1982)
8. Hofbauer, J. & K. Sigmund. *The theory of evolution and dynamical systems*, Cambridge University Press. (1988)
9. Schuster, P. & K. Sigmund. *Ber.Bunsenges.Phys.Chem.* **89** (1985) 668
10. Hofbauer, J. & P. Schuster. *Dynamics of Linear and Nonlinear Autocatalysis and Competition* in Stochastic Phenomena and Chaotic Behavior in Complex Systems, Springer-Verlag, Berlin. (1984)
11. Skulachev, V.P. *Antonie van Leeuwenhoek,* **65** (1994) 271-284.
12. Streissler, C. *Autocatalytic Networks Under Diffusion.* Doctoral thesis, (Universität Wien, 1992)
13. Stalder, P.F., W. Schnabl, C.V. Forst & P. Schuster *Bull. Math. Biol.* **57** (1995) 21-61.
14. Eigen, M. & P. Schuster. *The hypercycle — a principle of natural self-organization.* Springer-Verlag, Berlin. 1979
15. Eigen, M., J. Mc Caskill & P. Schuster. *Adv. Chem. Phys.* **75** (1989) 149-263
16. Bresch, C., U. Niestert & D. Harnash. *J.Theor.Biol.* **85** (1980) 399.
17. Nuño, J.C., M.A. Andrade & F. Montero. *Bull.Math.Biol.* **55** (1993) 417
18. Turing. A.M. *Philos.Trans.R.Soc.Lon.Ser. B* **237** (1952) 37
19. Cross, M.C. & P.C. Hohenberg. *Rev.Mod.Phys.* **65** (1993) 851
20. Boerlijst, M.C. & P. Hogeweg. *Physica D* **48** (1991) 17
21. Cronhjort, M.B. & C. Blomberg *J. Theor. Biol.* **169** (1994) 31-449.
22. Chacón, P. & Nuño, J.C.; *Physica D* **81** (1995) 398
23. Kaneko, K. *Physica D* **37** (1989) 60
24. Ponnamperuma, C., Sagan C. & Mariner, R. *Nature* **199** (1963) 222-238
25. Skulachev, V.P. *Molecular mechanisms in bioenergetics* (L. Ernster, ed.), Cap. 2. 37-73, Elsevier, Amsterdam (1994)

26. Luisi, P.L. & F.J. Varela. *Origins of Life and Evolution of the Biosphere* **19** (1989) 633-643
27. Andrade, M.A., J.C. Nuño, F. Morán, F. Montero & G.J. Mpitsos. *Physica D.* **63** (1993) 21
28. Schnabl, W., P.F. Stadler, C. Forst & P. Schuster. *Physica D* **48** (1991) 65

Evolutionary Dynamics and Optimization

Neutral Networks as Model-Landscapes for RNA Secondary-Structure Folding-Landscapes

Christian V. Forst*, Christian Reidys, and Jacqueline Weber

*Mailing Address:
Institut für Molekulare Biotechnologie
Beutenbergstraße 11, PF 100 813, D-07708 Jena, Germany
Phone: **49 (3641) 65 6459 Fax: **49 (3641) 65 6450
E-Mail: chris@imb-jena.de

Abstract

We view the folding of RNA-sequences as a map that assigns a pattern of base pairings to each sequence, known as secondary structure. These preimages can be constructed as random graphs (i.e. the *neutral networks associated to the structure s*).

By interpreting the secondary structure as biological information we can formulate the so called *Error Threshold of Shapes* as an extension of Eigen's *et al.* concept of an error threshold in the single peak landscape [5]. Analogue to the approach of Derrida & Peliti [3] for a flat landscape we investigate the spatial distribution of the population on the neutral network.

On the one hand this model of a *single shape landscape* allows the derivation of analytical results, on the other hand the concept gives rise to study various scenarios by means of simulations, e.g. the interaction of two different networks [29]. It turns out that the *intersection* of two sets of compatible sequences (with respect to the pair of secondary structures) plays a key role in the search for "fitter" secondary structures.

1. Introduction

The first theory of biological evolution was presented last century by Charles Darwin (1859) in his famous book *The Origin of Species.* It is based on two fundamental principles, *natural selection* and *erroneous reproduction* i.e. *mutation.* The first principle leads to the concept *survival of the fittest* and the second one to *diversity*, where fitness is an inherited characteristic property of a species and can basically be identified with their *reproduction rate.*

Au contraire to Darwin's theory of evolution the role of stochastic processes has been stated. Wright [31, 32] saw an important role for genetic drift in evolution in improving the "evolutionary search capacity" of the whole population. He saw genetic drift merely as a process that *could* improve evolutionary search whereas Kimura proposed that the *majority of changes* in evolution at the molecular level were the results of random drift of genotypes [18, 19]. The neutral theory of Kimura does not assume that selection plays *no* role but denies that any appreciable fraction of molecular change is caused by selective forces. Over the last few decades however there has been a shift of emphasis in the study of evolution. Instead of focusing on the differences in the selective value of mutants and on population genetics, interest has moved to evolution though natural selection as an optimization algorithm on complex fitness landscapes. However, for a short moment let us return to Darwin and his minimal requirements for adaption:

- a population of object that are capable of replication,
- a sufficiently large number of variance of those objects,
- occasional variations which are inheritable, and
- restricted proliferation which is constrained by limited resources.

In this paper we restrict ourselves to RNA, the possibly simplest entities that do actually fulfill all the four requirements listed above. We realize the fundamental dichotomy of genotypic legislative by RNA and the phenotypic executive is manifested by RNA secondary structures. In this context the mapping from RNA sequences to secondary structures is of central importance, since fitness is evaluated on the level of structures. This mapping induces naturally a *landscape* on the RNA sequences independent of any possible evaluation of RNA structures [27]. Following the approach in [23] we can construct these sequence structure maps by random graphs. By omitting any empirical parameter of RNA-melting experiments we obtain the so called *neutral networks* of sequences which each fold into one single structure. It can be shown that these neutral networks and the transitions between them are "essential" structural elements in the RNA-folding landscape [24]. These landscapes combine both in the first view contradicting approaches on biological evolution; Darwins survival of the fittest and Kimuras neutral random drift.

2. Realistic Landscapes

2.1. Fitness Landscapes and the Molecular Quasispecies

In this contribution we consider the most simple example of Darwinian evolution,namely a population $\mathcal{P}$ of haploid individuals competing for a common resource.

Following the work of Eigen [4, 5] we consider a population of RNA sequences of fixed length n in a stirred flow reactor whose total RNA population fluctuates around a constant capacity N. The definition of the overall replication rate of a sequence together with the constrained population size specifies our selection criterion.

In the limit of infinite populations its evolution is described by the *quasispecies equation*

$$\dot{c}_x = \sum_y Q_{xy} A_y c_y - c_x \Phi \tag{1}$$

where c_x denotes the concentration of genotype x, A_x is the replication rate of this genotype, and Q is the matrix of mutation probabilities, Q_{xy} being the probability for a parent of type y to have an off-spring of type x. The replication rates considered as a function of the genotypes x form a fitness landscape[1] [31] over the sequence space [6]. The total population is kept constant by a flux Φ compensating the production of new offsprings. The model mimics the asynchronous serial transfer technique [20].

As in the laboratory our RNA populations are tiny compared to the size of the sequence space. This fact forces a description in terms of stochastic chemical reaction kinetics. Two methods are appropriate to model stochastic processes:

- Gillespie [14] has described an algorithm for simulating the complete master equation of the chemical reaction network. We have used the implementation by Fontana *et al.* [10] for all computer simulations reported here. An individual sequence I_k can undergo two stochastic reaction events: either I_k is removed by the dilution flow, or it replicates with an average rate A_k that corresponds to the reaction rate constant in equ. (1). When an individual sequence is replicated, each base is copied with fidelity q. The overall model mimics the asynchronous serial transfer technique [20].
- While giving an complete description of the dynamics Gillespies algorithm does not allow for a detailed mathematical analysis. Therefore we approximate the quasispecies model by a birth-death process, following the lines of Nowak & Schuster [21] and Derrida & Peliti [3]. All analytical results presented in this contribution are based on this approach.

[1] For a recent review on fitness landscapes see, e.g., the contribution by Schuster and Stadler in the proceedings volume to the Telluride meeting 1993 [25]

In general all rate and equilibrium constants of the replication process and hence also the over-all rate of RNA synthesis are functions of the 3D-structure.

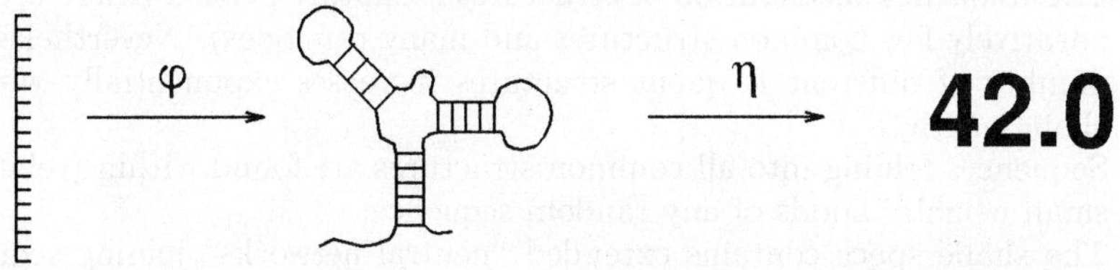

Fig. 1. Mapping of a genotype into its functional representation. The process is partitioned into two phases: The first phase is the complex mapping φ of sequences into secondary structures (phenotypes). Here neutrality plays a crucial rôle; in the second phase we omit the building of the spatial 3D-structure and evaluating its function. We assign arbitrarily a fitness-value to each phenotype by the mapping η.

This suggests to decompose the computation of the fitness into two steps: First we construct the shape of the RNA (phenotype) from its sequence (genotype), and then we consider the evaluation of this phenotype by its environment (figure 1). The effect of this composition is that we are left with two hopefully simpler problems, namely (1) to model the relation between sequences and structures in the special case of RNA, and (2) to devise a sensible model for the evaluation of there structures. The *combinatory map* of RNA secondary structures, i.e., the map assigning a shape $\varphi(x)$ to each sequence in the sequence space $\mathcal{C}$ will be discussed in the next section.

Formally, we consider the evaluation η assigning a numerical fitness value to each shape in the *shape space* $\mathcal{S}$. As even less is known in general about structure-function relations than about sequence-structure relations we will use the most simple model for the evaluation η. We assign arbitrary fitness-values $\eta(s_i)$ to specially chosen shapes s_i and a fitness-value to the background $\eta(\beta)$ (i.e. the remaining shapes) with the condition $\eta(s_i) > \eta(\beta)$ for all i.

Tying things together we are considering a fitness landscape of the form

$$f(x) = \eta(\varphi(x)). \tag{2}$$

2.2. The Combinatory Map of RNA Secondary Structures

Having defined the evaluation η of the structures we now turn to the sequence-structure relation φ. The phenotype of an RNA sequence is modeled by its minimum free energy (*MFE*) secondary structure.

The evidence compiled in a list of references [8, 9, 11, 15, 24, 26] shows that the combinatory map of RNA secondary structures has the following basic properties:

(1) Sequences folding into one and the same structure are distributed randomly in the set of "compatible sequences", which will be discussed below in detail.
(2) The frequency distribution of structures is sharply peaked (there are comparatively few common structures and many rare ones). Nevertheless, the number of different frequent structures increases exponentially with the chain length.
(3) Sequences folding into all common structures are found within (relatively) small neighborhoods of any random sequence.
(4) The shape space contains extended "neutral networks" joining sequences with identical structures. "Neutral paths" percolate the set of compatible sequences.
(5) There is a large fraction of neutrality, that is, a substantial fraction of all mutations leave the secondary structure completely unchanged (see figure 2).

These features are robust.

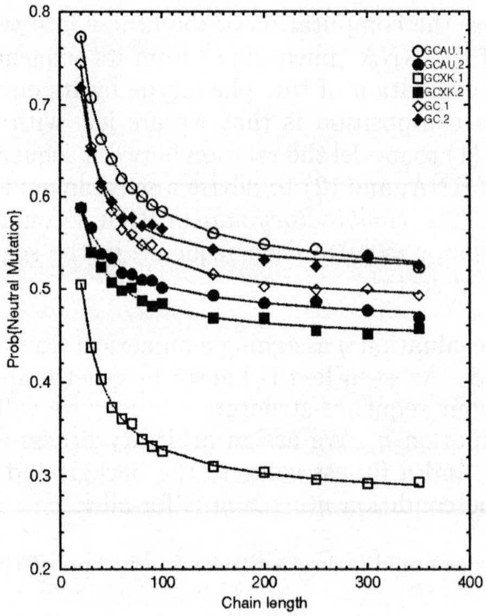

Fig. 2. Frequency of neutral mutations (λ_u and λ_p resp. — see section 2.3), counted separately for single base exchanges in unpaired regions (open symbols) and base pair exchanges (full symbols) for different alphabets.

A sequence x is said to be *compatible* to a secondary structure s if the nucleotides x_i and x_j at sequence positions i and j can pair whenever (i, j) is a base pair in s. Note that this condition does by no means imply that x_i and x_j will actually form a base pair in the structure $\varphi(x)$ obtained by some folding algorithm. The set of all sequences compatible with a secondary structure s will be denoted by $\mathbf{C}[s]$. There are two types of neighbors to sequence $x \in \mathbf{C}[s]$: each mutation in a position k which is unpaired in the secondary structure s leads again to a sequence compatible with s, while point mutations in the paired regions of s will in general produce sequences that are not compatible with s. This problem can be overcome by modifying the notion of neighborhood. If we allow the exchange base pairs instead of single nucleotides in the paired regions of s we always end up with sequences compatible with s. This definition of neighborhood allows us to view $x \in \mathbf{C}[s]$ as a graph. It can be shown [23] that this graph is the direct product of two generalized hypercubes

$$\mathbf{C}[s] = \mathcal{Q}_{\alpha}^{n_u} \times \mathcal{Q}_{\beta}^{n_p} \tag{3}$$

where n_u is the number of unpaired positions in s, α is the number of different nucleotides, i.e., $\alpha = 4$ in the case of natural RNAs, n_p is the number of base *pairs* in s, and β is the number of different *types* of base pairs that can be formed by the α different nucleotides; for natural RNAs we have $\beta = 6$. The sequence length is $n = n_u + 2n_p$.

2.3. A Random Graph Construction

Folding RNA sequences into their secondary structures is computationally quite expensive. It is desirable, therefore, to construct a simple random model for the sequence structure map φ with the same five properties that have been observed for RNA. Reidys *et al.* [23] have investigated random subgraphs of the hypercubes with the result that their approach is in fact able to explain the known facts about the combinatory map of RNA secondary structures.

2.3.1. A Mathematical Concept

We consider two closely related models. Consider a hypercube $\mathcal{Q}_{\alpha}^{n}$. We construct a random subgraph Γ'_{λ} by selecting each edge of $\mathcal{Q}_{\alpha}^{n}$ independently with probability λ. From Γ'_{λ} we obtain the induced subgraph $\Gamma^{\lambda} = \mathcal{Q}_{\alpha}^{n}[\Gamma'_{\lambda}]$ by adding all edges between neighboring sequences that have not been assigned already by the random process.[2] The probability λ is simply the (average) fraction of neutral neighbors.

The main result about these random subgraph models is that there is a critical value λ^* such that the subgraph Γ_{λ} is dense in $\mathcal{Q}_{\alpha}^{n}$ and connected (i.e., for any two vertices Γ_{λ} there is path in Γ_{λ} that connects them) whenever $\lambda > \lambda^*$. Explicitly it has been shown [23] that

$$\lambda^* = 1 - \sqrt[1-\alpha]{\alpha}\,. \tag{4}$$

Density and connectivity of the neutral network Γ result in percolating neutral paths.

[2] Alternatively, one could draw vertices from $\mathcal{Q}^{n}$ and consider corresponding the induced subgraph. Both random subgraph models have essentially the same properties.

2.3.2. Modeling Generic Fitness Landscapes

The model formulated above does not take into account that there are in general different probabilities for the two classes neutral mutations, $\lambda_u \neq \lambda_p$ for the *unpaired* and *paired* parts of the secondary structure, respectively. Using that the "graph of the compatible sequences" is a direct product of two hypercubes this limitation can be overcome by considering the direct product of two random graphs, one in each of the two hypercubes:

$$\begin{array}{ccc} & \mathcal{Q}_\alpha^{n_u} \times \mathcal{Q}_\beta^{n_p} & \\ & \uparrow \iota & \\ & \Gamma_{\lambda_u} \times \Gamma_{\lambda_p} & \\ \overset{in}{\nearrow} & & \overset{in}{\nwarrow} \\ \Gamma_n^{\lambda_u} & & \Gamma_{\lambda_p} \end{array} \qquad (5)$$

This model inherits its properties from the basic random subgraph model on the two hypercubes. In particular $\Gamma = \Gamma_{\lambda_u} \times \Gamma_{\lambda_p}$ is dense and connected if both "components" Γ_{λ_u} and Γ_{λ_p} are dense and connected. From now on we will only refer to this model for deducing our results in this paper.

A neutral network induces in a natural way a *fitness landscape* f_Γ on the complete sequence space $\mathcal{Q}_\alpha^n$:

$$f(x) = \left\{ \begin{matrix} 1 & \text{if} & x \notin \Gamma \\ \sigma & \text{if} & x \in \Gamma \end{matrix} \right\}. \qquad (6)$$

with $\sigma > 1$. We call f_Γ a *single shape* landscape in contrast to the single peak landscapes discussed for instance in [4, 5]. The two degenerated cases $\lambda_u = \lambda_p = 0$ and $\lambda_u = \lambda_p = 1$ are referred to the single peak landscape (Γ consists of a single sequence) and the *flat landscape* resp. In the following we will exploit the analogy between single peak and single shape landscapes quite extensively.

Summarizing the above discussion we claim that a single shape landscape is a much more realistic approximation of real fitness landscapes than a single-peak landscape or a spin glass like model landscape, since all these approaches lack what we think is the most important feature of biomolecular landscapes: *a high degree of neutrality.*

In chapter 5 we present a canonical generalization of the single-shape landscape to the more realistic *multi-shape* landscape. Transitions between two neutral networks are studied.

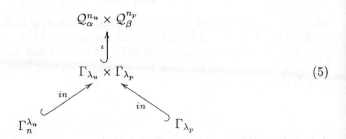

2.4. The Birth and Death Process Model

Let us now return to the dynamic behavior of a population $\mathcal{P}$ on such a landscape. Obviously f_Γ induces a *bipartition* of the population $\mathcal{P}$ into the subpopulation $\mathcal{P}_\mu$ on the network Γ and the remaining part $\mathcal{P}_\nu$ of inferior individuals. We call the elements of $\mathcal{P}_\mu$ *masters* (because they have superior fitness) and those of $\mathcal{P}_\nu$ *nonmasters.*

We will describe the evolution of $\mathcal{P}$ in $\mathcal{Q}_\alpha^n$ in terms of a *birth-death model* [17] with *constant* population size. At each step two individuals are chosen randomly; the first choice is subject to error-prone replication while the second choice is removed from the population [21]. The stochastic process is specified by the following probabilities:
$W^\Gamma_{\mu,\mu}$ is the probability that the offspring of a master is again a master;
$W^\Gamma_{\mu,\nu}$ is the probability that the offspring of a master is a non-master;
$W^\Gamma_{\nu,\mu}$ is the probability that the offspring of a non-master is a master; and
$W^\Gamma_{\nu,\nu}$ is the probability that the offspring of a non-master is again a non-master.

In general these probabilities will depend on the details of the surrounding of each particular sequence, namely on the number of neutral neighbors. It is possible, however, to show [23] that the fraction of neutral neighbors obeys a Gaussian distribution which approaches a δ-distribution in the limes of long chains. The same behavior was found numerically for RNA secondary structures. Hence we can assume that the number of neutral neighbors is the same for all masters, namely $n\lambda_u$ and $n\lambda_p$ for the two classes of neighbors. Consequently the probabilities $W^\Gamma_{\mu,\mu}$, $W^\Gamma_{\mu,\nu}$, $W^\Gamma_{\nu,\nu}$, and $W^\Gamma_{\nu,\mu}$ are independent of the particular sequence.

We consider each replication-deletion event as one single event per time-step. The consequence of this assumption is that depending on the individual fitness the equidistant time-step Δt in *reactor-time* results in different time-intervals per replication-round $\Delta\tilde{t}$ in physical time $\tilde{t}$. I.e. master replicate σ-times faster than nonmaster yielding in σ-times more individuals per replicated master than per nonmaster per physical time step $\Delta\tilde{t}$. This difference between physical time t and population-dependent *reactor time* $\tilde{t}$ has to be taken into account by calculating the probabilities for the replication-deletion events.

Analogously to the mutation-probabilities W^Γ we setup four probabilities P:
$P_{\mu,\mu}$ is the probability for choosing a master for replication and deletion;
$P_{\mu,\nu}$ is the probability for choosing a master for replication and a nonmaster for deletion;
$P_{\nu,\mu}$ is the probability for choosing a nonmaster for replication and a master for deletion;
$P_{\nu,\nu}$ is the probability for choosing a nonmaster for replication and deletion.

For the so called *birth-* and *death-*probabilities we obtain $\mathbf{P}_{k,k+1} = P_{\mu,\nu}W^\Gamma_{\mu,\mu} + P_{\nu,\nu}W^\Gamma_{\nu,\mu}$ and $\mathbf{P}_{k,k-1} = P_{\mu,\mu}W^\Gamma_{\mu,\nu} + P_{\nu,\mu}W^\Gamma_{\nu,\nu}$ resp.

After some lengthy calculations [22] we are able to compute the stationary distribution $\boldsymbol{\mu}_p$ of the birth-death process. According to [7, 17] $\boldsymbol{\mu}_p$ is given by $\boldsymbol{\mu}_p(k) = \pi_p(k)/\sum_k \pi_p(k)$ Then the stationary distribution is completely determined by

$$\pi_p(k) = \frac{W^{\Gamma}_{\nu,\mu}}{\mathbf{P}_{k,k-1}} \frac{B(N, C_2)}{(k+C_1)\,B(1+C_1,k)\;B(N-(k-1),C_2)} \cdot \left[\frac{\sigma\,W^{\Gamma}_{\mu,\mu} - W^{\Gamma}_{\nu,\mu}}{W^{\Gamma}_{\nu,\nu} - \sigma\,W^{\Gamma}_{\mu,\nu}}\right]^{k-1} \tag{7}$$

where $B(x, y)$ is the Beta-function. Λ_i and C_i $(i = 1, 2)$ are defined as follows:

$$\Lambda_1 \stackrel{\text{def}}{=} [\sigma\,W^{\Gamma}_{\mu,\mu} - W^{\Gamma}_{\nu,\mu}] \quad \Lambda_2 \stackrel{\text{def}}{=} [W^{\Gamma}_{\nu,\nu} - \sigma\,W^{\Gamma}_{\mu,\nu}]$$

$$C_1 \stackrel{\text{def}}{=} \frac{(N-1)\,W^{\Gamma}_{\nu,\mu}}{\Lambda_1}, \text{ and } \quad C_2 \stackrel{\text{def}}{=} \frac{(N-1)\,\sigma\,W^{\Gamma}_{\mu,\nu}}{\Lambda_2}.$$

For the dynamics in physical times holds

$$\frac{W^{\Gamma}_{\nu,\mu}}{\mathbf{P}_{k,k-1}} = \frac{W^{\Gamma}_{\nu,\nu}((\sigma-1)(k-1)+N)}{k\,[\sigma(k-1)\,W^{\Gamma}_{\mu,\nu} + (N-k)W^{\Gamma}_{\nu,\nu}]}$$

3. Diffusion on "Neutral" Landscapes

In general, "diffusion" can be understood as movement of the *barycenter* of the population in the high-dimensional sequence-space via point-mutations. The barycenter $M(t)$ of a population at time t is a real valued consensus vector specifying the fraction $x_{i\alpha}(t)$ of each nucleotide $\alpha \in \{A,U,G,C\}$ at every position i.

3.1. Diffusion in Sequence Space

Let us assume again that a secondary structure $s \in S_n$ and its corresponding neutral network Γ are fixed. In this section we study the *spatial distribution* of the strings on the network i.e. the spatial distribution of $\mathcal{P}_\mu$. Here we understand spatial distribution as distribution in *Hamming distances*. For this purpose we introduce the random variable $\hat{d}^{\mu}_{\Gamma}$ that monitors the pair distances in the population $\mathcal{P}$. The shape of the distribution of $\hat{d}^{\mu}_{\Gamma}$ is basically determined by the following factors.

- the distribution of the random variable $\hat{Z}_\mu$ whose states are the number of offspring.

- the *structure* of the neutral network Γ, given by the basic parameters for the construction of the random graph, $\{\lambda_u, \lambda_p, n_u, n_p\}$.
- the single digit error rate p for the replication-deletion process.

We will assume in the sequel that $|\,\mathcal{P}_\mu\,| = \mathbf{E}[\hat{X}_p]$, in other words the number of strings located on the neutral network is *constant.* Taking into consideration the genealogies along the lines of Derrida & Peliti [3] we can express the probability of having different ancestors in *all* i previous generations:

$$w_i \approx e^{-\mathbf{V}[\hat{Z}]\; i/(\mathbf{E}[\hat{X}_p]-1)}\,, \tag{8}$$

where $\hat{Z}$ describes the number of offspring produced by a master-string viewed as a random variable ($\mathbf{E}[\,.\,]$ and $\mathbf{V}[\,.\,]$ denote expectation value and variance resp.).

Following Reidys *et al.* [22] we consider then *random walks* on the neutral network Γ. For this purpose we introduce the probability $\varphi_\Gamma(t,h)$ of traveling a Hamming distance h on Γ by a random walk lasting t generations.

In this section we restrict ourselves to alphabets consisting of complementary bases that admit only complementary base pairs (consider for example $\{\mathbf{G},\mathbf{C}\}$ or $\{\mathbf{G},\mathbf{C},\mathbf{X},\mathbf{K}\}$). We consider *moves* as point-events, i.e. each move occurs at precisely one time step Δt. By use of the regularity assumption, we obtain the *infinitesimal error rates* (for unpaired and paired digits), $\lambda_u\, p\,\Delta t$ and $\lambda_p\, p^2\,\Delta t$.

Arbitrarily we set

$$\wp_u(t) \overset{\text{def}}{=\!=} \frac{\alpha-1}{\alpha}\,(1-e^{-\frac{\alpha}{\alpha-1}\lambda_u p\, t}) \text{ and } \wp_p(t) \overset{\text{def}}{=\!=} \frac{\beta-1}{\beta}\,(1-e^{-\frac{\beta}{\beta-1}\lambda_p p^2\, t})\,. \tag{9}$$

Combining the information on the genealogies and the random walks allows us to compute the distribution of $\hat{d}^\mu_\Gamma$ and leads to the main result in this section. For an alphabet consisting of complementary bases with pair alphabet $\mathcal{B}$ we have

$$\begin{aligned}\boldsymbol{\mu}\{\,\hat{d}^\mu_\Gamma = h\,\} = \mathbf{V}[\hat{Z}] \sum_{h_u+2h_p=h} \int_0^\infty & B(n_u, \wp_u(2[\mathbf{E}[\hat{X}_p]-1]\,\tau), h_u) \\ & B(n_p, \wp_p(2[\mathbf{E}[\hat{X}_p]-1]\,\tau), h_p)\; e^{-\mathbf{V}[\hat{Z}]\,\tau}\, d\tau\,,\end{aligned} \tag{10}$$

where $\wp_u((2[\mathbf{E}[\hat{X}_p]-1]\,\tau), h_u), \wp_p((2[\mathbf{E}[\hat{X}_p]-1]\,\tau), h_p)$ are defined above.

Next we turn to the average distance between the populations $\mathcal{P}(t)$ and $\mathcal{P}(t')$, where $t \geq t'$ are arbitrary times. Then we mean by

$$\begin{aligned}\mathrm{dist}(\mathcal{P}(t),\mathcal{P}(t')) &\overset{\text{def}}{=\!=} \frac{1}{\mathbf{E}[\hat{X}_p]^2} \sum_{\substack{v\in\mathcal{P}_\mu(t)\\ v'\in\mathcal{P}_\mu(t')}} d(v,v') \\ \mathrm{avdist}(\mathcal{P}(t),\Delta t) &\overset{\text{def}}{=\!=} \langle \mathrm{dist}(\mathcal{P}(t'),\mathcal{P}(t'+\Delta t))\rangle_{t'}\,.\end{aligned} \tag{11}$$

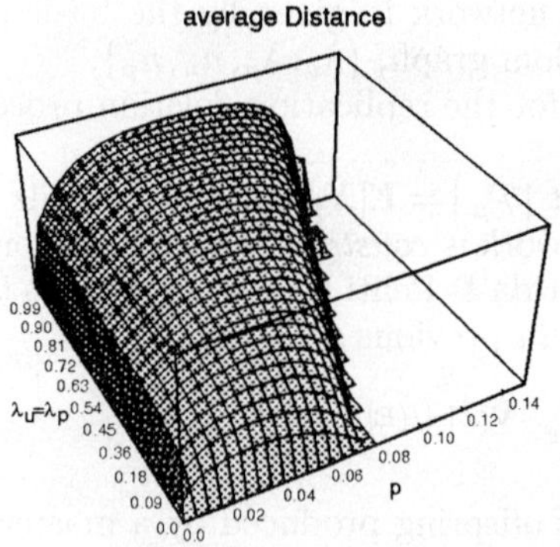

Fig. 3. The average pair distance $\mathbf{E}[\hat{d}^{\mu}_{\Gamma}]$ of master fraction of the population $\mathcal{P}$ on the neutral network Γ in the long time limes. We assume that $\lambda = \lambda_u = \lambda_p$. The distance is plotted as function of the single digit error rate p and the fraction of neutral neighbors for the paired and unpaired digits, λ. We observe that for wide parameter ranges the average pair distance of $\mathcal{P}_\mu$ is plateau-like. In particular the average pair distance becomes 0 at the shape-error threshold.

where $\langle\,.\,\rangle_{t'}$ denotes the *time average*. For binary alphabets with complementary base pairs it is shown in [22] that in the limes of infinite chain length

$$\begin{aligned}&\text{avdist}[\mathcal{P}_\mu(t), \Delta t] \sim \\ &n_u/2 \left[\frac{\chi_u}{\chi_u + \mathbf{V}[\hat{Z}]}\right] [1 - e^{-2\lambda_u\, p\, \Delta t}] + n_p \left[\frac{\chi_p}{\chi_p + \mathbf{V}[\hat{Z}]}\right] [1 - e^{-2\lambda_p\, p^2\, \Delta t}]\end{aligned} \tag{12}$$

(see figure 3). Now we study the displacement of the *barycenter* of the population $\mathcal{P}_\mu$. For this purpose it is convenient to write the complementary digits v_i of the sequence $x = (x_1, ..., x_n)$ as -1 and 1 respectively. We write $x \cdot x' \stackrel{\text{def}}{=\!=} \sum_{i=1}^{n} x_i\, x'_i$.

The *barycenter* of the fraction of masters $\mathcal{P}_\mu \subset \mathcal{P}$ where $|\,\mathcal{P}_\mu\,| = \mathbf{E}[\hat{X}_p]$, denoted by $M^\mu(t)$, is

$$M^\mu(t) \stackrel{\text{def}}{=\!=} \frac{1}{\mathbf{E}[\hat{X}_p]} \sum_{v \in \mathcal{P}_\mu} v\,. \tag{13}$$

We can compute the resulting *diffusion-coefficient* D of the barycenter $M^\mu(t)$ in the long time limes for a population $\mathcal{P}$ replicating on a neutral network Γ with constant master fraction (implying a constant mean fitness $\overline{\sigma} = \frac{(\sigma-1)\mathbf{E}[\hat{X}_p]+N}{N}$). Explicitly the diffusion coefficient is given by

$$\begin{aligned}&\frac{1}{\Delta t} \langle [M(t+\Delta t) - M(t)]^2 \rangle_{t'} \approx \\ &2\,\lambda_u\, n_u\, p \left[\frac{\chi_u}{\chi_u + \mathbf{V}[\hat{Z}]}\right] + 4\,\lambda_p n_p p^2 \left[\frac{\chi_p}{\chi_p + \mathbf{V}[\hat{Z}]}\right].\end{aligned} \tag{14}$$

and

$$\frac{1}{\Delta\tilde{t}}\langle[M(t+\Delta t)-M(t)]^2\rangle_{t'} = \overline{\sigma}\,\frac{1}{\Delta t}\langle[M(t+\Delta t)-M(t)]^2\rangle_{t'}\,,$$

where $\chi_u = 4\,\lambda_u\,p\,(\mathbf{E}[\hat{X}_p]-1)$ and $\chi_p = 4\,\lambda_p\,p^2\,(\mathbf{E}[\hat{X}_p]-1)$.

3.2. Mutational Buffering

We can now compare the analytical distributions of $\hat{d}^{\mu}_{\Gamma}$ with our simulations done in the case of *binary* alphabets (see figure 4).

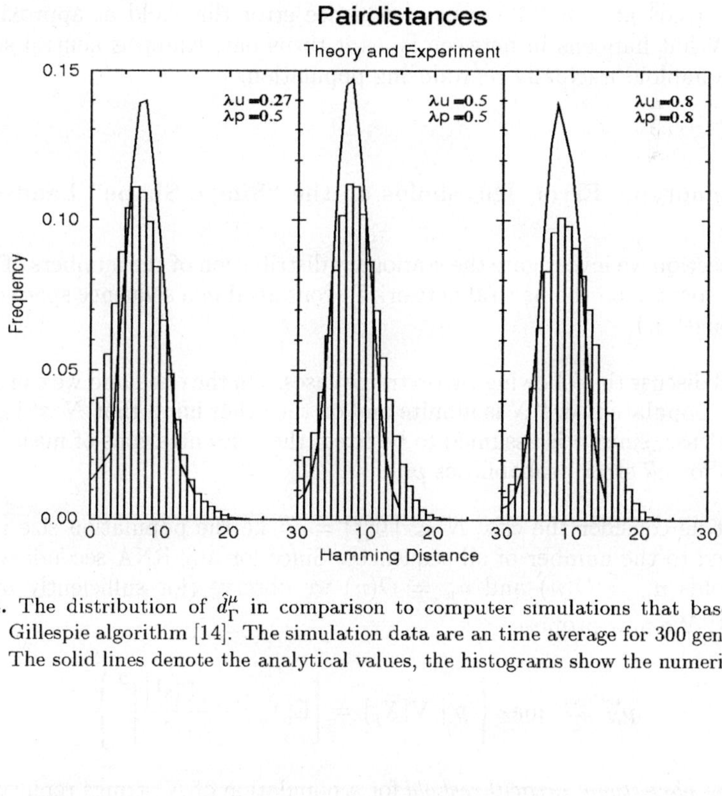

Fig. 4. The distribution of $\hat{d}^{\mu}_{\Gamma}$ in comparison to computer simulations that base on the Gillespie algorithm [14]. The simulation data are an time average for 300 generations. The solid lines denote the analytical values, the histograms show the numerical data.

The difference between the experimental and theoretical density curves is due to an effect known as *buffering* [16]. In the neutral networks a population is located preferably at vertices with higher degrees i.e.

$$v \in \mathrm{v}[\Gamma] : \delta_v \gg \lambda_u\,n_u + \lambda_p\,n_p\,.$$

For binary alphabets in particular the expected distance of pairs (v, v') with $\delta_v, \delta_{v'} \gg \lambda_u\,n_u + \lambda_p\,n_p$ is $n/2$, since the distance sequence of the Boolean hypercube is given by $\binom{n}{k}$. Therefore we observe a shift to higher pair distances in the population as the theory predicts for regular neutral networks.

4. Phenotypic Error Threshold

4.1. Genotypic Error Threshold

We must distinguish between an error threshold with regard to the genotype (sequence) population and a different error threshold, at higher error rates, with regard to the induced phenotype (structure) population marking the beginning of drift in structure space. *That* is when the population can no more preserve the phenotypic information and optimization breaks down. In the present case this occurred[3] at $p \approx 0.1$ versus a sequence error threshold at approximately $p = 0$. What happens in between is, as it turns out, Kimuras neutral scenario [19] in a haploid asexually reproducing population.

4.2. Phenotypic Error Thresholds on the "Single Shape" Landscape

In this section we investigate the stationary distribution of the numbers of strings that are located on the neutral network Γ (contained in a sequence space of fixed chain length n).

We shall discuss the following two extreme cases. On the one hand we can assume that the population size N is infinite and on the other hand that $N \ll |\mathcal{Q}_\alpha^n|$. In the first case, since n is assumed to be fixed, the concentrations of masters c_μ is *nonzero* for *all* error probabilities p.

Next let us consider the case $N \ll |\mathcal{Q}_\alpha^n| = \alpha^n$ i.e the population size is small compared to the number of all sequences. Since for any RNA secondary structures holds $n_p = O(n)$ and $n_u = O(n)$ we observe (for sufficiently large n) $\frac{|\Gamma|}{\alpha^n} \ll 1$. We now propose

$$p_N^* \stackrel{\text{def}}{=} \max\left\{ p \mid \mathbf{V}[\hat{X}_p] = \left[\mathbf{E}[\hat{X}_p] - \frac{|\Gamma[s]|}{\alpha^n}\right]^2 \right\} \tag{15}$$

to be the *phenotypic error-threshold* for a population of N strings replicating on a neutral network Γ. p_∞^* is further the *error threshold of the secondary structure* s. We immediately inspect that the above mentioned criterion generalizes the one used in the case of infinite population size in the single peak landscape of Eigen *et al.* [4], where p_∞^* is the solution of $c_\mu(p^*) = 1/\alpha^n$.

Let us discuss now the case of infinite population size. In this situation we can apply a completely deterministic ansatz solving a (well-known) rate equation

[3] depending on the fraction of neutral neighbors and relative superiority between individuals of different phenotype; a detailed study can be found i.e. in [12]

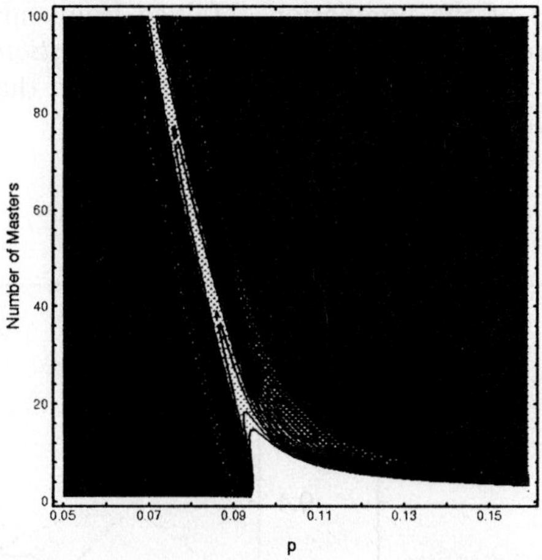

Fig. 5. For a regular neutral network Γ with parameters $\lambda_u = 0.5$ and $\lambda_p = 0.5$ we plot the distribution of $\hat{X}_p$ i.e. the number of masters of $\mathcal{P}$.

for the corresponding *concentrations* of master c_μ and nonmaster vertices c_ν, respectively. Assuming $W^{\Gamma}_{\nu,\mu} \approx 0$, i.e. neglecting back-flow mutations [4] and $\frac{|\Gamma|}{\alpha^n} \approx 0$, we derive

$$c_\mu \approx \left[\frac{\sigma W^{\tilde{\Gamma}}_{\mu,\mu} - (1 + W^{\tilde{\Gamma}}_{\nu,\mu})}{\sigma - 1}\right], \quad W^{\tilde{\Gamma}}_{\mu,\mu} \approx 1/\sigma \iff c_\mu = 0\,. \tag{16}$$

Using the threshold criterion of equ. (15) we can localize the error thresholds numerically for some population sizes and different Neutral-Network-landscapes.[4] with $\sigma = 10$ as superiority. The deterministic threshold values are obtained by solving $W^{\Gamma}_{\mu,\mu} \approx 1/\sigma$ (equ. (16)) for p (table 1).

Table 1: Theoretical and numerical Error Thresholds (for $\sigma = 10$)

λ_u	λ_p	Theory $N = \infty$	Theory $N = 1000$	Experiment $N = 1000$
0.1	0.1	0.079	0.071	0.065
0.27	0.5	0.081	0.08	0.0854
0.5	0.5	0.105	0.095	0.095
0.8	0.8	0.118	0.116	0.11

Finally we end this section by plotting the densities of the i-th incompatible

[4] The calculations are done with *Mathematica* [30].

classes $\mathbf{C}_i[s]$ (see figure 6) of the population obtained from our simulations[5]. We observe that at the error threshold there is a *sharp transition* from a population that is localized on the neutral network to a population that is uniformly distributed in sequence space.

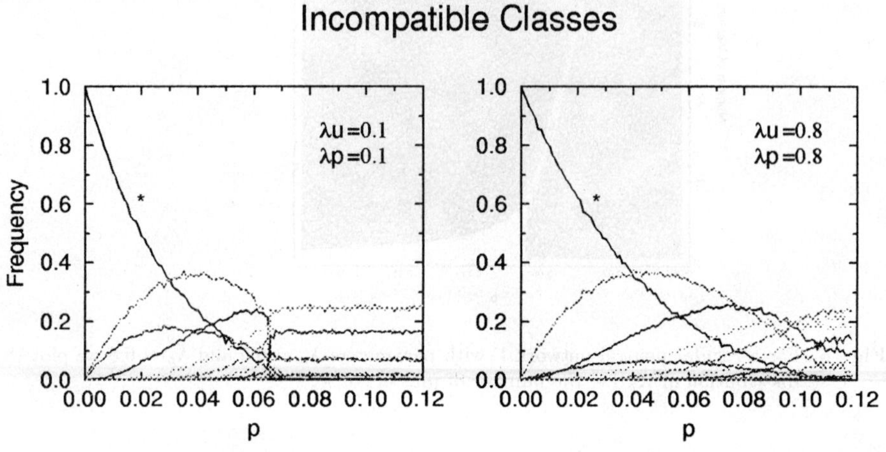

Fig. 6. In this figure we plot the error-classes in incompatible distances $\mathbf{C}_i[s]$ for certain Neutral-Network-landscapes. The underlying population size for the Gillespie simulation is $N = 1000$. The error-class $*$ denotes the number of masters i.e. the number of strings that are localized on the neutral network.

5. Transitions between Neutral Networks

Each neutral network is contained in the *set of compatible sequences* i.e. the set of sequences that could fold into one particular structure s. Each two sets of compatible sequences with respect to the pair of secondary structures have a nonempty *intersection*. This fact and the mathematical modeling of neutral networks as random graphs imply that the upper bound for the Hamming distance between two neutral networks is *four*. It turns out that the intersection is of particular relevance for *transitions* of finite populations of erroneously replicating strings between neutral neutral networks. In other words the intersection plays a key role in the search for "fitter" secondary structures.

It has been proven in [23] that the intersection is always nonempty. The intersection is constructed explicitly by using an algebraic representation of secondary

[5] In difference to the ansatz of constant population size, (the basic assumption for the birth-death model), the simulations are obtained by use of the Gillespie algorithm [14].

structures. As already proposed in [23] each secondary structure s can be interpreted as an element in S_n by use of the mapping

$$\pi\colon \mathcal{S} \to S_n, \quad s \mapsto \pi(s) \stackrel{\text{def}}{=} \prod_{i=1}^{n_p(s)} (x_i, x_i').$$

Here (x_i, x_i') is a base pair in s and $n_p(s)$ is the number of pairs in s. Clearly $\pi(s)$ is an *involution*, i.e. $\pi(s)\pi(s) = id$.

Using the fact that any two involutions $\imath$, $\imath'$ form a *dihedral group* $D_m \stackrel{\text{def}}{=} \langle \imath, \imath' \rangle$ for secondary structures s and s' this leads to the mapping

$$\jmath\colon \mathcal{S} \times \mathcal{S} \to \{D_m < S_n\}, \quad \jmath(s, s') \stackrel{\text{def}}{=} \langle \pi(s), \pi(s') \rangle.$$

In fact the operation of $\langle \pi(s), \pi(s') \rangle$ and especially the corresponding cycle decomposition is closely related to the structure of the intersection $\mathbf{I}[s, s']$.

An arbitrary cycle z is given by a sequence of positions where predecessor and successor are determined by the pairs in s and s'. We distinguish between two types of cycles: *open* and *closed* ones.

5.1. Size of the Intersection

This knowledge enables us to determine the size of the intersection. For alphabets $\mathcal{K}$ with complementary base pairs, e.g. $\mathcal{A} = \{G, C, X, K\}$ with corresponding base pairings $\mathcal{B} = \{GC, CG, XK, KX\}$, and $\alpha = |\mathcal{A}|$ we obtain

$$|\, \mathbf{I}[s, s'] \,| = \alpha^{n_1 + n_2 + \ldots + n_r},$$

where n_i is the number of cycles of length i. If we consider the physical alphabet $\mathcal{A} = \{G, C, A, U\}$ with corresponding pair alphabet $\mathcal{B} = \{GC, CG, AU, UA, GU, UG\}$ we obtain

$$|\, \mathbf{I}[s, s'] \,| = \prod_{i=1} (\alpha_i^{(o)})^{n_i^o} (\alpha_{2i}^{(c)})^{n_{2i}^c}$$

where n_i^c is the number of closed cycles with length i and n_i^o is the number of open cycles of length i and $\alpha_\nu^{(o)}$ and $\alpha_\nu^{(c)}$ are the numbers of all possible configurations for an open cycle of length ν or a closed cycle of length ν with

$$\alpha_\nu^{(o)} = \frac{2}{\sqrt{5}} \left[\left(\frac{2}{-1+\sqrt{5}} \right)^{\nu+2} - \left(\frac{2}{-1-\sqrt{5}} \right)^{\nu+2} \right] \quad \text{and}$$

$$\alpha_\nu^{(c)} = \frac{4}{\sqrt{5}} \left[\left(\frac{2}{-1+\sqrt{5}} \right)^{\nu-1} - \left(\frac{2}{-1-\sqrt{5}} \right)^{\nu-1} \right]$$

5.2. Structure of the Intersection

Definition 1. *Let s and s' be two secondary structures. The graph $\mathcal{I}[s,s']$ has the vertex set $\mathbf{I}[s,s']$. Two sequences $x,y \in \mathbf{I}[s,s']$ are neighbors, e.g. $\{x,y\} \in \mathbf{e}[\mathbf{I}[s,s']]$, if and only if x,y are neighbors in $\mathcal{C}[s]$ and $\mathcal{C}[s']$*

That means the intersection graph can be directly embedded in the graph structure of the sets of compatible sequences [23]. Note that the common unpaired positions in s and s' are the elements in the cycles of length 1. The common pairs of s and s' are represented by the closed cycles of length 2. Thus there exist two scenarios

(1) There are only open cycles of length 1 and closed cycles of length 2, then $\mathcal{I}[s,s']$ is a connected graph.

(2) There is at least one cycle of length greater than or equal 3 or one open cycle of length 2, then $\mathcal{I}[s,s']$ decomposes into components of equal size $((\alpha_1^{(o)})^{n_1^o} \cdot (\alpha_2^{(c)})^{n_2^c})$. The components are connected by paths in $\mathcal{C}[s]$ and $\mathcal{C}[s']$.

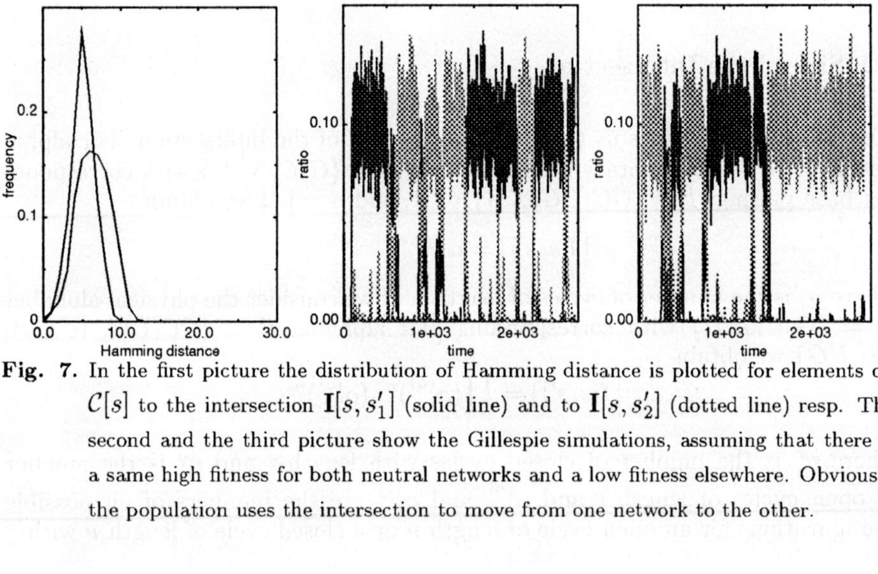

Fig. 7. In the first picture the distribution of Hamming distance is plotted for elements on $\mathcal{C}[s]$ to the intersection $\mathbf{I}[s,s'_1]$ (solid line) and to $\mathbf{I}[s,s'_2]$ (dotted line) resp. The second and the third picture show the Gillespie simulations, assuming that there is a same high fitness for both neutral networks and a low fitness elsewhere. Obviously the population uses the intersection to move from one network to the other.

5.3. Numerical Results

Suppose there are given two pairs of structures (s,s'_1) and (s,s'_2). We assume all λ values to be equal and an action probability of 1/2 on the intersection (figure 7).

The numerical results confirm the basic assumption of the neutral theory of Motoo Kimura [19]. The fixation of phenotypes is a consequence of a stochastic process.

6. Conclusions

Doing evolutive optimization on RNA secondary-structure folding landscapes is somehow different to optimization on typical rugged fitness landscapes. There are no local optima in the naive sense, but rather extended labyrinths of connected equivalent sequences which somewhere touch or come close to labyrinths of better sequences [13, 29]. What looks like punctuated equilibria in one projection (phenotype), presents itself as relentless and extensive change in another projection such as genetic makeup. Seen from this perspective the replicator concept that views genes as the sole unit of selection [2] may need an overhaul, since phenotypes are here to stay much longer than genes [1, 28].

Additional constraints at the sequence and the structure level may severely restrict the extent of neutral networks. However, there is no doubt about the evolutionary implications, should it turn out that RNA structures capable of performing biochemically interesting tasks do form neutral networks in sequence space or can be accessed from such networks. Given present day *in vitro* evolution techniques, these issues are within reach of experimental investigation.

References

[1] L. W. Buss. *The Evolution of Individuality.* Princeton University Press, Princeton, 1987.

[2] R. Dawkins. Replicator selection and the extended phenotype. *Zeitschrift für Tierpsychologie*, 47:61–76, 1978.

[3] B. Derrida and L. Peliti. Evolution in a flat fitness landscape. *Bull. Math. Biol.*, 53:355–382, 1991.

[4] M. Eigen. Selforganization of matter and the evolution of biological macromolecules. *Die Naturwissenschaften*, 10:465–523, 1971.

[5] M. Eigen, J. McCaskill, and P. Schuster. The molecular Quasispecies. *Adv. Chem. Phys.*, 75:149 – 263, 1989.

[6] M. Eigen and P. Schuster. *The Hypercycle: a principle of natural self-organization.* Springer, Berlin, 1979 (ZBP:234.

[7] W. Feller. *An Introduction to Probability Theory and its Applications*, volume I and II. John Wiley, New York, London, Sydney, 1966.

[8] W. Fontana, T. Griesmacher, W. Schnabl, P. Stadler, and P. Schuster. Statistics of landscapes based on free energies, replication and degredation rate constants of RNA secondary structures. *Monatshefte der Chemie*, 122:795–819, 1991.

[9] W. Fontana, D. A. M. Konings, P. F. Stadler, and P. Schuster. Statistics of RNA secondary structures. *Biopolymers*, 33:1389–1404, 1993.

[10] W. Fontana, W. Schnabl, and P. Schuster. Physical aspects of evolutionary optimization and adaption. *Physical Review A*, 40(6):3301–3321, Sep. 1989.

[11] W. Fontana, P. F. Stadler, E. G. Bornberg-Bauer, T. Griesmacher, I. L. Hofacker, M. Tacker, P. Tarazona, E. D. Weinberger, and P. Schuster. RNA folding and combinatory landscapes. *Phys. Rev. E*, 47(3):2083 – 2099, March 1993.

[12] C. V. Forst, C. Reidys, and P. Schuster. Error thresholds, diffusion, and neutral networks. *Artificial Life*, 1995. in prep.

[13] C. V. Forst, J. Weber, C. Reidys, and P. Schuster. Transitions and evolutive optimization in Multi Shape landscapes. in prep., 1995.

[14] D. Gillespie. Exact stochastic simulation of coupled chemical reactions. *J. Chem. Phys.*, 81:2340–2361, 1977.

[15] I. L. Hofacker, W. Fontana, P. F. Stadler, S. Bonhoeffer, M. Tacker, and P. Schuster. Fast folding and comparison of RNA secondary structures. *Monatshefte f. Chemie*, 125(2):167–188, 1994.

[16] M. Huynen, P. F. Stadler, and W. Fontana. Evolutionary dynamics of RNA and the neutral theory. *Nature*, 1994. submitted.

[17] S. Karlin and H. M. Taylor. *A first course in stochastic processes*. Academic Press, second edition, 1975.

[18] M. Kimura. Evolutionary rate at the molecular level. *Nature*, 217:624 – 626, 1968.

[19] M. Kimura. *The Neutral Theory of Molecular Evolution*. Cambridge Univ. Press, Cambridge, UK, 1983.

[20] D. R. Mills, R. L. Peterson, and S. Spiegelman. An extracellular darwinian experiment with a self-duplicating nucleic acid molecule. *Proc. Nat. Acad. Sci., USA*, 58:217–224, 1967.

[21] M. Nowak and P. Schuster. Error tresholds of replication in finite populations, mutation frequencies and the onset of Muller's ratchet. *Journal of theoretical Biology*, 137:375–395, 1989.

[22] C. Reidys, C. V. Forst, and P. Schuster. Replication on neutral networks of RNA secondary structures. *Bull. Math. Biol.*, 1995. submitted.

[23] C. Reidys, P. F. Stadler, and P. Schuster. Generic properties of combinatory maps and neutral networks of RNA secondary structures. *Bulletin of Math. Biol.*, 1995. submitted.

[24] P. Schuster, W. Fontana, P. F. Stadler, and I. L. Hofacker. From sequences to shapes and back: A case study in RNA secondary structures. *Proc.Roy.Soc.(London)B*, 255:279–284, 1994.

[25] P. Schuster and P. F. Stadler. Landscapes: Complex optimization problems and biomolecular structures. *Computers Chem.*, 18:295 – 324, 1994.

[26] M. Tacker, W. Fontana, P. Stadler, and P. Schuster. Statistics of RNA melting kinetics. *Eur. Biophys. J.*, 23(1):29 – 38, 1994.

[27] M. Tacker, P. F. Stadler, E. G. Bornberg-Bauer, I. L. Hofacker, and P. Schuster. Robust Properties of RNA Secondary Structure Folding Algorithms. *in preparation*, 1993.

[28] G. P. Wagner. What has survived of Darwin's theory? *Evolutionary trends in plants*, 4(2):71–73, 1990.

[29] J. Weber, C. Reidys, and P. Schuster. Evolutionary optimization on neutral networks of RNA secondary structures. in preparation, 1995.

[30] S. Wolfram. *Mathematica: a system for doing mathematics by computer.* Addison-Wesley, second edition, 1991.

[31] S. Wright. The roles of mutation, inbreeding, crossbreeeding and selection in evolution. In D. F. Jones, editor, *int. Proceedings of the Sixth International Congress on Genetics*, volume 1, pages 356–366, 1932.

[32] S. Wright. Random drift and the shifting balance theory of evolution. In K. Kojima, editor, *Mathematical Topics in Population Genetics*, pages 1 – 31. Springer Verlag, Berlin, 1970.

Population Evolution in a Single Peak Fitness Landscape How High are the Clouds?

Glenn Woodcock and Paul G. Higgs

Department of Physics, University of Sheffield, Hounsfield Road, Sheffield S3 7RH, UK.

Abstract: A theory for evolution of molecular sequences must take into account that a population consists of a finite number of individuals with related sequences. Such a population will not behave in the deterministic way expected for an infinite population, nor will it behave as in adaptive walk models, where the whole of the population is represented by a single sequence. Here we study a model for evolution of population in a fitness landscape with a single fitness peak. This landscape is simple enough for finite size population effects to be studied in detail. Each of the N individuals in the population is represented by a sequence of L genes which may either be advantageous or disadvantageous. The fitness of an individual with k disadvantageous genes is $w_k = (1-s)^k$, where s determines the strength of selection. In the limit L tends to infinity the model reduces to the problem of Muller's Ratchet: the population moves away from the fitness peak at a constant rate due to the accumulation of disadvantageous mutations. For finite length sequences, a population placed initially at the fitness peak will evolve away from the peak until a balance is reached between mutation and selection. From then on the population will wander through a spherical shell in sequence space at a constant mean Hamming distance $\langle k \rangle$ from the optimum sequence. This has been likened to the idea of a cloud layer hanging below the mountain peak. We give an approximate theory for the way $\langle k \rangle$ depends on N, L, s, and the mutation rate u. This is found to agree well with numerical simulation.

1. Introduction

The ideas of fitness landscapes and sequence spaces in models of evolution are now familiar (Eigen et al, 1989; Kauffman, 1993; Fontana et al, 1993). We may be considering the space of all possible proteins of length L composed of 20 types of amino acids, or the space of all possible length L sequences of DNA composed of four types of bases, or a chromosome with L loci, where a large number of alternative alleles may exist for each locus. The fitness landscape determines the multiplication rate of each sequence, which is either the mean number of offspring of an individual with a given gene sequence in a biological population, or the replication rate of a given chemical sequence in a model for molecular evolution.

If the population size is assumed to be infinite then deterministic equations can be obtained for the relative frequencies of sequences with different fitnesses within the population (Eigen et al, 1989; Schuster, 1986; Tarazona, 1992; Higgs, 1994). If the mutation rate (or replication error rate) is not too large, then the population tends to cluster about the fittest sequence. The distribution of frequencies converges to a stationary state known as the quasi-species. For simple landscapes the quasi-species distribution can be calculated analytically. Other studies have emphasized the ruggedness of fitness landscapes, and have modelled evolution as an adaptive walk (Kauffman and Levin, 1987 ; Macken et al. 1991; Flybjerg and Lautrup, 1992). The population is represented by a single point in sequence space which moves due to mutations onto neighbouring fitter sequences until a local optimum is reached, whereupon no further evolution is possible.

Both these types of model neglect the important features that the population is of finite size and that stochastic effects may be extremely large. One cannot assume that there is a finite concentration of copies of each sequence, since the number of possible sequences of a given length increases exponentially with the length, and may be far larger than the total number of individuals in the population. Neither can one assume that the population is just a single point in sequence space. The real population is a cluster of related sequences in a given region of sequence space. It is important that there is a range of sequences and of fitnesses within the population, otherwise natural selection has nothing to work on. If the landscape is flat (i.e. neutral evolution) the population will wander at random through sequence space (Derrida and Peliti, 1991; Higgs and Derrida, 1991, 1992). If the landscape is not flat then selection will tend to drag the population towards regions of higher than average fitness. However, natural selection is rather inefficient. It is by no means true that a population always evolves relentlessly uphill towards the nearest local fitness maximum, as in the adaptive walk models. The fitness of a population can often decrease due to stochastic effects.

The archetypal model which demonstrates this is Muller's ratchet (Haigh, 1978; Stefan et al, 1993; Lynch et al. 1993; Gabriel et al, 1993; Higgs and Woodcock, 1995). Here one considers a gene sequence of effectively infinite length, initially composed of favourable genes. Unfavourable mutations occur at rate U which each reduce the fitness of the individual by a factor (1-s). Although selection acts against these unfavourable mutations, it is powerless to stop them accumulating, and the fitness therefore decreases indefinitely until the population is no longer viable. Lynch et al, 1993, call this "mutational meltdown". The Muller's ratchet model assumes that all mutations are bad, and that there is no possibility of back mutation. This is entirely reasonable if the sequence is very long and is already very close to an optimum. If the sequence has only a moderate fitness which is not close to a fitness peak then there is a considerable chance of a mutation leading to an increase in fitness. Evolution will thus lead toward higher fitness sequences, but will never manage to get right to the top of the fitness peaks, since Muller's ratchet will set in. The population will evolve toward a steady state with constant fitness, where there is a balance between selection and unfavourable mutations. This steady state is a dynamic one in which the sequence can continue to evolve even though the mean fitness remains constant.

It is useful to borrow an image from Kauffman (1993, chapter 3). If the fitness landscape is viewed as a mountain range, then the population is likely to be found hanging like a layer of cloud below the mountain peaks but above ground level. The central question which we address in this article is how high are the clouds.

2. Analysis of the Model in the Steady State

We suppose that each individual has a sequence of L genes, and that each gene may be either of two possible alleles. Each individual will be represented by a sequence $\sigma_1\sigma_2...\sigma_L$ where each of the σ_i may be +1 or -1. A +1 represents a favourable allele having a relative fitness of 1, and a -1 represents an unfavourable allele having a relative fitness 1-s. The fitness of an individual with k unfavourable alleles is $(1-s)^k$, i.e. we have assumed that the contributions to the fitness from different loci are independent, and therefore multiplicative. The fitness of a sequence is only dependent on the number of -1 genes in the sequence and not on the position of the genes within the sequence.

In our model there are N individuals in the population at each generation. Individuals reproduce asexually with a reproduction rate proportional to their fitness. For each individual in a new generation, an individual is selected to be its parent from the previous generation with a probability proportional to the fitness of the parent. In this way the mean number of offspring of a given parent individual i is $w_i/\langle w\rangle$, where w_i is the fitness of individual i, and $\langle w\rangle$ is the mean fitness in the parent generation. After choosing the parent for each of the new individuals the gene sequences are copied from the parents to the offspring with a small probability u of mutation occurring at each gene. Hence, $\sigma_i^{offspring} = \sigma_i^{parent}$ with probability 1-u, and $\sigma_i^{offspring} = -\sigma_i^{parent}$ with probability u, where typically u << 1.

When simulating this model we began by setting all individuals to be identical to the optimum sequence ($\sigma_i = 1$ for all i). Initially almost all mutations are unfavourable, therefore the population moves away from the optimum sequence, and the number k of -1 genes in the sequence increases. As k increases the chance of a favourable mutation increases. After a certain time a steady state is reached where the occurrence of new unfavourable mutations is balanced by the action of selection plus the occurrence of favourable mutations. Figure 1 shows the mean number $\overline{\langle k \rangle}$ of unfavourable genes per individual in the steady state as a function of u. This is just the mean Hamming distance of the population from the optimum sequence. Note that two separate averages are necessary here : the angular brackets indicate an average over all individuals in the population at one moment in time, and the overbar indicates a time average over many generations after the steady state has been reached.

The expected value of $\overline{\langle k \rangle}$ is known in several limits. Firstly, in the neutral evolution limit, where s = 0, all sequences have an equal probability of occurring. On average half the genes will be -1, and therefore $\overline{\langle k \rangle} = L/2$. Secondly, if s is non-zero, and $u \to 0$ only the optimum sequence will remain in the population, so that $\overline{\langle k \rangle} \to 0$. If $u \to 1/2$, on the other hand, the offspring sequences will have no correlation with their parent sequences, and again $\overline{\langle k \rangle} \to L/2$. In general we know that $\overline{\langle k \rangle}$ is an increasing function of u and a decreasing function of s, and we would like a theory to predict $\overline{\langle k \rangle}$ for any values of u and s.

The other important variable in the problem is the population size N. Selection can only work if there is a range of fitnesses in the population. If N is small all the sequences will be very similar, since they can all be traced back to a common ancestor at a time of order N generations in the past. The spread of fitnesses within the population will therefore also be small for small N, and hence selection will be less effective. We therefore expect that $\overline{\langle k \rangle}$ will increase if we decrease the size of the population. In figure 1 we show $\overline{\langle k \rangle}$ as a function of u for three different population sizes, and in figure 2 we show $\overline{\langle k \rangle}$ as a function of N for three different values of L. In this figure u has been chosen so that uL = 1.0 in each case. When $N \to \infty$ the problem becomes a deterministic one. The fraction C_k of the population having k unfavourable genes can be calculated exactly (Woodcock and Higgs, 1995), and is found to be

$$C_k = \binom{L}{k} a^k (1-a)^{L-k}, \qquad (1)$$

$$\text{where } a = \frac{1}{2}\left((1-u+2u/s) - \sqrt{(1-u+2u/s)^2 - 4u/s}\right). \qquad (2)$$

Examples of stationary C_k distributions in other fitness landscapes are given by Higgs, 1994.

From (1), the mean Hamming distance from the optimum sequence is $\overline{\langle k \rangle} = \sum kC_k = aL$, and the mean fitness is $W = (1-as)^L$. If we suppose that s << 1, and u << 1, but that u/s may be of order 1, then the u terms in (2) are negligible, and a is just a function of u/s. If, in addition, u/s << 1, then $a \approx u/s$. If we take the limit u << 1, and L >> 1, keeping U =uL constant, then (2) becomes a Poisson distribution with $\overline{\langle k \rangle} =$

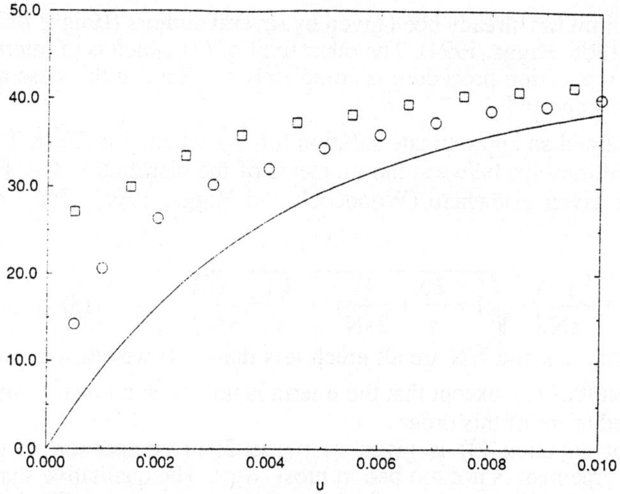

Figure 1. Mean Hamming distance $\overline{\langle k \rangle}$ from the optimum sequence as a function of mutation rate u, for L = 100 and s = 0.01. The solid line shows the exact result for the infinite population. Symbols indicate simulation results. Circles N = 384, Squares N = 96.

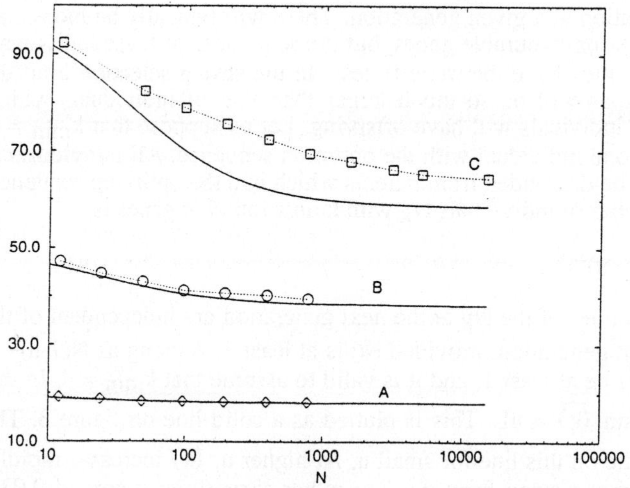

Figure 2. Mean Hamming distance $\overline{\langle k \rangle}$ from the optimum sequence as a function of population size N, for U = uL = 1.0 and s = 0.01. Solid lines show the result of an approximate theory, and symbols show simulation results. (A) L = 40, (B) L = 100, (C) L = 200.

U/s. The solution in this limit has already been given by several authors (Haigh, 1978; Kimura and Maruyama, 1966; Higgs, 1994). The other limit of (2) which is of interest is when u = 1/2, i.e. the replication procedure is completely random. In this case a = 1/2 independent of s, as is expected.

We have also obtained an approximate solution for $\overline{\langle k \rangle}$ when N is finite. This was done by deriving realtionships between the moments of the distribution C_k. The mathematical details are given elsewhere (Woodcock and Higgs, 1995). The final result is

$$\frac{\overline{\langle k \rangle}}{L} = \frac{1}{2}\left(\left(1+\frac{2u}{s}+\frac{1}{2sN}\right)-\sqrt{(1+\frac{2u}{s}+\frac{1}{2sN})^2-\frac{4u}{s}-\frac{1}{sN}}\right). \qquad (3)$$

Here we have assumed that u, s and 1/N are all much less than 1. If we take the limit $N \to \infty$, we obtain the result of (2), except that the u term is not given correctly, since we have already neglected terms of this order.

The prediction of equation (3) is given in figure 2, in comparison to the simulation results. The agreement is not too bad in most cases. The qualitative shape of the curve is given correctly. We see from (3) that a finite population will behave like an infinite one only if sN >> 1. If sN is of order one then $\overline{\langle k \rangle}$ is larger for the finite population than for the infinite one, and the mean fitness will be lower. This confirms what we stated qualitatively above: selection is less effective in small size populations, and the mean fitness decreases as the population size decreases.

In deriving the above approximation it was assumed that selection was weak, i.e. s << 1. Figure 3 shows the simulation results of $\overline{\langle k \rangle}$ as a function of u for larger s values. It can be seen that as $s \to 1$, the curves tend to a limiting form. Suppose that k_{min} is the number of unfavourable genes on the sequence which has the highest fitness within the population at a given generation. There will typically be more than one individual with k_{min} unfavourable genes, but the sequences of these individuals may differ, even though they have the same fitness. In the strong selection limit the fitness of these individuals will be so much larger than that of individuals with a higher k that only these indiviuals will have offspring. Let us suppose that $k_{min} = 0$, i.e. that there is at least one individual with the optimum sequence. All individuals at the next generation will be descended from parents which had the optimum sequence. Hence the expected number of individuals N_k with k unfavourable genes is

$$N_k = N\binom{L}{k}u^k(1-u)^{L-k} \ . \qquad (4)$$

Note that the expected values of the N_k at the next generation are independent of the value of N_0 at the parent generation, provided N_0 is at least 1. As long as $N(1-u)^L >$ 1, N_0 will almost always be at least 1, and it is valid to assume that $k_{min} = 0$. In this case from (4.1) we obtain $\overline{\langle k \rangle} = uL$. This is plotted as a solid line on figure 3. The data for s close to 1 do lie on this line for small u. At higher u, $\overline{\langle k \rangle}$ increases rapidly, and the data appear to move away from the line rather abruptly at u around 0.03 - 0.035. We would expect this to happen at $u = u_c$, where $N(1-u_c)^L = 1$. For these parameter values this formula gives $u_c = 0.045$. For $u > u_c$ the optimum sequence is not always present in the population. This argument is very similar to that given by Bonhoeffer and Stadler,1993, in their discussion of error thresholds in finite populations. For finite populations, however, there is no true singularity in the curve of $\overline{\langle k \rangle}$ versus u, and hence u_c is not defined precisely. We prefer only to use the term error threshold for infinite population models where there is a singularity (eg. Eigen et

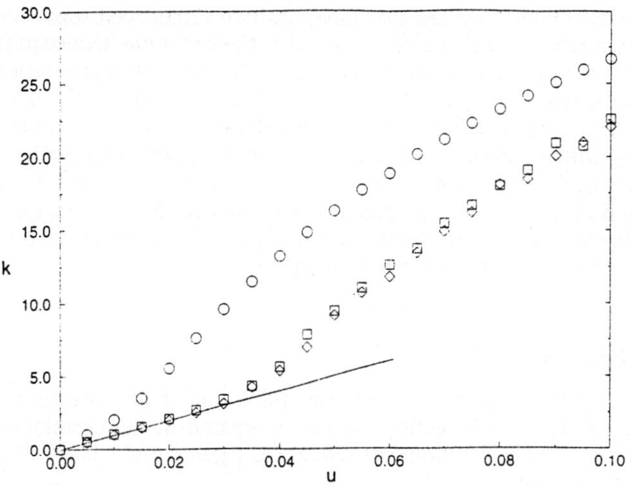

Figure 3. Mean Hamming distance $\overline{\langle k \rangle}$ from the optimum sequence as a function of mutation rate u, for N = 100, and L = 100. Symbols indicate simulation results. Circles s = 0.5, Squares s = 0.95, Diamonds s = 0.99966. The curves tend to a limit as s tends to 1. The limiting curve lies on the line $\overline{\langle k \rangle}$ = uL for u less than $u_c \approx 0.03$, indicating that the population is localized close to the optimum sequence for $u \leq u_c$.

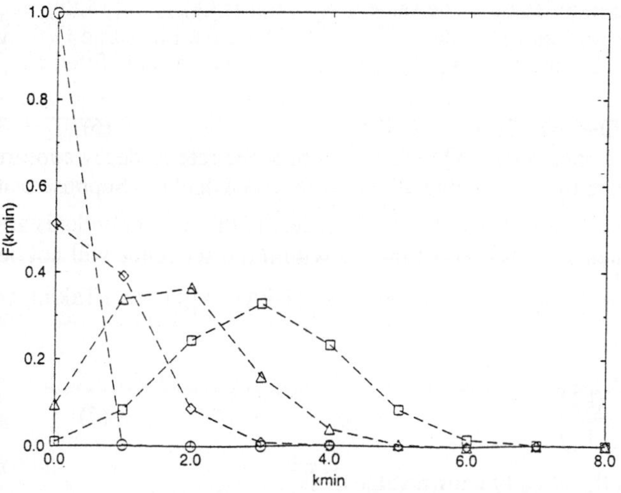

Figure 4. The probability $F(k_{min})$ that the fittest individual in the population has k_{min} unfavourable genes, calculated for N = 100 and L = 100, in the limit s tends to 1. Circles u = 0.025, Diamonds u = 0.035, Triangles u = 0.04, Squares u = 0.045. At u = 0.025 the optimum sequence is almost always present, and at u = 0.045 it is almost never present.

al, 1989; Higgs, 1994). Nevertheless, we see that there are two rather well separated types of behaviour in this model with finite N. At small u the optimum sequence is almost always present, and at larger u the fittest sequence varies from one generation to the next, and is usually not the optimum sequence. We have also calculated (Woodcock and Higgs, 1995) the probability $F(k_{min})$ that the fittest individual at any given generation has k_{min} unfavourable genes. This function is shown in figure 4 for four values of u close to u_c. We see that for u = 0.025 the optimum sequence is almost always present, and at u = 0.045 it is almost never present. For intermediate values the optimum sequence is present some of the time. The population thus gradually escapes from the optimum over a range of u values.

3. Dynamics in Sequence Space

Suppose that $\sigma_i^\alpha(t)$ represents the ith gene on individual α at generation t (which may be either +1 or -1). In this section we are interested in the correlation between gene sequences at a given generation and sequences t generations later. This is measured by the overlap function

$$Q(t) = \overline{\frac{1}{N^2L}\sum_\alpha\sum_\beta\sum_i \sigma_i^\alpha(t')\sigma_i^\beta(t+t')} \tag{5}$$

Here, the sum over α represents a sum over all individuals at a time t', and the sum over β represents a sum over all individuals at a time t'+t. The overbar indicates a time average over all t'. We assume that the population has reached the steady state and that the correlation between two generations only depends on the time t between them.

In the neutral limit, s = 0, it is easy to show that $Q(t) = Q_0\exp(-2ut)$, where the mean overlap between two individuals at the same generation is $Q_0 = 1/(1+4uN)$ (see Derrida and Peliti, 1991; Higgs and Derrida, 1991,1992). We have measured Q(t) by simulation for non zero s. Figure 5 shows that the results may be well fitted by a function of the form

$$Q(t) = Q_\infty + (Q_0 - Q_\infty)\exp(-2ut). \tag{6}$$

It can be seen that although both Q_0 and Q_∞ change with s, the rate of decay appears to be equal to 2u independent of s. The value of Q_∞ is easy to calculate. Suppose that individuals α and β have k_α and k_β -1s in their sequences. If the two individuals are widely separated in time then the positions of the -1s within the sequence will not be correlated, so that $q^{\alpha\beta} = \frac{1}{L^2}(k_\alpha k_\beta + (L-k_\alpha)(L-k_\beta) - k_\alpha(1-k_\beta) - k_\beta(1-k_\alpha))$.Taking an average of this equation gives

$$Q_\infty = 1 - 4\frac{\overline{\langle k\rangle}}{L}\left(1 - \frac{\overline{\langle k\rangle}}{L}\right). \tag{7}$$

Thus Q_∞ depends only on the already known value of $\overline{\langle k\rangle}$.

When the population is in the steady state, most of the individuals have k close to $\overline{\langle k\rangle}$. The population is thus to be found within a spherical shell at Hamming distance $\overline{\langle k\rangle}$ from the optimum sequence. Since the population is finite, and since all individuals are related by common ancestry, not all of the sequences at this Hamming distance from the optimum will be found. The whole of the population will be clustered together on "one side" of the optimum. The population will wander around the shell at random. Q_∞ is just the overlap between two randomly chosen points on

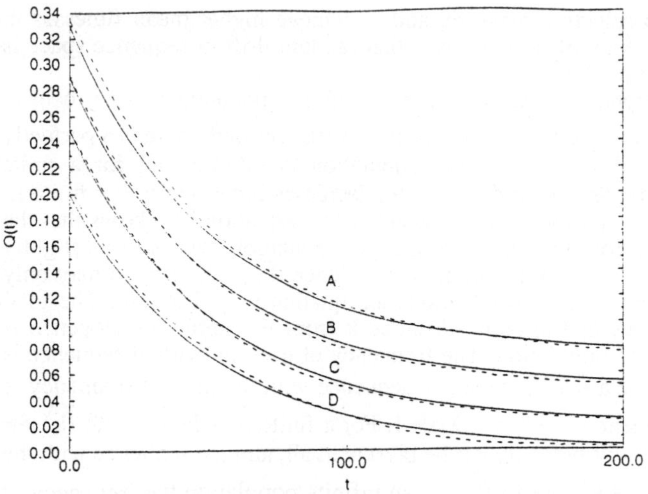

Figure 5. The overlap function Q(t) shown for N = 100, L = 100, and u = 0.01. Dashed lines show simulation results, and solid lines show best fits of the data to the exponential formula given in the text. (A) s = 0.02, (B) s = 0.014, (C) s = 0.01, (D) s = 0.0.

this spherical shell. We have not found a simple argument for Q_0. This is the mean overlap between two individuals at the same generation. As s is increased the probability that two individuals have a common ancestor in a recent generation increases, and the mean time since any two individuals had a common ancestor decreases. This means that Q_0 will increase, as is seen in the simulations. The problem can be solved exactly for the neutral case (Derrida and Peliti, 1991; Higgs and Derrida, 1991,1992; Higgs, 1995), but when selection is present an analytical solution is much more difficult. We have studied the distribution of times since the existence of common ancestors in more detail for the problem of Muller's ratchet (Higgs and Woodcock, 1995), and we expect a similar type of behaviour in this problem.

We have seen that the rate of the exponential decay in (6) appears to be independent of s. In other words, the rate of evolution within the spherical shell in sequence space does not depend on the radius of the shell. This result was somewhat unexpected, and we have no proof as why this is the case. However, we discuss elsewhere (Woodcock and Higgs, 1995) a simplified toy model where Q(t) can be calculated exactly, and shown to be of the form (6).

4. Discussion

When discussing the theory of evolution in a given fitness landscape it is important to remember that evolution works with populations of a finite number of individuals. At any one time the population "sees" only a small fraction of the fitness landscape. Natural selection can only act upon the sequences which are present in the population. Since all sequences in an asexual population can eventually be traced back to a single common ancestor, all the sequences are related to each other. The whole population is clustered together in a particular region of sequence space. The cluster increases in size as the population size increases, so that there is a larger spread of fitness values in a larger population. Larger populations therefore tend to respond more rapidly to selective pressure, and to achieve higher mean fitnesses in the steady state. Another way of saying this is that random drift in sequence space is less important for larger populations.

We wish to emphasize that finite and infinite populations can behave qualitatively differently. For the case $L \rightarrow \infty$ with $U = uL$ constant, there is a perfectly stable concentration distribution C_k if the population is infinite, but for a finite population Muller's ratchet sets in, and the fitness decreases indefinitely. For finite L, both finite and infinite populations reach a steady state. Equation (3) shows that the mean Hamming distance from the origin for a finite population may only be slightly higher than for an infinite one if sN is quite large. Hence the mean fitness may only be slightly lower. However, the two situations are qualitatively different. The total frequency of all sequences at Hamming distance k from the optimum sequence is given in (1) for the infinite population. The frequency of each individual sequence is therefore $a^k(1-a)^{L-k}$. Since $a < 1/2$, the single sequence with the highest frequency is always the optimum sequence, even if $\overline{\langle k \rangle} >> 1$. For a finite population with $\overline{\langle k \rangle} >> 1$, the optimum sequence will typically not be present at all, and the sequence with the highest frequency will have k close to $\overline{\langle k \rangle}$. In an infinite population the frequency of each sequence remains constant once a steady state is reached. For a finite population the steady state is dynamic, and evolution of the sequence continues to occur. In the single peak landscape the population becomes confined to a sperical shell in sequence space at Hamming distance close to $\overline{\langle k \rangle}$ from the optimum. The behaviour of the overlap function Q(t) shows that the population wanders at random around this spherical shell.

For this model we have also studied (Woodcock and Higgs, 1995) the rate of approach of $<k>$ to its steady state value. This makes the link with our previous work on Muller's ratchet (Higgs and Woodcock, 1995). The fraction of mutations which are favourable depends on the distance of the population from the fitness peak. A population initially very close to the optimum will move away from it, whilst a population initially far from the optimum will approach it. For some parameter values it is better to have as small a mutation rate as possible, whilst in other cases there is an optimum value of U for which the rate of increase of fitness is largest.

In finite populations, stochastic effects must be taken into account. This makes the problem much more difficult to treat mathematically than for infinite populations. We have used a very simple smooth single-peaked landscape here, so that at least an approximate analytical theory can be given. This is at least a starting point for a theory of the evolution of finite populations in rugged fitness landscapes. To obtain a mathematical theory of this behaviour is an important goal for future work.

References

Bonhoeffer, S. and Stadler, P.F. (1993) Error thresholds on correlated fitness landscapes. *J. Theor. Biol.* **164**, 359-72.

Derrida, B. and Peliti, L. (1991) Evolution in a flat fitness landscape. *Bull. Math. Biol.* **53**, 355-382.

Eigen, M., McCaskill, J. and Schuster, P. (1989) The molecular quasispecies. *Adv. Chem. Phys.* **75**, 149-263.

Flyvbjerg, H. and Lautrup, B. (1992) Evolution in a rugged fitness landscape. *Phys. Rev. A.* **46**, 6714-18.

Fontana, W., Stadler, P.F., Bornberg-Bauer, E.G., Griesmacher, T., Hofacker, I.L., Tacker, M., Tarazona, P., Weinberger, E.D., and Schuster, P. (1993) RNA folding and combinatory landscapes. *Phys. Rev. E* **47**, 2083-99.

Gabriel, W. Lynch,M. and Bürger, R. (1993) Muller's Ratchet and Mutational Meltdowns. *Evolution* **47**, 1744-57.
Haigh, J. (1978) The accumulation of deleterious genes in a population - Muller's ratchet. *Theor. Pop. Biol.* **14** , 251-267.
Higgs, P.G. (1994) Error thresholds and stationary mutant distributions in multi-locus diploid genetics models. *Genet. Res. (Camb)* **63**, 63-78.
Higgs, P.G. (1995) Frequency Distributions in Population Genetics Parallel those in Statistical Physics. *Phys. Rev. E.* **51**, 95-101.
Higgs, P.G. and Derrida, B. (1991) Stochastic models for species formation in evolving populations. *J. Phys. A (Math. & Gen.)* **24**, L985-L991.
Higgs, P.G. and Derrida, B. (1992) Genetic distance and species formation in evolving populations. *J. Mol. Evol.* **35**, 454-465.
Higgs, P.G. and Woodcock, G. (1995) The accumulation of mutations in asexual populations and the structure of genealogical trees in the presence of selection. *J. Math. Biol.* (in press).
Kauffman, S.A. (1993) *The Origins of Order.* Oxford University Press.
Kauffman, S.A. and Levin, S. (1987) Towards a general theory of adaptive walks on rugged landscapes. *J. Theor. Biol.* **128**, 11-45.
Lynch, M., Bürger, R., Butcher, D. and Gabriel, W. (1993) The mutational meltdown in asexual populations. *J. Hered.* **84**, 339-344.
Macken, C.A., Hagan, P.S. and Perelson, A.S. (1991) Evolutionary walks on rugged landscapes. *SIAM J. Appl. Math.* **51**, 799-827.
Schuster, P. (1986) Dynamics of molecular evolution. *Physica D* **22**, 100-119.
Stephan, W., Chao, L. and Smale, J.G. (1993) The advance of Muller's ratchet in a haploid asexual population : approximate solutions based on diffusion theory. *Genet. Res. (Camb)* **61**, 225-231.
Tarazona, P. (1992) Error thresholds for molecular quasispecies as phase transitions: From simple landscapes to spin-glass models. *Phys. Rev. A.* **45**, 6038-50.
Woodcock, G. and Higgs, P.G. (1995) Population evolution on a single peak fitness landscape (in press).

Replicators Don't![*]

Barry McMullin

School of Electronic Engineering
Dublin City University
Dublin 9
Ireland

Phone: +353 1 704 5432
Fax: +353 1 704 5508
E-mail: McMullin@EEng.DCU.IE
WWW: http://www.eeng.dcu.ie/~mcmullin/home.html

Abstract: Replicators don't. Replicate, that is. This is the shocking conclusion to which I have been forced by my attempt to figure out what *precisely* Richard Dawkins means by the term "replicator". Actually, it seems that Dawkins uses the term in at least two fundamentally different ways; but according to Dawkins' own specification of the problem which the "replicator" concept was intended to solve (namely, what entities can qualify as things that evolutionary adaptations are "good for") then "replicators" turn out to be a special form of *lineage* (what I shall term a *similarity lineage*); and these, in turn, do not actually "replicate" (in Dawkins' sense of the term) at all! Does this matter to the research programme of Artificial Life? Well yes, I believe it does. Dawkins has explicitly argued that there are principled reasons why Darwinian evolution, *in any medium whatsoever*, must rely on the participation of "replicators". Within limits I am inclined to agree. But it follows that, if we wish to realize artificial Darwinism, we had better be clear what a replicator actually is—and all the more so if it turns out that it doesn't...

1 Introduction

The notion that "replicators" play a uniquely distinguished role in biological evolution has been championed by Richard Dawkins (1976; 1989). Furthermore, Dawkins has argued that this idea can be generalised in a way which makes it applicable to any properly Darwinian evolutionary process, at least if that process gives rise to a growth in adaptive complexity (Dawkins 1983). It is evident, therefore, that if Dawkins' analysis is correct, it has profound implications for any attempt to realise a growth of adaptive complexity in *artificial* systems by Darwinian means.

This paper is concerned with trying to clarify just what, exactly, a Dawkinsian "replicator" might be. This has implications both for the specific field of Artificial Life, but also for the general debate in evolutionary biology about "units of selection". I provide a reformulation which, I claim, captures the valid core of Dawkins' insight, while, at the same time, avoiding certain confusions and misconceptions which might otherwise be read into his views.

2 Replicating Confusion

There has been considerable ambiguity, if not downright confusion, in the literature of evolutionary biology regarding specific technical usage of the terms *replication* and *replicator*. As

[*]Derived from material first presented in a series of three rather indigestible essays, previously available only in the form of internal technical reports (McMullin 1992a; 1992b; 1992c).

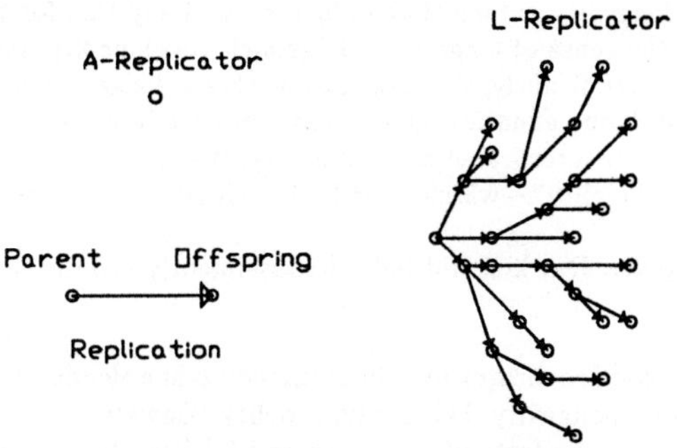

Figure 1: A-replicator vs. L-replicator

far as I am aware, the abstract, technical, idea of a *replicator* was first introduced by Dawkins (1976; 1978a). Hull subsequently elaborated the idea (Hull 1980; 1981), and Dawkins has since extended his own analysis somewhat further (Dawkins 1982a).

I shall argue that "replicator" has sometimes been used to refer to *actors*—individuals which are capable of replication[1]—and sometimes to refer to *lineages* formed as a *result* of such replication (Fig. 1). I suggest that the distinction between these two kinds of entities is actually very significant—and that it is only the latter (lineage) interpretation which can do the job Dawkins wants done.

> *Where necessary in the following, I shall explicitly distinguish references to actors with an* **A-** *prefix, and references to lineages with an* **L-** *prefix.*

The ambiguity of usage can be illustrated by considering the concept of replicator *longevity*. In *The Selfish Gene*, Dawkins first defines longevity as relating to the lifetime of an individual replicator, i.e. an actor or A-replicator:

> Certain molecules [supposed primordial replicators], once formed, would be less likely than others to break up again. These types would become relatively numerous in the soup, not only as a direct logical consequence of their 'longevity', but also because they would have a long time available for making copies of themselves. Replicators of high longevity would therefore tend to become more numerous and, other things being equal, there would have been an 'evolutionary trend' towards greater longevity in the population of molecules.
>
> Dawkins (1976, p. 18)

He goes on to argue that there would be an overall trend toward the evolution of "varieties" (?) of replicator with high "longevity/fecundity/copying-fidelity" (Dawkins 1976, p. 19). Thus, by "longevity" he must evidently mean the lifetime of individual actors, or A-longevity.

Somewhat later in the same source, Dawkins specifies that "Copying fidelity is another way of saying longevity-in-the-form-of-copies *and I shall abbreviate this simply to longevity*" (Dawkins 1976, p. 30, emphasis added). Now *this* version of longevity evidently refers to a replicator viewed as a lineage ("in-the-form-of-copies"); so this is L-longevity.

So far, any confusion is latent: as long as we remember that Dawkins is using "longevity" in two quite different ways, and judge his meaning from the context, it should not cause too much trouble. In particular, we might reasonably suppose that the slogan "longevity/fecundity/fidelity" will always refer to actors, not to lineages—i.e. the "longevity"

[1]See (McMullin 1992a, Section 3); *cf.* "Darwinian actor" (Gould 1982).

in question will be A-longevity rather than L-longevity. I say this for two distinct reasons. Firstly, A-longevity is the sense of longevity with which Dawkins first introduced the slogan. But secondly, and more significantly, Dawkins claims that L-longevity is effectively equivalent to copying *fidelity* (a dubious equation in any case, but let it stand). It follows that, if the longevity in the slogan were interpreted as L-longevity, the slogan would become synonymous with "fidelity/fecundity/fidelity"—which is at least redundant and confusing, if not actually incoherent.

Unfortunately, however, Dawkins did indeed subsequently use the slogan in precisely this confusing way:

> The qualities of a good replicator may be summed up in a slogan reminiscent of the French Revolution: Longevity, Fecundity, Fidelity [Dawkins 1976; 1978b]. Genes are capable of prodigious feats of fecundity and fidelity. *In the form of copies of itself, a single gene may persist for a hundred million individual lifetimes.*
>
> Dawkins (1978a, p. 68, emphasis added)

So we have the slogan, which I have just argued *must* imply A-replicator, followed immediately by an elaboration that obviously implies L-replicator. Given Dawkins' own confusion here, it is hardly surprising that Hull then compounded the error further:

> According to Dawkins [Dawkins 1978a, p. 68], the qualities of a good replicator may be summed up in a slogan reminiscent of the French Revolution: Longevity, Fecundity, Fidelity. As striking as this slogan is, it can easily be misunderstood. The fidelity which Dawkins is talking about is copying-fidelity, and the relevant longevity is longevity-in-the-form-of-copies [Dawkins 1976, p. 19, p. 30].
>
> Hull (1981, p. 31)

Hull's last citation here refers to the two locations in *The Selfish Gene* (1976 edition), which I have already identified above, where "longevity" was defined—but he omits to mention that these are two different, *and incompatible,* definitions!

Hull is certainly correct that Dawkins' slogan may be easily "misunderstood". On my view, both he (and Dawkins himself) have suffered from just such misunderstanding. The interpretation Hull gives here is not the "correct" one—i.e. that which accompanied the original formulation of the slogan in (Dawkins 1976, p. 19), and which referred to A-replicators—but the confusing and redundant one which refers to L-replicators (Dawkins 1976, p. 30; 1978a, p. 68).

Hull also uses Dawkins' slogan "longevity, fecundity and fidelity" in another paper (Hull 1980, p. 317), but, on this occasion, citing only (Dawkins 1978a) as the source. Again, Hull goes on to specify that the "relevant longevity concerns the retention of structure through descent" (i.e. L-longevity). Again, he does not comment on the fact that (according to Dawkins) this version of longevity is synonymous with (Dawkins' version of) fidelity, and is therefore redundant.

The problem is further compounded by Dawkins (1982a, p. 84) where he quotes, at length, from (Hull 1980), and specifically endorses Hull's interpretation of longevity in this context—thus reinforcing the confusion he himself originated in (Dawkins 1978a).

To summarise, while it seems that Dawkins typically uses "replicator" in the sense of a lineage or L-replicator rather than an actor or A-replicator, he also alternates between the two usages—sometimes with quite bewildering speed. Thus, we have the following two comments (quoted from consecutive paragraphs):

> A *germ line replicator* (which may be active or passive) is a replicator that is potentially the ancestor of an indefinitely long line of descendant replicators...
> [*Evidently this refers to actors.*]

> ... But whether it succeeds in practice or not, any germ line replicator is *potentially immortal.* It 'aspires' to immortality but in practice is in danger of failing. [*Yet now we must be talking about lineages.*]
>
> Dawkins (1982a, p. 83)

This confusion between A-replicator and L-replicator is counterpointed (presumably with unconscious irony) by Dawkins' approving remark that *Hull* (1980; 1981) is "particularly clear about the logical status of the lineage, and about its distinction from the replicator and the interactor" (Dawkins 1982a, p. 100).

In conclusion: the point of this rather laborious discussion has been to establish that, though Dawkins and Hull make considerable use of the term "replicator", their usage is quite generally ambiguous as between A-replicator and L-replicator, and calls for very careful interpretation. The recognition of this fact is an essential prerequisite for the analysis of the "unit of selection" controversy which follows.

3 The Elusive *Unit of Selection*

The question I now propose to address is this:

> *What entities, or kinds of entities, can qualify as the "units of selection" in Darwinian evolution?*

This issue has received extended consideration, as summarised by, for example, Hull (1981). Not unnaturally, this discussion has been carried out mainly in the particular context of biological evolution, but Hull remarks that the issue is so fundamental that it deserves to be called 'metaphysical'. Similarly, Dawkins has suggested that the question is not simply one of empirically deciding which real entities function as units of selection, but rather is a "dispute about about what we ought to *mean* when we talk about a unit of selection" (Dawkins 1982b).

Needless to say, given the intense scholarly attention which has been devoted to this issue, I do not claim to have a definitive resolution to offer. However, I shall at least try to offer a clear target for further criticism.

In particular, I embrace Dawkins' view that the substantive question here is to elucidate "the nature of the entity for whose benefit adaptations may be said to exist" (Dawkins 1982a, p. 81).[2] This, whatever it is, is the entity which I shall identify as the "unit of selection".

To make the question even more precise, I am interested in whether the unit of selection is an actor or a lineage. I shall suggest that, in certain, simple, cases, there may be very little difference between the two; and that this explains why the two possibilities are commonly confused. However, I shall conclude that, in the *general* case, the two possibilities are very different; and that the *correct* candidate is, quite unambiguously, a particular kind of lineage.

If "adaptation" evolves at all as a result of Darwinian processes[3] then it does so in the form of a change (or accumulation of changes) which, in each case, allows a new *lineage* to selectively displace an old one because it has higher fitness. I take it that there is no substantive dispute about this. Darwinian selection is clearly not a process of "displacement" of one *actor* by another. Similarly, while selection is more usually described in terms of "populations" displacing each other, these so-called "populations" are, in fact, not just any old groupings of actors (no matter how alike), but groupings *which are established by virtue of common ancestry*—which is to say, *lineages.*

---

[2]By implication, we are here talking exclusively about "Darwinian adaptations": that is, "adaptations" brought about by Darwinian evolution.

[3]There is, of course, never any *guarantee* that Darwinian evolution will lead to increases in "adaptation"—unless one wishes to interpret "adaptation" in such a way as to fall victim to one of the infamous Darwinian tautologies (McMullin 1992b, Sections 3.1, 5.1).

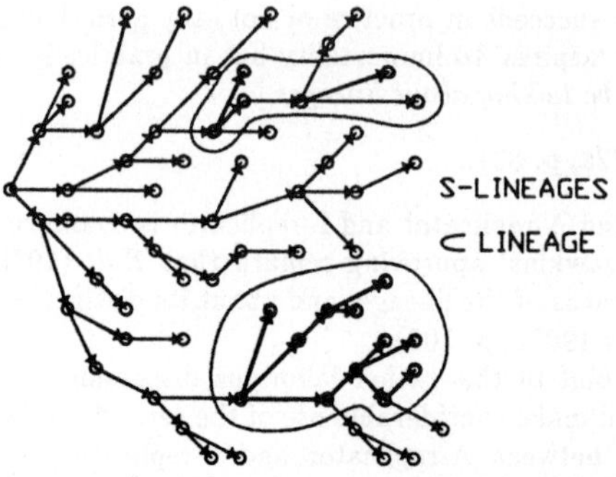

Figure 2: An S-lineage is a subset of a lineage.

By *lineage* here I mean simply the set of *all* descendents of a given *founder* (as already indicated in Fig. 1). Granted, it is only certain *kinds* of lineages which can participate in selection; these are lineages which are characterised by some selectively important attribute (or "trait"). I shall distinguish these by the term *Similarity Lineage* or S-lineage. Briefly, consider some actor which (through mutation or some other mechanism) is different from its parent(s) in some way which is potentially selectively significant. Then I call that actor an *S-founder*—founder of an S-lineage. The S-lineage is the set of all (and only) those descendents of the S-founder which are "similar" to the S-founder in respect of this trait or character (whether "expressed" or just as "carriers"). See Fig. 2.

The S-lineage concept relates to at least some of the common biological definitions of "species"—but can also potentially map onto both higher and lower taxonomic categories! A fundamental distinction from any conventional "species" concept is that S-lineages can, and generally will, *intersect*: a particular individual actor may be a member of numerous, distinct, S-lineages—even S-lineages which are competing with each other. Clearly, the concept is a moderately complex one, which I cannot elaborate fully here. A more detailed discussion is available in (McMullin 1992a, Section 5).

In any case, it follows from this that "adaptations" which are the subject of Darwinian selection must be such as to favour an *S-lineage* relative to its competitors. This immediately suggests the view that adaptations can be viewed as Darwinian *only* to the extent that they are beneficial to the S-lineage exhibiting them (at least for the duration of the selection episode): i.e. that S-lineages are the "units of selection". However, this needs to be considered in somewhat more depth.

In the simplest case, fitness can be improved by effects which are uniform and consistent for all members of an S-lineage. This kind of effect is particularly easy to envisage, given that all actors in a given S-lineage are already expected to be "similar" to a greater or lesser extent, simply by virtue of their common membership of that putative S-lineage.

Ultimately, this effect would be manifested in one of just two possible ways: if the actors uniformly[4] live longer than the actors of the competing S-lineage(s), then the mortality (expected death rate) of the S-lineage will be less than its competitors; or if the actors all uniformly have more offspring (per unit time) then the fecundity (expected birth rate) of the S-lineage will be greater than its competitors. In either case, the size of the S-lineage would be expected to increase relative to its competitor(s); and if the difference were large enough, and maintained long enough, the S-lineage would displace its competitors essentially deterministically—which is to say, an episode of Darwinian selective displacement would occur.

This way of thinking leads to associating fitness with individual actors, at least as a form of shorthand. This is always legitimate in a certain abstruse, formal, sense. The members of

[4]Note that I use the word in a probabilistic sense (McMullin 1992a, Section 6.1).

an S-lineage are guaranteed to be "similar" minimally in the sense that they *are* all members of that S-lineage. So any attribute of an S-lineage (such as fitness) can be imputed to the putative "Similarity Class" (S-Class) which also characterises all members of the S-lineage, and thus, implicitly, to single actors, considered as "exemplars" of their S-class.[5]

To put it in a slightly different way, any single actor is necessarily a member of its *own* S-lineage(s). Thus, it can surely favour its own S-lineage(s) by having as many offspring as it can (involving some kind of trade-off between living as long as it can while procreating as fast as it can). Again, this viewpoint estimates or measures fitness by reference to the activities or attributes of any single "typical" member of the S-class of actors making up a particular S-lineage.

The problem with this point of view is, however, that it is grossly simplistic. There may well be no such thing as a "typical" actor of a given S-lineage. In the simplest case, the selectively important "trait" may not be "expressed" at all in certain members of the S-lineage—but they *can* pass it on, and thus do indeed contribute to the selective dynamics of the S-lineage as a whole. But more generally, it is perfectly possible for the selectively important "trait" to be "expressed" in very different ways by different actors of the S-lineage. Thus, actors of the same S-lineage may be very different in a wide range of characteristics or attributes; or, for that matter, any single actor may engage in a wide variety of different behaviours, or manifest different traits, in different times and circumstances; we require only that these differences not be such as to establish two distinct, competing, lineages with distinct fitnesses relative to the particular selective competition we are studying (otherwise, of course, we would not be dealing with a single coherent S-lineage). The point is that the thing which is selectively important may not be, in any useful sense, a "trait" of a single actor at all, but rather is a "trait" of some groups of actors, or perhaps of the S-lineage as a whole.

Indeed, the fitness of an S-lineage may actually critically rely on the very *distribution of variations* in certain characteristics among the actors making it up. In that case, no single actor, *qua* actor, would even allow a determination of (S-lineage) fitness—except, perhaps, in the extremely contrived sense that we might, in theory, be able to somehow deduce or determine from that one actor what would be the "typical" distribution of variations within an S-lineage which it could found.

We can envisage a range of different ways in which variation in the characteristics of actors could be beneficial to an S-lineage. For example:

1. If there is a diversity of resources or habitats, whose availability varies, then an S-lineage which keeps its options open—distributes its actors over these resources or habitats—may well do better than competitors which lack this flexibility (whose actors are more uniform). This can work without any assumption of interaction or mutual recognition between the actors making up a single S-lineage.

2. Alternatively, if there *is* any mechanism whereby an actor can identify other actors with which it shares an S-lineage (with some degree of probability) then there may be other opportunities for *cooperation* between actors, perhaps involving specialisation or division of labour.

In general, both these possibilities could equally be based on dedicating certain actors to certain tasks on a lifetime basis, or on the use of a structured life cycle with different specialisations on the part of a single actor at different stages in its cycle; and, of course, the two possibilities could both be exploited to greater or lesser extents in a single S-lineage.

The notion of an actor being so constituted (in structure and/or behaviour) that it functions for the benefit of other actors, at a cost to itself (especially as measured in terms of the number of offspring it produces), is technically referred to as *altruism*, and has historically

[5] I do sincerely regret this further proliferation of prefixes; but it seems to be in the nature of the problems at hand to demand unusually precise vocabulary. If you doubt this, I can only point again at the terminological confusions documented in section 2.

been viewed as somewhat problematic for Darwinian theory. However, we can now see that this derives from a blinkered view of selection: the view that supposes that an actor can benefit its S-lineage(s) (improve the latter's fitness) *only* through direct "benefit" to itself. In particular, the second scenario mentioned above, where an S-lineage can benefit from mutual co-operation between its members, clearly allows the possibility of altruism (on the part of actors to other actors in a single lineage) being positively *favoured* by a Darwinian selection process.

I have suggested that altruism could be favoured where there is a mechanism for actors to recognise other actors sharing a single S-lineage. It seems to me that this is a *necessary* condition. By definition, altruism involves costs to the actors engaging in it. This can conceivably provide a net benefit to the S-lineage *only* if most of the benefit of the altruism is retained within the S-lineage. To put it another way, altruism is in permanent danger of being subverted or exploited. If one S-lineage contains actors which are unconditionally altruistic, and another is otherwise similar (and thus competing) but its actors succeed in restricting their altruism (even with limited effectiveness) to members of their own S-lineage, then the latter S-lineage clearly gets more benefits from altruism than its competitor, and will be favoured by selection.

Note that this argument would not generally go though if we considered a putative S-lineage which abandoned altruism entirely. Such an S-lineage could *initially* grow at the expense of the original S-lineage of altruistic actors. But recall our basic hypothesis that the altruism of the actors was indeed beneficial to the S-lineage. The new S-lineage, of entirely selfish actors, would, as it becomes more numerous, start losing this benefit. There is no general answer as to the final outcome of such a process, but it certainly does not follow that the original S-lineage would necessarily be eliminated.[6] It would be equally possible that another different S-lineage would arise which *restricted* its altruism: this would then generally be capable of selectively eliminating *both* of the others.

So altruism, if it arises, is expected to rely, to a greater or lesser extent, on mutual recognition between the actors of a single S-lineage. There are two basic mechanisms which suggest themselves for such recognition:

1. An actor's own "close" relatives (parents, sibling, offspring) are likely to be members (and, to a lesser extent, their close relatives in turn).

2. An actor may attempt to "recognise" other members of some class (which will reliably identify an S-lineage).

The former corresponds to what is conventionally referred to as "kin selection" in biological Darwinism, (Dawkins 1979). The latter is not generally explicitly emphasized in biology, though Dawkins has considered a variety of mechanisms which might come under this general heading (Dawkins 1982a, Chapter 8). Notwithstanding this apparent lack of emphasis among biologists, I speculate that benefits of altruism (in the broadest possible sense) mediated by recognition of class and/or recognition of close kin may be quite significant for Darwinian evolution generally. I say this simply on the basis that mutual recognition of this sort is clearly, in the general case, a difficult problem that calls for complex sensorimotor co-ordination, i.e. the utilisation of relatively complex anticipatory models of the world. Benefits (for S-lineages), deriving from more and more effective recognition and discrimination, seem to provide one plausible general basis for a sustained correlation between fitness and knowledge (or "adaptive complexity"), and thus for an extended period of growth of knowledge through evolutionary time. However, intriguing as this possibility may be, it is of limited relevance to my purposes here, and I shall not pursue it further.

To return to the central idea of this section, which is the notion of the S-lineage as the unit of selection, the significance of this point can be put at its most stark by asking what

[6]That is, it is not generally the case that such a process would meet all the requirements for selective displacement to go through; in particular, it may not meet the requirement of a *consistent* selective bias (independent of relative S-lineage size).

is the "selective" value *to an actor* of procreation itself? Actually, there is none, and yet everything else that might an adaptation of an actor is typically identified in terms of its contribution to procreation. Conversely, there is no problem when we think of the structure and behaviour of actors, insofar as it has been adapted by Darwinian evolution, as being consistently for the benefit of the S-lineage (procreation of at least some of the constituent actors is an absolutely necessary condition for the continued existence of *any* S-lineage). Which is simply to say that S-lineages which have come about through Darwinian evolution, are expected to be characterised by naked and exclusive *selfishness*; altruism on the part of an S-lineage (as opposed to the part of its component actors) could only be classified as pathological in a Darwinian context.

The essential point is that while the adaptations of an S-lineage *may* be clearly manifested in "adaptations" of the actors, this is an incidental effect, which may or may not occur, rather than being an intrinsic feature of Darwinian evolution in general. The general and reliable feature is the selfishness of the S-lineages themselves. The failure to realise this is manifested in the attempts to introduce entirely counter-intuitive modifications to the notion of "fitness" (of an actor) when it is discovered (as, for example, in the case of kin selected altruism) that, without such a correction, entities of lower "fitness" are apparently favoured by selection. This is the general idea of so-called "inclusive fitness". Dawkins has rightly pointed out that that is an inspired, but ultimately misguided, stratagem (Dawkins 1978a): the supposed difficulty simply does not arise when one realises that the actors are not the appropriate entities to ground the analysis in the first place—the *S-lineages* are.

So I consider that the unit of selection is an S-lineage; and that the implication of this is that a given phenomenon can be considered as an outcome of Darwinian evolution if and only if it is consistent with an unconditional selfishness on the part of the S-lineage possessing or exhibiting it. This, in short, is the doctrine of *The Selfish S-lineage.*

4 Interlude: Genic Selectionism

The biological literature on the Units of Selection issue also addresses another question, quite different from the one I have just chosen to answer. This second question might be phrased as follows:

> *Is there a uniquely distinguished "level" of organisation (biological or otherwise) which characterises entities that can qualify as Darwinian actors?*

I suggest that this is precisely the question which Hull set out to answer; whereas it is not clear to me whether Dawkins has ever consciously recognised this second question—not, at least, as a *separate* question. However he does *seem* to suggest an answer (implicitly or otherwise)—namely that *genes* are uniquely distinguished as the only biological entities which should properly be said to be actors. I believe that that claim is quite mistaken (but also that it is questionable whether Dawkins ever really intended to make it).

Specifically, to the extent that we interpret "gene" as a material part of an organism (a fragment of DNA, say), then I accept that such genes *might* be usefully considered as actors (in "suitable" circumstances); but I reject absolutely the idea that they are *uniquely* qualified for this role. I suggest that there are always alternative candidate actors (particularly, but not exclusively, organisms); alternatives which will be *equivalent* in the precise sense of yielding a selection dynamics which is either identical, or differs only by a bijective transformation of the state variables. Preferences among these different candidates therefore arise *only* from pragmatic considerations relating to the particular circumstances in which the theory is being applied.

It is, of course, no accident that "replicators" or "genes" turn up in two quite different senses in the two distinct questions regarding "units" of selection. For Dawkins, genes are the prototypical examples of replicators—and his conflation of both actor and S-lineage into the single term "replicator" is more or less mirrored in his usage of the term "gene". Again,

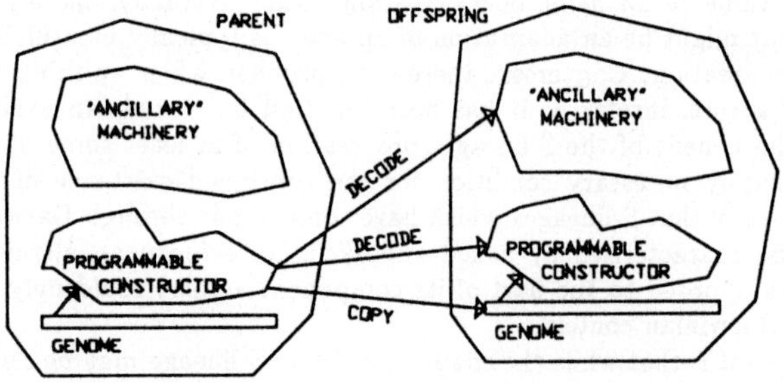

Figure 3: "Genetic" actor organisation.

we should try to distinguish between (at least) actor-genes or A-genes, and lineage-genes or L-genes. The former may (or may not) play the role of actors, and, correspondingly, the latter may (or may not) play the role of S-lineages. Dawkins effectively argues, correctly in my view, that only S-lineages can qualify as units of selection. But he also equates this with a claim that only "genes" (or "replicators") can qualify as units of selection; I can accept this also, if it is read in the sense of "only L-genes *(as opposed to A-genes)*" or "only L-replicators *(as opposed to A-replicators)*". But Dawkins seems to go on to parlay this into a claim that only "genes"—now specifically meaning A-genes, as opposed to (say) "organisms"—can qualify as Darwinian *actors*. I believe this to be definitely in error.

To repeat: I reject the idea that a certain level of biological organisation—the "genes" or the "DNA" or the "genetic material" or "replicators"—uniquely fulfills the role of Darwinian actor in terrestrial evolution. I refer to this flawed idea as *genic* selectionism—as opposed to *organismic* selectionism (Wright 1980; McMullin 1992b). My position is the pluralist one that genic and organismic selection, properly viewed, are not competitors or rivals, but merely alternative, formally interchangeable, descriptions of the same underlying biological reality; that, indeed, there may exist an indefinite number of other, similarly equivalent, descriptions; but that, *whichever* of these viewpoints may be adopted, there will be a crucial distinction between actors and S-lineages, with only the latter being properly regarded as units of selection, or entities for whose benefit (Darwinian) adaptations may be said to exist. This latter question, of actor versus S-lineage as the unit of selection, thus cuts at right angles to the question of gene versus organism as Darwinian actor; confounding these two questions leads only to confusion and error.

I may note that I do hold that there is *something* special, in evolutionary terms, about the "gene" level; or, more precisely, about that kind of actor organisation which involves a functional separation into a relatively passive information carrier or "genotype", and a relatively active information processor or "phenotype" (Fig. 3). This is, of course, the von Neumann architecture for self-reproduction (Von Neumann 1951; 1966a; 1966b), and it exhibits features which are crucially important for the possibility of a sustained Darwinian growth of complexity. Strangely, that this is so, and that this is precisely why von Neumann proposed the architecture he did, is still not generally appreciated even now, almost half a century after his seminal investigations.

Unfortunately, a properly detailed analysis and critique either of Dawkins' (alleged) genic selectionism, or of von Neumann's general theory of automaton evolution, would take me much too far afield in the current context. The interested reader (!) can find these discussions fully elaborated in (McMullin 1992c) and (McMullin 1992d, Chapter 4), respectively.

5 So: Does it or Doesn't it?

So, does the Dawkinsian replicator actually replicate?

Well, in my terms, actors clearly replicate (that is part of their definition). S-lineages, on the other hand, certainly *need not* do so. Of course, to have a *sustained* evolutionary process, new S-lineages have to come regularly into existence; and one can certainly describe this in terms of such new S-lineages being "offspring" of previous S-lineages. In some sense then, perhaps S-lineages *do* "replicate"—but they do *not* "replicate" in Dawkins' specific technical sense. The whole point of calling something a "new" S-lineage is that it is selectively *different* from, and typically competing with, its precursor(s); whereas, for Dawkins, the very definition of "replication" is that it *preserves* whatever is selectively significant.

Regretfully then, I am forced to the conclusion that Dawkinsian replicators don't, in fact, replicate! Strange but true.

This is not, of course, the whole story. To make my point here, I too have indulged in simplification. For example, I have glossed over the conceptual difficulties which arise because (under sexual forms of procreation) lineages, including S-lineages, need not be disjoint. Perhaps more seriously, it must be acknowledged that utterly new and distinct *kinds* of actors may sometimes emerge; and that the distinct evolutionary systems which result might interact in significant ways, which would render any explanation phrased in terms of only a single kind of actor quite inadequate; and, indeed, that this *may* already have been a significant phenomenon in biological evolution to date. I have in mind here the kind of hierarchical Darwinian theory described, for example, by Gould (1982), where an S-lineage in one evolutionary framework may actually function as an actor in another. In terms of such an hierarchical theory, my purpose has been restricted to the attempted clarification of Darwinian theory within *one* hierarchical level; but I do not imply, and do not suppose, that such a single level theory exhausts the scope of Darwinism.

6 Conclusion

In summary, my claims are:

1. The "replicator" terminology, as promulgated by Dawkins and Hull, suffers from significant, and confusing, internal contradictions. It would be best abandoned.

2. The *units of selection* in Darwinian evolution are lineages rather than actors.

3. More precisely, the units of selection are *S-lineages*. Unfortunately, these are somewhat complex and even nebulous entities. In particular, they are *not* simply groups of actors which share some overt, inherited, "trait"; nor are they simply "species" in any conventional biological sense. I suspect that S-lineages can only usefully be identified in a *relational* sense, in the context of a particular episode of Darwinian selection. That is, we identify an S-lineage not so much by any similarity among its members, as by the competively significant difference between two (or more) S-lineages.

4. Actors *may* have a von Neumann-like genetic architecture. This is generally the case for contemporary terrestrial organisms. In this (special) case, an S-lineage can typically be identified or marked by a particular segment of the genetic description, shared by all members of the S-lineage. This makes it tempting to regard this genetic "tag" (or "gene") as *being* the S-lineage, or to regard this genetic level of description as uniquely preferable for describing Darwinian selection dynamics ("genic selectionism"). *This temptation should be resisted.* A particular genetic tag can only identify a selectively important "trait" *in the context* of the actors which express or translate it; to put it another way, the "meaning" or "decoding" of a particular tag can vary (in principle by an arbitrary amount) from one actor to another, even in the "same" lineage; furthermore, in principle, the same "trait" can be correlated, independently, with multiple distinct genetic "tags".

5. The logic of Darwinian evolution requires replication with some kind of "heritability" so that coherent S-lineages can form and compete. However, it does *not* require any particular kind of hereditary mechanism, and, in particular, does not require that the component actors have a well-formed von Neumann style genetic architecture.

One reader of an earlier draft of this paper wondered what (if any) pragmatic implications it had for actually designing artificial evolutionary systems. I can only answer for myself of course. The most important implication for me is that, once the mystique of "genic" selectionism is dispelled, we can usefully ask how it is that von-Neumann style genetic organisation can spontaneously come into existence at all. As long as "replicators" (in Dawkins' narrow sense of fragments of genetic material—or even lineages thereof) are considered as necessary *pre-requisites* for Darwinian evolution, we can hardly even formulate this question, never mind answer it. But once we separate the notion of genetic organisation from the notion of the units-of-selection, we can envisage a form of ALife investigation in which we do not already "wire in" a genetic architecture, but ask how it could arise and be refined—by a process of Darwinian (but not yet "genetic"!) evolution. This is, of course, a fundamentally different direction from "conventional" GA or evolutionary strategy investigations. I suggest that it is a worthwhile alternative to pursue.

Acknowledgements

Perhaps contrary to appearances, it has not been my intention here to be merely polemical. Indeed, I should say that I have the greatest respect for the work of Richard Dawkins—it is precisely because that is so that I have ventured to expend so much attention on it. My hope is that, by expressing my views as clearly, and even trenchantly, as possible, I will facilitate subsequent criticism (and, no doubt, correction) of these views.

This paper arises from an ongoing attack on the problem of realising the spontaneous growth of Artificial Knowledge by Darwinian (or any other!) means. In this pursuit, I have benefited greatly from discussions with colleagues, particularly Noel Murphy in DCU, and John Kelly of University College Dublin. I am indebted to the School of Electronic Engineering in DCU (particularly through the agency of its Head, Charles McCorkell) for continuing encouragement. The final version of the paper has been significantly improved by the incisive comments of the ECAL '95 reviewers, to whom I am very grateful. All errors remain, of course, my own responsibility. Financial support for this work has been generously provided by Expert Associates Limited.

References

Burks, Arthur W. (ed). 1966. *Theory of Self-Reproducing Automata [by] John von Neumann.* Urbana: University of Illinois Press.

Dawkins, Richard. 1976. *The Selfish Gene.* Oxford: Oxford University Press. See also Dawkins (1989).

Dawkins, Richard. 1978a. Replicator Selection and the Extended Phenotype. *Zeitschrift für Tierpsychologie*, **47**, 61–76.

Dawkins, Richard. 1978b. The Value-judgements of Evolution. *In:* Dempster, M. A. H., & McFarland, D. J. (eds), *Animal Economics.* London: Academic Press. As cited by Dawkins (1978a).

Dawkins, Richard. 1979. Twelve Misunderstandings of Kin Selection. *Zeitschrift für Tierpsychologie*, **51**, 184–200.

Dawkins, Richard. 1982a. *The Extended Phenotype.* Oxford: W.H. Freeman and Company.

Dawkins, Richard. 1982b. Replicators and Vehicles. *Chap. 3, pages 45–64 of:* King's College Sociobiology Group (ed), *Current Problems in Sociobiology.* Cambridge: Cambridge University Press.

Dawkins, Richard. 1983. Universal Darwinism. *Chap. 20, pages 403–425 of:* Bendall, D. S. (ed), *Evolution from Molecules to Men.* Cambridge: Cambridge University Press.

Dawkins, Richard. 1989. *The Selfish Gene.* New edn. Oxford: Oxford University Press. This is a significantly revised edition of (Dawkins 1976), with endnotes and two new chapters added.

Gould, Stephen Jay. 1982. Darwinism and the Expansion of Evolutionary Theory. *Science,* **216**(23 April), 380–387.

Hull, David L. 1980. Individuality and Selection. *Annual Review of Ecology and Systematics,* **11**, 311–332.

Hull, David L. 1981. Units of Evolution: A Metaphysical Essay. *Chap. 1, pages 23–44 of:* Jensen, U. J., & Harré, R. (eds), *The Philosophy of Evolution.* Harvester Studies in Philosophy. Brighton, Sussex: The Harvester Press Limited. General Editor: Margaret A. Boden.

McMullin, Barry. 1992a (Mar.). *Essays on Darwinism. 1: Ontological Foundations.* Technical Report bmcm9201. School of Electronic Engineering, Dublin City University, Dublin 9, Ireland.
URL (Directory): ftp://ftp.eeng.dcu.ie/pub/autonomy/bmcm9201/.

McMullin, Barry. 1992b (Apr.). *Essays on Darwinism. 2: Organismic Darwinism.* Technical Report bmcm9202. School of Electronic Engineering, Dublin City University, Dublin 9, Ireland.
URL (Directory): ftp://ftp.eeng.dcu.ie/pub/autonomy/bmcm9202/.

McMullin, Barry. 1992c (May). *Essays on Darwinism. 3: Genic and Organismic Selection.* Technical Report bmcm9203. School of Electronic Engineering, Dublin City University, Dublin 9, Ireland.
URL (Directory): ftp://ftp.eeng.dcu.ie/pub/autonomy/bmcm9203/.

McMullin, Finbarr (Barry) Vincent. 1992d. *Artificial Knowledge: An Evolutionary Approach.* Ph.D. thesis, Ollscoil na hÉireann, The National University of Ireland, University College Dublin, Department of Computer Science.
URL (Directory): ftp://ftp.eeng.dcu.ie/pub/autonomy/bmcm_phd/.

Taub, A. H. (ed). 1961. *John von Neumann: Collected Works. Volume V: Design of Computers, Theory of Automata and Numerical Analysis.* Oxford: Pergamon Press.

von Neumann, John. 1951. The General and Logical Theory of Automata. *Chap. 9, pages 288–328 of:* (Taub 1961). First published 1951 as *pages 1–41 of:* L. Jeffress, A. (ed), *Cerebral Mechanisms in Behavior—The Hixon Symposium,* New York: John Wiley.

von Neumann, John. 1966a. Theory and Organization of Complicated Automata. *Pages 29–87 (Part One) of:* (Burks 1966). Based on transcripts of lectures delivered at the University of Illinois, in December 1949. Edited for publication by A.W. Burks.

von Neumann, John. 1966b. The Theory of Automata: Construction, Reproduction, Homogeneity. *Pages 89–250 of:* (Burks 1966). Based on an unfinished manuscript by von Neumann. Edited for publication by A.W. Burks.

Wright, Sewall. 1980. Genic and Organismic Selection. *Evolution,* **34**(5), 825–843.

RNA viruses: a bridge between life and artificial life

Andrés Moya[1], Esteban Domingo[2] and John J. Holland[3]

[1] Departamento de Genética and Servicio de Bioinformática, Facultad de Biología, Universidad de Valencia "Estudio General". c/ Dr. Moliner, 50. 46100 Burjassot, Valencia, Spain.
[2] Centro de Biología Molecular "Severo Ochoa", CSIC-UAM, Madrid, Spain.
[3] Department of Biology and Institute for Molecular Genetics. University of California, San Diego, La Jolla, California 92093. USA.

Abstract. RNA viruses can be an adequate bridge between life and artificial life. Under experimental conditions the parameters that in last instance are responsible for the evolution of replicons resembling primitive life forms can be easily studied. One year of a RNA virus evolving may be equivalent to one million years of an evolving DNA-based entity. High mutation rates as well as very short life cycles permit the capability of observing evolutionary effects in the lifetime of a human observer. Another important feature of RNA viruses, functionally related to its mutation rate, is the genome length, which ranges between 3 and 30 Kb, probably the shortest lengths with the highest estimated mutation rates that do not undergo error catastrophe when replicating. The evolutionary biology, that is to say the evolution of structural and functional properties, of RNA viruses can be probably simulated better than other non-RNA based life entities. The hypotheses underlying artificial life programmes could also be tested by experimental evolution of RNA viruses. Simplicity and rapid evolvability of RNA viruses are the basis for our proposal.

1 Darwin's analogy, RNA viruses and artifical life

Natural selection is the core of the neo-darwinian theory of evolution. Darwin found a deep analogy between the way breeders utilized appropiate artificial crosses to direct natural species towards new interesting varieties, and the way natural selection was acting on live forms to promote evolution. Manmade evolution is extremely rapid when compared with the products of natural evolution. However, Darwin reasoned that gradual changes acting during long periods of time, could ultimately shape the huge number of biological species and the variety of emergent biological properties that are present now or have previously disappeared. It was only a matter of time. Evidence for natural selection is mostly indirect, i.e., it is obtained when looking in the natural history repertoire and comparing patterns of change of a given character in populations and/or species. Then it is hypothesized that natural selection is responsible for the pattern observed. There is not enough time for a human observer or even for a limited set of human generations, to witness the emergence of, for instance, a new species, or the functional change from an ancestral property into a new function. There are two reasons that are responsible of the delay in observing the effects of natural selection. First, the genetic material of all known cell-based organisms is DNA, a very important stabilizing evolutionary event that must have followed a primitive RNA or RNA-like world. Second, the high fidelity of DNA replication catalyzed by cellular DNA polymerases. Accordingly, great evolutionary inventions triggered by natural selection acting gradually will be observed only after long periods of time. The question now is: can we observe and document great evolutionary events in our lifetime? The answer largely depends on the type of biological entity under consideration. Due to biological properties than will be discussed below, RNA viruses are excellent candidates for accelerated evolution (recent review in Morse [1]). Under appropiate experimental designs, RNA viruses can be used for direct demonstration of natural selection effects, and other parameters responsible for the evolutionary change. Darwin's analogy can now be taken "de novo" to demonstrate the power of evolution, in this case showing where viral evolution can go.

The basis of a potential bridge between life and artificial life offered by RNA structures, and more specifically by those that are replicating under high mutation pressures (see below), has already been developed by several authors utilizing different experimental (see [2], [3] and [4]) or theoretical (see [5] and [6]) approaches. Here, we would add to previous findings the idea that population and evolutionary biology of RNA viruses may represent an adequate bridge between life and artificial life studies, basically due to the powerful combination of experimental evolution in the laboratory together with computer simulations of simple replicons utilizing hypotheses, dynamics, quantitative parameters and population numbers as close as possible to those operating under experimental conditions.

2 Molecular variation and current theories on molecular evolution

Hubby and Lewontin [7] gave a list of criteria for the type of experimental technique needed to determine how much heterozygosity (i.e., genetic variability) there was per locus in a population. According to these authors a technique needed to provide the relevant information must satisfy the following four criteria:

Phenotypic differentiation. Phenotypic differences caused by allelic substitutions at single loci must be detectable in single individuals.

Genic distinction. Allelic substitutions at one locus must be distinguishable from substitutions at other loci.

Allelic substitution. A substantial proportion of (ideally all) allelic substitutions must be distinguishable from each other.

Genome sampling. Loci studied must be an unbiased sample of the genome with respect to the physiological effects and degree of variation.

These are experimental criteria needed for a proper validation of current theories of molecular evolution at population level. Current hypotheses on the mechanisms of evolution can be divided into two major groups. The first is centered around the neutral theory of molecular evolution [8], and states that the overwhelming majority of evolutionary changes at the molecular level are caused by random fixation of selectively neutral, or almost neutral, mutations under continuous mutation pressure. The second could generically be grouped in the context of positive selection, and would integrate such different but related views as the neo-darwinian theory of evolution (see [9] and [10]) or the quasi-species theory of molecular evolution [5]. Much of the controversy that has been created during the last thirty years in relation with major parameters (for instance positive darwinian selection versus random drift) ultimately responsible for molecular evolution come from the fact that there is problem in satisfying the four above mentioned criteria. The advent of molecular techniques have allowed us to have direct access to nucleotide sequences and, consequently, have supplied the best coverage of at least three out of four of the experimental criteria until now. Genome complexity, however, puts serious limitations on genome sampling. It is difficult to analyze genomes formed by thousands of genes, and particularly in a population context which attempts to relate mutational contributions to fitness of their carriers. It is not a problem caused by the molecular biology protocols; it is a problem posed by the organisms which are experimentally employed. At this point we can make a distinction between those organisms based on DNA and those based on RNA. Due to higher fidelity in replicatory machinery, evolutionary novelties are less probable in DNA than in RNA based organisms. A second major difference is genome complexity. RNA entities bear a small number of genes, whereas those of DNA organisms average several orders of magnitude higher. RNA genomes, in particular RNA viruses, not only conforms to the first three criteria of Hubby and Lewontin [7], but also the fourth one, and additionally one more that could be added to the list, that is:

Observable experimental evolution. In the case of RNA viruses, the time required for allelic substitutions should be detected by a human observer. Following Li and Graur [11] we define evolution at the molecular level as the process of progressive allelic fixations

3 Basic features of RNA viruses

The RNA viruses constitute the most abundant group of parasites of animals and plants (for a review see [12]); a very high percentage of the viruses that infect differentiated organisms are RNA viruses. Due to high mutation rates they have a great potential for variation that contribute to rapid adaptation. However the extent of variability and, consequently, adaptability is subject to important constraints (see [5] and [13]). These authors have shown that there is a crucial functional relationship between genomic size and accuracy of the replication machinery for meaningful information to be transmitted to the progeny. Those genomes that violate the relationship enter into the error catrastophe, and nucleotide sequences loose their information and become essentially random. The mutation rates operating on viral RNA during replication or reverse transcription range between 10^{-3} to 10^{-5} substitutions per nucleotide and replication round (for a review of the topic see [12] and [14]). Due to this high substitution rate and to avoid entrance into error catastrophe [5] the genomic length of RNA viruses must be very limited, and indeed all know RNA viruses have a genome size ranging from 3000 to 30000 nucleotides.

High mutation rates, coupled with extremely rapid replication cycles, are features that distinguish the genetics of RNA viruses from cellular DNA genetics. A direct consequence of the high mutation-rates operating during RNA replication is that mutant genomes are continuously being generated (see [3] and [15]). It means that in any infected cell, that amplifies one RNA molecule into 10^3 to 10^5 progeny molecules, many single and multiple mutants arise. Most of the new mutants will be maintained at low frequency and eventually will be eliminated by negative selection, even when they are highly adapted to the environment. In contrast, if the virus replicates in an environment different from that in which its parental genomes were used to replicate, the probability for newly arising mutants to become fixed will be greater [16]. These features endow RNA genomes not only with extreme adaptability but also with rapid evolutionary divergence, thus fitting the additional criterion five mentioned above.

4 Population genetics of RNA viruses

Several important concepts and parameters of population genetics have recently been evaluated with RNA viruses: positive darwinian evolution ([17] [18]), Muller's ratchet ([19] [20] [21]), the competitive exclusion principle ([22] [23]) and the Red Queen hypothesis ([23] [24]). As examples, the last three concepts will be summarized.

4.1 Muller's ratchet

This predicts than when mutation rates are high and a significant proportion of mutations are deleterious, a kind of irreversible ratchet mechanism will gradually decrease the mean fitness of small populations of asexual organisms. Experimental support for Muller's ratchet is scarce. It has been obtained in protozoa [25] and in a tripartite RNA bacteriophage [20]. We now have evidence that Muller's ratchet can operate in vesicular stomatitis virus (VSV), a nonsegmented nonrecombining pathogenic RNA virus of animals and humans ([21] [26] [27] [28] [29]) We did genetic bottleneck passages (plaque-to-plaque transfers) of VSV and then quantitated relative fitness of the bottleneck clones by allowing direct replicative competition in mixed infections in cell culture. We documented variable fitness drops (some severe) following 20 or more plaque-to-plaque transfers of VSV. Two additional remarks to emphasize the extreme genetic and biological variability of RNA virus populations. First, in some clones no fitness changes were observed. Second, the most regular and severe fitness losses ocurred during virus passages on a new host cell type. The relevance of these findings for population biology is clear: whenever genetic bottlenecks (of RNA viruses) occur, enhanced biological differences among viral subpopulations may result.

4.2 Competitive exclusion principle

In classical population biology this principle states that in the absence of niche differentiation, one competing species will always eliminate or exclude the other. The very high error frequency and rapid evolution during RNA genome replication may render improbable the prolonged coexistence of two genetically distinct viral subpopulations. Varied genomes are constantly generated that may have different replicative abilities and/or negative effects on competing virus populations. Clarke et al. [23] showed that two virus populations of approximately equal fitness can coexist through numerous generations during prolonged replication in the same environmet; however, stochastic changes in the population balance did eventually occur, and their impact on the competition was sudden and decisive: one population completely excluded the other.

4.3 Red Queen's hypothesis

Van Valen [24] proposed a hypothesis according to which each species is competing in a zero-sum game against others; each game is a dynamic equilibrium between competing species where no species can ever win and new adversaries grinningly replace the losers. Clarke et al. [23] showed with competing populations of VSV that the fact that one population eventually displaces its competitor should not obscure the fact that for very long series of virus transfers, during which immense numbers of viruses were produced, two competing virus populations maintained nearly equal relative fitness by constant running (fitness gain) to stay in the same place relative to its competitor.

Because of their high mutation rates, rapid replication, large population sizes, and controlled host cell environments, RNA viruses are excellent candidates for examining other evolutionary hypothesis. "One year of RNA evolution equals one million year of DNA evolution" is not a metaphor. Consequently, theories as important as the neutral theory, punctuated equilibrium, evolutionary stasis, etc. could be adequately tested using RNA viruses as biological model systems.

5 RNA viruses: a bridge between life and artificial life.

The concept of artifical life encompasses the following ideas [30]:

The biology of the possible. Artificial life deals with life as it could be.

Synthetic method. To synthesize life-resembling processes or behavior in computers or other media.

Real (artificial) life. Artificial life is the study of human created systems that exhibit behavior characteristic of natural, living systems.

All life is form. One can therefore ignore the material and instead abstract from it the logic that governs the process.

Bottom-up construction. The synthesis of artifical life take place best via a principle of computer-based information processing called "bottom-up programming".

Parallel processing. The principle of information processing in artificial life is based on a massive parallelism that occurs in real life.

Allowance for emergence. The must interesting examples of artifical life exhibit "emergent behavior".

What can biologists offer to those who work with artificial life? In our view, one possibility is to offer experimental results from those biological entities that, due to their structural and functional simplicity, more easily capture the qualities of life, or at least some of them. Artificial life would not be merely a metaphor for life if parallel experiments with simple "living" entities (i.e. RNA viruses , etc.) and computers or other devices were performed utilizing similar sets of working hypotheses. Interestingly, the seven concepts mentioned above that characterize artificial life, also characterize observable properties of RNA viruses. For instance, RNA viruses probably are the biological entities that generate new non-previously explored genomes in the adaptive landscape in the shortest absolute time. Probably better than any other type of genome, the RNA genomes represent the biology of the "possible" in time periods meaningful to a human observer. The search for new adaptive viral genome sequences, particularly due to the speed with with they are achieved, very much resemble a parallel process. The fact that the adaptive "sequence" in each case is actually a quasispecies "cloud" of related sequences in sequence space ([31], pp 79-86) contributes to, rather than detracts from, the parallel processing analogy. The finding which result from experiments (or natural events) inteliying RNA virus adaptive searches might sometimes be spectacular. A new virus can resemble the emergence of a completely new entity, with invasive and pathogenic properties that could be very far from those of its inmediate ancestors. With appropiate cautions, RNA viruses represent in real time (i.e., human time) the biology of the possible in evolution, the action of parallel processing in the search for optimal adaptive situations, a search that can culminate in the emergence of new viral "species". These are three important concepts that are at the core of artificial life programme.

High viral mutation rates, short replication rounds and extremely high population numbers can determine a rapid attainment of a new adaptive strategy. The RNA virus search strategy also parallels some strategies that have been used by computer experts to solve specific problems. As Langton [32] mentioned, biologists are much more interested in the mechanisms (i.e., processes) underlying evolutionary dynamics than in the final product, and this is also a major purpose for practitioners of artificial life.

Acknowledgements

This work has been supported by grant PB91-0051 from DGICYT (Spain) to AM and ED, by Fundación Ramón Areces and by grant AI14627 from NIH (USA) to JH. We also thanks the critical reading of A. Latorre, E. Barrio, A. Bracho and S.F. Elena.

References

1. Morse, S.S. (ed.): The Evolutionary Biology of Viruses. Raven Press, New York (1994)
2. Mills, D.R., Peterson, R.L., Spiegelman, S.: An extracellular darwinian experiment with a self-duplicating nucleic acid molecule. Proc. Natl. Acad. Sci. USA **58** (1967) 217-224.
3. Domingo, E., Sabo, D., Taniguchi, T., Weissmann C.: Nucleotide sequence heterogeneity of an RNA phage population. Cell **13** (1978) 735-744
4. Biebricher, C.K.: Darwinian selection of self-replicating RNA molecules. Evol. Biol. **16** (1983) 1-52
5. Eigen, M., Briebicher, C.: Sequence space and quasispecies distribution. pp. 211-245. Edit. by E. Domingo, J. Holland, P. Ahlquist in RNA Genetics (Vol. 3). CRC Press, Boca Ratón, California (1988)
6. Schuster, P.: Complex optimization in an artificial RNA world. Edit. by C.G. Langton, C. Taylor, J. Doyne Farmer, S. Rasmussen in Artitifical Life II, pp. 277-291. Addison-Wesley, Redwood City, CA (1992)
7. Hubby J.L., Lewontin, R.C.: A molecular approach to the study of genic heterozigosity in natural populations. I. The number of alleles at different loci in Drosophila pseudoobscura. Genetics **54** (1966) 577-594
8. Kimura, M.: The Neutral Theory of Molecular Evolution. Cambridge University Press, Cambridge (1983)
9. Dobzhansky, T.: Genetics of the Evolutionary Process. Columbia University Press, New York (1970)
10. Golding, B.: Non-Neutral Evolution. Chapman and Hall, New York (1994)
11. Li, W.H., Graur, D.: Fundamentals of Molecular Evolution. Sinauer, Sunderland, MA (1991)
12. Domingo, E., Holland, J.: Mutation rates and rapid evolution of RNA viruses. pp. 161-184. Edit. by S.S. Morse in The Evolutionary Biology of Viruses. Raven Press, New York (1994)
13. Eigen, M., Schuster, P.: The Hypercycle. A Principle of Natural Self-organization. Springer-Verlag, Berlin (1979)
14. Drake, J.: Rates of spontaneous mutation among RNA viruses. Proc. Natl. Acad. Sci. USA **90** (1993) 4171-4175
15. Temin, H.: Is HIV unique or merely different) J. AIDS **2** (1989) 1-9
16. Martínez, M.A., Carrillo, C., González-Candelas, F., Moya, A., Domingo, E., Sobrino, F. Fitness alteration of foot-and-mouth disease virus mutants: measurement of adaptability of viral sequences. J. Virol. **65** (1991) 3954-3957
17. Fitch, W., Leiter, J.M.E., Li, X., Palese, P.: Positive darwinian evolution in human influenza A viruses. Proc. Natl. Acad. Sci. USA **88** (1991) 4270-4274
18. Moya, A., Rodriguez-Cerezo, E., García-Arenal, F.: Genetic structure of natural populations of the plant RNA virus tobacco mild green mosaic virus. Mol. Biol. Evol. **10** (1993) 449-456
19. Muller, H.J.: The relation of recombination to mutational advance. Mutat. Res. **1** (1964) 2-9
20. Chao, L.: Fitness of RNA virus decreased by Muller's ratchet. Nature **348** (1990) 454-455
21. Duarte, E., Clarke, D., Moya, A., Domingo, E., Holland, J.: Rapid fitness losses in mammalian RNA virus clones due to Muller's ratchet. Proc. Natl. Acad. Sci. USA **89** (1992) 6015-6019
22. Gause, G.F.: The Struggle for Existence. Dover, New York (1971)
23. Clarke, D., Duarte, E., Elena, S., Moya, A., Domingo, E., Holland, J.: The red queen reigns in the kingdom of RNA viruses. Proc. Natl. Acad. Sci. USA **91** (1994) 4821-4824
24. Van Valen, L.: A new evolutionary law. Evol. Theory **1** (1973) 1-30

25. Bell, G.: Sex and Death in Protozoa: The History of an Obsession. Cambridge University Press, Cambridge (1988)
26. Clarke, D., Duarte, E., Moya, A., Elena, S.F., Domingo, E., Holland, J.: Genetic bottleneck and population passages cause profound fitness differences in RNA viruses. J. Virol. **67** (1993) 222-228
27. Duarte, E., Clarke, D., Moya, A., Elena, S.F., Domingo, E.,Holland, J.: Many-trillionfold amplifications of single RNA virus particles fails to overcome the Muller's ratchet effect. J. Virol. **67** (1993) 3620-3623
28. Duarte, E., Novella, I., Ledesma, S., Clarke, D., Moya, A., Elena, S.F., Domingo, E., Holland, J.: Subclonal components of consensus fitness in an RNA virus clone. J. Virol. **68** (1994) 4295-4301
29. Elena, S., González-Candelas, F., Novella, I., Duarte, I., Clarke, D., Domingo, E., Holland, J., Moya, A.: Evolution of fitness in experimental populations of vesicular stomatitis virus. Submitted
30. Emmeche, C.: The Garden in the Machine. Princeton University Press, NJ (1994)
31. Eigen, M.: Steps Towards Life. A Perspective on Evolution. Oxford University Press, Oxford (1992)
32. Langton, C.G.: Introduction. Edit. by C.G. Langton, C. Taylor, J. Doyne Farmer, S. Rasmussen in Artificial Life II, pp. 3-23. Addison-Wesley, Redwood City, CA (1992)

Complexity Analysis of a Self-Organizing vs. a Template-Directed System

Gad Yagil

Dept. of Cell Biology, The Weizmann Institute, Rehovot, Israel 76100.

Abstract: The structural biocomplexity of two viral structures is evaluated: that of the small RNA tobacco mosaic virus (TMV) and that of the larger dsDNA bacteriophage T4. Tobacco mosaic virus was chosen as a paradigm of a self-organizing biostructure, while the T4 represents biostructures where genome directed instructions are essential for the achievement of the correct virion structure. A large difference in complexity values is found: C = 4 for the TMV virion versus C = 117 for the tail part of the T4 virion. The considerable difference in these values indicates a correlation between the structural biocomplexity as defined and the pattern coding requirements of these organisms. It is proposed to utilize complexity analysis for the evaluation of expected genomic contribution to structural specification: the higher the complexity of a structure, the more genomic directions, of known or unknown nature, are likely to be required. The elements of biological complexity evaluation (Yagil, 1985; 1993b) are briefly summarized. A quantitative measure of order (0.93 for the T4 tail fiber) is an additional outcome of the formalism employed.

1. T4 vs. TMV

A major issue concerning pattern formation in living entities is whether biological structures are formed solely by spontaneous, self-organizing processes or whether additional, genome encoded signals are required for the specification of biostructures. In a classical demonstration of a self-organizing process, Fraenkel-Conrat (1963) dissociated the small RNA virus TMV (Tobacco Mosaic Virus) into its two components, RNA and coat protein, and showed that when these two components are remixed, fully infective virus particles are regained.

A different picture emerges when more complicated viruses such as the double stranded DNA bacteriophages are examined. Bacteriophage T4, for example, is composed of two principal parts, a head and a tail part. The tail structure is made of a sheath, a baseplate and 6 fibers attached to the base plate (see Figure), arranged in a close to perfect hexagonal symmetry. Bacteriophage T4 has a genome of about 166000 nucleotides (168699 nt in the t4 phage database as of 6/1994), compared with 6394 nt for TMV virus, and codes for at least 130 genes (Mosig and Eiserling, 1988). The correct expression of at least 49 of these genes is required for the assembly of the intact virion structure. Most of these genes code for the proteins composing the structure, but at least four genes are not found in the final structure: gene 57 codes for an enzymatic activity required for tail fiber assembly; gene 51, although essential, does not code for an identified protein and its role, is so far obscure; the product of gene 63 ("gp63") is responsible for chaperoning the correct joining of long tail fibers to the tail part (Wood

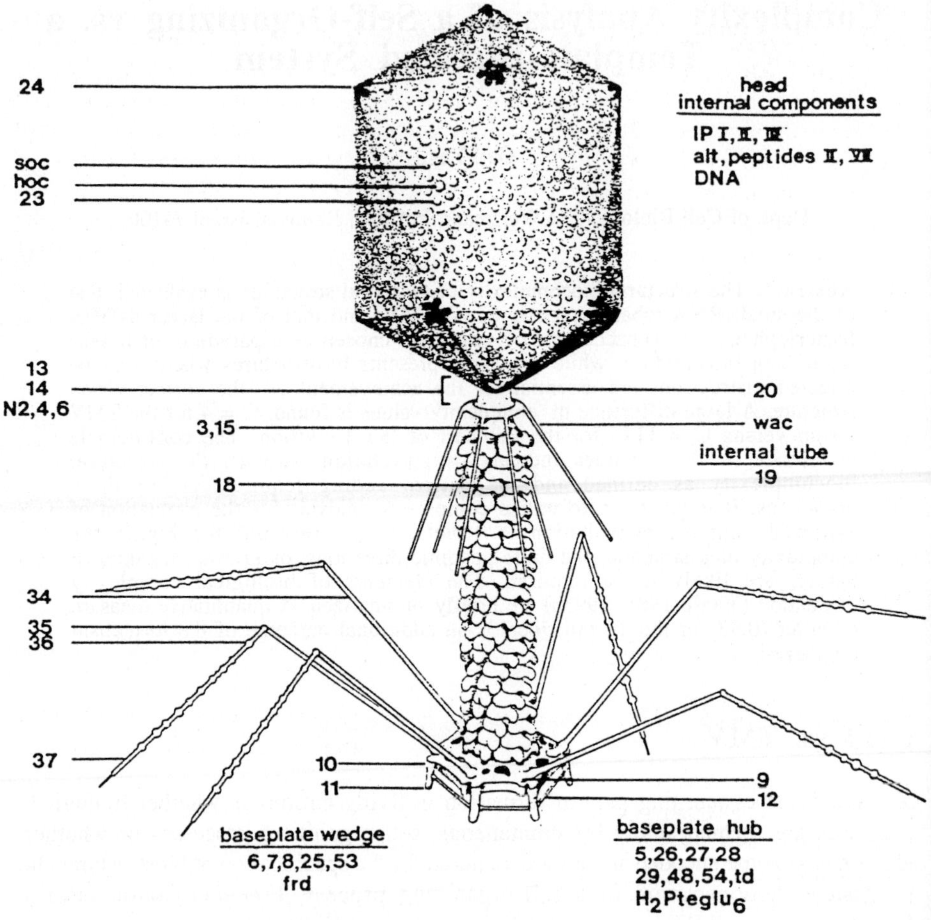

Bacteriophage T4. From Eiserling, 1983, loc. cit.

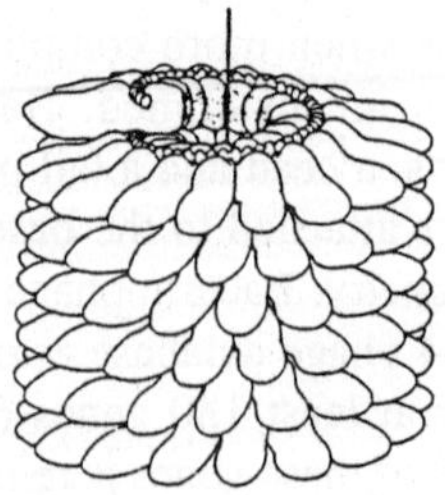

The TMV Virus.

and Crowther, 1983); gene 38 directs the correct assembly of tail fiber components (see Table 1 and Casjens and Hendrix, 1988). Gp29 and/or gp48 are proposed to serve as a tape or jig determining the correct length of the tail tube (Berget and King, 1983; Casjens and Hendrix, 1988). Other gene products have a role in directing the correct assembly of the head structure: gene 21 codes for a protease needed to cleave some of the head proteins before they can be properly assembled. Gp22 appears in the prohead to serve as a scaffold for the assembly of a correct head shell; only its degradation products appear in the final head structure (Black and Showe, 1983). It is clear that a simplistic principle of self-assembly is not applicable to bacteriophage T4. Rather, specific genomic instructions, some via yet unknown molecular pathways, have to be implemented before an infective virion particle can be produced in the infected cell. Direction by Such genomic signals may be dominating in higher organisms.

In this paper we shall apply a previously formulated theory of biocomplexity (Yagil, 1985; 1993a,b) to compare the structural complexity of the two virions - the self-organizing TMV virion with the instruction-directed T4 phage. It will be shown that a great difference in the numerical complexities of the two structures exists, leading to the proposition that self-organization can work well mainly for structures of low complexity, while template generated instructions are required for the generation of the more highly complex structures.

2. Structural Biocomplexity

Structural biocomplexity has been defined as *the length of the shortest list of numerical and symbolic specifications necessary to describe a given structure* (Yagil, 1985). This definition can be regarded as a form of the Kolmogorov algorithmic complexity (cf. Li and Vitanji, 1991) adapted to real, physical structures. The length of the list of specifications describing a physical system is determined by the number of regularities (i.e. identical numerical specifications) in a system - the more regularities that can be identified in the elements of the system (e.g. in the phage particles), the shorter is the specification list required. In quantitative form (Yagil, 1985):

$$\mathbf{C} = \Sigma_k \, [c(k)/k] - c' \qquad (1)$$

where **C** designs the structural complexity of the system; c(k) the number of point coordinates sharing a k fold regularity and c' the 5-6 coordinates necessary to place a rigid structure in the external world. We have previously shown how complexity **C** is evaluated for small molecules such as methane, ethane and adenine (1985), for proteins (1993a) and for hydrocarbon isomers (1993b). Here we shall apply the procedure to evaluate, the structural complexities of the virion structure of TMV and of the bacteriophage T4 tail structure.

The first step in the analysis is to decide upon the hierarchical level of interest, e.g. whether virion complexity is to be evaluated in terms of its structural proteins, in terms of their composing amino acids, or possibly in terms of the atoms constituting each amino acid. The decision is an arbitrary one and depends on the kind of information needed. When coding capacity of a genome is of interest, complexity in terms of gene products, i.e proteins, is called for. A simple level rule (Yagil, in preparation) connects the complexities of the different hierarchical levels.

Table 1. The Long Tail Fibers of Phage T4 Virion - Specification Table[a]

No.	n (i=0,n-1)	ε (protein)	r (nm)	ϕf	θ	Function
1	6 x 2[b]	gp34	695/2[c,e]	$\Phi_{34} + i\pi/3$	Any_1	Proximal part, main
2	6	gp35	R_{35}[d]	$\Phi_{35} + i\pi/3$	Any_2	Distal part
3	6 x 2[b]	gp36	690/2[d]	$\Phi_{36} + i\pi/3$	Any_2	Distal part
4	6 x 2[b]	gp37	690/2[d]	$\Phi_{37} + i\pi/3$	Any_2	Distal part, main
-	0	gp38	-	-	-	catalytic
-	0	gp63	-	-	-	catalytic
Specifications:	4		4+3[b]	4+3	0+3	

C = 21

Comments to Tables 1 and 2:

a. Source: Wood and Crowther, 1983.

b. These proteins are di, tri or tetrameric. The extra 3 or 4 coordinates are for the positions of the monomers within multimers, assuming symmetric arrangement.

c. A coordinate transformation to an origin at the edge of the base plate, e.g. at the center of gp9 applies. A transformation with previously defined parameters does not add to the complexity (Rule 6 ,Yagil, 1985).

d. The origin is here on gp34. A transformation with the parameters for gp34 applies.

e. Numerical values are given where known, to illustrate that complexity analysis is based on observed values. The numbers were extracted from journal figures, and can differ somewhat from the measured values.

f. Some of the ϕ coordinates may have the same value because of the hexagonal symmetry, but no firm data are available.

g. Sources: 1. Eiserling, F.A., Compreh. Virol. 13: p.558 (1979). 2. Berget, P. and King, J. (1983). 3. Casjens and Hendrix, (1988).

h. The numbers are for the hexagon form are from fig 4, Crowther et al., (1977).

j. From Amos and Klug (1977).

k. note that helically arranged gp18, gp19 need two numerical z specifications each.

m.See Kikuchi and King, 1975.

n. Converts gp12,34,37 (King and Laemmli, 1973).

p. the coordinates in small print are for the position of monomers within multimeres.

The next two steps are to set up a *specification table* and to choose a coordinate system. In the specification table all components of the system and their coordinates are listed (cf. e.g. Table 1). Each protein component is to be specified by four coordinates: One symbolic coordinate (ε) for its "type" or "color" (for H_2O, for instance, this would be: ε_1=O; $\varepsilon_{2,3}$=H) and three numerical coordinates for its position within the virion structure (the polar coordinates for H_2O would be: $r_1,\phi_1,\theta_1 = 0$; $r_{2,3} = r_{CH}$; $\phi_2 = 0$; $\phi_3 = \pi$; $\theta_{2,3} = \pm 52.25°$). A unique way to find the coordinate system which gives the shortest list of specifications can not be offered at present and several coordinate systems may have to be examined. For the analyses of the T4 tail and of TMV a cylindrical coordinate system (r,ϕ,z) turns out to be the most suitable one, because of the hexagonal and helical symmetries in their structures. For the tail fibers (Table 1), the polar system (r,ϕ,θ) is the more suitable one.

In the fourth step it should be determined whether each of the listed coordinates is a random or an ordered one. An *ordered* coordinate is one for which a fixed numerical value applies to all members of an ensemble (e.g. for all phage particles in a test tube), while a *disordered*,or *random* coordinate is one which takes a different value in each member of the ensemble. For instance, each tail fiber of T4 has two subparts the length of which is constant ("ordered coordinates") but have variable angles, depending on the extent each fiber is drawn in or extended (see Figure). These angles are "disordered coordinates", because they take a different value in every phage in an ensemble of phage particles and are assigned values of "Any" in Table 1. The angles of free rotating amino acids chains are other examples of random coordinates, while neighboring carbon - carbon distances are ordered features of these molecules. A coordinate is ordered only when it takes up the same value in each particle in ensemble. The distinction between disordered (random) and ordered coordinates is important, because only ordered coordinates, which can be assigned defined numerical values, contribute to the complexity of the system. For disordered coordinates it can not be stated whether they are complex or simply arranged; they have therefore to be excluded from the analysis. This is a major feature of the approach presented here.

As a bonus, data like those in Table 1 enable the assignment of a numerical value to the degree of order in a system in a quite natural way. The *degree of order* Ω has been defined (Yagil,1985, p.19) as *the ratio of ordered to total coordinates*. The tail fiber components have altogether 168 coordinates (7 proteins of 6 units each, 4 coordinates each); of these 12 are disordered (2 angles of 6 proteins), leaving 156 ordered coordinates. The ratio of ordered to total coordinates is therefore 156/168, yielding a degree of order of $\Omega = 0.93$ for the tail fibers. There are no "Any" coordinates in Table 2, consequently $\Omega = 1$ for the main tail structure.

3. The Complexity of the T4 Tail

The final step is to determine the minimal number of coordinates required to describe the ordered part of the system. This is done in Tables 1-2 for the tail fibers and the main tail structures of bacteriophage T4. The tail fiber is composed of the four gene products listed in Table 1 (Wood and Crowther, 1983); the two additional genes listed must be active for correct tail assembly but their products are absent from the final structure. Twenty two T4 gene products are listed in the specification table of the main tail structure (Table 2), divided into a sheath and tube, baseplate hub and six baseplate arms (or "wedges"). Almost all of the proteins listed are present in multiple units, up to 144 copies for the sheath and tube protein units (Berget and King, 1983). A numerical value has been entered whenever an experimental value was found in the literature. Almost all proteins listed share a six fold radial symmetry around the main axis of the phage, chosen here as the z axis. The z coordinates are the distances along the z axis to an arbitrarily chosen origin, e.g. to the base of the sheath structure. The radial arrangement of most units implies that only one r value need be listed for each protein, reducing the number of specifications needed for each six-fold subunit by 5. The radial symmetry dictates also the ϕ values for each component, so that one can write $\phi=2\pi i/6$ when six protein units are found in the structure. A slightly more complicated relation, involving a second numerical value, is necessary to specify the ϕ coordinate of the tube and sheath proteins subunits (gp18, gp19), because of the helical displacement of successive rings.

Table 2: The Tail Part of Phage T4 Virion - Specification Table [g]

No.	n (i=0,n-1)	ε (protein)	r (nm)	ϕf	z (nm)	Location and Function.
1	144	gp18	12.0 [e]	$\Phi_{18} + 17i$ [e]	$Z_{18} + 4.1\,\mathrm{int}(i/6)$ [e]	Sheath [j]
2	144	gp19	4.5	$\Phi_{19} + 17i$	$Z_{19} + 4.1\,\mathrm{int}(i/6)$	Tube [j]
3	1-6	gp15	R_{15}	$\Phi_{15}+i\pi/3$	Z_{15}	Sheath cap
4	1-6	gp3	R_3	$\Phi_3 + i\pi/3$	Z_3	Tube cap
5	6	gp8	0	$\Phi_{48}+i\pi/3$	Z_{48}	Sheath initiation, Jig?
6	6	gp54	R_{54}	$\Phi_{54}+i\pi/3$	Z_{54}	Tube initiation
8	6	gp29	R_{29}	$\Phi_{29}+i\pi/3$	Z_{29}	Hub center, fol.synt.,tape[m]
7	6	gp5	R_5	$\Phi_5 + i\pi/3$	Z_5	Hub, lysozyme
9	6	gp27	R_{27}	$\Phi_{27}+i\pi/3$	Z_{27}	Hub
10	3[e]	gp26	R_{26}	$\Phi_{26}+2i\pi/3$	Z_{26}	Hub assemb,folate synthase.
11	3[e]	gp28	R_{28}	$\Phi_{28}+2i\pi/3$	Z_{28}	Hub assemb, pteroyl-hexaglut. synthase.
12	6	frd	R_{frd}	$\Phi_{frd}+i\pi/3$	Z_{frd}	Hub, dhfolate reductase
13	3	td	R_{td}	$\Phi_{td} +2i\pi/3$	Z_{td}	Hub centr.,dT synthase.
14	12	gp6	R_6	$\Phi_6+i\pi/6$	Z_6	Arm, main inner
15	6	gp7	R_7	$\Phi_7+i\pi/3$	Z_7	Arm, to hub
16	6	gp8	R_8	$\Phi_8+i\pi/3$	Z_8	Arm, main inner
17	6 x 4[b,e]	gp9	15.3[h]+ r_9[p]	$\Phi_9+i\pi/3+ \phi_9$[p]	$Z_9 + z_9$[p]	Arm, long fiber att. site
18	6 x 2[b,e]	gp10	18.0[h]+r_{10}	$\Phi_{10}+i\pi/3+\phi_{10}$	$Z_{10}+ z_{10}$	Arm, spike or vertex
19	6 x 2[b,e]	gp11	19.1[h]+r_{11}	$\Phi_{11}+i\pi/3+\phi_{11}$	$Z_{11}+ z_{11}$	Arm, spike knob
20	6 x 3[b,e]	gp12	19.1[h]+r_{12}	$\Phi_{12}+i\pi/3+\phi_{12}$	$Z_{12}+ z_{12}$	Arm, spike fiber
21	6	gp53	R_{53}	$\Phi_{53}+i\pi/3$	Z_{53}	Arm, hexamer joiner?
22	6	gp25	R_{25}	$\Phi_{25}+i\pi/3$	Z_{25}	Arm, lysozyme
-	(6)	gp51	-	-	-	Hub assem,folate synt.
-	-	gp57	-	-	-	Not in capsid, catalytic[n]
Specifications:		22	22+4[b]	22+4[b]	24[k]+4[b]	

C = 102 - 6 = 96

The total number of specifications for each coordinate is listed at the bottom of the ε,r,ϕ,z columns in the tables. The total number for all four coordinates is the *structural biocomplexity* of the tail part of the virion, in terms of constituent gene products (proteins). The complexities which result are **C** = 21 for the tail fibers and **C** = 102 - 6 for the main tail structure (6 is subtracted for c' in eq. (1), i.e. for placement in external world). The structural biocomplexity of the complete tail can thus be assigned the value of **C** = 117. This is about half of the complexity of the entire virion, head and neck structures included. The full details of these parts will be described elsewhere. Structural complexity assumes necessarily larger values when evaluated in terms of lower hierarchical levels. The next lower level is that of the amino acids composing each protein. The total MW of the coded tail proteins is 1061 kD. If we take the average MW of an amino acids as 130, we obtain 8162 separate amino acids; each amino acid involves four specifications (few regularities on this level), which yields an overall minimal complexity of **C** = 8162x4 +14 -22 = 32640 (14 for multimers and helical arrangements; minus 22 for the 22 gene products, now respecified in terms of their component amino acids).

Table 3 Complexity of the Tobacco Mosaic Virus (TMV) Shell

No	n (i=0,n-1)	ε (protein)	r	φ	z
1	2100	Coat protein	R_1	$2\pi i/16 1/3$	330 i/2100
Specifications:		1	1	1	1

C = 4

Source : Caspar et al., *Adv.protein chemistry*, 18: 37-88

Comment: In the present analysis proteins were considered as point elements; the analysis can also be performed with the proteins as rigid bodies. Three additional coordinates for the orientation of each of the bodies within the virion would be needed. For an analysis of the complete virion (including the RNA part) as rigid bodies see Table 6, Yagil, 1985. RNA has been left out here for comparison with the unfilled T4 structure.

4. The Complexity of the TMV Virion

The value of **C** = 117 is to be compared with the complexity of the Tobacco Mosaic Virion. (Table 3). TMV has only a single protein component ("ε = coat protein") of n = 2100 units, arranged in 330 helical turns. The three helical parameters (R_1, $2\pi i/16 1/3$ and 330i/2100) are sufficient to describe the position of each subunit within the structure. The structural complexity of TMV coat, in terms of this single component, is therefore just **C** = 4. The complexity of the entire T4 is thus more than 50-fold higher than that of TMV. This large increase in complexity parallels the transition from a strictly self-organizing system to one which is heavily instruction directed. The comparison of the two organisms illustrates the utility of complexity analysis in predicting the type of organization to be expected.

5. Remarks and Conclusion

A more detailed discussion of the assumptions made in this analysis can be found in the three previous publications cited. These publications discuss also the relation between complexity as used here and common thermodynamic and information-theoretic quantities. In brief, biocomplexity can be regarded as an zero point entropy, which exists already at 0° Kelvin, and does not contribute, by the 3rd law, to the thermodynamically measured entropy of the system. The DNA and RNA of the viruses, really the heaviest contributors to the complexity of the two viruses (166000 nt and 6892 nt), have not been included in the present analysis; this has been done in order to highlight the structural contribution to biocomplexity. The contribution of nucleotide base sequence is nevertheless decisive in determining both shape and function of the viruses. This is a major point, as the increase in structural and functional complexity during evolution can hardly be imagined without the parallel increase in template complexity (certain aspects of brain function may be an exception). It is the emergence of templates capable of specifying organizing instructions which confers to living entities their high complexities, with all their manifestations. The complexities reached by the genetic templates remain unmatched in the inanimate world.

The difference between the two simple semi-organisms thus highlights the limitation of self-organization in producing complex biological structures. The requirement for template-specified directions should by no means be regarded as a conceptual hurdle - all highly evolved organisms have a vast repertoire of DNA with ample space for yet unaccounted functions (Hood, 1993). A search for novel organizational signals encoded in DNA, via small ORFs or otherwise, could be rewarding. The main object of this presentation is to illustrate, by way of an example, the relation between biocomplexity and morphogenetic elements and to point at the potential of complexity analysis to shed light on the emergence of the intricate patterns of life.

References

1. Amos, L.A. and Klug, A. (1977). Three dimensional image reconstruction of the contractile tail of the T4 bacteriophage. *J. Mol. Biol*. **99**, 51-73.

2. Berget, P. and King, J. (1983). T4 tail morphogenesis. In: *"Bacteriophage T4"* (C. Mathews, E. Kutter, G. Mosig and P. Berget, Eds.), ASM publications, Washington, pp. 246-258.

3. Black, L. and Showe, M. (1983). Morphogenesis of the T4 head. In: *"Bacteriophage T4"* (C. Mathews, E.Kutter, G.Mosig and P. Berget, Eds.) ASM publications, Washington, pp. 219-245.

4. Casjens, S. and Hendrix, R. (1988). Control mechanisms in dsDNA bacteriophage assembly. In: *"The Bacteriophages"*, Calendar, R., Ed. Plenum Press N.Y. pp 15-92.

5. Crowther, R.A., Lenk, E.V., Kikuchi, Y. and King, J. (1977) Molecular reorganization in the hexagon to star transition of the baseplate of bacteriophage T4. *J. Mol. Biol.* **116,** 489 - 623.

6. Eiserling ,F. (1983) Structure of the T4 virion. In: *"Bacteriophage T4"* (C. Mathews, E.Kutter, G.Mosig and P. Berget, Eds.) ASM publications, Washington, pp. 1-24..

7. Fraenkel-Conrat, H. (1963). *"Design and function on the threshold of life"*, Academic press, N.Y.

8. Hood, L., Koop, B.F., Rowen, L. and Wang, K. (1993). Human and mouse T cell loci; The importance of comparative large-scale DNA sequence analyses. *Cold Spring Harbor Symposia*, **57**, 339-348.

9. Kikuchi, Y. and King, J. (1975). Genetic Control of bacteriophage T4 baseplate morphogenesis. I. Sequential assembly of the major precursor, in vivo and in vitro. *J. Mol. Biol.* **99**, 645-716.

10. King, J. and Laemmli, U. K. (1973). Bacteriophage T4 tail assembly: structural proteins and their genetic identification. *J. Mol. Biol.* **75,** 315-337.

11. Li, M. and Vitanyi, P. (1993). *"An introduction to Kolmogorov complexity and its applications"*. Springer Verlag, New York, Inc.

12. Mosig, G. and Eiserling, F. (1988). Phage T4 structure and metabolism. In *"The Bacteriophages"*. Calendar, R. Ed., Vol. **2,** p.521- 606.

13. Wood, H.B. and Crowther, R.A. (1983). Long tail fibers: Genes, proteins, assembly and structure. In: *"Bacteriophage T4"* (C. Mathews, E. Kutter, G. Mosig and P. Berget, Eds. ASM publications, Washington, pp. 259 - 269.

14. Yagil, G. (1985). On the structural complexity of simple biosystems. *J. Theor.Biol.,* **112,** 1-23.

15. Yagil, G. (1993a). Complexity analysis of a protein molecule. In: *"Mathematics applied to Biology and medicine"*, J. Demongeot and V. Capasso, Eds., Wuerz publishing, pp. 305 - 313.

16. Yagil, G. (1993b). On the structural complexity of templated systems. in: *"1992 lectures in complex systems"* , L. Nadel and D. Stein, Eds., The Santa Fe Institute and Addison-Wesley, N.Y.

Tile Automaton for Evolution of Metabolism

Tomoyuki Yamamoto and Kunihiko Kaneko

Department of Pure and Applied Sciences,
University of Tokyo, 3-8-1 Komaba, Tokyo 153, Japan
E-mail: yamamoto@complex.c.u-tokyo.ac.jp, kaneko@complex.c.u-tokyo.ac.jp

Abstract. A new model, "Tile Automaton" ("Tile" for short)is proposed for the evolution of metabolism, where the importance of dynamical many-body relationships and spatial effects of shapes are stressed. Like the computer game "Tetris", our automaton consists of tiles with a variety of sizes, and shapes, moving in a 2-dimensional space with some velocity coded by a real number. By the collision, tiles react and change their shapes, according to the reaction rules given by the overlapped length by the collision. With this model, self-organization of ever-creating sets are obtained from a small number of simple-shaped tiles. Although a self-reproduction in the strict sense is not seen, we have found that a set of tiles is organized working as a unit of replication, self-maintenance and a source of novel patterns. The origin, maintenance, and evolution of complexity are discussed.
(key words: chaos-induced diversity, chaos-induced novelty scenario, prebiotic evolution)

1 Introduction

One of the most important aims of "Artificial Life(ALife)" studies lies in the understanding of the origin of life, which remains difficult and unresolved over long years. Restricting to chemical evolution, several models have been proposed, but the complete answer has not yet been obtained.

Difficulty of the problem lies not only in the huge number of involved degrees of freedoms–almost $O(10^{20})$ molecules. There is complexity of reaction process itself. Here, *complexity* means not only combinatorial complexity of chemicals, but also dynamical aspects of reaction. Chemical reactions occuring within cells are parallel, and dependent on the environment. Here we should note that the environment is provided by other components inside cells in addition to external parameters. Thus it is important to see how the interacting elements mutually play the role of *environment* with each other, even without introducing external environmental constraints.

As some ALife studies have shown, it is important to create a virtual world which has several essential features of the real world with the emergence of life-like behavior. Here we started from an inhomogeneous, parallel and many-body reaction process to obtain a self-organizing, reproducing and self-maintaining system.

There are many models for the evolution of metabolism. The Hypercycle[1] by Eigen and Schuster is thought to be one of the most successful models. However, this theory allows only for one self-replicating cycle. This model can explain the existence of a metabolic pathway, but it is not possible to discuss the evolution of variety, novelty, and stability of metabolic networks. Algorithmic Chemistry[2] by Fontana and Buss captures the essence of metabolism by adopting λ-calculus for reactors. They have succeeded in finding self-replicating organizations, self-maintaining organizations and self-maintaining metaorganizations by applying different boundary conditions respectively. However, only two-body reactions are considered there. The run of each "calculation" standing for reaction is completed before the next collision, in other words the relaxation is always fast enough. Thus, in the model one could hardly discuss an intermediate state, a catalyzed reaction or any other environment dependent process, which we think strongly important.

Conway's game of Life[3, 4] is a similar model to our Tile model to be discussed. Its process is parallel, and admits spatial aspects. However, the information retained in a cell is homogeneous in space, and does not allow for elements to have "inner states". So it may be impossible to discuss complex relationalship between objects in the model.

Taking into account of the above consideration, we propose a novel model including the following features:

- SHAPE : each tile consists of sets of connected cells. Any shape is allowed as long as it is connected as one body.
- MOTION : each tile moves in a parallel manner, with its own real-valued velocity.

We note that all tiles are movable units, in contrast with Conway's Life, where only few special patterns such as gliders can exist as movable units. By introducing the above two features, it is possible to take into account of the interference between "shapes" and "motion". In other words, with the change of its shape, a tile changes its boundary condition for the reaction and motion. This interaction allows much complexity of patterns.

2 Model

In this section, we describe our Tile model in more detail. The world of Tile consists of the following components. Beginning with the fundamental level, it is given by

1. WORLD : 2-dimensional plane, basis of cells, with a defined boundary condition
2. CELL : with a real-valued position coordinate and velocity. It occupies a unit square of the world. Coordinates are rounded to an integer when the reaction rules are applied.
3. TILE : defined as connected cells, moving as a unit.

The rules of reaction consist of one basic rule and two sub rules.

- the basic rule (the inversion rule):
  When tiles collide with others (where overlapping is not allowed), we invert the states of nearest neighbor cells of the colliding side. This area is called *reacting zone*. Momentum of eliminated cells is equally divided to created cells, so that the momentum is preserved. Note that the number of cells is not preserved. This process is shown in fig.1. After inverting the reacting zone, we connect contacting cells to form a tile. When the number of pairs of connecting cells is more than one, the rule means just to join the reacting zone. See also fig.3

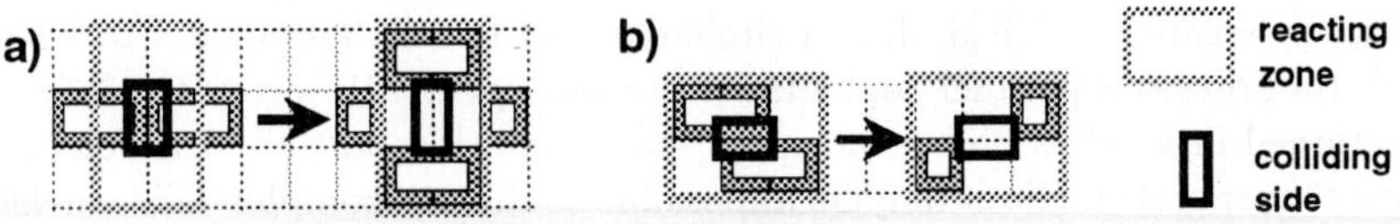

Fig. 1. Basic rule. Cells within a reacting zone are inverted. Note that the number of cells is not preserved.

- the joint rule:
  The new cells created by the above rule may contact with others which did not originally join the reaction. In such case, we connect them to form a tile. See fig.2.

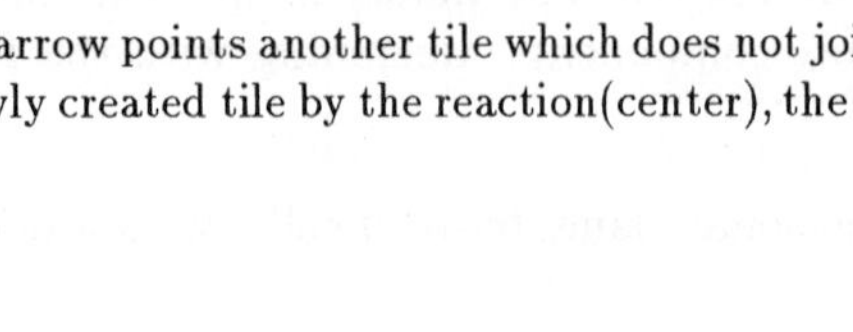

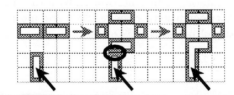

Fig. 2. Joint rule. An arrow points another tile which does not join by the reaction(left). If it collides with a newly created tile by the reaction(center), the collided tiles are joined together(right)

- the rate rule:
  This rule decides whether we apply the above basic reaction rule or not. We assume that the above reaction rules are applied only when the ratio of cells occupying the reacting zone falls within a predefined interval of values. See fig.3

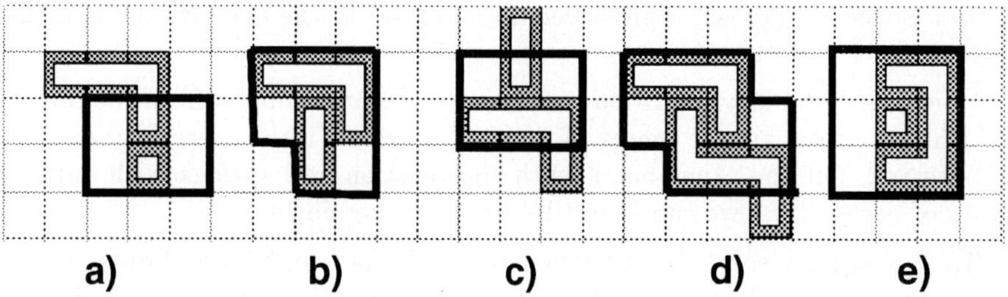

Fig. 3. Several possible cases of two-tile collisions. According to their relative positions, the colliding side and the reacting zone vary. Rates of cells to the reacting zone are: a: 2/6, b:6/8, c:4/6, d:8/10, e:6/9.

The inversion rule does not allow simple reproduction. We should note that shapes of tiles are strongly mixed by these rules.

On the other hand, the rules of motion are basically determined by Newtonian mechanics – by adopting molecular dynamics-like simulations. It consists of

- parallel motion
  All tiles moves parallelly unless their destination is occupied by other tiles.
- interaction between cells
  When the distance between two cells are smaller than 2.0, there is an interaction between the two. Here we assume harmonic interaction, in other words, we put a linear spring (with the natural length $x0$, and the spring constant k).
- reflection
  When the contacting tiles do not react according to the rate rule, these tiles reflect each other with the elastic collision. Thus the momentum and energy are preserved by this non-reactive collision.

For technical problems for simulations, we further impose few additional rules.

- friction
  To avoid tiles with too high velocities, we impose friction term for the tiles with velocities exceeding V_f. Using constant ν, the friction is defined as $\dot{\boldsymbol{v}} = -\nu(v - V_f)\boldsymbol{v}/v$ for $v > V_f$.

There are two reasons to put this interaction by the spring. One is rather technical, which is just to avoid overlapping of tiles. A spring brings about the repulsion to the tiles in the neighborhood. The other is more "chemical". Without this spring, the energy is totally conserved, since we have adopted Newtonian mechanics. By this spring, there appears flow of the energy when the two tiles react. At the reaction, as cells appear or disappear, springs are created or deleted. So the energy is not preserved. In other words, the reaction can be either exothermic or endothermic.

When the boundary condition is open, the tiles are dropped from the boundary. Thus the total kinetic energy of tiles will decrease, if the energy/momentum is conserved. On the other hand, with the reaction and springs with suitable values of parameters, we can keep tiles from slowing down.

To summarize, there are parameters r_h,r_l(upper and lower limits of rate rule), $x0$, k(natural length and constant of spring), V_f and ν (for friction) to specify our system. We use either a periodic or open boundary condition. As we mentioned, tiles just disappear when they go across the boundary, for the latter boundary condition. The system size is given by $x_s \cdot y_s$. The initial conditions are characterized by the parameters n_0 (number of cells), x_{ic}, y_{ic} (proportion of the area where cells are scattered), and the seed of random generator used to put cells randomly on the world. We set the position and velocity cell-by-cell using the random number generator.

The most significant parameter besides those for initial conditions is r_h. In the world of Tile, the rule of reaction is only the inversion. Thus cells' proportion to reacting zone corresponds to the growth rate of cells. Below r_h = 0.5, the decrease of cell is suppressed – either the growth or preservation is allowed. Here, we usually set r_h to 0.55 ~ 0.6, where typical interesting behaviors are observed, since cells grow in a non-trivial manner.

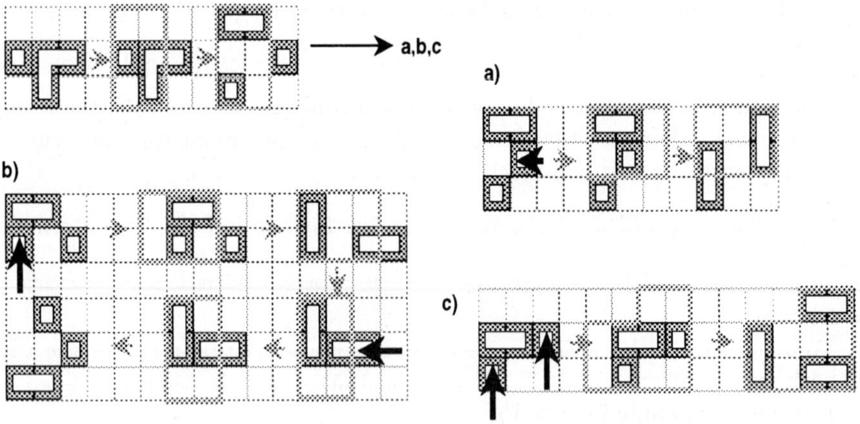

Fig. 4. Three different paths of reaction are shown. The reaction starts from two tiles(upper row). This reaction leads to three tiles. Depending on the initial condition and the interaction with other tiles(which are not shown in these figures), their velocities differ, which alters the pathways of the reaction.

3 Result

3.1 overview of Tile

Before showing the results of the actual run, we briefly sketch how the complexity is created by Tile. A schematic transition and its branching are shown in fig.4. On the uppermost row, two tiles (a square and upside-down L) collide to create three tiles. Depending on their initial velocity and interaction with other tiles through potentials, their motions can differ. The difference of motions causes different transition of shapes, and there are at least three resulting patterns as are shown in a, b and c of fig.4. In actual runs, due to the interaction, their transition can be more complex.

Results of some runs are shown figs.5-6. Two samples are shown, which we call world209 and world204, according to the initial number of cells. Open boundary condition is adopted. The evolution of world209 is shown on fig.5. Note that the initial shapes of tiles are simple. Even at time 100, their variation of shapes are rich. Although there are small number of tiles at this time, they start growing. To keep the world rich, the tiles must keep creating new tiles when an open boundary condition is adopted, since they continuously drop out of the boundary. The number of tiles get small when a set of active cluster disappears. From time 500, the growth gets stable. There is open-ended creation of tiles at least up to time 4000.

On the other hand, tiles fail to survive in the world204. Fig.6 shows the evolution of world204, which is completely different from that of world209. The difference of initial conditions is just the absence of only 5 cells, which eliminates 2 unit-square tiles and change the shapes of 3 tiles. Positions and velocities of all cells are same except the absence of 5 cells. From time 100 to 300, the diversity of tiles is not so low as that of world209. However, tiles fail to form an active core of creation later. All tiles drop out around time 800, although very few reactions occur from time 500, after which the tiles slowly move out of the world. The smaller number of initial cells in the world204 is not the reason of the death. Indeed, we have seen a continuous creation for the world203, and the death of the world286. We note that even difference of one cell can cause a completely different outcome.

Even classification of existent tiles and their transition require at least NP-hard problem since tiles change their shapes and motion according to their relative positions among them. Furthermore, the collision between the tiles can show deterministic chaos as in the billiard system, where tiny difference in initial positions and velocities are enhanced to show the macroscopic difference, and thus the later paths of reactions may totally differ.

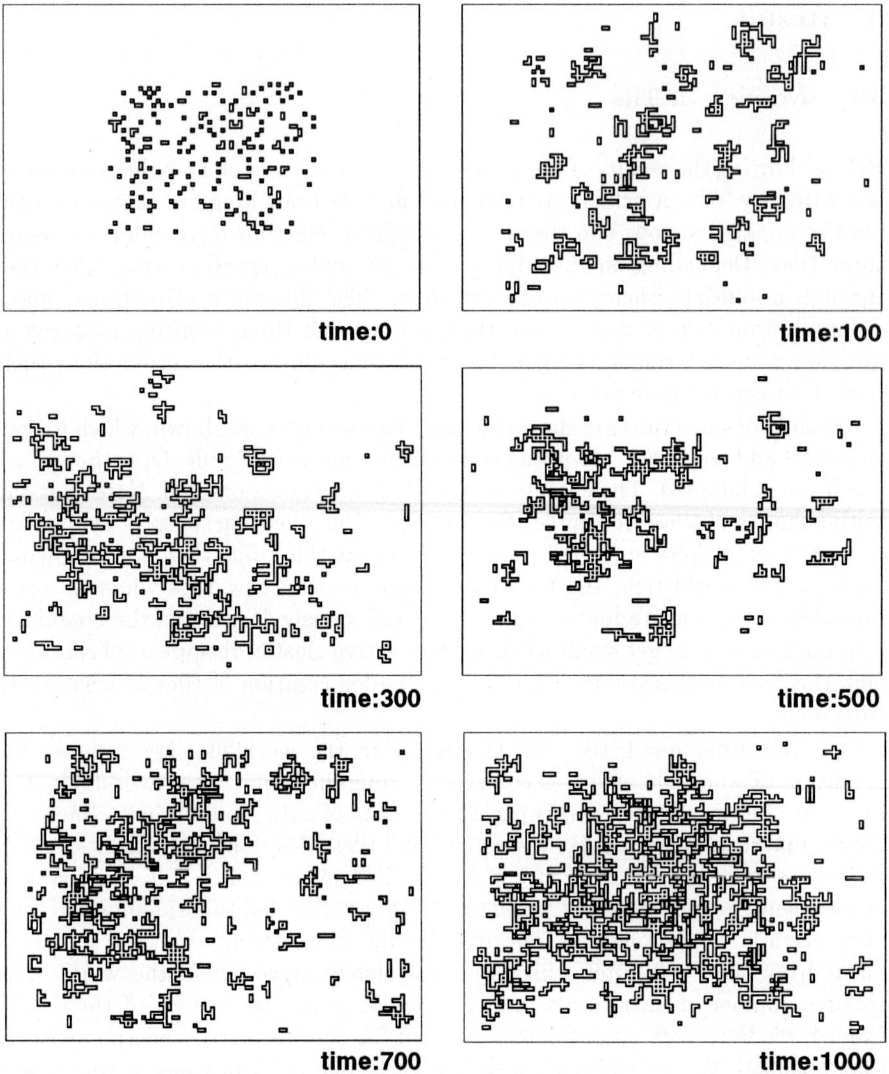

Fig. 5. The evolution of world209. Started with 209 cells, 162 tiles. The shapes of tiles are simple at the initial condition. System size is 96x72. Boundary is open. The parameters are:
$r_h = 0.55, r_l = 0, x0 = 2.0, k = 0.1,\ x_{ic} = y_{ic} = 0.5,\ V_f = 0.7,\ \nu = 0.7$

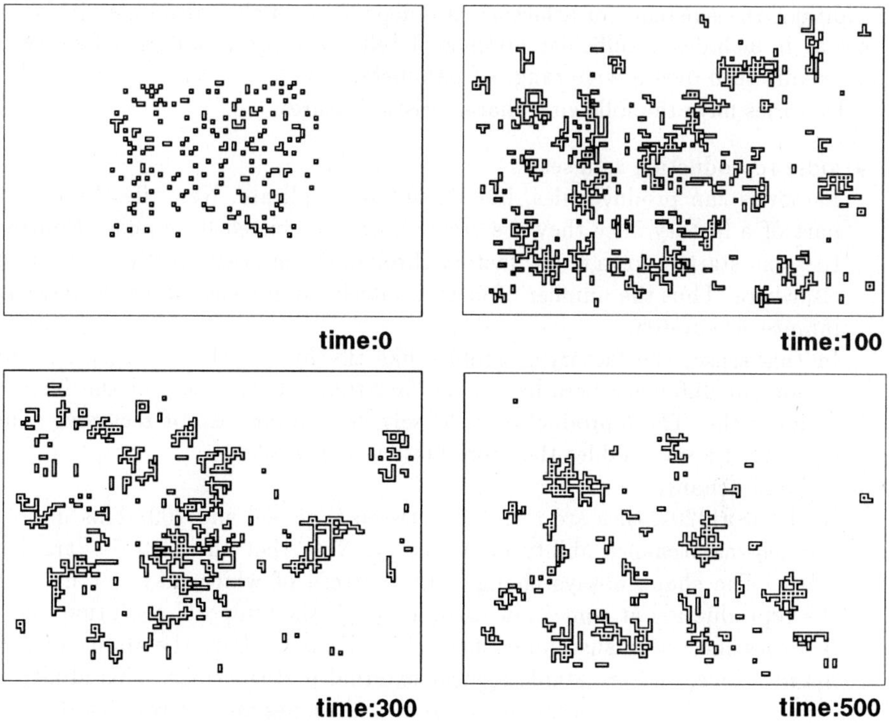

Fig. 6. The evolution of the world204. Started with 204 cells, 160 tiles. Parameters are as same as world209. The difference from world209 is absence of 5 cells at the initial condition. For existing cells, their positions and velocities are exactly same.

3.2 "Factory"

The most remarkable result of our Tile is the formation of a "factory". We call "factory" for processes which keep creating new tiles, as is seen in a set of tiles in world209. Though the factory itself does not self-replicate, but it grows by itself. It emits new tiles to outside, but the number of tiles within it does not decrease or grow. We note that the boundary of the factory is not fixed, but changes through the internal reaction. Thus it is sometimes hard to tell exactly which tiles belong to a factory, but as in the actual run, one can recognize it as a unit as already seen in the examples of the evolution of a factory in figs.5-6.

The evolution of a factory is irregular, and tiles with new shapes are created successively. More than one factories can coexist at one world if the system size is large. In the simulations we made so far with the size up to 192x144, factories finally join, to form a unified big factory at a later stage.

The formation of a factory is seen for most initial conditions with enough cells. Although the structure of a factory itself strongly depends on the initial condition, the existence of a factory is independent of the initial condition, as long as it includes a sufficient number of cells. The formation of a factory is universally seen over a wide range of parameters.

Factories have the following characteristic features.

- loose reproduction as a set
  Factories can produce tiles, but do not self-replicate. New tiles become a part of a factory once they are created, or are emitted to outside. Emitted tiles can start to form a new factory through the interaction with other tiles elsewhere. Thus the number of factories itself can increase when we have an infinite size system.
  In this sense, the factory resembles like the hypercycle by Eigen[1]. The important difference here lies in the fact that the evolution of the factory is not cyclic. The reproduction is loosely defined, and is not recursive. The factory, as a set, enables the reproduction of new tiles.
- self-maintenance
  In the world209, tiles are restored between time 500 and 700. This implies the self-maintenance ability of a factory. Note that the factories are not static. The shape always changes, the pattern of which may not fit with the reproduction at some time step. Through the temporal evolution, most factories succeed in sustaining themselves. Factories have the ability of self-maintenance, and are stable against external perturbations. This ability is expected since factories are not a fixed, and shapes are not restricted.
- open-ended complexity
  It should be noted that the factories seem to create tiles with novel shapes for ever. Such lasting increase of complexity is not expected in a finite world with a digital information, since the information contained therein is finite. For example, Conway's Life, when simulated in a finite lattice, will fall into periodic states. In our Tile, deterministic chaos in the Newtonian motion can lead to complexity with a real number. We believe that the open-ended production is originated through chaotic motion. In fig.7, histograms of mass of tiles are shown. Though classification of tiles is done only by mass, one can see that variety of tiles increases with the evolution.

3.3 contribution of sub rules

It must be useful to see how each rule contributes to the behavior of Tile. Of course it is not possible to decompose each process, but we may make a rough survey of the role of each rule, by comparing our results with the simulations of the pseudo-Tile world which lacks one of the rules.

- joint rule
  This rule is most important. Except this rule tiles could hardly grow, or only formation of a cluster of small tiles is possible. The variation of tiles is very small here, and the complexity is rather small.

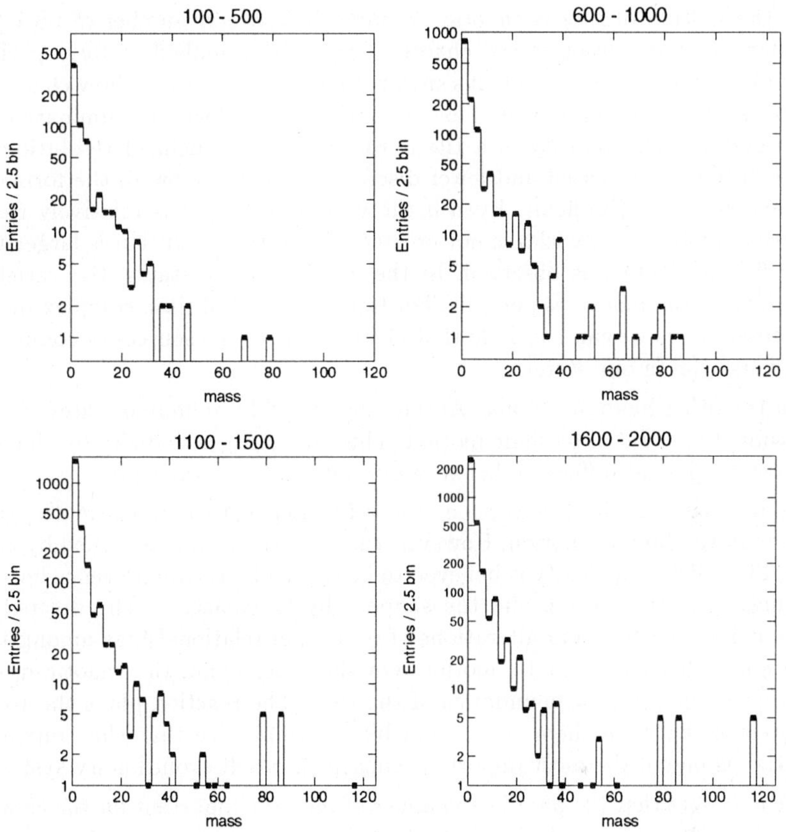

Fig. 7. World209's histograms of mass of tiles. Each plot is total sum of 5 snapshots with interval of 100 time units(not normalized). The time range is shown on the top of each graph. Here, shapes are not taken into account.

- rate rule
  This rule suppresses the reaction. Under this rule, tiles can contact with each other, which makes "mechanical motion" of tiles possible: the functional motion such as "gear" or "switch". Therefore, a cluster of tiles can keep its own memory. Although the effect of memory is not explicitly shown, it may have some contribution to Tile's complexity and the formation of a factory.

4 Discussion

What makes the tiles to grow the complexity? We think it is due to the interference between shape changes and chaotic motion of tiles. Within the context of the shape-to-shape reaction, combinatorial complexity exists. On the other hand, the motion of tiles is chaotic. Assume that several number of tiles form a cluster. If their motion is not chaotic, they will be unified or form periodic motion in finite time. Chaos breaks such periodic motion. To see how chaos contributes to Tile's complexity, we have discretized the velocity to eliminate chaos in our system. The velocity of a tile is rounded with a defined resolution. As the resolution is decreased and fewer discrete states are allowed, the formation of a factory is more difficult. Even if the variety of shapes is relatively rich, a sufficient number of new tiles is not created. When the resolution is larger than $\Delta v = 0.2$, no factory is observed. In the world of finite states, the variety is limited, since the motion is periodic. For lasting growth of tiles, complex motion is required. Thus we can conclude that chaos is one of the sources of complexity and diversity[5] in our system.

On the other hand, a Cellular-Automaton would have limited states. So it is impossible to get rid of periodic motion. This "breaking periodicity by chaos" is the most significant difference between our Tile and Conway's Life.

As is shown in the last section, the reflection of two tiles lead to strong chaos as in the billiard motion. However, the evolution is not governed by chaos alone. The Tile's complexity is believed to be caused by the interference between two contexts – the motion and the shape – by the reaction. The diversity of shapes is induced by rapid alterations of positional relationalships among tiles, which comes from the chaotic motion. On the other hand, the chaotic motion is originated in complex boundaries of the tiles. The reaction plays the role of the interface between them. We would like to emphasize that the "mixing of contexts" is one of the most important concept in a self-evolutionary system.

We also note that the parallel dynamics of tiles are important for the creation and maintenance of complexity. Indeed at the preliminary stage of the present work we started the simulation using a serial process, that is the successive motion of one tile at one time. However it did not work well. In this serial modeling, we have found only the formation of trivial clusters which prevent from reactions of tiles. Factories are never observed. We believe that the serial modeling failed because the correct ordering of the update of the tiles are never attained to keep the relationships among many tiles, while in the parallel motion such relationships are sustained. The formation of a factory is the result of parallel processes.

To sum up we have proposed the Tile Automaton in which the continuous reproduction of novel tiles and growing complexity from a finite world is attained through the formation of a self-maintaining factory. The reaction-induced interference of the motion and the shape is discussed as the origin of complexity. In fact, it is extremely hard to describe our dynamical, parallel and many-body process logically, for the logic itself is a serial process with static nature. Since all biochemical processes are parallel, there should be a gap between logic and physical systems. We think it is very important to create a dynamically complex virtual world as a thinking tool. That is why we have introduced an abstract model with the ability to grow complexity, rather than making a model to imitate realistic biological phenomena.

With this Tile Automaton, we have obtained a non-trivial self-evolutionary behaviors through the features of chaos-induced diversity, parallel real-time process and the mixing of contexts. We think that these behaviors are not model-specific so that our scenario of the evolution obtained here will be useful for the real biological evolution.

At the beginning of life, the formation of an exactly recursive cycle must be very difficult, which should be weak against perturbations. The loose reproduction here, is believed to be more robust, and indeed we have seen this formation of factories universally for most parameters. Thus it may be natural to conclude that this loose reproduction by a factory starts first, and the exact self-reproduction is evolved later[6, 7]. The metabolism is sustained by diversity from the beginning. Our factory provides an example of such loose reproduction with diversity. Thus our Tile will give a metaphor model for the origin of life and metabolism.

acknowledgement

The authors would like to thank Drs. T. Yomo, S. Sasa and M. Sasai for discussions. This work is partially supported by Grant-in-Aids (No. 05836006, 06302085 and 06-4028) for Scientific Research from the Ministry of Education, Science and Culture of Japan. One of authors (TY) is supported by research fellowship from Japan Society for Promotion of Science.

References

1. M. Eigen and P. Schuster. *The Hypercycle: A Principle of Natural Self-Organization.* Springer-Verlag, Berlin, 1979.
2. Walter Fontana. Algorithmic chemistory. In C. G. Langton, C. Taylor, J. D. Farmer, and S. Rasmussen, editors, *Artificial Life II*, page 159. Addison-Wesley, 1992.
3. E. Berlekamp, J.H. Conway, and R. Guy. *Winning Ways for Your Mathematical Plays.* Academic Press, New York, 1982.
4. M. Gardner. Mathematical Games: The fantastic combinations of john conway's new solitaire game 'Life'. *Scientific American*, 223(4):120–123, October 1970.
5. Kunihiko Kaneko. Chaos as a source of complexity and diversity in evolution. *Artificial Life*, 1(1/2):163–177, 1994.
6. F. Dyson. *Origins of Life.* Cambridge University Press, Cambridge, 1985.
7. A.G. Cairns-Smith. *Genetic Takeover and the Mineral Origins of Life.* Cambridge University Press, Cambridge, 1982.

Tracking the red Queen: Measurements of Adaptive Progress in Co-Evolutionary Simulations*

Dave Cliff[1] and Geoffrey F. Miller[2]

[1] School of Cognitive and Computing Sciences, University of Sussex
BRIGHTON BN1 9QH, U.K.
Phone +44 1273 678754, Fax +44 1273 671320, Email davec@cogs.susx.ac.uk
[2] Department of Psychology, University of Nottingham,
NOTTINGHAM NG7 2RD, U.K.
Phone +44 115 9515364, Fax +44 115 9515324, Email gfm@psyc.nott.ac.uk

Abstract. Co-evolution can give rise to the "Red Queen effect", where interacting populations alter each other's fitness landscapes. The Red Queen effect significantly complicates any measurement of co-evolutionary progress, introducing *fitness ambiguities* where improvements in performance of co-evolved individuals can appear as a decline or stasis in the usual measures of evolutionary progress. Unfortunately, no appropriate measures of fitness given the Red Queen effect have been developed in artificial life, theoretical biology, population dynamics, or evolutionary genetics. We propose a set of appropriate performance measures based on both genetic and behavioral data, and illustrate their use in a simulation of co-evolution between genetically specified continuous-time noisy recurrent neural networks which generate pursuit and evasion behaviors in autonomous agents.

1 Introduction

Some biologists have suggested that the 'Red Queen effect' arising from co-evolutionary arms races has been a prime source of evolutionary innovations and adaptations [19, 5, 16]. The Red Queen was a living chess piece in Lewis Carroll's *Through the Looking Glass*, who ran perpetually without getting very far because the landscape kept up with her. Similarly, in co-evolution between predators and prey, hosts and parasites, males and females, or competitors within a species, traits in organisms evolve against traits in competitor organisms that are themselves evolving: each lineage's fitness landscape changes perpetually. Adaptive advantage is continually eroded under co-evolution.

Or so the theory goes. But does sustained competition really lead to smooth, directional evolutionary progress, or to noisy, unreliable, fits and starts, or to endless cycling through different evolutionarily unstable strategies? How important is tight co-evolution among two or a few competing lineages, versus diffuse co-evolution among many? These issues are critical to the debate between those who view evolution as a smoothly running engine of adaptation (e.g. [5]), and other theorists who view it as a more contingent history of genetic drift, ad hoc modification, and developmental limitation (e.g. [9]). The Red Queen question is a microcosm of the ancient debate over the links between evolution, life, teleology, and progress.

Testing the significance of the Red Queen has proven difficult. The fossil record provides only ambiguous evidence of co-evolutionary progress [7], and fossils may not reveal the bodily and behavioral innovations that are important in most co-evolutionary scenarios. Simple population genetics models may over-estimate the smoothness of co-evolution by neglecting phylogenetic and developmental constraints that keep lineages stuck in local optima while their competitors surge ahead. Comparative studies across extant species reveal what adaptations exist, but not whether they were acquired through tight, synchronized co-evolution.

Evolutionary computer simulations are ideal for investigating co-evolution. They allow much more complex genotypes, phenotypes, behaviors, and interactions than population-genetic models or evolutionary game theory. And they allow researchers to make detailed measurements during and after co-evolution, revealing much more than could be inferred from fossils or comparative studies.

This paper focuses on developing measurement tools for such simulations. We are concerned with methods for measuring co-evolutionary 'progress', both to check that the evolutionary simulation is working properly, and to illuminate issues in theoretical biology. The difficulty in most interesting cases is that, unlike most genetic algorithms research, there is no pre-determined 'fitness landscape' against which progress can be measured. Lineages may evolve against each other with respect to certain domains of competition, but there may be no single correct solution (e.g. no single optimal stable strategy) for each domain.

The remainder of this paper is structured as follows. Section 2 reviews the the goals and methods of our experiments with simulated co-evolution. Section 3 then discusses the need for monitoring techniques in greater detail, and Section 4 describes several of the techniques we have developed.[3]

2 Co-evolution of Pursuit and Evasion

2.1 The Red Queen Effect in Co-evolutionary Simulations

Our interest in measuring co-evolutionary progress arises from our major ongoing research project: co-evolving things with eyes, brains, and wheels, that chase each other around. Or, more technically, using artificial co-evolution to develop neural-network sensory-motor architectures for controlling robot-like autonomous virtual pursuers which chase autonomous virtual evaders around a 2-dimensional (2-D) space, generating pursuit and/or evasion strategies on the basis of simulated visual input.

[3] An expanded version of this paper, with full illustrations, is available [4].

We have argued for the importance of studying these pursuit-evasion contests in previous papers [13, 12]. For the purposes of this methodological paper, we can simply note that pursuit-evasion contests offer a prime scenario for studying co-evolutionary dynamics. Pursuit-evasion contests are common in nature: predators pursue prey and prey evade predators, until the prey get eaten or the predator gets tired and abandons the chase. The success of strategies for pursuit and evasion are often mutually coupled: if a new strategy confers extra fitness on individuals, then that strategy should spread through the population, and its increased frequency makes it more likely that the opponent population will adapt to counteract it, thereby reducing its fitness benefits. It is this co-evolutionary coupling which underlies the Red Queen effect: the fitness landscape of one population is affected by the current strategies of any opponent populations; and the movements of one population over a fitness landscape can significantly alter the fitness landscapes of the other populations. Despite having fixed fitness *functions* for determining the reproductive success of individuals, the fitness *landscapes* will vary over time as adaptations in one population warp and shift and deform the fitness landscape of the other.

2.2 Simulation Methods

Brief details of our simulation system were given in [13], and full details are available in [3]. We will give here here only enough detail to establish the context for discussion of the problems of measuring co-evolutionary progress.

The simulation uses a conventional generational (as opposed to steady-state) genetic algorithm (GA). There are two separate populations which compete and co-evolve against each other: one undergoes selection for pursuit behaviors, the other for evasion. Each population is spatially distributed with local mating and local replacement. That is, each individual in the population is assigned a spatial location on a 2-D grid (with toroidal wrap-around at the edges). When a new generation is bred, each individual is only allowed to breed with other individuals from nearby grid locations, and the offspring is also placed in a nearby grid location. In principle, this spatial structuring of the population should allow for the emergence and maintenance of somewhat distinct *subpopulations*, that shade into each other across "clines".

Each individual has a genotype which is a string of approximately 1600 bits. A relatively complex 'morphogenesis' process translates the genotype into the agent's body morphology: the body has simple effectors, visual sensors, and a recurrent continuous-time dynamic artificial neural network whose parameters (connectivity, weights, thresholds, time-constants, etc) are determined by the genotype. Rather than using variable-length genotypes (e.g. [2]) to allow evolutionary control of the number of artificial neurons in the network, sequences of the fixed-length genotype are ignored in morphogenesis unless they are preceded by an appropriate 'marker sequence' which enables their expression: thus some portions of each genotype may be *active* (i.e. expressed in morphogenesis) while others may be *inactive* 'junk DNA'. Reproduction uses mutation, a stochastic multipoint crossover operator, and a translocation operator: see [3]. Elitism is used in the breeding phase (i.e. the newly bred population receives an unadulterated copy of the most fit genotype from the previous generation, located at the same grid position as before).

Reproductive success is determined by fitness, and fitness is evaluated for each individual by taking the mean score of a number of noisy trials with differing initial conditions (i.e. individual positions and orientations). In each trial a pursuer and an evader are given fixed amounts of energy which is expended in movement. They compete until one of three termination conditions is met: (1) if there is a collision between the pursuer and the evader; (2) if both contestants have run out of energy and drifted to a halt; or (3) 15 seconds of simulated time have elapsed.[4] Significant noise affects the simulated sensors and effectors, and in the activities of the artificial neural units. For efficiency, we use the same technique as Sims [18], where each individual's fitness is evaluated only in trials against the elite (i.e. highest-scoring) individual from the previous generation of the opponent population: we refer to this technique as LEO (Last Elite Opponent) evaluation. At the end of each trial, the individual under evaluation is given a score. In the experiments discussed below, the score for evaders is simply the amount of simulated time before the trial ended; the score for pursuers is a temporal integral of the instantaneous rate of approach (which encourages the pursuer to approach the evader), plus a 'bonus' reward awarded if a collision occurs. The differences in the scoring techniques mean that the contests are not zero-sum. See [3] for further discussion of this and other fitness scoring methods.

At the end of each generation, various statistics are calculated in order to monitor progress. The genotype of the elite of each generation is saved for use in the LEO contests of the next generation, and acts as a representative of the population for several of our monitoring procedures. Because starting conditions can vary and behavior is noisy, there is uncertainty as to whether the genotype ranked as the elite really is the best in the population or is an unexceptional individual that was lucky in evaluations. We reduce this uncertainty, using the standard statistical techniques such as blocking and stratified sampling across initial conditions. Thus, it is *highly* unlikely that the same poor genotype will be incorrectly identified as a legitimate elite over several successive generations.

3 The Need for New Measurement Techniques

3.1 Instantaneous Fitness Tells Us Little

During a simulation experiment, the fitness values of each individual are available at the end of each generation. These values have to be calculated and stored for use in the breeding phase of the generational GA. These values usually form the basis of monitoring progress in non-co-evolutionary applications where the fitness landscape is fixed in advance. In such non-co-evolutionary scenarios, progress can be monitored by plotting summary statistics (such as the mean or best) of the

[4] The simulation approximates continuous time using Euler integration techniques at a temporal resolution of 100 steps per simulated second.

fitnesses of the population. A successful simulation experiment can be expected to show a gradual increase in a relevant population fitness measure. The fitness measure will asymptote as the population converges on one or more fitness peaks, signifying an end to the evolutionary search. In unsuccessful experiments (e.g. where the GA parameters have been poorly set), fitness may never increase above that of the initial random population, or it may increase and then hold at a comparatively low value (indicating convergence to a suboptimal local fitness peak), or it may climb to a high value and then subsequently fall away (indicating convergence followed by genetic drift, caused by an excessive mutation rate).

However, in a co-evolutionary scenario, the Red Queen effect makes it hard to monitor progress by taking instantaneous measures of fitness at the end of each generation. Because fitnesses are defined relative to a co-evolving set of traits in other individuals, the fitness landscape(s) for the co-evolving individuals can vary dynamically. Hence periods of comparative stasis in instantaneous measures could signify a corresponding evolutionary stasis or could disguise a period of tightly-coupled co-evolution where adaptive changes in one population (which would register as increases in instantaneous fitness if the opponent population was held fixed) are matched by adaptive counter-changes in the opponent population (thereby holding the instantaneous fitness measures of both populations close to the values exhibited prior to the change and counter-change in strategies). Similarly, if the instantaneous measures decrease over time, this may represent either a setback in progress due to genetic drift, or co-evolutionary progress where *both* populations have adapted to make the pursuit-evasion contests significantly more difficult for their opponents.

As illustration, Figures 1 and 2 show instantaneous fitness measures for pursuer and evader populations in an experiment lasting 700 generations. If each population got better, the lines should go up. Although the data for the pursuer population shows a steady climb in fitness over the first 200 generations, the mean fitness value at generation 205 is never improved upon, and the mean fitnesses over generations 500 to 600 are roughly the same as those over generations 50 to 100, which could be interpreted as a lack of progress in the intervening 400 generations. Superficially, the data for the evader population is even worse: mean population fitness is at its highest at the start of the experiment, showing a steady decline over the first 300 generations, followed by 400 generations of roughly constant fitness at a level around 70% of that exhibited at the start. Given that both populations are selected for maximizing fitness, these data could indicate that something is seriously wrong, with progress either not occurring or not being maintained. As we shall see in the following sections, real progress *is* occurring in this experiment, but other monitoring techniques are required to demonstrate this.

We use the term *fitness ambiguities* to refer to such cases where qualitative trends in time-series of instantaneous fitness measures could feasibly be interpreted as either continuing progress or as a breakdown of the co-evolutionary process. Fitness ambiguities introduce two problems:

First, how do we know when to terminate an experiment that is failing due to

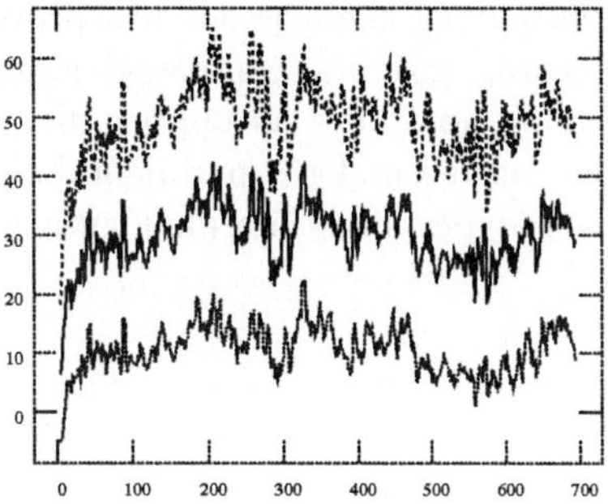

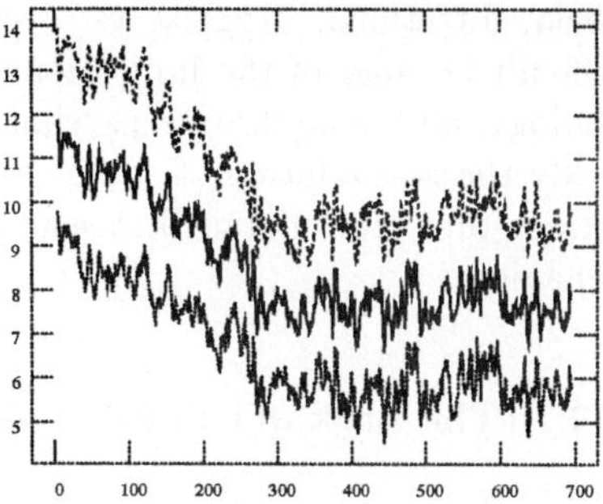

Fig. 1. Instantaneous fitness measures for pursuer population over 700 generations of co-evolution. Horizontal axis is generation number, vertical is fitness. Solid line is mean population fitness: dashed and dotted lines are mean plus and minus one standard deviation, respectively. Data smoothed by calculating rolling average over preceding five generations. See text for discussion.

Fig. 2. Instantaneous fitness measures for evader population over 700 generations of co-evolution: format as for Figure 1. The range of fitness values is different from that of the pursuers, because of the difference in evaluation functions. See text for discussion.

bugs or poor choice of parameters? The importance of this problem can be appreciated when one considers the computational requirements of our co-evolutionary simulation experiments. In the experimental regime described in Section 2.2, running on an unladen Sun SPARC20, with two populations each of size 100, using LEO competitions with 9 trials per individual, our fully optimized C code manages to evaluate each individual in an average of about 10 seconds (evaluation time can vary greatly because of the multiple termination conditions described in Section 2.2). So one complete generation takes roughly 35 minutes, and a single experiment of 1000 generations takes a little over three weeks of continuous computation. As one of our aims is to study the effects of varying the experimental conditions (e.g. different evaluation functions, genetic encoding techniques, settings of the constants governing motion, or environmental circumstances such as worlds with obstacles or boundaries), it would be nice to be able to disambiguate fitness ambiguities at the earliest possible opportunity. In short, we need to know whether to kill a pointless experiment or allow hidden progress to continue.

The second problem stems from our concern to establish an informative and reliable characterization of the co-evolutionary dynamics exhibited by our experiments. The possibility of ambiguities in time-series of instantaneous fitness measures makes such characterization impossible without further analysis. Yet we believe that our experiments offer an opportunity to empirically explore issues of progress that are keenly debated in the evolutionary biology literature: notions of teleology, diffuse vs. tight co-evolution, smooth vs. punctuated equilibria, directional progress vs. cycling, etc. Resolving these issues has proven difficult because of the limitations of fossil records, genetics, comparative morphology, and comparative psychology: our simulations offer an opportunity to study the co-evolution of complex autonomous agents under experimental conditions that allow detailed measurements of genotypes, body morphologies, and behaviors.

3.2 (The Lack of) Related Work

These two problems have led us to explore techniques that allow us to reliably monitor the sometimes hidden co-evolutionary progress in our simulations. To our surprise, we found that no applicable techniques had been developed in the fields of artificial life, theoretical biology, behavioral ecology, or evolutionary genetics. The complexity of our simulations violates many of the simplifying assumptions on which theoretical studies of evolution, co-evolution, and population dynamics, are founded. Furthermore, although other artificial life research has employed co-evolution to develop autonomous agent architectures (e.g. [18, 15]), such work has concentrated mainly on the end results, rather than on the dynamics of the co-evolutionary process.

Co-evolving pursuit and evasion strategies may appear related to the long-established literature on theoretical modeling of predator-prey population dynamics. Yet all such work with which we are familiar, from the well-known deterministic Lotka-Volterra equations to the more recent spatially-distributed stochastic population models (see e.g. [14]), depends on monitoring fluctuations in the sizes of two competing populations. As the population size is constant in our experiments, this large body of theoretical work is of little use to us.

Studies in ethology and behavioral ecology (e.g. [11]), although acknowledging the importance of predator-prey arms races, focus on the functions of current behavior rather than the dynamics of co-evolution. Macroevolutionary theory (e.g. [6]) typically treats co-evolution as a phenomenon that is hard to observe outside the fossil record.

Finally, evolutionary genetics and almost all research in either theoretical biology or artificial life which could be relevant (e.g. [1]) studies (co-)evolution at the level of discrete genes for particular traits. W. Hamilton and his associates (e.g. [17]) have developed techniques for visualizing and analyzing simplified co-evolutionary systems. However, the genetic encoding used in our simulations is sufficiently complex that there is no clear method of identifying a gene for a given trait: sequences of bits in a genotype may affect the connectivity of the artificial neural network, or may determine a parameter for one of the artificial neurons, but the observable behavior of the phenotype is a complex emergent property of the network interacting with the environment, which itself includes another network (controlling the opponent). In fact, because the space of possible behaviors for our artificial agents is continuous in both time and space, a precise definition of what constitutes a 'trait' is problematic. Although Kauffman's work on co-evolution in NKC fitness landscapes (e.g. [10, Chapter 6]) appears relevant, it is not clear how to determine N, K, or C for our simulations,[5] particularly as all three factors could vary dynamically as co-evolution progresses.

These issues have led us to conclude that our work is exploring largely uncharted territory: we are attempting to gain the insight offered by theoretical analyses in an artificial co-evolutionary system sufficiently complex that no established theoretical analysis tools are applicable. In the remainder of this paper we describe analysis and monitoring techniques we have developed to fill this gap.

4 Measurement Techniques

4.1 Ancestral Opponent Contests

In an earlier paper [13], we noted that one possible technique for monitoring co-evolutionary progress is to evaluate an individual I from generation g against representatives of I's opponent population from each previous generation $g - \Delta g : \Delta g \in \{0, 1, \ldots, g\}$. That is, I is entered into contests with the 'ancestors' of its current opponent. Current individuals should do well against outdated opponents; the more ancient the opponent, the better they should do. More technically, if progress has occured then we might expect that I's fitness will increase with Δg: the fitness scores for I will be positively correlated with Δg over some time-scale which we will refer to as the 'evolutionary time-lag' τ; I should in general do better in competitions against opponents drawn from earlier opponent generations, because co-evolutionary adaptation in I's population should have rendered these strategies less effective.

However, we needn't expect performance against ancestral opponents to improve all the way back. For example, ancient ancestors may have had tricks that more recent ancestral opponents have lost. So there are no strong reasons for expecting the positive correlation to be extended indefinitely, nor even for expecting the correlation to be monotonically increasing. For example: while I may reasonably be expected to do better in contests with individuals drawn from the opponent population at (say) generation $g - 10$ than in contests with opponents from generation g, it could be that when I competes with individuals from opponent generation $g - 100$ it fares much worse than it does against opponents from generation g. This statement may appear counterintuitive, but there are at least two possible explanations for such a result: the limits of 'evolutionary memory', and the possibility of cyclical trajectories through strategy space over evolutionary time-scales.

- Co-evolutionary adaptation in a population P_1 (e.g. pursuers) can be expected to render strategies from recent generations of an opponent population P_2 (e.g. evaders) less effective. But it is feasible that the genetic changes selected in P_1 to combat these recent P_2 strategies eliminates or reduces phenotypic traits in P_1 that contributed to counteracting P_2 strategies earlier

[5] N is the number of traits coded on a genotype, K the number of epistatically linked traits within a genotype, and C the number of epistatically linked traits in a co-evolving (opponent) species.

in evolutionary time. Such displacements will not reduce the fitness of P_1 individuals if the distant P_2 strategies are no longer employed in the current or recent generations of P_2. In a reciprocal manner, the P_2 population is less likely to have retained the genetic material responsible for the distant strategies if P_1's subsequent counter-adaptations rendered them ineffective. If bounds are imposed by limited resources or developmental constraints, displacement of out-dated genetic material is likely to form the basis of continuing adaptation. Even if such limitations are not significant, the escalating arms race may render distant P_2 strategies obsolete, so that they and their P_1 counter-strategies fade away through neutral mutations. Either way, the ultimate result is that current P_1 individuals fare badly when pitted against P_2 individuals from generations sufficiently distant that they are beyond the 'evolutionary memory' of the P_1 genomes.

- Cycling between strategies is possible if there is an intransitive dominance relationship between strategies. For example, suppose one attempted to co-evolve two populations that compete by playing each other at the childrens' game "rock-paper-scissors"[6] where each individual in the population is limited to one genetically determined choice which it uses in all contests in its lifetime (i.e., in game-theoretic terms, only pure strategies are allowed). More generally, consider a co-evolutionary competition between two populations P_1 and P_2 and let $P_i(f) \succ P_j(g)$ denote the fact that individuals from population P_i at generation f generally win competitions against individuals from population P_j at generation g; assume that $P_1(g) \succ P_2(g - \Delta g_a)$ and $P_1(g - \Delta g_a) \succ P_2(g - \Delta g_b)$ with $\Delta g_a < \Delta g_b$. A transitive dominance hierarchy would exist if $P_1(g) \succ P_2(g - \Delta g_b)$, but an intransitive dominance cycle could be established if $P_2(g - \Delta g_b) \succ P_1(g)$. LEO contests would tend to make cycles especially likely.

Of course, it is possible that particular co-evolutionary systems will not exhibit either of these two phenomena, but it is crucial to appreciate that, in general, co-evolutionary systems such as ours should not be expected to exhibit continuous progressive adaptation with each generation 'improving' on previous generations, toward some optimal state. Our qualitative notion of an evolutionary time-lag τ (which may itself vary over evolutionary time) serves to emphasize that the results from ancestral opponent contests need to be judged with care.

Figures 3 and 4 show results from ancestral opponent contests for the elite pursuer and elite evader from generation 700 of the run illustrated in Figures 1 and 2. In both cases there is a general trend towards higher fitness as Δg increases[7] (note that Δg decreases from left to right: in this case $\Delta g = 700$

[6] This is a two-player game where each player simultaneously announces a choice of 'scissors', 'paper', or 'rock': unless there is a tie, the player with the dominant choice wins. The dominance relationships are: scissors cut the paper; paper smothers the rock; and rock breaks the scissors.

[7] These data could perhaps be better characterized as periods of relative stasis either side of a period of significant improvement over generations 200 to 350: we return to discussion of these data in Section 4.2.

at $g = 0$ and $\Delta g = 0$ at $g = 700$). These ancestral fitness data support the claim made in Section 3.1 that the data presented in Figures 1 and 2 mask some underlying co-evolutionary progress.

Of course Figures 3 and 4 only compare generation 700 elites against their ancestral opponents. What about the corresponding data for all the previous elites against *their* ancestral opponents? To show that we would need as many graphs as there have been generations. However there is an efficient way to display all such comparisons: by representing data from Figures 3 and 4 as a row of intensities with high scores darker and lower scores lighter. Clearly for generation 1 we would have only one previous generation from which to draw ancestral opponents, yielding a single cell. As we compare elites from each successive generation against all of their ancestral opponents the rows will get longer and longer, and we can stack them one above the other, so the top row represents all the data from a single plot like Figures 3 or 4. We refer to the fitness data from these tests as CIAO data (from Current Individual vs. Ancestral Opponents).

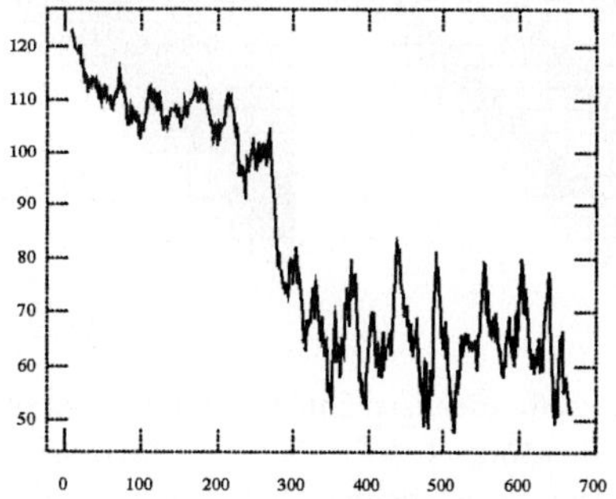

Fig. 3. Generation 700 elite pursuer scores better against earlier evaders. Graph shows pursuer ancestral fitness scores for the run illustrated in Figures 1 and 2. Horizontal axis is generation number g, from which an ancestral opponent is drawn; vertical axis is fitness scored by elite pursuer from generation 700 in contest with elite evader from generation g. Data smoothed by calculating rolling average over preceding ten generations. See text for discussion.

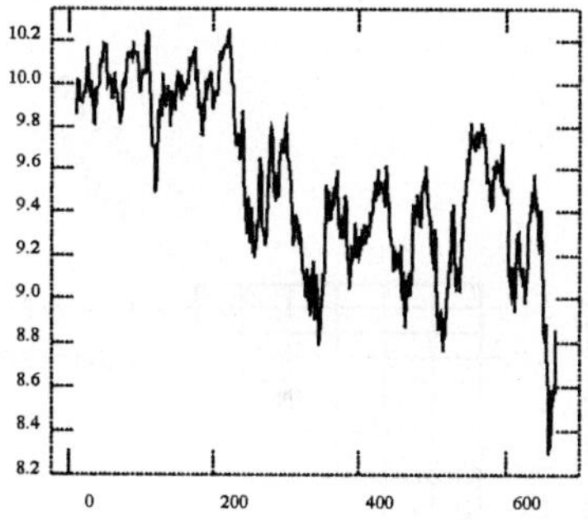

Fig. 4. Generation 700 elite evader scores better against earlier pursuers. Graph shows evader ancestral fitness scores for the run illustrated in Figures 1 and 2. Horizontal axis is generation number g, vertical is fitness scored by elite evader from generation 700 in contest with elite pursuer from generation g. Smoothing as for Figure 3. See text for discussion.

Figure 5 shows a simplified schematic of the CIAO display format, for visualizing the results of an experiment where two populations P_1 and P_2 have co-evolved. Essentially, the format is a triangle formed by stacking successive 2-D data-sets such as those shown in Figures 3 and 4. In order to spatially compress the data, the fitness scores for P_2 determine the darkness of each cell on the grid: darker cells signify higher scores. Thus the top row of cells is the ancestral fitness data for P_2 at generation 5, the one below that is the ancestral fitness data for P_2 at generation 4, and so on. The cells along the diagonal edge therefore represent the score of the elite of P_2 from generation g in contest with the elite from P_1 at the same generation, and the shading in the *next* diagonal line represents the scores from the elite of P_2 at generation g in contest with the elite of $P1$ from generation $g-1$: a clear parallel with LEO contests; although these are the instantaneous fitness scores of the elites, they should be in close agreement with the instantaneous population-average fitness data such as Figures 1 and 2.

The utility of this CIAO display format is indicated by Figure 6, which shows idealizations of the patterns that would be present if the co-evolutionary process was affected by limited evolutionary memory or by cyclic trajectories through strategy space.

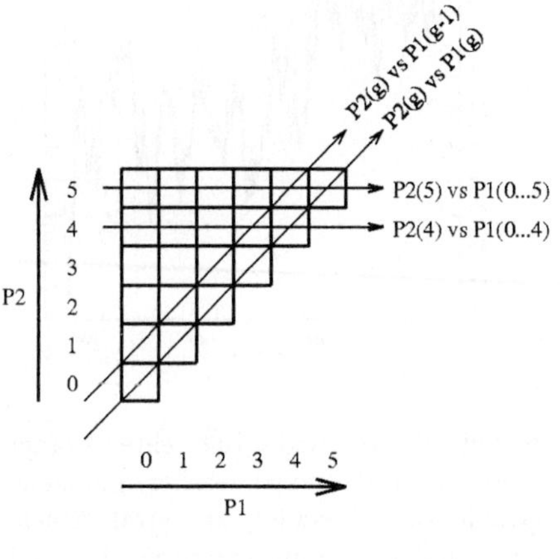

Fig. 5. Simplified Schematic of the CIAO fitness plots: the cells would be shaded to reflect the scores of individuals from P_2, with darker shades indicating higher fitness. See text for further details.

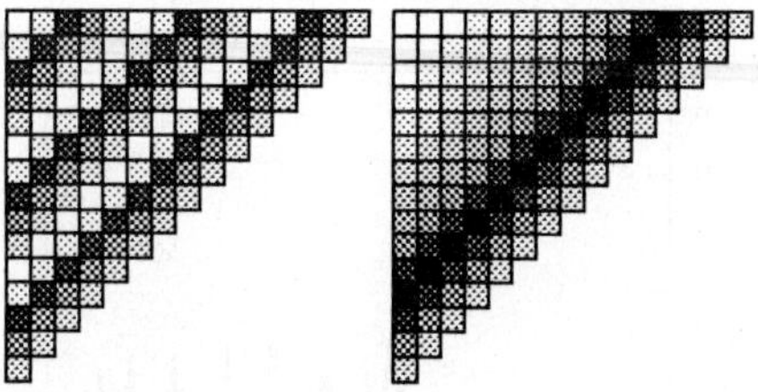

Fig. 6. Idealised illustration of patterns indicating intransitive dominance cycling (left, where current elites do well against opponents from 3 or 4 generations ago but not so well against those from generations further back); and limited evolutionary memory (right, where current elites do well against opponents from 3, 8, or 13 generations ago but not so well against generations inbetween). The presence of straight diagonal bands of intensity indicates a constant τ: if τ varies, the bands would follow curves.

The gray-scales in a CIAO plot could be set by normalizing all the scores in each data-set to the range of the entire data-set. Thus the darkest cell(s) in the figures would represent the highest scores in the entire data-set, and the brightest the worst. An alternative method of setting the gray-scales is to normalize all scores in each row in the image to the range of the data *in that row*. The effect on the image is similar to histogram-equalization used for contrast enhancement in image processing (e.g. [8]), but here the adjustment of gray-scales on each row has a natural interpretation: the darkest cell in each row signifies the highest score for the elite in P_2 of the generation plotted on that row, and the lightest the worst. Figures 7 and 8 show CIAO data from the experiment discussed above, with gray-scales normalized across rows.

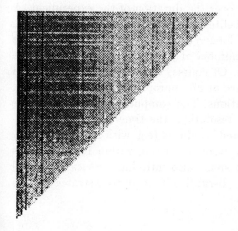

Fig. 7. Pursuer CIAO fitness scores for the run illustrated in Figures 1 and 2. Evader generations 0 to 700 run left to right in steps of 5, pursuer generations 0 to 700 run bottom to top in steps of 5. Darker cells represent higher fitness scores, gray-levels normalized to data range across each row. See text for discussion.

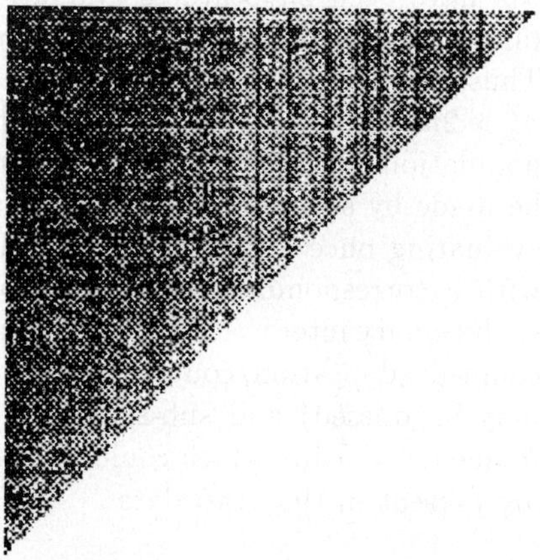

Fig. 8. Evader CIAO scores for the run illustrated in Figures 1 and 2. Pursuer generations 0 to 700 run left to right in steps of 5, evader generations 0 to 700 run bottom to top in steps of 5. Intensity range as for Figure 7. See text for discussion.

From these CIAO plots it is possible to give some account of the co-evolutionary dynamics of this particular experiment. Figures 7 and 8 both have the greatest density of (lighter) low-score cells towards the diagonal edge, with more of the (darker) high-score cells towards the left-hand edge. This indicates that in general there has been continuous progress in *both* populations, and that neither of the patterns shown in Figure 6 have occurred in the first 700 generations of this experiment.

In the pursuer scores there appears to be a significant improvement around generations 220 to 300: elite pursuers from all subsequent generations do well against the first 220 generations of elite evaders; moderately well against those in the range 220 to 300, and then fare fairly badly against evaders from generations 300 to 700. A similar pattern is revealed in the evader score, although there is evidence that the evaders improve slightly around generations 550 to 700.

Examination of CIAO data can clearly help identify major changes in the relative fitnesses of elites from the two populations, and helps to characterize the dynamics of the co-evolutionary experiment to date. Not surprisingly, these benefits come at a (computational) cost: the number of evaluations required for a complete CIAO analysis of a particular simulation experiment can easily exceed the number of evaluations in the experiment itself. Consider a co-evolutionary simulation lasting n_g generations, with two populations each of size p. Then there will have been $2n_gp$ fitness evaluations over the duration of the experiment. A CIAO analysis of the resulting two sets of elite individuals requires n_g^2 evaluations. Thus, the full CIAO analysis will take longer than the experiment itself once $n_g^2 > 2n_gp$ (i.e. $n_g > 2p$), or when the number of generations exceeds twice the population size, as it does in this run. Of course, computational savings can be made by sub-sampling the CIAO space at an appropriate resolution Δg, e.g. evaluating once every $\Delta g = 10$ generations. The computational savings come with a corresponding loss of temporal resolution: the time-scale at which co-evolutionary interactions can be monitored is reduced (e.g. with $\Delta g = 10$, tightly coupled adaptation/counter-adaptation events occurring within 10 generations may be missed) and sub-sampling the space also introduces issues of spatial frequency aliasing which could be highly disruptive if intransitive strategy cycles are present in the CIAO data.

4.2 Genetic Distance Measures

Although the CIAO data indicates co-evolutionary interaction and progress at the phenotype level (i.e. by showing the fitness values resulting from evaluations of behavioral performance), it gives no indication of the corresponding dynamics at the genotype level; yet for a complete account of the co-evolutionary process, such genetic analysis is necessary. To this end, we have developed a set of simple monitoring procedures which gives good indication of significant co-evolutionary interactions at the genetic level.

Furthermore, because these genetic analysis techniques do not require (computationally intensive) evaluation of phenotypes, relevant data can be calculated on-line during the progress of an experiment without incurring a significant processing overhead. Therefore these techniques offer the advantage that they can be used to monitor progress in an experiment while it is running, and hence allow for identification of experiments which should be terminated due to lack of progress. Two techniques are introduced below: "elite bitmaps" and "ancestral Hamming maps"; a third technique, "consensus distance plots", is described in [4]. All three techniques analyze one population in isolation from any other co-evolving populations: the intention is to identify periods of significant genetic change which can clarify features present in the CIAO data. For brevity, we illustrate these techniques with examples generated using only the pursuer population from the experiment analyzed in the previous sections: in practice it is necessary to separately apply these monitoring techniques to both the pursuer *and* the evader populations: see [4] for illustration of the results of applying these techniques to the evader population. Another technique we are exploring is *chronospeciation analysis*: testing whether successful offspring result from crossing individuals of generation g with individuals from generation $g - \Delta g$. If not, 'chronospeciation' has occured: current individuals can no longer breed with their ancestors, indicating that significant evolutionary change has occurred. The chronospecies concept in theoretical biology can't be tested very easily in nature, but simulations allow us to perform these cross-generational breeding experiments.

Elite Bitmaps At the end of each generation in our experiments we record the genotype of the elite individual in each population. Figure 9 illustrates the genotypes for the elite pursuers over the 700 generations of the experiment. This 'elite bitmap' shows the elite genotype at each generation, stacked horizontally with the earliest generation at the top and the latest at the bottom. Such raw genotype data reveals some qualitative structures, three of which are worth attention:

- There are clear vertical bands of varying extent: these bands correspond to bits in the elite genotypes which were largely unchanged over a series of generations. The horizontal extent of the band is governed by the length of the gene sequence which is constant between elites of successive generations, and the vertical extent indicates the number of generations during which this sequence was maintained in the elite.
- There are several 'noisy' areas where no banding is present: In general these areas correspond to sequences on the elite genotypes which have no impact at the behavioral level: either because they are not expressed at the morphogenesis stage, or because the morphological features governed by these sequences has a negligible effect on the behavior of the individual. In either case, a high degree of genetic variance in such sequences can be expected on the elite genotypes, yielding little or no correlation between successive generations and hence no clear vertical banding in the bitmap.
- There are horizontal 'faultlines' at various locations. Where banding either commences, ceases, or continues but with a different pattern of bits in the band. These faultlines indicate significant changes in the genetic profile of the population elite. The horizontal extent of the faultline indicates the degree of change. If the faultline extends across the entire genotype, then a new genetic 'strain' of elite has emerged from the underlying population. Faultlines with a more limited horizontal extent indicate more gradual changes in the genetic constitution of the elite. Some faultlines will be caused by the translocation operator employed in our GA. If the faultline initiates a new pattern of banding, or marks a transition from 'noise' to banding, then it indicates a major change to the affected sequence of genotype which is retained over successive generations. If the new banding pattern fades into noise, then the sequence could be 'hitch-hiking': that is, the sequence does not itself contribute to the fitness of the elite, but is retained by virtue of its presence in a genotype which has *other* sequences that confer sufficiently high fitness to maintain the genotype as the elite.

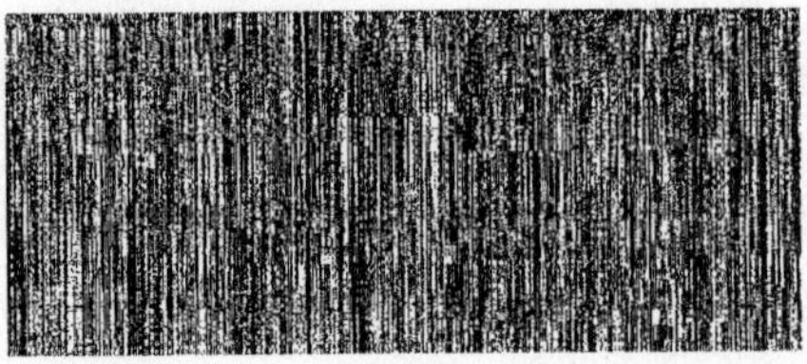

Fig. 9. Bitmap for elite pursuer genotypes. 1600-bit genotypes stacked horizontally, with generation 0 at the top and 700 at the bottom. Dark pixels represent a '0' in the genotype; light pixels represent '1'.

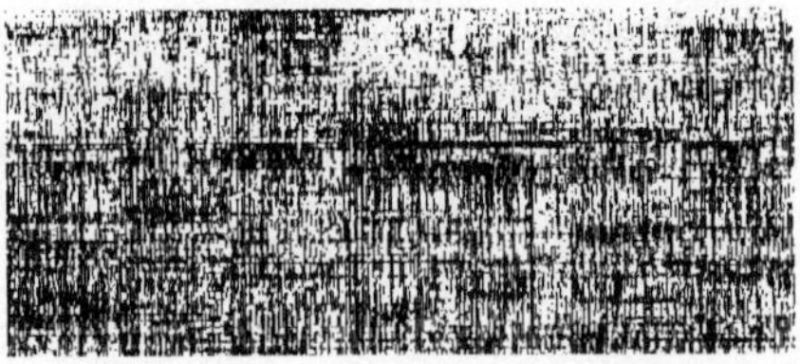

Fig. 10. Result of image-processing on the bitmap of Figure 10. Dark pixels indicate bits which are elements of vertical bands lasting for 9 generations or longer. See text for discussion.

These qualitative phenomena can be identified in a more objective manner by applying elementary image processing operations to the elite bitmap: convolving the bitmap with appropriate masks allows for the automatic highlighting of areas of banding, faulting, and noise. Figure 10 shows the result of convolving the elite bitmap with a simple one-dimensional mask which highlights vertical bands of 9 generations or more. Darker areas in the image indicate the presence of banding, and lighter areas indicate noise (or bands lasting less than 9 generations). The horizontal faultlines are also more prominent in the processed bitmap: Around generations 250 to 300 there is a clear group of faultlines with a large horizontal extent (indicating change in large sequences on the elite genotype). Several of these faultlines are followed by periods of strong banding (i.e. dark areas on the processed bitmap).

Ancestral Hamming Maps While the (processed) elite bitmap can help identify genotype sequences that are retained over successive generations and instances of significant change marked by faultlines, it is essentially a qualitative technique. To make meaningful comparative statements, quantitative measures of genotype-sequence retention and change are required.

In particular, it can be useful to quantify the degree to which a given elite genotype shares genetic material with the elites in the preceding generations. An obvious measure to use is the Hamming distance between the two bit-string genotypes. For brevity, we will use the following notation: let $E(g)$ denote the elite individual in a population at generation g; let $G(E(g))$ denote the genotype for $E(g)$; and let $H(f,g)$ denote the Hamming distance between $G(E(f))$ and $G(E(g))$ (note that if $f = g$ then $H(f,g) = 0$). In a manner similar to the CIAO plots, we can determine the Hamming distances $H(g, g - \Delta g) : \Delta g \in \{1, \ldots, g\}$. That is, the Hamming distances between $G(E(g))$ and the genotypes of the elites from each preceding generation of the same population. This "ancestral Hamming data" can be plotted as intensities on a 2-D grid, resulting in an *ancestral Hamming map*. Figure 11 shows a schematic ancestral Hamming map, and Figure 12 shows idealized qualitative patterns that indicate particular features of the underlying evolutionary dynamics. Further quantitative analysis of the ancestral Hamming data can be guided by searching for such qualitative features:

- If there is constant slight change in the genetic constitution of the elite over successive generations, then $H(g, g - \Delta g)$ should fall off smoothly as Δg increases, and all horizontal lines of cells in ancestral Hamming map will have roughly the same H (see the left-hand example in Figure 12).
- If a new elite genotype occurs at generation g and is sufficiently fit with respect to the opponent population that it or its immediate descendants also form the elite for subsequent generations, then this will show in the ancestral Hamming map as a 'wedge' of low-H cells on the grid: see the center example in Figure 12.
- Ancestral Hamming maps can give a useful indication of the genetic constitution of the underlying spatially distributed population in a GA. Because a spatially distributed population is capable in principle of sustaining separate 'subpopulations', as described in Section 2.2, there is no reason to expect that $H(g, g - \Delta g)$ will always decrease smoothly as Δg increases. In particular, it is possible that $H(g, g - \Delta g)$ is high for some small Δg, but low for a larger value of Δg (see the right-hand example in Figure 12). Such a situation would indicate that $E(g)$, is more strongly related to the earlier elites than to the more recent elites, and that the more recent elites came from a different 'subpopulation'. If ever $G(E(g))$ has a comparatively high H for all $G(E(g - \Delta g))$, then the indication is that $E(g)$ is a member of a subpopulation which shares comparatively little genetic material with the previous elites, and so g is the first generation in which members of that subpopulation have attained sufficient fitness to be selected as the elite.

Figure 13 shows an ancestral Hamming map for elites of the pursuer population introduced in Figure 1. For roughly the first 220 generations there is little evidence of any significant structure. Around generations 220 and 260 there are the first two of a number of dark 'wedges', indicating the emergence of genotype sequences which remain (partially) present in the population elite for many generations: even the column at generation $g = 400$ shows some darkening (i.e. lower H) around $\Delta g = 140$ and $\Delta g = 180$; this clearly (and quantitatively) indicates that the changes in genotype at $g = 220$ and $g = 260$ were retained for many subsequent generations: a fact that wasn't particularly clear in the processed bitmap of Figure 10.

Consensus Distance Plots Although monitoring genetic change in the elite genotypes is a valuable source of information for characterizing (co-)evolutionary dynamics, it is important to ensure that changes in the elite genotype are reflective of changes in the underlying population of genotypes. When the evaluation process is noisy or uncertain, there is always a possibility that changes in the elite genotype between one generation and the next are stochastic evaluation artifacts, rather than significant evolutionary events. To disambiguate the two,

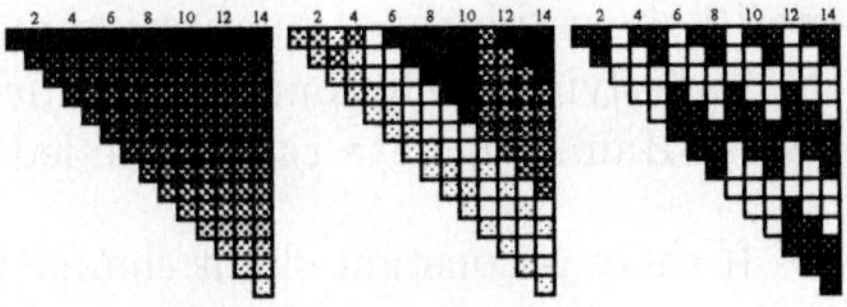

Fig. 12. Idealised illustration of patterns in Ancestral Hamming Maps, for three cases of generation 0 to 14. Left: steady genetic change (either through retention of adaptive mutations or through neutral genetic drift); the elite of each generation shares much genetic material with the elite of the previous generation (i.e. $H(g, g-1)$ is always low), but accumulated genetic change results in more distant ancestral elites having much less shared genetic material (i.e. $H(g, g-d)$: $d \gtrapprox 10$ is always high). Center: at generation 5 a new elite genotype appears that is largely dissimilar from all previous genotypes. The $g = 5$ elite genotype (or its descendants) remains as the elite until generation 10, when another new genotype, sharing much material with the $g = 5$ elite, becomes the elite for the remainder of the generations shown on the map. Right: two converged subpopulations alternate in their role as the elite, with little or no genetic material shared between the subpopulations.

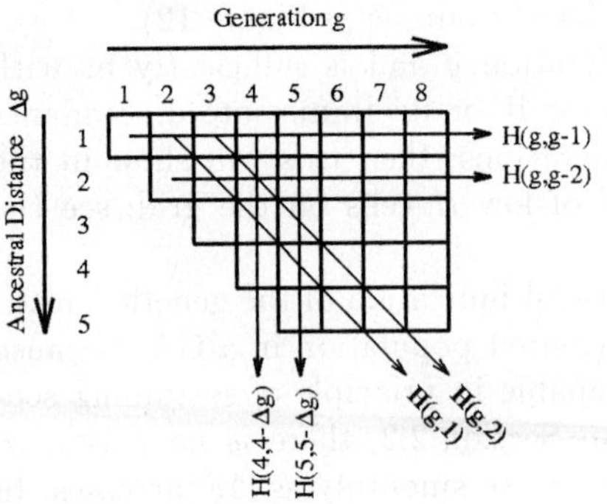

Fig. 11. Simplified schematic of Ancestral Hamming Maps. $H(f, g)$ denotes Hamming distance H between $G(E(f))$ and $G(E(g))$. Grid-cells would be shaded to reflect values of H (lower H given darker shading). Horizontal lines of cells indicate $H(g, g - \Delta g)$ for constant Δg while g varies. Vertical lines of cells indicate $H(g, g - \Delta g)$ for constant g while Δg varies. Diagonal lines of cells indicate $H(g, f)$ for constant f while g varies. This map terminates at $\Delta g = 5$, but could have been continued to $\Delta g = 8$.

it is necessary to monitor a representative population statistic. We have found that significant evolutionary events in our experiments can be identified by monitoring, at each generation, the distribution of Hamming distances from the population's *consensus sequence* to each individual genotype in the population. The consensus sequence for the population can be thought of as the "average" genotype in the population. The rationale for these plots, and illustration of their use on data from the above experiment, are given in [4].

5 Conclusion

Co-evolutionary simulations for developing advanced artificial autonomous agents are so complex that new techniques for monitoring progress are required. Al-

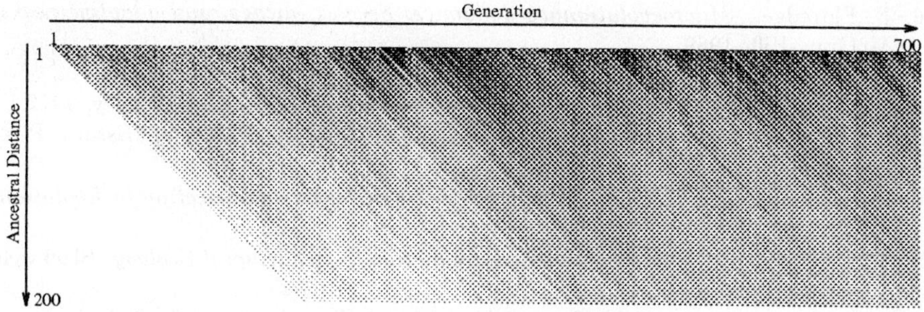

Fig. 13. Ancestral Hamming Map for elite pursuers. Generations 1 to 700 run from left to right. Ancestral distance (Δg) increases from top to bottom: top edge indicates Hamming distance from elite of generation $g-1$: bottom edge indicates Hamming distance from elite of generation $g-200$. Intensities indicate Hamming distance as a percentage of genome length: intensity increases linearly from black (0% distant) to white (> 50% distant). Note that the Hamming distance between two randomly generated bit-strings of the same length will be 50% of the length (on the average).

though this paper concentrated on our ongoing work in evolving pursuit and evasion strategies, we believe that both the problems and the solutions we have identified are general: it is likely that fitness ambiguities will occur in any co-evolutionary situation, and that these ambiguities can be resolved by combining CIAO tests with the various genetic analysis techniques described in Section 4.2. Thus, researchers interested in monitoring or analyzing evolutionary activity in either real or artificial systems where the fitness landscapes change over time (either through co-evolution or because the non-biotic environment is dynamic) should find use for the analysis methods we have described here. The open-ended nature of co-evolutionary simulations makes it difficult to detect the Red Queen's presence: the techniques developed in this paper now let us track her protean manifestations.

References

1. M. Bedau and N. Packard. Measurement of evolutionary activity, teleology, and life. In C. Langton, C. Taylor, J. D. Farmer, and S. Rasmussen, eds, *Artificial Life II*, pp.431–461. Addison Wesley, 1992.
2. D. Cliff, I. Harvey, P. Husbands. Explorations in evolutionary robotics. *Adapt. Behav.*, 2(1):71–108, 1993.
3. D. Cliff and G. F. Miller. Co-evolution of pursuit and evasion II: simulation methods and results. COGS Technical Report CSRP377, University of Sussex, 1995.
4. D. Cliff and G. F. Miller. Tracking the Red Queen: Measurements of co-evolutionary progress in open-ended simulations. COGS Technical Report CSRP363, University of Sussex, 1995.
5. R. Dawkins. *The Blind Watchmaker.* Longman, Essex, 1986.

6. N. Eldredge. *Macroevolutionary dynamics: Species, niches, and adaptive peaks.* McGraw-Hill, 1989.
7. D. J. Futuyama and M. Slatkin, editors. *Coevolution.* Sinauer, 1983.
8. R. C. Gonzalez and P. Wintz. *Digital Image Processing.* Addison-Wesley, 1977.
9. S. J. Gould. *Wonderful Life: The Burgess Shale and the Nature of History.* Penguin, 1989.
10. S. Kauffman. *The Origins of Order: Self-Organization and Selection in Evolution.* OUP, 1993.
11. J. R. Krebs and N. B. Davies. *An Introduction to Behaviuoral Ecology.* Blackwell Scientific, 1993.
12. G. F. Miller and D. Cliff. Co-evolution of pursuit and evasion I: Biological and game-theoretic foundations. Technical Report CSRP311, University of Sussex School of Cognitive and Computing Sciences, 1994.
13. G. F. Miller and D. Cliff. Protean behavior in dynamic games: Arguments for the co-evolution of pursuit-evasion tactics. In D. Cliff, P. Husbands, J.-A. Meyer, and S. Wilson, editors, *Proc. Third Int. Conf. Simulation Adaptive Behavior (SAB94)*, pages 411–420. M.I.T. Press Bradford Books, 1994.
14. E. Renshaw. *Modelling Biological Populations in Space and Time.* Cambridge University Press, 1991.
15. C. Reynolds. Competition, coevolution, and the game of tag. In R. Brooks and P. Maes, editors, *Artificial Life IV*, pages 59–69. M.I.T. Press Bradford Books, 1994.
16. M. Ridley. *The Red Queen: Sex and the evolution of human nature.* Viking, London, 1993.
17. J. Segers and W. D. Hamilton. Parasites and sex. In R. E. Michod and B. R. Levin, editors, *The evolution of sex: some current ideas*, pages 176–193. Sinauer, Sunderland, MA, 1988.
18. K. Sims. Evolving 3D morphology and behavior by competition. In R. Brooks and P. Maes, editors, *Artificial Life IV*, pages 28–39. M.I.T. Press Bradford Books, 1994.
19. L. van Valen. A new evolutionary law. *Evolutionary Theory*, 1:1–30, 1973.

The Coevolution of Mutation Rates

Carlo Maley

545 Technology Sq., NE43-803
Massachusetts Institute of Technology*
Cambridge, MA 02139, USA
email: cmaley@ai.mit.edu

Abstract. In order to better understand life, it is helpful to look beyond the envelop of life as we know it. A simple model of coevolution was implemented with the addition of genes for longevity and mutation rate in the individuals. This made it possible for a lineage to evolve to be immortal. It also allowed the evolution of no mutation or extremely high mutation rates. The model shows that when the individuals interact in a sort of zero-sum game, the lineages maintain relatively high mutation rates. However, when individuals engage in interactions that have greater consequences for one individual in the interaction than the other, lineages tend to evolve relatively low mutation rates. This model suggests that different genes may have evolved different mutation rates as adaptations to the varying pressures of interactions with other genes.

1 The Possibilities of Life

> ...we badly need a comparative biology. So far, we have been able to study only one evolving system and we cannot wait for interstellar flight to provide us with a second. If we want to discover generalizations about evolving systems, we will have to look at artificial ones.
> - J. Maynard Smith[39]

We only have a single example of the evolution of life. With a sample of one, it is difficult, if not impossible, to distinguish between historical and systematic constraints on the evolution of life. Artificial Life represents an emerging complementary approach to understanding biology.[43, 24, 26, 25] It is an attempt to expand the size of the sample of living systems we have to study. Artificial Life also commonly includes the synthesis of artificial systems designed to address specific questions in biology that are intractable under more traditional laboratory techniques. The principle weakness in the Artificial Life approach is the sometimes tenuous connection between the artificial systems and biological systems. This weakness stands in contrast to the high degree of experimental control over the artificial systems. An artificial evolving system can be restarted under identical initial conditions, excepting the random number generator seed,

* This work was supported in part by the generosity of the MIT AI Lab. I am greatful to M. Donoghue, P. Goss, K. Rice and L. King for their comments and support.

slightly different initial conditions, or different system parameters. Furthermore, the experimenter has access to all of the information in the system, and so can avoid many of the difficulties inherent in ecological research. The fact that Artificial Life research depends on models of life means that the conclusions must be predicated on the acceptance of the model as a reasonable description of life. Even given such acceptance of the realism of the model, the results generally can only be qualitative rather than quantitative. Given these limitations, Artificial Life helps us to understand both the constraints and possibilities of life.[30, 44, 22, 23, 7, 12]

Lewontin suggested that when we think about what life might be, rather than what we see before us, there are two dimensions that can be considered: mortality and heredity.[28] The life with which we are most familiar is mortal and reproduces with high fidelity. But why not mortal organisms that pass on little information to their offspring? For that matter, why not immortal organisms with varying degrees of hereditary transmission of information? This paper is an attempt to address those questions. Neither mortality nor heredity need be dichotomous qualities. Mortality might range over varying degrees of longevity. Heredity can be addressed through varying rates of mutation. At the extremes there is either no mutation and so perfect replication, or the mutation rate is so high that the offspring's genetic information is near random.

1.1 A Naive Model

Organisms described only by genes for longevity and mutation rates would be difficult to model on a computer. Assuming a constant rate of reproduction, the computer's memory would fill up with near immortal creatures. Once a lineage struck upon a low mutation rate combined with a high longevity, this growth would become exponential. Of course, in the real world, such growth would be limited by the available resources. The computer's memory in this case is an appropriate analogy for just such a limited resource. So if the model contained a fixed maximum carrying capacity, then an organism could only successfully reproduce if another died. Death, in this naive model, only comes from the longevity gene. Once an immortal organism evolved it would be there for good. Given non-zero mutation rates and a reasonable sized population, immortal organisms would continue to arise. In fact, the only selective pressure in this naive model is a pressure against mortality. Eventually, the entire carrying capacity of the environment would be filled with immortal organisms and evolution would come to a halt. An implementation of this naive model has confirmed this prediction.

1.2 Genetic Algorithms

A significant amount of work on genetic algorithms has focused on finding a mutation rate that will allow the population of solutions to converge on the best solution in the shortest amount of time.[4, 3, 33, 10, 17, 16] However, others point out that the optimal mutation rate depends on how a solution is encoded as a genome[42] and the choice of fitness function to evaluate a solution.[3] Bäck goes as far as to include a gene for mutation rate in each solution in the population.[3] This allows a "schedule" of mutation rates over time. Early on, mutation rates tend to be high. As the population approaches the optimal solution, mutation rates descend. Scaling down mutation rates over time is a common improvement technique in genetic algorithms.[4, 10, 16]

Traditional genetic algorithms employ an unrealistically static concept of fitness. The "problem" that the lineages are trying to solve in a genetic algorithm does not change over time. On the other hand, the naive model misses the point of environmental carrying capacity in biology. Rather than reproduction filling a viable hole in the population, the biological world follows the more Malthusian dynamic of explosive reproduction curtailed by starvation, disease, and conflict. If we allow the organisms to interact with each other, then we open up the possibility for competition, exploitation, and even "cooperation." With interactions, there is no longer any single, optimal solution.[2] Would the evolution of mutation rates and longevity change under coevolutionary pressures? The following model provides a tool for addressing that question.

2 The Model

Each organism in a population is explicitly represented in this configuration model.[11, 31, 20] Time is discrete and is organized by "generations." During each generation all of the organisms are allowed to interact in parallel with the other organisms near by in the environment. The form of these interactions is highly abstract, though not as simplified as other models of ecosystems[23, 22, 5], and have been inspired by the ECHO model.[18, 21, 19]

2.1 An Organism

An organism consists of four genes: *offense, defense, longevity* and *mutation.* Both the *offense* and *defense* genes consist of eight bits. These are treated solely as bit patterns. They determine the results of an interaction with another organism through matching of the bit patterns as detailed below in Section 2.2.

The genes for *longevity* and *mutation* are integers over the range 0 to 100, inclusive. The gene for *longevity* codes the percent likelihood that the organism will survive in a generation. Thus an organism with a 0 for a *longevity* gene would not survive past one generation. An organism with 100 for a *longevity* gene would be immortal.[3]

The *mutation* gene codes for the mutation rate of all the genes in the organism. Mutation only occurs during reproduction when the parental genes are passed on and then modified in the offspring. Mutation is different for the integer

[2] For a discussion of optimality see [27, 40]. For examples of coevolutionary models see [37, 18, 29].

[3] Such an "immortal" organism could still fall prey to competition, described in Sections 2.2 and 2.3.

and bit pattern genes. In the case of the integer genes, *longevity* and *mutation* itself, the genes of the offspring differ randomly from the parental genes along a Poisson distribution centered on the parent's gene, as defined in (1).

$$\frac{(\lambda^{x+\lambda-parental\,gene})e^{-\lambda}}{(x+\lambda-parental\,gene)!} \tag{1}$$

Here λ is the mutation gene, x is a potential offspring's gene, and *parental gene* is the gene of the parent that is being transmitted to its offspring. This gives a Poisson distribution around the parental gene. The offspring's gene is truncated at 0 and 100. When the mutation rate is low, the offspring's gene is likely to be very close to the parent's gene. However, when mutation rate is high, the offspring's gene may vary widely from the parent's gene. Figure 1 depicts three different distributions. When the mutation rate gene is zero, the offspring's genes are identical to the parent's genes.

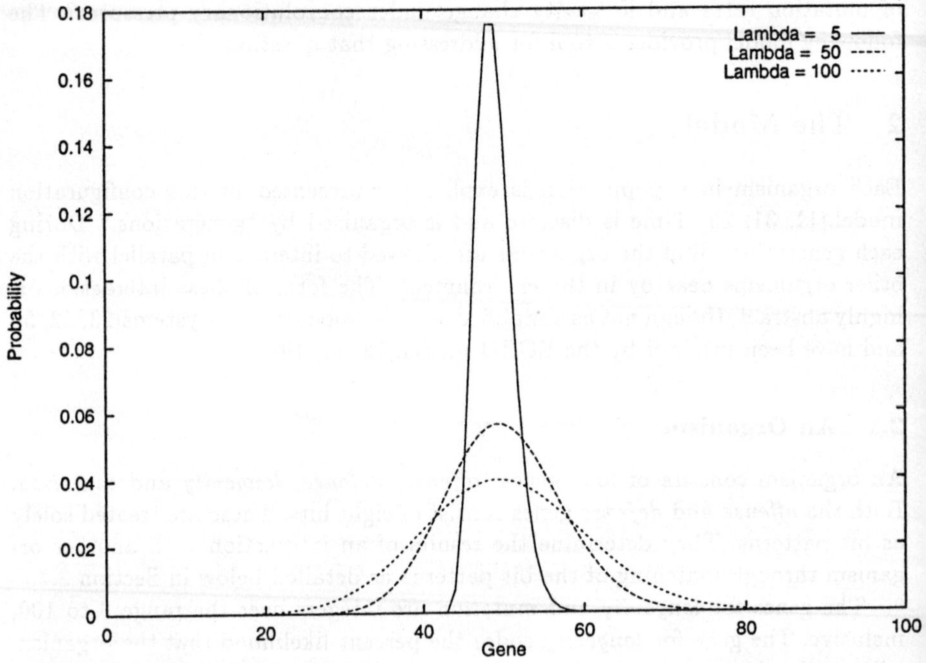

Fig. 1. For a gene with a value of 50, the genes of its offspring are distributed around it depending on the λ of the poisson distribution. This figure shows poisson distributions for $\lambda = 5$, 50 and 100.

Mutation of a bit pattern gene is scaled down via (2). The λ in (2), ranging from 0 to 8, was again used in a Poisson distribution. Here λ is the average number of bits that are flipped in an 8-bit gene during reproduction. This scaling was chosen to give both the full range of possibilities from 0 to 8 and to allow evolution to act on a fine grain of resolution of different mutation rates when the rates were low. Figure 2 shows the scaling curve.

$$\lambda = (mutation/50)^3 \qquad (2)$$

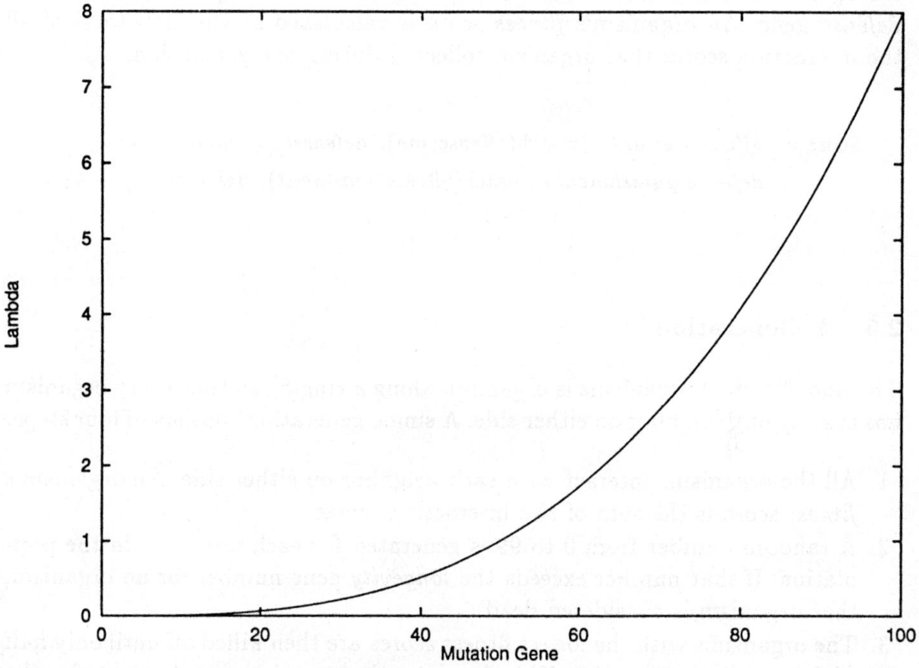

Fig. 2. The mutation gene is scaled from a range of 0 to 100 to a range of 0 to 8 by this curve. This scaling has the advantage of a fine grain resolution of different mutation rates when the mutation gene is low.

2.2 An Interaction

To model an ecosystem, interactions must at the least allow for the full scope of coevolutionary relationships, from mutualism to competition. More rigorously, an interaction can have a positive, negative, or neutral effect on either of the participants in the interaction. A model ecosystem should minimally be able to simulate all eight of these possible interactions. Comparison of bit patterns is one realization of this minimal requirement.[18] In this model, an interaction between two organisms, X and Y, consists of a comparison of X's *offense* gene bit pattern versus Y's *defense* gene bit pattern, and X's *defense* gene bit pattern versus Y's *offense* gene bit pattern.

A comparison is scored by subtracting four from the number of bits in the two genes that match. This is the reverse of the hamming distance. Subtracting 4 gives a score spread from -4 to 4. An organism with an *offense* gene of 11111111 would match 7 bits against an organism with a *defense* gene of 01111111 and so score 3 points. A full interaction is scored by subtracting the *defense-punishment* parameter times the comparison of X's *defense* gene and Y's *offense* gene from the *offense-reward* parameter times the comparison of X's *offense* gene and Y's *defense* gene. An organism's *fitness* score is calculated as the sum total of all the interaction scores that organism collected during one generation.

$$\begin{aligned}\text{Score} = \; & \textit{offense-reward} * (\text{match}(\text{offense}(\textit{me}),\ \text{defense}(\textit{opponent})) - 4) - \\ & \textit{defense-punishment} * (\text{match}(\text{offense}(\textit{opponent}),\ \text{defense}(\textit{me})) - 4) \; .\end{aligned}$$

2.3 A Generation

The population of organisms is organized along a ring[5], so that every organism has exactly one neighbor on either side. A single generation consists of four steps.

1. All the organisms interact with each neighbor on either side. An organism's *fitness* score is the sum of two interaction scores.
2. A random number from 0 to 99 is generated for each organism in the population. If that number exceeds the *longevity* gene number for an organism, that organism is considered dead.
3. The organisms with the lowest fitness scores are then killed off until only half of the population remains. This simulates the organisms producing twice the carrying capacity of the environment each generation.
4. The remaining half of the population is allowed to reproduce, according to their *mutation* genes, to fill the empty positions on the ring.

Reproduction is asexual. The new born offspring receive the genes of a single parent, modified by that parent's *mutation* rate gene. Furthermore, reproduction is fair. Each surviving organism gets a turn to reproduce, by order in the ring. The position of the last organism to reproduce in a generation is saved so that reproduction in the next generation starts with the following organism on the ring. Only living organisms are allowed to reproduce.

2.4 Missing Complexities

The model includes many simplifying assumptions. Interactions are highly stylized. The fitness payoffs for the interactions are simple, discrete, linear functions. The environment has a fixed carrying capacity and the organism's reproduction rates are assumed to be both identical and high enough to fill the environment beyond the carrying capacity. Organisms reproduce asexually. Perhaps most limiting is the absence of any life histories for the organisms. This means that an organism that has survived many generations has no *a priori* competitive advantage over a new born organism. It also means that longevity, and its complement, senescence, cannot be examined in the context of age dependent genetic effects.

2.5 Running the Model

When the model is started, parameters are input for population size, the number of generations to simulate, the number of neighbors on each side with which an organism will interact, the *offense-reward* and *defense-punishment* multipliers. All of the genes in the initial population are generated randomly.[4] During the course of the simulation statistics are gathered each generation on the average fitness, minimum fitness, maximum fitness, average longevity, average mutation rate, average age, and average number of offspring. Finally, at the end of a simulation, the resulting population is saved in a file for later examination.

3 The Results

In all of the following results the model was run with a population of 200 organisms for 2000 generations and an interaction radius of one. The independent variables in the experiment were the *offense-reward* and *defense-punishment* parameters.

3.1 Longevity

There are two separate selective pressures in the model, senescence and competition. The first depends solely on the longevity genes. It is perhaps not surprising then that the longevity genes of the organisms approach 100, or immortality. However, the population generally does not reach fixation on the maximum. To test if these results were significant, a null model was implemented. All interactions were removed from the null model and instead of killing off the organisms with the lowest fitness scores, half of the population was killed at random. The results showed that longevity of the population does not fixate at 100 because mutation continually introduces organisms with longevities below 100. There was no statistically significant difference between the evolution of longevity in the coevolutionary model and the null model.

3.2 Mutation Rates

The evolution of mutation rates differed dramatically depending on the settings of the *offense-reward* and *defense-punishment* parameters. The central results are detailed in Table 1. When the average mutation rates are compared for the condition where *offense-reward* = *defense-punishment* = 1 against the condition where the *offense-reward* is twice the *defense-punishment*, the later condition results in significantly lower mutation rates ($t = -4.0175$, $df = 9.075$, p-value $= 0.0015$).[5] As can be seen, the mutation rates often dropped to fixation

[4] Knuth's pseudo-random number generator was used, with a seed taken from the computer's clock.

[5] A Welch modified two-sample t-test was used with the alternative hypothesis that the second condition resulted in lower mutation rates.

at a zero mutation rate. The same is true if the *defense-punishment* is twice the *offense-reward*. The effect becomes even more dramatic as one of the two parameters is weighted more than the other. In all the extreme cases, where the *offense-reward* was weighted 100 times as much as the *defense-punishment* or when one of the two parameters was 0, mutation rates always dropped to 0.[6]

Table 1. These are the average mutation rates for a population of 200 after 2000 generations. In the headings of the columns, "O" and "D" refer to the *offense-reward* and *defense-punishment* multiplier that were used for that trial.

trial	*O=1, D=1*	*O=2, D=1*	*O=1, D=2*	*O=100, D=1*	*O=1, D=0*	*O=0, D=1*
1	53.525	61.170	0	0	0	0
2	47.575	0	0	0	0	0
3	51.310	0	50.275	0	0	0
4	67.455	31.980	0	0	0	0
5	66.485	0	22.805	0	0	0
6	71.195	33.480	55.465	0	0	0
7	67.470	34.890	24.380	0	0	0
8	66.725	45.505	0	0	0	0

4 Discussion

There seems to be nothing of interest in the evolution of longevity in this model. The results of the null model clearly argue that the only reason that longevity does not reach its maximum is that the mutation rates of the organisms constantly introduce a certain amount of genetic load. The central issue in the evolution of longevity is the attempt to distinguish between two theories for the phenomenon of senescence.[36, 14, 6, 41, 15] Senescence may be caused by a life history trade off. Genes might persist that enhanced survival and reproduction in an organism's early life at the expense of impairment in later life. The competing theory asserts that senescence is caused by deleterious mutations that are only expressed in old age, and so are not easily weeded out of the population. That is, senescence is just a form of genetic load. Clearly, without an implementation of life histories in the organisms, it is impossible to distinguish between these theories.

4.1 Game Theory Analysis of Mutation Rates

Game theory[38, 13] provides a useful tool for analyzing the evolution of mutation rates in the model. Why should an inequality in the *offense-reward* and

[6] In the two trials when one of the two parameters was zero, the population reached fixation at a zero mutation rate in less than 150 generations.

defense-punishment parameters lead to the evolution of low mutation rates? The players in these games are lineages. Each organism in the environment interacts with its two neighbors. The outcome of this interaction is fixed over the life time of the pair. As long as they both survive, the interaction will not change. There is no selection of alternate strategies within the life time of an organism. However, when an organism reproduces, its *offense* and *defense* genes may mutate. A mutation in either of these genes changes the way in which the new organism interacts with other organisms. So a lineage may fix on a certain form of interaction by evolving a zero mutation rate, or it may continually change its interactions by maintaining a non-zero mutation rate.

We can analyze the payoff matrices for a lineage mutating or not mutating in interaction with other lineages. To simplify the analysis, it helps to consider the payoffs around the extremes of interactions, with either a perfect match or mismatch between the *offense* and *defense* genes. This is not an unreasonable approach, because these extremes are often observed to be equilibria in the model. It also helps to assume that the lineages are coevolving over time. Of course, there is no guarantee that the next generation in a lineage will interact with the same "player." However, as a small number of lineages come to dominate the environment, the participants in the games will stabilize. The explanatory value of a game theoretic analysis appears to be robust to these simplifications.

In all of the following matrices, it is helpful to remember that a perfect match scores positive (e.g., 4) fitness points and a perfect mismatch scores negative (e.g., -4) points. The maximum payoff for an interaction, if *offense-reward* = *defense-punishment* = 1, would be $4 - (-4) = 8$, when the organism's *offense* gene perfectly matches its neighbor's *defense* gene, and the organism's *defense* gene perfectly mismatches the neighbor's *offense* gene. Let us begin with those parameter settings. If we consider the case where the two players perfectly match each other, in Table 2, then a mutation in an *offense* gene will always lower that player's payoff and a mutation in a *defense* gene will always raise that player's payoff. The interactions are completely symmetric. The first number in each paired entry of a matrix is the score for player 1 and the second number is the score for player 2. At the opposite extreme, the two players perfectly mismatch.

Table 2. *Offense-reward* = 1 and *defense-punishment* = 1. When lineages 1 and 2 perfectly match, a mutation in the *defense* gene produces these scores. This is a zero-sum game.

Lineage 1	*Lineage 2*: No Mutation	Mutation
No Mutation	0, 0	-1, 1
Mutation	1, -1	0, 0

Then a mutation in a player's *defense* gene will always lower its score. If we consider the mutations in the *offense* genes, then the exact same payoff matrix applies. A mutation in the player's *offense* gene will always raise its score. Under this condition, the benefits of mutation in one gene is always matched by a penalty for mutation in the other gene. Thus, when *offense-reward* = *defense-punishment* = 1 there is no selective pressure on mutation rates. The result is a population of lineages doing a sort of slow random walk in their mutation genes. The larger the population, the slower this walk.

Now consider what happens if *offense-reward* = 2 and *defense-punishment* = 1. If the players initially have a perfect match then the payoff matrix for mutating the *defense* gene is shown in Table 3. This is a Prisoner's Dilemma payoff matrix.[2, 1, 34, 32, 8, 9, 35, 45] A lineage scores better by mutating, or "defecting" in the Prisoner's Dilemma terminology, regardless of what its opponent does. However, mutual "cooperation" scores slightly better for both players than mutual "defection." Note that it is the lineages, and not the individuals, that are "cooperating" and "defecting." Thus, counter-intuitively, a pair of lineages might end up "cooperating" in a parasitic relationship. This only means that both lineages are maintaining the parasitic relationship, rather than changing the nature of their interaction.

The table for a mutation in the *offense* gene reverses each pair of scores in Table 3. Thus there is pressure against mutation in the *offense* gene and mutual "cooperation" still scores better than mutual "defection." Thus, on balance, there is pressure towards "cooperation," or zero mutation rates. A pair of well matched organisms with low mutation rates will thrive in a sea of organisms with high mutation rates. An historical accident could lead to the spread of "cooperation" and even fixation on a zero mutation rate. If we consider the

Table 3. *Offense-reward* = 2 and *defense-punishment* = 1. When lineages 1 and 2 perfectly match, a mutation in the *defense* gene produces these scores. This is a Prisoner's Dilemma game.

	Lineage 2	
Lineage 1	No Mutation	Mutation
No Mutation	4, 4	2, 5
Mutation	5, 2	3, 3

opposite extreme where the players are perfectly mismatched, as in Table 4, mutual "defection" scores better than mutual "cooperation." As long as the lineages generally mismatch they naturally evolve towards relatively high mutation rates. Once the lineages tend to match each other, the prisoner's dilemma payoff matrix emerges with its reward for mutual "cooperation." However, the difference in payoff between mutual "cooperation" and mutual "defection" is minimal. It still requires an adjacent "cooperating" pair of lineages to let "cooperation" spread. This explains why some of the trials fixate at zero mutation rates and others do not. Those that do not fixate at a zero mutation rate still have lower average mutation rates than the symmetrically weighted trials discussed above. Reversing the parameters does not change the analysis significantly. If we let

Table 4. *Offense-reward* = 2 and *defense-punishment* = 1. When lineages 1 and 2 perfectly mismatch, a mutation in the *offense* gene produces these scores.

Lineage 1 \ *Lineage 2*	No Mutation	Mutation
No Mutation	-4, -4	-5, -2
Mutation	-2, -5	-3, -3

offense-reward = 1 and *defense-punishment* = 2, mutual benefit can be gained when the players mismatch. The payoff matrix in this case is identical to Table 3. Similarly, in the case that the genes match, the payoff matrix is identical to Table 4. Reversing the two parameters results in an opposite but equivalent set of pressures. So these trials ought to behave similarly to the trials where *offense-reward* = 2 and *defense-punishment* = 1 except that "cooperating" organisms must mismatch rather than match. This is, in fact, what is observed in the populations that fixated on zero mutation rates.

When we increase the differences between the two parameters, the interactions become more lopsided, but the prisoner's dilemma remains. Table 5 shows the payoff matrix for *offense-reward* = 100 and *defense-punishment* = 1. It still benefits a lineage to mutate its *defense* gene. However, the advantage of such a mutation is only a single point, whereas the disadvantage if both lineages mutate is 99 points. The analysis of the mismatching case is also just a caricature of the Table 4. When the differences in the parameters are exaggerated like this, there

Table 5. *Offense-reward* = 100 and *defense-punishment* = 1. When lineages 1 and 2 perfectly match, a mutation in the *defense* gene produces these scores.

Lineage 1 \ *Lineage 2*	No Mutation	Mutation
No Mutation	396, 396	296, 397
Mutation	397, 296	297, 297

is even greater evolutionary pressure for the lineages to reduce their mismatches as well as to maintain "cooperation." Correspondingly, all of the trials run under these conditions fixated on zero mutation rates.

The limit of this trend towards asymmetric interactions is to set one of the parameters to zero. When *offense-reward* = 1 and *defense-punishment* = 0, there is an overall bias towards "cooperation." The payoff matrix of Table 6 makes it clear that there is no benefit to mutating the *defense* gene. When the pairs of payoffs are reversed in the table, it becomes clear that there is a pressure to prevent mutations in the *offense* gene. In total, mutual "cooperation" scores higher than mutual "defection" when the lineages perfectly match. All of the experimental runs under these conditions fixated at zero mutation rates. When the lineages mismatch, as in Table 7, there is pressure on the lineages to mutate to reduce mismatches and then to stop mutating when a match is achieved. Under this analysis, it is not surprising that occasionally a pair of species fixated at a zero mutation rate when they had a good but not perfect match. As above, the case where *offense-reward* = 0 and *defense-punishment* = 1 follows an exactly equivalent analysis.

Table 6. *Offense-reward* = 1 and *defense-punishment* = 0. When lineages 1 and 2 perfectly match, a mutation in the *defense* gene produces these scores.

	Lineage 2	
Lineage 1	No Mutation	Mutation
No Mutation	4, 4	3, 4
Mutation	4, 3	3, 3

Table 7. *Offense-reward* = 1 and *defense-punishment* = 0. When lineages 1 and 2 perfectly mismatch, a mutation in the *offense* gene produces these scores.

	Lineage 2	
Lineage 1	No Mutation	Mutation
No Mutation	-4, -4	-4, -3
Mutation	-3, -4	-3, -3

4.2 Implications for Biology

It is now clear that the results of the model follow the evolutionary pressures that are applied to the lineages of organisms in the model. But what has this got to do with biology? The dramatic distinction in the model is between the case where *offense-reward* = *defense-punishment* and the cases where *offense-reward* ≠ *defense-punishment*. This distinction roughly corresponds to zero-sum games versus non-zero-sum, or mixed-motive games. In more biological terms, the first case corresponds to interactions where the consequences are roughly the same for each participant. Intraspecific competition might be a common example. Under this form of interaction, the model predicts that the participants will evolve relatively high mutation rates. The second case corresponds to an interaction where the consequences are significantly different for the participants. Here the model predicts that the participant lineages will settle down into a mutualistic association with relatively low mutation rates. The dinner principle may be a familiar example. The cost/benefit analysis is dramatically different for predator and prey, as long as the predator is not on the brink of starvation. The preceding results suggest that such an imbalance in consequences should put the breaks on predator-prey genetic arms races.

Mutation should be understood under this theory as phenotypic mutation. Any heritable, causal factor that affects the rate of change in a phenotypic trait should be considered as a possible adaptation to the coevolutionary pressures on that trait. In a zero-sum game, the evolution of a relatively high "mutation rate" might be realized through the evolution of a greater sensitivity of the relevant trait to environmental conditions during development. This would produce a lower correlation between the trait of the parent and the trait of the offspring. On the other hand, the evolution of a high phenotypic mutation rate might be realized as a "hot spot" in the genes that influence the development of the relevant traits. Presumably, developmental flexibility is more likely to result in an adaptive response to environmental variation than genetic mutation.

These results suggest that different forms of interactions put different evolutionary pressures on the mutation rates of the participants. This leads to the prediction that different expressed genes should have evolved different intrinsic mutation rates as adaptations to their environment of interactions. We should expect to find structures in the cell and genome for selectively allowing some genes to mutate faster than others. This may mean actively boosting the mutation rates. A possible example might involve the selective methylization of cytosine into 5-methylcytosine in a gene. Conversely, it may simply mean that some genes are the subject of better error correction than others.

References

1. R. Axelrod. *The Evolution of Cooperation.* Basic Books, New York, NY, 1984.
2. R. Axelrod and W. D. Hamilton. The evolution of cooperation. *Science*, 211:1390–1396, 1981.
3. T. Bäck. The interaction of mutation rate, selection, and self-adaptation within a genetic algorithm. In R. Männer and B. Manderick, editors, *Parallel Problem Solving from Nature, 2*, pages 85–94. Elsevier Science Pubishers, Amsterdam, 1992.
4. T. Bäck. Optimal mutation rates in genetic search. In S. Forrest, editor, *Proceedings of the Fifth International Conference on Genetic Algorithms*, pages 2–8, San Mateo, CA, 1993. Morgan Kaufmann Publishers.
5. P. Bak and K. Sneppen. Punctuated equilibrium and criticality in a simple model of evolution. *Physical Review Letters*, 71(24):4083–4086, 1993.

6. G. Bell. Evolutionary and nonevolutionary theories of senescence. *American Naturalist*, 124:600–603, 1984.
7. M. Boerlijst and P. Hogeweg. Spiral wave structure in prebiotic evolution: Hypercycles stable against parasites. *Physica*, 48D:17–28, 1991.
8. R. Boyd. Mistakes allow evolutionary stability in the repeated prisoner's dilemma game. *Journal of Theoretical Biology*, 136:47–56, 1989.
9. R. Boyd and J. P. Lorberbaum. No pure strategy is evolutionarily stable in the repeated prisoner's dilemma game. *Nature*, 327:58–59, 1987.
10. M. F. Bramlette. Initialization, mutation and selection methods in genetic algorithms for function optimization. In K. Belew and B. Booker, editors, *Proceedings of the Fourth International Conference on Genetic Algorithms*, page 100=107, San Mateo, CA, 1991. Morgan Kaufmann Publishers.
11. H. Caswell and A. M. John. From the individual to the population in demographic models. In D. Deangelis and L. Gross, editors, *Individual-based Models and Approaches in Ecology*, pages 36–61, New York, 1992. Chapman and Hill.
12. R. Collins and D. Jefferson. The evolution of sexual selection and female choice. In F. J. Varela and P. Bourgine, editors, *Toward a Practice of Autonomous Systems: Proceedings of the First European Conference on Artificial Life*, pages 327–336. MIT Press, 1992.
13. A. M. Colman. *Game Theory and Experimental Games.* Pergamon Press Inc., Elmsford, NY, 1982.
14. V. J. Cristofalo. An overview of the theories of biological aging. In J.E. Birren and V.L. Bengtson, editors, *Emergent Theories of Aging*, pages 118–126. Springer Publishing Company, Inc., New York, NY, 1988.
15. R. G. Cutler. Evolutionary biology of senescence. In J.A. Behnke, C.E. Ellicott, and G. Moment, editors, *The Biology of Aging.*, pages 311–359. Plenum Press, New York, NY, 1978.
16. Terence C. Fogarty. Varying the probability of mutation in the genetic algorithm. In *Proceedings of the Third International Conference on Genetic Algorithms*, pages 104–109, 1989.
17. J. Hesser and R. Männer. Towards an optimal mutation probability for genetic algorithms. In H.-P. Schwefel and R. Männer, editors, *Parallel Problem Solving from Nature*, pages 23–32. Springer-Verlag, New York, 1990.
18. J. H. Holland. *Adaptation in Natural and Artificial Systems.* MIT Press, Cambridge, MA, 1992.
19. J. H. Holland. Echoing emergence: Objectives, rough definitions, and speculations for echo-class models. Technical Report 93-04-023, Santa Fe Institute, 1993.
20. M. Huston, D. DeAngelis, and W. Post. New computer models unify ecological theory. *Bioscience*, 38(10):682–691, 1988.
21. T. Jones and S. Forrest. An introduction to sfi echo. Technical Report 93-12-074, The Santa Fe Institute, 1993.
22. S. A. Kauffman. *The Origins of Order.* Oxford University Press, Oxford, UK, 1993.
23. S.A. Kauffman and S. Johnsen. Coevolution to the edge of chaos: Coupled fitness landscapes, poised states, and coevolutionary avalanches. *Journal of Theoretical Biology*, 149:467–505, 1991.
24. C. G. Langton, editor. *Artificial Life*, Reading, MA, 1989. Addison-Wesley.
25. C. G. Langton, editor. *Artificial Life III*, Reading, MA, 1994. Addison-Wesley.
26. C. G. Langton, C. E. Taylor, J. D. Farmer, and S. Rasmussen, editors. *Artificial Life II*, Reading, MA, 1992. Addison-Wesley.

27. R. C. Lewontin. Fitness, survival, and optimality. In D. J. Horn, G. R. Stairs, and R. D. Mitchell, editors, *Analysis of Ecological Systems*, pages 3–22. Ohio State University Press, Columbus, OH, 1979.
28. R. C. Lewontin, October 1993. Personal communication.
29. K. Lindgren and M. G. Nordahl. Cooperation and community structure in artificial ecosystems. *Artificial Life*, 1:15–38, 1994.
30. C. C. Maley. A model of the effects of dispersal distance on the evolution of virulence in parasites. In R. Brooks and P. Maes, editors, *Artificial Life IV*, pages 152–159, Cambridge, MA, 1994. MIT Press.
31. C. C. Maley and H. Caswell. Implementing i-state configuration models for population dynamics: An object-oriented programming approach. *Ecological Modelling*, 68:75–89, 1993.
32. M. Milinski. Cooperation wins and stays. *Nature*, 364:12–13, 1993.
33. H. Mühlenbein. How genetic algorithms really work i: Mutation and hillclimbing. In R. Männer and B. Manderick, editors, *Parallel Problem Solving from Nature, 2*, pages 15–26. Elsevier Science Pubishers, Amsterdam, 1992.
34. M. Nowak and K. Sigmund. A strategy of win-stay, lose-shift that outperforms tit-for-tat in the prisoner's dilemma game. *Nature*, 364:56–58, 1993.
35. M. A. Nowak and K. Sigmund. Tit for tat in heterogeneous populations. *Nature*, 355:250–253, 1992.
36. L. Partridge and N.H. Barton. Optimality, mutation and the evolution of aging. *Nature*, 362:305–311, 1993.
37. T. S. Ray. An approach to the synthesis of life. In C. G. Langton, C. Taylor, J. D. Farmer, and S. Rasmussen, editors, *Artificial Life II*, pages 371–408, Reading, MA, 1992. Addison-Wesley.
38. J. Maynard Smith. *Evolution and the Theory of Games*. Cambridge University Press, Cambridge, UK, 1982.
39. J. Maynard Smith. Byte-sized evolution. *Nature*, 355:772–773, 1992.
40. J. Maynard Smith. Optimization theory in evolution. In E. Sober, editor, *Conceptual Issues in Evolutionary Biology*, pages 91–118. MIT Press, Cambridge, MA, 1994.
41. T. M. Sonneborn. The origin, evolution, nature, and causes of aging. In J.A. Behnke, C.E. Ellicott, and G. Moment, editors, *The Biology of Aging.*, pages 361–374. Plenum Press, New York, NY, 1978.
42. D. M. Tate and A. E. Smith. Expected allele coverage and the role of mutation in genetic algorithms. In S. Forrest, editor, *Proceedings of the Fifth International Conference on Genetic Algorithms*, pages 31–37, San Mateo, CA, 1993. Morgan Kaufmann Publishers.
43. C. Taylor and D. Jefferson. Artificial life as a tool for biological inquiry. *Artificial Life*, 1:1–14, 1994.
44. J. W. Valentine and T. D. Walker. Diversity trends within a model taxonomic hierarchy. *Physica*, 22D:31–42, 1986.
45. G. S. Wilkinson. Reciprocal food sharing in the vampire bat. *Nature*, 308:181–184, 1984.

Coevolution of machines and tapes

Takashi Ikegami * and Takashi Hashimoto **

Institute of Physics, College of Arts and Sciences, University of Tokyo,
Komaba 3-8-1,Meguro-ku, Tokyo 153, Japan

Abstract. A problem of self-referential paradox and self-reproduction is discussed in a network model of machines and tapes. A tape consists of a bit string, encoding function of a machine. Tapes are replicated when it is attached by an adequate machine. Generally, a tape is replicated but it may be different from the original one. In this paper, external noise evolves diversity in a system. New reaction pathway induced by external noise will be reproduced deterministically by an emerging autocatalytic network. Hence it will remain stable after external noise is turned off. Low external noise develops a minimal self-replicative loop. When external noise is elevated, a more complex network evolves, where a core structure emerges. Tapes in a core network can be bifurcated into either a RNA-like or a DNA-like tape with respect to its usage in an autocatalytic loop.

1 Introduction

Origin of life is often attributed to the emergence of self-reproductive properties. People believe that both proteins and RNAs are necessary entities for self-reproduction. Taking RNAs as software, we need hardware to implement the software; i.e., proteins. In the same way, taking protein as hardware, we need some software to drive it; i.e., RNAs.

John von Neumann first proposed an automaton model for the self-reproduction problem [1]. In his abstract modeling, a fundamental problem is clear. To copy something, we first have to observe it. However, observation in some cases disturbs the object.

In addition to the problem of observation and copying, the self-referential problem known as Richard's problem occurs for self-copying. A self-reproducing automaton should interfere with itself for replication of itself. This generates a self-referential paradox [2]. To avoid the difficulty, von Neumann separates a machine from its description tape as well as proteins and RNA/DNAs. In Neumann's model, he defines a tape as a pattern of stationary states as a pattern of active states. If no external disturbance is possible, replication scenario is perfect. However, when other machines come to interact or if external noise causes error actions, replication becomes unstable.

---

* E-mail address : ikeg@sacral.c.u-tokyo.ac.jp

** E-mail address : toshiwo@sacral.c.u-tokyo.ac.jp

How to encode a machine function on a tape and how to make it stable under noisy environment is our main concern in this paper.

We propose the additional problem of self-replication, as is also stressed by Neumann himself. That is a problem of how to evolve more complex machines from simple self-replicating machines. Random updating of automaton states brings mutations into Neumann's automaton model. If mutations occur in a machine itself, a machine may change its function. But such mutations should not be copied onto the next generation due to Central Dogma. If mutations occur in a tape, it will be copied onto the next generation. Especially neutral mutations will be accumulated in future innovations of machine functions.

In this paper, we will study the evolution of machines and description tapes influenced by random external noise [3, 4]. While a machine reads a tape, both probabilistic and deterministic errors are assumed to occur. The probabilistic error is caused by external noise and is called a **passive error**. On the other hand, the deterministic error is caused by machine action and is named an **active error**.

In low external noise regime, perfect replicating network composed of one or two machines(s) is evolved. In high noise regime, a complex autocatalytic network sustaining deterministic mutation evolves. Self-replication not as an individual but as a network now becomes important. In the other autocatalytic models [5, 6], reproduction is discussed only by machines or by tapes. However, the important fact is that no machine can reproduce itself without the coding tapes and *vice versa*. Because of this restriction, autocatalytic loops of machines only or of tape only will be suppressed. By considering both machines and tapes, we will discuss the dynamics of coding of machines in the present model.

2 Modeling

Our system consists of two different objects, tapes and machines.

A tape has a bit string of a circular form. A machine consists of 3 different parts, a head, a tail and a transition table. Each head and tail is expressed by a 4 bit string, whose pattern will be compared with binary patterns of tapes. A transition table consists of 4 mappings; $\{\{(\sigma^m, \sigma^t) \rightarrow (\sigma'^m, \sigma'^t)\}$, where σ^m and σ^t represents a current binary state of machine and tape. A tape and machine state will change to (σ'^m, σ'^t), respectively depending on a current state of machine and tape.

Introducing an ensemble of tapes and machines, we carry out the machine-tape reaction process as follows:

(1) Interaction of machines and tapes

Machine $\mathbf{M_i}$ reads tape $\mathbf{T_j}$ iff tape $\mathbf{T_j}$ has a head $\mathbf{h_i}$ and tail $\mathbf{t_i}$ in a different site of the pattern. The site from $\mathbf{h_i}$ to $\mathbf{t_i}$ will be called the reading frame.

$$\mathrm{M_i} + \mathrm{T_j} \rightarrow \mathrm{M_k} + \mathrm{T_l} \tag{1}$$

Then machine $\mathbf{M_i}$ reads a tape $\mathbf{T_j}$ and rewrites from the site $\mathbf{h_i}$ to $\mathbf{t_i}$ according to the transition tables. A half population of machine starts to read a tape with the internal state 1 and the other half does with the state 0. As the result, it generates a new set of machine $\mathbf{M_k}$ and tape $\mathbf{T_l}$ per each interaction.

During the read/write process, both probabilistic and deterministic errors are assumed to occur. The probabilistic error is caused by external noise and is called a **passive error**. On the other hand, the deterministic error is caused by machine action, and is named an **active error**. We call it error since it does not replicate a tape but actively rewrites it. A rewritten tape can be taken as a mis-copy of the original tape. The rate of passive error is measured by the bit flip rate per bit. The active error is measured by the rewriting rate per a length of reading frame when a machine reads a tape. Namely, it is given by,

$$\mu_A = \frac{w}{L}, \tag{2}$$

where a symbol w denotes the number of rewritten bits and L denotes a length of reading frame.

(2) Translation of tapes:

Not only bits of a reading frame, but every bit of tape is repeatedly picked up to construct a new machine from a first site of the reading frame. If a length of a tape is not enough, the same bit is used for coding several different part of a new machine. In the present model, we use a fixed length of 7-bit tapes with 16-bit machines. A first 8 bits are mapped onto head and tail parts in order. The next 8 bits are mapped onto a transition table. In order to cover 16 bits by 7 bits, several bits are multiply used. This complicated mapping from bits of tape onto machine function is assumed to reflect the nonlinear mapping from one-dimensional DNA to 3-dimensional protein folding structure.

(3) Population dynamics:

We presume two abstract spaces of the size N for tapes and machines. A total m machines and t tapes distribute over the respective space. By iterating the following procedures, we simulate the machine/tape reacting system.

1. Compute concentration of tapes (f_m) and machines (f_t) by dividing the population number of each object by the capacity size N.

$$f_m^i = \frac{m^i}{N}, f_t^j = \frac{t^j}{N}. \tag{3}$$

2. Make $f_{ij}N$ numbers of new machines and tapes from the reaction of machine i and tape j. The frequency of reaction f_{ij} is given by,

$$f_{ij} = \frac{c_{ij} f_i^m f_j^t}{\sum_{k,l} f_k^m f_l^t}. \tag{4}$$

The coefficient c_{ij} takes a positive constant $c_m (= c_t)$, if a machine i can read tape j. Otherwise it takes zero value.

3. Remove d_m % of old machines and d_t % of old tapes.

4. Put the new machines and tapes back in each space. Hence the population of machine i and tape j of the next generation becomes,

$$m_i' = (1 - d_m)m_i + \sum_{k+j \to i} f_{kj}N, \tag{5}$$

$$t_j' = (1 - d_t)t_j + \sum_{k+i \to j} f_{ki}N. \tag{6}$$

   It should be noted here that each machine has its unique description tape but the inverse is not true. Generally each tape encodes several machines depending on which site the translation starts in.
5. Taking an integer part of the above population, we obtain the actual population of the next generation. The machine or tape whose frequency is lower than ϵ is removed from the population. Not to go beyond the total population size N, the coefficient should satisfy the relation $c_{m/t} \leq d_{m/t}$.

(4)Effect of external noise:

Out of $f_{ij}N$ new tapes, a number of tapes as well as machines, are erroneously generated by external noise. It is assumed that the rate of error depends on the reading frame. Namely, the rate of error replication by external noise is given by,

$$w = 1 - (1 - \mu_p)^L, \tag{7}$$

where the symbol L is the length of reading frame. We use Monte Carlo methods to get the mutant objects. At most $wf_{ij}N$ mutant populations are obtained by randomly flipping the bit within a reading frame.

(5)Reading tapes:

Each tape has a source where an attached machine starts to search for the head and tail pattern. Starting from the site, patterns are searched for in the clockwise direction of a circular tape. When a head pattern is found, a tail pattern starts to be searched in the clockwise direction. The site of source can be updated randomly when the tape is exploited and newly generated. Note that every translational invariant tape has the same source.

3 Destabilization of a minimal self-replicating loop

About 10 randomly selected machines with 2 or 3 tapes are prepared as the initial configuration. A machine without description tape is unstable and smoothly removed from the system. Hence an initial configuration which does not include any description tapes of the initial machines will die out if there is no mutation process. External noise may produce description tapes by mistake.

Even without external noise, a machine can acquire its description tape by the other machines as a normal product. To sustain the tape, we have to have the description tape of the machine which generates the tape of the first machine. In order to make it stable, a successive reproducing process should form a closed loop; each machine on the loop reproduces a machine for the next position.

We will see how the replicating networks evolve by changing the amount of external noise. Examples of temporal evolution of population of machines and tapes are shown below(Fig.1).

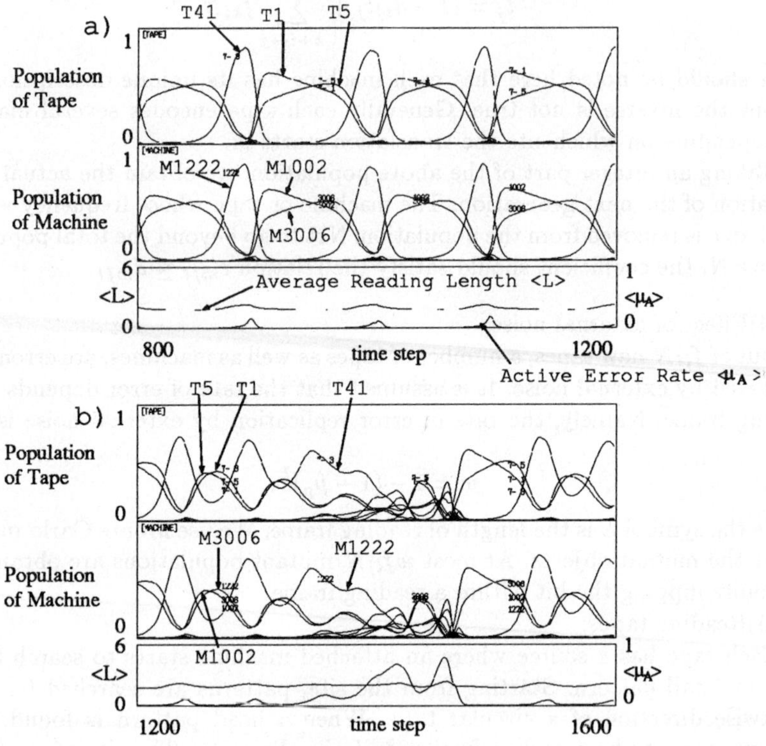

Fig. 1. Temporal evolution of population of machines(the top column)and tapes(the middle column). The bottom column displays temporal evolution of averaged active error rates and the averaged length of the reading frame. The parameters of population dynamics are $c_{m/t} = d_{m/t} = 0.6$ in the influence of, a) lower external noise (= 0.04) and b) higher external noise (= 0.055). Both start from the same initial states.

By introducing a lower amount of external noise in a system, we see a minimal autocatalytic loop evolve. In this example, a machine $\mathbf{M_{1002}}$ reads a tape $\mathbf{T_1}$ to replicate the same tape and its own machine. The number attached to tapes and machines are hexadecimal number converted from its binary representation. Many initial configurations reach this minimal autocatalytic system for a lower noise regime.

A system with the minimal autocatalytic loop is said to be metastable since it remains stable after turning off external noise. However the minimal loop is destabilized by increasing external noise. Translation under external noise generates many machines, most of which rewrites existing tapes. Increasing of parasite machines destabilizes the original self-replicating loop. In Fig-1-a), a parasitic machine $\mathbf{M_{1222}}$ invades the network with its description tape $\mathbf{T_{41}}$.

A greater variety of machines and tapes induces unstable oscillation in Fig.1-b). An original self-replicating pair becomes unstable if too many parasitic machines attach to it. Population of each machine and tape show unstable oscillation in time. An oscillation of large amplitude is caused by the original self-replicating loop. Other oscillations are caused by parasitic machines and tapes.

This temporal oscillation spontaneously crashes by exhausting the self-replicating loop. The system restarts by having the minimal self-replicating loop, otherwise it is exploited.

It should be noted that a non-zero active error rate begins to oscillate in time in Fig.1-b). A low rate of active errors means that the network has more individual self-replication; each tape is self-replicated without error. But a high active error suggests that many machines read tapes and producing different machines with different tapes from the original ones. Namely, failure of self-replication is reflected in the amplitude of active errors.

The minimal loop shows zero active error as being depicted in Fig.1-a). Intermittent bursts of active error rates are caused by the parasite machine $\mathbf{M_{1222}}$ and the tape $\mathbf{T_{41}}$.

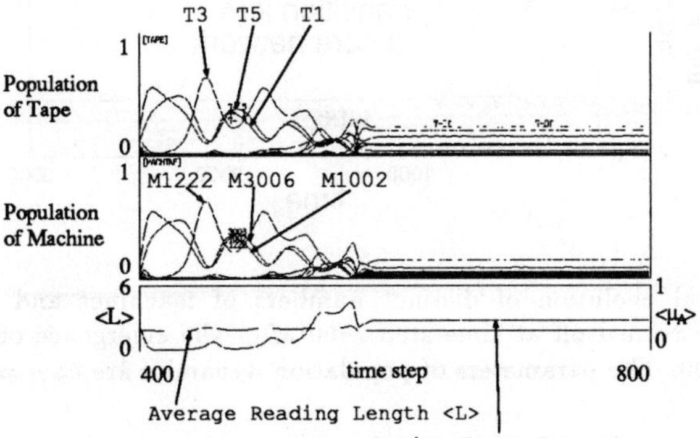

Fig. 2. A spontaneous transition into a core network of a fixed point state. It displays temporal evolution of population of machines(the top column)and tapes(the middle column), and averaged active error rates and the averaged length of the reading frame(the bottom column). The rate of external noise is set at 0.07. The parameters of population dynamics are $c_{m/t} = d_{m/t} = 0.6$.

4 Emergence of Core network

In the region of high external noise ($\mu_p \geq 0.05$), a stable structure seems to evolve. Unstable oscillation in population amplitude as we see in Fig.1-b) is spontaneously stabilized around time step 600 in Fig.2. At the same time, the active error rate is sustained at the high level.

If we turn off external noise after time step 600, the numbers of machines and tapes remain stable. But if the noise is removed before time step 600, it will back to a minimal self-replicating loop. We call a network which acquired an implicit stable structure a core network.

Fig.3 shows temporal evolution of machines and tapes after turning off external noise. The amplitude of total number of machines and tapes become less rugged by compared with that before a transition. Generally, machines disappears from the system when turning off the noise. Then a true core network is left in the system.

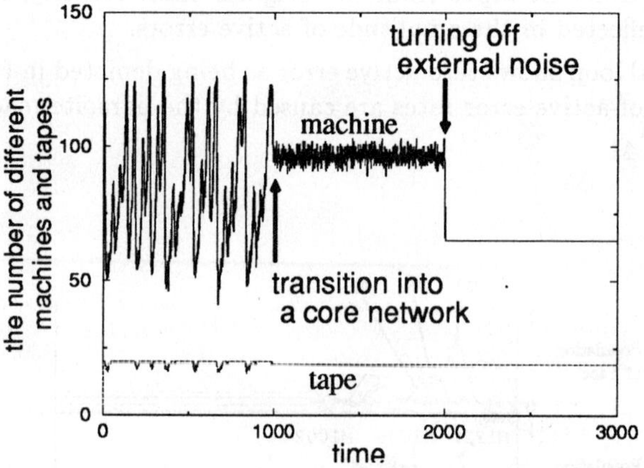

Fig. 3. Temporal evolution of distinct numbers of machines and tapes. External noise(= 0.08) is turned off at time step 2000 after the emergence of a core network at time step 1100. The parameters of population dynamics are $c_{m/t} = d_{m/t} = 0.6$.

A core network is not necessarily a fixed point state. It may start to oscillate after turning off the noise. Phenomenologically, core nets can be divided into distinguishable states with respect to the dynamical states of population. 1) Core networks with change in the number of machines but not of tapes. 2) Core networks without change in the number of machines or of tapes. 3) Core networks with change in both numbers. Note that a core network without any change in the number of machines and tapes may have temporal oscillation in population amplitude. Emergence of such temporal oscillation depends on the sustained network topology.

A minimal self-replicating loop(Fig.1-a)) is also an example of core net. It cannot sustain an active error by definition. All the examples of complex core networks which appear at mid external noise ($0.05 \leq \epsilon < 0.1$) have high active error rates. By increasing the amount of external noise more, an attainable core net again loses complexity, becoming a minimal self-replicating loop.

In Fig.4, we depict an active error rate of core nets as a function of external noise, which are attained by turning off external noise at time step 2000.

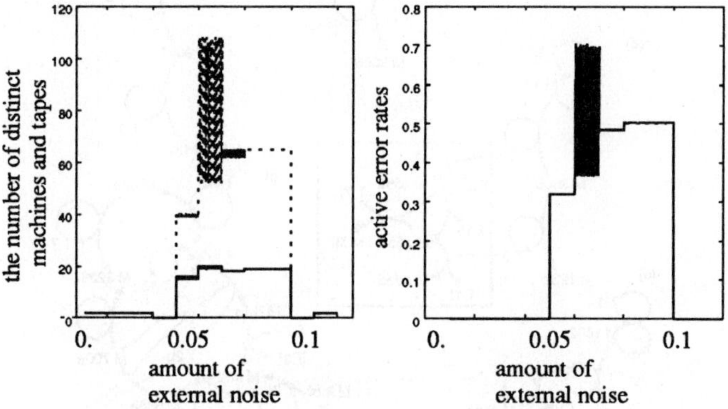

Fig. 4. Temporal evolution of the distinct numbers of machines and tapes(left figure) and those of active error rates(right figure) during time step [2500, 3000] are depicted as a function of external noise. The parameters of population dynamics are $c_{m/t} = d_{m/t} = 0.6$.

This diagram depends on the initial configuration of machines and tapes. Some initial configurations never attain any core networks. However the general tendency is that complex core networks emerge in he mid range of ($0.05 \leq \epsilon < 0.1$) of external noise. Namely, there exist upper and lower bound on external noise to evolve rich core networks. Core networks with a fixed point state are more attainable than ones with oscillating states. A core network at external noise of 0.05 contains a rather smaller number of machines than the other core networks. It has been found that this network uses more 0-rich tapes, which contain a bit of state 0 more than that of 1, in autocatalytic loops. Whereas the other core networks in a fixed point state use more 1-rich tapes, which contain a bit of state 1 more.

The other important criterion to distinguish network states is to investigate the embedded autocatalytic loops. Since Eigen and Schuster's pioneering work [6], notion of autocatalyticy has been known as a useful razor. We use this notion to dissect core networks in the next section.

5 Embedded autocatalytic loops

Machines to be sustained in a system should possess their description tapes. Evolution from a simple self-replicating loop to complex ones are depicted in Fig.5, where we express a system by actual machine-tape reaction graphs.

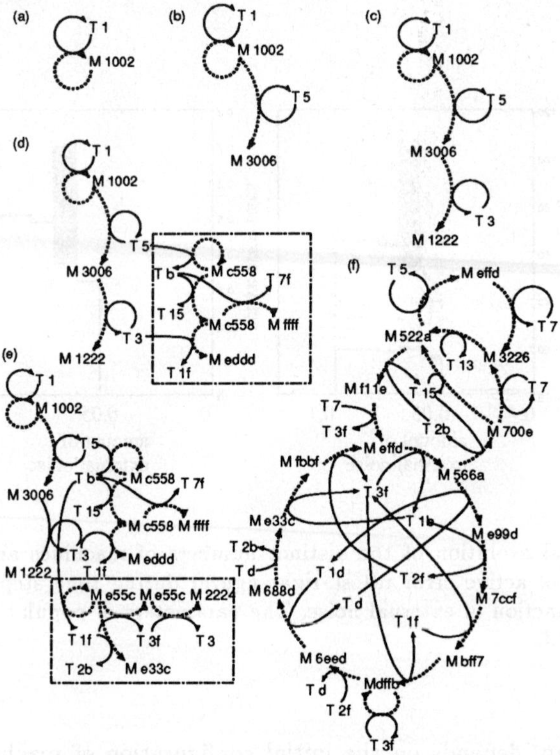

Fig. 5. Evolution from a simple self-replicating loop to a complex network a) An initially existing a minimal self-replicating loop, where a machine M_{1002} replicates itself by reading the tape T_1. A net a) will be successively exploited by M_{3006} with T_5 (b)then by M_{1222} with T_3 (c). The net which is depicted on d) is a net c) with a parasitic network(shown in a box) hanging on it. This is named a parasitic net, since as a network it depends on a net c) for the input tapes T_5 and T_3 to maintain its structure. In a network e), a hyper-parasite(shown in a box) hanging on a parasite net in d) is depicted. Embedded autocatalytic loops which are found in the core network after turning off external noise are depicted in f). We can see four independent double autocatalytic loops here. Two are of the Eigen-Schuster type and two are of the double loop types.

In Fig.5-a), a machine M_{1002} copies itself by reading its description tape T_1. This minimal self-replicating loop exemplifies Eigen-Schuster's autocatalytic network type. Namely, a tape is self-replicated without error. In high noise regime, the minimal loop is gradually destabilized by side chains. Fig.5 c) and d) show a successive appearance of parasitic networks. A structure obtained by subtracting a net c) from a net d) gives a parasite network. This parasite network depends two input tapes(T_3 and T_5) on a network c). Similarly, a structure obtained by e) minus d) gives a hyper parasite, which depends a tape T_b on the first parasite network. Without a host network c), successive parasite networks will be extinguished.

A finally established core network is a combination of several loops which are autocatalytic. With respect to machine sets only, a closed loop is defined as a chain of k machines where a machine M_j generates a machine M_{j+1}, the final machine M_k generating M_1. If a tape set needed to sustain the machine loop is self-reproduced by the machine loop, we say that a loop satisfies an autocatalytic condition.

As special case, the autocatalytic condition is locally satisfied; each tape is self-replicated. A network of this type is called the Eigen-Schuster type. Since each tape self-replicates, no active error exists in this network type. However, such autocatalytic condition is not locally satisfied in general. In the noisy environment a more general network which globally satisfies the condition will be expected. Difference between two possible network types are depicted in Fig.6.

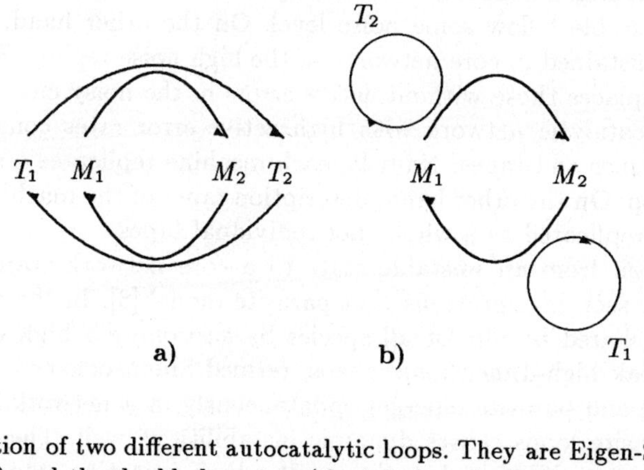

Fig. 6. Illustration of two different autocatalytic loops. They are Eigen-Schuster type(the right figure) and the double loop type(the left figure). Tapes are replicated without errors in Eigen-Schuster type. In the double type, both machines and tapes are replicated into the other.

In the general case of Fig.6-b), not only machines but also tapes form a loop structure. Hence we call the expected autocatalytic networks emerging in high noise regime a double autocatalytic network. In the low noise regime, a network can sustain its structure by Eigen-Schuster type. But for the high noise regime, it is difficult to maintain prefect replication. It then switches to the double loop structure. Since Eigen-Schuster type only allows replication without error, we call this type a DNA-like replicating system. On the other hand, a double network type edits tapes, which we call a RNA-like transcribing system.

Compared with core nets in a fixed point state, we find that cores in oscillatory states change the number of double loops in time. Details will be discussed elsewhere [7].

6 discussions

We have studied the coevolution of machines and their description tapes. We note that a minimal self-replicating loop, composed of one machine and one tape, emerges under influence of external noise.

When external noise is elevated, a minimal self-replicating loop is exploited by parasitic networks. Population of machines and tapes then begins to show unstable oscillations. In the mid regime of external noise, a system will evolve into a stable network against external noise spontaneously. When a network attains this state, it is stably sustained even after external noise is removed. We therefore call the network acquiring such noise stability a "core network". The core network consists of many autocatalytic loops, being a fixed point or oscillatory state.

A minimal self-replicating loop has no replication errors, and corresponding to the simplest Eigen-Schuster's autocatalytic network. Such autocatalytic network is only stable below some noise level. On the other hand, a high active error rate is sustained in core networks in the high noise regime. Self-replication with errors replaces those without active errors in the noisy environment.

This autocatalytic network with high active error rates consists of double loops of machines and tapes. Namely, each machine replicates a successive machine on a loop. On the other hand, description tapes of the machines composing of a loop are replicated as a whole, not individual tapes.

A transition from an unstable state to a core network state is similar to what we have seen in a previous host parasite model [8]. In the model, chaotic instability is shared by almost all species by sustaining a high mutation rate, leading to weak high-dimensional chaos, termed "homeochaos". In this paper, a role of host and parasite emerges spontaneously in a network. Interaction of host and parasite loops causes dynamic instability as well. The core structure suppresses the instability and as the result a high mutation rate is sustained in the present model.

The idea of introducing two different errors provides a new interpretation of self-replication. A network absorbs external noise as an active error of machine function. Namely, passive error caused by external noise is replaced by a deterministic error.

In real biological systems, DNA is replicated individually without active errors. RNA is also a mere copy of DNA. However, we have an example [9] which may be related to our double autocatalytic network. The authors showed that

DNA of macronucleus is generated from DNA of micronuclear in Oxytricha nova. DNA of micronuclear is transcribed once into RNA. Then the exons are completely rearranged and reverse transcribed into the DNA of macronucleus. If editing and reverse transcribing RNA is more stable than replicating DNA itself, an active error as editing will be favored in some genetic systems.

In order to enhance a desirable mutation, we have to generate mutations not by chance but by a deterministic process. In order to suppress the undesirable mutation, we have to change our coding structure, for example by replacing unstable description tape with more stable tapes. However it is difficult to chose stable coding generally. An optimal coding of a machine may lead to the worst coding of the other. This causes a temporal oscillation even in a core network. Note that self-organization of an autocatalytic network implies self-organization of a coding system [10]. How to improve coding cannot be determined locally but should be done in a network context. Alternation of coding driven by external noise has also been reported recently [11]. In their systems, a multi coding system is developed in the high noise regime.

Acknowledgment

One of the authors(T.I.) would like to thank Pauline Hogeweg for stimulating and critical discussions on this work, and her bioinformatic group for their heartful courtesy while he was staying in Utrecht. His stay was supported by NWO (Netherlands Scientic Research Foundation). This work is partially supported by Grant-in-Aids (No.06854014) for Scientific Research from the Japanese Ministry of Education, Science and Culture.

References

1. J.von Neumann, *Theory of Self-reproduction*(1968)
2. This is a main theme discussed by D.Hofstadter in his book, *Gödel, Escher, Bach: An eternal Golden Braid*(Penguin Books, 1980).
3. M. Eigen, Sci. Am. (1993) July, 32. Also see P.Schuster, Physica 22D(1986) 100.
4. See also, M.Andrade et al. Physica D 63 (1993) 21.
5. See e.g., D.J.Farmer, S.S. Kauffman, N.H.Packard and A.S.Perelson, Ann.Rev. NY Acad.Sci. 504(1987)118.
6. M.Eigen and P.Schuster, *Hypercycle*(Springer-Verlag, 1979).
7. T.Ikegami and T.Hashimoto, to be submitted.
8. See e.g., K.Kaneko and T.Ikegami, Physica D 56 (1992) 406:; T.Ikegami and K.Kaneko, CHAOS 2 (1992) 397.
9. A.F.Greslin, Proc.Natl.Acad.Sci. USA 86(1989)6264.
10. P.R.Wills, J.Theor.Biol. 162(1993)267.
11. P.Hogeweg and B.Hesper, Computers Chem. 16 (1992) 171.

Incremental Co-evolution of Organisms: A New Approach for Optimization and Discovery of Strategies

Hugues Juillé

Brandeis University, Computer Science Department
Volen Center for Complex Systems, Waltham, MA 02254-9110, USA
hugues@cs.brandeis.edu

Abstract. In the field of optimization and machine learning techniques, some very efficient and promising tools like Genetic Algorithms (GAs) and Hill-Climbing have been designed. In this same field, the Evolving Non-Determinism (END) model presented in this paper proposes an inventive way to explore the space of states that, using the simulated "incremental" co-evolution of some organisms, remedies some drawbacks of these previous techniques and even allow this model to outperform them on some difficult problems.
This new model has been applied to the *sorting network problem*, a reference problem that challenged many computer scientists, and an original one-player game named *Solitaire*. For the first problem, the END model has been able to build from "scratch" some sorting networks as good as the best known for the 16-input problem. It even improved by one comparator a 25 years old result for the 13-input problem. For the Solitaire game, END evolved a strategy comparable to a human designed strategy.

Keywords: Evolutionary optimization, Simulated co-evolution, Sorting networks.

1 Introduction

Simulation of the rules of life is an attractive and intuitive approach for the design of optimization tools or for machine learning. The evolution of a population of organisms is ruled by some operators that allow the exploration of the state space and a selection mechanism that works with respect to an objective function. Then, this simulated evolution may allow the emergence of specialized organisms or some complex behaviors. Genetic Algorithms are a perfect example of such a simulated evolution.
The aim of this paper is to describe an inventive model, called Evolving Non-Determinism (END), which proposes a new approach to explore the space of states. In fact, the END model can be seen as many simple competing organisms that interact locally one with each other and that become more and more complex with time, evolving using specialization. At each generation, they are evaluated according to a fitness function that allows most promising organisms to survive selection and to spread over.

The END model is extremely different from other well-known techniques: Genetic Algorithms (GAs), Hill-Climbing or Simulated Annealing (SA) in the way the information about the landscape (or the topology) of the space of states is used. This allows the END model to outperform these techniques in the case of problems for which there is only little information about the gradient of the landscape or for which state representation is problematic. Indeed, the main drawback of GAs is that crossover and mutation operators used to evolve genes may create genes that correspond to an invalid solution. For some problems, this drawback can be such that the population size has to be very important to counterbalance this undesirable property. For Hill-Climbing and SA, some operators have also to be designed in order to evolve solutions by finding some new solutions in their neighborhood. The design of such operators may be very elaborate for some problems. Unlike these techniques, the END model doesn't care about solution representation or local neighboring solutions since solutions are generated incrementally and are always valid. In fact, we shall see that the space of states can be represented by a tree and that a solution is a path from the root of this tree to a leaf.

The END model has been applied on two difficult "real-life" optimization problems. Encouraging results described in this paper, let us expect that a broader field of applications can be tackled by this model. The first problem is the follow-up of an established problem for which several approaches ([1, 5, 10]) have been used to try to improve some 25 years old results concerning sorting networks [8]. Actually, this problem was also at the origin of an early paper [13] in which GAs were used to try to replicate Hillis's experiment ([5]) for the 16-input problem and in which some ideas of the END model were presented. The second problem is a very simple one-player game for which the player tries to find a strategy to get a score as large as possible.

This paper is organized as follows: Principles and parameters of the model are presented in details in Section 2 along with a comparison with other optimization techniques. Results for the two real-life problems are described in Section 3. Section 4 presents a summary and possible future research.

2 Evolving Non-Determinism

2.1 Principles

2.1.1 Description of Organisms

The END model simulates the co-evolution of a population of N organisms. These organisms live in a grid world, wrap-around, for which there are as many slots as organisms. Each slot is occupied by an organism and every organism works as a non-deterministic Turing machine since it makes random decisions.

2.1.2 Representation of the Space of States

The class of problems we propose to study is such that the space of states can be represented by a *tree of solutions*. The leaves of this tree correspond to all possible solutions to the problem and any solution is uniquely defined by a path from the root to the leaf to which this solution is attached. Such a path can be described by an ordered sequence of choices, each choice corresponding to the node that has been picked while incrementally generating the path. A problem for which the space of states can be represented by such a tree is called an *incremental problem* in the following of the paper.

Moreover, we assume the following properties for the tree of solutions:

- A *fitness* can be assigned to each leaf (and therefore to each solution),
- For each internal node, children nodes are *correct and fair*. That is, no leaf corresponds to a non-valid solution and there are no useless choice in the description of a solution.

In the following of the paper, we shall also use the *partial solution* term to specify the first choices (or the *prefix*) of the description of a solution.

2.1.3 Incremental Co-evolution of Organisms

The simulated co-evolution is a sequence of competition rounds. At each round, each organism generates a path from the root to a leaf in the tree of solutions associated to the problem at hand. This path is built incrementally, choosing uniformly randomly a child for each node encountered.

Then, each organism is seen as the member of a *species* and a *fitness* is assigned to it:

- The prefix of the solution an organism generated defines the species to which it belongs. The length of this prefix is defined by a parameter called *commitment degree*. The value of this parameter equals 1 at the beginning of the simulated co-evolution and its increasement is managed thanks to a global strategy.
- The fitness is defined as the fitness of the solution this organism has generated.

According to the value of the fitness, a selection is performed in the population of organisms. This selection is operated as follows:

- At each slot, the fitness of the occupying organism is compared to the fitness of organisms in the neighborhood.
- Then, the organism with the highest fitness in this neighborhood is copied into the slot providing this fitness is better than the one of the organism currently in the slot. If several organisms have the same fitness, one of them is picked randomly.
- Otherwise, the slot occupant remains unchanged (and has eventually been copied into slots of its neighborhood).

The idea of this selection is that an organism with a higher fitness probably made some better choices for the first choices of its solution. Clearly, this simulated co-evolution of competing species allows such organisms to duplicate themselves and to take the place of worse organisms.

Now that the selection is completed, for the next round, every organism builds another solution. However, organisms keep the same prefix as their previous solution and build a new solution beginning with this prefix. This process can be seen as a backtrack in the tree of solutions and the incremental building of a new random path from this point. Therefore, the species organisms belong to doesn't change. Then, another round of selection is performed.

Figure 1 describes how such a population of competing species evolves. This simulation was performed in a 1024 slots world. There are 10 species competing; each species is represented by a number from 0 to 9. The problem tackled in this simulation is the sorting of integers: $\{0,\ldots,9\}$ and the fitness is defined as the number of ascents in the sequence generated by organisms. This problem is used in [6] to analyze performance of the END model. Looking at this simulated co-evolution, we can see that, as rounds go off, the species represented by '0' takes more and more importance and it almost dominates all the other species after 16 rounds.

It is easy to understand that, proceeding in that way, organisms corresponding to best first choices are theoretically stronger than others during the selection step and therefore duplicate more often. So, their species take more and more importance in the total population. If we let this scenario run indefinitely, we can expect a species to overcome all other species and to be the only one in the population.

That is why the *commitment degree* parameter has been introduced. Increasement of the commitment degree corresponds to the specialization of organisms since it defines the length of the prefix used to identify species of organisms.

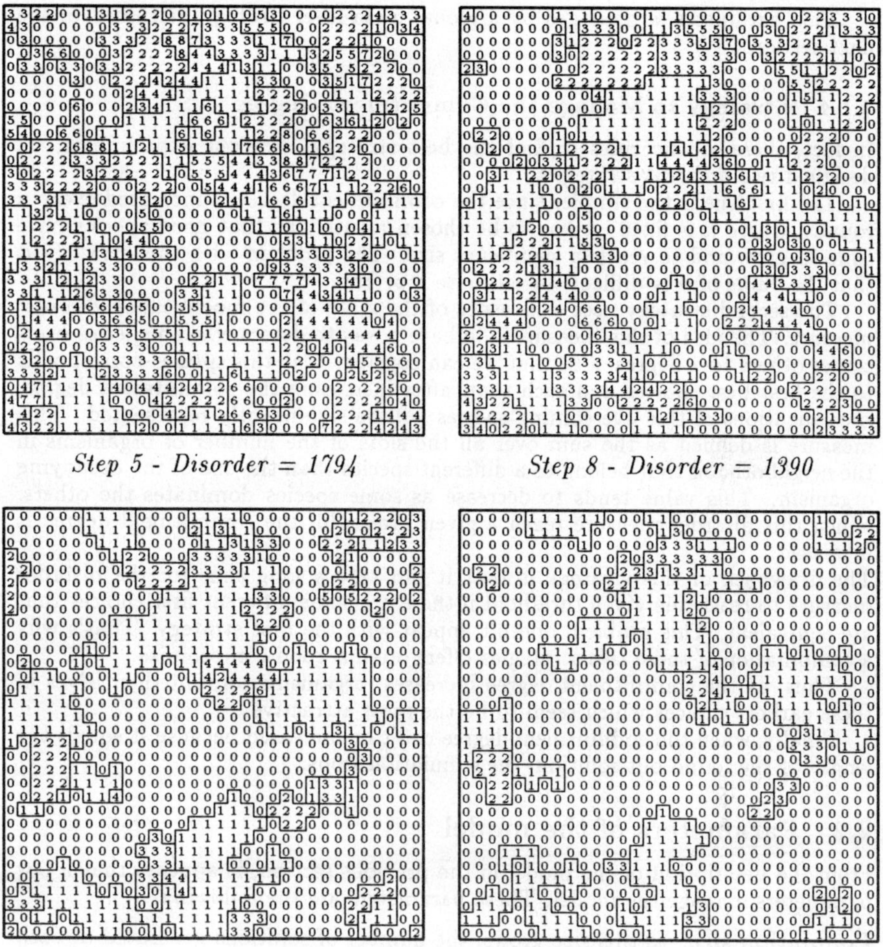

Step 5 - Disorder = 1794 *Step 8 - Disorder = 1390*

Step 12 - Disorder = 1046 *Step 16 - Disorder = 812*

Figure 1: *Simulation of the co-evolution of 10 competing species in a 1024 slots world.*

This parameter is managed by a global meta-level strategy discussed in the next section.

2.1.4 Strategy to Manage the Commitment Degree

In fact, there are no rules to find the best strategy. For our experiments, we designed two different strategies.

The first one is the simpler since the commitment degree is increased every n rounds, where n is fixed. n has to be chosen astutely so that good species have time to grow and to reach a significant size. The problem with this strategy is that the value of n is difficult to estimate *a priori*.

The second strategy uses a measure of the state of the model called *disorder measure*. The idea of this measure is to have a way to detect when a species overcomes others to a degree such that we can consider that this species corresponds to the first best choices and such that, after increasing the commitment degree, specializations of this overcoming species will be significantly represented. This measure is defined as the sum over all the slots of the number of organisms in the neighborhood that belong to a different species than the one of the occupying organism. This value tends to decrease as some species dominates the others. When this disorder measure reach a given threshold, the commitment degree is increased.
The drawback of this strategy is that it can take a long time for the disorder measure to reach the given threshold if there are some different first choices that are equivalent. This problem doesn't appear with the first strategy. That's why a combination of these two strategies offers often a good compromise.

Thus, as the commitment degree increases, organisms commit themselves in these earliest choices which seem to be the most attractive.

Finally, when the commitment degree can't be increased because it equals the length of the found out solution, the simulation stops.

2.2 Parameters of the model

From the description of the model in the previous section, it appears that a few parameters manage the model. These parameters are the following:

Population size: as this size grows, the number of solutions generated at each round increases. The size of the "sample" is more important and it is more likely that species corresponding to optimal solutions don't disappear because of a too small number of representatives.

Neighborhood used for selection: Unformally, the intuition behind this parameter is the following: If this neighborhood is large then we can expect that when a good solution is found out, it propagates more easily in the population. Therefore, the convergence to a good solution can be fast. However, a large neighborhood may forbid the discovery of a better solution since it shrinks the space of explored solutions. It is not the case when a small neighborhood is used. But, the drawback of the later case is that convergence is slower.

Management of the commitment degree evolution: the strategy used to schedule increasements of the commitment degree is very important. Indeed, it must be such that the number of representatives of "interesting" species at the next round is sufficient to expect them to overcome other species with a high probability.

All these parameters are analyzed experimentally in [6].

2.3 Comparison to other Optimization Techniques

As we have already said, our model doesn't exploit information about local gradient of the landscape. Indeed, once a solution is built, only first choices of this solution are used for the next round. That means that we don't try to get some better solution in the neighborhood of this solution, using some gradient information as it is the case for Simulated Annealing or Hill-Climbing.
This remark doesn't apply to Genetic Algorithms. However, the drawback of GAs are the operators used to evolve in the landscape: cross-over or mutation may make very difficult the search of an optimal solution if the new genotype doesn't correspond to a valid solution. In the case of some problems, like the two presented in this paper, this drawback is not negligible and good efficiency can only be reached if an extremely large population of genes is used.

To understand a little more easily how the END model works, let us make the following analogy: Children of the root of the tree of solutions can be seen as a partition of the space of states, each child (or species) corresponding to a particular sub-set of this partition. Then, the selection process allows species that generate a better solution on average (regarding the fitness) to dominate other species. Such species correspond to the domains of the space of states for which the mean value for the fitness is larger. Therefore, at this stage, details and gradient of the landscape of the space of states are not considered. Then, as the commitment degree increases, each domain is partitioned into smaller subdomains and, therefore, details of the landscape take more and more importance. Schraudolph and Belew ([12]) implemented a similar idea for GAs by tracking the convergence of the population to restrict subsequent search using a *zoom operator*.

Of course, it is easy to define a landscape such that this strategy doesn't work. For example, define a fitness such that the optimal correspond to a peak located in a region with a very low fitness and for which another region, far from this optimal peak, has a high average value. This strategy will be inclined to find out a local optimum in the region of high average fitness.
As we shall see later in this paper, the space of states for the sorting network problem has this kind of landscape.

However, the END model also has the ability to maintain a certain *degree of diversity*. This degree of diversity is directly related to the strategy used to manage the evolution of the commitment degree and it allows the model to not converge too quickly.

Finally, an analogy may also be done between the END model and a classical Artificial Intelligence heuristic search technique: *Beam Search* [2]. In this technique, a number of nearly optimal alternatives (the beam) are examined in parallel and some heuristic rules are used to focus only on promising alternatives, therefore keeping the size of the beam as small as possible. This technique proceeds like the END model in the sense that the search space is described by a directed graph in which each node is a state and each edge represents the application of an operator that tranforms a state into a successor state. A solution is a path from an initial state to a goal state.
Therefore, the problem modeling is very similar for the two techniques. However, the main difference is that the END model doesn't use some heuristic rules but only a fitness function that quantifies the degree of quality of a solution. Beam search uses heuristic rules to prune the set of alternatives at each step of the incremental building of a path to a goal state. This pruning is performed by the END model during selection but it is performed in an *auto-adaptive* way: no heuristic is provided to the model.
In fact, the END model has two important advantages compared to such an approach:

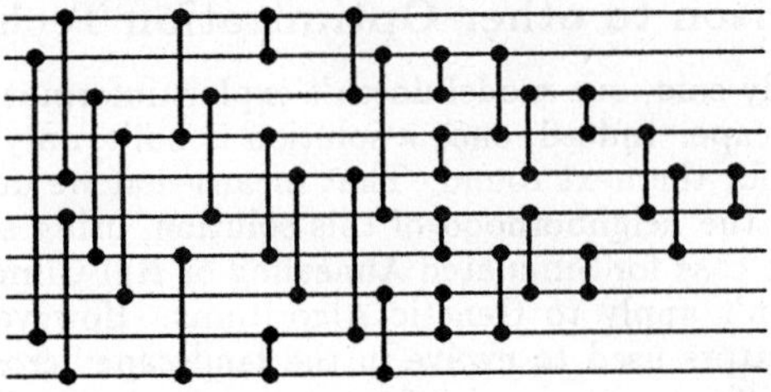

Figure 2: *A 10-input sorting network using 29 comparators and 9 parallel steps.*

- It doesn't need *a priori* knowledge of the topology of the space of states (no heuristics).
- A strategy is evolved by the model itself to find out an optimum, by eliminating "unpromising" species very soon and by maintaining diversity to keep attractive species. Therefore, an auto-adaptive behavior emerges while the model is searching for the optimum.

3 Results for Two Difficult Real-Life Problems

3.1 Sorting Networks

3.1.1 Presentation

An *oblivious comparison-based algorithm* is such that the sequence of comparisons performed is the same for all inputs of any given size. This kind of algorithm received much attention since they allow an implementation as circuits: comparison-swap can be hard-wired. Such an oblivious comparison-based algorithm for sorting n values is called an n-input *sorting network* (a survey of sorting networks research is in [8]).
There is a convenient graphical representation of sorting networks as the one in figure 2 which is a 10-input sorting network (from [8]). Each horizontal line represents an input of the sorting network and each connection between two lines represents a *comparator* which compares the two elements and exchange them if the one on the upper line is larger than the one on the lower line. The input of the sorting network is on the left of the representation. Elements at the output are sorted and the larger element is on the bottom line.
Performance of a sorting network can be measured in two different ways:

1. Its *depth* which is defined as the number of parallel steps that it takes to sort any input, given that in one step disjoint comparison-swap operations can take place simultaneously. Current upper and lower bounds are provided in [10]. Table 1 presents these current bounds on depth for $n \leq 16$.
2. Its *length*, that is the number of comparison-swap used. Optimal sorting networks for $n \leq 8$ are known exactly and are presented in [8] along with the most efficient sorting networks to date for $9 \leq n \leq 16$. Table 2 presents these results.
   The 16-input sorting network has been the most challenging one. Knuth [8] recounts its history as follows. First, in 1962, Bose and Nelson discovered a method for constructing sorting networks that used 65 comparisons and conjectured that it was best possible. Two years later, R. W. Floyd and D.

inputs	1	2	3	4	5	6	7	8	9	10	11	12	13	14	15	16
Upper	0	1	3	3	5	5	6	6	7	7	8	8	9	9	9	9
Lower	0	1	3	3	5	5	6	6	7	7	7	7	7	7	7	7

Table 1: *Current upper and lower bounds on the depth of n-input sorting networks.*

E. Knuth, and independently K. E. Batcher, found a new way to approach the problem and designed a sorting networks using 63 comparisons. Then, a 62-comparator sorting network was found by G. Shapiro in 1969, soon to be followed by M W. Green's 60 comparator network (see [4] and [8]).

inputs	1	2	3	4	5	6	7	8
comparators	0	1	3	5	9	12	16	19
inputs	9	10	11	12	13	14	15	16
comparators	25	29	35	39	45*	51	56	60

Table 2: *Best upper bounds currently known for length of sorting networks. Previously, the best known upper bound for the 13-input problem was* 46

3.1.2 Previous works

Most of the previous works on sorting networks focussed on the 16-input problem.

The first person who used optimization technics to design sorting networks is W. Daniel Hillis. In [5] and [9], he used GAs and then co-evolution to find a 61-comparator, only one more sorting exchange than the construction of Green. However, Hillis considerably reduced the size of the search space since he initialized genes with the first 32 comparators of Green's network. Indeed, since the pattern of the last 28 comparators of Green's construction is not really intuitive, one can think that a better solution exists with the same initial 32 comparators.

In a previous paper, overviewing the END model [13], a 60 comparator sorting network was found out with the same initial conditions as Hillis, that is as good as the best known. Two attempts were also done on the 15-input problem and the model was able to find two 56 comparators sorting networks, again as good as the best known, with no initial conditions.

Ryan [11] applied a Genetic Programming approach to the problem of 9-input sorting networks. Kim Kinnear also used GP, but in the area of adaptive sorting algorithms, to find an algorithm to sort n elements in $O(n^2)$ time [7].

Ian Parberry ([10]) worked on the optimal depth problem and found out optimal values for 9 and 10-input sorting networks using an efficient exhaustive search executed on a supercomputer.

More recently, Gary L. Drescher ([3]) designed a novel fitness function for evolving sorting networks, using GAs. His approach finds quickly and reliably 60-comparator networks, as compact as the best known, again starting with the first 32 comparators of Green's construction.

3.1.3 Non-Deterministic Sorting Network Algorithm

The aim of this non-deterministic algorithm is to generate incrementally and randomly some valid sorting networks. Each organism runs this algorithm (see

```
Begin with an empty or a partial network
DO BEGIN
    Compute the set of significant comparators
    IF (set of significant comparators IS NOT empty)
        Pick one of these comparators randomly and add
        it to the existing partial network
    END_IF
UNTIL (set of significant comparators is empty)
/* A valid and fair sorting network has been generated */
```

Figure 3: Non-deterministic algorithm run by each organism.

figure 3), making some random choices until a valid sorting network is found. A run of this algorithm corresponds to the incremental construction of a path in the tree representing all valid and fair sorting networks; that is, valid sorting networks with no useless comparators.
Valid sorting networks are built using the *zero-one principle*:

Zero-one principle: If a network with n input lines sorts all 2^n sequences of 0's and 1's into nondecreasing order, it will sort any arbitrary sequence of n numbers into nondecreasing order.

Therefore, we only consider all 2^n possible inputs instead of the $n!$ permutations of n distinct numbers to incrementally build a sorting network.

The fitness of a sorting network is defined as its length. However, for the selection process, ties are broken using the depth of the sorting networks. In that way, we also generate some efficient sorting networks regarding the number of parallel steps.

3.1.4 Results

Experiments were performed on a Maspar MP-2 parallel computer. The configuration of our system is composed of $4K$ processors elements (PEs). In the maximal configuration a MP-2 system has $16K$ PEs. Each PE has a 1.8 Mips processor, forty 32-bit registers and 64 kilobytes of RAM. The PEs are arranged in a two-dimensional matrix.
This architecture is particularly well-adapted for our model since it is designed as a two-dimensional (grid) world. Each PE simulates one organism if we want to study a $4K$ population, but it can also simulate more organisms if we want a larger population.

Results for the 16-input problem initialized with the first 32 comparators of Green's sorting network will not be presented. The last version of the model is able to evolve a sorting network as good as the best known with a $4K$ population size and a success rate of almost 100% within 5 to 10 minutes. This time performance is comparable to Drescher's results ([3]) who used a 64-node CM-5 computer. Actually, the interesting comparison between his GAs approach and the END model is that:

- GAs evolved a population of 2^{19} $(= 524,288)$ sorting networks (compared to a 4,096 population size for END),
- 29 to 100 generations are enough for GAs to find the optimum but 150 to 200 generations are often required for END.

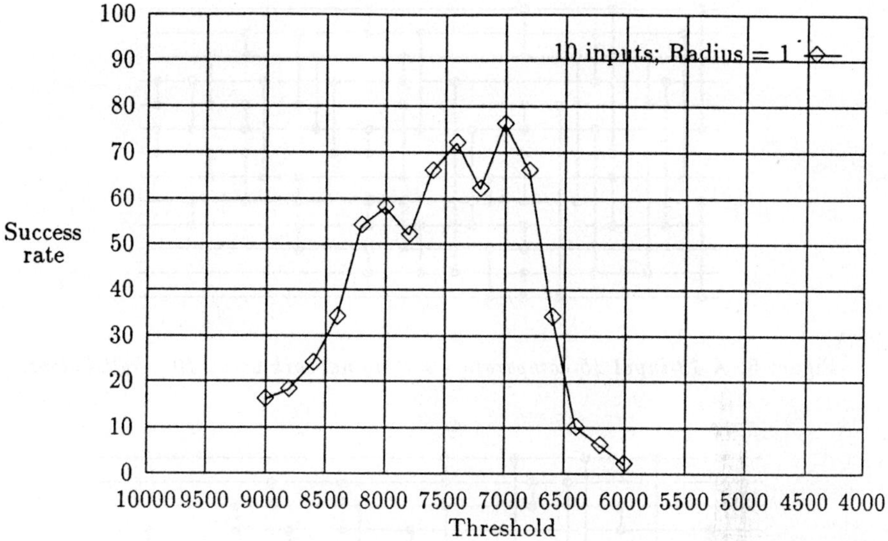

Figure 4: *Success rate for the 10-input sorting network problem versus the threshold value for the disorder measure strategy.*

In the following, only results for evolved sorting networks from scratch (*ie:* with no initialization) are described. In order to increase the probability of finding good sorting networks, we used a $64K$ population size, each processor of the Maspar simulating 16 organisms.

Before presenting results, let us come back to a previous observation. As it has been said in section 2.3, the landscape of the state space for the sorting network problem is such that optimal solutions don't correspond to the species which perform well for lower value of the commitment degree. That means that if we let the model evolve until there are only a few species remaining then the probability that the best solution be found out is very low. This behavior can be oberved in figure 4 which shows the evolution of the success rate as the value of the threshold decreases. First the success rate increases but, once the diversity degree is forced to be lower by decreasing the threshold value (the disorder measure strategy is used), the success rate also decreases.

This explains why one needs to maintain diversity to be able to find out an optimal solution. And of course, the larger the population size, the higher the probability of success.

We have been interested to tackle two different sorting network problems:

1. The 13-input problem because of the large gap between the best known result (46 comparators) and the 12-input one (39 comparators).
2. The 16-input problem because it is the most challenging one.

For the 13-input sorting network problem, we ran the END model 6 times with different values for the selection neighborhood radius (between 1 and 5) and the threshold. Each run was completed in about 8 hours. For 2 runs, we got a sorting network better than the best known. That is, the END model discovered two sorting networks using only 45 comparators, one comparator less than the best current known. Moreover, those two sorting networks use 10 parallel steps

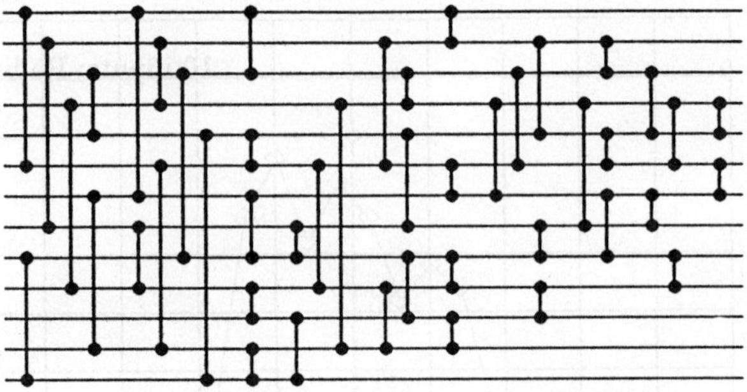

Figure 5: *A 13-input 45-comparators sorting network using 10 parallel steps.*

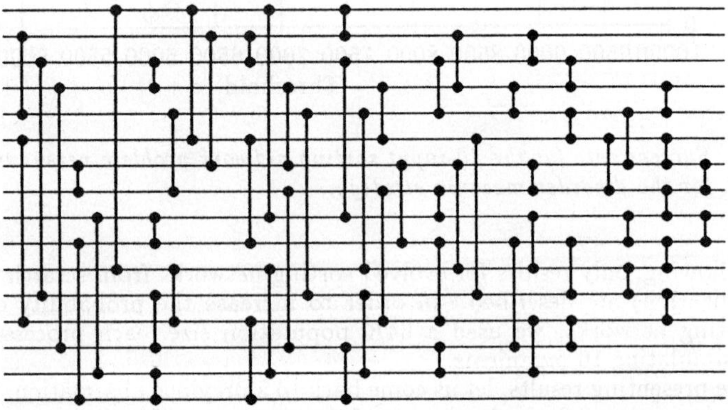

Figure 6: *A 16-input 60-comparators sorting network using 10 parallel steps.*

which is very good since to get smaller delay time one often has to add one or two extra comparator modules ([8]) and the best known delay for the 13-input problem is 9. Figure 5 presents one of these two 13-input sorting networks.

For the 16-input sorting network problem, we ran the END model 3 times. Each run was completed in a period of 48 to 72 hours[1]. For 2 runs, a 60 comparator sorting network was found out, each of them using 10 parallel steps. This is as good as the best human-built sorting network designed by M. W. Green. Those sorting networks were entirely designed from scratch by the END model (*ie:* there was no initial comparators as it was the case the previous times this problem was tackled using computers). Figure 6 presents one of these two 16-input sorting networks.

[1] A new algorithm that doesn't use the zero-one principle but that maintains the set of unsorted vectors using a list of masks has been recently implemented. This algorithm allowed us to divide execution time for the 16-input problem by a factor of about seven. Now, we can get reliable results within an execution time of 12 hours for a run. Using the maximal configuration for the Maspar, a run would take about 3 hours.

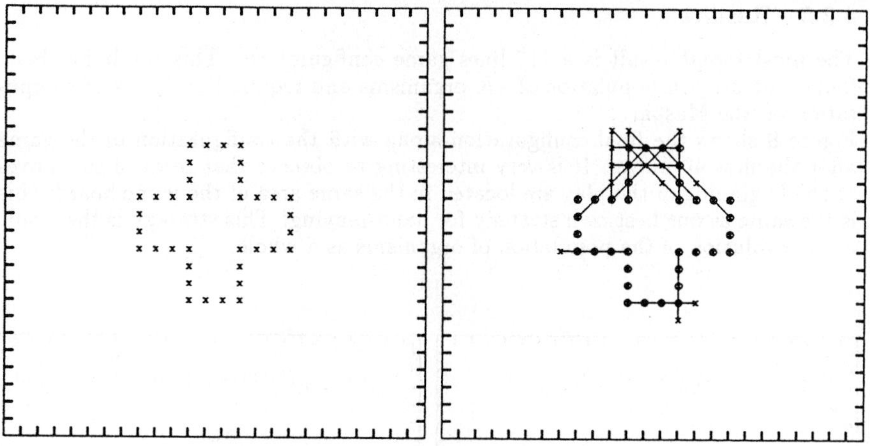

Figure 7: *The initial configuration and a possible configuration after 13 moves for the Solitaire game. For a clearer picture, the grid layout is not drawn but is represented by the rules on the border.*

3.2 A One-Player Game: Solitaire

3.2.1 Presentation of the game

This second problem is an original one and no one has published about it. Therefore, it is not possible to make some comparison. However, it is an interesting problem since it shows how difficult the modelling of a problem can be for other classical optimization techniques.

To play this game, you only need a piece of paper with a grid layout and a pencil. First, the player draws a fixed initial pattern composed of crosses, like the left picture in figure 7. The rule is that a cross can be added at an intersection of the grid layout if and only if it allows the drawing of a line composed of 5 crosses that do not overlap another line. This line may however be the continuation of another line or may cross another line. That is, the new line can share at most one common cross with another line. This new line can be drawn vertically, horizontally or diagonally.
The right picture in figure 7 shows a possible configuration of this game after a few moves. Crosses of the initial pattern are circled to be identified more easily. Now, the goal of this game is simply to draw as many lines as possible!

If this game is played by hand, one can see that a good strategy is difficult to elaborate. After a few plays, a score of 70-80 lines is relatively common. However, to reach 90 lines is less obvious and a score greater than 100 lines is exceptional.

Moreover, it can be proved that the maximum number of moves for this game is finite. However, no tight upper bound has been established for this optimum.

This game is interesting because it is a typical example of a problem wich seems impossible to code using an optimization technique like GAs or Hill-climbing. However, using an incremental approach, the description of solutions to this problem becomes obvious.

3.2.2 Results

The most recent result is a 117 lines game configuration. This result has been found out using a population of $4K$ organisms and required 30 hours of computation on the Maspar.
Figure 8 shows the final configuration along with the configuration of the game after the first 40 moves. It is very interesting to observe that most of the moves at the beginning of the play are located in the same area of the game board: this is the same as our best own strategy for hand-playing! This strategy is the result of the evolution of the population of organisms as a whole.

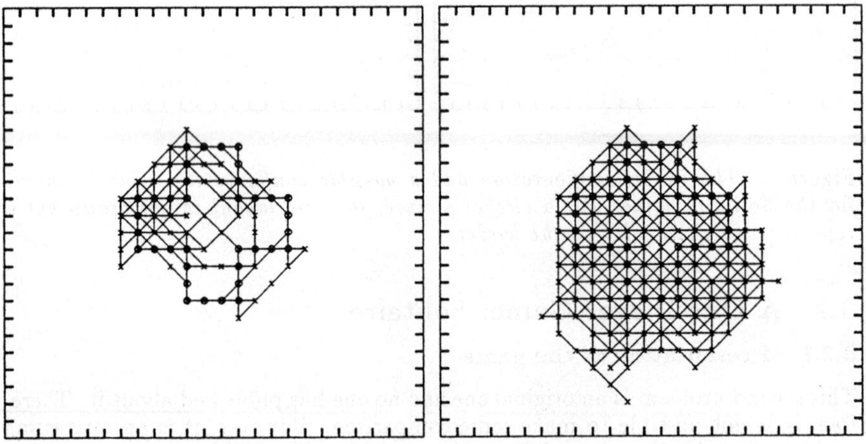

Figure 8: *The best configuration found out for the Solitaire game, using 117 lines (on the right); and the configuration for this best play after 40 moves (on the left).*

4 Conclusion

This paper presented an inventive and very promising technique. By using a new approach for the search in the state space and by constructing solutions incrementally, this model can outperform actual techniques. However, it should be noticed that when the topology of the search space for a given problem is well-known and its gradient is appropriate for Hill-Climbing or SA, these techniques are more efficient than the END model regarding execution time. Indeed, the approach of the END model is a statistical one which progressively takes into account details of the landscape; no gradient information is used. Moreover, analysis of the sorting network problem also reveals that the topology of the sub-space of valid networks makes the use of crossover and mutation critical. Therefore, it is difficult to take advantage of the "building block" mechanism exploited by GAs.

The aim of this paper was to present the parameters of the model. We focussed only on some elementary operators for selection and solution generating in order to identify clearly their importance for the efficiency of the model. The model described in this paper can be easily enhanced with some new features like:

- Allowing the use of some heuristics for solution generating to reduce the number of potential extensions at each node of the tree of solutions.
- Managing a local memory for each organism that would memorize its "past" and would allow learning.
- Each organism could look for a local optimum before selection rounds (using SA for example). When possible, this technique allows a faster convergence.

Moreover, we are currently working to replace the global strategy for the commitment degree management by a local strategy that would be managed by the model itself.

A very interesting advantage of the END model is that, intrinsically, it is highly parallelizable and its performance is related almost linearly to the size of the parallel system used.

Finally, the very encouraging results let us think about improving this approach and applying it on even more challenging problems like:

- multi-player games,
- problem solving,
- mechanical discovery of heuristics or theorems,

for which each organism would be an elementary and naive game player or problem solver. At each step, organisms would take a decision randomly but, as the result of the evolution of the population, some "high-level strategies" could be expected. This last idea needs however to be worked on.

5 Acknowledgments

I would like to thank Jordan Pollack, Patrick Tufts, Martin Cohn and Jacques Cohen for their advice and discussions. I would like also to thank Roger Gallier who challenged me one day with the Solitaire problem. Thanks also to the NSF whose grant allowed the Brandeis Computer Science to buy a Maspar computer. Finally, I want to thank my wife, Anne, for the moral support she provided me while I was working on this project and her constant curiosity.

References

[1] Richard Belew and Thomas Kammeyer: *Evolving Aesthetic Sorting Networks using Developmental Grammars.* In Proceedings of the Fifth International Conference of Genetic Algorithms.

[2] Roberto Bisiani: *Beam Search.* In Encyclopedia of Artificial Intelligence, Vol. 2, Second Edition, John Wiley & Sons, 1992.

[3] Gary L. Drescher: *Evolution of 16-Number Sorting Networks Revisited.* Submitted.

[4] Milton W. Green: *Some Improvements in Nonadaptive Sorting Algorithms.* Stanford Research Institute. Menlo Park, California.

[5] W. Daniel Hillis: *Co-Evolving Parasites Improve Simulated Evolution as an Optimization Procedure.* In Artificial Life II, Langton, et al, Eds. Addison Wesley, 1992, pp. 313-324.

[6] Hugues Juillé: *Evolving Non-Determinism: An Inventive and Efficient Tool for Optimization and Discovery of Strategies.* Draft paper, Computer Science Departement, Brandeis University, 27 pp.

[7] Kim Kinnear: *Generality and Difficulty in GP: Evolving a Sort.* In Proceedings of the Fifth International Conference on Genetic Algorithms, S. Forrest; Morgan Kaufmann Publishers, 1993.

[8] Donald E. Knuth: *The Art of Computer Programming: Volume 3 - Sorting and Searching.* Addison Wesley, 1973.

[9] Steven Levy: *Artificial Life: the Quest for a New Creation.*Pantheon Books, 1992.

[10] Ian Parberry: *A Computer-Assisted Optimal Depth Lower Bound for Nine-Input Sorting Networks.* In Mathematical Systems Theory, No 24, 1991, pp. 101-116.

[11] Conor Ryan: *Pygmies and Civil Servants.* In Advances in Genetic Programming, Kim Kinnear, Ed. MIT Press, 1994.

[12] Nicol N. Schraudolph and Richard K. Belew: *Dynamic Parameter Encoding for Genetic Algorithms.* In Machine Learning, Vol. 9, 1992, pp. 9-21.

[13] Patrick Tufts and Hugues Juillé: *Evolving Non-Deterministic Algorithms for Efficient Sorting Networks.* Submitted.

Symbiosis and Co-evolution in Animats

Chisato Numaoka

Sony Computer Science Laboratory Inc.
Takanawa Muse Building, 3-14-13 Higashi-gotanda
Shinagawa-ku, Tokyo, 141 Japan
TEL: +81-3-5448-4380 FAX: +81-3-5448-4273
E-mail: chisato@csl.sony.co.jp

Abstract. This paper investigates the dynamics of co-evolution of various types of animats. For this purpose, we use a problem we call the Blind Hunger Dilemma. This problem investigates the emergence of collective behavior concerned with shared resource. Our model of co-evolution is inspired by Margulis's idea of parasitism and symbiosis which she applies to the origin of the current eukaryotic cell. Originally, the environment we used has only one type of animat. By the invasion of "bacteria," animats will develop new sensing functionality. Through many generations, they co-evolve in the environment. Our results shows that animats that contributed to the improvement of mean fitness do not necessarily dominate the population. In this paper, we discuss the reasons within the context of biological study.

1 Introduction

Natural living things have evolved over 4 billion years by increasing biological variation and by natural selection. This is a well-known fact that many biologists have tried to substantiate from various viewpoints since Charles Darwin wrote his book "On the Origin of Species by Means of Natural Selection or the Preservation of Favoured Races in the Struggle for Life" in 1859, although there still remain controversial points.

Symbiosis. Lynn Margulis has the idea that parasitism and symbiosis were the driving forces in the evolution of cellular complexity. Her enthusiasm to prove the assumption biologically seems to have almost succeeded. Her assumption was that the main internal structure of eukaryotic cells did not originate within the cells but are descended from independent living creatures which invaded the cells from outside like carriers of an infectious disease [5]. The point is that the invading creatures and their hosts gradually evolved into a relationship of mutual dependence so that the erstwhile disease organism became by degrees a chronic parasite, a symbiotic partner and finally an indispensable part of the substance of the host.

According to the result of recent biological studies including Jeon's observation of the symbiotic interactions between amoeba and x-bacteria [3], it has been virtually possible to prove that mitochondria and plastids originate from organisms, which may be bacteria, living outside of the host cell. This is the ultimate type of endosymbiosis. Mitochondria and plastids cannot exist out of the host cell even if their chromosome has the information to create some important proteins that they need. They share their metabolic cycle with the host cell. Therefore they are no longer separate organisms, although the host cell had experienced a big jump in its evolutionary process with this invasion by the ancestors of mitochondria and plastids.

Co-evolution. Despite what the origin of current eukaryotic cells was, we suspect that the invaded cells would have to coexist with other creatures including the ancestors of current eukaryotic cells or invading bacteria. In order for the invaded cells to survive in such an environment, proper environmental conditions would be required. One of the conditions is that the cell's genes are grouped since the cells probably did not have a system for replicating genes accurately, as Sonea mentioned [10]. Another condition is that cells would have to be able to protect themselves against another invasion as securely as possible in order to establish a parasite-host relationship. This coexistence results in co-evolution of those which share their nutrition or nutritious sources.

Our approach. In this paper, we attempt to investigate how a property that a creature has accidentally acquired, in the way that ancestors of the current eukaryotic cell have acquired new functionality by combining with a parasitic bacteria in a metabolic level, will be preserved while they coexist with other types of organisms. In this papar, we do not look at the problem of how this closed relation between a parasite and a host cell has been established. What we are concerned with is the conditions needed to preserve creatures that acquired a new property by chance if it is effective for them to adapt to their environment.

Even if a property is useful, when the population of the organisms that acquired the property is small and there is another creature completely dominating the population, it is very difficult for this minority population which has acquired the new property to survive in such an environment. Therefore, the first condition must be that they are able to increase their population.

In order to investigate the dynamics of co-evolution of the original creatures and those that accidentally, by creating symbiotic relationships, acquired new functionality, we use a problem which we call the "Blind Hunger Dilemma" [8]. The problem was proposed for the purpose of investigating emergent collective behavior involved in the exclusive use of shared critical resources. Using this problem, we can investigate the dynamics of co-evolution of animats of different types sharing their nutritious source and a working environment which is limited in size.

We provided an animat that has four types of sensors; one for detecting the strength of a globally-distributed potential such as sound, smell, or light, one for pressure, one for forces of attraction, and one for viscous forces. In the previous studies [9], we supposed that every animat has all these sensing capabilities.

In this paper, we assume that there are a collection of animats that have only sensing capabilities to detect the strength of a globally-distributed potential and pressure. In addition, we assume that two types of bacteria exist and that they will invade the animats at some fixed rate. One type of bacteria will add the capability of detecting forces of attraction to host animats. The other type of bacteria will provide the capability of detecting viscous forces to host animats. We also assume that host animats will loose these capabilities by means of a mutation that results in fatal damage to part of a chromosome that originated from the parasitic bacteria, or by means of invasion by other types of animats which insert their chromosome into the invaded animat.

Organization of this paper. Again, what we investigate in this paper is the dynamics of co-evolution of four types of animat; one that is not invaded, one that has acquired the capability to detect forces of attraction, one that has acquired a capability to detect viscous forces, and one that has been invaded by both types of bacteria. The next section revisits the Blind Hunger Dilemma for demonstrating what was investigated with this problem. In Section 3, we define what an animat is and describe the experiment performed. We show the results of the experiment with some discussion in Section 4.

2 The Blind Hunger Dilemma

In [8], we introduced a problem that we call the Blind Hunger Dilemma. Here is an overview of the problem. There is a single energy supply base in an environment where many animats live. The energy base provides a limited number of facilities at each of which a single animat can dock to replenish its energy. The energy base emits a sound or a smell by which animats can recognize its location.

Animats, when they are hungry, go directly to the energy base and replenish their energy if they can reach one of the facilities. The problem happens when many animats simultaneously approach the energy base. While animats are replenishing, other animats will swarm around those replenishing animats.

As a result, even if they finish replenishment, they have no space to leave the energy base since they are surrounded by other swarming animats. On the other hand, hungry animats cannot go closer to the base since there are re-charged animats that cannot move away. This is the problematic situation that we call the Blind Hunger Dilemma (See Fig. 1 (b)). Provided there is an appropriate environment with queues or some locking mechanisms, this type of problem will never happen. It can also be avoided, if animats have developed or learned a certain ordered pattern of behavior to avoid as many collisions as possible (See Fig. 1 (a)).

As described in [9], this problem is concerned with the exclusive use of shared resources, the issue of locality with respect to the experiences of animats, real-time problem solving, and tasks that cannot be delegated to other animats.

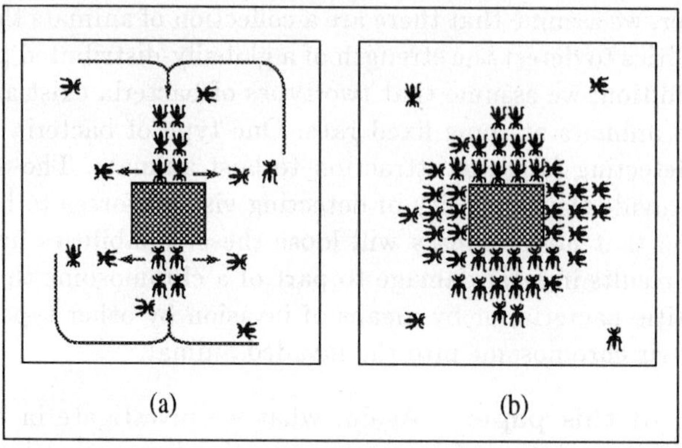

Fig. 1. Energy Replenishment

In a sense, conflict concerned with shared resources is the destiny of all species in ecological systems. In order to survive this conflict, various species have repeatedly evolved. Although the Blind Hunger Dilemma was proposed in the context of artificial systems concerned with shared resources, we believe that this problem is still effective even in the context of natural systems. In this paper, by using this problem as an experimental instrument, we will investigate the dynamics of co-evolution coupled with the notion of symbiosis.

3 Co-evolution of Animats

3.1 Animats

We will omit details about how animats choose the next direction to move in and explain only the part related to the later sections. We recommend reading [9] for those who are interested in the details of the definition of animats in our study.

The environment where the animats live is equipped with a single energy base around which are provided 8 facilities in 8 directions at each of which a single robot can dock. We suppose that the energy base emits a special smell. The energy base produces its own potential field which represents its smell and the animats have an ability to sense the strength of this potential field at any position.

The directions in which animats can move are quantized to 8; North, North-East, East, South-East, South, South-West, West, and North-West. Animats decide the next direction to move in by sensing the global force and some local forces. In the case of our animat, the source of the global force is the energy supply base.

In this study, we are concerned with the co-evolution of animats which differ in their capability to sense local forces. Potentially, they can sense the following three types of forces applied in eight directions. Note that, in the following explanation, K_P, K_A, and K_V are constants and $f_i^P(x,y,t)$, $f_i^A(x,y,t)$, and $f_i^V(x,y,t)$ are particular functions. The details of local forces are described in [9].

The first force is a component of inward pressure, $F_i^P(x,y,t) = K_P f_i^P(x,y,t)$. When an animat faces another animat and both try to exchange their position, each animat experiences a component of pressure from the other animat. In this case, they will both decrease the probability to move in that direction. When surrounded by many animats, this animat will experience the sum of these components as an inward pressure acting on itself.

The second force is an attraction, $F_i^A(x,y,t) = K_A f_i^A(x,y,t)$. Suppose at time $t-1$ an animat is surrounded by 7 animats and only the north direction is open for the animat to move. At time t, the animat located at the center moved to the north and the other 7 animats did not move. In this case, these 7 animats experience a force of attraction so that they are attracted to the central location.

The third force is viscous, $F_i^V(x,y,t) = K_V f_i^V(x,y,t)$. Under this force, an animat is encouraged to move parallel to an adjacent animat. For example, an animat will experience a viscous force on its west side if another animat, adjacent to its west side, decides to move north, and an animat in the central location experiences a viscous force on its west side toward the north.

In our study, we begin with a situation where all animats have, with respect to local forces, only a capability to sense inward pressure. As mentioned in Section 1, as a result of the invasion of a type of bacteria, animats may have other sensing capabilities withdrawn from their potential. In the next section, we will introduce a model of co-evolution and explain our experiments.

3.2 A Model of Co-evolution

Invasion by adjacent animats. Our model is inspired by the way that viruses invade a cell by inserting chromosome into the cell and clone themselves using various chemical materials inside the cell. We suppose that, if the amount of energy becomes lower than a certain level, one of the adjacent animats will invade the weakened animat with a certain probability. We can think of this as follows: if an animat turns out to be weakened, all adjacent animats will try to invade. However, these invasions have a certain probability of succeeding. Once they have invaded, their chromosomes pooled inside the invaded animat together with its original chromosome. Then, from within this pool, only one chromosome is chosen and others are discarded.

Bacteria. We suppose that there are two types of bacteria that can withdraw one of the potential sensing capabilities from a host animat by invading and confirming a symbiotic relationship with the host animat. This invasion will occur at a certain rate.

Chromosomes. The behavior pattern of our animats is characterized by two parameters α and β. α and β are, by definition, $\frac{K_A}{K_P}$ and $\frac{K_V}{K_P}$, respectively. Parameters K_P, K_A, and K_V are those used in the force definitions in the previous section. Animats generate two basic random floating values in the range 0.0 to 1.0 for α and β. We encode α and β to an 8 bit sequence "gene" and combine two genes as a "chromosome". Every animat has a parameter K_P whose value is fixed at 3000.0. The values of the other two parameters K_A and K_V are calculated in terms of α, β, and K_P.

Animats have a single chromosome by themselves. The chromosome is expressed as an n-bit sequence in which 16 bits are used for two genes representing the two parameters K_A and K_B.

Invasion mechanism. If an animat feels hungry, it has to get energy within a certain time. Otherwise, the animat will die for lack of energy. When an animat is weakened, it may be invaded by a single animat which inserts its chromosome into the weakened animat. There are two possibilities with which adjacent animat will invade; either the one which has the best fitness invades, or adjacent animats invade in probability in proportion to their fitness. We call the former *elite preservation* and the latter *roulette selection.* We call these two *invasion types.*

Fitness function. Fitness is calculated by using the following fitness function:

$$Fitness = \#Reprenishments - < \#Collisions > . \quad (1)$$

Here, $\#Reprenishments$ is the number of replenishment experienced during the lifetime of an animat with a chromosome. $< \#Collisions >$ is the mean number of collisions during its lifetime.

Recreation of a new chromosome. As described above, one adjacent animat will invade by replacing the chromosome of the invaded animat with the invader's chromosome. This invasion occurs in a certain rate. We defined an *invasion rate* (henceforth IR) as the probability that any one adjacent animat succeeds in copying its own chromosome to a newly born animat by reusing the body of the invaded animat. While it is copied, a certain rate of mutation is applied to every bit of the chromosome. We call this the *mutation rate* (henceforth, MR).

3.3 Experiment

Simulation Environment. Experiments were done in a 15×15 lattice space with 100 animats, which move one step at a time in one of the 8 directions around them. We made simulations for both elite preservation and roulette selection by means of varying IR and MR. We chose four values 0.0, 0.5, 0.9, and 1.0 for IR and a value 0.002 for MR. By arranging a single energy base at (7, 7), 300,000 steps were conducted for every simulation with different combinations of invasion type, IR, and MR.

We can have four types of animats: NEITHER, ONLY-ALPHA, ONLY-BETA, and BOTH. NEITHER type animats are sensitive only to inward pressure. ONLY-ALPHA type animats have another sensor for the attraction force. ONLY-BETA type animats become sensitive to the viscous force. BOTH type animats provide sensitivity to all three forces: inward pressure, attraction force, and viscous force.

Experiment 1: Co-evolution of Four Types of Animats. In this experiment, we investigated the dynamics of co-evolution starting from a situation where only animats of NEITHER type exist in the environment.

Animats have a single chromosome by themselves. The chromosome is expressed as an 18-bit integer sequence; The top two bits indicate which type of bacteria made a symbiotic relationship. The rest of the 16 bits are used for the two parameters of the animat.

Experiment 2: Co-evolution in light of Invasion Types. In this experiment, we investigated the effect of co-evolution around invasion types for co-evolution of four types of animats. We started simulations from a situation where only animats of NEITHER type exist in the environment and where they have a roulette selection as their invasion type.

Beside the above 18 bits, the chromosome has a further two high-order bits to represent the invasion type, elite preservation or roulette selection. As a result, a chromosome is composed of a 20-bit sequence.

4 Results and Discussion

4.1 The Results of Experiment 1

We show two examples of the results; one is for the elite preservation type of invasion (Fig. 2) and the other is for roulette selection (Fig. 3). In fact, we were not able to find any evident distinctions between the two invasion types with respect to their evolutionary dynamics. Two general characteristics irrespective of invasion type came out of Experiment 1. Firstly, if the period during which animats of NEITHER type dominate the population lasts longer, neither animats of ONLY-ALPHA or BOTH type can increase their population. Rather, animats of NEITHER and ONLY-BETA will alternately dominate the population. Secondly, in the case of roulette selection, we always observed the tendency for BOTH type animats to dominate the population after ONLY-ALPHA type animats dominate the population. The same observation was also made in the case of elite preservation when domination by the NEITHER type terminated quickly. We could not confirm the reason why this happens, and it requires further study.

4.2 The Results of Experiment 2

Fig. 4 shows one set of results for Experiment 2. We started this experiment from a situation where all animats have the roulette selection type of invasion. Since

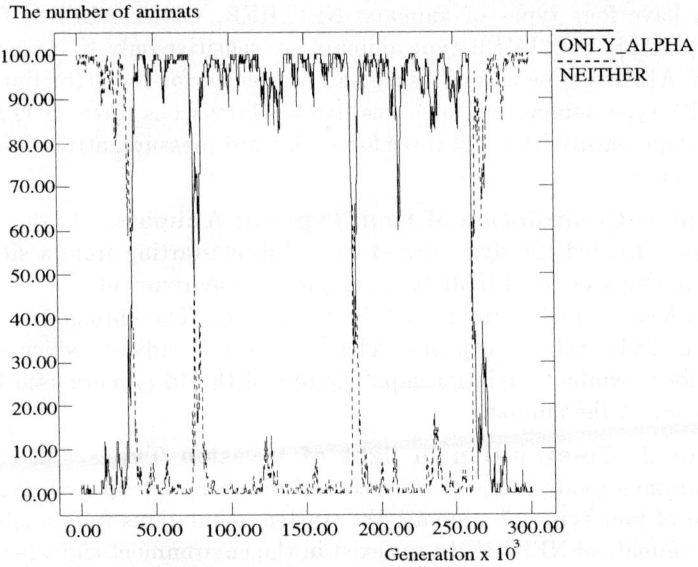

Fig. 2. A process of co-evolution (1): This is the result of a simulation where the invasion type is elite preservation, $IR = 0.5$, and $MR = 0.002$. In this figure, the numbers of only two types of animats, NEITHER and ONLY-ALPHA, are depicted since these two are alternately dominating the population.

there is no clear and general distinction between the evolutionary dynamics of the two invasion types, there is no reason for animats with the elite preservation invasion type to dominate the population. Nevertheless, one fact that we found is that elite preservation tends to dominate the population while animats of ONLY-ALPHA and BOTH types are alternately dominating the population.

Our conjecture is as follows. We observed the following three facts. At first, animats of ONLY-ALPHA or BOTH type directly contribute to raising the mean fitness. Secondly, the elite preservation invasion type directly contributes to preserving a chromosome of animats that have achieved a higher degree of fitness. Thirdly, when animats of roulette selection invasion type are dominating the population, there is a tendency for ONLY-ALPHA type animats and BOTH type animats to alternately dominate the population as described in the results of Experiment 2. As a result, the period during which animats of these two types are dominant is relatively long. From these observations, we conjecture that the elite preservation tends to appear while animats of ONLY-ALPHA or BOTH type dominate in order to preserve better chromosomes. Again, this is also a hypothesis. We need to confirm this fact by investigating in detail.

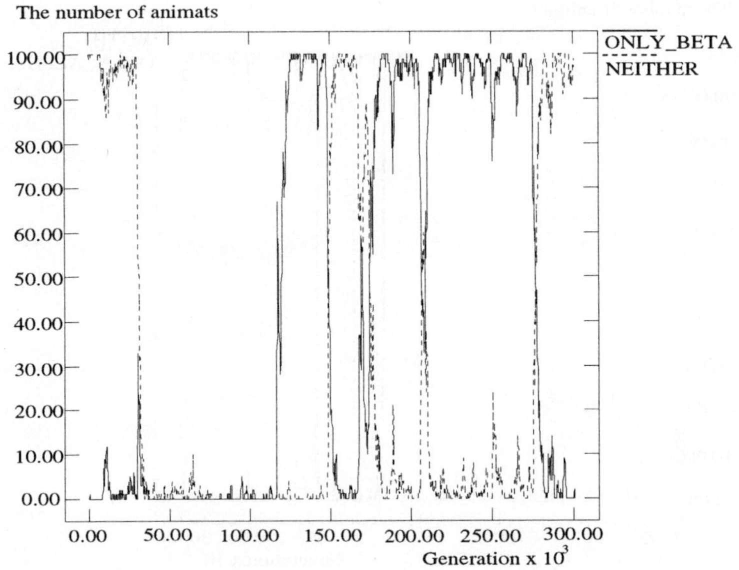

Fig. 3. A process of co-evolution: This is the result of a simulation where the invasion type is roulette selection, $IR = 0.5$, and $MR = 0.002$. In this figure, the numbers of only two types of animats, NEITHER and ONLY-BETA, are depicted since these two are alternately dominating the population.

4.3 The Direction of Evolution

In our experiment, we observed that, if a population of animats of either ONLY-ALPHA or BOTH type is dominant, the mean fitness is raised. Therefore, the direction of evolution seems to tend towards increasing the population of either of the above types. Nevertheless, it was still very difficult for these types of animats to increase their population in our experiment. There are a couple of conceivable explanations.

Population domination by animats of one type. The first explanation is that, in a situation where animats of either NEITHER or ONLY-BETA type are dominant, a small number of animats of either ONLY-ALPHA or BOTH type cannot survive. NEITHER type animats are only sensitive to inward pressure from other animats. This type of animat is very close to ALL-C animats in our previous work [8], which always attempt to avoid a collision with other animats. As described in [8], ALL-C animats have good survival properties. They can reach the energy base with almost equal opportunity although their waiting time is longer than the ideal case. Therefore, as long as they dominate the animat population, other types of animats find it difficult to survive.

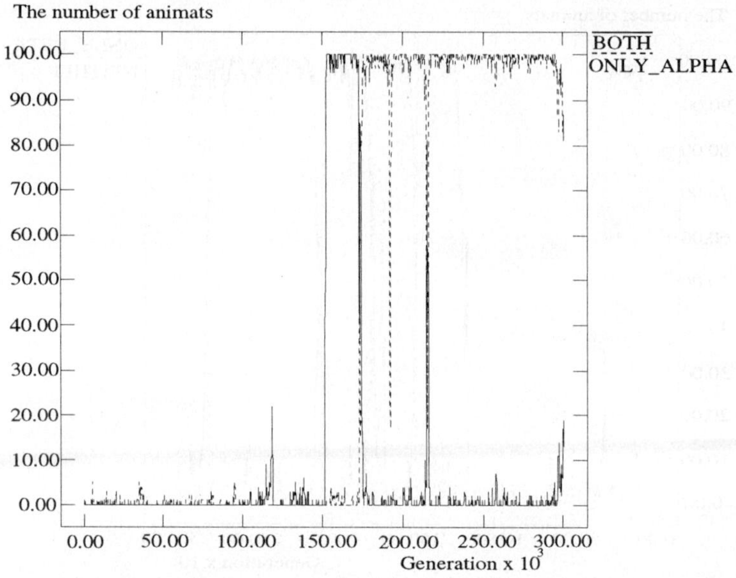

Fig. 4. A process of co-evolution with change of invasion type: This is the result of a simulation where $IR = 1.0$, and $MR = 0.002$. In this figure, the numbers of only two types of animats, BOTH and ONLY-ALPHA, are depicted since these two are alternately and reciprocally dominating the population. This tendency of reciprocal domination is typical of Experiment 2 and the same tendency was observed in the cases of $IR = 0.9$ and 0.5 as well.

Two parameters as introns. The second explanation may be related to the neutral mutation in a so-called *intron*. In the neutral theory of molecular evolution [4], Kimura claims that some mutated genes can spread through a collection of individuals even if they are not as well disposed to natural selection. His theory says that mutation occurs ubiquitously at the level of amino acid. Although the probability of causing mutation is the same for every amino acid, many mutated genes are not exposed to natural selection. These genes are located in introns. Introns do not function in coding for protein synthesis. As long as it is not involved in protein synthesis, it is not evaluated at the phenotype level. Natural selection always applies to the phenotype. Therefore, introns are neutral to natural selection.

Two genes coding parameters α and β work just like introns for animats of NEITHER type. If MR is greater than 0, these genes will flip the value of every bit (0 or 1) with a probability of MR. These genes become effective if a corresponding bacterium invades an animat of NEITHER type. The same thing applies to animats of either ONLY-ALPHA or ONLY-BETA type. For ONLY-ALPHA animats, the β gene is the intron whereas, for ONLY-BETA animats, the α gene is the intron.

When a type of bacteria invades an animat as a symbiont, these introns become effective and, if the genes that were introns have a higher fitness value, they will contribute to the survival of the animat in that environment. In contrast, if the value is not appropriate, the chances of the animat surviving will be decreased. This hypothesis is related to so-called *genetic drift* [1].

These are the explanations that we believe to be most plausible, although they are hypothesises and we have not as yet conducted any experiments to confirm them. This is a subject that we need to address using suitable experiments.

5 Conclusion

In this paper, using a problem called the Blind Hunger Dilemma, we have investigated the dynamics of the co-evolution of animats of different types, the differences the animats acquired by having a symbiotic relationship with either or both of two types of " bacteria."

Margulis's theory of symbiosis inspired us to conduct this research. However, we have ignored a very critical point; that is the process of establishing a metabolic cycle between a symbiont (bacterium) and an ancestor of the eukaryotic cell. This is because our focus in this paper is on the co-evolution of animats of different types. We are using the metaphor of symbiosis as a mechanism to produce a new functionality of animats that, we assume, emerge in the metabolic cycle together with a symbiont.

Margulis supposes that a rapid increase of O_2 molecules from 0.0001 % to 21 % [2] in the atmosphere of the earth (biosphere) 2,200 million years ago would have triggered the establishment of symbiotic relations with a kind of bacteria inside an ancestor of eukaryotic cells. It is believed that the bacteria invented aerobic or oxygen-using respiration [6]. In fact, to establish such a symbiotic relationship with such bacteria would have caused a great improvement in fitness of the ancestor of the eukaryotic cell. Even in our case, ONLY-ALPHA or BOTH type animats, which are dominant in the population, improve fitness.

We admit the claim that our way of study may be ad hoc and it lacks justifications supported by biologically certified facts. Nevertheless, we believe that the results of our experiment can show some truths of natural evolution, especially co-evolution, among species (i.e. animats of different types) which share nutritious sources (in our case, the energy base). The activity of ecological systems is well known (James Lovelock's Gaia hypothesis is an example to know their activity). The important factor in ecological systems is the relationship between species in light of their nutritious sources.

---

[1] As a related matter, we may view a relationship as *epistasis* between an invasion of either bacterium and a gene related to the invasion. This is because an invasion of either bacterium determines whether its related gene (α or β) is used or not. From this viewpoint, the length of chromosomes varies depending on invasions of bacteria. This is the same characteristic as SAGA (Species Adaptation Genetic Algorithms) demonstrates [2].

[2] This percentage of O_2 is the same as what we have now.

Many researches into co-evolution view the evolutionary process as a zero-sum game as in pursuit-evasion contests [7]. However, as Lynn Margulis and Dorion Sagan also point out, evolution is not a bloody struggle in which only the strong survive. This is a misunderstanding of natural selection, which the 19th century philosopher Spencer mistakenly named "survival of the fittest".

We already have a good example of a non zero-sum game, the Prisoner's Dilemma [1]. We can view the Blind Hunger Dilemma as another example of a non zero-sum game. In this game, animats do not differentiate between players. Furthermore, a game is implicitly and asynchronously played. This paper is just a beginning of the study of symbiotically triggered co-evolution. Many things have yet to be investigated. These are the subject of future work.

References

1. Robert Axelrod. *The Evolution of Cooperation.* Basic Books Inc., 1984.
2. Inman Harvey. Species Adaptation Genetic Algorithms: A Basis for a Continuing SAGA. In Francisco J. Varela and Paul Bourgine, editors, *Proceedings of the First European Conference on Artificial Life: Toward a Practice of Autonomous Systems*, pp. 346–354. The MIT Press/Elsevier, 1992.
3. Kwang W. Jeon. Amoeba and x-Bacteria: Symbiont Acquisition and Possible Species Change. In Lynn Margulis and René Fester, editors, *Symbiosis as a Source of Evolutionary Innovation*, pp. 118–131. The MIT Press/Elsevier, 1991.
4. Motoo Kimura. *The Neutral Theory of Molecular Evolution.* Cambridge University Press, 1983.
5. Lynn Margulis. *Symbiosis in Cell Evolution.* Freeman, 1981.
6. Lynn Margulis and Dorion Sagan. *Microcosmos.* A Touchstone Book, 1986.
7. Geoffrey F. Miller and Dave Cliff. Protean Behavior in Dynamic Games: Arguments for the co-evolution of pursuit-evasion tactics. In Jean-Arcady Meyer and Stewart W. Wilson, editors, *Proceedings of the Third International Conference on Simulation of Adaptive Behavior: From animals to animats 3*, pp. 411–419. The MIT Press/Elsevier, 1994.
8. Chisato Numaoka. Blind Hunger Dilemma: An Emergent Collective Behavior from Conflicts. In *Proceedings of the From Perception to Action Conference (PerAc'94)*. IEEE Computer Society Press, September 1994.
9. Chisato Numaoka. Introducing Blind Hunger Dilemma: Agents' Properties and Performance. In *Proceedings of the First International Conference on Multiagent Systems (ICMAS'95)*. AAAI Press, June 1994.
10. Sorin Sonea. Bacterial Evolution without Speciation. In Lynn Margulis and René Fester, editors, *Symbiosis as a Source of Evolutionary Innovation*, pp. 95–105. The MIT Press/Elsevier, 1991.

Artificial Endosymbiosis

Lawrence Bull and Terence C Fogarty
email {l_bull, tcf}@btc.uwe.ac.uk
Faculty of Computer Studies and Mathematics

A G Pipe
email ag-pipe@csd.uwe.ac.uk
Faculty of Engineering

University of the West of England,
Bristol, BS16 1QY, England

Abstract. Symbiosis is the phenomenon in which organisms of different species live together in close association, resulting in a raised level of fitness for one or more of the organisms. Endosymbiosis is the name given to symbiotic relationships in which partners are contained within a host partner. In this paper we use a simulated model of coevolution to examine endosymbiosis and its effect on the evolutionary performance of the partners involved. We are then able to suggest the conditions under which endosymbioses are more likely to occur and why; we find they emerge between organisms within a window of their respective "chaotic gas regimes" and hence that the association represents a more stable state for the partners. An endosymbiosis' effect on its other ecological partners' evolution is also examined. The results are used as grounds for allowing endosymbioses to emerge within artificial coevolutionary multi-agent systems.

1 Introduction

Symbioses are commonplace in the natural world and it is therefore argued that the phenomenon is of great evolutionary significance (e.g. [24]). In this paper we look at a particular form of symbiosis, that symbiologists term "endosymbiosis", the symbiotic association in which partners exist within a host partner, and compare its evolutionary progress to the equivalent association where the partners do not become so closely integrated. We use a version of Kauffman and Johnsen's [14] genetics-based NKC model, which allows the systematic alteration of various aspects of a coevolving environment, to show that the effective unification of organisms via this "megamutation" [8] will take place under certain conditions. Kauffman and Johnsen used this model to examine the dynamics of heterogeneous coevolution. Initially in our model potential endosymbionts exist as cooperating symbionts evolving within their respective separate populations, i.e. as heterogeneous species. During simulations we apply a megamutation operator to the members of these separate populations, causing them to become hereditary endosymbionts, forming their own sub-population, from within which they then evolve. The members of this sub-population compete against each other and members of the original cooperating symbiotic species for existence. Our results indicate that the succesful formation of an endosymbiotic sub-population occurs when the partners are, to use Packard's [27] terminology, away from their ideal liquid regimes - their edges of chaos - and are within their respective chaotic gas regimes. Endosymbiosis can therefore be seen as a mechanism by which organisms can stabilise the environment in which they evolve. We also examine the effect organisms forming endosymbiotic associations has on the other species with which they coevolve and find that it also helps the evolutionary performance of these ecologically coupled partners.

The paper is arranged as follows: section 2 describes symbiosis and endosymbiosis. Section 3 describes the simulated model of coevolution used in this work. Section 4 contains the results of our examination of endosymbiosis and in section 5 we discuss their relevance within the context of artificial evolutionary multi-agent systems in general.

2 Symbiosis

"[Symbiosis] is an extremely widespread phenomenon. A very significiant proportion, probably most, of the world's biomass depends upon it - for example the dominant organisms of all grasslands and forests ... and corals ... are mutualists" [2].

In 1879 the German biologist Anton De Bary introduced the term symbiosis to describe the living together of differently named organisms. Although De Bary used the term to include parasitism, where one organism benefits to the detriment of the other(s) involved, it is more commonly used to describe associations in which none of the partners are adversely affected. There are two categories of the thus defined symbiotic relationships: commensalism, where accrued benefit is one-sided and in which the other partner is neither harmed nor benefited; and mutualism, where all species benefit from the relationship. The strength of the association is categorised as either facultive, where the partner gains benefit but is not dependent upon the other(s), or obligate, where the partner is totally dependent upon the other(s). Here the term benefit is normally taken to mean a net benefit resulting in an overall improvement in an organisms food source, growth/reproduction rate, or conduciveness of its environment (see [31] for an alternative view) - i.e. an increase in its fitness [19] - as some interactions of the association may be costly to a symbiont. However some symbiologists stipulate that the association must persist for all, or a significiant part, of an organisms life span for it to be termed a symbiosis [22]. In all cases the forming of a symbiotic association, known as integration, has been described as resulting in "the display of structures, functions, etc., which are more than and different from, those of which the participants are capable as individuals" [20]. "The value of symbiosis is defined by the fact that, upon entering into an association, an organism becomes better adapted to the environment because of the use it makes of peculiarities already possesed by its partner" [12].

The most "intimate" [7] of symbiotic associations is termed endosymbiosis, in which one of the partners, the "host", incorporates the other(s) internally (purely external associations are termed ectosymbioses). Endosymbionts can occur within their host's cells (intracellular), or outside them (extracellular). Extracellular endosymbionts can be either between the cells of host tissue (intercellular), or in an internal cavity, such as the gut. Intracellular endosymbionts are usually enclosed by a host membrane (there may only be a single membrane separating host and endosymbiont cytoplasm). Burns [5] suggests that "most mutualisms exist as a solution to unpredictable environment variability" and the evolution of endosymbiotic associations can therefore be viewed as the logical outcome of such a process - he cites how anaerobic microrganisms "endosymbiotic in the cytoplasm of aerobes ... do not experience the present day atmosphere, which from their perspective is polluted with oxygen". Law [17] however suggests endosymbionts still indirectly experience environmental factors.

A large number of endosymbiotic associations are hereditary, where the host's endosymbiont(s) is(are) passed directly to offspring. The mechanism for this perpetuation ranges from transmission in the egg cytoplasm (transovarial transmission), e.g. in insects, to offspring ingesting the endosymbiont(s) shortly after birth, e.g. when cows lick their calves they pass on their rumen ciliates. From a genetic point of view, this mechanism of transferring information from one generation to the next is "twice as efficient in the case of symbiosis than Mendelian heredity, since half the [hosts] (the females) transmit [endosymbionts] to all progeny" [25]. Indeed the apparent correspondence between (meiotic) sexual reproduction and symbiosis to form new individuals has been highlighted by Margulis; "symbiosis and meiotic sexuality entail the formation of new individuals that carry genes from more than a single parent. In organisms that develop from sexual fusion of cells from two parents, these parents share very recent common ancestors; partners in symbioses have more distant ancestors"[24].

If symbiosis is viewed as organisms acquiring the properties of others, then endosymbiotic relationships particulary may be perceived as the direct inheritance of acquired characteristics - i.e. as a Lamarckian-like process, though this could also be said of all symbioses to some extent. Nardon and Grenier[25] point out how the acquisition of a symbiont fits well into the paradigm of natural selection, something first recognised by the Russian Kozo-Polyansky [16], but that "this evolutionary mechanism has been underrated by neo-Darwinian biologists, since it corresponds to Lamarckian conceptions of evolution ... We designate by neo-Lamarckianism a theory of evolution that takes into account the possibility of acquiring exogeneous characteristics ... [for] What is an [endosymbiont], from a genetic point of view ? A new cluster of genes !" [25]. This has also been referred to as the evolution of an "interspecies supraorganism about as well integrated as parts of an individual organism ... with selection operating on the system as a functional whole"[1]. Margulis has postulated that "the higher taxa (kingdoms or phyla) have evolved by acquisition of symbionts that have become hereditary ... [others] suspect lower taxa to have originated in the same way"[23].

We now describe the model used in the paper to examine this evolutionary mechanism.

3 Simulated Coevolution

3.1 The NKC Model

Kauffman and Johnsen [14] introduced the NKC model to allow the genetics-based study of various aspects of coevolution. In their model an individual is represented by a genome of N (binary) genes, each of which depends epistatically upon K other genes in its genome. Thus increasing K, with respect to N, increases the epistatic linkage, increasing the ruggedness of the fitness landscapes by increasing the number of fitness peaks, which increases the steepness of the sides of fitness peaks and decreases their typical heights. "This decrease reflects conflicting constraints which arise when epistatic linkages increase"[14]. Each gene is also said to depend upon C traits in each of the other species with which it interacts. The adaptive moves by one species may deform the fitness landscape(s) of its partner(s). Altering C, with respect to N, changes how dramatically adaptive moves by each species deform the landscape(s) of its partner(s). Therefore high values of C correspond to the obligate symbionts described above and low values to facultive dependencies; with this model we can adjust the strength of the symbionts' association from barely facultive through to being completely obligate in a systematic way.

The model assumes all inter and intragenome interactions are so complex that it is only appropriate to assign random values to their effects on fitness. Therefore for each of the possible K+C interactions, a table of $2_{(K+C+1)}$ fitnesses is created for each gene, with all entries in the range 0.0 to 1.0, such that there is one fitness for each combination of traits. The fitness contribution of each gene is found from its table. These fitnesses are then summed and normalised by N to give the selective fitness of the total genome (the reader is referred to [13] for full details of both the NK and NKC models).

Kauffman and Johnsen use populations of one individual (said to represent a homogeneous species) and allele mutation to evolve them in turn. That is, if a given mutant is found to be fitter than its parent in the current context of the other species, that species as a whole moves to the genetic configuration represented by the mutant. This is repeated for all species over a number of generations. In this paper we apply a generational genetic algorithm (GA) [9] to their model, slightly altering some aspects; the species evaluate and evolve at the same time and do so within populations of many individuals. This allows the appropriate association between the symbionts to emerge. However this does not appear to cause the loss of any of the dynamics Kauffman and Johnsen report. They show how both inter (C) and intragenome (K) epistasis affects a coevolving system, particulary in the attainment of Nash equilibria ("a combination of actions by a set of agents such that, for each agent, granted that the other agents do not alter their own actions, its action is optimal"[13]). We will return to their results later.

3.2 Genetic Algorithm

In this paper there are three species (A, B, and E), two of which (A and B) are said to be in a beneficial symbiotic relationship and have the potential to form an hereditary endosymbiosis. The third species (E) is said to represent the other species with which the symbionts coevolve - their ecologically coupled partners. For A and B there are three sub-populations (figure 1). There are the two sub-populations for the symbionts living cooperatively and evolving separately, each receiving there own separate measure of fitness, with these being initially set to size P (the sizes of these two populations are always the same as each other). The other sub-population is of hereditary endosymbionts, consisting of individuals carrying the genomes of both symbionts, where they are treated as "an interspecies supraorganism"[1] for selection by receiving a combined fitness. The size of this sub-population is initially set to zero, but during the course of a simulation individuals can move from the other sub-populations to this, and back, via a megamutation operator (hence there is always effectively P sets of the two genomes overall). Over evolutionary time the most appropriate configuration of the two symbionts of the model - be it an endosymbiosis or to stay as two separate populations of cooperating species - will emerge rather than being prescribed a priori; the hereditary endosymbiotic version of a given two-species symbiosis can emerge and compete against the separate but cooperating version of the association for population "space".

Initially the two symbiotic species exist in separate populations, both of size P, where they are paired simply by taking the corresponding members from their respective populations (that is speciesA[x] is paired with speciesB[x], where 0<x<P). A corresponding member of the environmental third species (speciesE[x]) is also chosen.

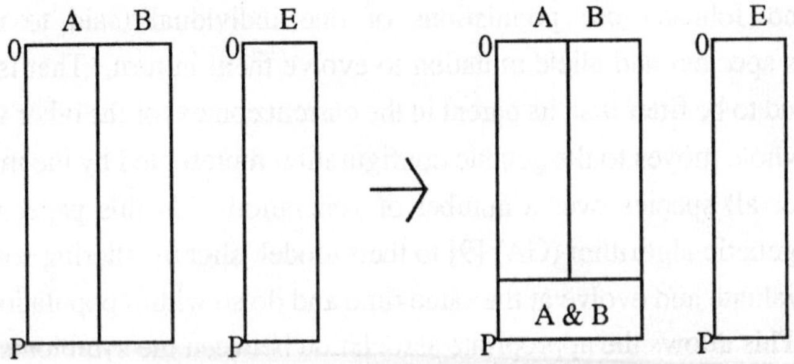

Figure 1. During evolutionary time the population space of the symbiotic species A and B can be invaded by their equivalent endosymbiotic association (A&B).

Individuals are then evaluated on their given NKC function and awarded their own separate measure of fitness. A generational GA is then run on the total population space of the two symbionts. At the end of this process, and each succeeding evaluation-selection-generation of individuals, a check is made to see if the symbionts form/ dissolve a more intimate association. The formation of a general symbiosis in nature is termed recognition [30] and "may include specific chemical interactions, tolerance or suppresion of host defences, and metabolic, morphological or behavioural interactions" [30]. Weisbuch and Duchateau have used differential equations to determine the recognition rates under which mutualism can emerge. They surmise that mutualism "can be established when a recognition mechanism allows the host to discriminate among parasites and bona fide symbionts"[32], i.e. when the differential rejection rates are large. For our purposes the formation of an endosymbiosis is established by simply testing a probability p_e. Satisfaction of this probability means both partners form an hereditary endosymbiosis, where they recieve a combined measure of fitness for selection and where an individual consists of two genomes - the speciesA and speciesB members. From then on they do not evolve within seperate populations, but are treated as one by the GA , having been joined via this megamutation operator. The same operator is also applied to the population of hereditary endosymbionts, working in reverse, to stop the drift of members away from the population of separate symbionts; separating is just as likely as joining. Evaluation of the two endosymbiotic sets of genes is the same as when they were separate; the speciesA genome is evaluated using the NKC fitness tables for speciesA and the speciesB genome with the speciesB tables.

The size of these competing symbiotic sub-populations (endo and separate) is controlled by a version of the roulette wheel proportional selection used in many genetic algorithms. Here the total fitness ratings of all individuals in all symbiotic sub-populations are summed at the end of each generation. Then a random number is chosen between zero and this total fitness. Starting with any given sub-population the scores of all individuals are again summed until this value is equal to or greater than the random number chosen. The following production of an offspring uses allele mutation on the selected single parent. This whole process is repeated until P new pairs of individuals have been created, from whichever sub-populations. It may have been noted that the cooperating separate species receive individual fitnesses, meaning they are on average half as likely to be chosen compared with the endosymbionts through the method of selection. Therefore we say that the separate individuals sub-population appears as one, with the combined fitnesses of speciesA and of speciesB during the summing processes. Whenever this sub-population is chosen a pair of separate symbionts are always created, proportionally, from their own populations. The environmental species (E) also evolves asexually within its own population via a generational GA, using proportional selection and allele mutation. This population allows us to examine the effect endosymbioses have on their other ecological partners. The process is repeated for a given number of evolutionary generations.

We now use the model described above to examine the evolutionary performance of endosymbiosis (the reader is also referred to [3] for a full investigation), with all results presented being averaged performance over fifty runs.

4 Endosymbiosis

4.1 Simulation

We implement our version of Kauffman and Johnsen's [14] NKC model using three species, each evolving separately in populations of one hundred (P=100) individuals. Three genome lengths are used throughout, N=8, 12 and 24 - all experiments are repeated for the various N. We introduce a second value of interdependence for the symbionts to the third species (Ce); we can alter both the interdependence between the symbionts (C) and the interdependence both symbionts have with their environment (Ce). This rate of environmental interdependence is the same for the symbionts whether or not they are joined. The per bit allele mutation rate of the GA (p_m) is set at 0.001 and the megamutation operator (p_e) is set at 1/(5P) - that is on average two sub-population members will alter their association once in five generations (it is suggested that the probability is comparitively low in nature with respect to the allele "micromutation"[18]). All species have the same K value.

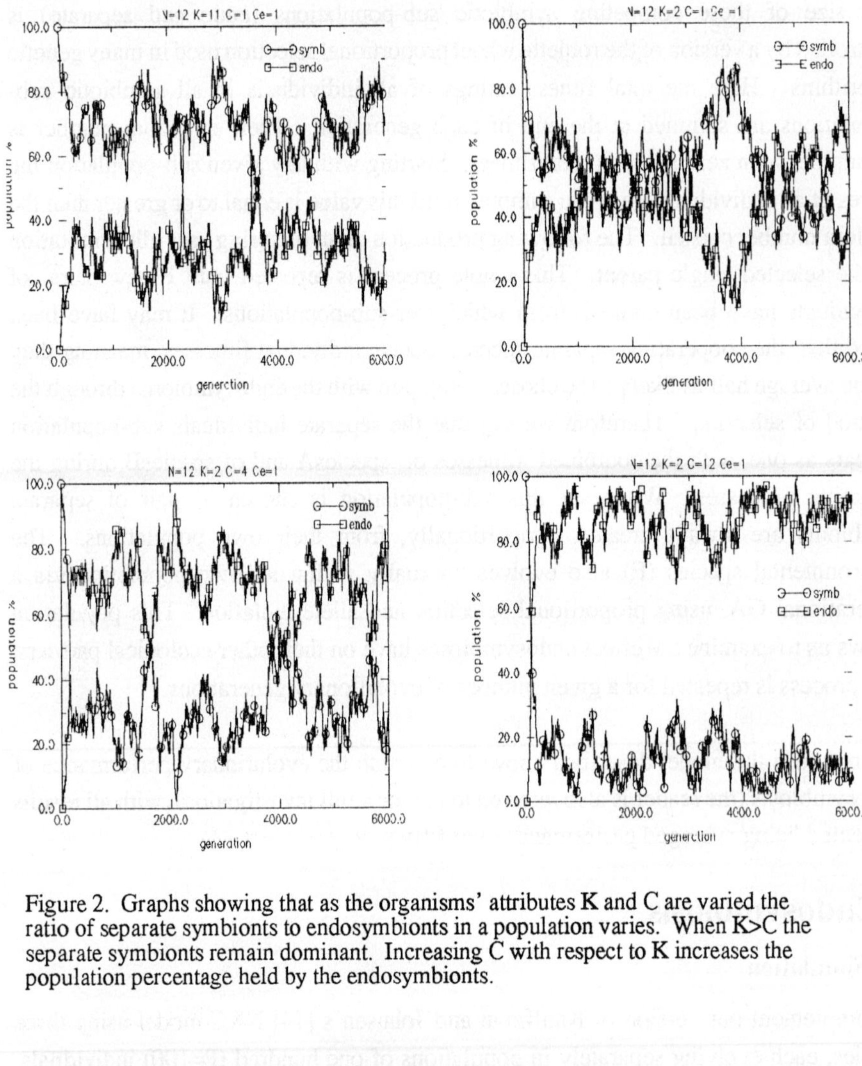

Figure 2. Graphs showing that as the organisms' attributes K and C are varied the ratio of separate symbionts to endosymbionts in a population varies. When K>C the separate symbionts remain dominant. Increasing C with respect to K increases the population percentage held by the endosymbionts.

A number of experiments were carried out in which the values of K, C and Ce were varied. We find that for various combinations of these values different sub-populations become dominant within the model.

For Ce=1 (i.e. low Ce) and any K we find that increasing C increases the population percentage of the hereditary endosymbionts, such that when C>K there are more endosymbionts than separate cooperating symbionts. The greater C is in relation to K, the bigger the difference in population percentage (figures 2&3). Conversely when K>C the cooperating separate symbiotic species are dominant (i.e. >50%), though they never dominate as much as the endosymbionts do. This shows that as the degree of dependency increases between symbionts, forming an hereditary endosymbiosis becomes increasingly beneficial to them. Also note from figures 2&3 that the difference between K and C must be large for either association to become significantly dominant (>80%). Kauffman and Johnsen state that K=C corresponds to the coevolving partners' optimum edge of chaos regime - their liquid state. It can be seen from our results that in these models the edge of chaos has quite a large basin of attraction towards coexistence of both associations; conditions must stray considerably before the population will diverge significantly either way. Correspondingly in experiments where the megamutation operator was disabled we found that overall performance actually drops in terms of attained fitness levels for all values of C; the search process is helped by having both associations coexist.

Increasing Ce does not appear to alter this general result, however from figures 3&4 it can be seen that an increase allows a higher percentage of endosymbionts to exist for lower C than before. Conversely for higher C a lower percentage of endosymbionts are found than for when Ce is low. That is, if the symbionts become increasingly dependent upon other aspects within their environment, their forming an endosymbiosis is beneficial only up to some point. As this dependence increases (i.e. when Ce is high), the benefits of joining appear to decrease; if the symbionts' or endosymbionts'

N=8 Ce=1

C \ K	1	3	5	7
1	51	49	43	34
3	73	60	54	45
5	85	70	69	49
7	87	80	70	70

N=8 Ce=3

C \ K	1	3	5	7
1	60	71	41	33
3	77	82	55	39
5	78	77	55	40
7	86	80	73	

N=8 Ce=5

C \ K	1	3	5	7
1	56	55	47	35
3	64	65	66	41
5	80	66	50	
7	77	82		

Figure 3. Showing the effects of varying K, C, and Ce on the population percentage held by the endosymbiotic association of a given symbiosis.

dependence upon other species within their environment increases, the benefits of either association are decreased and the system returns to its edge of chaos-like constitution. Burns' [5] suggestion that some endosymbionts no longer experience environmental factors may be a way in which nature has dealt with this problem, i.e. the effects of Ce are lost on endosymbionts.

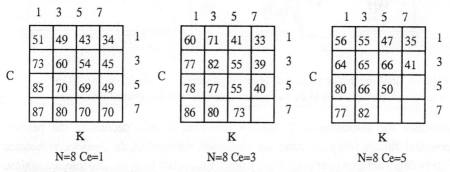

We have also examined the consequences of varying the other parameters in the model - p_e, p_m, and P. Briefly:

Increasing the megamutation rate p_e (e.g. $p_e = 1/P$) improves the separate symbionts' population share and fitness when the endosymbiosis would normally dominate because there is a greater movement of members between the two different associations; there is a greater flow of the results of the endosymbionts' better searching back in to the separate symbionts' gene pool. The reverse is true when p_e is decreased (e.g. $p_e=1/(10P)$) under the same conditions, with the endosymbionts dominating more easily. Natural symbioses and endosymbioses are often seen to act in a positive feedback loop [15] in which evolved beneficial adaption to the symbiosis further

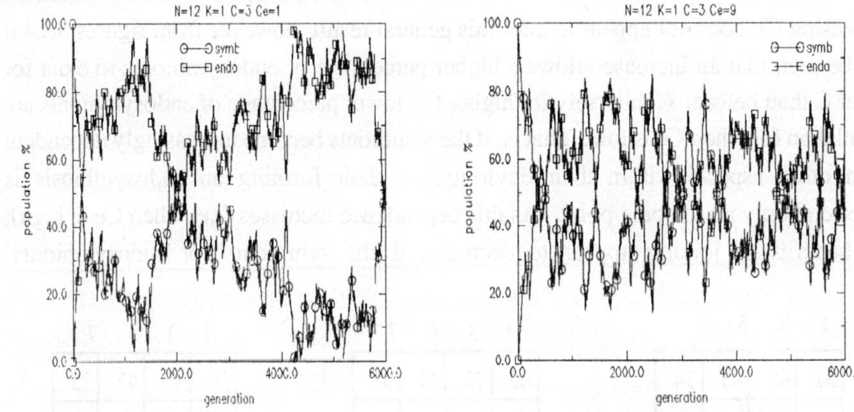

Figure 4. Showing that as the amount of interdependence with the environment (Ce) is increased, the difference in the two associations' performance gradually decreases.

increases the association's chances of selection, thereby decreasing the partners potential fitness when separate; the symbionts engaged in the association become increasingly obligate over evolutionary time. Redundant traits are also selected against, which may include traits critical to the partners' separated existence, exaggerating this unifying process [6]. Again it appears that under these circumstances nature has sided with endosymbiosis by making the dissolution of an association potentially harder over time, though as Nealson [26] points out the formation, perpetuation, and development of an hereditary endosymbiosis can be genetically costly, which could be seen to balance this process.

A higher allele mutation rate (e.g. $p_m = 0.01$) is coped with much better by the endosymbionts, but their best and mean fitness level performances drop slightly when compared to those with p_m=0.001. The separate symbionts' performance drops dramatically here. At a lower rate (e.g. p_m = 0.0001) the separate symbionts perform better in terms of population share, attaining a 50% share in more C>K situations, with there being no difference in the fitness levels of the two associations. Law [17] highlights the natural analogy of this by noting how obligate symbionts benefit from having a slower rate of genetic change because they need to stay similar to their parents to keep an association beneficial (also mentioned by [11]). In both cases megamutation again appears to help the mean performance of the separate symbionts. The endosymbionts effectively represent one organism of genome length 2N, so the higher allele mutation rate does not represent such large jumps in their (larger) search space compared to the separate symbionts; the mutation driven jumps are less likely to be to points beyond the correlation length of their fitness landscape for various amounts of epistasis below some limt (as K approaches N-1) [see 13].

With larger populations (e.g. P=200) we find that the hereditary endosymbionts do better for the same values of C>K in terms of population share and mean fitness. For smaller P (e.g. P=50) the reverse is true with the separate cooperating symbionts managing to hold larger percentages of the population. Overall performance of the fittest members in the populations is apparently unaffected by a change in population size. This correlates with the above findings from altering the mutation rates in that an increase in search potential, that is an increase in population size or an increase in the search mechanism rate (mutation), helps the endosymbionts with their larger search space.

On examining the evolutionay progress of the third distinct species, speciesE, said to represent the symbionts' ecological partners, we find that in models where C>K, and therefore the endosymbiosis becomes dominant, speciesE's mean and best performance are actually increased. That is, the formation of an endosymbiosis also appears to help the symbionts' ecology as a whole.

4.2 Discussion

From these simulations it can be seen that as the interdependence (C) between symbionts increases hereditary endosymbiosis becomes increasingly dominant within the overall population; the further the partners are into their chaotic gas regimes, the more likely they are to form an endosymbiosis.

Kauffman and Johnsen [14] investigated the effects interdependence has upon separate coevolving partners, finding that:

4A: When the partners are least interdependent (C is low) the time taken for the system as a whole to reach a Nash equilibrium is low, fitness during the period of pre-Nash oscillations is close to the equilibrium fitness and the oscillating fitness is high.

4B: When the degree of interdependence is increased, eventually to being obligate (C=N), they find the reverse is true with an increase in the time taken for equilibrium to be reached, with fitness during the oscillations being below the equilibrium fitness and where this oscillating fitness is low.

The results are attributed to the fact that as the amount of interdependence is increased, the effect of each partner on the others' fitness landscape increases; the higher C, the more landscape movement. This represents an extension to Kauffman's single organism NK model, in which he found that:

4C: Increasing epistasis (K) with respect to N increases the ruggedness of a fitness landscape, but where the height of the optima decreases.

We can use these findings to suggest underlying reasons for our result above, that as interdependence increases between symbionts the relative performance of an endosymbiotic association increases. Hereditary endosymbionts represent two or more genomes being carried together effectively as one genome, as Nardon and Grenier [25] have pointed out (see section2), and as a consequence the partners' intergenome epistasis becomes intragenome epistasis for a larger genome. The effects of intragenome epistasis are very different from intergenome epistasis, as can be seen from 4A - 4C above. That is the combined landscape of an hereditary endosymbiosis does not move due to the symbionts' interdependence C - but is more rugged, and larger, than that of the individual partners living separately(4C). There will still be movement from any environmental interdependences (Ce) of course, but now their internal epistasis has increased and they have (potentially - depending on Ce) moved into their combined form's solid state (K>C). This can be seen as a genetics-level analogy of Burns' [5] suggestion that closer associations form between symbionts as a response to environmental unpredictability (section2).

Thus for separate partners when C is low the amount of possible landscape movement is likewise low. From 4A it can be seen that separately coevolving symbionts oscillate with a high fitness under these circumstances. Therefore any hereditary endosymbionts must find at least as high a fitness in their larger, more stable, more rugged fitness landscape in order to survive in succeeding generations. They are also hindered by the fact that there is little time for them to search their larger space - even if the fitness level they find is comparable to that of separate individuals (which is more likely to be a local optimum for them anyway) - before the separate symbionts reach a higher Nash equilibrium. Thus the evolutionary benefits of an hereditary endosymbiosis in such cases are low because the separate symbionts can quickly find a high fitness level and then quickly reach equilibrium.

When C is high (as in obligate symbioses) the hereditary endosymbionts' combined landscape will be more rugged, containing many low optima (4C). In this case the oscillatory fitness of the coevolving separate symbionts is low (4B) - they are well into their gas regimes (C>K). Any hereditary endosymbionts will easily find optima in their more stable but rugged landscape and if that peak is comparable to, or better than, the low oscillatory fitness of the separate partners, they will survive into the succeeding generations. The time taken for the separate symbionts to reach equilibrium is longer here (4B), which also allows the larger endosymbionts greater searching time.

Kauffman [13] states that as K approaches its maximum (N-1), an organism experiences a "complexity catastrophe" whereby achievable fitness actually decreases as the height

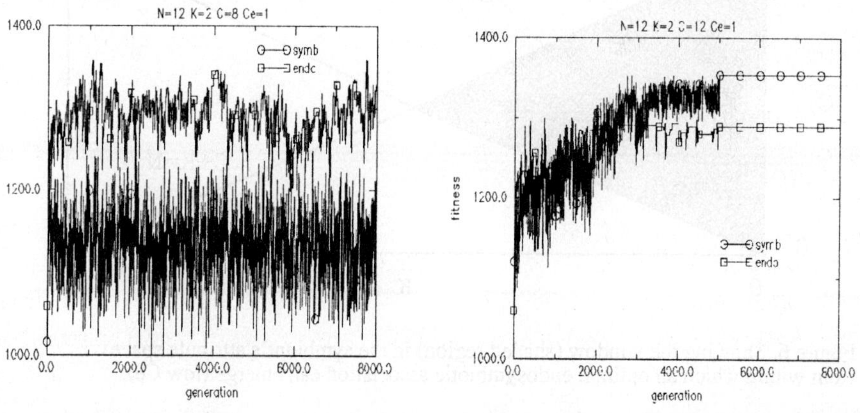

Figure 5. Graphs showing that a limit exists on how far into their respective gas regimes symbionts can be for an endosymbiosis to represent the optimal strategy.

of optima in its fitness landscape falls to the mean height (0.5). Intuitively therefore the formation of an hereditary endosymbiosis will not always represent an optimum strategy for the partners over a longer time scale. When C is high, the resulting K of the endosymbiotic supraorganism will also be high; intergenome interdependence becomes intragenome epistasis as described above. We ran the same experiments on a version of Kauffman and Johnsen's [14] NKC model and found that this was indeed the case (figure 5). Here the two separate symbiotic species were represented by two individuals, as in Kauffman and Johnsen's work, and the initial endosymbiotic version of the association was represented by an individual carrying both their initial genomes; in these simulations the existence of an endosymbiotic association was pre-determined and there was no gene transfer between the two associations during the simulations. The environmental species was also included (as a single individual). All species were evolved using Kauffman and Johnsen's [14] "fittest mutant" strategy, except that this was done concurrently as in our population-based GA models. We found that for the endosymbiotic version of the symbiosis to represent the most fit association over longer time scales a "window" of symbiotic organism attributes exists. When C is larger than K (when C>K) and these two values are less than the genome length (C+K<N), the endosymbiosis represents an evolutionarily fitter association (figure 6). This window of attributes represents an area covering half the organisms respective chaotic gas regimes. Forming an endosymbiosis moves the combined partners into a more solid regime as the effects of their interdepence are lost and they become a supraorganism

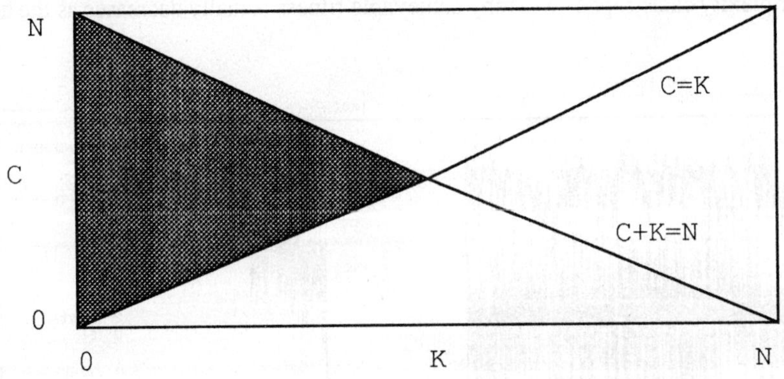

Figure 6. Showing the window (shaded region) in the symbiont's attribute space from within which an optimal endosymbiotic association can emerge (low Ce).

with high epistasis. If this new epistasis is large with respect to the new genome length (K>N/2) it is more efficient in the long run for the symbionts to stay separated. That is for highly dependent symbionts (i.e. those in their chaotic gas regimes), it is better to stay separated if joining via endosymbiosis will make them a supraorganism well into its solid regime. We would expect to find the same kind of dynamics occuring within our population-based GA models if they were left to run for much longer. Our results certainly match those using Kauffman and Johnsen's model over a shorter time scale.

In our models we also found that when the endosymbiosis became dominant within a system the performance of its ecologically coupled partners was increased. That is the supraorganism with a high K value helped the other species (speciesE in our models). Kauffman and Johnsen [14] also reported that when the amount of epistasis in any one species in an ecology is increased, the performance of the other coupled species is increased. In a more detailed study they found that for N=24 the optimum value of K for the coevolving partners coupled ecosystem was around 8 to 10 (just less than N/2) and that the presence of any species with this optimal value of K helped all coupled species, whether or not they had an optimum value of K themselves. Our finding that a window on the amount of epistasis within an endosymbiotic supraorganism exists, such that settings outside of it mean the association is sub-optimal, would appear to concur with their findings. This seems particularly true as we found a limit on K+C for the case of separate symbionts which must be less than N/2 of the resulting supraorganism. These results all seem to indicate that the formation of an endosymbiosis can help the symbionts' ecology as a whole.

5 Conclusion

In this paper we have shown that endosymbioses can represent an evolutionarily more efficient association within cooperative multi-species environments under the appropriate conditions. It has been seen that a window of organism attributes appears to exist from within which an optimum association can emerge, which in turn can help the symbionts' coupled ecological partners and hence their overall environment reach a more optimal state.

Previously (e.g. [4]) we have presented an approach with the aim of enabling the use of the genetic algorithm on complex problems. In that work we proposed that a given system is viewed as an ecology [10] or an endosymbiotic supraorganism and that a divisional approach be used to identify its constituent sub-systems (species); the system is broken into a number of smaller sub-systems a priori. In large and complex systems this can be difficult as the task often does not lend itself to formal analysis. Each sub-system is then represented by a coevolving agent (in our work these are Pittsburgh-style classifier systems [29]) able to communicate with any other sub-system, thus allowing for any interdependence between them. We have shown that such an approach can give an increase in performance when compared to using an equivalent single agent evolved with a single GA.

The research presented in this paper allows us to refine the approach by considering the amount of interdependence between the identified sub-systems and how best to assign any available global and local measures of fitness to them. The results above suggest that for various amounts of inter and intragenome dependence it can be better for some coevolving species (sub-systems) to be treated as a functional whole, rather than separately. If inappropriate divisions are made in a given system's search space, due to the complexity of the task, its evolutionary performance can be adversely effected. By adding our process of endosymbiosis to such systems, via a megamutation operator, the most efficient sub-system configuration is able to emerge through evolutionary pressure (recently Seredynski [28] has also shown that giving dependent agents a shared fitness can prove more efficient). It is hoped that in this way the use of coevolving communicating multi-agent systems, for simulation, modelling and other applications, will become an effective method for the solution of complex problems using genetics-based approaches.

6 Acknowledgements

The authors acknowledge the support of HP labs through their European Equipment grants programme. This work was supported under EPSRC grant no GR/J71687.

7 References

1. *Allee W C, Emerson A E, Schmidt K P, Park T & Park O* (1949), "Principles of Animal Ecology", Saunders Company.

2. *Begon M, Harper J L & Townsend C R* (1986), "Ecology", Blackwell.

3. *Bull L* (1995), PhD Thesis, UWE Bristol.

4. *Bull L & Fogarty T C* (1994), " Parallel Evolution of Communicating Classifier Systems", Proceedings of the IEEE World Congress on Computational Intelligence, IEEE.

5. *Burns T P* (1993), "Discussion: Mutualism as Pattern and Process in Ecosystem Organisation", Mutualism and Community Organisation, Oxford University Press.

6. *Cook C B* (1985), "Equilibrium Populations and Long-term Stability of Mutualistic Algae and Invertebrate Hosts", The Biology of Mutualism, Croom-Helm.

7. *Ehrman L* (1983), "Endosymbiosis", Coevolution, Sinauer Associates.

8. *Haynes R H* (1992), "Modes of Mutation and Repair in Evolutionary Rhythms", Symbiosis as a Source of Evolutionary Innovation, MIT Press.

9. *Holland J H* (1975), "Adaption in Natural and Artificial Systems", Univ. of Michigan Press, Ann Arbor.

10. *Hubermann B A* (1988), "The Ecology of Computation", North-Holland.

11. *Ikegami T & Kaneko K* (1991), "Computer Symbiosis - Emergence of Symbiotic Behaviour Through Evolution", Emergent Computation, MIT Press.

12. *Janzen D H* (1985), "The Natural History of Mutualisms", The Biology of Mutualism, Croom-Helm.

13. *Kauffman S A* (1993), "The Origins of Order: Self-organisation and Selection in Evolution", Oxford University Press.

14. *Kauffman S A & Johnsen S* (1989), "Co-evolution to the Edge of Chaos: Coupled Fitness Landscapes, Poised States and Co-evolutionary Avalanches", Artificial Life II, Addison-Wesley.

15. *Keeler K H* (1985), "Cost: Benefit Models of Mutualism", Biology of Mutualism, Croom-Helm.

16. *Kozo-Polyansky B M* (1924), "A New Principle of Biology", Puchina.

17. *Law R* (1985), "Evolution in a Mutualistic Environment", The Biology of Mutualism, Croom-Helm.

18. *Law R* (1991), "The Symbiotic Phenotype: Origins and Evolution", Symbiosis as a Source of Evolutionary Innovation, MIT Press.

19. *Law R & Lewis D H* (1983), "Biotic Environments and the Maintenance of Sex - some evidence from mutualistic symbioses", Biological Journal of the Linnean Society.

20. *Lewis D H* (1985), "Symbiosis and Mutualism: Crisp Concepts and Soggy Semantics", The Biology of Mutualism, Croom-Helm.

21. *Margulis L* (1970), "Origin of Eukaryotic Cells", Yale University Press.

22. *Margulis L* (1976), "A Review: the Genetic and Evolutionary Consequences of Symbiosis", Experimental Parasitology, 39.

23. *Margulis L* (1991), " Symbiogenesis and Symbionticism", Symbiosis as a Source of Evolutionary Innovation, MIT Press.

24. *Margulis L* (1992), "Symbiosis in Cell Evolution", W H Freeman and Company.

25. *Nardon P & Grenier M* (1991), "Serial Endosymbiosis Theory and Weevil Evolution: The Role of Symbiosis", Symbiosis as a Source of Evolutionary Innovation, MIT Press.

26. *Nealson K H* (1991), "Luminous Bacteria Symbiotic with Entomopathogenic Nematodes", Symbiosis as a Source of Evolutionary Innovation, MIT Press.

27. *Packard N* (1988), "Adaption to the Edge of Chaos", Complexity in Biologic Modelling.

28. *Seredynski F* (1994), "Loosely Coupled Distributed Genetic Algorithms", Parallel Problem Solving Nature - PPSN III, Springer-Verlag.

29. *Smith S F* (1980), "A Learning System Based on Genetic Adaptive Algorithms", PhD Dissertation, University of Pittsburgh.

30. *Smith D C & Douglas A E* (1987), "The Biology of Symbiosis", Arnold Edward.

31. *Templeton A R & Gilbert L E* (1985), "Population Genetics and the Coevolution of Mutualism", The Biology of Mutualism, Croom-Helm.

32. *Weisbuch G & Duchateau G* (1993), "Emergence of Mutualism - Application of a Differential Model to Endosymbiosis", Second European Conference on Artificial Life, Springer-Verlag.

Mathematical Analysis of Evolutionary Process

Tetsuya Maeshiro and Masayuki Kimura

School of Information Science, Japan Advanced Institute of Science and Technology,
15 Asahidai, Nomi, Ishikawa, 923-12 Japan
phone: +81-761-51-1225, +81-761-51-1699 ext.1381
fax: +81-761-51-1116
e-mail: maeshiro@jaist.ac.jp, kimura@jaist.ac.jp

Abstract. This paper proposes an analytic framework for the analysis of evolutionary mechanisms at genetic coding level, attempting to provide more detailed description than population genetics. It gives an estimated sequence after T replications in an environment given the initial genetic sequence. We assume that there is a principle obeyed by evolutionary mechanisms at genetic sequence level, such that some law, called action, is suboptimal. We propose such an action for haploid, asexual type living lives with replications involving only point mutations, as a function of fitness and the probability of change in sequence, so the evolutionary process is not a simple hill-climbing. Our method provides an intuitive view on evolution of genetic sequences, and it may be a powerful analysis tool when we need to treat directly the genetic sequence. It is useful for the analysis of real or artificial life such as genetic algorithms. This is a report of a work in progress, and we present the background, development, connection with population genetics, and some possible extensions of our work.

Keywords. evolutionary process / genetic sequence / path integral / population genetics / fitness landscape

1 Introduction

Population genetics has given important insights into evolution of life since its formulation at the beginning of the century by Haldane, Fisher and Wright [1, 2]. With the advance of molecular biology, some innovative theories appeared, for instance the neutral theory by Kimura [7], which opened an understanding of evolutionary mechanisms at molecular level, and the molecular evolutionary genetics by Nei [11]. However, population genetics is basically the study of change in proportions of each phenotype in a population, and although we can predict how the composition of phenotypes in a population changes, it is inappropriate to study the change in genetic sequence, giving little insights on how a genetic sequence changes, or evolves through the time.

The possibility to track the genetic information, typically DNA, is very useful in cases like creation of phylogenetic trees, artificial protein synthesis imitating evolution, and analysis of genetic data in simulations, among others.

In this paper, we give some mathematical frameworks that allow the prediction of a genetic sequence after T replications in a given environment. A strong assumption we make is the existence of the principle of least action in evolutionary mechanism at genetic sequence level. Here, we are treating only the haploid type with asexual reproduction involving point mutations.

2 Genetic Sequence

The gene is the fundamental information source when studying evolution of life. Although some genetic informations are implicit and therefore understandable only when their functions are activated at phenotype level, the source information is the gene.

The study on evolutionary mechanisms is a difficult task for many reasons, mainly: (i) the biological evolutionary process takes time interval much longer than the average life-span of human beings, limiting the availability of experimental data; (ii) the main information source of the past living creatures are fossil, which give only a partial information. The genetic information is not directly accessible, and the possibility of obtaining such an information is very remote, if not impossible. Although there have been some efforts in this direction, obtained informations would be incomplete.

If we have a correct analytic framework and genetic informations of living-forms at present time, it would be possible to predict roughly the future and estimate the evolutionary process in the past which lead to the origin of today's lives. For this purpose, it is not enough to handle the information of phenotype level, and there is a need for an analytic framework which allows the treatment of the information of genetic coding level.

Another fact that requires the treatment of genetic coding level is many to 1 correspondence between DNA sequence and the protein sequence, which are the genotype and the phenotype, respectively. Three bases, called triplets, are translated into one amino acid, and since there are 4 types of nucleic acid and 20 types of amino acid, some of 64 triplets are coded into an amino acid, for instance, 6 triplets are coded as arginine, 2 as histidine and 1 as methionine. It means that the distance between two arbitrary amino acids measured in number of mutated nucleic acids is not a fixed value. For instance, the number of necessary mutations between arginine and histidine is one, two or all three.

Therefore, it is simpler and more natural to treat directly the genetic sequence and assign the same mutation rate for all positions in the sequence, and assign the informational content value for amino acids of synthesized protein sequence. The informational content indicates the degree of importance of given site for the function of the protein, and can be interpreted as the number of amino acids that can be mutated without degenerating the function.

Genotype, Phenotype and Fitness Landscape. Here, we treat genotype simply as general terminology for genetic sequence, i.e. the genetic information, and phenotype as a protein sequence, i.e. the functional form.

Even though the changing unit is the genotype, the evaluation unit is the phenotype, whose fitness is given by some fitness landscape.

To each genetic sequence, one or more fitness values can be assigned, and a surface representing the fitness called fitness landscape in $m+1$-dimensional space, where m is the length of sequences, can be drawn, as in figure 1, where $m = 2$. The evolution can be interpreted as an optimization process where the creatures seek for maximum values of fitness, climbing hills.

3 Principle of Least Action and Path Integral

3.1 Principle of Least Action

In physics, it is well known that in many cases Nature seeks for such a way that some parameters or a combination of them are the least possible. For instance, the average kinetic energy less the average potential energy is minimum for the path of a body moving in a gravitational field.

This principle can be stated as

$$\text{Action} = S = \int_{t_a}^{t_b} \mathcal{L}\, dt \quad \longrightarrow \text{min} \tag{1}$$

where $\mathcal{L}$ is the Lagrangian. In the case of motion in a potential field, the Lagrangian becomes

$$\mathcal{L} = \frac{m}{2}\left(\frac{dx}{dt}\right)^2 - V(x,t) \tag{2}$$

where x and m are the coordinate and the mass of the body, respectively, the first term on the right hand side of equation (2) is the kinetic energy, and the second term is the potential energy.

The least action S_{min} is obtained when the body moves on the path $\boldsymbol{x}(t)$ satisfying the following differential equation,

$$-m\frac{d^2\boldsymbol{x}}{dt^2} - \nabla V(\boldsymbol{x},t) = 0 \tag{3}$$

which is $\boldsymbol{F} = ma$, the Newton's law.

3.2 Path Integral

The path integral was invented by Feynman [3], and is one of formalization of quantum mechanics. Its main idea is to calculate all possible paths that a particle can take between arbitrary coordinates and sum, or integrate over the amount of contribution of each path, a value defined as the phase proportional to the action of the path.

The probability that a particle reaches the position x_b at instant t_b from position x_a at instant t_a is $P(b,a)$, which is the absolute value of amplitude $K(b,a)$ squared, i.e., $P(b,a) = |K(b,a)|^2$. $K(b,a)$ is the sum of contributions from each path $\phi[x(t)]$,

$$K(b,a) = \sum_{\text{all paths}} \phi[x(t)] \tag{4}$$

and the contribution of a path $\phi[x(t)]$ is the phase proportional to an action S.

$$\phi[x(t)] = (\text{const}) \times \exp\left(\frac{i}{\hbar}S[x(t)]\right) \tag{5}$$

where S is given by equation (1), i is a unit of imaginary number, or $i = \sqrt{-1}$, $\hbar = h/2\pi$ where h is Planck's constant, and const is introduced to normalize $\phi[x(t)]$.

Let $\psi(\boldsymbol{x},t)$ be the state of a system. Then, $K(b,a)$ can be rewritten as $K(\boldsymbol{x}_2,t_2;\boldsymbol{x}_1,t_1)$, which is the probability that the state of the system changes to $\psi(\boldsymbol{x}_2,t_2)$ from $\psi(\boldsymbol{x}_1,t_1)$

$$\psi(\boldsymbol{x}_2,t_2) = \int_{-\infty}^{+\infty} d\boldsymbol{x}_1 K(\boldsymbol{x}_2,t_2;\boldsymbol{x}_1,t_1)\psi(\boldsymbol{x}_1,t_1) \tag{6}$$

The key idea is to take into account all the paths, assign them some weight value as a function of some action, and integrate them to calculate the probability that the state of the system changes from $\psi(\boldsymbol{x}_1,t_1)$ to $\psi(\boldsymbol{x}_2,t_2)$.

Now, we apply the idea to our problem to estimate how the genetic sequences change, to calculate the probability that a genetic sequence $\boldsymbol{x}_a$ at instant t_a become $\boldsymbol{x}_b$ at instant t_b.

4 Mathematical Framework

4.1 Basic Concepts

Let $\boldsymbol{x} = (x_1, x_2, \ldots, x_m)$ be a genetic sequence of length m, and n_e the number of possible types in each site x_i, so there are ${n_e}^m$ possible sequences. For instance, in DNA sequence, $n_e = 4$, as there are four types of nucleic acids: adenine, thymine, cytosine and guanine. μ is the mutation rate for each site of genetic sequence, and $n_\mu(t)$ is the number of mutations in a replication at instant t, whose expected value is $\mathrm{E}[n_\mu(t)] = \mu \cdot m$.

Let $W(\boldsymbol{x}, t)$, $0 \leq W \leq 1$, be the fitness of sequence $\boldsymbol{x}$ at instant t. W is the fitness landscape with dimension $m + 1$.

Now, suppose $\boldsymbol{x}_a$ and $\boldsymbol{x}_b$ are the genetic sequences at instants t_a and t_b, respectively. We want to calculate the probability that $\boldsymbol{x}_a$ becomes $\boldsymbol{x}_b$ after $T = (t_b - t_a)/\varepsilon$ generations, or replications, where ε is the duration of a generation. A certain number of mutations governed by mutation rate μ occur in every replication. The idea to calculate this probability is to count all possible paths between $\boldsymbol{x}_a$ and $\boldsymbol{x}_b$, assigning some probability value for each path.

Here we suppose that the mutation rate is constant for mutations between all element types on every site, and the generations do not overlap.

Let

$$v = \frac{1}{T} \sum_{t=t_a}^{t_b} n_\mu(t) \tag{7}$$

be the velocity of the sequence change, which is the average number of mutations per replication.

The probability associated with path between two adjacent generations is given by the number of point mutations. The probability due to mutation p^M that the sequence $\boldsymbol{x}_i$ is generated from $\boldsymbol{x}_j$ is

$$p_{ij}^M = \mu^{\nu_{ij}} (1 - \mu)^{m - \nu_{ij}} \frac{1}{(n_e - 1)^{\nu_{ij}}} \tag{8}$$

where ν_{ij} is the number of point mutations observed in a replication of $\boldsymbol{x}_i$ to generate $\boldsymbol{x}_j$

$$\nu_{ij} = m - \sum_{k=1}^{m} \delta(x_i^k - x_j^k) \tag{9}$$

which is the difference between sequence length and the number of non-modified. The number of identical sites between two sequences is calculated by

$$\delta(x_i^k - x_j^k) = \begin{cases} 1 & \text{if } x_i^k \text{ and } x_j^k \text{ are identical} \\ 0 & \text{otherwise} \end{cases} \tag{10}$$

where x_i^k is the kth site of sequence $\boldsymbol{x}_i$.

Let us call the probability p_{ij}^M the path probability between sequences $\boldsymbol{x}_i$ and $\boldsymbol{x}_j$.

Let us include the effect of other creatures of population. Even if the fitness of the offspring is higher than the parent, the path probability, which corresponds to the survival probability, will be low if the mean fitness of population is high. On the other hand, the offspring may survive with great chance if the mean fitness is low. So, the probability due to the influence of the rest of the population p^P is given by

$$p_i^P = \frac{W(\boldsymbol{x}_i, t) q_i}{\overline{w}} \tag{11}$$

where q_i is the frequency of sequence $\boldsymbol{x}_i$ given by

$$q_i = \sum_j^N \delta(\boldsymbol{x}_i - \boldsymbol{x}_j) \tag{12}$$

where N is the population size, and $\overline{w}$ is the mean fitness of the population given by

$$\overline{w} = \sum_k^N W(\boldsymbol{x}_k, t) q_k \tag{13}$$

where N is the population size.

Now, we have to take into account the shape of the fitness landscape $W(\boldsymbol{x}, t)$ to calculate the probability of paths, but its shape in biological world is unknown, although there has been great deal of activity in determining the properties of the landscape [12, 14], also some attempts to propose a model, such as NK model by Kauffman [5], and their analysis [9, 13].

Assume the fitness landscape $W(\boldsymbol{x}, t)$ is defined in some way. We have to find the Lagrangian that gives the least action to the change in genetic sequence. Some assumptions are made here: (i) the principle of least action is valid; (ii) the mutated gene's fitness value can be lower than the parent's gene.

Then, S is given by

$$S = \sum_{t_a}^{t_b} \mathcal{L} \quad \longrightarrow \text{max} \tag{14}$$

S should be maximized, but the principle is the same since the first order variation should be zero, although it is called principle of least action.

The contribution $\phi[\boldsymbol{x}(t)]$ of a path $\boldsymbol{x}(t)$ between $\boldsymbol{x}_a$ at t_a and $\boldsymbol{x}_b$ at t_b to the total probability of the change from $\boldsymbol{x}_a$ to $\boldsymbol{x}_b$ is

$$\phi[\boldsymbol{x}(t)] = A \cdot S[\boldsymbol{x}(t)] \tag{15}$$

where A is determined to normalize $\phi[\boldsymbol{x}(t)]$.

The total probability of the change from $\boldsymbol{x}_a$ to $\boldsymbol{x}_b$ is the sum of contributions over all possible paths from $\boldsymbol{x}_a$ at t_a to $\boldsymbol{x}_b$ at t_b

$$P(b, a) = \sum_{\text{all paths}} \phi[\boldsymbol{x}(t)] \tag{16}$$

Also, let ε be the duration of a generation given by $\varepsilon = t_b - t_a$, then $P(b,a)$ can be discretized and rewritten as $P(i+1,i)$, which is the probability that offspring $\boldsymbol{x}_{i+1}$ at t_{i+1} is replicated from $\boldsymbol{x}_i$ at t_i. The sum over all possible $\boldsymbol{x}_i$

$$\sum_{\boldsymbol{x}_i} P(i+1,i) \tag{17}$$

is the probability that $\boldsymbol{x}_{i+1}$ exists at t_{i+1}.

Then, for general a and b, $(t_b - t_a)/\varepsilon = T$ generations, $P(b,a)$ is

$$P(b,a) = \sum_{\boldsymbol{x}_{T-1}} \sum_{\boldsymbol{x}_{T-2}} \cdots \sum_{\boldsymbol{x}_1} P(b,T{-}1)P(T{-}1,T{-}2)\cdots P(1,a) \tag{18}$$

The summations (16) and (18) cannot be approximated to continuous form because the value of a site changes to any other value with equal probability under mutation, and the amount of change is not the usual distance in calculus or physics, so the derivative looses its sense, and conventional methods of calculus of variation cannot be applied. Therefore, the equations should be solved in discrete domain.

Visualization of Genetic Paths. There are two ways to visualize the evolution of genetic sequence. Here, we assume there is no change in genetic sequence length.

One is to imagine an m-dimensional surface which is the fitness landscape, and each sequence $\boldsymbol{x}$ is a point on that surface. The sequences "walk" on the surface of fitness landscape with time (figure 1).

Another way is to imagine a two-dimensional graph, and take vertical axis as the sequences and horizontal axis as time (figure 2). The sequences have length m and are mapped into one-dimensional space, ordered in some criteria. A probability is assigned to arrows between sequences of adjacent generations, and both axes are discrete.

4.2 Haploid, Asexual Replication

Here, we apply the idea to a special case, a population of haploid, asexual replication involving only point mutations. Therefore, there are neither homozygotes and heterozygotes as in the high level living forms such as eukaryotes, nor gene modifying mechanisms such as crossover, deletion and insertion.

Some assumptions are made in order to ease the interpretation. The first one we make is very small mutation rate, $\mu \ll 1$, very long gene sequence, $m \gg 1$. Then, the path probability due to mutation given by eq. (8) can be approximated by gamma function

$$p_{ij}^M = \frac{\mu^{\nu_{ij}}(1-\mu)^{m-\nu_{ij}}}{(n_e-1)^{\nu_{ij}}} \simeq \frac{e^{-\mu m}(\mu m)^{\nu_{ij}}}{\nu_{ij}!\, C_m^{\nu_{ij}}\, (n_e-1)^{\nu_{ij}}} \tag{19}$$

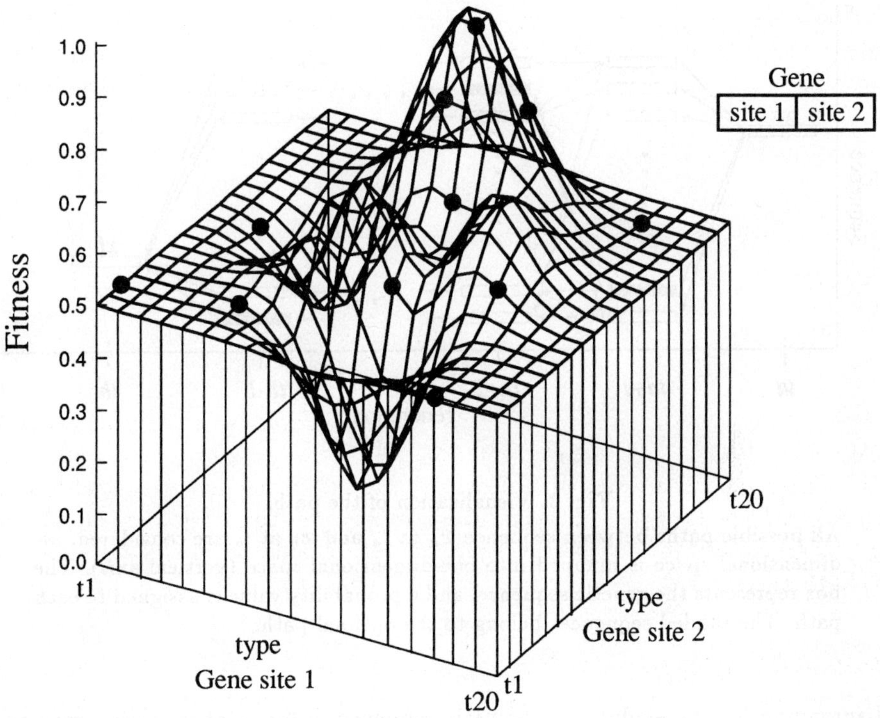

Fig. 1. An example of fitness landscape for gene with length 2 ($m = 2$, $n_e = 20$, $m^{n_e} = 2^{20}$ possible sequences). Each site can be any of 20 types $t_1 \dots t_{20}$. Circles are the genes, and they "walk" on the surface.

where $C_m^{\nu_{ij}}$ is the combination of ν_{ij} out of m

$$C_m^{\nu_{ij}} = \frac{m!}{\nu_{ij}!(m - \nu_{ij})!} \tag{20}$$

Also, the expected number of mutations per replications μm is 1, i.e., at most one site is mutated in a replication, so that $\nu_{ij} = 0$ or 1. Then, the path probability due to mutation is

$$\begin{aligned} p_{ij}^M &= e^{-\mu m}\left[\left(1 - \frac{\mu}{n_e - 1}\right)\delta(\boldsymbol{x}_i - \boldsymbol{x}_j) + \frac{\mu}{n_e - 1}\right] \\ &= e^{-\mu m}[(1 - \mu')\delta(\boldsymbol{x}_i - \boldsymbol{x}_j) + \mu'] \end{aligned} \tag{21}$$

where $\mu' = \mu/(n_e - 1)$.

Kimura suggested that genetic parameters are adjusted so that the total genetic load, introduced by Muller [10], is minimum [6, 7]. Here, we propose the

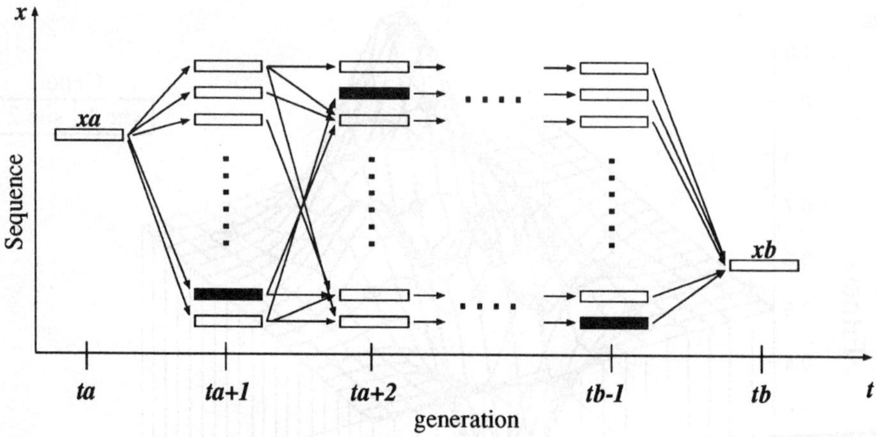

Fig. 2. Visualization of the paths
All possible paths between sequence $\boldsymbol{x}_a$ at t_a and $\boldsymbol{x}_b$ at t_b are considered. m-dimensional space is mapped into one-dimensional space (vertical axis). The box represents the genetic sequence, and a probability value is assigned to each path. The shaded sequences belong to the optimal path.

Lagrangian for the evolution in genetic sequence in terms of mutation rate and fitness landscape

$$\mathcal{L} = p_{i+1,i}\ \alpha W(\boldsymbol{x}_i) \tag{22}$$

where α is a constant, $p_{i+1,i}$ is the probability that the genetic sequence $\boldsymbol{x}_i$ changes to $\boldsymbol{x}_{i+1}$, and $W(\boldsymbol{x}_i)$ is the fitness value of sequence $\boldsymbol{x}_i$.

The principle introduced in the eq. (22) is to maximize the total fitness on the path, i.e., to keep as long as possible with maximum fitness, and this value is the expected value of total fitness gained through the path. The fitness W at any point is finite value, so the sum of this Lagrangian, S, always converges.

The probability $p_{i+1,i}$, which is the combination of probability p^M by mutation and p^P by selection, is included in the Lagrangian given by eq.(22) because the direction that allows the fastest increase in fitness may be difficult to reach due to the low probability of such a change in a replication. So, the principle of least action introduced here states that the sequences may take changes with relatively high probability that allow fast increase in fitness. In other words, the best path may be not an optimal one, but a suboptimal. It is different from a Lagrangian containing only the fitness value, even its summation is multiplied later with the total transition probability of the optimal path, since the criteria of the optimization is different. Therefore, the Lagrangian (22) is not a simple hill-climbing.

For computation of action S, we exclude the influence from the rest of population p^P from $p_{i+1,i}$ for the ease of analysis, so $p = p^M$ and the eq. (14)

becomes

$$S = \sum_{t_a}^{t_b} p_{j,i}\alpha W = \sum_{t_a}^{t_b} p_{j,i}^M \alpha W$$
$$= \sum_{t_a}^{t_b} e^{-\mu m}[(1-\mu')\delta(\boldsymbol{x}_i - \boldsymbol{x}_j) + \mu']\alpha W \tag{23}$$

The probability between genetic sequence $\boldsymbol{x}_i$ at generation i and its offspring $\boldsymbol{x}_{i+1}$ at generation $i+1$ is given by

$$P(i+1, i) = Ae^{-\mu m}[(1-\mu')\delta(\boldsymbol{x}_{i+1} - \boldsymbol{x}_i) + \mu']\alpha W \tag{24}$$

Then, for general a and b, $(t_b - t_a)/\varepsilon = T$ generations, $P(b, a)$ is

$$P(b,a) = \sum_{\boldsymbol{x}_{T-1}}^{M} \sum_{\boldsymbol{x}_{T-2}}^{M} \cdots \sum_{\boldsymbol{x}_1}^{M} A^T \sum_{j=1}^{T} e^{-\mu m}[(1-\mu')\delta(\boldsymbol{x}_{i+1} - \boldsymbol{x}_i) + \mu']\alpha W \tag{25}$$

where $M = n_e^m$, the number of possible sequences.

The above equation indicates that the most probable path to be taken is with smallest number of moves possible, since mutation rate μ is small, and when some move is performed through mutation, the destination has highest fitness value among other candidates. However, the optimal path is not necessarily unique since it depends on the shape of fitness landscape W. If the fitness landscape presents many peaks, as Kauffman suggests [5], there are many paths, and this is reasonable, meaning that life seeks not for an optimal path, but for suboptimal paths.

4.3 Relation with Population Genetics

The equations of population genetics for a haploid, asexual population are

$$p_i(t+\varepsilon) = \sum_j \frac{p_j(t) w_j}{\overline{w}} \mu_{ij} \tag{26}$$

where $\overline{w}$ is the mean fitness of population,

$$\overline{w} = \sum_k^N w_k p_k(t) \tag{27}$$

where μ_{ij} is the mutation rate between allele A_i and A_j. $p_i(t+\varepsilon)$ is the frequency of allele A_i in the next generation under natural selection and mutation.

Equation (26) is equal to the probability that allele A_i will exist on generation $t+\varepsilon$. If the correspondence between genetic sequence and allele is one to one, the probability is given by equation (17),

$$p_i(t+\varepsilon) = \sum_{\boldsymbol{x}_j} P(i,j) \tag{28}$$

where the summation is done over all genetic sequences of population.

However, the connection is not strict, and more work is necessary in this direction.

5 Discussion

The crucial point is that we do not know the shape of fitness landscape of protein, DNA, or any other biological elements, but works in this field have been intensifying.

Although our method is equivalent to the population genetics and allows more detailed analysis, it is not useful for analysis of change in proportion of phenotypes in population. Its main purpose is to provide a framework to treat directly the genetic sequence, and it is appropriate for the problems involving changes in genetic sequence like crossover and change in sequence length like insertion and deletion. Also, it is relatively simple to handle introns and exons. The number of genes can also be increased, and the mapping from genetic sequence to phenotype can be introduced by a conversion table.

The principle proposed in this paper is that the genetic sequence cannot just seek for the direction where the increase in fitness is fastest, but it also has to take into account the easiness of change in such direction. So the optimal path becomes quite different from the case considering only the shape of the fitness landscape and giving the transition probability as the weight to the path. We believe our formulation is more realistic, and the optimization method with transition probability may also have many applications in engineering.

An important and interesting question is: "Does the DNA sequence evolves following the principle of least action?", or, "Is there such a principle in evolutionary process?". In another words, "Do genes evolve through some specific path or take any one arbitrary?". Even the answer is no, our method is still valid, and it means the life does not seek for optimal. Although it still has to be proved, we believe the evolution at genetic sequence level takes suboptimal paths, allowing many paths to be taken between two sequences.

This method is expandable in many directions, and evolutionary mechanisms at molecular level can be incorporated naturally since it treats directly the genetic sequence.

6 Conclusion

In this paper, we proposed an analytic framework for the analysis of evolutionary mechanisms at genetic coding level, and presented the background idea based on path integral, mathematical frameworks, and its relation with population genetics. Advantages and problems of the method were discussed.

Although this work is still in initial stage and many problems have to be solved, it provides an intuitive way to analyze the evolution at genetic sequence level. It may be a powerful analysis tool when we need the description of genetic sequence, for instance analysis of genes in genetic algorithms or other algorithms based on evolution.

Computer simulations are also being done to check the adequacy of the method.

References

1. Crow, J. F.: *Basic Concepts in Population, Quantitative, and Evolutionary Genetics.* W. H. Freeman and Co. (1986)
2. Crow, J. F. and Kimura, M.: *An Introduction to Population Genetics Thory.* Harper & Row (1970)
3. Feynman, R. P., and Hibbs, A. R.: *Quantum Mechanics and Path Integrals.* McGraw-Hill (1965)
4. Gillespie, J. H: *The Causes of Molecular Evolution*, Oxford Univ. Press (1991)
5. Kauffman, S. A.: *The Origins of Order*, Oxford Univ. Press (1993)
6. Kimura, M.: Optimum mutation rate and degree of dominance as determined by the principle of minimum genetic load. *Journal of Genetics* **57** (1960)
7. Kimura, M.: On the evolutionary adjustment of spontaneous mutation rates. *Genet. Res.* **9** (1967)
8. Kimura, M.: *The Neutral Theory of Molecular Evolution*, Cambridge Univ. Press (1983)
9. Macken, C. A. and Perelson, A. S: Protein evolution on rugged landscapes. *Proc. Natl. Acad. Sci. USA.* **86** (1989)
10. Muller, H. J.: Our load of mutations. *American Journal of Human Genetics.* **2** (1950)
11. Nei, M.: *Molecular Evolutionary Genetics*, Columbis Univ. Press (1987)
12. Nishikawa, K., Ishino, S., Takenaka, H., Norioka, N., Hirai, T., Yao, T. and Seto, Y.: Constructing a protein mutant database. *Protein Engineering.* **7** (1993)
13. Perelson, A. S. and Kauffman, S. A.: *Molecular Evolution on Rugged Landscapes: Proteins, RNA and the Immune System.* Addison-Wesley (1989)
14. Reidhaar-Olson, G. F. and Sauer, R. T.: Combinatorial cassette mutagenesis as a probe of the informational content of protein sequences. *Science.* **241** (1988)

The Evolution of Hierarchical Representations

Franz Oppacher and Dwight Deugo
Intelligent Systems Group, School of Computer Science
Carleton University, Ottawa, Canada, K1S 5B6
e-mail: oppacher@scs.carleton.ca, dwightdeugo@scs.carleton.ca

Abstract. In the areas of Genetic Algorithms and Artificial Life, genetic material is often represented by fixed-length chromosomes. The simplification of a fixed size, sequential sequence of genes is in accord with the 'principle of meaningful building blocks'. The principle suggests that epistatically related genes should be positioned close to one another. However, in situations in which gene dependency information cannot be determined a priori, a Genetic Algorithm that uses static list-structured chromosomes will often not work. The problem of determining gene dependencies is itself a search problem, and seems well suited for the application of a Genetic Algorithm. In this paper, we propose a Genetic Algorithm that evolves a hierarchical representation in which gene dependencies and values of a chromosome coevolve.

1 The Evolution of Hierarchies

In the areas of Genetic Algorithms (GAs), Artificial Life (AL) and Animats, genetic material is often represented as a chromosome (see [5] and [9]) containing a fixed number of genes or features which can assume certain values, called alleles. The collection of alleles provides the genetic information required for constructing a phenotype or for determining a possible solution to a given problem. Under this representation, the chromosome is a static structure: every gene's locus is fixed. To simplify the encoding further, GAs often use alleles of 0 and 1 (see [5], [2], [8]).

The simplification of a fixed size, sequential sequence of genes is in accord with the 'principle of meaningful building blocks':

> *'The user should select codings so that short, low-order schemata are relevant to the underlying problem and relatively unrelated to schemata over other fixed positions.' [5]*

The above principle suggests that epistatically related genes should be positioned very close to one another so that crossover will be less likely to disrupt the building blocks. However, this is only possible if one knows a priori what the gene dependencies (building blocks) are, and if one knows a priori how many genes are required to encode the problem. It is, however, all but impossible to know the gene dependencies a priori without detailed knowledge of the fitness calculation. Therefore, a coding that does not violate the above principle is unlikely.

In situations in which gene dependency information cannot be determined a priori, a GA that use a static, list chromosome structure will often not work. For example, a GA will not work if it uses a fitness function that requires two specific genes to be located beside one another for optimal fitness, and it does not provide an operator for this to occur in the chromosome, such as inversion. One of the problems with a unary operator such as inversion is that the new chromosome it produces does not gain from the experiences of other chromosomes. There are situations, however, where GAs work even though their gene dependencies cannot be determined a priori (see [6], [4]). These successes may be due to a lack of dependencies or to the fact that the encoding happened to reflect them.

The problem of determining gene dependencies is itself a search problem, and seems well suited for the application of a GA. In this paper, we propose a self-organizing GA, and, after describing four different chromosome representations, show that the best one for it to use to coevolve the organization and contents (gene dependencies and values) of a chromosome is a hierarchy.

2 Hierarchical Chromosomes

Is there any benefit in evolving the organization of the chromosome? As stated in the first section, in all but the simplest domains it is an impossible task to determine a problem's gene dependencies a priori. Therefore, if the fitness of the chromosome depends on the positioning of its genes, then to achieve the best results one must allow the genes to move within the chromosome so that their adjacency relations reflect their dependencies, or else be satisfied with suboptimal results. The benefits are obvious. Allowing the chromosome to determine its own gene dependencies without relying on static, hand-coded ones, makes it a much more flexible and adaptable structure.

Can one coevolve the organization of a chromosome, its components, and its contents? Exactly what is meant by a part of a chromosome? How does one measure a part's fitness? Are a chromosome and its parts processed separately or simultaneously? What should the representation of the chromosome be?

To help answer these questions, we construct and discuss four different GAs. Each one uses a different chromosome representation, but all share a common fitness function. The first GA uses a chromosome constructed as a linked list, and is ordered with the correct gene dependencies. The second GA uses a chromosome constructed as a linked list, but uses a random gene ordering. In these two organizations, the gene at the head of the list is known as the root gene. The third GA uses a chromosome constructed as a random, rooted tree, with genes represented by the nodes of the tree and their dependencies represented by its links. The tree is considered complete: every one of the chromosome's genes is present in the tree. The fourth GA uses a chromosome similar to the third's organization but, because the GA does not require that all genes be present in the tree, it is considered a subtree. In each representation, a gene's value can be either 0 or 1.

The fitness function used by all organizations depends on the ordering of a chromosome's genes and their values, and is described as follows:

```
Fitness
    "self refers to the chromosome"
    ^ fitness := (self rootGene GeneFitness) / (2 * numberOfGenesInChromosome).

GeneFitness
    | sum |
    "self refers to the gene"
    sum := (self geneNumber) - (self parentGene  geneNumber).
    "Only reward a gene if its geneNumber is greater than its parent. The maximum
        fitness contribution of a gene is 2, 1 for being in the correct order - with
        respect to the parent gene - and 1 for a gene value of 1"
    (sum >= 0)
        ifTrue: [sum := 1 + self GeneAllele]
        ifFalse: [sum := self GeneAllele].
    self linkedGenes do: [:attachedGene |
        sum := sum + attachedGene GeneFitness].
    ^ sum
```

The fitness of a chromosome is calculated as the fitness of its genes divided by two times the number of genes in the chromosome. The reason for the division is to establish a fitness value between zero and one, since the maximum fitness value of the genes is two times the number of genes in the chromosome.

The recursive function GeneFitness works through a chromosome's linked genes (lists or trees) starting at the root gene. If the gene number of a gene's predecessor, or parent (in the case of a tree) is less than it[1], a value of 1 is added to the fitness total. Next, the gene's allele value is added to the total. Therefore, the contribution of each gene to the overall fitness can be either 0, 1, or 2.

Correct Order	Gene Value	Fitness Contribution
yes	0	1
yes	1	2
no	0	0
no	1	1

The fittest chromosome will have all alleles set to 1 and its genes ordered from the smallest to the largest gene number. The fitness function imposes a dependency on the chromosome: to receive the maximum fitness, a chromosome's genes must be in the correct order. Since this dependency is unknown to the chromosome, the chromosome must discover it and the correct allele settings, to achieve maximum fitness.

As an example of calculating a chromosome's fitness, consider the following correctly ordered, linked list, four gene example. The first gene is located at position 1, the second at position 2, and so on. The gene's alleles are as follows: 0101. In this case, the function GeneFitness would return 6: 1 + 2 + 1 + 2. Therefore, the fitness of the chromosome is 0.75 (6/8).

Gene Number:	1 <-	2 <-	3 <-	4	
Gene Allele:	0	1	0	1	
Fitness Contribution:	1	2	1	2	GeneFitness Total: 6

As another example of calculating a chromosome's fitness, consider the following randomly ordered, linked list, four gene example. In this example, the positions of the genes are randomly assigned in the linked list. In this case, the function GeneFitness would return 5: 1 + 2 + 0 + 2. Therefore, the fitness of the chromosome is 0.625 (5/8).

Gene Number:	3 <-	4 <-	1 <-	2	
Gene Allele:	0	1	0	1	
Fitness Contribution:	1	2	0	2	GeneFitness Total: 5

In this example, gene number 1 is penalized since its parent gene's gene number - 4 - is greater than its own.

As a final example, consider the following tree, four gene example. Gene 2 is the root gene, with allele 1, and has two children: gene 1 with allele 0, and gene 3 with allele 0. Gene 1 has one child: gene 4 with allele 1.

---

[1] The root gene's parent's gene number is the same as the root.

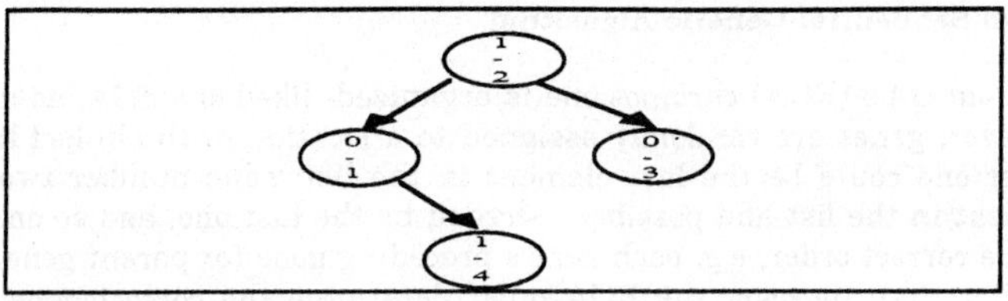

Figure 1. Example Tree Chromosome

In this case, the function GeneFitness would return 5: 0 + 2 + 1 + 2. Therefore, the fitness of the chromosome is 0.625 (5/8).

Gene Number:	4 ->	1 ->	2 <-	3	
Gene Allele:	1	0	1	0	
Fitness Contribution:	2	0	2	1	GeneFitness Total: 5

In this example, gene number 1 is penalized since its parent gene's gene number - 2 - is greater than its own.

The initial fitness of a subtree is computed like that of a tree. However, because a subtree does not have all the genes of the complete chromosome, its fitness value must be made to reflect this fact. This fitness calculation is discussed in section 2.4.

In the following subsections, we describe the results of using each of the above representations in a GA, and describe any new operators required to support them. For each representation, the GA uses the same gene dependent fitness function described above.

2.1 Sequential Genetic Algorithm

The sequential GA's (SGA) chromosome is organized as a linked list of genes. Gene number one is the first element in the list, gene number two is the second element in the list, and is linked to the first one, and so on. As the genes are already in the correct order, e.g. each gene's preceding gene (or parent gene) number is smaller than it own, all the SGA must do is determine the correct alleles for the genes - 0 or 1.

The SGA uses the two-point crossover operator described by Syswerda in [11] for combining chromosomes. One important feature of this operator, especially for the SGA, is that its application does not cause any change in the gene ordering of the resulting chromosome. The operator's application results in only an exchange of gene values, while maintaining gene ordering.

The fact that the SGA must only determine the correct gene values reduces the fitness function to that of the linear function $f(x) = c / n$. The value of the function is simply the number of 1 bits, c, in the input chromosome x divided by the total number of bits, n. This function presents no difficulty, since one hill climb from anywhere in the space will find the maximum value (see [1]).

The results in figure 5 prove this to be true. Figure 5 shows the median test run (out of 11) for the SGA to evolve the correct values for chromosomes of sizes of 5, 10, 20, and 40. The population sizes used by the SGA are a linear function of the size of the chromosome, defined as follows: PopulationSize(x) = 2 * size of x. As the graph indicates, the SGA has no problem in finding the best gene setting and quickly converges to the correct solution for varying chromosome sizes. This result is to be expected and establishes a baseline for the remaining GA test runs using different chromosome representations. However, remember that the SGA did not have to establish the gene ordering, because that was done a priori.

2.2 Random Sequential Genetic Algorithm

The random GA's (RGA) chromosome is organized, liked the SGA, as a linked list of genes. However, genes are randomly assigned to a location in the linked list. Therefore, gene number one could be the last element in the list, gene number two could be the middle element in the list and possibly preceded by the last one, and so on. As the genes are not in the correct order, e.g. each gene's preceding gene (or parent gene) number may not be smaller than its own, the RGA must determine the correct gene ordering and values.

Like the SGA, the RGA uses the two-point crossover operator. However, this causes a problem for the RGA. For example, given the following two chromosomes, A and B (with genes encoded as tuples: [value:gene-number], e.g. 0:3 means gene 3 has value 0), and crossover points 2 and 5, the application of the two-point crossover produces two chromosomes A' and B' as follows:

A:	1:1	1:3	1:5	1:7	1:9	1:2	1:4	1:6	1:8	1:10
B:	0:7	0:6	0:5	0:1	0:9	0:3	0:10	0: 4	0:8	0:2
A':	1:1	0:6	0:5	0:1	0:9	1:2	1:4	1:6	1:8	1:10
B':	0:7	1:3	1:5	1:7	1:9	0:3	0:10	0: 4	0:8	0:2

A good feature of this operator for the RGA is that its application causes a change in the gene orderings of the resulting chromosomes. Since the chromosome's gene ordering is randomly initialized, and only 1 in $n!$ gene orderings is correct, the RGA requires an operator that will alter the gene ordering. However, the operator's problematic feature is that its application results in both a duplication and loss of genes in the resulting chromosomes. For example, chromosome A' has genes 1 and 6 present twice, and for both values of 0 and 1. Also, genes 3 and 7 are missing from it. Without some other feature of the RGA to correct these problems, the RGA seems doomed to fail. This was in fact the end result. None of our test runs ever converged to the correct answer; therefore, these results have been omitted from figure 5. The main problem was that after a few generations, the population was filled with chromosomes with missing genes. Since mutation only changed the genes values and not the gene numbers, the missing genes could not return to the population. After this point, there was no way that the RGA could solve the problem.

Since it was next to impossible for the RGA to produce the correct chromosome using two-point crossover and mutation - there was always a remote chance that the RGA could do it, but it never happened in our experiments - the only other way for the RGA to produce it was to stumble upon it in the very first generation. The probability of this happening is $1 / 2^n * n!$, i.e. $n!$ different gene orderings, each with 2^n different allele settings. We conclude that the RGA is not a successful approach for coevolving the organization and values of a chromosome.

2.3 Tree Genetic Algorithm

The tree GA's (TGA) chromosome is organized as a rooted tree of genes. Like the RGA, genes are randomly placed in the tree, organized initially as a binary tree. Therefore, gene number one could be a leaf node, gene number two could be an internal node, the last gene could be the root node, and so on. As the genes are not in the correct order, e.g. each gene's parent gene number may not be smaller than its own, the TGA must determine the correct gene ordering and values.

Although initially constructed as binary tree, the tree will become a n-ary tree as soon as crossover is applied. The TGA uses a new crossover operator called the Tree Crossover Operator (TreCoP). The TreCoP creates one new tree from two chromosomes. First a node (gene) is selected in the original tree. The selection process is weighted to select its root node with probability 0.5, and all other nodes with equal probability. Next, a node in the mate's tree is selected. This selection process is weighted to select a child node of its root with probability 0.5, and all other nodes with equal probability. Next, the original's subtree is replaced by the mate's subtree, unless the original subtree selected is the entire tree, in which case the mate's subtree becomes the new rooted tree. Then, because crossover may cause genes to be duplicated or missing from the resulting tree, the duplicate genes are removed from it and the missing ones added.

The reason for using the weighted selection functions is simple. They result in a constant pulling of nodes towards the root of the tree, which is desirable for reorganizing the nodes in the tree. Since the original tree's selected subtree is often the root (at least 50% of the time), the mate tree's selected subtree will replace it. Since the mate's subtree is often a child of the root (at least 50% of the time), and since the trees tend to be similar as the population converges to a solution, this results in elevating a child to the status of the root of the tree.

What is also desirable about this operator is that the tree it produces closely resembles the two parent trees. To do this, care must be taken when removing duplicate genes and restoring missing ones. The procedure of removing duplicate genes is done in a depth first manner. Once a duplicate gene is detected, it is removed from the tree and its children added as the children of the removed gene's parent. For example, removing gene 2 from the left tree of figure 2 results in the right tree of the figure.

If the gene being removed is at the root of the tree, one of its children is selected at random to become the new root. The remaining children of the old root are then added as the children of the new root. For example, removing gene 1 from the left tree of figure 3 results in the right tree of the figure.

Missing genes are added into the new tree in the same location as they were located in the original tree[2]. For example, if gene 2 was a child of gene 3 in the original tree, and it is missing from the resulting tree, it is added as a child of gene 3 in the resulting tree. If this is not possible because the parent gene - gene 3 - is not present in the resulting tree, the gene is randomly added to one of the leaf nodes. By ordering the missing genes in the way they are visited by a breadth first search of the original tree, this problem is mostly avoided. The reason is that if two dependent genes are missing from the resulting tree, the parent gene is always added first, ensuring that the child can always be attached to it and henceforth in its proper location.

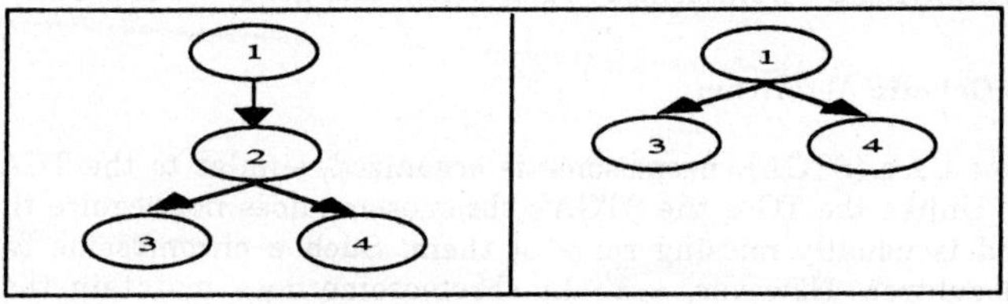

Figure 2 Removing a Gene

[2] Only those genes missing from the original tree, not the mate's tree, are added.

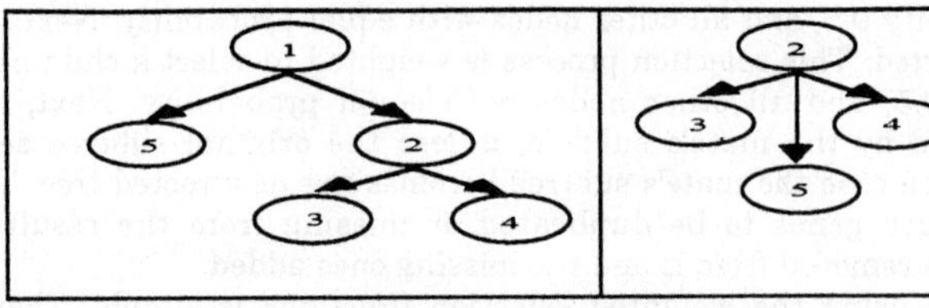

Figure 3 Removing the Root Gene

The TreCoP and its two supporting procedures for adding missing genes and removing duplicate ones attempt to maintain the same spirit of organization in the resulting chromosome as was found in the parent chromosomes, while also managing to change and reorganize the genes into a combination of the two. Since a TGA's chromosome ordering is randomly initialized, it is important that the TreCoP search for good gene orderings. However, it must not mix the existing chromosomes so much as to lose any information that has been gained from past explorations. The operator must simultaneously exploit old gene orderings while exploring new ones - the hallmark of GAs.

Does the TGA work? Does it locate the best gene ordering and values? The answer is yes. Figure 5 again shows the median test run (out of 11) for the TGA to coevolve the correct gene ordering and values for chromosomes of size 5, 10, 20, and 40. Although not as fast as the SGA, the TGA has no problem converging to a solution for varying chromosome sizes. The extra time required for the TGA to converge to a result, compared to the SGA, was to be expected. It must coevolve the organization and values of the genes. The SGA only evolves the values.

To get a rough estimate on the number of different trees in the TGA search space, consider the case of oriented trees. Oriented trees are not treated as different when they differ only in the respective ordering of subtree nodes, which is the case in the TGA's chromosome. For oriented trees with n labeled vertices (genes), there are n^{n-1} distinct oriented trees (see [7]). In addition to this space, the TGA searches the space of 2^n different gene values. Therefore, the TGA searches a space of size approximately: $2^n * n^{n-1}$.

Even with this enormous search space, the TGA works well, and is our first viable approach to coevolving a chromosome's organization and genes values. Can we do better?

2.4 Subtree Genetic Algorithm

The subtree GA's (STGA) chromosome is organized, similar to the TGA, as a rooted tree of genes. Unlike the TGA, the STGA's chromosome does not require the presence of all genes, and is usually missing some of them. Such a chromosome is known as a chromosomal subtree. However, a STGA chromosome does maintain that no gene is duplicated in a subtree.

The chromosome is subjected to a new crossover operator called the SubTree Crossover Operator (STreCoP). The code for the STreCoP is identical to that of the TreCoP, but its behavior is slightly different in one respect. As in the TreCoP, missing genes are added to the resulting tree in the same location they were located in the original tree. However, this procedure adds only those genes that are in the original tree, not the mate's tree, and missing from the resulting tree. This feature makes little difference in the TreCoP, since the original tree has all of its genes, so must its offspring. However, since now the original tree may not have all of its genes, those missing in it will also be missing in resulting tree, unless some of them were crossed over into the resulting tree from the mate's tree.

A minor difference between the TGA and the STGA is in the way the STGA measures fitness. Since STGA will have both complete and incomplete chromosomes of different sizes, a method is required to establish the fitness of each organization relative to the others.

The fitness of a chromosome is initially calculated as described at the beginning of section 2: the fitness of its genes divided by two times the current number of genes[3] in the chromosome. This method guarantees that all chromosomes have a fitness value between 0 and 1. Next, the fittest chromosome is selected from the STGA's population and its size - the number of nodes - is used to rank the remaining chromosomes relative to it. It is not the case that the best chromosome is always the largest. For example, a two node chromosome with both genes set to 1 and arranged in the correct order will have a fitness of 1. A ten node chromosome with all genes set to zero and in the wrong order will have a fitness of 0. Using the fittest chromosome's size ('relativeSize' below), we calculate the fitness of all other chromosomes as follows:

```
RelativeFitness: relativeSize
    " self refers to the chromosome "
    (self size > relativeSize)
    ifTrue: [
        "If my chromosome size is greater than the best chromosome's size, then
            decrease my fitness by an amount proportional to the ratio of the sizes"
        ^ fitness := self fitness * relativeSize / self size]
    ifFalse: [
            "If my chromosome size is less than the best chromosome's size, then
                decrease my fitness by an amount proportional to the ratio of the sizes"
            ^ fitness :=  self fitness *  self size / relativeSize]
```

The fitness of a chromosome - referred to as 'self' - is changed proportionally to the ratio of its size and the size of the fittest chromosome. Chromosomes that have more or less nodes than the best chromosome will see their fitness values decreased. Chromosomes that have the same number of nodes will see no change in their fitness. The effect of this fitness function is to keep the STGA working at one level - size of chromosome - until good subtrees - building blocks - are produced. These subtrees are then the material for the construction of the next level. Once a level is established, i.e. the chromosomes at that level have superior fitness, the STGA's effort is concentrated there, developing the subtrees for the next level, and so on.

For example, consider the following median run of a STGA using chromosomes with 10 genes. From left to right, the columns indicate the generation number, the fitness of the best individual in the population, the median fitness value of the population of twenty chromosomes, the fitness of the worst individual, and the fitness, chromosome organization, and gene values of the second best individual in the population (7/5:1 indicates that gene 7 is a child of gene 5 and has a value of 1).

0	1.0	0.36	0.01	[1.0]	5/5:1, 7/5:1
1	1.0	0.40	0.01	[0.66]	4/4:1, 5/4:1
2	1.0	0.65	0.12	[1.0]	5/5:1, 6/5:1, 7/5:1
3	1.0	0.59	0.15	[0.75]	3/3:1, 5/3:1, 10/5:1
4	1.0	0.63	0.28	[1.0]	1/1:1, 3/1:1, 9/1:1, 10/1:1
5	1.0	0.61	0.33	[1.0]	1/1:1, 3/1:1, 9/1:1, 10/1:1
6	1.0	0.59	0.26	[0.84]	1/1:1, 9/1:1, 2/1:1, 8/1:1, 3/8:1, 10/8:1
7	1.0	0.61	0.25	[0.84]	1/1:1, 3/1:1, 5/3:1, 7/5:1, 6/1:1, 4/6:1
8	1.0	0.59	0.22	[0.85]	1/1:1, 3/1:1, 5/3:1, 7/5:1, 2/1:1, 4/2:1

[3] In the TGA the number of genes in the chromosome is always the same. However, in the STGA this number may vary.

9	1.0	0.69	0.48	[0.86]	1/1:1, 3/1:1, 8/1:1, 4/1:1, 7/4:1, 5/7:1, 9/5:1
10	1.0	0.66	0.34	[1.0]	1/1:1, 3/1:1, 5/3:1, 7/5:1, 2/1:1, 10/2:1, 4/2:1
11	1.0	0.73	0.40	[1.0]	1/1:1, 3/1:1, 5/3:1, 7/5:1, 9/5:1, 2/1:1, 4/2:1
12	1.0	0.71	0.44	[1.0]	1/1:1, 3/1:1, 5/3:1, 7/5:1, 9/5:1, 2/1:1, 4/2:1
13	1.0	0.61	0.32	[1.0]	1/1:1, 3/1:1, 5/3:1, 7/5:1, 9/5:1, 2/1:1, 4/2:1
14	1.0	0.58	0.25	[0.76]	3/3:1, 5/3:1, 7/5:1, 10/5:1, 9/5:1, 8/3:1, 2/8:1, 4/2:1
15	1.0	0.60	0.34	[0.86]	1/1:1, 2/1:1, 4/2:1, 10/1:0, 3/1:1, 5/3:1, 7/5:1
16	1.0	0.62	0.34	[1.0]	1/1:1, 3/1:1, 5/3:1, 7/5:1, 9/5:1, 2/1:1, 4/2:1
17	1.0	0.58	0.40	[0.86]	1/1:1, 3/1:1, 5/3:1, 7/5:1, 9/5:0, 2/1:1, 4/2:1
18	1.0	0.57	0.34	[1.0]	1/1:1, 3/1:1, 5/3:1, 7/5:1, 9/5:1, 2/1:1, 4/2:1
19	1.0	0.73	0.40	[1.0]	1/1:1, 3/1:1, 5/3:1, 7/5:1, 9/5:1, 2/1:1, 4/2:1
20	1.0	0.71	0.44	[0.87]	1/1:1, 3/1:1, 5/3:1, 7/5:1, 6/5:1, 2/1:1, 4/2:1
21	1.0	0.68	0.44	[0.87]	1/1:1, 3/1:1, 5/3:1, 7/5:1, 9/5:1, 2/1:1, 6/2:1
22	1.0	0.68	0.39	[1.0]	1/1:1, 3/1:1, 5/3:1, 9/5:1, 10/5:1, 2/1:1, 8/2:1, 4/2:1
23	1.0	0.75	0.53	[1.0]	1/1:1, 2/1:1, 8/2:1, 10/2:1, 3/1:1, 5/3:1, 7/5:1, 6/5:1
24	1.0	0.73	0.46	[1.0]	1/1:1, 2/1:1, 8/2:1, 10/2:1, 6/1:1, 4/1:1, 5/4:1, 7/5:1
25	1.0	0.71	0.31	[1.0]	1/1:1, 3/1:1, 5/3:1, 7/5:1, 6/5:1, 2/1:1, 8/2:1, 4/2:1
26	1.0	0.66	0.25	[0.90]	1/1:1, 3/1:1, 5/3:1, 10/5:1, 7/5:1, 2/1:1,9/2:0,6/2:1,4/2:1, 8/2:1

Looking down the right hand side of the above STGA run, one can see how the size of the second best chromosome changes. Once the STGA produces a good chromosome of a particular size, it remains at that level until more good ones of the same size can be produced for the construction of chromosomes for the next level. Initially, this occurs quickly. The reason is that there are very few subtrees to consider at the initial levels. However, at the final levels, as the subtrees become larger, the switching of levels is not as fast because of the increased number of subtrees that have to be considered before proceeding to the next level. In the above test run, it took twenty-six generations for the STGA to produce the complete chromosome shown in figure 4: each gene in the correct position, and each with the correct gene value of 1.

Does the STGA work? Does it locate the best gene ordering and values? The answer is obviously yes. The surprising result is that it works faster than the TGA. Figure 5 again shows the median test run (out of 11) for the STGA to coevolve the correct gene ordering and values for chromosomes of sizes 5, 10, 20, and 40. Although not as fast the SGA, the STGA has no problem in finding the best gene setting and quickly converges to a solution for varying chromosome sizes. The extra time required for the STGA to converge to a result is to be expected. It must coevolve both the organization and values of the genes. The SGA only evolves the values.

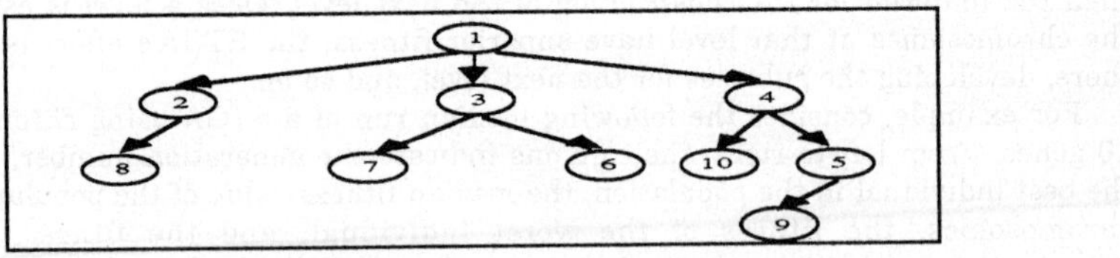

Figure 4 Complete Chromosome

We achieved similar results from our research into the Traveling Salesman Problem (TSP) in [3]. In that work, we developed an efficient schema-based representation and operators for solving the TSP. A schema was used as an explicit part of the representation rather than just a theoretical construct. In that domain, we found that the direct use of schemata provided a much more natural representation and made the application of genetic operators easier. For example, by using schemata incomplete tours were permitted, and, because operators did not have to form complete tours, exchange and variation of genetic material was simple. Empirical results showed that the new representation and operators quickly provide good results.

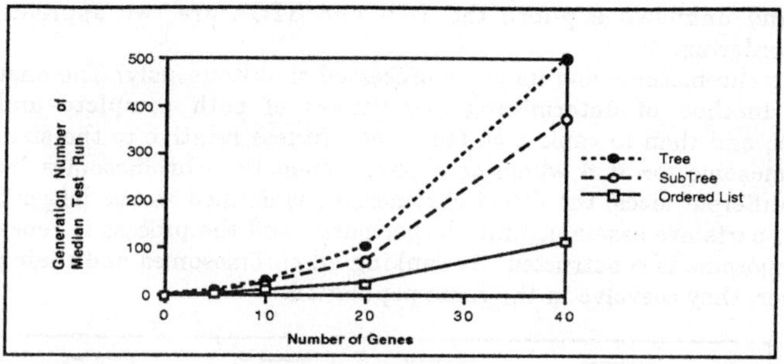

Figure 5 Gene Order Dependent GAs

In that paper, we conjectured that explicit schema processing would reduce the complexity of problems in other domains as well. Others also believe that 'the time required for a complex system, containing *k* elementary components, say, to evolve by a process of natural selection from those components is very much shorter if the system is itself comprised of one or more layers of stable component subsystems than if its elementary parts are its only stable components' (see [10]). Viewing a subtree as a schema (a stable component), the results of the STGA backs up this conjecture and belief.

The surprising result is that the STGA works even better than the TGA, and is our second, and best, approach to coevolving a chromosome's organization and gene values.

3 Conclusion

Can one coevolve the organization of a chromosome, its components, and their contents? The answer is yes. Both the TGA and the STGA manage to do both. These approaches are not as fast as the SGA, but then the SGA already has the correct gene ordering. The TGA and the SGA must discover this for themselves, and this activity requires time.

Exactly what is meant by a part of a chromosome? Both the TGA and the STGA use a chromosome structured as a rooted tree. Therefore, a part of their chromosome is nothing more than a subtree.

A subtree is analogous to a GA's building block: a compact way to express similarities among chromosomes. However, rather than imposing a strict ordering on all of the genes in the chromosome, as a schema does, a subtree only imposes an ordering on those genes for which the values are known. For this reason, the number of complete chromosomes represented by a subtree is more than that represented by a simple schema. Take for example the simple case of a schema of order n-1, with only one unknown gene value, e.g. 10*. This schema represents only two strings: 101, and 100. When the genes are positioned as nodes in an n-ary tree, the missing gene can be attached to any one of the existing n-1 nodes. Therefore, if we are using binary genes, the one subtree represents 2(n-1) different possible trees. In our example, the subtree (building block) would be part of four other possible trees, shown in figure 6.

The difference between the number of actual strings represented by a schema and by a subtree increases as the number of missing genes increases. Therefore, if the power of a GA's search is related to the number of schemata in a population, then the use of subtrees, or even trees, in a strongly epistatic environment, should enhance a GA's performance.

How does one measure the fitness of a subtree? As with a tree, a subtree's fitness is calculated as a function of the gene values. If this function is dependent on the ordering of the genes, and unknown a priori, the TGA and STGA are two approaches that will discover the ordering.

How are a chromosome and its parts processed simultaneously? The answer is to first establish a method of determining the fitness of both complete and incomplete chromosomes, and then to rank each (sub)tree's fitness relative to the size of the fittest known chromosome, be it a whole or a part. When two chromosomes have the same fitness, but different sizes, the fittest chromosome is defined as the larger of the two. In this manner, parts are assembled into larger parts, and the process is repeated until the desired chromosome is constructed. By ranking all chromosomes and their parts relative to one another, they coevolve in the same population.

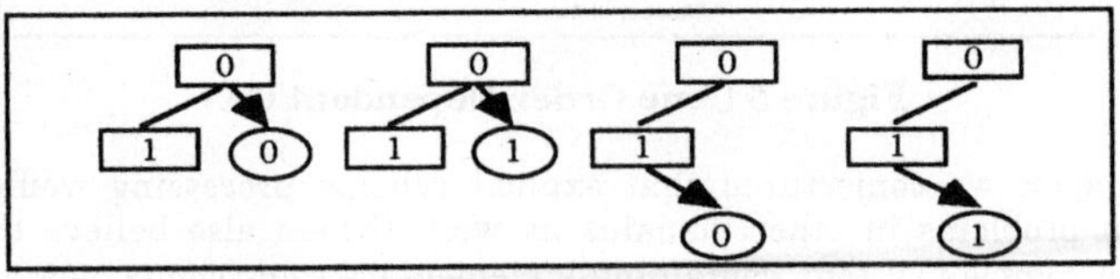

Figure 6 Subtree Building Block

What should the structure of the chromosome be? When the gene dependencies are known a priori, a simple list provides the best structure, as demonstrated by the SGA. However, when the gene dependencies are unknown and a list structure is used, as in the RGA, the approach fails. The structure to use in the latter case is a tree, as demonstrated by the TGA; a tree can be used to coevolve both the organization and content of the chromosome. However, if fitness values can be determined for the parts of a tree, the best structure for a chromosome is a subtree that grows into complete trees, as demonstrated by the STGA.

Is there any benefit to evolving the organization of the chromosome? The answer is yes. Since in most complex systems it is an impossible task to determine all gene dependencies a priori, the only way genetic methods will succeed is if these dependencies can be determined by the GA itself. Since GAs are good at searching, it seems an obvious choice for them to evolve their own organization!

4 References

1. Ackley, D.H., An Empirical Study of Bit Vector Function Optimization. In Lawrence Davis (Ed.), Genetic Algorithms and Simulated Annealing, Morgan Kaufmann, 170-204, 1987.
2. Calloway, D.L., Using a Genetic Algorithm to Design Binary Phase-Only Filters for Pattern Recognition, Proceedings of the fourth International Conference on Genetic Algorithms, Morgan Kaufmann, 422-427, 1991.
3. Deugo, D.L. & Oppacher, F., Explicitly Schema-Based Genetic Algorithms. Proceedings of the Ninth Biennial Conference of the Canadian Society for Computational Studies of Intelligence, Morgan Kaufmann, 46-53, 1992.
4. Gabbert, P.S., Brown, D.E., Huntley, C.L., Markowicz, B.P., and Sappington, D.E., A System for Learning Routes and Schedules with Genetic Algorithms, Proceedings of the fourth International Conference on Genetic Algorithms, Morgan Kaufmann, 430-436, 1991.
5. Goldberg, D.E., Genetic Algorithms in Search, Optimization, and Machine Learning, Addison-Wesley, 1989.

6. Grefenstette, J.J., A System for Learning Control Strategies with Genetic Algorithms, Proceedings of the third International Conference on Genetic Algorithms, Morgan Kaufmann, 183-190, 1989.
7. Knuth, D.E., The Art of Computer Programming: Fundamental Algorithms. V:1, Addison-Wesley Publishing Company, 391, 1973.
8. Nakano, R., Conventional Genetic Algorithm for Job Shop Problem, Proceedings of the fourth International Conference on Genetic Algorithms, Morgan Kaufmann, 474-479, 1991.
9. Rizki, M.M. & Conrad, M., Computing The Theory of Evolution, Physica D, **22**, 83-89, 1986.
10. Simon, Herbert, A., The Organization of Complex Systems, In H.H. Pattee (Ed.) Hierarchy Theory, 7, 1973.
11. Syswerda, G., Uniform Crossover in Genetic Algorithms, Proceedings of the Third International Conference on Genetic Algorithms, Morgan Kaufmann, 2-9, 1989.

3. Adaptive and Cognitive Systems

Adaptation and the Modular Design of Organisms

Günter P. Wagner

Center of Computational Ecology
Yale University, New Haven, CT 06511, USA

Abstract. In this paper the implications of the theory of evolutionary computation for evolutionary biology are explored. It is claimed that the concept of "representations" is particularly useful to understand the evolution of complex adaptation and the origin of the modular design of higher organisms. Modularity improves the adaptability of complex adaptive systems, but arises most likely as a side effect of adaptive evolution rather than being an adaptation itself.

1 Introduction

Adaptations are the result of natural selection on spontaneously arising, heritable variation. Little doubt remains on these essentials 150 years after the publication of Darwin's and Wallace's original proposals. Natural selection is a universal principle in every system where self-replicating entities show variation, heritability and differential reproduction (Lewontin, 1970). This universality of natural selection allowed an unexpected triumph of Darwinism in computer science where the principles of biological adaptations are increasingly used to solve computational problems. The rapidly growing arena of evolutionary algorithms includes genetic algorithms (Holland, 1992), evolutionary strategies (Rechenberg, 1973) and genetic programming (Koza, 1992). These applications show, in an artificial reality, the efficacy of the Darwinian process to solve even the most intricate optimization problems. In principle, this success should lay to rest the worry whether selection on spontaneous variation is able to explain the origin of complex adaptations, as for instance the notorious vertebrate eye (Darwin, 1859). However, evolutionary computer science not only demonstrates the problem solving power of the Darwinian method, but brings into focus the largely unsolved real problem of the origin of complex adaptations.

The process of adaptation can only proceed to the extent as favourable mutations occur. This is usually not a major concern to biologists, because they study the endproducts of evolution, and their very existence is powerful evidence that the favourable mutations have occured at a sufficient rate. However, the computer scientists, who want to solve engineering problems with evolutionary algorithms, start with inferior designs and want to improve them. For them the time it takes to actually obtain the improvement is money. In studying how quickly an improvement can be obtained it was discovered that the mutation/selection process is not universally effective. This is not an entirely new insight (see for instance Eden (1967), Bossert (1967), Simon (1965) or Bremermann (1966)), and it is also easy to see why this is the case. There is no way to improve the performance of a conventional computer program by randomly exchanging letters in the source code. However, Darwinian "evolution" of computer programs is indeed possible, as shown by the Tierra program of Thom Ray (1992) and the genetic programming methods (Koza, 1992). Hence, the Darwinian solution of optimization problems is possible if and only if the problem is "coded" in a way that makes the mutation-selection procedure an effective one. This fact is known as the "representation problem:" how to code a problem such that random variation and selection can lead to a solution? The representation problem is about the likelihood to obtain an improvement by mutation and/or recombination.

For biology the "representation problem" has some unsettling implications. If, as evolutionary biology asserts, all adaptations are the result of mutation and selection, organisms have to be evolvable. But how and why did an evolvable genetic representation of the phenotype originate in the first place? Is there an evolution of genetic representations of the phenotype? What are the evolutionary forces that shape genetic representations? The thesis of this essay is that the concept of *genetic representations of the phenotype* and its evolution forms a link among seemingly unrelated problems of evolutionary biology: the role of epistasis in adaptation, genetic canalization, developmental constraints, developmental and morphological integration, the evolution of complex adaptations, the biological basis of homology and perhaps the origin of body plans. Further it will be suggested that computer science and biology experience a phase of substantive convergence of interests, which makes computerscience the strongest allie in the attac on many of the still unsolved problems in evolutionary biology.

2 Variation and Variability

To accommodate a discussion of genetic representations in the language of evolutionary biology it is essential to clearly distinguish between *variation* and *variability*, even if these terms are often used synonymous in the scientific literature. The term *variation* refers to the actually present differences among the individuals in a population or a sample, or between the species in a clade. Variation can be directly observed as a property of a collection of items. In contrast *variability* is a term that describes the potential or the propensity to vary. Variability thus belongs to the group of "dispositional" concepts, like solubility (Goodman, 1955). Solubility does not describe an actual state of a substance, say sodium chloride, but its expected behavior if brought into contact with a sufficient amount of solvent, for instance water. Similarly, variability of a phenotypic trait describes its propensity to change in response to environmental and genetic influences. The representation problem is thus about the variability of the phenotype and not directly about the genetic variation within populations. And the concept of developmental constraints (sensu Maynard-Smith et al. 1985) is about the limits of variability of traits caused by the way they are "coded" in the genome.

As a directly observable property, variation is comparatively easy to measure. Genetic variation in a population is measured by the heterozygosity or the degree of polymorphism. Quantitative phenotypic variation is measured by the phenotypic, genetic and environmental variance or any other statistical measure of dispersion. In contrast, variability is much harder to measure. Genetic variability at the molecular level is measured as mutation rate. Genetic variability of quantitative phenotypic traits is measured by the mutational variance V_m, the average additive genetic variance produced per generation by mutations, or in the case of more than one trait, by the mutational covariance matrix, **M**. Each of these quantities requires elaborate experimental designs to be estimated.

The relationship between variation and variability is conditional. Clearly, if there is variation in a character it has to be variable, but the reverse is not true. Therefore the study of natural variation can give hints on the pattern of variability as for instance the study of osteological variation suggests the existence of constraints (Alberch, 1983; Rienesl and Wagner, 1992), but it is at best a surrogate of variability.

The genetic variance of a trait, i.e. the raw material of evolution, is a fairly ephemeral property. It depends on the complement of genes currently segregating in the population, the effect of the alleles present and their frequencies. Whenever an allele changes its frequency or gets fixed the genetic variance of the character may change (Bürger Wagner Stettinger, 1989; Turelli, 1988). The same is true for genetic correlations, which not only depend on the alleles segregating but also on the linkage disequilibrium among them (Bulmer, 1980). On the other hand the genetic variability of a character is a property of the genome. It remains the same as long as the complement of loci and the mutation rate is the same and as long as no epistatic mutations have been substituted. However, variability is under genetic control and may thus evolve.

Evidence that variability of phenotypic traits is under genetic control comes from research on the phenomenon of "canalization." The term was first introduced by Waddington (1957) to describe the tendency of development to produce clearly distinguished tissue and organ types. However the concept had only limited impact on developmental biology, but became important in quantitative genetics. It describes the fact that mutant phenotypes often show much more variation than the wild type phenotype. Some of this variation is genetic variation which was "suppressed" in the wild type genetic background (for a recent review, see Scharloo, 1991). Selection experiments suggested that the sensitivity to genetic variation of a trait can be decreased by artificial stabilizing selection (Rendel, 1967; Scharloo, 1988). Most recently it has been shown that the average effect of P-element induced mutations on life history traits in *Drosophila* is negatively correlated with the influence on fitness of the trait. The stronger the impact on fitness the smaller the average effect of a new mutation (Stearns et al., 1995).

Evidence for genetic control over phenotypic variability is of capital interest to evolutionary theory (Sharloo, 1991). This literature shows that evolution by fixation of spontaneously generated variation does not just happen, but that evolution can also change the "rules" under which heritable variation is produced, i.e. the variability of the traits itself can evolve. Some characters that were variable can become fixed (Riedl, 1975, Stebbins, 1974), while others may become integrated into a tightly coupled complex of characters (Stearns, 1993) or others may gain variability after a developmental constraint was broken (Vermeij, 1970). Perhaps Schmalhausen was the first to clearly see the theoretical implications of a genetic control of variability (Schmalhausen, 1949). His key observations of abundant epistatic effects among mutations is just another way of observing that the genetic variability of a trait is under genetic control. Per definition, epistasis is the influence of the locus at one locus on the effects of alleles at other loci. It thus reflects the fact that the expression of genetic variation is under the influence of other genes.

Population genetics has been developed to understand the dynamics of genetic *variation*. However, the issue here is the evolution of the *variability* of characters. So the question is how to describe the variability of a trait and its evolution in population genetic terms in order to link the theory of evolvability to the existing apparatus of evolutionary theory. Genetic variability of a character is determined by two factors: the rate of mutation of genes influencing the character and the effect of the mutations on the state of the character. Mutation rate is a standard parameter in population genetic model and there is also a theory on the selection forces acting on mutation rate (Altenberg and Feldman, 1987). The effect of a mutation can either be arbitrarily assigned to individual alleles, or described as the variance of the distribution of mutational effects. Mathematically, the relationship between the genotype and the phenotype is a function *f*, which assigns to each genotype *G* the average phenotype *P* (averaged over so-called 'environmental' variation).

$$G \xrightarrow{f} P$$

The idea of a genotype phenotype mapping function has been used in quantitative genetics, for instance in the study of genetic canalization (Rendel, 1967, Scharloo, 1987), multivariate mutation selection balance (Wagner, 1989a) and the study of epistatic effects (Gimelfarb, 1989, Wagner et al., 1994) and in evolutionary algorithms (for instance: Altenberg, 1994; Banzhaff, 1994; Schwefel, 1981). The genotype-phenotype mapping function describes how genetic variation is translated into phenotypic variation and is thus a way of describing how the phenotype is represented in the genotype. The evolution of genetic representations can thus be modeled as the influence of selection on the genotype-phenotype mapping function.

3 Complex Adaptation: When are they Possible?

The digression on variability and its genetic control sets the stage to consider the issued of evolvability in a biological context. If the expression of genetic variation is itself under genetic control, then it is conceivable that species evolve "strategies" of how to allocate the phenotypic effects of genetic mutations. But what exactly is evolvability and what influences its degree?

In the field of evolutionary algorithms, it is essential to understand which algorithm will effectively improve by random variation and selection. Therefore most of the theory of evolvability has been developed in the context of evolutionary algorithm theory (Jones and Rawlins, 1993; Altenberg. 1994). A comprehensive concept of evolvability was published recently by Lee Altenberg (1994). It is based on a generalization of Price's covariance theorem of natural selection (Price, 1969).

Evolvability Theorem (Altenberg, 1994): *the probability that a population generates individuals fitter than any existing is*

$$\overline{F}(w_{\max})' = \alpha\left\{R(w_{\max}) + \overline{\beta}_u(w_{\max}) + Cov\left[\beta_{jk}(w_{\max}), \tfrac{w_j w_k}{\overline{w}^2}\right]\right\}$$

where a is the maximal rate at which new genotypes are generated by mutation and/or recombination. *R(wmax)* is the probability that a random sampling of genotypes yields a genotype with fitness larger than the currently best, $\overline{\beta}_u(w_{\max})$ is the average search bias, the probability that the mutation and/or recombination of the given genotype produces better genotypes and $Cov\left[\beta_{jk}(w_{\max}), \frac{w_j w_k}{\overline{w}}\right]$ is the covariance between the current fitness of genotypes and the probability to produce better ones by mutation and or recombination (for details see Altenberg, 1994).

In short this theorem says that evolvability depends on two main factors: the rate of production of genetic variation by what ever means, and a correlation between the current fitness of genotypes and the likelihood to obtain even better genotypes from the already good ones. This theorem defines what the intuitive notion of a "smooth" adaptive landscape suggests: it is easy to evolve by natural selection if the adaptive landscape is smooth, which means that the better genotypes are found in the mutational "neighborhood" of the good genotypes.

Another way of expressing this result is that adaptations are possible if their improvement can be achieved in a cumulative or stepwise fashion. This has been known for some time, but what are the structural features that make stepwise improvement possible? The key feature is the covariance term in Altenberg's theorem. On average, further improvements must not compromise past achievements. This is the essence of the so-called "building block hypothesis" to explain the performance of Genetic Algorithms (Holland, 1992; Forrest and Mitchell, 1993). In terms of the representation of phenotype in the "genomic code" this implies that independent biological functions shall be coded independently so that their improvement can be also be realized with minimal interference. This leads to the concept of modularity underlying the various explanations of complex adaptations offered by biologists.

4 Modularity of Development

Evolution of complex adaptation requires a match between the functional relationships of the phenotypic characters and their genetic representation. This was clearly expressed by Rupert Riedl (1975) in his thesis of the "imitatory epigenotype." If the epigenetic regulation of gene expression "imitates" the functional organization of the traits then the improvement by mutation and selection is facilitated. Riedl predicts that the evolution of the genetic representation of phenotypic characters tends to favor those representations which imitate the functional organization of the characters. Imitation means that complexes of functionally related characters shall be "coded" as developmentally integrated characters but coded independently of functionally distinct character complexes.

Independent genetic representation of functionally distinct character complexes can be described as modularity of the genotype-phenotype mapping functions. A modular representation of two character complexes C1 and C2 is given if pleiotropic effects of the genes are more frequent among the members of a character complex than among members of different complexes (see Fig. 1).

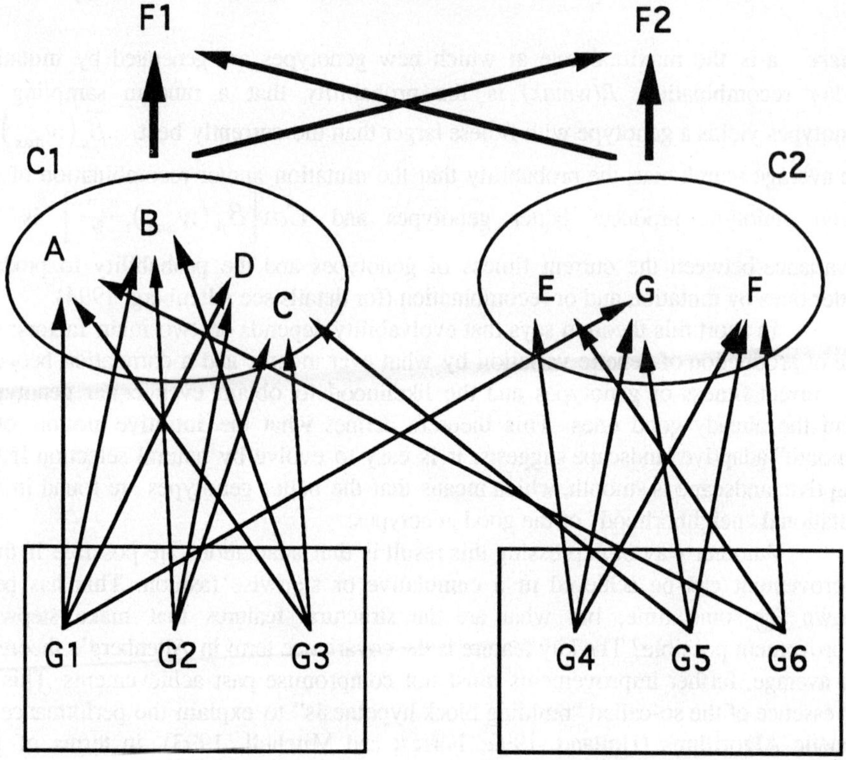

Fig. 1: Example of a modular representation of the character complexes C1={A, B, C, D} and C2={E, F, G} which serve to functions F1 and F2. Each character complex has a primary function, F1 for C1 and F2 for C2. Only weak influences exist of C1 on F2 and vice versa. The genetic representation is modular because the pleiotropic effects of the genes M1={G1, G2, G3} have primarily pleiotropic effects on the characters in C1 and M2={G4, G5, G6} on the characters in complex C2. There are more pleiotropic effects on the characters within each complex than between them.

The concept of modularity was clearly expressed by John Bonner in his concept of gene nets (Bonner, 1988):

"I will call [...] a 'gene net' [...] a grouping of a network of gene actions and their products into discrete units during the course of development." (P174) "This general principle of the grouping of gene products and their subsequent reactions into gene nets becomes increasingly prevalent as organisms become more complex. This not only was helpful and probably necessary for the success of the process of development, *but it also means that genetic change can occur in one of these gene nets without influencing the others, thereby much increasing its chance of being viable. The grouping leads to a limiting of pleiotropy and provides a way in which complex developing organisms can change in evolution.*" (p. 175, emphasis by GPW)

The idea that development is organized into semi-autonomous processes is actually much older, dating back to the beginnings of developmental biology and was summarized under the term "dissociability" by Needham (1933). Needham pointed out that even if development is a perfectly integrated process its component parts can be disentangled experimentally: growth can occur without differentiation and nuclear division without cell division and so on. The evolutionary importance of this fact was emphasized by Gould (1977, p 234) who suggested that dissociability is the developmental prerequisite for heterochronic change (see also Raff and Kaufman, 1983, p 150; Raff, 1995).

The existence of semi-autonomous units of the phenotype might be particularly important in connection with sexual reproduction (Stearns, 1993). Sexual reproduction rearranges genetic variation in every generation which creates the problem of maintaining functional phenotypic units intact. Stabilizing the development of functionally related character complexes allows to recombine integrated traits rather than true "random" variation.

The fact that the morphological phenotype can be decomposed into basic organizational units, the homologues of comparative anatomy, has also been explained in terms of modularity. It has been suggested that properly identified homologues are developmentally and genetically individualized parts of the organisms (Wagner, 1989b,c). The biological significance of these semi-autonomous units is their possible role as adaptive "building blocks" (Wagner, 1995).

5 The Evolution of Modularity

Even if the fact and importance of modularity has a long historical tradition, there is little understanding of how modularity has originated. Is it an inherent property of organisms and thus not the result of evolution in the Darwinian sense or is it the result of selection shaping the genotype-phenotype mapping function? Is modularity the result of integrating disconnected parts or, on the contrary, the result of parcellation of primarily integrated parts. Parcellation, a process which produces modularity from an integrated whole, consists in the differential suppression of pleiotropic effects among characters belonging to different complexes and the selective maintenance and augmentation of pleiotropic effects among the members of the same complex (Fig 2).

The first possibility, that modularity is a primitive property of all living beings, is unlikely. As much as the evolution of higher organisms consists in the acquisition of modular parts, like specialized organs the origin of modularity is most likely the result of evolutionary modification.

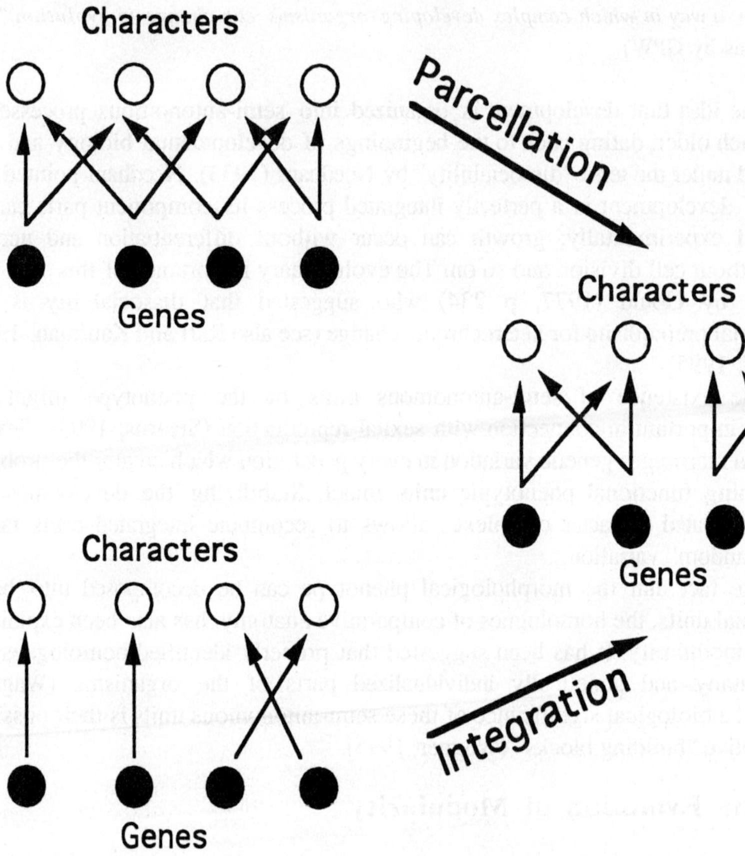

Fig. 2: Two ways of obtaining modularity. Parcellation consists of a differential suppression of pleiotropic effects between groups of characters and the maintenance and/or augmentation of pleiotropic effects among characters from the same group. Modularity through integration consists in the selective acquisition of pleiotropy among characters from the same group.

As to the 'direction' of evolution, integration or parcellation of modules (Fig. 2), the most prevalent direction seems to be parcellation, at least among metazoan animals. The origin of metazoans is the integration of conspecific unicellular individuals into a higher level unit (see Buss, 1987). Each of these units consists of cells which have the same genotype and only secondarily organize in specialized cell populations and anatomically separated organs. A very frequent mode of morphological innovation is the differentiation of repeated elements (Müller and Wagner, 1991; Weiss, 1990), for instance the differentiation of metameric segments at the origin of insects (see for instance Akam, 1989). The specialized organs acquire developmental autonomy in the course of phylogeny (Bonner, 1988). Hence, the origin of differentiated, complex animals appears to be dominated by the process of parcellation rather than secondary integration, even if integration certainly occurs, for instance in symbiotic integration of cells of different origin (mitochondria and plastids for instance).

Provided that modularity is most likely the derived state in the phylogeny of animals and is perhaps the result of parcellation rather than integration, the question arises of how parcellation has been caused by natural selection. Perhaps the most common and long lasting form of selection experienced by any species is stabilizing selection (Endler, 1986). However, stabilizing selection alone is the least likely candidate for causing parcellation. Stabilizing selection on all characters simultaneously favors suppression of all mutational effects (Wagner, submitted). It is thus unlikely to lead to modularity.

Two other mechanisms have been proposed: selection for adaptation rate (Rechenberg, 1973; Riedl, 1975) and "constructional" selection (Altenberg, 1994). Selection for adaptation rate assumes that modular or otherwise favorable representations of the phenotype will get selected because they are able to respond more quickly to directional selection. This is indeed the case and can happen without group selection (Wagner, 1981). However, the problem is that selection for adaptation rate requires high degrees of linkage disequilibrium (Wagner and Bürger, 1985) and is only effective in the absence of recombination. The reason is that recombination during sexual reproduction leads to a mixing of genotypes and thereby eradicates the adaptive advantages achieved by genotypes with a better genetic representation (Wagner, unpublished).

The second proposal suggests that modularity evolves during genome growth by gene duplication (Altenberg, 1994). This highly original proposal suggests that duplications of genes with fewer pleiotropic effects are more likely to be viable than duplications of genes with many pleiotropic effects. Indeed simulations have shown that such a process would lead to modular organizations, provided that gene duplications are associated with direct effects on fitness (Altenberg, 1994).

Another possibility of sufficient generality is that the combination of directional and stabilizing selection leads to the differential suppression of pleiotropic effects (Wagner, in preparation). This proposal assumes that adaptation to environmental perturbations includes directional selection on one or a few functions or character complexes (mosaic evolution). It implies that directional selection on adaptively challenged character complexes occurs simultaneously with stabilizing selection on all the other characters. This combination of selection forces creates strong selection for suppressing exactly those pleiotropic effects which connect the characters under different selection regimes (directional and stabilizing). Simulation studies show that the selection coefficient of a gene suppressing pleiotropy among adapting (= under directional selection) and non-adapting characters (= under stabilizing selection) can be up to 0.3 depending on the intensity of directional selection and the strength of stabilizing selection (Wagner, in preparation). However, it is not yet clear what the necessary conditions are under which this process is a likely explanation of modularity, and whether these conditions are realized in nature. More research into the population genetic theory of genotype-phenotype mapping functions is necessary to assess the plausibility of this and the other scenarios to explain the evolution of modularity. More knowledge of the developmental and evolutionary processes underlying the origin of modular parts of organisms is required to understand the significance of modularity.

6 Conclusions

To understand the conditions under which mutation, recombination and selection can lead to complex adaptations is of importance for evolutionary biology as well as its applications in computer science (evolutionary algorithms). The central idea uniting these two fields is the insight that the genetic representation of the phenotype determines its evolvability (the so-called representation problem). A recurrent theme in the biological literature is the concept of modularity, the fact that higher organisms are composed of semi-autonomous units (gene nets, Bonner (1988); dissociability, Needham (1933) and Gould (1977); individuality, Wagner (1989b,c); selfmaintaining organizations, Fontana and Buss (1994); developmental modules, Raff, 1995). However, even if the fact and the importance of modularity seems to be widely appreciated, there is little understanding of what selective forces have generated genetic and developmental modularity.

Acknowledgments: The author wants to thank the following individuals for stimulating discussions on the topic of the paper: Lee Altenberg, Leo Buss, Jim Cheverud, Junhyong Kim, Manfred Laubichler, Bernhard Misof, Christian Pazmandi, Andreas Wagner. The financial support by the Yale Institute for Biospheric Studies and NSF grant BIR-9400642 is gratefully acknowledged. This is contribution CCE-#25 of the Yale Center for Computational Ecology.

References

Akam, M., I. Dawson and G. Tear 1988. Homeotic genes and the control of segment diversity. Development 104: 123-133.

Alberch, P. 1983. Morphological variation in the neotropical salamander genus Bolitoglossa. Evolution 37: 906-919.

Altenberg, L. 1994. The evolution of evolvability. in press in Advances in Genetic Programming. J. K. E. Kinnear, ed. Cambridge, MIT Press.

Altenberg, L. and M. W. Feldman 1987. Selection, generalized transmission, and the evolution of modifier genes. Genetics 117: 559-572.

Banzhaf, W. 1994. Genotype-phenotype mapping and neutral variation - A case study in genetic programming. in Parallel Problem Solving from Nature - PPSN III. Y. Davidor, H.-P. Schwefel and R. Männer, ed. Berlin, Springer.

Bonner, J. T. 1988. The Evolution of Complexity. Princeton, NJ., Princeton University Press.

Bossert, W. 1967. Mathematical optimization: are there abstract limits on natural selection? in Mathematical Challanges to the Neo-Darwinian intepretation of evolution. P. S. Moorhead and M. Kaplan, ed. Philadelphia, Wistar Inst. Press.

Bremermann, H. J., M. Rogson and S. Salaff 1966. Global properties of evolution processes. in Natural Automata and Useful Simulations. H. H. Pattee, ed. Washington, DC, Macmillan Press.

Bulmer, M. G. 1980. The Mathematical Theory of Quantitative Genetics. Oxford, Clarendon Press.

Bürger, R., G. P. Wagner and F. Stettinger 1989. How much heritable variation can be maintained in finite populations by mutation-selection balance? Evolution 43: 1748-1766.

Buss, L. W. 1987. The Evolution of Individuality. New York, Columbia University Press.

Darwin, C. R. 1859. The Origin of Species. London, John Murray.

Eden, M. 1967. Inadequacies of neo-darwinian evolution as a scientific theory. in Mathematical Challanges to the Neo-Darwinian intepretation of evolution. P. Moorhead and M. Kaplan, ed. Philadelphia, Wistar Inst. Press.

Endler, J. A. 1986. Natural Selection in the Wild. Princeton, New Jersey, Princeton University Press.

Fontana, W., and L. W. Buss 1994. The arrival of the fittest. Bull. Math. Biol., 56: 1-64.

Forrest, S. and M. Mitchell 1993. Towards a stronger building-block hypothesis: effects of relative building-block fitness on GA performance. 109-126 in Foundations of Genetic Algorithms. C. D. Whitley, ed. Palo Alto, Morgon Kaufman.

Frazzetta, T. H. 1975. Complex Adaptations in Evolving Populations. Sunderland, MA., Sinauer Ass. Inc.

Gimelfarb, A. 1989. Genotypic variation for a quantitative character maintained under stabilizing selection without mutation: Epistasis. Genetics 123: 217-227.

Goodman, N. 1955. Fact, Fiction, Forecast. Indianapolis, Hackett Publ. Co.

Gould, S. J. 1977. Ontogeny and Phylogeny. Cambridge, MA., Harvard University Press.

Holland, J. H. 1992. Adaptation in Natural and Artificial Systems. Cambridge, MA, MIT Press.

Jones, T. and G. J. E. Rawlins 1993. Reverse hillcliming, genetic algorithms and the busy beaver problem. 70-75 in Proceedings of the Fifth International Conference on Genetic Algorithms. S. Forrest, ed. San Mateo, CA, Morgan Kaufmann.

Koza, J. R. 1992. Genetic Programming: On the Programming of Computers by Means of Natural Selection. Cambridge, MA., MIT Press.

Lewontin, R. C. 1970. The units of selection. Ann. Rev. Ecol. System. 1: 1-18.

Maynard-Smith, J., R. Burian, S. Kauffman, P. Alberch, J. Campell, B. Goodwin, R. Lande, D. Raup and L. Wolpert 1985. Developmental constraints and evolution. Quart. Rev. Biol. 60: 265-287.

Müller, G. B. and G. P. Wagner 1991. Novelty in Evolution: Restructuring the Concept. Annu. Rev. Ecol. Syst. 22: 229-256.

Needham, J. 1933. On the dissociability of the fundamental processes in ontogenesis. Biol. Rev. 8: 180-223.

Price, G. R. 1969. Selection and covariance. Nature 227: 520-521.

Raff, R. A. 1983. Embryos, Genes, and Evolution. New York, Macmillan Publishing Co.

Raff, R. A. In press. The shape of life. University of Chicago Press, Chicago, IL.

Ray, T. S. 1992. An approach to the synthesis of life. in Artificial Life II. C. G. Langton, C. Taylor, J. D. Farmer and S. Rasmussen, ed. Santa Fe, NM, Santa Fe Institute.

Rechenberg, I. 1973. Evolutionsstrategie. Stuttgart, Friedrich Frommann Verlag.

Rendel, J. M. 1967. Canalization and Gene Control. New York, Logos Press, Academic Press.

Riedl, R. 1975. Die Ordnung des Lebendigen. Systembedingungen der Evolution. Hamburg und Berlin, Verlag Paul Parey.

Rienesl, J. and G. P. Wagner 1992. Constancy and change of basipodial variation patterns: a comparative study of crested and marbled newts - Triturus cristatus, Triturus marmoratus - and their natural hybrids. J. Evol. Biol. 5: 307-324.

Scharloo, W. 1988. Selection on morphological patterns. 230-520 in Population Genetics and Evolution. G. de Jong, ed. Berlin, Spinger Verl.

Scharloo, W. 1987. Constraints in selection response. 125-149 in Genetic Constraints on Adaptive Evolution. V. Loeschke, ed. Berlin, Spinger Verl.

Scharloo, W. 1991. Canalization: Genetic and developmental aspects. Ann. Rev. Ecol. Syst. 22: 65-93.

Schmalhausen, I. I. 1949. Factors of Evolution. The theory of stabilizing selection. Chicago and London, University of Chicago Press.

Schwefel, H.-P. 1981. Numerical Optimization of Computer Models. Chichester, Wiley.

Simon, H. A. 1965. The architecture of complexity. General Systems 10: 63-73.

Stearns, S. C. 1993. The evolutionary links between fixed and variable traits. Acta Paleont. Polonica 38: 1-17.

Stearns, S. C., M. Kaiser and T. J. Kawecki 1995. The differential canalization of fitness components against environmental perturbations in Drosophila melanogaster. J. Evol. Biol. submitted:

Stebbins, G. L. 1974. Flowering Pants. Evolution Above the Species Level. Cambridge, MA, Belknap Press.

Turelli, M. 1988. Phenotypic evolution, constant covariances, and the maintenance of additive variance. Evolution 42: 1342-1347.

Vermeij, G. J. 1970. Adaptive versatility and skeleton construction. Amer. Nat. 104: 253-260.

Waddington, C. H. 1957. The Strategy of the Genes. New York, MacMillan Co.

Wagner, A., G. P. Wagner and P. Similion 1994. Epistasis can facilitate the evolution of reproductive isolation by peak shifts: a two-locus two-allele model. Genctics 138: 533-545.

Wagner, G. P. 1981. Feedback selection and the evolution of modifiers. Acta Biotheoretica 30: 79-102.

Wagner, G. P. 1989a. Multivariate mutation-selection balance with constrained pleiotropic effects. Genetics 122: 223-234.

Wagner, G. P. 1989b. The origin of morphological characters and the biological basis of homology. Evolution 43: 1157-1171.

Wagner, G. P. 1989c. The biological homology concept. Ann. Rev. Ecol. Syst. 20: 51-69.

Wagner, G. P. 1995. The biological role of homologues: A building block hypothesis. N. Jb. Geol. Paläont. Abh. 19: 279-288.

Wagner, G. P. and R. Bürger 1985. On the evolution of dominance modifiers II: a non-epuilibrium approach to the evolution of genetic systems. J. theor. Biol. 113: 475-500.

Weiss, K. M. 1990. Duplication with variation: Metameric logic in evolution from genes to morphology. Yearbook of Physical Anthropology 33: 1-23.

A Theory of Differentiation with Dynamic Clustering

Kunihiko Kaneko and Tetsuya Yomo

[1] Department of Pure and Applied Sciences,
University of Tokyo, Komaba, Meguro-ku, Tokyo 153, JAPAN
[2] Department of Biotechnology, Faculty of Engineering
Osaka University, 2-1 Suita, Osaka 565, JAPAN

Abstract. A novel theory for cell differentiation is proposed, based on simulations with interacting artificial cells which have metabolic networks within, and divide into two when the final product is accumulated. Results of simulations with coupled chemical networks and division process lead to the following scenario of the differentiation: Up to some numbers of cells, divisions bring about almost identical cells with synchronized metabolic oscillations. As the number is increased the oscillations lose the synchrony, leading to groups of cells with different phases of oscillations. At later stage this differentiation is fixed in time, and cells spilt into groups with different chemical constituents spontaneously, which are transmitted to daughter cells by cell divisions. Hierarchical differentiation, origin of stem cells, and anomalous differentiation by transplantations are also discussed with relevance to real biological experimental results.
(Keywords: differentiation, metabolic network, cell division, clustering, open chaos)

1 Introduction

It is often believed that the cell differentiation is completely determined by genes, with some regulatory networks among them[1]. Since genes interact with proteins and other chemicals, however, the differentiation process is not so simple. Indeed, there are some experiments, which cast a question to this widely accepted picture on the cell differentiation: As reported in [2], E. Coli cells with identical genes may split into several groups with different enzymatic activities. Rubin [3], in a series of papers, has shown that the tumor formation strongly depends on the history of the cultivation of cells over several generations, which are not explained by mutations.

In the previous paper[4] we have proposed a novel mechanism which potentially explains the spontaneous cell differentiation based on cellular interactions. The background of this theory lies in recent developments of the clustering theory of globally coupled chaotic elements [5], where chaos leads to the differentiation of identical elements through interaction among them. The relevance of dynamic change of relationships among elements to biological networks has been discussed [6].

In the present paper we extend the previous model[4] to show how cells are differentiated successively into different types. Here we adopt autocatalytic metabolic reaction networks in each cell, while interactions among cells are considered through the medium contacting with cells. We have explicitly included the cell division process, which leads to the increase of the number of cells. Thus the number of equations, consequently the degrees of freedoms of our model increase with time, and our problem provides an example of open chaos discussed earlier [4, 6]. Through our simulations it is shown that the cells lose totipotent ability, as the cells divide, in consistency with well-known fact in the cellular biology. It should be noted that the chemical composition of a cell is inherited by its daughter cells, without imposing any genetic constraints. Furthermore, emergence of stem cells and hierarchical differentiation of cellular types are also discussed.

In our model cells interact through a well stirred medium, and no spatial variation is included. Our results show that differentiation starts by a dynamic, rather than spatial, mechanism in contrast with Turing instability. Indeed our dynamic scenario is consistent with the experimental reports of the differentiation in a well stirred medium [2].

2 Model

The biochemical mechanisms of the cell growth and division are complicated, which include a variety of catalytic reactions. The reaction occurs both at the levels of inter- and intra- cells. Here we study a class of models which captures the metabolic reaction and cellular interactions.

Our model for cell society consists of

- Metabolic Reaction Network within each Cell : Intra-cellular Dynamics
- Interaction with Other Cells through Media: Inter-cellular Dynamics
- Cell Division

The basic structure is same as the previous model [4], although the present model includes metabolic network rather than a simple set of reactions, to cope with the complexity in a real cellular system.

(A) Metabolic Reaction

First we adopt a set of some chemicals' concentrations as dynamical variables in each cell, and also those in the medium surrounding the cells. We use the following variables; a set of concentrations of chemical substrates $x_i^{(m)}(t)$, the concentration of m-th chemical species at the i-th cell, at time t. The corresponding concentration of the species in the medium is denoted as $X^{(m)}(t)$. We assume that the medium is well stirred, and neglect the spatial variation of the concentration. Furthermore we regard the chemical species $x^{(0)}$ (or $X^{(0)}$ in the media) as playing the role of the source for other substrates.

The metabolic reactions are usually catalyzed by enzymes, which are inductive and are again synthesized with the aids of other chemicals $x^{(j)}$. Assuming that the dynamics for enzymes is faster, we adiabatically solve the reaction equations of enzyme concentrations, to represent the concentration by those of the substrates ($x^{(j)}$) corresponding to the synthesis [4]. For simplicity we assume that this synthetic reaction is linear in $x^{(j)}$, and adopt the Michaels-Mentens type reaction. Here we use the notation $Con(m,\ell,j)$ which takes the value 1 when there is a metabolic path from the chemical m to ℓ catalyzed by the chemical j, and takes 0 otherwise. In other words, a metabolic path $x^{(m)} \rightarrow x^{(\ell)}$ produces $x^{(\ell)}$, with the aid of the chemical j when $Con(m,\ell,j) = 1$. Here the choice of connected paths depends on each chemical ℓ, and generally there can be several paths for the production of the chemical ℓ. Thus the reaction from the chemical m to ℓ aided by the chemical j leads to the term $e_1 x_i^{(j)}(t) x_i^{(m)}(t)/(1+x_i^{(m)}(t)/x_M)$, where x_M is a parameter for the Michaels-Mentens' form. The coefficients for chemical reactions are taken to be identical (e_1) for all the paths.

In addition, we assume that there is a path to the final product, from all $x^{(k)}$, leading to a linear decay of $x^{(k)}$, with a coefficient δ. Summing up all these processes, we obtain the following contribution of the metabolic network to the growth of $x_i^{(\ell)}$ (i.e., $dx_i^{(\ell)}(t)/dt$);

$$Met_i^{(\ell)}(t) = e_1 x_i^{(0)}(t) x_i^{(\ell)}(t) + \sum_{m,j} Con(m,\ell,j) e_1 x_i^{(j)}(t) x_i^{(m)}(t)/(1+x_i^{(m)}(t)/x_M)$$

$$- \sum_{m',j'} Con(\ell,m',j') e_1 x_i^{(\ell)}(t) x_i^{(j')}(t)/(1+x_i^{(\ell)}(t)/x_M) - \delta x_i^{(\ell)}(t), \qquad (1)$$

where we note that the two terms with $\sum Con(\cdots)$ represent metabolic paths coming into ℓ and out of ℓ respectively.

When $m = \ell$, the reaction is regarded as autocatalytic, in the sense that there is a positive feedback to generate the chemical k. (In general, it is natural to assume that a set of chemicals work as an autocatalytic set.) Later we will study the case only with autocatalytic reactions, in a more detail.

(B) Active Transport and Diffusion through Membrane

A cell takes chemicals from the surrounding medium. Thus cells interact with each other indirectly through the medium. It is expected that the rates of chemicals transported into a cell are proportional to their concentrations outside. Further we assume that this transport rate also depends on the internal state of a cell. Since the transport here requires energy [1], the transport rate depends on the activities of a cell. To be specific, we choose the following form;

$$Transp_i^{(m)}(t) = \left(\sum_{k=1} x_i^{(k)}(t)\right) X^{(m)}(t) \qquad (2)$$

The summation ($\sum_{k=1} x_i^{(k)}(t)$) is introduced here to mean that a cell with more chemicals is more active. We choose the above bi-linear form for simplicity, although a nonlinear dependence on $\sum_{k=1} x_i^{(k)}(t)$ with a positive feedback effect leads to qualitatively similar results. Besides the above active transport, the chemicals spread out through the membrane by a normal diffusion process written as

$$Diff_i^{(m)}(t) = D(X^{(m)}(t) - x_i^{(m)}(t)) \tag{3}$$

Combining the processes (A) and (B), the dynamics for the source chemical $x_i^{(0)}(t)$ and $x_i^{(m)}(t)$ are given by

$$dx_i^{(0)}(t)/dt = -e_1 x_i^{(0)}(t) \sum_\ell x_i^\ell(t) + Transp_i^{(0)}(t) + Diff_i^{(0)}(t), \tag{4}$$

$$dx_i^{\ell)}(t)/dt = Met_i^{(\ell)}(t) + Transp_i^{(\ell)}(t) + Diff_i^{(\ell)}(t), \tag{5}$$

Since the present processes are just the transportation of chemicals through membranes, the sum of the chemicals must be conserved. The dynamics of the chemicals in the medium is then obtained by converting the sign, i.e.,

$$dX^{(m)}(t)/dt = -\sum_{i=1}^{N} \{Transp_i^{(m)}(t) + Diff_i^{(m)}(t)\}, \tag{6}$$

where N is the number of cells, which can change in time by cell divisions.

Since the chemicals in the medium can be consumed with the flow to the cells, we need some flow of chemicals (nutrition) into the medium from the outside. Here only the source chemical X^0 is supplied by a flow into the medium. By denoting the external concentration of the chemicals by $\overline{X^0}$ and its flow rate per volume of the medium by f, the dynamics of the source chemical in the medium is written as

$$dX^{(0)}(t)/dt = f(\overline{X^0} - X^0) - \sum_{i=1}^{N} \{Transp_i^{(0)}(t) + Diff_i^{(0)}(t)\}. \tag{7}$$

(C) Cell Division

Through chemical processes, cells can replicate. For the division, accumulation of some products is required. In our model the final product, generated from all chemical species, is assumed to act as the chemical for the cell division. (This final product can be regarded as DNA).

$$\int_{t_0(i)}^{T} dt \sum_k \delta \times x_i^{(k)}(t) > R \tag{8}$$

is satisfied, where R is the threshold for the cell replication. Here again, choices of some other division conditions can give qualitatively similar results as those to be discussed. We note that the division condition satisfies an integral form representing the accumulation.

When a cell divides, two almost identical cells are formed. The chemicals $x_i^{(m)}$ are almost equally distributed. "Almost" here means that each cell after a division has $(\frac{1}{2} + \epsilon)x_i^{(m)}$ and $(\frac{1}{2} - \epsilon)x_i^{(m)}$ respectively with a small "noise" ϵ, a random number with small amplitude, say over $[-10^{-3}, 10^{-3}]$. We should note that this inclusion of imbalance is not essential to our differentiation. Indeed any tiny difference is amplified to yield a macroscopic differentiation. It should be noted that for simplicity the volume of a cell is approximated to be constant except for a short span for the division. During the short span for the division, the volume is twice and thus the concentration is made half in the above process.

3 Few remarks on internal dynamics

Before presenting the dynamics of cell society, let us briefly describe the nature of metabolic reaction given by eq. (1). Roughly speaking, the dynamics strongly depends on the number of autocatalytic paths. If the number is large, only few chemicals are activated, and all other chemicals' concentrations vanish. In this case no metabolic paths are active, since the ongoing reaction is just the source chemical $0 \rightarrow x^{(k)} \rightarrow$ the final product, without any reactions $x^{(k)} \rightarrow x^{(\ell)}$. On the other hand, when the number of autocatalytic paths is small, many chemicals are generated, but their concentrations do not oscillate and are fixed in time. When the number of autocatalytic paths is medium, non-trivial metabolic reactions appear. Some, (not necessarily all) chemicals are activated. The concentrations of chemicals oscillate in time, which often shows a switching-like behavior: That is, chemicals switch between low and high values successively. This type of switching behavior is also seen in the randomly connected Lotka-Volterra equations as saddle-connection-type dynamics [7].

In the present paper we discuss cases with a medium number of autocatalytic paths, since they lead to non-trivial metabolic oscillations.

4 Proposed Scenario on cell differentiation

We have carried out several simulations of our model with $k = 8$, 16 or 64, with a variety of randomly chosen metabolic networks with connections from 2 to 6 per chemicals. Through these simulations, we propose the following scenario of the cell differentiation. Here we describe our scenario together with numerical results for a given network, although simulations with a variety of metabolic networks support the scenario rather well.

(1)Metabolic Oscillation of chemicals and Synchronized Division

Chemical concentrations within each cell oscillate in time by the metabolic reaction process, which provides the basis of the following differentiation process. Up to some number of cells, the oscillations are coherent, and all cells have almost same concentrations of chemicals. Accordingly, the cells divide almost simultaneously, and the number of cells increase as 1,2,4,8. It is interesting to note that cells in most real organisms are not differentiated up to some number of divisions.

(2) Clustering by Phases of Oscillations

As the division process proceeds, the metabolic oscillation starts to lose its synchrony. Cells often separate into several groups with distinct phases of oscillations, while the synchrony is preserved within each group of cells. Thus the differentiation sets in. At this stage, however, the differentiation is not yet fixed. In other words, only the phases of oscillation are different by cells, but the temporal averages of chemicals, measured over some periods of oscillations, are almost identical by cells. As has been discussed[4], this temporal clustering corresponds to the time sharing for resources, since the ability to get them depends on the chemical activities of cells. In Fig.1a), the temporal averages and snapshot values of some chemicals are plotted in the order of the birth time of cells. Here the averages are almost identical, but the snapshot chemical values start to show slightly different values. The difference here, indeed, is a trigger to the fixed differentiation at the next stage.

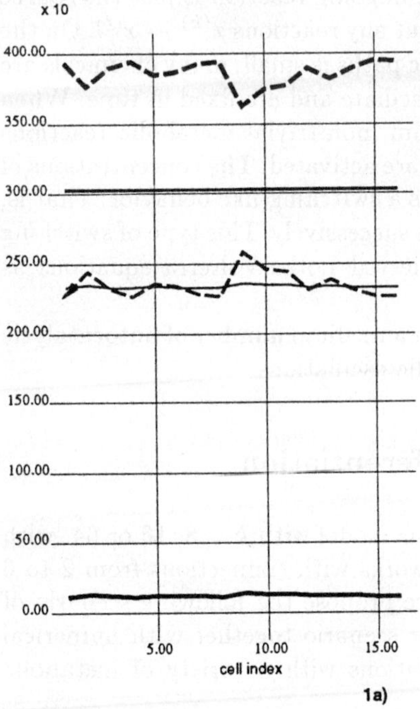

1a)

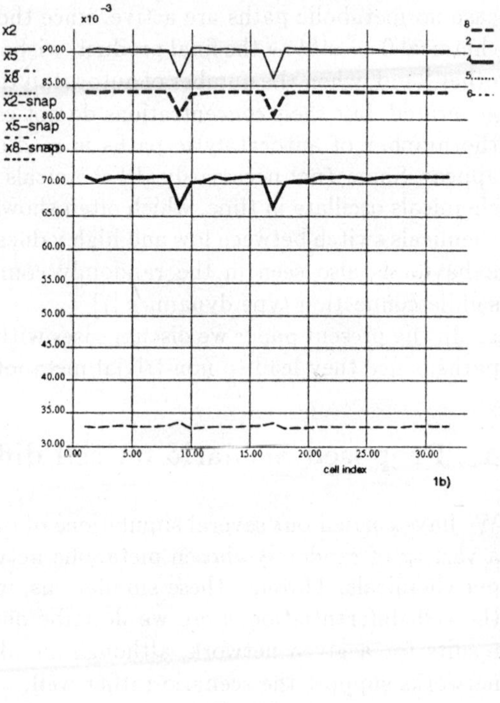

1b)

(3) Fixed Differentiation

After some divisions of cells (for example, at the stage of 32 cells) differences in chemicals start to be fixed by cells. When we measure the average chemicals over periods of oscillations, amounts of chemicals as well as the ratios of chemicals differ by cells. Thus cells with different chemical compositions are generated. This cells differentiate not only with respect to the strength of activities [4], but also to the compositions of chemicals. In Fig.1b)-e), we have plotted the

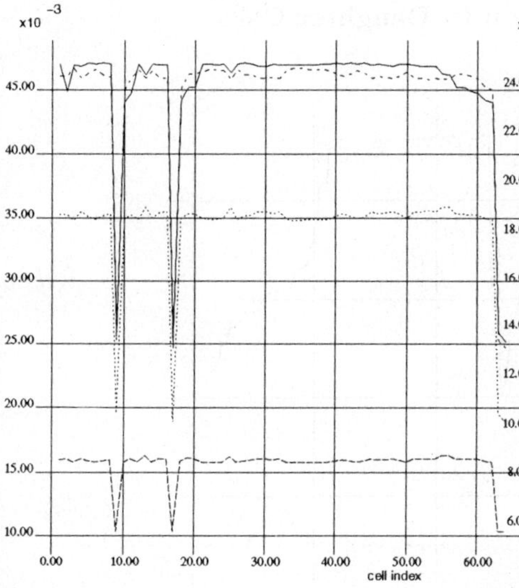

1c)

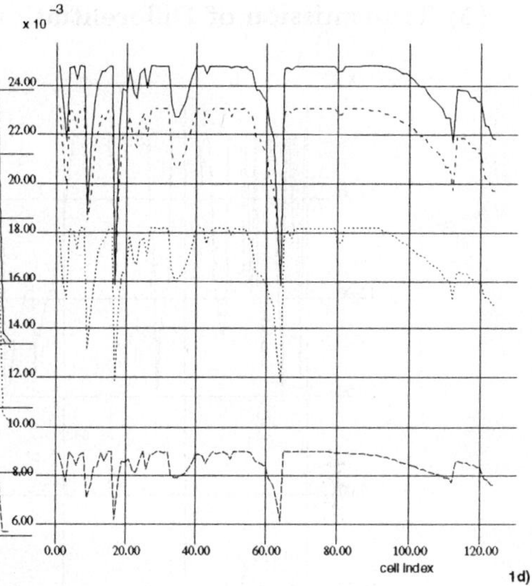

1d)

averages of chemicals for different temporal regimes. Distinct two groups of cells are created when the cell number is 32 in the figure. (see the item (6) for the further differentiation at a later stage).

It is noted that the phase difference still remains by each group. Thus there are two levels of differences by cells, one for the change of phases of the metabolic oscillations, and the other for the fixed differentiation. Indeed this two-level differentiation is the origin of the hierarchical organization, since the phase difference within each group leads to further fixed differentiations later. It is also interesting to note that the phase difference is by "analogue" means, while the fixed differentiation of averages is discrete, where cells are separated "digitally".

(4)Separation of inherent time scales

When differentiated, two groups of cells show distinct orbits, which lie in different regimes in the k-dimensional phase space $\{x(j)\}$. In Fig.2, we have plotted overlaid orbits of two groups of cells. In each group, the oscillation phases are different by cells but the orbits fall in the same attractor, while the difference of orbits between the two groups is clearly discernible.

Another important feature here is the differentiation of the oscillation frequency. One group of cells oscillates faster than the other group. Typically cells with lower activities oscillate in time more slowly with smaller amplitudes. Thus inherent time scales differ by cells, and the division speeds of cells are also differentiated. Indeed, one group of cells divides faster than the other group of cells.

(5) Transmission of Differentiation to Daughter Cells

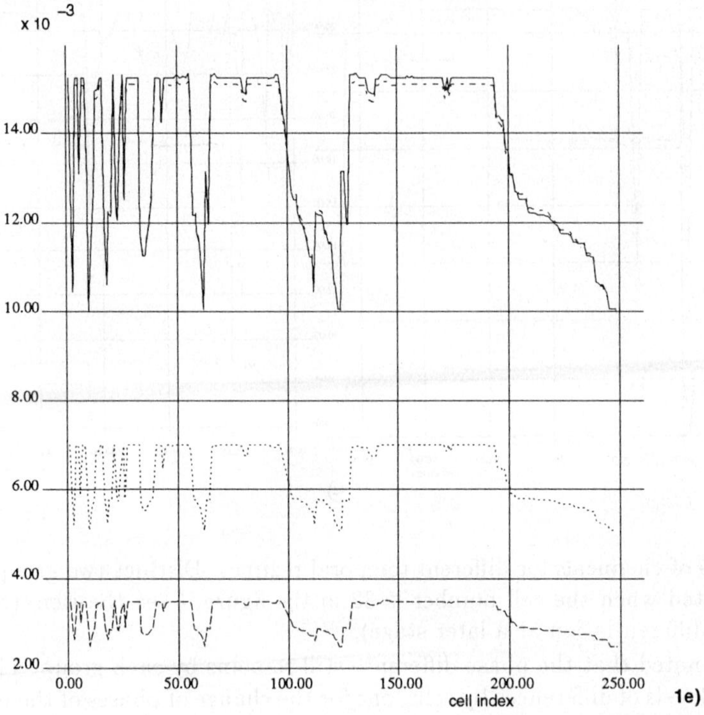

Fig. 1. Average chemical concentrations of $x^k(i)$. The cell index is defined in the order that the cell is born. Throughout the figures we use the parameters $e_1 = 1$, $s = 40$, $D = 0.02, f = 0.5$, $\delta = 0.2$, and $R = 100$, with the metabolic network of $k = 8$ chemicals, with 4 randomly chosen autocatalytic paths, although a variety of networks with the connection number from 2 to 4 lead to similar patterns of differentiation. The average is taken over the time steps while the cell number remains 16 (Fig.1a), 32 (Fig.1b), 64 (Fig.1c), 124 (Fig.1d), and 248 (Fig.1e). For reference, the snapshot values of two chemicals are also overlaid in Fig.1a).

After the fixed differentiation, the characteristic chemical compositions of each group of cells are inherited by their daughter cells. Daughters of a given type of cells keep the same character. Indeed, the cells with weaker activities in Fig.1, are successive daughters of an "ancestor" cell with such low activity. In other words, when the system enters into this stage, a cell loses totipotency. By using the term in the cell biology [1], we call that the determination of a cell has occurred at this stage, since daughters of one type of cells preserve its type.

It is important to note that the chemical characters are "inherited" just through the initial conditions of chemicals after the division, without any external implementation for a genetic transmission. The almost "digital" distinction of chemical characters, noted previously, is relevant to their preservation into

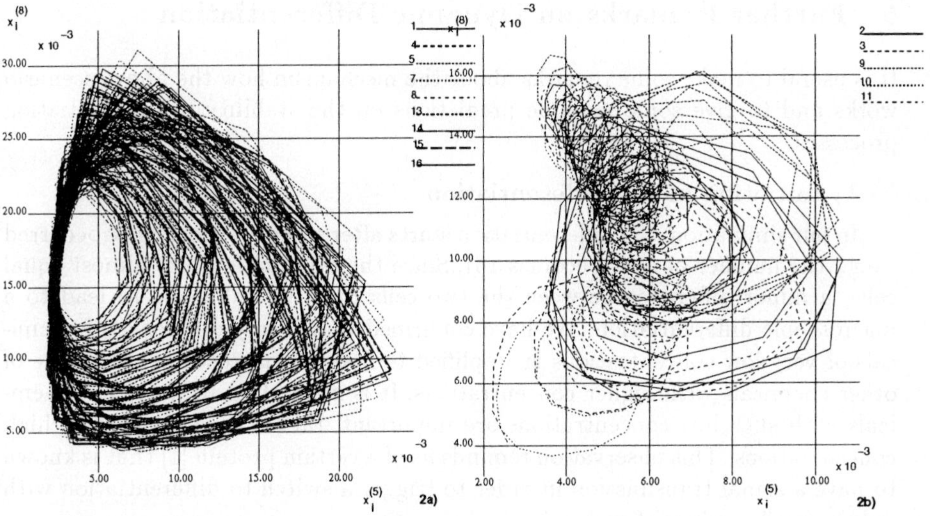

Fig. 2. Orbits of metabolic oscillations. Plotted are $(x_i^5(t), x_i^8(t))$ for time steps 1500 to 2000. (The piecewisely-discontinuos character in the orbit is just an artifact in the plot, since the data are sampled by 0.5 second due to the limited memory of data space, and the orbit, of course, shows a smooth change). In Fig. 2a), overlaid are orbits for the cell index 1,4,5,7,8,13,14,15,16, which have same average chemical characters as seen in Fig.1. Note that the oscillation phase differs by cells. The orbits of the cell 2,3,9,10,11 are plotted in Fig.2b), where the difference between the "strong" type cell in Fig.2a) is clearly discernible (also note the difference of scales).

daughter cells, since analogue differences may easily be disturbed by a possible noise at the division process.

(6) Hierarchical Differentiation

Further differentiation and determination of cells proceed successively in time. New types of cells are generated hierarchically. For example after two types of cells are differentiated (let us call them type-A and type-B cells for the moment) at the first stage, the type-A cell is differentiated into A1 and A2. Here the difference between A1 and A2 (for example that of chemicals or the frequency) is smaller than that between A and B. (see the two levels of "stronger" chemicals in Fig.1c)d)). Once this differentiation occurs, this character is fixed again, and after some time, such characters are determined by its daughter cells. With the cell divisions, this hierarchical determination of cells successively continues. For example daughters of type-B cells differentiate into B1 and B2 at a later stage. Examples of these successive determinations can be seen in Fig 1d)-e).

Since the daughters of A (B) cells can be either A1 (B1) or A2 (B2) cells, the A cell is regarded as the stem cell over the later A1 and A2 cells. The daughters of A1 cells, on the other hand, remain to be A1-type cells, and may further differentiate into cells with smaller differences (say A11 and A12 cells). In this case, the A1 cell can again be a stem cell over a narrower group of cells.

5 Further Remarks on Dynamic Differentiation

It is useful to make some remarks about the mechanism how the above scenario works and to give some possible predictions on the stability of differentiation processes.

1) Initiation of the differentiation

In our simulation, the differentiation starts after some divisions have occurred (e.g., the number of cells becomes 16). Since the division leads to almost equal cells, a minor difference between the two cells must be enhanced to lead to a macroscopic differentiation. We have confirmed that a tiny difference of chemicals of very low concentrations is amplified to make a macroscopic difference of other chemicals with higher concentrations. It is interesting to note that chemicals with such low concentrations are important, rather than those with high concentrations. This observation reminds us of a certain protein [1] that is known to have a signal transmission in order to trigger a switch to differentiation with only a small number of molecules.

2) Stability of our scenario

It should be noted that our scenario, although based on the chaotic instability, is rather robust against changes of initial conditions. Of course, which cell becomes one given type can depend on the initial conditions. The number distribution for each type of cells, on the other hand, is stable against a wide change of initial conditions. Still, if we start from a totally different kind of initial conditions, such as those with many identical cells, the results can change as is discussed later.

3) Memory as the inherited initial condition

The differentiation in our scenario is originated in the interaction among cells, but the chemical characters of a cell are later memorized as the initial condition after the division. The differentiation with the interaction mechanism is reversible, while the latter mechanism leads to the determination. It is interesting to note that the determination is not implemented in the model in advance, but emerges spontaneously at some stage when the cell number N exceeds some value.

In the natural course of the differentiation and in our simulations in §4, however, it is not possible to separate the memory in the inherited initial condition from the interaction with other cells. To see the tolerance of the memory as the inherited initial condition, one of the most effective methods is to pick up a determined cell and transplant it within a variety of surrounding cells, that are not seen in the "normal" course of the differentiation and development. Let us discuss some results of this "transplantation" experiments.

4) Transplantation of cells

In real biological experiments on the differentiation, some "artificial" initial conditions are adopted by transplantations of some types of cells. To check the validity of our scenario and to see the tolerance of the memory in the inherited initial condition, we have made several numerical experiments taking a variety of "artificial" initial conditions. As the initial conditions we choose determined cells at a later stage and mix them with undifferentiated cells at the earlier stage, to make the following observations.

i) Starting from few determined cells of the same type in addition to undifferentiated cells

The former group of cells keep its type, whose offsprings remain to be of the same type. Thus the determination is preserved, and the memory in the inherited initial conditions is robust against the change of cell interactions. The undifferentiated cells, on the other hand, start to differentiate to form many types of cells.

ii) Starting only from few differentiated cells of the same type

We have found either that the cells lose the non-trivial metabolic reaction, or that they start to differentiate again to generate different types of cells. Here the trivial reaction means that only one or two chemicals take non-zero values, without any ongoing reaction among the chemicals $\{x^j\}$ (i.e., the only direct path of the *source* $x^0 \rightarrow x^m \rightarrow$ *the final product* is active). This type of cells with trivial reactions divides faster since the metabolic pathway is direct. Taking also into account the fact that these cells destruct the chemical order sustained in the cell society, one may regard them as tumor cells. The formation of these tumor cells depends on initial densities of determined cells, which may be compared with the experiments by Rubin [3].

6 Discussions and Biological Implications

To sum up we have proposed a novel scenario on the cell differentiation, based on the interacting cells with the metabolic oscillations and the clustering of coupled oscillator systems[5]. The model, without any external mechanism, leads to successive spontaneous differentiations, which are transmitted to daughter cells. It is interesting to note that a variety of experimental results can be understood from this point of view, such as the loss of totipotency, the origin of stem cells, the hierarchical differentiation, the separation of division speeds by the differentiation, the germ-line segregation, and the importance of chemicals of low concentrations for the trigger to the differentiation.

In the present paper we have not included a cell death process. When it is included, our model leads to a stationary distribution of differentiated cells at a later stage, when the cell number reaches its maximum[8]. Local interactions in space are not included either, to focus on the dynamic clustering process. The inclusion of a local interaction is rather straightforward, and indeed some simulations in a 2-dimensional space lead to the spatial organization of differentiated cells as well as the developmental process of some forms[8].

Our results also provide novel perspectives to the dynamics of many interacting agents, generally. In particular we have succeeded in showing a mechanism of *division of labors* through the differentiation, as the segregation into active and inactive groups. It is interesting to extend the idea of the present paper to economics and sociology, and to discuss the origin of differentiation, diversity, and complexity there.

So far our results are rather universal as long as individual dynamics allows for some oscillations, and will be theoretically grounded by the studies of globally coupled dynamical systems, in particular, the spontaneous differentiation as the clustering [5]. Existence of a variety of chemicals in our problem leads to the "dual" clustering, both for the cell index and the chemical species. Construction of a minimal model with the dual differentiation will be an interesting problem as a dynamical systems theory, in future. Besides this viewpoint of coupled dynamical systems, it should be noted that our system is "open-ended" in the sense that the degrees of freedom increase with the cell division, where the notion of "open chaos" [6] will be useful to analyze the mechanism of the cell differentiations.

The authors would like to thank Chris Langton and Howard Gutowitz for useful discussions. This work is partially supported by a Grant-in-Aid for Scientific Research from the Ministry of Education, Science, and Culture of Japan. The authors would like to thank Chris Langton and the members of Santa Fe Institute for their hospitality during their stay, while the paper is completed.

References

1. see e.g., B. Alberts, D.Bray, J. Lewis, M. Raff, K. Roberts, and J.D. Watson, *The Molecular Biology of the Cell*, 1983
2. E. Ko, T.Yomo, I. Urabe, Physica 75D (1994) 84
3. A. Yao and H. Rubin,, Proc. Nat. Acad. Sci. 91 (1994) 7712; M. Chow, A. Yao, and H. Rubin, ibid, 91(1994) 599; H. Rubin, ibid, 91(1994) 1039; 91(1994) 6619
4. K. Kaneko and T. Yomo, Physica 75 D (1994), 89
5. K. Kaneko, Phys. Rev. Lett. 63 (1989) 219; 65 (1990) 1391; Physica 41 D (1990) 137; Physica 54 D (1991) 5
6. K. Kaneko, Physica D (1994), 55-73
   K. Kaneko, Artificial Life 1, 1994, 163-177
7. S. Sasa and T. Chawanya, submitted to Phys. Rev. Lett.
8. K. Kaneko and T. Yomo, " Dynamic Theory of Differentiation and Development I,II", in preparation

Cell Differentiation and Neurogenesis in Evolutionary Large Scale Chaos

Hiroaki Kitano
Sony Computer Science Laboratory
3-14-13 Higashi-Gotanda, Shinagawa
Tokyo 141 Japan
kitano@csl.sony.co.jp

Abstract. This paper reports dynamic phenomena analogous to cell differentiation. The model used in this paper is based on evolutionary large scale chaos, a large numbers of coupled chaotic elements whose logistic map is evolutionary acquired. The logistic map function itself may change dynamically according to chemical concentration in each cell by regulating its gene. Each individual starts from a single cell which can cause cell division. successfully evolved individual can form a cell cluster of the substantial size. It can form a neural network by growing axons from cells differentiated to neurite. We have observed that even such a simple dynamical system, phenomena similar to cell differentiation takes place, and creates characteristic patterns which may be observed in actual biological systems. Some of temporal patterns of axon growth observed were similar to actual growth patterns.

1 Introduction

This paper presents a simple model of cell differentiation and neurogenesis, which is subject to evolution, development, and learning. The model is an extension of our previous model of evolution of metabolism [Kitano, 1994]. A major extension enables the modeling of axon growth. This is the first model to incorporate aspects of evolution, metabolism, morphogenesis, neurogenesis, neural networks, and learning. Refinement and variants of the model are potentially useful for wide range of applications and basic research.

Our interest is to examine whether a simple dynamical system exhibits spatio-temporal patterns which are analogues to some of patterns in actual biological systems. Our hypothesis is that a dynamical system composed of large numbers of simple and coupled chaotic elements has parameter regions which exhibits spatio-temporal patterns similar to these in actual biological systems. The other hypothesis is that, given the appropriate fitness measure, evolutionary process can be used to acquire logistic map function to create such a spatio-temporal patterns. Since the logistic map function is created by reading chromosome, this means that gene to cause interesting spatio-temporal can be acquired.

These issues are obviously scientific questions, rather than engineering questions. Although a degree of engineering impact is anticipated during the course of this project, our primary concern is the scientific issues. To be more specific, we are interested in finding a general mathematical model to describe the dynamics of a range of biological phenomena, in particular, morphogenesis, neurogenesis, and brain activities. Unlike traditional biology, which emphasizes an analytical and divide-and-conquer approach, our approach is synthetic and holistic. This reflects our belief that highly complex and nonlinear systems can not be understood by an analytical and divide-and-conquer method, the dominant scientific approach to date. We advocate a synthetic and holistic approach that uses massive computing power. We argue that complex and nonlinear systems can be only understood by actually replicating then on other media and by mimicking their behavior in various situations. By replicating complex systems on computer, we hope to be able to collect the data necessary to analyse these phenomena. Unfortunately, collecting all necessary data on-line and in real-time is not possible with currently available technologies. Hopefully, 30 years from now, we will be able to develop extremely innovative devices to simultaneously monitor cell metabolism and other biological phenomena at more than 1,000 points in real-time. Once we have reached this stage, we will be able to critically test the validity of the computational model.

Although there is an enormous desire to advance the science of biology, research has only just started, hence our models are very premature. The model presented in this paper is by no means an exception to this. It is still very premature and incomplete. There are ad-hoc assumptions, omissions of known biological facts, and rough abstractions. However, even such a simplified model gives us an insight into many important factors, such as the likely computational cost of the modeling, the biological research necessary to enable the modeling, necessary implementation techniques, and so forth.

2 The Model

2.1 Overview

First, our model is a continuous model, in the sense that there are no distinct processing stages, such as an evolution stage, development stage, or training stage. While evolution acts as the outermost loop, other processes are performed concurrently without changing to the state of the model. For example, in our model, axons grow while cell metabolism are being computed. While some synapses are changing their connection strengths, some axons are still growing, and some cells are undergoing differentiation.

Our model acquires metabolic reactions within each cell. Fig. 1 shows the overall organization of the model. A genetic algorithm is used to evolve a population of individuals. The chromosome in each cell determine the metabolic reactions in that cell. Metabolic reactions are computed using a continuous value discrete time simulation. The production of certain chemicals is related to the synthesis of DNA. When the synthesis of DNA reaches a certain level, cell division occurs place. After a certain number of simulation cycles, each individual will consist of a number of cells, some which have already differentiated. When this model is applied to neurogenesis, the model should be extended to simulate the formation of networks.

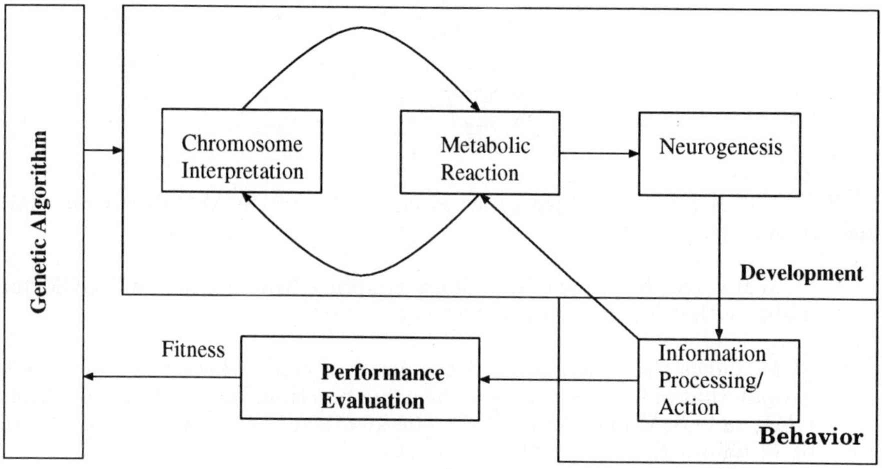

Fig. 1. Overview of the model

2.2 Evolutionary Super-Coupled Map

Mathematically, the model uses a new type of coupled map. A coupled map is a mathematical formalization of large-scale chaos, where a system consists of a large number of elements, each of which exhibits chaotic behavior. In a coupled map lattice (CML), the new state of a cell is determined from the state of the cell and its neighbors. Equation 1 formalizes this model in the case of a one-dimensional cell array.

$$x_i(t+1) = [1-e]f(x_i(t)) + \frac{e}{2}[f(x_{i-1}(t)) + f(x_{i+1}(t))] \quad (1)$$

The other extreme is the global coupled map (GCM), described by equation 2.

$$x_i(t+1) = [1-e]f(x_i(t)) + \frac{e}{N}\sum_{j=1}^{N} f(x_j(t)) \quad (2)$$

In this case, the new state of a cell is determined from the state of the cell and a global variable (e.g. the average energy level of the system). Both CML and GCM were proposed by Kaneko [Kaneko, 1992].

In this paper, we introduces a new map, the *super-coupled map (SCM)* whose one-dimensional cell array formalization is described in equation 3.

$$x_i(t+1) = [1-e-g]f_i(x_i(t))$$

$$+\frac{e}{2}[f_{i-1}(x_{i-1}(t)) + f_{i+1}(x_{i+1}(t))]$$
$$+\frac{g}{N}\sum_{j=1}^{N} f_j(x_j(t)) \tag{3}$$

Fig. 2 illustrates the three types of coupled map. SCM differs from CML and GCM in two points:

1. SCM involves both local and global coupling, where CML and GCM formalize either local or global coupling.

2. SCM allows the functions to differ between cells, where CML and GCM assume that all cells encompass the same function. In the above equations, CML and GCM allow only $f(x)$ to be identical in every cell of the system. SCM allows $f_i(x)$ to be unique in each cell.

These differences are essential to the modeling of biological systems, because, in actual biological system, (1) each cell is influenced by both the local and global chemical concentration, and (2) each cell may adopt a different state of gene regulation and expression.

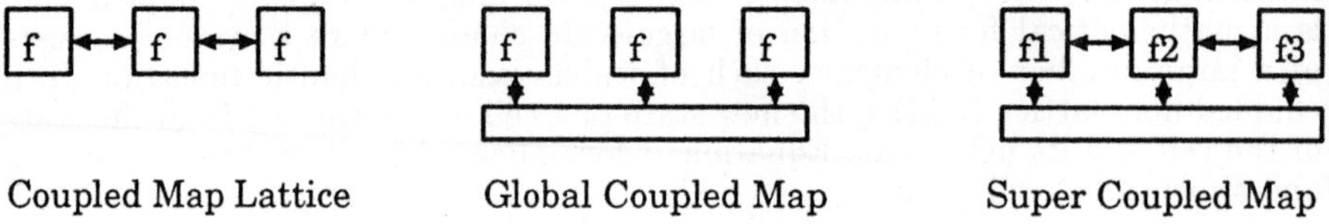

Fig. 2. Three coupled maps

An evolutionary super-coupled map (ESCM) further augments SCM by incorporating evolutionary computation. In ESCM, map functions and the coupling of elements are determined by using the genetic information acquired through evolution. In the model presented in this paper, only map functions can be determined genetically, while the coupling between elements is determined through a developmental process, guided by genetic codes in each element.

2.3 Evolution

A genetic algorithm was used to simulate the evolution process. We used a fixed population, proportional reproduction strategy combined with elitest reproduction, adaptive mutation, and two-point crossover. the chromosome is a binary representation used to encode a sequence of fragments, each of which represents a reaction rule. The length of chromosome used in the experiments presented in this paper was 1,000 bits, and the population size was 10. Fitness was measured from the weighted sum of the total DNA produced, the number of cells, and the total length of the axons. Ideally, fitness should reflect the behavior of an individual in the environment, rather than static structural features.

2.4 Metabolic Reactions

One of the central issues related to this model is how to simulate metabolic reactions in each cell. Chemicals and enzymes are represented by bit strings of a given length. The existence of a bit string of length n, representing chemicals and enzymes, means that there are 2^n possible chemicals and enzymes in the model. The chromosome represents a set of rules governing the metabolic reactions. As shown in Fig. 3, a chromosome is a set of fragments, each of which represents a metabolic rule.

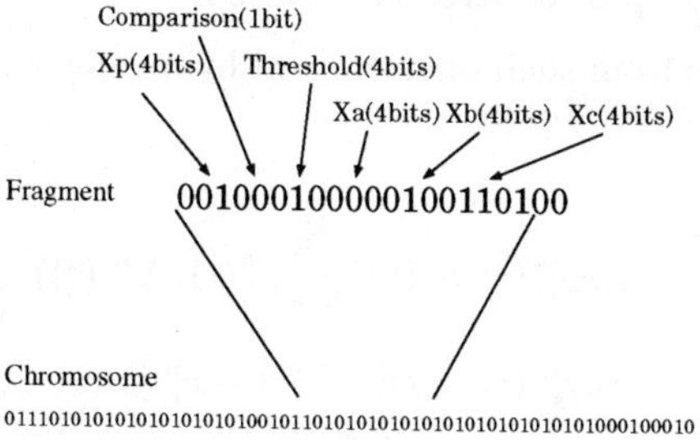

Fig. 3. Chromosome and Fragment

For example, the fragment 000000010000100010010 is interpreted as;

$$x^0 < 0.5 \Rightarrow x^1 \rightarrow E^{1,2}(x^1 \rightarrow x^2) \tag{4}$$

This reads as; if the concentration of x^0 within a cell is less than 0.5, then activate a metabolic reaction which creates enzyme $E^{1,2}$ from x^1. Enzyme $E^{1,2}$ acts as a catalyst for reaction $x^1 \rightarrow x^2$.

During each cycle, these rules are applied to test whether the cell's chemical concentration triggers the activation or suppression of a metabolic reaction. Thus, the metabolic reactions vary dynamically during the cell's life span. In an actual implementation, a chromosome is decoded into a look-up table, with metabolic reactions being represented as a matrix, enabling efficient matching.

Once the metabolic reactions have been determined by testing a rule set derived from the chromosome, the amount of each chemical in each cell can be determined from the following equations:

$$x_i^m(t+1) = x_i^m(t) + \frac{dx_i^m(t)}{dt} \tag{5}$$

$$\begin{aligned}\frac{dx_i^m(t)}{dt} = &\sum_{k=1}^{M} m(E^{k,m}x_i^k) - \sum_{k=1}^{M} m(E^{m,k}x_i^m) \\ &+Transp_i^m(t) + Diff_i^m(t) \\ &+LocTransp_i^m(t) + LocDiff_i^m(t)\end{aligned} \tag{6}$$

$$m(S) = \frac{V_{max}S}{K_m + S} \tag{7}$$

where x_i^m is the chemical concentration of chemical m in the i-th cell. The first term is the production of x_i^m from other chemicals, while the second term is the production of other chemicals using x_i^m. When there is no enzyme (e.g. $E^{k,m} = 0$), no reaction occurs. This modeling is inspired by Kaneko and Yomo's model [Kaneko and Yomo, 1994]. $m(S)$ is the Michaelis-Menten equation, which provides the reaction speed of the enzyme-substrate complex. V_{max} and K_m are constants.

2.5 Active Transport and Diffusion

Cells take in chemicals from their environment by means of active transport and diffusion through their membrane.

$$Transp_i^m(t) = P\, t(\sum_{k=1}^{M} x_i^k(t), X^m(t)) \tag{8}$$

$$Diff_i^m(t) = D[X^m(t) - x_i^m] \tag{9}$$

$$t(a, x) = \frac{a \times x}{0.5x + a} \tag{10}$$

where X^m is the chemical concentration of chemical m in the medium, and P and D are constants. In this equation, active transport corresponds to the amount of chemicals in the cell. Diffusion correlates to the difference in the chemical concentration inside the cell and that of the medium. The speed of active transport is limited according to the density of the chemicals in the medium and the activation level of each cell.

2.6 Cell Division

Cell division takes place when a DNA synthesis process reaches a certain level, R. In this paper, we assume that DNA synthesis correlates with chemical x^0 by a factor γ. This is represented by the following equation:

$$\int_{t_{0(i)}}^{T} \gamma x_i^0(t)dt > R \tag{11}$$

When cell division takes place, the chemicals in a cell are shared between two cells with a certain level of fructuation.

2.7 Cell Death

Cell death takes place when the metabolism of the cell fall to below level S_{min}, or rises above level S_{max}.

$$\sum_{k=1}^{M} x_i^{(k)}(t) < S_{min} \tag{12}$$

$$\sum_{k=1}^{M} x_i^{(k)}(t) > S_{max} \tag{13}$$

2.8 Interaction with Neighbor Cells

Interaction between neighboring cells is achieved through active transport and diffusion limited to adjacent, or near-by, cells. The method is to completely describe cell-to-cell interaction. The diffusion and active transport equation will be:

$$LocDiff_i^m(t) = DL \sum_{j}^{N} C(i,j)[X_{i,j}^m - x_i^m] \tag{14}$$

$$LocTransp_i^m(t) = P \sum_{j=0}^{N} t(\sum_{k=1}^{M} x_i^k(t), X_{i,j}^m) \tag{15}$$

where $X_{i,j}^m$ is the medium between cell i and j, and DL is a constant. $C(i,j)$ returns 1 when cell i and j are in contact. Otherwise, it returns 0.

2.9 Neurogenesis

2.9.1 Nerve Growth Factor

This model assumes several NGF types. The type of NGF emitted by a cell is determined by density of certain specified chemicals, the chemical having densest concentration being selected as the NGF type of that cell. When the axon follows the NGF of other cells, the NGF followed by the axon of the cell is also determined by the chemical density within the cell. For example, let us assume that x_n, x_{n+1}, x_{n+2}, and x_{n+3} are specified as the chemicals which determine the NGF type, and $x_{n+m}, x_{n+m+1}, x_{n+m+2}$, and x_{n+m+3} are specified as the chemicals which determine the target NGF type. n and m are constants. The target NGF (TNGF) type defines the NGF which the axon of the cell should follow. Let us suppose that x_{n+1} is the densest of the four chemicals used to define the NGF type of the cell, the cell's NGF type being defined as NGF_1. Next, when x_{n+m+3} is the densest of the four chemicals which define TNGF, then the TNGF type will be 3. In this case, the axon growing from this cell will follow an NGF gradient of NGF_3. The density of the NGF emitted from a cell depends on the cell itself, and is determined by application of the diffusion equation.

2.9.2 Axon Growth

The growth of an axon starts once the accumulation of a specified chemical in a cell has reached a certain level, and the nerve growth factor (NGF) has reached a certain density. The direction in which the axon grows is determined by the gradient of the NGF. Since the two-dimsional space in the model is quantized using orthogonal coordinates, the possible directions of growth are restricted to up, down, left, and right. The growth cone of the axon checks the density gradient in each of the four directions, the axon growing along the steepest NGF gradient. The speed of axon growth is regulated by the maximum growth speed and by the level of activity of the cell:

$$\frac{dL}{dt} = \frac{T_{max}A}{C+A} \tag{16}$$

2.9.3 Synaptic Connection

When the growth cone reaches a cell, a synaptic connection is created. Whenever a synapse is created, an initial connection weight is assigned. The weight is assigned randomly, but a a regulated range. If $x_E > x_I$, the weight is assigned a positive value, corresponding to an excitory connection. Otherwise, a negative weight is assigned, creating an inhibitory connection. This connection weight, however, is subject to modulation through learning.

2.10 Parameters

Unless otherwise stated, the following parameters are used in experiments presented in this paper; $P = 0.1$, $D = 0.1$, $PL = 0.1$, $DL = 0.1$, $R = 10.0$, $F = 0.1$, $S_{min} = 3.0$, $S_{max} = 100.0$, $\bar{X} = 15.0$, $\delta = 0.1$, $V_{max} = 10.0$, $K_m = 5.0$, and Time Step = 0.1.

3 Experimental Results

Since the process of evolution and basic characteristic was reported in [Kitano, 1994], this section focus of neurogenesis and differentiation. Further details are provided in [Kitano, 1995].

A chromosome of the best individual of the 150-th generation was taken to analyse the development stage of the individual. Fig. 4 shows a 2-D view of the development process of the neural network for this individual. In the early stages of development, there are a number of cell types, the chemical balance of each cell changing drastically within a few cell cycles. However, after certain number of cell divisions (generally when an individual consists of more than 30 cells), the cell types tend to be stabilized. We have observed that, at this stage, cells are differentiated into at least two types. The first type exists at the outer edge of a cluster, while the second type exists near the center of a cluster. This phenomenon were observed in almost all cases where a cell cluster grew over a certain size.

It should be noted that simple differentiation can be observed in Fig. 4. The NFG types of the cells near the center and near the surface of the cluster are different. A network structure is a mixture of local and long-distance connections. Long-distance connections are generally started from cells closer to the surface, extending to a cell near the center of the cluster. Coaxial patterns of cell differentiation can be best observed in the Fig 5.

Changing the local diffusion parameter drastically changes behavior of the development. At DL = 0.05, activity of cells are converged into coherent mode. Also, it shows drastic increase in axon length. At DL = 0.005, activity level converge into few clusters. This is due to local diffusion. At DL = 0.0005, no cluster

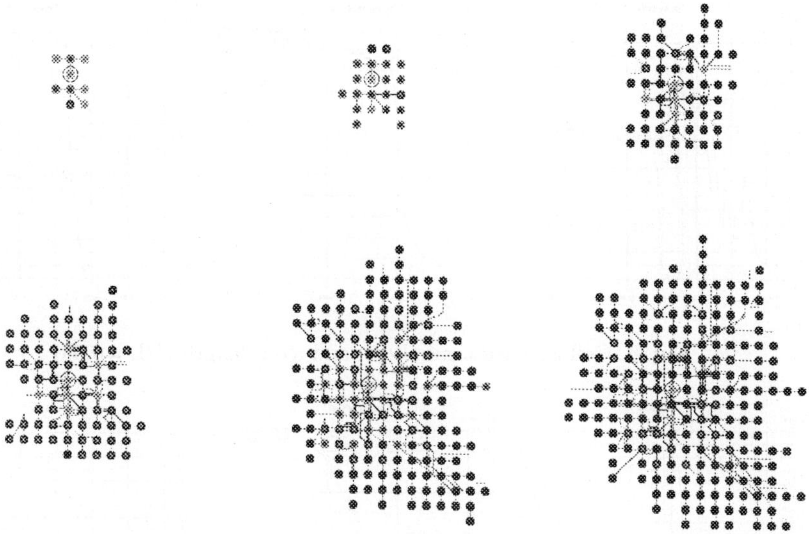

Fig. 4. Growth of Neural Network

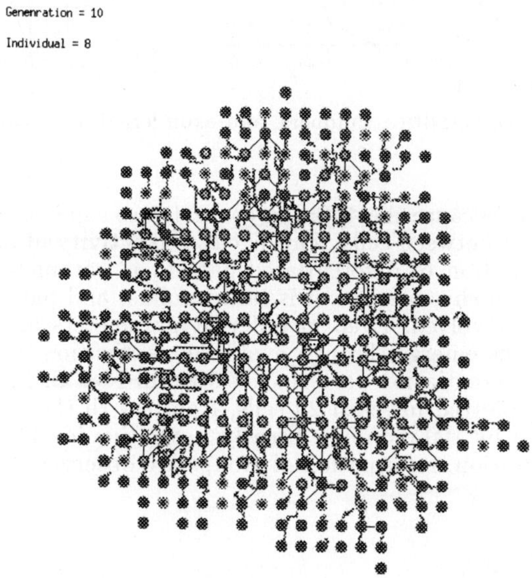

Fig. 5. Differentiated cell clusters

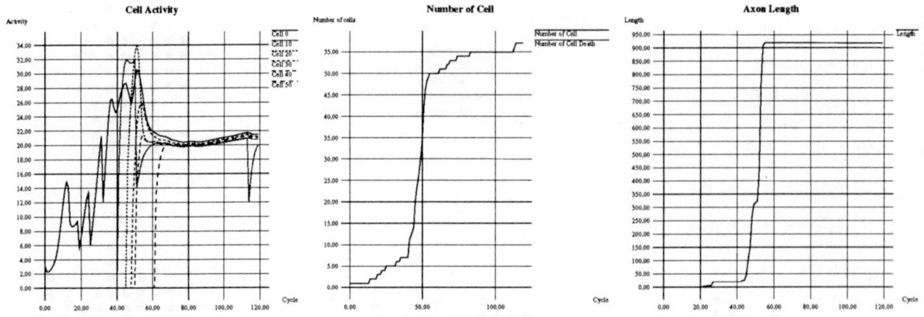

Fig. 6. Cell activities, numbers, and axon length (DL =0.05)

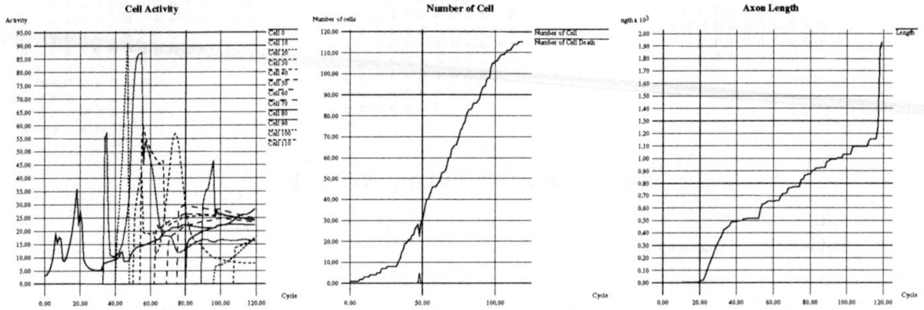

Fig. 7. Cell activities, numbers, and axon length (DL =0.005)

is observed. This is because the diffusion contact is so small that virtually no interaction tool place between cells. This is close to activity of independent cells. Although the formation of differentiated pattern is parameter sensitive when simulated on a given chromosome, it is rather robust and parameter insensitive when evolution is involved. Even in different parameters are set, similar differentiation patterns emerge through evolution. Activation levels of cells in an individual with differentiation patterns is shown in Fig. 9. Fig. 9(a) is activation level of cells plotted for each 25 cells, from (b) to (f) are activation level of cells partitioned by 100 intervals. Clearly, activation levels are clustered into two groups. In addition, reformation of groups are observed.

4 Summary

Several interesting phenomena were revealed by the experiments. First, the experimental results consistently demonstrate the possibility of the differentiation of cell types through a developmental process. In this paper, cell type refers to the type of chemical balance in each cell, which reflects expression and regulation of gene.

The second point we noticed was that the speed of the change in the chemical balance in each cell slowed down as the size of the cell cluster grew. Once, when

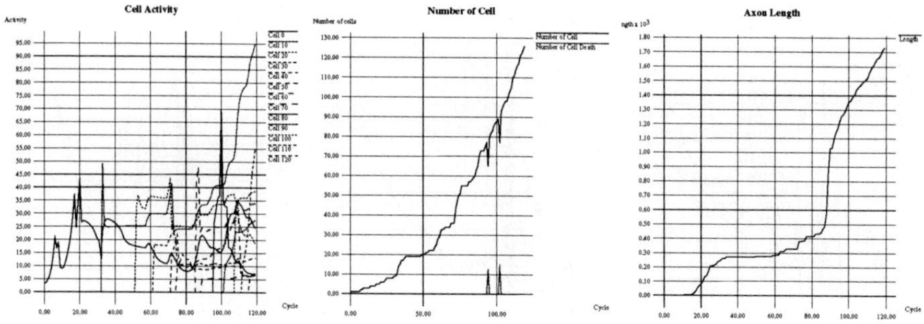

Fig. 8. Cell activities, numbers, and axon length (DL =0.0005)

possible, differentiation started, the chemical balance of each cell stabilized.

In terms of the growth of nerve systems, our experiments indicate that there are several stages of active axon growth. This can be clearly seen in Fig. 6 7 and 8. To date, we have not analyzed this phenomenon, from the point of view of whether there is a biological correspondence in any aspect of neurogenesis.

Third, experimental results implies that interaction between cells are very important to form global behavior of the system. With large diffusion constant, differentiation is less likely to take place. At the same time, with very small diffusion constant, interaction do not take place so that no coherent behavior takes place. An interesting region exists within certain window of parameters that exhibits clustering of cells. Further research goal is to identify characteristics of this region on what dynamical property it may entail.

Finally, relation ship between number of cells and length of axon was found to be non-trivial. Increase in number of cell and total axon length has certain time delay. This relationship changes when there are clusters or not. More detailed analysis is necessary to make any conclusion on this phenomena.

ESCM is a new formalization of large-scale chaos with an evolutionary feature. The model based on this idea incorporates evolution, metabolism, and morphogenesis. This is a major advance from existing models [Fleischer and Barr, 1994; Lindenmayer, 1968] A series of experiments demonstrated that complex neural networks can be formed using the model. Possible cell differentiation was observed. In addition, we discovered emergent differentiation patterns which were widely observed in subsequent, independent experiments. This implies the existence of universal dynamics in the development of biological systems.

References

[Fleischer and Barr, 1994] Fleischer, K. and Barr, A., "A Simulation Testbed for the Study of Multicellular Development: The Multiple Mechanism of Morphogenesis," *Artificial Life III,* Addison Wesley, 1994.

[Kaneko and Yomo, 1994] Kaneko, K. and Yomo, T., "Cell Division, Differentiation and Dynamic Clustering," *Physica D,* 1994. (to appear)

[Kaneko, 1992] Kaneko, K., "Overview of coupled map lattices," *Chaos,* 2 (3), 1992.

[Kitano, 1994] Kitano, H., "Evolution of Metabolism for Morphogenesis," *Proc. of Alife-IV,* 1994.

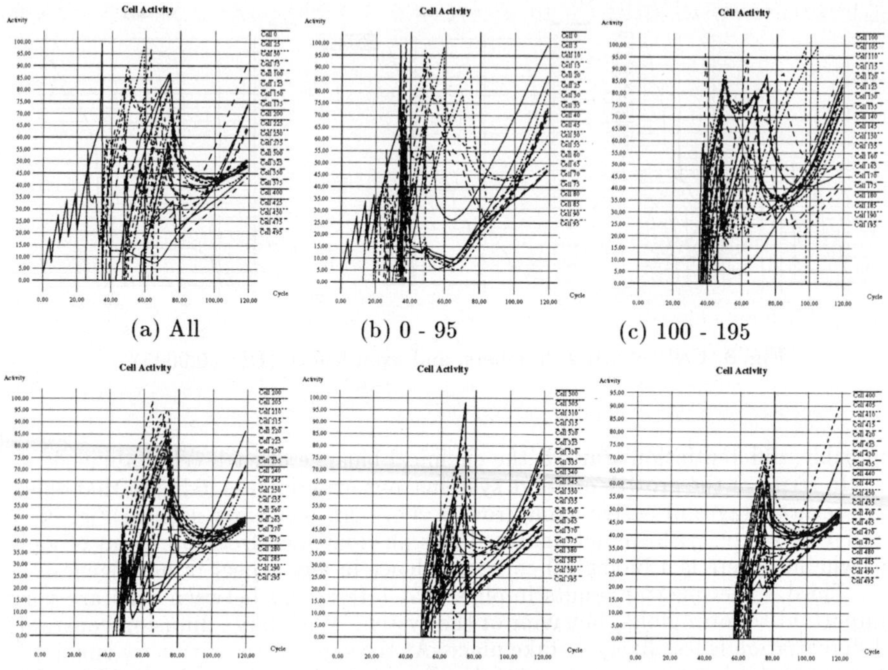

(a) All (b) 0 - 95 (c) 100 - 195

(d) 200 - 295 (e) 300 - 395 (f) 400 - 495

Fig. 9. Cell activities for differentiated individual

[Kitano, 1995] Kitano, H., "A Simple Model of Neurogenesis and Cell Differentiation based on Evolutionary Large Scale Chaos," *Artificial Life,* Vol. 2, No. 1, 1995.

[Lindenmayer, 1968] Lindenmayer, A., "Mathematical Models for Cellular Interactions in Development," *J. theor. Biol.,* 18, 280-299, 1968.

Evolving Artificial Neural Networks that Develop in Time

Stefano Nolfi and Domenico Parisi

Institute of Psychology, National Research Council
15, Viale Marx, 00137, Rome, Italy
e-mail: stefano@kant.irmkant.rm.cnr.it - domenico@kant.irmkant.rm.cnr.it

Abstract. Although recently there has been an increasing interest in studing genetically-based development using Artificial Life models, the mapping of the genetic information into the phenotype is usually modeled as an abstract process that takes place instantaneously, i.e. before the creature starts to interact with the external world and is tested for fitness. In this paper we show that the temporal dimension of development has important consequences. By analyzing the results of simulations with temporally developing neural netwoks we found that evolution, by favouring the reproduction of organisms which are efficient at all epochs of their life, selects genotypes which dictate early maturation of functional neural structure but not of nonfunctional structure. In addition, we found that development in time forces evolution to be conservative with characters that mature in the first phases of development while it allows evolution to play more freely with characters that mature later in development. Finally, characters that mature in the first phases of development tend to be phylogenetically older than characters that mature later.

1 Introduction

Development in natural organisms is a temporal phenomenon. There is not a single phenotype that is the final result of the genotype-to-phenotype mapping - the mature or adult form of the phenotypical individual - but the individual passes through a sequence of phenotypical forms from the initial cell (the egg) to the adult form. Therefore, the information contained in the genotype is reflected not only in the initial state of the organism (at birth or conception) but also - and crucially - in the succession of changes that occur in the organism throughout the organism's life (although these changes may be more pronounced during the organism's developmental age).

Such developmental issues have potentially important evolutionary consequences. Because in many animals (including humans) the gametic cells are sequestered from the embryology affecting the rest of the somatic cells, the "Weissmann doctrine" has often been invoked to decouple developmental issues from population genetics. Like all other characters, however, the developmental process itself evolved and the wide range of developmental "strategies" employed across the various phyla points to important interactions between development and evolution (Buss, 1987). For example, the effective in utero insulation of the mammalian fetus in some species has allowed the developmental process much more flexibility during this stage than in species (frogs, for example) where the juvenile's viability is costantly being tested. Gould has been a particularly vocal advocate of ways in which apparently minor "heterochronic" mutations affecting the relative timing of subprocesses of development can have drammatic consequences (Gould, 1977). At a more local level, features of a species' evolved "life history strategy" such as when individuals reach reproductive maturity and the evolutionary consequences of senescence must also be considered part of any developmental account.

Recently, there has been an increasing interest in studing development using Artificial Life models. Wilson (1987), Kitano (1990), and Belew (1993) have proposed models that include a process resembling the cellular duplication and differentiation process. Nolfi and Parisi (in press) have proposed a model in which the connectivity of the nervous system of artificial creatures grows in a way that has similarities to neural development in natural organism. Cangelosi, Parisi, and Nolfi (1994) and Dellaert and Beer (1994) have proposed still more complex models in which both the cellular duplication and differentiation process and the neural development process are simulated. In Dellart and Beer, in particular, the molecular level is also simulated (even if, of course, in a very simplified way) with a boolean network that emulates genetic regulatory processes (see also Kauffman, 1969). However, in all these works and, as far as we know, in all published works in Artificial Life, the mapping of the genetic information into the phenotype is viewed as an abstract process that takes place instantaneously, i.e. before the creature starts to interact with the external world and is tested for fitness.

Several published works that use neural networks to represent the nervous system of artificial creatures describe models that entail some form of learning that produce after-birth changes in evolving creatures (e.g. Parisi, Nolfi, and Cecconi, 1991; Ackley and Litmann, 1991; Nolfi, Elman, and Parisi, 1994). However learning is esogenous change, i.e. change caused by the interaction of the individual with the external environment. Development is endogeneous change, i.e. change due to genetically inherited information which is already "inside" the organism. We are aware of the fact that learning and development are interconnected processes and that it may be difficult to separate them in real biological organisms. However, the possibility to study development without learning in artificial organisms is an advantage if one wants to understand the influence of development on evolution. (For a model that incorporates both development in time and learning, cf. Nolfi, Miglino, and Parisi, 1994).

If development is realized as a temporal succession of phenotypic forms a number of important research questions that cannot be posed with nontemporal mapping models are open to research with simulations.

The fact that development occurs in time implies that all the successive phenotypical forms which constitutes development may be subject to evaluation in terms of fitness. While in nontemporal mapping schemes it is only the single adult form whose fitness is measured and which determines the individual's reproductive chances, if one adopts a temporal genotype/phenotype mapping each successive developmental form must demonstrate its value in terms of fitness. In fact, the global fitness of an individual is likely to be a complex function of the separate fitnesses of each of the successive developmental forms of the individual. On one side, a particular developmental form has the role of preparing in the best possibile way the future forms. On the other side, an intermediate developmental form is not only a way-station to the final adult form but it is something which must possess properties that allow it to survive and to contribute independently to the global fitness of the organism. Therefore, evolution won't shape only the final adult form of a particular organism but also all the intermediate developmental forms. And it will shape the intermediate developmental forms by taking both their role as way-stations to the adult form and their intrinsic fitness into consideration. Therefore, a first important question that can be asked is how evolution shapes the particular sequence of developmental forms that characterizes a particular population of organisms. Is there acceleration or retardation of developmental changes from one generation to the next? What is the role of heterochrony, i.e. the differential speed of development of the various parts and traits of an organism, in determining the final adult form?

But it is not only the case that evolution shapes development. Development also constrains evolution. Since an individual is a particular succession of developmental forms, evolution cannot just change a single developmental form in isolation. The possible change in fitness caused by the new developmental form depends on the entire succession of forms in which the changed form is included. Hence, the entire succession of developmental forms of an organism will constrain the evolutionary changes that will be retained or discarded. A second research question is then how the particular succession of forms that characterizes the development of an organism constrains the types of evolutionary changes that can occur at the population level. Since evolution shapes the genotype and the genotype controls development, evolution is crucial to understand development. But since, development constrains evolution, development is crucial to understand evolution.

A final question is a classical one: What is the relationship between evolutionary changes and developmental changes? Are there similarities between the two? Does development (ontogeny) recapitulates evolution (phylogeny)?

2 The model

Let us begin by assuming that our ultimate goal is to create an organism (O) which is able to find and eat food in its environment. O includes two components: (a) a phenotypical neural network which controls O's behavior in the environment, and (b) some genetic material (genotype) encoding developmental instructions which generate a certain number of neurons and control the growing and branching process of the neurons' axons (Nolfi and Parisi, in press). The result of this growing process is a neural network that represents the nervous system of the corresponding O. The architecture and connection weights of such a network determine the way in which O responds to environmental stimuli, i.e. its behavior. O's behavior determines its fitness, i.e. its reproductive chances, through O's interaction with the environment to which it is exposed.

Each O lives in a simulated environment which is a two-dimensional square divided up into cells. At any particular moment O occupies one of these cells. A number of food elements are randomly distributed in the environment with each food element occupying a single cell. O has a facing direction and a rudimentary sensory system that allows it to receive the angle (relative to where O is currently facing) and the distance of the nearest food element as input. O is also equipped with a simple motor system that provides it with the possibility, in a single action, to turn any angle from 90 degrees left to 90 degrees right and then move from 0 to 5 cells forward. Finally, when O happens to step on a food cell, it eats the food element which disappears.

O's genotype is represented as a string of 0 and 1. It has a fixed length (1600 bits) and is divided up into 40 blocks, each block corresponding to a single neuron. Since some blocks may not be expressed, the mature phenotypical network may include less than 40 neurons. Each block encodes developmental instructions specifying a number of properties of the corresponding neuron (Fig. 1).

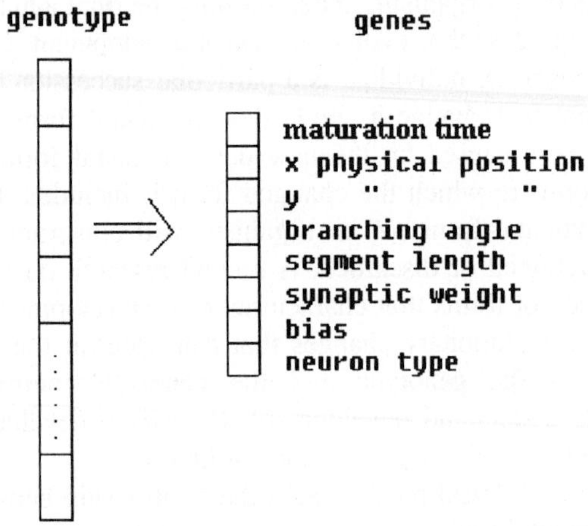

Fig. 1. O's genotype. Each block contains a set of developmental instructions for the corresponding neuron. The developmental instructions specify when during ontogeny the block will be expressed (this is the 'maturation time gene'), the x and y spatial coordinates of the neuron in the nervous system's bidimensional space, the angle of branching and the length of the branching segments, the weight of the connections departing from the neuron (same weight for all connections), the activation bias of the neuron, the type of the neuron: sensory, internal, or motor.

When a growing axonal branch of a particular neuron reaches another neuron a connection between the two neurons is established. Figure 2 shows the growth of axons resulting from a random genetic string on the left and the resulting neural network on the right. Functional neurons and connections, i.e. neurons and connections that actually map input into output and therefore can influence the O's behavior, are represented with larger circles and thick lines.

The functional network determines O's behavior through the interaction with the environment. At each time step, O receives a pattern of activation values on its sensory neurons encoding the position of the nearest food element relative to O. Such an input determines, through a spreading activation process, the activation value of the internal and output neurons. These last neurons determine O's motor reaction to the current input, i.e. O's behavior.

Since different blocks in an individual's genotype may contain different 'maturation time genes', different neurons will grow their neurites at different times during the life of the individual. The branching axons will establish connections that can vary from one epoch to the successive epoch of life because new connections can be added. Therefore, the behavior of the individual in the environment will be controlled by a succession of different phenotypical forms of its nervous system and, as a consequence, the behavior itself will be different in various epochs of life.

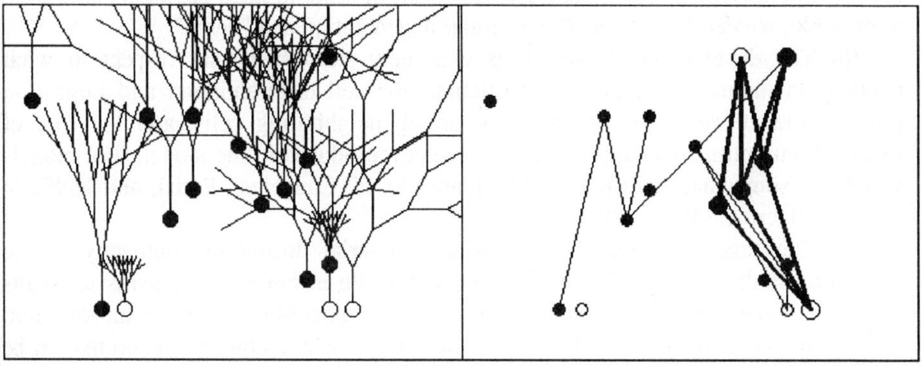

Fig. 2. Development of a neural network from a randomly generated genetic string. The growing and branching process of axons is shown on the left side of the Figure. The resulting neural network after removal of nonconnecting branches is shown on the right side. Isolated or nonfunctional neurons and connections are represented by small circles and thin lines. Functional neurons and connections are represented as large circles and thick lines. The bottom layer contains sensory neurons, the upper layer motor neurons, and the remaining layers internal or hidden neurons. Filled and empty circles denote different types of encoded information.

In order to obtain adapted Os, i.e. Os that exhibit an efficient food collecting behavior, we simulate an evolutionary process using a genetic algorithm. (Holland, 1975). We ran 20 simulations each starting with 100 different randomly generated genotypes. This is generation 0 (G0). Os of G0 are allowed to develop and "live" for 8 epochs, each epoch consisting of 250 actions in 5 different environments (50 actions each). The environment is a grid of 40x40 cells with 10 pieces of food randomly distributed in it. Os are placed in individual copies of these environments, i.e. they live and develop in isolation. At the end of their life (2000 actions) they are allowed to reproduce. However, only the 20 individuals which have accumulated the most food during their life reproduce by generating 5 copies of their genotype. These 20x5=100 new Os constitute the next generation (G1). Random mutations are introduced in the copying process resulting in possible changes in the phenotypic network and/or in the time of development of different subparts of the phenotypic network. The organisms of G1 are also allowed to live for 2000 cycles. The process continues for 1000 generations.

The neural development model is inspired by biology but it does not pretend to be a realistic model of neural development. It is simplified in several respects and in particular:

(a) It contemplates only growth and no degeneration or disappearence of parts of the developing neural structure. One could easily include in the model some form of "death" of units or connections, either programmed in the genotype or as a consequence of time past since the appearence of the particular unit or connection or because of experience and (in)activity. However, in order to keep the model as simple as possible, we decided to not allow cellular degeneration and "death".

(b) The developmental process is extremely simplified with respect to what happens in natural organisms. In particular, the cell differentiation and migration process, the effects of the interactions among neighboring cells, and the role of extracellular entities and trophic factors are not reproduced in the model. (For models which are somewhat more biologically plausible, cf. Cangelosi, Parisi, and Nolfi, in press; Dellaert and Beer, 1994).

(c) The time of appeareance of neural material during the ontogeny of the individual is directly encoded in the genotype while in natural organisms it results from an interaction between genetic products and possibly exogenous information. This implies that in our model neurons and groups of interconneted neurons can be anticipated or postponed in ontogeny independently of each other. This in turn implies that they cannot assume interphene functions. A character has an interphene function when in addition to having an adaptive value with respect to the external environment it assumes an adaptive value as an intermediate stage which is necessary to induce subsequent developmental characters (cf. Mayr, 1994). We choose to use this type of implausible coding both to keep the model simple and to be able to track diacronic changes phylogenetically.

(d) In our model individual fitness, i.e. the individual's reproductive chances, is calculated by summing the number of food elements collected by the individual during each epoch of life. This implies that the individual's ability to find food at different phases of the individual's ontogeny has the same impact on fitness. In natural organisms the situation is more complex and it can be different in different species. In most species the first phases of ontogeny appear to be the most important ones because if individuals are not able to collect enough energy or to escape from predators they can easily die. In other (e.g. mammal) species, with strong parental care behaviors, the survival in the first phases of ontogeny is facilitated and almost independent from the ability of the individual. As a consequence, one can hypotesize that the ability in the first phases of ontogeny weights less on the reproductive chances of the individual.

3 Simulation results

If we look at how fitness changes phylogenetically and ontogenetically we observe an increase in both cases (Fig. 3 and 4). If we count the number of food elements our Os are able to collect during their life across the 1000 generations, we see that Os of successive generations are increasingly able to approach food elements. Figure 3 shows the fitness value of the Os of the winning lineage for each of the 1000 generations. (Average of 20 different simulations.) The winning lineage is the lineage of the best individual of the last generation. The lineage is constituted by this individual, by its (single) parent in the preceding generation, by its grand parent, its grandgrand parent, etc., until the originator of the lineage in the first generation is reached. The evolutionary increase in fitness implies that generation after generation Os are able to adapt their architecture and synaptic weights to the evolutionary task. The number of food elements available in the entire lifetime of an O is 400. On average the best Os of the last generation reach 57% of these food elements, with one of the 20 simulations reaching 95%.

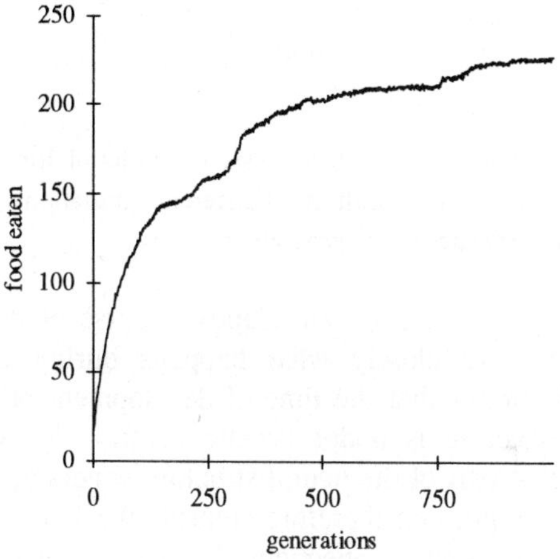

Fig. 3. Fitness (= total number of food elements eaten during life) of the Os of the winning lineage. The graph represents the average result of 20 simulations starting from initial populations of different randomly generated genotypes.

Figure 4 shows the number of food elements Os are able to eat in different epochs of their life. The amount of food eaten increases during life, especially in the early epochs. This implies that genetically controlled developmental changes in neural structure induce behavioral changes that lead to progressively more effective behaviors with increasing age. In other words, selective reproduction and mutations in successive generations of individuals results in the evolutionary emergence of developmentally adapted genotypes.

This might in part be explained as a necessary by-product of our model of development. Our model contemplates neural growth but no neural degeneration or programmed cell death. Hence, development tends to be a one-way progressive process with addition but no subtraction of neural structure. However, even the addition of newly developed neural structure could in principle have negative effects on fitness. Apparently, the evolutionary process in our simulations is able to rule out this possibility.

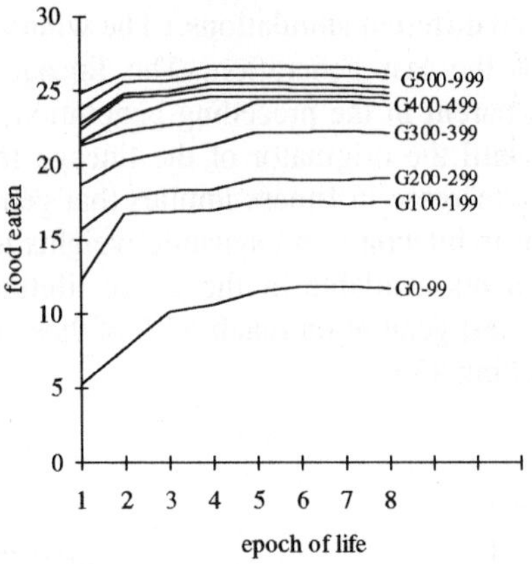

Fig. 4. Number of food elements eaten in successive epochs of life by Os of the winning lineage. The graph shows the average result of 20 different simulations. Each curve represents the average performance of 100 successive generations.

But the role of evolution in selecting an adaptive course of development becomes evident if we analyze more closely what happens during development in our organisms. Our model implies that the time of development of both functional and nonfunctional neural structure is under genetic control. As will be recalled, the functional structure is that part of the neural structure generated during development which maps input into output and therefore controls the individual's behavior. The nonfunctional structure (units and connections) is also generated during development but is not part of the input/output pathways and therefore has no role in controlling behavior. We can ask how evolution shapes the temporal dimension of development both generally and with respect to these two components of neural structure. In other words, is there a general tendency to evolutionarily anticipate or posticipate the generation of neural structure during development and are there differences in this tendency between functional and nonfunctional structure?

An answer to this questions is in Figure 5 which shows how the average epoch of life in which functional and nonfunctional units are generated changes evolutionarily. (Functional units are units that are part of the functional network while nonfunctional units are part of the nonfunctional neural structure.) Functional units tend to develop earlier in life than nonfunctional units from the very first generations on and the effect of evolution is to increase the average difference in the epoch of development for the two types of units. More specifically, there is an evolutionary tendency for units that are part of the functional network to be anticipated in their epoch of development across generations.

A similar conclusion is reached by counting the number of neutral and non-neutral mutations in the genotype that cause posticipation or anticipation in the time of development of individual neurons (cf. Fig. 6). Neutral mutations are those mutations which do not translate into a change (increase or decrease) in fitness. Non-neutral mutations are those that cause either an increase (adaptive mutations) or a decrease (maladaptive mutations) in fitness. Non-neutral anticipations (i.e. mutations affecting the 'maturation time genes' that determine a modification in the individual's fitness and cause an anticipation in the time of development of neurons during ontogeny) largely outnumber posticipations. On the contrary for neutral mutations, posticipations and anticipations do not differ significantly. (Notice that in the winning lineage non-neutral mutations are likely to be adaptive rather than maladaptive.)

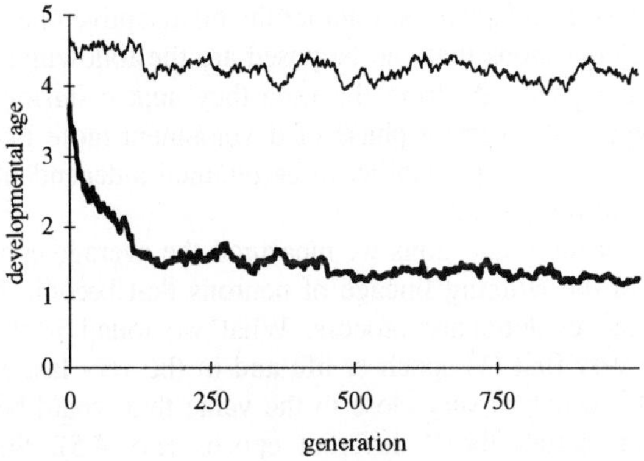

Fig. 5. Average epoch of development of functional and nonfunctional units in the individuals of the winning lineage across 1000 generations. Average results of 20 different simulations.

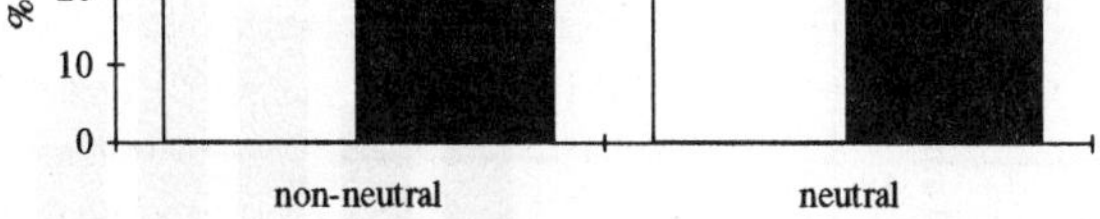

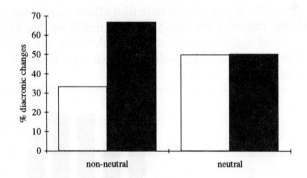

Fig. 6. Percentage of posticipations (empty bars) and anticipations (full bars) for non-neutral and neutral mutations in the 1000 individuals of the winning lineage. Average results of 20 different simulations.

These results show that evolution selects genotypes which dictate early maturation of functional neural structure but not of nonfunctional structure. Selective reproduction based on fitness appears to be the force that causes this developmental anticipation of newly evolved structures by favouring the reproduction of Os which are efficient at all epochs of their life and, therefore, as early as possible during lifetime. It is clear that anticipation pressures only apply to functional neurons (i.e. to neurons that contribute to determining the individual's behavior) and, among them, only to neurons that produce an increase in performance. Functional neurons that decrease performance are likely postponed. Thus, in simulations with artificial organisms that develop in time evolution can follow two adaptive routes: (a) it can select individuals with adaptive characters overall (cf. Fig. 4), or (b) it can anticipate adaptive characters and posticipate maladaptive ones. The fact that retained adaptive anticipations outnumber posticipations can be interpreted as a consequence of (a) which ensures that among non-neutral characters the adaptive outnumber the maladaptive ones.

Other interesting questions that can be posed are the following: Are evolutionary novelties retained independently from the time they appear during ontogeny or are novelties which appear at a certain phase of development more like to be retained? And have mutations the same probability to be retained independently from the time in which they affect development?

In order to answer these questions we measured the average epoch of appearence in the individuals of the winning lineage of neurons that become functional for the first time during the evolutionary process. What we found is that novelties may appear both in the very first (1) epoch of life and in the very last epoch (8) with an average value of 4.4 which is very close to the value that would be obtained with a uniform distribution across the 8 different epochs (i.e. 4.5). So it appears that novelties may be retained whatever the epoch in development in which they emerge.

Although evolutionary novelties can be retained independently from their epoch of appearence during development, the probability that adaptive mutations are retained depends on the time of appearence (cf. Fig. 7). In particular, neurons that appear in the later epochs of life are more likely to receive mutations than neurons that appear in the first epochs of life. Mutations that have their effects early in life are more likely to affect greatly the fitness of the individual than mutations that have their effects later in life. But most mutations have negative rather than beneficial effects. As a consequence, it is very unlikely that individuals that have received mutations with effects early in life will leave offspring.

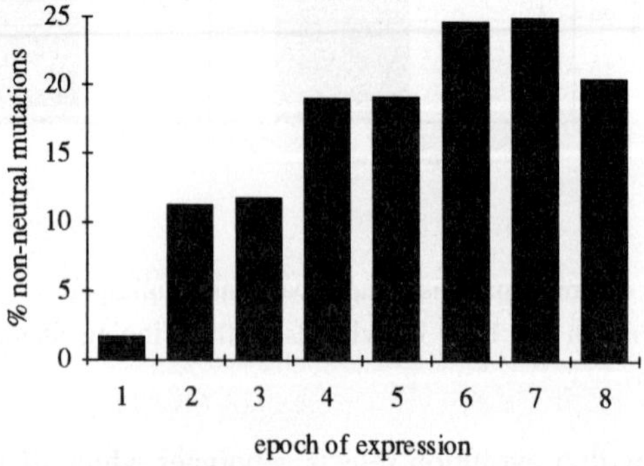

Fig. 7. Probability that a neuron of an individual of the winning lineage was subject to a non-neutral mutation for different epochs of appearence of the neuron. Average results of 20 different simulations.

The neurons that mature in the early phases of development in addition to being the most important in determining a fit behavior and the less mutated are also the oldest phylogenetically. As Figure 8 shows, neurons that appear in the first epochs of life are significantly older on average (i.e. have become part of the functional network a greater number of generations before) than neurons that mature in the last part of an individual's lifetime.

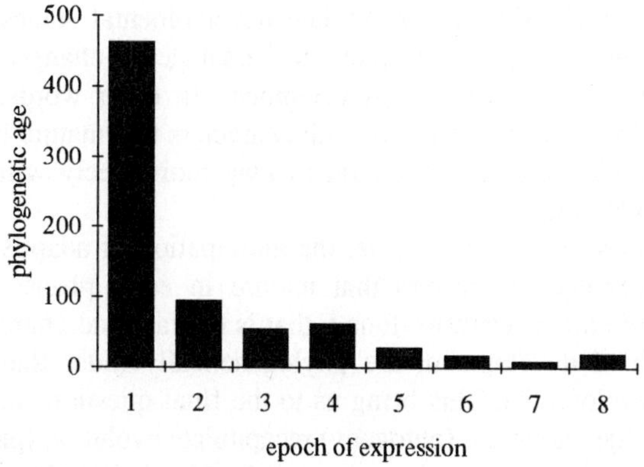

Fig. 8. Average number of generations since when neurons that appear at different epochs of life have become functional. Average result of 20 different simulations.

The fact that the anticipation process applies only to functional neural structure combined with the fact that only neurons that are preserved for many generations have the possibility to be anticipated (the probability for an individual to receive a mutation that anticipates a particolar neuron being very small) ensures that the earlier a neuron or a neural module appears in development the higher the probability that such a neuron or neural model is useful and important in determining a fit behavior and, as a consequence, the higher the probability that it will be preserved in successive generations.

4 Discussion

We have shown that to develop in time has important consequences. In particular, our results support the claim we made in the introduction that evolution shapes development and development constrains evolution.

Evolution shapes development by anticipating the maturation of adaptive characters but not of neutral ones (and of maladaptive ones). By favouring individuals that become efficient as soon as possible selective reproduction appears to be the force that causes this developmental anticipation (for a similar explanation see Muller, 1864). This implies that anticipations in non-neutral characters will outnumber posticipations at least if, as seems to be the case in our simulations, evolution is able to select individuals in which adaptive characters outnumber maladaptive ones. This also implies that the fact that individuals develop in time allows evolution to follow two adaptive routes: (a) to select individuals with adaptive characters overall; (b) to anticipate adaptive characters and posticipate maladaptive one.

Development constrains evolution by allowing an easy retention of mutations affecting characters that mature in later stages of development while it remains more difficult to retain mutations with earlier effects. Therefore, although we did not find any tendency for novelties to appear in late developmental stages, it appears that characters that mature late are more likely to be subject to changes and innovations than characters that mature early in development. In other words, development in time forces evolution to be conservative with characters that mature in the first phases of development while it allows evolution to play more freely with characters that mature later in development.

The combination of these two facts: the anticipation of adaptive characters and the tendency to preserve characters that mature in early phases of development, probably explain the other result we found, that is the fact that characters that mature in the first epoch of development are phylogenetically older than characters that mature later in development. This bring us to the final question we discussed in the introduction: Does development (ontogeny) recapitulate evolution (phylogeny)?

The recapitulation theory as was proposed by Haeckel (1966), i.e. the idea that characters that develop in an individual from conception to maturity repeat the evolutionary history of the individual's species, involves two fundamental assumptions: terminal addition and anticipation. Terminal addition means that novelties can only be introduced at the end of the developmental process. Anticipation means that the length of an ancestral ontogeny is continuously shortened during the subsequent evolution of the lineage. This assumption means that phylogenetically older characters which first appeared at a certain stage of development often develop at an earlier stage during the ontogeny of individuals of successive generations while more recent phylogenetic characters develop later. This theory is false because it has been empirically shown that acceleration is not general or equal for all characters, that new characters can be introduced at any stage of ontogeny (hence, even earlier than older characters), and that development can be retarded as well as accelerated (cf. Gould, 1977).

However, if one considers, as Gould does, recapitulation not as an absolute law but as a simple tendency which is the result of a more general process, the evolutionary alteration of developmental times and rates to produce acceleration and retardation in the ontogenetic development of specific characters, one can conclude that recapitulation phenomena can be observed if there is some tendency to accelerate development and to preserve characters that have been accelerated. Our results suggest that in individuals that develop in time both requirements may be satisfied to a certain extent and therefore recapitulations phenomena are actually observed in our simulations. This does not imply that development in time is the only cause of recapitulation. There are in fact at least two other causes that have to do with development per se and that do not concern its temporal character. The first cause has to do with the recursive nature of the cellular duplication process (Belew, 1993;

Cangelosi, Parisi, and Nolfi, 1994). Mutations that affect the first cellular duplication phases have in general a huge impact on the resulting phenotype and therefore can be retained very rarely. The second possible cause of recapitulation concerns the fact that characters that mature in early phases of development even if they loose their adaptive function can have an interphene functions, i.e. they can assume an adaptive value as an intermediate stage which is necessary to reach subsequent developmental stages (Mayr, 1994). Both facts may force evolution to be conservative with characters that mature in the early phases of development. This may explain why Dellaer and Beer (1994) also observe some recapitulation phenomena with their model in which individuals develop istantaneously.

Finally, we have shown that the anticipation tendency only applies to the functional characters of individuals, i.e. to characters that imply a direct advantage in the reproductive chances of the individual. In fact, one can expect that, as already observed by Mehnert and Massart at the end of the last century (Menhert, 1898; Massart, 1894), characters which must function first develop first.

However, the fact that there is no evolutionary anticipation of neutral characters during development should not be taken as implying that the nonfunctional neural structure - which is generated under genetic control exactly like the functional structure - is irrelevant from the point of view of the development of the individual. Miglino, Nolfi and Parisi (in press) have argued that although the nonfunctional neural structure has no role in determining the behavior of the individual, it can have an important role in the evolution of the population of which the individual is a member. Nonfunctional neural structure which has accumulated evolutionarily because of neutral mutations can become suddenly functional because of a further mutation and perhaps cause a change in fitness (Loomis, 1988). Without the already existing nonfunctional structure the mutation might be unable to cause the change in fitness.

The same reasoning applies at the individual level if the individual develops in the course of its life. At any given epoch (age) the nonfunctional structure which has developed in the previous epochs of the individual's life has no role in determining the behavior of the individual at that epoch. But a further step in development can make (part of) this nonfunctional structure functional and cause a change in the fitness collected by the individual in the next epochs of life. The new developmentally emerging functional structure and the resulting change in fitness might never be realized unless the previous stages in development had not caused the emergence of the critical nonfunctional structure.

In both cases, a static and a dynamic view of the phenomena concerned offer different perspectives on the role of nonfunctional stucture. If an organism is viewed statically as a nonchanging entity, the nonfunctional structure which is generated inside the organism under genetic control appears to be useless. Only at the dynamic level of evolutionary change in the population, the nonfunctional structure can have a role and a meaning, if not for the individual, for the population to which the individual belongs. Furthermore, if the individual itself is viewed dynamically as a developing entity, then the nonfunctional structure has a role and a meaning also for the individual. The behavior that the individual exhibits in a particular epoch of its life can be dependent on the nonfunctional structure which has developed in previous epochs.

In fact, the various results we have described concerning the evolutionary and developmental changes in the functional and in the nonfunctional neural structure allow us to draw a picture in which development tends to have two opposite functions from an evolutionary point of view. On one side, development tends to be conservative and to guarantee a certain amount of fitness to an individual. This is obtained by anticipating the development of functional neural structure and by sheltering this structure from the effects of disrupting mutations. Mutations that negatively affect the initial functional neural structure tend to be quickly eliminated so that there is no trace of these mutations in the best individual of the next generation. A few mutations can affect the changes in functional neural structure that occur during development but the individual is relatively sheltered from the possible negative effects of these mutations because there is little change in functional neural structure after birth and because most of the individual's fitness is collected using the stable functional structure already present at birth. Those mutations that turn out to have a positive effect on fitness are retained and tend to be anticipated in development in the successive generations.

However, after the initial generations most retained mutations are those that affect the nonfunctional neural structure. These mutations may happen to be retained because they are neutral. By affecting the nonfunctional neural structure, they do not change the behavior of the individual and, therefore, its fitness. This aspect of development may represent an important tool to explore novelty at disposal of evolution. Many evolutionary changes in nonfunctional neural structure can be retained because they do not affect fitness. At a certain point in evolution, due to some further mutation, one of these changes is carried over from the nonfunctional to the functional neural structure. If the change is maladaptive it does not affect fitness too much because it happens later in life. If it turns out to be favourable from the point of view of fitness, the change is retained because it is adaptive and it may be anticipated in development later in evolution. Hence, development appears to be a flexible tool in the hands of evolution with both a conservative role and a role to explore evolutionary innovation.

References

Ackley, D.E. and Littman, M.L. (1991). Interactions between learning and evolution. In C.G. Langton, J.D. Farmer, S. Rasmussen, and C.E. Taylor (eds.) *Proceedings of the Second Conference on Artificial Life*. Addison-Wesley: Reading, MA.

Belew, R.K. (1993). Interposing an ontogenetic model between Genetic Algorithms and Neural Networks. In J. Cowan (ed.), *Advances in Neural Information Processing (NIPS5)*, San Mateo, CA, Morgan Kaufmann.

Buss, R. (1987). *The evolution of individuality*. Princeton University Press.

Cangelosi, A., Nolfi, S. Parisi, D. (1994). Cell division and migration in a 'genotype' for neural networks. *Network*, **5**, pp.497-515.

Dellaert, F., and Beer, R.D. (1994). Toward an evolvable model of development for autonomous agent synthesis. In R. Brooks and P. Maes (eds.) *Artificial Life IV: Proceeding of the Fourth International Workshop on the Synthesys and Simulation of Living Systems*, MIT Press: Cambridge,MA.

Gould, S.J. (1977). *Ontogeny and Phylogeny*. Harward University Press: Cambridge,MA.

Haeckel, E. (1866). *Generelle morphologie der organismen*. Georg Reimer:Berlin.

Holland, J.J. (1975). *Adaptation in natural and artificial systems*. University of Michigan Press: Ann Arbor, Michigan.

Kauffman, S. (1969). Metabolic stability and epigenesis in randomly constructed genetic nets. *Journal of Theoretical Biology*, pp.437-467.

Kitano, H. (1990). Designing neural networks using genetic algorithms with graph generation system. *Complex Systems*, **4**, pp. 461-476.

Loomis W.F. (1988). *Four billion years: an essay on the evolution of genes and organisms*. Sinaver Associates, Inc.: Sunderland, MA.

Mayr E. (1994). Recapitulation reinterpreted: the somatic program. *The Quartely Review of Biology*, **69**, pp.223-232.

Massart, J. (1894). Le recapitulation et l'innovation en embryologie vegetale. *Bull. Soc. Roy. Bot. Belgique,* **33**, pp.150-247.

Menhert, E. (1898). *Biomechanik*. Gustav Fisher, Jena.

McKinney, M.L., McNamara, K.J. (1991). *Heterochrony*. Plenum: New York.

Miglino, O., Nolfi, S., Parisi, D. (in press). Discontinuity in evolution: how different levels of organization imply pre-adaptation. In R.K. Belew, and M. Mitchell (eds.) *Plastic Individuals in Evolving Populations*. SFI Series, Addison-Wesley.

Muller, F. (1864). Fur Darwin. In A. Moller (ed.) *Werke, Briefe und Leben*. Gustav Fischer: Jena, pp. 200-263.

Nolfi, S., Elman, J.L., Parisi, D. (1994). Learning and Evolution in Neural Networks. *Adaptive Behavior*, **1**, pp.5-28.

Nolfi, S. Parisi, D. (in press). "Genotypes" for Neural Networks. In M. A. Arbib (ed.) *The Handbook of Brain Theory and Neural Networks*. Bradford Books, MIT Press.

Nolfi, S., Miglino, O., Parisi, (1994). Phenotypic Plasticity in Evolving Neural Networks. In: D. P. Gaussier and J-D. Nicoud (Eds.) *Proceedings of the International Conference From Perception to Action*, Los Alamitos, IEEE Press: CA.

Parisi, D., Nolfi, S., Cecconi, F. (1991). Learning, Behavior and Evolution. In: Varela, F, Bourgine, P. *Toward a pratice of autonomous systems*. MIT Press.

Wilson, S.W., 1987, The genetic algorithm and biological development. In *Proceedings of the Second International Conference on Genetic Algorithms*. Erlbaum: Hillsdale, NJ.

Contextual Genetic Algorithms: Evolving Developmental Rules

Luis Mateus Rocha
Department of Systems Science and Industrial Engineering, Watson School of Engineering and Applied Science, SUNY-Binghamton, Binghamton NY 13902, USA
e-mail: ba05099@bingsuns.cc.binghamton.edu

Abstract. A genetic algorithm scheme with a stochastic genotype/phenotype relation is proposed. The mechanisms responsible for this intermediate level of uncertainty, are inspired by the biological system of RNA editing found in a variety of organisms. In biological systems, RNA editing represents a significant and potentially regulatory step in gene expression. The artificial algorithm here presented, will propose the evolution of such regulatory steps as an aid to the modeling of differentiated development of artificial organisms according to environmental, contextual, constraints. This mechanism of genetic string editing will then be utilized in the definition of a genetic algorithm scheme, with good scaling and evolutionary properties, in which phenotypes are represented by mathematical structures based on fuzzy set and evidence theories.

1. Introduction

The essence of GA's lies on the separation of the description of a solution (e.g. a machine) from the solution itself: variation is applied solely to the descriptions, while the respective solutions are evaluated, and the whole selected according to this evaluation [14]. A genetic algorithm "is primarily concerned with producing variants having a high probability of success in the environment" [19, page 35]. Nonetheless, one important difference between evolutionary computation and biological genetic systems, lies precisely on the connection between descriptions and solutions, between signifier and signified. In genetic algorithms the relation between the two is linear and direct: one description, one solution. While in the biological genetic systems there exists a multitude of processes, taking place between the transcription of a description and its expression, responsible for the establishment of an uncertain relation between signifier and signified, that is, a one-to-many relation.

> "The proteins encoded by [DNA] are [...] oxymorphic: their individual shapes are precisely unpredictable. So long as this is true, the genomic language, like our own languages, will not have a logical link between signifier and signified. This will not prevent its being read or understood; rather, it will assure that DNA remains a language expressing as full a range of meanings through arbitrary signifiers as any other language." [26, p. 70]

In other words, the same genotype will not always produce the same phenotype; rather, many phenotypes can be produced by one genotype depending on changes in the environmental context. If the effects of changing environmental contexts affecting gene expression within an individual can be harnessed and used to it's selective advantage in a changing environment, then we can say that such an individual has achieved a degree of control over its own genetic expression. It is the objective of this paper to propose a computational scheme which may be able to achieve this degree of control. It will be further suggested, that the modeling of biological development may be linked precisely to GA's capable of evolving this extra degree of control.

To establish this one-to-many relationship between descriptions and solutions in GA's, I will propose an extra mechanism inspired by the edition of RNA in biological genetic systems. Section 2 will introduce some of the known mechanisms of RNA Editing. Section 3 will introduce semiotic model offering a theoretical framework for RNA editing. Section 4 will propose computational counterparts to RNA editing. Section 5 will discuss the utilization of such mechanisms regarding the problem of development. Finally, Section 6 will present one particular algorithm, which utilizes fuzzy set and evidence theory to introduce even higher levels of uncertainty to description/solution relations.

2. RNA Editing

The discovery of messenger RNA (mRNA) molecules containing information not coded in DNA, first persuaded researchers in molecular biology that some mechanism in the cell might be responsible for posttranscriptional alteration of genetic information; this mechanism was called 'RNA Editing' [2, 1986]. "It was coined to illustrate that the alterations of the RNA sequence (i) occur in the protein-coding region and (ii) are most likely the result of a posttranscriptional event" [3, page 16]. The term is used to identify any mechanism which will produce mRNA molecules with information not specifically encoded in DNA. Usually we will have insertion or deletion of particular bases (e.g. uridine), or some sort of base conversion (e.g. adenosine → guanisine).

The most famous RNA editing system is that of the African Trypanosomes [3; 36]. The mitochondrial DNA of this parasite, responsible for sleeping sickness, "consists of several dozen large loops called maxicircles and thousands of smaller ones called minicircles." [27, page 132] At first, the minicircles were assumed to contain no genetic information, while maxicircles were known to encode mitochondrial rRNA. However, the maxicircles were found to possess strange sequence features such as genes without translational initiation and termination codons, frame shifted genes, etc. Furthermore, observation of mRNA's showed that many of them were significantly different than the maxicircles from which they had been transcribed. These facts suggested that mRNA's were edited posttranscriptionally.

It was later recognized that this editing was performed by *guide RNA's* (gRNA's) coded mostly by the minicircles, the strands of DNA previously assumed to contain no useful information [37; 4]. In this particular genetic system, gRNA's operate by inserting, and sometimes deleting, uridines. To appreciate the effect of this edition consider figure 1. The first example [3, p. 14] shows a massive uridine insertion (lowercase u's); the aminoacid sequence that would be obtained prior to any edition is shown on top of the base sequence, and the aminoacid sequence obtained after edition is shown in the gray box. The second example shows how potentially the insertion of a single uridine can change dramatically the aminoacid sequence obtained; in this case, a termination codon is introduced.

It is unclear how exactly gRNA's insert uridines into mRNA's; basically, the shorter gRNA strings base-pair with stretches of mRNA, and at some point will insert a number of uridines [33]. An interesting aspect of the gRNA/mRNA base-pairing is that it is more general than the Watson-Crick base-pairing found in DNA and RNA, it is more ambiguous since "uracils in mRNA can be specified by either guanine or adenine in gRNA" [36, page 36]

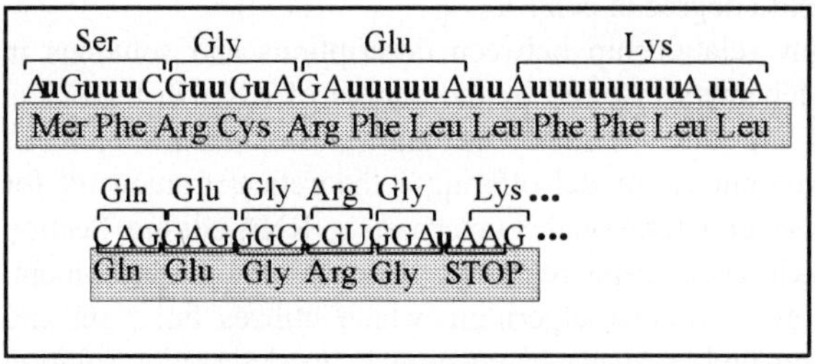

Figure 1: U-insertion in Trypanosomes

But even if the precise mechanisms of RNA editing are not yet know, its importance is unquestionable, since it has the power to dramatically alter gene expression: "cells with different mixes of [editing mechanisms] may edit a transcript from the same gene differently, thereby making different proteins from the same opened gene." [26, page 78] (one-to-many relations). It is important to retain that a mRNA can be edited in different degrees precisely according to the concentrations of editing operators it encounters. Thus, at the same time, several different proteins coded by the same gene may coexist, if all (or some) of the mRNA's obtained from the same gene, but edited differently, are meaningful to the translation mechanism.

If the concentrations of editing operators can be linked to environmental contexts, the concentrations of different proteins obtained may be selected accordingly, and thus evolve a system which is able to respond to environmental changes without changes in the major part of its genetic information (genome size optimization). One gene, different contexts, different proteins. This may be precisely what the trypanosome parasites have achieved: control over gene expression during different parts of their complex life cycles.

> "Space is clearly not a problem for mammalian nuclear DNA, so the [previous] rationale is not so obvious for the [editing mechanisms of mammals]. Also there, however, we see one gene encoding two proteins. In mammalian genomes, gene duplication followed by separate evolution of the two copies would be a more obvious way of producing closely related proteins in regulatable amounts. RNA editing, however, does provide the opportunity to introduce highly specific, local changes into only some of the molecules. [...] It could be reasoned that somehow this would be more difficult to achieve via gene duplication, since independently accumulating mutations would make it harder to keep the remainder of the two sequences identical" [3, p. 22]

Thus, RNA editing may be more than just a system responsible for the introduction of uncertainty (one-to-many relations), but also, and paradoxically, a system that may allow the evolution of different proteins constrained by the same genetic string. In other words, even though one gene may produce different mRNA's (and thus proteins), the latter are not allowed heritable variation since they are always constrained by the gene from which they are edited, and which is ultimately selected and transmitted to the offspring of the organism. We can see RNA Editing, especially in the case of gRNA's, as a case of co-adaption of two distinct systems: the stored genetic information (e.g. maxicircles) and the contextual editors (e.g. minicircles), also stored in DNA, but independent and meaningless to the larger semantic loop of the genetic code.

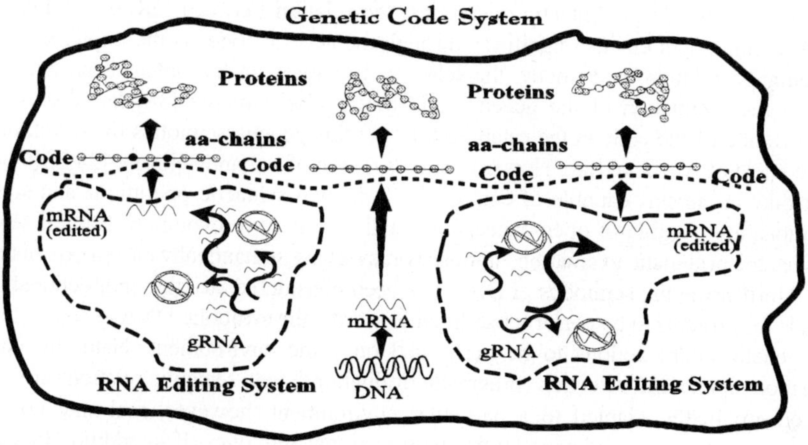

Figure 2: Co-adaptation of the RNA Editing and Genetic Code Systems

The dependent evolution of one gene and several contexts, as expressed by Rob Benne in the previous quote, may allow the introduction of highly specific, local (contextual) changes, more effectively than the independent evolution of several genes. If all of the different expressions were allowed different genes, they would evolve separately not only increasing the size of the genome, but also, possibly, making it harder to maintain coherent, multicellular, phenotypes as well as coherent developmental processes. For instance, the editing of several genes of the *Trypanosoma Brucei* is developmentally regulated [36] which may be of evolutionary advantage for these parasites [35]. Though in the course of evolution editing was partially or completely eliminated in many lineages of eukaryotic organisms containing mithocondria, by reverse transcription of partially edited mRNA's, it may be useful for the development of parasitic adaptations as is the case of the developmental regulation of editing in *T. Brucei* [35].

The role of RNA editing in the development of multicellular organisms has also been shown to be important, Lomeli at al [21] have discovered that the extent of RNA editing affecting a type of receptor channels responsible for the mediation of excitatory postsynaptic currents in the central nervous system, increases in rat brain development. As a consequence, the kinetic aspects of these channels will differ according to the time of their creation in the brain's developmental process.

3. A Theoretical Model: Evolving Semiotics

Semiotics concerns the study of signs/symbols in three basic dimensions: syntactics (rule-based operations between signs within the sign system), semantics (relationship between signs and the world external to the sign system), and pragmatics (evaluation of the sign system regarding the goals of their users) [23]. The importance of this triadic relationship in any sign system has been repeatedly stressed by many in the context of biology and genetics [39; 24, 25; 26]; in particular, Peter Cariani [6] has presented an excellent discussion of the subject. We can understand the semiotics of the genetic system if we consider all processes taking place before translation (from transcription to RNA editing) as the set of syntactic operations; the relation between mRNA (signifier) and folded amino acid chains (signified), through the genetic code, as the implementation of a semantic relation; and finally, the selective pressures on the obtained proteins as the pragmatic evaluation of the genetic sign system. Jon Umerez [38] has discussed the importance of the code in the establishment of this genetic semiotics by developing, in the context of Artificial Life, Howard Pattee's notion of semantic closure [24,25]: the idea that only organisms capable of controlling their own syntactic operations and semantic relations are capable of open-ended functional creativity or evolution. Natural selection defines the pragmatic evaluations imposed on evolving semantically closed organisms.

Until now, the semiotics of DNA has been considered strictly unidirectional: DNA stands for proteins to be constructed. In other words, the symbolic DNA encodes (through the genetic code) actions to be performed on some environment. Naturally, through variation and natural selection (pragmatic evaluations) new semantic relations are created which are better adapted to a particular environment, however, real-time contextual measurements are not allowed by this unidirectional semiotics. If in addition to symbols standing for actions to be performed, the genetic sign system is also allowed a second type of symbols standing for environmental, contextual, measurements, then a richer semiotics can be created which may have selective advantage in rapidly changing environments, or in complicated, context dependent, developmental processes.

Figure 3 depicts such a sign system. The top plane contains two different types of symbols which are combined in different ways (symbolic operations). *Type 1* symbols stand for actions through a *code* ϕ (e.g. the genetic code) and *type 2* symbols stand for measurements through a different *code* γ which is being hypothesized here. The evidence presented in section 2 refers to genetic systems in which RNA Editing is used in different amounts according to different contexts (namely, different stages of a developmental process). We can think of DNA as a set of symbolic descriptions based on two types of symbols: *type 1* symbols will be expressed in mRNA molecules and will stand for actions to be performed; *type 2* symbols will be expressed in gRNA molecules (or other editing mechanisms) and will stand for contextual observables. RNA editing can be seen as a set of symbolic operations performed with symbols of both types, resulting in symbols of *type 1* to be translated into actions by the genetic code.

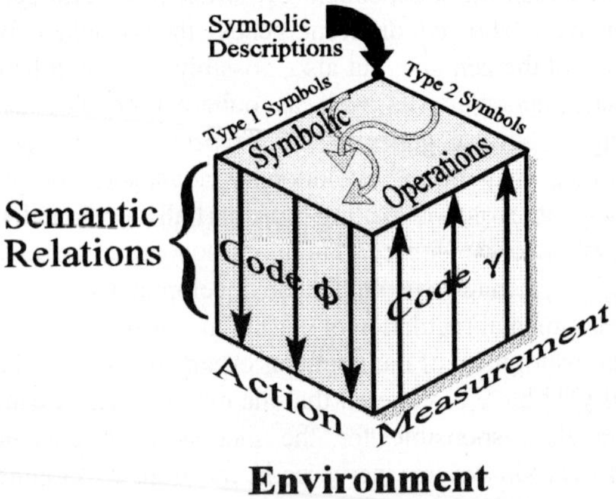

Figure 3: DNA Semiotics with two symbol types

Notice that *code* γ is proposed here as an abstraction referring to the set of mechanisms which will vary the concentration of editing agents *(type 2* symbols) according to environmental context. It is **not** expected to function as a proper genetic code. This issue has been dealt with in [29, 30] in the context of evolutionary systems and second order cybernetics.

4. Artificial Genetic Editing

In GA's, genes are substituted by strings of symbols taken from a binary vocabulary $V = \{0, 1\}$ and called ***V-strings***. The genotype of an individual, referred to as its *symbolic description*, is the set of V-strings necessary to produce a phenotype or *solution alternative* [12]. The translation of symbolic descriptions into the space of solutions is performed by invariant formal rules which define a code for a particular application. In the following, a symbolic description is comprised of only one V-string.

Definition 1. V is a vocabulary with two symbols: $V = \{0,1\}$.

Definition 2. S is a V-string of dimension n : $S = s_1 s_2 s_3 \dots s_n$, $s_i \in V$, $I = 1, 2, \dots, n$. Let $\mathbf{S}^n$ denote the power set of V-strings of dimension n.

Definition 3. $P(g) = \{S_i \mid I = 1, \dots, n_p\}$, is a population of n_p V-strings at generation g.

Definition 4. $\mathbf{X} = X_1 \times X_2 \times \dots \times X_d$ is a space of solutions, of dimension d, for a particular problem. X_i is the universal set of a relevant variable x_i, $I = 1, 2, \dots, d$. ϕ maps V-Strings S into solution alternatives $\mathbf{x}$. $\phi : \mathbf{S}^n \to \mathbf{X} \mid \phi(S) = \mathbf{x} \in \mathbf{X}$. This mapping establishes the translation rules between symbolic descriptions and solution alternatives: the code.

An individual is composed of a symbolic description, $S \in \mathbf{S}^n$, and a solution alternative, $x \in \mathbf{X}$. But the relation between S and x is not a result of direct application of the mapping ϕ. Before S is translated into the space of solutions, it will possibly be altered through interaction with a different sort of string.

Definition 5. U is a vocabulary with three symbols: $U = \{0,1,*\}$.

Definition 6. E is a string of length m over the vocabulary U, or a U-string of dimension m:$E = e_1 e_2 e_3 \dots e_m$, $e_i \in U$, $I = 1, 2, \dots, m$. Let $\mathbf{E}^m$ denote the power set of U-strings of dimension m.

These U-strings will function as the editing agents of the population of V-strings. The length of U-strings is supposed much smaller than that of the V-strings: $m << n$, usually an order of magnitude. Maintaining the analogy with the RNA editing system of the Trypanosomes, V-strings can be referred to as *maxistrings*, and U-strings as *ministrings*. Here I will assume that the editing agents are constant, that is, the structure of the ministrings will be maintained through the successive generations of $\mathbf{P}$.

Definition 7. Let $\mathcal{E}$ denote a finite family (ordered set) of l U-strings: $\mathcal{E} = \{E_1, \dots, E_l\}$.

Definition 8. For each family of U-strings, $\mathcal{E}$, there exists an associated family of mappings $\mathcal{F} = \{f_1, f_2, \dots, f_l\}$. Each mapping f_i associates its respective U-string in $\mathcal{E}$ with a V-string, and produces another V-string: f_i: $\mathbf{E}^m \times \mathbf{S}^n \to \mathbf{S}^n$. The associated pair $(\mathcal{E}, \mathcal{F}) = \{(E_1, f_1), (E_2, f_2), \dots, (E_l, f_l)\}$ is called a family of editors.

In other words, each editing ministring will have a function which is also dependent on the maxistring to be edited. This function will result in an edited maxistring, and thus specifies how a particular ministring edits maxistrings: when the ministrings match a portion of a maxistring, a number of symbols from the V vocabulary is inserted into or deleted from the (V-)maxistring. To introduce the sort of ambiguity the guanine-uracil base pairing allows the gRNA/mRNA duplex, the U includes an extra symbol '$*$' , matching both '1' or '0' in V. Ministrings match more than one subsequence of maxistrings.

Definition 9. A U-string $E \in \mathbf{E}^m$, *matches* a substring, of size m, of a V-string, $S \in \mathbf{S}^n$, at position k if:

$$\exists_{k \mid 1 \le k \le n} : \begin{cases} s_{k+i} = 1 \text{ and } e_i = (1 \vee *) \\ s_{k+i} = 0 \text{ and } e_i = (0 \vee *) \end{cases} \forall_{i = 1, 2, \dots, m}$$

Example of a family of mappings ***f*****:** $E^m \times S^n \rightarrow S^n$. $\mathcal{F} = \{Add\_1(E, S), Del\_1(E, S)\}$. ***Add_1*** will add the symbol '1' at position $k+m+1$ if E matches S at position k; all string symbols in S from position $k+m+1$ to n-1 are shifted one position to the right (the symbol at position n is lost). ***Del_1*** will instead delete the symbol '1', if it is present at position $k+m+1$ when E matches S at position k; the string symbols are shifted in the inverse direction (the symbol at position n is randomly selected from V).

Definition 10. Let the *concentration* of a family of editors ($\mathcal{E}$, $\mathcal{F}$) be defined by $\mathcal{C} = \{v_1, v_2, ..., v_l\}$, where v_i represents the average number of editors (E_i, f) per V-string of a population $\boldsymbol{P}$. If n_p is the number of V-strings in $\boldsymbol{P}$, then there will be $v_i\, n_p$ editors (E_i, f) randomly distributed by the n_p V-strings of $\boldsymbol{P}(g)$.

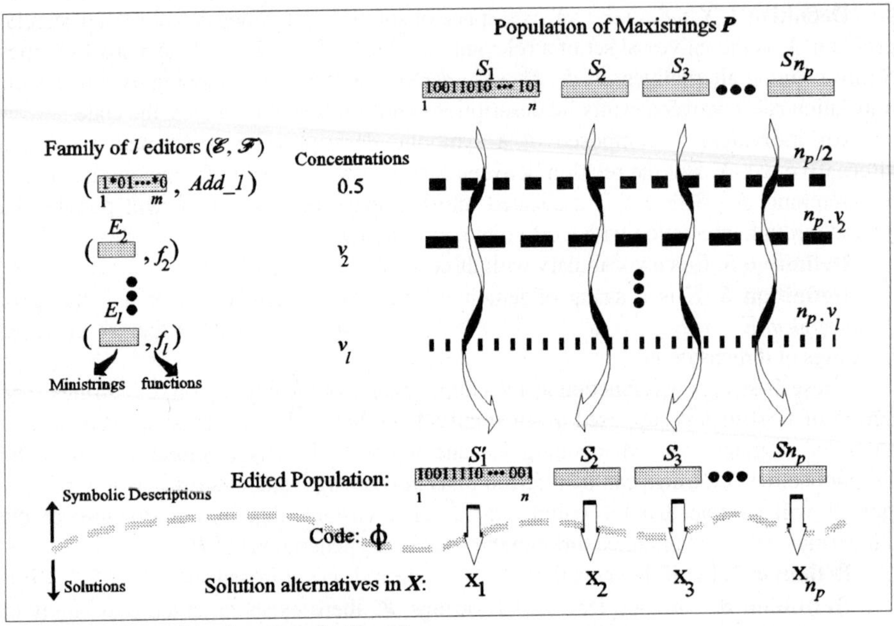

Figure 4: String Editing in a Genetic Algorithm

Figure 4 shows the operational layout of this genetic algorithm with string editing. Generally, we have a population $\boldsymbol{P}$ of n_p maxistrings, and a family of l editors with different concentrations. Before the maxistrings can be translated into the space of solutions $\boldsymbol{X}$, by the mapping ϕ, they must "pass" through successive layers of editors, present in different concentrations. At each generation, the same number of editors (given by the concentrations) is randomly distributed over these layers. Thus, in the example of figure 4, editor 1 (E_1, *Add_1*) with a concentration of 0.5, will have n_p /2 copies of itself randomly distributed by the n_p positions of its layer; there will be on average 0.5 of such editors 'meeting' each maxistring. When an editor meets a maxistring, and its ministring matches some subsequence of the maxistring, the editor's function is applied and the maxistring is altered.

5. Context and Development

5.1 Context

In biological genetic systems, RNA editing regulates gene expression; somehow, organisms have used the edition of mRNA molecules to their advantage, perhaps by linking it to environmental context. If a particular external event has the effect of changing the concentrations of editing agents in some genetic system, then those genes which are able to produce fit phenotypes in the different contexts will be selected. Notice that changing environmental context will not merely affect the concentration of editing agents, but also, potentially, the fitness landscape of the genetic system. Thus, the ability to link changes in the environment with internal parameters such as concentrations of editing agents, gives organisms an adaptive advantage as gene expression can become contextually regulated. The idea is the introduction of the second kind of semantic relation leading to a second type of symbol described in section 3. The editing strings are now more than symbolic constraints, but are also semantically related to context variation through a (postulated) code γ.

Figure 5 shows precisely this kind of coupling between environmental context and the regulating effects of editor concentrations. Notice, at the bottom of the figure, the dependence of the fitness landscape of the solution alternative space X, on environmental context. When the context changes, not only are the symbolic descriptions edited differently, but the solution alternatives are also evaluated differently. The inclusion of this extra level of semantic relations and pragmatic evaluations establishes the kind of genetic semiotics described in section 3.

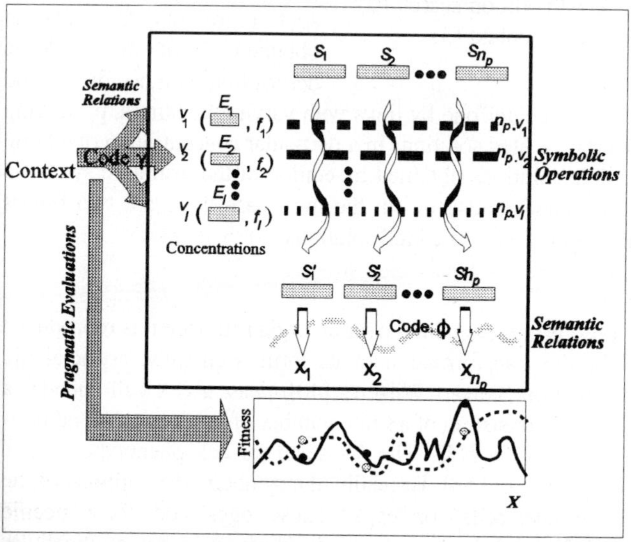

Figure 5: GA with editing parameters linked to context

Consider now two sets of concentrations C_1 and C_2 of our family of editors ($\mathcal{E}$, $\mathcal{F}$) linked respectively to two evaluation functions, *fitness1* and *fitness2*. When the first context is at play, we obtain a population of solution alternatives $\mathcal{X}_1$ which will be evaluated by *fitness1*; alternatively, when the second context is at play, $\mathcal{X}_2$ is evaluated by *fitness2*. Notice that both $\mathcal{X}_1$ and $\mathcal{X}_2$ are produced from the same population of symbolic descriptions $\boldsymbol{P}$. Those symbolic descriptions in $\boldsymbol{P}$ which tend to produce fit solution alternatives in $\mathcal{X}_1$ and $\mathcal{X}_2$ (evaluated by *fitness1* and *fitness2* respectively) will have a higher probability of being selected. This result will of course be dependent on the timing and sequence of application of contexts: if contexts are alternated rapidly, then it will be possible to have symbolic descriptions, with a high probability of selection in the population, which produce fit solutions in only one of the contexts; if contexts are maintained a bit longer before alternating, those symbolic descriptions that tend to produce fit solutions in both contexts will have a higher probability of selection; if the contexts are maintained too long, however, it will be more difficult to evolve symbolic descriptions able to survive in both contexts. These results follow Richard Levins [20, chapter 2] strategies of adaptation.

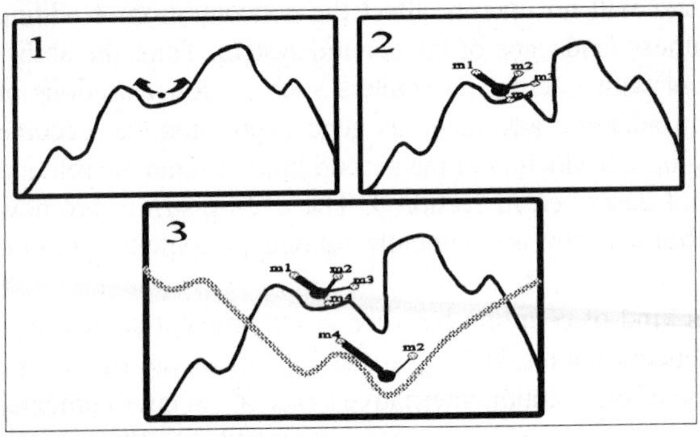

Figure 6: GA search (1); GA with edition search (2); Search in GA with edition linked to context (3)

Figure 6 shows the different searches of traditional GA's, GA's with edition, and GA's with edition linked to environmental context. In the first case, one solution alternative, directly obtained from a symbolic description, is evaluated in a fitness landscape. In the second case, a set of possible solution alternatives, where the dark spot represents the solution obtained from a symbolic description with no edition, and the lighter spots, connected to the center one by links with varying thickness, portraying the relative probabilities of certain edited solutions in a particular concentration of editing agents, represent all the possible solutions obtained by edition of the primitive symbolic description, is evaluated in a fitness landscape. In the third case there are two fitness landscapes evaluating the **different** clouds of edited solutions.

5.2 Development

Development refers to those processes taking over an organism once it is reproduced and which are responsible for the transformation of its form. Generally, artificial life models of development are based on Stewart Wilson's [40] ideas: a GA will encode "a *production system program* (*PSP*) consisting of a finite number of production (condition-action) rules [...] of the form: $X + K_i \Rightarrow K_j K_k$. The *K*'s stand for cell phenotypes and *X* represents the local context". [40, page 159]. Basically, the symbolic descriptions of the GA code for a population of "mother cells", or "eggs". These "eggs" code for a specific PSP (a set of production rules) dictating how the "cell" develops into some multicellular aggregate, which is then evaluated for its fitness. The more fit aggregates will have the symbolic description of its "egg" reproducing with a larger probability in the population. These ideas have been used mostly to generate neural networks [16; 1; 13] or more generally sensorimotor control systems [for a good overview see 15]. Recently, the idea of encoding metabolic cycles in a genetic algorithm [17], represented by boolean networks for instance [11], which will then in certain conditions effect developmental steps has also been proposed. This approach aims at an increasing self-organization of the developmental PSP's.

Developmental cycles have been argued to offer an expanded universe of solution alternatives, that is, rather than precisely encoding a fixed number of parameters, more general rules are encoded which will themselves organize, and search a larger universe of alternatives. Thus development cycles come as a necessary solution for design problems affected by scaling constraints (such as neural networks). By the same token, we can expect developmental cycles in artificial life models to come up with more complicated morphologies arising through the interaction of several developmental rules (PSP's) rather than direct encoding. Basically, the evolutionary advantage of these PSP's is the definition of a smaller search space which is then amplified through development into more complicated morphologies. We can also think that this reduced search space is more amenable for evolution since lower dimensionality spaces will have more valleys; if more morphology details have to be encoded then dimensionality is increased and the search becomes more difficult (see [9] for a discussion of these topis). Related to this is Conrad's tradeoff principle between structural programmability and evolvability [7, 10].

The several approaches vary in many ways, for instance, on the degree of context allowed in production rules of the various PSP's (e.g. how rules are applied depending on a cell's neighbors). Nevertheless, in all of them, the symbolic description-solution space relation is always certain. The production rules are primitives of the representational system and encoded in a one-to-one manner in the symbolic descriptions of the genetic algorithm. The more self-organizing approaches of Kitano [17] and Dellaert and Beer [11], seem to offer a way out for this one-to-one correspondence, but the wiring of the boolean networks (or metabolic cycles) is still encoded in the genetic algorithm in a one-to-one manner. The metabolic networks will then reach some state corresponding to a particular developmental rule; however, this correspondence, established by a **second** set of semantic relations, a simulation code (see section 6), is also completely certain. These systems are very powerful, and offer very interesting and sound approaches to modeling developmental cycles in artificial organisms, however, they do not aim at the understanding of how and why developmental stages arise in the first place through internal regulation of genetic expression.

If the editing system above is able to evolve developmental stages triggered by the internal control of the expression of symbolic descriptions, then we are moving towards utilizing the principle of natural selection not only at the level of the individual, but also at levels internal to the individual, namely through the evolution of semantic referents, for contextual information, in the genetic system.

> "More specific to GA's is the central question of representation. [...] The choice of system primitives (in the case of GA's, the features that comprise the genotype) is a decision that cannot be automated." [22, page 281]

The direct engineering of a relationship between descriptions and solutions allows only what Peter Cariani [5] has referred to as syntactic emergence, that is, the inability of a formal system to change its primitives and create new observables, and therefore respond with open-ended evolution. The kind of automation that Mitchell and Forrest refer above, would amount to the evolution of the semantic relationship between symbolic descriptions and solution alternatives itself, the representation issues above, and would therefore shed some light on the problem of the origin of symbols.

This is not what is pursued here, the direct semantic relationship of the GA will be given by the mapping ϕ which is predefined from the beginning. The choice of primitives for this mapping, the code, is permanent. What can be utilized as a source of contextual input, is the editing system of the GA's presented above. Remember once again that this system is independent of code ϕ, and is therefore only taking place at the syntactic level (symbolic descriptions) of the GA. However, the symbolic descriptions can be made to evolve with the editing constraints, tied up to environmental context, which become referents for this context. In other words, the aim is to evolve the contextual semantic relations for type 2 symbols described in section 3. Thus, it is possible to evolve context referents for the rules of a PSP, rather than predefine them from the start, provided different sets of concentrations of editors are linked to different fitness functions. Also, since the solutions of the same symbolic descriptions in the various contexts are not allowed independent evolution, as only the "mother descriptions" are reproduced, the evolved rules will be more related than if evolved independently (with distinct descriptions), and have therefore the potential to evolve more coherent PSP's with shorter symbolic descriptions.

6. Physical Simulations and Fuzzy Developmental Rules

In artificial life, it is important to distinguish between the code of the GA (the mapping between symbolic descriptions and solution alternatives, ϕ, or genetic code) and the code of a simulation. The latter refers to all the physical characteristics the modeler attributes to the solution alternatives of his or hers simulation. It is important to realize that this code is external to the GA and does not affect its search. Often, these distinctions are blurred in artificial life and evolutionary computation precisely because traditional GA's, due to their one-to-one mapping between symbolic descriptions and solution alternatives, do not distinguish between the two, or metaphorically, do not distinguish genotype from phenotype.

Naturally, in a computational realm all material aspects must be simulated and therefore a semantic relation is imposed which refers the simulation's symbols to the physical characteristics we desire to model. It is important to keep this in mind especially in the simulation of developmental cycles since these are defined on two stages: first the GA searches for a particular developmental program, and then this program is executed. The first stage depends on the GA's code (ϕ), independently from the physical attributes of the simulation, while the second stage executes the program according to some simulated physics defined by the simulation code, from now on referred to as code β.

If we are to utilize contextual GA's to tackle the problem of development, the primitives of the solution alternatives obtained should naturally code for all the characteristics needed to form the rules of a PSP, namely, phenotypic characteristics such as "cell thickness" as well as orders such as "divide in two", etc. However, there will be no coding of rules themselves, in particular, the context in which a rule should be applied, will not be a semantic primitive, but allowed to evolve from the coupling of the editing system of the GA to the external contexts. This is by no means achieved, or easy to achieve, it indicates a proposed research direction necessary to attain true evolution of development cycles in artificial life models.

6.1 Fuzzy Sets as Uncertain Physical States

Fuzzy sets may be ideal mathematical structures to characterize some simulated physical dynamics. For instance, the stable states of metabolic networks used for the definition of developmental cycles referred above [17, 11] can be represented by a fuzzy set in which the nodes of the network and their activation states are the elements of the set and their membership degrees respectively. More generally, the elements of a fuzzy set can refer to some desired physical attributes (through the simulation code β) while their membership degrees can describe the degree to which such physical attributes are present in a certain situation. In the context of developmental cycles, certain actions will be taken when certain elements have membership degree beyond a specified value.

To allow for a better representation of uncertainty, that is, if we desire the physical characteristics to observe in addition to fuzziness the two other recognized forms of uncertainty — nonspecificity and conflict [18] — then a more complicated set structure can be used. This structure is referred to as an *Evidence Set* [28, 31,32] and is based on the extension of fuzzy sets by utilizing *Evidence Theory* [34]. Basically, this structure formalizes the membership degree of an element in a set, with a finite number of weighted subintervals of [0,1]. A degree of membership in [0, 1] captures uncertainty in the form of *fuzziness*, an interval of membership introduces *nonspecificity*, and finally several competing intervals introduce *conflict*. The measurement of uncertainty in set structures is discussed in [31].

Evidence sets can be obtained through the operation of simpler fuzzy sets. Several operations for evidence sets have been defined in [28, 32]. Consider now a string of fuzzy sets, defined on some universal set K, and operations amongst them together with parenthesis which group the operations in the string in different ways: $S = F_1 \otimes ((F_2 \odot F_3) \cdots F_{n-2}) \ominus F_{n-1} \oplus F_n$. Consider further that these fuzzy sets, F, are picked from a finite, small, family of possible fuzzy set shapes, and the operations are likewise picked from a small family of operations. Finally, a number of parenthesis is somehow randomly distributed over the string. Once a string is generated, it must be parsed in order to obtain an evidence set: parenthesis will have to be matched and operations performed. If a right (left) parenthesis is not matched all the fuzzy sets and operations to its left (right) are discarded. Thus, from an original string with n fuzzy sets, after parsing, we will obtain strings with 0 to n fuzzy sets.

Returning to our GA's, consider now that the edited strings obtained will code (through ϕ) to such a string of fuzzy sets and operations. In other words, the solution alternatives of the GA will be fuzzy set strings which will be parsed and operated into evidence sets whose elements (of K) refer to some simulated physics through code β. Since fuzzy sets capture only one form of uncertainty (fuzziness) and evidence sets capture three (fuzziness, nonspecificity, and conflict), we can metaphorically say that the fuzzy set strings "fold" from a one dimensional into a three dimensional uncertainty state. Figure 7 presents a scheme of this process.

To make things more general, the fuzzy sets, F_i, define only shapes of membership as seen in figure 7. These shapes are then positioned and stretched over some pre-defined portions of the universal set K. This is a very important point since it eliminates any scaling problem of whatever physical attributes we wish to simulate. To explain this better, I must be a bit more formal. Consider that the universal set K of our fuzzy sets is divided into octants (eight portions of K). A fuzzy set shape can now be associated with a particular octant as well as with some width stretching over a number of octants. If we have eight possible fuzzy set shapes we only need 3 bits of information to express the shape, plus 3 bits to position it in an octant, and finally 2 extra bits can specify 4 possible widths for stretching the shape over K. This way, a fuzzy set can be specified by only 8 bits: 1 byte. Likewise for the fuzzy operations and parenthesis. 8 different operations are possible (3 bits). If we specify that an operation will carry with it a left or a right parenthesis one fourth of the time, we need 2 bits for each parenthesis (4 bits). With an unused bit, an operation with parenthesis can be specified by one byte. A string with 8

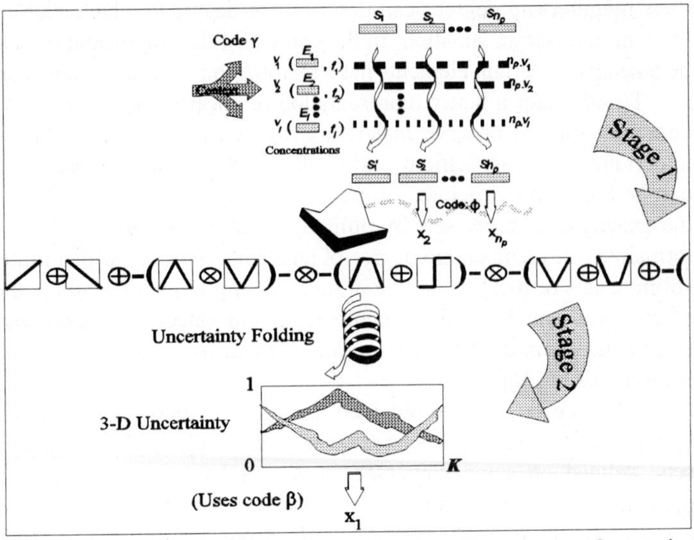

Figure 7: Two stage contextual GA coding for 3-D uncertainty fuzzy rules

fuzzy sets and 8 operations can be described by 16 bytes (128 bit long string), for **any** finite cardinality universal set *K*.

Hence, the definition of the solution alternatives of the contextual GA in terms of fuzzy set strings is independent of the size of a particular physical simulation, that is, of the number of physical characteristics of our artificial organisms. Whatever the number of these characteristics, whatever the cardinality of *K*, the search space of the GA will be the same, namely, the one defined by the 128 bit long strings coding (through φ) into fuzzy set strings. Nevertheless, and naturally, the size of *K* is relevant for other aspects of the simulation external to the GA. A larger *K*, will mean that the definition of an organism (by an evidence set) will require a finer tuning of the composing fuzzy set string, which may take longer for the GA to reach. In any case, the search space will remain constant, only more elaborate searches will be required.

So far the fuzzy set strings have been shown to increase the uncertainty description of a simulation, as well as to allow for a good scaling management. But they possess yet another important evolutionary advantage: a buffering mechanism for genetic mutation. Michael Conrad [1990] has developed the notion of genetic buffering as an important requirement for evolvability. Though mutation is required for evolution, it is also important that certain shapes may be resilient to changes which may potentially destroy an important physical functionality. As discussed above, the fuzzy set strings will be parsed according to its parenthesis. Consider the following parsing situation:

$$F_1 \oplus F_2 \otimes (F_3 \odot \ldots F_7 \oplus F_8 \mapsto F_1 \oplus F_2$$

all the fuzzy sets and operations to the right of the unmatched left parenthesis are discarded. This means that any bit to the left of the second fuzzy set is free to mutate without any effect on the final organism, except those few bits which may cause a matching parenthesis to occur to the left of F_2.

As a final note, crossover was not considered in this model precisely to not disrupt this kind of genetic buffering. Also, since eventually this kind of buffering is transcended, usually with a dramatic change of form (a string with two functional fuzzy sets can suddenly become a string with, say, eight functional fuzzy sets), crossover seems to be unnecessary as a source of more variety.

7. Final Remarks

The most important characteristic of all the mechanisms here introduced is related to a conflict between introducing more variety and constraining, or buffering, this variety. Contextual editing allows for a variety, a cloud, of solution alternatives to effect the genetic search, however, this variety of alternatives is not allowed independent evolution and is constrained to an original symbolic description (which can be edited in different ways) ultimately reproduced. On another level, the uncertain fuzzy set strings, though introducing a large amount of variety in their parsing and uncertainty folding, also observe the kind of genetic buffering described earlier. It is believed that the right amounts of variety and constraint lie at the core of evolvability [8,9]. Only the implementation and testing of the proposed model will tell if it has the right amount of both. In addition, the inclusion of context in the genetic algorithm, or the limited evolution of a semantic relationship between editing mechanisms and contexts (the type 2 symbols in a semiotic relation), opens the way for the study of the emergence of developmental cycles triggered by contextual constraints.

References

1. Belew, R.K. [1993]."Interposing an Ontogenic Model Between Genetic Algorithms and Neural Networks." In: *Adv. in neural information processing*. J. Cowan (Ed.). Morgan Kaufmann.
2. Benne, R. Et al ."Major transcript of the frameshifted coxII gene from trypanosome mitochondria contains four nucleotides that are not encoded in the DNA." *Cell* N. 46, 819-826
3. Benne, Rob (Ed.) [1993]. *RNA Editing: The Alteration of Protein Coding Sequences of RNA*. Ellis Horwood.
4. Blum, B, N. Bakalara, and L. Simpson [1990]."A model for RNA editing in kinetoplast mitochondria: "guide" RNA molecules transcribed from manicircle DNA provide the edited information." In: *Cell* No. 60, pp. 189-198.
5. Cariani, Peter [1991]."Some Epistemological Implications of Devices which Construct their own Sensors and Effectors." In: *Towards a Practice of Autonomous Systems, Proc. First European Workshop on Artificial Life*. F. Varela and P. Bourgine (Eds.). MIT Press. pp 484-493.
6. Cariani, Peter [1995]."Towards an evolutionary semiotics: the role of symbols in organisms and adaptive devices." In: *Proceedings of the International Seminar on Evolutionary Systems, Vienna 1995*. Stanley Salthe and Gertrudis Van de Vijver (eds). In Press.
7. Conrad, Michael [1974]."The limits of biological simulation." *J. The. Biology* Vol. 45, 585-590.
8. Conrad, Michael [1983]. *Adaptability*. Plenum Press.
9. Conrad, Michael [1990]."The geometry of evolutions." In: *BioSystems* Vol. 24, pp. 61-81.
10. Conrad, Michael [1993]."Adaptability theory as a guide for interfacing computers and human society." In: *Systems Research* Vol. 10, No. 4, pp. 3-23.
11. Dellaert, F. and R.D. Beer [1994]."Toward an evolvable model of development for autonomous agent synthesis." In: *Artificial Life IV: Proceedings of the Fourth International Workshop on the Synthesis and Simulation of Living Systems*. R. Brooks and P. Maes (Eds.). MIT Press.
12. Goldberg, D. E. [1989]. *Genetic Algorithms in Search, Optimization, and Machine Learning*. Addison-Wesley.
13. Gruau, Frédéric [1992]."Genetic Sythesis of boolean Neural Networks with a cell rewriting developmental process." In: *Proceedings of the International Workshop on Combinations of Genetic Algorithms and Neural Networks*. Whitley, L.D. and J.D. Schaffer. IEEE. pp 55-74.
14. Holland, John H. [1975]. *Adaptation in Natural and Artificial Systems*. U. of Michigan Press.
15. Husbands, P., I. Harvey, D. Cliff, and G. Miller [1994]."The use of genetic algorithms for the development of sensorimotor control systems." In: *(retrieved from the Internet)* .

16. Kitano, H. [1990]."Designing Networks using Genetic Algorithms with Graph Generation System." In: *Complex Systems* Vol. 4, pp 461-476.
17. Kitano, Hiroaki [1994]."Evolution of Metabolism for Morphogenesis." In: *Artificial Life IV: proc. 4th international workshop on the synthesis and simulation of living systems*. R. Brooks and P. Maes (Eds.). MIT Press.
18. Klir, George J. [1993]."Developments in uncertainty-based information." In: *Advances in Computers*. M. Yovits (Ed.). Vol. 36, pp 255-332.
19. Langton, Christopher G. [1989]."Artificial Life." In: *Artificial Life: SFI Studies in the Sciences of Complexity*. C.G. Langton (Ed.). Addison-Wesley. pp. 221-249.
20. Levins, Richard [1968]. *Evolution in Changing Environments: Some Theoretical Explorations.* Princeton Univ. Press.
21. Lomeli, H. et al [1994]."Control of kinetic properties of AMPA receptor channels by nuclear RNA Editing." In: *Science* Vol. 266. pp. 1709-1713.
22. Mitchell, M. and S. Forrest [1994]."Genetic algorithms and Artificial Life." In: *Artificial Life* Vol. 1, pp 267-289.
23. Morris, Charles W. [1946]. *Signs, Language, and Behavior.* G. Braziller, New York.
24. Pattee, Howard H. [1982]."Cell psychology: an evolutionary approach to the symbol-matter problem." In: *Cognition and Brain Theory* Vol. 5, no. 4, pp 325-341.
25. Pattee, Howard H. [1995]."Evolving self-seference: matter, symbols, and semantic closure." *Comm. and Cog. - AI* Vol. 12, nos 1-2 pp 9-27.
26. Pollack, R. [1994]. *Signs of Life: The Language and Meanings of DNA.* Houghton Mifflin.
27. Rennie, J. [1993]."DNA's New Twists." In: *Scientific American* March 1993.
28. Rocha, Luis M. [1994]."Cognitive Categorization revisited: extending interval valued fuzzy sets as simulation tools for concept combination." In: *Proc. of the 1994 Int. Conference of NAFIPS/IFIS/NASA*. . IEEE. pp 400-404.
29. Rocha, Luis M. [1995]."Contextual Genetic Algorithms and an Evolutionary Semiotics." In: *Proceedings of the Int. Seminar on Evolutionary Systems, Vienna 1995*. Stanley Salthe and Gertrudis Van de Vijver (eds.). In Press.
30. Rocha, Luis M. [1995]."Eigen-states and symbols." In: *Heinz von Foerster Festschrift (Special Issue)*. Ranulph Glanville (Ed.). *Systems Research* Vol. 12, No 3. (In Press).
31. Rocha, Luis M. [1995]."Computing Uncertainty in Interval Based Sets." In: *Advances in Interval Computation*. Vladik Kreinovich (ed.). Kluwer. (In Press).
32. Rocha, Luis M. [1995]."Evidence (interval based) Sets: Modelling Subjective Categories." In: *International Journal of General Systems* In Press.
33. Seiwert, S. D. and K. Stuart [1994]."RNA Editing: transfer of genetic information from gRNA to precursor mRNA in Vitro." In: *Science* Vol. 266 pp. 114-116.
34. Shafer, Glenn [1976]. *A Mathematical Theory of Evidence.* Princeton Unversity Press.
35. Simpson, L. and D. Maslov [1994]."RNA Editing and the evolution of parasites.". *Science* Vol. 264, pp. 1870-1871.
36. Stuart, K. [1993]."RNA Editing in mitochondria of african trypanosomes." In: *RNA Editing: the Alteration of Protein Coding Sequences of RNA*. R. Benne (Ed.). Ellis Horwood. 26-52.
37. Sturn, N.R., and L. Simpson [1990]."Kinetoplast DNA minicircles encode guide RNA's for editing of cytochrome oxidase subunit III mRNA." In: *Cell* No. 61, pp. 879-884.
38. Umerez, Jon [1995]."Semantic Closure: A guiding notion to ground Artificial Life." In: *Proc. of the ECAL 1995*
39. Waddington, C.H. [1972]. *Biology and the History of the Future.* Endinburgh University Press.
40. Wilson, S.W. [1988]."The genetic algorithm and simulated evolution." In: *Artificial Life: SFI Studies in the Sciences of Complexity*. C.G. Langton (Ed.). Addison-Wesley. pp. 157-166.

Can Development Be Designed? What We May Learn from the Cog Project

Julie C. Rutkowska

School of Cognitive & Computing Sciences, University of Sussex
Brighton, BN1 9QH, UK
Email: julier@cogs.susx.ac.uk

Abstract. Neither design nor evolutionary approaches to building behavior-based robots feature a role for development in the genesis of behavioral organization. However, the new Cog Project aims to build a humanoid robot that will display behavioral abilities observed in human infants; and proposes making use of ideas from evolution and developmental psychology in its design. This paper provisionally evaluates this work from a developmental perspective, to show how developmental study may offer not only a source of phenomena for modelling but also a method that contributes to our understanding of self-organization. Cog's design methodology confronts problems with selection and interpretation of component behaviors, and how these may be better understood through appropriate developmental study is illustrated. Cog's design principles are shown to exhibit interesting convergences with infant mechanisms, based on the significance of emergent functionality and the action- as opposed to representation-based nature of initial and outcome mechanisms, but analysis of infants suggests a more constructive view of ability is required.

1 Routes to Understanding Autonomous Agents

Artificial Intelligence's new behavior-based robotics is unified by commitment to understanding intelligent systems in terms of specifics of their physical embodiment, their sensorimotor coupling with the environment, and the organizational possibilities of the situatedness to which these properties give rise. There is less agreement as to whether a historical process must be a necessary component of the construction of a system that is to become capable of survival in our normal environment. Engineering methods are at the heart of a 'design' approach to building robots, attempting to pre-specify component behaviors that are required and the mechanisms through which they can be implemented. Brooks's insect-like Creatures (1986, 1990, 1991a & b), based on a subsumption architecture with layered control, provide elegant and successful examples of this strategy. Exponents of evolutionary robotics see this kind of hand-design as simply too hard to be feasible at any but a toy scale, however ingenious the experimenter. Instead, processes inspired by evolution are exploited to automate design. For example, notions like mutation, recombination and selection have been employed to evolve sensorimotor controllers in recurrent dynamical artificial neural networks by repeatedly evaluating and 'breeding' sets of (initially randomly generated) networks, thus arriving at a maximally adaptive genotype-like structure (Cliff, Harvey & Husbands, 1992, 1994; Harvey, Husbands & Cliff, 1992, 1994).

The overall role that a historical dimension plays in these approaches to autonomous organization is more complex than it may initially seem, at least as far as evolution is concerned. Superficially, they might appear to correspond to the distinction between "implementation by design" and "implementation by evolutionary strategies" that Varela (1988) identifies as one criterion of a theoretical shift from traditional symbol-manipulating cognitivism, focussing on the subject's representation-based activities in a 'pre-given' world, to an enactive framework, in which the subject's activities serve to construct a domain of interactions through which a world is 'brought forth'.

On closer inspection, the gap is smaller than it seems. For example, both approaches purport to reject traditional representationalist assumptions (in favour of pragmatic 'wiring' considerations in the former case, and dynamical systems analysis in the latter). Furthermore, Brooks's Creatures are deliberately engineered incrementally, while current evolutionary robotics pre-specifies behavioral outcomes insofar as selection is achieved through evaluation procedures that depend on objective, task-oriented fitness functions. Both approaches differ from autopoietic notions, which focus on how systems generate and maintain their own organization (i.e. are 'self-producing') and characterize evolution in terms of 'natural drift' rather than progressive adaptation to the environment (Maturana & Varela, 1988; Varela, 1988, 1993).

Neither approach features an explicit role for the individual historical dimension, development, in the genesis of behavioral organization. However, the ambitious new Cog Project aims to build a humanoid robot that will display behavioral abilities observed in the first couple of years of human life (Brooks & Stein, 1993; Brooks, 1994). Cog is not intended to be a model of human development. Nevertheless, it aims at biological relevance (Brooks, 1994); and proposes making use of ideas from both evolution and developmental psychology (Brooks & Stein, 1993). This paper aims to evaluate these proposals from a developmental perspective. Its emphasis is on going beyond the use of developmental study as a source of phenomena for modelling, to consider how it can provide a method that contributes to our understanding of self-organization.

2 From Creatures to Cog

The Cog Project is at an early stage, so any assessment must also be of a provisional nature. It aims to extend the methodology of generating complex seeming abilities through hard-wired networks of simple sensorimotor coordinations, each capable of engaging in independent interaction with the environment. As far as possible, such 'task-achieving behaviors' are to be added incrementally, with each of these layers being tested and debugged before attempting to build in another, continuing an analogy to evolution (Brooks & Stein, 1994, p.7). The initial schema for Cog includes far more layers than have so far been built into any Creature, offering an increased challenge for undertanding behavioral coherence (Brooks, 1994). These layers begin with apparently basic abilities such as visual following and sound localization, and cumulate in complex ones such as generic object recognition and protolanguage. For example, layers most closely related to the 'development' of prehension are: body stability, leaning, resting; bring hands to midline; own hand tracking; hand linking; batting static objects; grasping & transfer; and body & arm reaching.

Cog "will get a continuous large and rich stream of input data of which it must make sense, relating it to past experiences and future possibilities in the world" (Brooks & Stein, 1993, p.6), but it appears to follow its Creature predecessors in lacking a clearly defined role for endogenous organizing mechanisms. Attaining novel abilities by building in additional layers is the consistent focus, in preference to more dynamic notions of designing the system so that novel abilities might 'come for free' as far as the design process is concerned, emerging from what happens when implemented layers are allowed to 'run' in the environment. For example, when it is assumed that functional behavior layers for correlating hearing and vision should serve as a usable basis for discrimination between 'interesting' events and background noise, this is 'use' more by the human designer than autonomously by Cog as part of an internal process of self-organization.

2.1 Cog, Evolution and Development

Basing layered control on an analogy with evolutionary development (Brooks & Stein, 1993) could be seen as implying a view of the evolution-development relationship based on the notion of terminal addition. The implications of this view are illustrated by a model that views Piagetian stages of intellectual development, from sensorimotor abilities to abstract thought, in terms of evolutionary selection pressures operating on genotypes that make possible specific behavioral adaptations (Parker, 1985; Parker & Gibson, 1979). Cross-species observations of relative attainments on developmental stages of ability are used as evidence that ontogeny recapitulates phylogeny, based on a series of terminal additions of new structures or stages through evolution. The fundamental argument is that "more intelligent species achieve their greater intelligence not by altering early developmental processes, but by adding later stages of intelligence to the end of the developmental cycle" (Gibson, 1981, p.52).

This view contrasts strongly with the epigenetic position favoured by Piaget (e.g. 1971), the origins of whose developmental stages it purports to explain, insofar as he considers stages as evidence for levels of knowledge that are neither additive nor genetically predetermined but the product of developmental processes that operate from the very outset. While Piagetian stages and developmental processes are contentious, there are good grounds for sharing his dissatisfaction with proposals for genetically pre-specified additive/sequential behavioral outcomes that arise from this way of using orthodox neo-Darwinian theory to frame an account of development. Attributing a privileged role to genes in the determination of development, commonly enshrined in the 'genetic program' notion, has appropriately been criticized as denying the very development that it seeks to explain (Oyama, 1985). Genes can be thought of as inputting certain parameter values into a developmental process involving a system of multiple variables and relations, but they do not define the organizing principles of that process. Those depend on the dynamics of the developing system as a whole (Goodwin, 1993). Observations of human acquisition of everyday, apparently universal sensorimotor abilities suggest processes capable of flexible outcomes that strain the notion of genetic predetermination. For example, infants who are raised with sparse adult interaction may not walk, even by around 3 years of age. Instead, they acquire 'scooting' (sitting while using arms to pull the body along), a behavior whose form is surely not pre-programmed (Dennis, 1960).

Brooks and Stein (1993) acknowledge that the analogy between layered control and evolution is 'simplistic and crude', and seem unlikely to wish to characterize development in the (over)simple additive, genetically bound terms sketched above. Their current models, however, appear compatible with that direction. The evolutionary approach to constructing robots is sensitive to the fact that its current models locate the form of individual performance too exclusively 'in the genes', and both approaches agree on serious reservations about viewing real evolution as an orthodox process of optimisation with contemporary animals seen as solutions to problems posed in their species' distant evolutionary past (Brooks, 1994; Cliff, Harvey & Husbands, 1994). However, just how new sensorimotor coordinations emerge in development, if not through an essentially additive sequence of gene-behavior mappings, remains an open question. Some developmental ideas about how transactions between phylogenetically determined initial mechanisms and the environment may guide ontogenetic change are sketched in the following section of this paper.

2.2 Using Developmental Observations

Constructing Cog by design entails pre-selection of behaviors into which its abilities will be decomposed. Its planned layers thus embody an implicit developmental theory, to the extent that they highlight a restricted range of the behaviors that have been studied by developmental psychologists, and provisionally order them so that earlier layers are expected to aid the implementation and operation of later ones.

Brooks's work emphasises the difficulty of achieving an effective behavioral decomposition of abilities, and how we may frequently be misled into thinking that our observers' discriminations map straightforwardly onto demarcations in our subjects' mechanisms. Certainly, the selection and interpretation of behaviors for such a design plan raise closely related problems. On the one hand, behaviors that are necessary to the developmental sequence may be missed. In the case of prehension, for example, arm raising behavior is commonly found after batting objects is observed but before top-level reaching appears; while the infant intently fixates the object, the arm and hand are raised in its direction and held at the horizontal, often with signs of considerable effort. This behavior is not included in the Cog schema, although it may play a more significant role in the emergence of reaching and grasping than, say, hand linking. Equally, of course, it may not. Deciding which is the case depends on interpretation of the behavior concerned; that requires a hypothesis, or at least a hunch, as to what mechanisms are involved, which clearly affects plausible implementation strategies.

The difficulty of such interpretation as far as individual behaviors are concerned can be illustrated easily, without considering potential Cog layers that have controversial psychological connotations (e.g. that dedicated to 'multiple drafts emergence' associated with Dennett's ideas about consciousness). What is the significance, for example, of the readily observable behavior of bringing hands to the midline? There is no developmental consensus as to why infants exhibit midline activity (Rutkowska,

1994b). Possible interpretations range from initially out-of-sequence fine motor movements that will eventually be used for manipulating grasped objects; to stress reduction in the case of hands brought to the mouth; and a side-effect of the mechanics of failed early reaching attempts. Likewise, is batting at objects a form of ballistic reaching, superseded by reaching with visual feedback, or is it an attempt to palpate an extended surface that is perceived as too large for grasping?

A prime source of such difficulties is methodological. Much mainstream developmental psychology does not itself employ a particularly developmental method. It tends to concentrate on relatively isolated behaviors; is preoccupied with 'when' in the subject's chronology those behaviors appear; and with judgments of 'success/failure' that treat behavior as a mere criterion of some other, supposedly underlying ability. And it generally works backwards from possibly erroneous assumptions about outcome behaviors to processes of acquisition (Rutkowska, 1993). Faced with the question 'How do you get from **A** to **B**?', it concentrates too exclusively on the nature of **B**. An essential change of focus is needed from initially asking 'Can this system do **B**?' to the more fundamental question 'What is this system doing?' This change of orientation is facilitated by employing a more genuinely developmental method.

The strategy involves three things that enrich, and often alter, our understanding of the nature of **B**:

- Taking behavior seriously by looking at the patterns through which it achieves succeed/fail outcomes in terms of the observer's criteria.
- Working forwards by taking seriously the idea that you can't get from **A** to **B** unless you start from a good idea of what **A** is.
- Observing changing behaviors in a domain of activity, using the relative position of a behavior within a sequence to constrain its interpretation.

3 Behavioral Interpretation Through Development

The developmental strategy can be illustrated by looking at infants' changing performance on a simple visual tracking task that presents them with a moving object, part of whose path is hidden by an occluder (Rutkowska, 1993, 1994a & c). Their looking behavior is generally assumed to index knowledge of the object and of its motion ('success' = look to exit as/before the object reappears; 'failure' = look elsewhere). Even very young infants will sometimes succeed in 'anticipating' the object's emergence from behind the occluder in an operational sense, by looking at the exit side as or before the object comes back into view. Should we therefore conclude, depending on theoretical preference, that infants come equipped with visual procedures for solving the problem of object search or 'believe' that objects continue to exist while out of sight? Considering the details of this behavior in the context of others displayed by 3-, 6- and 9-month-old infants makes such interpretations extremely implausible. Three aspects of the data are notable:

- The behavior pattern of fixations and head and eye movements that sometimes leads 3-month-olds to be looking at the object's reappearance point before it comes into view is quite different from the pattern through which 9-month-olds attain the same outcome. While 3-month-olds simply continue tracking as the object disappears from view, sometimes tracking as far as the reappearance point, 9-month-olds characteristically pause as the object disappears from view, then make a single head and eye movement to the reappearance side of the occluder, which they fixate until the object returns to view.

- Although 3-month-olds' continued tracking has the appearance of functional search for the disappeared object, its frequency declines rather than increasing with age. Nor is it simply replaced by a corresponding increase in the 'entry-exit' fixation pattern found in 9-month-olds, despite infants getting faster and faster at turning to refixate the reappearing object, from wherever they do happen to be looking, as it comes into peripheral vision. 6-month-old subjects exhibit less of either form of 'successful' anticipation than 3- or 9-month-olds, demonstrating the kind of U-curve that characterizes many instances of development.
- What does increase are behavior patterns involving attention to the object's disappearance point. The one most characteristic of 6-month-olds can be described as backtracking: as the object disappears, the infant continues tracking, but then turns head and eyes sharply back to fixate the object's disappearance point. This is a strange observation as far as attempts to interpret backtracking in isolation are concerned, since those generally assume the infant must have noticed some change in the reappearing object and be looking back to the disappearance point where the original object was last seen. Here, however, a single object moving at constant speed is involved, and is generally still out of view when the infant turns back.

These and other aspects of the data suggest the observer-labelled tracking task is not initially a problem with the goal of 'find the object' as far as the infant's viewpoint is concerned. 3-month-olds' behavior is not wired up to search for objects that move and disappear, but they are initially equipped with a preadapted behavioral procedure for tracking visible object movement. Their continued tracking is no more than a failure to alter ongoing behavior when environmental circumstances change. They fail to do so because the recurrent visual pattern (kinetic occlusion) that marks the moving object's disappearance has yet to become salient to them. It may be available preattentively, at the sensory process level, but has yet to be usable at the level of action through coordination with behavior. If this is the case, we would expect 3-month-olds to do one of two things when the object moves out of sight: nothing, i.e. look away or 'lose interest'; or what they are already doing, i.e. continue tracking. These prove to be the behavior patterns that are most characteristic of that age. 6-month-olds' backtracking can be seen as indicating the beginnings of attention to kinetic occlusion, which develops further with behavior patterns such as intently fixating the object's disappearance point during the entire period that it is out of sight. They are not seeking a changed/missing object. Only the 9-month-olds' coordination of attention to kinetic occlusion with turning to look to the opposite side of the occluder marks the beginnings of search from both the observer's and the infant's viewpoints.

Infants, then, are not initially trying to 'do the task' as an observer sees it. Their changing performance is more akin to task construction than task solution. Through repeated sensorimotor interactions with the environment, they come to construct the problem of search for missing objects through their experience of finding them. Even at this everyday level, development may be seen as an enactive process, in Varela's (1988) terms, insofar as its processes appear to be directed at problem-definition rather than problem-solving.

4 Action-Based Task Construction

The broad issue of how novel abilities are constructed can be crudely divided into three questions, each of which allows additional evaluation of the architectural principles and design methodology of the Cog Project from a developmental perspective: What are initial mechanisms like? What is the process through which they change like? And what do they change to?

4.1 Emergent Functionality in Scaffolding Development

A key point of rapprochement between design and developmental methodologies involves the notion of emergent functionality, through which complex abilities may result from the independent interaction of more basic components with the environment. Emergent functionality is central to the Cog Project's attempt to maintain behavioral organization through layered control, and it may be developmentally advantageous in two ways, at least as far as the early stages of acquiring novel abilities are concerned (cf. Rutkowska, 1994a & c).

Firstly, emergent functionality could support an initial organization of independent sensorimotor coordinations, such as the visual following featured in the preceding section, that is neither a *tabula rasa* nor a blanket prewired solution to problems that will be encountered. This would offer preadaptation without rigid predetermination. Interactions between preadapted abilities of such a system and the environment in which it finds itself could enable it to 'tune in' sensorimotor coordinations, and sequences of such coordinations, that prove viable in the individual's experience. Novel coordinations (e.g. locomotion by scooting) would not be precluded in case of altered environmental conditions and/or properties of the subject (e.g. physical-motor disability).

Secondly, within the developmental process, the phenomenon of scaffolding can be viewed as a form of supervised learning in which emergence of function is temporarily engineered to establish the developmental space within which viable patterns of activity can be stabilized. Scaffolding, as originally viewed in social terms, marks the process through which more able humans manipulate the infant's transactions with the environment so as to foster novel abilities (e.g. Valsiner, 1987; Wood, Bruner & Ross, 1976).

The process begins with sensory and motor processes that are not coordinated by the infant but are set in alignment with the environment by adults. For example, if an infant's head is moved to look at someone leaving a room and simultaneously his/her hand is moved up and down, whatever the infant is doing, initially s/he is not waving goodbye. Key features are: customizing or simplifying the environment; reducing the number of degrees in the target task; directing attention by marking critical attributes; and enabling repeated experience of the end, outcome or goal of an activity that the infant would be unable to seek voluntarily. This sets up the possibility of serendipitous learning by the infant, that is of an accidental (i.e. unplanned) yet fortunate discovery of possibilities for effective action, in which the balance of behavioral control shifts from the environment to the subject.

The ubiquitous nature of such phenomena has been seen as evidence for all aspects of human development being socially and culturally guided, but adults may be exploiting and directing inbuilt processes that also operate in infant's spontaneous interactions with the environment. For example, in the previous section's account of the development of visual tracking, initial serial ordering of behaviors emerged from ongoing interaction with the environment; it was not governed by a goal or plan directed at finding the disappeared object. Spatio-temporal properties of the infant's interactions with the environment supported recurrent sequences of sensory and motor processes, most notably attention to kinetic occlusion followed by turning to refixate (and hence to experience 'finding') the reappearing object. In principle, such processes may share the main properties of social scaffolding, provided attention can be limited through processing restrictions such as spatiotemporal constraints (for a relevant simulation of an attention mechanism in the context of sensorimotor learning see Foner & Maes, 1994).

The notion of scaffolding begins to provide a way out of problems faced by traditional AI's view of learning. This tends to see it in terms of adaptive change that enables a system to do a task better next time round; and which is unnaturally difficult unless the subject knows the goal in advance (Mitchell, 1983). Such assumptions make it difficult to see where novel abilities and goals might come from. It is notable that a robot system such as Darwin III, which is purported to exhibit self-organization in the absence of supervision and with unbiased internal connectivity in place of inbuilt sensorimotor structure, relies heavily on designer-coded 'value shemes' that evaluate the outcomes of its behavior (Reeke, Finkel, Sporns & Edelman, 1990). These intrinsic value schemes (e.g. getting visual stimuli onto the fovea = good; making contact with bumpy objects = bad; hand in region of foveated object = good) share non-trivial properties with traditional internal goals (and with externally specified fitness functions), hence encounter similar problems. As a means of ensuring recurrent experience of novel viable activities, coupling emergent functionality with scaffolding may offer a better characterization of constraints on infant's changing behavior.

4.2 Is Development Additive?

The kinds of behavioral developments that characterize the first year of life appear to involve more than straightforward addition of novel sensorimotor coordinations. In many domains of activity, there appears to be a move away from a reactive mode, which is essential to the basic operation of layered control in Brooks's robots, to increasingly anticipatory functioning and what might be called 'nascent plans'. In the case of visual search, for example, the infant develops from attention to kinetic occlusion plus turning to fixate the reappearing object to attending to kinetic occlusion then turning *in order to* re-fixate the reappearing object.

An explanation of the novel mechanisms that underlie such changes need not invoke qualitative change in the form of concepts or mentally represented goals controlling behavior. It will need to account for an extended time-scale of coupling between the subject's activities and the environment; and the changing functionality of sensorimotor components that is illustrated by eye and head movements initially associated with fixation coming to be used also for re-fixation. Traditional computational explanations could permit new and old 'programs' to invoke common lower-level movement primitives. It is, however, an interesting empirical question as to whether, and how, an intricately hardwired subsumption architecture could generate such phenomena.

4.3 What Do 'Internal World Models' Model?

Meyer & Guillot (1994) refer to anticipatory triples of the form *'if, in sensory circumstance 1, I do behavior B, I shall get sensory circumstance 2'* as 'internal world models'. While agreeing that such mechanisms underlie a form of nascent planning, it is worth emphasizing that, to the extent that they 'model' anything, it is constraints on effective action rather than an external 'world' in which action takes place. Developmental psychology and cognitive science have become relatively fixated on the notion that model-like internal representations of the environment underlie intelligent functioning, a notion that is not endorsed here. Nor are representations featured in Brooks's (1991b) foundational assumptions about intelligence, although it has been argued that his robots may in fact use internal representations and require them for further progress to be made (e.g. Clark & Toribio, 1994).

I doubt this conclusion as far as both infants and robots (whether humanoid or otherwise) are concerned. This is not because there is no interpretation of 'representation' that can be mapped onto their functioning, but because the notion of an internal representation attempts to demarcate one component of a complex subject-environment system, and to give it a privileged status in the genesis of organization. In doing so, it limits attempts to deepen understanding of how that organization is achieved.

Insofar as it works to establish selective correspondence(s) between subject and environment, action maps onto some perfectly good treatments of representation and the establishment of meaning, which are equally applicable to human infants, to Brooks's Creatures and, by extrapolation, to Cog (Rutkowska, 1994a & c). What these action-based, process-oriented approaches share is a scale of analysis that spans sensory and motor processes and their functional coordination in the environment, unlike traditional preoccupations with representation as a substitution for the environment, which locates it firmly 'in the head'. A typical direction is the situation semantics notion of 'attunement to constraints' (Israel, 1988), which allows infant sensory-motor coordinations, such as reaching towards or avoiding things, and the task-achieving behaviors of Brooks's robots (e.g. 1991b) to be thought of as underlying human and robot subjects' representation and understanding of constraints on acting in the world, and their ability to satisfy them. Also significant are Dretske's (1988) notion of a 'natural system of representation', which can be applied to cases of infant action, as when very simple directionally selective elements acquire the function of indicating that something is approaching from the way they are 'wired' with avoidance behaviors; and to robots, as when a sonar pattern associated with free space acquires the function of indicating a place to visit when wired into a task-achieving behavior that successfully embeds the robot in its environment. Also applicable is Varela's (1993) view of meaning emerging from processes that establish domains of interaction between a 'self' and its environment, for which the CNS's sensory-, motor- and inter-neurons are only one specialist adaptation for achieving closure, a reflexive interlinking of subject and environment processes that supports construction of neurocognitive identity.

As far as such systems are concerned:

- Explicit internal models of an objective external environment are unnecessary to adaptive behavior, provided embedded sensory and motor processes are taken as the scale of analysis. The usefulness (efficiency) of sensory processes lies not in how exhaustively they enable the subject to model an object or an event but in how successfully they limit and allow for possibilities for action.
- No 'bit' of action mechanisms is 'the' internal representation. The capacity for successfully locking onto the environment and anticipating the consequences of activity within it is distributed across the operation of perceptual and behavioral processes. For example, infants' knowledge of invisible object movement is embedded in the way the disappearance event becomes involved in determining future head and eye movements towards the reappearance point.
- The notion of 'representation', when viewed in terms of action-based mechanisms, does no explanatory work in the sense of being a (more or less localizable) functional component of action. Representation, whether by selective correspondence or by substitution, is one vantage point onto processes that are grounded in an action system spanning subject and environment. It makes equally little sense to consider this a central/internal phenomenon or an external one.

When it comes to considering the environmental contribution to the interaction of subject and environment, it seems that Brooks's methodology does not fully follow through the implications of this systematic, action-based approach. Whereas traditional 'representation' was appropriately dismissed, its opposite number, 'information', is invoked in vestigial but non-trivial allusions to animals having sensors that "extract just the right information about the here and now around them" (Brooks, 1991a). Despite a subjective focus on embodiment and action, a notion of an objective environment containing information for action is brought in, albeit information that is selected with the subject's action requirements in mind. This can raise inappropriate assumptions about the subject's 'access' to the world.

In the context of challenging model-like representations, work in behavior-based robotics has often alluded approvingly to Gibson's (1979) theory of direct perception and the notion of 'affordances' (invariant combinations of properties that specify what things are 'for' for a given subject). It is often unclear just how far this flirtation with ecological psychology's methodological realism is intended to go. Buying wholeheartedly into the view that there is an organized world of objects and events, independent of the subject's activity, and that this is the world that is perceived and known, risks simply substituting a discredited notion of internal representations about the world for equally dubious external information about the world as 'the' foundation for organization. This can only be counterproductive as far as establishing a genuinely systematic account of adaptive behavior is concerned (cf. Bersini,

1992). It can make for easier questions for robot design and for development (though they may not be appropriate questions). To understand a subject's world, we do not need to ask how organization is constructed, only about what objects and events they are sensitive to, and about what information specifies those things. Likewise, development is not a constructive process but one of coming to perceive better through the 'education of attention' to things that are already 'out there'.

If we reject the information pickup/recovery metaphors for perception, and assume the environment may not come ready populated with the categories of objects and events about which language-using scientists and philosophers routinely speak, there may still be no need to be pessimistic about how we might talk about the environment. If the aim is to understand emergent organization within an environment, as opposed to what is 'really out there' beyond the subject and the subject's activities, behavior-based robotics suggests some interesting possibilities.

One route comes from evolutionary robotics' focus on the adaptation of simple robots in simple environments (e.g. Cliff, Husbands & Harvey, 1992). The power of this method lies in its potential for 'reverse engineering' to clarify what sensorimotor control structures the robot evolves for itself, in place of checking its success/failure in acquiring the categories of its creators. A further direction involves turning the issue of what sensors are doing away from the heavily entrenched notion that they provide environmental information (whether for direct action control or further processing). Dynamical systems theory may support a promising alternative if it can develop the notion that sensors are not measurement devices. It suggests that sensor signals need not 'encode information' specifying particular states of a robot in its environment, it is enough that they "vary in some way that depends upon the dynamics of the robot-environment interaction" (Smithers, 1994, p.70), delivering what might be considered sensorimotor invariants. This kind of re-think may provide an original route into genuinely enactive, mutual notions of organization.

5 Conclusion

Not only the successes but also the failures of the Cog Project will offer significant opportunities to clarify our thinking about development. As far as principles underlying the design of Cog are concerned, it starts out exhibiting some interesting convergences with the human infants whose behavioral abilities it aims to model, revolving around the significance of emergent functionality and the action- as opposed to representation-based nature of both initial and outcome mechanisms. As far as the design methodology is concerned, both selection and interpretation of Cog's component behaviors pose difficulties, which might be eased by developmental data (of which there is less than one would like) that takes seriously behavior and the process of development. Along the lines of evolutionary processes enabling the construction of structures that may be too hard to achieve by hand design, exploiting developmental change may enable appropriate interpretation of behaviors and acquisition possibilities that are too hard to achieve purely through rational analysis. An ideal aim for Cog would be to reduce the design element in favour of carving out a greater role for an internally driven contribution to its own 'development' through interaction with the social and physical environment. This would offer a unique opportunity to clarify how both the outcome and the process of development are grounded in effective action.

Acknowledgements

Thanks to the ECAL95 referees for their useful comments on this paper.

References

1. Bersini, H. (1992). Animat's I. In F.J. Varela & P. Bourgine (Eds.), *Towards a Practice of Autonomous Systems: Proceedings of the First European Conference on Artificial Life.* Cambridge, MA: MIT Press/Bradford Books.
2. Brooks, R. (1986) A robust layered control system for a mobile robot. *IEEE Journal of Robotics and Automation,* **RA 2,** 14-23.
3. Brooks, R. (1990) Elephants don't play chess. In P. Maes (ed.) *Designing Autonomous Agents.* Bradford/M.I.T. Press.
4. Brooks, R.A. (1991a) Intelligence without reasoning. *Proceedings of the Twelfth International Joint Conference on Artificial Intelligence.*
5. Brooks, R. (1991b) Intelligence without representation. *Artificial Intelligence,* **47.** 139-160.
6. Brooks, R.A. (1994) Coherent behavior from many adaptive processes. In D. Cliff, P. Husbands, J.-A. Meyer & S.W. Wilson (Eds.) *Animals to Animats 3: Proceedings of the Third International Conference on Simulation of Adaptive Behavior.* Cambridge, MA: MIT Press/Bradford Books.
7. Brooks, R.A. and Stein, A. (1993) Building brains for bodies. MIT AI Laboratory Memo No. 1439.
8. Clark, A. and Toribo, J. (1994) Doing without representing? University of Sussex, Cognitive Science Research Paper, Serial No. CSRP 310.
9. Cliff, D., Harvey, I. and Husbands, P. (1992) Incremental evolution of neural network architectures for adaptive behavior. University of Sussex, Cognitive Science Research Paper No.256.
10. Cliff, D., Harvey, I. and Husbands, P. (1994) General visual robot controller networks via artificial vision. University of Sussex, Cognitive Science Research Paper No.318.
11. Dennis, W. (1960) Causes of retardation among institutional children: Iran. *Journal of Genetic Psychology,* **56** 77-86.
12. Dretske, F.I. (1988) *Explaining Behavior.* Cambridge, MA: MIT Press/Bradford Books.
13. Foner L. and Maes, P. (1994) Paying attention to what's important: Using focus of attention to improve unsupervised learning. In D. Cliff, P. Husbands, J.-A. Meyer & S.W. Wilson (Eds.) *Animals to Animats 3: Proceedings of the Third International Conference on Simulation of Adaptive Behavior.* Cambridge, MA: MIT Press/Bradford Books.
14. Gibson, J.J. (1979) *The Ecological Approach to Visual Perception.* Boston MA: Houghton-Mifflin.
15. Gibson, K.R. (1981) Comparative neuro-ontogeny. In G. Butterworth (ed.) *Infancy and Epistemology.* Brighton: Harvester.
16. Goodwin, B. (1993) Development as a robust natural process. In W. Stein & F.J. Varela (eds.) *Thinking About Biology.* Reading, MA: Addison-Wesley.
17. Harvey, I., Husbands, P. and Cliff, D. (1994) Issues in evolutionary robotics. University of Sussex, Cognitive Science Research Paper No. 219.

18. Harvey, I., Husbands, P. and Cliff, D. (1994) Seeing the light: Artificial evolution, real vision. In D. Cliff, P. Husbands, J.-A. Meyer & S.W. Wilson (Eds.) *Animals to Animats 3: Proceedings of the Third International Conference on Simulation of Adaptive Behavior.* Cambridge, MA: MIT Press/Bradford Books.
19. Israel, D. (1988) Bogdan on Information. *Mind and Language,* **3** 123-140.
20. Maturana, H. and Varela, F.J. (1988) The Tree of Knowledge. Boston & London: Shambhala.
21. Meyer, J.-A. & Guillot (1994) From SAB90 to SAB94: Four years of animat research. In D. Cliff, P. Husbands, J.-A. Meyer & S.W. Wilson (Eds.) *Animals to Animats 3: Proceedings of the Third International Conference on Simulation of Adaptive Behavior.* Cambridge, MA: MIT Press/Bradford Books.
22. Mitchell, T.M. (1983) Learning and Problem Solving. *Proceedings of the Eighth International Joint Conference on Artificial Intelligence.*
23. Oyama, S. (1985) *The Ontogeny of Information.* Cambridge: Cambridge University Press.
24. Parker, S.T. (1985) Higher intelligence as an adaptation for social and technological strategies in early Homo Sapiens. In G. Butterworth, J.C. Rutkowska & M. Scaife (eds.) *Evolution and Developmental Psychology.* New York: St. Martin's Press
25. Parker, S.T. and Gibson, K.R. (1979) A developmental model for the evolution of language and intelligence in early hominids. *Behavioral and Brain Sciences,* **2** 367-407.
26. Piaget, J. (1971) *Biology and Knowledge.* Edinburgh: Edinburgh University Press.
27. Reeke, G.N., Finkel, L.H., Sporns, O. and Edelman, G.M. (1990) Synthetic neural modeling: A multilevel approach to the analysis of brain complexity. In G.M. Edelman, W.E. Gall & W.M. Cowan (eds.) *Signal and Sense: Local and Global Order in Perceptual Maps.* New York: Wiley-Liss.
28. Rutkowska, J.C. (1994a) Emergent functionality in human infants. In D. Cliff, P. Husbands, J.-A. Meyer & S.W. Wilson (Eds.) *Animals to Animats 3: Proceedings of the Third International Conference on Simulation of Adaptive Behavior.* Cambridge, MA: MIT Press/Bradford Books.
29. Rutkowska, J.C. (1994b) Prehension intention from 12 to 22 weeks. Presented at the IXth International Conference on Infant Studies. Paris, 2-5 June.
30. Rutkowska, J.C. (1994c). Scaling up sensorimotor systems: Constraints from human infancy. *Adaptive Behavior,* **2** 349-373.
31. Smithers, T. (1994) On why better robots make it harder. In D. Cliff, P. Husbands, J.-A. Meyer and S.W. Wilson (eds.) *From Animals to Animats 3.* Cambridge, MA: MIT Press/Bradford Books.
32. Valsiner, J. (1987) *Culture and the Development of Children's Action.* Chichester: Wiley.
33. Varela, F.J. (1988). *Cognitive Science: A Cartography of Current Ideas.* Author's unpublished translation of F.J. Varela (1989). *Connaitre – Les Sciences Cognitives: Tendances et Perspectives.* Paris: Editions du Seuil.
34. Varela, F.J. (1993). Organism: A meshwork of selfless selves. Second European Conference on Artificial Life. Brussels, 24-26 May.
35. Wood, D., Bruner, J.S. and Ross, G. (1976) The role of tutoring in problem-solving. *Journal of Child Psychology and Psychiatry,* **17** 89-100.

Emergent Organization of Interspecies Communication in Q-Learning Artificial Organisms

N.Ono, T.Ohira, and A.T.Rahmani

Department of Information Science and Intelligent Systems
Faculty of Engineering, University of Tokushima
2-1 Minami-Josanjima, Tokushima 770, JAPAN
E-mail: ono@is.tokushima-u.ac.jp

Abstract. Recently in the fields of artificial life and robotics, several researchers have attempted to let the populations of artificial organisms or real robots synthesize some intra- or inter-species cooperative relationships. Here taking the *symbiont-finding* problem as such a problem domain, we show how multiple populations of Q-learning artificial organisms synthesize symbiotic behavior needed to achieve their goals effectively. Optimality of thus synthesized behavior and its organization processes are also analyzed.

1 Introduction

Recently, by employing various learning algorithms, several researchers in the fields of artificial life and robotics have attempted to let the populations of artificial organisms or real robots synthesize intra- or inter-species cooperative relationships among them (e.g., [8][9][3][4][6]). In this paper, we consider the *symbiont-finding* problem, a learning task inspired by natural symbiosis, which needs a mutual understanding by two distinct artificial organisms to achieve their goals. The main feature of the problem is existence of two parallel learning subsystems which have to agree on common communication protocols to succeed in accomplishing their tasks. By representing each of these organisms as a Q-learning agent[7], we attempt to let the organisms spontaneously synthesize common communication protocols. Based on a Q-learning algorithm and a simple information-exchange operator, called *unification*, these organisms gradually organize optimal communication protocols needed to achieve their goals effectively.

2 Symbiont-Finding Problem

The primary task domain we have treated is *symbiont-finding* in populations of artificial organisms. The environment is defined as a 100 by 100 grid where edges are continued toroidally. Each cell is either empty or occupied by an organism. Two species of organisms, called *flower* and *bee* organisms respectively, are operating in the same environment. We randomly place 200 flower organisms and 200 bee organisms into the environment, so the population density for each species is 2%. A typical environment is shown in Fig.1. A flower can sense the position and direction of bees nearby. The direction of a bee has significant meanings for flowers as well as its position, while that of a flower is not for bees because they can not sense such information.

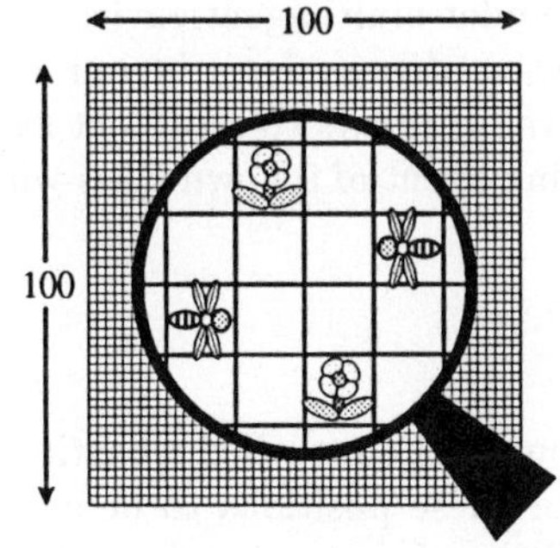

Fig. 1. A typical environment showing the distribution of organisms.

Flowers and bees possess a common goal, i.e., to repeatedly make a contact with each other and to eventually establish a kind of *symbiotic* relationship between them just as some *entomophilous* flowers and their *symbiont* bees do.

Flowers can not move but they can look at their surroundings and when detecting a bee in a nearby position within its *visual field* then producing a sound which is represented as a signal pattern. They have a repertoire of eight such signals. When they do not detect any bee, they just keep silent. On the other hand, bees are totally blind but they can hear the sound which has been emitted by the nearest flower in their *auditory field*. Upon receiving a signal they interpret it as one of four possible actions they can take (MOVE_FORWARD, TURN_RIGHT, TURN_LEFT, STAND_STILL). If no signal is heard then they just take an action as dictated by their current decision policies.

At each time step, at first all flower organisms scan their nearby positions, and accordingly generate their signal patterns. Then bee organisms detecting these signals make their moves. If a contact happens between a bee and a flower, i.e., the bee gets to the flower's position, then they both receive reinforcement signals from environment, and at the same time these two symbiont organisms are moved to new random locations in the environment.

Our primary concern is whether these two species of organisms are able to spontaneously organize optimal *interspecies* communication protocols. Without organizing such protocols, these organisms are not able to accomplish their goals effectively because their function is substantially limited as explained above. To accomplish this, each bee, through interaction with flowers, has to learn how to appropriately respond to signals from them, while each flower has to learn how to optimally guide blind bees.

3 Implementation

In our preliminary experiments, we decided to represent each of our organisms as a Q-learning agent, mainly aimed at the evaluation of learning capabilities of the Q-learning algorithm under multi-agent environments.

As there are two species of organisms in our simulated environment, we consider two sets of Q-learning agents to represent each class. Each organism is associated with a Q-learning agent of its own class which models the organism's behavior.

3.1 Representation

There are 24 positions around a flower which constitute its visual field. We have to consider an ordering for these positions to be scanned by the flowers. If we discriminate these 24 positions from each other, then for the 24 positions and 4 directions, the Q-learning agent has to learn about 96 states to deal with all possible situations. Instead if we employ the partitioning of these positions ($R1, R2, \ldots, R8$) as shown in Fig.2, we can reduce the number of states which must be treated to 32.

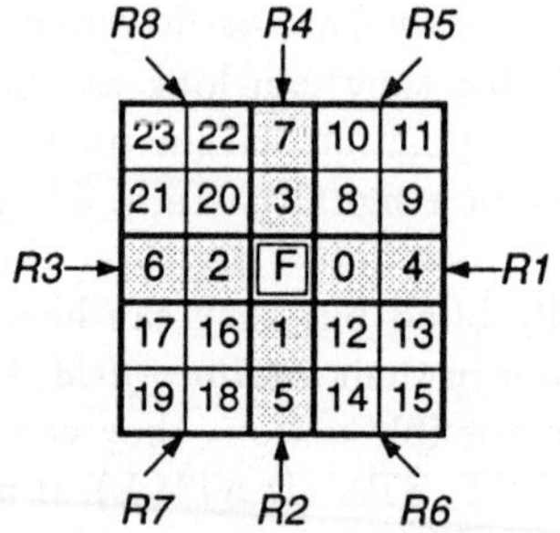

Fig. 2. An appropriate partitioning of a flower's visual field. These 24 positions surrounding a flower represented by "F" are classified into 8 regions, $R1$–$R8$.

A flower can detect only the nearest bee even when there are multiple ones in their nearby positions. It concerns bees only and ignores the existence of other flowers in their nearby positions. A bee can detect only the largest signal, i.e., that emitted by the nearest flower. Thus, the possible states for a bee organism are 8 signal patterns emitted by the nearest flower and a special input indicating that no signal is detected, while those for a flower organism are 32 visual patterns (there are 8 regions and 4 directions for detected bee) and a special pattern indicating that no bee is detected.

3.2 Q-Learning Algorithm

We use a standard one-step Q-learning algorithm[7]. At each time step, an organism receives sensory input x (i.e. its state) from the environment and accordingly determines its action a. Then it takes that action, receiving immediate reward r if available, and it moves to the new state y. With each state-action pair (x, a) of an organism, Q-value $Q(x, a)$ is associated and it is updated by:

$$Q(x,a) \leftarrow (1-\alpha)Q(x,a) + \alpha(r + \gamma \max_{b \in A(y)} Q(y,b))$$

where α is learning rate, γ is discounting factor of rewards, and A gives possible actions for each state.

We select the action by a probabilistic selection algorithm based on Boltzmann distribution[5]. Given a current input x, each possible action $a_i \in A(x)$ is selected with a probability:

$$p(a_i|x) = \frac{e^{Q(x,a_i)/T}}{\sum_k e^{Q(x,a_k)/T}}$$

where parameter T is temperature which controls exploration rate at the learning.

When making a contact, the flower and bee involved receive 10.0 reward immediately. Our organisms receive no reward in any other cases. A relatively low temperature value of 0.1 was selected. We fixed this value but regularly changed the value of α from 0.9 to 0.1 and that of γ from 0.7 to 0.99 respectively[1]. Each of Q-values was set to zero initially.

3.3 Unification Mechanism

Apart from Q-learning algorithm, we use a *unification* operator[4][3] to encourage coordinated behavior of these organisms.

Unfortunately, if we let all of our reinforcement-learning organisms independently explore their decision policies, they can not agree on a single communication protocol. In order for our organisms successfully communicate with each other, they should agree on a single communication language, but there exist a variety of such languages that are equally useful and it is difficult for independently reinforcement-learning organisms to do so. So long as such a communication language has not been established, it is impossible for our organisms to identify optimal decision policies.

[1] Although in Q-learning algorithms, the value of discounting factor γ is usually supposed to be fixed over iterations, it is regularly changed in our experiments. With fixed one, however, we could also get a comparable result, e.g., by setting and fixing γ to 0.98 and by regularly changing T from 1.0 to 0.1 and α from 0.9 to 0.1 respectively.

To allow our organisms to agree on a single communication language, we decided to let them share their experiences locally. At some fixed intervals, e.g., every 100 steps, each flower is allowed to scan its surroundings and if it can find another flower in a nearby position, then their decision policy tables are averaged pairwise. The same procedure is also performed for each pair of bees mutually locating in nearby positions. We call this a *unification* operator, because it will allow our organisms to unify and coordinate their behavior with each other.

4 Results

To obtain a performance baseline for comparing our results with, we performed a number of experiments with both agents for bees and flowers initialized with optimal policy tables. So to see how our learning system works, we compare our results with the results we get from the above simulations.

Fig.3 shows the performance when we use a moving average of the number of steps to symbionts' contact as the performance measure.

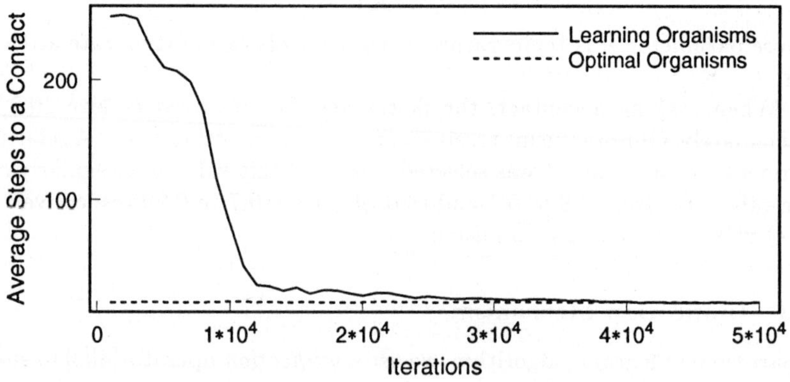

Fig. 3. Bee organisms learn how to reduce average steps to a contact.

Another performance measure we considered was the percentage of number of correct steps taken by bees. We count the number of correct steps to contact when a bee is within the visual field of a flower and report the result every 2000 iterations. Fig.4 shows the result.

Again this is compared with the same measure obtained from experiments with optimal organisms mentioned above. The reason that even the optimal system does not reach a 100% correct performance is that there are many misleading interactions among these organisms. For example, it is possible that more than one bee be within the visual field of a flower, but the flower can only detect the

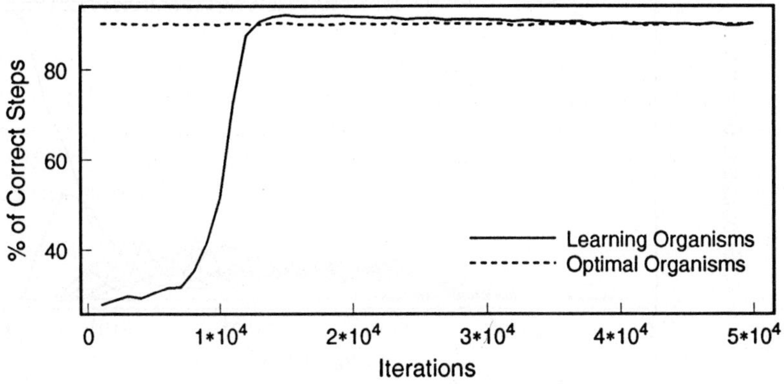

Fig. 4. Performance of Q-learning Organisms.

nearest one and send signal appropriate for that one's position and orientation. But the same signal is received by the other bee which may not be a good one in its case. This shows how noisy our environment may be.

To analyze what happens during the course of mutual learning process by our organisms, we have investigated typical transitions of dominative flowers' and bees' policies. Gradually, our flower and bee organisms start employing some of deterministic policies and eventually only a small number of such policies come to be dominative. To clarify this, at every 500 learning steps, we classified each population of our organisms into its subpopulations, each consisting of organisms employing the same deterministic policy, ignoring those that have not acquired any deterministic policies. Fig.5 shows a typical development of flowers' and bees' subpopulations, each consisting of organisms employing the same policy. In this case, a single common policy has dominated both of flowers' and bees' populations.

5 An Extension: Symbiont-Discrimination Problem

Encouraged by the successful results shown above, we extended the symbiont-finding problem as follows. Let us call flowers and bees in symbiont-finding problem *white* organisms. Suppose that another group of organisms called *gray* organisms are operating in the same environment. Gray organisms are further classified into two different species of organisms, called *gray-flowers* and *gray-bees* respectively, and each subgroup consists of 200 individuals. Flowers and bees belonging to the same group wish to repeatedly make a contact with each other and to eventually establish a symbiotic relationship between them just as in the symbiont-finding problem.

All of white- and gray-flowers are functionally equivalent to each other as well

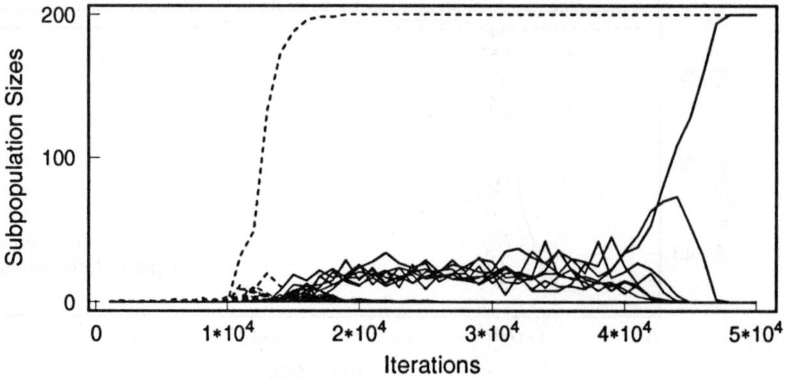

Fig. 5. Typical transition of dominative bees' and flowers' policies (depicted with dashed and solid lines respectively). Sizes of the subpopulations of bees and flowers, each employing the same policy, are shown as functions of time.

as all of white- and gray-bees. White- and gray-organisms share signal patterns they can emit and receive. They have a repertoire of sixteen such patterns in common. Just as in the symbiont-finding problem, at each time step, all of flowers generate their signal patterns first and then all of bees take their moves. A bee since totally blind can not discriminate which group of a flower is currently emitting a signal to it, while a flower can discriminate which group of a bee is currently located in its visual field. When detecting both groups of bees at a time, a flower attempt to emit a signal toward the nearest one belonging to the same group, though it may be also received by the other bees and thereby mislead them. A flower also select and emit some of signals, even when the same group of bees are not detected.

Here the primary question is whether each of white- and gray-groups is able to spontaneously synthesize an optimal communication protocol internally. Without synthesizing such protocols, these two groups are not able to accomplish their goals effectively just as flowers and bees in the symbiont-finding problem. The communication protocol organized for one group should not only be optimal for the group but also minimally interfere with that organized for the other group. This is not an easy learning task because white- and gray-flowers have in common their repertoire of signal patterns. Through interaction with other organisms, each bee has to learn how to appropriately respond to signals from flowers of the same group, while ignoring signals from irrelevant flowers. On the other hand, each flower has to learn how to optimally guide bees of the same group without misleading irrelevant bees.

We performed a series of experiments regarding this problem by employing the same implementation scheme as that explained above and observed that almost optimal communication protocols are steadily synthesized by each group and thus synthesized protocols minimally interfere with each other, by appropriately setting learning parameters. Eventually all of white- and gray-bees come to repeatedly get to their corresponding flowers' positions quite effectively, guided by relevant signals and ignoring irrelevant ones.

6 Related Works

The above-mentioned symbiont-finding problem is a modified version of *mate-finding* problem by Werner and Dyer[8]. In mate-finding problem, what we call flowers and bees are supposed to be *female* and *male* organisms of the same species respectively and hence they have their genome structures in common. When a mating happens between a male and a female, i.e., the male gets to the female's position, then two offsprings are produced, a male and a female, and these two offsprings replace old organisms randomly selected from the population. At the same time the parents as well as their offsprings are moved to new random locations in the environment. In symbiont-finding problem, on the other hand, flowers and bees are totally different species of organisms and do not share genome structures with each other. Thus, if a flower and a bee make a contact, they do not reproduce. Instead, they both receive reinforcement signals from environment and then they are placed to new locations.

Females and males in mate-finding problem attempt to genetically combine and transmit their own successful policies to their descendants over generations through their reproduction and genetic operators, while flowers and bees in symbiont-finding problem attempt to combine and transmit their acquired successful policies to other organisms of the same species respectively through periodical information-exchanging operations. Especially in symbiont-finding problem, more successful policies are expected to more effectively infect other organisms and accordingly alter their behavior; in this sense those policies propagated among flowers and bees might correspond to *memes* suggested by Dawkins[1].

According to the results reported in [8], not all of females and males implemented as recurrent neural networks agree on a single optimal communication protocol unlike our flowers and bees, and consequently their overall performance is not optimal, though Werner and Dyer did not explicitly attempt to let them behave uniformly. The co-evolutionary process by females and males in [8] also seems different from the mutual-learning process by our flowers and bees. At an early stage of the co-evolution, those males that usually go straight take over the population; then those males appear that turn appropriately when in the same row or column as a nearby female. On the other hand, our mutually-learning flowers and bees have never exhibited such phenomena in any runs of our simulation under our parameter settings; there have been no such stages that most of bees in simply keep moving forward irrespective of the signals they receive.

Using an interval estimation reinforcement-learning algorithm[2], Yanco and Stein [9] attempted to let two real robots develop a simple communication language needed to accomplish their coordinated task effectively, but the learning task they addressed is relatively simple. While our reinforcement-learning artificial organisms attempt to synthesize not only a communication language but also an optimal communication protocol under *delayed* reinforcement, their robots simply attempt to synthesize a communication language and moreover they can receive some beneficial reinforcement at every iteration.

Our results presented here are strongly dependent upon our preliminary ones[4][3] where each of flowers and bees was represented as an independent learning classifier system. Independently of these results, Tan[6] attempted to let a small number of Q-learning agents effectively learn by averaging their policy-tables periodically. This policy-sharing operator is equivalent to our unification one, but Tan applied it mainly to accelerate the learning by individual agents. In our case, on the other hand, we applied this operator mainly to combine and eventually unify organisms' policies, since to let all members of each population behave uniformly is crucial point in our problem domain.

7 Concluding Remarks

Recently, by employing various learning algorithms, several researchers in the fields of artificial life and robotics have attempted to let the populations of artificial organisms synthesize cooperative relationships among them. In this paper, aimed at the evaluation of learning capabilities of Q-learning algorithm under such multi-agent problem domains, we have attempted to let multiple distinct populations of Q-learning artificial organisms spontaneously synthesize optimal interspecies communication protocols. Our results have suggested that how effectively Q-learning algorithm can solve the temporal credit assignment problem which is essential to any learning autonomous systems and that it can work in conjunction with our simple information-exchanging operator even in multi-agent environments.

References

1. Dawkins, R.: The Selfish Gene. New Edition. Oxford University Press (1976)
2. Kaebling, L.P.: Learning in Embedded Systems. The MIT Press (1993)
3. Ono, N., Rahmani, A.T.: Self-Organization of Communication in Distributed Learning Classifier Systems. Albrecht,R.F.*et al.*(Eds.): Artificial Neural Nets and Genetic Algorithms. Springer-Verlag Wien New York (1993) 361–367
4. Rahmani, A.T., Ono, N.: Co-Evolution of Communication in Artificial Organisms. Proc. 12th Intl. Workshop on Distributed Artificial Intelligence (1993) 281–293
5. Sutton, R.S.: Integrated Architectures for Learning, Planning, and Reacting Based on Approximating Dynamic Programming, Proc. 7th Intl. Conf. on Machine Learning (1990) 216-224
6. Tan, M.: Multi-agent Reinforcement Learning: Independent vs. Cooperative Agents. Proc. 10th Intl. Conf. on Machine Learning (1993) 330–337
7. Watkins, C.J.C.H.: Learning With Delayed Rewards. Ph.D.Thesis. Cambridge University (1989)

8. Werner, G.M., Dyer, M.G.: Evolution of Communication in Artificial Organisms. Langton, C.G. *et al.*(Eds.): Artificial Life II. Addison-Wesley (1991) 330–337
9. Yanco, H., Stein, L.A.: An Adaptive Communication Protocol for Cooperating Mobile Robots. From Animals to Animats **2**. The MIT Press (1992) 478–485

Self and Nonseld Revisited: Lessons from Modelling the Immune Network

Jorge Carneiro and John Stewart

Immunobiology Unit, Pasteur Institute,
25 rue du Dr Roux, 75724 Paris Cedex 15, France

Abstract. In this paper we present a new model for the mechanism underlying what is traditionally known in immunology as the "self-nonself" distinction. It turns out that in operational terms, the distinction effected by this model of the immune system is between a sufficiently numerous set of antigens present from the start of the ontogeny of the system on the one hand, and isolated antigens first introduced after the system has reached maturity on the other. The coincidence between this "founder versus late" distinction and the traditional "somatic self-foreign pathogen" one is essentially contingent, an example of the purely opportunistic tinkering characteristic of biological organization in general. We conclude that the so-called "self-nonself" distinction in immunology is a misleading misnomer. This raises the question as to what would genuinely count as a "self-nonself" distinction, a fundamental question for biology in general and Artificial Life in particular.

1 Introduction

Over the years, immunology has established a substantial reputation as the "Science of Self-Nonself Discrimination" (Klein 1982, Bernard et al 1990). In this paper, we wish to take seriously the project of applying the metaphors of cognitive science to the biology of immunology; this will lead us to the conclusion that the concepts of "self versus nonself" are less straightforward than they seem at first sight, and constitute in fact a problematic issue of general importance for Artificial Life.

The fundamental problem with the notion of "self versus nonself" in immunology was clearly stated almost 20 years ago by Vaz and Varela (1978). They pointed out that an autopoietic subject can only perceive items that exist in its own cognitive repertoire : in the words of the Portuguese poet

Pessoa, "what we see/ is not what we see/ but what we are". Since the subject constructs its own cognitive repertoire, everything that it perceives is, in a fundamental sense, necessarily "self"; if an item were truly "nonself" it would simply not be perceived at all. It follows that no subject can literally make a "self-nonself" distinction; hence the emblematic title of their classic article, "Self and non-sense".

In light of this, let us recall why immunologists talk of "self versus nonself" in the first place. Historically, immunology was born of the attempt to put the late 19th-century discoveries of Pasteur and Koch concerning vaccination on a scientific footing. Thus, the mainstay of classical immunology is the "immune response" : if a foreign entity is introduced into a mammal, the organism produces "antibodies" which specifically recognize the entity and lead to its destruction. Generically, entities which provoke an immune response are known as "antigens"; the prototypical examples are bacteria and viruses, and in such cases the immune response serves the function of defending the organism against pathogenic microbes. However, any foreign molecule whatsoever - from molecules synthesized by human chemists that have never existed before, to tissues from other species or even other individuals of the same species - can function as an antigen and provoke an immune response. Immunologists express this by saying that the repertoire of the immune system is "complete" - there are virtually no molecules (of sufficient size) that are not recognized.

This inevitably raises the question of a possible immune response against the organism's own tissues. With relatively rare exceptions which are pathological and cause "autoimmune disease", the empirical fact is that such responses do not occur in practice. More than that, an immune response of this sort would destroy the organism, and would thus be hopelessly dysteleological; this was clearly recognized from the earliest days of Immunology by Paul Ehrlich, who coined the term *horror autoxicus* to express the notion that such responses *cannot* occur. In both principle and in fact, the immune system must and does make a distinction between foreign antigens which provoke immune responses, and antigens from the organism's own body which do not. This is the distinction which immunologists - naturally but as we shall see somewhat carelessly - have labelled the "self-nonself" distinction.

Since the immune system as classically conceived destroys whatever it recognizes, the upshot is that the system must recognize "nonself" but ignore "self". This would seem to be in flat contradiction with Vaz and Varela's contention that a cognitive system recognizes only "self", and can but ignore "nonself". The resolution to the apparent paradox lies in Maturana's dictum :

"anything said is said by an observer" (Maturana & Varela 1980). The question is, in each case, *who* is making the distinction? When immunologists speak of the "self-nonself" distinction, they are of course speaking from their privileged vantage-point as human scientists. This gives them the ressources to make a clear categorical distinction - in their own cognitive domain - between molecules which belong to a particular individual (mouse or man), and those molecules which comes from outside the organism. Vaz and Varela, by contrast, refer to the immune system itself and to distinctions that can exist within *its* cognitive domain. The immune system, of course, is not a human scientist and does not have the same ressources; consequently, it cannot (and does not) make the *same* distinction.

This brings into focus a question that is somewhat obscured by the common-places of classical immunology : what is the so-called "immune system" doing *in its own terms?* This question can only be answered by an adequate knowledge of the actual mechanisms underlying its operation. In biology, the invocation of teleology or final causes is not acceptable as a substitute for the identification of immediate efficient causes. Thus, the fact that the immune system must (and apparently does) make a distinction that immunologists, after the event, can label from their viewpoint "self-nonself" is no substitute for a proper scientific account of the mechanism whereby this is achieved. Since there is absolutely no systematic physico-chemical difference between "'self" and "'nonself" molecules, the capacity of the immune system to make a distinction that can be so described is truly remarkable; and it is perhaps not entirely surprising, although it is certainly an occasion for modesty, that a century of immunological science has, so far, failed to provide a satisfactory account of how this is achieved.

2 Classical Theories

Antibodies are produced by a special class of cells, the B-lymphocytes. It is now well established that antibody diversity is essentially localized in a "variable region" of the molecule, and results from a process of somatic DNA rearrangement which occurs early in B-cell differentiation; all the cells in the same cell-line or *clone* inherit the same rearrangement and produce the same antibody, but the potential diversity is so great that each clone produces a unique antibody with a characteristic specificity. The classical *clonal selection theory,* due to Burnet (1959), holds that when an antigen enters the system, the pre-existing clones that happen to recognize it are stimulated both to proliferate and to secrete antibodies. On this basis, the avoidance of autoimmunity is accounted for by the *clonal deletion theory* : clones which recognize somatic "self" antigens of the organism are supposedly deleted from the B-cell repertoire.

This theory is in line with the general classical idea that the immune system ignores the "self", but it is not entirely satisfactory for a number of reasons. Firstly, there is experimental evidence that the avoidance of autoimmunity is not simply a negative, passive process but is rather a positive, active process denoted by the term "tolerance". When irradiated mice are reconstituted with a mixed population of lymphocytes from two donors, one of whom is "tolerant" and the other "autoimmune", the recipient animal is tolerant and not autoimmune (Sakaguchi et al 1994). This result is difficult to explain on the deletion theory, which would predict that tolerance should be recessive and not dominant as is the case. Secondly, there is the awkward fact that B-lymphocytes are produced in large numbers throughout life; moreover, the generation of diversity by somatic DNA rearrangement is essentially a random process. If the recognition of somatic "self" is a non-event, as the deletion theory would have it, then a B-lymphocyte freshly produced by the bone-marrow would have to distinguish between two non-events : (i) its reactivity with somatic "self", which should lead to its deletion; and (ii) its reactivity with a future foreign antigen, in which case it should remain in the repertoire so as to be able to proliferate and give rise to an immune response. It is really difficult to imagine how any cognitive agent, unless equipped with the reflexive linguistic capacities and powers of abstract reasoning of human beings, could possibly distinguish between two non-events.

3 Network Models

A possible way forward originated with the idiotypic network theory first proposed by Jerne (1974). Jerne's argument was primarily theoretical : if the antibody repertoire is indeed complete, then each antibody should in principle be recognized by other antibodies. In other words, the antibodies should function as a set of antigens for each other, thus giving rise to a highly connected self-activating network. In this perspective, the term "antibody" becomes something of a misnomer, both because the same molecules are both "antigens" and "antibodies", and because they do not necessarily lead to the destruction of everything that they recognize. Such molecules are therefore often referred to by the somewhat more neutral term "immunoglobulin" (Ig); specific interactions between immunoglobulins are known as "idiotypic" interactions.

The initial attempts to exploit the concept of an idiotypic network, in the years immediately following Jerne's suggestion, were largely centered on the classical phenomena of immune responses, vaccination and immunological "memory"; the results were disappointing, and led to an eclipse of the network theories. Recently, however, there has been a regain of interest in so-called "second generation networks" (Varela & Coutinho 1991) characterized by a focus on positive self-recognition and tolerance. Briefly, these studies have shown that if a model immune system is organized as a highly connected network, then it can couple with antigens in a self-regulating tolerant mode; this corresponds to what Coutinho et al have termed the "Central Immune System" or CIS. Conversely, if the model system has little or no supra-clonal organization, then it will couple with antigens in an explosive immune-response mode; this corresponds to the "Peripheral Immune System" or PIS.

These recent studies may well represent a big step in the right direction; but with respect to the "self versus nonself" question they suffer from a serious defect. To date, none of these models provides a satisfactory account as to how the CIS/PIS distinction might come about. The comprehensive study by Detours et al (1994) shows that "tolerance" can be achieved if but only if the parameters governing the dynamics are such that a stable fixed point exists. This dynamic regime is qualitatively the same as that observed with a single compartment (Neumann & Weisbuch 1992). In this case, the whole shape space is partitioned into zones of two types: those in which the clones are present in relatively low concentration, receive a relatively high field and are coupled to "tolerized" antigens; and those in which the clones are present in relatively high concentration, receive a low field and are couples to "vaccinated" antigens. In this configuration, 50% of the clones are "tolerant", the other 50% being "vaccinated"; moreover, "tolerance" is strictly identical to the "vaccinated" condition for the complementary idiotype, and vice versa. This does not seem satisfactory as a model for natural tolerance. More generally, the problem is that if the parameter settings are such that the model system reliably couples with antigens in the tolerant CIS mode, then this mode of operation percolates to the entire repertoire; the PIS (defined as unactivated clones not coupled to the network) is reduced to zero and the model system can no longer produce any immune responses. In other words, no models to date have succeeded in simulating a co-existence of the CIS and PIS modes. Consequently, these models can have no account of "the difference that makes a difference" (Bateson 1972); they cannot answer the question as to what the immune system may actually be doing when it makes the distinction that immunologists call "self-nonself".

A cardinal principle in biomathematical modelling is that of maximal simplification; the reason is that if a model is too complicated, then it may be almost as difficult to understand the functioning of the model as it is to understand the actual biological system itself, so that even if (miraculously) the model behaves the same way as the real system, nothing has been gained. The attitude of resolute simplification adopted by the "second generation " models is thus largely justified. However, we have come to the conclusion that if the aim is to illuminate the "self-nonself" question, these models may actually have been *too* simplified. In particular, it may be necessary to take into account the fact that mammalian immune systems are based on a dual recognition mechanism, composed of a B-cell and a T-cell compartment.

4 B-Cells And T-Cells

Ever since the early days at the beginning of the century, immunology has been divided into two branches: the "humoural" branch concerned with antibodies (immunoglobulins) and other soluble factors; and the "cellular" branch which includes the ameoba-like macrophages that actually engulf and digest bacteria and other foreign bodies. It is now known that this division corresponds to a distinction between B-lymphocytes and T-lymphocytes. The variable-region molecules of B-lymphocytes are the immunoglobulins, which exist in two forms: integrated in the cell membrane where they function as receptors (m-Ig); or secreted as free molecules (the classical "antibodies"). Immunoglobulins, free or membrane-bound, interact directly with antigens via stereochemical affinity. The variable-region molecules of T-lymphocytes are simply known as "T-cell receptors" (or TCR for short), and exist only as membrane-bound receptors. The interaction of TCRs with antigens is indirect. The antigen must first be engulfed and digested by an "antigen presenting cell" (typically non-specific macrophages, but in general APC for short), which then "presents" peptide fragments of the antigen on its cell surface in the context of a special "MHC" molecule. The TCR then interacts with the "MHC + peptide" complex.

There has been, and to some extent still is, a tendency to separate these two branches of immunology. In fact, however, a typical immune response requires co-operation between T-cells and B-cells: the latter are only stimulated to proliferation and immunoglobulin secretion if they receive "help" from T-cells. The usual condition for "help" is that the B-cell, which is itself an excellent APC, should present "MHC + peptide fragments" that are recognized by the TCR of the T-cell. There is thus a dual recognition system: antigens must be recognized both in their native 3-dimensional form by immunoglobulins, and as MHC-presented peptide fragments by TCRs. Concomitantly, full B-cell activation requires *both* that their m-Ig receptors be cross-linked by binding with a ligand (which is typically an antigen but which can also be an idiotypic immunoglobulin); *and* that molecules on the surface of the B-cell be recognized by the TCRs of activated T-cells.

To date, all published second-generation network models have made the simplifying assumption that the T-cell compartment can be neglected. Biologically, this corresponds to the explicit postulate (Varela et al 1988) that T-cell help is never a significant limiting factor for B-cell activation. It is this assumption that we now consider may have been an over-simplification. We therefore present a new model which specifically includes B-T cell co-operation.

5 A New Model

The variables of this new model distinguish B-cell clones B_i ($i=1...N_B$) and secreted free immunoglobulin F_i as before (Varela et al 1988), but now also include a set of T-cell clones T_k ($k=1...N_T$). T-cell activation occurs through TCR-recognition of MHC-presented antigen fragments. B-cell activation requires both M-Ig binding by ligand and T-cell help, as explained in the previous section. B-cell and T-cell clones can be added to or removed from the system by a metadynamical process of recruitment, as in the previous models.

Under these conditions, a single antigen rather straightforwardly gives rise to an unlimited immune response. (Figure 1). A T-cell clone with TCR complementary to MHC-presented antigen will be recruited; once recruited, it will be activated and grow exponentially. A B-cell clone whose Ig is sterically complementary to the antigen will be recruited; it will itself digest and present the antigen; it will thus receive help from the activated T-cell, and will also grow exponentially. It is to be noted that in this configuration, the Ig is complementary to the whole antigen, whereas the TCR is complementary to the MHC-presented peptide; in general, these two "forms" of the antigen are quite unrelated, so that there is no relation between the TCR and the Ig.

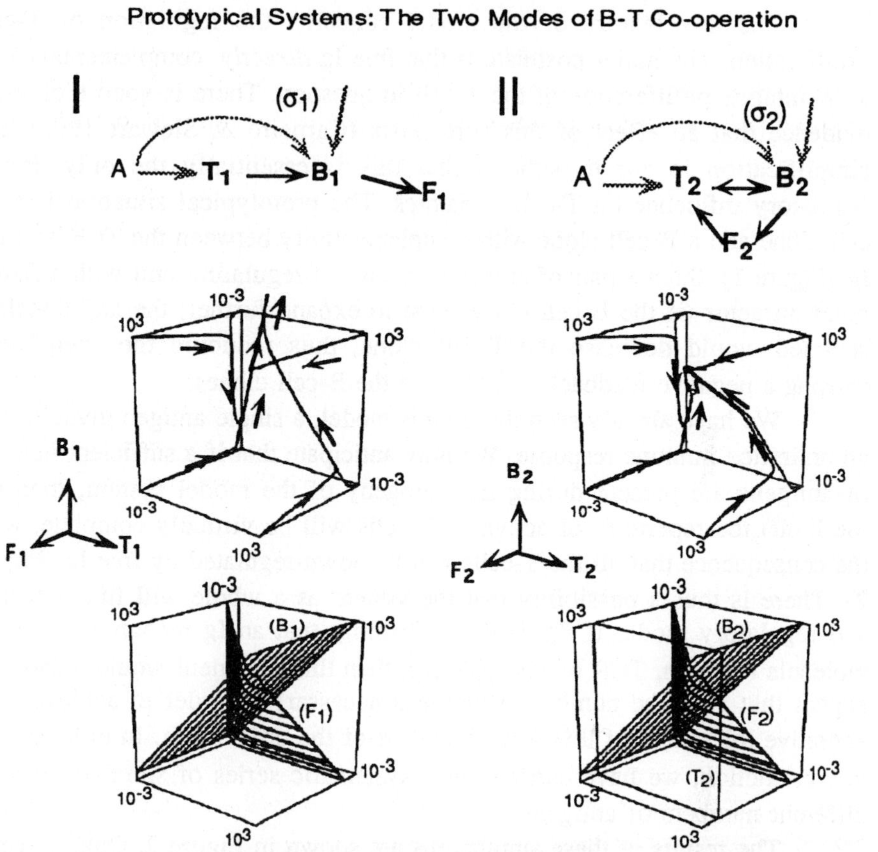

Fig. 1. Two prototypical modes of B-T co-operation are depicted. I- The antigen (A) specific B-cell clone (B_1) and T-cell clone (T_1) are stimulated and their co-operation is mediated by antigen presentation. The immunoglobulins (F_1) produced by stimulated B_1 do not recognize directly T_1, which are therefore not controlled and grow exponentially driving the system to explosion. II - The B-cell clone (B_2) and antigen specific T-cell clone (T_2) are stimulated but their co-operation is now mediated by mutual recognition. The immunoglobulins (F_2) produced by stimulated B_2 recognize directly T_1, and inhibit its growth, defining a negative feed-back loop that renders the system stable. The two situations are illustrated by diagrams and phase-space representations [(T_1,B_1,F_1) and (T_2,B_2,F_2)] of simple mathematical models implementing the two situations. In the middle: several trajectories of the two models are plotted, leading (I) to explosion and (II) to a fixed-point attractor. Bottom panel: the nullclines for the system components: note the absence of a nullcline for T-cells in model I. The phase-space representations were obtained using the package GRIND by R.J.De Boer.

The key feature of the model concerns the regulation of T-cell proliferation. The major postulate is that free Ig *directly* complementary to a TCR inhibits proliferation of the T-cell in question. There is good biological evidence that an effect of this sort exists (Carneiro & Stewart 1995); the simplification is our hypothesis that this is essentially the only down-regulatory influence on T-cell dynamics. The prototypical situation is a T-cell clone and a B-cell clone with complementarity between the TCR and the Ig (Figure 1). Such a pair of clones forms a self-regulating unit with a fixed-point attractor (if the B-cell clone were to expand further, the additional Ig secreted would decrease the T-cell clone, thus reducing the "help" and causing a negative feedback reduction in the B-cell clone).

We have already seen that in this model, a single antigen gives rise to an unlimited immune response. We may anticipate that if a sufficient number of antigens are present during the ontogeny of the model system, then (in the limit) the repertoire of activated B-cells will be virtually complete, with the consequence that all the T-cells will be down-regulated by free Ig (Figure 2). There is thus a possibility that the system as a whole will function in a self-regulatory mode. If P_R is the probability that an Ig recognizes another molecule (antigen, TCR or idiotypic Ig), then this argument would lead us to expect that N_A, the number of antigens necessary in order to achieve "co-operative tolerance" of this sort, should be of the order $1/P_R$. In order to test this prediction, we have carried out a systematic series of simulations with different numbers of antigens.

The results of these simulations are shown in Figure 2. Qualitatively, the expectations are fulfilled. For a small number of antigens, unlimited immune responses are produced; as the number of antigens increases, more and more of the T-cell clones are controlled, and for a sufficient number of antigens the model reliably enters into a globally "tolerant" mode. The details of the ontogenetic process are however more complex than the simple qualitative expectation outlined above, and merit closer description. In what follows, we shall concentrate on cases in which tolerance to the initial set of antigens is achieved.

Firstly, in the initial phase of the ontogeny there is a "cascade" in the recruitment of B-cell clones, many of which are activated by idiotypic interactions (Figure 2). As a result, the number of activated B-cell clones, N_B, is an order of magnitude greater than N_A. Since, as may be expected, the condition for global tolerance is that N_B should be of the order of $1/P_R$, the critical value of the product $N_A.P_R$ necessary to achieve tolerance is of the order of 0.1.

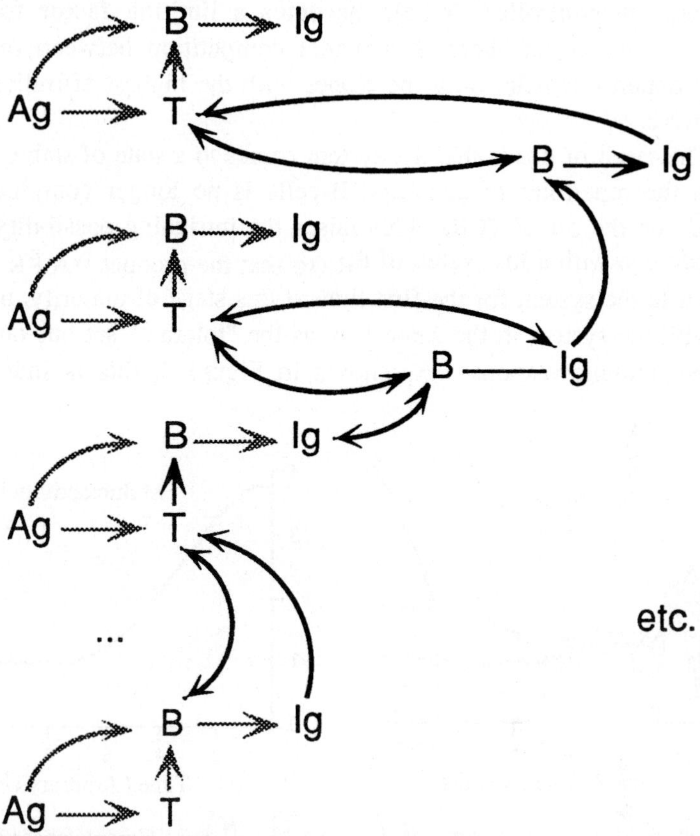

Fig. 2. Scheme of the development of a B-T lymphoid network. Available antigens will drive antigen specific B- and T-clones. The activation of this first 'layer' of B-clones is ensured by co-operation with the antigen driven T-clones mediated by antigen presentation. Subsequent 'layers' of B-clones can be activated and therefore recruited into the network, if and only if, they are specific both for the immunoglobulins (Ig) produced by other B-clones and for the T-clones which are already activated. Since the newly recruited B-clones have the potential to control the T-clones that sustain them (see fig.1-II), the process of their co-operative recruitment can therefore lead to stability of the entire system, when all the antigen driven T-clones are controlled by feed-back loops.

The second unexpected result is that after passing through a maximum at the point in time when the T-cell clones all come under control, there is a second phase of the ontogeny during which N_B decreases and falls to a fraction of its maximum value. The reason for this is that when the T-cell clones are controlled, T-help becomes a limiting factor for B-cell proliferation and hence there is stringent competition between the B-cell clones to obtain this help; only the clones with the highest affinities for the TCRs survive.

The result of this is that the system comes to a state of stable maturity in which the repertoire of activated B-cells is no longer complete but is "focussed" on the set of TCRs. This raises the intriguing possibility that an isolated antigen with a low value of P_R (so that the product $N_A.P_R$ is 0.01), introduced to the system for the first time at this stage of maturity, might not couple with the system in the same way as the "tolerant" set but rather give rise to an immune response. As shown in Figure 3, this is indeed what happens.

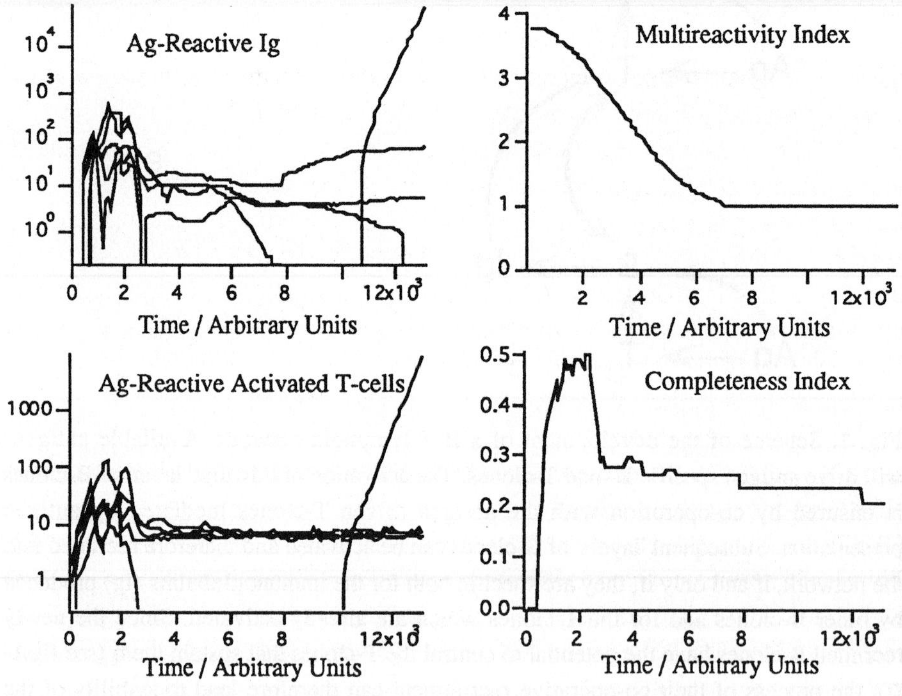

Fig. 3. Co-existence of two modes of coupling with antigen. The 4 graphs are time-plots of the following quantities during a typical computer simulation of the model: concentration of antigen specific Ig-molecules (top-left), number of antigen specific T-cells (bottom-left), multi-reactivity index (top-right), and completeness index (bottom-right). A discrete shape-space formalism with one center of symmetry and dimension (25x25) was used to generate the interaction coefficients between the model components. The "multireactivity index" is proportional to the parameter P_R. "Completeness index"

is the fraction of shape-space which is under the influence of the network and is a result emerging from the simulation. The number of founder antigens was 5 with $N_A.P_R$=0.3; one new antigen was introduced at time=10000 with $N_A.P_R$=0.01.

6 Discussion

The new model of the immune network that we have presented here has the property that it is able to couple simultaneously with some antigens in a tolerant CIS mode, and with other antigens in an immune response PIS mode. Moreover, there is a systematic distinction between the two classes of antigen. Operationally, the model system makes a distinction between a sufficiently numerous set of "founder antigens" that are present from the start of the ontogeny of the system, and isolated antigens that are introduced to the system for the first time at a stage of ontogenetical maturity. The next question is whether this is acceptable as a model for natural tolerance and for what is classically known as the "self-nonself" distinction.

The answer to this question would seem to be positive. The antigens that an immunologist identifies as belonging to the body of the organism are indeed numerous and present from the start of the ontogeny of the immune system in the late pre-natal period. By contrast, the antigens that an immunologist identifies as "foreign" are indeed introduced to the system one by one at a later stage. In the post-natal period, young mammals are not "immunologically competent"; they only produce full-fledged "immune responses" to "foreign" antigens when they reach the young adult stage.

In fact the notion that the so-called "self-nonself" distinction is actually a distinction between "founder" antigens and later arrivals is not at all a new idea in immunology. It was proposed 40 years ago by Medawar (1957), and has been corroborated by a variety of experimental evidence. Thus, additional antigens introduced into the fetus pre-natally (either experimentally or, in the case of dizygotic twins in cattle, naturally) will be recognized as "self" by the immune system. Conversely, tissues that genetically and anatomically form part of the body but are isolated from the circulation, such as the eye-lens and spermatozoa, will be treated as "nonself" by the immune system if introduced into the blood stream in the adult.

This notion, that the "self-nonself" distinction is actually a distinction between "founder" versus "late" antigens, is so far from being a complete innovation that it is even incorporated in some versions of the classical clonal deletion theory (in other versions, the "clonal deletion" takes place throughout life in the thymus). The present model does however have a number of distinctively novel features. Firstly, it provides a specific (if as yet hypothetical) mechanism whereby the "imprinting" with "founder" antigens could come about. Secondly, tolerance is thereby explained as an active, dominant process rather than a negative, recessive one. Thirdly, the set of founder antigens are recognized collectively and co-operatively as a "Gestalt". It is because these antigens are connected into a coherent set that the idiotypic network can "focus" around the TCRs that recognize them, and thus release the greater part of the potential repertoire for the PIS and possible immune responses.

A question which arises here is whether an idiotypic network which is focussed ont the inhibitory regulation of TCRs is still a full network in the sense of the "second generation networks" as defined by Varela & Coutinho (1991) which have been at the centre of recent work in Theoretical Immunology. We may note that each regulatory idiotype recognizes a TCR which in turn recognizes a MHC-presented antigenic peptide; it is therefore an "internal image" of the peptide in a functional sense (which does not necessarily imply that it has similar 3-dimensional structure). Now since in general there is no relationship between the 3-dimensional native antigen and the corresponding MHC-presented peptide, the B-clones that produce the regulatory idiotypes will only be sustained if they are engaged in idiotypic Ig-Ig interactions in the network. Another way of stating this is that the essentially arbitrary set of "founder" antigens will only form a coherent "Gestalt" if they are linked via idiotypic interactions. We conclude that although the repertoire of the idiotypic network is no longer complete, being focussed around the set of TCRs to be regulated, our model is still fundamentally a network model.

7 Conclusions

We are now in a position to return to the question raised in the introduction. What is the relationship between the "founder versus late" distinction as proposed by our model, and the "self versus nonself" distinction of classical immunology?

The main point we wish to make is that there *de facto* a strong and reliable correlation between the two distinctions; but that in an important sense, this correlation is essentially contingent. The molecules that a human biologist identifies as belonging to the somatic body of a mammal are indeed, with rare exceptions, present from the inception of the ontogeny of the immune system. Interestingly, these molecules are also numerous; if the mechanism proposed here is correct, this fact would also be significant.

Conversely, molecules that enter the body fluids for the first time in young adult life do by and large correspond to what an immunologist would call "nonself".

However, this does not mean that the founder antigens are identified by the immune system as "self" in any *intrinsic* sense. To be sure, the founder antigens form a coherent gestalt which is "imprinted" in the cognitive repertoire of the system with a special status; but when the newly-hatched chick is "imprinted" by the first animate object it sees, and to which it henceforth attaches a special status, we do not therefore say that the mother-surrogate is identified by the bird as "self". Likewise, it seems to us that there is no good reason to maintain that the distinctions effected by immune system can be properly labelled "self versus nonself".

Our general conclusion is thus that what at first sight seems to be a "self-nonself" distinction turns out, when the actual mechanism underlying the distinction is properly elucidated, to be a different distinction that does not involve "self" in any intrinsic sense. This raises the question as to what would constitute an authentic establishment of "self" as a biological category. This is indeed one of the fundamental questions of biology (Buss 1987), and it is of great importance for Artificial Life if the field is to amount to more than engineering and computer games. We obviously cannot answer that question here, but we do hope to have raised it in a useful way.

Acknowledgements

J.Carneiro acknowledges the financial support of 'Junta Nacional de Investigação Científica e Tecnológica - Programa Ciência', Lisbon (grants BD/2319/92-ID).

References

Bateson G. (1972). *Steps to an ecology of mind.* New York, Chandler.

Bernard J., Bessis M. & Debru C. (Eds) (1990). *Soi et non-soi.* Paris, Editions du Seuil.

Burnet F.M. (1959). *The clonal selection theory of acquired immunity.* London, Cambridge University Press.

Buss L.W. (1987). *The Evolution of Individuality.* Princeton, Princeton University Press.

Carneiro J. & Stewart J. (1995). *Journal of Theoretical Biology* (submitted).

Detours V., Bersini H., Stewart J. & Varela F. (1994). Development of an Idiotypic Network in Shape Space. *Journal of Theoretical Biology* **170**, 401-414.

Jerne N.K. (1974). Towards a network theory of the immune system. *Annales d'Immunologie (Institut Pasteur)* **125C**, 373-89.

Klein J. (1982). *Immunology : the science of self-nonself discrimination.* New York, Wiley.

Maturana H.R. & Varela F.V. (1980). *Autopoiesis and cognition : the realization of the living.* Dordrecht, Reidel.

Medawar P.B. (1957). *The Uniqueness of the Individual.* London, Methuen.

Neumann A.U. & Weisbuch G. (1992). Window automata analysis of population dynamics in the immune system. *Bull. math. Biol.* **81**, 645-670.

Sakaguchi S & Sakaguchi N. (1994). Thymus, T cells and autoimmunity. In A. Coutinho & M.D. Kazatchkine Eds., *Autoimmunity: physiology and disease,* pp. 203-27. New York, Wiley-Liss.

Varela F.J., Coutinho A., Dupire B. & Vaz N.M. (1988). Cognitive networks: immune, neural and otherwise. In A.S.Perelson Ed., *Theoretical Immunology. Vol II,* pp. 359-374. New York, Addison Wesley.

Varela F.J. & Coutinho A. (1991). Second generation immune networks. *Immunology Today* **12**, 159-66.

Vaz N.M. & Varela F.V. (1978). Self and non-sense : an organism-centered approach to immunology. *Medical Hypothesis* **4**, 231-67.

On Formation of Structures

Jari Vaario*[1] and Katsunori Shimohara[2]

[1] Nara Women's University, Faculty of Science,
Department of Information and Computer Science,
Kita-Uoya, Nishi-Machi, Nara 630 JAPAN
Phone/Fax: +81-742-20-3443, e-mail: jari@ics.nara-wu.ac.jp
[2] ATR Human Information Processing Research Laboratories
2-2 Hikari-dai, Seika-cho, Soraku-gun, Kyoto 619-02 JAPAN
Phone: +81-7749-5-1070, fax: +81-7749-5-1008, e-mail: katsu@hip.atr.co.jp

Abstract. This paper describes a method for a computer exploration of formation of structures based on the network of autonomous units. This method has a biological correspondence with morphogenetic processes. The interactions in the network of autonomous units are modeled by two kinds of forces: *repulsive* and *attractive* forces. When and what kind of forces are active at each unit is based on genetic information and environmental factors. Genetic information enables the use of evolutionary algorithms to evolve the interactions and thus to create new structures. Environmental factors provide the needed restrictions for the space of possible structures. Depending on what meaning is given to the units, the system is capable of simulating various kinds of emergent phenomena. For example, in the case where units are interpreted as cells, where the repulsive and attractive forces represent collision and adhesion forces, a formation of multicellular organism can be achieved.

1 Introduction

Recently Artificial Life has become a popular topic. Programs that show some kind of behavior found also in natural living systems are classified under artificial life. Thus computer viruses, genetic algorithms, evolutionary systems *etc.*, have been placed under the artificial life concept.

Although there is no problem with classifying the above like systems under artificial life, the focus of this paper is on another stricter view of artificial living systems. The approach is to create a virtual reality that models artificial living systems including individual development as well as evolution. The main motive is not to simulate biological living systems, but to model processes that create them. These processes are then used to create artificial 'living' systems, that, hopefully, are significant from the engineering point of view (for example, intelligent autonomous systems, or software agents). Another motivation is that these processes can provide some new ideas for the concept of emergence.

---

* During this work the first author has been affiliated at ATR Human Information Processing Research Laboratories

This paper gives an overview of the biological processes that are thought to be significant in the creation of living systems. Our emphasis in these processes is on *synthesis*, *emergence* and *local control*. With synthesis we want to emphasize that systems must be constructed gradually as a part of the environment. The process will generate a system in the sense that the final system is constructed based on simple rules that do not explicitly define the final system. The final system is a result of the execution of these simple rules that collectively give rise to it. This means that the whole system is constructed based on local control, rather than global control.

Through the synthesis process the systems are created based on environmental effects as well as genetic information. This provides them a direct and immediate capability to react to environmental changes. This response can be viewed as adaptation. To find a general adaptation mechanism through the synthesis process, *i.e.*, how the local changes effect the global structure or behavior, is one of our main research goals.

In this paper the biological adaptations are overviewed after a general overview of artificial life. This is followed by a description of emergent concept. The simulation system is viewed with an emphasis on the environmental model, rather than on the interpretation mechanism of genetic information. Some simple examples are used to demonstrate the environmental model of simulation system. The organism level behaviors, including neural network formation, are described in other papers (see, for example, [11, 13]) being omitted here. Also the focus of this paper is on the self-organization rather than evolution. Thus no evolutionary simulation results are reported here.

2 Our View of Artificial Life

With Artificial Life we understand the research of life-like behaviors where the emphasis is on the mechanisms of creating the behavior by *synthesis* rather than analysis (of existing biological behavior). Thus having a *mechanism* for creating life-like behaviors will provide 1) better understanding of existing life, especially what are the essential properties of life (*scientific purpose*), and 2) generally adaptive artificial systems (*engineering purpose*).

The basic idea is to study biological systems in order to understand the essence of their behaviors. This is the *life* part in artificial life. The *artificial* part describes how the life-like systems could be constructed by humans. The media of this construction could be computer (software), silicon, etc. (hardware), or chemical molecules (wetware). The media could also be very theoretical including only some philosophical thinking how everything could be constructed at the conceptual level. Our interest is mainly on the software synthesis without forgetting the philosophical basis of it.

There is not yet common agreement about what is *the* necessary and sufficient mechanism(s) to create life-like behaviors. The problem is that life shows a large variety of behaviors that are difficult to cover under a single method. There are two usually used approaches (that I want to emphasize here) "competing" with each others: 1) *evolution* and 2) *self-organization*.

The former approach considers evolution through *natural selection* (survival of the fittest) as the necessary and sufficient mechanism. What happens during the individual development process is not important for evolution. This idea has been already challenged in biology, but it still exists strong in artificial life community. The examples of this approach are *genetic programming*, *evolutionary systems*, *Tierra*, etc.

The latter approach considers self-organization as the origin of natural systems. Even evolution itself is a result of self-organization. Therefore the study of artificial life should start from the principles behind self-organization. *Autopoiesis* is an example of how a system can create biological structures and maintain their stability [7]. Other examples are the various interaction networks (*e.g.*, immunologically inspired networks) that show great adaptability to environmental changes [6].

The biggest problem with the self-organization approach is to create a purposeful behavior, whereas the evolutionary approach (where explicitly given goals define the purpose) does not have the same capability for creativity. This could be solved by combining two approaches creating a system where self-organization provides the micro dynamics, and evolution provides the macro dynamics.

Figure 1 brings together different artificial life research approaches. The bottom-up approach is a *self-organization* process based on *environment* and *genetic information* resulting in an *emergent structure*. This is a collection of basic elements (for example cells) interacting with each other and thus it is also called *collectionism*. From the biological point of view these structures are called the phenotype, or if the environmental effect is emphasized *phenocopies* [8], or *phenotypic variations* [10]. From these structures the *evolutionary feedback* is generated presenting the top-down approach. It is divided into a mechanism of *natural selection* that acts as a *filter of phenotypic variations* and a mechanism of *production of genetic variations*, *i.e.*, *genovariations* [10]. The evolutionary feedback mechanisms are also called *macro-mechanisms*, and the self-organization based construction mechanisms as *micro-mechanisms*.

3 Overview of Biological Adaptations

Our research aims at creating artificial systems (agents) capable of intelligent behavior where intelligence is viewed as adaptability to the environment. Because of this purpose we define the life of these systems similar to the biological systems.

The systems consist of a *genotype*, that generates a phenotype; a *phenotype*, that results in behavior; a *behavior* that leads to reproduction, or death; and *reproduction* that produces a new genotype closing the loop. Between each transition from one stage to the next, there is a process that is affected by the environment. Biological adaptation is defined by these processes: *morphogenesis*, *ontogenesis*, *natural selection*, and *genetic variations*. The general concept is shown in Figure 2.

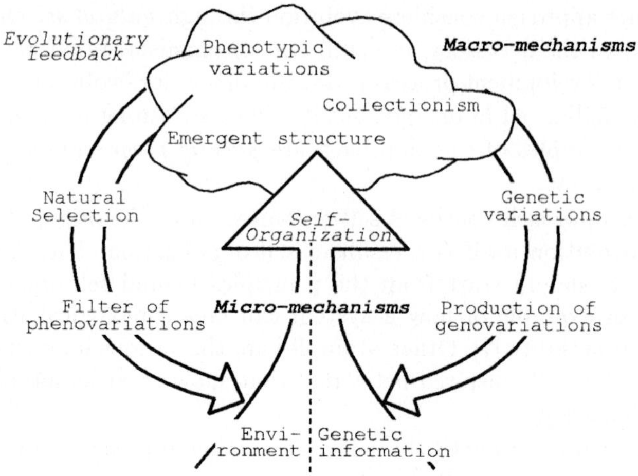

Fig. 1. A general concept for the artificial life research approaches.

The morphogenesis and ontogenesis are processes of individual, *i.e.*, *developmental* processes. In the same way natural selection and genetic variations, even happening for individuals, are usually observed at the population level being *evolutionary* processes.

It has become obvious that individual development plays an important role in evolution as well [4, 10]. For all these processes the environment is a necessary element. Usually the environment is replaced by a well defined interface with a possible parameter space. However, the systems should also be given a capability to make use of different environmental features, *i.e.*, evolve their perception space and the possible actions. This is illustrated in Figure 3.

The problem for modeling an intelligent system is on the principles used thus far. We want to argue that a self-organization process is needed to allow intelligent systems to emerge, rather than to define the system explicitly. The emergent process is needed when the final state cannot be defined. That is the case of intelligence. In general, for intelligent systems, the final state is a result of continuous adaptation process. For this we need a mechanism that is capable not only of dynamic modification, but also of dynamic creation including the interface to the environment. This paper concentrates on this dynamic creation through a self-organization process as follows.

4 Emergence of Structures

The approach selected here is based with autonomous elements which are sensitive to (and capable of creating) environmental forces. Each element is considered as an independent process having its own instruction set, *i.e.*, *action rules*. The network of these elements is the result of execution of these local

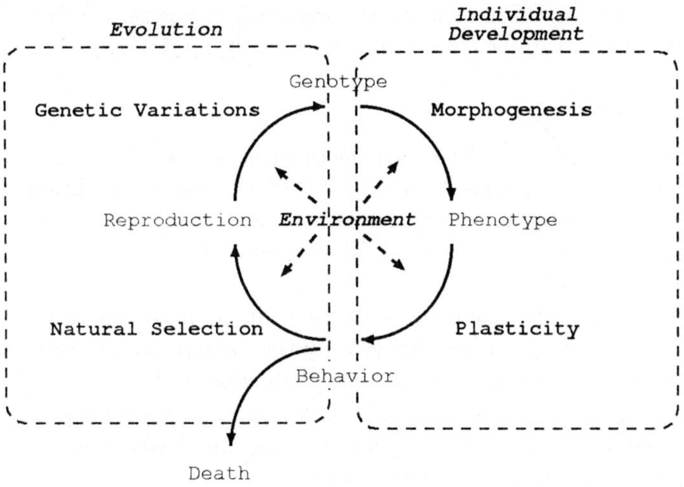

Fig. 2. A general concept for biological adaptation.

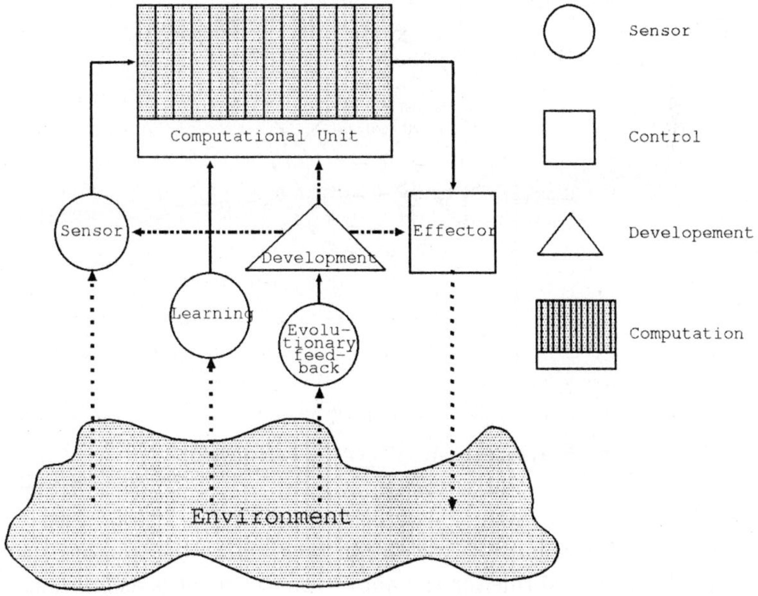

Fig. 3. The concept of a general evolvable autonomous system. (Adapted from [2])

rules in a common environment. In the following the general principle is given with a conceptual example (that is inspired partly by [5]).

4.1 Basic Element

The main idea is to define a basic element as a composition of action rules. The basic element is capable of modifing itself based on the conditions of internal and external (surrounding environment) states. These actions rules could be thought to be an emergent result of lower level actions rules, althought the lowest level has to be defined.

In addition to these action rules, we need an environment to create the dynamics. The above action rules change only the passive components of the basic elements. How these changes cause some behavior is the responsibility of the environment. Thus we have to assume as a minimal requirement the existance of environmental forces (*gradient fields*) to cause the basic elements to flow according to their sensitivity to a particular field.

In Figure 4 we give an explicit definition for a basic element that exists in the environment and is capable of creating a gradient field to which the other element is sensitive.

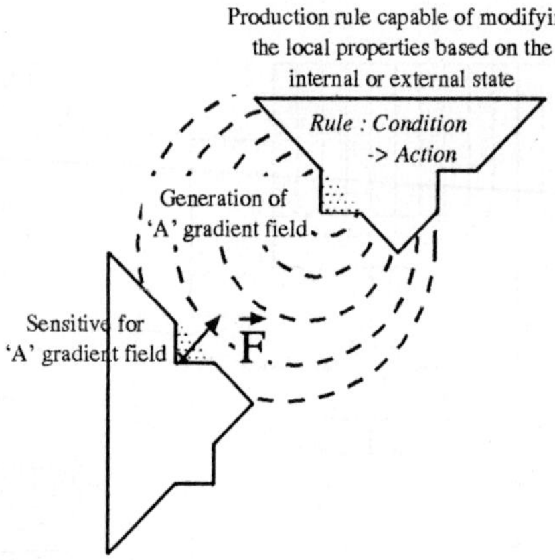

Fig. 4. A basic element of emergence

The behavior of such basic elements can be given by the format of a simple production rule. A logical rule for the above basic element is as follows.

if ... **then** *activate gradient field 'A'*
if ... **then** *be sensitive to gradient field 'A'*

Note that at this first stage we do not need any condition, since the activation could be always on.

4.2 Network Formation

When several basic elements exist in the environment each creating a gradient field(s) and being sensitive to other gradient field a self-assembling of a structure results. This structure can be considered as an interaction network of the basic elements.

This is demonstrated in Figure 5. A few elements are located in the environment creating a network of elements interacting through the physical forces. As a result of these interactions a structure is self-assembled.

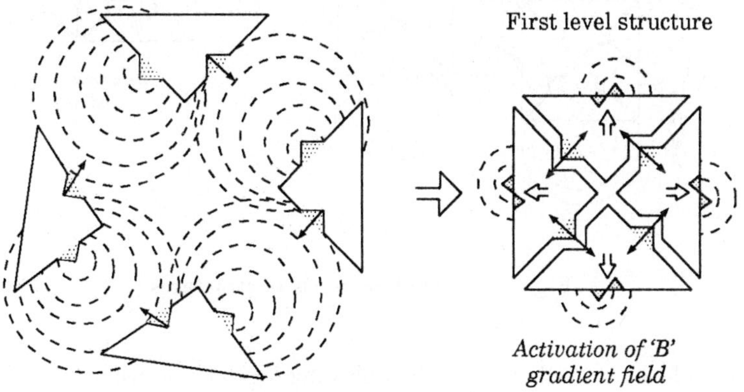

Fig. 5. A network of basic elements

The establishment of the network of basic elements can be verified by a simple logical condition.

if *connected to another unit on both sides*
then *activate gradient field 'B' at the free edge ...*

This condition can be used to initiate an additional behavior like activating the free edge. This in turn causes the next level of self-organization as described below.

4.3 Formation of Multilevel Networks

If an action rule is capable of detecting the existence of neighbor cells, it can change the internal value of a basic element in a such a way that an additional attractor appears (Figure 5). Now, the first level structure can attract similar structures creating a next level structure (Figure 6). Once again when the basic element detects a new neighbor, it modifies its internal state. This is recognized by the neighbor cells, and results in a 'turn off' of the first level effect, and causing the second level structure to become inactive for any further modifications, *i.e.*, a stabile structure has been reached.

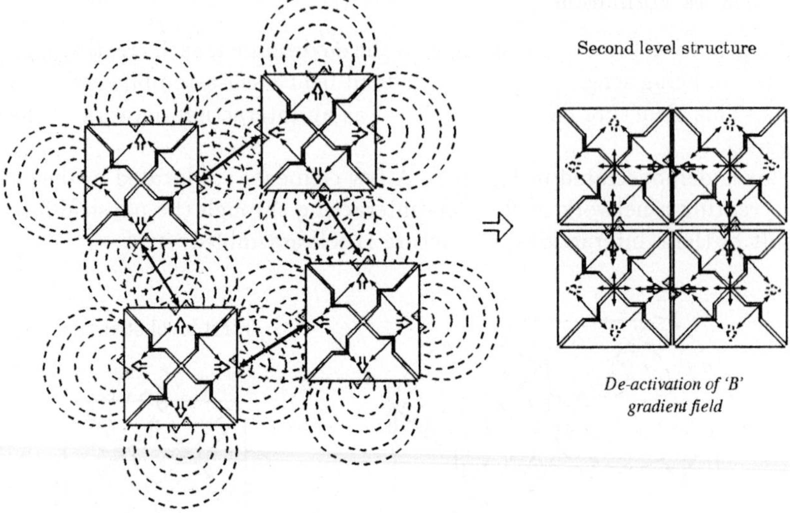

Fig. 6. The multiple levels of structures

This could be described with the following logical condition.

if (*not connected to another unit* **and**
opposite neighbor connected to another unit)
then *de-activate 'B' gradient field* ...

4.4 Environmental Effect

In the above principle the effect of environmental factors is straightforward. Each environmental factor is modeled by similar gradient fields effecting to the self-organization process. For example, the second level structure in Figure 6 could be a chain, if a global environmental field favors elements to move only in a specific direction.

4.5 Biological Correspondence

The basic elements in the above examples can easily be thought to be cells, that forms tissues and more complicated organisms. A cellular structure can be formed through multiple cellular divisions that occur in a particular direction. The physical forces are collision and adhesion forces. Collision forces keep the cells away from each other while the adhesion forces pull them together. A cell has itself a capability to change its surface in order to adjust its adhesion properties.

5 The Simulation Method

The basic idea in modeling the above self-organization of structures is to separate the environmental model from the model of basic elements. This is the main difference to cellular automaton (CA), where the environment is modeled as a part of transition rules of each CA's cells. Similiarly, the difference to the lattice models is that the interactions are not direct between each element, but between the elements and the environment.

The basic element model consists of production rules that operate on the internal and external states. The environment model must be able to model the physical forces to move basic elements.

5.1 Top-down vs. Bottom-up

Figure 7 gives an overview of the hierarchy of the production rules. The environment level consists of organisms and production rules that describe the physical interactions between the organisms. The organism level consists of cells and production rules that describe the physical interactions between the cells. The initial state consists of at least one cell belonging to an organism and the environment. As a result of execution the cell will divide and form a mature organism. After this a seed cell can be ejected meaning that a new organism is created.

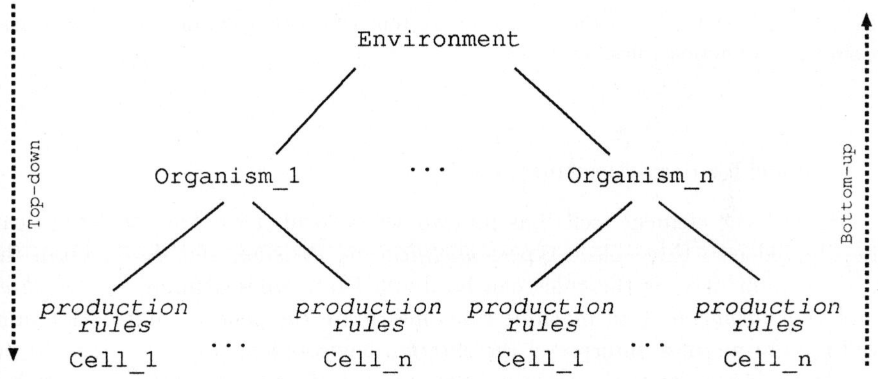

Fig. 7. The modeling principle: top-down or bottom-up direction.

A slightly different view of the above could be reached by reversing the direction of the arrows. Now the independent cells are executing their production rules resulting in the formation of organisms. However, because now we do not have the production rules for the organism and environmental levels, we need a separate environmental model. An initial state consists of a single cell, that divides and forms an organism.

5.2 Overview of Implementation

The implementation consists of a parser to create the initial state for modeling. The life cycle of each basic element consists of the evaluation of production rules, that modify (including division and death) the basic elements. The environmental simulation consists of the simulation of physical forces as will be explained later. The graphical interface is used only for illustration. The overview of these are shown in Figure 8.

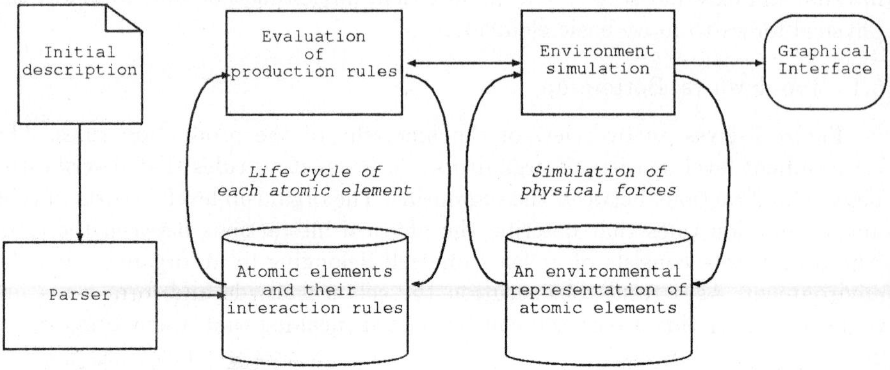

Fig. 8. Basic structure of the simulation system where the environment is separated from the production rules.

5.3 Model for Basic Elements

Each basic element (cell) has its own set of local rules. The local rules can be divided into three basic types: *modification*, *division*, and *death*. From the logical point of view the rules can be divided into rules defining *cell machine* and *cell membrine*. Cell machine rules interprete the genetic information, and cell membrine rules 'interprete' the external environment [14].

At an initial stage the system might consist of only one basic element that through division multiplies itself. On the other hand, these division rules might be ignored by giving enough basic elements as an initial condition.

The format of local rules follow the basic concept of production rules of Lindenmayer-systems [9]. The basic characteristics of rewriting systems are preserved. However, the parameters (of parametric L-systems) are replaced by attributes (key = value). Multilevel rules are also allowed through a sub-symbol concept where each symbol (letter) can have sub-objects with their own production rules. (See details in [12].)

5.4 Model for Environment

The basic idea is that environment simulates some forces that move the basic elements based on their properties. The properties are modified by the basic elements themselves (or by some other factors) depending on their state and surrounding environment. These forces could be defined by the concept of gradient fields. This approach is more efficient than an analytical approach used, for example, by Fleischer and Barr in [3].

Using gradient fields to model the physical forces has various advantages over the analytical approach. First, the basic concept is the same for all kinds of forces, which implies a simple implementation. Second, the calculation of the gradient field value at a given location is quite straightforward and fast. Third, the forces acting on multi-part bodies are the sum of forces of the sub-parts. This is especially important in our work where the shape is dynamic.

Each element itself can create a gradient field. While the element with a gradient field can move we have a system where dynamic gradient fields can superimpose (Figure 9). The used gradient fields are round shaped, but other shapes could be also considered.

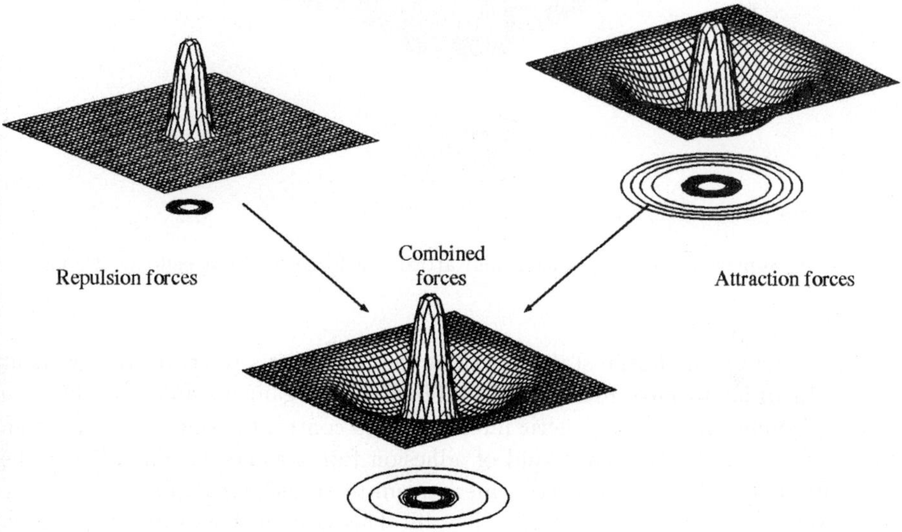

Fig. 9. Superposing repulsion and attraction forces in 2D space with fields of round shape.

The combined objects are created by attraction forces. For example, eight basic elements attract each other, creating the organization shown in Figure 11. The forces applied to the created system, is a superposition of the forces of each part. This is illustrated in Figure 10.

The form in Figure 10 is very unstabile, because of the attraction valley in the center of cells. In simulations rounding errors caused a situation where every second cell collapsed into a valley. This is demonstrated in Figure 11.

Some stabile forms are illustrated in Figure 12. The forces create patterns

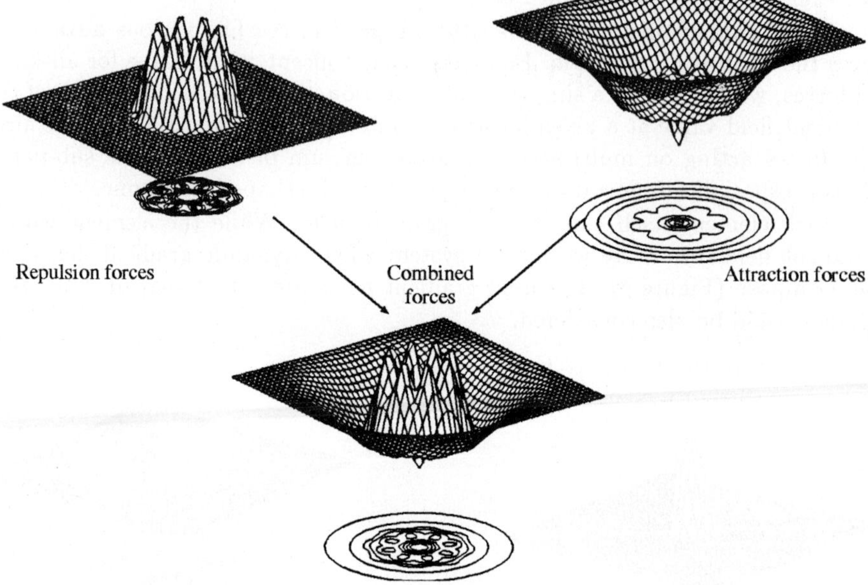

Fig. 10. Superposition of repulsion and attraction forces of eight cells in 2D space.

that depend on the initial state as well as the parameters to create the gradient fields. In order to have meaningful patterns, *i.e.*, organisms able to behave in the environment, we need genetic information to control the cell activities: when to divide, when to die, what kind of adhesion forces to create and when to be influenced. In addition we need a mechanism to transfer information from the cell's immediate surroundings into the cell. This is used, for example, to check whether there are any neighboring cells.

6 Life Like Structures

In Figure 13 eight adhesion fields are used to create stability of a given form. This form is a result of repeated divisions and cell differentiations as described in. The adhesion field is created by a cell based on the genetic information and the historical information of cell lineage.

The example does not yet include any signal propagation and thus it represents only the creation of form. Some cells could differentiate into sensors (a group of cells numbered from 30 to 35; and cells 16-19, 27, 28), some as muscle cells (a group of cells 2-4, 8, 10, 12, 20; and cells 5-7, 9, 11, 15, 26), and some as neuron cells (1, 13, 14, 22-24, 32, 33).

Similar to our previous research [11, 13] the neurons grow the connections to the sensors and to the muscle cells. The signals received from the environment

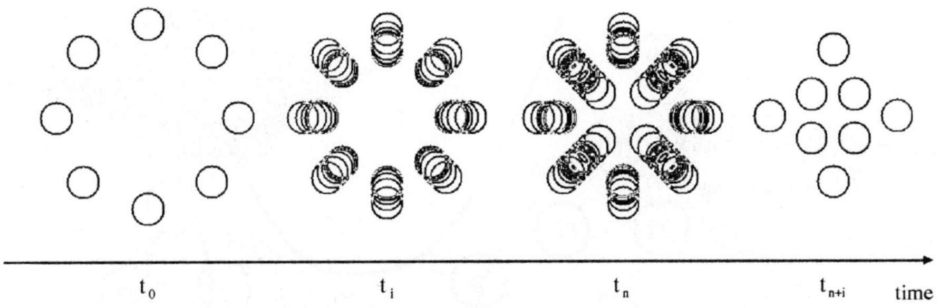

Fig. 11. Collapse of a circular structure of eight cells into a diamond form. (t_0 initial state; t_i gradual collapse; t_n sudden collapse of every second cell into the attraction valley; t_{n+i} the final stabile state.)

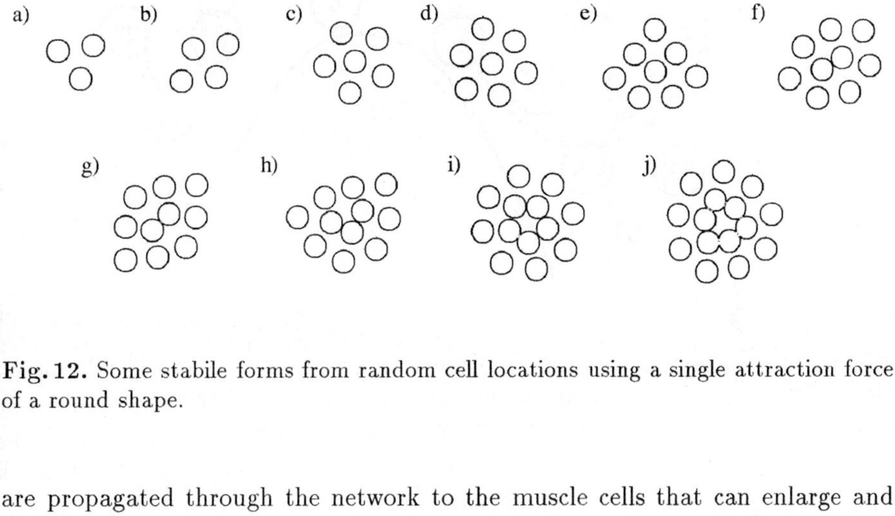

Fig. 12. Some stabile forms from random cell locations using a single attraction force of a round shape.

are propagated through the network to the muscle cells that can enlarge and shrink causing the whole structure to move in the environment. Our previous work demonstrate that already through simple network structures quite complex behavior could be achieved.

7 Conclusion

This paper described briefly how the creation of forms can be simulated to consist of cooperative influences of environment and genetic information.

The idea of using simulation to describe biological phenomena is not new. However, the described method of using synthesis to create forms that are able to behave in the simulated world is characteristic of artificial life implementations.

Direct applications of the field still lie in the future. However, the main contribution of the field is on widening the perspective of simulation and exploring new methods for the simulation of adaptive behavior.

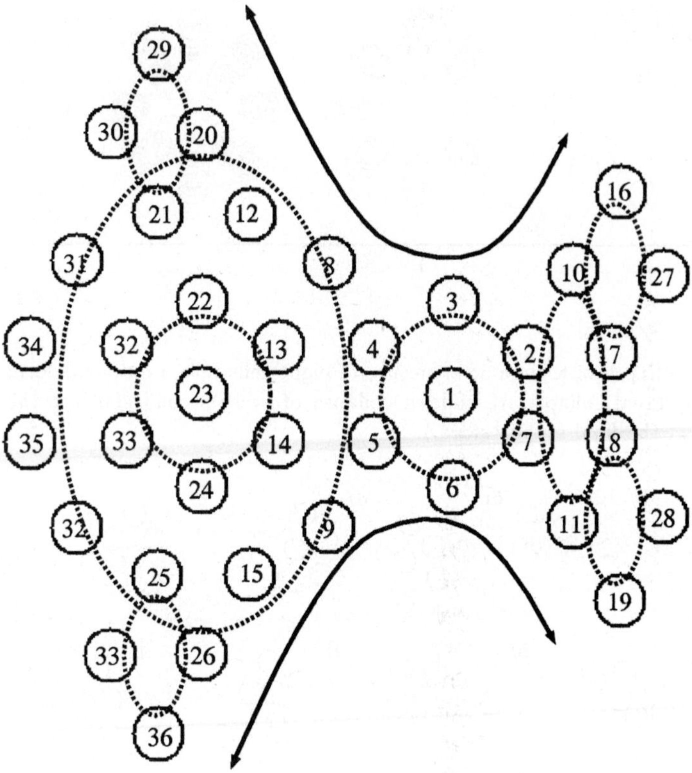

Fig. 13. A formed structure. The stability creating adhesion groups are shown by ellipses. The muscle extension and flexion (thought to be created by the expansion and contraction of single muscle cells) shown by two directional arrows.

Acknowledgment

Thanks are due to Hugo de Garis for comments during preparation of this manuscript.

References

1. Rodney Brooks and Pattie Maes, editors. *Artificial Life IV*. The MIT Press, 1994.
2. Peter Cariani. Emergence and artificial life. In Christopher G. Langton, Charles Taylor, J. Doyne Farmer, and Steen Rasmussen, editors, *Artificial Life II*, pages 775–798. Addison-Wesley Publishing Company, 1992.
3. Kurt Fleischer and Alan H. Barr. A simulation testbed for the study of multicellular development: The multiple mechanisms of morphogenesis. In Christopher G. Langton, editor, *Artificial Life III*. Addison-Wesley, 1994.
4. Gilbert Gottlieb. *Individual Development and Evolution*. Oxford University Press, 1992.

5. Kazuo Hosokawa, Isao Shimoyama, and Hirofumi Miura. Dynamics of self-assembling systems — Analogy with chemical kinetics —. In Brooks and Maes [1].
6. Stuart A. Kauffman. *The Origins of Order - Self-organization and selection in evolution.* Oxford University Press, New York, Oxford, 1993.
7. H. R. Maturana and F. J. Varela. *Autopoiesis and Cognition: The Realization of the Living.* Reidel, 1980.
8. Jean Piaget. *Adaptation and Intelligence — Organic Selection and Phenocopy.* The University of Chicago Press, 1974. (reprint 1980).
9. Premyslaw Prusinkiewicz and Aristid Lindenmayer. *The Algorithmic Beauty of Plants.* Springer-Verlag, 1990.
10. David S. Thaler. The evolution of genetic intelligence. *Science*, 264:224–225, 1994.
11. Jari Vaario. *An Emergent Modeling Method for Artificial Neural Networks.* PhD thesis, The University of Tokyo, 1993.
12. Jari Vaario. Artificial life as constructivist AI. *Journal of SICE (Society of Instrument and Control Engineers)*, 33(1):65—71, 1994.
13. Jari Vaario. From evolutionary computation to computational evolution. *Informatica*, 18(4):417—434, 1994.
14. Jari Vaario. Modeling adaptative self-organization. In Brooks and Maes [1].

4. Artificial Worlds

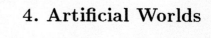

Learning in the Active Mode

Domenico Parisi and Federico Cecconi

Institute of Psychology, National Research Council,
Viale Marx 15, 00137 Rome, Italy

Abstract. The paper distinguishes between two different modes of learning by neural networks. Traditional networks learn in the passive mode by incorporating in their internal structure the regularities present in the input and teaching input they passively receive from outside. Networks that live in a physical environment (ecological networks) can learn in the active mode by acting on the environment and learning to predict what changes in the environment or in their relation to the environment are caused by their actions. Being able to predict the consequences of one's own actions is useful when one wants to cause desired consequences with these actions. The paper contrasts learning to predict the consequences of one's actions with learning to predict environmental changes that are independent from the network's actions. It then discusses how perceptually 'hidden' properties of the environment such as the weight of objects are better learned in the active rather than in the passive mode and how learning in the active mode can be particularly useful in a social environment and in learning by imitating others. Learning in the active mode appears to be a crucial component of the human adaptive pattern and is tightly linked to another component of this pattern, i.e., the human tendency to modify the external environment rather than adapt to the environment as it is.

1 Causing changes in the environment or in one's relation to the environment

Imagine a simulated organism that lives in a simulated physical environment. The behavior of the organism is controlled by a neural network. The network's input units encode information about the current state of the environment in the vicinity of the organism and the organism responds with motor actions encoded in the network's output units.

When the organism executes a motor action, the action is likely to cause changes that can be roughly classified in three categories. The organism's motor action can cause changes (a) inside the organism's body, (b) in the relation of the organism's body to the external environment, (c) in the external environment itself. In many cases more than one type of changes will be caused by the same action. We will not consider here the first category of changes, i.e., those that are restricted to the organism's body, but will dedicate our attention to the changes that concern the relation of the organism's body to the external environment without changing the environment itself and the changes that do affect the external environment.

An example of the former type of changes is when the organism rotates its body or some part of its body (say, its head). The environment remains unchanged but the relation of the organism to the environment changes. An example of an organism changing the environment itself is when the organism displaces an object by pushing the object away using one of its forelimbs (say, its arm and hand). In both cases the changes caused by the organism's motor actions can have consequences for the sensory input the organism is receiving from the environment. Consider the case in which the organism rotates its body or head or eyes. If at time T1 the organism was perceiving an object as located 90 degrees on its left, at time T2, after the motor action of turning, say, 90 degrees to the left has been executed by the organism, the object will be perceived by the organism as lying in front of itself. Or consider the case in which the organism launches or pushes an object away. In this case, too, the action has consequences for the sensory input to the organism. Prior to the action of pushing the object the organism was perceiving the object at a certain distance from itself. After pushing the object with the hand, the object will be perceived by the organism as lying at a greater distance.

The motor actions of the organism may not involve gross displacements of the body of the organism or of body parts such as the head or limbs. For example, by moving its phonoarticulatory motor organs an organism can cause changes in the external environment under the form of sound waves. These changes in the environment caused by the organism's motor actions are perceived by the organism itself as auditory stimuli. Hence, the motor actions of the organism have consequences for the organism's sensory input.

We conclude that neural networks that live in a (simulated) physical environment (ecological neural networks; cf. Parisi, Cecconi, and Nolfi, 1990) have a tendency to influence their own input with their motor output. They cause changes in the external environment or, at least, in the relation of their body to the external environment and, by so doing, they at least partially determine their future inputs.

2 Predicting the changes in sensory input resulting from one's own actions

As already noted, some of the changes caused by an organism either in the external environment or in its relation to the external environment can have consequences for the sensory input the organism is receiving from the environment. Now, imagine an ecological network that has the following capacity: the network can predict the changes in sensory input that will result from its own motor actions. The network is given a description of a planned but not actually (physically) executed motor action as input, and the it is able to generate as output a description of the sensory input that will result when the motor action is physically executed. (Jordan and Rumelhart, 1992, discuss this problem under the name of learning a forward model but their discussion is restricted to learning while we discuss prediction learning in an evolutionary framework.)

This prediction ability can be demonstrated in both cases we have distinguished above. If the organism's motor action changes the relation between the organism and the environment without changing the environment, the network may be able to predict the new input that will result from its changed relation to the environment. For example, if an object is perceived as being located 90 degrees to the left and the network is given a description of a planned action of turning 90 degrees to the left, the network may be able to generate a correct description (prediction) of where the object will be perceived to be located after the action has been physically executed: the object will be in front of the organism. Similarly in the case of an action that changes some aspect of the external environment. If an object is perceived as lying, say, 50 centimeters from the organism and the organism is given a description of a planned action of pushing the object away with a given force, the organism may be able to generate a prediction that the object will be located, say, 80 centimeters from itself after the action of pushing the object with that force has been executed. Or, in the case of an organism producing some movement of its phono-articulatory organs, the organism can predict the particular sound it will hear after actually executing a particular phono-articulatory movement.

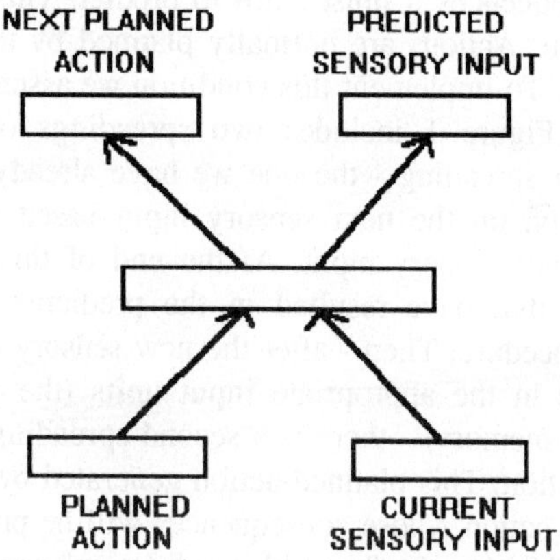

Fig. 1. Neural network that learns to predict the sensory consequences of planned actions

Neural networks can learn to predict the sensory consequences of motor actions using the backpropagation procedure (cf. Jordan and Rumelhart, 1992). The network's input units encode two types of information: (a) a description of a planned motor action, and (b) a description of the current sensory input. Activation spreads through the network and it reaches the output units. The output units encode a description (prediction) of the sensory input that will result when the planned motor action will be physically executed. Then, the planned action is actually executed by the organism (by the algorithm controlling the whole simulation) and the body (or body part) of the organism is physically moved. The algorithm computes the new sensory input and it uses this sensory input as teaching input to change the network's connection weights. At the beginning of the simulation the predictions generated by the network are likely to be wrong (large discrepancy/error between predicted sensory input and actual sensory input). However, after a certain number of learning cycles the discrepancy/error will approach zero. The network has learned to predict what sensory input will result as a consequence of physically executing planned actions. An architecture for a network which is to learn to predict the results of motor actions is shown in Figure 1.

The architecture of Figure 1 is appropriate for the case in which the next sensory input depends on both the planned motor action and the current sensory input. For example, if I am about to turn 90 degrees to the left, the next sensory input from an object near me will depend both on this planned action and on the current sensory input from the object. In other cases the next sensory input appears to depend only on the action I am about to do. For example, the auditory stimulus I will hear in a moment depends almost exclusively on the phono-articulatory action I am planning to do. Hence, in these cases the network architecture would not include the "current sensory input" portion of the input.

One important variable in learning to predict the sensory input resulting from motor actions is who is deciding what the next action is going to be. There can be various possibilities. The network itself can generate the planned motor actions whose sensory consequences it must learn to predict. This is the most ecologically appropriate condition. Actions are normally planned by the same organism that is executing the action. To implement this condition we assume that each activity cycle of the network of Figure 1 includes two spreadings of activation through the network. In the first spreading - the one we have already described - the network generates a prediction on the next sensory input based on the currently planned action and the current sensory input. At the end of this spreading the network's connection weights that have resulted in the prediction are corrected using the backpropagation procedure. Then - after the new sensory input has been substituted to the previous one in the appropriate input units (the preceding planned action remains as a sort of memory) - there is a second spreading of activation that results in a new planned action. This planned action generated by the network itself is used as the new planned action whose consequences will be predicted by the network in the next cycle. (Notice that the "next planned action" output is ignored in the first spreading and the "prediction" output is ignored in the second spreading.)

We can compare this condition in which the network generates its own actions with two other conditions in which the motor actions are not decided by the network. In one condition the planned motor actions are randomly generated. In the other condition the planned motor actions are generated by another network.

The networks live in an environment that contains food elements. At the beginning of learning the networks already know how to approach the food elements. This ability is the result of applying a genetic algorithm to a population of networks. The networks used in the experiment on learning to predict are evolved members of the population. They tend to respond to sensory information concerning the position of the nearest food element by reducing the distance from the food element. When a network happens to step on a food element, the food element disappears (it is eaten). The details of the simulations are the following.

Each organism lives alone in an environment which is a bidimensional grid of 10x10 cells. The organism occupies one cell and it has a facing direction. Its motor repertoire includes four possible actions: moving one cell ahead, turning 90 degrees to the right or to the left, do nothing. The four motor actions are encoded as binary patterns (11, 10, 01, 00) in the planned action input units of the network of Figure 1. The motor output units' continuous activation values are thresholded to either 1 or 0.

The environment contains 10 randomly distributed food elements with each food element occupying a single cell. A new set of 10 elements periodically replaces the preceding set to compensate for the food eaten. At any given time the organism is informed by its senses about the location of the nearest food element. The position of the food element is encoded in the two sensory input units as (a) angle of the food element with respect to the organism's facing direction, and (b) distance of the food element from the organism. The same encoding is also used for the two prediction output units.

We have trained a set of 50 different networks in each of three conditions. In one condition a network generates its own actions and it learns to predict the sensory changes resulting from these actions. In a second condition the network is trained using a randomly generated file of actions. In a third condition the network is trained using a file of actions generated by another network. (Notice that while in the first and third condition the current sensory input results from the previous action, in the second condition it is randomly generated together with the action.)

The results of learning to predict in the three conditions are shown in Figure 2.

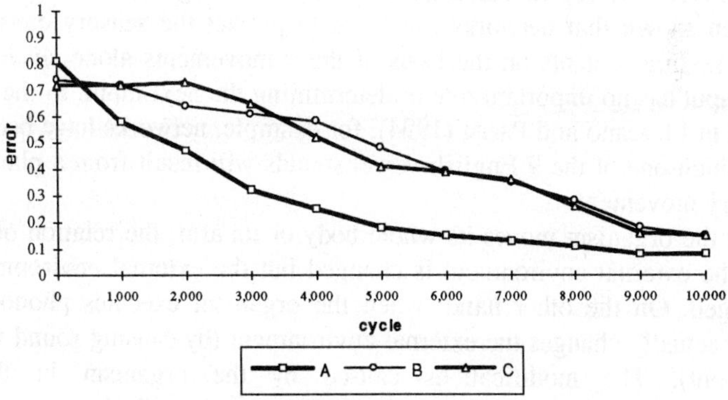

Fig. 2. Average prediction error across 10,000 learning cycles for 50 networks that (A) generate their own actions, (B) use randomly generated actions, (C) use the actions generated by another network

Figure 2 contains two results. It first establishes the fact that networks can learn to predict the sensory consequences of planned motor actions that cause changes in the relation between the network and the environment. Secondly, Figure 2 shows that learning to predict the sensory consequences of motor actions is facilitated if the motor actions are generated by the same neural network that is learning to predict. Ecological networks self-select their input, i.e. they restrict the sensory input to which they are exposed to some preferred sub-set of all possible inputs (Nolfi and Parisi, 1993). This may explain why networks that are trained using their own actions (a sub-set of all possible inputs) learn better than networks that are trained using randomly generated actions (therefore, sampling all the possible inputs). However, networks that are trained with their own actions learn better than networks that are trained using the actions generated by another network. In both cases, the networks must learn to predict the consequences of a sub-set of all possible inputs but, apparently, it is easier for a network to learn to predict the consequences of planned motor actions if the network that learns to predict is the same network that decides the motor actions.

The ability of networks to learn to predict the sensory consequences of their own actions has also been demonstrated with other types of tasks. A neural network can learn to predict the position of the endpoint (hand) of its 2-segment arm on the basis of sensory information about the current position of the hand and a planned movement of the arm. Or one and the same network can learn to predict at the same time (a) the position of an object with respect to its own body on the basis of sensory information about the the current position of the object and a planned movement of its whole body, and (b) the position of its hand on the basis of sensory information about the current position of the hand and a planned movement of its arm (Cecconi and Parisi, 1991).

In all these cases predictions are generated on the basis of two inputs: (a) a planned movement, (b) the current sensory input (cf. Figure 1). In other simulations it has been shown that networks can learn to predict the sensory consequences of their planned movements on the basis of these movements alone since the current sensory input has no important role in determining the next input. In the simulations described in Floreano and Parisi (1994), for example, networks have been trained to predict which one of the 9 English vowel sounds will result from a planned phono-articulatory movement.

When the organism moves its whole body or its arm, the relation of its body or hand to the external environment is changed but the external environment itself is not changed. On the other hand, when the organism executes phono-articulatory actions it actually changes the external environment (by causing sound waves in the environment). The modifications caused by the organism in the external environment are even more obvious when the organism displaces an object in the environment with its actions, e.g., by pushing the object away with its hand. In other simulations (Cecconi and Parisi, unpublished data) we have shown that neural networks can learn to predict how much an object will be removed from the organism after a planned pushing-away movement with a given force is physically executed by the organism. The neural network is given a description of (a) the current position and characteristics of the object and (b) the force of the pushing-away movement as input, and it learns to generate a prediction of the new position of the object as output.

3 Predicting the consequences of one's own actions and causing useful consequences with these actions

In the preceding Section we have shown that neural networks can learn to predict the sensory consequences of planned motor actions. We believe that real organisms, especially the more advanced ones, actually possess a capacity to predict the consequences of their own actions and that they may learn this capacity using a procedure not very different from the backpropagation procedure. The backpropagation procedure is often criticized because it unrealistically assumes that organisms have "teachers" constantly available to them that know the correct responses to inputs and are capable and willing to describe these correct responses to them. This assumption is implausible in most cases in which the backpropragation procedure is used to teach neural networks various abilities. In the case of prediction learning in ecological neural networks, however, the assumption seems to be perfectly realistic since it is the environment itself and the organism/environment interactions that automatically provide the organism with the appropriate teaching input for backpropagation learning. When a planned motor action is physically executed by the organism, the resulting sensory input to the organism is the teaching input for the organism's prediction learning. Therefore, the backpropagation procedure is not behaviorally and ecologically unrealistic in the case of prediction learning (although, of course, it may still be implausible neurophysiologically).

Why should organisms possess prediction abilities? What is the functional significance, the usefulness of possessing these abilities? By itself, being able to predict the consequences of one's own actions should not increase the reproductive chances of an individual. So, why should prediction abilities emerge in organisms?

To answer these questions we should refer to what is the fundamental problem faced by organisms during their life in the environment. Organisms must be able to generate the appropriate motor actions in response to the various sensory inputs arriving from the environment. They cannot just randomly select an action in response to a particular input. An organism that would do so would not survive for long and would not have offspring. Organisms must know what particular action is appropriate for each particular sensory input. We saw that the motor output (action) of an ecological neural network tends to cause changes either in the organism's relation to the environment or in the environment itself. We can define an "appropriate" reponse to a particular sensory input as a motor action that causes some "desired" change in the organism's relation to the environment or in the environment itself. "Desired" changes in turn are defined as those changes that allow organisms causing them to have more survival and reproductive chances.

Our hypothesis is that possessing an ability to predict the sensory consequences of one's own actions is beneficial with respect to the capacity to respond to sensory input with the appropriate actions, i.e. with actions that cause desired changes - and this is why organisms possess prediction abilities. The sensory consequences of one's own actions tend to be correlated with the physical changes caused by the organism's actions. Hence, organisms which are able to predict the sensory consequences of their actions are at the same time able to predict the changes caused by these actions.

Our hypothesis is that organisms that can predict the changes caused by their actions will be more able to cause desired changes with these actions. And organisms that have more sophisticated prediction abilities will tend to be organisms that have a more sophisticated repertoire of behaviors in comparison to organisms with limited or nonexistent prediction abilities.

That knowing (being able to predict) what changes are caused by what actions can be useful in the moment the organism must select the action that causes a particular desired change appears to be intuitively plausible. In various simulations we have shown that this is actually the case.

In the simulations described in the preceding Section in which neural networks learned to predict the changing position of food elements as a consequence of their actions, the individual networks started their prediction learning by already incorporating in their connection weights the ability to approach food. In other words, they were already able to cause desired changes with their actions. The desired changes were changes that reduced the distance between the network and the food elements. This is in agreement with our definition of "desired changes" as changes that increase the reproductive chances of the organism causing them. In fact, these networks were members of evolving populations of neural networks in which each individual network reproduces as a function of its fitness. The fitness of each individual is the number of food elements eaten during the individual's life. (Remember that our networks 'eat' the food element when they step on a food cell.) The individuals with most fitness reproduce (agamically) by generating copies of their matrix of connection weights (with random mutations to some of the weights). The initial generation is composed of neural networks with random connection weights. The ability to eat of these networks is almost nonexistent. However, after a certain number of generations selective reproduction and mutations result in networks possessing weight matrices that incorporate a high level of eating ability. An individual belonging to one of the later generations is clearly attracted by food and it will eat most of the food present in the environment.

These networks evolve without learning to predict during their life. They learn to predict at the end of evolution, i.e., after they have already evolved an ability to eat. The succession is: first, evolve an ability to eat (more generally, to cause desired changes); second, learn to predict the changes caused by your own actions. But neural networks can evolve at the population level an ability to cause desired changes and, at the same time, they can learn at the individual level to predict the changes caused by their actions. During its life, while the network is tested to determine its ability to eat and, therefore, its reproductive chances (fitness), the network can learn to predict the changes caused by its actions. The network architecture is the same as that of Figure 1.

If learning to predict the changes caused by one's own actions benefits an individual's ability to cause desired changes with these actions, we should expect individual networks that learn to predict during their life to increase their eating ability in successive epochs of life. This is what is observed in simulations in which a population of neural networks reproduces on the basis of eating ability and at the same time the individual networks learn to predict how food position changes with their actions during their life (Nolfi, Elman, and Parisi, 1994). Figure 3 shows the number of food elements eaten in successive epochs of life by networks of successive generations in a population of networks that learn to predict during life.

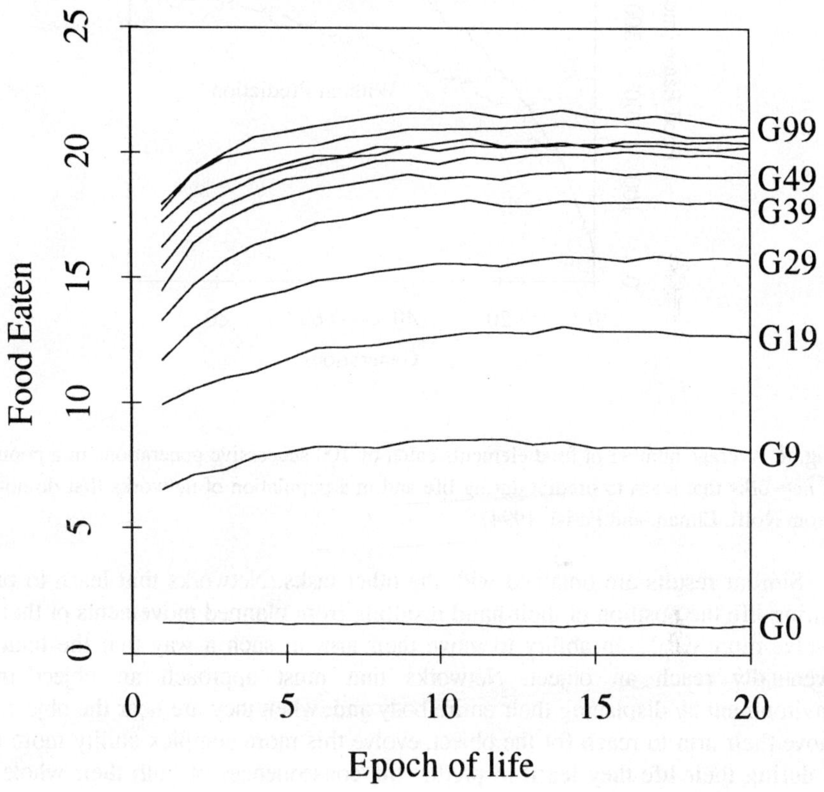

Fig. 3. Number of food elements eaten in successive epochs of life by networks of successive generations in a population of networks that learn to predict during life (from Nolfi, Elman, and Parisi, 1994)

At the same time the beneficial effects of learning to predict during life tend to emerge at the population level also. Figure 4 shows the evolution of the average eating ability across 100 generations in a population of networks that learn during life and in a population of networks that do not learn.

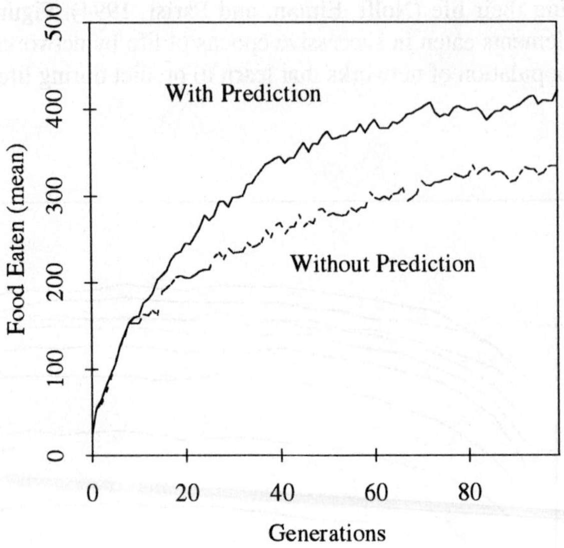

Fig. 4. Average number of food elements eaten by 100 successive generations in a population of networks that learn to predict during life and in a population of networks that do not learn (from Nolfi, Elman, and Parisi, 1994)

Similar results are obtained with the other tasks. Networks that learn to predict during life the position of their hand resulting from planned movements of their arm evolve more easily an ability to move their arm in such a way that the hand will eventually reach an object. Networks that must approach an object in the environment by displacing their entire body and, when they are near the object, must move their arm to reach for the object, evolve this more complex ability more easily if during their life they learn to predict the consequences of both their whole-body movements and their arm movements (Cecconi and Parisi, 1991). Finally, networks that must launch an object to a desired distance by pushing the object away from them with a given force evolve this ability more easily if they learn during life to predict where the object will end up when launched with actions of varying forces (Cecconi and Parisi, unpublished data).

We conclude that (some) organisms possess (learn, develop) an ability to predict the consequences of their own actions because possessing this ability helps them cause desired consequences with their actions. Individuals that tend to learn to predict the sensory consequences of their actions have more reproductive chances than individuals that do not have this tendency.

If the ability to predict the consequences of one's actions is useful from the point of view of producing desired consequences with these actions, it would seem to be more appropriate to first acquire the ability to predict the consequences of one's actions and then to acquire an ability to produce desired consequences with these actions. In a new set of simulations we have reversed the succession of events and we had a population of networks with random initial weights first learn to predict the changes in food position resulting from their actions and then evolve, in successive generations, the ability to eat. We found that populations of networks that evolve an ability to eat starting from an initial generation that has learned to predict how food position changes as a consequence of actions have a better evolutionary increase in eating ability than populations that do not initially learn to predict (cf. Figure 5). (Notice that in these simulations, unlike the previous simulations, the networks do not learn to predict during life. It is only the first generation that, prior to evolution, learns to predict.)

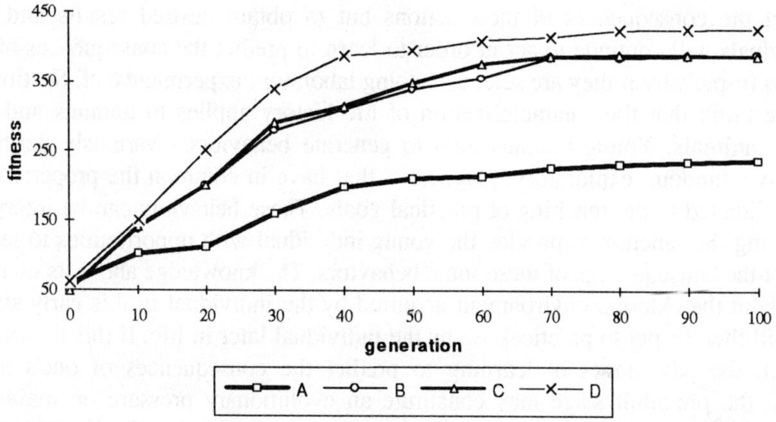

Fig. 5. Average number of food elements eaten by 100 successive generations of a population whose initial generation has learned to predict prior to evolution (D) and of a population which never learns to predict (A). Two additional evolutionary curves are shown: of a population that learns on the basis of a randomly generated teaching input (B) and of a population that learns an auto-association task (C)

In Figure 5 two additional evolutionary curves of the food eating ability are shown. Curve (B) refers to a population that learns prior to evolution on the basis of a randomly generated teaching input (random learning). Curve (C) refers to a population that learns an auto-association task. As we already found for random learning during life (Parisi, Nolfi, and Cecconi, 1992), apparently any type of learning (changes in weights) has a beneficial effect on the evolution of the eating ability but learning to predict the consequences of actions has a stronger effect.

In most organisms both the ability to predict the changes resulting from one's actions and the ability to cause desired changes with these actions are not genetically inherited and already present at birth (congenital) but they develop during life on the basis of an interaction of genetically inherited information and environmental information. However, if being able to predict the consequences of one's actions is useful in order to cause desired changes with these actions, we should expect the lifetime of organisms to be divided up into two successive periods. In the initial period the individual learns to predict the changes resulting from its own actions and in the subsequent period it learns to cause desired changes with these actions. In other words, when it is young the individual will be particularly dedicated to learning to predict the changes of its own actions and it will be less preoccupied with obtaining desired results with its actions. (Of course, this is only possible if other individuals will take care of the individual during this initial stage.) As the individual grows up, it will be more and more concerned with obtaining desired results with its actions (including the taking care of younger individuals) and less with learning to predict the consequences of its actions. Of course, the separation is unlikely to be very sharp. Young individuals will also sometimes act not to learn to predict the consequences of their actions but to obtain desired results and adult individuals will continue to act in order to learn to predict the consequences of their actions (especially if they are scientists doing laboratory experiments; cf. Section 5).

We think that this characterization of life history applies to humans and other higher animals. Young humans tend to generate behaviors - variously defined as reflexive, random, exploratory, play, etc. - that have in common the property of not being directed to the reaching of practical goals. These behaviors can be interpreted as having the function to provide the young individual with opportunities to learn to predict the consequences of these same behaviors. The knowledge about its own body and about the external environment acquired by the individual in this early stage of life will then be put to practical use by the individual later in life. If this reasoning is correct, the advantages of learning to predict the consequences of one's actions during the pre-adult stage may constitute an evolutionary pressure on maintaning such a stage as part of life history. (On the need to clearly identify the evolutionary advantages of an immaturity stage to compensate for its costs, cf. Stearns, 1992; Cecconi and Parisi, 1995; Cecconi, Menczer, and Belew, 1995.)

There is some evidence that early behavior by human infants can be interpreted as generated in order to learn to predict its sensory consequences. For example, newborn babies that are experimentally allowed to see only one of their two arms move the arm they can see more than the arm they cannot see (van der Meer, van der Weel, and Lee, 1995). If we assume that they are moving their arms in order to learn to predict the visual consequences of their arm movements, it makes sense for them to move a particular arm if they can verify their predictions about the visual consequences of their movements but it is useless to move an arm that is not seen. (Arm movements have also somatosensory consequences and predictions about these consequences can be verified even in unseen arms. However, predictions about the somatosensory consequences of arm movements may have already been learned earlier in life - the subjects of the van der Meer et al's experiment were 18-day-old babied - or babies may be born with a congenital ability to make these predictions.)

Another example is the spontaneous phono-articulatory movements of babies during their first semester of life (babbling). It can be hypothesized that these phono-articulatory movements are generated with the function to learn to predict the sounds resulting from them (cf. Section 2). In the second semester of life the network that has learned to predict the sounds resulting from one's own phono-articulatory movements is incorporated in a larger network that learns to imitate the sounds heard in the child's environment (produced by the phono-articulatory movements of adults) on the basis of the previously acquired ability to predict the sounds resulting from one's own phono-articulatory movements (cf. Floreano and Parisi, 1994; Parisi, 1995).

It is also possible to hypothesize a general developmental sequence in learning to predict the changes resulting from one's own actions. The sequence follows a direction from the organism to the external environment. The organism first learns to predict the changes inside its own body resulting from its own actions, with no reference to the external environment. (We have ignored these changes. Cf. Section 1.) Then the organism learns to predict the changes in the relation of its own body to the external environment. And, finally, the organism learns to predict the changes caused by its own actions in the external environment. Since, as we will argue in the next Section, learning to predict the changes caused by one's own actions is crucial for acquiring knowledge about reality, organisms would first acquire knowledge about their own body and then they would gradually acquire knowledge more detached from their own body and more concerned with the external, independent environment. Furthermore, simpler organisms may be restricted to predictions about changes in their own body or in the relation of their body to the external environment and, therefore, to knowledge centered on their own body, while only humans appear able to acquire an enormous quantity of organized knowledge about the external environment.

5 Acquiring knowledge about the environment by observing how the environment reacts to our actions

All organisms must somehow 'know' their environment or, more generally, reality (which includes their own body besides the external environement) in order to survive and leave descendants. Their knowledge of the environment is demonstrated by their ability to generate appropriate responses to stimuli. Since appropriate responses are motor actions that cause desired changes in the environment (cf. our definitions in Section 3 above), no organism would be able to generate appropriate reponses unless the organism possesses some knowledge of its environment.

A knowledge of the environment can be acquired passively, i.e. by extracting the regularities present in the inputs from the environment or, as we will see in the next Section, by predicting future inputs based on past inputs. This is what classical neural network models help us explain. The environment sends its stimuli to the organism according to its own structure and the organism tries to recover the structure of the environment from the stimuli. We think there are intrinsic limitations in the type (amount, complexity, sophistication, etc.) of knowledge that can be acquired in this purely passive way, although many simple organisms can be actually restricted to a passive mode of knowledge acquisition.

There is another route to knowledge acquisition, however. This is the active mode. In the active mode knowledge is acquired more through the actions generated by the organism than through the stimuli arriving to the organism. In the passive mode the environment sends stimuli to the organism and the organism reacts to these stimuli. In the active mode it is the organism that 'stimulates' the environment with its actions and the environment reacts to these 'stimulations' on the part of the organism. By observing how the environment reacts to its stimulations the organism learns about the structure of the environment. (According to Elizabeth Bates (personal communication), Peter Green interpreted Piaget as suggesting exactly this inversion of the classical S-R paradigm.)

In both modes the problem for the organism is to learn about the structure of the environment. However, in the passive mode the organism is restricted to learning about the environment by analyzing the stimuli the environment 'decides' to send to the organism. In the active mode, the organism itself decides which stimuli the environment will send to the organism. This has two consequences. First, in the active mode more inputs can be gathered from the environment, especially inputs that would not be accessible at all if one were restricted to the passive mode. Second, since in the active mode the arrival of inputs from the environment is under the control of the organism, the process of knowledge acquisition ceases to be purely inductive to become in some sense hypothetical-deductive. (Notice that making a prediction and then checking if the prediction is correct is analogous to making an hypothesis, deducing some particular consequence from the hypothesis, and then verifying if the consequence is realized. See below.)

Learning about the environment in the active mode appears to be a much more powerful method for acquiring knowledge than learning in the passive mode. The obvious evidence in favor of this statement is constituted by humans. Humans know more of the environment than any other species and humans are specialized for learning about the environment in the active mode. They have developed a whole set of powerful methods for interrogating the environment instead than just listening to what the environment has to say spontaneously. First, to a much greater extent than any other species humans with their behavior change the external environment and not only their own body or their body's relation to the external environment. Hence, they can collect information on the external environment and not only on their own body. Second, humans not only 'stimulate' the external environment with their bare hands, so to speak, but they construct tools for acting on the environment with the purpose to collect information about the environment. Third, humans have developed the experimental scientific method which is a systematization of the active method of collecting information about the environment. Fourth, humans have the tendency to turn their communicative behavior into a tool for collecting information about the environment. Humans do not wait passively that some conspecific communicates to them some useful information, as other animals do, but they ask for information.

This species-specific adaptation of humans has both genetic and socio-cultural causes. Genetically, humans appear to have both morphological and behavioral specializations that predispose them to learn in the active mode. They have complete bipedalism that allows them to use their forelimbs (hands) exclusively for manipulation. They have a prolonged period in their life history which is prevalently dedicated to learning to predict the changes caused by their actions. Etc. But socio-cultural causes are equally important. The most important product of socio-cultural evolution from the point of view of learning in the active mode is the experimental method of science. In a scientific experiment what the experimenter does is to manipulate variables to observe the effects of these manipulations. But to manipulate variables is to execute actions that are intended to cause predicted changes in some phenomenon or object in order to verify the goodness of these predictions. In the laboratory the learning procedure is backpropagation. The discrepancy between prediction and observation is used to change our ideas about the phenomena under study and, therefore, ultimately the connection weights in our brains. The special powerfulness of the experimental method as an active mode of learning is due to the fact that phenomena are observed in the laboratory under controlled conditions. The controlled conditions of the laboratory mean that the experimenter is trying to rule out other variables (other actions) as causes of the predicted changes in order to refine and increase the reliability of his/her predictions and, consequently, of his/her knowledge of the external environment. (Of course, the experimental method is so powerful for other reasons also: observations are precise and quantitative, instruments and not only 'bare hands' are used to manipulate and observe reality, the knowledge acquired is socially accumulated and checked, etc.)

The enormous advances in our knowledge of the external environment that have been made possible by the introduction of the experimental method in the XVI century in contrast to the much slower progress of science prior to the XVI century testify to the greater knowledge-acquiring power of learning about the external environment through acting on the environment, predicting the consequences of our actions, and verifying the goodness of our predictions, as compared with learning about the external environment by just passively noting the regularities that the environmental input presents to us. Furthermore, the fact that the scientific knowledge about the external environment acquired in the laboratory is being increasingly translated into an increased ability to cause desired changes in the environment through technology testifies to the ultimate functional significance of the human tendency to learn to predict the changes caused by their own actions in order to increase their capacity to cause desired changes (cf. Section 4).

More generally, it is the uniquely human attitude to learn about the environment in the active mode that explains the uniquely human adaptation of changing the external environment so radically that humans can be said to live not in a natural environment but in a human-made environment. Notice that we are not claiming that humans first developed an attitude to learn about the environment in the active mode and then, because of the much deeper, more extended, and more detailed knowledge about the environment that this attitude made it possible to collect, they started using this knowledge to change the external environment in ways that suited their goals. Even if the actual course in human evolution may be difficult to decipher, it is more likely that the two developments co-evolved and reinforced each other during the course of human biological and cultural evolution. Efficiently collecting information about the environment in the active mode made it possible to change the environment in useful ways and, at the same time, acting on the environment to change it to suit our goals represented a context in which it was possible to collect much information about the environment in the active mode.

6 Predicting the changes caused by one's own actions vs predicting intrinsic changes

Let us come back to prediction in neural networks. What a predicting network predicts are changes in its input. These changes, however, can be the result of the network's own actions (the output of the network) or they can be the consequences of intrinsic changes in the environment that are independent of the actions of the network. An ability to predict future states of the environment can be useful to prepare to these future states and this ability is present in some form in all organisms. However, the knowledge of the environment that can be obtained if one is restricted to predicting intrinsic environmental changes appears to be limited and it cannot lead to an adaptation that includes changing the external environment as a crucial component. On the other hand, if the ability to predict intrinsic environmental changes is found together with the ability to predict the changes resulting from one's own actions, as in humans, this arrangement may produce a rich and powerful knowledge of the environment and may enable organisms to change in radical ways the environment to suit their goals.

The ability to predict intrinsic environmental changes reduces to the ability to predict the next environmental input given the current input and perhaps some memory of the preceding inputs. (Remember that, in contrast to this, predicting the changes which are caused by one's own actions is predicting the next input given the current input and a planned action. In some cases, as in predicting the sounds resulting from planned phono-articulatory movements, the prediction may be based on the planned action only, ignoring the external environment.) Elman's networks that learn to predict the next word in a sentence given the current word and some 'trace' of the preceding context as input, are an example of this type of prediction ability (Elman, 1990; 1991a). One can imagine that the external environment of the network changes for independent reasons word after word as a spoken sentence is heard by the network, and the network learns to anticipate each time the next word.

The ability to predict the next word in a sentence increases with the length of previous context that the network is able to store and utilize at the time of prediction. In Elman's simulations the previous context is represented by some 'context' units which replicate the activation pattern of the network's hidden units. This stored activation pattern is input to the network's hidden units in the next cycle together with the sensory input constituted by the current word. The storage mechanism, therefore, maintains a cumulative trace of the past. Elman (1991b) has shown that how distant the memory trace reaches in the past is a crucial factor in allowing the network to learn to predict the next word in syntactically complex sentences with extended syntactic dependencies.

This dependence of the ability to predict the future on the length of the record of the past is quite understandable considered that the ability to predict intrinsic environmental changes exploits the temporal regularities present in the environment and is based on the capacity of the network to discover these regularities during learning. (There is a whole literature on networks that learn to capture the temporal regularities in the input. Cf, e.g., Weigand, Rumelhart, and Huberman, 1991.) Since the interesting temporal regularities tend to extend beyond two successive environmental events, in order to be able to predict future events a network must refer not just to the preceding environmental event but to an entire sequence of previous events.

The situation with respect to the ability to predict the consequences of one's own actions is in principle different. A particular prediction is based on the currently planned action and the current sensory input but it normally does not need to go back further in the past. Hence, networks that must acquire an ability to predict the consequences of their own actions do not normally need context units or other special memory mechanisms to store the past. All they need is a description of the currently planned action and of the current sensory input. On this basis, they can generate a prediction on what the next sensory input will be. Of course, as already suggested, an ability to predict what states of the environment are followed by what states (i.e. to predict intrinsic changes) can be very useful in association with an ability to predict the consequences of one's own actions. If I can predict that my action will have such and such consequences and, given the intrinsic nature of the environment, these consequences will be followed by such and such further consequences, and so on, this may be very useful information in choosing what action to actually generate. But a human-like adaptation that includes changing the environment as one of its major features necessarily includes a well-developed ability to predict the consequences of one's own actions and cannot emerge simply from an ability to predict the intrinsic changes that occur in the environment.

The distinction between predicting the intrinsic changes in the environment and predicting the changes resulting from one's own actions reflects the distinction between classical neural networks and ecological networks or networks viewed in an Artificial life perspective. Classical networks do not live in an environment. Their 'environment' is the set of input patterns and, possibly, teaching input patterns provided by the researcher to the network (McClelland, 1989). But a classical network has no influence on its own input and it cannot change the environment because there is no independent environment to be changed. Therefore, a classical network can only acquire an ability to predict future inputs given past and present inputs. This is knowledge acquisition in the passive mode.

On the other hand, since ecological networks live in an independent environment they can change the environment with their output (motor actions) and, by so doing, they can influence their subsequent input. It is therefore only by using ecological networks that one can study the acquisition of an ability to predict the consequences of the outputs generated by the network itself and the emergence of an adaptation that includes modifying the physical environment as one of its crucial components.

7 Some properties of the environment can be learned in the active mode much better than in the passive mode

An organism that learns in the active mode can acquire knowledge about the environment it is unlikely to acquire if the organism is restricted to the passive mode. Let us come back to our simulations on launching objects and predicting where an object will end up when launched with a given force. It is obvious that different objects will end up at different distances from the organism when launched with the same force. In other words, if an organism wants to predict where a given object will end up when it is launched with a given force, the organism must take the properties of the particular object into consideration. Now, some properties of objects can be directly perceived because the environment spontaneously exposes these properties to the sensory organs of the organism. Hence, even an organism that is restricted to the passive mode of knowledge acquisition can access these properties of objects. However, other properties of objects are hidden in them and cannot be discovered by an organism restricted to the passive mode. For example, the size of an object or the particular texture of the object's surface can be directly perceived. They can be directly accessed by the organism through the input the environment makes available to the organism. However, the weight of an object cannot be directly accessed through the input made available by the environment. To know the weight of an object an organism must act on the object. It must lift the object or push the object or launch the object. The weight of an object can be known by the organism by observing the sensory input resulting from these actions on the part of the organism. Weight is a property of objects that can only be known if the organism is not restricted to the passive mode but it learns about reality by acting on reality and by learning to predict the consequences of its actions.

Ecological neural networks can know the weight of objects. In our simulations (Cecconi and Parisi, unpublished data) the network is informed about two directly accessible properties of an object: (a) the size of the object (two input units encode the width and height of bidimensional quadrangular objects), and (b) the surface texture of the object (there are four possible textures corresponding to four different materials objects can be made of, with each material having a different specific weight). Furthermore, the network's input units encode the force with which the network is planning to launch the object (planned action). The network's output units encode a prediction of where (i.e., at what distance from the organism) the object will end up when the planned launching action is physically executed. As we have already said, neural networks can learn to make this sort of predictions and, furthermore, networks that acquire this prediction ability during their life tend to evolve at the population level a better capacity to launch objects at desired distances.

Networks that can predict where an object with given directly accessible properties (size and texture of the object) will end up when launched with a given force can be said to know the weight of the object - a property of the object that is not directly accessible to the organism's senses. They can also be said to know the specific weight of different materials objects can be made of - another property of the environment hidden to the senses. These hidden properties of the environment can in principle be known also by organisms that are restricted to the passive mode of knowledge acquisition - as demonstrated by the fact that even populations of networks that do not learn to predict where launched objects will end up can evolve an ability to launch objects to desired distances. Hidden properties such as the weight of objects and the specific weight of different materials can be implicitly deduced from accessible properties of objects such as their size and external texture. However, hidden properties can be more easily known by organisms if they learn about the environment in the active mode as shown by the results of our simulations according to which populations of networks that learn to predict the consequences of their launchings are better able to obtain desired results with these launchings.

There are many important properties of the environment that can be known more easily - or even at all - if the networks learn by acting on the environment and do not wait passively that the environment makes available the necessary information. Take the properties of objects such as the quantity of substance that remain invariant under reversible actions. An organism can know that the quantity of a substance remains constant if it observes that the perceptual properties of the substance change when it moves the substance from one container to another container with a different shape but the original perceptual properties are restored when it moves the substance back to the first container. Being able to predict the restoration of the original perceptual properties when the substance is put back in the original container means that the organism understand what is the quantity of the substance. Or take the numerosity of a set of objects. The perceptual properties of the set of objects can change (e.g., the objects can occupy more or less space) but the numerosity of the set remains constant provided the organism can put each object in one-to-one correspondence with a reference set. The organism demonstrates its understanding of the numerosity of the set of objects if it is able to predict that the correspondence will be maintained even if the perceptual properties of the set have changed. (Of course, an individual can learn about invariant properties even when it is another individual who is doing the actions of moving some substance from one container to another and back or is putting the objects in correspondence with a reference set. On learning in the active move in a social environment, cf. the next Section.)

8 Predicting the consequences of one's actions in a social environment

In the preceding Sections we have considered learning to predict in an environment that contains inanimate objects only. But imagine a network that lives in a social environment, that is, an environment containing other networks (conspecifics). In this Section we want to explore some implications of learning to predict the consequences of one's actions in a social environment.

One important consequence of living in a social environment is that the motor actions of one network can cause changes in the environment of another network. The motor actions of the first individual can be directly perceived by the second individual or the first individual can cause changes in the external environment and these changes are perceived by the second individual. For example, the first individual can move its body or body parts and these movements become encoded in the visual or tactile input units of the other individual. Or the first individual can move its phono-articulatory organs and these movements produce sound waves that cause auditory inputs for the other individual. Or the first individual can displace or otherwise change some objects in the external environment and these changes in the objects of the external environment are perceived by the other individual.

The changes in sensory input to the other individual that are caused by our actions may provoke an action on the part of the other individual. And this action, in turn, can cause changes that are perceived by us. Hence, we may develop an ability to predict the sensory consequences for us of our actions when these consequences are mediated by the actions of another individual through the causal processes we have described. We can apply our usual scheme here. The neural network predicts its next sensory input on the basis of (a) a planned action on its part, and (b) the current sensory input from the environment (which in this case includes the conspecific). The only peculiarity here is that the next sensory input is mediated by the behavior of the other individual.

To be able to predict the conquences of one's actions in a social environment can have important implications for the emergence of more sophisticated social behaviors and more complex social organization. As we have argued, we mainly know the environment by verifying our predictions on how the environment will react to our actions. Therefore, we may know the 'mind' of other individuals by verifying our predictions on how they will react to our actions addressed to them (or even not addressed to them). (This view of how we know the 'mind' of others is similar to the "simulation" view which has been proposed by Gordon (1986; 1992) and Goldman (1989) in opposition to the view that we have an implicit "theory" of mind (cf., e.g., Wellman, 1990)).

Another implication of living in a social environment derives from the peculiar nature of conspecifics as 'objects' existing in the environment. Conspecifics are remarkably similar to the predicting individual we are considering. They are both morphologically and behaviorally similar to our organism. Hence, if the organism can discern the similarities between its own actions and the actions of conspecifics, it can generalize both ways. Imagine that our individual is already able to predict that some particular action will be followed by some particular consequence in its own case, that is, when it is the author of the action. Then, when the organism perceives a conspecific which is engaging in the same action, it can predict that the same consequence will follow. Conversely, if our organism has learned to predict a particular consequence when a conspecific is engaged in a particular action (a case of prediction of intrinsic change, cf. Section 6), it can generalize to its own case and predict the same consequence when it is planning the same action.

Since the social environment, unlike the nonsocial environment, is populated by entities that are similar to us, an organism living in a social environment can learn about the environment vicariously, that is, by learning to predict the consequences of the actions of other individuals, and not only the consequences of its own actions. Hence, a social environment acts as a multiplier of experience, and of the knowledge extracted from this experience, for all individuals living in the social environment.

Of course, generalizing from my actions to the actions of others and viceversa requires an ability to discover similarities in nonidentical experiences (inputs). The sensory consequences of an action if the action is executed by the organism itself or by a conspecific in many cases are not identical. An action planned by an organism is not identical for the organism to an action in which another individual is perceived to be engaged. We are assuming that organisms (and neural networks as models of organisms) have an intrinsic tendency to generalize. However, there are limits to this generalizing ability and different species may be more or less able to generalize in different cases. We believe that humans are particularly able to generalize in the social domain (perhaps only because they have a large primate brain) and this may explain their tendency to learn by imitating others, as we will see in the next Section.

9 Predicting the sensory consequences of actions and learning by imitating others

A simple neural network model of learning by imitating other individuals that already know the behavior one is trying to learn is the following (Hutchins and Hazelhurst, in press; Denaro and Parisi, 1994). Imagine a neural network that already can associate the appropriate outputs in response to a set of inputs. This is the 'model'. There is a second network that has random connection weights and, therefore, it cannot associate the appropriate outputs to the set of inputs. This is the 'imitator'. The imitator learns by imitating the model. When the same input is encoded in the input units of both the model and the imitator, both the model and the imitator respond by generating some particular output. The two outputs will tend to be initially different, for obvious reasons. The imitator uses the output of the model as its teaching input. In other words, it compares its own output with the model's output and uses the discrepancy (error) to modify its connection weights. After a certain number of learning cycles, the behavior (output) of the imitator will approximate the behavior of the model in response to the set of inputs.

Notice what is required for this model of learning by imitating others to be effective. First, the inputs to the two networks, the model network and the imitator network, must be functionally the same. "Functionally" means that the two inputs may be physically different but they must be considered as the same by the imitator in the sense that the same output is to be associated to both inputs. For example, if I perceive an object from a given perspective and my model perceives the same object from its perspective, the two perspectives as physical sensory inputs are likely to be different but I must be able to consider them as sufficiently identical to learn from the way in which my model responds to its input.

But what is of more specific interest here is that our model of imitation contemplates the use of the output of the model as teaching input for the imitator. In order to compare the output of the model with its own output the two outputs must be encoding the same thing for the imitator. However, while the output of the imitator encodes a motor action, the output of the model is not directly accessible to the imitator as a motor action but only as the sensory consequences for the imitator of the motor action of the model. For example, if an imitator is trying to generate the same phono-articulatory movements that are generated by a model, the output of the imitator (a phono-articulatory movement) cannot be directly compared with the output of the model (another phono-articulatory movement) but only with the sensory consequences of the output of the model (the sound caused by the model's phono-articulatory movement). In order to compare comparable things (and generate an error for the backpropagation procedure) the solution is to compare the sensory consequences (sound) of the imitator's output with the sensory consequences (another sound) of the model's output (cf. Section 3 and Parisi, 1995).

But if this is how imitation can work, it is first necessary for the imitator to learn to predict the sensory consequences of its own actions so that it can then compare the sensory consequences of its own actions with the sensory consequences of the model's actions. Being able to learn to predict the sensory consequences of one's own actions, therefore, appears to be preliminary for learning by imitating others. We suspect that the great development of the ability to imitate in humans, in comparison to other animals including non-human primates (Visalberghi and Fragaszy, 1990), can be explained with the great development of their ability to learn to predict the sensory consequences of their actions since the ability to learn to predict is a prerequisite for learning to imitate (at the sophisticated level exhibited by humans).

10 Conclusion

We have argued that a more sophisticated knowledge of the environment can only be acquired if an organism is not restricted to learning about the environment in the passive mode, that is, by noting the regularities present in the environmental inputs, but it learns in the active mode, that is, it learns by acting on the environment and noting the consequences of its actions for the inputs arriving from the environment. In practice, this means that the organism learns to predict the sensory consequences of its planned actions. Many important properties of the environment can be discovered while the organism is acquiring sophisticated prediction abilities. Knowledge of these properties can then be used to react to environmental inputs with responses that increase the reproductive chances of the organism and to change the environment to suit the organism's goals. The adaptive pattern of humans appears to be typically based on (a) sophisticated prediction abilities, mainly developed during a prolonged early life stage specifically dedicated to learning to predict the consequences of actions, (b) a pronounced tendency to change the external environment which co-evolves with the prediction abilities, and (c) the use of the prediction abilities to learn by imitating the actions of others.

Classical connectionist networks do not live in an independent environment and, therefore, cannot act on the environment or on the relation between themselves and the environment in order to learn the predict the consequences of their actions. Therefore, classical connectionism is more appropriate for studying learning in the passive mode. To study learning in the active mode it is necessary to view neural networks in the perspective of Artificial Life. Artificial Life Neural Networks (ALNNs) are ecological networks, i.e. they live in a physical (simulated) environment and they have a physical (simulated) body. We have shown in the present paper how ALNNs can be used to study the acquisition of the ability to predict the consequences of one's own actions in an environment that contains inanimate objects and also conspecifics and to study how the ability to predict can be useful for attaining goals in the environment and for increasing one's reproductive chances.

References

Cecconi, F., Menczer, F. and Belew, R.K. Learning and the evolution of age at maturity. Submitted to Adaptive Behavior, 1994.

Cecconi, F. and Parisi, D. Evolving organisms that can reach for objects. In J-A. Meyer and S.W. Wilson (eds.) From Animals to Animats. Carmbridge, Mass., MIT Press, 1991.

Cecconi, F. and Parisi, D. Learning during reproductive immaturity in evolving populations of neural networks. Submitted to Journal of Theoretical Biology, 1994.

Denaro, D. and Parisi, D. Imitation and cultural transmission in populations of neural networks. Institute of Psychology, CNR, Rome, 1994.

Elman, J.L. Finding structure in time. Cognitive Science, 1990, 14, 179-211.

Elman, J.L. Distributed representations, simple recurrent networks, and grammatical structure. Machine Learning, 1991a, 7, 195-225.

Elman, J.L. Incremental learning, or The importance of starting small. In Proceedings of the 13th Annula Conference of the Cognitive Science Society. Hillsdale, N.J., Erlbaum, 1991b.

Floreano, D. and Parisi, D. A connectionist account of language development: imitation and naming. Institute of Psychology, CNR, Rome, 1994.

Goldman, A.I. Interpretation psychologized. Mind and Language, 1989, 4, 161-185.

Gordon, R.M. Folk psychology as simulation. Mind and Language, 1986, 1, 158-171.

Gordon, R.M. The simulation theory: objections and misconceptions. Mind and Language, 1992, 7, 11-34.

Hutchins, E. and Hazelhurst, B. How to invent a lexicon: the development of shared synbols in interaction. In N. Gilbert and R. Conte (eds.) Artificial Societies: the Computer Simulation of Social Life. London, UCL, in press.

Jordan, M.I. and Rumelhart, D.E. Forward models: supervised learning with a distal teacher. Cognitive Science, 1992, 16, 307-354.

McClelland, J.L. Parallel distributed processing: implications for cognition and development. In R.G.M. Morris (ed.) Parallel Distributed Processing: Implications for Psychology and Neurobiology. Oxford, Clarendon Press, 1989.

Nolfi, S., Elman, J.L. and Parisi, D. Learning and evolution in neural networks. Adaptive Behavior, 1994, 3, 5-28.

Nolfi, S. and Parisi, D. Self-selection of stimuli for improving performance. In G.A. Bekey (ed.) Neural Networks and Robotics. New York, Kluwer Academics, 1993.

Parisi, D. An Artificial Life approach to language. Submitted to Brain and Language, 1995.

Parisi, D., Cecconi, F. and Nolfi, F. Econets: neural networks that learn in an environment. Network, 1990, 1,149-168.

Parisi, D., Nolfi, S. and Cecconi, F. Learning, behavior, and evolution. In F.J. Varela and P. Bourgine (eds.) Toward a Practice of Autonomous Systems. Camrbidge, Mass., MIT Press, 1992.

Rumelhart, D.E. Toward a microstructural account of huamn reasoning. In S. Vosniadou and A. Ortony (eds.) Similarity and Analogical Reasoning. Cambridge, Cambridge University Press, 1989.

Stearns, S.C. The Evolution of Life Histories. New York, Oxford University Press, 1992.

van der Meer, A.L.H., van der Weel, F.R., and Lee, D.N. The functional significance of arm movements in neonates. Science, 1995, 267, 693-695.

Visalberghi, E. and Fragaszy, D. Do monkeys ape? In S. Parker and K. Gibson (eds.) "Language" and intelligence in monkeys and apes. Cambridge, Cambridge University Press, 1990.

Weigand, S.A., Rumelhart, D.E. and Huberman, B.A. Generalization by weight-elimination with application to forecasting. In R.P. Lippman, J.E. Moody, and D.S. Touretsky (eds.) Advances in Neural System Processing 3, San Mateo, Cal., Morgan Kaufmann, 1991.

Wellman, H.M. The Child's Theory of Mind. Cambridge, Mass., MIT Press, 1990.

Learning Subjective "Cognitive Maps" in the Presence of Sensory-Motor Errors

A. G. Pipe[1], B. Carse[1], T. C. Fogarty[2], A. Winfield[1]

1 Intelligent Autonomous Systems Laboratory, Faculty of Engineering
2 Faculty of Computer Science & Mathematics

University of the West of England
Coldharbour Lane, Frenchay, Bristol BS16 1QY
United Kingdom

email {ag-pipe, b-carse, a-winfie}@uwe.ac.uk, tcf@btc.uwe.ac.uk

Abstract

In this paper we present a new version of our previous work on a maze learning animat. Its sensory/motor capabilities have been extended and modified so that they are more biologically plausible than before. The animat's learning architecture is based around a hybrid RBF Neural Network/Evolutionary Strategy implementation of an Adaptive Heuristic Critic. We conduct experiments in which the animat either acquires persistent but undetectable internal errors in its sensory equipment, or operates in an environment where undetectable factors influence motor actions. We also observe the effects of random sensory errors on the usefulness of the information which the animat acquires. Through interactions with its environment the animat learns a subjective "cognitive map" which is a fusion of the features in its surroundings, the path to a goal state, and the errors/environmental influences which it cannot directly detect. We find that despite the subjective nature of the map it remains useful under quite high levels of error/distortion in our experiments.

1 Background

Much recent animat research has featured an emphasis on behaviour based systems, for details the reader is referred to numerous examples from [ECAL 1991 & 1993] and [SAB 1992 & 1994]. This approach has often concentrated upon building or evolving systems with multiple Stimulus-Response (SR) modules interacting simultaneously with the environment, and each other, to produce useful overall behaviour in some context. Real or simulated animats based upon this approach have proved to be very successful in tackling problems with modest processing resources, an achievement which had largely eluded the earlier endeavours of the classical AI and robotics communities.

A major feature of this approach is the absence of an explicit "world model" or "cognitive map"[1]. There is a valid argument, based in evidence from the natural world, to support the view that a large amount of animat functionality can be achieved this way. We would not wish to disagree with this view. However there is a similarly large body of evidence, also from the natural world, that under many circumstances the actions of even simple animals (such as foraging wasps and bees) are derived from something other than basic behaviour based mechanisms. An argument for this is developed in [Gallistel 1990]. If higher animals such as birds and rats are considered then this argument becomes even more forceful [also Gallistel 1990].

We would not wish to argue that some other "cognitive"[2] level of processing *precludes* behaviour based mechanisms, rather that they are insufficient alone to account for the manner in which even simple animals have been observed to achieve their full range of capabilities, let alone those of more complex organisms such as birds and rodents. This has been well summarised in [Roitblat 1994]. In this paper we do not intend to discuss this point further, our aim here is to briefly argue for the relevance of investigating a computational model by which an animat may learn and use an explicit "cognitive map" which can be related to capabilities found in the animal kingdom.

It could be argued that amongst the many stumbling blocks of the classical AI approach to animat-type problems which used "world modelling" was the attempt to establish an accurate and *objective* map of the environment. Our animat learns an approximate, or "fuzzy"[3], *subjective* map of the world, the nature of which is affected (ie. distorted or transformed in some way) by factors which the agent cannot directly observe. These may be either internal to the agent or external, ie. in the environment. The subsequent usefulness of the map thus obtained is dependent on both the extent and persistence of these factors. We argue that in a large number of cases the model remains useful and is broadly similar to those found and used in nature.

---

1 *In some evolutionary neural network based approaches [eg. Cliff 1994] it could be argued that if appropriate evolutionary pressure were applied elements of a "cognitive map" might become represented in a distributed fashion across the weights and connectivity of the network without being separated out as a distinct entity.*

2 *The meaning of this word is too disputed to use it without the use of quotes.*

3 *Clearly we borrow this term from Fuzzy Logic, there are certain similarities between the RBF Neural Networks used here and Fuzzy Logic Systems (FLSs), especially with respect to their local generalizing characteristics in the input space.*

A pair of simple analogies may be illustrative here, the former an internal source of error, the latter external.

An animat might develop a persistent error in its distance measuring equipment. A human subject with normal vision who puts on someone else's spectacles is in a similar situation, for example objects might appear closer than they actually are. However the subject would still able to navigate around a house, albeit in a rather cautious fashion. Whilst doing this the subject could still build up a useful internal map of the building provided that the distortions were not too great. If on the other hand objects appear further away than they actually are then after some time spent in simple distance calibrating experiments with the environment the subject would be able to adapt their internal space modelling processes to these new conditions. Any spatially oriented perception-action sequences learnt whilst in either of these situations will be useful until the spectacles are removed.

An animat may not be capable of measuring inclination of the surface over which it moves, however large inclinations might produce undetectable "slip" in a wheeled robot. Though most humans are capable of sensing inclination and are clearly not "wheeled" we still commonly perceive a distant place to be further away than it really is due to a slight uphill gradient on the route. However transferring this inaccurate information to others is usually not problematic since they would also be subjected to the same uphill gradient.

2 Related Work

As mentioned frequently in the "animat literature" there are many researchers active in this area, spread across a wide range of disciplines, from Biology and Ethology through Psychology to Control Engineering and Robotics. The following brief review is therefore intended to be representative rather than exhaustive.

Temporal Difference (TD) reinforcement learning algorithms [Barto et al 1989] such as the Adaptive Heuristic Critic (AHC) [Sutton 1984] [Werbos 1992] and Q-learning [Watkins 1989], both "on-line" approximations to Dynamic Programming [Barto et al 1991], have shown great promise when applied to 2-dimensional maze problems where an agent, such as a mobile robot, attempts to establish an efficient path to a goal state by interaction with its environment. In many problems presented to date the area covered by the maze is divided into a number of states, usually on an equally spaced 2-dimensional grid [Lin 1993], [Roberts 1993] & [Sutton 1991].

Other strongly related work includes the use of a Classifier System [Pipe 1 1994], [Booker et al 1989] in the creation and reinforcement of rules for navigation in a "grid world" [Roberts 1993]. However another important body of work using a rule-based approach does not use a "grid world" approach. The *Samuel* architecture [Grefenstette 1991] is perhaps the most representative of these.

Here we develop our previously reported learning architecture [Pipe 2 & 3 1994], which is strongly related to the Adaptive Heuristic Critic and Q-learning work mentioned above, but is also based on a fully continuous model of the environment and actions therein. Of course all the examples above share the common feature of learning an explicit "world model" or "cognitive map" which is isomorphic with features in the environment.

3 Introduction

Our animat uses a combination of primitive "piloting" and "dead reckoning"[4] to navigate from a start position to a goal position around static obstacles. Both of these techniques are commonly used in this way in the animal kingdom [Gallistel 1990].

The animat builds up, via interactions between itself and the environment, an indirect isomorphic mapping between a combination of a number of relevant external & internal factors and the weight vector of a Radial Basis (RBF) Function Neural Network. The following factors are combined in the resultant weight vector;

a) the physical features in the environment,
b) the "best" path from start to goal found to date,
c) any perception or action errors/distortions which are not directly detectable by the animat but which affect its progress.

The Neural Network contains a mapping between perceived[5] positions in the environment and the "value" or "fitness" of those positions in terms of eventual completion of the task, ie. reaching the goal position. It is the animat's "world model"[6] or "cognitive map".

This mapping is intentionally approximate, and often incomplete. It is a geocentric map but it is essentially *subjective* in nature. Even when converged item a) may well be incomplete, and subjective, since only those obstacles which required investigation

---

4 *Like Gallistel we borrow from nautical terminology here.*

5 *We use the word "perceived" here since the experiments we have conducted result in errors building up between the position which the animat perceives itself to be in and its actual position.*

6 *There is a plethora of terminology to describe the purpose of our "world model" here. In the Evolutionary Algorithm community it is the "fitness function". In the animat community it is the "cognitive" or "world map". Within the Temporal Difference learning parlance it is the major part of the "Critic", ie. the "V-function" or "Utility" function. Because this is a hybrid architecture used in an animat we use all these terms interchangeably here.*

in order to identify a path to the goal will be investigated. Item b) can be found from the converged model by following a path of gradually increasing fitness until the goal position is attained. Item c) is incorporated via distortions of the perceived positions of items a) and b) with respect to their actual positions.

The learning architecture for the animat is derived largely from our earlier related work [Pipe 2 & 3 1994], an overview of which is given below. In the remainder of this paper we go on to describe the experiments which were conducted and report the results which were obtained. This is followed by a discussion of the major points of this paper, conclusions and suggestions for further work.

4 The Animat

We work in simulation here as we did previously. Although clearly lacking the richness and diversity of the real world, simulation does allow careful control over the introduction of errors, inaccuracies and environmentally induced distortions so that their effects may be observed in isolation.

Though the learning architecture is broadly similar to our previous work [Pipe 2 & 3 1994], there are some significant changes in the animat's features and the knowledge which it can obtain from its environment.

Firstly all movements and positions known to the animat are relative rather than absolute as in our previous architectures. We are interested here in investigating the effects of persistent errors in the perceptions and actions of the animat. If sensing and movement are taken relative to the current perceived location of the animat then this allows the build up of errors in the model, clearly a more plausible situation.

Secondly our previous animat was "blind", it knew only its absolute position and, where relevant, the fact that it had bumped into an obstacle. The set of sensors which the simulated animat now possesses are;

a) shaft encoders on each wheel,
b) a compass,
c) an ultrasonic distance sensor which can be rotated to any position,
d) a "bump stop" which skirts the animat's perimeter.

Items a) and b) together allow simple calculation of a "dead reckoning" movement estimate, relative to the current position and orientation. Items b) and c) together allow estimates to be made of the distance to an obstacle in a given direction relative to the present position and orientation. Note that there is no simulated vision system for detecting visual cues such as colour or patterns on the walls or obstacles of the maze, this is a little reminiscent of some of Tolman's experiments with rats [Tolman 1946] wherein he deliberately tried to exclude this as a source of information by careful maze design.

Before making a movement the animat "looks around" using its compass, distance measuring equipment and internal "world model" to assess the value of various positions it could move to in a straight line. Once it has established what it believes to be the best position it makes a "dead reckoning" movement using only its shaft encoders for feedback. When sensing or movement errors are present item d) clearly allows for the detection of an unexpected collision, allowing subsequent update of the animat's model with the perceived position of the collision. This process is then repeated from the new position until the goal is attained.

Below we give an overview of the hybrid learning architecture which drives these processes, and updates the animat's model of its environment after each movement. Figure 1 illustrates the main features.

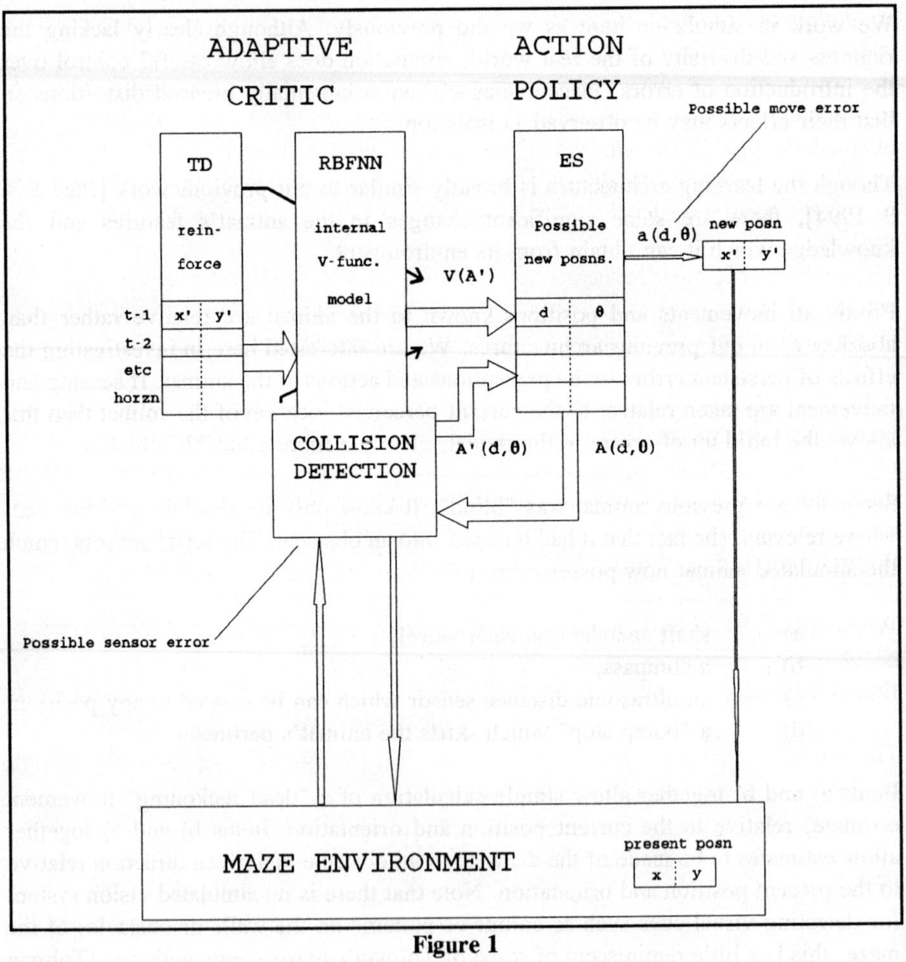

Figure 1

The two principle components of the Adaptive Heuristic Critic (AHC) architecture used here are the "action policy" and "critic" parts. Here an Evolutionary Strategy (ES), is used directly as the main exploring part of the agent, ie. it forms the "action policy" for each movement in the maze. We use an Evolutionary Strategy (ES) here to search the RBF Neural Network "world model" for good movements because it is a good real-valued function optimizer which assumes nothing about the inherent features of the function to be optimized. The Radial Basis Function (RBF) Neural Network, used to store the knowledge gained from exploration, ie. the "cognitive map", forms the main part of the "critic". We use a Temporal Differences (TD) learning algorithm to update the shape of this map as the search progresses.[7]

Before each movement the Evolutionary Strategy (ES) is restarted with a fresh population. Each population member consists of two parts, a distance d to move and an angle from geographical north θ. The fitness function used to rate ES population members is supplied by the RBF Neural Network, ie. the "cognitive map". Before this evaluation takes place however the animat's distance measuring equipment is used to determine whether a collision with an obstacle would occur if this trajectory were pursued, if this is the case then two operations are performed. Firstly the ES population member is modified accordingly. Secondly the "world model" is modified to reflect a low fitness level centred around the location of the projected collision site.

This process continues through multiple ES generations until either the population has converged or a maximum number of generations has been reached. The highest rated population member is then used to make a movement to the next position in the maze.

After this a Temporal Difference learning algorithm is executed on the RBF Neural Network to change the shape of "cognitive map". Changes are made in the regions of the maze surrounding the movements executed so far in the current trial by distributing a discounted reward or punishment back to them according to a simple variant of the standard TD algorithm [Sutton 1991]. This is "daisy-chained" back through time until a "horizon" of backward time steps is reached.

When sensing or movement errors are present clearly an unexpected collision could occur during movement. If this occurs the "cognitive map" is updated with a low fitness centred at the perceived position of the collision[8].

The processes described above are then repeated from the new position in the maze until the goal position is reached. This completes one trial. The RBF Neural Network

---

7 *ie. the V-function or Utility function to use AHC parlance.*

8 *Derived from the previous position estimate plus the orientation and number of wheel revolutions.*

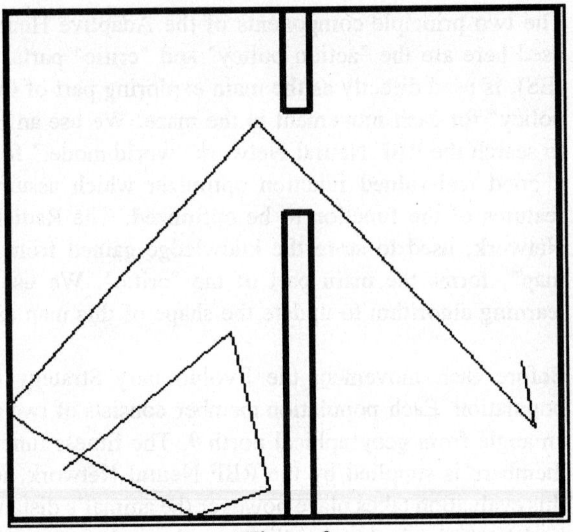

Figure 2

is thus used as an Adaptive Heuristic Critic (AHC) of the ES's every attempt to explore the maze. It learns a "world map" or V-function which reflects the value of being in those regions of the maze surrounding the positions which have so far been visited, refining the accuracy of the function after every movement in every trial of the maze.

5 Experiments & Results

We have shown elsewhere [Pipe 2 & 3 1994] that simulated animats based on this learning architecture can model and solve static mazes of quite high complexity. In the same work we also looked at the convergence performance of the algorithm over repeated trials and the quality of the converged solution. Here we wished to focus on the effects of introducing selected types of persistent error, inaccuracy and random errors.

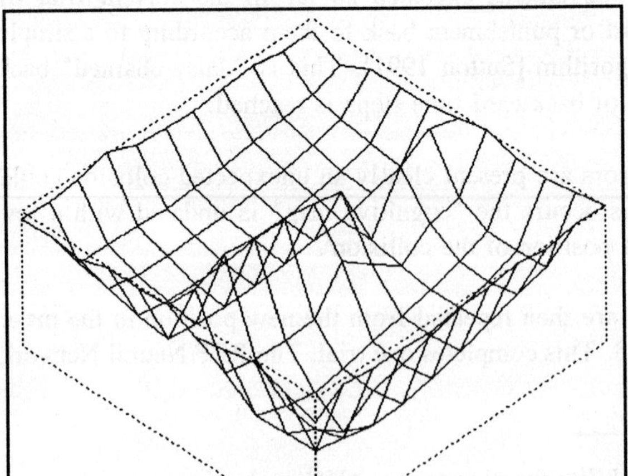

Figure 3

We therefore chose the simplest maze we could whilst retaining a worthwhile problem to solve. This was done to focus attention on the effects of the errors rather than other interactions and to ease visual interpretation of the results presented in this paper. The maze is shown in figure 2 along with an example of an initial route taken from the start position on the left side to the goal on the right.

Since we were not interested here in convergence behaviour or quality of the best route for such a simple problem we have not presented results for any of these[9]. Rather our interest is in the acquired "world model" or "cognitive map". Figure 3 shows a typical map acquired after completing the first trial of the maze, ie. the trajectory shown in figure 2. The map has been inverted and scaled for clarity, obstacles and "bad" places appear as peaks whilst "good" places appear as valleys in the function. Representations of the maze walls, obstacles, goal position and the stopping places on the route can all be observed with a little effort.

This model was generated whilst the animat was moving and sensing in a "perfect" error-free world.

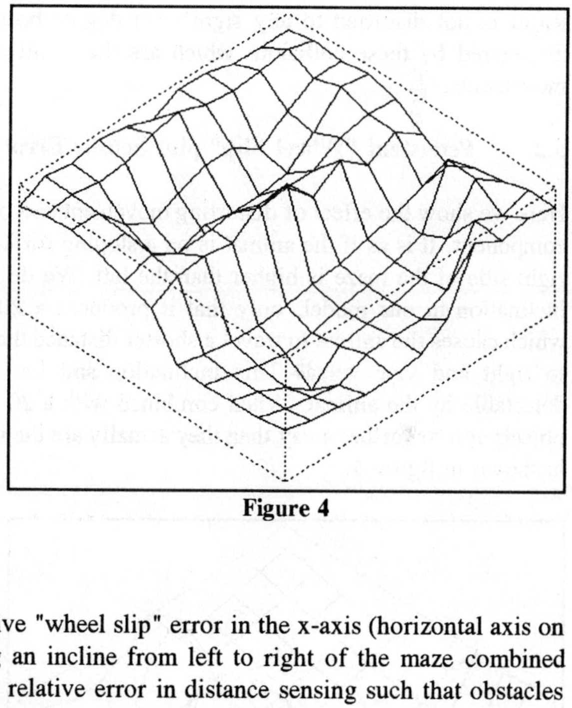

Figure 4

Though we have experimented with many forms of persistent and random errors in both sensing and movement, we picked results from three different types of induced error/inaccuracy for discussion here because they are broadly representative;

a) a persistent 20% relative error in the distance sensing measurement such that obstacles appear to the agent to be closer than they actually are,

b) a persistent 20% relative "wheel slip" error in the x-axis (horizontal axis on figure 2) representing an incline from left to right of the maze combined with a persistent 20% relative error in distance sensing such that obstacles appear further away than they actually are,

c) a random +/- relative error in distance sensing measurement.

5.1 Persistent Sensor Error

Figure 4 shows a typical map generated during the solution of the maze with this error present. When compared with figure 3 it is clear to see the effects on the

---

9 *For this maze all runs we conducted converged to two or three step move sequences from start to goal.*

animat's view of its environment. In the areas which it has investigated the walls and obstacles have apparently "closed in" on it. Not surprisingly this produces movements in the environment which are more constrained, of shorter average length and avoid walls and obstacles at a greater distance than in the error-free situation. The mapping has been so distorted in the area of the gap between the obstacles that it no longer appears to exist. However the animat is still able to solve the maze because as it approaches the gap it finds that it can pass through. This effect does however slow down the speed with which the gap is discovered since it must find it by "blind" search.

If we reverse the error so that objects appear further away than they actually are then a rather "bold" search results. Although the animat collides many times the map learnt is not distorted to any significant degree because most object positions are discovered by these collisions which are the result of error-free "dead-reckoning" movements.

5.2 Persistent "Wheel Slip" plus Sensor Error

Here we show the effect of distorting movements according to the size of their x-axis component. It is as if the animat is on a sloping surface which is inclined so that the right side of the maze is higher than the left. We do not need to know this angle of inclination in our model, only that it produces a 20% relative "wheel slip" error which causes the animat to travel a shorter distance than expected when travelling left to right and vice versa. This inclination and the "wheel slip" are not directly detectable by the animat. When combined with a 20% relative sensor error making objects appear further away than they actually are the resulting internal representation is shown in figure 5.

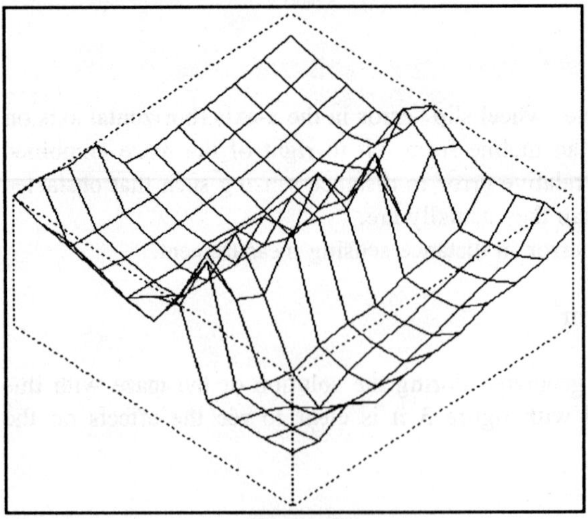

Figure 5

The errors in left-to-right movements and the sensor errors balance each other out. This means that the maze can be solved quite adequately even though the obstacles now appear shifted and thickened towards the right-side of the map. Further, the learnt information is valuable on subsequent trials of the maze because, in the animat's subjectively interpreted world, the obstacles *are* that far away when viewed from a distance.

Each one of these effects on their own do not produce significant distortions in the "cognitive map" since whichever is the "true" measure predominates. If we reverse both of these errors so that the maze is downhill to the right and distance sensing makes objects appear closer than they really the result is a similar, though more pronounced, effect to that observed with the sensor error alone as described in section 5.1.

5.3 Random Sensor Error

We introduced a random relative error into each occurrence of distance sensing. Selecting randomly between a +15% or -15% relative error still produced results which allowed the animat to converge on a stable path, though this took typically three to five times as many runs through the maze compared with the error-free case. If the magnitude of the relative error was increased to +/- 30% then the first run through the maze produced the map shown in figure 6. When compared with figure 3 it can be seen that at this level of error the effects on the internal model are quite pronounced, as one would expect, and indeed the signal to noise ratio became so deteriorated that the information contained in the "world model" was largely unusable by the animat on subsequent runs. Under these circumstances the animat reverts to a more or less random traversal of the maze, eventually arriving at the goal. The acquired information is so unreliable that full convergence to a short path on subsequent trials is not possible.

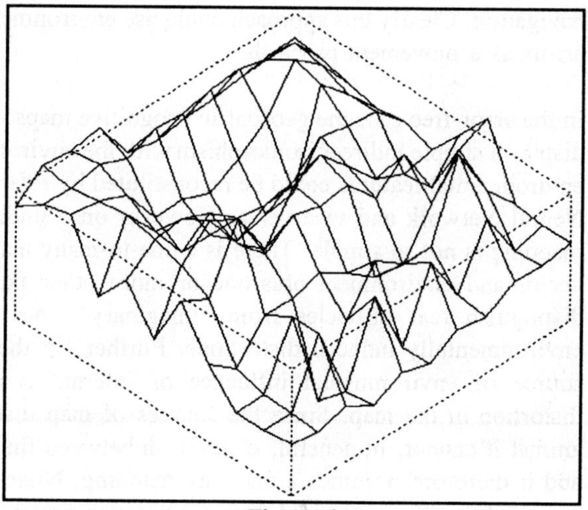

Figure 6

6 Discussion

The above results show that persistent errors within the agent or undetectable environmental factors which affect the animat's motion can be incorporated into a subjective internal model of the environmental features along with a reasonable approximation to the best path through them to some goal position from any start point in the maze. The animat is still able to acquire useful information which can be used to improve its performance over successive trials.

Not surprisingly the introduction of error, especially random error, is detrimental to the animat's learning performance, although for this simple maze quite high levels could be tolerated. In real systems there would in all likelihood be a mixture of persistent and random errors, in our experiments we have shown that the total error can be quite large especially if the persistent element is the major contributor.

In these experiments we use pure "dead reckoning" for all movements once they have started, clearly this is far more susceptible to the errors discussed above than the more biologically plausible model of continuous "piloting" feedback during navigation. Clearly this approach could use environmental cues to minimize trajectory errors as a movement proceeds.

In the error-free case the geocentric "cognitive maps" learnt by this simulated animat display a simple indirect isomorphism with the environment, ie. exact positions of the environmental features could be reconstituted by reference to the weights of the RBF Neural Network and vice versa. However once undetectable errors are present the mapping is not so simple. There is a one-to-many mapping between the RBF weight vector and environment plus one or more other factors. The animat is unable to distinguish real obstacles from "imaginary" ones caused by internal errors or environmentally induced distortions. Further, in the general case, more than one source of environmental influence or internal error could give rise to similar distortion in the map. Since the sources of map distortion are undetectable by the animat it cannot, in general, distinguish between the effects of the sources of error and it therefore becomes a one-way mapping. None of this is a problem of course provided that the sources of distortion remain present. If removed, or changed, then the animat would need to adapt the map on-line. However we argue that unless the changes were large there is still likely to be considerable worth in modifying the old map rather than starting afresh.

7 Conclusion and Further Work

We have taken our previous work [Pipe 2 & 3 1994] with a simulated animat based on a hybrid Adaptive Heuristic Critic learning architecture and extended it so that it is more biologically plausible. This has allowed us to make a study of the effects of internal errors in the sensory/motor equipment and external influences from the environment, neither of which are directly detectable by the animat. We find that in many cases the acquired subjective "cognitive map" remains useful even though it may be transformed by high levels of persistent error/distortion. Not surprisingly the usefulness of the map is reduced when sensing or movement errors are introduced, though quite high levels may be tolerated in the simple maze experiments which were conducted here.

Once internal errors or external undetectable influencing factors are introduced the "cognitive map" is no longer isomorphic in a straightforward way with the features of the environment because the differences between "real" and "imaginary" obstacles cannot be simply resolved by the animat, further there is more than one possible cause for any given distortion in the map. This is not problematic in the majority of cases provided that the source(s) of distortion in the learnt map are persistent. We argue that this is quite likely to be the case for the examples described in this paper. Even if they are removed or changed then there is likely still to be considerable useful information contained in the old map.

In our further work we will be repeating these experiments on a real mobile robot which is at present under construction. We would also like to pursue the possibility of making either continuous or periodic "piloting" checks on environmental features whilst a movement is taking place since this could greatly reduce the effects of sensory and "dead reckoning" errors in mobile animats of the type described in this paper.

References

Barto A. G., Bradtke S. J., Singh S. P., 1991, 'Real-Time Learning and Control using Asynchronous Dynamic Programming', Dept. of Computer Science, University of Massachusetts, USA, Tech. Report 91-57

Barto A. G., Sutton R. S., Watkins C. J. C. H., 1989, 'Learning and Sequential Decision Making', COINS Technical Report 89-95

Booker L. B., Goldberg D. E., Holland J. H., 1989, 'Classifier Systems and Genetic Algorithms', Artificial Intelligence 40, pp.235-282

Cliff D., Harvey I., Husbands P., 1993, 'Explorations in Evolutionary Robotics', Journal of Adaptive Behaviour, 2(1), pp.71-104

ECAL I, 1991, 'Towards a Practice of Autonomous Systems', Proceedings of the First & Second European Conference on Artificial Life, Eds. Varela F. J., Bourgine P., MIT Press

ECAL II, 1993, Proceedings of the Second European Conference on Artificial Life, MIT Press

Gallistel C. R., 1990, 'The Organization of Learning', MIT Press

Grefenstette J. J., 1991, 'Lamarckian Learning in Multi-agent Environments', Proceedings of the Fourth International Conference on Genetic Algorithms, Morgan-Kaufmann, pp.303-310

Lin L., PhD thesis, 1993, 'Reinforcement Learning for Robots using Neural Networks', School of Computer Science, Carnegie Mellon University Pittsburgh, USA

Pipe A. G. 1, Carse B., 1994, 'A Comparison between Two Architectures for Searching and Learning in Maze Problems', Selected papers from AISB Workshop in Evolutionary Computation, Springer-Verlag Lecture Notes in Computer Science #865, pp.238-249

Pipe A. G. 2, Fogarty T. C., Winfield A., 1994, 'A Hybrid Architecture for Learning Continuous Environmental Models in Maze Problems', From Animals to Animats 3, Proceedings of third International Conference on Simulation of Adaptive Behaviour, Eds. Cliff D., Husbands P., Meyer J-A., Wilson S. W., MIT Press, pp.198-205

Pipe A. G. 3, Fogarty T. C., Winfield A., 1994, 'Hybrid Adaptive Heuristic Critic Architectures for Learning in Mazes with Continuous Search Spaces', Parallel Problem Solving from Nature (PPSNIII), Proceedings of the third International Conference on Evolutionary Computation, Springer-Verlag Lecture Notes in Computer Science #866, pp.482-491

Roitblat H. L., 1994, 'Mechanism and Process in Animal Behaviour: Models of Animals, Animals as Models', From Animals to Animats 3, Proceedings of third International Conference on Simulation of Adaptive Behaviour, Eds. Cliff D., Husbands P., Meyer J-A., Wilson S. W., MIT Press, pp.12-21

Roberts G., 1993, 'Dynamic Planning for Classifier Systems', Proceedings of the 5th International Conference on Genetic Algorithms, pp.231-237

SAB92, From Animals to Animats 2, Proceedings of the Seconds International Conference on Simulation of Adaptive Behaviour, Eds. Meyer J-A., Roitblat H. L., Wilson S. W., MIT Press

SAB94, From Animals to Animats 3, Proceedings of third International Conference on Simulation of Adaptive Behaviour, Eds. Cliff D., Husbands P., Meyer J-A., Wilson S. W., MIT Press

Sutton R. S., 1984, PhD thesis 'Temporal Credit Assignment in Reinforcement Learning', University of Massachusetts, Dept. of computer and Information Science

Sutton R. S., 1991, 'Reinforcement Learning Architectures for Animats', From Animals to Animats, pp288-296, Editors Meyer, J., Wilson, S., MIT Press

Tolman E. C., Ritchie B. F., Kalish D., 1946, 'Studies in Spatial Learning I. Orientation and the Short-Cut', Journal of Experimental Psychology #36, pp.13-24

Watkins C. J. C. H., 1989, PhD thesis 'Learning from Delayed Rewards', King's College, Cambridge.

Werbos, P. J., 1992, 'Approximate Dynamic Programming for Real-Time Control and Neural Modelling', Handbook of Intelligent Control: Neural, Fuzzy, and Adaptive Approaches, Van Nostrand Reinhold, Ed. White D. A., Sofge D. A.

Specialization under Social Conditions in Shared Environments

Henrik Hautop Lund

[1] DAIMI,University of Aarhus, Ny Munkegade, 8000 Aarhus C., Denmark
[2] Institute of Psychology, National Research Council, Viale Marx 15, 00137 Rome, Italy
E-mail: henrik@caio.irmkant.rm.cnr.it

Abstract. Specialist and generalist behaviors in populations of artificial neural networks are studied. A genetic algorithm is used to simulate evolution processes, and thereby to develop neural network control systems that exhibit specialist or generalist behaviors according to the fitness formula. With evolvable fitness formulae the evaluation measure is let free to evolve, and we obtain a co-evolution of the expressed behavior and the individual evolvable fitness formula. The use of evolvable fitness formulae lets us work in a dynamic fitness landscape, opposed to most work, that traditionally applies to static fitness landscapes, only. The role of competition in specialization is studied by letting the individuals live under social conditions in the same, shared environment and directly compete with each other. We find, that competition can act to provide population diversification in populations of organisms with individual evolvable fitness formulae.

1 Introduction

Within an animal taxon there is often striking variability in the degree to which members restrict themselves to hosts or resources of a certain kind (e.g. Brooker and Brooker, 1990, Rothstein, 1990, Rothschild and Clay, 1961). Some species will be *generalists* and use a very large number of different food sources or hosts. An example of a generalist is the hen flea (*Ceratophyllus gallinae*) that has been recorded from birds of over 65 species in Britain (Rothschild and Clay, 1961). Other species act like *specialists* and use very few different food sources or hosts, like the related flea species (*C. rossittensis*) that is found on the crow (*Corvis corone*) only.
The preponderance of specialized forms in nature presents an evolutionary puzzle (Futuyma, 1991), since some individuals within populations of specialists will be unable to obtain sufficient resources, implying that there are individual advantages associated with capabilities to increase or maintain resource breadth. Some specialist lineages are derived from generalists, and the other way around, some generalist lineages are derived from specialists (Bernays and Wcislo, 1994).
The environments influence on this evolution is significant, and it has been shown that individuals that develop in abnormal environments often express abnormal behavior (Santibanez-H. and Lindemann, 1986) or structure (Matsuda, 1987). The environment can be expressed as the availability of resources, the degree of interspecific competition, the degree of intraspecific competition, etc. We will study the environments influence on evolution of specialist and generalist behaviors in populations of artificial neural networks. Further we study the individual organisms ingestion capabilities as a trait of the individual organism that co-evolve with either the specialist or the generalist behavior.

In section 2 we present an artificial life model of a population of organisms that live in an environment with different kinds of food sources. Each single organism will be represented by an artificial neural network. In section 3 we expand the model with evolvable fitness formulae as a trait of organisms. With this model we do simulations in section 4 with both fixed fitness formulae and evolvable fitness formulae under social conditions in a shared environment. Finally, in section 5, we discuss the results, and outline some future directions of research.

2 Model

We let a population of organisms live in an environment that contains three different food types, let us call them A, B, and C. The population consists of 100 organisms, each represented by an artificial neural network. Each neural network has 5 input units, 9 hidden units, and 2 output units, as shown on Fig. 1.

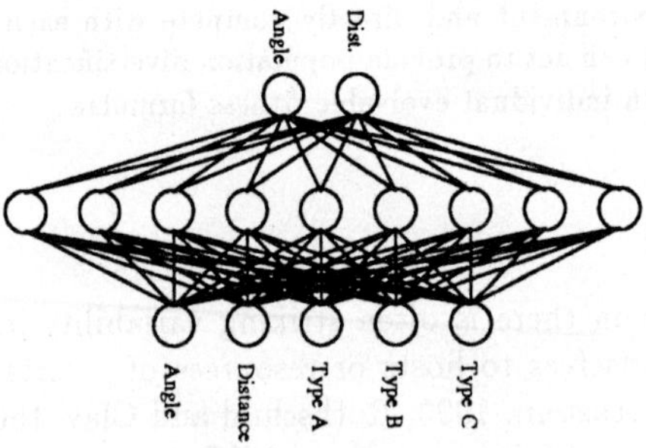

Fig. 1. Each of the 100 organisms in a population is represented by a neural network with 5 input units that encode sensory informations from the environment, 9 hidden units, and 2 output units that encode motor actions in the environment.

The 5 input units represent sensory input from the environment. The first unit gives the angle to the nearest food element in the environment, the second unit the distance to the nearest food element in the environment, and the last three units encode the type of the nearest food element in the environment. Both angle and distance information is mapped in the 0-1 interval. Each of the last three input units is assigned to one of the three types, A, B, and C. The unit corresponding to the type of the currently nearest element takes a value of 1, while the other two units take a value of 0. Information will flow through the 9 hidden units, where it is thresholded to 0 or 1 in each unit, up to the 2 output units that encode motor actions in the environment. The first output unit encodes the angle to turn, which can be any of the 360 degrees, and the second output unit encodes the distance to move forward after the turn. The distance to move forward can be from 0 to 5. Equipped with this kind of sensory and motor system, the organisms will be able to sense food elements and move around in the environment in order to approach

and capture food elements. When an organism happens to step on a food element, the food element will be ingested and disappear from the environment.
The 100 organisms are placed under social conditions in the same, shared environment, and thereby directly compete for the food elements. The environment consists of 5 different worlds of size 100*100, and each world will initially contain a random distribution of 100 food elements of type A, 100 food elements of type B, and 100 food elements of type C. The lifespan of the organisms will be 20 epochs of 50 actions in 5 different worlds, giving a total lifespan of 5000 actions. Thereby, the simulation differs from traditional simulations with ecological neural networks (e.g., Parisi et al., 1990), where artificial neural networks are let to live in each their environment with no contact with each other, why there is no direct competition on the behavioral level, but only in the selection where the most fit are selected to reproduce. In the present simulation, on the other hand, there will be a direct competition among the organisms. We call this type of environment in which organisms live together a *shared environment* .
We simulate natural evolution by the use of a genetic algorithm (Holland, 1975, Goldberg, 1989), that will search on the individual neural network connection weights. The genetic algorithm used to simulate the evolution process is shown on Fig. 2.

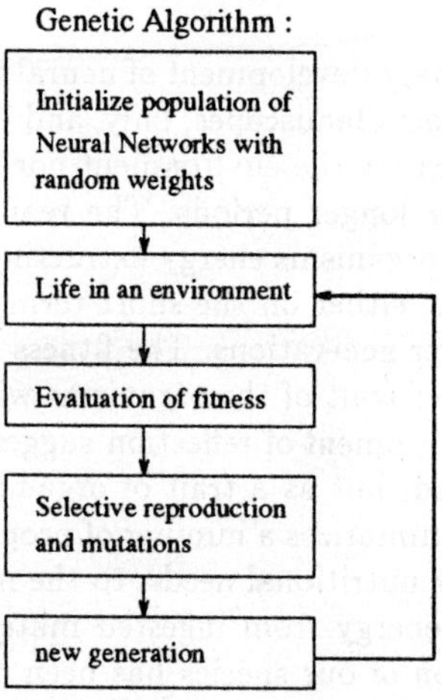

Fig. 2. The genetic algorithm used to search for neural networks to control the individual agents behavior.

Initially, we construct a population of 100 artificial neural networks with random connection weights. These 100 organisms are then let to live in parallel for their whole lifespan of 5000 actions in their proper environment. After these 5000 actions of each of the 100 organisms we do truncation selection of the best 20%, i.e. the 20 best organisms are selected to do agamic (non-sexual) reproduction. These 20 organisms will make 5 copies each, making a new population of 100 organisms.

The new 100 organisms will then have mutations applied to 10% of their connection weights. Again, we can let the new population live in the environment, evaluate each organisms performance, select and make reproduction of a new generation. In this way the cycle will go on for a number of generations, and we have constructed an evolution process. We expect progress on the evaluation measure, since we select the best organisms to reproduce, and the new generation will therefore be successors of the best organisms, only, yet differing to some degree from these, due to the mutations.

3 Evolvable Fitness Formula

In the preceding section with the introduction of the model using a genetic algorithm, it was not explained how the fitness, upon which selection is based, should be measured. The individual organisms should generate offspring as a function of the degree that they satisfy a criterion of fitness, or a *fitness formula* (Lund and Parisi, 1994a). For example, the organisms may reproduce in proportion to the number of food elements they are able to capture during life. The fitness formula in this case is number of food elements captured. The fitness formula of organisms living in another environment may be number of food elements eaten of type A and type B, minus number of food elements eaten of type C, that in this case, will act poisonous to the organisms.

Almost all work on evolutionary development of neural network architecture considers development in static fitness landscapes, only, and often in static environments, only. Yet, one can neither expect the environment nor the fitness landscape of real systems to remain static for longer periods. The real environment changes every moment, and e.g. biological organisms energy extracting abilities, upon which their fitness is based, changes too, either on the short term, during their life, or on the long term, evolutionarily over generations. The fitness formula should therefore be let free to evolve as any other trait of the organisms without any necessary control from the researcher, since a moment of reflection suggests, that in biological reality the fitness formula isn't fixed, but as a trait of organisms, is evolvable. More precisely, the fitness formula summarizes a number of properties of a particular species of organisms related to their nutritional needs, to the mechanisms and processes in their bodies which extract energy from ingested materials, etc. (The relation between food and the evolution of our species has been outlined by Harris and Ross (1987)). Like all other traits of organisms, the fitness formula can evolve. In fact, we can interpret the fitness formula of a particular species of organisms as an inherited property of that species of organisms. If the fitness formula varies (slightly) from one individual to another, is inherited, and is subject to random mutation, we can study its evolution in a population of organisms as that of any other trait.

In one of the presented simulations each organism will have an individual evolvable fitness formula. The individuals will have different energy extracting abilities as expressed in the individual evolvable fitness formula. The fitness formula is inherited from parent to child, but is subject to random mutation in the range of -0.1 to 0.1 on one of the three fitness values. In this type of simulations, there will therefore be selected for both behavior and fitness formula, and the fitness formula becomes evolvable as any other trait of organisms. As described above, the rationale for this interpretation of the fitness formula is that the quantity of energy that an individual organism extracts from a particular element type present in the environment is a function of the properties of the organism. The body of an organism includes mechanisms for processing ingested elements and extracting energy from these ingested elements. The quantity of energy which is extracted in each particular case (and which ultimately determines the fitness of the individual, i.e. its reproductive chances) depends on the particular properties of the element ingested but also on the particular properties of the individual organism. Since the relevant properties of organisms are not identical from one individual to another (and from one species of organisms to another species), the fitness formula which describes the quantity of energy (what we have called the energy value) of each type of element becomes a property of the individual organism which can be inherited exactly like all other properties. In this case, the overall fitness, F_i , of an organism, i , can therefore be described by (1)

$$F_i = \sharp A_i f_i^A + \sharp B_i f_i^B + \sharp C_i f_i^C \tag{1}$$

where $\sharp A_i$, $\sharp B_i$, and $\sharp C_i$ are the number of food elements ingested of each type by the ith organism, and f_i^A , f_i^B , and f_i^C are the respective fitness value from the individual fitness formula of organism i.

Lund and Parisi (1994a) showed how fixed fitness formulae and sensory apparatus of populations of organisms shape the behavior which emerges in a particular environment. With evolvable fitness formula it is possible to study the role of co-evolution of the fitness formula and of behavior given a particular sensory apparatus, for example one can observe how organisms with either a generalist behavior (i.e. organisms which tend to eat of all kinds of food distributed in the environment) or a specialist behavior (i.e. organisms which tend to eat only some, and not all, kinds of food distributed in the environment) emerges, along with the evolution of a fitness formula that either assigns equal values to all food types or assigns high values only to the preferred type(s).

The concept of ***limited evolvable fitness formulae*** should also be considered. In a limited evolvable fitness formula there is a limit on the energy, that an organism can extract with its internal processing mechanism (fitness formula) from a given food source. Although an evolving population of organisms can change its internal mechanism for extracting energy from food, there are likely to be various constraints and limits on this evolutionary process of change. A limited evolvable fitness formula limits the maximum quantity of energy that can be extracted from a given food source. When the limit has been reached, any mutation can only decrease this quantity. Simulations with limited evolvable fitness formulae result in very abrupt behavioral changes of organisms (Lund and Parisi, 1994a, Lund, 1994). It can generally be concluded that when the fitness formula evolves, but there are limits on the possible energy value of single food types, the behavioral strategy which emerges initially is specialist but then this strategy is replaced by progressively more generalist strategies. The organisms first prefer the single food type that has the highest energy value but when the energy value that can be extracted from this food type has reached its maximum value, they include other food types that can provide increasing quantities of energy.

In a *dynamic environment*, e.g. an environment where food elements are depleted due to some environmental catastrophe such as pollution, the evolved system will have to self-adapt to the changed situation. With an evolvable fitness formula the system is allowed to change the fitness landscape in order to cope with the changed environment. With changes in the environment on the evolutionary time scale, it has been found, that populations of artificial neural networks with evolvable fitness formulae show pre-adaptations to environments different from those in which they evolved (Lund and Parisi, 1994b). In the next sections we will study the influences of changes in the environment on the short term time scale due to interactions of competing organisms.

4 Simulations in Shared Environments

A population of artificial neural networks is placed in a shared environment. The individual organisms will live under social conditions in the same, shared environment, and thereby directly compete for food elements on each time step. It therefore happens that an organism with intention to approach a specific food element does not sense that food element in the next time step, not because it has moved away from the position of the food element, but rather because another organism already has reached the food element and ingested it. In an isolated environment an ecological neural network, as the ones used in the present simulations, will have the possibility to self-select its' sensory inputs (Nolfi and Parisi, 1993). This is the case in simulations where the organisms with evolvable fitness formulae live in an isolated environment, too, as shown in (Lund and Parisi, 1994b). Based on their movements the organisms decreased the possibility to sense unpreferred food types by maximum displacement away from such food elements, and increased the possibility to sense preferred food types by approaching these food elements. Since the organisms lived in an isolated environment, there would be no disturbing factors influencing the self-selection of sensory input. Yet, this is not true in shared environments. Here food elements will disappear from the environment not only due to the organisms own actions (upon which the self-selection of sensory input is based), but also due to the actions of other organisms in the environment. Therefore, the organisms might have to use other behavioral strategies than was the case in the isolated environment. Especially, one can imagine that the organisms with evolvable fitness formulae will have a rather difficult task, since it has been found in isolated environments, that the whole population quickly converges to a specialist behavior (Lund and Mayoh, 1995). In a shared environment there will therefore be a hard competition for food elements of the preferred food type.

4.1 Fixed fitness formula

We let a population of organisms live in a shared environment. All organisms have the same fixed fitness formula that assigns 1 energy unit to each food element of each of the three food types. The energy extracting mechanism that lets the individual extract 1 energy unit from each ingested food element will remain unchanged

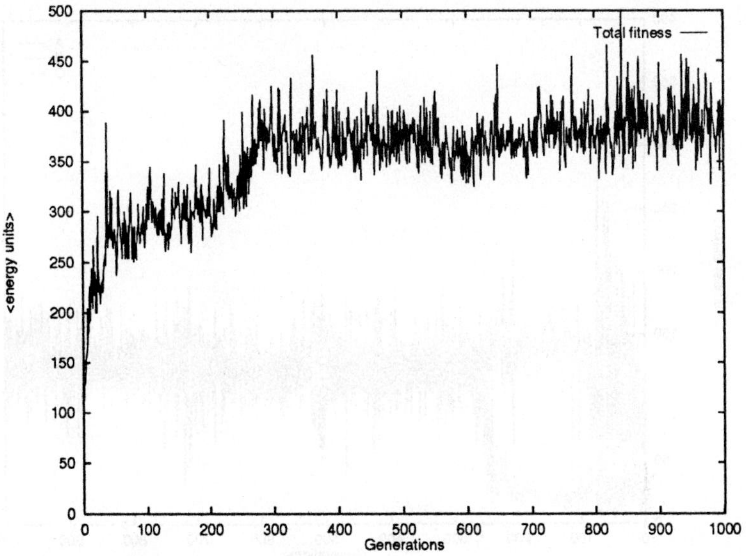

Fig. 3. The total fitness of the best organism of each generation with a fixed fitness formula.

over the whole evolution process. The only thing that changes over the evolution process is the connection weights of the individual neural network that decides the behavior expressed by the individual organism.

The results are as follows. On Fig. 3 we see, that the total fitness of the best organism in each generation increases fast to a high level, and in the end of the evolution process, the best organism will obtain fitness of approximately 400 energy units. Since the fitness does not increase much in the last 700 generations, it seems that the organisms have found an optimal strategy.

The increase in total fitness from 300 energy units to 400 energy units at around generation 300 can be explained by looking at the behavioral strategy of the organisms, that is shown on Fig. 4. It shows the number of food elements of each type ingested by the best organism of each generation. From generation 100 to generation 300 the best organism will eat a high number of food elements of type A and type C, and nearly none of type B, why we call these organisms two-type AC *specialists*. Yet, just before generation 300 the organisms with two-type AC specialist behavior are replaced by organisms that will eat approximately the same number of food elements of each of the three food types distributed in the environment. Therefore, we say that the organisms are *generalists*. The generalists will totally eat more food elements than the specialists, since they ingest food elements of all three food types, why we observe the increase in total fitness around generation 300. From generation 300, the behavioral strategy of the best organisms will not change, why there will be only minor changes in the total fitness.

Figure 4 only shows the behavioral strategy of the best organism of each generation, and hence, from the figure we can not say anything about the whole population. We therefore make a *population diversification analysis* of the behavior of each of the

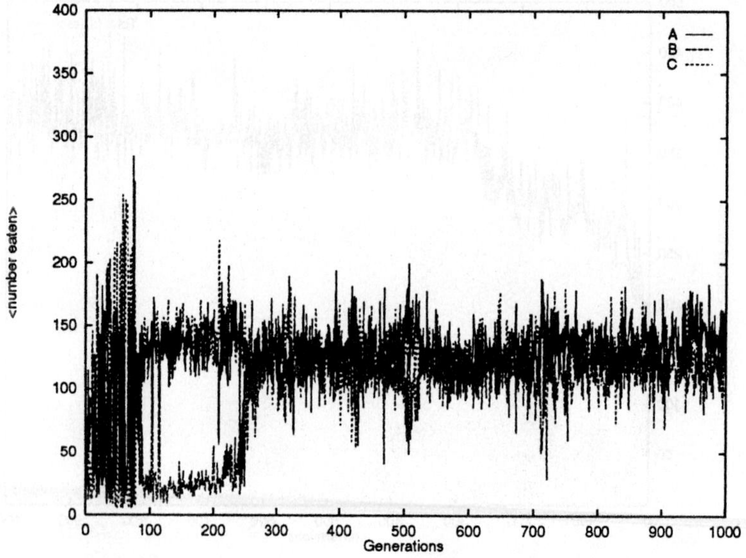

Fig. 4. The number of food elements eaten of each type of the best organism of each generation. During the last 700 generations, the best organism of the population will always behave as a generalist.

100 organisms in the population, to see whether they act as generalists or specialists. We classify organisms that ingest more than 25% type A food elements, more than 25% type B food elements, and more than 25% type C food elements of the total number of ingested food elements as generalists. Organisms that eat less than 25% type A food elements, but more than 25% type B food elements, and more than 25% type C food elements of the total number of ingested food elements are classified as two-type specialists, i.e. BC specialists. Similarly, we classify AB specialists and AC specialists. Finally, organisms that eat more than 25% of type A food elements, but less than 25% type B food elements, and less than 25% type C food elements of the total number of ingested food elements are classified as one-type specialists, i.e. A specialists. B specialists and C specialists will be classified in a similar way. By choosing a 25% limit instead of the, at first sight, more obvious 33% limit, a monopolization of specialist strategies is obtained. The results of the population diversification analysis are shown on Fig. 5. In fact, we see that not all members of the population are generalists. In most of the evolution process, only about 50% - 60% are generalists, while the remaining part of organisms in the population are different kinds of specialists, i.e. there will be at least one food type of which they will eat less than 25% of the total number of ingested food elements. However, from generation 100 to generation 300, the picture is totally different. The population of artificial neural networks will consist of mainly AC specialists, that occupy up to 70% of the population.

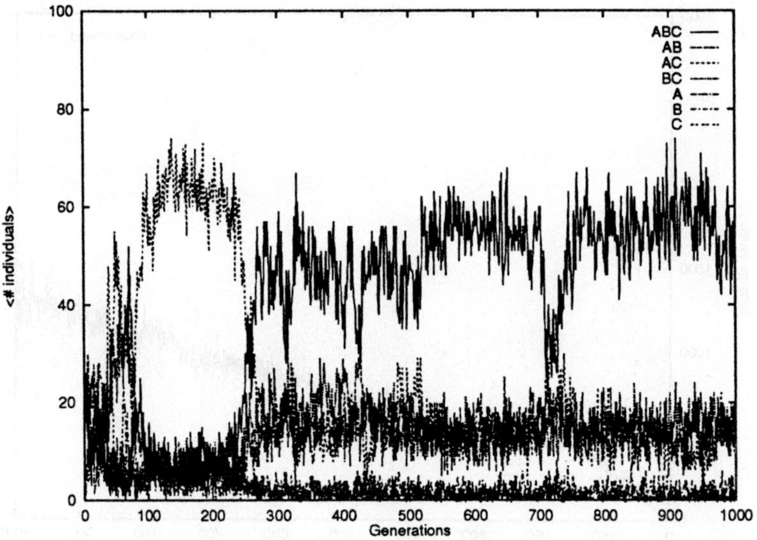

Fig. 5. The distribution of behavioral strategies over the population of 100 organisms in each generation. In the first 300 generation, the population will mainly consist of AC specialist, while generalists will dominate the rest of the evolution process, though only 50%-60% of the population.

4.2 Evolvable fitness formula

The results of a simulation with a population of organisms with individual evolvable fitness formulae in a shared environment are shown on Fig. 6, 7, 8, and 9. The fitness of the best organism, shown on Fig. 6, increases over the whole evolution process, as opposed to the preceding simulation, where the fitness reached a stable state after approximately 300 generations. Yet, the number of food elements captured during life reaches a stable state at around generation 200-300, from which point on it will not increase, but rather decrease with very small portions, as shown on Fig. 7.

The behavioral strategy is also different from the preceding simulation where generalists that ate approximately the same number of food elements of all three food types emerged. In the present simulation, the best organisms will initially be very different, and use different behavioral strategies, but later on specialists emerge. These specialists eat nothing else but food elements of type C, of which they will eat a bit less than 150, while eating approximately 25 food elements of both type A and type B. We call type C the *preferred* food type, and type A and type B the *unpreferred* food types. If an organism senses a food element of an unpreferred type, it will go away from the food element with a maximum movement, in order to increase its' chances to sense a food element of the preferred type in the next time step. This is done until the organism finally senses a food element of the preferred food type. Then the organism will try to approach the food element with slower movements, in order to reach and ingest the food element. Yet, as noted above, the food element may disappear as an effect of the social conditions under which the organisms live. In fact, it has been shown, that organisms with evolvable fitness formulae that live

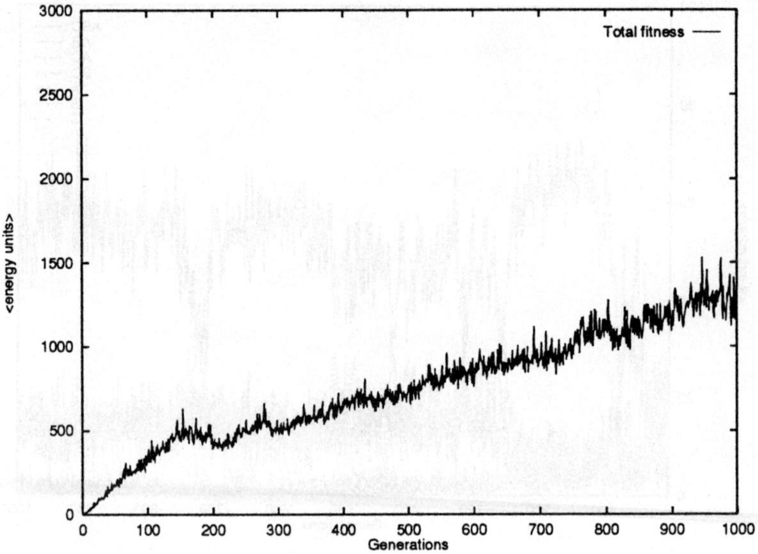

Fig. 6. The total fitness of the best organism of each generation with an evolvable fitness formula.

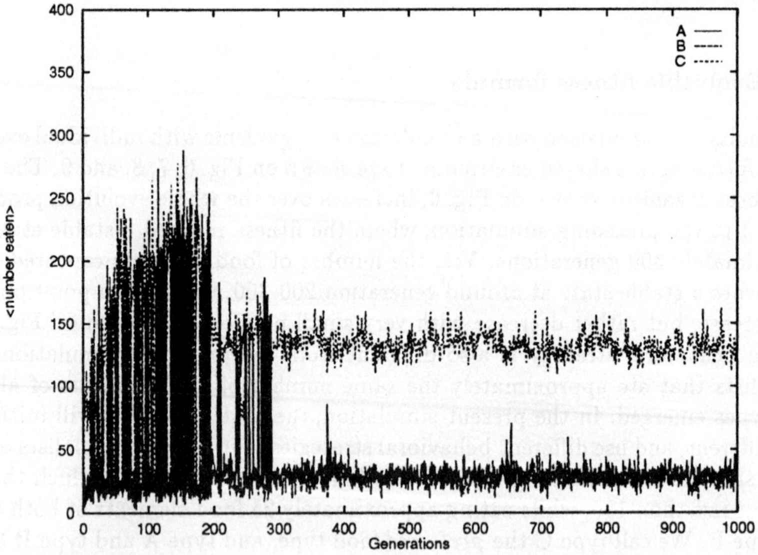

Fig. 7. The number of food elements eaten of each type of the best organism of each generation. After generation 300 the best most fit organism is always a C specialist.

in individual, isolated environments will reach a significantly higher number of food elements of the preferred type (Lund and Parisi, 1994a). The convergence of the whole population to a single specialist behavioral strategy may cause the decrease

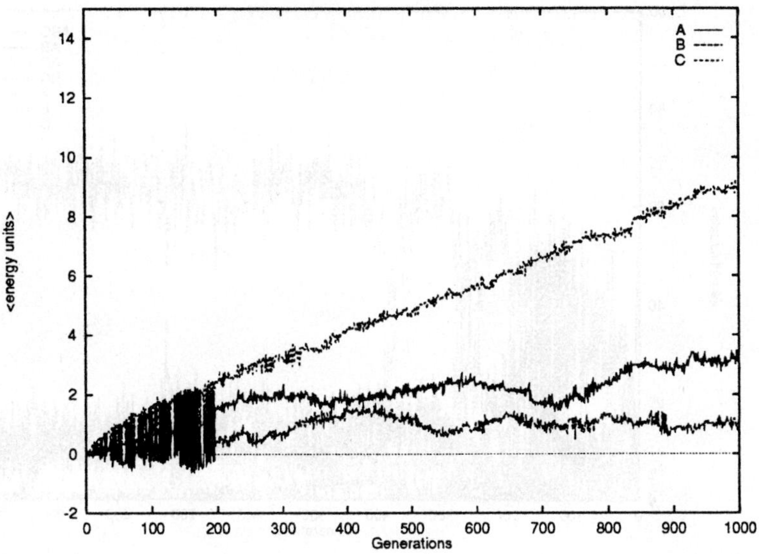

Fig. 8. The individual evolvable fitness formula of the best organism of each generation. The organisms will be able to extract high energy from food elements of type C, and much lower energy from type A and type B food elements.

in number of reached food elements, but the other way around, the competition may also cause population diversification. We will study this in a moment, but let us first try to explain the constant increase in total fitness.

The description of behavioral strategy could not explain the constant increase in fitness over the whole evolution process, since organisms eat a constant number of food elements from generation 200-300 on, but the fitness continues to increase. Instead we turn to analyze the fitness formula, which can explain the phenomena. The individual evolvable fitness formula of the best organism of each generation is shown on Fig. 8. We notice, that the fitness formula co-evolve with the behavior. The specialist organisms will obtain increasingly more and more energy from food elements of the preferred food type, C, while they will obtain only a small amount of energy from food elements of the two unpreferred food types, A and B. It is the constant increase in energy obtained from each single food elements of the preferred food type, C, that is the reason for the increase in net fitness of organisms over the whole evolution process. Since the energy extracted from food elements of the two unpreferred food types is not very high compared to the energy extracted from food elements of the preferred food type (respectively a 1/3 and a 1/10 in the end of the evolution process), the behavior to avoid these food elements with a maximum displacement in the environment is reasonable, since it enables the organisms to sense food elements of the preferred type more often. Yet, what might happen, is that the preferred food element is ingested by another, competing organism, before the organism itself is able to reach the food element, meaning that the organism can not do a perfect self-selection of sensory input, as in an individual, isolated environment.

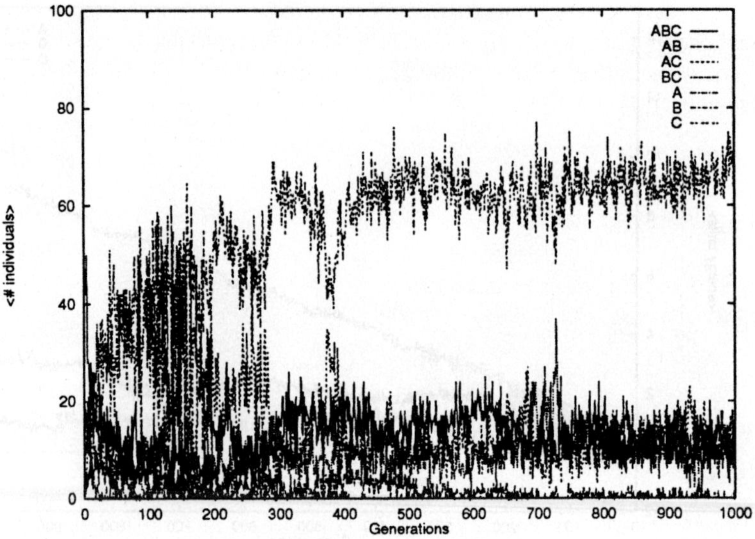

Fig. 9. The distribution of behavioral strategies over the population of 100 organisms in each generation. Only 60%-70% of the organisms in the population are C specialists.

When we turn to look at the whole population on Fig. 9, we observe diversification in the population of organisms with evolvable fitness formula, that live under social conditions in a shared environment. In the first many generations of the evolution process, there will be a lot of different behavioral strategies present in the population, and even in the end of the evolution process, no more than 60%-70% of the population will be C specialists, while the rest 30%-40% of the population will use different strategies. This is highly in contrast with populations of organisms with evolvable fitness formula, that live in isolated, individual environments. In isolated, individual environments, all organisms of a population will converge quickly to the specialist behavior (Lund and Mayoh, 1995), while we here, under social conditions, observe population diversification. In other words, from having a population with one single strategy in an isolated environment, we obtain population diversification when transferring the population to a shared environment.

5 Conclusion

We have studied how specialist and generalist behaviors may emerge in populations of artificial neural networks that live under social conditions in shared environments. We got evidence that specialist behavior may emerge even with the fixed fitness formula at some evolutionary stages under competition in the shared environment. Yet, the specialist behavior was later replaced by a generalist behavior. Further, when the fitness formula is let free to evolve as a trait of organisms, a specialist behavior tends to evolve together with the co-evolution of a fitness formula that let the organism extract high energy from food elements of the preferred food type, only. In isolated environments all organisms behave equally as specialists of the same type (Lund and Mayoh, 1995), while competition in the shared environment causes population diversification, so that not all organisms of the population behave equally.

Acknowledgement

The author would like to thank Domenico Parisi, C.N.R., and Charles Taylor, UCLA for useful discussions on the concept of evolvable fitness formulae.

References

- Bernays, E. A. and Wcislo, W. T. (1994). Sensory Capabilities, Information Processing, and Resource Specialization. The Quarterly Review of Biology, 69, 2, pp. 187-204.
- Brooker, L. C. and Brooker, M. G. (1990). Why are cuckoos host specific? Oikos, 57, pp. 301-309.
- Futuyma, D. J. (1991). Evolution of host specificity in herbivorous insects: genetic, ecological, and phylogenetic aspects. In P. W. Price, T. M. Lewinsohn, G. W. Fernandes, and W. W. Benson (eds.) Plant-Animal Interactions, pp. 431-454, John Wiley & Sons, New York.
- Goldberg, D. E. (1989). Genetic Algorithms in Search, Optimization and Machine Learning, Addison-Wesley, Reading.
- Harris, M. and Ross, E.B. (1987). Food and Evolution. Toward a theory of human food habits. Philadelphia, Temple University Press.
- Holland, J. J. (1975). Adaptation in natural and artificial systems. University of Michigan Press, Ann Arbor, Michigan.
- Lund, H. H. (1994). Adaptive Approaches Towards Better GA Performance in Dynamic Fitness Landscapes. DAIMI PB-487, Aarhus University, Aarhus.
- Lund, H. H. and Mayoh, B. H. (1995). Specialization in Populations of Artificial Neural Networks. In *Proceedings of 5.th Scandinavian Conference on Artificial Intelligence (SCAI'95)*, IOS Press, Amsterdam.
- Lund, H. H. and Parisi, D. (1994a). Simulations with an Evolvable Fitness Formula. Technical Report PCIA-1-94, C.N.R., Rome.
- Lund, H. H. and Parisi, D. (1994b). Pre-adaptation in Populations of Neural Networks Evolving in a Changing Environment. C.N.R., Rome. Submitted to *Artificial Life*.
- Matsuda, R. (1987). Animal Evolution in Changing Environments. John Wiley & Sons, New York.
- Nolfi, S., and Parisi, D. (1993). Self-selection of input stimuli for improving performance. In C. A. Bekey (ed.) *Neural Networks and Robotics*. Kluwer Academic Press.
- Parisi, D., Cecconi, F., and Nolfi, S. (1990). ECONETS: neural networks that learn in an environment. Network, 1, pp. 149-168.
- Rothschild, M., and Clay, T. (1961). Fleas, Flukes, and Cuckoos. Arrow Books Ltd., London.
- Rothstein, S. I. (1990). A model system for coevolution: avian brood parasitism. Annual Rev. Ecol. Syst., 21, pp. 481-508.
- Santibanez-H., G., and Lindemann, M. (1986). Introduction to the Physiopathology of Neurotic States. VEB George, Leipzig.

Iterated Prisoner's Dilemma with Choice and Refusal of Partners: Evolutionary Results

E. Ann Stanley[1], Dan Ashlock[1], and Mark D. Smucker[2] *

[1] Department of Mathematics, Iowa State University, Ames, IA 50010 USA (e-mail: stanley@iastate.edu, danwell@iastate.edu)

[2] Department of Computer Sciences, University of Wisconsin, 1210 W. Dayton St., Madison, WI 53706, USA (e-mail: smucker@cs.wisc.edu)

Abstract. In a series of papers we have examined what happens when individuals make very calculated choices of partners, based on past interaction histories [17, 1, 16]. In Iterated Prisoner's Dilemma with Choice and Refusal (IPD/CR), players use expected payoffs, which are based on the play history between the players plus an initial expectation, to assess the relative desirability of potential partners and refuse play with those judged to be intolerable. We have primarily studied this model using evolved populations of finite state machines. In each generation, individual behaviors generate social networks of interacting players. Here we provide an overview of our previous evolutionary results, and include some preliminary results on the impact of increasing the population size and including more randomness into the partner selection procedure.

1 Introduction

Prisoner's dilemma is often used as a model for situations where two entities have the possibility of either "cooperating" for the common good or "defecting" for personal gain[3, 2]. The entities may be two nations, with cooperation being temporary compliance with a treaty, say on air pollution levels, and defection being temporary noncompliance. If both nations comply, they both gain cleaner air, but if only one does not comply it gets both cleaner air and cheaper operating costs, and if neither complies they both gain dirty air. Or they might be two individuals working on a computer code together, with compliance being putting time into the work and being truthful about the state of the code and tests on the code, and noncompliance being not working, or being dishonest about results. In some situations, such as the air pollution example, the agents involved have no control over the identity of their partners: neighboring nations have to deal with each other. But in many social situations, individuals or groups can use

* Special thanks to Leigh Tesfatsion, who helped develop the IPD/CR model and with whom we have had many thoughtful discussions. Partial support came from an Iowa State Research Grant funded under DHHS Grant # 2SO7RR07034-26 and from the Los Alamos Center for Nonlinear Studies. Many thanks to the Iowa State Physics and Astronomy Department, especially the Gamma Ray Astronomy research group, for providing computer facilities and office space to Mark Smucker.

knowledge and experience to select partners. If a fellow researcher has proved difficult to work with or unreliable in the past, one will avoid him or her in the future. While most studies of iterated prisoner's dilemma (IPD) assume that individual players have no control over which opponents they play, a number of researchers have begun looking at different ways that partner selection can occur, and its impact on the emergence of cooperation. Experiments have shown that people who are given the option of playing or not are more likely to choose to play if they are themselves planning to cooperate. As well, more cooperative individuals are more likely to anticipate that others will be cooperative[13]. In a tribal situation, defecting individuals may be ostracized by the tribe[7]. Or it may simply be that some agents refuse those who have ever defected against anyone[15, 9, 4]. In the N–person prisoner's dilemma, it may be that individuals can change groups if they don't like the size of their group[5]. Individuals may use inherited tags to select partners[14, 11]. The use of expected payoffs in Iterated Prisoner's Dilemma with Choice and Refusal (IPD/CR) is meant to capture the idea that players attempt to select partners rationally, using some degree of anticipation, even though they do not know their partners' strategies and payoffs[17].

In order to study IPD/CR, we evolve populations of players, whose genetic code specifies their strategy for iterated prisoner's dilemma, and sometimes also one of their parameters for choice and refusal. The social networks formed by populations of these players are studied using combinatorial graph theory. Our research into this model [17, 1, 16] has shown that when average interaction lengths are comparable full cooperation is much more likely to evolve under our choice and refusal mechanism than under conditions of either random partner selection or round-robin matching. However, this is not the whole story, since there are competing pressures on the population. Choice allows cooperators to select each other, but it can also allow defectors to select cooperators. Refusal allows cooperators to protect themselves from repeated attacks, but not necessarily from occasional defections, and refusal can also lead to populations, which we call wallflower ecologies, where play stops forever.

In all previous studies we have used thirty players. Here we show that the results with sixty players are very similar, at least for the baseline choice of parameters. We then present preliminary results exploring what happens if partner selection and refusal allows for some uncertainty in the use of the expected payoffs.

2 The Original IPD/CR Model

Each run of IPD/CR consists of a series of generations of players who are evolved via a genetic algorithm, starting from a randomly chosen initial population. Each generation of players goes through I iterations, in each of which they select and refuse partners for games of prisoner's dilemma.

Prisoner's dilemma is a two-player game in which both players make simultaneous moves of either cooperate or defect. If both players cooperate, they each get a payoff of C, and if they both defect they each get a payoff of D. If one player defects and the other player cooperates, the defector gets the highest payoff, H, and the cooperator gets the lowest payoff, L. The payoffs obey the relations $L < D < C < H$ and $(H + L)/2 < C$. In all of the simulations described here, we use $L = 0, D = 1, C = 3, H = 5$.

In order to determine who to select and refuse as partners, each player n maintains an expected payoff $\pi(m|n)$ of every other player m, which it uses to rank them. In addition to ranking them, there is a minimum tolerance level, τ. Player m is intolerable to player n if $\pi(m|n) < \tau$. Expected payoffs for the entire population are initialized to π_0 at the start of each new generation, so that all players initially look identical to each other. When n receives a payoff from an opponent m, n updates its expected payoff from m via the assignment rule

$$\pi(m|n) \leftarrow \omega\pi(m|n) + (1 - \omega)U ,$$

where U denotes the payoff n received from m and ω is the memory weight.

Every iteration, each player makes an offer of game play to the player which corresponds to its top expected payoff, so long as that player is tolerable. If more than one player has the same expected payoff, random draws break the ties. If all other players are considered intolerable by a player, then that player receives a wallflower payoff of W.

For each chosen opponent, a prisoner's dilemma game is played between the player and the opponent if the opponent does not reject/refuse the offer of play. Players reject all offers from intolerable players and accept all others. If two players choose each other, only one game is played between the pair and not two. If a player's offer of play is rejected, that player receives a rejection payoff R from the rejector. The rejector does not receive a payoff from the chooser.

The players use a deterministic finite state machine to determine their moves for iterated prisoner's dilemma with each of the other players. In iterated prisoner's dilemma two players repeatedly play each other and can use the history of their interactions to determine their sequences of moves. The restriction on the payoffs, $(H + L)/2 < C$, prevents two IPD players from obtaining an average payoff greater than cooperation by alternating cooperation and defection. The finite state machines are a method of coding this use of history in a way which allows a variable amount of past moves to be used to determine the next move[12]. A player stores a unique machine state for every other member of the population that is updated each time they play. In our simulations we generally use either 16 state Moore[12] or 16 state Mealy machines[1] with both giving similar results.

2.1 Evolution

The genetic material of each player is coded as a bit-string, which specifies its finite state machine. In the cases where a choice and refusal parameter is evolved, this parameter is coded with additional bits. A genetic algorithm [8, 6] is used to co-evolve the population of players. For the experiments described here each generation usually has thirty individuals. A player's *fitness* at the end of the I iterations is the average of all payoffs it received, including wallflower and rejection payoffs. At the end of a generation, the top twenty individuals as ranked by fitness, the elite, are chosen to survive to the next generation. Individuals of equal fitness have equal probability of surviving. From the twenty elite individuals, ten individuals are chosen with replacement via fitness biased selection to pair up, mate and fill up with their children the ten openings. An individual can mate with itself. When two individuals mate, their bit-strings are subjected to one point crossover and the resulting two children are then subjected to a mutation operator (see [1, 16] for details). For the next generation, all memories of previous plays are removed, so that their expected payoffs are reinitialized to $\pi_0 = 3$, and their finite state machines are reinitialized to their starting states.

Our experiments have shown that when a successful mixture of players evolves, the genetic algorithm quickly creates populations which are fairly genetically homogeneous. The populations maintain low average ages, between two and a half and five generations[16].

3 Understanding Population Behaviors

Despite the simple formulation of IPD/CR, the population behavior which results is complex. This complexity is a desirable feature of the model, which seeks to capture some of the richness and variability of the real world. To understand our results, we look at them from several perspectives. At the global level, we plot the population statistics of average payoff, and the number of each kind of payoff. At the other extreme, we examine each individual's finite state machine or the sequence of plays between pairs of individuals. At an intermediate, and perhaps most interesting level, we examine the social networks [10, 16].

In IPD/CR, the choices individuals make based on their opinion of other individual players build up networks of interacting players, and the results of these interactions lead to the global statistics for the population. We study the social networks two ways: by observing play dynamically as it occurs and taking *snapshots*, and by using the *significant play graph* defined below.

Dynamic visualization is done as follows. At the beginning of each generation, the twenty elite individuals are placed counter clockwise around a circle, starting at 3 o'clock, according to their rank from the previous generation. The ten children are placed randomly into the bottom of the circle. Individuals are represented by open circles, except that individuals receiving a rejection payoff are represented by squares, and those who have found everyone else intolerable are solid circles. Players move toward each other when they play PD. Rejected players move away from the rejector. When two players interact, a line is drawn between them. Five iterations are overlapped at a time. One of these *snapshots* is shown at the left in Figure 1.

At the end of each generation, the individuals in the system are arranged back on their starting circle, the individuals' fitnesses are listed next to them, and the

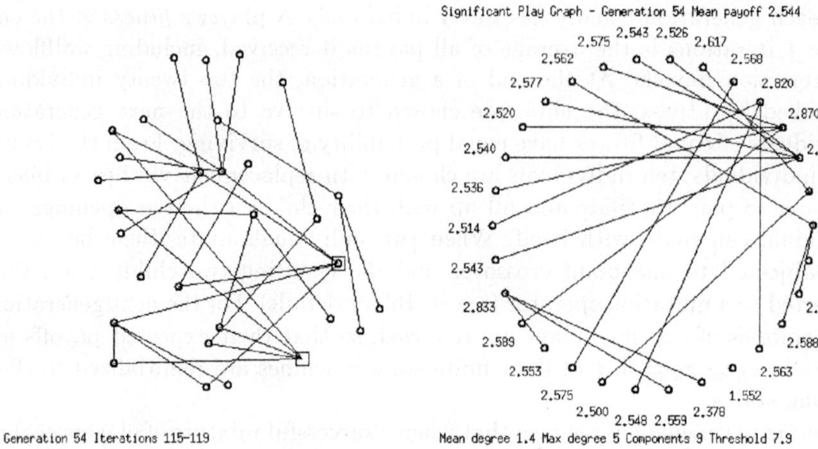

Fig. 1. A sample snapshot (left) and a significant play graph (right) from a rare run in which three genetically different subpopulations each continued to have offspring for over 70 generations. In the snapshot, lines indicate all play offers which have occured in the past five generations. Boxes indicate players being refused. This snapshot, and the resultant significant play graph are visually very similar to the stars networks described below. However, the stars are smaller and tend to be connected to each other more frequently. The networks for all 70 generations looked very similar to this after the first few iterations. The significant play graph on the right shows that in general the players who did well in the last generation also did well here. Positions indicate relative rankings in the previous generation, and numbers indicate fitnesses this generation.

significant play graph is drawn, as on the right-hand side of in Figure 1. The significant play graph is intended to capture the pattern of significant contacts between individuals. To obtain it we start from a fully connected combinatorial graph (or graph for short) which consists of a set of *vertices*, which represent individuals, and the set of all *edges* which connect two vertices. The edges are assigned a weight which is the number of times the individuals at the vertices played prisoner's dilemma with each other during the generation. A hypothetical play graph is shown on the left in Figure 2.

Next the edges of this graph which do not represent significant levels of game activity are zeroed out and the other edges are set equal to one. Determining what represents significant play is the key to obtaining useful graphs (see [16]). For the present paper, an edge is considered significant if it is greater than a threshold value. The threshold is the mean edge weight, μ, if the standard deviation, σ, of the edge weights is greater than μ, then the threshold is $\mu - 2\sigma$, unless $\mu - 2\sigma < 0$, in which case the threshold is set to one. After applying the computed threshold to the edge weights, the nonzero edges determine the significant play graph. This somewhat ad hoc method, illustrated in Figure 2, gives results which are consistent with our visual observations, but more work needs to be done on this issue.

To a certain extent the structure of the significant play graph can be as-

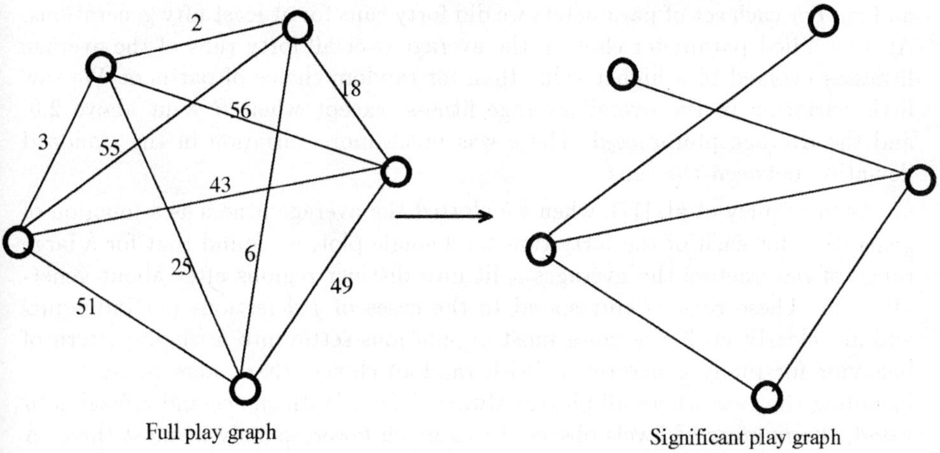

Fig. 2. The algorithm for generating the significant play graph of a hypothetical population of five players (represented by circles). On the left is the full play graph, with the edge weights along each edge indicating the number of PD games between the individuals connected by the edge. For these weights, $\mu = 30.8$ and $\sigma = 22.4$. Thus $\sigma < \mu$, and the first procedure described in the text, namely that all edges are zeroed out if their weight is less than μ, is applied to get the significant play graph shown on the right. Only the nonzero edges, all of which have a weight of one, are shown.

certained from its *average degree*, *maximum degree*, and number of *connected components*. The degree of a node is equal to the number of edges to which it is connected. A graph is connected if all vertices are reachable by all other vertices via a walk along the edges which connect the vertices. If a graph is not connected, then it is composed of two or more connected components. Each connected component is completely disjoint from every other connected component. The average degree represents the average number of significant relationships per individual. The maximum degree can be used in combination with the average degree to distinguish populations with central players who are more popular than other players from those without them. The number of components can be used to identify populations in which players have clustered into separate groups, or populations that have ostracized individuals, since an ostracized individual often forms its own component.

4 Average Fitness and Population Behaviors

In Ashlock et al.[1] we studied the sensitivity of the average fitnesses of IPD/CR populations to changes in the model parameters. In all cases, each generation had 30 players who played 150 iterations of IPD/CR before their fitnesses were determined. We started with a standard set of parameters ($W = \tau = 1.6$, $\pi_0 = 3.0$, $\omega = 0.7$, and $R = 1.0$), and varied the parameters across their ranges one at a time, holding $W = \tau$. We also did a set of runs simultaneously varying both τ

and π_0. For each set of parameters we did forty runs for at least fifty generations. At all studied parameter choices, the average over all forty runs of the average fitnesses evolved to a higher value than for random choice of partners. We saw little variation in the overall average fitness, except when R went above 2.0, and the average plummeted. There was much more variation in the standard deviation between the runs.

As in Stanley et al. [17], when we plotted the average fitness as a function of generation for each of the forty runs on a single plot, we found that for a large range of parameters the averages split into distinct regions after about generation 25. These regions correspond to the cases of j defections per individual and are clearly visible because most populations settle into a single pattern of behavior for many generations. With random choice, there were many levels, including the case where all players always defect. With choice and refusal activated, the number of levels observed was much fewer, usually at most three. A standard parameter run is shown in Figure 3. The most common levels observed were full cooperation, one defection, and always defect (which quickly leads to a situation where all players refuse to play anyone, and end up with mostly wallflower payoffs). Occasionally, the two–defection level was observed. As well, other fitness levels sometimes occurred and persisted, but these were rare. At some parameter values, the average fitness plots were very noisy-looking, and no levels were seen, but in these cases all runs had relatively high values compared to the case of always defect seen in the random choice plot.

The presence of wallflower ecologies is easily predicted; they occur either when any early defection against a cooperator causes the cooperator to reject the defector or when rewards for rejection are high. Thus these ecologies are observed at high τ, low ω, and high R. They are also sometimes observed at low values of π_0, but they are rare since early cooperations wash out the impact of π_0. The condition on early defections ensures that wallflower ecologies and the multiple-defection levels are not observed at the same time unless R is larger than 2.0.

An intriguing phenomenon are those ecologies whose average fitness jumps above the always cooperate level of 3.0. An average fitness above 3.0 is made possible by the fact that not all players play the same number of games, and each player's fitness is its average payoff. These ecologies can never maintain this level for very long before dropping below 3.0. These peaks often indicate the presence of the "Raquel and the Bobs" populations described in the next section.

Note that no ecology appears to be completely stable to its own offspring. If watched for long enough, it appears that all ecologies eventually will leave their current payoff level, even those at the 3.0 mutual cooperation level.

When we added one parameter to the genetics of the individual, either τ or ω, and allowed that parameter to evolve [1], the mean fitness of the population was lower than many of the fixed parameter cases, including the standard parameter case. On average, ω dropped to 0.4 and τ rose to 2.1. These values implied that wallflower ecologies were very common, which explains the lowered average

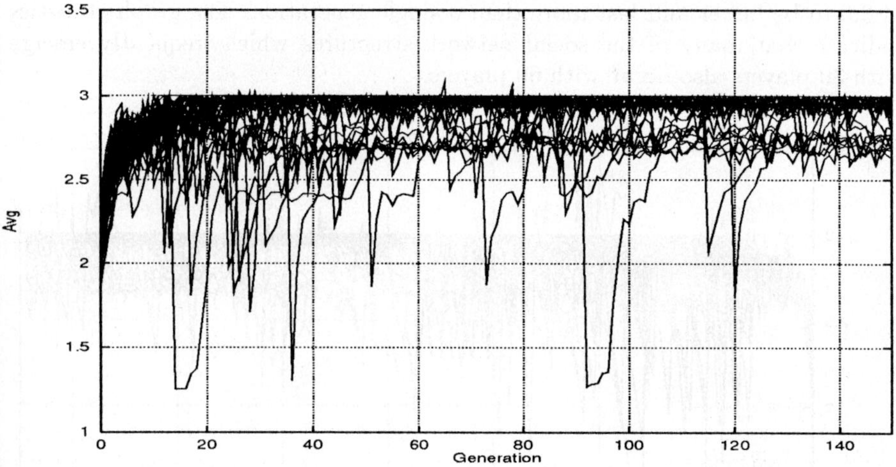

Fig. 3. The average fitness over the population for each of forty runs as a function of generation number. This series of runs is at the baseline parameter values, with $N = 30$, $I = 150$, $W = \tau = 1.6$, $\pi_0 = 3.0$, $\omega = 0.7$, and $R = 1.0$. Players are 16-state Moore machines [12]. The populations appear to split by about generation 25 into two fitness regions, with the upper region being one in which players essentially always cooperate with each other, and the lower region being one in which players on average defect exactly once. Here we also see one population which persists for a long time at a lower fitness level, with more defections.

fitness. Once again, the fitnesses of the populations were concentrated in three narrow regions, near the always cooperate, one–defection, and always defect (wallflower) levels.

The type of finite state machine (Moore or Mealy) affects the simulation very little, although there is a little more randomness with the Moore machines (evolution operates slightly differently on their bit-strings). The sorting algorithm used to rank individuals is more important, because individuals often end up with identical fitnesses at the end of a generation, it is important to use a randomized sort which does not have any age biases.

What happens if we change the number of players to sixty? In doing this, we wish to keep average payoffs roughly the same for similar behaviors. In particular, we want to keep the average for the single-defection level at about the same place. Thus we need to roughly keep the ratio N/I constant. With 60 players and 300 iterations we expect to have very similar results as with 30 and 150 iterations. However, with 60 players there is more genetic material available in the initial population, and it seems likely that there is a greater chance that complex game strategies and social networks can emerge. This could lead to more variability in fitnesses. On the other hand, stochastic models tend to be less noisy as population sizes increase. Preliminary studies at the standard parameter values indicates that the results are very similar (see Figure 4), although the larger population size allows the jumps in fitness above the cooperative payoff of 3.0 to be larger and last more than a single generation. The graph statistics indicate that many of the social network structures which frequently emerge with 30 players also occur with 60 players.

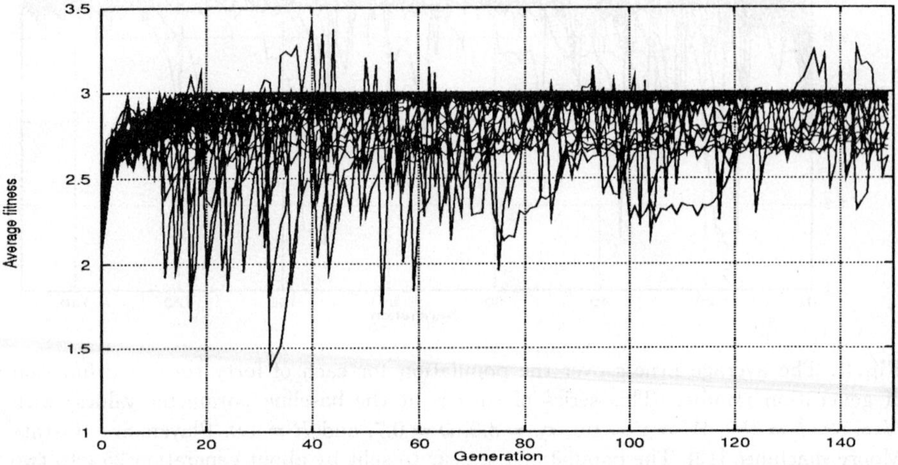

Fig. 4. The average fitness over the population for each of forty runs as a function of generation number. Here $N = 60$, there are 40 elite each generation, and $I = 300$. All other parameters are the same as in Figure 3.

5 Local Behavior and Social Networks

The average payoff gives only a gross sense of the interactions within a given ecology. When we looked more closely at individual behaviors, and the resultant social networks, we discovered that the ecologies within the different payoff levels were not all the same. Moreover, most of the ecologies had highly structured social networks and social systems, which remained constant for many generations. These populations generally contain either one or two subpopulations whose members are fairly similar, at least in their expressed behavior.

Ecologies lying near the 3.0 level consist of individuals who always or nearly always cooperate with each other. They have fully connected significant play graphs. These individuals may differ greatly in the unexpressed part of their bit-string, and thus differ in their response to more defecting individuals and in their probability of generating such an individual as an offspring, but they are identical in their behavior toward each other.

Similarly, wallflower ecologies tend to consist of individuals who always defect with each other until play refusal occurs, usually leading to a fully disconnected significant play graph. In a few rare cases a run has evolved into an ecology with individuals who cooperate once or twice with each other before going into the always-defect, wallflower behavior.

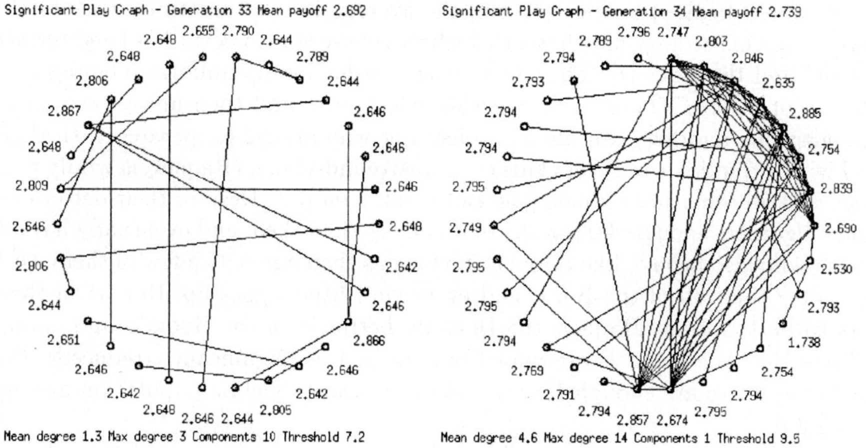

Fig. 5. Significant play graphs from latching (left) and connected centers (right) populations. In the latching population, each individual defects on the first move, and then cooperates afterward. This means that each individual prefers the individual it has played the most times and thus always chooses from the same set of individuals after an initial set of explorations. In the connected centers population, the individuals with many connections (the centers) are nice to each other and never defect first. The ones with a single connection (the outsides) defect once against all others and then cooperate. However, the centers retaliate against their defection by defecting once and then cooperating. This causes the outside individuals to latch onto the first center individual that is nice to them. The center individuals with the most outsiders latched on do the best.

The intermediate ecologies vary greatly. Near the one-defection payoff there are at least three different behaviors which occur fairly frequently, each of which has distinct-looking social networks. Full details of these are given in [16]. Very briefly, there are ecologies where each individual defects exactly once, on the same move (usually on the first move) and otherwise always cooperates. This leads all individuals to choose to play the same individual over and over (we term this latching) and creates small, disconnected components (see Figure 5). The other two behaviors each have two subpopulations. In one of them, the interactions lead to large disconnected groups with one individual in the group chosen every time by all the other individuals. These appear visually as disconnected stars with one individual at the center, similar to those seen in Figure 1. The individuals in the center are members of one subpopulation, and the individuals on the outside are a behaviorally different subpopulation. Individuals in each subpopulation are dependent on the other subpopulation to score highly, since they defect fairly frequently with those like themselves. The other population behavior with two subpopulations has a single component in the significant play graph, with one fully connected subpopulation whose members always cooperate with each other and another subpopulation whose members each latch onto a single individual in the other subpopulation (see Figure 5).

As noted in the previous section, the average fitness sometimes rises briefly above 3.0. One ecological behavior which can cause this to occur has been termed Raquel and the Bobs [1]. This starts with a behaviorally uniform latching population of "Bobs." These players, who defect once and then always cooperate amongst one another, can have an offspring who always cooperates with them and with a duplicate of itself. This cooperative individual (Raquel) is wildly popular, and all other individuals (the Bobs) latch onto it. Despite their defections, Raquel garners enough 3.0 payoffs to do better than them and eventually has offspring who also behave like Raquels. So long as there are only a few of them, they do better than all of the Bobs, and everyone's fitness goes up. But when there gets to be too many Raquels the Bobs do better than the Raquels and elitism ensures that the Raquels are wiped out (or at least significantly reduced). Occasionally there are enough Raquels at their peak to yield a population average above 3.0.

The two-defection and lower fitness levels also contain a wide variety of population behaviors such as "cyclic" populations, whose members all express the same play behavior with each other; namely, they have a repeated pattern of cooperations and defections with each other. Cyclic populations have shifting social networks, since after players mutually defect they are likely to choose someone else on the next interaction.

6 A More Random Partner Selection Procedure

In all previous studies of IPD/CR, we assumed that the partner selection procedure was deterministic, unless there was more than one potential partner with the highest expected payoff in which case a random number was drawn to choose the partner. In some cases, players ended up distinguishing between potential partners with very tiny differences in expected payoffs. Such perfect measurement of past actions seems unrealistic. Since it appears that some of the interesting social behaviors we observe may occur only because of this perfect discernment, we have begun to explore a variation in which individuals may make some errors in their partner choices.

We continue to assume that individuals can perfectly calculate expected payoffs, however, they cannot perfectly rank individuals or compare to the tolerance threshold, τ. Instead, players use probability distributions to do these rankings and comparisons. The probability that player n will choose a partner at all is given by

$$\frac{1}{2}(1 + \tanh \frac{\pi_{max} - \tau}{\alpha}) ,$$

where $\pi_{max}(n) = \max_m\{\pi(m|n)\}$ is player n's largest expected payoff of the current iteration and α is a tunable parameter. Once it is determined that a partner will be chosen, the probability that the selected individual will be individual m is $p(m|n) = \frac{f(m|n)}{\sum_{i\neq n} f(i|n)}$, where

$$f(m|n) = \delta(\pi(m|n) - \pi_{max}) + \epsilon \exp(\frac{-(\pi(m|n) - \pi_{max})^2}{B}) ,$$

and where $\delta(\pi(m|n) - \pi_{max})$ is 1 if $\pi(m|n) = \pi_{max}$ and 0 otherwise. The probability of accepting an offer from player m is similar to the probability of choosing anyone at all, with $\pi(m|n)$ replacing π_{max}, or

$$\frac{1}{2}(1 + \tanh \frac{\pi(m|n) - \tau}{\alpha}) .$$

Here α is taken to be the same in both tanh functions. As $\alpha \to 0$, $\frac{1}{2}(1 + \tanh \frac{(x-\tau)}{\alpha})$ goes to a step function which jumps from 0 to 1 as x passes τ, and therefore the decision whether or not to choose any partner as well as the refusal of an offer from a partner becomes identical to the original procedure. Similarly, as either ϵ or $B \to 0$, the partner choice algorithm returns to the original procedure that all those with $\pi(m|n) = \pi_{max}(n)$ are equally likely, and all others are not chosen.

In this modified model, wallflower payoffs are given to each player who neither makes nor accepts any offers in the iteration. Note that it is possible for players in this model to receive a wallflower payoff in one iteration and make or accept an offer in the next iteration. As well, a players may refuse an offer from a player to which it has made an offer, though they will still play a PD game unless they both refuse each other.

We anticipated that even small amounts of noise in this system could dramatically change the behavior of IPD/CR. To our surprise, in our preliminary studies of this new choice and refusal system, we continue to see the same average fitness levels as before, even for what might seem like fairly large values of α, B and ϵ (0.3, 0.09, and 0.5, respectively). The mean fitness remains substantially better than random choice at those levels. Although we have not yet studied the social networks, we see evidence for many of the behaviors we expected to disappear, such as Raquel and the Bobs.

As α increases beyond this, refusals become more and more common, until up to 50% of offers are refused and 50% of the time players do not choose any partners. This substantially decreases the mean fitness of the populations, especially since the average number of games played per individual is small and defection becomes an attractive behavior, although cooperation remains a common evolutionary outcome.

7 Conclusion

With IPD/CR we have explored the impact of individual behaviors on global social characteristics. Evolution often leads to populations with highly structured social networks. Allowing individuals to select and refuse partners greatly increases the level of cooperation which evolves when averaged over many simulations with different random seeds (as compared with random partner selection). However, it does not ensure that every individual run is more cooperative. In fact, populations which defect until no individuals can tolerate anyone else are a fairly common outcome. The most interesting social networks occur in populations with small but nonzero levels of defection. Increasing the population size seems to decrease the likelihood that defection evolves.

References

1. Dan Ashlock, Mark D. Smucker, E. Ann Stanley, and Leigh Tesfatsion. Preferential partner selection in an evolutionary study of prisoner's dilemma. *BioSystems*. To appear.
2. Robert Axelrod and Lisa D'Ambrosio. Annotated bibliography on the evolution of cooperation. University of Michigan, Obtainable via ftp from host www.ipps.lsa.umich.edu in the directory /ipps/papers/coop as Evol_of_Coop_Bibliography.txt, October 1994.
3. Robert Axelrod and Douglas Dion. The further evolution of cooperation. *Science*, 242:1385–1390, 1988.
4. John Batali and Philip Kitcher. Evolutionary dynamics of alturistic behavior in optional and compulsory versions of the iterated prisoner's dilemma. In Rodney A. Brooks and Pattie Maes, editors, *Artificial Life IV*, pages 343–348. MIT Press, 1994.
5. Natalie S. Glance and Bernardo A. Huberman. Organizational fluidity and sustainable cooperation. In K. Carley and M. Prietula, editors, *Computational Organization Theory*. Lawrence Erlbaum Associates, 1993.
6. David E. Goldberg. *Genetic Algorithms in Search, Optimization, and Machine Learning*. Addison-Wesley Publishing Company, Inc., Reading, MA, 1989.
7. D. Hirshleifer and E. Rasmusen. Cooperation in a repeated prisoners' dilemma with ostracism. *Journal of Economic Behavior and Organization*, 12:87–106, 1989.
8. John H. Holland. *Adaptation in Natural and Artificial Systems*. MIT Press, 1992.
9. Philip Kitcher. Evolution of altruism in repeated optional games. Working Paper, Department of Philosophy, University of California at San Diego, July 1992.
10. David Knoke. *Network Analysis*. Sage Publications, 1982.
11. Kristian Lindgren and Mats G. Nordahl. Artificial food webs. In Christopher G. Langton, editor, *Artificial Life III*, pages 73–103. Addison-Wesley, 1994. SFI Studies in the Sciences of Complexity, Proc. Vol. XVII.
12. John H. Miller. The coevolution of automata in the repeated prisoner's dilemma. A Working Paper from the SFI Economics Research Program 89-003, Santa Fe Institute and Carnegie-Mellon University, Santa Fe, NM, July 1989.
13. J. M. Orbell and R. M. Dawes. Social welfare, cooperator's advantage, and the option of not playing the game. *American Sociological Review*, 1993.
14. Rick Riolo, 1994. Personal Communication.
15. Rudolf Schuessler. Exit threats and cooperation under anonymity. *Journal of Conflict Resolution*, 33:728–749, 1989.
16. Mark D. Smucker, E. Ann Stanley, and Dan Ashlock. Analyzing social network structures in the iterated prisoner's dilemma with choice and refusal. Technical Report CS-TR-94-1259, University of Wisconsin-Madison, December 1994.
17. E. Ann Stanley, Dan Ashlock, and Leigh Tesfatsion. Iterated prisoner's dilemma with choice and refusal of partners. In Christopher G. Langton, editor, *Artificial Life III*, pages 131–176. Addison-Wesley, 1994. SFI Studies in the Sciences of Complexity, Proc. Vol. XVII.

Abundance-Distributions in Artificial Life and Stochastic Models: "Age and Area" Revisited

Chris Adami, C. Titus Brown and Michael R. Haggerty

W.K. Kellogg Radiation Laboratory
California Institute of Technology
Pasadena, CA 91125

Abstract. Using an artificial system of self-replicating strings, we show a correlation between the age of a genotype and its abundance that reflects a punctuated rather than gradual picture of evolution, as suggested long ago by Willis. In support of this correlation, we measure genotype abundance distributions and find universal coefficients. Finally, we propose a simple stochastic model which describes the dynamics of equilibrium periods and which correctly predicts most of the observed distributions.

1 Introduction

Species-abundance distributions have played an important role in our understanding of the process of evolution on the one hand, and in the field of ecology on the other. Early on, Willis [1] remarked that the frequency distribution of species within genera is markedly concave, i.e. there are many genera with very few species but only a few with very many species. In fact, it was Willis' objective to disprove Darwin's claim that new species arise through the "survival of the fittest," in a scenario where whole genera adapt gradually. In such a scenario, there is no correlation between the age of a genus and the number of extant species. A genus with many species may be old, or it may have adapted as a whole, gradually, until the differences are so large that a new genus is formed. Then, even though the genus would be considerd young [2], it would show a lot of variation. Willis, on the contrary, believed that mutation acts on the individual, and that any "young" genus will have, on average, less speciation then an "old" one.

This view, of the origin of species *per saltum*, i.e. the creation of new genera by extremely rare mutations that trigger an avalanche of speciation, as opposed to the gradual adaptation of species through natural selection, was taken up by Yule [3] in a remarkable paper. He developed a mathematical theory of evolution based on this picture, which matched the species-abundance curves obtained by Willis with high accuracy. The theory proposed was simple. For one ancestral genus, he assumes a certain probability for a "specific" mutation which creates a new species, as well as a (smaller) probability for a mutation that gives rise to a new genus: a "generic" mutation. Iterating this process, he predicts (in the limit of large number of species and infinite evolutionary time) a distribution for the number of genera N_g with n species

$$N_g(n) \sim \frac{1}{n^{1+1/\rho}}\,, \qquad (1.1)$$

where $\rho \geq 1$ is the ratio between the probability for a "specific" to a "generic" mutation. This allowed him to fit most observed distributions with a parameter $\rho \approx 2$, which fit the species-abundance relations available to him (see Fig. 1).

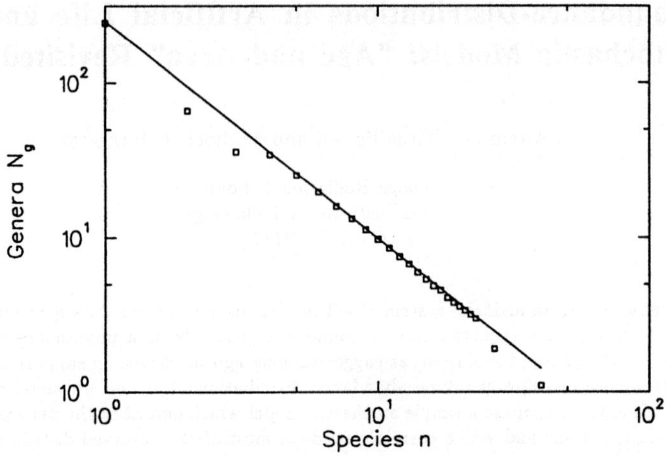

Fig. 1. Number N_g of genera (binned) of *Leguminosoe* with n species, compiled by Willis from "Dictionary of Flowering plants," as quoted by Yule [3] (errors are unavailable). The solid line is Yule's model-fit (infinite time) for $\rho = 2.457$, i.e. $D_s = 1.4$. The "finite time" fit raises the power to $D_s = 1.5$.

Furthermore, he concluded from his model that there is indeed, as postulated by Willis, a relationship between age and "size" of a genus, and therefore that evolution proceeds by leaps and bounds, rather than by gradual Darwinism. Seventy years later we have much sympathy with this interpretation, since punctuated equilibrium as promoted by Gould and Eldredge (see, for example, [4]) has become the standard evolutionary picture.

Ecologists, on the other hand, are interested in the abundance-distribution of species in a specified area, to understand the mechanisms that govern influx of new species and the extinction of others. The connection between these apparantly unrelated distributions is established via the age/"size" correlation of species or genera. Should such a correlation exist, the ecological distribution reflects the dynamics of competition rather than a distribution of resources, and those species taking up more space in the area studied are simply the older ones. In the following, we would like to argue that, not only is there such a relationship in the simple Artificial Life system that we investigate, but that most species-abundance distributions arise from very simple underlying dynamics governed by random processes. We suggest that "taxon"-abundance curves, independent of the placement in the hierarchy of taxonomic groups, should show universal power laws with coefficient (-3/2), based on the model presented in section 3. In fact, there is some evidence supporting this point of view from the work of Burlando [14], who compiled abundance-distributions from taxonomic data and the fossil record, and observed power-laws with coefficients consistent with the one obtained here.

Previously [8, 9], it has been suggested that such power-laws can be understood if living systems are in a self-organized critical state. In this paper, we proceed from a different point of view. We measure, in an Artificial Life system, the most basic of the abundance distributions, that of genotypes. We try to establish that the power-law obtained there is roughly universal, by studying its dependence on system size and mutation rates. We establish that the power-law is observed for the "instantaneous" (i.e. ecological) distribution, by taking snapshots of it in time, and the "integrated" (i.e. historical) distribution, which determines the "size" of a specific genotype. We use the latter to establish the "Age-Area" relation. In the next step, we present a simple model that reproduces most of the features observed in genotype-abundance distributions, but which in principle can explain the gross features of abundance distributions of *any* taxon, instantaneous and integrated, simply by relabelling the probabilities.

2 Abundance-Distributions and Artificial Life

Several models for the species-abundance curve have been proposed, and fitted to data obtained from a variety of fauna and flora. Some, like MacArthur's "broken-stick" model [11], relate the species-abundance distribution to the distribution of resources and niches, while others invoke simple branching and mutation models (without competition) such as Yule's, predicting geometric series (and power laws in the limit of a large number of species). Others include the effects of competition (such as [12] and much later [13]), and obtain frequency distributions ranging from exponential to power-law.[1]

Here, we take a novel approach to the determination of species-abundance relations, made possible by the advent of pioneering Artificial Life systems such as the tierra system, developed by Ray [5], and the avida system [10], developed by our group at Caltech. Keeping in mind the caveats mentioned in the Introduction, we investigate genotype frequency distributions from a system of self-replicating bit-strings subject to mutation and survival of the fittest. Each codon (from an alphabet of 20) codes for an instruction in a special machine-language for simple "programs" running on virtual CPUs. The language permits programs that self-replicate, and the chemistry of self-replication is thus substituted by the execution of the program (see [5, 7, 10] for more details on the tierra and avida systems). The aspects of the statistical mechanics of self-replication of the bit-strings have already been investigated with one approach [9], and will be investigated with another in this paper; such theoretical understanding allows us to make predictions and test them against the experimental results obtained with the Artificial Life systems. How much this artificial system resembles the global behavior of populations of self-replicating RNA will become known once such natural systems become available.

2.1 Dynamics of Self-Replicating Strings

Any string in the population is characterized by its specific sequence of instructions, which is termed the *genotype* of the string. If a genotype replicates accurately, we can associate a *replication rate* ϵ (number of offspring per unit time) with it, and the string then competes with neighboring strings for the placement of offspring. In avida, strings (or "cells") are arranged on a two-dimensional grid of fixed size and the total number of cells is constant throughout the run. When a new cell is spawned, the offspring replaces the oldest of the nine cells in the neighborhood of its parent. This mechanism of placing offspring in nearest-neighbor sites (thereby removing potentially competing cells) constitutes the only significant method of interaction between cells in this system, and results in the dissipative transport of information contained in the genome throughout the population (see [10] for details on this system).

[1] see [13] for a brief review of abundance-rank relations.

Strings are subjected to Poisson-random "cosmic ray" mutations that replace instructions randomly at an average rate R (mutations per site per unit time) such that the probability for a string of length ℓ to be hit by a mutation is $R\ell$. In the first approximation, then, genotypes are governed by the following "kinetic equation," which models the non-stochastic aspects of the time development of the occupation number (or frequency) $n_i(t)$ of genotype i:

$$n_i(t+1) - n_i(t) = (\epsilon_i - \langle\epsilon\rangle - R\ell)n_i(t) + C, \tag{2.2}$$

where $\epsilon_i n_i$ is the average number of cells of genotype i born per unit time, and $\langle\epsilon\rangle n_i$ is the average number that die (when cells of different genotypes replicate into their spot). The flux term C models cells mutated into genotype i from "mutationally close" genotypes j, and can be neglected in most situations.

In that case, and when the average replication rate $\langle\epsilon\rangle$ is roughly constant, Eq. (2.2) describes exponential growth or decline

$$n_i(t) = n_i(0)e^{\gamma_i t} \tag{2.3}$$

where γ_i is the growth factor $\epsilon_i - \langle\epsilon\rangle - R\ell$ of genotype i. Any genotype with $\gamma < 0$ experiences exponential decline and is soon pushed into extinction. A newly created genotype with a replication rate better than the old average, on the other hand, will experience exponential growth, quickly taking over the soup and temporarily reducing the diversity of the population until mutations restore it (see [8, 9] for a more complete description of the dynamics). In practice, we find that the system spends most of its time not in either exponential regime, but rather in equilibria where most genotypes have $\gamma_i \approx 0$. It is this third regime which turns out to dictate most of the properties of genotype distributions; the properties of this regime will be understood below through the use of the DL model.

2.2 Results

The empirical data which will be presented below were produced in a number of avida simulations. We have measured the number of genotypes $n_i(t)$ for each living genotype at different times (every ten updates), effectively taking "snapshots" of the genotype-abundance distribution, and averaged the 3,000 snapshots for each of twenty runs for populations of sizes 20x20 and 40x40, as well as 3,000 snapshots for each of five runs of size 80x80, all at an intermediate mutation rate $R = 40 \times 10^{-5}$ mutations per genome site per update.

The distributions $N_g(n)$ of genotypes with n living copies are shown in Fig. 2a as log-log plots. The slope of a straight line in a log-log plot determines the exponent in a power law, and we obtain from these measurements distribution functions of the form

$$N_g(n) \sim \frac{1}{n^{D_s}} \tag{2.4}$$

with D_s between 1.45 and 1.85 for the different sizes.

The peculiar rise in the distribution close to the maximum population size is due to the finite size of the lattice. The genotypes accumulating there are in fact the few ones that enjoy exponential growth after an invention that gives them an edge over all extant genotypes.

We are also interested in the dependence of the distribution function on the mutation rate. For the 20x20 system, we have measured the distribution function at half and twice the mutation rate used in Fig. 2a; the results can be seen in Fig. 2b. The power-law exponents D_s for the three mutation rates fall between 1.45 and 1.64.

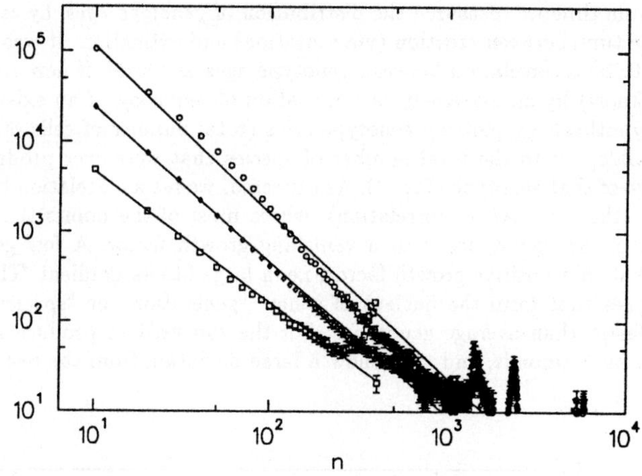

Fig. 2a. Genotype-abundance distributions (number of genotypes with n living copies), for three different population sizes in avida: 20x20 (squares), 40x40 (diamonds), and 80x80 (circles). The straight lines are least-squares fit to the data with power-law coefficients of 1.45, 1.7, and 1.84 respectively.

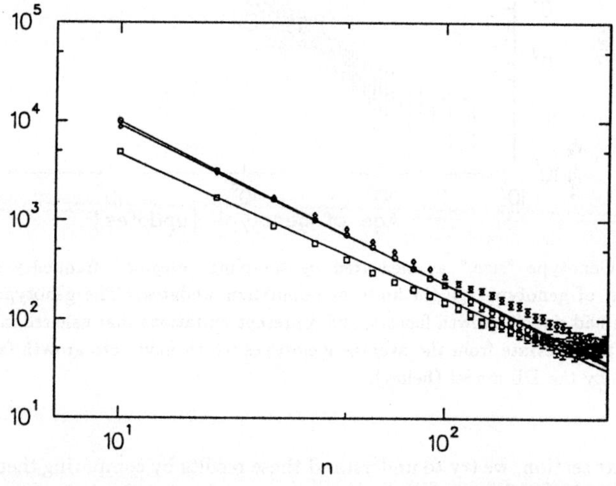

Fig. 2b. Genotype-abundance distributions (number of genotypes with n living copies), for three different mutation rates in avida (small: squares, medium: diamonds, high: circles) and a population size 20x20. The straight lines are least-squares fit to the data with power-law coefficients 1.45, 1.60, and 1.64 respectively.

At the same time we measured the distribution of *genotype ages*, by recording, for each genotype, the time between creation (via mutation) and extinction. It was Willis' idea that there ought to be a correlation between genotype ages and sizes, if new species are created (as we now know) by an extremely rare mutation of *one* copy of an existing one. We can check this hypothesis by plotting genotype sizes (total number of cells of a genotype that ever lived, analogous to the total number of species that were ever produced by a genus) versus the age of that genotype (Fig. 3). As expected, we see a correlation between genotype size and age (the "Age-Area" correlation), where most of the points that lie close to the diagonal are in fact genotypes with a vanishing growth factor. A few genotypes in each run start out with a positive growth factor, i.e. a large fitness gradient. These are precisely those genotypes that form the nucleus of a new "generation" or "species," and therefore grow much faster than average genotypes. For the run used to produce Fig. 3, they have been marked by diamonds, and they show a large deviation from the rest of the Age-Area curve.

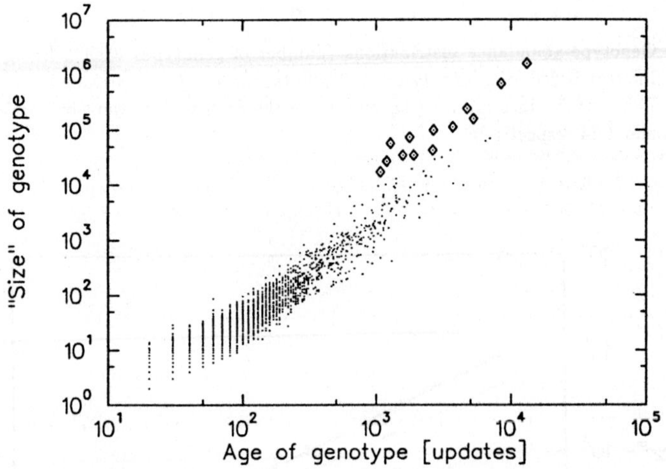

Fig. 3. Genotype "size," as measured by sampling *integrated* frequency over time, as a function of genotype age (in units of population updates). The genotypes marked by diamonds had sizable growth factors, and represent mutations that ushered in a new epoch. As such, they deviate from the average genotypes which have zero growth factors and are described by the DL model (below).

In the next section, we try to understand these results by comparing them to the simplest model that can be used to describe species or genotype distributions.

3 The DL Model

Analysis of genotype frequency curves as a function of time as obtained from the avida system reveals that these curves share many characteristics with a simple random walk. While of course this cannot be the whole story (for example, it lacks the dynamics of interaction between genotypes present in avida), it turns out nonetheless that this simple "dumb luck" model of equilibrium genotype dynamics reproduces much of the universal behavior seen in the full avida model.

In the DL model, a species is born with a single member, when an individual of another species suffers a mutation. The new species is characterized by a rate of growth, R_g, and a rate of shrinkage, R_s, which are assumed to be constant. We rule out unbounded exponential growth by making the additional assumption that $R_g \leq R_s$.

From that point on, the population $n(t)$ of the species increases or decreases with randomly occurring single births or deaths, given for a small time Δt by the probabilities

$$P_g \equiv P(\text{one birth}) = R_g n(t)\Delta t, \tag{3.5}$$

$$P_s \equiv P(\text{one death}) = R_s n(t)\Delta t. \tag{3.6}$$

Both births and deaths are necessarily assumed to be proportional to the number of individuals alive at that time (as in (2.2)). The first time that $n(t)$ becomes zero again, the species is extinct. The key point is that the number of individuals is always an integer, and changes in discrete steps according to the Poisson process described. The population $n(t)$ thus takes a random walk—prosperity or plague determined by dumb luck, modified only by an overall species fitness parameter (R_g/R_s). As it turns out, this model seems to describe most of the equilibrium population dynamics of species in avida, where creature interactions are relatively unimportant and mixing takes place on a shorter time scale than evolution.

Trivially, the long-time behavior of a species in this model is exponential decline whenever $R_g < R_s$, and that behavior ensures that every species will eventually become extinct. However, we are interested in the dynamics in the regime of low population counts, where the stochasticity of the model plays the dominant role. Therefore we have written a simulation of genotype dynamics following the DL model to compare with avida data. Some typical population time series of this simulation are shown in Fig. 4.

3.1 Comparison with a positive random walk

The classical random walk problem is similar to the DL model, and studying it will give us some insight into the behavior of our system. In a random walk, there are constant probabilities P_g and P_s (with $P_g + P_s = 1$) of taking unit steps in the positive or negative directions, respectively, at each unit of time. If we choose variable time steps of

$$\Delta t = \frac{1}{n(t)(R_g + R_s)}\,, \tag{3.7}$$

in the equations above, then the proportionality of growth and shrinkage rates to n is attained. If we further constrain the random walk to positive values (extinction being identified with the first time the walk returns to zero), the models are alike in their essential characteristics.

For the positive random walk, the probability of ending up at position n after exactly N steps can be calculated analytically; it is

$$P(N,n) = \begin{cases} \frac{2n}{P_g N}\binom{N}{(N+n)/2} P_g^{(N+n)/2} P_s^{(N-n)/2} & \text{for } n > 0 \\ \frac{1}{2(N-1)P_g}\binom{N}{N/2} P_g^{N/2} P_s^{N/2} & \text{for } n = 0, \end{cases} \tag{3.8}$$

where $N + n$ must be even. Although the expression does not hold rigorously for the DL model, it is a good approximation and will help illuminate several points.

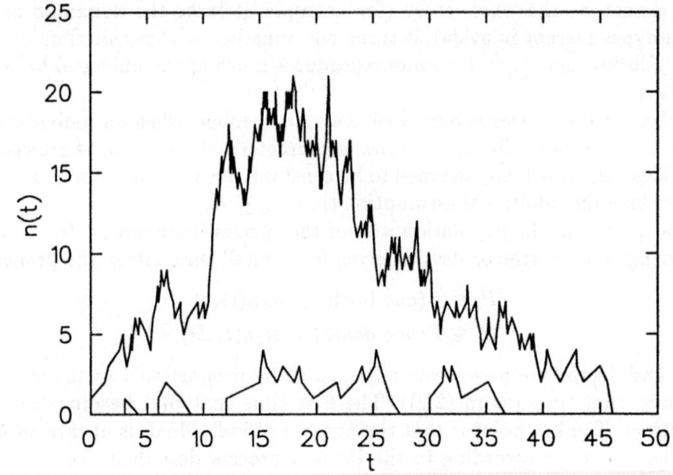

Fig. 4. Time-series of 5 independently run DL simulations of genotype abundances. Each species has the same fitness $R_g/R_s = 0.999$, so the surprising success of the one can only be attributed to "dumb luck."

3.2 Total historical genotype populations

The positive random walk and DL models make identical predictions for the distribution of the total historical genotype population (the total number of individuals of a genotype that have ever been born). Since each time step coincides with either a single birth or a single death, the sum of births and deaths equals the number of time steps. Since the genotype starts and ends with zero members, moreover, the number of births and the number of deaths must be equal. The probability that a genotype has a total number N_b births during its whole history is therefore given by $P(2N_b, 0)$ from Eq. (3.8)—for large N_b, approximately

$$P(2N_b, 0) \approx \frac{1}{P_g 4\sqrt{\pi}} \frac{1}{N_b^{3/2}} (4P_g P_s)^{N_b}. \tag{3.9}$$

This expression has two interesting limits. First, for creatures with low fitness ($P_g \ll P_s \approx 1$), there is an exponential suppression $(4P_g)^{N_b}$ of large historical populations (see Fig. 5). This is nothing more than the tendency of an unfit species towards exponential decline. This is one reason why unfit creatures do not have a big effect in observed data.

In a realistic avida run, however, mutations that are not fatal are usually neutral, and thus the most important case is $P_g \approx P_s \approx \frac{1}{2}$. In this limit, the exponential factor is negligible, and the distribution is dominated by the power law $N_b^{-3/2}$. This is indeed the behavior seen in the DL model simulations (see Fig. 5), avida population data (Figs. 2a,b), and also in many studies of biological species abundances [3].

The universal exponent $(-3/2)$ relies only on the assumption that populations of species are noninteracting; it is a property of any species which evolves stochastically. In particular, it does not depend on the mutation rate (rate of creation of new genotypes), in contrast to the model of Yule. Thus, the fact that his fits to observational data produce an exponent of

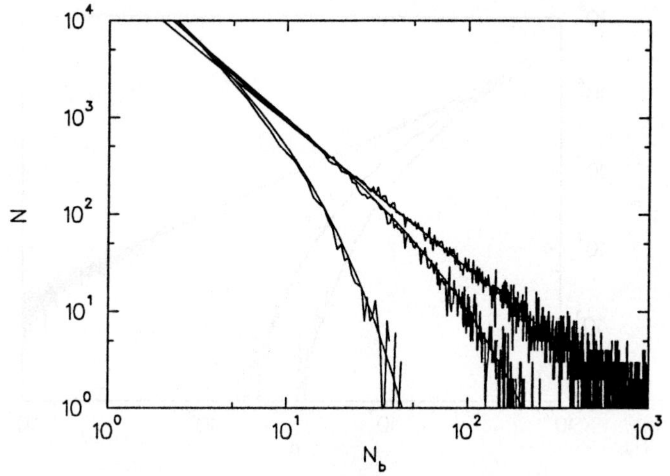

Fig. 5. Distribution of total number N of random DL walkers with N_b total births, with fitness parameter $\rho = R_g/R_s = 0.999$ (upper curve), 0.8 (middle curve), and 0.5 (lower curve). The distributions are fit using the parameters obtained by the theoretical estimate $N \propto N_b^{-3/2}[4\rho/(1+\rho)^2]^{N_b}$ above. This models the "integrated," or "historical," distributions.

roughly ($-3/2$) should not be seen as a property of the biological systems that he studied, but rather as a universal property accidentally uncovered in his analysis of the data.

3.3 Instantaneous abundance distribution

The distribution of instantaneous abundances $n(t)$ for the positive random walk is found to be roughly exponential: $P(n) \propto \exp(-\alpha n)$. In this case the connection to the DL model can be made exactly: since its time steps are dilated by a factor of $n(t)$ during periods of higher abundances, those points are systematically oversampled by the same factor, and the instantaneous abundance distribution for the DL model is roughly

$$P(n) \propto \frac{e^{-\alpha n}}{n} \tag{3.10}$$

(see Fig. 6a).

This distribution is heavily influenced by one factor present in avida but not included in the naive DL model: the effect of genotypes with high fitness ($P_g > P_s$), which appear occasionally and take over a large fraction of the soup. While neutral genotypes spend exponentially little time at high abundances, high fitness genotypes spend much of their time there. Therefore, when comparing avida data to theory, we omitted in each case the contribution of any genotype which ever filled more than 10% of the soup; this removes most of the high-fitness genotypes. (The cut also removes a significant number of neutral growth rate genotypes that happen to cross that threshold, but it does so equally for both data sets.) The result is that theory matches avida data quite well, as seen in Fig. 6b.

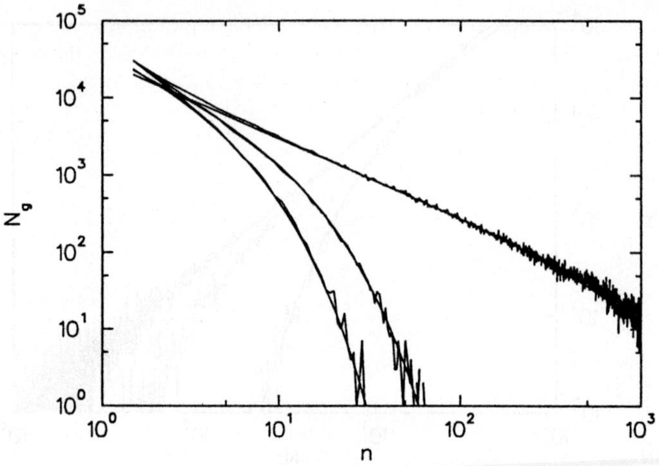

Fig. 6a. Genotype abundance distribution in a DL model obtained by sampling genotype frequencies (such as in Fig. 4) every 10 time units ("instantaneous," or "ecological" distribution). The upper curve was obtained for a simulation of genotypes with $\rho = P_g/P_s = 0.999$ (almost neutral growth rates); the middle curve, $\rho = 0.9$; and the lowest curve, $\rho = 0.8$ (significantly inferior). The solid lines are fits to the data, of the form $N_g(t) \sim \exp -\alpha n/n$, with α parameters 0.001, 0.11, and 0.24 respectively.

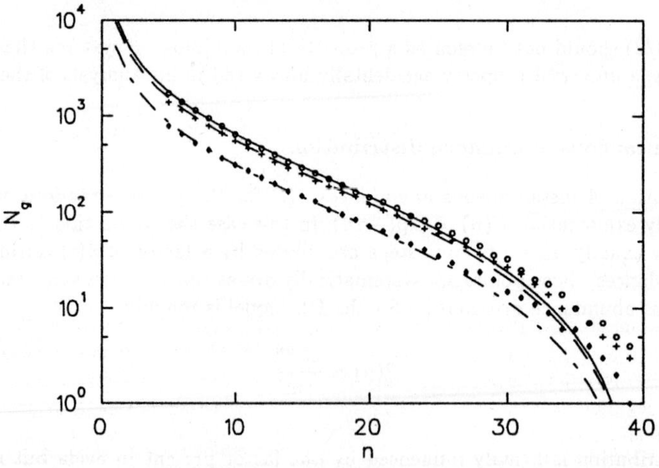

Fig. 6b. Genotype abundance distribution (ecological) from avida, omitting the "leading," fast replicating (high-fitness) creatures and compensating for that skewing on the neutral fitness abundance distribution. We have plotted here the genotype abundance distribution for genotypes with less than 40 members, in a 40 × 40 population, for small (diamonds), medium (crosses), and large (circles) mutation rates, semi-logarithmically. The short-dashed, long-dashed, and solid lines are the theoretical estimates from the DL model, with $R_s = R_g$ and no other free parameters except the normalization.

3.4 Age-area in the DL model

One can also make an age-area plot, comparable to Fig. 3, with data from the DL model simulation (see Fig. 7). Again, the data show that larger genotypes are older ones, consistent with evolution by infrequent jumps as opposed to gradual evolution. Comparison with Fig. 3 also highlights the fact that the vast majority of genotypes are neutrally fit, and explained well by the DL model. The few that lie above the "neutral" curve are the exceptional genotypes which have discovered some competitive advantage—and they are as unimportant to most statistical measures of diversity as they are fundamental to the process of evolution.

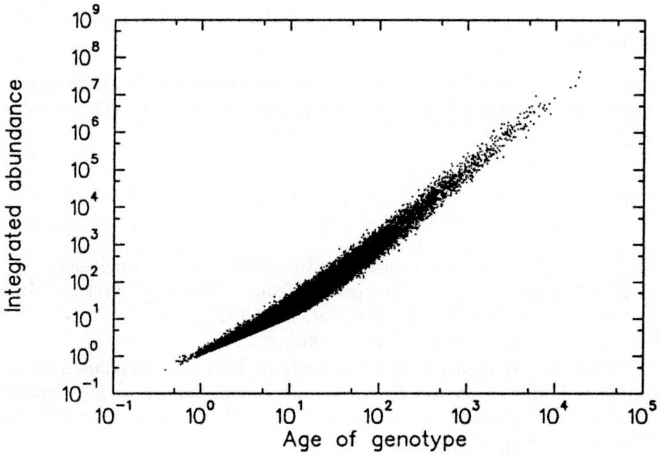

Fig. 7. Genotype age *vs.* integrated abundance for the DL model (analogous to Fig. 3).

4 Conclusions

Measuring abundance distributions in any living system is notoriously difficult, as witnessed by the limited amount of data in the literature, and reflected in a continuing uncertainty as to the nature of the distributions. The advent of Artificial Life systems analogous to simple natural systems has introduced the ability to study the old unanswered questions but with new methods and tools.

With Artificial Life it is now possible to gather huge amounts of population data from simple artificial evolving systems. From a simulation of an artificial world, not only the current ecology, but also the complete "paleontological" history can be recorded and examined in full detail. Parameters such as mutation rates and carrying capacities can also be adjusted easily.

In this paper we have re-examined some of the old questions by measuring both abundance distributions and the area-age relation in an Artificial Life system. Although it is currently not possible to distinguish species in these RNA-type systems, we are able to produce genotype abundance distributions showing universal behavior, and to speculate that the genotype distributions are the very foundation of the ubiquitous (-3/2) power-laws in species abundance curves obtained already in 1922 by Willis, and later throughout the taxonomic system and the paleontological record by Burlando.

Finally, we used our experience with the avida system to develop the DL stochastic model of population dynamics. This surprisingly simple model explains most of the dynamics of equilibrium periods, which in turn occupy the vast majority of the time of the simulations. It also provides accurate quantitative predictions for effects seen both in avida and in nature. In the future, we expect to refine the DL model to more accurately incorporate the spatial and finite-size effects present in the avida system, in order to isolate the unique features of an adaptive, evolving system.

Acknowledgements

We would like to thank C. Ofria for discussions and collaboration in the design of avida. This research was supported in part by NSF grant PHY90-13248. C.A. acknowledges a Caltech Division Fellowship.

References

1. Willis, J.C. (1922). *Age and Area.* Cambridge University Press, Cambridge.
2. Darwin, C (1859). *On the Origin of Species by Means of Natural Selection.* Murray, London.
3. Yule, G.U. (1924), Proc. Roy. Soc. London Ser. B **213**, 21.
4. Gould, S.J. and Eldredge, N. (1993) Nature **366**, 223.
5. Ray, T.S. (1992). An Approach to the Synthesis of Life. In *Artificial Life II*: Proc. of an Interdisciplinary Workshop on the Synthesis and Simulation of Living Systems, (Santa Fe Institute Studies in the Sciences of Complexity, Proc. Vol 10., edited by C.G. Langton et al., Addison-Wesley, Reading, MA, p. 371.
6. Ray, T.S. (1994). An Evolutionary Approach to Synthetic Biology. *Artificial Life* **1**, 195.
7. Adami, C. (1995). Learning and Complexity in Genetic Auto-Adaptive Systems. Physica **D** **80**, 154.
8. Adami, C. (1994). Self-Organized Criticality in Living Systems. KRL preprint MAP-167, Caltech.
9. Adami, C. (1994). On Modelling Life. In *Artificial Life IV*, proceedings of the Fourth International Workshop on the Synthesis and Simulation of Living Systems, R.A. Brooks and P. Maes eds. (MIT Press, Cambridge, MA,), p. 269; *Artificial Life* **1**, in print.
10. Adami, C. and Brown, C.T. (1994). Evolutionary Learning in the 2D Artificial Life System "Avida." In *Artificial Life IV*: Proc. of the Fourth International Workshop on the Synthesis and Simulation of Living Systems, R.A. Brooks and P. Maes eds. (MIT Press, Cambridge, MA), p. 377.
11. MacArthur, R.H. (1957). Proc. Nat. Acad. Sci. **43**, 293.
12. Motomura, I. (1932). J. Zool. Soc. Japan **44**, 379.
13. Teramoto, E., Shigesada, N., and Kawasaki, K. (1984). Species-Abundance Relation and Diversity. In: *Mathematical Ecology*: Proceedings of the Autumn Course held at the International Centre for Theoretical Physics, Trieste (Italy), Nov. 29 - Dec. 10 1982, Lecture Notes in Biomathematics, Springer Verlag.
14. Burlando, B. (1990). The Fractal Dimension of Taxonomic Systems. *J. theor. Biol.* **146**, 99; (1993); The Fractal Geometry of Evolution. *J. theor. Biol.* **163**, 161.

Elements of a Theory of Simulation

Steen Rasmussen[1,2] and Christopher L. Barrett[1,2]

[1] Los Alamos National Laboratory, Los Alamos NM 87545, USA
[2] Santa Fe Institute, 1399 Hyde Park Road, Santa Fe NM 87501, USA

Abstract. Artificial Life and the more general area of Complex Systems does not have a unified theoretical framework although most theoretical work in these areas is based on simulation. This is primarily due to an insufficient representational power of the classical mathematical frameworks for the description of discrete dynamical systems of interacting objects with often complex internal states.
Unlike computation or the numerical analysis of differential equations, simulation does not have a well established conceptual and mathematical foundation. Simulation is an arguable unique union of modeling and computation. However, simulation also qualifies as a separate species of system representation with its own motivations, characteristics, and implications. This work outlines how simulation can be rooted in mathematics and shows which properties some of the elements of such a mathematical framework has.
The properties of simulation are described and analyzed in terms of properties of dynamical systems. It is shown how and why a simulation produces emergent behavior and why the analysis of the dynamics of the system being simulated always is an analysis of emergent phenomena. Indeed, the single fundamental class of properties of the natural world that simulation will open to new understanding, is that which occurs only in the dynamics produced by the interactions of the components of complex systems. Simulation offers a synthetic, formal framework for the experimental mathematics of representation and analysis of complex dynamical systems.
A notion of a universal simulator and the definition of simulatability is proposed. This allows a description of conditions under which simulations can distribute update functions over system components, thereby determining simulatability. The connection between the notion of simulatability and the notion of computability is defined and the concepts are distinguished. The basis of practical detection methods for determining effectively non-simulatable systems in practice is presented.
The conceptual framework is illustrated through examples from molecular self-assembly end engineering.

Keywords: simulatability, computability, dynamics, emergence, system representation, universal simulator

1 Introduction

It is typical in both science and engineering to be interested in properties or causal details of a phenomena for which we do not have an adequate, explicit model. For instance this is the case when we are investigating the complicated phenomena of life. In other less complicated situations a model of the phenomena may not yet be derived; indeed, the derivation may be the very question under investigation. Also, analytic solutions may not be tractable due to inherent system complexity.

In these situations it is increasingly common to resort to *synthetic* methods using computer simulation. In a certain sense then, the comments that follow may seem as merely common sense. However they lead, we believe, to genuinely important considerations that can be briefly summarized in the following paragraph.

The essence of synthetic methods is that a simulation is a mechanism which interacts many state transition models of individual subsystems (i.e., system components) and thereby *generates* system dynamical phenomena. There is nothing inherently explanatory in this stage of investigation, it is essentially representational. The issues are that the system dynamical representation is *implicit* and *constructive*: The relations that constitute the properties of interest are nowhere explicitly encoded in the simulated component subsystems, but rather *emerge* and become accessible to *observation* as a result of the collective effects of computed interactions among these subsystems. This implies that certain assumptions have been made concerning the status of the capabilities of simulation as a species of representation in relation to computability, observability, numerical stability, and many other issues.

In addition to these important and formalizable issues are the need to establish an elementary and general concept of the generative concept of emergence. That is to separate the dynamics of the simulation mechanism, which is itself an iterated system, from the simulated system, which itself is represented in the framework provided by the simulation system. Thus, there is a need to relate the concept of emergence and the concept of simulation. Some clarification of at least the basic form of these issues is the aim of this paper.

Consistent with what is summarized in the above paragraph, the general research, or indeed any application-specific, program of investigating or developing control strategies for complex dynamical phenomena using simulation is in four basic parts: (i) We must always be aware that a simulation is generating dynamical phenomena at a level which is *higher* than the level from which the elemental interactions are described. If we are to exploit simulation it is necessary to understand what this capability to produce hierarchies of emergent relations implies. (ii) We must have methods with which to identify the elements of the underlying system that create the phenomena of interest. (iii) It is then necessary to formulate models of the important underlying subsystems (those that define the elemental subsystems and the element-element or object-object interactions). (iv) Finally, we must create the framework in which the simulation of the subsystems in interaction is composed, and embody the system representation in that framework so that the phenomena of interest can be generated and analyzed.

Part (i) does not seem to have been treated in general terms of dynamics, which is odd, since it is the foundation of all simulation-based work in many disciplines. Part (ii) is problem specific, but general principles do exist such as those mentioned above that, in one form or another, occupy the attention of systems science. Part (iii) involves how an appropriate "level of aggregation" and useful "collective coordinates" can be chosen. These are not simple questions. Among other things, aggregation depends on which global system properties are of interest, how the component representations can be made as parsimoniously as possible, and what can be observed about the system behavior. Parts (iii) and (iv) can be combined to form a single, broader, question: "Given a system composed of many interacting subsystems, how does one formulate models of the subsystems and cause them to interact in a simulation environment that will generate appropriate global system dynamics?" Or the Artificial Life variant: "How can we generate life-like behaviors using low-level, local rules?" Over the years, a variety of proposals has been given in an attempt to answer this question, depending on the characteristics of the system under investigation, the kind of system properties of interest - as well as taste.

In large complex systems the system, in the sense of the generator of the dynamics of interest, is implicitly presumed to exist. For example, we presume that a solution of polymers with hydrophilic heads and hydrophobic heads in water is indeed a system. Furthermore, we assume that it is a dynamical system in the sense that the state of the system at time, t, and some "state transition rules" completely determine the state of the system at time, $t + \Delta t$. That is to say that the system has a "model". However, we doubt the possibility of writing down an explicit analytical expression of these dynamics in terms of all of the relevant state variables and parameters and, therefore, the model of that system. Moreover, we doubt the tractability of the solution of such a model even if one could be somehow defined.

However, we do not discount the possibility of modeling each of the relevant elemental subsystems in isolation. By "elemental" subsystems we mean as monomers and solvent molecules [11], or perhaps as vehicles, roadway segments, signals, and travel goals in a traffic system [3] [9]. We can imagine various specialist-practitioners to be able to define or, at least, hypothesize possible relevant ental subsystems and characterize local interaction rules. We can imagine modeling these individual subsystems and interaction rules well enough to define the state transition and interaction possibilities of a single class of subsystems that could be inherited by every instance of that class.

Given an object-entity-subsystem perspective, interactions can in general be viewed as *discrete events* among the subsystems undergoing local state changes and communicating these state changes to its neighbors in some space. That is to say that the interactions can be viewed as calculable by means of discrete event, object-oriented, simulation of collections of subsystems.

The concept of an *event driven simulation* contains the most general updating scheme for a simulation, since an event can either be externally or internally

[3] As in TRANSIMS, an ongoing, large scale Transportation ANalysis and SIMulation System project at the Los Alamos National Laboratory.

generated, and, as a special case, an event can also be *a time step*. Thus, a time stepped simulation is a special case of an event driven simulation, namely a discrete event simulation where the update is driven by the event that a clock-entity object produces as its internal state and transmits to all other objects. It is perhaps more accurate to say that any time stepped method can be simulated in some event driven method. It is, however, not the purpose of this paper to develop new update schemes or to review the extensive literature in this area, (e.g. Jefferson[6] and Lubachevsky[7]). We only mention the issue here to clarify our use of language and basic concepts as well as to set the stage for what we do intend to investigate. These issues will include the most general issues of the concept of object state update and the coordination, the scheduling, of updates of interacting object-subsystem.

The properties and consequences of a generalization of the simulation scheduling problem, together with the notion of hierarchies of emergent dynamical relations, seem to us to form the elementary foundations of simulation rooted in dynamics.

We have for the current presentation mostly focused on discrete space and discrete time systems (mappings), which, for the most part, are defined through interacting objects with some (minimal) internal state complexity. We have chosen to do so because the proposed framework is natural for such systems, but it should be noted that continuous space, continuous time, dynamical systems, equally well can be treated in the given framework. Moreover, discrete space, discrete time systems do not have any other general formal framework within which the dynamical properties can be generated and analyzed.

2 Simulation

This paper describes simulation as an *iterated mapping* of a (usually large and complicated) system. The simulated system is usually decomposed to a level where subsystems or system components are individually defined as encapsulated objects that calculate and communicate internal state. The simulation is an iterative system in which the simulated system is represented and its dynamics calculated. Thus the simulation and the simulated system are both dynamical systems and the interplay between the dynamics of the coordination of the simulation updates and the dynamics calculated in the time series of system states are essential issues.

In the above paragraph, we have distinguished four "systems" that comprise a simulation. We assume the existence of some Σ_R, a real or natural system in the world that we are interested in, $\Sigma_{(S_i \in M)}$, models of subsystems S_i of this system and rules that define interactions among the subsystems, Σ_S, a simulation of Σ_R involving $\Sigma_{(S_i \in M)}$ and some update functional U, and finally, Σ_C, a formal (and equivalent physical) computing machine on which Σ_S is implemented.

Definition: The *objects* (elements or subsystems) in a simulation are defined as

$$S_i = S_i(f_i, I_{ij}, x_i, t_i), i \text{ and } j = 1, ..., n, \tag{1}$$

where f_i is the representation of the dynamics of the ith object and where $I_{ij,j=1,\ldots,n}$ is the ith object's interaction rules with other objects j. Interaction and dynamics operate on x_i, the state of the ith object. t_i is the local object time coordinate.

Definition of $\Sigma_{(S_i \in M)}$: S_i is an element in the system $\Sigma_{(S_i \in M)}$; that is, S_i is a *model* of the ith element of the set of modeled system elements in M, $i = 1, ..., n$. Thus, the algorithmic part of S_i is equivalent to f_i and I_{ij}.

Definition of U: An object *update functional* U is the state transition

$$S_i(t_i + \Delta_i) \leftarrow S_i(t_i), i = 1, ..., n, \tag{2}$$

or

$$S_i(t_i + \Delta_i) = U(S_i(t_i)), i = 1, ..., n, \tag{3}$$

where U, the update functional, defines, organizes and executes the formal iterative procedure that prescribes the state transition.

Definition of Σ_S: A *simulation* is the *iteration* of object updates over the entire set of objects

$$\{S_i(t+1)\} \leftarrow \{S_i(t)\}, i = 1, ..., n. \tag{4}$$

or

$$\{S_i(t+1)\} = U(\{S_i(t)\}), i = 1, ..., n. \tag{5}$$

A valid update functional U also needs to be able to time align all objects at regular intervals or at a given time, perhaps at each update. Note that f_i together with I_{ij}, x_i and U *implicitly* defines the dynamical properties of the system. U can be viewed as the "active" part of Σ_S where as f_i, I_{ij} and x_i can be viewed as the "passive" parts of Σ_S.

Thus, the iteration of the dynamics of $\Sigma_{S_i \in M}$ using U constitutes a formalization of Σ_S, the simulation system.

Definition of Σ_C: The formal, or equivalent physical implementation, of the mechanisms of the iteration procedure that prescribe the interactions and consequent object state transitions and their storage. Σ_C is normally a physically and conceptually digital computer of some kind.

3 Emergence

Having defined n objects or structures $S_i^1 \in \Sigma_M$ and an update functional U at some level of description, say L^1, we now also introduce an observational function O^1 by which the objects can be inspected. Iterating Σ_M using U a new structure S^2 may be produced over time

$$S^2 \leftarrow U\{S_i^1(f_i, I_{ij}^1, x_i, t_i),\ O^1\}, i \text{ and } j = 1, ..., n. \tag{6}$$

This is what we call a *second order structure* occuring at level L^2. This new structure may be subjected to a possible new kind of observer O^2.

Definition: We define that a property P is *emergent* iff

$$P \in O^2(S^2),\ but\ P\ \notin O^1(S_i^1). \tag{7}$$

In this sense emergence depends essentially on the observer in use which may be *internal* or *external.* It should be noted that the generated, emergent properties may be *computable* or *non-computable.* For a comprehensive discussion of emergence we refer to [1].

This process can be iterated in a *cumulative,* not necessarily a *recursive,* way to form *higher order emergent structures* or *hyperstructures* of e.g. order N:

$$S^N \leftarrow U(S_i^{N-1}, O^{N-1}, S_k^{N-2}, O^{N-2}, \ldots) \tag{8}$$

Note that the definition of an observation function is no more - or just as - arbitrary as the definition of the objects and their interactions.

Examples of emergent properties could for instance be the dynamical properties of a polymer in solution or the properties of a congestion in a traffic system. The polymer as well as the congestion can be viewed as S^2 structures. A lower level L^1 description of the interactions will in the polymer example mean to describe the monomer-monomer interactions together with the monomer-solvent molecule interactions (S_i^1-S_j^1 interactions). In the example of traffic congestion it means to describe the vehicle-vehicle interactions together with the vehicle-roadway and -signal interactions (again S_i^1-S_j^1 interactions).

In these examples the S_i^1 interactions generate the S^2 phenomena, but the S^2 structures also have a *downward causal effect* on the S_i^1 structures. The polymer restricts the dynamics of the monomers, that it is made up of. The jam does the same, it also restricts the dynamics of the vehicles it is made up of.

However, emergent properties, as defined above, may not always have a downward causal effect. For example, the joint distribution of heads and tails generated from two independent coin flips is an emergent property of the system, but the distribution does obviously not have any influence on the dynamics of the coins.

A central question to ask here is: What is the *minimal* (or *critical*) *object complexity* needed to generate an emergent property of a given order in Σ_S? Complexity here refers to computational complexity, which may be defined through the time (or number of steps) or the capacity (memory), which at a minimum, is needed to generate the particular property [10].

4 Simulation and Emergence

It is in the general case very difficult, and perhaps in some cases even impossible, to come up with a direct, a priori description (a model) of the dynamics of the phenomena S^2 of interest in systems consisting of many, interacting elements with some internal complexity. In general it may, however, be possible to identify the level, say L^1, from which the phenomena of interest *emerges* and where it in a direct way is possible to describe the interactions or the dynamics of the elements or objects that generate S^2.

If we assume that a formal description of the object-object interactions is possible at level L^1 and that some observation mechanism O^2 exists so that properties of S^2 can be detected and their dynamics followed, then the situation is the following at level L^1: Explicit models S_i^1 exist to describe the dynamics of and the interactions between the n objects where the object's states depend on each other. However, a global state dynamics function F^1 may only *implicitly* exist at level L^1. F^1 is the global function that describes the system wide state changes caused by the object-object interactions described by the set of local f_is and I_{ij}s. The total system state $\chi^1(t)$ at level L^1 can be obtained via appropriate observational functions O^1 successively applied to each of the objects. Thus,

$$\chi^1(t) = \{x_1^1(t), ..., x_n^1(t)\}. \tag{9}$$

The state dynamics function F^1 is therefore always at least implicitly given at level L^1, since χ can be computed at any time. Thus the description of the L^1 dynamics is in principle known on the form

$$\chi^1(t+1) \leftarrow F^1(\chi^1(t)). \tag{10}$$

To actually *produce* the dynamics some update functional U is needed which is able to organize the update of the interacting set of objects in a consistent way. Assuming that some update functional U exists we have

$$\{S_1^1(t+1), ..., S_n^1(t+1)\} = U(\{S_1^1(t), ..., S_n^1(t)\}) \tag{11}$$

or

$$\chi^1(t+1) = U(F^1(\chi^1(t))). \tag{12}$$

Thereby the dynamics of system Σ_S can be *generated.*

From the above it is clear that whenever it is possible to define an update functional U that can organize the interactions of the objects defined at level L^1 through the set of models M, then the L^2 phenomenon of interest S^2 *emerges* and can be followed, applying the observation function O^2. Note that this is possible even without knowing F^1 explicitly. Thus, a recursive application of U to the objects *generates* S^2 and the dynamics of S^2 (which is a property P^2 of S^2) can then be followed by a recursive application of O^2.

The central point is that *a simulation is a representational mechanism that is distinguished by its capacity to generate relations that are not explicitly encoded.*

Recall that S^2 for instance could be a polymer described through monomer-monomer and monomer-solvent interactions. In that situation P^2 could be the polymer elasticity. S^2 could also be a traffic jam described through vehicle-vehicle and vehicle-roadway interactions and then P^2 could for instance be the lifetime of a jam.

Thus, we have

$$S^2 \leftarrow U(\{S_1^1, ..., S_n^1\}) \tag{13}$$

and

$$P^2 = O^2(S^2). \tag{14}$$

Note that S^2 in (13) is defined through the implicit (emergent) relations that are generated between the objects on the left hand side of (11).

Recall that we in principle would like to be able to follow the state dynamics of Σ_R through some Σ_M at level L^2 in a direct way. But this requires that the state variables $\{x_1^2(t), ..., x_m^2\} = \chi^2(t)$ together with the state dynamics function F^2 at level L^2 were known explicitly, thus that we could write

$$\chi^2(t+1) \leftarrow F^2(\chi^2(t)), \tag{15}$$

which expresses that the state dynamics can be derived from the current state by applying some F^2. Knowing F^2 would in principle also enable some update functional U^2 to produce the dynamics

$$\chi^2(t+1) = U^2(F^2(\chi^2(t))). \tag{16}$$

Since we assume that the system cannot *a priori* be described at level L^2, but that it can be described at level L^1 the dynamics at level L^2 can be *generated* by *simulating* the interactions of the objects $S_1^1, ..., S_n^1$ at level L^1. In other words: By simulating the interactions of S_i^1 at level L^1 the phenomena and relations of interest at level L^2 will *emerge*.

Note that simulation is a direct generative way to obtain knowledge of this kind of non-explicitly encoded (dynamical) relations and phenomena. *Simulation is therefore a natural method to study emergence.* The non-explicitly encoded relations may later explicitely be modeled in a closed form, but that is irrelevant.

In fact, science if full of descriptions of systems where we have both an L^1 and an L^2 description. Classical examples include the Statistical Mechanical (L^1) versus the Thermodynamical (L^2) description of matter as well as the Lattice Gas Automata for fluid particle dynamics (L^1) versus Navier Stokes equations for macroscopic fluid dynamics (L^2).

5 Simulatability

A major remaining issue concerning simulation is: Under which conditions does an update functional exist for a large number of different, interacting objects? It is obvious that the object interactions can be rather involved and thus difficult to "untangle" so that the objects actually can be updated.

Let $\varphi(S_q^1, ..., S_s^1)$ be a hierarchically distinct representation of a subset of the interacting objects $S_q^1, ..., S_s^1$. Thus, $\varphi(S_q^1, ..., S_s^1)$ defines a sub-aggregation (an aggregated model) of some of the objects.

Definition: If

$$U(\{S_1^1, ..., S_n^1\}) = \{U(S_i^1), U(S_j^1), ..., U(\varphi(S_q^1, .., S_s^1)), U(S_l^1), ..., U(S_p^1)\} \quad (17)$$

for some order of the objects, then the update U is ***distributable*** over the decomposition Σ_M of the system and each object and object aggregation can be updated independently of each other.

Note that if U is operating in a sequential manner the list on the right hand side of (17) is ordered. Thus the sequence in which the objects are updated matters. If U is operating in a parallel manner the order in which the objects are updated does not matter.

A simple example of a ***subaggregation*** in a simulation is a particle collision in a lattice gas automata [4] [5]. As long as the fluid particles do not interact they are updated independently of each other - they just propagate along the lattice. But when they collide they are aggregated and the individual particle velocities after the collision are given by a collision table which takes the incoming, colliding particle velocities as arguments.

The nature of the update functional U has a significant influence on the dynamics. The same model decomposition M will in general generate different dynamical properties if different update functionals U are applied. For instance, the elementary 1D, radius one cellular automata with rule 58 (00111010) will exhibit very different dynamics using a parallel or a random update respectively. Obviously, both the random and the parallel update ***distributes*** over any of the elementary rules on the 1D lattice.

It is clear that the representation of the objects and their interactions, M (the models of interactions at L^1) is crucial for whether a given update U can distribute or not. It is assumed that each object, given the (M, U) pair, individually can be updated when supplied with the appropriate state information from its communicating objects.

Thus, a system is ***simulatable*** iff there exists a pair (M, U) such that U ***distributes*** over $S_i \in M$.

The above follows directly from our assumptions and definitions. Since it is assumed that each object, or sub-aggregation of objects, given the (M, U) pair, can be updated individually and that problems can only occur due to the order and the organization of the object-object interactions. Since each of the objects and/or sub-aggregations can be updated independently in a system where the (M, U) pair allows the update to distribute the above follows.

A direct consequence of this is that if no distributable U exists for some sub-aggregation of the objects M which allows the update to distribute, then the system is non-simulatable. Conversely, if no M exists so that a given U can distribute then the system is also non-simulatable.

A situation may occur where the smallest sub-aggregation which can be updated independently is the system itself. In this situation the sub-aggregation defines a *model* for the whole system at level L^1.

An example of a *non-simulatable* systems consider a model polymer defined on a 2D lattice [11]. Assume that the polymer is embedded in some solvent (heat bath) and that we would like to update each of the monomers in parallel. To perform the update and thus generate a possible new (lattice) position for each of the monomers in the polymer the polymer should not break and it should respect its excluded volume. That means that each monomer requires information about every other monomer in the polymer to be able to resolve possible conflicts due to the no-break and the excluded volume constraints. Thus, the minimal complexity of the model M depends on the polymer length! Only allowing, say k steps, in the update cycle (in the model of monomer-monomer and solvent-monomer interactions) implies that polymers above a certain (finite) length are non-simulatable, because the update does not distribute over the objects, since the objects cannot be updated independently of each other after they have communicated with each other.

However, defining a update scheduling such that alternating monomers are updated in every first and every second part of the update cycle, polymers of any length can be simulated. The monomer models M becomes much simpler using such a two step parallel scheduling instead of using a strict parallel update [11].

6 A Universal Simulator

We define a universal simulator US as a machine that is able to resolve all causal dependencies among the objects S_i^1. Thus, a US can determine whether the system given the (M, U) pair is simulatable or not. Further it can give an appropriate order of updating for the objects if it is simulatable and detect where the problems are if the system is non-simulatable. Thus, the *scheduling problem* lives in the US. Since the causal dependencies is being done on-line it may be of particular, practical interest for event driven simulations.

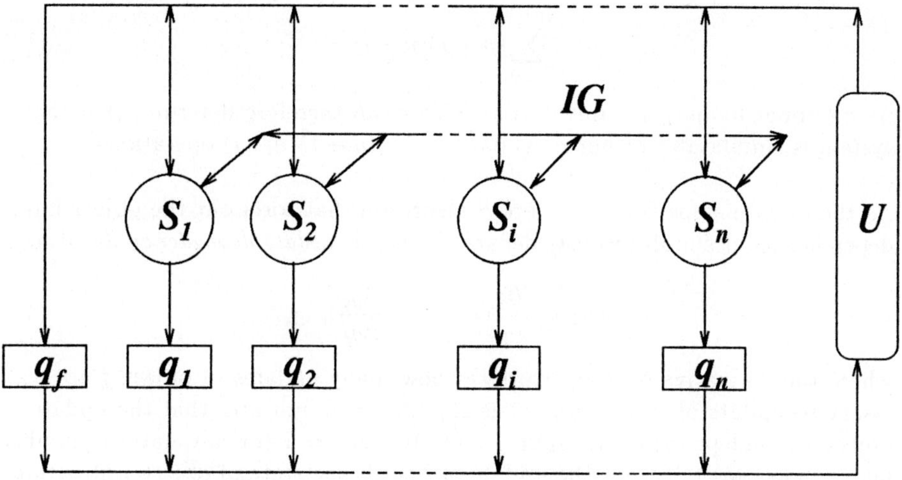

Fig. 1. *A universal simulator. The objects are denoted S_i^1, the q_i's denote counters associated with the objects, $\mathrm{q} = (q_1, ..., q_n)$, $Q = \sum_i(q_i)$ (the sum of all the individual object counters). q_f counts the failed update trails, the update functional is denoted U and the interaction graph is denoted IG.*

The structure of a US is defined in figure 6.

To perform a complete update of Σ_S each of the objects S_i^1 need to be updated at least once and all need to be aligned in time. Assume that we are at *time* 1. The dynamics of the machine is as follows: Start by attempting to update S_1^1. If S_1^1 does not depend on the state of any other object at *time* 1 it is updated and its associated counter q_1 is incremented by one. If, however, S_1^1 has dependencies (depends of the state of one or more of the other objects at *time* 1) the object S_1^1 is exited and q_f is incremented and the next object S_2^1 is attempted to be updated. If also S_2^1 has dependencies q_f is again incremented and object 3 is attempted to be updated and so forth until an updatable object, say S_i^1 is found.

Without any loss of generality assume that all objects have internal dynamics on the same time scale (no objects need to be updated with any smaller time resolution than any other) and that U is a discrete time update.

There are n ways to pick the first object to update and each of these objects have at most $(n-1)$ objects they can depend on. Thus an upper bound on the number of operations needed to find the first updatable object is $n(n-1)$, given that it exists. The second updatable object does at most require $(n-1)$ operations to find and it takes at most $(n-1)$ operations to check whether any of the other objects influence it. Thus, an upper bound on the number of operations it takes to update the whole system once is given by

$$\sum_{i=0}^{n-1}(n-1)(n-i). \tag{18}$$

As an upper bound, the universal simulator can therefore determine whether a system is simulatable or not in at most $\sum_{i=0}^{n-1}(n-1)(n-i)$ operations.

As the update steps through the US-algorithm that sorts out the object interdependencies it simultaneously defines the *object update dependency Jacobian.*

$$D\mathbf{q} = (\frac{\partial q_i}{\partial q_j})_{n\times n} \equiv (\frac{\Delta q_i}{\Delta q_j})_{n\times n} \tag{19}$$

where the derivative $\partial q_i/\partial q_j$ expresses how many updates of object j are necessary to update object i once. Thus $\Delta q_i/\Delta q_j = 0$ indicates that the update of object i is independent of object j and $\Delta q_i/\Delta q_j = 1$ (or any natural number larger than one) tells that the update of i needs the state of object j at current time before it can be updated. As a special case $\Delta q_i/\Delta q_j \equiv 1$ for $i = j$.

A system Σ_S is *simulatable* if $D\mathbf{q}$ is a diagonal matrix (unit matrix). This follows directly from the definition of the object update dependency Jacobian.

If $D\mathbf{q}$ contains sub-matrices on the diagonal and Σ_S is found non-simulatable it is an indication that Σ_S could become simulatable by the construction of sub-aggregations including the objects contained within each of the sub-matrices. Recall the definition of simulatability in section 5.

In the case of an upper diagonal object update dependency Jacobian a reorganization of the ordering of the updates can make a diagonal matrix.

Note that a parallel updatable system as well as a strictly sequential updatable system have the same structure in their update dependency matrices. If $D_i\mathbf{q}$ is a diagonal matrix it only insures that the system is simulatable and it gives a causal order of the object updates. Other orderings may exist.

Since $\partial Q/\partial q_i$, $Q = \sum_i(q_i)$, defines the relative computational load of the ith object the problem of *load balancing* also naturally lives in the US.

Also note that since the matrix $D\mathbf{q}$ is purely emperical, that is it evolves during the course of the simulation, the control problem associated with coordinating the simulation update is formalizable in terms of the trajectories of $D\mathbf{q}$ as a function of the update actions.

7 Achievable Simulatability

Suppose we are given a simulation system Σ_S with n objects implemented on some physical or formal machine Σ_C (a computer). Assume that we need to know whether this computer is able to handle the integration of the system for the time interval T_S (model time) within some pre-specified time interval T_R (real time). In other words; Σ_C needs to be updated T_M time units model time, within T_R time units real time.

Without any loss of generality we assume that Σ_C is a sequential machine. Further assume that s_i is the number of cpu cycles it takes to update the simplest object, S_i, and that τ_i is the (real) time of one cpu cycle. The minimal number of cpu cycles to update Σ_S is therefor ns_i and

$$\tau_{min} = ns_i\tau_i \tag{20}$$

therefore defines the minimal (real) time it takes to update the whole systems Σ_S once.

$$R_{max} = \frac{1}{\tau_{min}} \tag{21}$$

therefor defines the maximal (real) rate by which Σ_C can be updates. If we assume that the largest time increment (model time) allowed in each update of the system is Δt then

$$N = \frac{T_M}{\Delta t} \tag{22}$$

defines the minimal number of system (Σ_S) updates to be done to complete the task.

The minimal number of updates left to be done at (real) time t is therefore

$$Q_{min}(t) = ns_i(N - \frac{Q(t)}{n}) \tag{23}$$

where Q(t) is defined as the total number of updates performed by the machine at any given time t (recall the definition of Q in the universal simulator in section (6)).

Thus, an on-line (optimistic) estimate of whether the desired integration can be accomplished within the specified time frame of T_R time units, real time, can therefore be found through a comparison of the current, minimum update rate needed to complete the task

$$R_{needed}(t) = \int_0^t \frac{Q_{min}(\tau)}{T_R - \tau} d\tau \tag{24}$$

and the maximal, real time, rate by which Σ_C can be updates (recall (21)).

If $R_{needed} > R_{max}$ at any time t then the system Σ_S is not *achievably* simulatable on Σ_C.

This follows directly of the definition of R_{needed} and R_{max}.

8 Computability and Simulatability

Consider the iterated map

$$z(t+1) = z^2(t) + c, \tag{25}$$

where z and c are complex numbers (see for instance [8]). It can be shown that for most c the location of the closure of the unstable equilibrium set, the Julia set, for this mapping is *non-computable* where computation is defined over the real numbers [2] [3]. This means that it is *non-decidable* whether a given point in the complex plane is a member in the Julia set. Note that it is the observational function O^2, that decides membership in a set with certain properties, that is non-computable.

The above mapping is obviously simulatable and the Julia sets are also obviously emergent S^2 structures for this mapping. Thus, this example shows a simulatable system with non-computable emergent properties. The concepts of simulatability and computability are distinct for the purposes of the discussion here.

9 Conclusion

We have demonstrated that the foremost property of simulation is its ability to produce emergence. A simulation is an *emergence engine*. It is a representational mechanism that is distinguished by its capacity to generate relations that are not explicitly encoded. This ability enables us to study complicated dynamical properties which are otherwise intractable.

A system may be non-simulatable for certain (model, update) pairs pairs, but simulatable for other (model, update) pairs. Only when the update distributes over the ensemble of objects is a system simulatable.

We have defined and discussed the notion of a universal simulator and shown how the scheduling problem as well as the problem of load balancing naturally lives in this machine. Using the universal simulator we have shown that for any given set of model formulations of the interacting objects that constitute the system, together with a given update functional, it is possible in a finite number of operations to determine whether a system is simulatable or not. This is equivalent to a diagonal form of a corrosponding object update dependency Jacobian. This machine is also a useful practical device as an indicator for whether a system is (achievable) simulatable or not given a physical or formal machine (computer) where the simulation is implemented.

We have also demonstrated that a system may be simulatable, but have non-computable, emergent properties and thus the concepts of computability and simulatability are thus distinct for the purposes of the discussion here.

Acknowledgments

We would like to thank Nils Baas and Kai Nagel for constructive comments on earlier versions of the text as well as the many people at LANL and at SFI who have helped clarify our first ideas on the topic. It has been a long process.

References

1. N.A. Baas. Emergence, hierarchies and hyperstructures. *Artificial Life III, proceedings, ed. C.G. Langton, Santa Fe Institute Studies in the Sciences of Complexity /Addison-Wesley, New York*, XVII:515–537, 1994.
2. L. Blum. Lectures on a theory of computation and complexity over the reals (or an arbitrary ring). *Lectures in Complex Systems, SFI Studies in the Sciences of Complexity/ Addison-Wesley, New York*, II:1–47, 1989.
3. L. Blum, M. Shub, and S. Smale. On the theory of computation and complexity over the real numbers: Np completness, recursive functions and universal machines. *Bull. Amer. Math. Soc*, 21(1):1–46, 1989.
4. U. Frisch, B. Hasslacher, and Y. Pomeau. Lattice-gas automata for the navier-stokes equation. *Physical Review Letters*, Vol 56:1505–1508, 1986.
5. B. Hasslacher. Discrete fluids. *Los Alamos Science*, Special Issue (15):175–217, 1987.
6. D.R. Jefferson. Virtual time. *ACM Transactions on Programming Languages and Systems*, Vol 7, No. 3:404–425, 1985.
7. B. Lubachevsky and A.Weiss. An analysis of rollback-based simulation. *ACM Transactions on Programming Languages and Systems*, Vol. 1, No. 2:154–193, 1991.
8. B.B. Mandelbrot. *The fractal geometry of nature*. W.H. Freeman, New York, 1983.
9. K. Nagel and S. Rasmussen. Traffic at the edge of chaos. *Artificial Life IV, proceedings, eds. R.A. Brooks and P. Maes/MIT Press, Cambridge Massachusetts*, pages 222–235, 1994.
10. C.H. Papadimitriou. *Computational Complexity*. Addison-Wesley, 1994.
11. S. Rasmussen and J.R. Smith. Lattice polymer automata. *Ber. Bunsengs. Phys. Chem.*, 98 (No. 3):1185–1193, 1994.

To Simulate or Not to Simulate: A Problem of Minimising Functional Logical Depth

Gregory R. Mulhauser

Department of Philosophy, University of Glasgow, Scotland
scarab@udcf.gla.ac.uk

Abstract. I define a new measure of complexity called functional logical depth and apply the measure to specify the levels of description at which, for the purposes of creating artificial life, it may be most appropriate to simulate the temporal evolution of classes of real living organisms. According to the approach outlined here, when the functional logical depth of particular information transforming relationships in such a simulation is *minimised*, the more complex transforms which typify life appear from the simpler relationships as emergent properties. I situate this view within a context of computational convenience and evolutionary plausibility and discuss implications for developing genetic algorithms and for evaluating various arguments about levels of description in living systems.

1 Too Much, Too Little, or Just Right?

While many artificial life projects have little to do with directly imitating existing biological life, often biological life offers a guide as to what behaviours we should seek in artificial creations. In some cases, we do try explicitly to simulate the complex adaptive behaviours of particular classes of biological organisms. Yet the adaptive changes within real organisms may be described at many different levels of detail. At the lowest levels, living organisms are collections of elementary particles interacting according to the laws of physics, while at a higher level some of the most advanced ones may be understood as sophisticated agents with their own subjective awareness and characteristic psychology. We are here concerned with the problem of selecting the level or levels of description at which it is most useful in the context of artificial life to simulate a living organism. This problem appears clearly in an example contrasting the two extremes of highest and lowest levels of description.

At the lowest level, we might—if we had ridiculously capable computing resources or if we were interested only in the very simplest organisms—choose to simulate life by tracing the temporal evolution of an enormous collection of interacting particles. If our simulation was very good, we might even observe behaviour from our simulated collection of particles which resembled the behaviour of the collections of particles we call living organisms. But simulating a living system at this level offers little insight into the patterns of organisation or information transformation characteristic of living things. This kind of simulation traces *too much* internal complexity, complexity which may be largely irrelevant to what it means to be alive.

At the other extreme, we might take as inspiration John Searle's well known Chinese room example [19], in which all the input/output relationships necessary for one type of behaviour—conversing in the Chinese language—are captured by a an exhaustive rule book which tells what symbols should be returned as output in response to any possible input symbol. Whereas the simulation at the level of elementary particles captures every single interaction within an organism, whether relevant or not, a simulation along the lines of the Chinese room *ignores* every single interaction within an organism (unless, of course, the organism happens to have an exhaustive rule book inside). Such a simulation includes *too little* internal complexity of real organisms and probably ignores characteristics which are highly relevant to being alive. Both kinds of simulation might be able to converse with someone in Chinese perfectly well, yet most cognitive scientists and artificial life researchers would no doubt agree that both simulations have missed the point.

Problems which arise from trying to settle on one or more intermediate non-trivial levels at which organisms ought to be studied or simulated are at the heart of a range of debates in artificial life, artificial intelligence, cognitive science, and the philosophy of mind. They bear on arguments about the capabilities of connectionist models as compared to symbolic ones [10, 9, 22, 4, 7, 20, 3]. They also figure centrally in ongoing debates about the rôle of low level dynamics in cognitive state transitions [13, 11, 12, 14, 16, 18, 21], and they may be important in deciding on features of the genomes and fitness functions used in genetic algorithms development.

In the following, we discuss how the complexity of an input/output relationship may be quantified and how we might discern appropriate levels for simulation by minimising the complexity of certain features of simulations. We note that this approach makes sense from the standpoint of computational convenience and see how it is motivated by considerations from evolutionary biology. Before some closing thoughts, we also comment on the relevance for our purposes of the noncomputability of the standard complexity measures.

We begin with a rather lengthy sojourn into information theory.

2 Complexity Simplified

2.1 Standard Measures

Information theory offers at least two standard measures of complexity: the logical depth due to Charlie Bennett and the Kolmogorov/Chaitin/Solomonov, or KCS, definition of algorithmic complexity. The latter, also known as algorithmic information content, algorithmic randomness, or algorithmic entropy, is the length, in bits, of the most concise description (usually, the shortest program for a Universal Turing Machine) of a physical entity or bit string for a given level of accuracy. We apply to bit strings by evaluating the question of whether the string can be *compressed* in such a way that the compressed bit string together with its required decompression procedure can be specified significantly more succinctly than the original string.

The closely related logical depth measure [1, 2] is defined as the *execution time* of the shortest program for a universal computer which can generate a desired bit string or description of an object. (More precisely, it is the harmonic mean of all such programs.) The idea is that logically deep entities should contain internal evidence of having resulted from long computations or dynamical processes. Whereas wholly disordered strings can be generated quickly by long programs which just print them out, highly ordered strings might be generated more slowly by shorter programs. A slight improvement on this measure, called quantum logical depth, or Q-logical depth, is due to Deutsch. [8] Deutsch's Q-logical depth is keyed to the execution times of the shortest programs for his own Universal Quantum Computer. The key point of difference with ordinary logical depth stems from the suggestion that in Nature, random states are generated not by "long programs" but by short programs exploiting nondeterministic hardware. The quantum analogue of logical depth solves this minor problem by generating random sequences with very short and quickly executing programs. In this respect, however, Q-logical depth seems little different than logical depth measured, for instance, with respect to a wholly classical Bernoulli-Turing Machine. For present purposes, differences between the two are largely irrelevant.

The main shortcomings of both algorithmic information content, or KCS complexity, and the logical depth measures are first that they are typically applied to single bit strings, without regard for how those strings were generated, and second that there may be nearly arbitrary variation in the complexity of very similar strings. The first problem means we may comment only on the *outputs* of information transforming systems, but not on whatever processes instantiate them. Under the KCS measure, for instance, where complexity is related to the compressibility of a bit string, it does not matter whether a string was actually produced by tossing coins and noting heads or tails or by watching signals on the parity line of a digital computer with parity checked memory. Although coin tossing is a paradigmatic example of randomness and a digital computer is a paradigmatic example of order, the KCS measure of both strings may be similar. Likewise, although the logical depth measures are meant to extract information about how strings could have been generated, the shortest program to generate most strings is just a program which will print out that string, so in most cases logical depth will not differ markedly from some inverse relationship with algorithmic complexity.[1] In the parity checking example, although parity bits may actually be generated by the more computationally demanding application of XOR to successive bits of a byte, the shortest program which will reproduce a typical stream of parity bits is just one which prints them out.

The second problem with the two measures is the possibility for nearly arbitrary variation between two similar strings when one is specifiable succinctly as the initial condition of some dynamical system—which amounts to a compression system—and the other is not. One interesting example occurs when a chaotic dynamical system is interpreted in some way so as to generate bit strings: in this case, a single succinct initial condition may generate an infinite class of unique bit strings of length one to infinity. Many of these strings, highly compressible by means of the chaotic dynamical system, will be of intermediate algorithmic randomness and high logical depth (because the shortest program to generate them will specify the dynamical system and its initial condition, rather than the string itself). Yet there will be a class of other strings, very similar to these, which will be entirely algorithmically random and of low logical depth because they are *not* compressible with this dynamical systems method (and because the procedure for extracting them from similar strings which *are* so compressible cannot be specified succinctly).

For instance, consider digitised photographs of two very similar looking brothers. Although we might intuitively consider the two as of roughly similar complexity, it might be the case that bits in one digitised photograph could be specified from the

---

[1] This observation derives from the fact that most strings are not compressible (and thus cannot be generated by any significantly shorter program). Indeed, it can be shown easily that fewer than one in a million binary strings are compressible by more than 20 bits. While the overwhelming majority of strings have high algorithmic complexity and low logical depth, it is also true that logical depth appears to be at a maximum for intermediate values of algorithmic complexity and at a minimum when algorithmic complexity is either minimal or maximal.

output of some dynamical system with a very succinct initial condition, while the bits in the other could not (perhaps because the initial condition required many more bits to specify or because something like it couldn't be generated by that system at all). Thus, the complexity measures might return vastly different results for the two images. More importantly for our purposes, if we are interested in quantifying the complexity of an input/output system, it won't do to appeal to a complexity measure which indicates that sometimes the system may produce output of high algorithmic entropy or low logical depth and sometimes it may produce output of intermediate algorithmic entropy or high logical depth or whatever. This reveals little about the system *itself*. To get at the complexity of the system demands a new measure.

2.2 Functional Logical Depth

By focusing on the relationship between the input space and the output space of a system, the new measure of complexity overcomes some difficulties of the other two measures and accommodates new observations which they do not. Functional logical depth, or F-logical depth, is defined as the mean execution time, over all inputs, of the shortest Bernoulli-Turing machine description of the input/output relationship of the system in question. [18] That is, it is the average over all possible inputs of the length of time taken by the shortest program (or the harmonic mean of the average time taken by all such programs) to produce an output identical to that which the system in question would produce in response to the same inputs. (Hereafter, by 'complexity' we shall mean this 'functional logical depth' unless specified otherwise.[2]) I hope mathematicians everywhere will forgive the silliness of trying to express it in the following way, but in the simplest two dimensional case, the functional logical depth F of a relationship described by Bernoulli-Turing program S for a given level of precision P (i.e., how many bits we're looking for in the outputs), where "pseudo-function" E returns the execution time to produce an output of precision P for a single initial condition specified in x and y, might look something like:[3]

$$F(S,P)=\frac{\int_{y_0}^{y_b}\int_{x_0}^{x_a} E(S(x,y),P)\,dx\,dy}{(x_a-x_0)(y_b-y_0)}$$

Here we integrate over relevant x and y in the domain of the system, and dividing by the area of the domain space yields average execution time. Graphically, this is a process of scanning across an (x, y) plane and plotting a height along an E axis corresponding to the execution time to produce the output for each (x, y). The two

---

[2] Notice, incidentally, that F-logical depth is closely related to but not identical with the computational complexity of algorithm analysis, in which algorithms are ordered according to the number of steps which must be performed in their completion.

[3] This is really simplified beyond plausibility, with, for instance, an easy (x, y) domain over which we can integrate in this simple manner. Indeed, for some systems, both the (x, y) domain and the manifold in E might even be fractal. I've deliberately written it out in this entirely inaccurate way purely for illustrative purposes; shortly we observe that F-logical depth is not a computable value anyway.

integrals give the volume under this E manifold, which we divide by the area of the (x, y) domain to derive the average height of the manifold above the (x, y) domain.

A few minor issues deserve attention before we note more significant points about the measure. First, for nondeterministic systems, we aren't concerned that S produce outputs identical with that of the real system—how could it?—but only that its probability density function matches to within P.[4] Second, in comparing outputs, we won't be concerned with the time it might take otherwise output-identical systems to give their outputs, since for the purposes of F-logical depth, we measure the complexity of the *relationship* itself, not whether one system is quicker or slower in instantiating that relationship. Also, F-logical depth does not necessarily reflect the *internal* complexity of a particular process, but only the complexity of the input/output relationship of that process. So, any two processes which instantiate the same relationship are equivalent in terms of F-logical depth, regardless of their individual computational complexities. Finally, perhaps the clumsiness of the pseudo calculus will be excused in light of the fact that F-logical depth is no more a computable measure than KCS, logical depth, or Q-logical depth.[5] (All the complexity measures suffer from the shortcoming of undecidability: we can never know for sure of the shortest program for generating a string or instantiating a given input/output relationship.)

There are several things more worthwhile noticing about F-logical depth as a measure of complexity. First, the value is minimal where outputs are nondeterministically random or trivial. If there is no probability relation whatsoever between inputs and outputs, then for each input S can simply offer a scaled output from its random number generator and be done with it. Likewise, the F-logical depth is minimal if the relationship is trivial, since S might either just look up the appropriate output value in a table or output a constant value or whatever, according to the particulars of the trivial system in question. (This observation also reveals that F-logical depth is more meaningful for systems with continuous or with discrete but very large input spaces. For systems with discrete and small input spaces, it may always be quicker and shorter to exploit a simple lookup table than to undertake calculations more closely related to the real internal behaviour of the input/output system under consideration.)

At the other end of the complexity spectrum are the kinds of deep processes the products of which logical depth might have been able to pick out if the shortest program for generating most single strings weren't just one which printed the string. By defining the F-logical depth measure over all possible inputs, we have avoided one difficulty which arose for ordinary logical depth, namely, the problem of returning low complexity for those outputs which truly were created by long and complex processes but whose initial conditions in terms of those processes required very many bits to specify. Because F-logical depth requires only that we specify the dynamics of a mimicking system and not that we provide any particular set of initial conditions, functional logical depth is immune to such variations. Again by focusing on an overall input/output relationship, this measure also avoids the kind of variation in

---

4 The usefulness of F-logical depth distinctions depends on the value of P. At low precision, all processes look F-logically shallow.

5 We might, incidentally, measure a quantity similar to F-logical depth by applying the algorithmic complexity measure to a bit description of the relationship between a system's inputs and outputs. This approach does, however, entail certain difficulties which I believe make F-logical depth preferable.

complexity which occurs when, for instance, offering typewriters to a million generations of a million chimpanzees really does result in the creation of a highly ordered encyclopædia. F-logical depth just indicates the process *itself* is of low complexity and makes no claims about individual outputs.

Next we return to the question of levels of description and explore how F-logical depth may help us to choose the most appropriate levels for simulation.

3 What to Minimise...

3.1 Background Assumptions

The approach to complexity in levels of description which we will shortly explore is inspired by several assumptions. Most importantly, I take it that choosing levels of description for a living system is not merely a matter of coarse-graining the lowest level description of particles until we have thrown away enough detail to arrive at some higher level or levels where we describe arbitrary collections of such particles. Instead, levels of description and organisational features have a subtle reciprocal relationship. The patterns of organisation displayed by elementary quarks and leptons, for instance, suggest protons and neutrons and atoms at the next higher level of description, the patterns into which *those* patterns organise suggest the next higher level of molecules, and so forth. This is of course not always a straightforward progression, but in general it might proceed something like: elementary particle → atom → molecule...neuron → neuronal group → computational element...brain and so forth, where ellipses indicate we're skipping through several other levels. (What specific structures we place at each level are entirely irrelevant; I intend these only as an example.)

The second main assumption is that complex living organisms are typically composites of many different modular subsystems, all of which interact to some extent with at least some of the others and which contribute to the overall adaptive behaviour of the organism. Significantly, not all of these subsystems which might be relevant to understanding overall behaviour may appear at the same level of description. For instance, understanding a process of visual perception might require understanding systems described at the level of individual photoreceptors (in the retina), at the level of large neuronal groups (from the retina itself through the optic nerve) and repertoires (in the primary visual cortex), and so on. While describing the behaviour of individual neurons may be important at the stage of initial response to a light stimulus, individual neurons might be largely irrelevant to the collective properties of large repertoires in the primary visual cortex. That is, I take it that not all of the features of the inputs to particular subsystems are relevant for preserving the overall input/output relationship of an organism. Thus, any scheme for selecting levels for simulating a living organism must be able to accommodate simulation of different subsystems, or modules, at possibly different levels of description. It is to such a scheme that we turn our attention next.

3.2 Selecting Modules by Minimising Complexity

Our basic goal is to select levels of description so as to minimise simultaneously the functional logical depth of two sets of relationships. First, we seek to minimise the functional logical depth of the input/output relationships for each of the selected subsystems common to some class of living organisms we are wanting to simulate. At the same time, we would like to minimise the complexity of a relationship which takes a class of organisms and a set of subsystems as inputs and provides as output a description of an overall arrangement of the subsystems which would be appropriate for simulating the given organisms. Thus, returning to the vision example, we would like simultaneously to minimise the functional logical depth of the input/output relationships we are simulating at the level of the individual photoreceptor, the retina and optic nerve, and the primary visual cortex, *as well as* the relationship between these subsystems and the way they would have to be interconnected to mimic a given organism. (Obviously, the subsystems needn't be arranged serially, as in the vision example: they might be arranged instead in parallel or in some sophisticated combination.) Put nontechnically, the goal is to select levels of description to yield the simplest modules which can be linked together in the simplest ways. If the modules are selected to be too simple, specifying how they should be linked will be more complex; alternatively, if modules are selected to be very simple to link together, the relationships they themselves instantiate will be complex.

Because the quantities we would like to minimise are coupled, achieving minimal values for them simultaneously would not be a straightforward matter even if F-logical depth itself were computable. Moreover, there is no guarantee that a unique choice of subsystems will meet this requirement. For an oversimplified case of just two subsystems, we might imagine the relationship between the complexity of the two subsystems and that of the overall arrangement of those subsystems for some class of organisms as a manifold in 3-space. Points on this manifold which are closest to the origin then indicate the best minimisation of all complexities simultaneously.

Of course in practice the actual relationships between such values likely wouldn't be nearly so straightforward. Not only would it be unlikely that there existed a continuous manifold relating the arrangement complexity to that of the subsystems, but the number of subsystems might not even be fixed: decreasing complexity for one might require an extra be introduced, while increasing the complexity of another might obviate the need for a third. Likewise, there might be two or more non-identical sets, each of the same number of subsystems, and while the complexity of the subsystems in each might be the same, the complexity of specifications for overall simulations deriving from them might well be different. These problems notwithstanding, the theoretical value of this approach will emerge shortly. First, we evaluate in this context our two extreme examples of "bad" simulations.

3.3 Example "Bad" Simulations

First was the example of simulating every single particle in a living organism. In this case, the processing modules correspond to simple descriptions of how each particle interacts in accordance with the laws of physics. Since the laws of physics are fairly straightforward, the F-logical depth of these tiny subsystems is relatively low. But the complexity of an overall simulation is very high, because describing an overall simulation of an organism solely on the basis of descriptions of particle interactions requires specifying arrangements of trillions upon trillions of particles.

The Chinese room example offers a similarly imbalanced mix of complexity. In this case, the overall simulation is very simple to specify, because we need concern ourselves with only one processing module: the rule book. But it is difficult to conceive of any *single* Bernoulli-Turing Machine program, long or short, which would offer a very fast execution time for processing inputs from the infinite domain of well formed linguistic symbol strings. Thus it is a fairly safe bet that the rule book itself instantiates an input/output relationship of enormous complexity. (Note that execution time on a Bernoulli-Turing Machine with sequential memory access remains high even if the input/output relationship is instantiated with a simple lookup table.)

So what about the case of a "good" simulation, where we have managed to minimise to some extent the complexity both of the modular simulations and of their interconnections for an overall simulation? At what levels of description might these modules have been selected? We can make a number of interesting observations about this, and these observations set the stage for examining the rationale of the view on offer from the standpoints of computational convenience and evolutionary biology.

3.4 Complexity in a "Good" Simulation

We can best get an intuitive feel for complexities in a "good" simulation by first assuming that our subsystems have been selected at some intermediate level of description and then examining what features of the real organism have been preserved or ignored, starting at the lowest level and working up.

The first interesting thing to emerge is that any low level noise or stochastic variation which doesn't have a direct rôle in shaping the output of a given module may be ignored by that module. This is a result of the interrelationship between the complexities of the subsystems and their interconnection, and it emerges from a quick analysis. Given some type of low level variation in the inputs to a module, either the fine detail of that variation is functionally relevant to preserving the overall input/output relationship instantiated by an organism, or it is not. In the first case, the module could reduce its complexity and ignore the detail only at the expense of requiring that some other module or some feature of interconnections between modules worked so as to recover it. If the overall input/output relationship is to be maintained, there must be *some* module or arrangement of modules *somewhere* which acts to pick up this fine detail. If, on the other hand, the low level variations are not relevant to the overall input/output relationship, then the given module may reduce its complexity by ignoring them. (If low level variations are irrelevant to the outputs of a module *itself*, then F-logical depth doesn't distinguish between the complexity of a module which does something with them and one which ignores them.) This line of thought of course presents no difficulty for cases where outputs of an overall system are meant to include some low level variations which are not deterministically related to such variations in inputs, since they can simply be introduced with a random number generator at low cost to the complexity measure.

This line of analysis applies just as well when we consider the functional relevance of any other features of inputs, right up through higher levels of organisation. The view on offer indicates that subsystems should be selected not only at a level of description such that they ignore irrelevant low level variations, but they should also ignore *any* features of their inputs which are irrelevant to maintaining the overall input/output relationship of the organism. This is suggested by the exact same line of reasoning: a module may ignore a functionally relevant feature of its inputs only at the expense of forcing some other module to process it or forcing the arrangement of modules somehow to account for it. On the other hand, every module can lower its complexity by ignoring features of inputs which it otherwise might process but which turn out to be irrelevant for preserving the overall relationship at hand.

Thus, the view on offer automatically suggests a goal which it is safe to assume we would already have had anyway, namely, that our simulations of living organisms should concentrate just on those features of the organisms which are functionally relevant to their behaving the way they do. That the view on offer is compatible with what we would already have sought should not be too surprising, but it may be reassuring to know that so far it isn't suggesting anything explicitly at odds with existing goals. What this view on minimising complexity does do, however, is offer a way of narrowing the class of different ways of meeting this goal. The goal of "simulating only those features which are functionally relevant to preserving the particular behaviours" doesn't offer us much reason to prefer any of several non-identical ways of preserving the same overall input/output relationship. Such a goal doesn't by itself, for instance, offer us any immediately obvious reason to reject a Chinese room sort of simulation in which the input/output relationship is preserved and low level variations in inputs—along with everything else—are ignored.

Next we'll explore some considerations which motivate the view on offer, and along the way we'll see some other advantages of this view over simpler goals of just searching out functional relevance.

4 ...And Why

At first glance, it may seem counterintuitive to suggest that *minimising* the complexity of the relationships we've described is the best way to go about simulating living organisms. However, the approach has a number of advantages which go beyond simple compatibility with existing goals of focusing on functional relevance.

4.1 Computational Convenience

With respect to real world resources, perhaps the most obvious benefit of minimising the complexity both of the subsystems we would like to simulate and of the arrangement of these subsystems to form an overall simulation is that the strategy is computationally economical. Since our simulations don't actually run on Bernoulli-Turing Machines, the relationship between F-logical depth and computing resources isn't entirely straightforward, but generally speaking the lower the complexity of a simulation the less demanding it will be on computer resources.

Likewise, in line with our observations of the previous section, following the strategy here on offer can automatically reduce the capacity for "overfitting" in a simulation by suggesting detail which can be ignored by the individual processing modules. To be sure, this doesn't relieve the designer of the task of actually designing modules which are immune to overfitting, but it can suggest contexts in which such design efforts are called for or in which irrelevant details might simply be filtered out as a stopgap measure in the absence of more capable simulation modules.

But while computational considerations are important for simulations which have actually to be run on real computers, the more theoretically interesting advantages of the minimisation strategy appear with respect to evolutionary considerations.

4.2 Evolutionary Biology

First, it is important that genomes typically code just for structures and characteristics which are—or *were*, at some stage of phylogenetic progression—relevant to seeing an organism through ontogenetic development to reproductive age (although, of course, some characteristics which are irrelevant to survivability and reproduction but which are nonetheless by-products of characteristics which *are* relevant may also be indirectly coded). Also, within a species, some structures vary only in detail between individual examples of the species. That is, within a species, structures at certain levels of description are approximately the same regardless of the genotype of the particular organism. This suggests that these structures are *robustly* coded in the genome, either directly or indirectly, and that these structures have achieved a sort of structural stability or persistence—a resistance to slight variations introduced by mutation or recombination—through the course of phylogenetic development.

Now, the complexity (in a nontechnical sense) of the human genome, for instance, is far too low to account even for the full connectional complexity of the central nervous system. [5, 6] Thus, a great deal of functionally relevant complexity lurks in human phenotypes which is not coded in the genome and which thus emerges epigenetically and developmentally at least partly on the basis of structures which *are* coded. It makes sense to think that Nature will have selected for the simplest "building blocks" which can be coded robustly in a genome and which will promote rapid ontogenetic development and adaptation of the kinds of complex structures which cannot be directly coded but which are essential to the organism's survival and reproduction. For instance, it makes sense that Nature might specify the structure of neurons and their gross arrangements and capacities for adaptation which allow, say, the visual cortex to recognise patterns, but not that it would specify in detail the exact interconnections which instantiate any particular pattern recognition system.

With these general biological observations to hand, we are now in a position to see how our complexity minimisation strategy offers an avenue for mimicking the selection of particular structures by evolution and for understanding both our own simulations and Nature's creations in a broader theoretical context. First, we can immediately see that our strategy of choosing minimally complex subsystems whose interconnections can also be specified with minimal complexity is entirely compatible with our inferred natural "strategy" of robustly coding for relatively invariant structures which appear throughout a species and for coding arrangements of these structures which provide at least a minimally sufficient foundation for rapid ontogenetic development and adaptation. Nature's apparent preference for the simplest "building blocks" within a species corresponds to our preference for subsystems of lowest complexity, while Nature's apparent preference for arrangements which can be simply coded in a genome but yet provide a sufficient developmental foundation corresponds to our preference for low complexity in the arrangement of our subsystems. The minimisation strategy may help us mimic the emergent modularity of species differentiation by suggesting those very subsystems which have in fact developed a sort of structural stability through the course of phylogenetic progression and which can in fact be arranged simply to form a whole organism.

None of this suggests that real biological phenotypes do not instantiate what amount to some extraordinarily complex input/output systems which yield very sophisticated behaviours. But along the lines of our second biological observations above, it seems clear that Nature does not code *directly* for very complex behaviours but only for those structural characteristics which make possible complex adaptive behaviour in the *developing* organism. More precisely, Nature is liable to code for structural characteristics which enable the ontogenetic *emergence* of far more complex structures subserving sophisticated behaviours. We have learned the value of such an approach ourselves, for example, in developing artificial neural networks: for most applications, it is easier to specify basic network nodes and learning rules and then to allow networks to develop their own patterns of connection strengths in response to training data than to attempt to design complete networks ourselves. Not only is this method easier for the designer, but it supports the creation of more general networks which can be trained on a wide range of data sets. Given even the weakest similarity between artificial neural networks and their biological counterparts, the variability in stimuli faced by members of the same generation of the same species of an organism suggests it is not biologically plausible to suppose the phylogenetic development of most species of even moderate complexity would have pressed Nature to code for these more "application-specific" characteristics of neural networks.

But this is really just common sense: we know from everyday experience that, unlike Athena, who sprang fully grown from Zeus's head, higher organisms in Nature do not emerge spontaneously from the genotype with an entire repertoire of sophisticated adult behaviours at their disposal. Instead, they emerge with the "building blocks" with which to develop those behaviours, and it is these building blocks which, I suggest, are selected not only by the natural processes of evolution, but also by the strategy of minimising functional logical depth in simulations of real biological examples of millions of years of phylogenetic adaptation. The complexity minimisation strategy thus offers a broader theoretical framework for understanding both the selections of evolution and the selections of "good" simulations.

Within this framework, we can understand the general complexity of biological phenotypes and of appropriate simulations as a property which emerges from the right combination of simplicity in basic building blocks and their initial interconnection. Our framework offers a method for relieving the burden of designing systems to simulate the most complex overall behaviours by shifting their complexity onto the emergent dynamics of simple adaptive subsystems connected in simple ways. Just as Nature may not be able to reduce the overall complexity of an organism without compromising its survivability, we can't reduce *overall* complexity without compromising the quality of a simulation—but this method allows us simply to provide a basic adaptive substrate from which greater overall complexity can emerge.

4.3 Genetic Algorithms

This broad theoretical framework also offers straightforward hints about the kinds of characteristics which should be coded in the bit strings which represent genomes when artificial neural networks are developed with genetic algorithms. As I indicate in [17] on the basis of biological considerations, genetic algorithms needn't just create zero plasticity networks for particular applications: by coding for learning rules or other features and by tailoring fitness functions to favour characteristics such as the ability to adapt to variations in an environment, genetic algorithms may develop generalised networks more similar to those which appear in Nature. The complexity minimisation strategy and the theoretical framework it prompts suggest that fitness functions and genomes should accommodate simultaneous coding for characteristics of subsystems and of interconnections. And where we can find with the present method or others some likely candidates for subsystems (or their interconnections), genetic algorithms can be applied to find good interconnections (or subsystems).[6]

4.4 Levels of Description and Cognitive Transitions

The present framework also offers a pragmatic—but not philosophically decisive—angle on debates about symbolic or connectionist processing and appropriate levels of

---

6 Because F-logical depth is not computable, it would be possible to apply genetic algorithms directly to the complexity minimisation problem only in approximation.

description for analysing the capabilities of living systems. Complexity minimisation indicates levels appropriate for understanding genetically coded "building blocks" and perhaps for minimising the distortions of evolution which may be introduced by interpreting subsystems at incorrect levels as well as indicating levels appropriate for understanding what substrates may enable the emergence of the more complex adaptive behaviours characteristic of living organisms. It does not, however, offer any comment on philosophical questions of whether it is *possible* or even more efficient to instantiate cognitive or living systems with simulations at other levels of description. With respect to this framework, questions about the possibility or necessity of symbolic or connectionist processing are irrelevant except as they bear on complexity issues. This view only recommends choices of levels of description where functional relevance *can be* isolated (and perhaps *has been* isolated by Nature) and leaves open questions about where else it may be interpreted or how else it might be implemented either symbolically or with connectionist networks.

The framework also suggests that the dynamics of cognitive state transitions should be traced with respect to the indicated levels of description. Where such transitions turn out to be nondeterministic, we may anticipate either that the indeterminism is functionally irrelevant or that it reflects a systematic stochastic relationship between the inputs and outputs of the subsystem under scrutiny. Questions about what level or levels of simulation will preserve a deterministic relationship between inputs and outputs are on this view secondary to the complexity minimisation question, and levels of description with nondeterministic dynamics may be preferred over levels with deterministic dynamics in those cases where nondeterministic input/output relationships are of lower functional logical depth. Here again, the framework on offer does not offer a philosophically decisive approach to the problem, but it does offer pragmatic reasons for side-stepping it both in creating artificial living systems and in analysing biological ones.

4.5 Noncomputability

All this is fine for understanding real and artificial living systems within a broader theoretical framework, but some might object that because neither F-logical depth nor the other complexity measures we've described are computable, such a theoretical framework has little real application. By way of reply, we can concede that it is certainly true the complexity measures are noncomputable values, because we can never be sure of the shortest bit string which will generate a given pattern. Yet, we can still place a bound on such measures if we know any method at all of compressing a string, we can approximate the value of such measures on the basis of whatever compression algorithms we do have available, and—since the point of our subsystems, just like the point of the Bernoulli-Turing programs S(x, y) in the definition of F-logical depth, is to reproduce an input/output relationship—we can even compare directly the execution times of the actual programs we have available for simulation. That is, instead of calculating functional logical depth with respect to unknown programs S(x,y), we might calculate it with respect to the actual programs available. While this last method of course offers a view of levels of description which is coloured by the particulars of whatever hardware and software facilities are to hand, in the apparent absence of any other objective procedure for taking such decisions, we are almost certainly better off to adopt *some* method, especially where the method offers such a rich theoretical framework.

5 Conclusion

We have defined a new measure called functional logical depth which is specifically designed for comparing the complexity of input/output systems, and we have seen how to apply the measure to the problem of finding appropriate levels of description at which to simulate natural living organisms. The strategy of minimising the F-logical depth both of the subsystems into which we decompose an organism's structure and of the requisite interconnection of those subsystems is meant to offer a description of an organism which mimics that of an organism's genotype. In the genetic context, there is good reason to believe that simple adaptive subsystems common to a species are coded robustly in the genome and that more complex behaviours of phenotypes emerge ontogenetically from the dynamics within and between these subsystems. The present strategy is meant to "reverse engineer" this process by starting with an organism, the product of phylogenetic development, and finding the simple kinds of subsystems with which it would have made sense for Nature to produce it. In so doing, the complexity minimisation strategy also offers a theoretical framework within which to compare natural species differentiation and phylogenetic development and our own attempts at creating artificial life.

The theoretical framework carries interesting implications for genetic algorithms development as well as for levels of description debates concerning symbolic and connectionist processing and cognitive state transitions. Although we have not discussed such areas explicitly, the strategy may also bear on problems in the development of autonomous robots, on questions about organisms' modelling of the behaviour of other organisms as well as of their own, and even on recursion theoretic issues surrounding chaotic analogue systems which I have discussed elsewhere. [15, 18]

Overall, introducing the new measure of complexity and taking a position on the appropriate distribution of complexity in natural living organisms yields a rich theoretical framework with implications for a wide range of issues relevant to the project of developing artificial life.

6 Acknowledgements

I am grateful to the Gifford Committee, who now fund my research at the University of Glasgow, and to HM Government's Marshall Aid Commemoration Commission, who previously funded my research at the University of Edinburgh, some of which has figured directly in developing the present material. Thank you also to two anonymous ECAL referees, not all of whose comments could be fully addressed in the paper before the conference.

References

1. Bennett, C.H.: Information, dissipation, and the definition of organization. In Emerging Syntheses in Science, ed. David Pines. Reading, Massachusetts: Addison-Wesley. (1987)
2. Bennett, C.H.: How to define complexity in physics, and why. In Complexity, Entropy, and the Physics of Information, SFI Studies in the Sciences of Complexity VIII, ed. Wojciech H. Zurek, 137-148. Redwood City, California: Addison-Wesley. (1990)
3. Browne, A., Pilkington, J.: Variable binding in a neural network using a distributed representation. In European Symposium on Artificial Neural Networks 1994, ed. Michel Verleysen, 199-204. Brussels: D facto. (1994)

4. Chalmers, D.J.: Syntactic transformations on distributed representations. Connection science **2** (1990) 53-62.
5. Changeux, J.-P., Danchin, A.: Selective stabilization of developing synapses as a mechanism for the specification of neuronal networks. Nature **264** (1976) 705-711.
6. Changeux, J.-P., Heidmann, T, Patte, P.: Learning by selection. In The Biology of Learning, ed. P. Marler and H.S. Terrace, 115-137. New York: Springer-Verlag. (1984)
7. Chrisman, L.: Learning recursive distributed representation for holistic computation. Connection science **3** (1991) 345-365.
8. Deutsch, D.: Quantum theory, the Church-Turing principle and the universal quantum computer. Proceedings of the Royal Society of London **A400** (1985) 97-117.
9. Fodor, J.A., McLaughlin, B.P.: Connectionism and the problem of systematicity: Why Smolensky's solution did not work. Cognition **35** (1990) 183-204.
10. Fodor, J.A., Pylyshyn, Z.: Connectionism and cognitive architecture: A critical analysis. Cognition **28** (1988) 3-71.
11. Horgan, T., Tienson, J. (1993) Levels of description in nonclassical cognitive science. PHILOSOPHY **34**, Royal Institute of Philosophy Supplement, 159-188.
12. Horgan, T., Tienson, J.: A nonclassical framework for cognitive science. Synthese: special issue on connectionism and philosophy of mind, ed. A. Clark (to appear)
13. Marr, D.: Vision. New York: Freeman. (1982)
14. Mulhauser, G.R.: Chaotic dynamics and introspectively transparent brain processes. Presented 4 July 1993 at the Second Annual Conference of the European Society for Philosophy and Psychology in Sheffield, England. (1993)
15. Mulhauser, G.R.: Computability in chaotic analogue systems. Presented 21 July 1993 at the International Congress on Computer Systems and Applied Mathematics in St. Petersburg, Russia. (1993)
16. Mulhauser, G.R.: Cognitive transitions and the strange attractor: A reply to Peter Smith. Presented 7 September 1993 at the Fifth Joint Council Initiative Summer School in Cognitive Science and Human Computer Interaction in Edinburgh, Scotland. (1993)
17. Mulhauser, G.R.: Biologically plausible hybrid network design and motor control. In European Symposium on Artificial Neural Networks 1994, ed. Michel Verleysen, 79-84. Brussels: D facto. (1994)
18. Mulhauser, G.R.: Mind Out of Matter: Topics in the Physical Foundations of Consciousness and Cognition. Typescript, University of Glasgow. (1994)
19. Searle, J.: Minds, brains and programs. Behavioral and Brain Sciences **3** (1980) 417-457.
20. Sharkey, N.E.: The ghost of the hybrid: A study of uniquely connectionist representations. Artificial Intelligence and Simulation of Behaviour Quarterly **79** (1992) 10-16.
21. Smith, P.: Chaos: Explanation, Prediction & Randomness. Unpublished manuscript of Easter Term 1993 Cambridge Department of Philosophy lecture series. (also to appear from Cambridge University Press)
22. Smolensky, P.: Tensor product variable binding and the representation of symbolic structures in connectionist systems. Artificial Intelligence **46** (1990) 159-216.

Quasi-Uniform Computation-Universal Cellular Automata

Moshe Sipper
Department of Computer Science
Tel Aviv University
Tel Aviv 69978, Israel
e-mail: moshes@math.tau.ac.il

Abstract. Cellular automata (CA) are dynamical systems in which space and time are discrete, where each cell obeys the same rule and has a finite number of states. In this paper we study non-uniform CA, i.e. with non-uniform local interaction rules. Our focal point is the issue of universal computation, which has been proven for uniform automata using complicated designs embedded in cellular space. The computation-universal system presented here is simpler than previous ones, and is embedded in the minimal possible two-dimensional cellular space, namely 2-state, 5-neighbor (which is insufficient for universal computation in the uniform model). The space studied is *quasi*-uniform, meaning that a small number of rules is used (our final design consists of just two rules which is minimal), distributed such that most of the grid contains one rule except for an infinitely small region which contains the others. We maintain that such automata provide us with a *simple*, *general* model for studying Artificial Life phenomena.

1 Introduction

Cellular automata (CA) are dynamical systems in which space and time are discrete. The states of cells in a regular grid are updated synchronously according to a local interaction rule. Each cell obeys the same rule and has a finite (usually small) number of states [29]. The model was originally conceived by John von Neumann in the 1950's to provide a more realistic framework for studying the behavior of complex, extended systems [31]. Over the years it has been applied to the study of general phenomenological aspects of the world, including: communication, computation, construction, growth, reproduction, competition and evolution [4, 26, 29].

The CA model is both *general* and *simple*. Generality implies two things: (1) the model supports universal computation and (2) the basic units encode a general form of interaction rather than some specialized action. Simplicity implies that the basic units of interaction are modest in comparison to Turing machines. If we imagine a scale of complexity with Turing machines occupying the high end then simple machines are those that occupy the low end, e.g. finite state automatons. The CA model is one of the simplest, general models available.

From an Artificial Life (AL) perspective it has been noted that the main difficulty with the CA approach seems to lie with the extreme low-level representation of the interactions. CA's are programmed at the level of the local physics of the system and, therefore, higher-level cooperative structures are difficult to evolve in CA's [23]. One of our primary goals is to increase the "capacity" for AL modeling while preserving the essential features of the original CA, namely massive parallelism, locality of cellular interactions and simplicity of cells. Thus we attain a model that is close to CA with regards to generality and simplicity [24, 25]. In this paper we consider non-uniform CA, i.e. with non-uniform local interaction rules. Such automata function in the same way as uniform ones, the only difference being in the cellular rules which need not be identical for all cells.

This type of model has been investigated by others. [8] presents two generalizations of cellular automata, namely discrete neural networks and automata networks. These are compared to the original model from a computational point of view which considers the classes of problems such models can solve. The work of [30] discusses a one-dimensional CA in which a cell probabilistically selects one of two rules, at each time step. They showed that complex patterns appear characteristic of class IV behavior[1], however universal computation was not discussed (see also [9]). In [13, 22] adaptive stochastic cellular automata are considered which are essentially non-uniform automata whose rules are drawn from the same probability distribution function. Their approach focuses on

[1] CA class numbers are those introduced by [32].

the learning aspect where an automaton is trained to solve some problem (e.g. pole balancing). As noted above, our motivation is different than these works: we wish to study Artificial Life phenomena such as evolution, emergence and multi-cellularity in a simple, general model which operates at a higher level than CA [24, 25].

The focal point of this paper is the issue of universal computation in two-dimensional CA, namely the construction of machines, embedded in cellular space, whose computational power is equivalent to that of a universal Turing machine [10]. The first such machine was described by von Neumann, who used 29 state, 5-neighbor cells [31][2]. Codd provided a detailed description of a computer embedded in an 8-state, 5-neighbor cellular space, thus reducing the complexity of von Neumann's machine [5]. Banks described a 2-state and a 3-state automaton (both 5-neighbor) which support universal computation with an infinite and finite initial configuration[3], respectively [1]. A cellular space with a minimal number of states (two) and a 9-cell neighborhood proven to support universal computation (with finite initial configuration) involves the "game of life" rule [2]. One-dimensional CA have also been shown to support universal computation [28, 27]. A recent review of theoretical results is provided in [6].

Codd proved that there does not exist a computation-universal 2-state, 5-neighbor cellular automaton with finite initial configuration. His proof concerns the original uniform model [5]. We present a universal non-uniform system embedded in such space.

Section 2 presents the details of our construction, consisting of ten different cell rules, which are reduced to six in Section 3. Both sections involve an infinite initial configuration. Section 4 describes the implementation of a universal machine using a finite initial configuration. A quasi-uniform automaton is discussed in Section 5 where quasi-uniformity implies a small number of rules distributed such that most of the grid contains one rule except for an infinitely small region which contains the others. Our final implementation consists of just two rules which is minimal. A discussion of our results follows in Section 6 linking them with Artificial Life issues.

2 A universal 2-state, 5-neighbor non-uniform cellular automaton

In order to prove that a two-dimensional CA is computation-universal we proceed along the lines of [2, 21, 1, 5, 31] and implement the following components[4]:

1. Signals and signal pathways (wires). We must show how signals can be made to turn corners, to cross and to fan out.
2. A functionally complete set of logical gates. A set of operations is said to be ***functionally complete*** (or ***universal***) if and only if every switching function can be expressed entirely by means of operations from this set [11]. We shall use the $NAND$ function (gate) for this purpose (this gate comprises a functionally complete set and is used extensively in VLSI since transistor switches are inherently inverters [19]).
3. A clock that generates a stream of pulses at regular intervals.
4. Memory.

In the following sections we describe the implementations of the above components (note: the terms 'gate' and 'cell' are used interchangeably).

[2] The neighborhood consists of the cell itself together with its four immediate nondiagonal neighbors. This is also the neighborhood used throughout this paper.

[3] A configuration is an assignment of non-zero states to cells in the space.

[4] Another approach is one in which a row of cells simulates the squares on a Turing machine tape while at the same time simulating the head of the Turing machine. This has been applied to one-dimensional CA [28, 27, 1].

Name	Rule	Symbol	Name	Rule	Symbol
Right propagation cell	$*$ / x $*$ $*$ / $*$ $\mapsto x$	$\rightarrow$	Exclusive Or (*XOR*) cell (type *a*)	y / x $*$ $*$ / $*$ $\mapsto x \oplus y$	$\oplus$
Left propagation cell	$*$ / $*$ $*$ x / $*$ $\mapsto x$	$\leftarrow$	Exclusive Or (*XOR*) cell (type *b*)	x / $*$ $*$ y / $*$ $\mapsto x \oplus y$	$\oplus$
Up propagation cell	$*$ / $*$ $*$ $*$ / x $\mapsto x$	$\uparrow$	Exclusive Or (*XOR*) cell (type *c*)	$*$ / $*$ $*$ x / y $\mapsto x \oplus y$	$\oplus$
Down propagation cell	x / $*$ $*$ $*$ / $*$ $\mapsto x$	$\downarrow$	Exclusive Or (*XOR*) cell (type *d*)	$*$ / y $*$ $*$ / x $\mapsto x \oplus y$	$\oplus$
NAND Cell	y / x $*$ $*$ / $*$ $\mapsto x \mid y$	▮	No Change (NC) cell	$*$ / $*$ x $*$ / $*$ $\mapsto x$	

Each rule specifies the new state of the central cell.
'$*$' represents the set of states: $\{0, 1\}$.
$x, y \in \{0, 1\}$.
'$\oplus$' is the *XOR* function (modulo-2 addition), '$|$' is the *NAND* function.

Table 1. Cell types (rules)

2.1 Signals and wires

A wire in our system consists of a path of connected *propagation cells* each containing one of the propagation rules. A signal consists of a state, or stream of states being propagated along the wire. There are four propagation cell types (i.e. four different rules), one for each direction: right, left, up, down (Table 1). Figure 1a shows a wire constructed from propagation cells. Figures 1b - 1d demonstrate signal propagation along the wire. Note that all cells which are not part of the machine (i.e. its components) contain the *NC* (No Change) rule (Table 1) which simply preserves its initial state indefinitely.

A wire in our system possesses a distinct direction, a characteristic which is highly desirable as it simplifies signal propagation [5]. In most cases signals must propagate in one direction only, and should bi-directional propagation be required then two parallel wires in opposite directions may be used. We note in Figure 1 that wires support signal propagation across corners. Fan out of signals is also straightforward as evident in Figure 2.

The last problem we must address concerning signals is wire crossing (there are four possible crossings since wires may run in four directions). We first demonstrate that at least three gates (cells) are required for this operation. To see this note that one gate is insufficient since there are two bits of information, denoted x and y, whereas the intersection cell can only contain one bit:

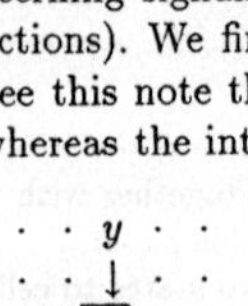

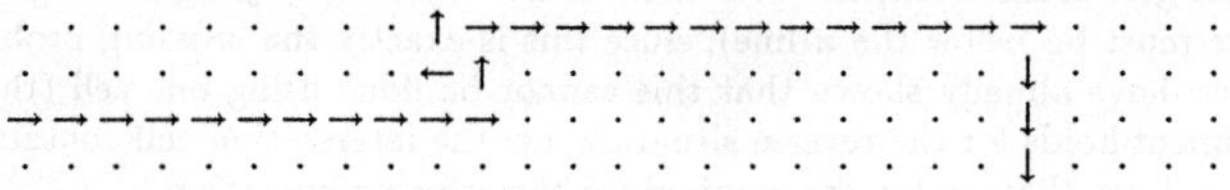

(a) cellular arrangement

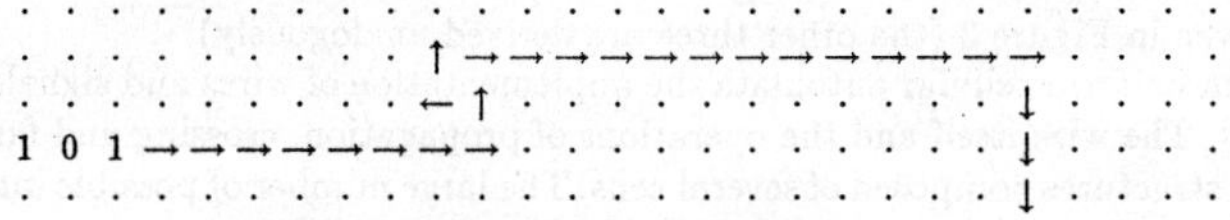

(b) propagation of signal: time = 0

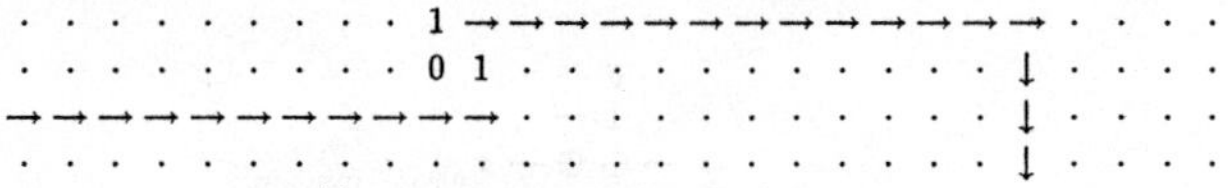

(c) propagation of signal: time = 11

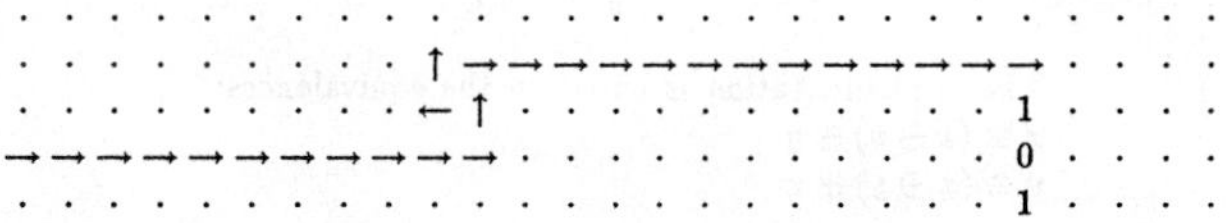

(d) propagation of signal: time = 27

Fig. 1. Signal propagation along a wire.

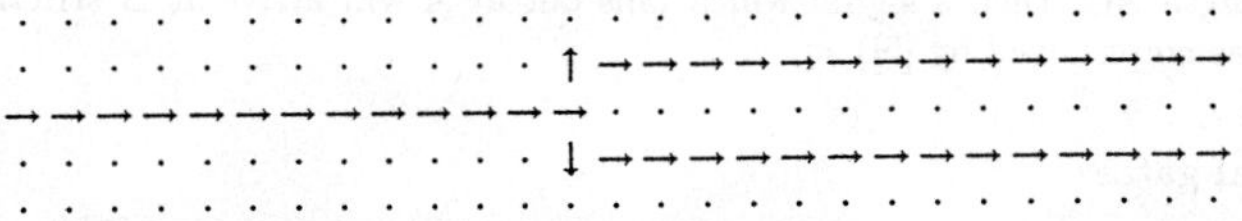

(a) two-way fan out

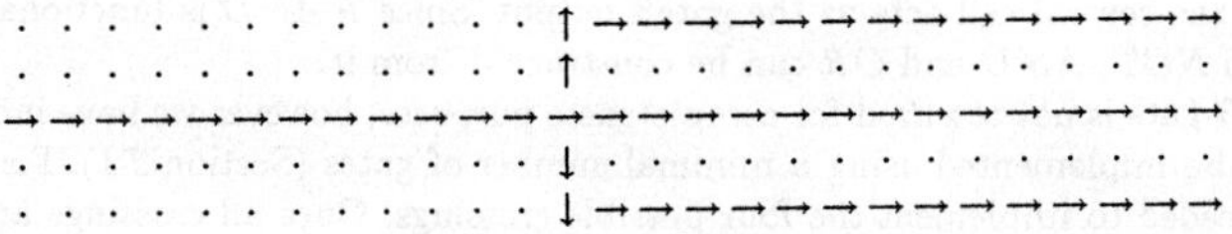

(b) three-way fan out

Fig. 2. Signal fan out.

Thus, either x crosses the y (vertical) line, in which case the y signal is lost, or conversely the x signal is lost. In case there are two gates then they must eventually contain x and y (otherwise information is lost). The situation is therefore as follows:

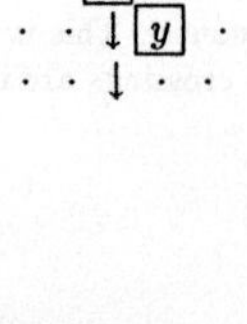

The x signal gets transferred, however there is no way for the y signal to get to its gate (note that the y gate must be below the x line), since this is exactly the crossing problem we are trying to solve, and we have already shown that this cannot be done using one cell (the remaining one). A similar argument holds for the reverse situation, i.e. the intersection cell contains y. We therefore conclude that at least three gates are required for the crossing operation.

Towards this end we have selected the XOR cells (Table 1) since wire crossing can be implemented using a minimal number of such gates (three). An implementation of one of the four possible crossings is given in Figure 3 (the other three are derived analogously).

Note that in uniform cellular automata the implementation of wires and signals is highly complicated [31, 5, 1]. The wire itself and the operations of propagation, crossing and fan out are attained using complex structures composed of several cells. The large number of possible interactions between these structures makes the design task arduous.

It is important to note the direct relationship between path length and time: if two paths branch out from a common point A to points B, C respectively, and if path length AB is strictly greater

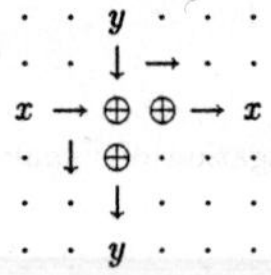

The implementation is based on the equivalences:
$x \equiv (x \oplus y) \oplus y$
$y \equiv (x \oplus y) \oplus x$

The XOR gates used are type a (see Table 1).

Fig. 3. Implementation of wire crossing.

than path length AC, then a signal which fans out at A will arrive at B strictly later than at C (this issue was emphasized by [5]).

2.2 Logical gates

Table 1 includes a 2-input, 1-output $NAND$ gate (cell), which forms a functionally complete set, thereby providing us with the second component discussed above. Two neighboring cells act as inputs while the central cell acts as the gate's output. Since $NAND$ is functionally complete other gates such as NOT, AND and OR can be constructed from it.

The XOR gate is not required for completeness purposes, however we have included it since wire crossing can be implemented using a minimal number of gates (Section 2.1). Four XOR cell types (rules) are needed to implement the four possible crossings. Once all crossings are possible we only need one $NAND$ gate since the two wire inputs can always be made to arrive at the two input cells of the gate's neighborhood[5].

2.3 Clock

The third component of our system is a clock that generates a stream of pulses at regular intervals. We implement this using a wire loop, i.e. a loop of propagation cells. Figure 4 presents the implementation. Note that any desired waveform can be produced by adjusting the size and content of the loop. The implementation of the clock is made simple due to the manner in which wires are constructed, i.e. as cellular arrangements. Thus it is possible to obtain such a closed loop, which proves highly useful in our case.

[5] Note that both signals must arrive synchronously. This is possible since delays can always be introduced by using, e.g., loops which are feasible once crossings are implemented.

3 Reducing the number of rules

In Section 2 we presented the components of a universal machine employing ten different cell rules (Table 1). This number may be reduced to six, by using a more complex wiring scheme, involving the implementation of the propagation cells using XOR gates (implementation is not shown due to lack of space).

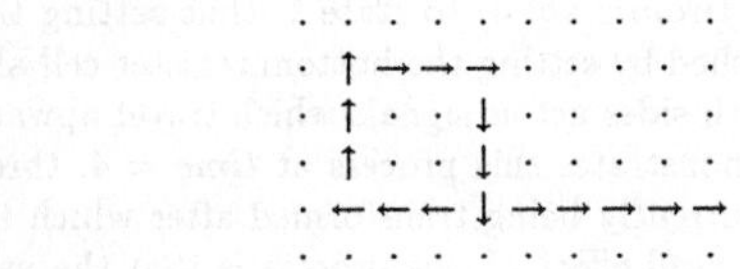

(a) cellular arrangement

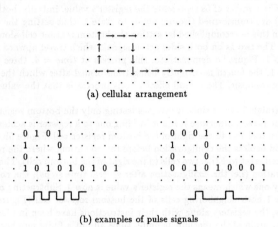

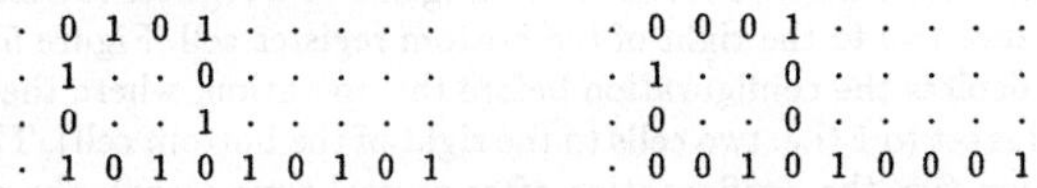

(b) examples of pulse signals

Fig. 4. Implementation of clock.

4 Implementing a universal machine using finite configurations

The components presented in the previous sections are sufficient in order to build a universal machine using an infinite initial configuration. Codd conjectured that an unbounded but boundable propagation is a necessary condition for computation universality and used this conjecture in a proof that there does not exist a 2-state, 5-neighbor universal cellular automaton with finite initial configuration [5]. Following his work universality was implemented by using more states or larger neighborhoods [1, 2, 21].

The problem with finite configurations involving the above components is that a computation may require an arbitrary amount of space and therefore some method must exist for increasing the information storage (memory) by arbitrarily large amounts. In order to prove universality we implement Minsky's two-register universal machine, which consists of [20]:

1. A programming unit (finite control).
2. Two potentially infinite registers.
3. The following set of instructions:
   - Increase the contents of a register by one.
   - Decrease the contents of a register by one.
   - Test whether the contents of a register equal zero.

The finite control unit may be realized using the components described in Sections 2 and 3. The major difficulty is the implementation of the infinite registers and the three operations associated with them. According to Codd's proof a single rule in 2-state, 5-neighbor cellular space is insufficient since, as noted above, unbounded but boundable propagations cannot be attained [5].

While other researchers have turned to cellular spaces with more states or larger neighborhoods, our approach is based on non-uniformity. As noted above the minimal number of rules needed to implement a register is two. Indeed, we have found two such rules: one which we denote the *background* rule, the other being Banks' rule [1] (rules are not shown due to lack of space). The implementation of a universal computer consists of a finite control unit, which occupies a finite part of the grid. All other cells contain the *background* rule except for two cellular columns, infinite in one direction, containing Banks' rule. These *register columns* originate at the upper part of the control unit and each one represents one register (Figure 5a depicts the basic machine configuration after initialization has taken place, see ahead). The cells in these columns are denoted register cells.

The three register instructions are implemented as follows: at any given moment a register column consists of an infinite[6] number of cells in state 1, and a finite number in state 0, occupying the bottom part of the column. The number of 0s represents the register's value. Initially, both register columns (i.e. all register cells) are transformed (from state 0) to state 1, thus setting the register's value to zero. For each column this is accomplished by setting the bottom register cell along with its left and right neighbors to 1. The two 1s on both sides act as signals which travel upward along the column, setting all its cells to 1. Figure 5a demonstrates this process at $time = 4$: three cells have already been transformed to 1, the fourth is currently being transformed after which the (dual) signal will continue its upward movement. The overall effect of this process is that the value of both registers is initialized to zero.

The zero test is straightforward since it involves testing only the bottom register cell: if its state is 1 the register's value is zero, otherwise it is not. Adding one to a register is achieved by setting to 1 the cell which is at distance two to the right of the bottom register cell. Figure 5b demonstrates this operation: the left grid depicts the configuration before the operation, where the register's value is 3 and the appropriate cell is set to 1 (i.e. two cells to the right of the bottom cell). The right grid depicts the effect of the operation (i.e. the configuration after several time steps): the column's number of zeros has increased by one which means the register's value is now 4. Subtracting one from a register is done by setting to 1 both neighboring cells of the bottom register cell (Figure 5c demonstrates this operation). Thus, the registers along with their instructions have been implemented.

The initial configuration of the machine is finite, since all but a finite number of cells are set to zero. The total number of rules needed to implement a universal computer equals the number of rules necessary for implementation of the finite control unit plus the two additional memory rules (background and Banks). Thus we need a total of 12 rules using our implementation of Section 2 and 8 rules using the implementation of Section 3.

5 A quasi-uniform cellular space

As noted in Section 3 by increasing the complexity of our machine a reduced set of rules (six) may be used to construct the finite control. We can go one step further and construct the finite control with only one rule, e.g. the Banks rule [1] (note that this increases the complexity of the basic operations). As noted in Section 4 the complicating issue is not the finite control but rather the infinite memory which cannot be implemented in uniform 2-state, 5-neighbor cellular space.

Our conclusion from the previous paragraph is that a universal computer can be implemented using only two rules: background and Banks. Thus we have implemented universal computation using the minimal number of rules.

The two rules necessary to implement universal computation are distributed unevenly. Most of the grid contains the background rule, except for an *infinitely small* region which contains the other one. By this we mean that although there is an infinite number of both rules in the grid, each (infinite) row contains an *infinite* number of background rules with only a *finite* number of the other. In fact, except for a finite region of the grid each row contains only two Banks rules and an infinite number of background rules. Hence we say that our cellular space is *quasi*-uniform.

[6] More precisely the number of cells in state 1 tends to infinity as time progresses, see ahead.

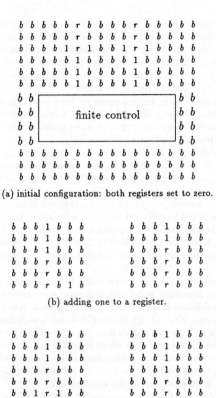

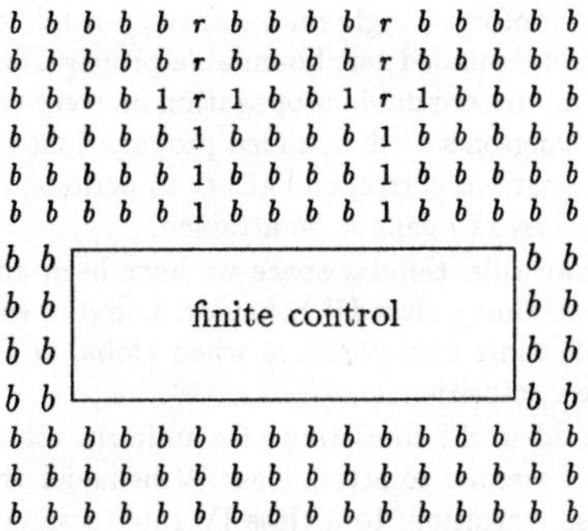

(a) initial configuration: both registers set to zero.

b	*b*	*b*	1	*b*	*b*	*b*		*b*	*b*	*b*	1	*b*	*b*	*b*
b	*b*	*b*	1	*b*	*b*	*b*		*b*	*b*	*b*	1	*b*	*b*	*b*
b	*b*	*b*	1	*b*	*b*	*b*		*b*	*b*	*b*	*r*	*b*	*b*	*b*
b	*b*	*b*	*r*	*b*	*b*	*b*		*b*	*b*	*b*	*r*	*b*	*b*	*b*
b	*b*	*b*	*r*	*b*	*b*	*b*		*b*	*b*	*b*	*r*	*b*	*b*	*b*
b	*b*	*b*	*r*	*b*	1	*b*		*b*	*b*	*b*	*r*	*b*	*b*	*b*

(b) adding one to a register.

b	*b*	*b*	1	*b*	*b*	*b*		*b*	*b*	*b*	1	*b*	*b*	*b*
b	*b*	*b*	1	*b*	*b*	*b*		*b*	*b*	*b*	1	*b*	*b*	*b*
b	*b*	*b*	1	*b*	*b*	*b*		*b*	*b*	*b*	1	*b*	*b*	*b*
b	*b*	*b*	*r*	*b*	*b*	*b*		*b*	*b*	*b*	1	*b*	*b*	*b*
b	*b*	*b*	*r*	*b*	*b*	*b*		*b*	*b*	*b*	*r*	*b*	*b*	*b*
b	*b*	1	*r*	1	*b*	*b*		*b*	*b*	*b*	*r*	*b*	*b*	*b*

(c) subtracting one from a register

b denotes a cell in state 0 containing the background rule, *r* denotes a cell in state 0 containing Banks' rule. In figures (b) and (c) the left grid shows the configuration before the operation, the right grid shows the configuration upon its completion (after several time steps). The bottom line represents the bottom register cell and its neighbors.

Fig. 5. Register operation.

6 Discussion

We presented a quasi-uniform 2-state, 5-neighbor cellular automaton capable of universal computation. Quasi-uniformity implies a small number of rules distributed such that most of the grid contains one rule except for an infinitely small region which contains the others. Three implementations were described in which the number of rules used is: 12, 8 and finally 2.

The following paragraphs provide a discussion of a speculative nature linking our result with that of Langton [12] (see also [14]). He addressed the following question: under what conditions can we expect a dynamics of information to emerge spontaneously and come to dominate the behavior of a physical system? This was studied in the context of CA where the question becomes: under what conditions can we expect a complex dynamics of information to emerge spontaneously and come to dominate the behavior of a CA? [12].

Langton showed that the rule space of CA consists of two primary regimes of rules, periodic and chaotic, separated by a transition regime. His main conclusion was that information processing can emerge spontaneously and come to dominate the dynamics of a physical system in the vicinity of a critical phase transition.

According to Codd's proof a uniform (single rule) 2-state, 5-neighbor cellular space is insufficient for universal computation since unbounded but boundable propagations cannot be attained [5]. Either every configuration yields an unboundable propagation or every configuration yields a bounded propagation. In the context of Langton's work bounded propagations correspond to fixed point rules (class I) and unboundable propagations correspond either to periodic rules (class II) or chaotic ones (class III). Complex behavior (class IV) cannot be attained.

By using a quasi-uniform, two rule, cellular space we have been able to achieve unbounded but boundable propagations, thus attaining class IV behavior. Langton suggested that the information dynamics which gave rise to life came into existence when global or local conditions brought some medium through a critical phase transition.

Imagine an information-based world consisting of a uniform cellular automaton, which is not within the class IV region. If we wanted to attain class IV behavior the *entire* space would have to "jump", i.e. go through a phase transition, to a class IV rule (assuming this is at all possible, e.g. in our case of a 2-state, 5-neighbor space it is not). However, as a conclusion of the work presented above we offer an alternative: a small perturbation may be enough to cause some infinitely small part of the world to change. This would be sufficient to induce a (possible) transition from class II or class III behavior to class IV behavior. Furthermore, such a change could be effected upon a very simple world (in our case 2-state, 5-neighbor). As noted by [3] "frozen accidents" play an important role in the evolutionary process. These accidents are mainly caused by external conditions, i.e. external relative to a given system's laws of functioning.

A (highly) tentative comparison may be drawn to the famous experiment performed by [17] (see also [18]) in which methane, ammonia, water and hydrogen, representing a possible atmosphere of the primitive Earth, were subjected to an electric spark for a week. After this period simple amino acids were found in the system. The analogy to our CA world is as follows: we start with a simple uniform world, consisting of a single rule, which does not support complex (class IV) behavior[7]. At some point, a "spark" causes a perturbation in which a small number of cells change their rule. This is all that is needed. Our central conclusion is that such an infinitely small change in our world can suffice to generate a phase transition such that class IV behavior becomes possible. Note that this can happen independently in other regions of the world as well.

While the above discussion has been of a speculative nature we may also draw some practical conclusions from our work. As noted in Section 1 the main difficulty with the CA approach seems to lie with the extreme low-level representation of the interactions. Essentially, we construct world models at the level of physics. By slightly changing the rules of the game (no pun intended) we can increase the "capacity" for AL modeling while preserving the main features of CA, namely massive parallelism, locality of cellular interactions and simplicity of cells. Thus, we obtain a simple, general model (Section 1) which allows us to evolve complex behavior with the ability to explore, in-depth, the inner workings of the evolutionary process.

We have already begun venturing along this path [24, 25]. In these works a CA derived model is explored in which the basic units are slightly more complex than those of CA. Furthermore, evolution proceeds not only in state space as in CA but also in rule space, i.e. rules evolve over time. We have observed several phenomena of interest involving emergence, evolution and multi-cellularity. Another example is that of *embryonics*, standing for embryological electronics [15, 16, 7]. This is a CA based approach in which three principles of natural organization are employed: multi-cellular organization, cellular differentiation and cellular division. Their intent is to create an architecture which is complex enough for (quasi) universal computation yet simple enough for physical implementation. The approach represents another attempt at confronting the aforementioned problem of CA, namely the low level of operation.

It is hoped that the development of such AL models will serve the two-fold goal of: (1) increasing our understanding of biology and (2) enhancing our understanding of artificial models, thereby providing us with the ability to improve their performance. AL research opens new doors providing us with novel opportunities to explore issues such as adaptation, evolution and emergence which are central both in natural environments as well as man-made ones.

[7] either due to inability of the cellular space to support such behavior at all or due to the rule being in the non class IV regions of rule space.

Acknowledgment

I am grateful to Daniel Mange for his suggestions which helped improve this paper.

References

1. E. R. Banks. Universality in cellular automata. In *IEEE 11th Annual Symposium on Switching and Automata Theory*, pages 194–215, Santa Monica, California, October 1970.
2. E. R. Berlekamp, J. H. Conway, and R. K. Guy. *Winning ways for your mathematical plays*, volume 2, chapter 25, pages 817–850. Academic Press, New York, 1982.
3. E. W. Bonabeau and G. Theraulaz. Why do we need artificial life? *Artificial Life Journal*, 1(3):303–325, 1994. The MIT Press, Cambridge, MA.
4. A. Burks, editor. *Essays on cellular automata*. University of Illinois Press, 1970.
5. E. F. Codd. *Cellular Automata*. Academic Press, New York, 1968.
6. K. Culik II, L. P. Hurd, and S. Yu. Computation theoretic aspects of cellular automata. *Physica D*, 45:357–378, 1990.
7. S. Durand, A. Stauffer, and D. Mange. Biodule: an introduction to digital biology. Technical report, LSL, Swiss Federal Institute of Technology, Lausanne, Switzerland, September 1994.
8. M. Garzon. Cellular automata and discrete neural networks. *Physica D*, 45:431–440, 1990.
9. H. Hartman and G. Y. Vichniac. Inhomogeneous cellular automata. In E. Bienenstock, F. Fogelman, and G. Weisbuch, editors, *Disordered Systems and Biological Organization*, pages 53–57. Springer-Verlag, 1986.
10. J. E. Hopcroft and J. D. Ullman. *Introduction to Automata Theory Languages and Computation*. Addison-Wesley, 1979.
11. Z. Kohavi. *Switching and Finite Automata Theory*. McGraw-Hill Book Company, 1970.
12. C. G. Langton. Life at the edge of chaos. In C. G. Langton, C. Taylor, J. D. Farmer, and S. Rasmussen, editors, *Artificial Life II*, volume X of *SFI Studies in the Sciences of Complexity*, pages 41–91, Redwood City, CA, 1992. Addison-Wesley.
13. Y. C. Lee, S. Qian, R. D. Jones, C. W. Barnes, G. W. Flake, M. K. O'Rourke, K. Lee, H. H. Chen, G. Z. Sun, Y. Q. Zhang, D. Chen, and C. L. Giles. Adaptive stochastic cellular automata: theory. *Physica D*, 45:159–180, 1990.
14. W. Li, N. H. Packard, and C. G. Langton. Transition phenomena in cellular automata rule space. *Physica D*, 45:77–94, 1990.
15. D. Mange and A. Stauffer. Introduction to embryonics: Towards new self-repairing and self-reproducing hardware based on biological-like properties. In N. M. Thalmann and D. Thalmann, editors, *Artificial Life and Virtual Reality*, pages 61–72, Chichester, England, 1994. John Wiley.
16. P. Marchal, C. Piguet, D. Mange, A. Stauffer, and S. Durand. Embryological development on silicon. In R. A. Brooks and P. Maes, editors, *Artificial Life IV*, pages 365–370, Cambridge, Massachusetts, 1994. The MIT Press.
17. S. L. Miller. A production of amino acids under possible primitive earth conditions. *Science*, 117:528–529, May 1953.
18. S. L. Miller and H. C. Urey. Organic compound synthesis on the primitive earth. *Science*, 130(3370):245–251, July 1959.
19. J. Millman and A. Grabel. *Microelectronics*. McGraw-Hill Book Company, second edition, 1987.
20. M. L. Minsky. *Computation: Finite and Infinite Machines*. Prentice-Hall, Englewood Cliffs, New Jersey, 1967.
21. F. Nourai and R. S. Kashef. A universal four-state cellular computer. *IEEE Transactions on Computers*, c-24(8):766–776, August 1975.
22. S. Qian, Y. C. Lee, R. D. Jones, C. W. Barnes, G. W. Flake, M. K. O'Rourke, K. Lee, H. H. Chen, G. Z. Sun, Y. Q. Zhang, D. Chen, and C. L. Giles. Adaptive stochastic cellular automata: applications. *Physica D*, 45:181–188, 1990.
23. S. Rasmussen, C. Knudsen, and R. Feldberg. Dynamics of programmable matter. In C. G. Langton, C. Taylor, J. D. Farmer, and S. Rasmussen, editors, *Artificial Life II*, volume X of *SFI Studies in the Sciences of Complexity*, pages 211–254, Redwood City, CA, 1992. Addison-Wesley.
24. M. Sipper. Non-uniform cellular automata: Evolution in rule space and formation of complex structures. In R. A. Brooks and P. Maes, editors, *Artificial Life IV*, pages 394–399, Cambridge, Massachusetts, 1994. The MIT Press.
25. M. Sipper. Studying artificial life using a simple, general cellular model. *Artificial Life Journal*, 2(1), 1995. The MIT Press, Cambridge, MA.

26. A. Smith. Cellular automata theory. Technical Report 2, Stanford Electronic Lab., Stanford University, 1969.
27. A. R. Smith. Simple computation-universal cellular spaces. *Journal of ACM*, 18:339–353, 1971.
28. A. R. Smith. Simple nontrivial self-reproducing machines. In C. G. Langton, C. Taylor, J. D. Farmer, and S. Rasmussen, editors, *Artificial Life II*, volume X of *SFI Studies in the Sciences of Complexity*, pages 709–725, Redwood City, CA, 1992. Addison-Wesley.
29. T. Toffoli and N. Margolus. *Cellular Automata Machines.* The MIT Press, Cambridge, Massachusetts, 1987.
30. G. Y. Vichniac, P. Tamayo, and H. Hartman. Annealed and quenched inhomogeneous cellular automata. *Journal of Statistical Physics*, 45:875–883, 1986.
31. J. von Neumann. *The Theory of Self-Reproducing Automata.* University of Illinois Press, Illinois, 1966. Edited and completed by A.W. Burks.
32. S. Wolfram. Universality and complexity in cellular automata. *Physica D*, 10:1–35, 1984.

A New Self-Reproducing Cellular Automaton Capable of Construction and Computation

Gianluca Tempesti

Logic Systems Laboratory, Swiss Federal Institute of Technology
INN-Ecublens, CH-1015 Lausanne, Switzerland

Abstract. We present a new self-reproducing cellular automaton capable of construction and computation beyond self-reproduction. Our automaton makes use of some of the concepts developed by Langton for his self-reproducing automaton, but provides the added advantage of being able to perform independent constructional and computational tasks alongside self-reproduction. Our automaton is capable, like Langton's automaton and with comparable complexity, of simple self-replication, but it also provides (at the cost, naturally, of increased complexity) the option of attaching to the automaton an executable program which will be duplicated and executed in each of the copies of the automaton. After describing in some detail the self-reproduction mechanism of our automaton, we provide a non-trivial example of its constructional capabilities.

1 Introduction

The history of self-reproducing cellular automata basically begins with John von Neumann's research in the field of complex self-reproducing machines. Advised by the mathematician Stan Ulam, he applied his concepts in the framework of a "cellular space", a two-dimensional grid of identical elements where each element (cell) is a finite state automaton whose next state is a function of its present state and of the present state of its 4 neighboring cells.

Within this framework, von Neumann was able to conceive a self-reproducing automaton endowed with the properties of both computational and constructional universality [1]. Unfortunately, the automaton was of such complexity that, further simplifications notwithstanding, even today's state-of-the-art computers lack the power to simulate it in its entirety.

The next significant event in the history of self-reproducing automata was the development of the automaton commonly referred to as "Langton's loop" [2]. By dropping the requirements of computational and constructional universality, Langton created an automaton capable of non-trivial self-replication, that is an automaton where the replication is actively directed by the automaton itself, rather than being a mere consequence of the transition rules.

The automaton we introduce seeks to go beyond Langton's loop, which is capable exclusively of duplicating itself, by adding computational and constructional capabilities to self-reproduction. In fact, while our automaton is based on the utilization of a "loop" similar to that of Langton's automaton, we have modified the self-reproducing mechanism so that it requires only a fraction of the data circulating in the loop to perform its task, thus making the remaining data available for other purposes.

In the next chapter, we will present an overview of the cellular automata mentioned above, and compare them with our own automaton. We will then describe in detail the operation of our automaton, and provide an example of its constructional capabilities.

2 Self-reproducing cellular automata

2.1 Von Neumann's automaton

Von Neumann's self-replicating cellular automaton was a result of the mathematician's interest in complex machines and their behavior [1]. His research led to the conclusion that the following characteristics should be present in a self-reproducing machine:

- Computational universality, that is the ability to operate as a universal Turing machine, and thus to execute any computational task.
- Constructional universality, that is the ability to construct any kind of configuration in the cellular space starting from a given description; self-reproduction is then a particular case of universal construction.

To implement these properties in a cellular automaton, von Neumann set out to design a universal constructor, i.e. an automaton capable of constructing, through the use of a "constructing arm", any configuration whose description can be stored on its input tape (Fig. 1). This universal constructor, therefore, is able, given its own description, to construct a copy of itself, thus achieving self-reproduction.

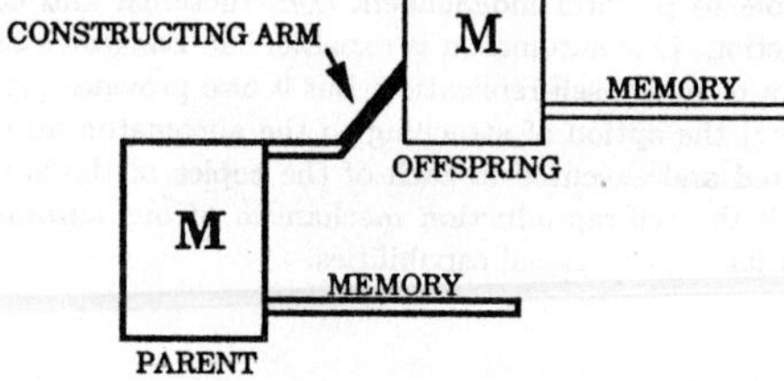

Fig. 1: Von Neumann's self-reproducing automaton

The automaton developed by von Neumann used tens of thousands of 29-state cells and a 5-cell neighborhood (the cell itself plus its four cardinal neighbors). Codd [3] and others managed to reduce the complexity of von Neumann's machine, but the automaton retains a level of complexity too high for simulation. In fact, while parts of the machine have been successfully simulated, the task of simulating the whole automaton remains virtually impossible given current technology.

2.2 Langton's loop

Langton's automaton [2] is based on one of the components of Codd's universal constructor, namely the "periodic emitter" [3]. The automaton (Fig. 2) is essentially a square loop, with internal and external sheaths, where the data neccssary for the construction of a duplicate loop circulate counterclockwise. Duplication is achieved by extending a constructing arm which will be forced to turn 90 degrees to the left at regular intervals corresponding to the size of one side of the loop. After three such turns the arm will have folded upon itself. When the new loop is closed the constructing arm will retract and the new loop will be active, that is will be able to reproduce itself as the original loop did. The original loop will then repeat the process by creating a second copy of itself in another direction, and finally "die" by losing the information within the loop. Given sufficient time, the automaton will replicate itself to fill the available space.

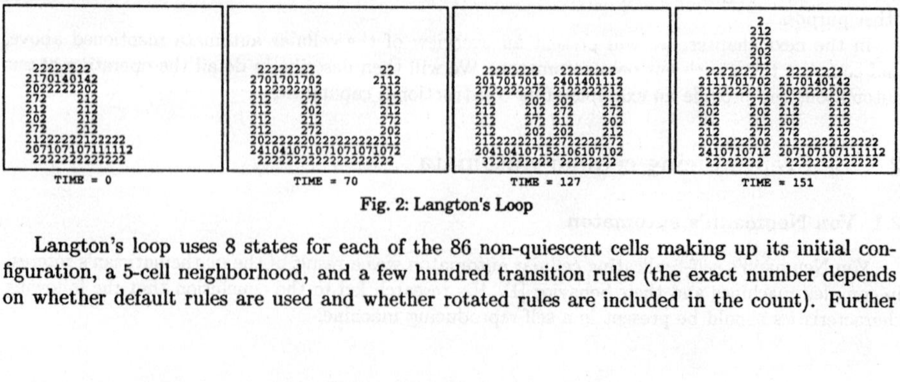

Fig. 2: Langton's Loop

Langton's loop uses 8 states for each of the 86 non-quiescent cells making up its initial configuration, a 5-cell neighborhood, and a few hundred transition rules (the exact number depends on whether default rules are used and whether rotated rules are included in the count). Further simplifications to the automaton were introduced by Byl [5], who eliminated the internal sheath and reduced the number of states per cell, the number of transition rules, and the number of non-quiescent cells in the initial configuration. Reggia et al. [6] managed to remove also the external sheath. Given their low complexity, at least relative to von Neumann's automaton, all of the mentioned automata have been thoroughly simulated.

2.3 The new automaton

Our automaton uses some of the concepts found in Langton's loop. In particular, we retain the concept of loop, which Langton himself derived from Codd's periodic emitter, to store the data dynamically. However, there are some substantial differences between our loop and Langton's automaton (Fig. 3):

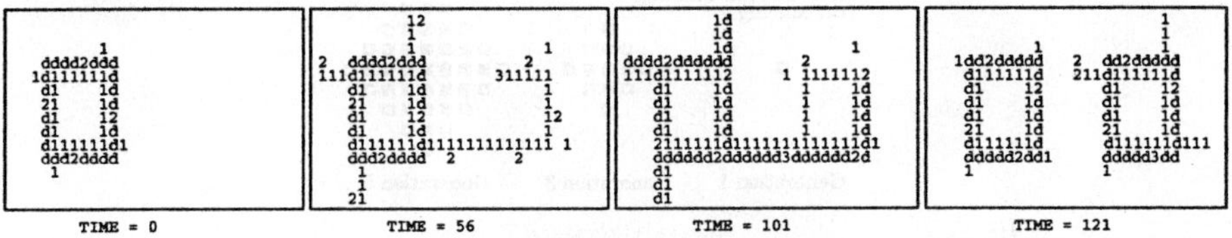

Fig. 3: Our Loop

- We use a 9-cell neighborhood (the cell itself plus its 8 neighbors).
- As in Byl's version of Langton's loop, we use only one sheath, but contrary to Byl, we retain the internal sheath and eliminate the external one. This allows us to let the data in the loop circulate without the need for leading or trailing states (the 0s in Langton's loop). In addition to the internal sheath, we have four "gate cells" (in the same state as the sheath) outside the loop at the four corners of the automaton. These cells are initially in the "open" position, and will shift to the "closed" position once the copy is accomplished.
- We extend four constructing arms in the four cardinal directions at the same time, and thus create four copies of the original automaton in the four directions in parallel. When the arm meets an automaton already in place where the copy should be (which happens for all but the original automaton), it simply retracts and puts the corresponding gate cell in the closed position.
- Rather than being directed to advance, our constructing arm advances by default. As a consequence, it is necessary only to direct it to turn at the appropriate moment. This is done by sending periodic "messengers" to the tip of the constructing arm, which advanced at a slower pace with respect to the messengers.
- The arm does not immediately construct the entire loop. Rather, it constructs a sheath of the same size as the original. Once the sheath is ready, the data circulating in the loop is duplicated and the copy is sent along the constructing arm to wrap around the new sheath. When the new loop is completed, the constructing arm retracts and shifts the corresponding gate cell to the closed position.
- As a consequence of the above, rather than using all of the data in the loop to direct the constructing arm, we use only four of the cells circulating in the loop to generate the messengers. Since the only operation performed on the remaining data cells is duplication, they do not have to be in any particular state. In particular, they can be used as a "program", i.e., a set of states with their own transition rules which will then be applied alongside the self-reproduction to execute some function.
- Unlike Langton's loop, our loop does not "die" once duplication is achieved, as the circulating data remains untouched by the self-reproduction process. Therefore, any program stored in the loop will be able to continue to execute. Also, it is possible to force the loop to try and

duplicate again in any of the four directions simply by shifting the corresponding cell back to the open position.

- When the duplicated loops arrive next to the border of the array, the constructing arm detects the border and retracts without attempting to duplicate the data. Thus, our automaton, unlike Langton's, does not crash when the duplication process reaches the edge of the cellular space.
- Because the reproduction process occurs in the four directions at the same time, the growth of the colony follows a symmetric pattern (Fig. 4), unlike the spiraling pattern of Langton's automaton.

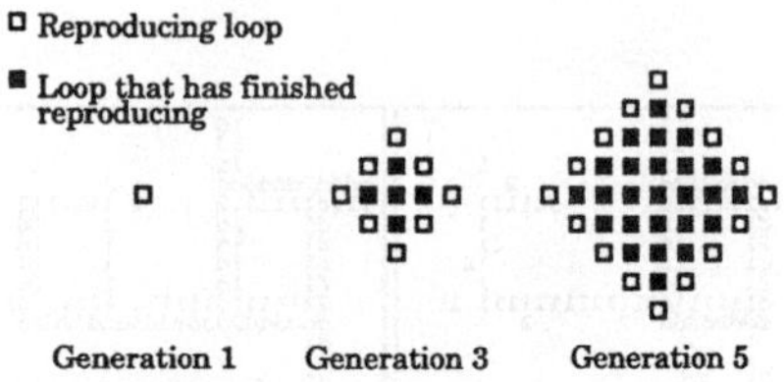

Fig. 4: Growth pattern

As should be obvious from the above, while our loop owes to von Neumann the concept of constructing arm and to Langton (and/or Codd) the basic loop structure, it is in fact a very different automaton, endowed with some of the properties of both.

As far as the complexity of the automaton is concerned, its estimation is more difficult than for Langton's loop, as it depends on the data circulating in the loop. The number of non-quiescent cells making up the initial configuration depends directly on the size of the circulating program. The more complex (i.e. the longer) the program, the larger the automaton. It should be noted, however, that the complexity of the self-reproduction process does not depend on the size of the loop. The number of states also depends on the complexity of the program. To the 5 "basic states" used for self-reproduction (see description below) must be added the "data states" (at least one) used in the program, which must be disjoint from the basic states. The number of transition rules is obviously a function of the number of data states: in the basic configuration (i.e., one data state), the automaton needs 692 rules (173 rules rotated in the four directions). By default, all cells remain in the same state.

The complexity of the basic configuration is therefore in the same order as that of Langton's and Byl's loops, with the proviso that it is likely to increase drastically if the data in the loop is used for some purpose. In fact, the number of rules in the automaton we have described grows as D^4, where D is the number of data states. A different version of the automaton limits the growth to D^3 (at the expense of some versatility), but the increase remains substantial.

In the next chapter we will describe in some detail the operation of the automaton in a small, basic configuration, and illustrate an example of a loop where a program has been included in the loop to demonstrate the construction capabilities of our automaton.

3 Description of the automaton

3.1 Cellular space and initial configuration

As for von Neumann's and Langton's automata, the ideal cellular space for our automaton is an infinite two-dimensional grid. Since we realize that a practical implementation of such a cellular space might prove difficult, we added some transition rules to handle the collision between the constructing arm and the border of the array. On meeting the border, the arm will retract without attempting to make a copy of the parent loop.

The cells of the array require five basic states and at least one data state (see Fig. 4 at time 0). State 0 is the "quiescent state" and is represented by a blank space in the figures. State 1 is the "sheath state", that is the state of the cells making up the sheath and the four gates. State 2 is the "activation state". The four cells in the loop directing the reproduction are in state 2, as are the messengers which will be used to command the constructing arm and the tip of the constructing arm itself for the first phase of construction, after which the tip of the arm will pass to state 3, the "construction state". State 3 will construct the sheath that will receive the copy of the loop, signal the parent loop that the sheath is ready, and lead the duplicated data to the new loop. State 4, the "destruction state", will destroy the constructing arm once the copy is ready. In addition to these states, we have labeled 'd' the data state, with the understanding that this one symbol might in fact represent any set of states not including states 0 to 4.

The initial configuration is in the form of a square loop wrapped around a sheath. The size of the loop is a variable that for our example have set to 8x8. The loop is a sequence of data states in which four cells in the activation state are placed at a distance from each other equal to the side of the loop. Near the four corners of the loop we have placed four cells in the sheath state. These are the gate cells, and the position they occupy signifies that the gates are open (that is, that the automaton should attempt to duplicate itself in all four directions).

3.2 Operation

Once the cellular space starts operating, the data starts turning around the loop. Nothing happens until the first 2 reaches a corner, where it finds the gate open. Since the gate is open, the 2 splits into two identical cells. One cell continues turning around the loop, while the second starts extending the arm (Fig. 5a). The arm advances by one cell each two time periods. Once the arm has started extending, each 2 that arrives to a corner will again split and one of the copies will start running along the arm, advancing by one cell per cycle (Fig. 5b). Since the arm is extending at half the speed of these messengers and the messengers are spaced 8 cells apart (the length of one side of the loop), the messengers will reach the tip of the arm at regular intervals corresponding to the length of one side of the loop.

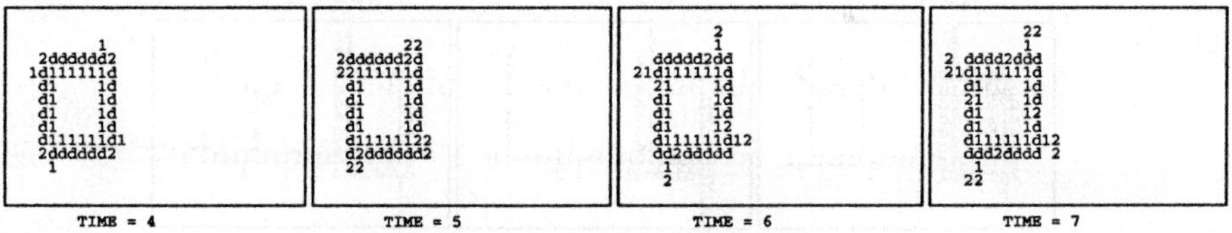

Fig. 5a: The constructing arm starts extending

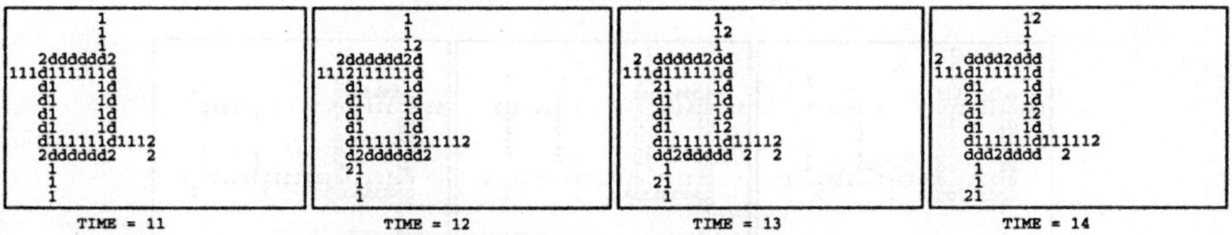

Fig. 5b: The first messenger leaves the loop

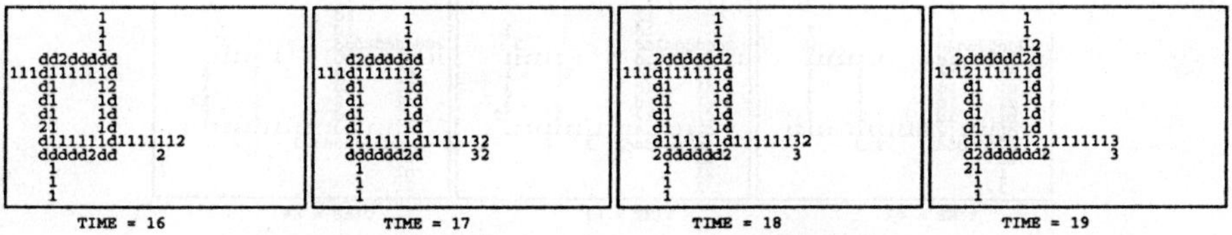

Fig. 5c: The first messenger reaches the tip of the constructing arm

When the first messenger reaches the tip of the arm, the tip, which was until then in state 2, passes to state 3 and continues to advance at the same speed (Fig. 5c). This transformation tells the arm that it has reached the location of the offspring loop and to start constructing the new sheath.

The next two messengers will force the tip of the arm to turn left (Fig. 5d), while the fourth will reach the tip as the arm is closing upon itself (Fig. 5e). It causes the sheath to close and then runs back along the arm to signal to the original loop that the new sheath is ready.

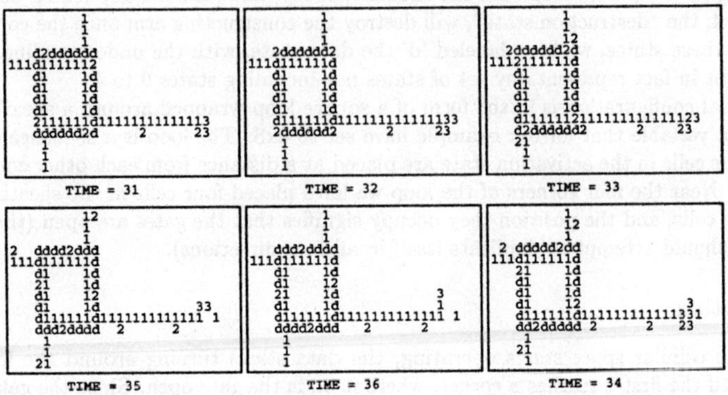

Fig. 5d: The second messenger forces the arm to turn to the left

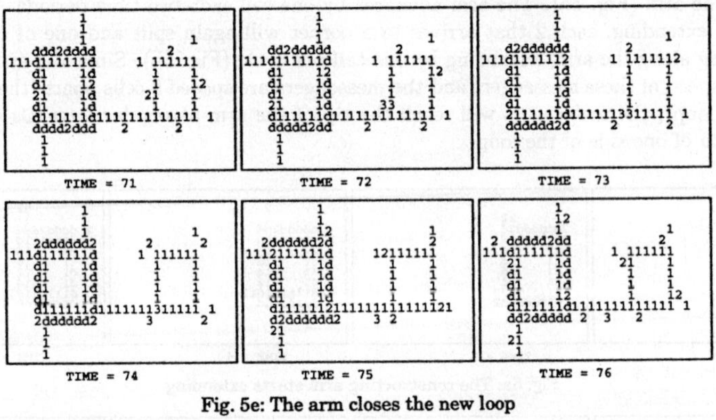

Fig. 5e: The arm closes the new loop

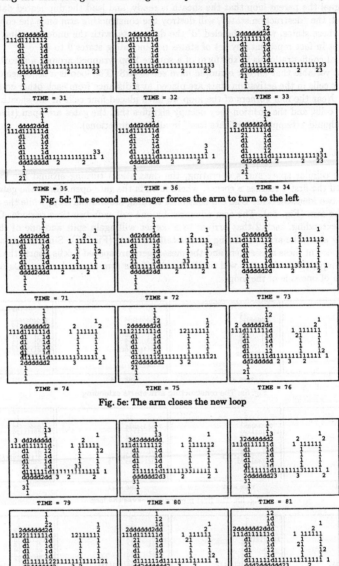

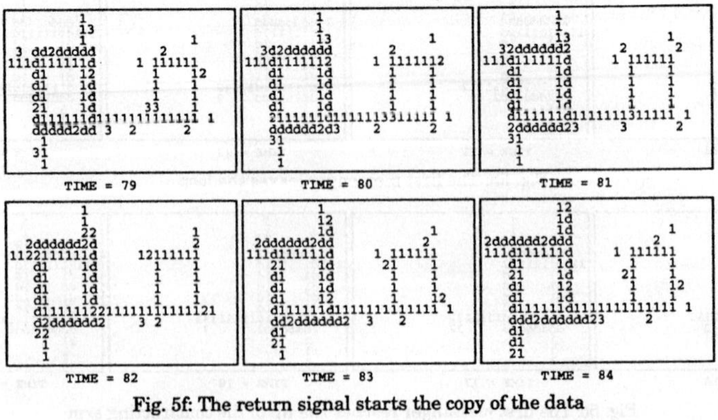

Fig. 5f: The return signal starts the copy of the data

Once the return signal arrives at the corner of the original loop, it waits for the next 2 to arrive (Fig. 5f). When the 2 sees the 3 waiting by the gate, again it splits, one copy staying around the loop, the other running along the arm. This time, however, rather than running along the arm in isolation as a messenger, it carries behind him a copy of the data in the loop. In the copy, the activation cells are temporarily switched off (set to state 3) until the new sheath is reached, where they will again become 2s and start their function.

Always followed by the data, it runs around the sheath until it has reached the junction where the arm folded upon itself (Fig. 5g). On reaching that spot, it closes the loop and sends a destruction signal (the 4) back along the arm. The signal will destroy the arm until it reaches the corner of the original loop, where it closes the gate (Fig. 5h).

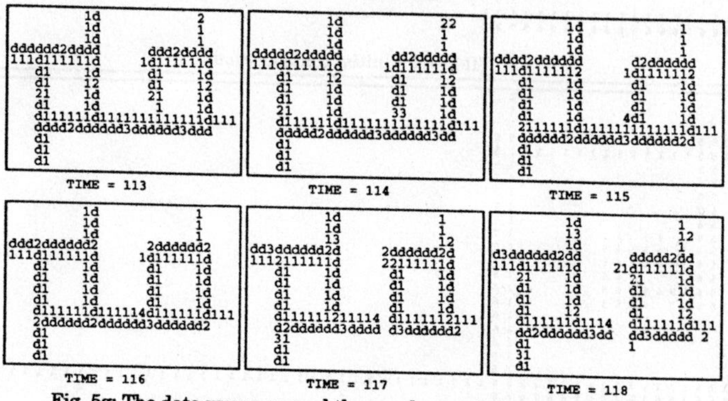

Fig. 5g: The data wraps around the new loop and the arm is destroyed

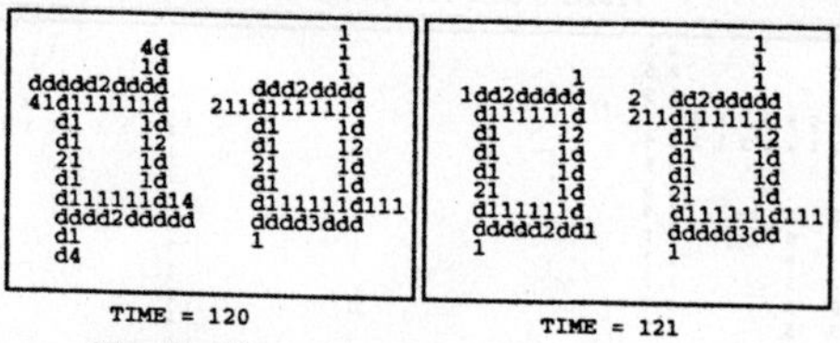

Fig. 5h: The original cell becomes inactive

Meanwhile, the new loop is already staring to reproduce itself in three of the four directions. One direction (down in the figures) is not necessary since another of the new loops will always get there first, and therefore its corresponding gate will be set to the closed position.

After 121 time periods the gates of the original automaton will be closed and it will enter an inactive state, with the understanding that it will be ready to reproduce itself again should the gates be opened.

3.3 Example

In fig. 6, we illustrate an example of how the data states can be used to carry out operations alongside self-reproduction. The operation in question is the construction of three letters, LSL (the acronym of Logic Systems Laboratory), in the empty space inside the loop. Obviously this is not a very useful operation from a practical point of view, but it is a non-trivial case of construction that should demonstrate some of the capabilities of the automaton.

For this example, we have used 5 data states, which have brought the number of transition rules to 35202. Of these, 326 are new rules which control the behavior of the program, and do not concern self-reproduction. The loop size is 20x20, and a full reproduction of a loop requires 321 time periods.

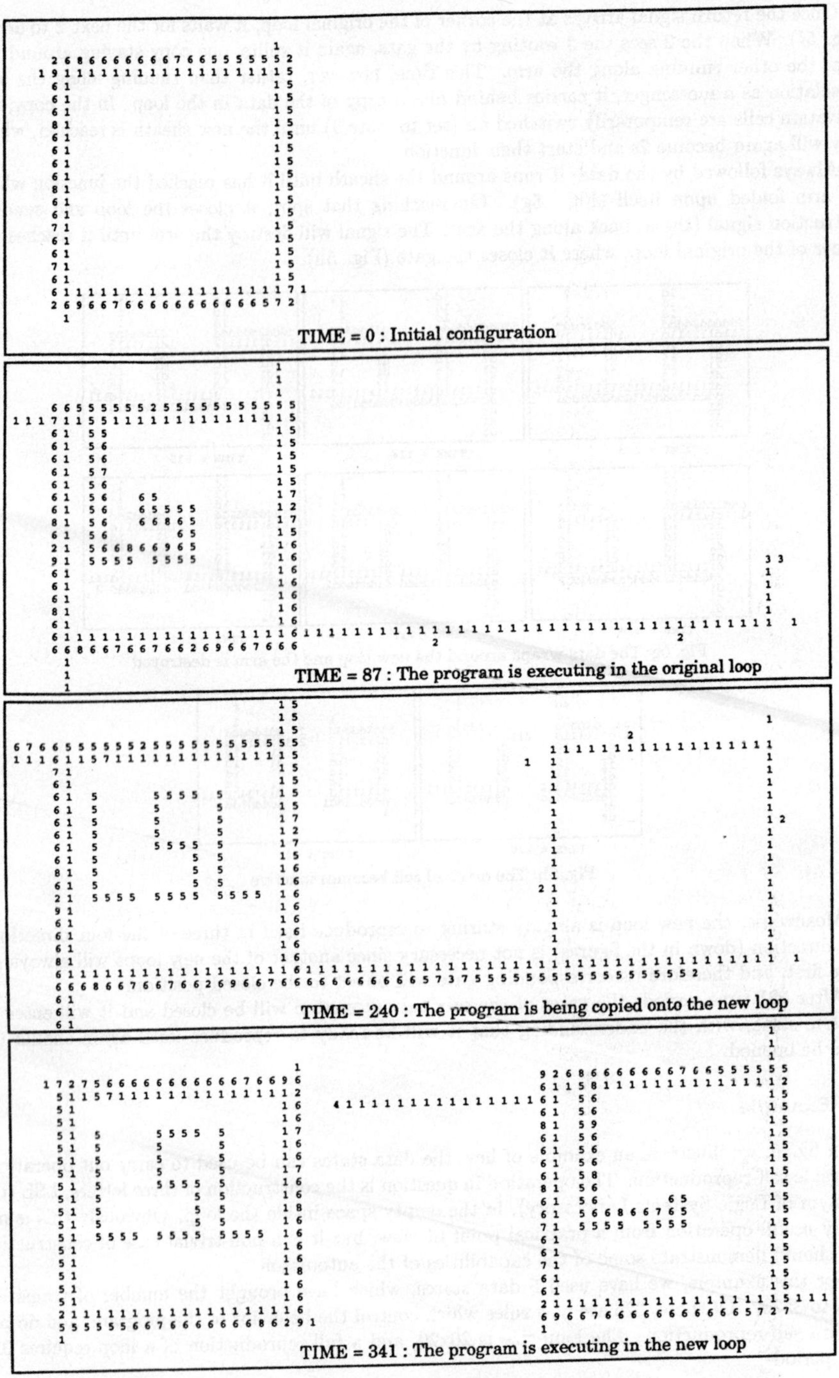

Fig. 6: An example of the capabilities of our automaton

The operation of the program is fairly straightforward. When a certain "initiation sequence" within the loop arrives to the top left corner of the loop, a "door" is opened in the internal sheath. The rest of the program, as it passes by the door in its rotation around. the loop, is duplicated and one of the copies enters the interior of the loop, where it is treated as a sequence of instructions which direct the construction of the three letters. The construction mechanism is somewhat similar to the method Langton used in his own loop, with single-cells instructions such as "turn left", "advance", etc. The construction ends when a "termination sequence" arrives at the door. At that stage, the door is closed and a flag is set in the sheath to warn that the program has already executed.

During the process of reproduction, the program is simply copied (as opposed to interpreted as in the interior of the cell) and arrives intact in the new loop, where it will execute again exactly as it did in the parent loop.

This is a simple demonstration of one way in which the data in the loop could be used as an executable program. Many other methods can be envisaged, and we are currently working on the development of other, hopefully more useful, programs. Our ultimate goal would be to be able to construct, using a program stored in one of our loops, a universal Turing machine. Some of the features of such machines render this a difficult task which will probably require a modification of the basic mechanisms of our automaton, but we are confident that such a construction is indeed feasible.

4 Conclusion

We have described a new self-reproducing cellular automaton which provides some of the features of both von Neumann's and Langton's automata. Our automaton is capable, like Langton's loop and with comparable complexity, of simple self-replication, but it also provides (at the cost, naturally, of increased complexity) the option of attaching to the automaton an executable program which will be duplicated and executed in each of the copies of the loop.

The example we illustrated, while not trivial in its complexity, is far from being a full demonstration of the capabilities of the automaton.

In particular, we have shown that our loop is capable of *some* construction beyond simple self-reproduction, and thus, trivially, of *some* computation. However, we have not yet fully investigated the limits of the automaton as a constructor. That is, we have not yet determined whether our machine is capable of constructional universality. Should that be the case, computational universality should follow trivially (since the automaton would be able to construct a universal Turing machine), and the properties outlined by von Neumann for self-reproducing machines would be met.

References

1. J. von Neumann, *The Theory of Self-Reproducing Automata*, A. W. Burks, ed., University of Illinois Press, Urbana, 1966.
2. C. G. Langton, "Self-Reproduction in Cellular Automata", *Physica 10D*, pp. 135-144, 1984.
3. E. F. Codd, *Cellular Automata*, Academic Press, New York, 1968.
4. R. A. Freitas and W. P. Gilbreath, "Advanced Automation for Space Missions", *NASA Report CP-2255*, 1982.
5. J. Byl, "Self-Reproduction in Small Cellular Automata", *Physica 34D*, pp. 295-299, 1989.
6. J. A. Reggia, S. A. Armentrout, H.-H. Chou, Y. Peng, "Simple Systems That Exhibit Self-Directed Replication", *Science*, Vol. 259, pp. 1282-1287, 26 February 1993.

Self-inspection based Reproduction in Cellular Automata*

Jesús Ibáñez, Daniel Anabitarte, Iker Azpeitia, Oscar Barrera, Arkaitz Barrutieta, Haritz Blanco and Francisco Echarte

Lengoaia eta Sistema Informatikoak Saila
Informatika Fakultatea
649 Posta Kutxa, 20080 Donostia, Spain
Phone: +34-43-218000, Fax: +34-43-219306, Email: jipibmaj@si.ehu.es

Abstract. In this work we analyze different existing cellular automata capable of reproducing, and conclude that reproduction in those models is attained as a fixed point of the more general operation of construction, instead of being founded as a higher level specialized mechanism. Thus reproduction becomes a very unstable property and this fact is an obstacle in the aim of studying it *together* with other Alife characteristics, mainly because of destructive interactions. We propose a cellular reproduction model based on self-inspection as an attempt to overcome at least partially these drawbacks. We argue that our model attains a better founded an more robust reproduction scheme, and briefly explore the possibilities of using it to model the emergence of reproducing structures, to study the interplay between autonomy and reproduction, and to use it as a basis for evolutionary experimental scenarios.

1 Introduction

The ability of living organisms to reproduce[1] has been considered for a long time to be one of their most characteristic features, and so it has become a main target in the Alife research program. Cellular automata probably constitute the field in which most remarkable results have been obtained in the search of reproductive structures in formal or computational environments. Not surprisingly, it is also the one in which the earliest efforts took place. Indeed, cellular automata were devised by John Von Neumann as abstract machines whose computation would deal with the same elements used in their "physical" construction, constituting his second attempt to achieve reproduction performed by purely informational objects. As Arthur Burks [2] quotes, Von Neumann didn't try to simulate reproduction of natural systems in its low-level details (he could hardly do so,

* This work was supported in part by the DGCYT of the Spanish Ministerio de Educación y Ciencia, grant n. PB920456 and by the Euskal Herriko Unibertsitatea (UPV/EHU), grant n. 003.230-HA160/94.

[1] Though widely spread in cellular automata literature, the term self- reproduction will be avoided throughout this paper for the sake of clarity; thus we will just use the words construction and reproduction

considering the degree of knowledge that biochemistry had by this date), but to abstract the problem in its logical form.

Some authors [9] consider that reproduction is secondary to self-production in the explanation of the logic of life. According to that reproduction wouldn't be a necessary precondition for the living organization. In any case, it seems that there is no controversy about the central role that it plays in the phylogenetical achieving of biological complexity, and that could be the reason why the main scientific interest in artificial reproduction is linked with artificial evolution.

On the other hand, now widespread approaches [5] to incorporate the power of Darwinian evolution in formal and computational environments neither use reproduction as an individual operation nor found it on lower level interactions. Researchers seem to have a dilemma between evolution and reproduction: either they choose to neglect the importance of the latter so that it even turns into a routine, global and futile step (as a counterpart for having free hands concerning the design of artificial selection forces and/or genetic operators), or either they stress upon it, with all the incompatibility problems that it supposes. In the case of cellular automata these obstacles seem to get magnified. Cellular configurations are extremely fragile units, and if we consider the restrictions imposed on a two-dimensional space, in which potentially destructive contact is mandatory in a population of reproducing entities, we have to admit that their interaction can be anything but unproblematic.

2 Reproduction in cellular automata

In his aim of building reproducing structures Von Neumann started by approaching the general problem of a cellular configuration (understood as the state of a finite, contiguous portion of the cellular space) designed to construct another configuration in an appropriate region of free cellular space neighbor to it. He defined a 29-state cellular space in which he could accomplish construction tasks by means of special configurations composed of two specific elements: a control unit and a construction unit. The control unit could send appropriate constructing instructions and thereby the construction unit would interpret them to build the desired cellular structure.

The next step should be the design of a universal constructing configuration, who would be able to build not just a specific cellular structure, but any desired configuration instead. To do so the constructor would be provided with an external cellular region called the tape unit, suitable of storing a description of any other configuration to be constructed. The control unit would read the content of the tape and translate it into instructions for the constructing unit. Finally the universal constructor would be extended with the ability of copying the contents of its own tape to put aside the newly constructed configuration. This way the reproducing cellular configuration would consist of the extended universal constructor along with a tape including its own description.

Though never simulated in actual computing environments, Von Neumann's reproducing cellular automaton and other cellular systems inspired on it [4] clearly established that universal construction capability ensures the existence of reproducing cellular automata, which can safely be obtained with minor modifications on those constructors. Anyway, Chris Langton [7] proved that the universality requirement was not a necessary condition for reproduction, though, as himself pointed out, its removal introduces another problem: that of avoiding trivial versions of reproduction. Langton established two criteria that any nontrivial reproducing structure should fulfill: a) the construction of the copied structure should be "actively guided by the configuration itself", and b) the configuration should treat its stored description in two manners: interpreted (translation of the information into effective constructing actions) and uninterpreted (transcription of that information so that the reproduced structure inherits the same functionality of its builder).

According to these criteria Langton devised his well-known sheathed reproducing loops. He used as primitive a particular class of configurations called data-paths that had been defined earlier by Codd [4]. A data-path is a linear structure that can contain signals flowing along it. Data-paths have constructing properties, since the signals contained in them can be constructing instructions. This way Langton achieves three features that will turn out very important for his purposes:

1. the necessity of reading the description from a static tape unit is avoided, since the description is contained into the structure itself in its active form (constructing instructions)
2. since the only relevant property of a data-path is its shape, the constructing instructions get enormously simplified with respect to Von Neumann's system: they only have to specify the direction in which the next constructing action is to take place
3. the control unit and the constructing unit become one, since the description of the structure to build is directly executable

Thereby any data-path can be constructed by another one with enough capacity to contain its constructing instructions. To achieve reproduction the main drawback is that the full description of any path is longer than itself, so Langton chooses a special looping path whose regularity (it is a perfect square) allows to compact its description to 1/4 of its original size, and whose cyclic structure allows to repeat the emission of the constructing instructions the eight times required, firstly to construct a copy of the loop (translation) and finally to transmit the instructions to the constructed loop, so that it can in turn reproduce (transcription).

3 Universality, construction and reproduction

Though the size, number of states and reproduction cycle can be further optimized without leaving the constructing principles [3] or slightly altering them [14], the reduction in complexity between Von Neumann's model and Langton's one is really astonishing, and it is undoubtedly due to the fact that the latter avoids the overwhelming task of designing a universal constructor[2]. Nevertheless, and despite this fundamental difference, both models share a common origin of the reproducing properties. First, either implicitly or explicitly, a class of interesting cellular configurations is defined. Second, a valid construction mechanism over that class is established. Third, a special configuration that reproduces (constructs itself) is designed. That is to say, *reproduction is achieved as a special case of construction.* The main difference between both models is the way in which this special case is constructed.

Though being a concept borrowed from Nature, we find an essential difference between the *logic* of reproduction in cellular automata and that of natural replication mechanisms. We have no evidence that in Nature reproduction arises as a particular case of a general constructing method. Note that the reproducing entities designed by Von Neumann, Langton and other authors could construct *something entirely different* from themselves just altering the internal description that they carry along. If we fall in the temptation of making analogies such as phenotype/constructor and genotype/description, we find that the relationship between genotype and phenotype would be absolutely contingent in both cases: reproduction becomes an accident in these systems.

On the contrary, as Von Neumann [18] himself pointed out, this is a very common phenomenon that arises in formal systems capable of self-reference, and is in particular very similar to the fixed point result in Recursive Function Theory, firstly expressed by Kleene and now commonly known as the Second Recursion Theorem [15]. One very well known consequence of this result is the existence of a self-descriptive Turing machine (as a fixed point of the general operation performed by Turing machines describing other Turing machines). Whenever we have formal systems whose capabilities include dealing with objects of the same kind (functions defined in terms of other functions, programs that use other programs as data, formulae that express properties of other formulae, etc.), we just need very weak conditions to establish fixed point results [8], that ensure the existence of such special objects (functions defined in terms of themselves, programs that act as if they had theimselves as input, formulae that express properties about themselves, etc.).

Sometimes a version of a universality property can be very useful in proving some of these results, but it is not the general case. Universality is a much more restricted concept, and it is always relative: actually Von Neumann's constructor is not *strictu sensu* universal, that is to say, *it is not able to construct any possible configuration* of the cellular space. Moreover, it has been proven the existence of non-constructible ones. The solution is, obviously, to restrict the universe to a special class of interesting cellular structures containing the constructor itself

[2] It wouldn't be possible anyway, because if we assume a finite number of states in the cellular system (as it is usually done), since any data-path must contain the description of what it is going to build, the size of that description is limited by the total length of the data-path, and so the number of different constructions is finite. That excludes the possibility of having any universal constructor among the data-paths.

[16]. We can conclude that Langton's configuration is much simpler than Von Neumann's because *it doesn't use* a universal constructor to design it, rather than because this universal constructor really *doesn't exist*.

Nevertheless, Barry McMullin's [10] objection against reproduction schemes not based upon universal constructors shouldn't be neglected. McMullin argues that Von Neumann's search for a universal constructor as a base for reproduction obeys a powerful reason: such a reproducer along with some more reasonable preconditions could allow the definition of a family of analogous reproducers anchored in it, with genetic similarity properties and with unbounded complexity. This way we could have something sounder than a mere artificial reproducing entity: instead, this reproduction scheme could serve as a solid basis to bring about artificial Darwinism (the ultimate reason to explain the interest in artificial reproduction). In other reproduction schemes we wouldn't have this for granted.

4 A model for reproduction in cellular automata based on self-inspection

When Langton establishes his second condition to avoid trivial cases of reproduction he is doing more than that: he is assuming that reproduction must be driven by the interpretation of a description of the object to be build that *must be independent* from the nature of the builder. We claim that nontrivial reproductive behaviors can be obtained without this independence precondition by self-inspection based mechanisms: the description of the object to be built is dynamically constructed at the same time it is being interpreted. As Richard Laing [6] pointed out and proved with his kinematic reproducers, there is no logical need for the presence of an explicit blueprint of the reproducing structure.

We present a model of reproducing cellular automata whose behavior relies in a higher order operation than mere construction. It has been developed under the program ACD (Automate Cellulaire Distribué), from the Laboratoire de Systèmes Logiques in Lausanne (Switzerland) on a Macintosh Quadra 840 AV. The total number of states is 16, though only 7 are essential (in the sense that the other 8 are automatically generated as by-products or to signal intermediate steps whenever appropriate).

The configuration has some principles very similar to those of Langton's [7] and Byl's [3] models: we have the data-paths, in which we distinguish the sheath (state $\star$) and stuffing (state $\bullet$). Several constructing signals can flow along them, and whenever a signal arrives to the open end of a path, it becomes executable and performs a particular construction operation. There are three of such signals: ahead (state $\circ$), left and right (not appearing in the figures). Each signal is preceded by an orientation mark (state $\diamond$) that indicates its propagation direction. Our reproducing structure is composed by a loop and a constructing arm, but *we are not restricted to square loops*, since in principle any shape can be reproduced. Figure 1 shows an example of a cycling data-path with arbitrary shape.

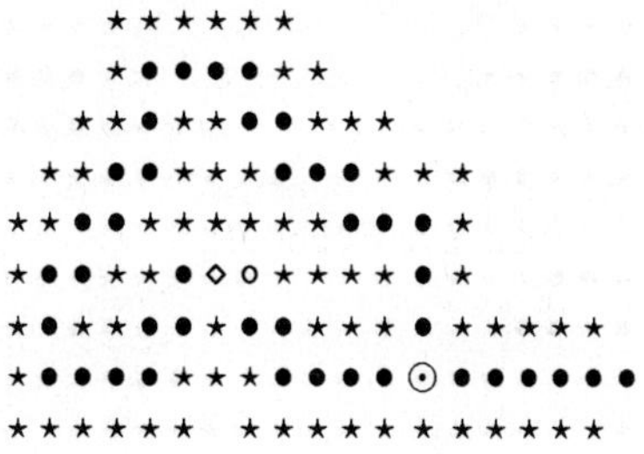

Fig. 1. Example of a reproducing loop

Note that corners must be defined as part of the path to assure proper transmission of signals in several special cases. The only thing needed to start reproduction (besides having free space to do so) is the inclusion of a special reading cell (state ⊙) that will travel along the path in which it is situated. Any signal traversing the path forces the reading cell to advance across it. Depending on the shape of the path that it follows, the reading state sends backwards appropriate constructing signals that, when arriving to the end of the construction arm, build a mirror image of the original loop. Also these signals are duplicated at the T-junction to continue firing the reading process. Figure 2 shows how reproduction has been partially accomplished (note that the position of the reading state corresponds to the most advanced position of the already constructed part of the new copy).

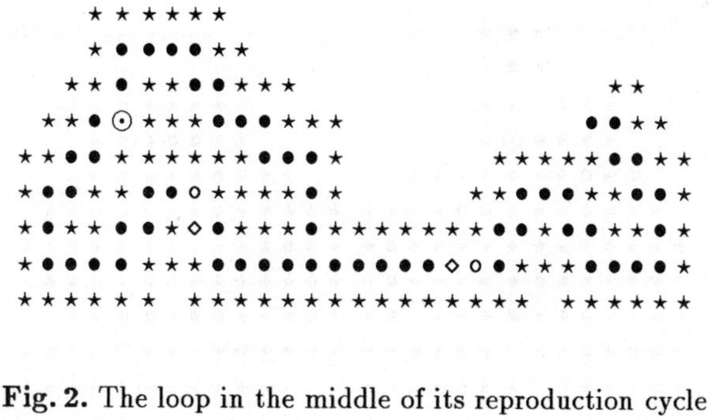

Fig. 2. The loop in the middle of its reproduction cycle

Once the copied version of the reproducing loop has completed its cyclic part, that is to say, once it has collided with itself (see Figure 3), the process is similar to that of Langton's model.

Appropriate states and transitions are provided to cut the umbilical cord and to open a new constructing arm both in the constructor and in the constructed

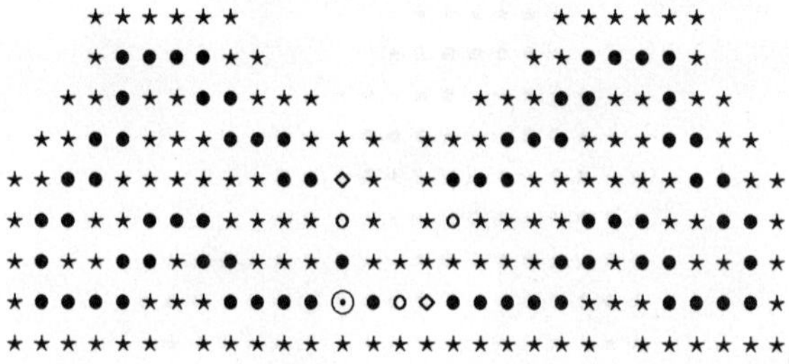

Fig. 3. The construction of the cyclic part of the loop is accomplished

units. But since we can have an arbitrary shape we cannot guarantee that this may be accomplished "next corner", as it happens with Langton's loops (there could be saw-teeth profiles, invaginations, etc.). For those cases there is a trial-and-error mechanism so that the system looks for an appropriate corner to open the arm. If it finds any obstacle, the arm retracts and another opening point is sought for. Whenever it is found, the arm is generated and the reproduction cycle is triggered for both the parent and the child structure, as it can be seen in Figure 4.

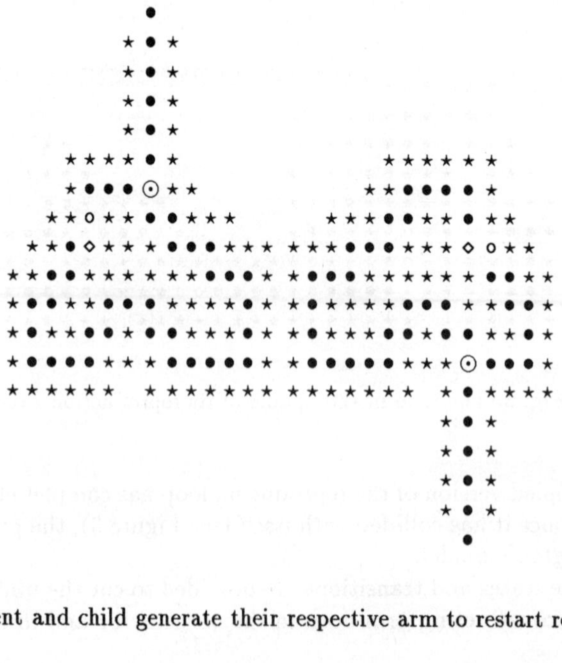

Fig. 4. Parent and child generate their respective arm to restart reproduction

Then each structure starts the reproduction in the chosen direction up to completion of its respective copy, and the process starts again. The maximum number of children that will eventually be generated will of course be four, but no direction will be abandoned without exploring it. Eventually the reproducing structures will start to fill the cellular space in all directions as they crowd in the way shown in Figure 5. Note that the timing of the different structures is not identical and some of them have got ahead. This is because of the different delays that each reproducing configuration has according to the particular distance that it must traverse between two consecutive branching points.

Fig. 5. After 5500 steps the loop has spread its descent in all directions

5 Genesis, autonomy and evolution

Our model is nothing but just an example of how easy might be the definition of self-inspection based reproduction mechanisms in cellular automata. In any case our principal aim in its design is to prototype a toolkit for testing the relationships, interactions and trade-offs between reproduction and some important problems that arise in Artificial Life[3]. We shall briefly state three of them, explaining our point of view about the guidelines for attempting to embody them in a class of models similar to ours.

First of all we consider the problem of explaining how reproducing abilities can emerge from lower level processes. In Langton's model constructing properties are founded and explainable in terms of the much simpler local interactions determined by the state-transition rules of the cellular automaton. In other words, the increase in complexity between the overall automaton dynamics (that deals with cells) and the specific construction dynamics (that deals with paths) is relatively small. However, it is very difficult to dynamically decompose reproduction in terms of lower level properties. Let's imagine an experiment in which the cellular space would be filled with reasonably biased randomly distributed states. Let's suppose the biasing oriented to promote the initial existence of some potentially constructing structures. The natural evolution of the system once freed to the interaction laws would undoubtedly tend to waste the constructing signals, thus enhancing the static structures, but at the same time impoverishing the informational patrimony of the system. To achieve reproduction we would need two conditions:

1. the spontaneous formation of loops
2. the filling of those loops with the precise signals to describe them, so to start reproduction[4]

Though any of these conditions cannot separately be discarded, the combination of both seems hard to achieve in an ordinary human/computer time scale. This unfeasibility of the spontaneous emerging of reproduction is due to the above expressed fact that in the model, reproduction is a very unlikely special case of construction.

At first glance it could seem that in our model, since we specialize reproduction as a separate and higher level feature from construction, it would be even harder to achieve spontaneous emerging of reproducing structures. We claim that this is not the case. It should be noted that, since we have essentially copied from Langton the construction mechanism, we have it also well founded in the (lower level) transition mechanism. But in addition we have that the self-inspection based reproduction operation results in:

- the significant increase in the probability of spontaneous formations of loops; if we only had constructing signals, the maximum number of them in any cellular space would be a fraction of the total length of existing paths, but now we have this limit as an upper bound of the number of self-inspecting

[3] It should be understood that we restrict ourselves to problems arising in Artificial Life *computer simulations.*

[4] We could weaken this condition, since it would be sufficient to have the description of another reproducing loop preserving some reproduction-compatibility restrictions with the host loop.

signals, each of which is able to produce as much constructing signals as path-cells it visits during its lifetime

- the concept of loop is wider, since we can admit almost any shape[5] in our model; moreover, non-reproducing loops (due to multiple breaking points in the loop) are nevertheless useful in avoiding the system's dying, since they are still non-stopping producers of constructing signals
- once formed, a loop just needs a reading signal inside to ensure that the transmitted information will correspond to its own description, thus drastically facilitating the triggering of spontaneous reproduction.

Roughly speaking, the setting of what we could call VENUS-like experiments [12] (in which we are working at present) implicitly poses the question of basing the "biochemical" properties of our system in explicit "chemical" features. It will be very important to tune appropriately the initial conditions as well as to enrich via new transitions the "inorganic" behavior of the states (for instance, it can be supposed a "polymerizing" tendency of the states that constitute the sheath of the data-paths).

The second problem that could have a chance of being studied along with reproduction, by means of a self-inspecting reproducing model, could be that concerning self-maintenance. The fragility of the cellular reproducing structures is its main drawback to put the reproducing properties to work in interacting scenarios. Our present solution tends to minimize the intercourse between different configurations. A reproducing structure will use its retractile arm whenever it encounters the frontier of any neighbor. In fact, the only exception to that rule takes place whenever the growing configuration needs to collide with itself in order to close the loop of its child. This self-recognizing process is performed thanks to some synchronization properties (if reproduction is worked out without errors the last extending instruction will meet a corresponding one travelling by the path to which it is approaching, and this circumstance will allow the opening of the latter).

We need to imagine different contexts in which the identity of a reproducing loop would be threatened. One of them could be to loosen the conditions in which a structure could damage another one. At the same time the system should be reinforced by means of transitions devoted to restore possible damages and to assure boundary formation. Probably this wouldn't be enough, and perhaps a good inspiration source could be the tessellation example of autopoiesis proposed by Francisco Varela [17]. That suggests that we should not neglect the possibility of using probabilistic transition functions to design structures capable of absorbing noise effects. It can be argued that all that can easily well be accomplished in a model like Langton's, but it should be noted that in our self- inspecting model the reproduction operation is at least in principle more robust facing external or noisy perturbations. If a particular signal (not the self-inspecting one) would

[5] Actually there are some minor exceptions to that law. In some degenerate cases either signal generation or transmission doesn't work appropiately, and thus reproduction cannot take place. We have preferred to leave aside these few cases rather than incrementing the number of special-purpose states.

decay due to any hazard, the only affected structure (and perhaps not fatally) by that fact would be the copy to be reproduced at that precise moment, while the original configuration would preserve its reproducing capabilities.

Finally we will make some comments about the possibility of using our self-inspection based reproductive scheme as a basis for artificial evolution. This attempt is not new: it has quite successfully been applied in other environments, and the most paradigmatic of them is probably Tierra system [13]. We must confess that the analogy with Tierra has been an encouraging guideline in our work, although cellular spaces lack many of the features that contribute to the interesting results by Thomas Ray, and in particular:

- all actions and interactions must take place by direct contact; this compulsory proximality not only yields that any action implies that the agent must physically move, but also the possibility of unavoidable obstacles to do so
- in particular, reproduction is bound to take place in the immediate vicinity of the configuration, with the subsequent fertility limitation for the reproducers; this is in sharp contrast with the hundreds of descendants that a Tierra creature can produce
- the writing protection facilities that shield Tierra creatures and forbids destructive interaction have no correspondence in a cellular space, and certainly life can be far more uncomfortable in the latter
- in a classical computational environment we can take advantage of an old powerful theory that sets which are the minimal operations to expect as much expressive and computational power as can be; moreover, we have a long tradition in designing instruction-sets so as to have a cultural sense of what can be useful, time saving or tricky facing the choice of the primitive operational tool kit of our system; that is not the case in cellular automata, where the magnitude of the choice possibilities is overwhelming even with a very reduced number of states, and where it is very difficult to evaluate the influence of a small change in the definition of the behavior of one particular state.

One of the main problems will be the stability of the reproducing entities in an environment in which the main goal is to make things grow around. Probably the previous study of the self-maintaining capabilities would be very important in designing more robust and self-repairing loops. Difficulties can arise if in one hand we need to prevent blind destructive power from growing structures, but in the other it becomes desirable to abandon the assumption of free surrounding space as a precondition for reproduction to enhance it in a wider range of conditions. Finally, it should be noted that an appropriate formalization of the variation mechanism is far from being self-evident, and it brings about again the problem of the primitives: should it be limited the number of possible states? does a mutation affect slightly the behavior of a particular cell, altering its transition function, or else the changes take place abruptly among a carefully chosen set of possible state values?. The study of possible answers to these questions opens interesting trends for future work.

An important fact concerning the possible advantages of the system proposed relies in the fact that reproduction is a separate feature from construction. As McMullin [10] quotes, if we want to meet the requirements of the original Von Neumann's project, by no means it suffices with a computational system containing reproducing units. On the contrary, it is fundamental for the chosen set of reproducers to be genetically connected in some sense. This means that they should be able to absorb mutation-like variations without abandoning their condition of reproducers. Of course, if a reproducer is nothing but a special kind of constructor, it seems probable that most of its genetic relatives (those entities close to it according to some mutation mechanism) will be no more reproducers. Fortunately, it doesn't seem to be the case in the proposed model, where there is a wide range of possibilities of provoking mutation-like alterations without loosing that condition.

6 Conclusion

The model that we present is the most general one among a series that we have developed based on the same ideas. Since its virtual interest lies in the possibility of using it to study the problems sketched in the previous section, there will be more specialized versions of it in the next future.

We think that the difficulty of these problems is increasing in order (the easiest being the study of origins while the evolutionary aspects probably carry out the more complex objections). Moreover, it seems likely that the previous study of the origins of reproduction might be very enlightening to face the problem of autonomy, and that before approaching the evolutionary behavior of the system it would be very helpful an attentive exploration of its self-maintaining properties.

In any case we don't claim that self-inspection is actually the best way to approach *separately* either genesis of life, autonomy or evolution, but we hope that it can be useful as a bench mark for enlightening the relationships and interactions between the former concepts and reproduction, or even as a step towards the study of the developmental aspects of evolution [1]. It is also worth to mention that self-inspection based reproduction allows the inheritance of phenotypical variations [6], thus supporting Lamarckian propagation of acquired characteristics. It has been noted [1] that, unlike in Nature, Lamarckian evolution can be a major tool to speed up evolutionary processes in artificial systems.

It can be argued that the removal from the cellular reproduction scheme of the genetic foundations is an excessive price to pay, since self-inspection avoids the possibility of genotype/phenotype distinctions (or at least reduces it drastically in time, since in some sense genotype is present in an ephemeral and discontinuous way during the migration of the construction signals). According to that, our model would intrinsically render less complex behavior than it could do otherwise. In any case, though it has been postulated [11] that complex systems require self-descriptions, the latter constitute by no means a sufficient condition to assure complexity. So far known cellular systems are very simple indeed, and we ought to admit that the exhibited reproducing capabilities have a closer scientific metaphor to the replicating molecule than to the reproducing organism.

Acknowledgments

We would like to thank to Eduardo Sánchez, from the Laboratoire de Systèmes Logiques in Laussanne (Switzerland) who kindly supplied the software used, and to Alvaro Moreno, from the Logika eta Zientzi-Filosofia Saila in Donostia (Spain) for his helpful comments and suggestions.

References

1. Ackley D.H. and Littman M.L.: A case for Lamarckian evolution. In *Artificial Life III*, edited by Langton, C.G. Addison Wesley (1994)
2. Burks A. W.: Von Neumann's reproducing automata. In *Essays on cellular automata*, edited by Burks A.W. University of Illinois Press (1970)
3. Byl J.: Reproduction in small cellular automata. Physica D **34** (1989) 295–299
4. Codd E. F.: Cellular automata. Academic Press (1968)
5. Goldberg, D.E.: Genetic algorithms in search, optimization and machine learning. Addison-Wesley (1989)
6. Laing R.: Automaton models of reproduction by self-inspection. Journal of theoretical Biology **66** (1977) 437–456
7. Langton C.G.: Reproduction in cellular automata. Physica D **10** (1984) 135–144
8. Manna Z. and Vuillemin J.: Fixpoint approach to the theory of computation. Communications of the ACM **15** (1972) 528–536.
9. Maturana H.R. and Varela F.J.: Autopoiesis and cognition: the realization of the living. Reidel (1980)
10. McMullin B.: Artificial Darwinism: the very idea! In *Autopoiesis and Perception*, Technical Report bmcm9401 of the School of Electronic Engineering at the Ollscoil Chathair Bhaile Átha Cliath / University of Dublin City (1994)
11. Pattee H.H.: Dynamic and linguistic modes of complex systems. International Journal on General Systems **3** (1977) 259–266
12. Rasmussen S., Knudsen C. and Feldberg R.: Dynamics of programmable matter. In *Artificial Life II*, edited by Langton C.G., Taylor C., Farmer J.D. and Rasmussen S. Addison Wesley (1992)
13. Ray T.S.: An approach to the synthesis of life. In *Artificial Life II*, edited by Langton C.G., Taylor C., Farmer J.D. and Rasmussen S. Addison Wesley (1992)
14. Reggia J.A., Armentrout S.L., Chou H.-H. and Peng Y.: Simple systems that exhibit self-directed replication. Science **259** (1993) 1282–1287
15. Rogers H.: Theory of recursive functions and effective computability. McGraw-Hill (1967)
16. Thatcher J.W.: Universality in the Von Neumann cellular model. In *Essays on cellular automata*, edited by Burks A.W. University of Illinois Press (1970)
17. Varela, F.J.: Principles of biological autonomy North Holland (1979)
18. Von Neumann, J.: Theory of reproducing automata, edited and completed by Burks, A.W. University of Illinois Press (1966).

5. Robotics and Emulation of Animal Behavior

Evaluation of Learning Performance of Situated Embodied Agents

Maja J Matarić
Volen Center for Complex Systems
Computer Science Department
Brandeis University
Waltham, MA 02254
tel: (617) 736-2708 fax: (617) 736-2741
email: maja@cs.brandeis.edu

Abstract

This paper discusses the complexities of designing and evaluating Alife learning systems using physical robots interacting in complex, dynamic environments. We use the learning data from two different implemented multi-robot learning systems to illustrate the difficulties with traditional "objective" methods of evaluation. We then describe the methods we used to evaluate the behavior of such learning systems in order to preserve its dynamics and demonstrate their effect on the performance.

1 Introduction

Adaptation is one of the hallmarks of biological systems, and one of the main challenges of Alife, a field striving to produce artificial systems which display properties of biological life. Adaptation is particularly difficult yet crucial in systems dealing with dynamic, uncertain environments, such as the physical world mobile robots interact with. The field of robotics has worked on achieving adaptive behavior through various approaches, from traditional adaptive control systems (Barto 1990), to more recent neural–networks (Pomerleau 1992, Millán 1994) and genetic algorithms (Floreano 1993, Steels 1994). Many of those approaches fall under the broad and, in the last decade in particular, exceedingly popular approach called "reinforcement learning."

Learning by reinforcement is widely spread in biology, and animals use it to effectively fine–tune old behaviors and acquire various new ones (McFarland 1985). However, studies in ethology and psychology show that learning curves of animals and artificial agents differ drastically, not only in speed to convergence, but their basic nature (McFarland 1987). In this paper we discuss some reasons underlying these fundamental differences.

2 Choice of Domain and Related Work

According to evolutionary theory, the complexity of the niche drives the adaptation of the species. Many fundamental problems with adaptation in artificial systems do not appear until the systems and their environment are sufficiently complex. Even the most benign physical environments offer abundant potential for complexity due to the dynamics interaction between the simplest of physical robots with the environment (Smithers 1994).

Another challenging and critical level of complexity is introduced when multiple agents interact in the physical world, and nature abounds with examples of such systems. Much research has addressed societies with large numbers of simple agents, usually referred to as "swarm intelligence", dealing with the design of communication protocols, social rules, and strategies for avoiding conflict and deadlock (Dario & Rucci 1993, Dudek, Jenkin, Milios & Wilkes 1993, Huang & Beni 1993), etc.

Many Alife researchers have focused on modeling colonies of ant–like agents (Corbara, Drogoul, Fresneau & Lalande 1993, Colorni, Dorigo & Maniezzo 1992, Drogous, Ferber, Corbara & Fresneau 1992, Travers 1988), and some have experimented with both live and simulated ants (Deneubourg, Goss, Franks, Sendova-Franks, Detrain & Chretien 1990, Deneubourg, Theraulax & Beckers 1992). Related work, including Assad & Packard (1992) and Hogeweg & Hesper (1985), demonstrated simulations of other simple organisms producing complex behaviors emerging from simple interactions, and the evolution of simple communication strategies in such systems (Werner & Dyer 1990, MacLennan 1990).

Until recently, work in Alife largely focused on systems with large numbers of extremely simple agents. In the last few years, several systems featuring multiple mobile robots have been used to implement and test Alife methodologies (Matarić 1992, Beckers, Holland & Deneubourg 1994). Experiments using such systems have brought to light many important issues and have brought into question a number of basic assumptions held by most of the research community. This paper focuses on those issues concerning adaptation with reinforcement learning, currently the most popular on–line learning method for autonomous agents.

3 Reinforcement Learning

Reinforcement learning (RL) is a class of learning methodologies in which the agent adapts based on external feedback it receives from the environment, and interprets as positive or negative scalar reinforcement. The agent's goal is to maximize reward and minimize punishment over time, by producing an effective mapping of states to actions called a *policy*. RL has been successfully demonstrated on a variety of domains, and has recently been attempted on situated, embodied agents. The following are basic assumptions integral to theoretical reinforcement learning which are challenged in complex, embodied adaptive robot domains:

- *Markovian Learning Models:* Most learning algorithms are based on Markovian models that assume the world and the agent can be represented as synchronized finite state automata, with the agent choosing actions and the world neatly changing states in response. In physical domains none of those assumptions hold true due to incomplete knowledge, limited and faulty perception, noise, and intrinsic uncertainty (Matarić 1994 b).

• *Probabilistic Learning Models:* Alternatives to Markovian models have been proposed. They feature probabilistic state transitions, but require world models that specify the probabilities, and which are unavailable for physical robot domains that are extremely difficult to characterize precisely. As was argued in Matarić (1994 a), producing such models requires observing the system long enough to establish sufficient statistics, thus long enough to learn an adaptive behavior.

• *Asymptotically Converging Algorithms:* Most successful traditional reinforcement learning algorithms are based on dynamic programming (Bellman 1957) and are favored because of their provable asymptotic convergence properties (Barto, Bradtke & Singh 1993). However, these algorithms traditionally trade built–in knowledge for large numbers of learning trials. The opposite appears to be true for animals, whose reinforcement learning curves are much steeper and converge with many fewers iterations (McFarland 1985). This is not surprising, as evolution favors faster adaptation to dynamically changing environments. Additionally, evolution employs "parallel search" for effective solutions through large populations where individuals are disposable. In contrast, for pragmatic reasons, physical robotic agents must conserve individuals. Furthermore, robot learning must minimize the number of required trials, even in learning behaviors that are, to date, relatively simple compared to any biological ones.

• *Optimal Policies:* Theoretical models allow for establishing a canonical optimal policy or behavior which can be used to evaluate the agent's performance. In physical environments, however, it is typically impossible to specify a single, objective optimal policy. While some behaviors are clearly better than others, many appear "good enough" or similar enough in performance relative to the goal.

Adaptive autonomous agents do not have a single well–defined *goal* in a traditional sense. Instead, like animals, they maintain large numbers of maintenance goals, including avoiding damage to themselves and staying charged. In addition, they pursue multiple achievement goals which change over time, such as finding food, shelter, and information. Achievement and maintenance goals are difficult to translate into a single monolithic evaluation function. In order to combine them, high–level goals must be represented in terms of low–level actions, a process that greatly enlarges the agent's search space, and thus requires even more learning trials (Matarić 1994 a). In worlds with continuous space, time, and goals, a single optimal strategy is difficult to define and translate all the way down to the action level.

• *Objective Evaluation Strategies:* Related to the challenges of defining an optimal policy, it is generally very difficult to find ways to objectively evaluate the performance of a physically embodied adaptive system. While it may appear quite simple to define evaluation criteria in terms of "time to convergence" and "energy spent", it is usually not helpful to measure absolute time in event–driven systems that can vary greatly from one–experiment to another, as is the case in

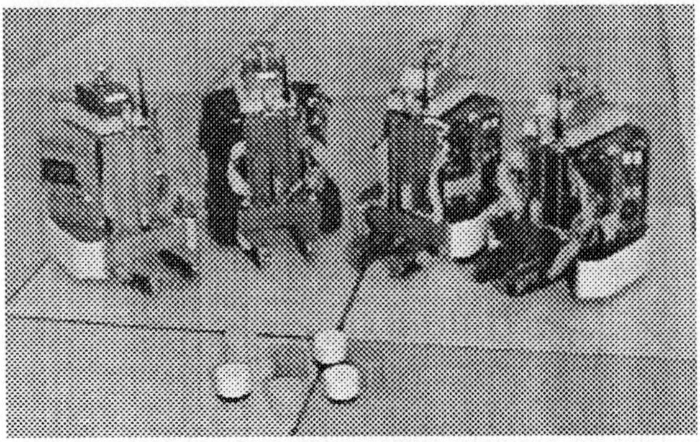

Figure 1: The mobile robots used to validate the group behavior and learning methodologies. These robots demonstrated learning to forage by using group safe–wandering, following, and resting behaviors.

multi–agent systems. Furthermore, simple measures of energy, such as battery power, can produce irrelevant results, since they hide important details about the environment and system dynamics that cannot be represented in the evaluation function. This and the other issues we have addressed will be illustrated on robot data in the following sections.

4 Experimental Environments

Our work was motivated by the challenge of learning through social interaction in which agents learn in parallel and independently, while the society as a whole dynamically adapts. We used two different multi–robot experimental environments. The first consists of four mobile robots capable of detecting and moving objects, and sensing and communicating with other robots (Figure 1). The second environment involves a pair of communicating six–legged robots capable of sensing an object in front using contact sensors (Figure 2). In both cases the robots are equipped with on–board power and computation, their sensory capabilities provide only local, partial information about the world, and their radio-based communication capabilities are local and undirected. The systems are programmed in the Behavior Language (Brooks 1990).

5 Learning in Cooperating Wheeled Robots

The following is a brief summary of the learning system we implemented (for more details please see Matarić (1994 a) and Matarić (1994 b)). The wheeled robots were used to implement a society that learned to forage based on two

Figure 2: Two Genghis-II six–legged robots were used in the experiments. The robots are 30 cm long and 15 cm high. The sensory suite of each robot consists of two frontal contact whiskers and an array of five pyroelectric sensors for detecting moving IR sources. The robots are powered by rechargeable batteries and controlled by five 68HC11 microprocessors.

forms of shaped reinforcement, heterogeneous reward functions and progress estimators, both of which accelerated learning. We used a simple learning algorithm that summed the reinforcement over time, and applied no reward discounting, since the policy to be learned was purely reactive (Matarić 1994 b).

The behavior space included the foraging subset of basic behaviors including *safe-wandering*, *dispersion*, and *homing*, and *resting*, augmented with *grasping*, *dropping*, and *resting*. The condition set was reduced to the power set of: *have-puck?*, *at-home?*, *night-time?*, and *near-intruder?*. Three learning algorithms were tested, and the performance of each was averaged over 20 trials and plotted based on the mean percent of correct policy learned (Figure 3).

In most complex situated domains the "optimal policy" is not readily available for comparison because it is usually difficult to establish the most effective solution, and the user is typically only intuitively aware of an effective class of solutions. Fortunately, we did have an optimal policy available for comparison and evaluation, from our previous foraging experiments in which we hard–wired different strategies and found the most effective one by hand. Using it to evaluate each of the three algorithms proves that our proposed algorithm outperforms the alternatives, but gives little insight into the reasons.

To interpret the results, we tabulated the data for each condition–behavior (state–action) pair and evaluated it based on three additional criteria: 1) time to convergence to a any solution, 2) stability of the solution, and 3) correctness

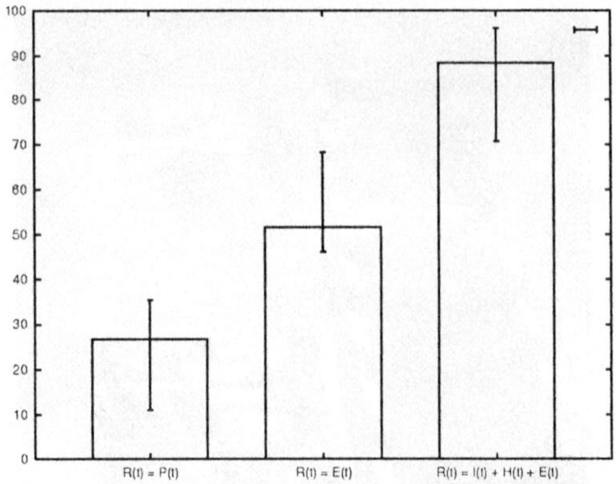

Figure 3: The performance of the three reinforcement strategies on learning to forage. The x-axis shows the three reinforcement strategies. The y-axis maps the percent of the correct policy the agents learned, averaged over twenty trials.

of the solution. We found that some learned conditions oscillated between two similar solutions one of which we had considered to be "correct" but both of which turned out to be equally effective for the robots. Only after an extensive qualitative analysis of each entry in the learning table we were able to produce an evaluation of the results, as described in Matarić (1994 b).

6 Learning in Cooperating Legged Robots

In the case of the legged robots, the learning problem consisted of finding the best strategy for cooperatively moving an oversized elongated box to a goal indicated with a light. The size of the box prohibited efficient individual solutions and required that the robots find a coordinated cooperative strategy (Matarić, Nilsson & Simsarian 1994). The robots could sense the box with their whiskers and the light with their five pyroelectric sensors. However, they could not tell the global position and orientation of the box or of the other robot. Thus, the task was challenging because it involved complex dynamics of the robot–to–box and robot–to–robot interactions. The robot–to-box interaction strategy was hard–wired: a low–level behavior kept the robots in contact with the box. If contact was lost, the robot turned toward the direction of last contact and proceeded to search for the box.

The robot–to–robot interaction strategy was learned. We set up the general cooperative box–pushing scenario so the robots took turns "having control", i.e. getting to decide what to do (based on the pyroelectric values), taking one of the four possible actions *left*, *right*, *forward*, and *stop*, and communicating to the other robot what it should do. After having control, the robot received reinforcement based on its alignment relative to the light: rewards were given

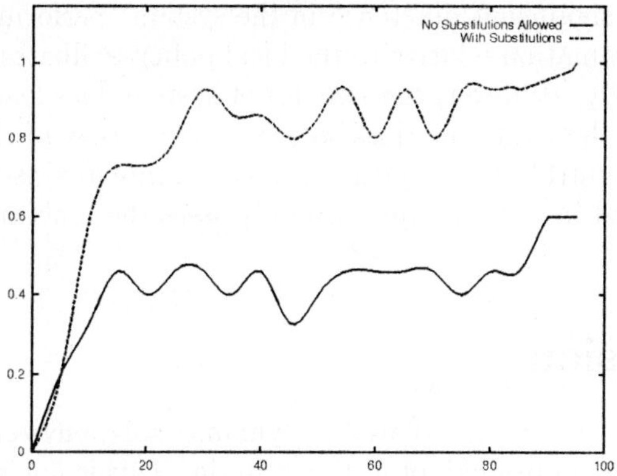

Figure 4: This graph illustrates the learning performance on the harder task, consisting of both *my–action* and *its–action*. The x–axis plots contiguous trials over a 50–minute run. The y–axis plots the percentage of the learned policy. The solid graph shows the percentage of correct values relative to the optimal policy. The dashed graph shows the percentage compared with a relaxed margin of error.

for keeping the light in the middle of the pyroelectric array, and punishment was given otherwise [1].

We worked on two different learning problems. In the easier one, the robots were given a hard–wired policy for *my–action*, i.e. what to do in each perceptual state, and the task consisted of learning *its–action* i.e. what to communicate to the other robot in order to keep the actions coordinated and the box moving toward the light. In the harder task, the robots had to learn both what to do and what to communicate in each situation.

Even more so than in the foraging example, the box–pushing task made it difficult to establish a unique optimal policy. We found that many combinations of cooperative actions would result in efficient box–pushing trajectories. While we were able to eliminate many ineffective combinations, we were left with a set of possibilities we found impossible to rank using any realistic criterion. Solely using the time required to deliver the box to the goal was insufficient due to a large variability across trials, caused by sensory and effector noise.

Consequently, we evaluated the results using two different policies. In the first case we compared the learned matrix to the actions that appeared to most quickly take the object to the goal based on an intuitive ideal designer–derived strategy. In the second case we relaxed the ideal policy to include substitute actions that appeared to be almost equally effective.

Figure 4 shows the comparison between the strict and relaxed evaluation

[1] If the light is centered then the robot and the box it is pushing are both moving directly toward the goal.

policies and the significant difference in the system "performance" that results. The objective evaluation relative to the ideal policy evaluates the system performance very poorly. However, the two–robot system does learn to cooperatively push the box to the light, and does so robustly regarless of the noise and errors in sensing and control. Consequently, a more context–sensitive evaluation approach is required in order to appropriately asses the functional behavior of the system.

7 Discussion

Embodied learning systems situated in dynamic, noisy environments are difficult to evaluate using traditional, objective criteria. This is not surprising considering that these criteria, along with models and methodologies we discussed earlier in this paper, have been adopted from learning work in very different domains. Objective, domain–independent evaluation can and has been used successfully in a great variety of domains. Traditional machine learning, for example, typically features more cleanly characterizable domains, such as game–theory, automata learning, and indictive learning, which allow for clean models of the agent(s), environment, and the learning algorithm, with provable convergence and performance properties. At the other end of the learning domain spectrum, work in real–time systems and adaptive controllers has also applied objective evaluation criteria such as minimization of time or energy.

However, work in Alife has already demonstrated that evaluation of complete, situated agents and creatures is a much more complex problem. In the domain of genetic learning, evaluation is equivalent to selecting the fitness function, a notoriously difficult problem in complex environments (Forrest 1989). Analogous problems arise in situated domains featuring creatures with multiple interacting goals. Both mirror the issues of strategy evaluation in biological systems whose utility functions can be extremely complex.

We believe that the ability to perform objective evaluation is another assumptions that follows from traditional agent models, but is being challenged by work in dynamic situated domains, just as it has been by researchers who study animal behavior in biology and ethology. As we move away from domains that can be usefully [2] modeled with standard techniques, we are forced to give up traditional notions of evaluation based on similarly rooted monolithic evaluation functions. In such domains, it becomes increasingly difficult and decreasingly useful to define a system's performance relative to a single easily measurable quantity, such as time or energy.

The performance of a complex system, such as a situated agent, results from the interaction dynamics between itself and the world in which it is situated, as well as the interaction dynamics between its behavior control components. For example, the multi–robot learning environment we have described feature interactions at many levels, including those between behaviors within each of the

[2] We consider a model useful if it has predictive power.

robots (low–level behaviors such as avoidance or maintaining contact with the box and higher–level behaviors such as homing and going to the light), between a robot and the objects in the world (pucks or the box), and between one or more robots. Since all of these interactions have an impact the overall "behavior" and "performance" of the system, it is difficult to combine them into a monolithic, numerical evaluation function. While it is certainly possible to abstract any of those interactions away in a higher–level system description, it has already been argued that such models are not realistic (Brooks 1991, Brooks & Matarić 1993).

The experiments described in this paper, as well as those by other researchers in Artificial Life, adaptive behavior, and mobile robotics, as well as neural modeling and other complex domains, demonstrate that a new approaches to both setting up the problems and evaluating the results are needed.

8 Conclusions

In this paper we have discussed some of the key issues in designing and evaluating adaptive Alife systems using physical robots interacting in complex, dynamic environments. We used the examples of two different multi–robot learning systems to illustrate the difficulties with traditional, objective, evaluation methods. Finally, we summarized the evaluation methods we used to properly describe the learning behavior and performance of two multi–robot learning systems.

In summary, as the complexity of the environment and the agent interaction increase, so do the challenges of the learning algorithms even for the simplest of tasks. Consequently, it is necessary to look beyond simple numerical recipes toward more detailed, context–dependent and qualitative approaches to task representation and evaluation in order to embrace rather than avoid the complexity of the physical multi–agent world.

Acknowledgements

The author wishes to thank Kristian Simsarian and Martin Nilsson for collaborating on the box–pushing project; Martin for writing and debugging the communication hardware and software, and Kristian for patiently running experiments and analyzing the pushing data. Thanks to both for many inspiring discussions about all topics from robot hardware to learning. Finally, thanks to the three anonymous reviewers for their objective evaluation and comments.

References

Assad, A. & Packard, N. (1992), Emergent colonization in an artificial ecology, in F. Varela & P. Bourgine (eds.) Toward A Practice of Autonomous Systems: Proceedings of the First European Conference on Artificial Life, MIT Press, pp. 143–152

Barto, A. G. (1990), Some learning tasks from a control perspective, Technical Report COINS TR 90–122, University of Massachusetts

Barto, A. G., Bradtke, S. J. & Singh, S. P. (1993), Learning to act using real-time dynamic programming, AI Journal

Beckers, R., Holland, O. E. & Deneubourg, J. L. (1994), From local actions to global tasks: Stigmergy and collective robotics, in R. Brooks & P. Maes (eds.) Artificial Life IV, Proceedings of the Fourth International Workshop on the Synthesis and Simulation of Living Systems, MIT Press

Bellman, R. E. (1957), Dynamic Programming, Princeton University Press, Princeton, New Jersey

Brooks, R. A. (1990), The behavior language; user's guide, Technical Report AIM-1127, MIT Artificial Intelligence Lab

Brooks, R. A. (1991), Intelligence without reason, in Proceedings, IJCAI-91

Brooks, R. A. & Matarić, M. J. (1993), Real robots, real learning problems, in Robot Learning, Kluwer Academic Press, pp. 193–213

Colorni, A., Dorigo, M. & Maniezzo, V. (1992), Distributed optimization by ant colonies, in F. Varela & P. Bourgine (eds.) Toward A Practice of Autonomous Systems: Proceedings of the First European Conference on Artificial Life, MIT Press, pp. 134–142

Corbara, B., Drogoul, A., Fresneau, D. & Lalande, S. (1993), Simulating the sociogenesis process in ant colonies with manta, in Toward A Practice of Autonomous Systems: Proceedings of the First European Conference on Artificial Life, pp. 224–235

Dario, P. & Rucci, M. (1993), An approach to disassembly problem in robotics, in IEEE/TSJ International Conference on Intelligent Robots and Systems, Yokohama, Japan, pp. 460–468

Deneubourg, J.-L. & Goss, S. (1989), Collective patterns and decision-making, in Ethology, Ecology and Evolution 1, pp. 295–311

Deneubourg, J. L., Goss, S., Franks, N., Sendova-Franks, A., Detrain, C. & Chretien, L. (1990), The dynamics of collective sorting, in From Animals to Animats: International Conference on Simulation of Adaptive Behavior, MIT Press, pp. 356–363

Deneubourg, J. L., Theraulax, G. & Beckers, R. (1992), Swarm-made architectures, in F. Varela & P. Bourgine (eds.) Toward A Practice of Autonomous Systems: Proceedings of the First European Conference on Artificial Life, MIT Press, pp. 123–133

Drogous, A., Ferber, J., Corbara, B. & Fresneau, D. (1992), A behavioral simulation model for the study of emergent social structures, in F. Varela & P. Bourgine (eds.) Toward A Practice of Autonomous Systems: Proceedings of the First European Conference on Artificial Life, MIT Press, pp. 161–170

Dudek, G., Jenkin, M., Milios, E. & Wilkes, D. (1993), A taxonomy for swarm robotics, in IEEE/TSJ International Conference on Intelligent Robots and Systems, Yokohama, Japan, pp. 441–447

Floreano, D. (1993), Patterns of interactions in shared environments, in Toward A Practice of Autonomous Systems: Proceedings of the First European Conference on Artificial Life, pp. 347–366

Forrest, S. (1989), Emergent Computation: Self-Organizing, Collective, and Cooperative Phenomena in Natural and Artificial Computing Networks, North Holland, Amsterdam

Fukuda, T., Sekiyama, K., Ueyama, T. & Arai, F. (1993), Efficient communication method in the cellular robotics system, in IEEE/TSJ International Conference on Intelligent Robots and Systems, Yokohama, Japan, pp. 1091–1096

Hogeweg, P. & Hesper, B. (1985), Socioinformatic processes: Mirror modelling methodology, Journal of Theoretical Biology 113, 311–330

Huang, Q. & Beni, G. (1993), Stationary waves in 2–dimensional cyclic swarms, in IEEE/TSJ International Conference on Intelligent Robots and Systems, Yokohama, Japan, pp. 433–440

MacLennan, B. J. (1990), Evolution of communication in a population of simple machines, Technical Report Computer Science Department Technical Report CS–90–104, University of Tennessee

Matarić, M. J. (1992), Designing emergent behaviors: From local interactions to collective intelligence, in J.-A. Meyer, H. Roitblat & S. Wilson (eds.) From Animals to Animats: International Conference on Simulation of Adaptive Behavior

Matarić, M. J. (1994 a), Interaction and intelligent behavior, Technical Report AI-TR-1495, MIT Artificial Intelligence Lab

Matarić, M. J. (1994 b), Reward functions for accelerated learning, in W. W. Cohen & H. Hirsh (eds.) Proceedings of the Eleventh International Conference on Machine Learning (ML-94), Morgan Kauffman Publishers, Inc., New Brunswick, NJ, pp. 181–189

Matarić, M. J., Nilsson, M. & Simsarian, K. T. (1994), Cooperative multi–robot box–pushing, in Proceedings, IROS-95, Pittsburgh, PA

McFarland, D. (1985), Animal Behavior, Benjamin Cummings

McFarland, D. (1987), The oxford companion to animal behavior, in Oxford, University Press

Millán, J. D. R. (1994), Learning reactive sequences from basic reflexes, in Proceedings, Simulation of Adaptive Behavior SAB-94, MIT Press, Brighton, England, pp. 266–274

Pomerleau, D. A. (1992), Neural Network Perception for Mobile Robotic Guidance, PhD thesis, Carnegie Mellon University, School of Computer Science

Smithers, T. (1994), On why better robots make it harder, in The Third International Conference on Simulation of Adaptive Behavior

Steels, L. (1994), Emergent functionality of robot behavior through on-line evolution, in R. Brooks & P. Maes (eds.) Artificial Life IV, Proceedings of the Fourth International Workshop on the Synthesis and Simulation of Living Systems, MIT Press

Travers, M. (1988), Animal construction kits, in C. Langton (ed.) Artificial Life, Addison–Wesley

Werner, G. M. & Dyer, M. G. (1990), Evolution of communication in artificial organisms, Technical Report UCLA–AI–90–06, University of California, Los Angeles

Seeing in the Dark with Artificial Bats

Kourosh Teimoorzadeh *

* Laboratoire d'Intelligence Artificielle Université Paris 8

93526 Saint-Denis Cedex 02 France

E-Mail : kouros@tao.ai.univ-paris8.fr

Abstract. We are providing a simulation model of the echolocation phenomenon and biological sonar of bats during night flight. Our simulations are based on stationary or mobile obstacle avoidance and prey recognition (moths) by the artificial bats. Echolocation is the navigation system adopted by bats, dolphins, killer whales, as well as the majority of autonomous mobile robots (AMR).

To begin with, we will give a detailed description of our model's sonar configuration. We will then present our artificial biotope[1] NetFreeFly inside which our bats manoeuvre in relation to a specific scenario. This biotope allows the study of bat behaviour during the phases of form recognition, obstacle avoidance, and prey capture. Lastly, we will describe the modelling of a memorisation circuit of sensory expressions based on feedback and association of sensory phenomena. This circuit reinforces the mental representation of acoustic images coming from the aural perception of a bat model.

Keywords : animal behavior; bat; echolocation; biological sonar; acoustic perception; adaptive sensing; mobile agents;

1. Introduction

The study of an animal's adaptation to its biotope offers an exploration ground for the comprehension of its behaviour. Our simulation of echolocation phenomenon in bats (chiropteres) permits us to understand the adaptation reactions of these animals to variations in their environment. Furthermore, our study on the autonomy and control phenomena of bats directly applies to the understanding of problems tied to navigation and intelligent control of AMRs during their movement in hostile environments[2] .

Thanks to their biological sonar, bats are capable of flying and hunting in total darkness and in heavy rain. Depending on the species and how an echo is specifically treated, there are two distinct types of echolocation (Simmons, 1967). One is based on the use of a wide band frequency in order to produce multidimensional acoustic images of the prey[3] . Whereas the other, using a narrow frequency band, is transmitted during detection of prey morphological characteristics and movement.

Sound echoes emitted by a bat provide it with essential information concerning its biotope such as:

---

1. We use the word biotope to refer to the life environmental of bats (*bio-topos*).

2. The life of an AMR depends on initially blurred information which can be changed during their treatment and resolution processes (Chatila, 1981; Cliff, Bullock, 1993; Meystel, 1982, 1984, 1990, 1991).

3. "*...They live in a world of echoes, and probably their brains can use echoes to do sommething akin to 'seeing' images might be like... These bats are like miniature spy planes, bristling with sophisticated instrumentation. Their brains are delicately tuned packages of miniaturized electronic wizardry, programmed with the elaborate software necessary to decode a world of echoes in real time.*" (The Blind Watchmaker, R. Dawkins, 1988, p. 24).

- The echo's delay and its Doppler shift[4] indicate the distance and relative speed of the prey respectively ;
- Rapid oscillations of the echo's frequency inform on the flapping of insect wings ;
- The echo's amplitude, correlated to the distance of the target, gives prey size ;
- The differences in amplitude and delay between the two ears indicate the prey's azimuth[5] ;
- Sonar interferences, formed in the outer ear, communicate target elevation (Suga, 1990).

Bats receive and transmit, in brief regular intervals, ultrasonic frequencies in a range of 50 to 90 KHz (Griffin, 1958, 1960, 1982). Synchronisation and delay time between a transmitted sound and its reception in the form of an echo make up an efficient system of object recognition and detection and prompts the animal to modify the direction of its flight. It is why these animals have significant influence on aerospace projects (Byrne, 1964; Webster, Brazier, 1965), concerning autopilots and explorer robots. The design of Chirps Radars (Klauder, Price, 1960) and obstacle detectors intended for blind persons such as Blind Guide (Dobelle, 1977; Kay, 1962, 1976) highlights the interest of the echolocation phenomena in scientific work.

For our part, in modelling the echolocation phenomenon, we are trying to elaborate simplified representations of what we have established from reality. The representations are partly based on experimental and zoological facts concerning bat behaviour[6]. We have generalised this information so that they correspond to the behaviours of an experimental prototype in various conditions in which the characteristics are partially known (Teimoorzadeh, 1993, 1994). Thus, our prototype Bat represents an artificial bat which can be subjected to measurements, calculations, and physical tests which cannot be applied to a real animal. It is placed in scenes composed of fixed and mobile objects (obstacles, prey, bats) which constitute its artificial biotope.

Our article is arranged as follows. First, we shall give a detailed description of the artificial bat's composition (receiver, transmitter, sensory surfaces of the aural cortex). We will then describe the organisation of our artificial biotope in relation to the scene and scenario. Lastly, we will present the simulation of an artificial bat's mental image during acoustic perception and a course of retroactive memorisation.

2. Modelling an artificial bat

We have managed to work-out the three essential phases of echolocation;

[1] The screech: transmission of a simple or composite sound (ultrasounds) depending on the species;

[2] Listening: echo reception and analysis of the sonar image obtained;

[3] Movement: avoidance of fixed or mobile obstacles.

Our Bat module simulates an artificial bat. Its basic structure is very simple and includes:

- Two ultrasonic receivers which simulate the bat's ears;
- An ultrasonic transmitter which simulates, depending on the species of bat studied, the mouth or nose[7] (Teimoorzadeh, 1993).

---

4. In flying bats, the echo frequency is higher than the emission frequency owing to Doppler shifts caused by the relative velocity between bat and target. For stationary targets, the relative velocity is dependant on the angle between flight path and target. the Doppler shifts decrease with the cosine of the increasing target angles (Schnitzler, Henson, 1967, p. 140).

5. The interaural differences in arrival time depend on the distance between the two ears. This distance is relatively short in bats (8 - 22 mm) (Schnitzler, Henson, 1967). Thus, Schnitzler think, the azimuth of target direction is encoded by differences in the phase and arrival times of echoes.

6. (Busnel & Fish, 1967) ; (Cahlender, 1964a-b) ; (Fenton, 1983, 1985) ; (Griffin, 1958, 1960, 1982) ; (Pye 1961, 1962, 1980, 1983a-b) ; (Simmon, 1974, 1979a-b) ; (Suga, 1965, 1975, 1977, 1989, 1990).

7. For example, the Vespertilio transmits by the mouth (Salvayer, 1980) which is kept open during flight. The frequency of their transmitted signal varies between 30 KHz and 70 KHz. The ears does not move but the animal moves its head to localise echoes.

2.1 Composition of the artificial ears

Contrary to the majority of AMRs (Rovisec, 1988), the receivers of our bat model are self-directed. We have introduced a dynamic control system into the ear movements which are independent of head movement (Figure 1).

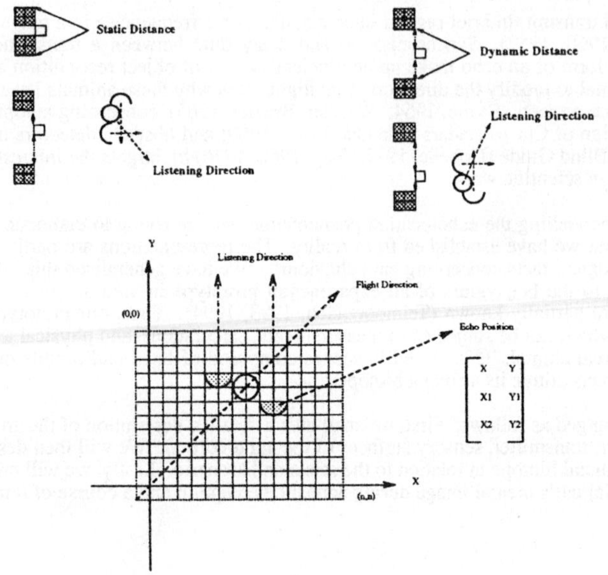

Figure 1. Modelling of head and ear movements of an artificial bat.

Our strategy is based on a multidimensional aural perception system which corresponds to the simultaneous transmission of several frames of signals in multiple directions, and multidimensional sweeping in order to capture their echoes. We feel that this strategy permits significant improvement in the AMR's field of action. Currently, the ears of our model are in the experimentation phase and are only operational in our artificial environment framework. However, they have characteristics and significant points in common with those of the natural model. For a model, ear sensitivity is proportional to its size. Each ear of the model differs from the other in its sensitivity just like a real bat (Webster, 1963, 1965). Thus, when several echoes are simultaneously detected by both ears, the intensity difference of the signal captured permits the animal to localise the relative source of the echo.

2.2 Modelling a screech

We have classified the bats according to their ability to transmit signals of constant or unique frequency (FC) or modulated and damped frequencies (FM) or even mixed frequencies (FC-FM). Our models are capable of transmitting mixed frequency signals of FC/FM type[8] . They produce screeches resembling iiiiiiu in French. The sound transmitted by a model is composed of several types of impulses: FC, FM, FC-FM. Thus, the signals generated are not pure, but composed of a fundamental frequency and several higher harmonics:

1. one FC1 frequency → constant fundamental frequency ;

8. As the species Pteronotus Parnellii, a bat of Central America (Suga, 1965, 1975, 1977, 1989, 1990).

2. three harmonics FC2, FC3, FC4 ;
3. one frequency FM1 → modulated fundamental frequency ;
4. three harmonics FM2, FM3, FM4 ;

In our simulations, the Transmitter and Impulse modules represent the screech and the ultrasonic transmitter of our models respectively. They are implemented by the merger of relative facts concerning behaviour of the various species of bats and sonar technologies. During each screech, the sounds emitted by a model are made up of several signals which last between 1 and 5 milliseconds. Thus, a model can transmit up to 60 signals per second.

2.3 Specialised artificial sensory units

In a prey-predator relationship, the signals which serve to guide a predator such as a bat[9] or alert a prey such as a moth, are not generally transmitted with communication in mind. We consider the aural system of bats as a chain of specialised sensory filters. Thus, stimuli were modelled in the form of an information frame exchanged between the different specialised sensory units (SSU). (Figure 2).

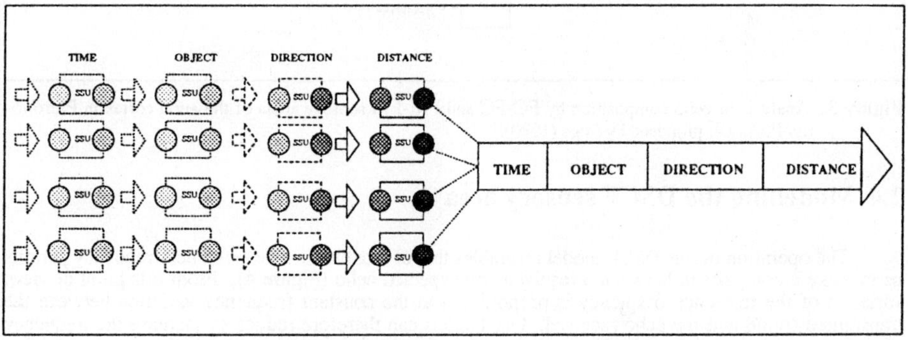

Figure 2. As stimuli advances in the SSU of our models, their contents are treated and renewed continuously with supplementary information.

We divided the bat's aural system into two distinct parts:

1. The peripheral aural system: composed of zones sensitive to echoes but insensitive to the animal's screeches. A distinguishing factor of these zones is their self-adjustment capability permitting them to detect weak signals in hostile (noisy) areas. These adjustments permit the capture of frequency oscillations due to flapping of insect wings.
2. The central aural system: composed of sensory areas specialised in the analysis and specific processing of sound: DSCF area (Doppler Shifted Constant Frequencies), FC-FC area, and FM-FM area (Figure 3).

9. Thus, we can consider the bat as a very efficient self-guided missile of which the fuel is provided by the target (Pye, 1961, 1962).

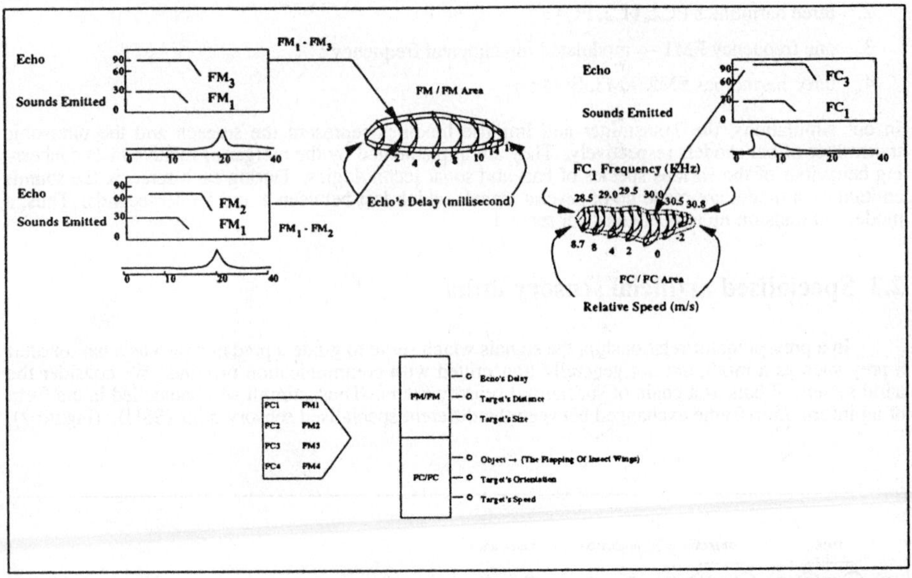

Figure 3. Analysis of echo composition by FC-FC and FM-FM sensory areas of the aural cortex in Pteronotus Parnellii proposed by Suga (1990).

2.4 Modelling the DSCF sensory area

The operation of our DSCF model resembles that of a balance. The reference frequency permits us to make a comparison between a received and expected echo (Figure 4). From this point of view, variation of the reference frequency is proportional to the constant frequency variation between the transmitted sound and the echo received. Our models can therefore reduce or increase the frequency of their transmissions when the harmonic reverberation FC2 does not correspond to expected value (Teimoorzadeh, 1993).

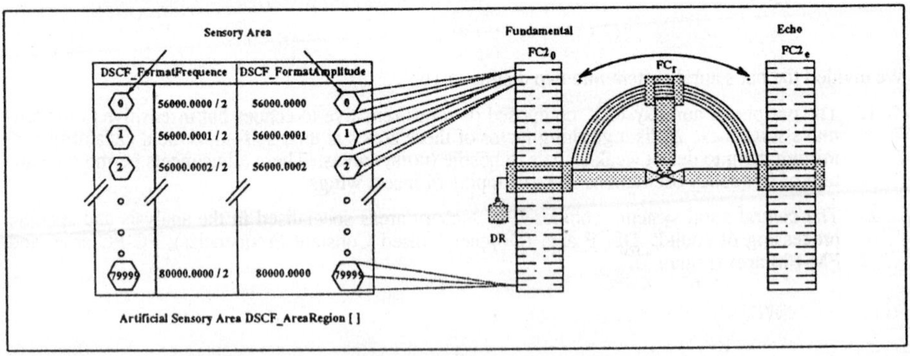

Figure 4. Modelling the DSCF area model as a balance.

Our DSCF-Area module simulates operation of the DSCF sensory area of the bat's aural cortex. This artificial area is sensitive to Doppler shift. It is composed of sensory regions laid out in columns. The columns are composed of neurons sensitive to frequency variations and amplitudes of signals captured (echoes) by the animal. Furthermore, these neurons are only sensitive to FC type frequencies between 60.60 KHz and 62.30 KHz (Suga, 90).

In order to increase sensitivity of this artificial region, we extended sensitivity from 56 KHz to 64 KHz, and frequency shift is calculated from 0.0001 Hz accordingly. Therefore our

DSCF_AreaRegion contains sensory zones sensitive to frequency variations between 56 KHz and 64 KHz with a 0.0001 Hz precision factor. According to Suga (1990), this strategy is incorporated by real bats in order to stabilise subsequent echoes of their transmissions according to more suitable reference frequency: DR → experimental value

$$FC_{Reference} = \left[FC2_{fundamental} - \left(\left(FC2_{echo} - FC2_{fundamental} \right) - DR_{Doppler-Rat=0.2} \right) \right]$$

2.5 Modelling the FC-FC sensory area

The FC-FC area is organised in a Cartesian coordinates system (CCS) in which each point represents a specific target's speed. According to J. A. Simmons (1974, 1979a-b) and N. Suga (1990), two distinct sensory areas exist inside the FC-FC area, and are referred to as the FC1-FC2 area and FC1-FC3 area. The neurons of these two sensory zones are only sensitive to frequency variations between the FC1 sounds transmitted and their return in the form of an echo FC2 and FC3. This process permits a real bat to determine the relative speed of its prey, using Doppler shifts between the signals transmitted and the signals received (Simmons, 1974, 1979a-b). In the artificial brain of our models the FC_FC_Format module simulates the topological organisation (CCS) of the specialised areas FC1-FC2 and FC1-FC3. The FC_FC_Format module is included in the FC1_FC2_Area and FC1_FC3_Area artificial areas which respectively simulate the specific operations of FC1_FC2 and FC1_FC3 areas. We can, therefore, by integrating the FC1_FC2_Area and FC1_FC3_Area modules in the artificial FC_FC_Area simulate FC-FC area sensitivity to Doppler shifts (Figure 5).

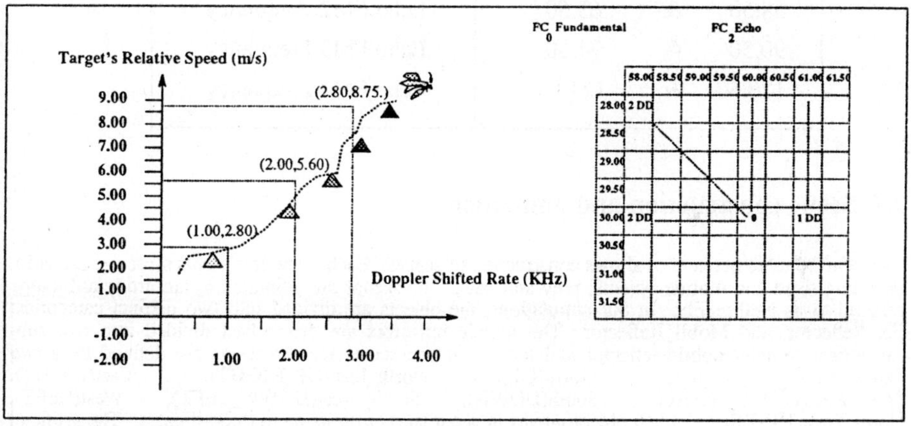

Figure 5. Modelling the FC-FC area.

Inside our artificial FC-FC area, a neuron only reacts to frequency transmission FC_0 and a Doppler shift equal to: [DS = ((FC_0 * 2) - FC_e)] . Our models can calculate the relative speed (RS) of a moving object as a function of two frequencies FC_0 and FC2 and the echo FC_e': DS = ($FC2_0$ - $FC2_e$ = and RS = (DS * R_S) . Here, R_S is an experimental value which represents the speed of 2.8 m/s for a shift of 1 KHz.

During our simulations, the neurons of two artificial zones FC1-FC2 and FC1-FC3, respond to coupling between a frequency of 28 and 30 KHz and frequencies in a range of 61 KHz ($FC2_0$) and 92 KHz (FC3) respectively (Fenton, 1983, 185; Suga, 1975, 1977, 1989, 1990). Furthermore, the frequencies FC2 and FC3 are slightly superior to the fundamental frequency. This means that they are not exactly the double or triple of the fundamental frequency transmitted. Therefore, in calculating the Doppler shift between the screech transmitted and the echo received, and artificial bat establishes the speed of its prey.

2.6 Modelling an FM-FM sensory area

Our FM_FM_Format module models the operation of this specialised region of bats' aural cortex. It simulates the specific processes carried out by this area[10]. This operation of this module is based on the implementation of a time delay (TD) and a relative reflection surface. TD corresponds to the return time of a fundamental sound FM_0 transmitted and a FM harmonic received in the format echo. As for the surface detected, its represents the relative size of a reflector object detected. The artificial area FM_FM_Area is integrated into the FM_FM_Format module and thus permits simulation of the FM_FM sensory area. The FM_FM_Area is arranged in columns. The nerve cells which make up these columns are stimulated by a target of given size and at a precise distance. Furthermore, the nerve cells only react to precise frequency modulations (Table 1). The speed variations perceptible by these cells are between 0.4 m/s and 18 m/s and correspond to distances which vary between 7 and 310 cm. Our models evaluate the distance of a reflector as follows:

$$TD = (\text{Arrival Time} - (T_0)/2) \text{ and } \text{Distance} = (\text{Echo's Delay} * D\_R)$$

where D_R represents an experimental coefficient which corresponds to a 1 millisecond delay.

Table 1: FM echoes recognised by an artificial bat.

FM Frequency (KHZ)	Origin
29.25 & 31.75	Fundamental Frequency
58.50 & 63.50	Echo FM2 Frequency
90.50 & 94.50	Echo FM3 Frequency
120.50 & 124.50	Echo FM4 Frequency

2.7 Scene construction and animation

Our graphic Scene tool allows construction of scenes. Each scene represents a precise scenario, including fixed and mobile objects, prey and bats. The scenes are animated by an artificial biotope referred to as NetFreeFly. In our simulations, the objects are divided into two distinct categories: Fix_Reflector, and Mobil_Reflector. The mobile reflectors are themselves divided into two sub-categories: simple mobile reflector and target. In one scene, the directions are defined by a two dimensional matrix: North(UP), North_East(UP_RIGHT), East(RIGHT), South_East(DOWN_RIGHT), South(DOWN), South_West(DOWN_LEFT), West(LEFT), North_West(UP_LEFT). Each object moves in accordance with its matrix coordinates. The artificial biotope NetFreeFly which simulates the daily life of bats permits the study of the behaviour of each model in a chosen context in order to observe its strategy to survive based on prey recognition and object avoidance (Figure 6).

10. Certain species (Pteronotus Parnellii) of bats transmit FM frequencies in order to measure the time elapsed between the transmission and the echo. This measurement then allows their distance with respect to the reflector object to be determined. According to Suga (1990), a delay of one millisecond corresponds to a distance of 17.3 centimetres when the air temperature is 25 degrees Celsius. Several bat species can discern distances to within 12 millimetres and detect delays to within 69 millionths of a second! This elapsed time calculation takes place in the FM-FM region of the aural cortex. The neurons which make up this area react only to one FM frequency and to a given echo return time.

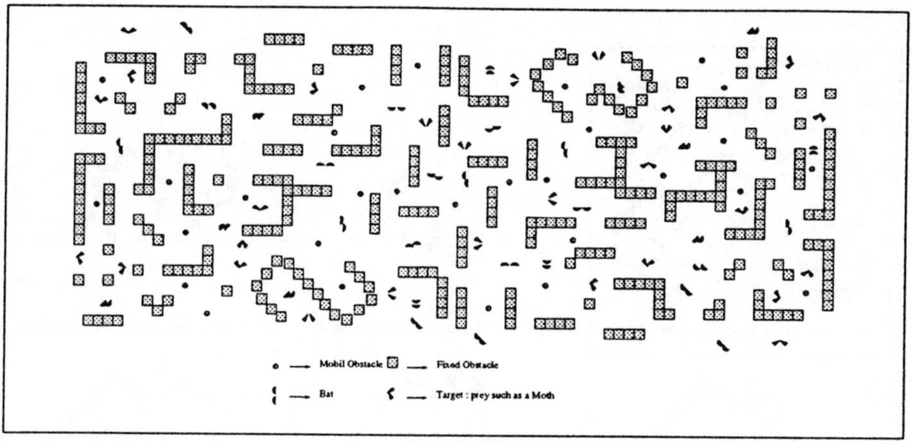

Figure 6. Captured image during operation of the artificial biotope NetFreeFly'.

3. Perspectives: artificial mental representation

What is a blind person's or a bat's mental image of an obstacle? If we had at our disposal a means of detection comparable or even more elaborated than those used by bats, would we be capable of recognising objects by using acoustic imagery? In order to be able to respond to these types of questions, we must establish a model which includes the entire process of stimulus and sensory expressions processing including their memorisation in the form of emotional experiences in all artificial bats. Our perspective consists in simulating an artificial bat's mental representation of an acoustic image (Figure 8), in accordance with the capabilities of a real bat to navigate in its biotope (Figure 7).

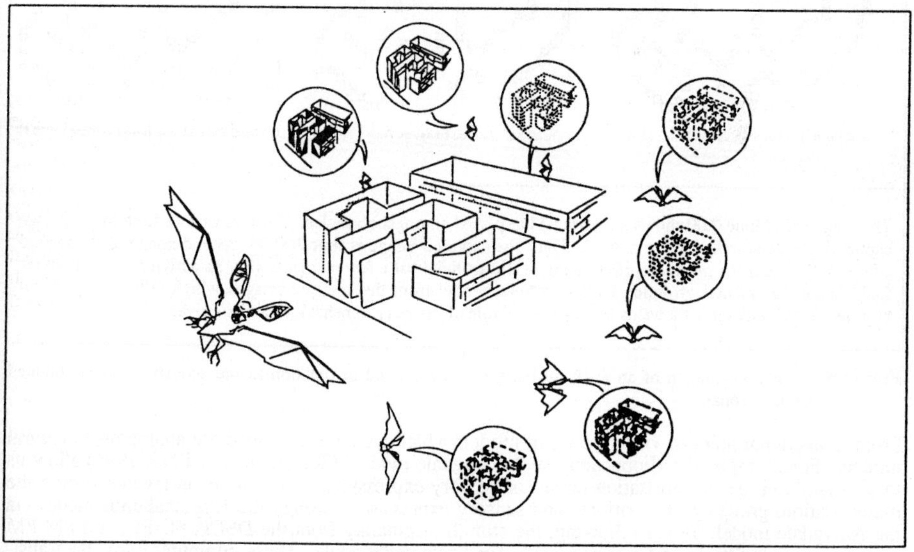

Figure 7. This figure illustrates the ability of a bat to navigate in its biotope with the help of its biological sonar and in function to its acoustic perception.

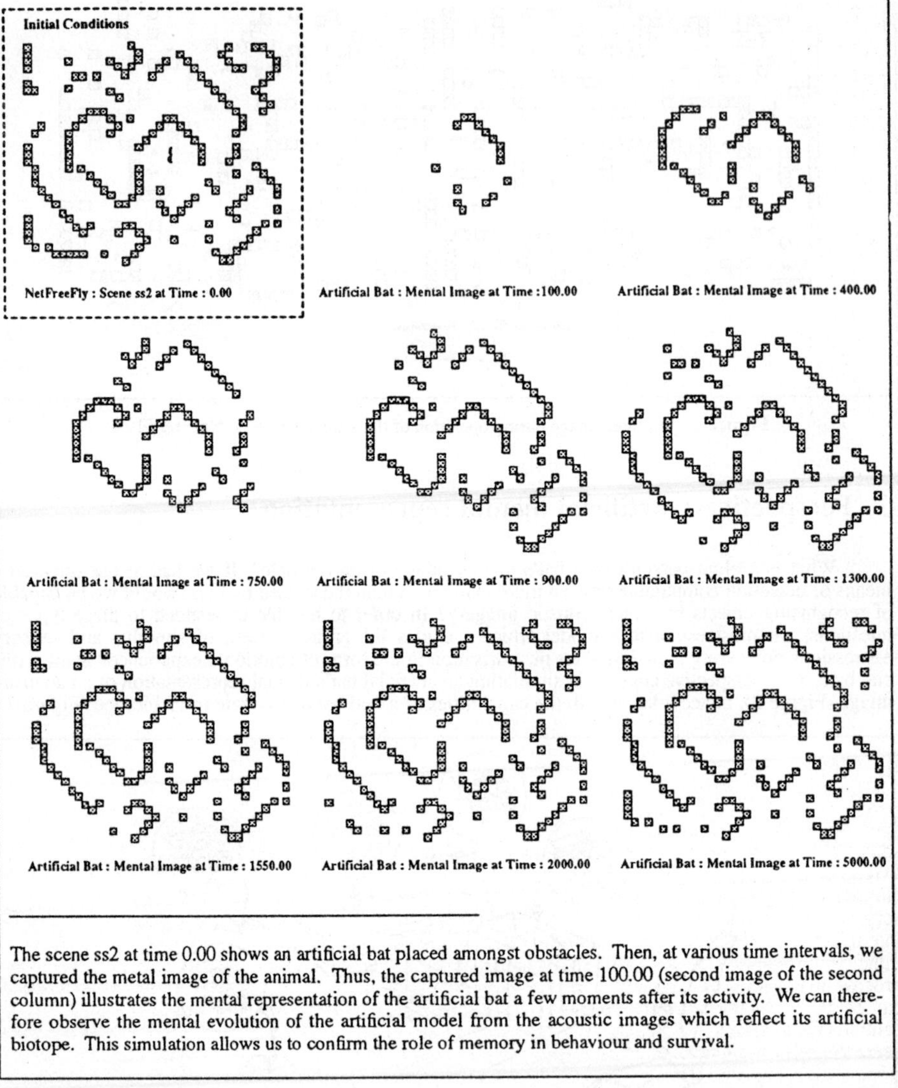

The scene ss2 at time 0.00 shows an artificial bat placed amongst obstacles. Then, at various time intervals, we captured the metal image of the animal. Thus, the captured image at time 100.00 (second image of the second column) illustrates the mental representation of the artificial bat a few moments after its activity. We can therefore observe the mental evolution of the artificial model from the acoustic images which reflect its artificial biotope. This simulation allows us to confirm the role of memory in behaviour and survival.

Figure 8. Mental evaluation of an artificial bat through time and in function to the acoustic images obtained from its sonar.

From a functional point of view, the set of models which we have described are analogous to natural models. For example, the Hippocampe and Amygdale models (Teimoorzadeh, 1993, 1994) allow us to simulate a virtual memorisation circuit of sensory expressions. This circuit intervenes during the memorisation, grouping of emotions, and sensory expressions through the Hypothalamus model via the Amygdale model. In the following, the stimuli originating from the DSCF, FC-FC, and FM-FM sensory areas are filtered and grouped inside our Thalamus module. These groupings allow the transition of stimuli in the form of sensory impressions and emotional recollection toward an artificial memory circuit. This strategy allows feedback of sensory stimuli in the closed circuit and also reinforce neuron representation of the sensory phenomenon. Therefore, the artificial Thalamus module deals with the transition of stimuli originating from sensory zones towards the Amygdale and Hypothalamus memory regions. (Figure 9).

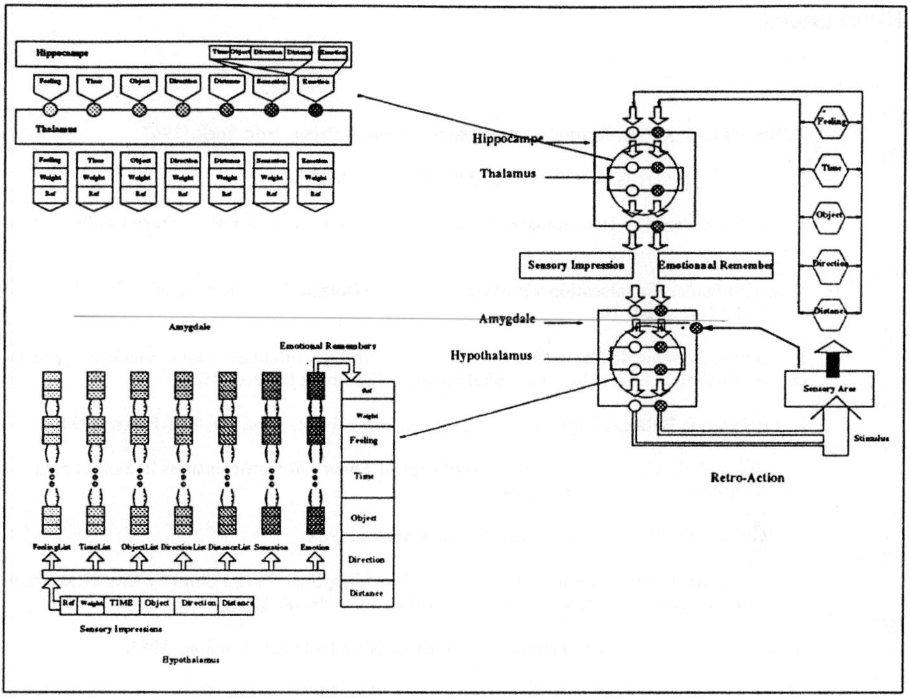

Figure 9. Modelling a retroactive memorisation circuit model.

4. Conclusion

We have presented our processing model of the phenomenon of echolocation and the biological sonar of the bat. We have described the specialised mechanisms of the bat's aural cortex: areas DSCF, FC-FC, FM-FM. We are especially interested in the study of natural signal processing by specialised sensory organs in bats. This natural processing calls upon complex mechanisms such as sensory and emotional memorisation. Problems tied to perception, form recognition and memory are interrelated in certain respects and cannot be studied separately (Crick, Koch, 1990). With this in mind, we considered the perception phenomena of the live animals as being part of the total cognitive process and in relation to memory. In modelling bat behaviour, we are trying to present solutions that can be adapted to autonomy problems of AMRs. Therefore, during AMR design, we want to avoid centralised decisional and behavioural determinism. Our methodological approach is based on a decentralised and modular view of AMR's sensory and decision making units. Therefore, mental representation of a sensory phenomena by the machine is not due to a pre-established functional scheme but to collaboration of autonomous sensory units. This implementation strategy is illustrated by the adaptation behaviour[11] and survival strategy[12] used by our artificial bats in a hostile environment.

Acknowledgments

This work was done under the supervision of Patrick Greussay and Jacqueline Signorini at the Artificial Intelligence Laboratory of Paris-8 University. Special thanks to Patrick Greussay, Jacqueline Signorini, and Pierre Chataigner for interesting discussing, reading earlier versions of this article, and providing useful comments.

11. Perception and acoustic orientation.

12. Obstacle avoidance by using biological sonar.

References

[1] R. Busnel and J. Fish, Animal Sonar Systems, Plenum Press, New York, 1967.

[2] J. Byrne, Bat Radar is Fine if you Just After Bugs, 1965.

[3] D. A. Cahlander, The Determination of Distance by Echolocation Bats, Nature London, 1964. (a)

[4] D. A. Cahlander, Echolocation with Wide-Band Waveforms: Bat Sonar Signals, Tech. Rep. Lincoln Lab MIT, 1964. (b)

[5] R. Chatila, Système de Navigation pour un Robot Mobile Autonome : modélisations et processus décisionnels, Thèse : Université Paul Sabatier, Toulouse France, 1981.

[6] D. Cliff and S. Bullock, Exploration in Evolutionary robotics, 2, p. 73, MIT Press, 1993.

[7] F. Crick and C. Koch, Towards a Neurobiological Theory of Consciousness in Seminars in the Neurosciences, vol. 2, p. 263, 1990.

[8] R. Dawkins, The Blind Watchmaker, Penguin Books, 1988.

[9] W. H. Dobelle, Current Status of Reaserch on Providing Sight to the Blind by Electrical stimulation of the Brain, p. 289, Visual Impairement and Blindness, 1977.

[10] M. Fenton, Just Bats Brock Fenton, University of Torontw Press, London, 1983.

[11] M. Fenton, Communication in the Chiropteres, p. 137, Bloomington : Indiana University Press, 1985.

[12] D. Griffin, Listening in the Dark: the Acoustic Orientation of Bats and Men, Yale University Press, New Haven, 1958.

[13] D. Griffin and D. Thompson, High Altitude Echolocation of Insects by Bats, p. 303, Behavior Ecol. Soc., 1982.

[14] D. Griffin, F. Webster, and C. Michael, The Echolocation of Flying Insect by Bats, p. 141, Animal Behavior, 1960.

[15] L. Kay, A Plausible Explanation of the Bat's Echolocation Acuity Animal Behavior, 10, p. 34, Animal Behavior, 1962.

[16] L. Kay, Auditory Perception and its Relation to Ultrasonic Blind Guidance Aids, 24, p. 309, Brit. I.R.E., 1962.

[17] L. Kay, A Plausible Explanation of the Bats Echo-Locating Acuity, 10, Animal Behaviour, 1962.

[18] L. Kay and M. A. Do, Resolution in an Artificially Generated Multiple Object Auditory Space Using New Auditory Sensations, 36, p. 9, Acoustica, 1977.

[19] J. Klauder and A. Price, The Theory and Design of Chirp Radars, p. 745, Bell Syst.Tech, 1960.

[20] A. Meystel, Autonomous Mobile Robots : Vehicles with Cognitive Control, World Scientific, Singapore, 1991.

[21] A. Meystel, On the Phenomenon of High Redundancy in Robotic Perception, 158, Springer-Verlag, 1990.

[22] A. Meystel, Planning in the Anthropomorphical Machine Intelligence, p. 648, Proc of the IEEE int. Conf. on Cybernetics and Society, Seattl WA, 1982.

[23] A. Meystel and R. Chavez, Structure of Intelligence for an Autonomous, Proc. of the IEEE Int. Conf. on Robotics, Atlanta GA, 1984.

[24] J. Pye, Echolocation Signals and Echoes in Air, Plenum (Animal Sonar Systems), New York, 1980.

[25] J. Pye, Techniques for Studying Ultrasound, p. 39, Academic Press, New York, 1983. (a)

[26] J. Pye, Echolocation and Counter Measures, Academic Press, New York, 1983. (b)

[27] J. Pye, Perception of Distance in Animal Echolocation, p. 362, Nature, London, 1961.

[28] Rovisec, International Conference on Robot Vision and Sensory Controls, IFS. Springer-Verlag, Zurich, 1988.

[29] H. Salvayre, Les Chauve-Souris, Balland, Paris, 1980.

[30] H. Schone, Spatial Orientation : The Spatial Control of Behavior in Animal and Man, Princeton University, N.J., 1984.

[31] J. A. Simmons, The Processing of Sonar Echoes by Bats, p. 695, Animal Sonar System, 1967.

[32] J. A. Simmons, Response of the Doppler Echolocation system in the bat, p. 672, Acoust.Soc.Amer, 1974.

[33] J. A. Simmons, Perception of Echo Phase in bat Sonar, p. 1336, Science, 1979. (b)

[34] J. A. Simmons and R. Stein, Acoustic Imaging in Bat Sonar: Echolocation and the Evolution of Echolocation, Comp. Physiol., 1979. (a)

[35] N. Suga, Further Studies on the Peripherical Auditory System of "CF-FM" Bats Specialized for the Fine Frequency Analysis of Doppler-Shifted Echoes, p. 207, Exp. Biol., 1977.

[36] N. Suga, Functional Properties of Auditory Neurones in the Cortex of Echolocation Bats, p. 671, Physiol., 1965.

[37] N. Suga, Parallel-Hierarchical Processing of Biosonar Information, Plenum Press (Animal Sonar Systems), New York, 1989.

[38] N. Suga, Peripherical Control of Acoustic Signal in the Auditory System of Echolocation Bats, p. 277, Exp. Biol., 1975.

[39] N. Suga, Le Système Sonar des Chauve-souris, Pour la Science, 1990.

[40] K. Teimoorzadeh, La modélisation du phénomène d'écholocation chez les chauve-souris, Mémoire de Maîtrise, Saint-Denis France, 1993.

[41] K. Teimoorzadeh, La modélisation d'une chauve-souris artificielle, Mémoire de DEA, Saint-Denis France, 1994.

[42] F. Webster and O. Brazier, Experimental Studies on Target Detection Evaluation and Interception by Echolocating bats, U.S.A.F Aerospace Medical Div., 1965.

[43] F. A. Webster, Some Acoustical Differences Between Bats and Men : Sensory Devices for the Blind , London, 1963.

Navigating with an Adaptive Light Compass

Dimitrios Lambrinos

AILab, Institute for Informatics
University of Zurich-Irchel
Winterthurerstrasse 190, CH - 8057 Zurich, Switzerland
E-mail: lambri@ifi.unizh.ch

Abstract. One of the fundamental abilities required in autonomous mobile agents is the one of homing. Natural agents like ants solve this problem by mainly using dead-reckoning mechanisms within an egocentric frame of reference. Here we present a biologically inspired orientation mechanism, an adaptive light compass, that was used for homing in "Myrmix", a mobile robot equipped with infrared and ambient light sensors. The control architecture is adaptive by using a self-organizing neural network. Herewith, the robot learns to associate signals coming from the light sensors with the corresponding motor actions. This approach is less computational than others, since apart from the length of the path travelled it is based on local rules. Preliminary results of experiments with this control architecture are reported and contrasted with a similar, but more computational, architecture introduced by [5].

1 Introduction

Many animals, such as insects use information from the sky to navigate. The kind of information and the way in which it is being used depends on the species. It has been found that insects such as the desert ant *Cataglyphis bicolor* and the Honey bee *Apis mellifera* use both the pattern of light polarization in the sky (polarized light compass), as well as the position of the sun (sun compass) [12], [10], [11]. Especially the polarized light is used for precise homing. In both species a specialized region (the POL area) of the eye is used for detecting and analysing the polarized sky-light. Inspired from the insects light compass we derived an orientation scheme, the adaptive light compass (ALC), that uses ambient light information to provide orientation cues.

A light compass is often used in combination with dead reckoning in animals. Models concerning dead reckoning suggested by biologists are often complex using goniometric relationships [6] or arithmetic means and angle estimations [7]. As an alternative, we study the behaviour steered by simple, local, action-based rules such as in [3].

2 The robot Myrmix

The mobile robot used in the experiments, named Myrmix (which stands for ant in Greek) is shown in figure 1. It is a $Khepera^{TM}$ robot, 55mm in diameter and 32mm high (weight 70g). It is equipped with 8 infra-red proximity sensors (6 in the front, 2 on the back) which can also be used to measure ambient light. The body of the robot is supported by two wheels, each controlled by a DC motor with an incremental encoder. The robot is controlled by a Motorola 68331 micro-controller. The robot can either be controlled from a workstation via a serial link, or run completely autonomously by having the control program down-loaded in the on-board RAM. In the experiments reported in this paper the robot was controlled from a workstation.

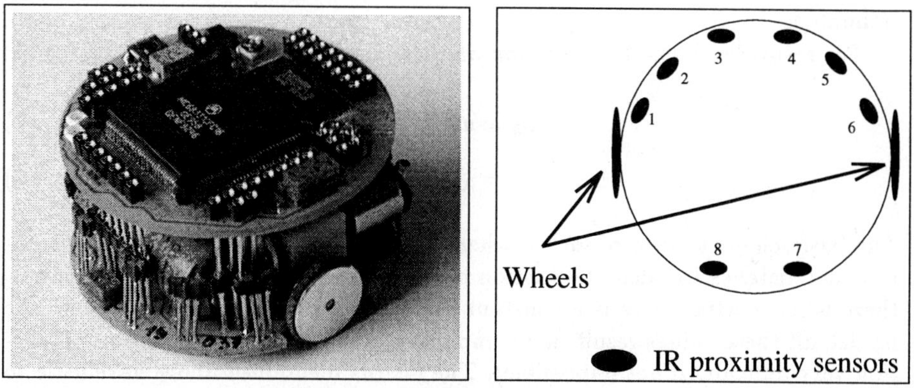

Fig. 1. The mobile robot Khepera.

3 Homing strategies

Probably the most representative example of goniometric models is the one proposed in [6]. This model was tested in a real robot (the Edinburgh R2) by [5]. According to this model the position and direction of the agent are calculated as follows:

$$dx = cos(\phi)v\Delta t \tag{1}$$

$$dy = sin(\phi)v\Delta t \tag{2}$$

where dx and dy is the displacement on the X and Y axis respectively, ϕ is the current heading of the robot, and v is the speed (assumed constant during Δt).

The cosine and the sine of the current direction in the case of R2 was measured by referring to a reference direction through five light sensors. In that case dx and dy where given by:

$$dx = \frac{S_1 - S_2}{|S_1 - S_2| + \left|\frac{S_2+S_3}{2} - S_0\right|} \tag{3}$$

$$dy = \frac{\frac{S_2+S_3}{2} - S_0}{|S_1 - S_4| + \left|\frac{S_2+S_3}{2} - S_0\right|} \tag{4}$$

where S0,..S4, were the signal values from the five light sensors around the robot. The speed and time terms where abstracted from the above equations under the assumption of being constant.

The current position of the robot is given by:

$$x(t+1) = x(t) + dx \tag{5}$$
$$y(t+t) = y(t) + dy \tag{6}$$

where the current direction is either measured (through some kind of sensor) or estimated.

For returning home the direction and the distance are given by:

$$\phi_h = atan(\frac{y}{x}) \tag{7}$$
$$\lambda_h = \sqrt{x^2 + y^2} \tag{8}$$

The basic characteristic of such a scheme is the Cartesian way of using orientation and distance to calculate position in each step. Velocity is assumed constant, there is no inertia, there is a constant light gradient,...etc. When added to the model, all these things result in a complexity that makes the estimation of position very difficult (if not impossible). The factors mentioned above have little to do with the agent itself or the environment alone. It is the interaction of these two that has to be taken into account [8]. The homing strategy that is used in Myrmix is based on simple rules and the complexity of the behaviour is left to the agent-environment interaction. The rules are:

1. Move straight ahead if no obstacles are encountered.
2. If an obstacle is encountered turn away.
3. After avoiding an obstacle, compensate the avoidance turn by a counter-turn.
4. Record the distance travelled.
5. After a fixed period of time turn 180° and travel back recorded distance.

The rules are partly biogically inspired and were already used in previous experiments (see [3]). Several animal species move mostly in a straight line when exploring and homing, making a full turn in between [11], are assumed to remember the distance travelled by the number of steps taken or by recording the rate of visual motion of the ground [11] and have been described to compensate a forced turn (for example when a path was blocked) by a turn in the opposite direction (e.g. [2]) leading to a parallel walk as shown in figure 2.

The underlying assumptions in [5] introduce at least one limitation on the design of the control architecture of the robot. Namely the lack of adaptivity to changing conditions in the world, such as changes of the light gradient. By contrast, Myrmix is adaptive to changing light conditions. Adaptivity occurs through the learning process of a self-organizing neural network, the adaptive light compass.

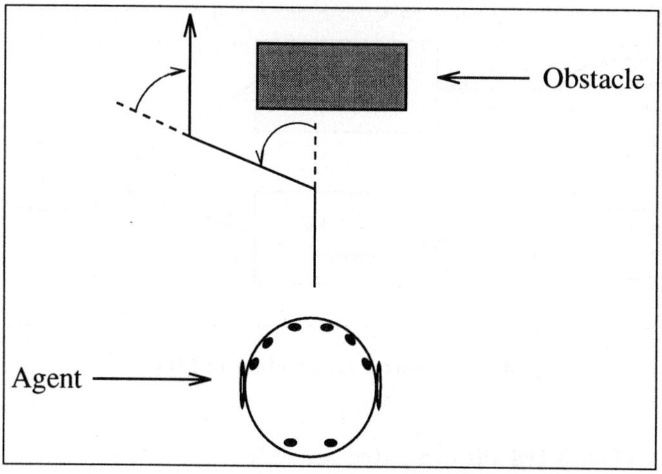

Fig. 2. The homing strategy of Myrmix. The robot goes straight until it is forced to turn because of an obstacle. Immediately afterwards it compensates the forced turn with a turn in the opposite direction and continues moving in direction parallel to the initial one.

4 The control architecture

The general design of the control architecture used in Myrmix is shown in figure 3: Two neural networks are responsible for the locomotion of the robot. A self organizing neural net (the Adaptive Light Compass) takes input from the ambient light sensors and outputs to the motors (the light source used in the experiments reported in this paper was our laboratory window). The task of this network is to keep track of the current direction and steer the robot in such a way that it always moves in one direction. This network can be realized as the implementation of the turn compensation rule 3).

The second network is responsible for the obstacle-avoidance behaviour of the robot. It is a network with fixed weights inspired by Braitenberg [1] that basically implements the avoidance reflexes of the robot.

4.1 The Adaptive Light Compass

The adaptive light compass (ALC) is based on a Kohonen self-organizing net [4]. Its structure is depicted in figure 4. ALC consists of a two-dimensional lattice of cells which all receive the same input. The input to the cells is an eight-dimensional vector $\bar{i}$ of the normalized signal values obtained from the ambient light sensors. Each cell in the lattice is connected to two units, the motor units. The output of these units is a two-dimensional vector $\bar{o}$ with elements that take values in the range [-1,1]. In order to enable ALC to directly steer the motors, the extended Kohonen model [9] has been used. Each cell of the lattice is associated with an incoming weight vector $\overline{w}^{in}$ and an outgoing weight vector $\overline{w}^{out}$. The

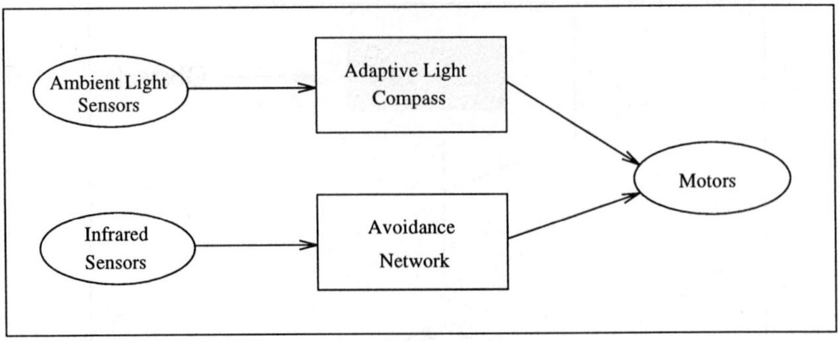

Fig. 3. The control architecture of Myrmix.

activation a_i of each cell i is computed as:

$$a_i = \sum_{j=1}^{8} w_{ij}^{in} \; i_j \tag{9}$$

The output of the net $\overline{o} = (o_l, o_r)$ is calculated as follows: each time an input vector $\overline{i}$ is presented to the net, the "winning" cell s whose weight vector $\overline{w}^{in}$ best matches the input vector $\overline{i}$ is determined by the condition:

$$\left\| \overline{i} - \overline{w}_s^{in} \right\| \leq \left\| \overline{i} - \overline{w}_r^{in} \right\| \; \forall r \in L \tag{10}$$

where r is an index on the cells of lattice L. The output of net $\overline{o}$ is then set to:

$$\overline{o} = \overline{w}_s^{out} \tag{11}$$

The extended Kohonen model allows the system to learn a control task (the turn compensation in our case) when a sequence of correct input-output pairs $(\overline{i}, \overline{o})$ are available. Here $\overline{i}$ denotes the input vector, that is the eight-dimensional vector of the normalized sensory signals, and $\overline{o}$ stands for the correct motor action associated with $\overline{i}$. The input-output pairs $(\overline{i}, \overline{o})$ are constructed as follows: every time the agent leaves its home it performs an "orientation movement" (usually a turn of 360 degrees) during which it collects data from the ambient light sensors and the associated direction ϕ as derived from the wheel encoders. The input vector $\overline{i}$ is constructed by normalizing the signal values from the ambient light sensors and the output vector $\overline{o}$ is then set to:

$$\overline{o} = (o_l, o_r) = \left(\frac{\phi}{\pi} - 0.5, -\frac{\phi}{\pi} + 0.5 \right) \tag{12}$$

where o_l, o_r is the speed of the left and right wheel respectively, normalized in the range [-1,1] (-1 stands for maximum speed backwards, and 1 stands for maximum speed forwards). As a result of this relation, the correct speed for the motors is set to be proportional to the angle that has to be compensated. In other words, if the robot has been forced to deviate from the original direction

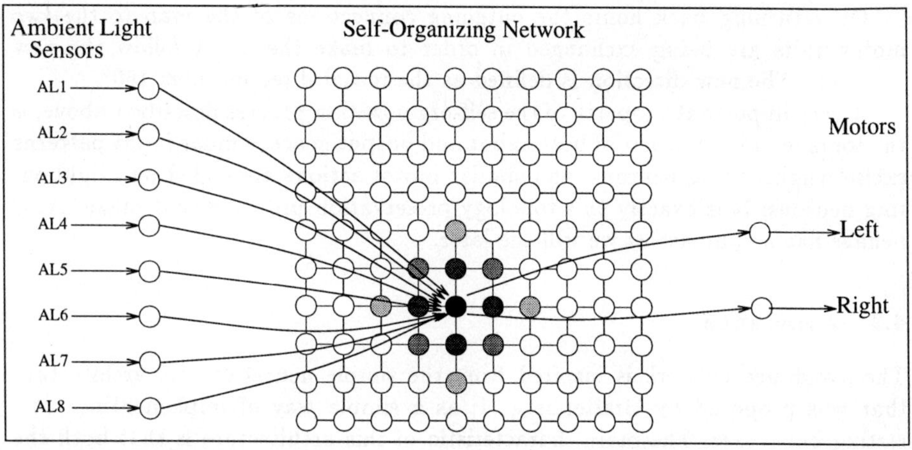

Fig. 4. The ALC consists of a self-organizing neural network. Each cell in the network is fully interconnected to both input and output layers. The input layer consists of an array of cells that receive their inputs directly from the ambient light sensors of the robot. The output layer consists of two cells corresponding to the left and right motors of the robot.

because of an avoidance movement, the contribution of the net to the motors will be proportional to the deviated angle.

After the data collection, training takes place in the following way: an input $\bar{i}$ is selected from the sequence and presented to the net, the best matching cell is then determined according to (Eq.10), the weights of all cells in the lattice are then updated according to:

$$\Delta\overline{w}_r^{in} = \overline{w}_r^{in} + a\ h_{rs}\ \left(\bar{i} - \overline{w}_r^{in}\right) \tag{13}$$

$$\Delta\overline{w}_r^{out} = \overline{w}_r^{out} + a'\ h'_{rs}\ \left(\bar{o} - \overline{w}_r^{out}\right) \tag{14}$$

where $\bar{o}$ is the output vector associated to $\bar{i}$, a is the learning rate, and h_{rs}, h'_{rs} are functions that depend on the distance of $||r - s||$, which is in our case the gaussian:

$$h_{rs} = e^{-\frac{||r-s||^2}{2\sigma(t)^2}} \tag{15}$$

$\sigma(t)$ determines the diameter of the neighbourhood where changes in weights are going to occur and is defined as a linear function of time. $\sigma(t)$ is set to a relatively high value initially and it declines to zero as training progresses:

$$\sigma(t) = 1.0 + (r_{max} - 1.0)\,\frac{t_{max} - t}{t_{max}} \tag{16}$$

r_{max} is the initial value of $\sigma(t)$ and t_{max} is the maximum number of training steps. The learning rate is also defined to be a function of time as follows:

$$\alpha(t) = \alpha_{max}\frac{t_{max} - t}{t_{max}} \tag{17}$$

On returning back home the outgoing connections of the map to the two motor units are being exchanged in order to make the agent follow the new direction. The new direction is defined as the initial direction plus 180°.

A very important property of the self-organization process described above, is the topology preservation of both input and output space: similar input patterns excite neighbouring neurons, and similar motor actions are coded on neighbouring neurons. It is exactly this topology preservation property that makes ALC behave like a compass as we will see later.

4.2 Avoidance

The avoidance network is inspired from the simple neural control architecture that was proposed by Braitenberg [1] as a simple way of implementing interesting behaviour. The main characteristic of this architecture is that both the motors and the sensors are modelled as formal neurons, with hard-wired links connecting them. Interesting behaviours can then be implemented by choosing an appropriate connectivity pattern for the links between sensors and motors. An example of such an architecture implementing the avoidance behaviour in our robot myrmix is shown in figure 5. The speed of each motor is set to the

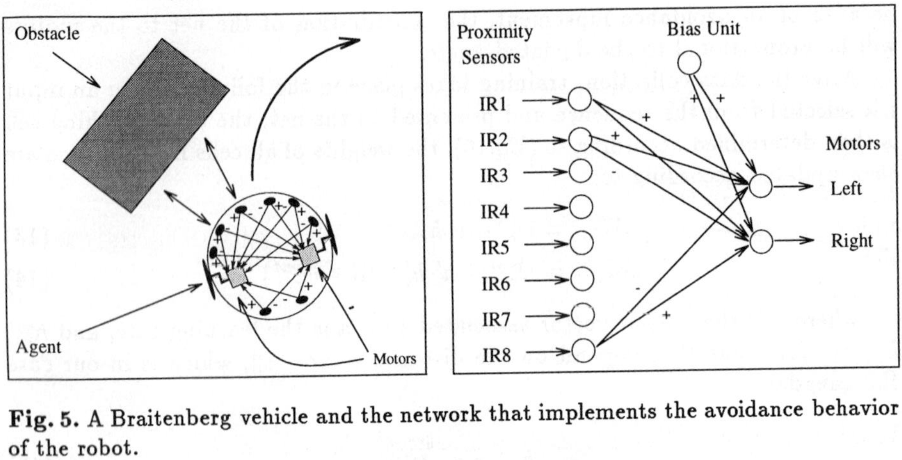

Fig. 5. A Braitenberg vehicle and the network that implements the avoidance behavior of the robot.

weighted sum of the signal coming from the sensors. As a result of the connectivity pattern, detecting an obstacle on the left will lead to large values for the sensors on the left side and thus increased speed for the left motor and decreased speed for the right motor.

The avoidance network is modelled exactly the same as a Breitenberg vehicle with the addition of a Bias unit. There are eight input units fully interconnected with two output nodes via excitatory and inhibitory links as shown in figure 5. Each input node in the network takes its input from the corresponding proximity sensor. Each output gets an activity in the range [-1,1]. The activity of each of

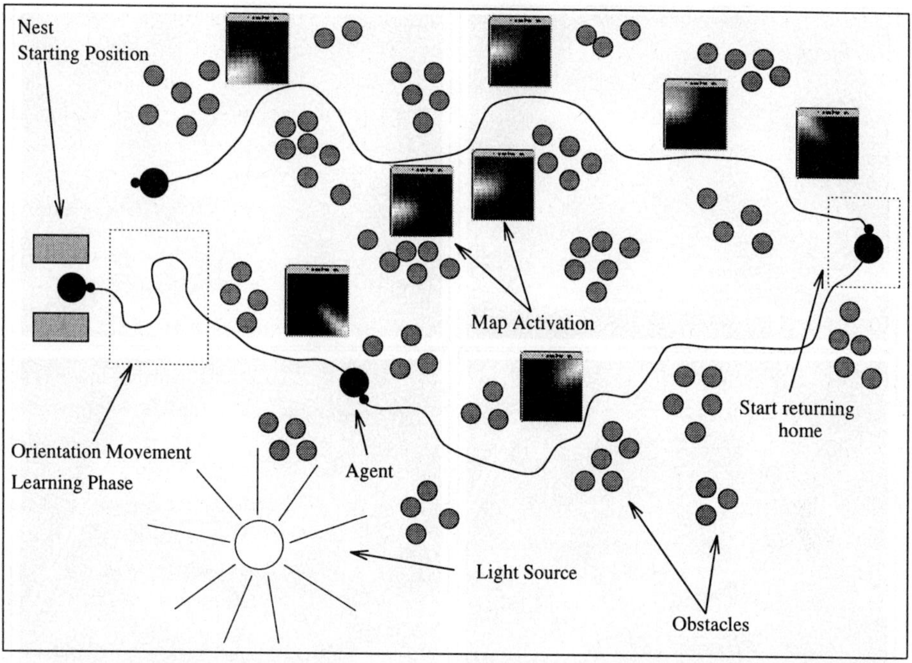

Fig. 6. A Typical path of the robot. The activity of the map is shown at specific points of the path corresponding to different orientations from the light source. Light areas on the map contain cells with high activation value while dark areas contain cells with low activation value.

these nodes translates in a linear fashion into speed for the wheels. The Bias unit is connected to both motors via excitatory links, and is responsible for the default forward motion of the robot. The outputs of both avoidance and ALC networks are directly fused to the motor units.

5 Experiments

In all experiments, the dimensions of the self-organizing network were 10x10. An initial learning rate α_{max} of 0.6 was used, and r_{max} was set to 10.

Figure 6 shows a typical path of the robot. The activation of the network is displayed at several points corresponding to different headings of the robot. It is interesting to see how the resulting activity on the net corresponds to different directional angles.

Maximum activity occurs at specific cells of the net for certain angles. Turning the robot manually 360 degrees results to the maximum activation area being shifted all around the net in a circular manner like a compass.

Due to the lack of an external device for deriving the real orientation of the robot and since information about the orientation of the robot that is derived from the wheel encoders is often unreliable and subject to cumulative errors, we

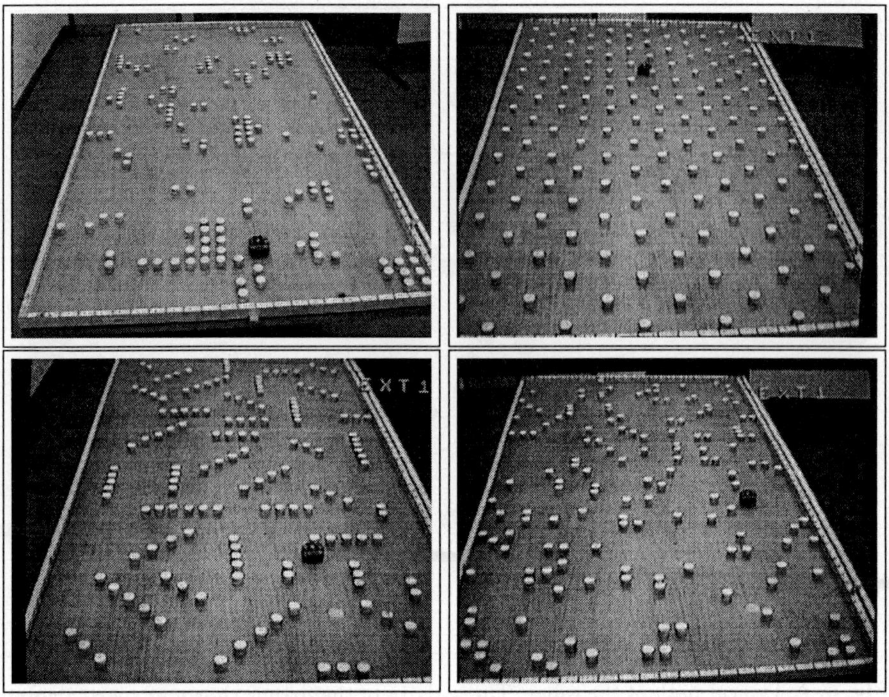

Fig. 7. The four different types of environment used in the experiments.

decided for the time being not to start with measurements on the precision of the isolated light compass but rather concentrate on the performance of the complete homing mechanism. However, it is in our intentions to further investigate the precision and the robustness of the light compass as a general directional tool. For this purpose we are currently developing an external device that is based on a CCD camera which will record both position and orientation of the robot during the exploration and homing trip.

Three types of environment were used as in [3]. In order to measure the performance of the homing mechanism an arena (120cm x 80cm large) is used in which we've placed 100 round obstacles of 2.5 cm in diameter and 2cm high. The obstacles were distributed in four different ways: regular, at random, clumped and with walls. Figure 7 shows the experimental setup for all four cases. For each environment, eight experiments were performed, two for each of the four different starting points. For each experiment, the robot was made to cover a distance of 110cm (twenty times its body diameter) during the outward journey, and then go back home. The length of the path was always 220cm. Starting points were at a distance of 15cm from the nearest wall. The homing error was measured as the distance between the start position and the return position.

In all experiments, sun-light coming from our laboratory window was used as the light source for the ambient light sensors of the robot. The initial directions of the robot were set at about 90° from the orientation of the window.

6 Results

The results are summarised in table 1 and graphically displayed in figure 8 for each type of environment.

Table 1. Descriptive statistics on the homing error. Split By: Distribution of obstacles.

Environment	Mean	Std. Dev.	Std. Error	Minimum	Maximum	Error
Regular	15.062	9.010	3.185	3.000	29.000	6.846
Clumped	14.062	4.851	1.715	6.000	18.600	6.391
Walls	14.375	6.217	2.198	4.500	22.400	6.534
Random	8.050	3.871	1.369	3.600	13.900	3.659
Total	12.888	6.629	1.172	3.000	29.000	5.858

As it can be seen the homing error is almost the same (around 15cm) for regular, clumped, and walls, but it drops significantly in the case of environments where the obstacles are randomly distributed. In an environment with randomly distributed obstacles the displacement error is more likely to be cancelled out during the return journey. Exactly the opposite happens in the case of environments with clusters and walls where a bias (a long wall, or a big cluster of obstacles) can drive the robot far away from the initial path which in turn results in large homing errors.

In the case of regularly distributed environments the robot was unable to compensate the deviated angles immediately after an obstacle was encountered, because the distance between two neighbouring obstacles was too small. An average of 2 avoidance movements in a row were performed before the agent compensated the deviated angle. It is expected that, if we change the distance between the obstacles the homing error will become lower.

Comparing the results with the ones reported for the R2 [5] one should consider the following factors:

1. The trip length of myrmix (220cm) was set to 40 times the diameter of the robot (5.5cm) compared to 16 in the case of the R2.
2. Myrmix was tested in four different types of environment instead of one in the case of the R2.

In the experiments with the R2, a 7.9% navigational error (an average of 44,9cm, with Std. Dev. being 22.6) is reported. With the 5.8% that we obtained with myrmix for all types of environment, and 3.6% in the case of environments with randomly distributed obstacles, myrmix does better than R2.

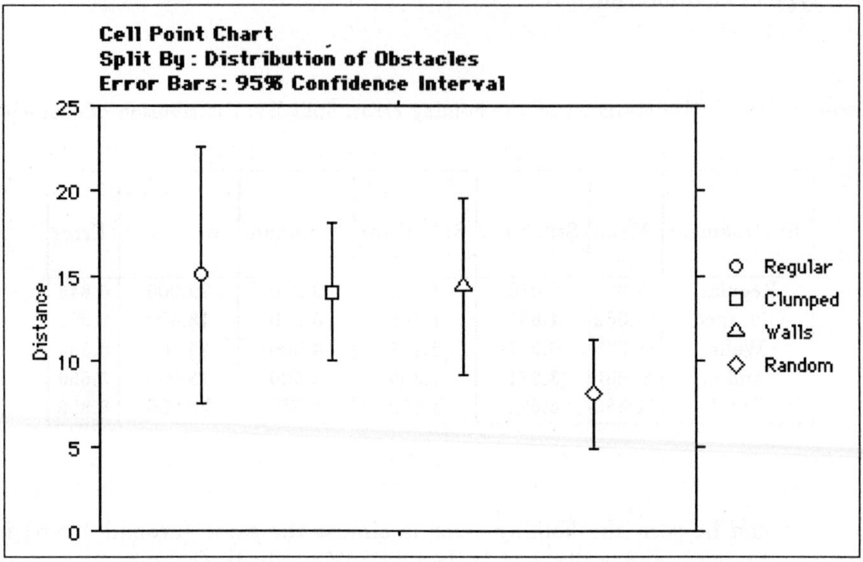

Fig. 8. Results from the experiments performed with the four types of environment.

7 Discussion

We have proposed a solution to one of the most crucial problems of navigation: that of returning to the starting position after being out for performing another task. Our solution differs from goniometric models like [5] in that it is not based on a Cartesian frame of reference but it is steered by mostly local rules. Therefore the proposed model is less computational than other models so far in the sense that there is no calculation of position in every step; the only information that is being encoded is the length of the path.

We solve the problem of changing environmental conditions like the light intensity and direction by making the control architecture adaptive. The architecture consists of two artificial neural networks that are directly coupled to the sensors and motors. A self-organizing map that steers the robot to a particular direction, and a simple completely reactive network that is responsible for the avoidance behaviour of the agent. Adaptation to changing light conditions occurs through the learning process of the self-organizing neural network at the beginning of every outward journey.

The proposed solution takes the environment-agent interaction into account by using only local, action-based rules. It is exactly this interaction that determines the homing precision.

It is in our future plans to incorporate the Adaptive Light Compass in a landmark-based navigation scheme. This type of navigation has been observed in insects like bees [11]. In such a scheme the agent could retrace the outward path using a series of landmarks memorised during the outward journey, where navigation between successive landmarks is done with ALC.

Acknowledgements

We are especially grateful to professor Rolf Pfeifer for stimulating this project. Thanks to Charlotte Hemelrijk, Rene te Boekhorst and Christian Scheier for looking through a former draft. This work is supported by the Swiss National Science foundation, grant number 20-40581.94 to D. L.

References

1. V. Braitenberg. *VEHICLES, Experiments in synthetic psychology.* MIT-Press, Cambridge, Massachusetts, 1984.
2. M. L. Burger. Zum mechanismus der gegenwendung nach mechanisch aufgezwungener richtungsänderung bei schizophyllum sabulosum (julidae, diplopoda). *Z. verg. Physiologie*, 71:219–254, 1971.
3. C. K. Hemelrijk and D. Lambrinos. Performance of two homing strategies in environments with differently distributed obstacles. In P. Gaussier and J. D. Nicoud, editors, *From Perception to Action*, Lausanne, 1994. IEEE Computer Society Press.
4. Teuvo Kohonen. *Self-Organization and Associative Memory.* Springer Series in Information Sciences. Springer Verlag, 3rd edition, 1989.
5. Bredan McGonigle and Ulrich Nehmzow. Robot navigation by light. In J. A. Mayer and S. W. Wilson, editors, *From animals to animats*, Cambridge, 1990. MIT Press.
6. H. Mittelstaedt. Analytical cybernetics of spider navigation. In F.G. Barth, editor, *Neurobiology of Arachnids*, pages 298–316. Springer, Berlin, 1985.
7. M. Müller and R. Wehner. Path integration in desert ants, cataglyphis fortis. In *Proc. Natl. Acad. Sci.*, number 85, pages 5287–5290, U.S.A, 1988.
8. R. Pfeifer and C. Scheier. From perception to action: The right direction? In P. Gaussier and J. D. Nicoud, editors, *From Perception to Action*, Lausanne, 1994. IEEE Computer Society Press.
9. Helge Ritter, Klaus Schulten, and Thomas Martinetz. *Neural Computation and Self-Organizing Maps.* Addison-Wesley, Reading, Massachusetts, 1992.
10. S. Rossel and R.Wehner. The bee's e-vector compass. In R. Memzel and A. Mercier, editors, *Neurobiology and Behaviour of Honeybees*, pages 76–93. Springer, Berlin, 1986.
11. R.Wehner. *Animal Homing*, chapter 3. Chapman and Hall, London, 1992.
12. R. Wehner and F. Raeber. Visual spatial memory in desert ants cataglyphis bicolor. *Experientia*, 35, 1979.

Collision Avoidance using an Egocentric Memory of Proximity

R. Zapata, P. Lépinay, P. Déplanques

Laboratoire d'Informatique, de Robotique et de Microélectronique
LIRMM – UM CNRS C9928 – Université de Montpellier II,
161 rue Ada, 34392 Montpellier cedex 5, FRANCE
Tel:(33) 67.41.85.60. Fax:(33) 67.41.85.00. E–mail: zapata@lirmm.fr

Abstract : This paper describes a formal model of a dynamic and egocentric memory used for predicting the evolution of a local function of proximity. This model is represented by a polar function able to trig the reactive behaviors (reflex actions for collision avoidance) of a planar mobile robot. It also addresses the implementation of this model with an artificial neural network first on a simulator that has been developed in order to test this model and also on a real holonomic autonomous robot.

1. Introduction

Artificial reflex actions for mobile robots can be defined as the ability to react when unscheduled events occur, for instance when unknown and dynamic obstacles can collide the robot. This problem can be solved by using a behavioral approach, consisting in directly relating inputs (stimuli) to outputs (actions) through state machines [1], [2], [3], [4] or by considering a sensor–based approach which feedbacks sensory information to the robot control loop [5], [6], [13].

The method developed here is a sensor–based approach. A polar function, representing the distances all around a moving planar robot, is used to trig its reactive behaviors (reflex actions for collision avoidance). The function creates a Deformable Virtual Zone (DVZ), protecting the robot, and whose deformations are due to the intrusion of proximity information in the robot space. The control action consists in reforming this zone. This model is described in paragraph 2.

Paragraph 3 describes a formal model of a dynamic and egocentric memory used for predicting the evolution of this function of proximity.

This model has been implemented with an neural network on a simulator and also on a real autonomous holonomic robot. The description of these experiments is the purpose of paragraph 5.

2. The Deformable Virtual Zone approach of collision avoidance

Reactive behaviors are based on the DVZ concept which was developed for wheeled robots and first described in [7] and [8]. In this section, we briefly recall this sensor–based control model.

For a mobile robot, we define the *Deformable Virtual Zone* as a state function, denoted by Ξ which is a vector[1] representing a deformable zone whose geometry characterizes the interaction between the robot and its environment. This zone depends on a vector π characterizing the motion capabilities of the robot (its translational and rotational velocities for instance). The *internal control* is a relation linking the undeformed zone Ξ_h to this vector π, namely $\Xi_h = \varrho(\pi)$. We assume that we can control π through its first derivative.

The complete evolution of this internal state is driven by a 2–fold input vector $u = \begin{pmatrix}\phi \\ \psi\end{pmatrix}$. Its first vectorial component $\phi = \dot{\pi}$ is due to the control module of the robot (internal control) which tends to minimize the deformation of the DVZ. The second one , ψ, is induced by the environment itself (intrusion of information) which is the DVZ deforming agent.

In the problem of collision avoidance, the robot/environment interaction can be described as a DVZ surrounding and protecting the robot from collision. The DVZ can be considered as the sum of 2 terms :

(1) $$\Xi = \Xi_h + \Delta.$$

with :

(2) $$\begin{cases} \Xi_h = \varrho(\pi) \\ \Delta = \alpha(\Xi_h, I) \end{cases}$$

where Δ represents the deformation of the state vector due to intrusion (denoted by I and sensed by proximity sensors) of information in the robot space (See Figure 1). The vector Δ is an explicit function of the undeformed DVZ and of the intrusion of information. We note $\mathfrak{R}_S$ the curve representing the range of the proximity sensors and $\mathfrak{R}(t)$ the curve representing the *proximity function* (distance Robot/environment with respect to the angular coordinate).

1. In the case of a small number n of sensors, Ξ is a n–dimensionnal vector. Otherwise (or in the case of a continuous sensor), Ξ is a polar function of proximity.

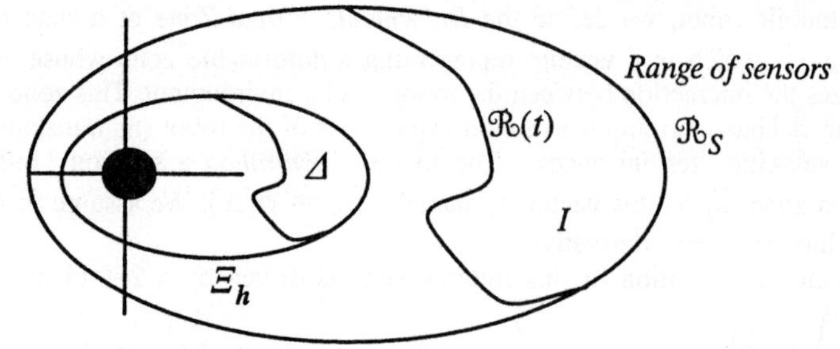

Figure 1: *Intrusion of information and deformation of the DVZ*

By differentiating equation (1)with respect to time and by letting $\psi = \dot{I}$ be the second control vector, we obtain the complete state equation of the system :

$$\text{(3)} \qquad \dot{\Xi} = \left(\frac{\partial a}{\partial \Xi_h}(\Xi_h, I) \times \varrho_\Xi{}'(\pi) + \varrho_\Xi{}'(\pi) \right)\phi + \frac{\partial a}{\partial I}(\Xi_h, I)\ \psi$$

This equation allows to control the robot reactions for collision avoidance : the environment modifies the undeformed DVZ (through the control vector ψ) and the control module of the robot tries to minimize this deformation (through the control vector ϕ). This problem can be seen as a member of the Differential Game Theory class of problems [9] and hence, be solved by using optimal control laws. Several other methods have been developed in this sense [10], [8],[11].

3. The concept of Egocentric Memory of Proximity

3. 1 The biological basis

In the hypothetical case of a robot able to sense distances in all directions of space, the curve $\Re(t)$ is directly determined at each instant. In the more realistic case of a small number of sensors, this curve can only be extrapolated, using the information obtained by the sensors and the robot motion.

Along an unsensed direction, the robot has to 'imagine' how varies the distance of the closest object in this direction. If this direction was previously sensed, the robot must keep alive this information and make it evolves in a dynamic and 'egocentric memory'. This principle is commonly used by living beings, able to move toward goals without seeing them (except at an initial time), only by measuring their own ego–motions (vestibular system). This basic principle was implemented to simulate the dynamic memory in vision systems [12]. Our aim is to model this principle for a memory of proximity and to use it for the automatic motion of artificial beings (a mobile robot moving in the plane) .

3. 2 Basic definitions

Let the mobile robot moving in the plane be kinematically characterized by the moving frame $(R) = (R, \vec{x}, \vec{y})$, referenced with respect to an absolute frame

$(A) = (O, \vec{X}, \vec{Y})$ by the coordinates of its center $R = \begin{pmatrix} X_R \\ Y_R \end{pmatrix}$ and by the angle $\beta = [OX\ Rx]$ (Figure 2); the curve $\mathfrak{R}(t)$, representing the proximity function, is defined by its polar equation in (R). This equation can be written (implicit form) :

(4) $g(r, \theta, t) = 0$

or (explicit form) :

(5) $r = \bar{g}(\theta, t)$

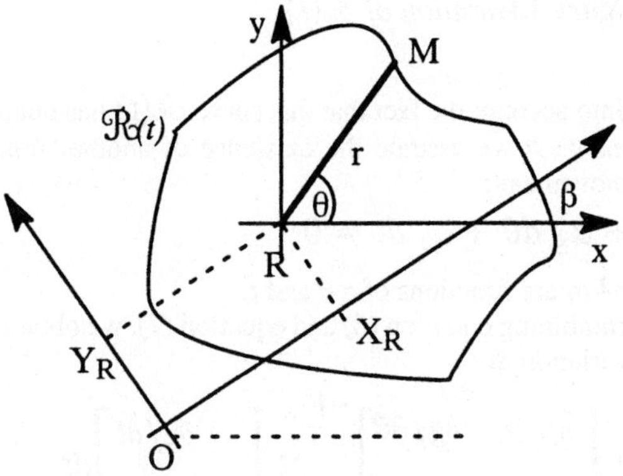

Figure 2: *The proximity function*

Between *t* and *t+dt*, the frame (R) is moving relatively to (A) and the curve $\mathfrak{R}(t)$ becomes $\mathfrak{R}(t + dt)$.

The basic idea here, is to predict this curve, by "measuring" its characteristics at previous times. This can be computed at the first order, by assuming that the "measures" of $\mathfrak{R}(t)$ and $\mathfrak{R}(t + dt)$ allow to predict the curve $\mathfrak{R}(t + 2dt)$ by a simple extrapolation.

Now, if this process is initialized by the measure of $\mathfrak{R}(t_0)$ and $\mathfrak{R}(t_0 + dt)$, it is possible to iteratively determine an approximation of $\mathfrak{R}(t)$ with $t \geq t_0$.

At each time, the predicted curve $\mathfrak{R}(t)$ can be considered as a dynamic and egocentric memory of the proximity function Robot/Environment.

We now derive the mathematical expression of this extrapolation. The differentiation of equation (4) gives:

(6) $$\frac{\partial g}{\partial r} dr + \frac{\partial g}{\partial \theta} d\theta + \frac{\partial g}{\partial t} dt = 0$$

If $M = \begin{pmatrix} r \\ \theta \end{pmatrix}$ is a point of $\mathfrak{R}(t)$ at time *t*, there exists a point $N = \begin{pmatrix} r + dr \\ \theta + d\theta \end{pmatrix}$ verifying equation (6) belonging to $\mathfrak{R}(t + dt)$ at time *t+dt*, and which can be considered as the successor of M (see Figure 3).

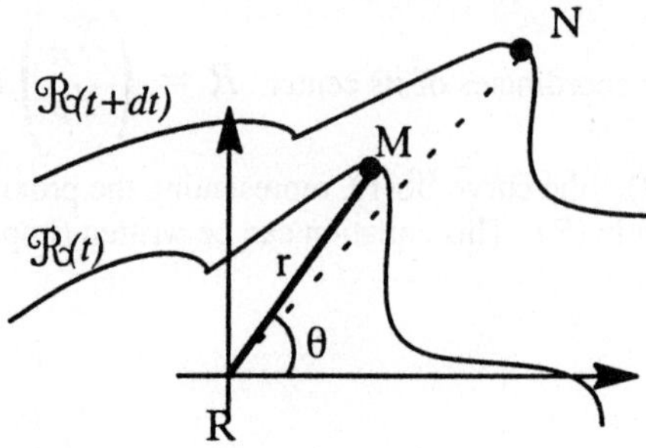

Figure 3:*Variation of* $\Re(t)$

In order to take into account the fact that the curve $\Re(t)$ has qualitatively the same shape between[2] t and $t+dt$ we assume the existence of another relation between the differentials of the movement :

(7) $$a_r\, dr + a_\theta\, d\theta + a_t\, dt = 0$$

where a_r , a_θ and a_t are functions of r, θ and t.

In this case, by combining equation (6) and equation (7), we obtain the variations dr and $d\theta$ due to the variation dt :

(8) $$\begin{bmatrix} dr \\ d\theta \end{bmatrix} = \begin{bmatrix} \partial g/\partial r & \partial g/\partial\theta \\ a_r & a_\theta \end{bmatrix}^{-1} \times \begin{bmatrix} -\partial g/\partial t \\ -a_t \end{bmatrix} dt$$

Measuring $\Re(t)$ and $\Re(t + dt)$ leads to an evaluation of $\begin{bmatrix} \frac{\partial g}{\partial r} & \frac{\partial g}{\partial \theta} & \frac{\partial g}{\partial t} \end{bmatrix}$ and hence, by asserting a model of the functions a_r , a_θ and a_t we can obtain a first–order approximation of the variation dr and $d\theta$ and hence, to predict the curve $\Re(t + 2dt)$. This algorithm is schematicaly represented in Figure 4.

2. This motion can be seen as a shape wave moving along the proximity curve.

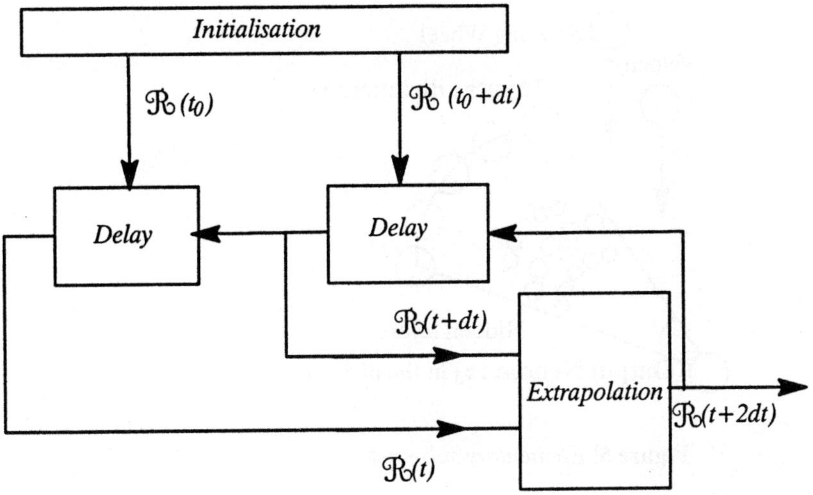

Figure 4:*Evolution of the egocentric memory*

4. Implementation

4. 1 Neuro–computing of the proximity function

Instead of analytically determining $\begin{bmatrix} \frac{\partial g}{\partial r} & \frac{\partial g}{\partial \theta} & \frac{\partial g}{\partial t} \end{bmatrix}$ and choosing the functions a_r, a_θ and a_t, it is possible to build a neural architecture in order to compute the variation *dr* in each direction of space. Actually, we can imagine a 'annular' neural network, whose external layer encodes the curve $\Re(t)$ and whose most internal layer (output layer) encodes the curve $\Re(t + 2dt)$. One of the internal layers is used to encode $\Re(t + dt)$ (by a delay of the external layer) and the other ones are used for the computation .

The neural net we have developed, is a feed–forward with back propagation network, composed of several sub–nets, each of them covering a small angular sector equal at least at $2d\theta_{max}$, which validates the first–order approximation (Figure 5).

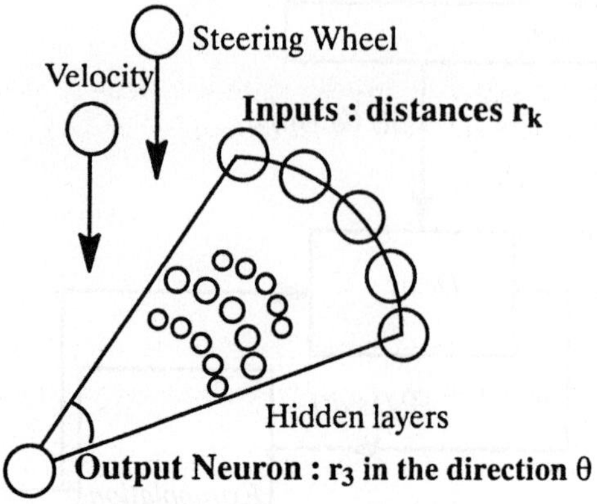

Figure 5: *Elementary sub−net*

Each sub–net allows the prediction of the proximity function in one direction. The sub–nets are "geometrically" arranged in order to cover all the robot surrounding space and to predict the curve $\mathfrak{R}(t)$ in this space.

It is clear that this system is not able to provide a long term memory and will diverge unless the egocentric memory of proximity is refreshed by real measures. Actually, the curve $\mathfrak{R}(t)$ is a combination of memorized parts (evolving in open loop) and of measured parts. A good and simple strategy consists in using the sensors to regularly refresh all parts of $\mathfrak{R}(t)$, and to let the other parts being imagined by the memory (Figure 6). This will be exposed in the paragraph 4.3. Another strategy would be to use the sensors in the most dangerous directions (fast evolution of $\mathfrak{R}(t)$).

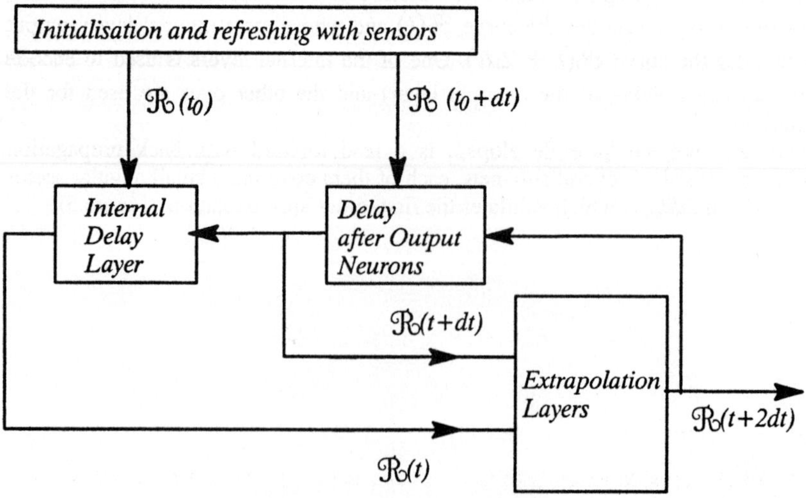

Figure 6:*Implementation with a neural network*

4. 2 Computing the DVZ

Let $r = R_{\max}$ the polar equation of $\mathcal{R}_S$. Using the polar representation $\bar{g}$ of $\mathcal{R}(t)$ (Equation (5)), we define the intrusion of information as the polar function :

(9) $$I(\theta) = \bar{g}(\theta, t) - R_{\max}$$

Next, we define the undeformed DVZ by :

(10) $$\Xi_h(\theta) = G_v(\theta)$$

where $G_v(\theta)$ is a polar function depending on the absolute velocity of the robot (for instance $G_v(\theta)$ can be the polar equation of an ellipse whose long axis is oriented by v and whose length is proportional to the module of v)[3].

At least, the deformation of the DVZ[4] is defined by (see the shaded region of Figure 7):

(11)

$$\Delta(\theta) = \begin{cases} I(\theta) + R_{\max} - \Xi_h(\theta) \text{ if } \| I(\theta) \| \geq \| R_{\max} - \Xi_h(\theta) \| \\ 0 \quad \textit{otherwise} \end{cases}$$

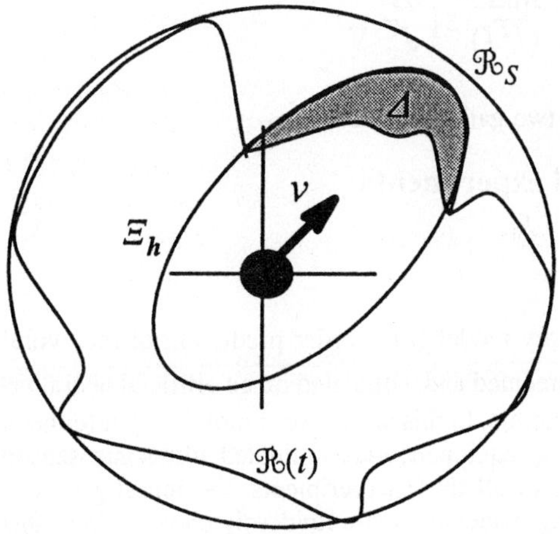

Figure 7: *The DVZ function of the robot's dynamics*

---

3. The robot 'looks' towards the direction it has to go, and the farest the fastest.
4. The undeformed DVZ acts as a dead zone.

4. 3 The collision avoidance

When the DVZ deforms (trigged by $\Delta(\theta) < 0$) the main strategy carrying out a reflex action, consists in changing the orientation of the main axis of $G_v(\theta)$ in order to modify or cancel this deformation or in decreasing the module of $G_v(\theta)$ ((the module of $\Delta(\theta)$) by acting on v. In our case, the control vector is $\phi = \begin{pmatrix} V \\ \lambda \end{pmatrix}$ where V is the module of v and λ its orientation

Equation (2) provides the variation of the DVZ deformation :

$$\text{(12)} \qquad \dot{\Delta} = \left(\frac{\partial \alpha}{\partial \Xi_h}(\Xi_h, I) \times \varrho_{\Xi}'(\pi) \right)\phi + \frac{\partial \alpha}{\partial I}(\Xi_h, I)\,\psi$$

which can be written $\dot{\Delta} = J\phi + K\,\psi$. This vector Δ is an m–dimensionnal vector (if we have m virtual sensors). The avoidance strategy consists in computing the "best" control vector ϕ to obtain a given variation $\nabla = J\phi$. This is done by pseudo–inverting this *mx2* matricial relation. The vector ∇ is heuristically computed with respect to the deformation vector. For instance :

$$\text{(13)} \qquad \begin{cases} \nabla = -M\Delta - N\dot{\Delta} \\ \phi_{best} = (J^T J)^{-1} J^T \nabla \end{cases}$$

where M and N are two gain matrices.

5. Simulation and experiments

5. 1 Simulation

The dynamic memory model (first–order prediction of the evolution of the curve $\mathfrak{R}(t)$) has been implemented and simulated on an artificial neural net, developed in C language on a PC compatible. In this simulator, a mobile robot (either a car–like vehicle or an holonomic vehicle equipped with simulated ultrasonic sensors) moves in an unknown environment. In all these experiments, the initial position of the robot, its velocity and the steering wheel angle are randomly chosen. From this initial state, the robot moves till it meets the boundary of its world. During this motion, the feed–forward process and the back propagation process are run at each sampling time. A preliminary learning phase is carried out in a stationary environment (to teach the neural network its ego–motion).

5. 2 Implementation on RATMOBILE

This model has also been implemented on a real holonomic mobile robot (RATMOBILE) able to sense its environment over a small angular sector (one to three ultrasonic sensors). The robot moves in corridors, and from time to time checks its environment by rotating while translating. Between these instants of measure, it moves using the egocentric memory as a virtual perceptive system (Figure 8).

Figure 8: *RATMOBILE the holonomic robot*

The general architecture of RATMOBILE consists of 2 levels, each of them being controlled by a single transputer board.The low level is based on a T222 Inmos™ transputer which is in charge of the management of all the sensors. The high level is based on a T805 Inmos™ transputer which computes the control laws of the 3 actuators, checks the state of the interaction Robot/Environment (detection of obstacles and generation of the avoiding trajectories).

The programs of RATMOBILE are developed in Parallel C™ on a PC compatible and then, downloaded to the inboard computer (Figure 9).

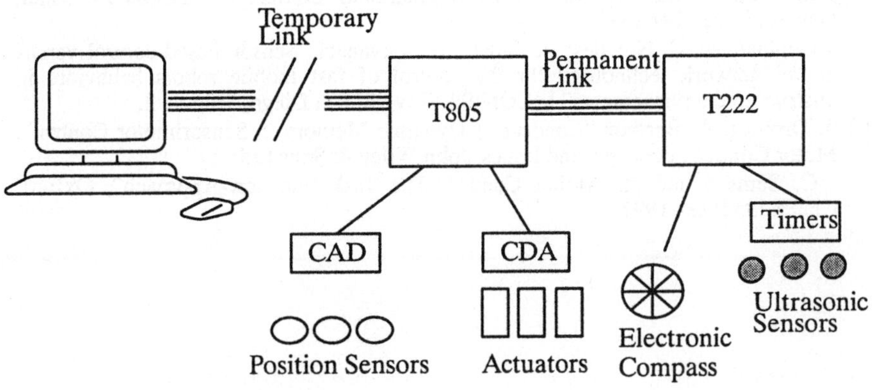

Figure 9:*Hardware architecture*

6. Conclusion

This paper has described a model of an egocentric memory used to predict the evolution of a function of proximity protecting a mobile robot. This model is represented by a polar function able to trig the reflex actions of this robot. It has also addressed the implementation of this model with an artificial neural network and the experiments that have been carried out on a simulator also on a real holonomic autonomous robot. The algorithms implementing the DVZ method can be seen as low level interactions with the robot environment and must be coupled with high level path planing procedure\s. We have implicitly assumed that these high levels of control were existing, in order to let the robot come back to its initial mission after a purely "reactive behavior".

7. References

1. R.A. Brooks “Robot Beings”, International Workshop on Intelligent Robots and Systems ’89, Sep 1989, Tsukuba, Japon .
2. T.L. Anderson “Autonomous Robots and Emergent behaviors : A Set of Primitives Behaviors for Mobile Robot Control”,IROS’90, Tsukuba 1990, Japan
3. M. Soldo “Reactive and Preplanned Control in a Mobile Robot”, Proc. IEEE Int. Conf. on Rob. and Aut.,Cincinnati, Ohio, USA, May 90
4. D.T. Lawton, R.C. Arkin, J.M. Cameron “Qualitative Spatial understanding and Reactive Control for Autonomous Robot”, IROS’90, Tsukuba 90, Japan
5. B. Espiau & *al* , "Nouvelle approche de la relation vision–commande en robotique", rapport INRIA RR–1172, 1990
6. Holenstein, E. Badreddin “Collision Avoidance in a Behavior–based Mobile Robot Design”, Proc. IEEE Int. Conf. on Rob. and Aut., Sacramento, California, April 1991.
7. R.Zapata "Quelques aspects topologiques de la Planification de Mouvements et des Actions Reflexes en Robitique Mobile", These d’etat, Université Montpellier II, July 91, France.
8. R.Zapata, P.Lépinay, P. Thompson "Reactive Behaviors of Fast Mobile Robots", Journal of Robotic Systems, January 1994
9. R. Isaacs “Differential Games”, John Wiley and Sons, Inc., New York 1965.
10. P. Lépinay, R. Zapata, B.Jouvencel : "Sensor–based Control of the Reactive Behaviors of Walking Machines", International Conference IECON’93, Maui, Hawaii, December 1993
11. P.Deplanques, C. Novales, R. Zapata, B. Jouvencel "Sensor–based control versus neural network technology for the control of fast mobile robots behaviors in unstructured environments" IECON’92, Nov 92, San Diego, Volume 2.
12. J. Droulez, A. Berthoz “Concept of Dynamic Memory in Sensorimotor Control”, Motor Control : concepts and Issues, John Wiley & Sons Ltd
13. C. Samson and al. “Robot Control: The Task Function Approach”, Oxford, Clarendon Press 1991.

A Useful Autonomous Vehicle with a Hierarchical Behavior Control

Luís Correia*, A. Steiger-Garção**

Universidade Nova de Lisboa - Faculdade de Ciências e Tecnologia
Quinta da Torre, 2825 Monte de Caparica, PORTUGAL
Phone: +351-1-295 3220, Fax: +351-1-295 5641 E-Mail: {lc,asg}@fct.unl.pt

Abstract

This paper describes a vehicle to perform a simple task using a decision control composed of a behavior hierarchy. First the model is presented. It is based on two fundamental components: the behavior module defining how each behavior element is implemented; and the blocker which is the basic component of the arbitration system, responsible for behavior selection. Behavior modules are composed of two submodules, one to produce actuator actions and other to produce priorities associated to actions. They also have three parameters whose values can be changed prior to execution, as a means of configuring a particular control architecture. Competition among behaviors is resolved by arbitration structures, each one consisting of a series of blockers working in a completely distributed way. The arbitration system can be automatically generated from the mere enumeration of the constituting behaviors. The development of the model in order to support behavior hierarchies is also presented as well as the possibility of using global variables. Then the prototype - a vehicle to clear away film boxes - is described in detail namely all the behaviors constituting its control. Finally some conclusions based on the obtained results are presented.

1 Introduction

The development of mobile robots as artificial beings has been inspired in several ways by animals. Either by the neural structure of animals, in robots controlled by artificial neural networks, or by the macroscopic manifestations of behavior segments, in behavior based autonomous vehicles [1].

---

* Dept. of Computer Science

** Dept. of Electrotechnical Engineering

Our approach has been focused on developing a model for a behavior based architecture, able to allow an easy specification of the autonomous vehicle's decision control [2, 3]. Regarding other research in this area [4-7] our model uses a distributed arbitration system to select active behaviors and introduces behavior fatigue as a general feature of every behavior module.

As was mentioned, inspiration for these approaches has drawn heavily on biology and animal behavior. Behavior based autonomous vehicles have thus been used mostly in performing behaviors not related with any specific mission, just enabling the vehicle to "live" in a particular environment. We are interested in exploring applications of behavior based vehicles in useful tasks. And in doing so we have developed a behavior architecture model where the user is relieved from defining behavior interconnections. Thus the arbitration system which selects active behaviors can be generated automatically from a simple enumeration of the behaviors to be included in each particular controller.

In the next section we present the behavior model used as the main component of the architecture - the set of all behaviors being responsible for the global control of the vehicle. The third section is dedicated to the presentation of the arbitration system, constituted by simple elements (blockers) which are used, as many as necessary, to support behavior selection in each particular architecture. In section 4 a generalization of the model is done, in a way that encompasses behavior hierarchies and global variables. Section 5 describes a new prototype developed using this model. It is a vehicle to find film boxes, grab them and put them near a wall. In section 6 some remarks are made about the theoretical model, where, besides conclusions to draw from prototype development, we defend the idea of considering behavior based approaches as a new programming paradigm.

2 Behavior Model

The main component of this architecture is the behavior module. Each instance is responsible for limited aspects of the performance of the Autonomous Vehicle (AV). From the interaction of all behavior instances emerges the global behavior of the AV. Seen that these modules will interact much in a mutual competition form, the behavior model must cope with the need to resolve competition situations among different modules. At each moment the control of the AV will have to consider external and internal stimulus and, in a behavior based architecture such as this one, select the behavior (or behaviors) that will take over the control of the vehicle.

Thus a behavior produces as outputs an activity value (which is effectively its priority) and an action value for the actuators, both of them being used as inputs to the arbitration system described in section 3. Our solution to accomplish this goal is to have a behavior divided into two sub-modules. Here we named them Action and Activity sub-modules, producing two independent Action and Activity outputs (fig.1).

The Action sub-module is responsible for producing the commands, to be delivered to the actuators, which will enable the VA to exhibit the specific aspects of behavior concerning its module. The Activity sub-module will output a value representing the urgency (or priority) the behavior has in expressing itself - taking into account the stimulus inputs it is receiving, its internal state and a user parameter. We can say that Action is the fundamental component of the behavior module, since it is the responsible for its actuation output. In case there was no other behavior competing there would be no need for the Activity sub-module.

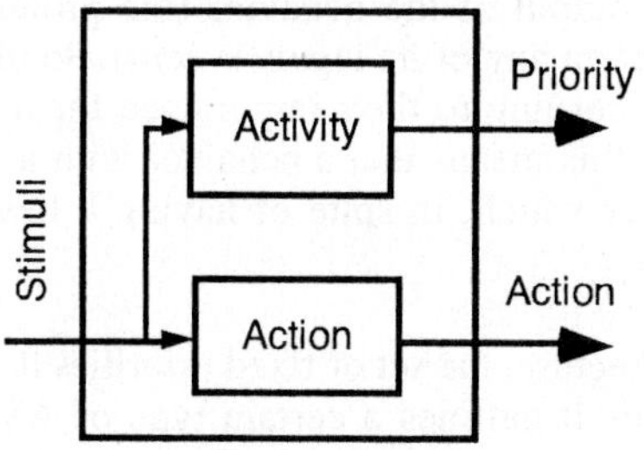

Fig. 1. Behavior model - abridged representation

In this way we separate the instantaneous action produced by each behavior module from the instantaneous priority it presents. This allows for a flexible behavior selection depending on the output priorities, unlike solutions in which this selection is predefined by hardwiring at the moment of the construction of the vehicle (e.g. [6]). Also this type of solution - with two separate activity and action outputs - is necessary if variable priorities are to be considered. In fact a model with only the action output underlies fixed priorities of the behaviors, with at most a commutation between an all active and an all inactive state for each behavior (step type of function in fig.2).

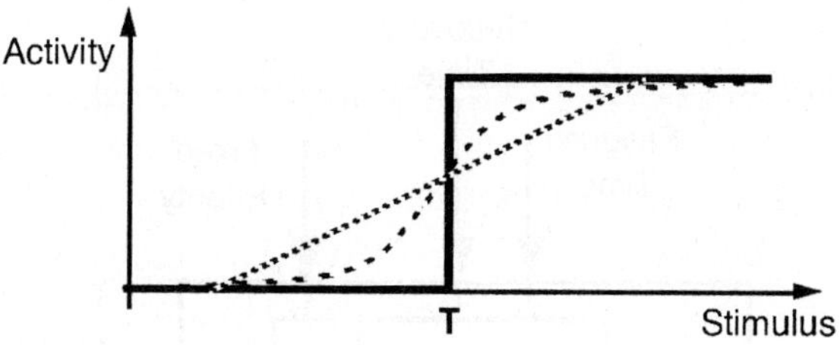

Fig. 2. Examples of behavior activity functions

In fig.2 are also depicted other forms of the Activity function, having shapes of continuous functions. All of them are monotonic increasing functions of the stimulus - we consider a domain (e.g.[0,1]) equal for all activities. This means that behaviors will be selected accordingly to the value of their input stimulus - the stronger the stimulus the higher the activity and therefore the behavior has more chances of being selected (however there is no need for all the activity functions to have the same shape). Naturally a vehicle limited to such a solution would have a completely stimulus driven character, meaning that the behavior with the stronger stimulus response would always be selected, disregarding any considerations on the relative importance of the behaviors.

Thus in order to allow the user to assign a relative emphasis to the different behaviors each of these modules has a *fixed priority* input parameter. It is combined with the variable priority, consisting of the activity value, by a product function, in order to produce a final priority output of the behavior (the product has the advantage of producing a zero result when any of its inputs is zero). So, the behaviors will start of with uneven priorities according to their importance for the global behavior of the AV. In runtime however this means that a behavior with a strong stimulus response may be beaten by another which, in spite of having a lower stimulus response, is more vital for the vehicle.

In a given behavior architecture, the set of fixed priorities defines the character of the vehicle, or in other words it outlines a certain type of AV giving it the necessary basis for the kind of task it is supposed to perform (see prototype in section 5). On the other hand the variable priority, generated internally by a behavior, defines the situation in which the vehicle is at that moment, as "seen" by that behavior.

A further feature exists in our behavior model which also draws on the fundamental properties of animal behavior. It is the *Fatigue* phenomenon - This constitutes the decrease of action intensity down to complete lack of response in presence of a continued or frequently repeated stimulus. The behavior subsequently needs a further recovery time before the action can be performed again. Fatigue is here defined by two parameters: *fatigue time* - time from beginning of action till beginning of decrease - and *recovery time* - taken from stimulus ceasing till action allowed again.

The external representation of the complete behavior model is as represented in fig.3.

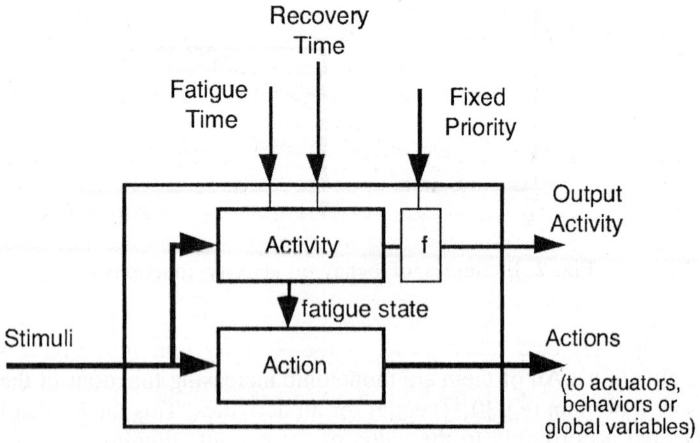

Fig. 3. Behavior model - complete representation (*f* - product function)

Why should this fatigue effect be included? From our point of view, it helps in solving a problem characteristic of decentralized architectures: deadlock (or impasse).

It usually happens as a consequence of a conflict between two behaviors, whose myopic views of the vehicle in the environment are unable to detect the situation. Other solution, based on watch-dog behaviors, has also been proposed in previous versions of the architecture [8, 9], however such an approach is more situation dependent. By using only the above results inspired on ethology research (fatigue and recovery times) we can obtain a general solution already quite efficient, being able to get around many deadlock situations. When one of the two conflicting behaviors eventually reaches a fatigue situation the other will gain control.

Fixed priority, fatigue and recovery times are all defined by the user as part of the behavior architecture of a particular AV. While the first of those is only relevant to the Activity module, the fatigue parameters affect both sub-modules.

In programming a vehicle we consider a coding phase, where the workings of a behavior module are coded, and a configuration phase, where a user selects behaviors to be used and specifies their parameters. Varying only the parameter values is enough to produce, with the same set of behavior modules, a different global behavior.

3 Arbitration Model

Taking into account the goals and restrictions of the control architecture of an AV - namely an emergent behavior solution, based on the interaction of independent behavior modules, with distributed decision regarding behavior selection and flexibility to endure easy design changes in the implementation of specific vehicles - we propose a general model in which modules direct interaction (inhibition, stimulation, or communication in general) is reduced to a minimum as an orientation guide-line. In fact, presently, in our model their only interaction is indirect through the world, using a common structure *(arbitration structure)* responsible for solving the behavior selection problem. Our aim is to minimize direct connections between behaviors - without hampering possibilities of overall performance. We feel that by unrestricting direct behavior interconnections the resulting architecture tends to present a very intricate structure which will condition the design of a vehicle.

These are only orientation principles. It is possible that there will be need for one behavior to feed another with some kind of input (see for example [8] and prototype described next). Although these will be considered exceptions and limited to the strictly necessary, the model should also be able to support them. On the other hand the use of internal sensors, for example, can be of interest in several cases as an alternative to direct behavior connections.

With an arbitration structure and a convenient model of the behaviors (already presented) it is possible to debug and adjust each behavior separately and then to integrate them in a particular vehicle and tune the global architecture - by varying the relative priorities and the other parameters of the different modules. This allows for an incremental development of a behavior architecture (successively adding new behaviors to an existent architecture).

A behavior based architecture for the control of an Autonomous Vehicle produces a global behavior as a result of the interaction of its constituents. Thus it is important for each one to act no more nor less than it needs. A behavior must be able to control whatever it needs, but it must not over-exert its actions since it may block other behaviors which would otherwise be able to express themselves. In this way we propose a decomposition of the arbiter into multiple structures (blocker structures), one for each independent actuator of the AV (see fig.4 where an architecture with five blockers in two blocker structures is represented) - e.g. in a car-like AV there will be two arbitration structures, one for propulsion and another for steering.

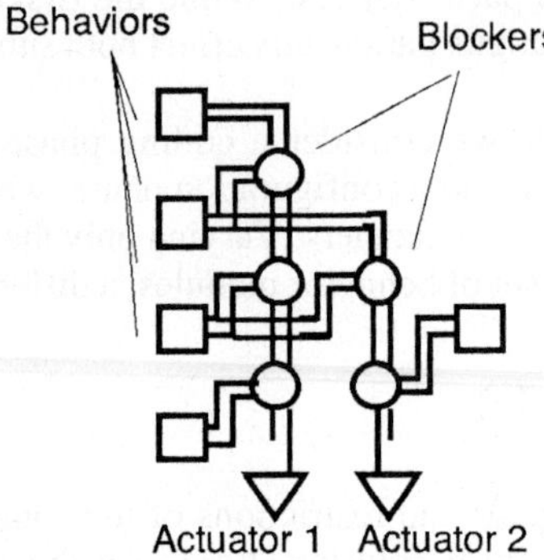

Fig. 4. Example of a control architecture with two blocker structures

The idea is to have each behavior using only actuators needed for its actions, freeing others which may then be controlled by *orthogonal* behaviors. By *orthogonal* we mean behaviors which produce independent activities, thus not interfering with one another. For instance the depth control in an Autonomous Underwater Vehicle (AUV) can usually be performed independently of the motion in the horizontal plane in many situations, or, in a land vehicle, direction can be controlled independently of forward movement [2].

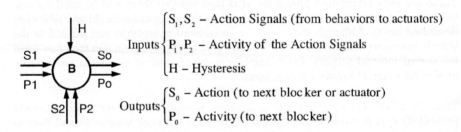

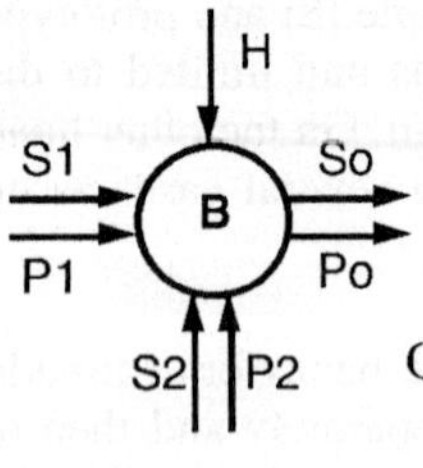

Inputs:
- S_1, S_2 – Action Signals (from behaviors to actuators)
- P_1, P_2 – Activity of the Action Signals
- H – Hysteresis

Outputs:
- S_0 – Action (to next blocker or actuator)
- P_0 – Activity (to next blocker)

Fig. 5. Detailed representation of a blocker

Also, from our point of view, a behavior module when controlling some actuator should not have its action changed by the action of other behavior, unless the later is more prioritary and in that case it will impose its own action. This means that the solution for the arbitration structure should not perform combination of behavior actions.

The constitution of the arbitration structure is completely based on blockers, represented (in a simplified form - without the hysteresis input) as circles in fig.4. Each blocker (see fig.5) solves competition among two behaviors with the following output functions' transitions (sgn(x) in equation (2.1) is the signal function):

$$S_0 = \begin{cases} \rightarrow 0 & \text{if} \quad |P_1 - P_2| \leq H \quad \text{and} \quad \operatorname{sgn}(S_1 S_2) \neq 1 \\ * \rightarrow S_2 & \text{if} \quad P_2 > P_1 + H \\ * \rightarrow S_1 & \text{if} \quad P_1 > P_2 + H \quad * \equiv \text{any output} \end{cases} \tag{2.1}$$

$$P_0 = \begin{cases} P_1 \rightarrow P_2 & \text{if} \quad P_2 > P_1 + H \\ P_2 \rightarrow P_1 & \text{if} \quad P_1 > P_2 + H \end{cases} \tag{2.2}$$

Due to hysteresis in the blockers the output functions' transitions are a more understandable representation than the output functions themselves - alternatively blockers can also be characterized as simple state machines with the above output transitions. Thus in any situation not specified in equations (2.1) and (2.2) there is no transition in the blocker outputs.

If not for hysteresis, a blocker has the ability to choose, among two behaviors, the one with higher priority. In doing this it passes on to the next blocker both the priority and the activity signal of the "wining" behavior. These inputs have to stand another comparison with the inputs from a third behavior and this process goes on till the last blocker of the corresponding structure, which outputs the action signal directly to the actuator

Hence behavior modules using multiple actuators will have each different actuation output connected to a distinct blocker structure (see fig.4). There will be n_i-*1* connections in each blocker structure, n_i being the number of behaviors using the corresponding actuator. This may amount to a large number of connections. However these connections can be computed automatically from the list of behaviors used in a particular architecture. Since the former are all of the same type the problem is rather simplified [10]. And in spite of the sequential character of the arbitration structure this is a completely distributed way of behavior selection.

The hysteresis input serves to prevent repeated commutation between two behaviors presenting a similar priority, and thus it contributes to stabilize the vehicle's global behavior. Without this input in the blockers the vehicle could often get into situations of repeated behavior selection, due to small fluctuations in sensors readings.

Another variation of this same problem, prone to occur in autonomous vehicles based on the interaction of modular behaviors, is the oscillation of the AV between two conflicting behaviors [11, 12]. In animals this problem is solved mostly with the mutual cancellation of both behaviors [13, 14], giving rise to the display of what ethologists call *displacement behaviors* (in [11] a solution for an animat which displays this type of behavior is also presented). In our model of the blocker (fig.5 and equations 2.1 and 2.2) the activity signal output is set to zero when two behaviors having a similar priority try to control the same actuator in opposite ways - *similar* here meaning that the two priorities differ less than the value of hysteresis. Thus blockers force a cancellation of equal priority conflicting behaviors, helping also to solve the oscillation problem.

We discussed this issue as if the two conflicting behaviors were physically connected to the same blocker but the reasoning is general since the blockers let go through the most prioritary of the inputs. Therefore whenever the two most prioritary behaviors in a blocker structure are conflicting their outputs will eventually be resolved at one unique blocker. The priority output in this case is not set to zero in order to still block other behaviors less prioritary than these two.

Conversely, when behaviors with similar priorities but not opposing in their actions compete at a blocker, the action output is kept as well as the priority since the behaviors are competing but they are not in conflict regarding the motion of that actuator.

Another consequence of both this solution and the assignment of a blocker structure for each actuator is that, in case of behavior conflict in one of the blocker structures, the vehicle may still be enabled to move actuators under other blocker structures. Sometimes this is as much as necessary to remove the conflict situation.

The structure thus defined combined with the behavior model allows for a high flexibility in the definition of the architecture for a particular AV. The fact that priorities are separated from the actions of the behaviors plus the continuous character of those same priorities, that can be dealt with gracefully by the blockers, produces a flexible control structure. By allowing continuous varying priorities the arbitration structure complies with behaviors whose instantaneous priorities change accordingly to their input stimulus.

4 Behavior Hierarchy and Global Variables

The use of a flat architecture, meaning that all behaviors lie at the same level, turns out to be rather limiting, especially if there is need to specify a behavior sequence - a case where we only want a behavior to be present if another behavior has been active previously. To fulfill that purpose the model is extended to allow *coordinator behaviors*, which enable or disable underling behaviors.

With the behavior model presented above, the solution for hierarchical control is quite straightforward. The action outputs of a coordinator behavior connect directly to the Fixed Priority inputs of its subordinate behaviors. Thus, by attributing null fixed priorities, the coordinator is capable of shutting down behaviors while selecting others, with some non-zero value of the fixed priority. Sequencing behaviors is done by successively selecting them (with a state-machine type of behavior). In a certain sense it is a form of inhibition that subordinate behaviors undergo. This type of hierarchical control is easily extended to further layers, allowing for a growing complexity of the global behavior of the vehicle.

Sharing of behaviors by coordinator modules is possible using the same type of arbitration structures as those presented above to access actuators. In this way, the activity output of a coordinator module will qualify its action outputs, both delivered to blocker structures, for purposes of accessing shared subordinate behaviors.

A further enhancement is achieved with the inclusion of global (internal status) variables. These can be used by any module in the architecture both for reading and for writing. In the later case again blocker structures are used, when necessary, to arbitrate access to variables shared, in writing, by multiple behaviors.

5 Prototype

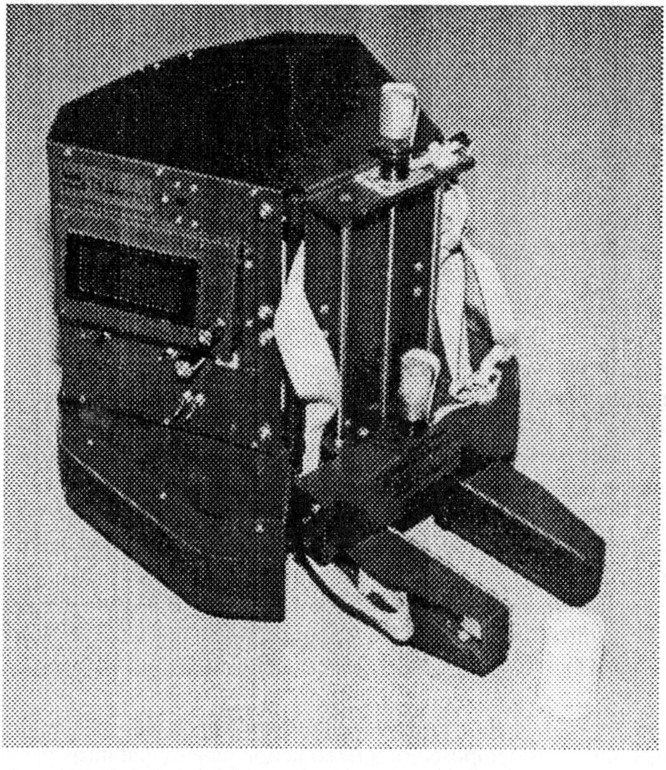

Fig. 6 - R-2 Vehicle (next to a film box)

A control architecture for a prototype based on an R-2 vehicle, from AAI (fig.6) was developed using the concepts described above. Concerning actuation, the vehicle has 4 wheels with differential motorization in the anterior ones and a gripper with one degree of freedom (up and down movement). It has 8 IR sensors, 2 of them at the end of the gripper fingers, and 7 contact sensors, 2 on the inside surface of both fingers and the other 5 around the vehicle. Also, there is a IR beam sensor inside the gripper, to detect, by beam interruption, objects between the fingers.

The task performed by the prototype is to collect photo-film boxes and deposit them next to a wall. This was an adaptation of the contest organized in the PerAc'94 conference. The original task of the contest was to have a robot collecting film boxes from a circle with a diameter of about one meter and stacking them in a pile under a light source. The winner would be the robot producing the highest pile.

Since we didn't have any sensors other than the original ones, it turned out not to be feasible for the R-2 to perform the contest task. The only sensors usable to identify a box and the pile are the 2 IR sensors at the end of the fingers. Since the gripper only has one degree of freedom, any position adjustment (except for height) has to recur to wheel movements, which are not precise enough to allow building a stack of boxes. Also the size of the vehicle (approximate length - 30cm) is too large to work inside a circle of 1m in diameter.

Thus we adapted the task, simplifying it somehow. The vehicle's typical environment is a corridor, where most objects are vertical planes which are easily detected by the vehicle's sensors. The working area however is unlimited, meaning that the vehicle doesn't have any representation whatsoever of the space it is working in. An arbitrary number of boxes is spread on the floor of the corridor in a random way. The vehicle should collect them, one at a time, and transport them to the neighborhood of a wall, where they should be deposited.

5.1 Behaviors

To perform this task we developed a hierarchical control architecture with two top coordinator behaviors and several behaviors below, some of them shared by the top coordinators (fig.7). We should stress that there would certainly be other ways of producing a control architecture with similar capabilities, depending on the programmers idea of the solution.

Since the task has two clearly separate phases, search for a box and deposit it, each of the two top coordinators is responsible for establishing the behaviors adequate to that subtask. Thus, for the search phase the top behavior is *Search*. It becomes active whenever the gripper is empty, as it happens when the vehicle drops a box. It activates the behaviors *Return, Wander, Detect* and *Grab*.

Return drives the vehicle back to the neighborhood of the place where the box it just dropped was found. It is composed of a fixed action pattern that turns the vehicle back (by 180°) and then a forward movement whose duration is defined by a global variable (*distance*). This behavior is triggered by a transition in the gripper contact sensors when the R-2 drops a box. It is the most prioritary behavior in this initial period of the search phase.

Then the vehicle will normally be under control of the *Wander* behavior. It will move about the environment, with sharp turns, in order to maintain the R-2 in a small area, stopping for a while as a result of fatigue. *Wander* has a low priority and thus it will only get selected when no other behavior under *Search* is active.

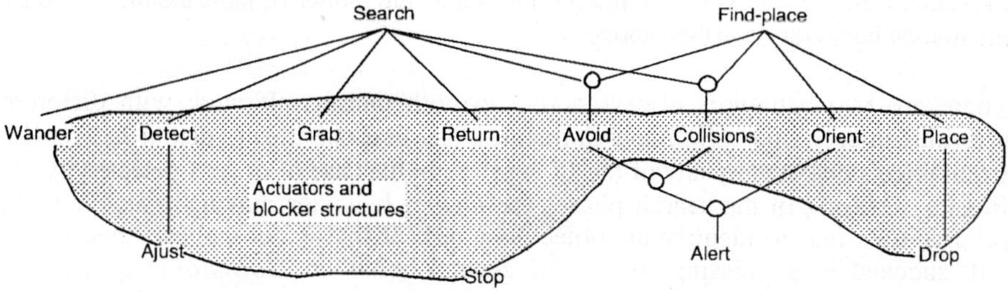

Fig. 7 - Behavior architecture of the box storer

While the vehicle is roaming the *Detect* behavior tries to detect a stimulus in any of the two front (finger) IR sensors. If that happens this behavior measures the height of the object to decide if it is a box or not. With a positive decision it activates the subordinate behavior *Adjust* which tries to align the vehicle with the detected box, so that the box will get positioned inside the gripper.

Under *Search* there is still another behavior, *Grab*. It just closes the gripper whenever it detects an object inside the fingers. This behavior is always active and thus it provides an important opportunistic character to the vehicle. Either as a consequence of a fine positioning of the vehicle, by the action of *Adjust*, or by a casual movement while wandering, anytime a box befalls inside the gripper it will be grabbed.

Fixed priorities produced by *Search* are ordered, in descending order, as follows: *Collisions, Grab, Adjust, Detect, Avoid, Return* and *Wander*.

The detection of an object held in the gripper, by reading the contact sensors of the fingers, triggers the other top level behavior, *Find-place*. This module selects (activates) the adequate behaviors to perform the deposit box phase. These are: *Orient* and *Place*.

In order to allow the user to select the wall where boxes should be deposited, the vehicle waits for some user intervention during a small interval right after grabbing a box. This behavior segment is produced by *Orient*. It produces a sound signal, through activation of the behavior *Alert*, indicating that the vehicle is waiting for a user intervention. By positioning an object (hands for instance) in front of any of the two lateral IR sensors the user can generate rotation movements of the vehicle orienting it towards the wall chosen. After this intervention the user can "tell" the vehicle to start moving in that direction by pushing the rear contact sensor. If after a while, in fact after the fatigue time specified for this behavior, there is no user action, the vehicle will start moving in the direction it is heading.

The *Place* behavior just drives the R-2 in a forward movement until a wall is detected. When that happens it stops the vehicle and increases the fixed priority of the *Drop* behavior, activating it. *Drop* only opens the gripper putting the box down at that position. This action will again trigger the *Search* behavior, as described above. It should be noticed that *Place* measures the time spent from the beginning of the movement till the wall is found and produces it as the global variable *distance* read by the *Return* behavior described above.

There are two behaviors always active in both phases, although with different priorities in each one: *Avoid* and *Collisions*. The first one tries to detour the vehicle from objects detected by any of the IR sensors. In the case of objects detected by the fingers' sensors, in the search phase, the *Detect* behavior, having a higher fixed priority tries first to identify the object as a box. If that is not the case then *Avoid* will succeed in expressing itself. The second behavior (*Collisions*) gets active whenever a collision is detected with any of the contact sensors around the R-2. It then executes the fixed action pattern corresponding to the detected sensor, in order to get away from the object.

The fixed priorities generated by *Find-place* for its subordinates are ordered, in descending order, as follows: *Collisions, Orient, Place* and *Avoid.*

The *Alert* behavior already referred only generates an intermittent buzzer sound when it has a non-zero fixed priority. It is shared by three behaviors and it exists with the purpose of alerting the user to important situations. We have already described one of these, when *Orient* is waiting for a user intervention. Another situation in which it is used is to signal the fatigue state in key behaviors (namely *Avoid* and *Collisions*). This was considered to be important so that the observer's attention can be drawn to a situation from which the vehicle is having difficulty in getting away. Human assistance may be necessary, for instance in clogged situations with no easy way out. However, if the user does nothing, after recovery both *Avoid* and *Collisions* will resume their evasive maneuvers.

Finally there is the *Stop* behavior. Notice that it is not subordinate to any other behavior. Also, it has no stimulus and therefore it is always active trying to stop the vehicle. Its fixed priority was defined as the lowest of all the behavior set, just the minimum necessary for it to actuate when there is no other behavior active, in what regards actuation of the locomotion motors. The inclusion of this behavior has two important advantages. One is in the debugging phase, since it works like defining the default movement of the vehicle - stopped. The second is that, in this way, behaviors only have to act when they have stimulus and thus they don't have to produce any action when their stimulus is absent (or after the duration of a fixed action pattern). They only have to set their output activity down to 0 in order to allow *Stop* to get control.

To support these 14 behaviors we needed 31 blockers, used in 7 different arbitration structures. These are one for each of the four vehicle actuators, one to share *Avoid*, another to share *Collisions* (with only one blocker each) and another to share *Alert* (with two blockers).

All the architecture was programmed using a subset of the Behavior Language supplied with R-2. This language is a subset of Lisp with primitives to handle concurrent processes (concurrence is actually simulated) and others to allow implementation of Brooks' inhibitors and suppressers in connections, as well as spreading activation [15, 16]. We only used basic Lisp instructions, primitives to simulate concurrence and simple connections between processes.

5.2 Operation

The global operation of the vehicle in performing its task is reasonable, taking into account the poor set of sensors available for doing so. In tests performed with 8 boxes spread out through an area of approximately $4m^2$ in a corridor (about 2 meters wide), the vehicle has been able to clear 6 boxes (storing them next to walls) in periods of 5 to 10 minutes. Since the search is random, it gradually takes more time in finding boxes as their number decreases.

No deadlock or impasse condition has been observed in normal operation of the vehicle, such that user assistance was needed. Some indecisions can be noticed due to behavior conflict but all have been promptly solved by the workings of the model. This is an important result for it has been pointed as one of he major drawbacks in behavior based control.

The fact that the working area is unlimited gave rise to a couple of situations where the R-2 would go straight along the corridor after grabbing a box. However due to the fatigue phenomenon in the *Place* behavior, the vehicle would stop after 3 to 4 meters and drop the can right there. Naturally this distance can be easily changed just by defining a different fatigue time. Notice that in these occurrences the *Return* behavior is still able to take the vehicle back to the neighborhood of the place where it found the last box.

In observing the vehicle running there are some particular displays worth mentioning.

With around 140 (simulated) processes, in spite of most of them being rather simple, the architecture execution shows some slowness, in form of slight delays in reactions to stimulus.

The IR sensors are not very discriminating, having only three usable values for positive detection of objects. Furthermore the intrinsic problems of IR sensors with surface color and polishing further deteriorate their use. Thus, in a light polished floor, the vehicle drops boxes at distances of up to 10cm of the walls, while in darker environments it may position them at less than 2cm. This issue has to do with the adaptation of the vehicle to the environment, but we purposely didn't want to produce an overadapted vehicle, in order to allow more flexibility even at expenses of a slight performance degradation.

As the wall identification is only the detection of an object ahead, sometimes it happens that the vehicle drops a box next to another one in the middle of the working space. However it has also been able to collect them later, either separately or together, correctly transporting and dropping them next to a wall.

It does also grab other objects not so similar to film boxes, such as a pack of cigarettes. The identification of a box is solely based on its height and thus any object with a similar height and not too wide (which is the case of cigarette packs when standing on the face with medium area) will be gladly stored away.

The odd phenomenon of *flying boxes* blocks the vehicle to a standstill. This happens whenever a vicious experimenter takes the object away before the R-2 closes the gripper. The cause of this is a simplification made to the *Grab* behavior (in order not to slow down the system too much the same simplification has also been made in *Alert* and *Stop*) where the processes to simulate fatigue and recovery were not included. However is should be stressed that this is not a situation the vehicle is prone to encounter during its normal operation and therefore we should not worry about evolving it in that direction.

6 Conclusions

The presented model, proved to be an interesting approach to the specification of decision control in autonomous vehicles for unstructured environments.

From the specification point of view the user has a simplified task since he only has to enumerate the behaviors used in a particular control architecture, from a library of available behaviors. The model allows then an automatic generation of the distributed arbitration system, which will handle, in run-time, the behavior selection problem. Exception is made for specification of behavior hierarchies.

But even for hierarchy specification as well as global variables, the model presents a very interesting unifying character that simplifies their integration in the basic model.

From the execution point of view the model has clearly shown, namely in the prototype, to have intrinsic characteristics adequate to solve some of the typical problems in behavior based control. Such is the case of the fatigue phenomenon which proved very useful even in a simple form as the one presented here. Its influence in eliminating deadlock and impasse problems due to behavior conflicts must be stressed.

Future work will be focused on the introduction of adaptation schemes in the model, so that a vehicle might be able to tune itself to its surroundings, adjusting itself to slightly different environments. With this purpose, the behavior model with only three external parameters (fixed priority, fatigue and recovery times) will possibly have to be enriched with other parameters specific to each particular behavior.

Acknowledgments

We are indebted to Sérgio Santos who participated in the development of the prototype and provided insightful comments on its implementation. The prototype was developed in cooperation with CRI center of UNINOVA institute. This work was partially supported by a grant from Junta Nacional de Investigação Científica e Tecnológica (Proc° 29702/INIC) for L.Correia.

References

[1] Meyer, J.-A. and A. Guillot. Simulation of Adaptive Behavior in Animats: Review and Prospect. in First International Conference on Simulation of Adaptive Behavior. 1991. Paris, September 24-28, 1990: MIT Press.

[2] Correia, L. and A. Steiger-Garção. *Behavior Based Architecture with Distributed Selection*. in *The Biology and Technology of Intelligent Autonomous Systems*. 1994. Trento, Italy, 1-12 March 1993: Springer-Verlag (in publication).

[3] Correia, L. and A. Steiger-Garção. *A Model of Hierarchical Behavior Control for an Autonomous Vehicle*. in *PerAc'94 From Perception to Action*. 1994. Lausanne, Switzerland, 7-9 Sep.: IEEE Computer Society Press.

[4] Maes, P., ed. *Designing Autonomous Agents*. 1990, MIT Press.

[5] Brooks, R.A., *A Robust Layered Control System For A Mobile Robot*. IEEE Journal of Robotics and Automation, 1986. **vol. RA-2**(No.1, March).

[6] Connell, J.H., *Minimalist Mobile Robotics*. 1990, Academic Press.

[7] Steels, L., *Building Agents out of Autonomous Behavior Systems*, in *The 'artificial life' route to 'artificial intelligence'. Building situated embodied agents*, L. Steels and R. Brooks, Editor. 1993, Lawrence Erlbaum.

[8] Correia, L. and A. Steiger-Garção. *An AUV Architecture and World Model*. in *Fifth International Conference on Advanced Robotics - Robots in Unstructured Environments*. 1991. Pisa - Italy, June 19-22: IEEE.

[9] Correia, L. and A. Steiger-Garção. *A Reactive Architecture and a World Model for an Autonomous Underwater Vehicle*. in *IEEE/RSJ International Workshop on Intelligent Robots and Systems '91*. 1991. Osaka-Japan, November 3-5: IEEE.

[10] Correia, L. *An AUV Simulator for Test and Development of a Behavior Based Architecture*. in *Workshop on Autonomous Underwater Vehicles*. 1994. Porto, Portugal, 1-3 September 1993: Springer-Verlag (to be published).

[11] Maes, P. *A Bottom-Up Mechanism for Behaviour Selection in an Artificial Creature*. in *First International Conference on Simulation of Adaptive Behavior*. 1991. MIT Press.

[12] Bellingham, J.G., *et al*. *Keeping Layered Control Simple*. in *Symposium on Autonomous Underwater Vehicle Technology*. 1990. Washington, DC, USA, 5-6 June.

[13] Slater, P.J.B., *An introduction to Ethology*. 1985, Cambridge University Press.

[14] McFarland, D., *Animal Behaviour*. 1985, Pitman.

[15] IS_Robotics, *The R-2 Wheeled Robot Manual*. 1992.

[16] Brooks, R.A., *The Behavior Language; User's Guide*. 1990, MIT - AI Lab.

Evolving Electronic Robot Controllers that Exploit Hardware Resources *

Adrian Thompson

School of Cognitive and Computing Sciences,
University of Sussex,
Brighton BN1 9QH,
England.
E-mail: adrianth@cogs.susx.ac.uk

Abstract. Artificial evolution can operate upon reconfigurable electronic circuits to produce efficient and powerful control systems for autonomous mobile robots. Evolving physical hardware instead of control systems simulated in software results in more than just a raw speed increase: it is possible to exploit the physical properties of the implementation (such as the semiconductor physics of integrated circuits) to obtain control circuits of unprecedented power. The space of these evolvable circuits is far larger than the space of solutions in which a human designer works, because to make design tractable, a more abstract view than that of detailed physics must be adopted. To allow circuits to be designed at this abstract level, constraints are applied to the design that limit how the natural dynamical behaviour of the components is reflected in the overall behaviour of the system. This paper reasons that these constraints can be removed when using artificial evolution, releasing huge potential even from small circuits. Experimental evidence is given for this argument, including the first reported evolution of a real hardware control system for a real robot.

Keywords: Evolvable Hardware, Evolutionary Robotics,
Physics of Computation.

1 Introduction — What is Evolvable Hardware?

Evolvable hardware [4, 10, 8, 9, 20, 24] is a reconfigurable electronic circuit, which can be changed by an adaptive process such as a genetic algorithm. This paper considers the evolution of hardware to control an autonomous mobile robot; initially by examining exactly what evolvable hardware is, and what its advantages over software systems are, and then by concentrating on one of these benefits: the exploitation of the physics of the implementation, and how this may be maximised. Finally, early results are presented for the first ever evolution of real hardware to control a real robot, which benefits from the change of perspective that I claim evolvable hardware justifies.

A type of commercially available VLSI chip called a Field Programmable Gate Array (FPGA) [25] will provide a good illustration of how hardware may be subject to adaptation, although many other evolvable hardware architectures (both analogue and digital) are possible. A typical FPGA consists of an array of hundreds of reconfigurable blocks that can perform a variety of digital logic functions, and a set of wires to which the inputs and outputs of the blocks can be connected (Figure 1). What logic functions are performed by the blocks, and how the wires are connected to the blocks and to each-other, can be thought of as being controlled by electronic switches (represented as dots in the diagram). The settings of these switches are determined by the contents of digital memory cells. For example, if a block could perform any one of the 2^4 boolean functions of two inputs, then four bits of this "configuration memory" would be needed to determine its behaviour. The blocks around the periphery of the array have special configuration switches to control how they are interfaced to the external connections of the chip (its pins).

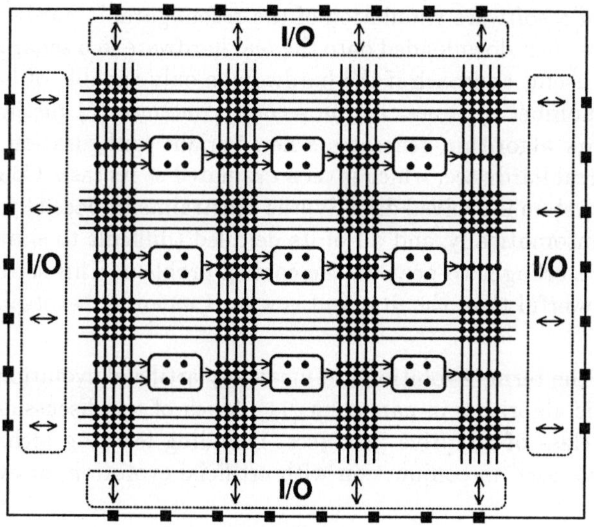

Fig. 1. Part of a simplified typical FPGA architecture.

The configuration memory of an FPGA can be conceptualised as its genotype, which determines what the blocks do and how they are wired together. If an FPGA is the control system of a robot, then artificial evolution can manipulate an encoding of the configuration memory (genotype) according to the robot behaviour induced by the corresponding circuit implemented on the FPGA: the hardware is evolvable. Commercial FPGAs have more features than I have mentioned, but the principles are the same.

There is often no clear distinction between software and hardware, however.

The contents of an FPGA's configuration memory can be thought of as causing it to instantiate a particular circuit by means of setting its configuration switches, but also the contents of the configuration memory might be viewed as the "program" of a parallel computer that happens to be called an FPGA. Which of these viewpoints is adopted is a matter of deciding on the most appropriate style of description for the particular system in question: a conventional personal computer could be envisaged as evolvable hardware, but it is usually more useful to think of it as a fixed piece of electronics that operates upon data and instructions held in a memory. An intimately linked consideration is whether the evolving system is viewed as a computational one (performing calculations or manipulating symbols), or whether it is seen from the much wider perspective of dynamical systems theory. Later, I shall advocate robot control systems that are appropriately described as non-computational dynamical systems evolving in hardware.

There are two ways in which artificial evolution may be applied to a reconfigurable hardware system. In the first (sometimes known as "*extrinsic* evolvable hardware"), an evolutionary algorithm produces a configuration based on the performance of a software simulation of the reconfigurable hardware. The final configuration is then downloaded onto the real hardware in a separate implementation step: a useful approach if the hardware is only capable of being reconfigured a small number of times. In the second ("intrinsic"[2]) method, each time the evolutionary algorithm generates a new variant configuration, it is used to configure the real hardware, which is then evaluated at its task. Consequently, an implemented system is evolved directly; the constraints imposed by the hardware are satisfied automatically, and all of its detailed (difficult to simulate) characteristics can be brought to bear on the control problem. This second method is much more powerful than the first and is what I mean by "evolvable hardware" in this paper.

I shall use the term "evolution" to mean any artificial evolution-like process, and the genetic algorithm in particular [11]. Much of the discussion also applies to a broader class of adaptive processes, including learning techniques, which may usefully be used in conjunction with artificial evolution, or even instead of it.

Next, Section 2 examines the motivations behind the evolution of hardware systems. One of the possible objectives — the maximal exploitation of hardware resources — is singled out as being particularly interesting; to facilitate this, a new approach is proposed. Section 3 consolidates some of the ideas put forward, by means of a simulation study using a highly abstract model of an FPGA. Finally, Section 4 describes a new evolvable hardware architecture, which is easy to build, but yet takes many of the important ideas on-board. The benefits of this architecture, and the underlying approach, are demonstrated by the real-world performance of this machine in controlling a robot.

---

[2] The terms *extrinsic* and *intrinsic* evolvable hardware are due to Hugo de Garis.

2 Why Evolve in Hardware?

Under what conditions is the use of evolvable hardware beneficial, when compared with systems evolved in software simulation? This section identifies three areas: the first two are related to the raw speed of hardware, while the third is more profound and suggests powerful new kinds of system that have not been seen before.

2.1 Speed of Operation

Currently, much of the research on evolved control systems for autonomous agents is centred around simulations of neuro-mimetic networks performed on a general purpose computer. As expertise in the evolution of complex systems advances, so the size and complexity of these networks will increase, with a corresponding decrease in the speed with which they can be simulated in this way. If the speed increase in general purpose computers does not keep pace with improvements in evolutionary techniques, then general purpose computers will have to be abandoned, and more specialised parallel machines used to perform the simulations. Initially, coarse-grain parallel multi-processor computers will be sufficient, but if the complexity of the networks increases still more, progressively finer grained parallelism will have to be utilised, until eventually there is one special purpose processor for each neural node. We would then have arrived at a reconfigurable hardware implementation for artificial neural networks — evolvable hardware.

The assumption that our ability to evolve control systems will outstrip the speed of simulating computers is not necessarily true. Nevertheless, it is interesting to observe that if our aspirations for building the most complex technically-possible brain-like systems are fulfilled, then the result will be an evolvable hardware implementation of the system. In that case, would it not be possible to evolve hardware *in its own right* (as an electronic circuit, and not as an implementation of anything else) and obtain an even more powerful system, better suited to silicon than other kinds of architectures like neural networks? I shall return to this question soon.

2.2 Accelerated Evolution

When evolving a robot control system to achieve some task, the time taken before a satisfactory controller is obtained is usually dominated by how long it takes to evaluate each variant controller's efficacy at that task (fitness evaluation). The genetic operations take a small amount of time in comparison.

The obvious way to evaluate a control system is to connect it to the robot and see how well it performs in the real world, in real time. Even for tasks that take a short length of time to perform (a minute or two), the large number of fitness evaluations normally required can make this highly time-consuming. Consequently, there is a case for interfacing the evolving hardware control system to a high-speed simulation of the robot and its environment, in order to accelerate the entire evolutionary process.

It is suggested by de Garis[4] that the environment simulation could be implemented in special purpose electronics situated next to the evolving hardware control system on a VLSI chip. The implementation of the simulator in hardware is made feasible by modern automatic synthesis techniques, which can derive a circuit from a textual description resembling a computer program. Implementing the environmental simulator in hardware rather than software makes it faster, but does not solve the problem that it is extremely difficult adequately to simulate the interactions between a control system and its environment, such that a control system evolved in the simulated world behaves in a satisfactory way in the real world. This is especially the case when vision is involved [2]. Nevertheless, it is possible that environment simulation in special purpose hardware will be an important tool as new techniques are developed [13, 23].

When a circuit that has been rapidly evolved for behaviour in a high-speed simulated world is ready for use in the *real* world, all of its dynamics that influence the robot's behaviour must be slowed down by the same factor by which the real world is slower than the simulation. (Imagine a controller that was evolved for a high-speed simulated world and was then let loose in the real world without being slowed down. Everything in the environment would then be happening slower than it "expected," and the motor signals produced would tend to be too fast for the robot's actuators and the world. It would probably no longer perform the task.) This means that the acceleration of evolution through the use of a high-speed simulated environment is at the cost of the efficiency of the control circuit produced. The final circuit cannot be making maximal use of the available hardware when it is operating in the real world, because it is capable of producing the same behaviour in a world that is running faster: the resources needed to allow for this could be being used for real-world performance. This may, however, be a sensible use of some of the high speed available from electronic hardware.

The fact that it must be possible to adjust the speed of all of the control circuit's dynamics that affect the behaviour of the robot restricts the set of control circuits that could be produced by evolution in a high-speed world. Either the time-scales of all of the semiconductor physics must be adjustable by large amounts (this is not practical), or the aspects of the control circuit's dynamics that make a difference to the robot must be restricted such that they are more easy to control. The latter can be arranged by restricting evolution to use predefined indivisible modules that have adjustable time-constants, or by applying the restriction of discrete-time dynamics through the use of an adjustable clock. Each of these possibilities diminishes the degree to which evolution can exploit the resources available from the reconfigurable hardware. Again, this may often be a sensible sacrifice to make for accelerated evolution, but the remainder of this paper deals with an interesting alternative.

2.3 Exploitation of Semiconductor Physics

Work with the "gantry robot" at the University of Sussex [7] suggests that it may be feasible to carry out artificial evolution in a real-world environment with fitness evaluations taking place in real time. If, in addition, evolution is given control over the reconfigurable hardware at the lowest possible level (for example the "configuration bits" of the FPGA mentioned in Section 1) to produce an electronic circuit as a type of control system in its own right (and not as an implementation of anything else), then circuits of unprecedented power and efficiency will be produced. These systems can exploit every facet of the characteristics of the reconfigurable hardware on which they are developed, because all of the natural real-time behaviours of its components (their physics) are allowed potentially to affect the robot's behaviour, and detailed control is provided to tune their collective action to achieve the task.

In nature, control systems are always adapted to the manner in which they are implemented, because their evolutionary success is determined by their effectiveness as real implemented systems. If engineering success is the objective, then implementations of neural networks (that evolved for an implementation allowing slow, highly unconstrained connections between slow units) may be a bad use of integrated circuits, which are very fast, but have severe constraints on interconnections (because of their planar nature). Perhaps it will be possible to evolve an architecture that is better suited to silicon than neural networks, but yet induces intelligent ("brain-like") behaviour into a robot. The fine-grained control over the hardware that is required implies a vast search-space for a system of any complexity, so techniques need to be developed cope with this. One possibility is the use of developmental genetic encoding schemes for genetic algorithms [12], which allow the evolution of re-usable building blocks (analogous to neurons?), permitting fine-grained tuning of the building blocks, but yet reducing the search space to systems built out of them.

The space of circuits of the type I propose is very much larger than the space of solutions available to a human designer. The designer works with high-level models of how components or higher-level building blocks behave and interact. Design constraints are adopted to prevent the imperfections of these models from affecting overall behaviour. One such constraint is the modularisation of the design into parts with simple, well defined interactions between them. Another is the use of a clock to prevent the natural dynamics of the components from affecting overall behaviour: the clock is used so that the components are given time to reach a steady state before their condition is allowed to influence the rest of the system. I suggest (with the aid of the empirical evidence presented in the following sections) that such constraints should be abolished whenever they are a limitation on the potentially useful behaviour of an evolving hardware system. A designer carefully avoids "glitches," "cross-talk," "transients" and "meta-stability," but all of these things could be *put to use* by artificial evolution.

Evolution could also put to use properties of the hardware that the designer could *never* know about. For instance, a circuit may evolve to rely on some internal time-delays of an integrated circuit that are not externally observable. Even if there is a silicon defect, the system could evolve to use whatever function the "faulty" part happened to perform. This raises a fundamental problem: a circuit that is evolved for a particular evolvable hardware system (a certain FPGA chip, for example) may not work on a different system that is nominally identical — no two silicon chips are the same. There are several ways in which this could be avoided. The circuits could be evolved to be robust to perturbations in some properties that vary from chip to chip (by altering the chip's temperature or power supply during evolution, for example). Evolution could be forced to produce building blocks that are repeatedly used in the circuit, and would therefore be insensitive to characteristics that varied across one chip. Evolution could evaluate a configuration on more than one piece of reconfigurable hardware when judging its quality (this can also be done using a single reconfigurable chip that can instantiate the same circuit in several different ways, e.g. by using an FPGA's rotational symmetry). Finally, it could be accepted that further adaptation will have to take place each time a configuration is transferred from one reconfigurable device to another [14, 15, 16, 17].

The next two sections of this paper will provide experimental evidence for the ideas I have put forward here. Firstly, I present a simulation of the evolution of an FPGA configuration, which demonstrates that evolution can produce circuits optimised for a particular implementation, and in the absence of modularisation and clocking constraints. Then I describe a real evolved hardware control system that controls a real robot, and was produced according to the above rationale, demonstrating its benefits.

3 A Millisecond Oscillator from Nanosecond Logic Gates

Abandoning the external clock can reap even more rewards than were mentioned above. A clocked digital system is a finite-state machine, whereas an unclocked (asynchronous) digital system is not. To describe the state of an unclocked circuit, the temporal relationships between its parts must be included. These are continuously variable analogue quantities, so the machine is not finite-state. This theoretical point gives a clue to a practical advantage: in an unclocked digital system, it is possible to perform analogue operations using the time dimension, even when the logic gates assume only binary values (see for example, the *pulse stream* technique [21, 22]).

In the previous section, I argued that when producing circuits by evolution rather than design, the use of a clock is often an *unnecessary* limitation on the way in which the natural dynamics of the components can be used to mediate robot behaviour. This is not always the case — electronic components usually operate on time-scales much smaller than would be useful to a robot; unless the system can evolve such that the overall behaviour of the components (when integrated into the sensorimotor feedback loop of the robot) is much slower than the behaviour of individual components, then a clock (perhaps of evolvable frequency) will be required to give control over the time-scales. (Sometimes, the use of a clock can *expand* the useful dynamics possible from the evolving circuit.)

A simulation has been performed to investigate whether a genetic algorithm can evolve a recurrent asynchronous network of high speed logic gates to produce behaviour on a time-scale that would be useful to a robot. The number of logic nodes available was fixed at 100, and the genotype determined which of the boolean functions of Table 1(a) was instantiated by each node (the nodes were analogous to the reconfigurable logic blocks of an FPGA), and how the nodes were connected (an input could be connected to the output of any node, without restriction). The linear bit-string genotype consisted of 101 segments (numbered 0..100 from left to right), each of which directly coded for the function of a node, and the sources of its inputs, as shown in Table 1(b). (Node 0 was a special "ground" node, the output of which was always clamped at logic zero.) This encoding is based on that used in [2]. The source of each input was specified by counting forwards/backwards along the genotype (according to the 'Direction' bit) a certain number of segments (given by the 'Length' field), either starting from one end of the string, or starting from the current segment (dictated by the 'Addressing Mode' bit). When counting along the genotype, if one end was reached, then counting continued from the other.

Name	Symbol
BUFFER	
NOT	
AND	
OR	
XOR	
NAND	
NOR	
NOT-XOR	

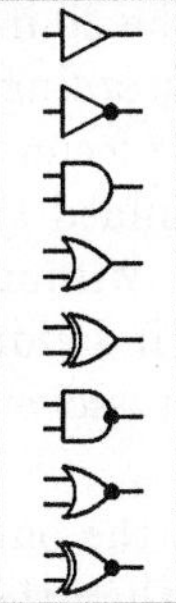

(a)

BITS	MEANING
0-4	Junk
5-7	Node Function
	POINTER TO FIRST INPUT
8	Direction
9	Addressing Mode
10-15	Length
	POINTER TO SECOND INPUT
16	Direction
17	Addressing Mode
18-23	Length

(b)

Table 1. (a) Node functions, (b) Genotype segment for one node.

At the start of the experiment, each node was assigned a real-valued propagation delay, selected uniformly randomly from the range 1.0 to 5.0 nanoseconds, and held to double precision accuracy. These delays were to be the input-output delays of the nodes during the entire experiment, no matter which functions the nodes performed. There were no delays on the interconnections. To commence a simulation of a network's behaviour, all of the outputs were set to logic zero. From that moment onwards, a standard asynchronous event-based logic simulation was performed [19], with real-valued time being held to double precision accuracy. An equivalent time-slicing simulation would have had a time-slice of 10^{-24} seconds, so the underlying synchrony of the simulating computer was only manifest at a time-scale 15 orders of magnitude smaller than the node delays, allowing the *asynchronous* dynamics of the network to be seen in the simulation. A low-pass filter mechanism meant that pulses shorter than 0.5ns never happened anywhere in the network.

The objective was for node number 100 to produce a square wave oscillation of 1kHz, which means alternately spending 0.5×10^{-3} seconds at logic '1' and at logic '0'. If k logic transitions were observed on the output of node 100 during the simulation, with the nth transition occurring at time t_n seconds, then the average error in the time spent at each level was calculated as :

$$\text{average error} = \frac{1}{k-1} \sum_{n=2}^{k} \left| (t_n - t_{n-1}) - 0.5 \times 10^{-3} \right| \qquad (1)$$

For the purpose of this equation, transitions were also assumed to occur at the very beginning and end of the trial, which lasted for 10ms (but took very much more wall-clock time to simulate). The fitness was simply the reciprocal of the average error. Networks that oscillated far too quickly or far too slowly (or not at all) had their evaluations aborted after less time than this, as soon as a good estimate of their fitness had been formed. The genetic algorithm used was a conventional generational one [5], but used elitism and linear rank-based selection. At each breeding cycle, the 5 least fit of the 30 individuals were killed off, and the 25 remaining individuals were ranked according to fitness, the fittest receiving a fecundity rating of 20.0, and the least fit a fecundity of 1.0. The linear function of rank defined by these end points determined the fecundity of those in-between. The fittest individual was copied once without mutation into the next generation, which was then filled by selecting individuals with probability proportional to their fecundity, with single-point crossover probability 0.7 and mutation rate 6.0×10^{-4} per bit.

The experiment succeeded. Figure 2 shows that the output after 40 generations was approximately $4\frac{1}{2}$ thousand times slower than the best of the random initial population, and was six orders of magnitude slower than the propagation delays of the nodes. In fact, fitness was still rising at generation 40 when the experiment was stopped. The final circuit (Figure 3) *was* exploiting the characteristics of its "implementation" — if the propagation delays were changed, it reverted to behaviour similar to that at the first generation. A spike-train, rather than the desired square-wave was produced, allowing the phenomenon of spike trains of slightly different frequencies beating together to produce a much lower frequency (but it is difficult to gain the massive reduction in frequency required and yet produce a regular output). The entire network contributes to the behaviour, and meaningful sub-networks could not be identified.

This simulation, although quite an unrealistic model of the evolution of a real FPGA configuration, has shown how evolution can assemble high speed components to produce behaviour on a time-scale that approaches that useful to a robot. It exploits the characteristics of the implementation, and does not require the imposition of spatial or temporal constraints such as modularisation or clocking. The style of solution adopted (the beating of spike trains) is an

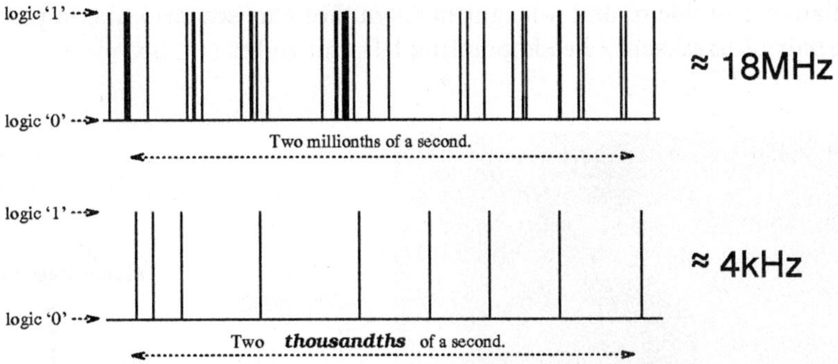

Fig. 2. Output of the evolving oscillator. (Top) Best of the initial random population of 30 individuals, (Bottom) best of generation 40. Note the different time axes. A visible line is drawn for every output spike, and in the lower picture each line represents a single spike.

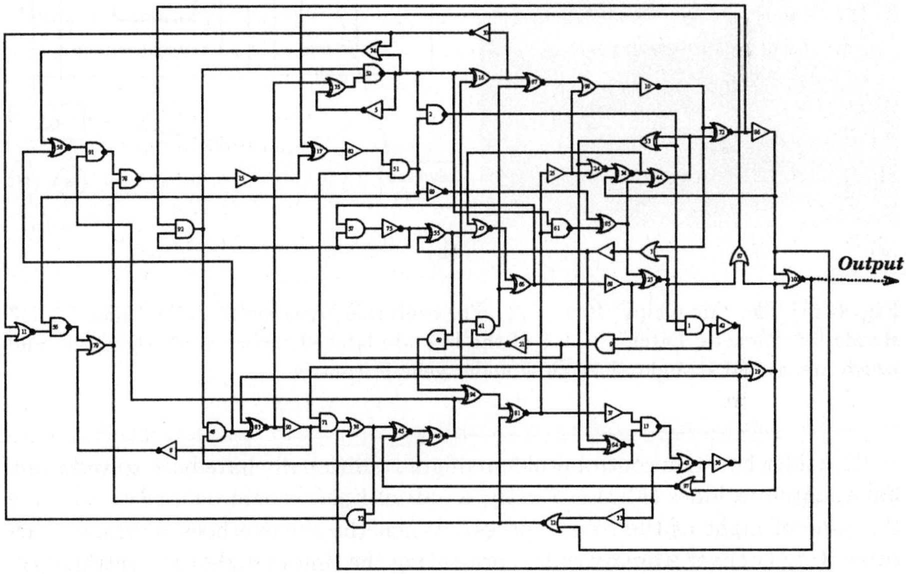

Fig. 3. The evolved 4kHz oscillator (unconnected gates removed, leaving the 68 shown).

analogue operation over the time axis, and would have been more difficult in a discrete time system.

4 A Real Evolved Hardware Robot Controller

In this experiment, a real hardware robot control system was evolved for wall-avoidance behaviour in an empty 2.9m×4.2m rectangular arena, using sonar time-of-flight sensing. The two-wheeled robot ("Mr Chips," Figure 4(a)) has a diameter of 46cm, and a height of 63cm. For this scenario, its only sensors were a pair of fixed sonar heads pointing left and right.

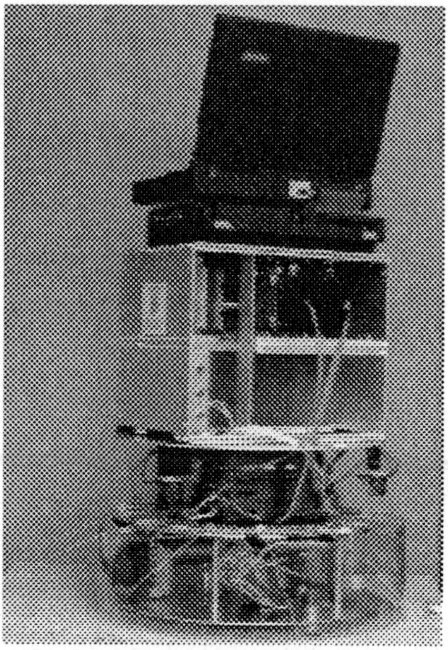

(a)

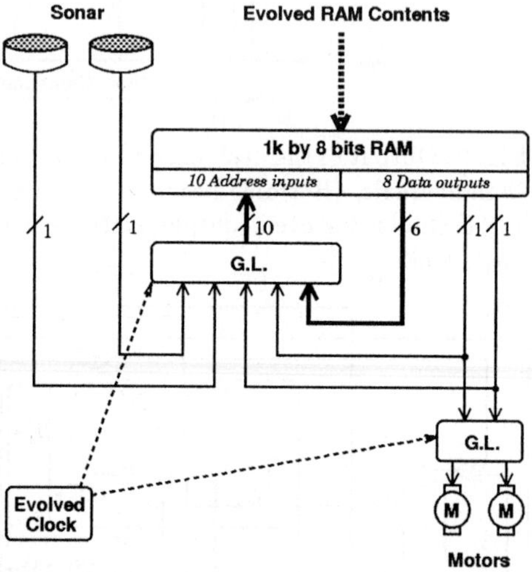

(b)

Fig. 4. (a) The "Mr Chips" Robot. (b) The evolvable Dynamic State Machine. "G.L." stands for "Genetic Latch": which of the bits are latched according to the clock, and which are passed straight through is under genetic control.

Consider how this control problem might traditionally have been solved using finite-state machines (FSMs). Firstly, a pair of FSMs would be used to measure the time of flight of the sonar "pings." When the sonar echoes returned, each timer would deliver a binary code representing the time of flight to a central "control" FSM, which would compute a motor response on the basis of its current internal state and these inputs. The control FSM would then deliver a binary code representing desired motor speed to each of a pair of pulse-width-modulation FSMs, which would drive the d.c. motors with the appropriate waveforms. Notice that there is a very strict sensory/control/motor functional decomposition inherent in this architecture, and the system is a computational one, in which the control FSM deals in binary codes and is totally divorced from the dynamics of the sensorimotor systems and the environment. It would be possible to evolve the control FSM in hardware, but there would be no benefit: *exactly* the same behaviour would be obtained from an implementation of the FSM in software. A hardware implementation would use a clocked register to hold the current state, and a random-access memory (RAM) chip with evolvable contents would hold the next-state and output variables corresponding to each present-state and input combination (this is the well known "direct-addressed ROM" implementation of an FSM [3]).

The hardware control system used in this experiment is superficially similar to an FSM, but fundamentally different. Call it a Dynamic State Machine (DSM). As its input, it directly takes the echo signals from the sonars. There is one wire from each sonar, on which pulses arrive: the lengths of these pulses are equal to the time of flight for that sonar. (The sonars fire, and the pulses begin, simultaneously, with a pulse repetition rate of 5Hz. The sonar firing cycle is completely asynchronous to the DSM.) The output of the DSM goes directly to the motor drivers; if the motors are to go at an intermediate speed, then the DSM must pulse them itself. Like the FSM, the DSM implementation is centred around an RAM chip with evolvable contents, but which of the input, state and output variables are clocked, and which are free running (asynchronous) is genetically determined. For those that *are* clocked, the clock frequency is also genetically determined. The full arrangement is shown in Figure 4(b).

The sensory/control/motor functional decomposition has now been removed, and the control system is intimately linked to the dynamics of the sensorimotor signals and the environment, with time now able to play an important role throughout the system. The possibility of mixing asynchronous state variables with state variables being clocked at an evolved frequency endows the system with a rich range of possible dynamical behaviour: its actions can immediately be influenced by the input signals, but at the same time it is able to keep a trace of previous stimuli and actions over a time-scale that is under evolutionary control. It is able to exploit special-purpose tight sensorimotor couplings because the temporal signals can quickly flow through the system, being influenced by, and in turn perturbing, the DSM on their way.

The presence of asynchronous state variables means that this is not a finite-state machine (their continuous-valued temporal relationships need to be included in a description of the machine's state). It would not be possible to simulate this machine in software, because the effects of the asynchronous variables and their interaction with the clocked ones depend upon the characteristics of the hardware: meta-stability and glitches will be rife, and the behaviour will depend upon physical properties of the implementation, such as propagation delays and meta-stability constants. Similarly, a designer would only be able to work within a small subset of the possible DSM configurations — the more predictable ones.

For the simple wall-avoidance behaviour, only the two state variables that also go to the motors were used — the others were disabled, and can be introduced incrementally as the difficulty of the task is increased. The genetic algorithm was the same as that described in the previous section, with the contents of the RAM (only 32 bits required for the machine with two state variables), the period of the clock (16 bits, giving a clock frequency from around 2Hz to several kHz) and the clocked/unclocked condition of each variable all being directly encoded onto the linear bit-string genotype. The population size was 30, probability of crossover 0.7, and the mutation rate was set to be approximately 1 bit per string. (It can be shown that this small DSM is statistically likely to visit all of its possible states: DSMs with more state variables are likely to visit a smaller fraction of their possible states, and the mutation rate needs to be higher, because many mutations will have no immediate phenotypic effect.) If the distance of the robot from the centre of the room in the x and y directions at time t was $c_x(t)$ and $c_y(t)$, then after an evaluation for T seconds, the robot's fitness was a discrete approximation to the integral:

$$\text{fitness} = \frac{1}{T}\int_0^T \left(e^{-k_x c_x(t)^2} + e^{-k_y c_y(t)^2} - s(t)\right) \mathrm{d}t \qquad (2)$$

k_x and k_y were chosen such that their respective Gaussian terms fell from their maximum values of 1.0 (when the robot was at the centre of the room) to a minimum of 0.1 when the robot was actually touching a wall in their respective directions. The function $s(t)$ has the value 1 when the robot is stationary, otherwise it is 0: this term is to encourage the robot always to keep moving. Each individual was evaluated for four trials of 30 seconds each, starting with different positions and orientations, the worst of the four scores being taken as its fitness [6]. For the final few generations, the evaluations were extended to 90 seconds, to find controllers that were not only good at moving away from walls, but also *staying* away from them.

For convenience, evolution took place with the robot in a kind of "virtual reality." The real evolving hardware controlled the real motors, but the wheels were just spinning in the air. The wheels' angular velocities were measured, and used by a real time simulation of the motor characteristics and robot dynamics to calculate how the robot would move. The sonar echo signals were then artificially synthesised and supplied in real time to the hardware DSM. Realistic levels of noise were included in the sensor and motor models, both of which were constructed by fitting curves to experimental measurements, including a probabilistic model for specular sonar reflections.

Figure 5 shows the excellent performance which was attained after 35 generations, with a good transfer from the virtual environment to the real world. The robot is drawn to scale at its starting position, with its initial heading indicated by the arrow; thereafter only the trajectory of the centre of the robot is drawn. The bottom-right picture is a photograph of behaviour in the real world, taken by double-exposing a picture of the robot at its starting position, with a long exposure of a light fixed on top of the robot, moving in the darkened arena. If started repeatedly from the same position in the real world, the robot follows a different trajectory each time (occasionally *very* different), because of real-world noise. The robot displays the same qualitative range of behaviours in the virtual world, and the bottom pictures of Figure 5 were deliberately chosen to illustrate this.

When it is remembered that the DSM receives the raw echo signals from the sonars and directly drives the motors (one of which happens to be more powerful than the other), with only two internal state variables, then this performance

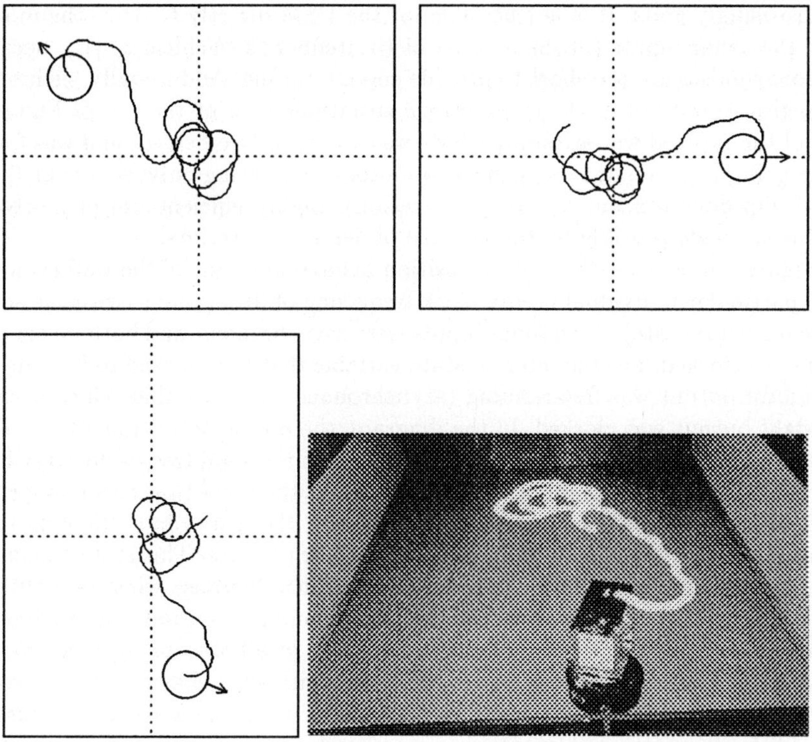

Fig. 5. Wall avoidance in virtual reality and (bottom right) in the real world, after 35 generations. The top pictures are of 90 seconds of behaviour, the bottom ones of 60.

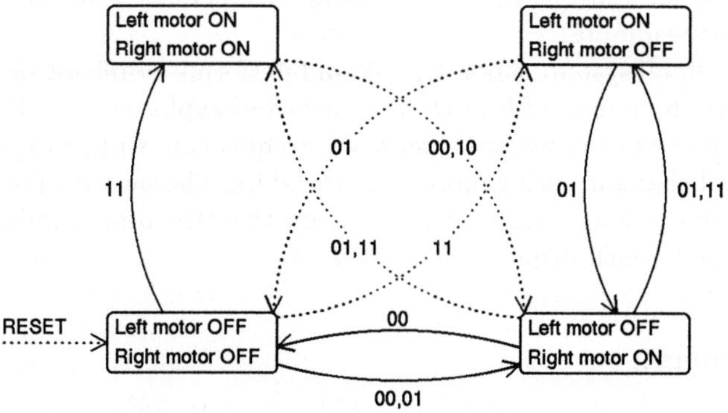

Fig. 6. A representation of one of the wall-avoiding DSMs. Asynchronous transitions are shown *dotted*, and synchronous transitions *solid*. The transitions are labelled with *(left, right)* sonar input combinations, and those causing no change of state are not shown. There is more to the behaviour than is seen immediately in this state-transition diagram, because it is not entirely a discrete-time system, and its dynamics are tightly coupled to those of the sonars and the rest of the environment.

in surprisingly good. It is not possible for the DSM directly to drive the motors from the sonar inputs (in the manner of Braitenberg's "Vehicle 2" [1]), because the sonar pulses are too short to provide enough torque. Additionally, such naïve strategies would fail in the symmetrical situations seen at the top of Figure 5. One of the evolved wall-avoiding DSMs was analysed (see below), and was found to be going from sonar echo signals to motor pulses using only 32 bits of RAM and 3 flip-flops (excluding clock generation): highly efficient use of hardware resources, made possible by the absence of design constraints.

Figure 6 illustrates the state-transition behaviour of one of the wall avoiders. This particular individual used a clock frequency of 9Hz (about twice the sonar pulse repetition rate). Both sonar inputs were asynchronous, and both motor outputs were clocked, but the internal state variable that was clocked to become the left motor output was free-running (asynchronous), whereas that which became the right output was clocked. In the diagram, the dotted state transitions occur as soon as their input combination is present, but the solid transitions only happen when their input combinations are present at the same time as a rising clock edge. Since both motor outputs are synchronous, the state can be thought of as being sampled by the clock to become the motor outputs. This state-transition representation is misleadingly simple in appearance, because when this DSM is coupled to the input waveforms from the sonars and its environment, its dynamics are subtle, and the strategy being used is not at all obvious. It is possible to convince oneself that the diagram is consistent with the behaviour, but it would have been very difficult to predict the behaviour from the diagram, because of the rich feedback through the environment and sensorimotor systems on which this machine seems to rely. The behaviour even involves a stochastic component, arising from the probabilities of the asynchronous echo inputs being present in certain combinations at the clocking instant, and the probability of the machine being in a certain state at that same instant (remember that one of the state variables is free-running).

Even this small system is non-trivial, and performs a difficult task with minimal resources, by means of its rich dynamics and exploitation of the real hardware. Only time will tell whether the DSM architecture will be capable of more sophisticated behaviour, using more state variables. The success of this particular architecture in the long term is less important than the powerful demonstration of the principles which inspired it, given here.

5 Conclusion

This paper has been a manifesto for an approach to evolving hardware for robot control. This approach aims fully to exploit the potential power of reconfigurable electronic hardware by relaxing the constraints on the spatial and temporal organisation of the system that are necessary for a human designer. The resulting systems may seem extraordinary to engineers at first, but experimental evidence has been presented to corroborate the claim that this may be a route to electronic control systems of unprecedented power and efficiency. A particular architecture

— the Dynamic State Machine — has been proposed as a reconfigurable system that is easier to build than the more sophisticated alternatives based around Field Programmable Gate Array (FPGA) technology, but yet contains the essential ingredients. Its successful use in the first reported evolved hardware control system for a real robot demonstrates the *viability* of the new framework, but the long term goal must be to evolve systems as hardware that cannot be made in any other way — it could be that this is possible even with currently available FPGAs.

6 Acknowledgements

This research is funded by a D.Phil. scholarship from the School of Cognitive and Computing Sciences, for which I am very grateful. Special thanks are also due to Phil Husbands, Dave Cliff and Inman Harvey for their kind, expert and tireless help.

References

1. Valentino Braitenberg. *Vehicles : Experiments in Synthetic Psychology.* MIT Press, 1984.
2. Dave Cliff, Inman Harvey, and Phil Husbands. Explorations in evolutionary robotics. *Adaptive Behaviour*, 2(1):73–110, 1993.
3. David J. Comer. *Digital Logic & State Machine Design.* Holt, Rinehart and Winston, 1984.
4. Hugo de Garis. Evolvable hardware: Genetic programming of a Darwin Machine. In C.R. Reeves R.F. Albrecht and N.C. Steele, editors, *Artificial Neural Nets and Genetic Algorithms - Proceedings of the International Conference in Innsbruck, Austria*, pages 441–449. Springer-Verlag, 1993.
5. David E. Goldberg. *Genetic Algorithms in Search, Optimisation & Machine Learning.* Addison Wesley, 1989.
6. I. Harvey, P. Husbands, and D. Cliff. Genetic convergence in a species of evolved robot control architectures. CSRP 267, School of Cognitive and Computing Sciences, University of Sussex, 1993.
7. Inman Harvey, Phil Husbands, and Dave Cliff. Seeing the light : Artificial evolution, real vision. In Dave Cliff, Philip Husbands, Jean-Arcady Meyer, and Stewart W. Wilson, editors, *From animals to animats 3: Proceedings of the third international conference on simulation of adaptive behaviour*, pages 392–401. MIT Press, 1994.
8. Hitoshi Hemmi, Jun'ichi Mizoguchi, and Katsunori Shimohara. Development and evolution of hardware behaviours. In Rodney Brooks and Pattie Maes, editors, *Artificial Life IV*, pages 317–376. MIT Press, 1994.
9. Tetsuya Higuchi, Hitoshi Iba, and Bernard Manderick. *Massively Parallel Artificial Intelligence*, chapter "Evolvable Hardware", pages 195–217. MIT Press, 1994. Edited by Hiroaki Kitano.
10. Tetsuya Higuchi, Tatsuya Niwa, Toshio Tanaka, Hitoshi Iba, Hugo de Garis, and Tatsumi Furuya. Evolving hardware with genetic learning: A first step towards building a Darwin Machine. In *Proceedings of the 2nd Int. Conf. on the Simulation of Adaptive Behaviour (SAB92).* MIT Press, 1993.

11. J. H. Holland. *Adaptation in Natural and Artificial Systems.* Ann Arbor: University of Michigan Press, 1975.
12. Philip Husbands, Inman Harvey, Dave Cliff, and Geoffrey Miller. The use of genetic algorithms for development of sensorimotor control systems. In P. Gaussier and J-D. Nicoud, editors, *From Perception to Action Conference*, pages 110–121. IEEE Computer Society Press, 1994.
13. Nick Jakobi, Phil Husbands, and Inman Harvey. Noise and the reality gap: The use of simulation in evolutionary robotics. In *To appear: Proceedings of the 3rd European Conference on Artificial Life (ECAL95)*, Granada, June 4-6 1995. Springer-Verlag.
14. D. Mange. Wetware as a bridge between computer engineering and biology. In *Proceedings of the 2nd European Conference on Artificial Life (ECAL93)*, pages 658–667, Brussels, May 24-26 1993.
15. Daniel Mange and André Stauffer. *Artificial Life and Virtual Reality*, chapter "Introduction to Embryonics: Towards new self-repairing and self-reproducing hardware based on biological-like properties", pages 61–72. John Wiley, Chichester, England, 1994.
16. P. Marchal, C. Piguet, D. Mange, A. Stauffer, and S. Durand. Achieving von Neumann's dream: Artificial life on silicon. In *Proc. of the IEEE International Conference on Neural Networks, icNN'94*, volume IV, pages 2321–2326, 1994.
17. P. Marchal, C. Piguet, D. Mange, A. Stauffer, and S. Durand. Embryological development on silicon. In Rodney Brooks and Pattie Maes, editors, *Artificial Life IV*, pages 365–366. MIT Press, 1994.
18. Carver A. Mead. *Analog VLSI and Neural Systems.* Addison Wesley, 1989.
19. Alexander Miczo. *Digital Logic Testing and Simulation.* Wiley New York, 1987.
20. Jun'ichi Mizoguchi, Hitoshi Hemmi, and Katsunori Shimohara. Production genetic algorithms for automated hardware design through an evolutionary process. In *IEEE Conference on Evolutionary Computation*, 1994.
21. A. F. Murray et al. Pulsed silicon neural networks - following the biological leader. In Ramacher and Rückert, editors, *VLSI Design of Neural Networks*, pages 103–123. Kluwer Academic Publishers, 1991.
22. Alan F. Murray. Analogue neural VLSI: Issues, trends and pulses. *Artificial Neural Networks*, (2):35–43, 1992.
23. Stefano Nolfi, Orazio Miglino, and Domenico Parisi. Phenotypic plasticity in evolving neural networks. In P. Gaussier and J-D. Nicoud, editors, *From Perception to Action Conference*, pages 146–157. IEEE Computer Society Press, 1994.
24. David P.M. Northmore and John G. Elias. Evolving synaptic connections for a silicon neuromorph. In *Proc of the 1st IEEE Conference on Evolutionary Computation, IEEE World Congress on Computational Intelligence*, volume 2, pages 753–758. IEEE, New York, 1994.
25. Trevor A. York. Survey of field programmable logic devices. *Microprocessors and Microsystems*, 17(7):371–381, 1993.

Classification as Sensory-Motor Coordination

A Case Study on Autonomous Agents

Christian Scheier and Rolf Pfeifer

AILab, Computer Science Departmement
University of Zurich-Irchel
Winterthurerstrasse 190, CH - 8057 Zurich, Switzerland
E-mail: scheier@ifi.unizh.ch

Abstract. In psychology classification is studied as a separate cognitive capacity. In the field of autonomous agents the robots are equipped with perceptual mechanisms for classifying objects in the environment, either by preprogramming or by some sorts of learning mechanisms. One of the well-known hard and fundamental problems is the one of perceptual aliasing, i.e. that the sensory stimulation caused by one and the same object varies enormously depending on distance from object, orientation, lighting conditions, etc. Efforts to solve this problem, say in classical computer vision, have only had limited success. In this paper we argue that classification cannot be viewed as a separate perceptual capacity of an agent but should be seen as a sensory-motor coordination which comes about through a self-organizing process. This implies that the whole organism is involved, not only sensors and neural circuitry. In this perspective, "action selection" becomes an integral part of classification. These ideas are illustrated with a case study of a robot that learns to distinguish between graspable and non-graspable pegs

Keywords: Classification - Sensory-Motor Coordination - Autonomous Agents

1 Introduction

If an agent is to function in the real world it must be able to classify objects in its environment: some things are eatable, others not, the nest is different from the rest of the world, etc. Some aspects of this capability are innate such as the taste of particular foods. But the location and perhaps the shape of these food items might vary strongly and cannot be predetermined genetically. Rather these distinctions have to be learned. The learning should be self-organizing, as has been argued extensively in the literature (e.g.[8],[13]).

In this paper we have applied this idea to a robot that learns to distinguish between objects it should collect and other things. Classification, as it is studied in cognitive psychology, but also in computer vision and robotics, is normally treated as an information processing question: the sensors receive a particular input which is processed and mapped onto a kind of internal representation. Often neural networks are used for this purpose. The problem encountered is that normally an enormous number of highly different sensory patterns should map onto the same representation. This problem has also been referred to as the problem of *perceptual aliasing* ([18]). Efforts to deal with this issue have only been successful to a limited extent. Recently, there have been more successful attempts to come to grips with the problem by incorporating motion as in "animate vision" (e.g.[1]).

The fundamental problem with the classical approach seems to be the isolation of a perceptual (sub-)system which is responsible for classifying objects in the environment. Recent studies in developmental psychology suggest that perception is not an information process which takes place on the sensory or input side, but rather a *sensory-motor coordination* (e.g.[2],[5],[14]). This idea has already been pointed out by [6] and we have argued elsewhere along similar lines ([12]). Viewing perception and classification as a sensory-motor coordination sheds new light on this difficult problem. Because it takes the agent's own actions into account some of the problems become much simpler.

Before turning to the case study one additional point must be discussed. Agents in the real world have always several things to do, like eating, drinking, grooming, or in the case of robots, they have to perform a task while avoiding obstacles and maintaining their battery charge. At least some of these tasks will be incompatible. For example, moving forward and moving backward are incompatible. A difficult problem has been to decide when an agent should do what. In the autonomous agents literature this issue has been referred to "action selection". It will be argued that classification cannot be separated out from other processes; this is why it is intrinsically linked to as "action selection". We will also discuss why this term is in fact misleading.

2 Learning to classify: a case study

2.1 The basic set-up

In this case study, an agent has to collect small pegs. Therefore it has to learn the distinction between graspable and non-graspable pegs. There is an environment with pegs of different sizes (see Figure 1). We have used the robot $Khepera^{TM}$. It is equipped with eight IR sensors, six in the front and two in the back. It has two wheels which are individually driven by DC motors. The "gripper" consists of a bent wire attached to the back of the robot, forming a kind of loop. With this "gripper" the agent can grasp certain pegs but not others, some are too big and do not fit into the wire loop.

Pegs inside the wire loop (the shaded object in Figure 1) can be detected by one of the rear IR sensors. The agent should collect the ones it can grasp. The simplest strategy is to try and grasp every object it encounters, and if the grasping is successful the agent brings the peg home, if not it moves on. But this would be highly inefficient. In order to make the collection process more efficient the distinction between graspable and non-graspable objects should be based on information which is more readily available. This information comes from both its sensors and from information stored about its own actions.

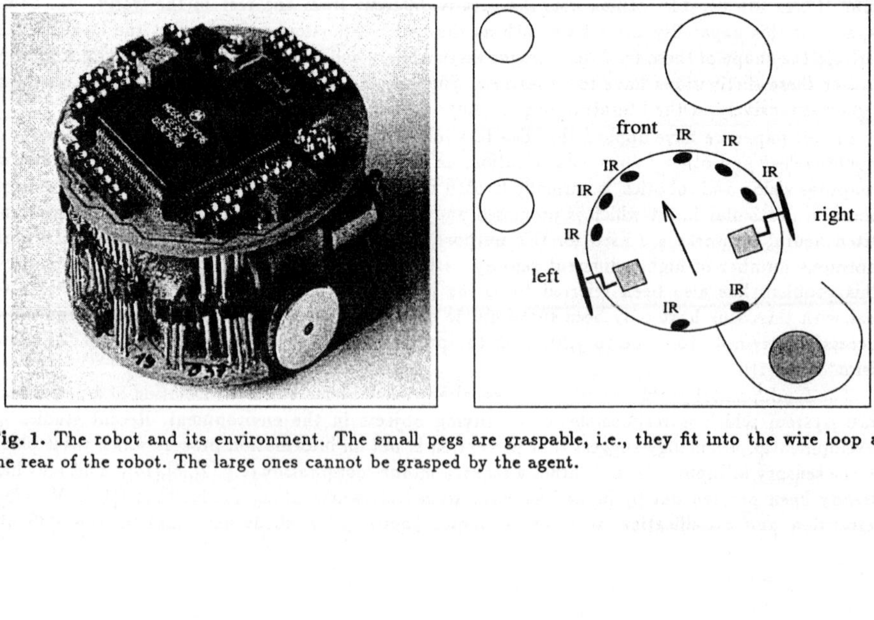

Fig. 1. The robot and its environment. The small pegs are graspable, i.e., they fit into the wire loop at the rear of the robot. The large ones cannot be grasped by the agent.

2.2 Sensory-motor coordination

As pointed out earlier, object recognition is normally seen as a process which happens on the sensory side. This leads to the problem of normalization of the sensory inputs. By contrast, we can view perception as a process which involves action as well. Perception then becomes a matter of sensory-coordination rather than information processing alone. We can exploit the fact that the agent is able to move around. Instead of looking at a particular (fixed) sensory pattern we let the agent move along an object. This is a reflex action: whenever it gets close to an object it will move along it at a fixed distance. Graspable objects have a particular size. If they are too large, they do not fit into the wire loop, if they are too small, they do not sufficiently stimulate the sensors.

The "retinal images" of a graspable object, i.e. the sensory patterns from the IR sensors, will differ widely, depending on the distance to the object and on the agent's direction. What we are now interested in is a sequence of sensory-motor states: on the one hand there is a sensory input, on the other hand there is a particular action the agent is involved in. It is this sequence that forms the basis of the perceptual process. This sequence will, on avarage, be within clear limits (given a certain statistical variation) for graspable objects. In this way we can use - in a sense - the agent's action to "normalize" the sensory stimulation.

2.3 The control architecture

In the real world, agents always have to do several things and at least some of them will be incompatible. To decide what the agent should do in a particular situation is one of the important functions of a control architecture. In the literature this problem has been called "action selection". The term is inappropriate because it introduces a particular bias on how the issue should be tackled (see below). A comprehensive review of "action selection" is given in [17].

There is a fundamental problem with most approaches. They explicitly or implicitly rely on the assumption that what is expressed behaviorally has an internal correspondence. If we see an agent following a wall we suspect a wall-following module (which is sometimes called a "behavioral layer"). But there is a frame-of-reference issue here. Behavior is by definition the result of a system-environment interaction and the internal mechanisms cannot be directly inferred from the behavior alone. Of course, there must be something within the organism which leads to this behavior. But it would be a mistake to infer that, if we want an agent to follow walls, we have to define a special module for wall-following. But this is precisely what is often done.

[15] proposed an alternative scheme which does not suffer from this problem. Instead of having modules (or behavioral layers) there are a number of simple processes which all run in parallel and continuously influence the agent's internal state. This is an idea which goes back to [3]. Let us illustrate the point directly with our case study.

2.4 The basic processes

There are the following processes: `move forward`, `move along object`, `avoid obstacle`, `grasp`, and `turn back`. This sounds very much like the traditional approaches. The main difference is the following. All the processes run all the time. The influence they exert on the behavior of the agent varies depending on the circumstances. So, under certain conditions they will have no influence and, in others, they will constitute the major influence; but they are not on or off. The basic architecture is shown in Figure 2.

There are two values (called quantities) associated with the motors, namely their speed. The processes continuously write onto these quantities, i.e. they add or subtract particular values. Thus, the activation of the speed quantities is a superposition of the output values from all processes:

$$\mathbf{s}(t) = (s_l(t), s_r(t)) = \left(\sum_{i=1}^{N} o_i^l(t), \sum_{i=1}^{N} o_i^r(t) \right) \tag{1}$$

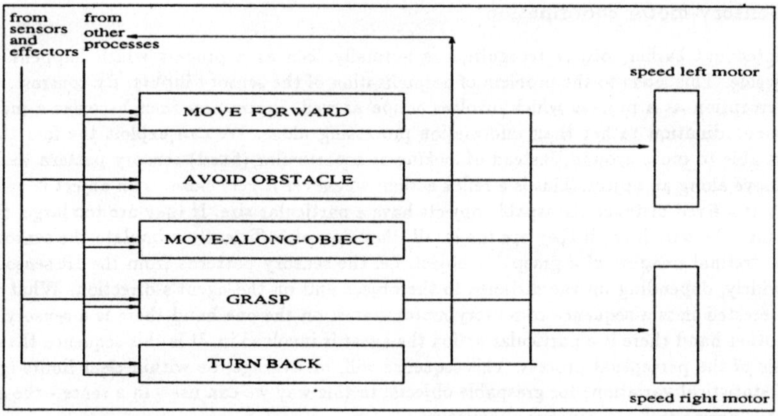

Fig. 2. Overview of the control architecture. The process all run in parallel. They receive input from the sensors and effectors, and from quantities of other processes. They all continuously add and substract to the values of the speed quantities which constitute the "effector system".

where s_l, s_r is the speed of the left and the right motor and o_i^l, o_i^l is the output of the i-th process to the speed quantity of the left and the right motor, respectively and N is the number of processes. What the agent does is determined by these speed quantities since the agent has only the two wheel motors as effectors. The agent does not care where the speed values for the motors come from: they can originate from any process writing onto the speed quantities. The equations for the various processes are as follows:

- `move forward`: This is a default process: if the other processes are not or only slightly activated `move forward` causes the robot to move straight ahead with a default speed:

$$o_{mf}^l(t) = o_{mf}^r(t) = \lambda \tag{2}$$

- `avoid obstacle`: This process increases its influence on the speed quantities as the robot gets close to an obstacle. Its output is the difference between the weighted sum of the signals coming from the three infrared sensors on the left and the right side, respectively (see also Figure 1) :

$$o_{ao}^l(t) = \sum_{i=1}^{3} \phi_i IR_i(t) - \sum_{i=4}^{6} \phi_i IR_i(t) \tag{3}$$

$$o_{ao}^r(t) = -\sum_{i=1}^{3} \phi_{6-i+1} IR_i(t) + \sum_{i=4}^{6} \phi_{6-i+1} IR_i(t) \tag{4}$$

  where ϕ_i is a parameter determining the influence of each corresponding infrared sensor IR_i on the output of the process. Obstacles on the right will lead to large values in the sensors on the right side and thus to an increase of the speed quantity asscociated with the left motor, s_l, and a decrease of the speed quantity s_r associated with the right motor.
- `move along object`: Whenever a sensor on the right or on the left of the agent is on, the agent tries to maintain this condition. This makes the agent move along anything which stimulates a lateral IR sensor. Together with the `avoid obstacle` process it will also start moving along obstacles it encounters head on - there is no need to provide explicitly for this behavior internally. The agent will first turn which leads to activation in one of the lateral IR sensors, which in turn causes the agent to move along the object. This process is implemented as follows:

$$o^l_{mao}(t) = \theta(-IR_1(t) + IR_6(t)), o^r_{mao}(t) = \theta(IR_1(t) - IR_6(t)) \tag{5}$$

where θ is a weighting parameter. Additionally, the `move along object` process continuously adds to a quantity Q which determines the strength of the influence of the grasp process onto the speed quantities (see below).

- `grasp`: This process is a function of the quantity Q. Q determines how strongly the motors are influenced by grasp. We give only the equations here. Because this is a core process we give it seperate treatment below:

$$o^l_g(t) = f(Q(t)) = \alpha Q(t), o^r_g(t) = f(Q(t)) = \beta Q(t) \tag{6}$$

where α and β are constants. Depending on the values of Q, the robot will turn and grasp the object. The quantity Q is computed as follows:

$$Q(t+1) = Q(t) + \Delta Q \tag{7}$$
$$\Delta Q = \varpi(t)\mathbf{w}(t) + \delta O_{mao}(t) - \gamma(t) \tag{8}$$

where $\varpi(t)$ is a 6-dimensional vector of angular velocities, δ is a constant, $\mathbf{w}(t)$ is a weight vector, and $\gamma(t)$ is a decay term (see below). Angular velocities are used to normalize the sensory input, as mentioned above. The key point is that, once the robot is moving along the object, they will be larger for small pegs than for large pegs. Angular velocities are approximated by taking the difference between the values of the wheel encoders WE of the left and the right motor, respectively, at two subsequent time steps:

$$h_i(t) = \frac{\Delta\varphi}{\Delta t} \tag{9}$$
$$= \frac{\varphi(t+1) - \varphi(t)}{(t+1) - t} \tag{10}$$
$$= \left|(WE^l(t+1) - WE^r(t+1)) - (WE^l(t) - WE^r(t))\right| \tag{11}$$
$$\omega_i(t) = f(h_i) \tag{12}$$

where f is the logistic function. The weights $\mathbf{w}(t)$ in (8) are updated according to a Hebbian rule combined with a value signal which provides the reinforcement (see next section for details). In addition to the weighted angular velocities, the weighted output of the `move along object` is added to Q:

$$O_{mao} = \frac{o^l_{mao} - o^r_{mao}}{\zeta} \tag{13}$$

where o^l_{mao}, o^r_{mao} is the output of the `move along object` and ζ is a constant. This leads to an increase in Q as the robot circles around an object. Finally, Q is decayed by $\gamma(t)$ which is computed by taking the normalized sum of the weights $\mathbf{w}(t)$.

- `turn back`: As the robot has "grasped" an object it has to turn back to face in the same direction as before the grasping. We have used the following equation:

$$o^l_{tb} = -\alpha IR_8, o^r_{tb} = -\beta IR_8, \tag{14}$$

where α and β are the same constants as in (6). If the robot is not able to grasp an object after 30 steps, i.e. if it is unsuccessfully trying to grasp a large peg, IR_8 is set to 1 which causes the robot to turn back. In either case Q is reset to zero after each grasping behavior.

2.5 Process fusion, tuning and self-organization

While designing the individual processes is quite straightforward, having the processes cooperate to achieve the desired behaviors is nontrivial: how much should a processes add to the various quantities. Similar to the sensor fusion problem of the classical approach or the behavior fusion problem in behavior based robotics (e.g. [4]) there is a process fusion problem in our approach. For the designer the difficulty is as follows. If one process works fine, say `move forward`, and a new one is added, say `avoid obstacle`, this new process will interfere with the existing ones since it also writes onto the same quantities (the speed values of the motors).

How should the parameters be chosen, how should the tuning be done? The solution is to use principles of self-organization. This implies that the agent must be embedded in a value system which guides the self-organized learning. (e.g. [9]). Since in our case study we are primarily interested in the grasping behavior we have used a simple Hebbian network to tune the output of the grasp process. The network is similar to an architecture we have used on earlier robots (e.g. [13]), but in the present approach it is embedded into a flexible control architecture. This part of the architecture is illustrated in Figure 3.

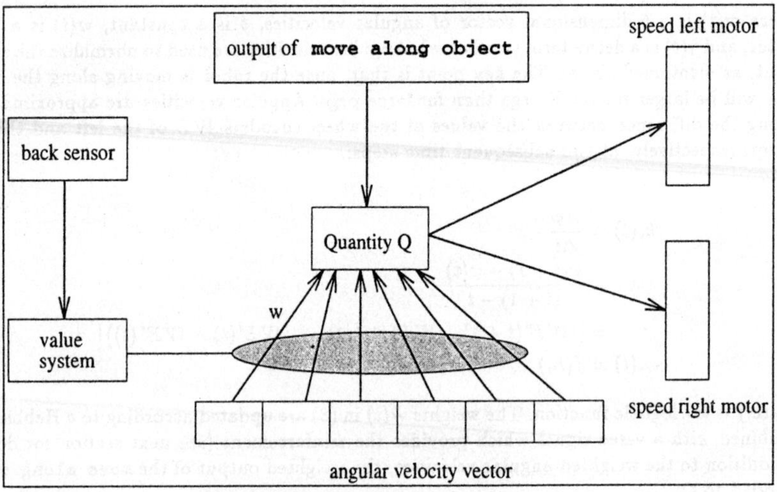

Fig. 3. Tuning of the grasp process. There is a vector of the angular velocities. It represents a particular sequence of sensory-motor states. This vector is characterisitc of the size of a particular object. If the grasp process becomes dominant and if this leads indeed to a grasping event (i.e., the peg being inside the loop) this vector is reinforced. Next time a similar vector is encountered, recognition will be faster.

In essence, the angular velocity vector is associated with the quantity if there is a reinforcement signal. Reinforcement is generated whenever the back sensor is on, i.e. the robot was able to grasp a small peg. In a pure reinforcement learning approach, the agent initially makes more or less random movements. If by chance it happens to do the right thing, i.e. something which leads to a reinforcement, it can learn and this particular sensory-motor coordination will be strengthened. But as is well-known in the reinforcement literature (e.g.[10], [16]), this would take much too long. An initial bias must be introduced: the agent must be constrained to explore the environment in a particular manner, not just at random.

In our example, the agent must be endowed by the designer with an "idea" of when to try and grasp an object. After learning has occurred this initial bias should be tuned such that the agent grasps small but not large pegs. As we mentioned earlier the agent is "normalizing" the sensory stimulation by approaching an object and circling around it. The latter is achieved by the `move along object` process. It continuously adds to the quantity Q . Thus Q only gets bigger if this process is active for a certain period of time. Q is monitored by the `grasp` process. It determines the strength of the influence of the `grasp` process on the speed quantities. As this influence gets high it will eventually cause the agent to make a grasping movement (turning on the spot). If the peg is caught in the wire loop the value system generates a reinforcement signal, which in turn enables the Hebbian learning to occur. The updating of the weights in (8) thus is as follows:

$$w_i(t+1) = w_i(t) + \Delta w_i(t) \tag{15}$$

$$\Delta w_i(t) = \frac{1}{n}\left[r(t)\left(\eta Q(t_b)\omega_i(t_b) - \epsilon\omega_i(t_b)\right)\right] \tag{16}$$

where n is the number of units in the velocity vector, η is the learning rate, $r(t)$ is the reinforcement signal triggered by the value system ($r(t) = IR_8$), t_b is the time step before the robot starts to grasp, $Q(t_b)$ is the quantity Q, ω_i is the value of the i-th element of the velocity vector ϖ, and ϵ is a decay parameter.

3 Results

Experiments were conducted on a flat arena (80cm x 50cm) with walls (8cm height) on each side. Small pegs were of 1.5cm in diameter and 2cm high, large pegs were of 3cm in diameter with the same height (2cm). The shape of both types of pegs was round (see Figure 1). There were 10 trials each lasting until the robot had encountered 50 small and 50 large pegs, respectively. Whenever the robot had tried to grasp a peg and learning had occurred, the peg was removed from the arena. The results of the experiments are depicted in Figures 4, 5 and 6. The trajectories were recorded with a video camera and then hand traced. Figure 4 shows a typical trajectory at the beginning of the learning process. The white square indicates the robot's starting locations and arrows indicate grasping behavior. Large and small filled circles indicate large and small pegs, respectively.

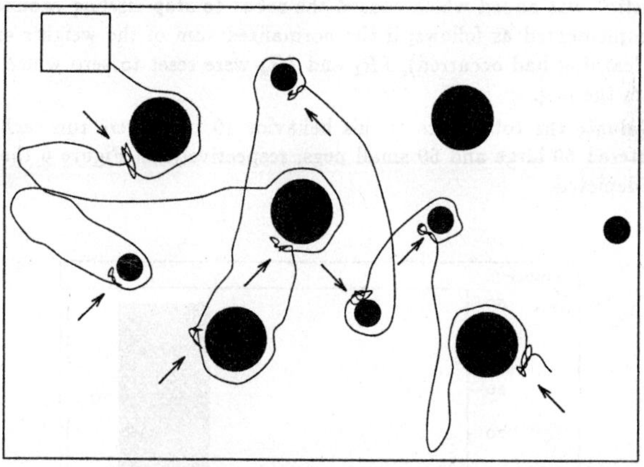

Fig. 4. A typical trajectory of the robot before learning.

The traces show smooth moving forward, avoiding, moving along object, grasping and turning back behavior. As can be seen, the `move along object` and `grasp` behavior are initially triggered by the bias and as a consequence the robot is not able to distinguish between large and small pegs. After a given number of steps of circling around a peg it tries to grasp the peg. As observers we can say that the robot is not able to distinguish between graspable (small) and non-graspable (large) pegs. In Figure 5 a typical trajectory of the robot after it has encountered a large number of pegs is shown. There is a clear distinction between the `move along object` and `grasp` behavior for large and small pegs, respectively.

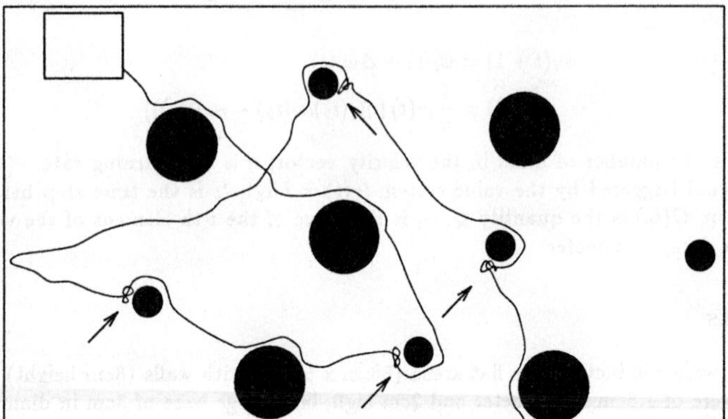

Fig. 5. A typical trajectory of the robot after it has encountered a large number of pegs.

If the robot reaches a small peg, it starts grasping it after a few steps of circling around it. In contrast, if the robot encounters a large object, it starts to circle around it without trying to grasp it. Since with the current setup the robot would circle around large pegs for a large number of steps, a "heuristic" was added which caused the robot to stop circling around large pegs. The "heuristic" was implemented as follows: if the normalized sum of the weights exceeded a certain threshold (i.e., if learning had occurred), IR_1 and IR_4 were reset to zero which caused the robot to move away from the peg.

In order to evaluate the robustness of this behavior 10 trials were run each lasting until the robot had encountered 50 large and 50 small pegs, respectively. In Figure 6 the mean number of grasped pegs are depicted.

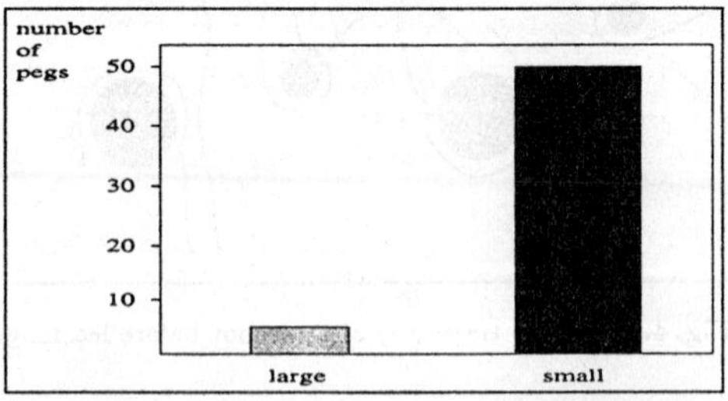

Fig. 6. Number of pegs grasped. Data are mean values over 10 trials. In each trial the robot encountered 50 pegs of each type.

As can be seen, there is a significant difference between the number of small and large pegs the robot grasps. While the robot grasped 100% of the small pegs it encountered it stopped grasping large pegs after learning. Over all trials, 10% of the large pegs were grasped. Large pegs were grasped in the beginning of each trial because of the initial bias mentioned above.

In sum, the results show the emergence of classification in our robot. This classification is based on a learned coordination of sensory data - used to move along an object - and of data taken from the motors, the angular velocities. In other words, robot classifies graspable and non-graspable pegs based on this sensory-motor coordination.

4 Discussion

4.1 Behavior and internal mechanisms

If the agent consistently shows a behavior like the one in Figure 5 we are inclined to say that it has learned to distinguish graspable from non-graspable objects. Note that we, as observers, attribute this capacity to the agent - there is no explicit internal representation of this distinction within the agent. Simply based on its behavior we say it has learned the distinction.

Note also that there is no "action selection" mechanism. What the agent does at a particular point in time is determined by the speed quantities associated with the motors. And they are the result of the values added to these quantities by all processes. So there is no "action" which is somehow selected and this is also why we do not like to use the term "action selection". So, at any point in time it is not possible to say what "action" the agent is performing.

This approach has some interesting consequences for reflexes. We mentioned initially that there are a number of basic reflexes like avoiding obstacles or moving along objects. Reflexes can be considered as part of the value system: the designer decides that they are good for the agent. But in our scheme reflexes are implemented just like any other process. They can therefore be overridden by other processes. In some cases it might be necessary to actually bump into objects (e.g. for pushing them). If the pushing process influences the speed values of the motors more strongly than the avoid process it will in fact bump into the object. The original process is still functioning, but it simply does not effect the behavior of the agent as much as the other processes. Typically, overriding a reflex should be hard, otherwise there would not be much sense in having reflexes. But there is no in principle difference between reflex processes and other processes. The former simply tend to become dominant very quickly under certain circumstances (e.g. an object approaching quickly).

Let us look at the interaction of `move along object` and `grasp` for a moment. If nothing interesting happens in the environment, i.e. not many other processes are active, the agent, as it is moving along an object, will still now and then try to grasp the object even if it does not match the learned sensory-motor coordination. But this is precisely what is required if the agent is to keep its capacity to come up with new classifications in different environments. Above, we argued that classification cannot be separated out as an invidiual module. Training the agent to grasp objects was done in the context of all other processes operating simultaneously.

In the context of a different set of processes it would tune itself differently. This is another example demonstrating why it is not sensible to preprogram a particular module for grasping but that it must be acquired in the system-environment interaction. Thus, classification must be seen within the context of a complete agent. It depends on the value system. Each process serves a particular value which can be explicit or implicit. Only if a process has adaptive value, i.e. relates to one of the system's basic values, it makes sense to have them in the first place. From an observer's perspective we can say that the basic values represent what is good for the agent. It is good for the agent to have pegs inside the wire loop. It does not matter that it knows nothing about it.

4.2 Sensory-motor coordination: a biological analogy

Using Dewey's metaphor of perception as sensory-motor coordination we have derived a control architecture which is based on a sensory-motor coordination rather than on sensory information only. There is an interesting biological analogy, namely retinotopic matching in Drosophila. Using a sophisticated experimental set-up, flies have been trained to recognize certain patterns ([7]). From their experiments the authors conclude: "... for recognizing a memorized visual pattern, flies have to shift the actual image to the same location in the visual field where the image had been presented during the storage process." (Dill et al., 1993, p. 752). In other words, the flies have to move such that they can get the image into exactly the right position on the retina, a process which beautifully illustrates the incorporation of action into a recognition process: there is a clear sensory-motor coordination.

Our agent does a very similar thing. Instead of positioning its body to have a particular sensory image it moves along an object and if this movement (which includes particular sensor readings) corresponds to the learned ones, it will grasp it, and we, as observers, would then say it has classified the object as graspable.

4.3 Limitations and further work

The main limitation of the current approach is the relatively small number of processes. However, we believe that by making each process "self-adaptive" (via intrinsic values and self-organizing processes) we can achieve scalability to many different processes. In a next step we will incorporate a home process which - based on an adaptive light compass ([11]) - will cause the agent to bring grasped pegs to a home base. Moreover, the `move along object` process will be improved in order to avoid the circling around large pegs observed in this study; the "heuristic" should no longer be required. Another important limitation concerns the simplicity of the effector system. At the moment there are only two DC motors with the corresponding speed variables. If a more sophisticated classification is required more complex effectors will be needed in order to have a more interesting sensory-motor coordination.

4.4 Conclusions

In this paper we have demonstrated three main points: First, viewing classification as a sensory-motor coordination rather than a process happening on the sensor-side can lead to significant simplifications. It enables the agent to reliably learn a classification of objects in the environment in a straightforward and simple way. Second the "action selection" problem "drops out" when using a control architecture that is conceived as a dynamical system rather than as a set of individual behaviors. And third, classification comes about via a process of self-organization which is at the same time used to tune the indivividual processes.

Acknowledgements

We are especially grateful to Dimitri Lambrinos and Ralf Salomon for looking through a former draft of this paper. This work is supported by the Swiss National Science foundation, grant number 20-40581.94 to C. S.

References

1. D. Ballard. Animate vision. *Artificial Intelligence*, 48:57–86, 1991.
2. H. Bloch and B. Bertenthal, editors. *Sensory-Motor Coordination and Development in Infancy and Early Childhood.* Kluwer Academic Publishers, 1990.
3. V. Braitenberg. *Vehicles.* Kluwer Academic Publishers, 1984.

4. R. Brooks. A robust layered control system for a mobile robot. *IEEE Journal of Robotics and Automation*, RA-2:14–23, September/October 1986.
5. G. Butterworth. Dynamic approaches to infant perception and action. In Z.B. Smith and E. Thelen, editors, *A Dynamic Systems Approach to Development*, pages 1–25. MIT Press, Massachusetts, 1993.
6. J. Dewey. The reflex arc concept in psychology. *Psychological Review*, (3):357–370, 1896.
7. M. Dill, R. Wolf, and M. Heisenberg. Visual pattern recognition in drosophila involves retinotopic matching. *Nature*, (365):751–753, 1993.
8. G. M. Edelman. *Neural Darwinism.* Basic Books, New York, 1987.
9. G. M. Edelman. *The Remembered Present.* Basic Books, New York, 1987.
10. Leslie Pack Kaelbling. *Learning in Embedded Systems.* PhD thesis, Department of Computer Science, Standford University, 1990.
11. D. Lambrinos. Navigating with an adaptive light compass. In *to appear in: Proceedings of the Third European Conference on Artificial Life ECAL95*, 1995.
12. R. Pfeifer and C. Scheier. From perception to action: the right direction? In *Proceedings "From Perception to Action" Conference*, pages 1–11, Los Alamitos, 1994. IEEE Computer Society Press.
13. R. Pfeifer and P. Verschure. The challenge of autonomous agents: Pitfalls and how to avoid them. In L. Steels and R. Brooks, editors, *The "Artificial Life" Route to "Artificial Intelligence"*. Erlbaum, Hillsdale, NJ, 1994.
14. M.A. Schmuckler. Perception-action coupling in infancy. In J.P. Geert and Z. Savelsbergh, editors, *The Development of Coordination in Infancy.* Elsevier, Amsterdam, 1993.
15. L. Steels. Building agents with autonomous behavior systems. In L. Steels and R. Brooks, editors, *The "Artificial Life" Route to "Artificial Intelligence"*. Erlbaum, Hillsdale, NJ, 1994.
16. S. Thrun. Efficient exploration in reinforcement learning. Technical report, School of Computer Science, Carnegie Mellon University, Pittsburgh, 1992.
17. T. Tyrell. *Computational Mechanisms for Action Selection.* PhD thesis, University of Edingburgh, 1990.
18. S.D. Whitehead and D.H. Ballard. Active pprception and reinforcement learning. In W. Porter and R.J. Mooney, editors, *Machine Learning. Proceedings of the Seventh International Conference*, pages 179–190, 1990.

High-pass filtered positive feedback for decentralized control of cooperation

Holk Cruse, Christian Bartling, Thomas Kindermann

Dept. of Biol. Cybernetics, Fac. of Biology, Univ. of Bielefeld, D- 33501 Bielefeld, FRG

Abstract. In a multilegged walking system, the legs, when in stance mode, have to cooperate to propel and support the body and, at the same time, to avoid unwanted forces across the body. As a simple method to control the joint movement, we propose to use local high-pass filtered positive feedback. This does not only make redundant the determination of equations for coordinate transformation, but is also robust against all kinds of geometrical changes within the mechanical system as, for example, changes in leg segment length, bending of segments, changes in orientation of the rotational axes (by accident or because the suspension is soft by design), or addition of further joints. This simplification is possible because, instead of applying abstract, explicit computation, the physical properties of the world are exploited. It permits extreme decentralization, namely control on the joint level, and therefore allows very fast computation. In order to provide height control of the body, one leg joint is subjected to a negative feedback controller.

1 Introduction

In behavioral and physiological experiments, systems are sometimes described to be based on positive feedback. In some cases this term is used in a way which may be misleading. For example, when a leg is actively moved, and the load it has to carry is experimentally increased, the motor output of the leg may be increased, too. This is sometimes described as assisting reflex or positive feedback [24,25]. However, this need not necessarily be based on a circuit which, in a logical sense, contains positive feedback. The description is, at first sight, only justified on a phenomenological level because the underlying circuit could also be a negative feedback controller or, in other words, a servosystem. In such a system, the actual leg position, due to the increased load, drags behind the desired position, thus giving rise to a greater error signal and a higher motor output. When such a negative feedback controller is involved, the leg would, in the case of any external disturbance, try to move in the direction of the desired value or setpoint. Of course, a system with positive feedback may also exist, and an experimental distinction between both circuits is possible. In contrast to a servocontroller, the movement of the leg would, under positive feedback control, follow every externally imposed movement and continue this movement, possibly with continuous acceleration. Positive feedback can, thus, lead to instability, if there are no mechanisms for limiting the output. Although positive feedback may, therefore, appear not to be an acceptable design principle, it is nevertheless successfully applied in a number of systems. On the molecular level, many autocatalytic systems exist which are, in fact, positive feedback systems (e.g. synthesis of pepsin from pepsinogen).

Interesting examples are also described on a much higher level, namely the behavior of social insects. In ants, for instance, path recruitment and foraging strategies are shown to be based on positive feedback [19]. In several cases, a positive feedback or an assisting reflex is described to occur as reflex reversal, where under special conditions, the usual negative feedback or resisting reflex changes sign (see [1], [3], [21], [20], [27], [30], and [31]). However, as mentioned above and already discussed by Zill (1985), this could also result from a servocontroller with negative feedback if, due to external conditions, the reference input is influenced appropriately.

Below, a problem will be described which, at first sight, appears to require a complex mathematical solution, but could be radically simplified by applying positive feedback. When a multilegged walking system, e. g. a six-legged insect or an artificial walking machine, moves its body forward, all the legs on the ground have to cooperate and move in such a way that no unwanted forces arise across the body. The movement of a leg is produced by movements of the individual joints. Therefore, the actual control task is to determine the movement of the individual joints such that the whole system cooperates adequately. In such walking systems, the calculation of the necessary joint angle changes is a rather nonlinear task, and the situation is even worse if the rotational axes of the joints are not perpendicular. It can be accomplished, however, using the classical methods of kinematics. The task becomes more difficult when we consider walking on soft ground since, in this case, one leg may move at a different speed compared to the other legs. A similar situation occurs when the system walks through curves. In any case, a central controller system seems to be required which determines how the joints have to act in such difficult situations. Even for the simple case of straight walking on a horizontal surface, this computation is only possible, if exact knowledge of the geometry of the system, i.e., the orientation of the joint axes and the lengths of the limbs, is available. If this knowledge is not (or only partly) available because of a limited resolution of sensory systems, or soft suspension (changes due to different loading), or because these values change during life, the task becomes very difficult if not unsolvable. However, as it is easily accomplished even by "simple" insects, there has to be a way not only to solve the problem, but to solve it fast enough so that on-line computation is possible even for the slow biological neuronal systems.

2 Local positive feedback

It was mentioned earlier by Gibson [22] that many values have not to be explicitly calculated, but are already available in the world. This means that, instead of an abstract calculation, one could directly exploit the physical properties of the world. This is very much the case for the problem discussed here. Imagine a multilegged system with unknown arrangements of joints and leg segment lengths standing on irregular ground. Now we activate one of these joints to perform a small movement. The task is to know what active movements should all the other joints perform in order to move the whole body. The answer is already given by physics because the active movement of the one selected joint elicits the exactly corresponding movements in all the other joints. Therefore, each joint has only to measure its own movement and use this signal to control its active movement. This means that positive feedback

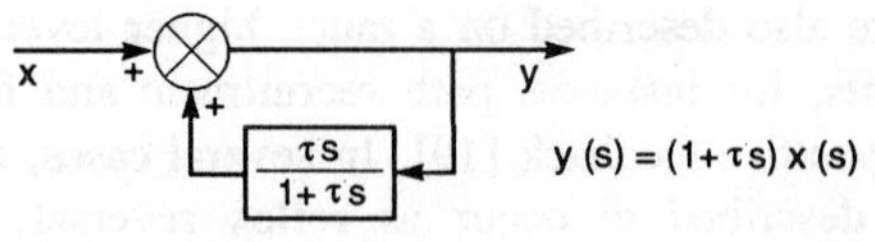

Fig. 1. A network with high-pass filtered positive feedback and its transfer function.

is required at each joint. According to this concept, the appropriate movement of all joints can thus be performed after movement has been started by any means, for example by an arbitrary start impulse in one joint, followed by local positive feedback on the joint control level. This is sufficient for a further continuation of the movement. This extremely simple solution involves three problems, though. One will be dealt with in the next paragraph. The second concerns the question of height control and will be considered in the next chapter. The latter concerns a special aspect of a more general problem, which will be treated in the final part of the Discussion.

The first problem is that simple positive feedback would lead to a continuous increase in the speed of the movement although an about constant speed is required. This problem can be easily solved by introducing a high-pass filter into the feedback loop or, in other words, feeding back not the angular position, but angular velocity. As shown in Fig.1, the transfer function of a system with high-pass filtered positive feedback corresponds to that of the sum of the input and the integrated input value. This means that a short velocity impulse given at the input leads, after the pulse, to a continuous constant velocity output value.

To show the behavior of such a positive feedback controller, we used a simple two-dimensional system. This "creature" consists of a body and two legs with two segments each (Fig.2a). Thus, there are four joints to be controlled. In order to be consistent with the 3D version of an insect leg (see [10]), these joints are called β and γ. The α joint, which is not considered here (see, however, Figs. 4 and 5), would move the leg perpendicular to the drawing plane of Fig.2.

As the physical part of the system is essential, its mechanics have to be simulated first. For this purpose, a special recurrent system is used which was originally proposed as a model for a multijoint arm (MMC net, see [17]). The model used here corresponds to an arm with five segments and five joints, but the positions of the shoulder and endpoint are fixed. Here, they correspond to those points where the two legs are fixed to the ground. These joints cannot be moved actively. Active movement is only possible for the four joints β_1, γ_1, β_2, and γ_2.

When three of the four active joints are given, the geometrical solution is determined. Since we applied positive feedback to all four leg joints (β_1, γ_1, β_2, γ_2), this corresponded to an overdetermined situation, and changing all four joints could lead to a geometrically impossible solution. The difference to a geometrically possible solution is small, however, because the movement starts from a geometrically realistic position, and the error is due only to the linearization of this nonlinear system. To cope with this problem, the simulation of the mechanical system is first constrained to search for a solution with all four joint angles being prescribed by using positive feedback control; but then the MMC system is released to find a geometrically

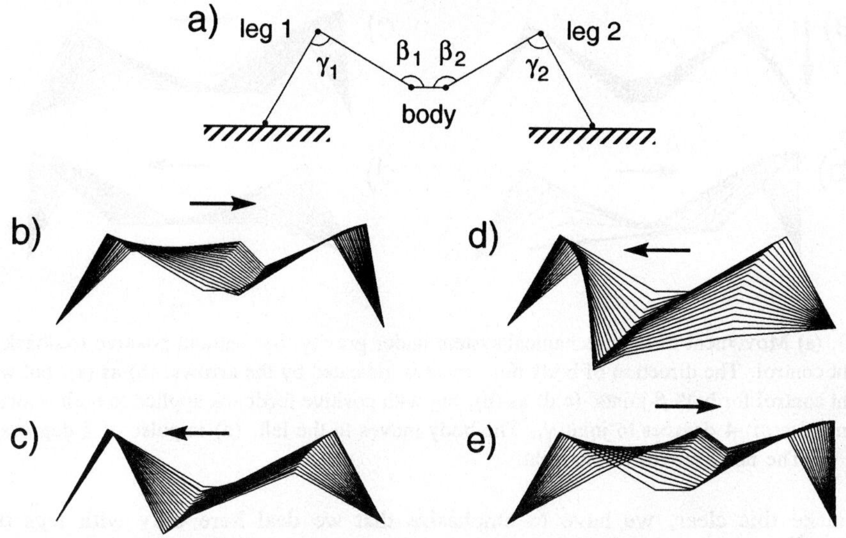

Fig. 2. (a) The mechanical system consisting of body and two legs with two segments each. The active joints are β_1 and γ_1 for the left leg, and β_2 and γ_2 for the right leg. The figure shows the starting position which was used for all simulations. (b-d) Movement under the control of a high-pass filtered positive feedback for each of the four active joints. The direction of body movement is indicated by the arrows. Starting impulse of -20 degrees (b) or of 20 degrees (c) applied to joint β_1. Starting impulse of -4 degrees (d) or of 4 degrees (e) applied to joint γ_1.

possible solution, which turns out very similar to that proposed by the strict positive feedback control. This second part simulates the elastic properties of the muscles. The smaller the incremental step size, i.e., the time resolution, chosen, the smaller the error.

Figs 2b,c show the behavior of the system, using this simulation of the mechanics and the introduction of high-pass filtered positive feedback at each joint, after an initial impulse of -20 degrees (Fig.2b) or 20 degrees (Fig.2c) applied to the β_1 joint. Figs 2d,e show the behavior of the system when the movement of angle γ_1 was initialized with an impulse of -4 degrees (Fig.2d) or 4 degrees (Fig.2e). As can be seen, the legs and the body move at about constant speed. Thus, both legs actively perform a proper movement, although no central controller exists. This independence of a central controller could be made more obvious by the following consideration: the system would immediately show the same performance, if any geometrical value of the system was changed, for example the length of a segment, or if a segment was bent. Similarly, it is possible to apply stronger changes to the mechanical system by using legs with three joints instead of two (not shown), for example.

However, the movements shown in Fig.2 seem unnatural in two ways. First, the body does not maintain a proper distance from the ground; this question will be considered in the following chapter. The other problem is that the legs move until the joints reach their mechanical limits. But this is not a problem for the walking system.

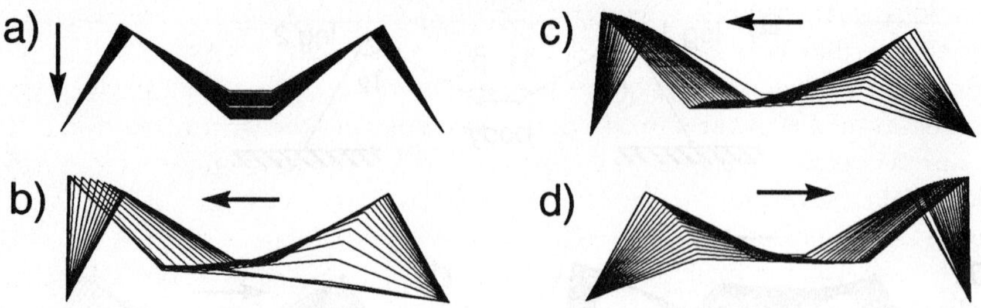

Fig.3. (a) Movement of the mechanical system under gravity, but without positive feedback or height control. The direction of body movement is indicated by the arrows. (b) as (a), but with height control for both β joints. (c,d) as (b), but with positive feedback applied to both γ joints: (c) impulse of -4 degrees to joint γ_1. The body moves to the left. (d) impulse of 4 degrees to joint γ_1. The body moves to the right.

To make this clear, we have to emphasize that we deal here only with legs that perform a stance movement. As was shown earlier (see [[2], [8]), a superior system is required for each leg to decide when stance movement is finished and swing has to be started (and vice versa). Simple solutions for this problem have already been presented, and we will, therefore, not discuss this question here (see Fig.4b and related text). At any rate, this superior system will force the leg to start a swing movement before it reaches the mechanical joint limits.

3 Height control

As mentioned, there is another problem. The simulations presented in Fig.2 did not take into account the effect of gravity. When the body is subject to gravity, this force pulls down the body, and positive feedback would even augment this effect. Thus, a simple positive feedback system would not allow body height to be maintained even without gravity (Fig.2), and would even less allow this under the influence of gravity. It has been shown for insects that body height is maintained during walking by each leg acting as an independent height controller which functions as proportional controller (see [5], [16]). Thus, no central height controller is necessary. For an individual leg, how can the requirement of maintaining a given height via negative feedback and the proposal of local positive feedback be fulfilled at the same time?

To solve that problem we assume that positive feedback is only provided for the γ joints (and in the 3D case, the α joints, see Figs. 4 and 5), but not for the β joints. For the β joint, we assume that it is subject to a negative feedback height controller. Using the knowledge of the segment lengths and the two joint angles β and γ, the height of the leg in a body-centered coordinate system can easily be determined. Comparison with a desired height value provides an error signal which can be used to change the β angle accordingly. The behavior of this system is shown in Fig.3. Fig.3a shows the mechanical system without positive feedback and without height controller, but under the effect of gravity. As expected, the body moves downward. In Fig.3b, height control influencing the two β joints is introduced. As can be seen, the system moves somewhat to the left, but maintains height much better. There is a kind of labile equilibrium because under gravity the mechanical system could be pulled downward to a small extent either to the right or the left side. The movement to the left results from a slight asymmetry in the simulation of the mechanical system. If we now add the positive feedback to the γ joints, and give an impulse of -4 deg to joint γ_1, the body moves to the left (Fig.3c). If the impulse amounts to 4 deg, the corresponding movement to the right can be observed (Fig.3d). Thus, as already shown when using only positive feedback (Fig.2), the legs move with adequate coordination of the joint movement, although no central controller exists. Furthermore, a sufficient control of body position is also possible in this case. We assume that this behavior will also be found when we consider the 3D case and, for this purpose, introduce an additional joint α as described below (Figs. 4 and 5).

In the following simulation, we applied the principle of positive feedback to a six-legged walking machine. Earlier versions of our walking machine (see [12]) contain a leg controller which has a simple selector net consisting of two input units (ground contact, GC, for termination of the swing movement or return stroke and posterior extreme position, PEP, for termination of stance movement or power stroke) and two internal neurons (PS for power stroke, RS for return stroke), which are provided with positive feedback (Fig.4b). This selector net controls the output of two subsystems, the swing net and the stance net. Both provide the angular velocity for the three leg joints α, β, and γ. The orientation of the angles is shown in Fig.4a. The swing net is very simple and contains only three output units (one for each joint), six input units, and one bias unit (see [10]). In order to produce a dynamic, i.e., time-varying output, the application of a recurrent network would, in principle, be appropriate. Therefore, we began by developing the swing net using a recurrent net. However, in this case, too, the physical properties of the world can be exploited. The net already contained recurrent connections via the outer world. Thus, for the internal part of the system, a feedforward net in its simplest form turned out to be sufficient to produce appropriate swing movements. According to the positive feedback control discussed here, we now can propose an extremely simplified version of the stance net. This is shown in Fig.4b. As discussed above, α and γ are controlled by the simple high-pass filtered positive feedback. The β angle is subject to a negative feedback controller which controls the ground distance of the leg as calculated by using the values of the three leg angles. To start the return stroke movement, a 'starting impulse' has to be applied to at least some joints. This is not shown in Fig.4. One possibility is to use the high-pass filtered output of the PS unit of the selector net. Alternatively, its unfiltered output might be multiplied by fixed weights and added to some joints, for example all α joints.

Figs 5a,b show simulations of a six-legged system under gravity. All six legs are standing on the ground. The axis of rotation of the basal joints are not orthogonal to the body coordinate system. (The values of the angles ϕ and ψ (Fig.4a) were given in [9].) According to the principles described above, the six β joints are controlled by negative feedback, whereas the six α joints and the six γ joints are subject to high-pass filtered positive feedback. In Fig.5a, a starting impulse of -3 degrees is applied to the two α joints of the front legs. As can be seen, the body moves forward and all the legs perform appropriate stance movements. In the walking insect, the angle of the

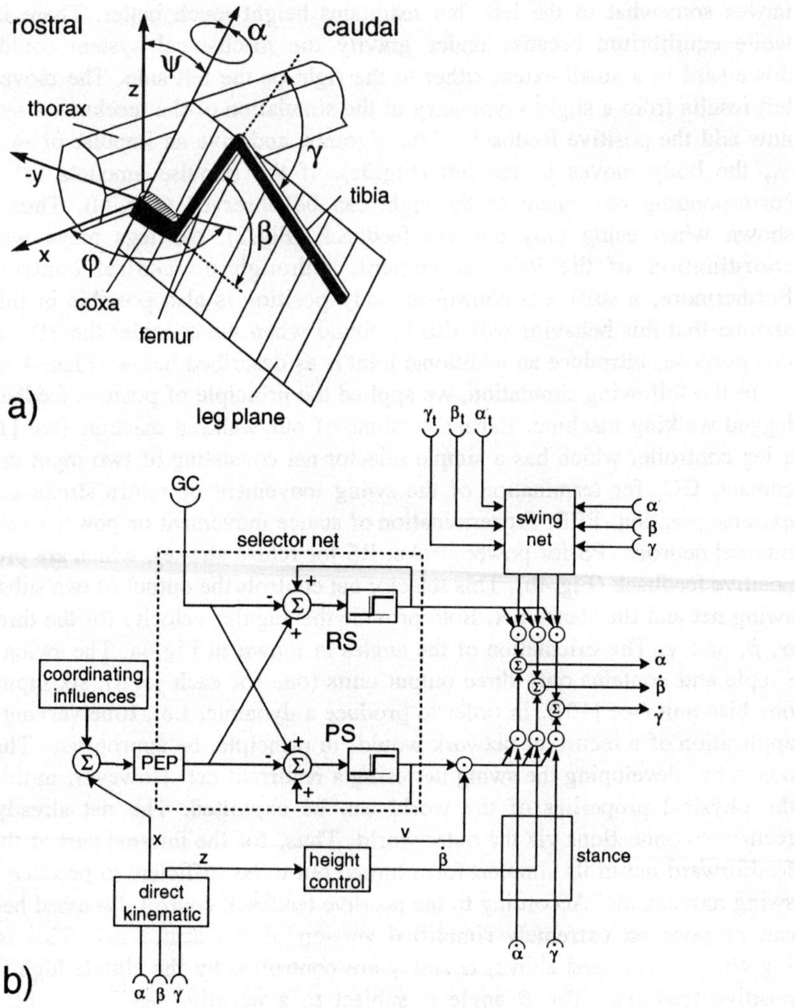

Fig.4. (a) The geometrical arrangement of the joints in an insect leg. (b) The leg controller as proposed earlier (Cruse et al. 1993a). However, the stance net is now extremely simplified compared to the earlier version.

γ joint of the middle leg increases during the first half of the stance and decreases during the second half. This sign reversal is also found in the simulation, in spite of the positive feedback. In the example of Fig.5b, the starting impulse of -7 degrees is applied only to the α joint of the right front leg. In this case, the system begins to negotiate a curve.

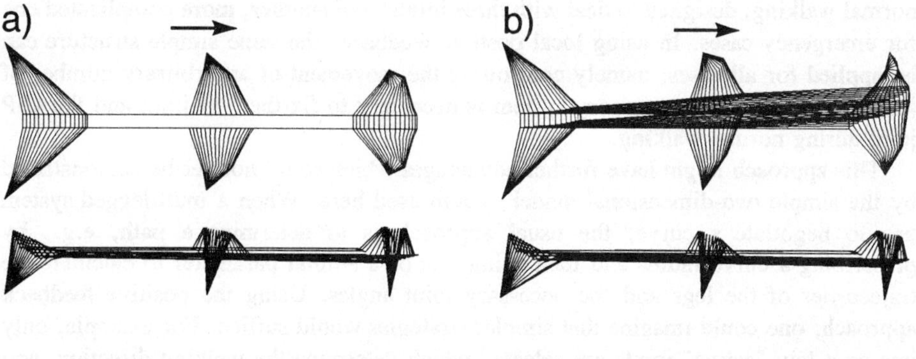

Fig.5. Movement of a six-legged system under gravity with negative feedback applied to all six β joints and positive feedback applied to all α joints and all γ joints. Movement direction is from left to right. Upper part: top view, lower part: side view, only right legs are shown. (a) The starting impulse of -3 degrees is applied to the α joints of both front legs. All legs perform a stance movement. (b) The starting impulse of -7 is applied only to the α joint of the right front leg. The system shows the beginning of a left turn.

4 Discussion

It could be shown that the problem of determining the movement of the joints of a multilegged system could be extremely simplified by using local high-pass filtered positive feedback. This does not only make redundant the determination of equations for coordinate transformation, but is also robust against all kinds of geometrical changes within the mechanical system as for example changes in leg segment length, bending of segments, changes in orientation of the rotational axes (by accident or because the suspension is soft by design), or addition of further joints. This simplification is possible, because instead of an application of abstract, explicit computation, the physical properties of the world are exploited. Thus, "the world" is used as "its own best model" (see [4]). Due to its extremely decentralized organization and the simple type of calculation required, this method furthermore allows for a very fast computation.

The main advantage of a positive feedback controller is its robustness. Changes in segment length or bending of a segment could occur by accident. Furthermore, changes of joint axis orientation can occur and have, for example, been found in the walking stick insect. The axis of the basal leg joint, the thoracic-coxal joint, shows systematical changes in orientation during the step (see [9]). Moreover, introduction of additional joints may occur in animals. In many crustaceans, the crayfish for example, two of six leg joints are usually not used, i.e., held fixed, during walking, namely the Ischiopodite-Meropodite joint and the Carpopodite-Propodite joint. Only under special circumstances, for instance when passing through a narrow gap, these leg joints may be actively bent. Therefore, the use of a traditional leg controller meant this either had to deal always with five joints, although only three of them were used most of the time, or, alternatively, two separate controllers had to be installed, one for normal walking, designed to deal with three joints, and another, more complicated one for emergency cases. In using local positive feedback, the same simple structure can be applied for all cases, namely controlling the movement of an arbitrary number of joints. Of course, a supervisory system is necessary to fix the I-M joint and the C-P joint during normal walking.

This approach might have further advantages which could not yet be demonstrated by the simple two-dimensional model system used here. When a multilegged system has to negotiate a curve, the usual approach is to determine a path, e.g., by prescribing a curve radius and then using this or a similar parameter to calculate the trajectories of the legs and the necessary joint angles. Using the positive feedback approach, one could imagine that simpler strategies would suffice. For example, only one or a few "active" joints are selected which determine the walking direction, and the other joints follow automatically (Fig. 5b). What detailed strategies are used by animals, however, remains to be investigated.

Such a positive feedback system, in exploiting the mechanical properties of the system, can also serve as a kind of averager which cancels inappropriate motor output of a single-joint controller because, by means of the mechanical coupling, such an error is corrected not only on the level of the movement, but, due to positive feedback, already on the level of the control signal. One might imagine that simple passive compliance, instead of positive feedback, in some joints could provide similar movements. However, in stick insects all leg joints produce active torques during walking (see [5]), so passive joint movement appears not to be part of the biological solution.

One major disadvantage we saw was that height control under different loading situations could not be performed by the simple positive feedback control. This problem was solved here by choosing one particular joint, i.e., the β joint, and subject it to classical negative feedback control. With respect to height control, decentralization in this system is only possible on the level of the leg, not on that of the individual joint. Moreover, robustness against changes of segment length or segment bending does not exist in this subtask. For stick insects, height control is shown to occur on the basis of the leg, but it is not known whether the actuator of this negative feedback system affects only one joint, as assumed here, or several joints (standing animal: [28], [14], walking animal: [5], [16], [26]). For the standing leg, all joints are subject to negative feedback (see [[2], [11]). Whether for the walking leg the separation assumed here can also be found in the biological system, is an open question.

To what extent can these assumptions be based on knowledge from biological systems? Positive feedback occurring in circuits situated within the brain, i.e., not including sensory signals, is discussed in terms of being used to drive sustained activity (see [23]). This basically corresponds to the simple circuit used in the model of [12] to sustain power stroke or return stroke activity in the walking leg (selector net, Fig.4b). In contrast to this kind of application of positive feedback, we here discuss the situation in which the feedback loop includes elements outside the neuronal system, i.e., provides information about the environment. Concerning this kind of positive feedback during walking, there are, at first sight, contradictory results. In the stick insect, negative feedback, controlling velocity or force, was shown to be active during stance (see [6], [7], [13], [15], [26], [29]) and during swing (see [18]). On the other hand, a series of investigations, starting with [1] and reviewed by Bässler [3], showed that at the beginning of the stance the so-called "active reaction" is found which has been interpreted as positive feedback: when the sense organ signals the leg joint as being flexed (which would, in this case, correspond to a stance movement), the excitation of the flexor muscles becomes stronger. Such reflex reversal was also found in cats (see [21]), and in crustaceans (see [20], [27], [30]). As mentioned in the Introduction, it is not clear whether these assistance reflexes are based on a negative or a positive feedback system.

Generally, however, a pure positive feedback is improbable. If positive feedback was the only mechanism, a walking animal which is momentarily pulled backward by an experimenter would then continue to walk backward. This was never observed. Therefore, some kind of supervisory system might be required, and positive feedback may only be active within a limited range. A possible solution enabling us to understand both results, i.e., those pointing to positive and those pointing to negative feedback, might be that a supervisory system consisting of nonlinear characteristics decides between both strategies depending on the amplitude or the velocity of the imposed movement the leg experiences. Results reported in [3] could be interpreted in the latter sense as, for very fast stimuli, he found a negative feedback effect for the active reaction.

Acknowledgments: We want to express our thanks to Ms. A. Baker for emending the English manuscript. The work was supported by DFG grant Cr 58/8-2, the Körber Foundation, and BMFT.

References

1. Bässler,U.: Reversal of a reflex to a single motor neurone in the stick insect Carausius morosus. Biol. Cybern. **24** (1976) 47-49
2. Bässler,U.: Neural basis of elementary behavior in stick insects. Springer, Berlin, Heidelberg, New York 1983
3. Bässler,U.: The femur-tibia control system of stick insects - a model system for the study of the neuronal basis of joint control. Brain Res. Reviews **18** (1993) 207-226
4. Brooks,R.A.: Intelligence without reason. IJCAI-91, Sydney, Australia, (1991) 569-595
5. Cruse,H.: The control of the body position in the stick insect (Carausius morosus), when walking over uneven surfaces. Biol.Cybern. **24** (1976) 25-33
6. Cruse,H.: Is the position of the femur-tibia joint under feedback control in the walking stick insect? I. Force measurements. J.exp.Biol. **92** (1981) 87-95
7. Cruse,H.: Which parameters control the leg movement of a walking insect? I. Velocity control during the stance phase. J.exp.Biol. **116** (1985) 343-355
8. Cruse,H.: The influence of load, position and velocity on the control of leg movement of a walking insect. In: M.Gewecke, G.Wendler (eds): Insect Locomotion, 19-26. Parey, Hamburg, Berlin 1985
9. Cruse H., Bartling,Ch.: Movement of joint angles in the legs of a walking insect, Carausius morosus. J. Insect Physiol. (in press)
10. Cruse,H., Bartling,Ch., Cymbalyuk,G., Dean,J., Dreifert,M.: A modular artificial neural net for controlling a six-legged walking system. Biol. Cybern. (in press)

11. Cruse,H., Dautenhahn,K., Schreiner,H.: Coactivation of leg reflexes in the stick insect. Biol. Cybern. **67** (1992) 369-375
12. Cruse, H., Müller-Wilm, U., Dean, J.: Artificial neural nets for controlling a 6-legged walking system. In: J.A. Meyer, H. Roitblat. S.Wilson (eds.) From animals to animats 2. 52-60. MIT Press 1993
13. Cruse,H., Pflüger,H.-J.: Is the position of the femur-tibia joint under feedback control in the walking stick insect? II. Electrophysiological recordings. J.exp.Biol. **92** (1981) 97-107
14. Cruse,H., Riemenschneider,D., Stammer,W.: Control of body position of a stick insect standing on uneven surfaces. Biol.Cybern. **61** (1989) 71-77
15. Cruse,H., Schmitz,J.: The control system of the femur-tibia joint in the standing leg of a walking stick insect Carausius morosus. J.exp.Biol. **102** (1983) 75-185
16. Cruse,H., Schmitz,J., Braun,U., Schweins,A.: Control of body height in a stick insect walking on a treadwheel. J.exp.Biol. **181** (1993) 141-155
17. Cruse,H., Steinkühler,U.: Solution of the direct and inverse kinematic problem by a unique algorithm using the mean of multiple computation method. Biol. Cybern. **69** (1993) 345-351
18. Dean,J.: Control of leg protraction in the stick insect: a targeted movement showing compensation for externally applied forces. J.Comp.Physiol A **155** (1984) 771-781
19. Deneubourg,J.L., Goss,S., Fresneau,D., Lachaud,J.P., Pasteels,J.M.: Self-organization mechanisms in ant societies (II): Learning during foraging and division of labor. In: From individual characteristics to collective organisation in social insects, J.M. Pasteels, J.L. Deneubourg (eds.). Experientia Supplementum **54** (1987) 177-196
20. DiCaprio Clarac,F.: Reversal of a walking leg reflex eliceted by a muscle receptor. J.exp.Biol. **90** (1981) 197-203
21. Forssberg,H., Grillner,S., Rossignol,S.: Phase dependent reflex reversal during walking in chronical spinal cats. Brain Res. **85** (1975) 103-107
22. Gibson,J.J.: The senses considered as perceptual systems. Boston, Houghton Mufflin 1966
23. Houk, J.C., Keifer,J., Barto, A.G.: Distributed motor commands in the limb premotor network. Trends in Neurosciences **16** (1993) 27-33
24. Pearson,K.G.: Central programming and reflex control of walking in the cockroach. J.exp.Biol. **56** (1972) 173-193
25. Pringle,J.W.S.: Proprioception in arthropods. In: The cell and the organism, J.A. Ramsey, V.B. Wigglesworth (eds.). pp. 256-282. Cambridge, Cambridge Univ. Press 1961
26. Schmitz, J.: Control of the leg joints in stick insects: differences in the reflex properties between the standing and the walking states. In: M.Gewecke, G. Wendler (eds.) Insect Locomotion, 27-32. Parey Hamburg Berlin 1985
27. Skorupski,P., Sillar,K.T.: Phase-dependent reversal of refelxes mediated by the thoracocoxal muscle receptor orgen in the crayfish, Pacifastacus leniusculus. J. Neurophysiol. **55** (1986) 689-695
28. Wendler,G.: Laufen und Stehen der Stabheuschrecke: Sinnesborsten in den Beingelenken als Glieder von Regelkreisen. Z.vergl.Physiol. **48** (1964) 198-250
29. Weiland, G., Koch,U.: Sensory feedback during acitve movements of stick insects. J.exp.Biol. **133** (1987) 137-156
30. Vedel,J.P.: The antennal motor system of the rock lobster: competitive occurrence of resistance and assistance reflex patterns originating from the same proprioceptor. J.exp.Biol. **87** (1980) 1-22
31. Zill,S.N.: Plasticity and proprioception in insects. II. Modes of reflex action of the locust metathoracic femoral chordotonal organ. J.exp.Biol. **116** (1985) 463-480

Learning and Adaptivity: Enhancing Reactive Behaviour Architectures in Real-World Interaction Systems

Miles Pebody[1]

Department Of Computer Science,
University College London,
Gower Street,
London WC1E 6BT.
UK

e-mail: M.Pebody@cs.ucl.ac.uk
Tel: +44 171 387 7050 ext. 3697
Fax: +44 171 387 1397.

Abstract. The success of the behaviour-based approach to designing robot control structures has largely been a result of the bottom-up development of a number of fast, tightly coupled control processes. These are specifically designed for a particular agent-environment situation. The onus is on the designers to provide the agent with a suitable response repertoire using their own knowledge of the proposed agent-environment interaction. A need for learning and adaptivity to be built into robot systems from the lowest levels is identified. An enhancement to basic reactive robot control architectures is proposed that enables processes to learn and subsequently adapt their input-output mapping. This results in both a local and global increase in robustness as well as a simplification of the design process. An implementation of the proposed mechanism is demonstrated in a real-world situated system: the control of an active laser scanning sensor.

1. Introduction

Robotics is one of the central themes of the Artificial Life field. It deals with the design, understanding and construction of artificial agents that interact with and in real-world environments. For a number of years the Behaviour-Based approach to robotics has been the main and most successful paradigm in system design. Many examples exist that demonstrate the benefits of the approach in terms of response time, robustness and reliability. [Brooks and Flynn 89] provides an interesting but by no means complete catalogue of such robots. There are also many examples of architectures and design frameworks that have been developed to support the Behaviour-Based design of robots, such as: (the first of all) the Subsumption Architecture [Brooks86], PDL [Steels93], AuRA [Arkin90], Spreading Activation Networks [Maes89] and ALFA [Gat et al 94]. One of the main characteristics of these methodologies is that control systems are built as a number of specially tailored (by the designer) reactive processes, each tightly coupled to the robot's environment. The processes are generally fast, simple and contain a minimum of explicit domain or internal state information. The structure of these control architectures generally leads to a highly specialised and tuned control solution to a particular agent-environment situation (which is perhaps the fundamental reason for their success). It may be argued that the designers have built their own knowledge of the domain into the structure of the agent control system providing the right process connections and constant values to elicit the right agent-environment interaction dynamics ([Brooks91] and [Smithers94]). This designer knowledge can be seen as a property of the agent architecture in the form of implicit domain knowledge.

The implicit domain knowledge is a designer oriented view of the agent control architecture. In terms of the agent it manifests as basic hardware structure (including fixed software structure) which forms the basis of its behavioural repertoire. The task of the robot designer then is to provide a maximally robust and effective control structure. However, as robot task requirements, environmental constraints and hardware become more complex the practicalities of utilising many simple designer-crafted processes will seriously restrict progress. It is easy to envisage task domains for which the designer or designers will be unable to grasp the intricacies of the full set of possible robot-environment interactions (the frame of reference problem [Pfeifer 93]).

[1]Funded by the Engineering and Physical Sciences Research Council and Sira Ltd.

Techniques of adaptive and learning control have been widely seen as a solution to this problem but in many cases have so far failed to be transferred from the idealistic domain of simulation. Examples that do deal with real-world implementations of learning and adaptation can be seen in [Kaebling93], [Mahadevan and Connell 92] and [Mataric94] where aspects of discrete-state reinforcement learning are experimented with at the behaviour selection and adaptation level. [Pfeifer and Verschure 93] details work on learning associations between complex sensors and actuators in the form of models of classical conditioning applied to individual behaviours. Also [Nehmzow and Smithers 90] looks at the use of self organising feature maps for landmark detection. Other work reported in [Clark, Arkin and Ram 92] looks at on-line tuning of parameters in reactive control architectures.

Many learning and adaptive control problems remain open with solutions to date having many limitations. Reinforcement learning tends to rely on the problem space being defined in terms of a Markovian model of synchronous, discrete state and time steps for both agent and environment. These characteristics are difficult to identify in the real world. Reinforcement learning also typically suffers from a bootstrap problem which is characterised by an initial period in which the agent builds up its interaction processes through environmental exploration and experimentation. During this time the agent's performance is unreliable at best. Other solutions, for example the classical conditioning and self-organisation models that do have useful bootstrap mechanisms, tend to be very specific and centralised. A complication here is that distributed hardware architectures are becoming the norm for robot control with different parts of the robot being locally controlled by their own microprocessors. Another issue then regarding the solution to providing learning and adaptive control for a robot agent is that the design architecture should be able to deal with physically distributed centres of computation which have limited bandwidth communications.

Given that learning and adaptation are necessary additions to Behaviour-Based robot control architectures this paper presents a partial solution to the following problems: i) that the techniques of reinforcement learning need to be grounded in the dynamics of real sensors and actuators from the bottom up, ii) adaptive and learning mechanisms need to be distributed around multiprocessing hardware architectures and iii) an agent must have a minimum level of ability while the learning and adaptive mechanisms are initialised. An adaptive Behaviour-Based control architecture is presented that is based on the subsumption architecture [Brooks 86]. The behaviour processes are augmented finite-state machines (AFSMs) and the work here has concentrated on enhancing these basically reactive processes with a mechanism that can learn by association and continuously adapt the input-output mapping of the rule element of the process. It has been found that the design task is simplified and the AFSM behaviour more robust. The adaptive architecture increases the amount of domain knowledge that the agent can acquire for itself. The agents are also less dependent on the initial control structure and parameters provided by the designer.

The next section outlines the associative learning mechanism that is added to the basic reactive rule element of a subsumption AFSM. This is followed by an example application in the form of the low-level control of an active sensory subsystem, with emphasis here on the need for increased attention to true bottom-up development of control architectures. That section then ends by outlining a brief experiment that compares the proposed adaptive architecture with a basic reactive version. Section 4 discusses the implementation and the results of these early experiments and section 5 draws conclusions from the work.

2. Adding Learning and Adaptivity to Reactive Systems

A common feature of many Behaviour-Based control architectures is the utilisation of a number of parallel processes repeatedly looping and executing condition statements of the form: "if *condition* do *action*". The success of the Behaviour-Based approach to robot control can be attributed largely to the skill of the designer in selecting the right conditions and action parameters for each behaviour process and the right interprocess connections. In order to automate this process the nature of the condition and action must be examined more closely.

The condition predicate part of the statement usually refers to a number of input values to the process that may originate either from a sensor or another internal process. It will typically be either a simple binary-valued variable or the binary result of some comparison of values, for example: *input_a > threshold.* As well as thresholds, minimum-maximum range tests are also common. The critical aspects of the condition when it comes to building in domain knowledge are: i) which inputs are relevant and should be included and ii) how these inputs are evaluated to generate a true or false result, for example at what value should the threshold be set?

The action part of the statement usually consists of a value that is output from the process. This may either be a direct output value to an actuator or it may be connected to an input of another behaviour process. This value is also often updated as part of the output process. The critical aspects of the action output are either i) the actual value output if it is locally constant or ii) the size of modification to the output.

It would appear then that some form of continuous parameter tuning is required to thresholds and action modification values as well as some form of active search of sensory input combinations to format the condition predicate. In this way the initial rule as specified by the designer may be adjusted and tuned as the agent gains experience in its environment.

In order to implement the automatic tuning of the parameters of an AFSM rule within the subsumption architecture it would be possible to add more AFSMs into the system control structure. Each of the critical parameters that contain the magical designed-in domain knowledge could be turned into a register that is updated, depending on the activities of the robot, by a higher-level AFSM. However this approach, whilst adding robustness to the agent architecture, involves a potentially exponential increase in the workload of the designer with each new AFSM requiring additional domain knowledge for its implementation. More conditions and constant settings will have to be tuned. It is also the case that this technique does not lend itself to on-line automatic methods. In this instance the learning process would involve the dynamic allocation of new AFSMs, connecting them with wires to lower level AFSM registers and selecting suitable rules that include both suitable condition predicates (with possible combinations of threshold and max-min classification values) and action outputs. Clearly a solution is required that can adapt around an existing predefined control structure.

The rule part of an AFSM may be regarded as the formulation of an input-output mapping, the input being some pattern of input state defined by the condition predicate and the output being an action that is associated with a particular condition. The problem of parameter selection and tuning of the input can be partially solved with the use of a connectionist or artificial neural network approach. Artificial neural networks are able to learn on-line and also have useful generalisation characteristics. Noisy and incomplete input patterns can still trigger a specific response that was originally associated with more complete data. Both problems (identified above) of selecting the inputs and tuning the parameters of the AFSM rule's condition predicate may be addressed in this way. Another way of seeing this arrangement is as an additional process learning a feedforward response based on a lower-level AFSM.

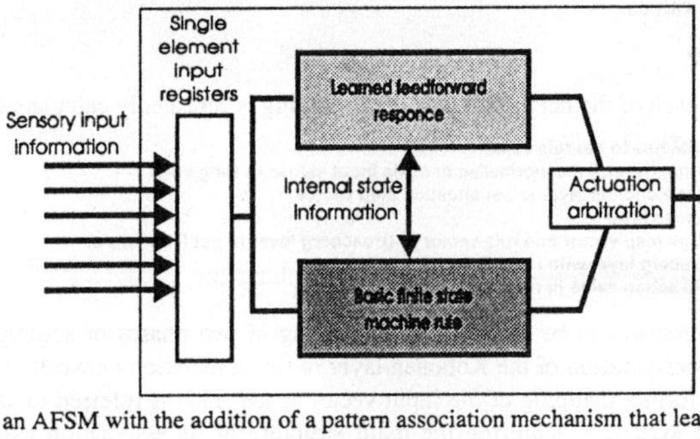

Figure 1. Enhancing an AFSM with the addition of a pattern association mechanism that learns the process's feedforward response.

Figure 1 shows the proposed format of an enhanced AFSM. The basic AFSM pattern is maintained (as is the loosely synchronised timing characteristic of the subsumption architecture). The rule acts both to bootstrap the association network and to ensure that the AFSM has a consistent minimum performance. The task of the designer is now to specify the basic rule (which may now be less detailed) using a few key input registers and then to specify a number of other auxiliary inputs that will correlate with the rule input-output mapping in some way that is too complex to detail explicitly. The idea is that the network learns an association between the array of inputs based on reinforcement from the rule activity. Since the network is continuously modified depending on changing environment situations a form of parameter tuning is provided. Additional inputs that enhance the selectivity of the rule provide an added robustness and redundancy in input state may be utilised in the case of missing or noisy input data.

Work presented in [Pfeifer and Verschure 93] has demonstrated that self-organising robot controllers can be developed from simple artificial neural networks. While this work focused on the issues surrounding classical conditioning [McFarland 85] it provided insight into the nature of continuous adaptive control when applied to real-world robots and more specifically how such adaptive mechanisms might be initialised (or bootstrapped) so as to ensure that the correct input-output association was acquired. The strategy presented in this paper for learning and adapting the input-output mapping of an AFSM is based strongly on these adaptive associative networks that are able to learn the associations of complex sensor states from a limited set of simple key inputs. Modifications and additions to the basic mechanisms include:

ii) The use of a two layer counterpropagation network [Wasserman 89] in place of the single layer model which from experimentation did not converge well for AFSMs with small numbers of real valued inputs (notwithstanding problems associated with linear seperability of input space).

ii) The framework of a single AFSM for implementation allows a physically distributed control architecture to be built using the enhanced adaptive AFSM processes as and where necessary.

iii) The output stage is different to that of the classical conditioning model in order to allow for the possible lack of any temporal sequencing of events (which is a critical factor of Pfeifer and Verschure's method).

2.1. Implementing a Two Layer Association Mapping Network

Figure 2 shows the addition of a counterpropagation network to a standard AFSM. This network algorithm was chosen in preference to the more commonly known multilayer perceptron and backpropagation training method primarily due to reported faster weight convergence times [Wasserman 89]. In addition the robustness of the Kohonen network algorithm has been shown in real-world robot applications. For example [Nehmzow and Smithers 90] reports on experiments with on-line acquisition of landmark feature maps in mobile robots using such techniques.

During every processing cycle of the AFSM (figure 2) the input registers are sampled. The key inputs are used to process the AFSM rule and also used in conjunction with the auxiliary inputs to create the associative network input vector. The rule output is used both to train the network and (in the early stages) to provide the main AFSM output, thus serving to reinforce and bootstrap the learning process.

The counterpropagation algorithm is a combination of two single-layer algorithms. The first layer is a Kohonen self-organising feature map [Kohonen 88] whilst the second is an outstar network [Grossberg 69], similar in many respects to a single-layer perceptron. The counterpropagation algorithm functions as a look-up table that is capable of generalisation allowing accurate reproduction of the output despite incomplete or partially incorrect input data. The training process associates input vectors with output vectors which may be either binary or continuous values.

2.1.1. The Main Algorithm

After randomly initialising each of the network layers the following is repeatedly calculated for each time step:

```
Process AFSM rule to get rule vector
Preprocess input vector (i.e. normalise or scale input values to range 0-1)
Apply input to Kohonen layer to get situation map vector
Update Kohonen layer
Apply situation map vector and rule vector to Grossberg layer to get final output
Update Grossberg layer with rule vector
Output AFSM action value (if required)
```

The life of the AFSM process can be described as consisting of two phases of activity resulting from the learning rates and update configuration of the Kohonen layer of the association network. Initially the Kohonen network must have time to form a mapping of the input vector space. This is referred to as the learning phase and is the period when the system is acquiring the main structure of its association mappings. The second phase is the adaptive phase. This is the normal operation mode of the network and takes over once the mappings in the Kohonen layer have become stable. It is recommended in [Wasserman 89] that to acquire statistically significant clustering of inputs a Kohonen network should be given 500 times the number of cells worth of training cycles. This is currently the time period used in the implementations described here. For the initial learning period it is helpful if the system is presented with as rich a set of input stimuli as possible since it is at this time that the system is maximally responsive to learning various input characteristics.

2.1.2. Basic Rule

The basic rule takes a number of key inputs and uses them to generate an output vector R which is scaled and then combined with the Grossberg output layer of the association mapping network.

2.1.3. Input Scaling

The association network input vector includes the key inputs as used by the rule and a set of extra auxiliary inputs. Real valued inputs to Kohonen networks are usually normalised before being applied. However, the standard normalisation procedure as provided in [Wasserman 89] causes the inputs to influence each other in terms of both overall magnitude and changes over time. To avoid this, a simple scaling factor was used for each input. The value of scaling was chosen as an approximate maximum value for each input type so that a scaled input would have a range between 0 and 1. For binary inputs this process is not required.

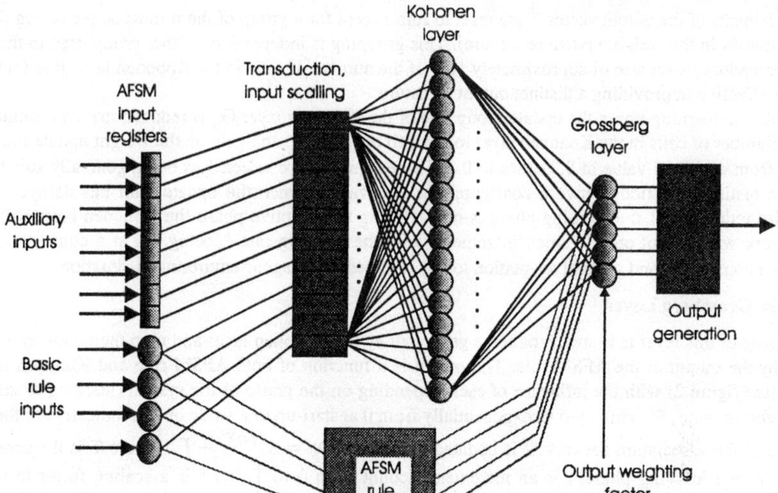

Figure 2. An AFSM rule with a counterpropagation network implementation of the association network. Basic input values are routed to both AFSM rule (light grey lines) and the network (black lines); auxiliary inputs are routed to the network only and are scaled into the range 0-1. In normal response the output of the rule and the output of the Grossberg layer are calculated with output bias being a function of the output weighting factor Ψ. The output generation converts the output vector from the Grossberg-Rule combination into an integer AFSM output value. During network weight update the rule output is used to calculate weight changes for the Grossberg layer.

2.1.4. The Kohonen Layer

Kohonen self organising networks have been shown to have useful mapping characteristics. The self-organisation of these networks involves the clustering of similar input vectors applied over time and convergence can be relatively fast depending on the degree of difference in the input vectors. However, the speed of convergence is also influenced by factors such as the number of cells in the network and the number of distinct clusterings that result from the input.

The input-stage layer in Figure 2 uses a one dimensional Kohonen network with a number of weight storage units in a vector K. The number used depends on the required associative mapping. From experimentation it was found that $4n^2$ units was generally suitable where n is the approximate number of input states to be differentiated. Each unit is a vector with the same number of elements as the input vector, each unit vector contains the weights associated with that unit and these are updated according to the following scheme (having been initialised to random values in the range of 0-1):

1. An input vector (X with elements x_i) is applied to each of the Kohonen layer units and a measure of appropriateness to the current input is made using a simple dot product of the two vectors. The most active unit is selected.
2. The weights (vector with elements k_{ji}) of the selected unit and an adjacent group of units of size G_u are updated according to the following where ξ_k controls the gain of the weight update and X is the input vector:
$$k_{ji}(t+1) = k_{ji}(t) + \xi_k.\left[x_i - k_{ji}(t)\right]$$
3. The update gain and the size of the update group are reduced and steps 1 and 2 repeated for other input vectors. The group size is reduced gradually until only one unit is selected for updating and the update gain reduced down to a value of 0.1

The calculation of the output (vector Y with y_j elements) of the Kohonen layer is made in the same basic way as many other artificial neural net algorithms:

$$y_j = \sum_{i=0}^{n} k_{ij} . x_j$$

All elements of the output vector Y are reset to zero except for a group of the n most active (using the same dot function as in the weight update procedure). This grouping is independent of the group used in the weight update procedure. A set size of approximately 10% of the number of cells in the Kohonen layer was found to be the most effective in providing a distinct output signature

During the learning phase the update group size of the Kohonen layer G_u is reduced from an initial size of half the number of cells in the Kohonen layer to one cell only and $\xi_{k,}$ the gain on the weight update function, is reduced from an initial value of 0.7 down to 0.1. These figures were selected as being generally suitable after extensive testing of various network configurations and values. Once the update gain has decayed and the group size reduced to 1 the learning phase is over. During the adaptive phase the Kohonen layer still adjusts the network weights but only in small increments with the learning rate $\xi_k$being left at a constant 0.1. This provides a continuous and gradual adaptation to any changes in the agent-environment situation.

2.1.5. The Grossberg Layer

The purpose of this layer is to match patterns generated by the Kohonen layer and map them onto an output as trained by the output of the AFSM rule. The output is a function of both AFSM rule and Kohonen mapping outputs (see figure 2) with the influence of each depending on the phase of the system (learning or adaptive). The weighting factor Ψ, which grows exponentially from 0 at start-up to a value of 1 is indicative of the output influence of the association network.Ψ is updated according to $\psi = e^{(f.t/T)} \div T$. Where T is the time period selected for the learning phase, t is an incremental count from 0 to T and f is a scaling factor to ensure a growth from zero to one in the time period, it is initialised as the natural logarithm of T. The AFSM output was calculated using the following:

$$z_j = \sum_{i=0}^{n} w_{ij} . y_j$$

$$out_j = (1-\psi).r_j + \psi.\frac{1}{1+e^{-z_j}}$$

The reason for introducing the weighting factor is that Pfeifer and Verschure's work was based on modelling the phenomenon of classical conditioning which relies on a temporal difference between the basic and associated reactions. It depended on the associated response occurring before the built-in response. The decay used here ensures that the associated action becomes the dominant output whether or not there is a timing factor involved.

The weight update function of the Grossberg layer was of the standard single-layer perceptron form with ξ_g, the learning rate constant, set to 0.2 in both learning and adaptive phases. This ensured that the layer responded quickly to continuing changes on the Kohonen layer during the ongoing adaptive phase. At all times the weights in the Grossberg layer were updated using the difference between the current output Z and the output from the AFSM rule R and the output vector Y from the Kohonen layer. The weights i of element j of the Grossberg layer were updated using the following:

$$g_{ij}(n+1) = g_{ij}(n) + \xi_g . y_i . [r_i - z_i]$$

3. Experiments in Embedded Adaptivity

In order to test the ideas so far outlined for enhancing the basic reactive Behaviour-Based control architectures, a real-world test-bed is required. Whilst many experiments of this nature take place in the domain of mobile robots the example here deals with the more restricted domain of the control of a sensory subsystem, part of a highly sensitive industrial laser scanning inspection device [IAL93]. This is in keeping with earlier mention of using a genuine bottom-up development approach. In this case the behaviour of a complex sensory subsystem must be taken care of before any higher-level functionality can be considered. First in this section the test-bed is outlined to give a general overview of the problem domain, then an example of a subsumption architecture control solution is given that incorporates an adaptive AFSM process.

3.1. The Laser Scanning Inspection System Experimental Test-Bed

The laser scanning sensor system works by illuminating a surface with a laser stripe and detecting reflected laser light with a number of photomultiplier light-sensitive devices. Photomultiplier tubes are light-detecting devices that have a huge dynamic range. Light falling on the tube's cathode end causes electrons to be emitted from a phosphorous layer on the inside surface. The electrons are accelerated down the tube by a number of extra-high-tension (EHT) voltage coils towards the anode end of the tube. The electrons striking the anode result in an output signal with a voltage level that is indicative of the intensity of the detected light. Figure 3 shows a typical signal as the laser scans across an object in ideal conditions. The control problem explored here is that of controlling a photomultiplier sensor channel so that it maintains an optimal signal output compensating for environmental influences.

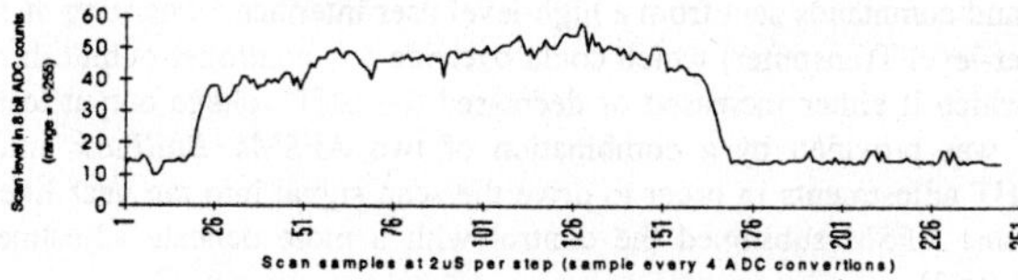

Figure 3. Ideal photomultiplier output signal showing single-surface scan with no major defects.

The dynamic range of the output signal of a photomultiplier may be controlled by varying the voltage on the EHT coils of the tube. The graph in figure 4 shows a typical photomultiplier response for a constant light level and three different sensor-surface distances. The gain on the X axis (EHT voltage level) was increased from zero until the output signal from the photomultiplier became saturated and no longer accurately reflected the detected light intensity. It can be seen that this response characteristic is highly non-linear and can be categorised into four main states: 1,2,3 and 4 on the graph. The regions change with different surface distance as well as other environmental effects such as changing ambient light levels and changing surface characteristics. The sensor controller must maintain a set-point operation in the state 2 region in order to provide a stable and usable light-intensity signal. Should the system be in any other state, then the sensor signal is unusable. Different control strategies are required to return to state 2 from states 1, 3 and 4. Consequently fast assessment of signal state is a fundamental aspect of the control process.

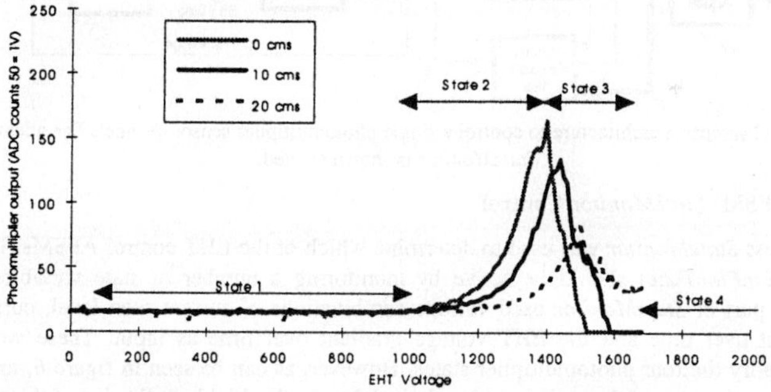

Figure 4. Response of the test-bed photomultiplier for constant light level and changing sensor-surface distance. State 1: Quiescent, State 2: Near linear response, State 3: Unstable fold-back, State 4: Zero level fold-back.

For each sensor channel, control software is executed on a dedicated Transputer microprocessor which is part of an interconnected network of such devices. A tool for building Transputer-based subsumption architectures has been developed. Augmented finite state machines are implemented as Transputer processes and behavioural levels as collections of Transputer processes. Subsumption wires use the Transputer channel protocol which makes the physical architecture transparent to the subsumption network architecture [Pebody94].

3.2. Illustrating the Use of an Adaptive AFSM

Figure 5 shows a two-layer subsumption control structure for a single photomultiplier channel. In this example which is of only a single sensor channel control, there is only one adaptive AFSM process: *StateMonitor*. The other AFSM processes are briefly described first.

The first level contains the sensor and actuator device driving AFSMs *EHTOutput* and *ScanAcquire* as well as the AFSMs *ScanPercentile*, *EdgeMonitor* and *MetaPixelGen*. The latter four AFSMs formed a small subsystem that processed the scan level input data. The *ScanAcquire* AFSM controlled the analogue to digital conversion of scan stripes and loaded the data into a vector of up to 1000 pixels. This vector was then accessed by the other AFSMs in this group. Each of these AFSMs was asynchronous; there was no handshaking or other form of synchronisation. The *EdgeMonitor* processed the scan stripe vector searching for large gradients in pixel value (each pixel containing a value representing the intensity of the light detected). The first positive gradient was taken as the left edge and the last negative gradient the right edge. The *ScanPercentile* AFSM calculated the percentile of the scan line between the left and right edges using a rolling average of the last *n* scan lines in order to filter out high-frequency signal noise. The AFSM *MetaPixelGen* assembled a vector of 16 large pixels each being the average level of 62 scan-stripe pixels. The *EHTOutput* AFSM arbitrated between controlled EHT output and commands sent from a high-level user interface (consisting of a network of higher-level AFSMs on a higher-level Transputer) which could override any controller output. It received as input an EHT delta value with which it either increased or decreased the EHT voltage output to the photomultiplier. Finally the EHT delta was provided by a combination of two AFSMs: *BallPark* and *NearLinFineTune*. *BallPark* made large EHT adjustments in order to drive the scan signal into the near linear state 2 and once here the *NearLinFineTune* AFSM subsumed the control with a more delicate adjustment to maintain the required scan signal (figure 3).

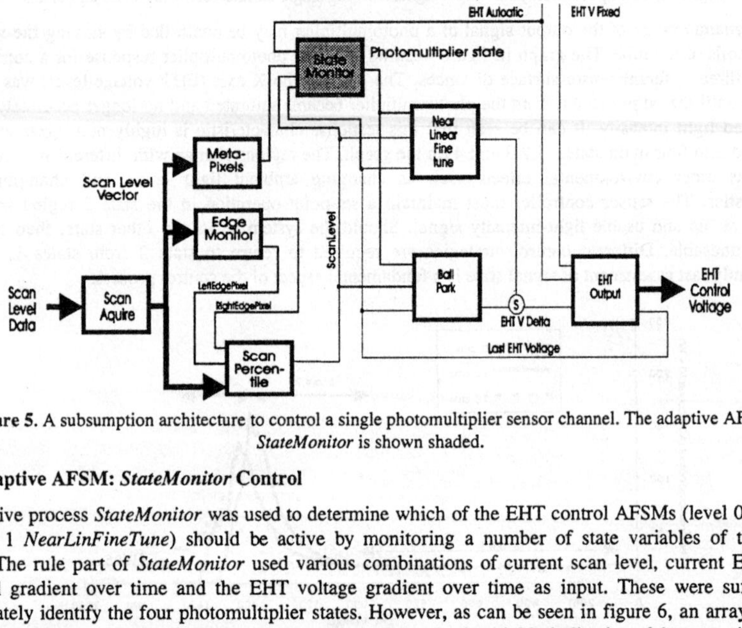

Figure 5. A subsumption architecture to control a single photomultiplier sensor channel. The adaptive AFSM *StateMonitor* is shown shaded.

3.2.1. Adaptive AFSM: *StateMonitor* Control

The adaptive process *StateMonitor* was used to determine which of the EHT control AFSMs (level 0 *BallPark* and level 1 *NearLinFineTune*) should be active by monitoring a number of state variables of the sensor channel. The rule part of *StateMonitor* used various combinations of current scan level, current EHT level, scan level gradient over time and the EHT voltage gradient over time as input. These were sufficient to approximately identify the four photomultiplier states. However, as can be seen in figure 6, an array of meta-pixels taken by averaging sets of the main scan-level pixel data is also highly indicative of the sensor's state but is difficult to characterise in terms of rule conditions. Thus for this adaptive AFSM the scan level, scan level gradient, EHT level and EHT-level gradient were used as basic rule inputs and the meta-pixel array used as a set of auxiliary inputs. A pseudo-coding of the AFSM rule follows:

```
If scan_level < ZEROFOLDBACKLEVEL{
        state = ZERO_FOLDBACK_STATE
}else{
        If QUIESCENT_HI_LEVEL > scan_level < QUIESCENT_LO_LEVEL {
                state = QUIESCENT_STATE
        }else{
                If scanLevel_gradient direction = EHT_level_gradient direction{
                        state = NEAR_LINEAR_STATE
                }else{
                        state = UNSTABLE_FOLDBACK_STATE
                }endif
        }endif
}endif
```

The rule output was an array of four binary values, one for each of the states. It was used to train the associative network and was combined with the output in the way described above in section 2. The 16 meta-pixels from the *MetaPixelGen* AFSM were combined with the four rule inputs to give an input array of 20 sensor readings. The transduction function for these inputs was set to ensure that the values ranged from 0 - 1 most of the time (in the case of the meta-pixels an approximate ranging was found to be sufficient; values slightly greater than 1 did not adversely effect the outcome). The output of the associative network was of the same format as the rule output, a 4 element array of binary values.

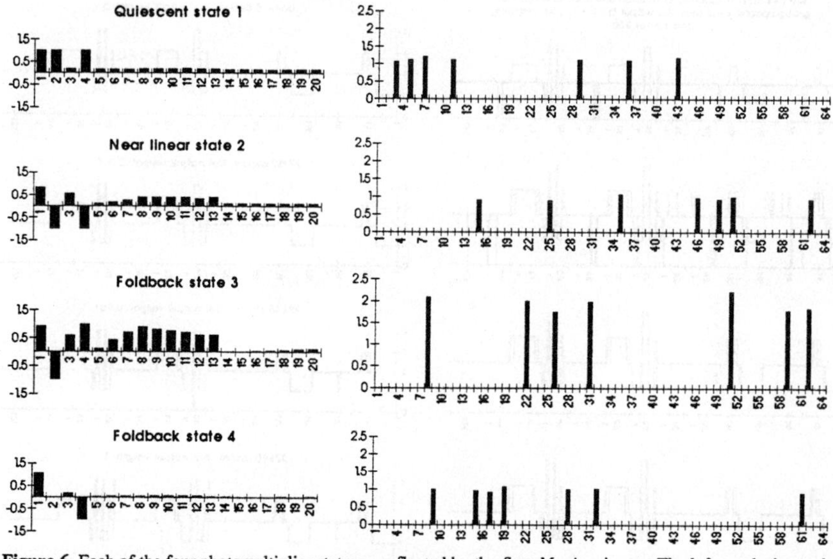

Figure 6. Each of the four photomultiplier states as reflected by the *StateMonitor* inputs. The left graph shows the StateMonitor input vector with the 4 basic inputs leftmost and 16 meta-pixel inputs. The right-hand graph is the mapping output of the Kohonen self-organising network.

3.3. Experiments

To compare the usefulness of the approach both in terms of ease of design and in actual effectiveness at the target task a series of experiments was conducted. The test-bed was set up with the sensor focused on a rotating surface providing a repeating pattern of changing laser light intensity that included dark areas of zero reflectivity (27% of total surface area). Each rotation took approximately 1000 AFSM processing cycles. Four tests were run: i) Using the AFSM rule only, ii) Using the combined rule and network output iii) A fine-tuned standard AFSM solution and iv) Testing of the robustness to missing inputs. The output of the rule, the network output (where appropriate) and the combined and weighted AFSM output was monitored. In addition the stability of the scan signal output was monitored as a global indication of the implementation's effectiveness.

4. Results and Discussion

Figure 7 shows a series of *StateMonitor* outputs for sensor transition over the test surface. The left column of graphs show the development of the association-network output in comparison with the AFSM rule output. During this test the extra input information of the meta-pixel array was not included and the network response was based on the same information as the rule only. It can be seen that the network output eventually develops a similar response to the rule but differs in small ways, particularly in the speed of state transition and filtering out of transitive state changes. The right-hand column shows the merged AFSM output compared to that of the rule output. Here it can be seen that as expected, in the early stages the rule output dominates. However by the end the output has become that of the network output and it can be seen clearly that the an association has caused the early indication of state 4 on two occasions.

Examination of the standard deviations of the sensor output variation for the four test runs, figure 8, would suggest that the response learned by the adaptive AFSM was sufficient at least in comparison with the other control solutions tested here and may even be seen to be slightly better. The results on tests for system robustness were most encouraging. The right-hand graph in figure 8 shows that the system performance degraded only slightly for significant degradation of AFSM input connections. In fact the degradation of performance resulted in a general slowing of the system's response speed to the changing external situation. In the tests run this resulted in a more stable signal due to the repeating nature of the test surface and it should be emphasised that in a more diverse environment the degradation of performance would (most likely) be more marked.

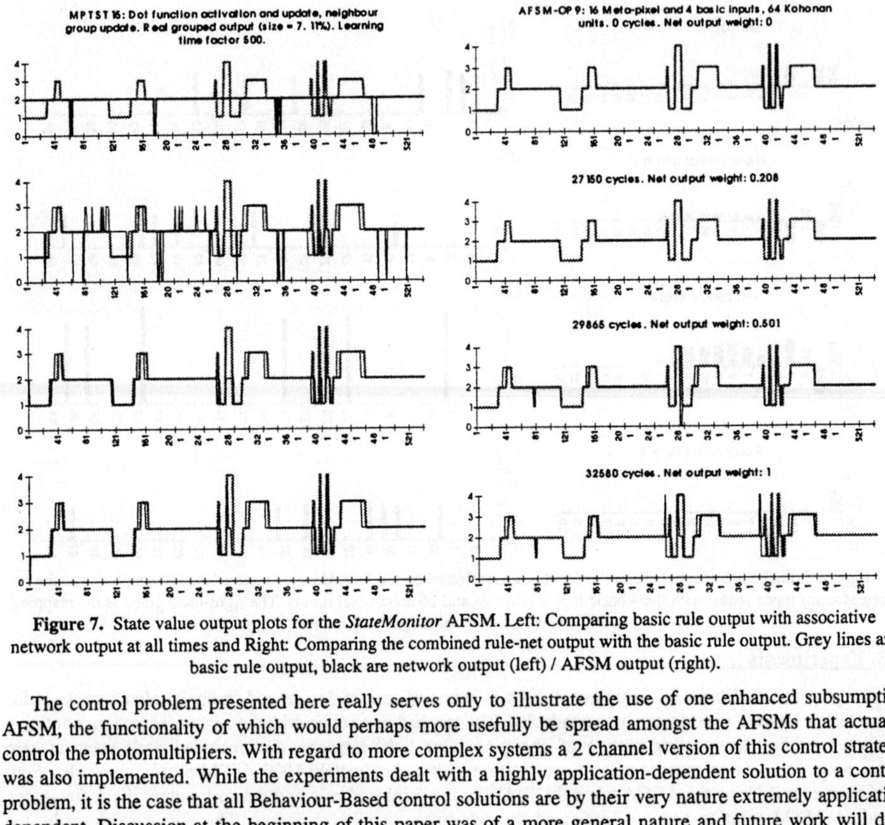

Figure 7. State value output plots for the *StateMonitor* AFSM. Left: Comparing basic rule output with associative network output at all times and Right: Comparing the combined rule-net output with the basic rule output. Grey lines are basic rule output, black are network output (left) / AFSM output (right).

The control problem presented here really serves only to illustrate the use of one enhanced subsumption AFSM, the functionality of which would perhaps more usefully be spread amongst the AFSMs that actually control the photomultipliers. With regard to more complex systems a 2 channel version of this control strategy was also implemented. While the experiments dealt with a highly application-dependent solution to a control problem, it is the case that all Behaviour-Based control solutions are by their very nature extremely application dependent. Discussion at the beginning of this paper was of a more general nature and future work will deal with, (amongst other aspects) a greater complexity of adaptive processes. One important issue that is not dealt with in these experiments is that of stability in a system with multiple adaptive AFSMs. This is currently being addressed in more experiments.

The problem of building domain knowledge into a Behaviour-Based architecture that was identified at the beginning of this paper has only been partially addressed by the associative mechanisms presented here. In particular, the solution only deals with a fixed number of AFSM inputs and future work will need to look at creating new connections dynamically, perhaps using strategies similar to those described in [Maes and Brooks 90] in which connections between competence modules of a spreading activation network are learned. The adaptive AFSMs presented here also do not deal with tuning of output actuation. The associative net is only able to learn mapping from inputs to the predefined (by the rule) output. This can be currently worked around by adding higher level AFSMs to tune this parameter. The addition of the associative mapped input-output makes this easier.

Another interesting and important issue that requires work is that of time and timing of processes and their interactions. The classical conditioning models discussed in [Pfeifer and Verschure 93] necessitate either an implicit or explicit time-based relationship between the built-in response and the associated response. The adaptive AFSM strategy here does not account for this and the associations are instantaneous. However, it is possible that inbuilt time information that is characteristic of subsumption AFSMs might be used as input to the associative network in the same way as any other input data. Related to this is another possible future direction to look at ways of externally influencing the variables (G_k, ξ_k, ξ_g and Ψ) that control the dynamics of the system initialisation so that the association learning process might be restarted to some degree should environmental situations necessitate a large change in system response repertoire.

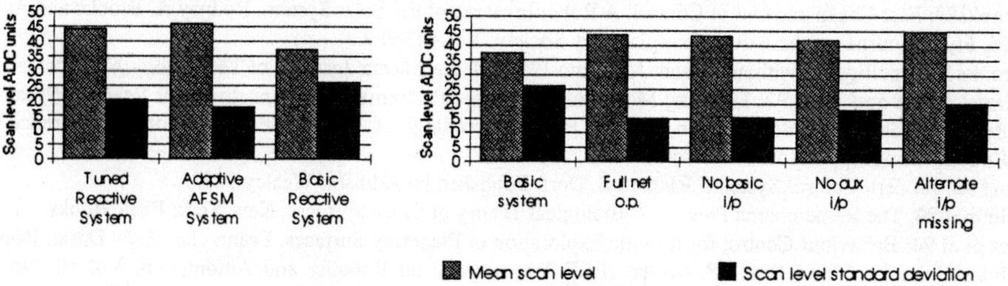

Figure 8. Comparing standard deviations and mean level of the sensor output. Left: for each of three controller configurations and right: for robustness to missing inputs. Tests were conducted on a segmented surface which meant that 100% reliability was not possible, hence mean scan levels only approach the optimum of 50 ADC counts and standard deviations indicate the nature of the recapture of the signal after a transition period. A lower standard deviation indicates that less time was spent searching for the new signal setting.

4.1. Related Work

The problem of providing artificial autonomous agents with sufficient domain knowledge to perform satisfactorily in real-world environments is widely acknowledged. Many areas of research are looking at using the various techniques of learning and adaptive control ranging from the classical adaptive parameter tuning of [Dorf 92] to areas of learning control [Harris 94] and reinforcement learning [Kaebling 93]. However, the work reported in this paper deals more specifically with the problem of grounding the learning and adaptive processes that deal with real sensor and actuator dynamics. Apart from the work already mentioned in [Pfeifer and Verschure 93] and [Nehmzow and Smithers 90] other work in this direction includes that reported in [Clark, Arkin and Ram 92] and [Gaussier and Zrehen 94]. These also deal with adaptivity with an architectural perspective to agent control.

5. Conclusions

The need for automating the application of domain knowledge to Behaviour-Based agent-control architecture development has been highlighted. This can either be regarded in terms of implicit domain knowledge acquired by the agent to function in its domain, or it can be seen as the on-line development of agent internal state. Whichever, the state of an agent is more than just a number of updateable variables in a control architecture: it includes all parts of the agent and environment, and as such mechanisms for adaptivity need to be built in at the lowest levels possible. The problems associated with designing more complex agent systems have been characterised as ones of providing sufficient mechanisms that are able to acquire domain knowledge in the form of control structure rather than building increasingly complex reactive architectures. To explore the nature of such mechanisms an enhancement to the finite-state machine processes of the subsumption architecture has been implemented and experimented with. It was shown that the basic rule of an AFSM may be used to bootstrap an associative learning device which in turn may allow other sources of information such as more complex sensory input to be incorporated into the rule's activation condition. The proposed strategy, being embedded in a Behaviour-Based architecture, proved to be a useful means of localising the learning and adaptive functionality required to initialise and maintain the interaction dynamics of a distributed control architecture. A means to build in adaptivity on a bottom-up basis was provided. Building such systems involves specifying key inputs and rules for the AFSMs which can be seen as a form of "genetically" defined set of start-up parameters or value scheme [Edelman 89]. The mechanism provided by the counterpropagation network can usefully accumulate input-output mappings that both enhance and generalise the action of the AFSM rule.

References

Arkin90: The Impact of Cybernetics on the Design of a Mobile Robot System: A Case Study. Ronald C. Arkin, IEEE Transactions on Systems, Man and Cybernetics, Vol 20, no 6, November/December 1990.

Brooks86: A Robust Layered Control System For A Mobile Robot, Rodney A. Brooks, IEEE Journal of robotics and automation. Vol. RA-2, No 1, March 1986.

Brooks89: Fast Cheap and Out of Control: A Robot Invasion of the Solar System. Rodney A. Brooks and Anita M. Flynn. Journal of the British Interplanetary Society. Vol 42 1989.

Brooks 91: Intelligence without reason. AI memo 1293, Massachusetts Institute of Technology AI laboratory.

Clark, Arkin and Ram 92: Learning Momentum: On-Line Performance Enhancement for Reactive Systems. Russell J. Clark, Ronald C. Arkin, Ashwin Ram. Proceedings of 1992 IEEE International Conference on Robotics and Automation. Nice, France.

Dorf 92: Modern Control Systems. Richard C. Dorf. Published by Addison-Wesley 1992.

Edelman89: The Remembered Present: A Biological Theory of Consciousness. New York: Basic Books.

Gat et al 94: Behaviour Control for Robotic Exploration of Planetary Surfaces. Erann Gat, Rajiv Desai, Robert Ivlev, John Loch and David P. Miller. IEEE Transactions on Robotics and Automation, Vol 10, No. 4, August 1994.

Gaussier and Zrehen 94: Complex Neural Architectures for Emerging Cognitive Abilities in an Autonomous System. Conference proceedings: From Perception to Action 94. IEEE Computer Society Press 1994.

Grossberg69: Some networks that can learn, remember and reproduce any number of complicated space-time patterns. Journal of Mathematics and Mechanics 19:53-91. 1969.

Harris94: Advances in Intelligent Control. Edited by C.J. Harris. Published by Taylor and Francis 1994.

IAL93: Automatic Inspection Saves Money and Improves Quality. Commercial publication, Image Automation Ltd, Kelvin House, Worsley Bridge Road, Sydenham, London, SE26 5BX., UK.

Kaebling93: Learning in Embedded Systems. Leslie Pack Kaebling, Bradford Books MIT Press 1993.

Kohonen88: Self-Organisation and Associative Memory. Springer-Verlag 1988.

Maes89: How To Do The Right Thing. Pattie Maes, Connection Science, Vol. 1, No 3, 1989.

Maes and Brooks 90: Learning to Coordinate Behaviours. Pattie Maes and Rodney A. Brooks. AAAI 1990 Conference Proceedings.

Mahadevan and Connell 92: Automatic Programming of Behaviour Based Robots Using Reinforcement Learning. Sridhar Mahadevan and Jonathan Connell, Artificial Intelligence 55, Elsevier Science Publishers 1992.

Mataric94: Reward Functions for Accelerated Learning. Maja J. Mataric. MIT Artificial Intelligence Laboratory. 545 Technology Square, Cambridge MA 02139 USA.

McFarland85: Animal Behaviour, section 2.3 Animal Learning. David McFarland. Publishers Longman Scientific and Technical. 1985.

Nehmzow and Smithers 90: Mapbuilding using Self-Organising Networks in "Really Useful Robots". Ulrich Nehmzow and Tim Smithers. University of Edinburgh Department of Artificial Intelligence Research Paper No. 489.

Pebody94: Acting to Sense: The lowest levels of a Subsumption Architecture. Miles Pebody. Conference proceedings: From Perception to Action 94. IEEE Computer Society Press 1994.

Pfeifer93: Studying Emotions: Fungus Eaters. Rolf Pfeifer. Proceedings 2nd European Conference of Artificial Life 1993.

Pfeifer and Verschure 93: Categorisation, Representations and the Dynamics of System-Environment Interaction: A case study in autonomous systems. Paul F.M.J. Verschure and Rolf Pfeifer. Proceeding of the Second International Conference on the Simulation of Adaptive Behaviour, Bradford books MIT Press 1993.

Smithers94: What the Dynamics of Adaptive Behaviour and Cognition Might Look Like in Agent-Environment Interaction Systems. Tim Smithers. Workshop proceedings: On the Role of Dynamics and Representation in Adaptive Behaviour and Cognition. 9 and 10 December, 1994, Hosted by University of the Basque Country (UPV/EHU) Faculty of Informatics.

Steels93: Building Agents Out Of Autonomous Behaviours. Luc Steels in The 'Artificial Life Route To Artificial Intelligence: Building Situated Embodied Agents. L Steels and R. Brooks (eds) 1993. New Haven: Lawrence Erlbaum Ass.

Wasserman89: Neural Computing: Theory and Practice. Philip D. Wasserman. Van Nostrand Reinhold. 1989.

Interactivism: A Functional Model of Representation for Behavior-Based Systems

Sunil Cherian and Wade Troxell

Department of Mechanical Engineering
Colorado State University, Ft. Collins, CO 80523, USA
E-mail: {sunilc, wade}@lance.colostate.edu

Abstract. In this paper we present the *interactive* model of knowledge representation and examine its relevance to behavior-based robot control. We show how the interactivist position that *representation* and *motivation* are different aspects of the same underlying ontology of interactive dynamic systems emerges within the framework of our *Behavior-Network* architecture. The behavior-network architecture consists of a collection of task-achieving behaviors that interact amongst themselves and with the environment generating global behavior that aid survival in its environment while carrying out various tasks. The representational and motivational aspects of the underlying control structure are implicit in the connectivity between the behavior modules and in the current activity levels over those connections. The interactivist stance enables us to address the issue of representation in behavior-based systems and suggest directions for future inquiry.

1 Introduction

The role of representations in building intelligent autonomous systems has become quite contentious in recent years. Opinions range from the view that intelligence constitutes the appropriate manipulation of symbolic representations that correspond to objects and events in the external world [14], to a flat rejection of internal representations in favor of system-environment dynamics as the basis for building intelligent systems [7]. In this paper we argue that interactivism, a functional explication of representation championed by Bickhard [2, 3, 4], emerges naturally within the framework of behavior-based systems.

2 Knowledge Representation

The concept of *internal representation* has been a central issue in Artificial Intelligence (AI) research. The standard account of representations consider internal representation to be the *common language produced by the various modules of the cognitive agent* [8]. Vision and language modules, inference and planning modules, and muscle and speech systems are examples of such modules. The same representation may be embodied differently, but the resulting implementations will be inter-translatable. Essentially all the implementations will be variants of the same abstract internal representation, and constitutes the knowledge at the disposal of the agent. All the inputs to the system are translated into some internal representation and stored in an explicit form. Some active process, generally one that follows specific rules, uses this internal representation for deductions, planning, etc. The results are also in the form of internal representations. This information is then decoded to form the output of the system.

Bickhard [2] calls these kinds of explicit models *picture models* of knowledge representation. The basic assumption is that knowledge of something must somehow constitute a picture of it. Picture models are based on two basic assumptions: (i) knowledge can be assimilated to structural representation (encodings) and (ii) representation can be assimilated to structural isomorphism.

Bickhard [2] argues that the assumption that structural representation constitutes the essence of knowledge is false and that the attempt to model the essence of representation as some form of structural isomorphism is impossible. He introduces the interactive model of knowledge representation as a viable alternative.

In the realm of robotics research, recognition of the limitations of picture models of knowledge representation has spawned several alternative approaches such as situated cognition [1], physical grounding [5], analogical representations [15], and emergent intelligence [13] for building intelligent systems. The dynamics of agent-environment interactions is the favored alternative to explicit representations as the appropriate basis for synthesizing intelligent systems. Such systems are now widely called behavior-based systems.

Although there are many behavior-based robot control practitioners, the nature of representations in behavior-based systems has not been investigated in detail except to show that explicit correspondence-based representations are not necessary for building competent autonomous systems. It is in this context, that we analyze the interactive model of knowledge representation and its relevance to behavior-based systems.

2.1 The Interactive Model

While picture models of knowledge representation focuses on *elements* of representation, the interactive model views representation as a functional aspect of certain sorts of system processing [4]. Action and interaction between the epistemic agent and its environment, rather than static encodings, are fundamental to the interactivist perspective. Control structures that allow the agent to interact effectively with its environment account for the knowledge at the disposal of the agent. The representation of knowledge is therefore assimilated to dynamic control structures as opposed to static encodings.

The interactive model assumes that knowledge consists of the ability to successfully interact with the environment. The success of an interaction is defined in terms of some sort of goal for the interaction. Bickhard [2] states that:

> knowledge consists of the ability to transform situations in the environment in accordance with goal criteria, and that ability is constituted within goal-directed and goal-competent control structures.

The goal itself need not be explicitly represented. It just needs to be some control relation that orients system processes.

The course of the interaction of a goal-directed system engaged in some interaction with its environment will depend mainly on two factors: (i) the internal organization of the system and (ii) the conditions and responses of the environment with which the system is interacting. Different system organizations will yield different types of interactions within the same environment. Similarly, for the same system organization, different environments will result in differing interactions.

The internal outcome of an interaction can be thought of as implicitly differentiating the current environment into classes that afford differing types of interactions. That is, internal outcomes of current interactions implicitly constitute indications of further interaction potentialities. These indications are therefore anticipatory in nature, and they can be detected to be in error during the course of future interactions. According to interactivism, indications of potential system interactions (indications that are capable of being in error), constitutes the foundational form of representation out of which all other forms of representation are constructed.

Clearly, this representation is not an encoding form of representation. The control structures that enable the system to interact with its environment do not contain any explicit information about what they represent as in the case of structural encodings. What they do have is the ability to indicate interaction potentialities, and to select future activity based on those indications.

2.2 Representational aspect of interactive control structures

The representational ability of interactive control structures hinges on a change in perspective of what it means to know something. A system can be said to know an invariant condition or feature in the environment if the system has the appropriate control structures that can engage the system in particular types of interactions whenever that condition or feature is encountered. The control structure can be said to represent a particular feature or object if the class or category that the control structure interacts with has only one instance [2]. Interactive control structures implicitly define and therefore represent the category of all those things that yield certain patterns of interactions. At the same time, different patterns of interactions can make a functional difference to the system only if they result in differences in resultant internal system states - internal states that affect the future course of activity.

A system engaged in interactions with its environment will continually have to make choices about which interaction to engage in since many interaction potentialities may exist simultaneously. Selection can be based upon indications of interaction potentialities and anticipated outcomes. Such indications constitute the representational aspect of interactive control structures.

Interactive representations can be in error in goal-directed systems if the indicated interaction potentialities do not eventuate. Due to the goal-directed nature of interactive systems, error is detectable by the system affording the modification of future behavior. Interactive representations are therefore capable of system detectable error [4]. The conception of "goal" in interactive goal-directed systems need not themselves involve representations. They only need to be internal functional switches that switch system processes between various alternatives. Therefore, "goals" within interactivism do not succumb to the circularity of requiring representations themselves.

2.3 Motivational aspect of interactive control structures

Within the interactive perspective the appropriate way to pose the question of motivation is: "What makes the system select, or what constitutes the system selecting, one thing to do rather than another?" [4]. This question differs from the standard question: "What makes the system do something rather than nothing?" Addressing the former question involves an explication of action selection in autonomous systems. Selection of activity may depend on indicated interaction potentialities and their relation to anticipated outcomes. This dependence constitutes a motivational aspect of interactive control structures.

To summarize, the interactive model presents a functional explication of representational and motivational phenomena that occur in interactive open systems. Representation and motivation are appropriately viewed as different aspects of the same underlying ontology of interactive dynamic systems. Indication of interaction potentialities and action selection based on those indications, among other things, are seen as properties of a common underlying interactive control structure.

3 Behavior-Based Robot Control

Control architectures that decompose the overall competence of an autonomous system into task-achieving behavior modules - emphasizing a bottom-up design strategy, have gained ground in recent years as a viable alternative to traditional top-down, functional controller decomposition. This trend involves building robots that break away from conventional models that emphasize accurate world modeling, correspondence based symbolic representations, declarative knowledge bases, centralized control architectures, etc. Control architectures such as subsumption architectures [5, 6], spreading activation networks [11], action networks, and behavior-based architectures [10, 12, 9] favor a distributed approach in which the controller is decomposed into behavior modules with tightly integrated sensing and action.

We view behavior-based control as an abstraction for structuring activity in some underlying computational medium such that the resultant dynamical coupling between controlled system processes (agent processes) and environmental processes converge in such a manner that desired tasks are achieved in the given environment. Since a behavior-based control strategy is an abstraction mechanism for controller design, it can be realized in various computational media. Although it places constraints on the underlying computational medium, it makes

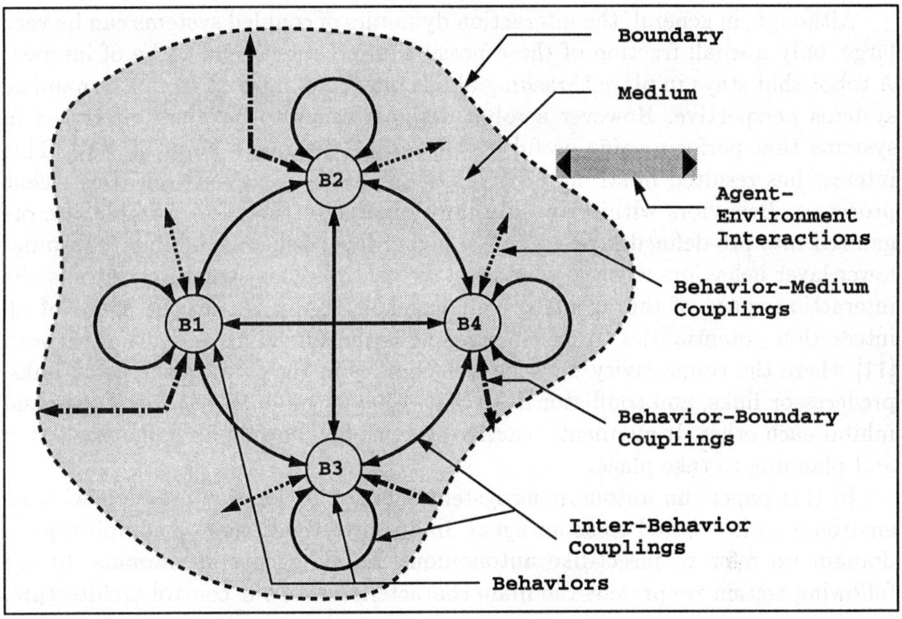

Fig. 1. Generalized representation of behavior modules and their coupling modalities.

no assumptions about the nature of the medium itself. A complete description of a physically realizable behavior-based system must provide descriptions of the underlying computational medium as well as component processes. This can be done without appealing to behavior-based control abstractions. Therefore, behavior-based control abstractions are not necessary for computation *per se*; however, they provide a rigorous framework for structuring as well as understanding the computation.

Figure 1 represents our view of a collection of task-achieving behaviors along with their coupling modalities. The collection interacts with the environment in order to achieve system goals. Each behavior-module is capable of achieving a unique task in the environment or maintaining critical internal system states within acceptable limits. The actions they take and the conditions they perceive modify the agent-environment interaction dynamics. Each behavior is capable of three types of couplings. (i) Behavior-boundary couplings establish information transfer from and to sensors and actuators. (ii) Inter-behavior couplings can be excitatory or inhibitory and enables the synergistic integration of multiple, possibly competing, behaviors. (iii) Behavior-medium couplings also facilitate coherent integration of multiple behaviors by providing a mechanism for utilizing global resources. While inter-behavior connections have a specific source and target, behavior-medium couplings connect individual behaviors with a global resource pool.

Although, in general, the interaction dynamics of coupled systems can be very large, only a small fraction of those possible interactions tend to be of interest. A robot that stays in place thrashing about may be of interest from a dynamical systems perspective. However a robot designer tends to be more interested in systems that perform some useful task from the designer's point of view. This interest has resulted in strong constraints on the coupling between constituent processes. Behaviors within the subsumption architecture, for example, are organized into pre-defined layers where a higher-layer behavior inhibits/subsumes lower layer behaviors when it wants to take control. This structure restricts the interaction space of the agent to a manageable region within the space of all interaction potentialities. Another example is the spreading activation network [11] where the connectivity between behaviors is in the form of successor links, predecessor links, and conflictor links that behavior modules use to activate and inhibit each other. It augments reactive systems by allowing for some prediction and planning to take place.

In this paper, an autonomous system engaged in some interaction with its environment is referred to as an *agent*. In our investigations in a computational domain we refer to insect-like autonomous mobile agents as *animats*. In the following section we present the main characteristics of our control architecture.

3.1 Behavior-Network Architecture

This architecture draws extensively on the working principles of behavior-based systems. The overall architecture consists of a collection of behavior modules that are capable of communicating with each other, interacting with a boundary layer, and with an inter-behavior medium. The boundary layer propagates sensor signals to various behaviors and merges actuation commands from the behaviors and transmits them to the final control elements. The inter-behavior medium enables global interactions between behavior modules. One of the primary motivations behind developing this architecture was to decouple the role of behavior-environment interactions from the interactions between behaviors. This decoupling allows us to pose questions about the competence of individual behaviors, and about the relationship between inter-behavior connectivity and system performance, independent of each other.

Each behavior-module is designed to achieve a specific task within the agent-environment system. A behavior is said to be *activated* if sensed conditions, internal or external to the agent, are conducive to carrying out its task. Influences between behaviors can either be *excitatory* or *inhibitory*. A behavior is said to be excited when the sum of excitatory/inhibitory inputs from other behaviors and self-excitation from itself exceeds some threshold. The behavior is inhibited if the sum is below some threshold. In order for a behavior to generate actuation commands that are actually passed on to the boundary layer and from there to final control elements, it must be both activated and excited. Each behavior also generates excitatory and inhibitory influences that are transmitted to other behaviors. These influences depend upon current activation and excitation

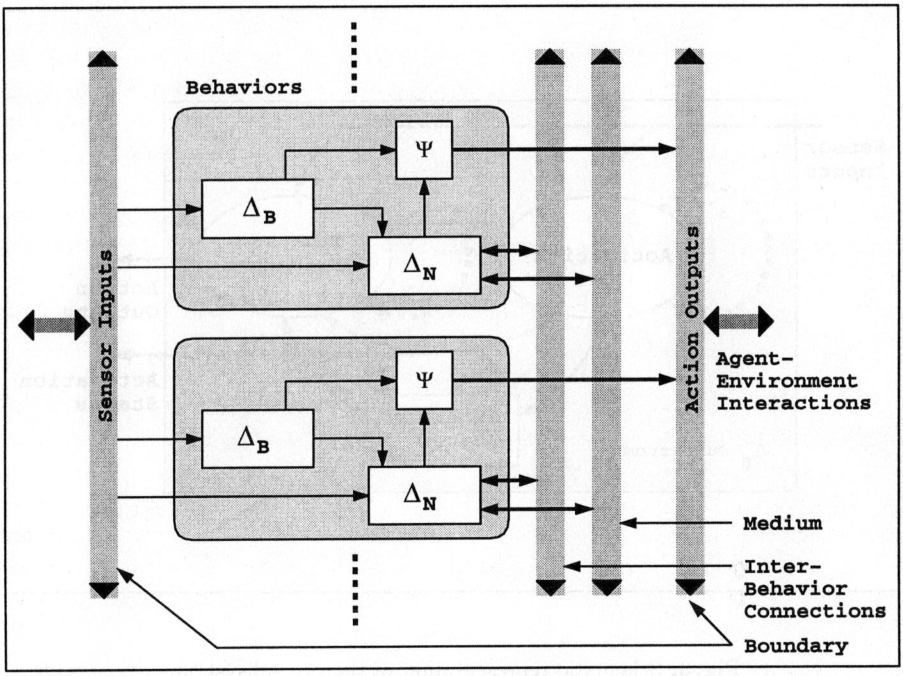

Fig. 2. Schematic representation of the Behavior Network Architecture.

conditions. If a behavior is excited, then influences that further the chances of its being able to execute are transmitted. If a behavior is already executing, it transmits influences to other behaviors such that it can continue execution until its task is achieved.

Each behavior module consists of three subsystems, the Δ_B, Δ_N, and Ψ subsystems. The Δ_B subsystem carries out behavior-environment interactions by handling task-specific processes within each behavior module. It receives sensory information from the boundary layer, outputs actuation commands for carrying out its task to the Ψ subsystem, and transmits its activation status to the Δ_N subsystem.

The Δ_N subsystem handles inter-behavior interactions. It receives excitation and inhibition from other behaviors and sends excitatory and inhibitory influences to other behaviors so as to promote its own chances of executing its task. The Ψ subsystem receives the actuation commands from the Δ_B subsystem and excitation status from the Δ_N subsystem. Actuation commands are passed on to the boundary layer if the behavior is excited and suppressed if it is not. The boundary layer combines actuation commands from all the behaviors and transmits the resultant commands to the final control elements.

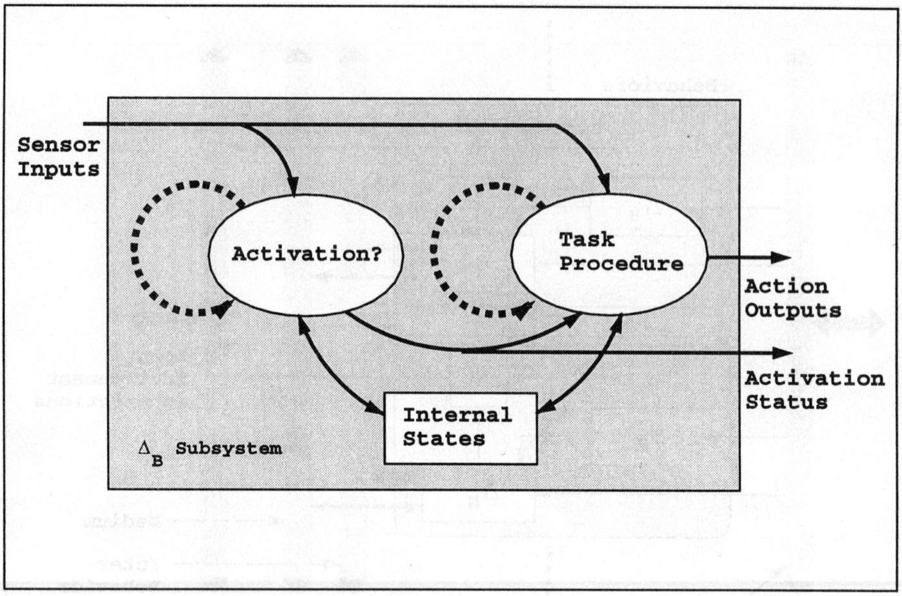

Fig. 3. Schematic representation of the Δ_B subsystem.

Figure 3 shows a schematic representation of the Δ_B subsystem. It continually computes the *activation* status of the behavior. The activation status is some sensed state of affairs in the system or environment conducive to carrying out its task. The activation status is transmitted to the task procedure and to the Δ_N subsystem. If the behavior is activated, the task procedure executes and actuation commands are send to the Ψ subsystem. The activation and task procedure can maintain some internal state information in order to aid system processes.

The Δ_N subsystem, Figure 4, deals with the synergistic interaction of the collection of behavior modules. The Δ_N subsystem receives excitatory, inhibitory, or no signals from other behaviors. In the simplest model these signals are summed and the excitation status is determined. The resultant excitation or inhibition is passed on to other behaviors differentially. The proportion of excitation or inhibition that is transmitted to other behaviors is determined by connection weights maintained by each Δ_N subsystem and by the activation status of the the corresponding Δ_B subsystem. For example, if the the Δ_B subsystem is not activated, while the Δ_N subsystem is excited, then, other behaviors that are likely to change the system/environment relationship in such a way that the behavior is likely to be activated are preferentially excited. Alternatively, if the behavior is activated and excited, behaviors that are likely to interfere with its task execution are inhibited. This preferential excitation and inhibition al-

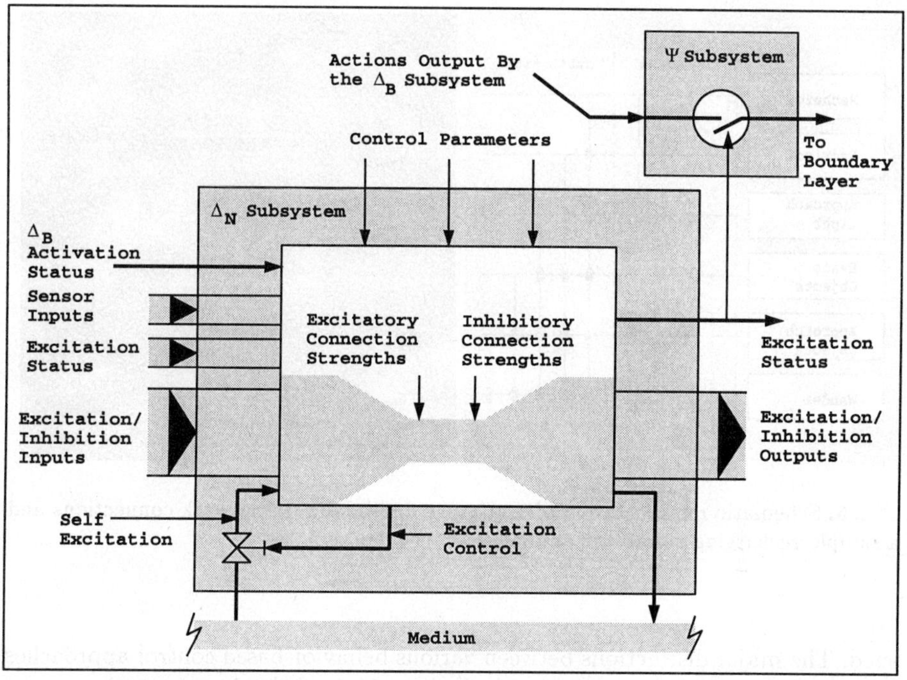

Fig. 4. Schematic representation of the Δ_N and Ψ subsystems.

low various functional relationships to be set up between behavior modules for achieving high level goals consisting of complex sequences of primitive behaviors.

The Δ_N subsystem also accepts sensor inputs and the excitation status of other behaviors. Sensor inputs are used to determine the level of self-excitation. For example, a sensed "hunger" level may determine the level of self-excitation of a "forage" behavior. Self-excitation levels can have complex dependencies on the exciting conditions. The excitation status of other behaviors can be used for modulating influences to those behaviors as well as for adaptation purposes. The control parameters of the Δ_N subsystem, shown in Figure 4, are used to set learning rate, decay rate, percentage connectivity, etc., in order to study the role of connectivity and adaptation in robust behavior.

The Ψ subsystem acts as a switch, selectively allowing actuator commands to pass through if the behavior is both activated and excited. Otherwise the actuator commands are suppressed.

The main point to be noted here concerns the connectivity between Δ_N subsystems in the behavior-network. The connections can either be established by an adaptive process or they can hand wired. In either case they serve the function of integrating various behaviors such that higher level goals can be coherently pur-

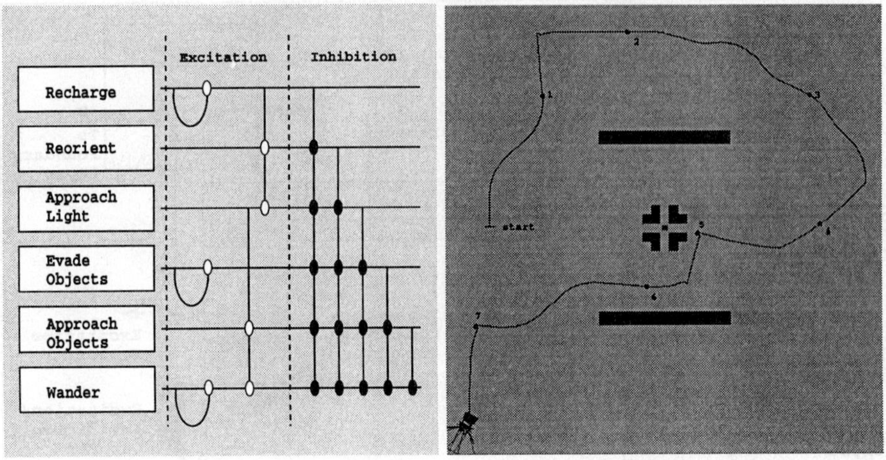

Fig. 5. Schematic representation of excitatory and inhibitory network connections and a sample recharging run of the animat.

sued. The major distinctions between various behavior-based control approaches can be said to reside in the nature and type of inter-behavior connections.

3.2 Example Application

In this section a sample implementation of the control architecture is presented. The behavior network architecture was used to control an animat within 'AnSim'. AnSim is a computational environment that models a 2D world that can be populated with animats. The contents of the world can be specified by the user and consist of blocks, lights, recharging stations, etc. Animats are built individually by the user with an arbitrary number of sensors such as tactile whiskers, infra-red obstacle detectors, light detectors, etc. Sensors provide only binary responses; i.e., either on or off. Actions that an animat can carry out are: move forward, move backwards, turn left, turn right, veer left, and veer right. The experiment we describe was done with an animat with two whiskers, three non-contact obstacle detectors, five light detectors, and a charge detector that uses the animat's whiskers as contact elements. A schematic representation of the controller and a sample run of the animat are shown in Figure 5. The behavior network that controls the animat consists of the behaviors: wander, approach objects, evade objects, approach light, reorient, and recharge.

Normally the animat wanders about in its environment merely avoiding bumping into objects. The recharge behavior becomes self-excited when the available energy of the animat falls below a predetermined threshold. The activation condition of the recharge behavior is the detection of the charging posts by the whiskers of the animat. If this activation condition is not present when the recharge behavior is excited it passes excitation to two other behaviors -

reorient and approach light. The approach light behavior, whose activation condition is the detection of a light source, is designed to guide the animat to the light beacon emitted by the recharging station. If the beacon is not detected in the current situation, approach light feeds excitation to the wander behavior and the approach objects behavior. The animat then wanders around until it detects an object. It then follows the boundary of the object by first avoiding and then approaching the object. On detecting the beacon, the approach light behavior stops feeding excitation to other behaviors and inhibits the other behaviors instead. The animat then approaches the light source and when it bumps into the recharging station, attempts to reorient until its whiskers enable it to recharge. As soon as charge is detected, the recharge behavior stops feeding exciting reorient and approach light behaviors and instead inhibits conflicting behaviors until recharging is complete. Once fully recharged, the recharge behavior stops feeding inhibitory influences to other behaviors and the animat reverts to its exploratory mode.

The network dynamics are easy to comprehend due to the fact that network connections are fairly limited. Feedback loops and self-inhibition are not allowed in the network. This restriction simplifies analysis of network dynamics. However, it does not affect the point we wish to make concerning interactive representation and motivation in behavior-based systems.

4 Representation and Motivation in the Behavior Network Architecture

Within the behavior-network architecture various conditions in the environment as well as within the agent result in the activation of different loci of control. Excitation and inhibition of behavior modules, either by self-excitation or through influences from other behaviors, is achieved through inter-behavior connections. Each behavior module keeps track of inhibitory and excitatory connections propagating to other modules from itself. Each behavior also determines the actual activity over connections to other behaviors based on its current activation and excitation status. Excitatory and inhibitory influences transmitted to other behaviors and actuation commands generated attempt to improve each behavior's chances of achieving its own task.

These facts leads to the following observations. The topology of network connectivity, at any time, can be viewed as indicative of interaction potentialities. For example, in Section 3.2, if the animat behavior "approach light" is excited but unable to execute since no light sources are currently visible, approach light passes excitation on to the "wander" behavior. This activity is based on the anticipation that executing the wander behavior will lead to conditions that enable approach light to execute. It does not matter, for the current argument, how the connections were determined to begin with - hand coded, acquired through some adaptive mechanism or otherwise. What is important is that the connections are indicative of interaction potentialities. Actual activity over inter-behavior connections is modulated by the activation status and excitation status of behavior modules. Since the interactive outcome predicted by the network connectivity can be determined to be true or false by the system itself through interaction, we can conclude that the behavior-network architecture is capable of displaying primitive forms of interactive representation.

Within the behavior-network architecture selection of activity depends on the status of each behavior during each control cycle. The status of the behaviors, in turn, depends on the indicated interaction potentialities. In the animat example in Section 3.2, the selection of a behavior is the outcome of the resultant excitation/inhibition between behaviors. The activation status of each behavior modulates the transmission of excitatory and inhibitory influences between behaviors. This dependence of selection of activity on indicated interaction potentialities closely relates to the interactivist notion of motivation presented in Section 2.3.

Indications of interaction potentialities, the predictive relationships between various behavior modes, and the selection of current activity are implicit in the dynamics of the overall control structure. This aspect of the behavior network architecture is in close agreement with the notion of representation and motivation within the interactive perspective.

5 Conclusion

According to interactivism, representation and motivation cannot be seen as synthesis primitives for building competent autonomous agents but as functional emergents of their underlying control structure. As Bickhard [4] points out, representation and motivation are to interactive control structures as a circle and a rectangle are to a cylinder. They represent aspects of the underlying ontology of interactive dynamic systems, best understood as *properties* of interactive control structures rather than as *elements* of the control structure. Within the behavior-network architecture, activation and excitation status of behavior modules and the activity levels along excitatory and inhibitory connections between behaviors, constitute indications of interaction potentialities that are anticipatory in nature. These indications can therefore be viewed as a primitive form of interactive representation. Since selection of ongoing activity implicitly depends upon these indications, the motivational aspect of interactive control structures is also seen to emerge within the behavior-network architecture.

Interactivism enables a re-integration of representational and motivational phenomena within the behavior-based control paradigm as opposed to their rejection as subjective notions dependent upon observer semantics.

In this paper, we were only able to scratch the surface of interactivism and its implications in the synthesis of autonomous systems. Goal-directedness, the possibility of system detectable error, interactive conceptions of communication, learning and language, and architectures based on modulations among oscillatory processes are included in the promise of interactivism. We believe that interactivism requires serious consideration as an epistemic foundation for the synthesis of intelligent autonomous systems.

References

1. Philip E. Agre and David Chapman. What are plans for? In Pattie Maes, editor, *Designing Autonomous Systems: Theory and Practice from Biology to Engineering and Back*, pages 17–34. MIT Press, 1990.
2. Mark H. Bickhard. *Cognition, Convention, and Communication.* Praeger Publishers, 1980.
3. Mark H. Bickhard. Representational content in humans and machines. *J. Expt. Theor. Artif. Intell.*, 5:285–333, 1993.
4. Mark H. Bickhard and Loren Terveen. *Foundational Issues in Artificial Intelligence and Cognitive Science - Impasse and Solution.* Elsevier Science Publishers, 1995.
5. Rodney A. Brooks. Elephants don't play chess. In Pattie Maes, editor, *Designing Autonomous Agents: Theory and Practice from Biology to Engineering and Back*, pages 3–15. MIT Press, 1990.
6. Rodney A. Brooks. Challenges for complete creature architectures. In Jean-Arcady Meyer and Stewart W. Wilson, editors, *From Animals to Animats: Proceedings of the First International Conference on Simulation of Adaptive Behavior*, pages 434–443. MIT Press, 1991.
7. Rodney A. Brooks. Intelligence without representation. *Artificial Intelligence*, 47(1-3):139–159, 1991.
8. Eugene Charniak and Drew McDermott. *Introduction to Artificial Intelligence.* Addison-Wesley, 1985.
9. Sunil Cherian and Wade O. Troxell. Intelligent behavior in machines emerging from a collection of interactive control structures. *Computational Intelligence.* Blackwell Publishers. Cambridge, Mass. and Oxford, UK. In Press.
10. Hillel J. Chiel, Randall D. Beer, Roger D. Quinn, and Kenneth S. Espenschied. Robustness of a distributed neural network controller for locomotion in a hexapod robot. *IEEE Transactions on Robotics and Automation*, 8(3):293–303, 1992.
11. Pattie Maes. Situated agents can have goals. In Pattie Maes, editor, *Designing Autonomous Agents: Theory and Practice from Biology to Engineering and Back*, pages 49–70. MIT Press, 1990.
12. Pattie Maes. Behavior-based artificial intelligence. In Jean-Arcady Meyer, Herbert L. Roitblat, and Stewart W. Wilson, editors, *From Animals to Animats: Proceedings of the Second International Conference on Simulation of Adaptive Behavior*, pages 2–10. MIT Press, 1993.
13. Chris Malcolm, Tim Smithers, and John Hallam. An emerging paradigm in robot architecture. In T. Kanade, F. C. A. Groen, and L. O. Hertzberger, editors, *Proceedings of the Second Intelligent Autonomous Systems Conference*, pages 284–293, Amsterdam, 1989.
14. Allen Newell and Herbert A. Simon. Computer science as empirical inquiry: Symbols and search. In *ACM Turing Award Lectures.* Communications of the ACM, 1976.
15. Luc Steels. Exploiting analogical representations. In Pattie Maes, editor, *Designing Autonomous Agents: Theory and Practice from Biology to Engineering and Back*, pages 71–88. MIT Press, 1990.

Noise and the Reality Gap: The Use of Simulation in Evolutionary Robotics

Nick Jakobi and Phil Husbands and Inman Harvey

School of Cognitive and Computing Sciences
University of Sussex
Brighton BN1 9QH, England
email: nickja or philh or inmanh@cogs.susx.ac.uk

Abstract. The pitfalls of naive robot simulations have been recognised for areas such as evolutionary robotics. It has been suggested that carefully validated simulations with a proper treatment of noise may overcome these problems. This paper reports the results of experiments intended to test some of these claims. A simulation was constructed of a two-wheeled Khepera robot with IR and ambient light sensors. This included detailed mathematical models of the robot-environment interaction dynamics with empirically determined parameters. Artificial evolution was used to develop recurrent dynamical network controllers for the simulated robot, for obstacle-avoidance and light-seeking tasks, using different levels of noise in the simulation. The evolved controllers were down-loaded onto the real robot and the correspondence between behaviour in simulation and in reality was tested. The level of correspondence varied according to how much noise was used in the simulation, with very good results achieved when realistic quantities were applied. It has been demonstrated that *it is* possible to develop successful robot controllers in simulation that generate almost identical behaviours in reality, at least for a particular class of robot-environment interaction dynamics.

Keywords: Evolutionary Robotics, Noise, High Fidelity Simulations, Artificial Evolution.

1 Introduction

A number of New-Wave roboticists have consistently warned of the dangers of working with over-simple unvalidated robot simulations [2, 1, 16]. Indeed, as Smithers has pointed out [16], the word simulation has been somewhat debased in the fields of AI, robotics, and animat research. Many so-called simulations are abstract computer models of imaginary robot-like entities, not carefully constructed models of *real* robots. Whereas these abstract models can be very useful in exploring some aspects of the problem of control in autonomous agents, great care must be taken in using them to draw conclusions about behaviour in the real world. Unless their limitations are recognised, they can lead to both the study of problems that do not exist in the real world, and the ignoring of problems that do [1]. Behaviours developed in a simulation worthy of the name must correspond closely to those achieved when the control system is down-loaded onto the real robot.

One area of New-Wave robotics where these issues may be particularly pertinent is evolutionary robotics [8]. This is the development of control systems (and potentially morphologies) for autonomous robots through the use of artificial evolution. Populations of robots evolve over many generations in an open-ended way under the influence of behaviour-based selection pressures. Two of the earliest papers on this topic both stressed the likelyhood of having to work largely in simulation to overcome the time consuming nature of doing all the evaluations in the real world [3, 8]. However, both discussed the potential problems with simulations and remarked on the great care that would have to be taken. In [3], Brooks was highly sceptical[1]:

> There is a real danger (in fact, a near certainty) that programs which work well on simulated robots will completely fail on real robots because of the differences in real world sensing and actuation – it is very hard to simulate the actual dynamics of the real world.

and later,

> ... [sensors] ... simply do not return clean accurate readings. At best they deliver a fuzzy approximation to what they are apparently measuring, and often they return something completely different.

But since the aim of evolutionary robotics is to produce working real robots, if simulations are to be used, these problems must be faced. The question is how. In [8] (with further elaborations in [9]) it is argued that:

- The simulation should be based on large quantities of carefully collected empirical data, and should be regularly validated.
- Appropriately profiled noise should be taken into account at all levels.
- The use of networks of adaptive noise tolerant units as the key elements of the control systems will help to 'soak up' discrepancies between the simulation and the real world.
- Noise added *in addition to* the empirically determined stochastic properties of the robot may help to cope with the inevitable deficiencies of the simulation by blurring them. A control system robust enough to cope with such an envelope-of-noise may handle the transfer from simulation to reality better than one that cannot deal with uncertainty over and above that inherent in the underlying simulation model.

This paper reports the results of experiments that were intended to explore the validity of some of these assertions and claims. Network-based control systems for generating simple behaviours were evolved in simulations of different

[1] It is likely that these comments were influenced by experiences with devices rather different to the robot used in the experiments described later. This issue is returned to in the Conclusions.

levels of fidelity and then down-loaded onto the real robot. Comparisons were made between behaviours in the simulations and in the real world.

In [8] it was argued that as the robot's sensory coupling with its environment becomes more complex, simulations would become extremely difficult to construct and would be slower than real time unless highly specialised hardware were available. This problem resulted in the development of the Sussex gantry-robot which allows evolution in the real world [7]. This issue is revisited in the present paper in the light of the experiments outlined above.

The next section discusses related work. Following that is a description of the robot simulation and then an outline of the experimental setup. After detailing the evolutionary techniques used, experimental results are presented and discussed. Finally conclusions are drawn.

2 Related Work

Recently there have been a number of reports on experiences with transferring control systems from simulation to reality. These have met with varying degrees of success. Mondada and Verschure [14] describe the development, through the use of a learning algorithm, of a network-based control system both in simulation and in reality. They used the Khepera robot, the same device involved in the study described later in this paper (see Section 3). Qualitatively similar behaviours were observed in simulation and in reality. However, behaviour on the real robot was significantly less robust and a far greater number of learning steps were needed to achieve reasonable results. This appears to be partly due to the fact that noise was not modelled in the simulation. Miglino, Nafasi and Taylor [13] evolved recurrent network controllers for a very crude computer model of a simple Lego robot. Not surprisingly the evolved controllers generated significantly different behaviours in the real robot. Nolfi, Miglino and Parisi evolved network-based controllers for a simulation of the Khepera robot (described in [15]). Behaviours developed in simulation did not transfer at all well to the robot, but if the GA run was continued for a few generations in the real world (using techniques similar to those described in [5]) successful robust controllers were obtained. This was probably due to the fact that the simulation was based on empirically sampled sensor readings. Sampling appears to have been too coarse, and possibly not enough readings were taken at each point to accurately determine the statistical properties of the sensors. We feel a better approach is to use large amounts of empirical data to set parameters and mappings in a continuous mathematical model of the robot-environment interactions. This technique seems to be vindicated by the results presented later in this paper. In [18] Yamauchi and Beer describe an experiment in which dynamical neural networks were evolved, in a manufacturer supplied simulation of a Nomad 200 robot, to solve a landmark recognition task using sonar. When the evolved controllers were used on the real robots, behaviours were very similar, although not quite as successful, as in the simulation. The simulator was probably very accurate and the highly dynamical networks used are likely to be good at 'soaking up' discrepancies between simulation and reality. Thompson had significant success (although there were some discrepancies) in transferring evolved hardware controllers developed in a semi-simulation (the robot's sonar-environment interaction was simulated, the real hardware and actuators were used) [17]. This was despite the fact that the robot involved is much more cumbersome and less 'clean' than devices such as Khepera, and the sonar simulation was rather crude (although noise was added to the underlying model).

3 The Robot Simulation

3.1 Khepera

The robot used in this project was the Khepera robot developed at E.P.F.L. in Lausanne, Switzerland. It has been specifically designed as a research tool allowing users to run their own programs and control algorithms on the powerful 36 MHz Motorola 68331 chip carried on board (see [11]). The robot is cylindrical with a diameter of 5.8 cm and a height of 3.0 cm. It has eight active IR proximity sensors mounted six on the front and two on the back. The receiver part of these sensors may also be used in a different mode to passively measure surrounding ambient light. The wheels are driven by extremely accurate stepper motors under P.I.D. control. Each motor incorporates a position counter and a speed of rotation sensor. It should be noted that it is probably far easier to build an accurate simulation of this sort of robot than it would be for many others.

3.2 The Simulation

The simulation of the Khepera was built using empirical information obtained from various experiments. The simulation is based on a spatially continuous, two dimensional model of the underlying real world physics and not on a look-up tabl approach as in [15]. This affords greater generality with respect to new environments and unmodelled situations although at some computational expense. The simulation is updated once every 100 simulated milliseconds: the rate at which the inputs and outputs of the neural network control architectures are processed. This results in relatively coarse time slicing, some of the effects of which may be moderated by noise.

Modelling Methodology The decision on which level of detail to pitch the simulation was arrived at intuitively. The idea was to build into the simulation all the important features of the Khepera's interaction with its environment without going into so much detail that computational requirements became excessive. After initial experimentation, it became clear which features were important to model (as fields in a state vector) and which could be left out. For example the distance from an IR sensor to the nearest object is an important feature model whereas the height of objects is not.

An idealised mathematical model was constructed by applying elementary physics and basic control theory to the interactions between these environmental variables. More specifically, generalised equations (with unassigned constants) were derived capable of producing values proportional to the ambient light intensities, the reflected infra red intensities and the wheel speeds.

After developing these general equations, several sets of experiments were performed to find actual sensor values and noise levels for specific settings of environmental variables. Using curve-fitting techniques it was then possible to produce mappings from the predictions of light intensities, wheel speeds and so on produced by the model to the actual sensor values observed during experimentation. As part of the same process values were also attributed to all unassigned constants to produce the set of equations (given below) that are used by the simulation to calculate specific values for the IR sensors, ambient light sensors and wheel speeds.

Modelling the Khepera's Movement, Motors and PID controllers All experiments performed on the Khepera's motors, PID controllers and general movement were carried out with the aid of the in built position and speed sensors. By connecting the Khepera to a host computer using the supplied serial cable, accurate statistics on the Khepera's current speed and position could be gathered while the robot was moving. In this way a profile of the Khepera's response to motor signals was calculated and mapped onto the model given below.

When one of the motors on the Khepera is set to run at a certain target speed V its actual speed, U, will in fact oscillate around this figure due to axle noise, irregularities on the ground surface and so on. Khepera uses a PID control algorithm [12] that ensures that U never varies significantly from V. It also has the important consequence that each wheel, over time, travels approximately the correct distance.

The PID algorithm changes the motor torque T according to the equation:

$$T \propto K_p(V - U) + K_i \int V - U\,dt + K_d \frac{d(V - U)}{dt}$$

where K_p K_i and K_d are the proportional, integral and derivative constants.

In the simulation, the integral and derivative terms could only be approximated due to the relatively coarse time slicing involved. The proportional error, P, at time t is calculated as $P_t = (V - U)/0.1$ since the simulation is updated every 0.1 seconds. The integral term, I_t, is the sum of the proportional terms over the last five time steps. The derivative term, D_t, is proportional to the force applied to the wheels on the last time step. It is calculated by dividing the change to the wheel speeds on the last time step, δv_{t-1}, by an empirically found constant of 50000, equivalent to the robot's mass. Once these terms have been calculated then the change to the wheel speed at time t, δv_t is calculated as:

$$\delta v_t = K_p \times P_t + K_i \times I_t + K_d \times D_t$$

The simulated robot's movement was found empirically to best match that of the Khepera when K_p K_i and K_d were set at 3800, 800 and 100 respectively. These values are, in fact, the same as those used by the PID controller on the Khepera (see [11]).

Static friction (the force that has to be overcome to start an object moving) is modelled by noisily thresholding the integral term. The path taken by each wheel during a simulation update is modelled as an arc of a circle.

Modelling the Khepera's Infra Red Sensors Ray tracing techniques are used to calculate values for the IR sensors (see [6]). Ten rays are used for each sensor arranged in an arc spanning 180°. If the distance from a particular sensor along a ray i to an object is d_i, then the sensor value is calculated as

$$I = \sum_{i=1}^{10} \cos \beta_i (a/d_i^2 + b)$$

where β_i is the angle at which the ray i leaves the sensor and a and b were set empirically (see above) to 3515.0 and -91.4 respectively.

3.3 The Ambient Light Sensors

In reality there are many factors that have a measurable effect on the ambient light sensor values. The model used in the simulation is, therefore, correspondingly complicated. Ray tracing to a depth of two rays (the IR sensors are calculated from a depth of one) the intensity of the ambient light at a sensor is calculated as a sum of direct illumination and reflection. Experiments using the ambient light sensors were carried out in an environment with one major light source (see Section 4). The light source (in reality a 60W desk lamp) is modelled by five point sources. For each point at which one of the rays leaving the sensor hits the wall, five lamp-rays are calculated between this point and the five point sources approximating the lamp. The number of unobstructed lamp-rays is used to calculate the brightness of the reflection. Similarly direct illumination of a sensor is calculated from the number of unobstructed lamp-rays between that sensor and the five point sources.

Because the light source is brighter in the middle than at the edges, lamp-rays originating from the centre play a greater role in determining sensor illumination than those at the edges. The direct illumination D of a sensor is calculated as:

$$D = (-4.15 + \sum_{i=1}^{5} L_i \times k_i)/d^2$$

where d is the average of the distances from the sensor to each point source. L_i is 0 if the lamp-ray from the sensor to point source number i is occluded and 1 otherwise. k_i are the empirically derived weights for each point source: 4.19 for the point source on each edge, 4.24 for the pair inside each of these and 8.24 for the point source at the centre.

If D_i is the direct illumination of ambient sensor i and D_{ij} is the direct illumination of the point at which ray j of length d_{ij} from sensor i hits an object, then the total illumination A_i of sensor i is calculated as:

$$A_i = D_i + \sum_{j=1}^{10} (D_{ij} \times \cos^2 \beta_{ij} \times 1.5)/d_{ij}^2$$

where β_{ij} is the angle at which ray j leaves sensor i.

Finally the simulated sensor value V is calculated from the total illumination A according to the empirically determined mapping:

$$V = 55.0 + 1/\sqrt{A}$$

4 The Experimental Setup

In order to test the assertions outlined in the introduction, two sets of evolutionary runs were carried out, each involving the evolution of a different behaviour.

In the first set of experiments, obstacle avoiding behaviours were sought. The task here was to move around the environment covering as much ground as possible without crashing into objects. The environment, shown in figure 1, consisted of a square arena with sides of length 50cm constructed from yellow painted wood and four grey cardboard cylinders with a radius of 4cm.

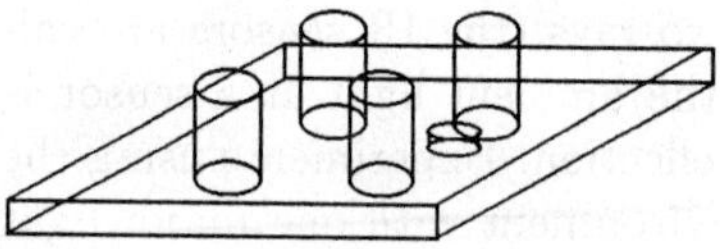

Fig. 1. The environment used in obstacle avoiding experiments

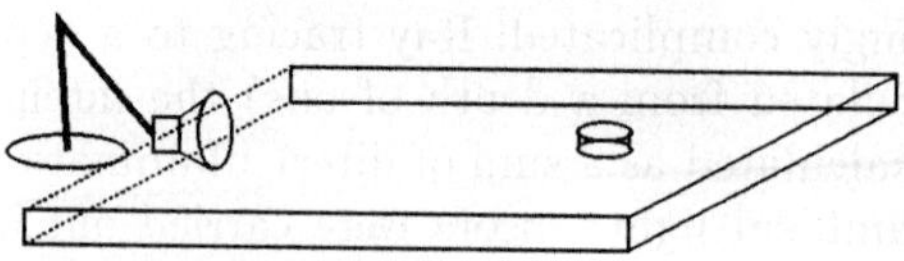

Fig. 2. The environment used in light-seeking experiments

In the second set of experiments an ordinary sixty watt desk lamp was placed at one end of a 110cm by 70cm arena again made from yellow painted wood (Figure 2). A light seeking behaviour was sought which would enable the robot to move towards the lamp when started from random orientations and positions at the other end of the arena.

Before describing the experiments in detail, the evolutionary techniques will be explained.

5 Evolutionary Machinery

5.1 The Genetic Algorithm

In order to try and avoid some of the problems of premature convergence associated with more traditional genetic algorithms, a distributed GA was used

[4]. The population was distributed over a two dimensional grid, and local selection was employed. Each member of the population breeds asynchronously with a mate chosen from its eight immediate neighbours on the grid. The mate is selected probabilistically using a linear fitness-rank-based distribution. The offspring replaces a member of the neighbourhood (this could potentially be either of its two parents) according to a linear inverse fitness-rank-based probability distribution. The genetic operators employed were mutation and crossover at rates of 0.05 (mutations per piece of information stored on the genotype) and 0.8 respectively. Neurons and links were also added and/or deleted from the offspring genotype (see below) on each breeding with probabilities of 0.05 and 0.1 respectively.

5.2 The Encoding Scheme

The encoding scheme is the way in which phenotypes (in this case neural nets) are encoded by genotypes (the structure on which the genetic algorithm operates). The most commonly used encoding schemes involve direct one to one mapping between genotype and phenotype. The genotype consists of a series of fields expressed in bits, real numbers or characters. Separate fields specify the characteristics of each neuron and the connections associated with it. Networks encoded using a direct scheme can suffer gross distortions when their genotypes are allowed to grow or shrink.

The encoding scheme used in this research was designed to resolve some of these problems. The size of the phenotype is under genetic control but the addition or deletion of neurons and connections has minimal effect on the structure of the original network. The main difference between the encoding scheme used here and more normal direct encoding methods is that while phenotype size is under evolutionary control genotype size stays fixed. Instead of using a series of fields that specify neuronal characteristics, each genotype is made up of a series of 'slots', each one of which *may or may not* define a particular neuron on the phenotype and the links associated with that neuron. Connections address neurons by the absolute address of the slot they are associated with. Provided the genotype does not run out of spare slots to store new neurons in, any addition or deletion of neurons has minimal consequences on the rest of the network. In the runs described below there were 30 slots per genotype, of which typically 10–12 were used.Full details of the scheme can be found in [10].

5.3 The Neural Networks

A form of arbitrarily recurrent dynamical network was used in this research. The activation function of each neuron is defined as a simple linear threshold unit with fixed slope and a genetically determined lower threshold. Connections had genetically set time delays and weights.

In general, the smaller the number of parameters there are that need to be set in order to define a particular neural network, the quicker most learning algorithms or search techniques will be in finding suitable values for solving a particular problem. For this reason network parameters were restricted to a small number of integer values. Connection weights and delays were restricted to the interval ± 4, where the unit of time delay is ten milliseconds. Activation thresholds were restricted to the interval ± 10 and neuron output values to the interval ± 10. All inputs and outputs of the network were scaled to the interval ± 10 in order to maximise network response and were updated every hundred milliseconds.

6 Experimental Results

Once a good underlying simulation model had been developed, it was found that in general, a neural network evolved in simulation evoked qualitatively similar behaviour on the real robot. During the entire period of this research there was never a negative instance in the sense of no similarity at all. The correspondence between simulated and situated behaviour turns out to be a matter of degree rather than binary valued. The following experiments were designed to inspect two factors that affect this correspondence: the nature of the behaviour itself and the level of noise present in the simulation.

For each of two behaviours, obstacle avoiding and light seeking, three sets of five evolutionary runs were performed, one set for each of three different noise levels. These three noise levels were set at zero noise, observed noise and double observed noise. Observed noise (on sensors, motors etc.) refers to a roughly Gaussian distribution with standard deviation equal to that empirically derived from experiments. Double observed noise refers to the same distribution with double the standard deviation.

Because of the stochastic nature of the evolutionary process thirty runs were performed in total in order to acquire some statistical support for the conclusions that may be drawn from them. After the runs were complete, the evolved behaviours were subjectively marked by the authors on their optimality (how close they came to the ideal strategy/behaviour) and the correspondence between behaviours in simulation and reality[2]. The results for both behaviours are displayed in Tables 1 and 2.

Figure 3 and 5 both contain pictures showing paths taken by the Khepera in the real world. These were made by applying image processing techniques to short films of the Khepera, moving around its environment, with a specially constructed black and white disk placed on its uppermost face. Each frame underwent convolution operations using D.O.G. [3] center-surround masks specifically designed to respond maximally to the white patches at the centre and front of the disk placed on the Khepera. The positions of the peaks in the resultant intensity arrays were then used to pinpoint precisely the position and orientation

[2] It is possible that some objective scoring system could be devised based on statistics of agent-environment interactions, but because of the nature of the problem it is not clear what this system would look like.

[3] A three dimensional mask constructed from the rotation of the difference between two Gaussian curves, of appropriate widths, around the vertical axis

of the Khepera in each frame. After processing an entire sequence, the lines, one per frame, joining the centre of the Khepera to its leading edge, were overlaid on the final frame to produce an image of the Khepera with a white 'tail' behind it. The pictures of paths taken by the simulated robot contain a 'tail' of the same form. These were constructed by failing to erase a line, plotted on each previous time step, also joining its centre to its leading edge.

6.1 Obstacle Avoidance

Nolfi et al. [15] were the first to try evolving behaviours in simulation for the Khepera robot. In their work, outlined in Section 2, they were attempting to evolve obstacle avoiding behaviours. As already discussed, they did not achieve close correspondence between simulation and reality.

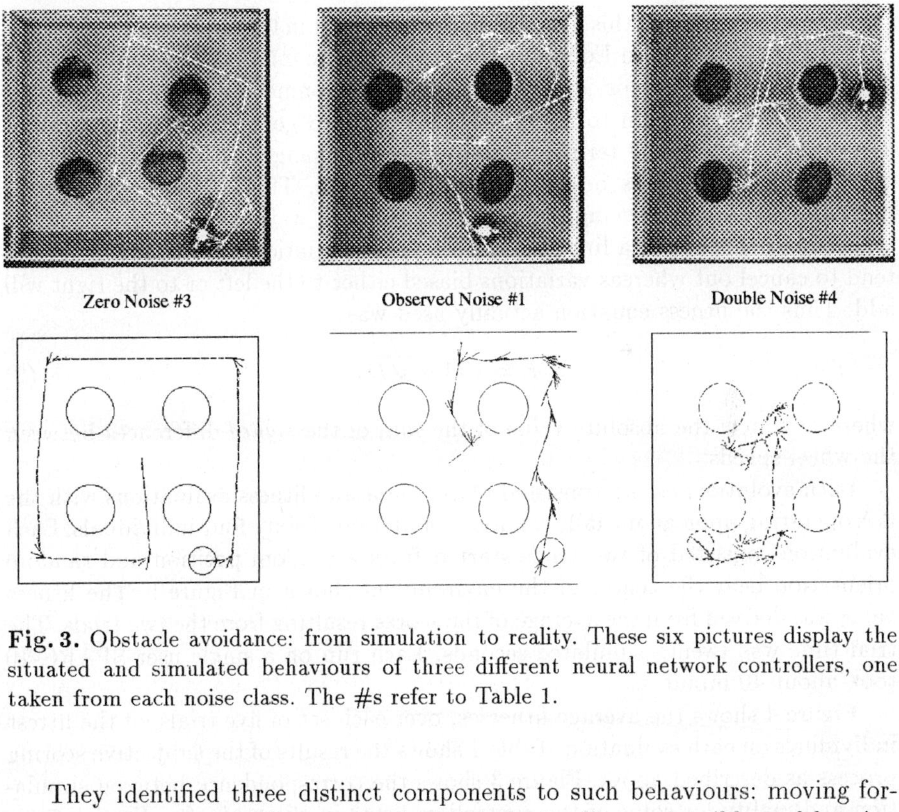

Fig. 3. Obstacle avoidance: from simulation to reality. These six pictures display the situated and simulated behaviours of three different neural network controllers, one taken from each noise class. The #s refer to Table 1.

They identified three distinct components to such behaviours: moving forwards as fast as possible, moving in as straight a line as possible and keeping as far away from objects as possible. The fitness function they employed reflects this. During a trial three normalised sums are calculated: V, the sum of the wheel speeds at each time step, D, the signed sum of the absolute differences between the speeds of the two wheels at each time step and I the sum of the

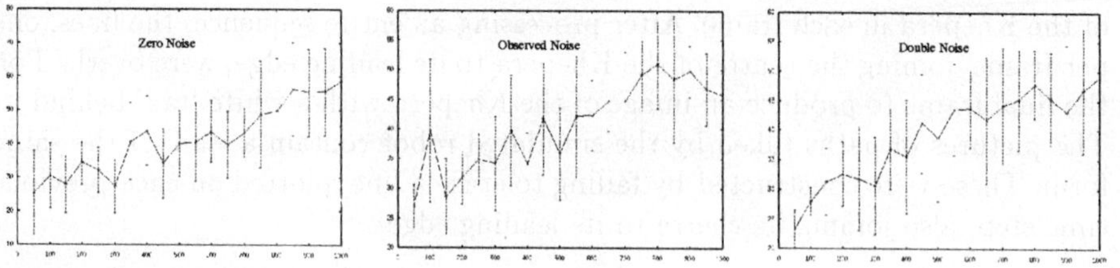

Fig. 4. Obstacle avoidance. The average fitnesses, over each set of five trials, of the fittest individuals on each evaluation in simulation.

largest of the eight sensor values at each time step. These are combined to give a score F according to Equation 1:

$$F = V(1 - \sqrt{D})(1 - I) \quad (1)$$

A slight variation of this fitness function was used in the experiments reported here. The $(1 - I)$ term in Equation 1 was found to be implicit if the environment is cluttered enough. This is because, in the environment used (Figure 1), the robot will have to learn to avoid objects if it is to go as fast and as straight as possible. Also the D term in Equation 1 was changed to the unsigned sum of the signed differences between the wheel speeds. This was thought to be a more effective way of forcing the robot to turn both ways to avoid objects while traveling in as straight a line as possible. Slight variations to left and right will tend to cancel out whereas variations biased either to the left or to the right will add. Thus the fitness equation actually used was:

$$F = V(1 - \sqrt{D}) \quad (2)$$

where D is now the absolute value of the sum of the *signed* differences between the wheel speeds.

Each evolutionary run consisted of one thousand fitness evaluations with the GA operating upon an initially random population of sixty four individuals. Each evaluation consisted of two trials started from a random position and random orientation near the centre of the environment shown in Figure 1. The fitness value was derived from the average of the scores resulting from the two trials. The trial time was twenty simulated seconds. Each run on a single user SPARC-10 took about 40 minutes.

Figure 4 shows the average fitnesses, over each set of five trials, of the fittest individuals on each evaluation. Table 1 shows the results of the subjective scoring process as described above. Figure 3 shows the correspondence between simulation and reality for some of the controllers listed in the table. See Section 7 for a discussion of these results.

6.2 Light Seeking

#	*type of behaviour*	*behaviour score*	*differences*	*correspondence score*
		Zero Noise		
1	one way turner	5	a little noisier	8
2	wall follower	3	qualitatively similar	7
3	one way turner	4	noisier	5
4	one way turner	3	a little noisier	7
5	one way looper	5	noisier	4
		average 4		average 6.2
		Normal Noise		
1	two way turner	8	a little noisier	8
2	one way looper	5	very similar	9
3	wall follower	4	a little noisier	7
4	wall follower	6	very similar	9
5	wall follower	7	a little noisier	7
		average 6		average 8
		Double Noise		
1	one way turner	4	a little less responsive	8
2	two way turner	8	slightly less noisy	8
3	one way looper	4	less responsive	5
4	one way turner	5	less responsive	6
5	wall follower	3	less responsive	7
		average 4.8		average 6.8

Table 1. Obstacle avoidance. This table shows the scores subjectively given by our panel of judges to the evolved neural networks on the basis of the quality of their behaviours in simulation and the correspondence between their behaviour in simulation and their behaviour in reality. All scores are out of a maximum of 10.

This behaviour is perhaps easier to evolve than obstacle avoidance as everything happens on a much slower scale. An obstacle avoider using short range IR sensors must turn the moment (or very soon afterwards) it senses an obstacle. A light seeking robot may employ a number of different strategies that will take it to the light eventually. It may react instantly to the light *or* take slow curving paths. The fitness landscape of such a task is therefore much smoother and light seeking behaviour emerges fairly early on in the evolutionary process.

The fitness function for this behaviour was simply calculated as the reciprocal of the sum of the squares of the distance from the light source at any particular time step. In other words if D_i is the distance to the light source at time step i, then the fitness F after n time steps is calculated as:

$$F = \frac{1}{\sum_{i=1}^{n} D_i^2} \tag{3}$$

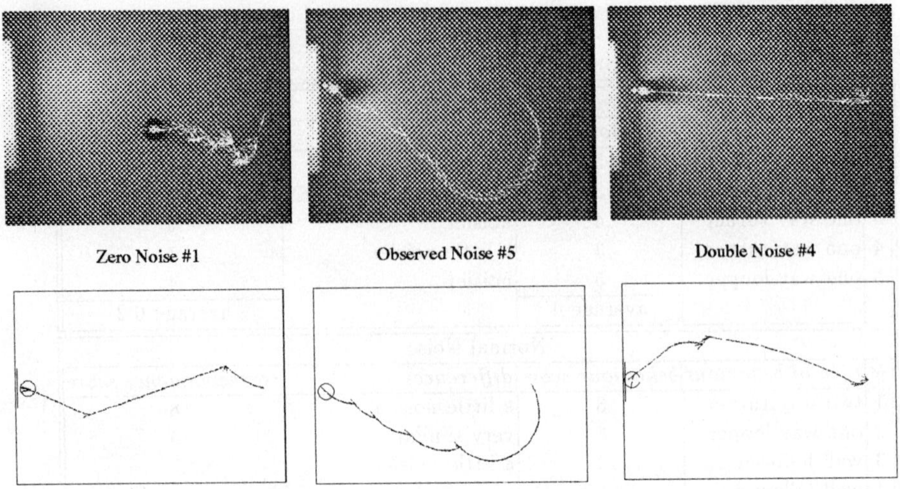

Fig. 5. Light seeking: from simulation to reality. These six pictures display the situated and simulated behaviours of three different neural network controllers, one taken from each noise class. The #s refer to Table 2.

Again, each evolutionary run consisted of one thousand fitness evaluations with an initially random population of sixty four individuals. A single evaluation consisted of two separate trials started from a fixed position and random orientation near the opposite end of the environment (shown in Figure 2) to the light source. The fitness value was the average of the two scores. The trial time was twenty simulated seconds. Each run on a single user SPARC-10 took about an hour.

Figure 6 shows the average fitnesses, over each set of five trials, of the fittest individuals on each evaluation. Table 2 shows the results of the subjective scoring process as described above. Figure 5 shows the correspondence between simulation and reality for some of the controllers listed in the table. See Section 7 for a discussion of these results.

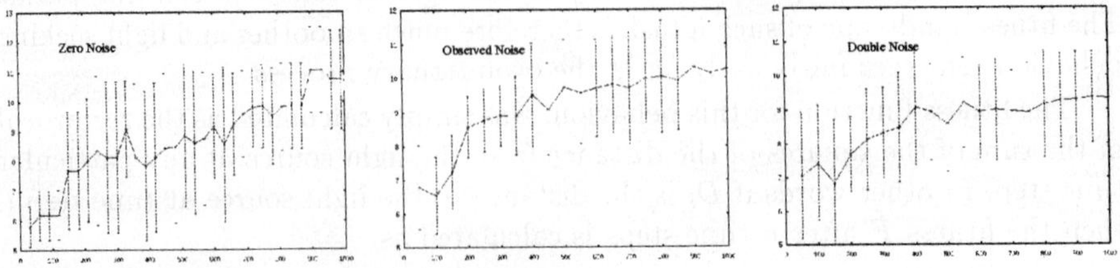

Fig. 6. Light seeking. The average fitnesses, over each set of five trials, of the fittest individuals on each evaluation in simulation.

Zero Noise				
#	*type of behaviour*	*behaviour score*	*differences*	*correspondence score*
1	three point turner	6	noisier	4
2	straight looper	4	loops more	6
3	two way turner	6	similar	8
4	two way turner	8	noisier	4
5	two way turner	8	a little noisier	6
		average 6.4		average 5.6
Normal Noise				
#	*type of behaviour*	*behaviour score*	*differences*	*correspondence score*
1	straight looper	5	a little noisier	8
2	two way turner	8	a little noisier	6
3	two way turner	9	very similar	9
4	two way turner	7	a little noisier	6
5	bobber	5	observably identical	10
		average 6.8		average 7.8
Double Noise				
#	*type of behaviour*	*behaviour score*	*differences*	*correspondence score*
1	two way turner	7	much noisier	2
2	bobber	5	less noisy	6
3	straight looper	3	less noisy	7
4	two way turner	8	less noisy	6
5	bobber	6	less noisy	6
		average 5.8		average 5.4

Table 2. Light seeking behaviours. This table shows the scores subjectively given by our panel of judges to the evolved neural networks on the basis of the quality of their behaviours in simulation and the correspondence between their behaviour in simulation and their behaviour in reality. All scores are out of a maximum of 10.

7 Discussion

7.1 The General Picture

The overall picture to be gleaned from a study of Tables 1 and 2 is perhaps not very surprising. In general, networks evolved in an environment that is less noisy than the real world will behave more noisily when downloaded onto the Khepera and, conversely, networks evolved in an environment that is noisier than the real world will behave less noisily when downloaded. Simulation to situation correspondence seems to be maximised when the noise levels of the simulation have similar amplitudes to those observed in reality. The behaviours shown in Figures 3 and 5 graphically illustrate this.

Noise also plays a part in determining the quality (see Tables 1 and 2) of behaviours that evolve. This was a score subjectively attributed to behaviours on the grounds of how robust and optimal they proved to be *in simulation* after repeated testing. Since a fitness evaluation only involves two trials, these scores do not necessarily correspond well to the fitness values ascribed by the GA. This correspondence could be improved by increasing the number of trials that make up a fitness evaluation, but at the expense of lengthening the time taken to perform an evolutionary run. The behaviour scores, then, show which noise levels give the best 'value for money' in terms of robustness against the number of trials that make up a fitness evaluation.

For both obstacle avoidance and light-seeking, the set of experiments running under observed noise obtained the highest average behaviour score. In a zero noise environment, brittle 'hit or miss' strategies tend to evolve which either score incredibly well or incredibly badly on each fitness trial, depending on their initial random starting positions. Although noise, in general, blurs the fitness landscape, reducing the possibility of 'hit or miss' strategies evolving (since they are far more likely to 'miss' rather than 'hit'), too much randomness in the environment, as in the double noise case, ensures that the same genotypes may again achieve very different scores on two otherwise identical fitness evaluations. A balance between these two cases seems to be achieved at the observed noise level. However, the fact that the particular level of noise that most favours the evolution of robust behaviours in simulation is also the level of noise that achieves the highest simulation to reality correspondence should probably not be regarded as anything other than a coincidence.

7.2 The Noise Level Has to Be Right

If the noise levels used in the simulation differ significantly from those present in reality, whole different classes of behaviours become available which, while acquiring high fitness scores in simulation, necessarily fail to work in reality. The zero noise behaviour shown in Figure 3 is one example. Evolution has taken advantage of the fact that, in a zero noise environment, the simulated robot will react identically in similar situations. By always turning through exactly ninety degrees, the simulated robot circumnavigates the room, ad infinitum, scoring well on the fitness test. In reality, the Khepera will never respond in exactly the same way twice. As can be seen from the picture, its behaviour is qualitatively similar, but the fact that its turns are never exactly ninety degrees means that it cannot settle down into a steady state of circumnavigation. Instead, because the neural network controlling it is 'blind' on one side, it ends up hitting objects and displaying far from optimal behaviour.

It is perhaps less obvious that evolution can also take advantage of *too much* noise in the simulation to produce networks that rely on the extra noise and are thus incapable of reproducing their behaviours in reality. The first double-noise light-seeker in Table 2 is one such example. Here a network that displayed near optimal (if noisy) light-seeking behaviour in simulation proved totally useless when downloaded onto the Khepera. It would jitter rapidly from side to side, approaching the light only very slowly, and occasionally backing right away from it into the rear wall. The neural network in question uses two sensors, one on either side, to find the light. Each sensor controls the wheel on the opposite side of the robot in a Braitenberg fashion, but they are so arranged that, when the Khepera exactly faces the light-source, they are both *only just* illuminated. Since each input of the neural network is divided into bands, they are never both illuminated sufficiently (in a low noise world) to provide positive input to the neural network at the same time, and thus the robot jitters on the spot. However, in the double noise environment, there is enough noise present to push each sensor input's value up from the lowest band every now and then, thus providing positive inputs from both sensors at the same time, driving the robot forwards.

The experimental results provided some (inconclusive) support for the envelope-of-noise conjecture mentioned in the Introduction – that inevitable deficiencies in the simulation could be blurred by noise. In this case, sensor noise *profiles* were not modelled, rather a simple distribution of noise at the right level was used. Also differences between individual sensors were not modelled.

8 Conclusions

It has been shown that it is possible to artificially evolve successful network-based control systems in simulation that generate almost identical behaviours in reality. However, great care must be taken in building the simulation and appropriate levels of noise *must* be included.

The robot-environment interactions modelled here are relatively simple. Difficulties in simulating interference between the IR and ambient light modes of the Khepera's sensors, suggest that the approach taken here rather quickly become less feasible as the interaction dynamics become more complex. We still feel that real-world evolution techniques such as those described in [7, 5] are necessary, at least for the time being, to deal with these more complex couplings. However, it does appear that simulations are not quite the dead-end some had suggested. For simpler cases at least, it has been shown that they can be made accurate enough. Their attractive qualities of speed and ease of data collection can then be made use of.

Acknowledgements

Nick Jakobi is supported by a COGS postgraduate bursary. Thanks to colleagues in the Evolutionary and Adaptive Systems Group for useful discussions. Special thanks to David Young and Bob Ives for help in conducting the experiments described in this paper.

References

1. R.A. Brooks. Intelligence without reason. In *Proceedings IJCAI-91*, pages 569–595. Morgan Kaufmann, 1991.
2. R.A. Brooks. Intelligence without representation. *Artificial Intelligence*, 47:139–159, 1991.

3. Rodney A. Brooks. Artificial life and real robots. In F. J. Varela and P. Bourgine, editors, *Proceedings of the First European Conference on Artificial Life*, pages 3–10. MIT Press/Bradford Books, Cambridge, MA, 1992.
4. R. Collins and D. Jefferson. Selection in massively parallel genetic algorithms. In R. K. Belew and L. B. Booker, editors, *Proceedings of the Fourth Intl. Conf. on Genetic Algorithms, ICGA-91*, pages 249–256. Morgan Kaufmann, 1991.
5. D. Floreano and F. Mondada. Automatic creation of an autonomous agent: Genetic evolution of a neural-network driven robot. In D. Cliff, P. Husbands, J.-A. Meyer, and S. Wilson, editors, *From Animals to Animats 3, Proc. of 3rd Intl. Conf. on Simulation of Adaptive Behavior, SAB'94*. MIT Press/Bradford Books, 1994.
6. A.S. Glasner, editor. *An Introduction To Ray Tracing*. Academic Press, London, 1989.
7. I. Harvey, P. Husbands, and D. Cliff. Seeing the light: Artificial evolution, real vision. In D. Cliff, P. Husbands, J.-A. Meyer, and S. Wilson, editors, *From Animals to Animats 3, Proc. of 3rd Intl. Conf. on Simulation of Adaptive Behavior, SAB'94*, pages 392–401. MIT Press/Bradford Books, 1994.
8. P. Husbands and I. Harvey. Evolution versus design: Controlling autonomous robots. In *Integrating Perception, Planning and Action, Proceedings of 3rd Annual Conference on Artificial Intelligence, Simulation and Planning*, pages 139–146. IEEE Press, 1992.
9. P. Husbands, I. Harvey, and D. Cliff. An evolutionary approach to situated AI. In A. Sloman et al., editor, *Proc. 9th bi-annual conference of the Society for the Study of Artificial Intelligence and the Simulation of Behaviour (AISB 93)*, pages 61–70. IOS Press, 1993.
10. N. Jakobi. Evolving sensorimotor control architectures in simulation for a real robot. Master's thesis, School of Cognitive and Computing Sciences, University of Sussex, 1994.
11. K-Team. Khepera users manual. EPFL,Lausanne, June 1993.
12. J. Kunt. *An Introduction to Control Theory*. Huddersfield and Wallstone, 1964.
13. O. Miglino, C. Nafasi, and C. Taylor. Selection for wandering behavior in a small robot. Technical Report UCLA-CRSP-94-01, Dept. Cognitive Science, UCLA, 1994.
14. F. Mondada and P. Verschure. Modeling system-environment interaction: The complementary roles of simulations and real world artifacts. In *Proceedings of Second European Conference on Artificial Life, ECAL93*, pages 808–817. Brussels, May 1993, 1993.
15. S. Nolfi, D. Floreano, O. Miglino, and F. Mondada. How to evolve autonomous robots: Different approaches in evolutionary robotics. In R. Brooks and P. Maes, editors, *Artificial Life IV*, pages 190–197. MIT Press/Bradford Books, 1994.
16. Tim Smithers. On why better robots make it harder. In D. Cliff, P. Husbands, J.-A. Meyer, and S. Wilson, editors, *From Animals to Animats 3, Proc. of 3rd Intl. Conf. on Simulation of Adaptive Behavior, SAB'94*, pages 54–72. MIT Press/Bradford Books, 1994.
17. A. Thompson. Evolving electronic robot controllers that exploit hardware resources. In *Submitted to ECAL'95*, 1995.
18. B. Yamauchi and R. Beer. Integrating reactive, sequential, and learning behavior using dynamical neural networks. In D. Cliff, P. Husbands, J.-A. Meyer, and S. Wilson, editors, *From Animals to Animats 3, Proc. of 3rd Intl. Conf. on Simulation of Adaptive Behavior, SAB'94*, pages 382–391. MIT Press/Bradford Books, 1994.

Essential Dynamical Structure in Learnable Autonomous Robots

Jun Tani
Sony Computer Science Laboratory Inc.
Takanawa Muse Building, 3-14-13 Higashi-gotanda,
Shinagawa-ku,Tokyo, 141 JAPAN
email: tani@csl.sony.co.jp
Fax +81-3-5448-4273
Tel +81-3-5448-4380

Abstract. This paper studies the essential dynamical structure that arises in two different classes of learning of the sensory-based navigation, namely *skill-based learning* and *model-based learning*. In *skill-based learning* a robot learns navigational skills for a fixed navigational task such as *homing*, while in *model-based learning* a robot learns a model of the environment, then conducts planning on the model to reach an arbitrary goal. We formulated that the former is achieved by learning the state-action map, and the latter does by learning the forward model of the environment, using recurrent neural learning scheme. The analysis of the dynamical structure from the coupling of the internal neural dynamics and the environment showed that generation of the global attractor is crucial for both learning cases. Experiments were conducted using a mobile robot with a laser range sensor, which verified our assertions in a simple obstacle environment.

1 Introduction

Recently, many have discussed how the knowledge should be represented internally for a mobile robot that navigates based on its local sensory inputs. Conventionally, the navigation problem has been approached in rather straightforward manner. A global representation formula is employed: a robot builds an environmental map, represented in global coordinates, by gathering geometrical information as it travels [2]. Although a variety of methodologies has been proposed in this context, potential problems still remain, especially in robot localization. The localization is not always robust enough in the noisy environments of the real-world since there exist gaps between the knowledge of the global map and the information provided by the local sensory inputs. The problem to consider is how the task knowledge can be represented as intrinsic [3] to the robot, and how such representations can be obtained through its behavioral experiences.

Others [7, 13] have developed an alternative approach based on landmark detection. In this approach, the robot acquires a graph-type representation of landmark types. This representation is equivalent to a finite state machine (FSM), as a topological modeling of the environment. In navigation, the robot can identify its topological position by anticipating the landmark types in the FSM representation. This scheme enables the robot to acquire the internal model of the obstacle environment by a local representation scheme. It is, however, considered that the representations by the FSM are still "parasitic" since symbols manipulated in the FSM are in the arbitrary shape regardless of their meaning in the physical world. A crucial gap exists between the actual physical systems defined in the metric space and their representation in the non-metric space, which makes the discussion of the structural stability of the whole system difficult.

This paper addresses the above problems by using the dynamical system's approach [1, 4, 10], expecting that this approach would provide other effective representational. The approach focuses on the fundamental dynamical structure that arises from coupling the internal and the environmental dynamics [1]. Here, the objective of learning is to adapt the internal dynamical function such that the resultant dynamical structure might generate the desired system behavior. The system's performance becomes structurally stable if the dynamical structure maintains a sufficiently large basin of attraction against possible perturbations. The most advantage of this approach is that we are able to conduct structural analyses of the system by fact that the internal representations, embedded into attractor dynamics, share the same metric space with the physical environment.

We investigate two classes of task learning, namely *skill-based learning* and *model-based learning*. The *skill-based learning* aims to ensure that a robot will acquire skills (represented as a state-action map) for a fixed navigational task, such as homing or cyclic routing, under the supervision of a trainer. In the *model-based learning* the robot learns the internal model of the environment rather than the direct state-action map so that the robot may adapt flexibly to different goal tasks. An important difference between two is that, in the former approach, the action is determined in a reactive way just by looking up the map while it is determined in a deliberative way through *mental simulation* of the model in the latter approach. We study how the state-action map or the environmental model can be represented by means of neural dynamical functions. Then we will explicate the conditions that each learning scheme should satisfy from the view point of dynamical systems.

2 Navigation Architecture

We review the navigation architecture [11] which is applied to the *YAMABICO* mobile robot [14]. *YAMABICO* can obtain the range image, covering a 160 degree arc in front of the robot, by a laser range finder in real-time.

In our formulation, maneuvering commands are generated as the output of a composite system consisting of two levels. The control level generates a collision-free, smooth trajectory using a variant of the potential field method [5]—i.e. the robot simply proceeds towards a particular potential hill in the range profile (direction toward an open space). The navigation level focuses on the topological changes in the range profile as the robot moves. As the robot moves through a given workspace, the profile gradually changes until another local peak appears when the robot reaches a branching point. At this moment of branching the navigation level decides whether to transfer the focus to the new local peak or to remain with the current one. The navigation level functions only at the branching point that appears in unconstructed environment. Hereafter, our discussions focus on how to determine the branching sequences.

3 Skill-Based Learning

The objective of *skill-based* learning is that the robot learns a fixed navigational task on the topological trajectory comprising branch points. We consider two specific tasks, namely homing and cyclic routing as examples. In the homing task, the robot has to travel back to a fixed branch point starting from an arbitrary position in the workspace. In the cyclic routing task, the robot have to travel into a fixed cyclic loop comprising branch points with starting from an arbitrary position.

3.1 Learning state-action map

The neural adaptation schemes are applied to the navigation level so that it can generate an adequate state-action map for a given task. Although some might consider that such map can be represented by using a layered feed-forward network with the inputs of the sensory image and the outputs of the motor command, this is not always true. The local sensory input does not always correspond uniquely to the true state of the robot (the sensory inputs could be the same for different robot positions). Therefore, there exists an ambiguity in determining the motor command solely from sensory inputs. This is a typical example of so-called non-Markovian problems which have been discussed by Lin and Mitchell [6]. In order to solve this ambiguity, a representation of contexts which are memories of past sensory sequences is required. For this purpose, a recurrent neural network (RNN) [4, 8] was employed since its recurrent context states could represent the memory of past sequences. The employed neural architecture is shown in Figure. 1. The sensory input p_n and the context units c_n determine the appropriate motor command x_{n+1}. The motor command x_n takes a binary value of 0 (staying at the current branch) or 1 (a transit to a new branch). The RNN learning of sensory-motor (p_n,x_{n+1}) sequences, sampled through the supervised training, can build the desired state-action map by self-organizing adequate internal representation in time.

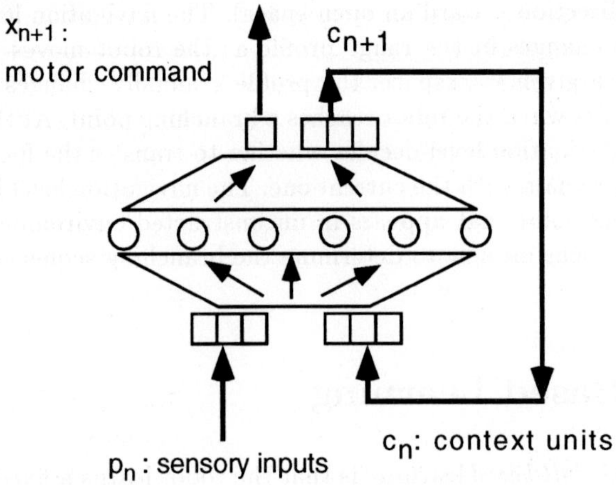

Fig. 1. Neural architecture for skill-based learning.

3.2 Embedding problem

The objective of the neural learning is to embed a task into certain global attractor dynamics which are generated from the coupling of the internal neural function and the environment. Figure 2 illustrates this idea. We define the in-

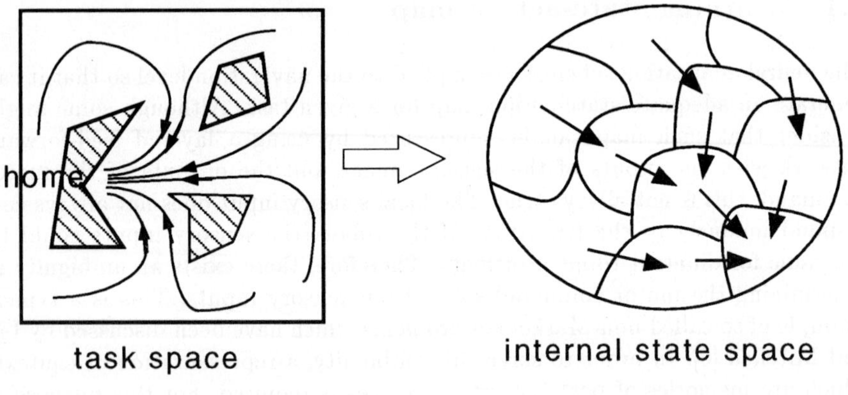

Fig. 2. The desired trajectories in the task space and its mapping to the internal state space.

ternal state of the robot by the state of the RNN. The internal dynamics, which are coupled with the environmental dynamics through the sensory-motor loop, evolve as the robot travels in the task space. We assume that the desired vector field in the task space forms a global attractor, such as a fixed point for a homing task or limit cycling for a cyclic routing task. All that the robot has to do is to follow this vector flow by means of its internal state-action map. This requires a condition: the vector field in the internal state space should be self-organized as being topologically equivalent to that in the task space in order that the internal state determine the action (motor command) uniquely. This is the embedding problem from the task space to the internal state space, and RNN learning can attain this, using various training trajectories. This analysis conjectured that the trajectories in the task space can always converge into the desired one as long as the task is embedded into the global attractor in the internal state space.

3.3 Experiment

An experiment of learning a cyclic routing task is presented. The assigned task is to repeat looping of a figure of '8' and '0' in sequence. In the training the robot moved by collision-free control, and branching of the navigation level was taught by the trainer. The employed RNN consists of three input units, eight hidden layer units, two context units and one output unit. The trainer guided the robot back to the target loop from arbitrary selected starting position. The travel of the training was repeated until the robot is assured of being capable of achieving the given task when started from an arbitrary position.

It was found that robot could achieve the task in a stable way after 10 times repetitions of the training. In the test travel, the robot was started from arbitrary initial positions, with setting the initial values of context units as random. Figure 3 shows examples of the test travels. The result appeared that the

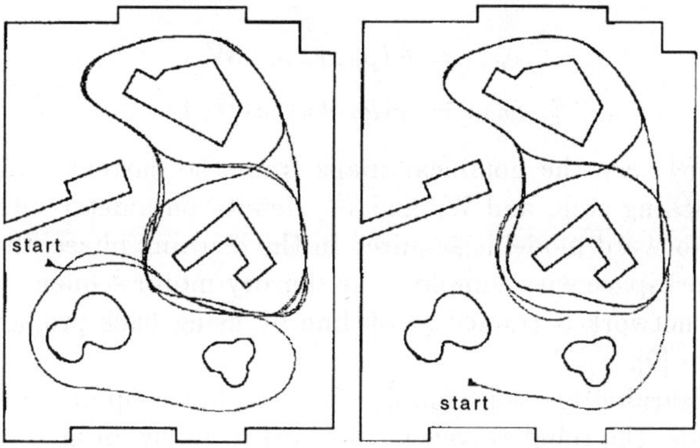

Fig. 3. Trace of test travels for cyclic routing.

robot can converge to the desired loop from any position in the workspace. Its convergence, however, takes a certain period depending on the case. The RNN initially cannot output normally until the context units catch up the context. As the robot moves around the workspace, encountering a sequence of known sensory input, the orbit in the internal state space starts to converge from the initial transient one. Noises affects the navigation performance remarkably. When miscellaneous noise such as mechanical, sensory and radio noise is present, the branching sometime become unstable. Thus, even after convergence, the robot could by chance go out the loop, perturbed by such noise. However, it always comes back to the loop after while. Although the actual navigation contains stochastic property in its local decisions, it can be said that the structure of convergence is quite stable in terms of the global attractor dynamics generated.

4 Model-Based Learning

In this learning, the main concern is how a robot can acquire the internal model as an intrinsic function which enables the *mental simulation* of its own actions in the obstacle environment. Here, we attempt to apply the scheme of forward modeling [4] to the problem.

4.1 Forward modeling

The objective is to build a forward model through which a robot can conduct lookahead prediction of the sensory input sequence (as the distal output) as a result of the given motor program (of the proximal input) in branching sequence. (Hereafter, the term "motor program" denotes a sequence of motor commands.) The objective forward model is embodied using a standard discrete time RNN architecture, as shown in Figure 4. The mapping function of the RNN can be written as;

$$\begin{aligned} c_{n+1} &= f_c(p_n, x_n, c_n, W_c) \\ \hat{p_{n+1}} &= f_p(p_n, x_n, c_n, W_p) \end{aligned} \qquad (1)$$

where f_c and f_p are the nonlinear maps from the current branching step to the next branching step, and W_c and W_p denote parameter sets of connective weights. The forward model is acquired in the learning phase; the robot travels around the workspace with sampling the sensory-motor sequence in the branching, then the network is trained as off-line by using back-propagation through time algorithm [9].

After the learning phase is completed, the robot is operated in the so-called open-loop mode: the robot travels in the workspace by an arbitrary motor program while conducting the one-step lookahead prediction (predicts next sensory input as the result of the current motor command). The RNN predicts the next sensory input $\hat{p_{n+1}}$ by inputting the current sensory input p_n and the current motor command x_n to the network. The RNN, in the beginning of the travel, cannot predict the next sensory input correctly since the initial context value is set randomly. However, the context value can get *situated* as the RNN continues to receive the sensory-motor sequence during the travel, then the RNN begins to predict correctly.

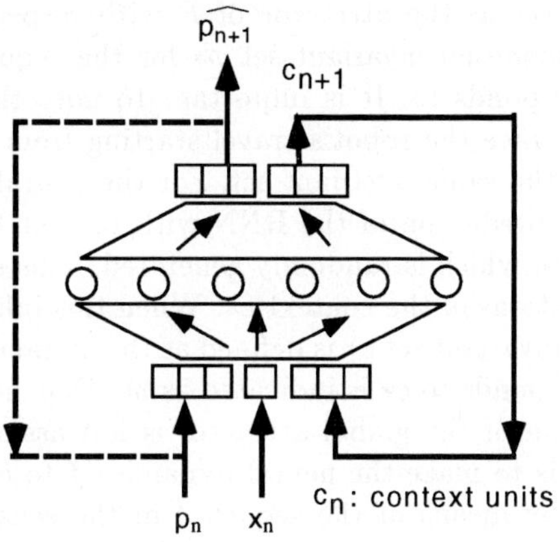

Fig. 4. Forward model by RNN architecture.

After the robot is *situated* to the environment, the RNN can be switched into the closed-loop mode with stopping the robot at a branch point. Now, a lookahead prediction of an arbitrary length for a given motor program can be made by copying the previous prediction of the sensory input to the current sensory input. (As indicated by a dotted line in Figure 4, the closed-loop for the sensory input is made.) Let us denote the motor program as $x*$. Then the lookahead prediction of the sensory input sequence $\hat{p}*$ can be obtained by recursively applying $x*$ to the RNN mapping function, with using the initial values of context units c_0 and the sensory input p_0 which have been obtained in the open-loop mode.

4.2 Dynamical mechanism of *situatedness*

This sub-section investigates the mechanism of *situatedness* by focusing on the coupling between the internal neural dynamics and the environmental dynamics.

First, we will define the term "attractor" for both of the environmental and the internal dynamics. Let us consider the environmental dynamics F. We consider an infinite length of randomly generated binary sequences (the motor program $x*$) to be fed into the robot. Let $s*$ be the resultant state transitions of the environmental state in the branching sequence. The environmental state s can be represented by the robot's position (including the orientation) upon branching. In the ideal case with no noise in the environment, the infinite travel of the robot forms an invariant set $\underline{s}*$, since the trajectory of the robot is limited to be in a subspace of the entire workspace after an initial transient period. We define this invariant set as the attractor of F with respect to the excitatory input $x*$. Also, we define an invariant set $\underline{p}*$ for the sequence of the sensory input which $\underline{s}*$ corresponds to. It is important to note that this attractor is the global attractor, since the robot's travel starting from any position in the workspace results in the same invariant set. For the neural dynamics f, let us consider a lookahead prediction of the RNN with respect to a motor program $x*$ of an infinite length which is randomly generated. This generates an infinite sequence of the transitions of the context $c*$. When this infinite sequence forms an invariant set, this invariant set $\underline{c}*$ is defined as the attractor of f. The sensory sequence which corresponds to $\underline{c}*$ is indicated as $\hat{\underline{p}*}$. Depending on the learning process, the generation of the global attractor is not assured for f. Since the objective of learning is to make the neural dynamics f to emulate the environmental dynamics F by means of the sequence of the sensory input, f in the limit of a learning process satisfies, for an arbitrary motor program $x*$, that:

$$\exists \underline{c}_0,\ \exists \underline{s}_0\ \Rightarrow\ \hat{\underline{p}*}\ =\ \underline{p}* \tag{2}$$

The conclusion here is that there is, at least, one attractor for f by which the lookahead prediction of the sensory input can be made correctly, as satisfying (2). Now let us consider the coupling of these two dynamics. In the open-loop mode, the RNN predicts the next sensory inputs $\hat{p_{n+1}}$ using the current sensory inputs p_n while the robot travels following the motor program $x*$. This coupling is schematically shown in Figure 5. In this coupling, it is conjectured that two sequences $p*$ and $\hat{p*}$ converge into the same sequence for all the initial states of s_0 and c_0 if f has been formed as global attractor dynamics. This implies that the internal dynamics, with arbitrary setting of the initial state, always become harmonized with the environmental dynamics and predict the sensory inputs correctly, as long as the internal model is embedded in the global attractor dynamics.

This feature of the entrainment of the internal dynamics by the environmental one assures an inherent robustness of the robot's behavior against temporal perturbations. The robot, during its travel, could lose its context if perturbed by noise. The robot, however, can get *situated* again by means of the entrainment as long as it continues to interact with the environment.

4.3 Experiment

4.4 Learning and lookahead prediction

We conducted experiments on the scheme using *YAMABICO*. The robot samples the data of the sensory-motor sequence while it wanders around the adopted workspace for a certain period, then it learns the forward model of the navigation level using the data obtained off-line. The adopted RNN architecture is three-layered having 10, 12 and 9 units for the input, hidden, and output layers respectively. It has four context units. After learning 193 sampled data, it

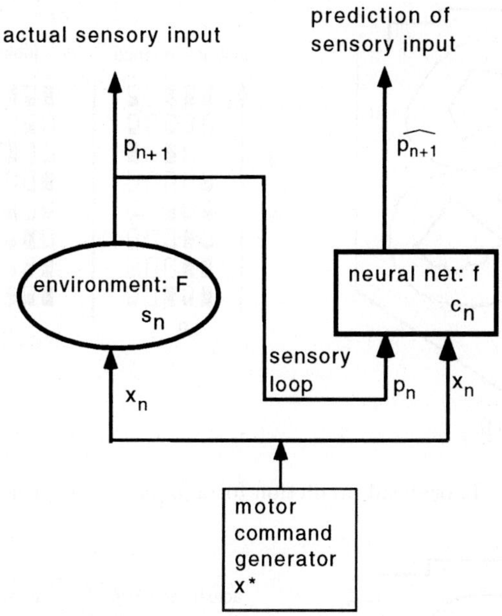

Fig. 5. Entrainment of internal dynamics by environment.

was observed that lookahead predictions became accurate except in cases with certain noise effects.

An example of a lookahead prediction test is shown in Figure 6. In (a) an arrow denotes the branching point where the robot conducted a lookahead prediction of a motor program given by 1100111 with switched to the closed-loop mode (after get *situated*). The robot, after conducting the predictions, traveled following the motor program, generating the trajectory of a "figure of eight", as shown. In (b) the left side shows the sensory input sequence, while the right side shows those of the lookahead, the motor program and its context values. The values are indicated by the bar heights. It can be seen that the lookahead for the sensory inputs agrees very well with the actual values.

We have stated that the global attractor provides an inherent robustness for context dependent navigation as a natural consequence of coupling between the internal and the environmental dynamical systems. The following experiment shown in Figure 7 demonstrates an example of auto-recovery from temporal perturbation. The robot traveled in the workspace while predicting the next sensory inputs with the RNN switched to the open-loop mode. During this travel, an additional obstacle was introduced. The upper part of Figure 7 shows the trajectory of the robot's travel; the lower part shows the comparison of the actual sensory inputs and corresponding one-step lookahead prediction. The branching sequence number is indexed beside the trajectory; this number

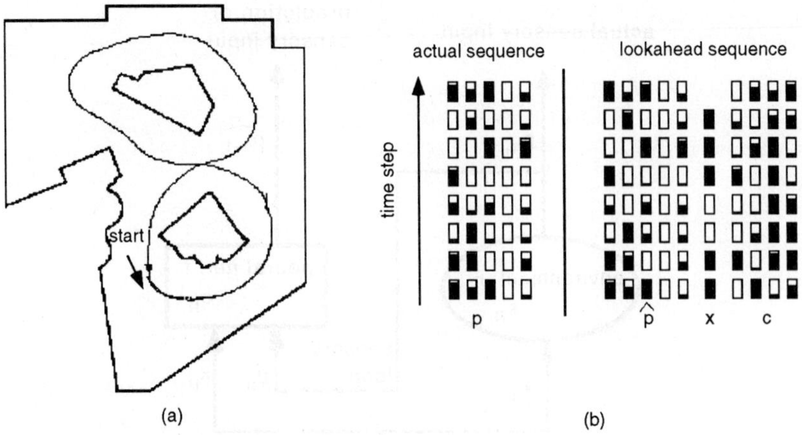

Fig. 6. Lookahead prediction for a given motor program.

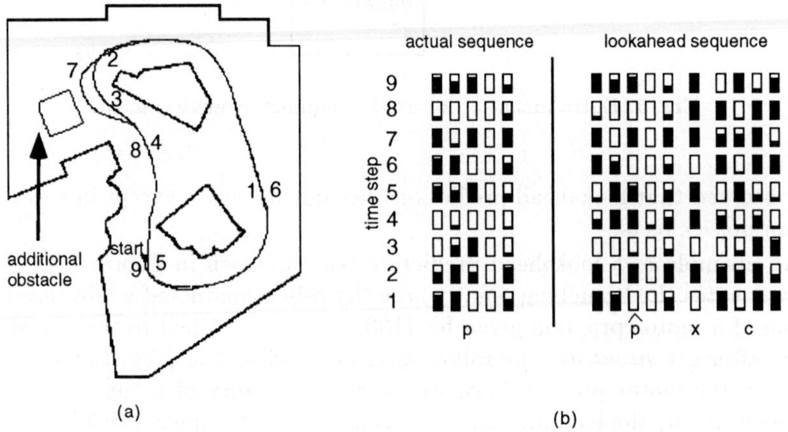

Fig. 7. Auto-recovery from an addition of an obstacle.

corresponds to the prediction sequence in the lower part of figure. The prediction starts to be incorrect once the robot passes the second branching point, as it encounters the unexpected obstacle. The robot, however, continues to travel and meanwhile the obstacle is removed. After the sixth branching point, as the lost context is recovered by means of the regular sensory feed, the prediction returns to the correct evaluation.

The repeated experiments showed that the mechanism of the auto-recovery is general. This implies that the learning of the RNN might have created the global attractor. In order to confirm this, we analyzed the dynamical structure self-organized in the RNN. The RNN, switched to the closed loop mode, was activated for two thousand forward steps using input sequences of random motor commands. The phase diagram was plotted as a two-dimensional projection using the activation state of two context units, excluding 100 points from the initial transient steps. Fig. 8(a) shows the resulting phase diagram, while (b) shows an enlargement of part of (a) in which a one-dimensional structure is seen. We repeated this several times with different initial values of the internal states,

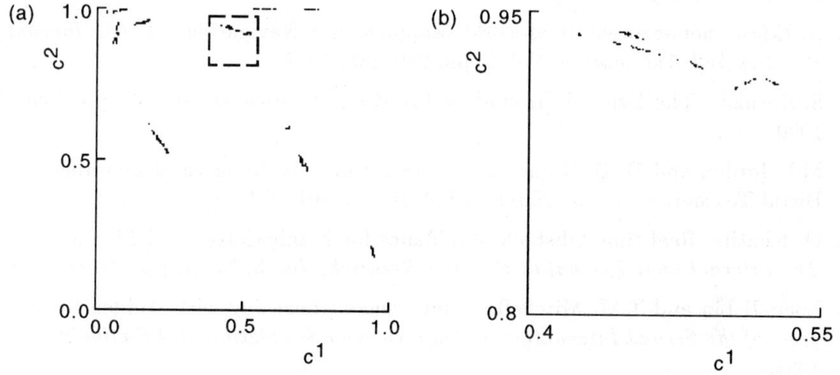

Fig. 8. Attractor observed in the internal dynamics.

and found that they all resulted in the same attractor structure. It confirmed that the internal dynamics are self-organized in the form of the global attractor dynamics. Although any theory has not been established to explain the creation of low-dimensional global attractor in the recurrent neural learning, its tendency is suggested in other numerical experiments of learning simple grammatical descriptions [8, 12].

5 Discussion and Conclusion

We have investigated the dynamical structure that arises in the coupling of the internal neural function and the environment, and have shown that the generation of the global attractor is essential for the embedding as well as the entrainment discussed in the *skill-based learning* and *model-based learning* respectively.

Our formulations have also clarified essential differences in their dynamical compositions. In the *skill-based* scheme, the dynamical structure arises only from the coupling of the internal and the environmental dynamics when the robot actually travels. In the *model-based* scheme, attractor dynamics exists in the internal dynamics even when it is decoupled from the environmental one. This decoupling allows the robot to have a "symbolic process" which accounts for its cognitive activities of *mental simulation* or planning. This symbolic process can be grounded to the physical world by means of *situatedness* when coupled with the environmental dynamics.

References

1. R.D. Beer. A Dynamical System's Perspective on Agent-Environment Interaction. *Artificial Intelligence*, in press.
2. A. Elfes. Sonar-based Real-world Mapping and Navigation. *IEEE Journal of Robotics and Automation*, Vol. 3, pp. 249–265, 1987.
3. S. Harnad. The Symbol Grounding Problem. *Physica D*, Vol. 42, pp. 335–346, 1990.
4. M.I. Jordan and D. E. Rumelhart. Forward models: Supervised Learning with a Distal Teacher. *Cognitive Science*, Vol. 16, pp. 307–354, 1992.
5. O. Khatib. Real-time Obstacle Avoidance for Manipulators and Mobile Robots. *The International Journal of Robotics Research*, Vol. 5, No. 1, pp. 90–98, 1986.
6. Long-Ji Lin and T.M. Mitchell. Reinforcement Learning with Hidden States. In *proc. of the Second International Conference on Simulation of Adaptive Behavior*, 1992.
7. M. Mataric. Integration of Representation into Goal-driven Behavior-based Robot. *IEEE Trans. Robotics and Automation*, Vol. 8, pp. 304–312, 1992.
8. J.B. Pollack. The Induction of Dynamical Recognizers. *Machine Learning*, Vol. 7, pp. 227–252, 1991.
9. D.E. Rumelhart, G.E. Hinton, and R.J. Williams. Learning Internal Representations by Error Propagation. In D.E. Rumelhart and J.L. Mclelland, editors, *Parallel Distributed Processing*. MIT Press, Cambridge, MA, 1986.
10. L. Steels. Mathematical Analysis of Behavior Systems. In *proc. of From Perception TO Action*, 1994.
11. J. Tani and N. Fukumura. Learning Goal-directed Sensory-based Navigation of a Mobile Robot. *Neural Networks*, Vol. 7, No. 3, pp. 553–563, 1994.
12. J. Tani and N. Fukumura. Embedding a grammatical description in deterministic chaos: an experiment in recurrent neural learning. *Biological Cybernetics*, in press.
13. B. M. Yamauchi and R. D. Beer. Spatial Learning for Navigation in Dynamic Environment. *IEEE Trans. on Systems, Man, and Cybernetics*, in press.
14. S. Yuta and J. Iijima. State Information Panel for Inter-Processor Communication in an Autonomous Mobile Robot Controller. In *proc. of the IEEE International Workshop on Intelligent Robots and Systems (IROS'90)*, 1990.

Optimizing the Performance of a Robot Society in Structured Environment through Genetic Algorithms

Mika Vainio, Torsten Schönberg, Aarne Halme, Peter Jakubik

Helsinki University of Technology, Automation Technology Laboratory,
Otakaari 5 A 02150 Espoo, FINLAND
mikav@automaatio.hut.fi

Abstract. Societies are formed as collaborative structures to execute tasks that are not possible or are otherwise difficult for individuals alone. There are many types of biological societies, but societies formed by machines or robots are still rare. The concept offers, however, interesting possibilities especially in applications where a long term fully autonomous operation is needed and/or the work to be done can be executed in a parallel way by a group of individuals. To explore these features a foraging society formed by several autonomous mobile robots working in structured environment was defined. The society members have restricted communication properties and their work is evaluated and controlled when needed by an outside operator. So far the problems for the society's functional optimization with a simulator have been caused by the large number of parameters with built-in cross relations and by the strong dependence on a given environment. To deal with these problems and to provide the society ways to adapt itself to changing circumstances Genetic Algorithms were applied. The concept was tested with simulations, which indicated the existence of adaptation both at the individual member level and at the higher society level.

1 Introduction

In Nature the survival of the fittest has always been the guiding line in the evolution process; organisms either adjust to changing environment conditions or they will become extinct. Besides mammals, with high level of intelligence, there are other truly complex living organisms, whose evolution has gone through another just as brilliant way. Instead of having one large unit, with superior intelligence, Nature has created systems where this intelligence has been distributed over many small units, with lower level of intelligence. Good examples of these kind of living forms are the social insects, e.g. ants and bees, which despite of their low level of intelligence have survived and found their niche (see [1] and [2]). These animals form seemingly chaotic structures, societies, that when studied more closely, show high level of distributed intelligence. Their ability to survive comes from the high redundancy and from the structure of the society. The structure allows modifications in the volume of the society; adding or deleting some members will not cause any extra reconfigurations. Redundancy comes directly from the high number of members in the society, destruction of some members will not become fatal for the society's operation. Considering these features, it is no wonder that some robot systems nowadays bear a great resemblance with these living structures. The work presented by Brooks in [3] acted as a kind of catalysts, and today terms like animat and behaviors belong to the robotics like servos and sensors. Cooperation between multiple mobile robots both in simulations and in real robots has been a target for a very lively research all over the world. Among the most popular research topics are definitely swarming [4]- [6], inter-robot communications [7] - [10], emerging and evolving group behaviors [11] - [15], and self-organization [16], [17].

In this paper the cooperation between autonomous mobile robots is studied through the essential "hidden" structures, which form and control the operation of a model robot society presented in [18]. This model society is formed by multiple small-scaled autonomous mobile robots, which forage in an initially unknown environment. Communication is locally restricted with a very simple protocol. The society operator is connected to the system through a base station, which operates as a "root" for the society's distributed sensing (i.e. information is "injected" to the system through it and all incoming data from society members is collected through it). It is also the place, where the collected objects are brought. So far the work has been done mostly with a simulator and with a prototype of the society's member. The previous simulations have expressed clearly the importance of choosing the right values for the society's so called operational parameters. These parameters in the selected domain include e.g. the number of members in the society, their sensor configurations and the range of communication. All these parameters have significant influence on the overall performance of the society and thus make the optimization difficult. On the basis of these experiences other means to improve the society's performance were searched for. The robot society's main principles were derived from Nature, accordingly it seemed a feasible solution to choose Genetic Algorithms (GAs) for this task. GAs are stochastic optimization algorithms, which are based on a model of the natural evolution. Holland's pioneering work presented in [19] together with Goldberg's presentation in [20] provide a comprehensive introduction to GAs. Here GAs were used to optimize the performance of the individual members as well as the performance of the whole robot society. Several references can be found, where GAs have been used to improve or evolve the operation of a single autonomous mobile robot (see e.g. [21] and [22]). These cases often include a neural network -based controllers as stated by Husbands et al. in [23]. These controllers' structures and/or weights are evolved with GAs. The goal here was not to develop a general controller, rather to give the robot society the ability to adjust itself to a changing environment and thus improve its performance. The adaptation could be observed at two levels, i.e. at the individual member level and at the whole society level. It was accomplished by giving the members the ability to vary their operational parameters.

A method for optimizing the operation of the cooperative autonomous mobile robots with the aid of Genetic Algorithms is presented. Both the robot society and the applied GA algorithm are explained in more detail and simulations are included to verify the feasibility of the concept.

2 The Model Robot Society

Basically every society is a group of individuals, called members, with information and control structures. All members of a society need not to be similar. Members having same properties can form clusters or classes. The information structure defines how information is spread among the members and how an individual member communicates with the other members of the society. The control structure defines the way that the society affects to its members. Because all working power is produced by the members the control structure controls the task execution of the society. The practical goal of the robot society concept is to construct a kind of "distributed robot" or "group robot", which can execute tasks which are defined by the user or the "society controller", like in the case of a conventional industrial robot. This means that the behaviour of the society must be controllable from outside and that the society must thus have information connection to the controller. However, it is important from a practical point of view that this connection is not built to every member of the society, but rather to its information system. This is because a society may include a large number of members that are located in places where a communication system is difficult to build. Basically the communication in a society is done on member to member bases.

The robot society concept is illustrated with a society formed by several autonomous mobile robots. The society's task is to forage in an initially unknown area. The society has two types of members: working units that do the actual foraging and energy units which carry energy from the base station and act as work coordinators. The working units execute following tasks: random searching, moving towards a given destination, obstacle avoidance and object collecting. Small objects ("stones") are picked up and carried to the base station. The base station serves also as the interface with the user. The operator can send commands through it and the information collected by the members can be communicated further back to the operator. The robots avoid obstacles and each others in a simple reactive way, and they navigate with a dead reckoning / beacons based system. Communication between the members takes place locally by using a broadcasting principle, where only members within a certain radius can communicate. A set of simple messages can be sent and received. These include e.g. location coordinates, stones found, need help, detecting nearby units, and change operation mode. The workers have a limited energy resources which are refilled from the energy units when needed. Two working units can also equalize their energy resources when they meet (not used in these simulations). Thus the society can also distribute energy like it distributes information. When the energy level of a working unit drops under $threshold_{upper}$ it tries to navigate to the last known location of a energy unit for refilling. If the working unit's energy level goes below a $threshold_{lower}$ before it finds a energy unit, it will switch its systems off and "die".

A graphic simulator has been constructed to study the basic properties of such model society [24]. It has been implemented on a PC with the ACTOR object oriented programming language. Simulations have maximum length, but if all stones are collected before that or all working units are dead the simulation will be terminated. The simulator provides a way to optimize society's structures before they are implemented to real world robot societies. The outlook of the simulator environment is shown in Fig.1.

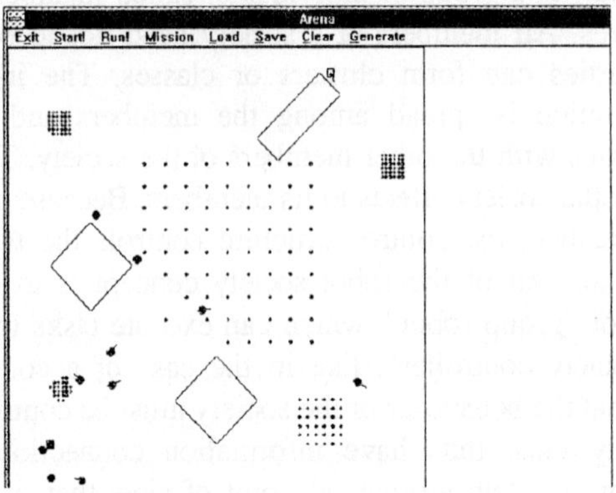

Fig. 1. The graphic society simulator

For the physical robot society implementation a small-scale autonomous mobile robot called HUTMAN (Helsinki University of Technology's Mobile Autonomous Navigator) was built. It provided the needed test-bench for developing suitable structures for the model society's members. The chassis of the robot has a diameter of 15 cm and the robot is 14 cm high. HUTMAN, showed in Fig. 2, uses transputers, infrared communication links, narrow beam ultrasonic sensors and has bumpers for collision detection.

Fig. 2. Robot society member, HUTMAN

It has two driven wheels, both controlled by a stepper motor, and four balancing swivel castors. Robot's navigation system uses Kalman filter for fusing the information from separate sub-systems. To give HUTMAN the ability to accomplish the stone collecting task, a simple one degree of freedom gripper was applied to it. The robot is completely autonomous; it can operate approximately 40 minutes without energy refuelling and the processing takes place only in the processors located onboard. More detailed description of the robot can be found in [25].

3 The Genetic Approach

The task for the model robot society was defined to be foraging. The idea was to illustrate how the interaction between a dynamic structured environment and the model robot society works, i.e. how the society members influence to the environment and vice versa. To succeed in the changing environment the society needs to have some kind of adaptability both at the member level and at the whole society level. This adaptation was expected by applying Genetic Algorithms to the society operation.

GAs have been widely implemented for autonomous mobile robot applications especially during the last few years. Beside the traditional GAs, the Genetic Programming (GP) paradigm presented by Koza in [26] has also been successfully used in several cases. The concept where real computer programs evolve from scratch to accomplish quite demanding tasks, is more than tempting. The choice here was though the traditional approach, where society members' chromosomes were represented as fixed length bit strings.

3.1 The Chromosome

From the member's several operational parameters, only two significant were chosen to be represented in the genotype; the range of the communication and the range of the sensors. Other possible parameters would have been e.g. the number and location of the sensors, the energy threshold values, the turning rate, etc. In the following simulations these other parameters are fixed based on a priori knowledge gained by extensive testing. The sensors' minimum range was 20 cm (robot's one step in simulation, and the minimum range for the ultrasonic sensors in real robots) and the maximum value was around 50 cm. So the range to be coded was 30. This required the first five bits of the chromosome. The communication should be able to cover the whole working area (a grid of size 1000*1000). This value was represented with the last 8 bits due to a scaling. So the total length of the chromosome was fixed to 13 bits.

3.2 The Fitness Functions

The GA approach was used for two goals; to improve the operation of the individual member and furthermore to improve the whole society's performance. The actual "tools" for these optimization are the fitness functions F_m (member's) and F_s

(society's). When the optimization of the foraging done by a society or by a single member is in concern, the foundations of the fitness functions should of course be in this foraging task. A term, *cooperation*, was added to the member's fitness function to give some kind of account for how well an individual member is doing as a part of a society. If the task is considered from a real world point of view, the number of collisions should also be limited to the minimum. This added one term more. So the member's fitness function (F_m) to be minimized has the following structure:

$$F_m = foraging + cooperation + collisions \tag{1}$$

In this equation, *foraging*, evaluates member's success in foraging. It is based on an expected average value for the member's performance (i.e. the estimated number of collected stones), which is formulated as follows:

$$ens = \frac{tns}{nm} \tag{2}$$

where
tns = total number of stones in the environment
nm = number of members in the society

The actual *foraging* is defined as:

$$foraging = \frac{ens - cs}{ens} \tag{3}$$

where
cs = collected stones

Thus term *foraging* indicates how well the member is doing in its stone collecting task relative to the other members in the society. This part can have a negative value, when the member is able to pick more than its "fair" share.

Term *cooperation* approximates the society's cooperation level. The members are able to work together with the aid of the restricted communication. This cooperation level is measured based on the ratio between the received messages (*tnrm*) and the results (*sf*). In other words:

$$cooperation = \frac{tnrm - sf}{tnrm} \tag{4}$$

where

$tnrm$ = total number of received messages from the other members about possible stones ($tnrm_{min} = 1$)
sf = stones found based on $tnrm$

The last term, *collisions*, counts the collisions. It is represented as follows:

$$collisions = \frac{1 - e^{-slope \cdot NC}}{1 + e^{-slope \cdot NC}} \quad (5)$$

where

NC = number of collisions

$slope$ = small positive coefficient

The parameter, *slope*, changes the "weight" of the collisions term. The larger this parameter is the bigger influence it will naturally have on the total value of the member's fitness function.

The society's fitness (F_S), which is to be maximized, has the following form:

$$F_s = tncs \cdot \frac{mts}{uts} \quad (6)$$

where

$tncs$ = total number of collected stones during one generation

uts = used time steps for collecting all stones

mts = maximum number of time steps

Term (mts/uts) will increase the "raw" fitness ($tncs$), if the society has been able to accomplish the foraging task before the maximum simulation time.

3.3 The Algorithm

An overall description of the applied GA algorithm is presented in Table 1. It includes basic reproduction operators (i.e. one-point crossover, mutation and asexual reproduction). Elitism was used to confirm that the best individual will always survive to the next generation. If the population size is very small, this kind of strategy can easily lead to a premature convergence. But here the search space was quite small and hard selection (70 % of the population consists of new individuals and thus quarantees the diversity of the population) was used, so this did not cause any major problems. After all the goal was to find near-optimal solutions in non-stationary environments.

Table 1

1. INITIALIZE

- *Define maximum number of generations GEN_{max}*
- *Set Generation.number:= 0*
- *Create the initial population of 10 members*

2. RUN ONE GENERATION
- *Generation.number:= Generation.number + 1*

3. EVALUATE THE SOCIETY FITNESS F_s
- *Save F_s*
- *IF*

  Generation.number = GEN_{max}

  Plot F_s values; STOP

  ELSE

  CONTINUE

4. EVALUATE EACH MEMBERS FITNESS FUNCTION VALUE F_m

5. ORDER THE MEMBERS BASED ON THEIR F_m VALUE

6. COPY THE BEST MEMBER DIRECTLY TO THE NEXT GENERATION'S POPULATION

7. FROM FIVE BEST CHOOSE RANDOMLY TWO

8. USE ONE-POINT CROSSOVER WITH PROBABILITY $P_{crossover} = 1.0$

9. MUTATE THE TWO NEW OFFSPRINGS WITH PROBABILITY $P_{mutation} = 1/chromosome.lenght$

10. ADD THE TWO MUTATED OFFSPRINGS TO THE NEXT GENERATION'S POPULATION

11. GENERATE RANDOMLY 7 NEW MEMBERS TO THE NEXT GENERATION'S POPULATION

12. GO TO STEP 2

4 Experiments and Results

Tests were performed to verify the usefulness of the presented GA for the performance of a robot society. As mentioned earlier the GAs were used for two tasks:

- To make it possible for an individual member to adjust its sensor and communication ranges according to what is needed in dynamic environments (member level adaptation).
- To provide the whole society the means to improve its overall performance (society level adaptation)

4.1 Member Level Adaptation

The use of GAs for low level (individual) adaptation was tested with a simple method. Instead of usual (5000 simulation steps) generations, one run was split to 50 separate cases (100 steps per each). Each of these short runs was dealt as one generation and the presented algorithm was applied to them. The idea was to show that the sensor and communication values vary accordingly what is needed for a good operation in the dynamic environment.

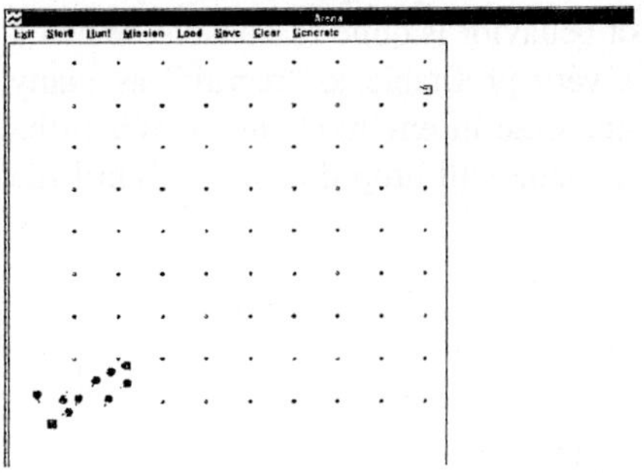

Fig. 3. Environment 0 with evenly placed stones

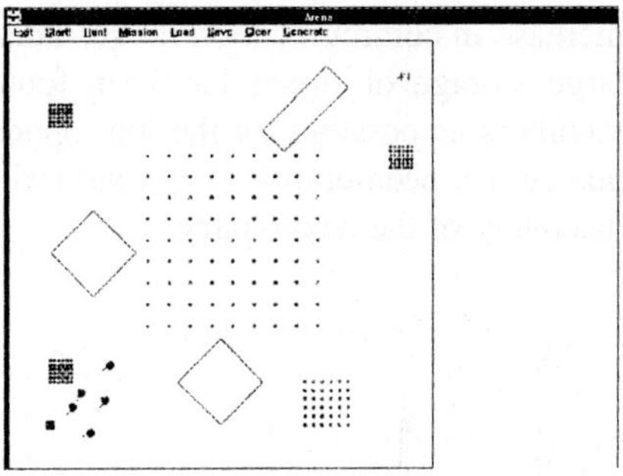

Fig. 4. Environment 1 with stone quarries

The two test cases (Figs. 3,4) were chosen to clarify some quite unexpected environment related behaviors, such as why the communication and the sensor ranges do not always reach their maximum values, even though it would seem to be the best operation strategy.

Fig. 5 and Fig. 6 show the results from the environment 0. For both the communication and the sensor ranges the values started with large values and significant fluctuations. After 30 generations these values dropped drastically.

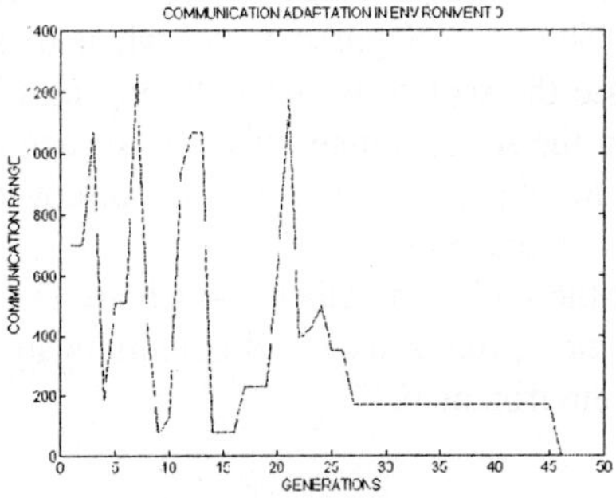

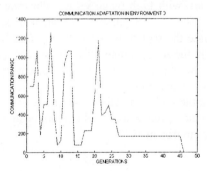

Fig. 5. The communication range adaptation in environment 0

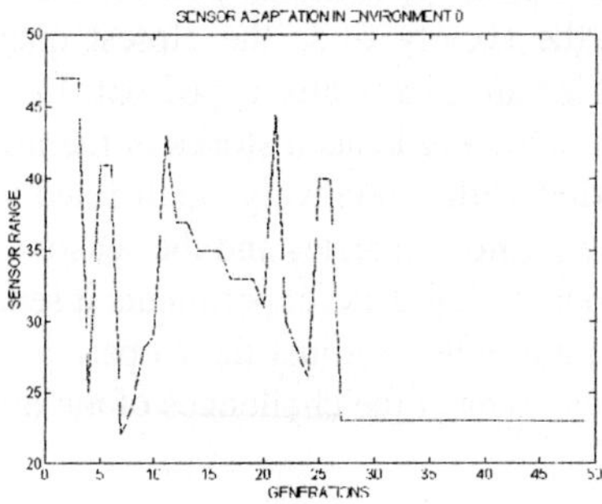

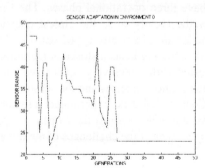

Fig. 6. The sensor range adaptation in environment 0

This kind of behavior in sensor and communication ranges was naturally due to the placement of the stones. In this environment the stones were situated evenly. The reason members with big communication did not do well, was that they were repeatedly receiving messages from the other members about found stones. Because there were no stone quarries and the stones were located sparsely, these individuals usually did unnecessary trips to these places. The evolving of sensor range is based on the obstacle avoidance and stone collecting behaviors. These behaviors determine that short range values will work better in this kind of environment, where the members operate inside a sparse stone "lattice."

The results from the environment 1 are shown in Fig.7 and Fig 8. Fig. 7 illustrates how the communication range changes according to some specific environmental conditions, i.e. the discovery of a stone quarry leads always to an increase in communication range. This kind of behavior is quite reasonable. When a large storage of stones has been found, it is very profitable to "recruit" as many members as possible for the job (opposite to the case in environment 0). When the quarry has been emptied, it is natural that the value will drop downwards until the discovery of the next quarry.

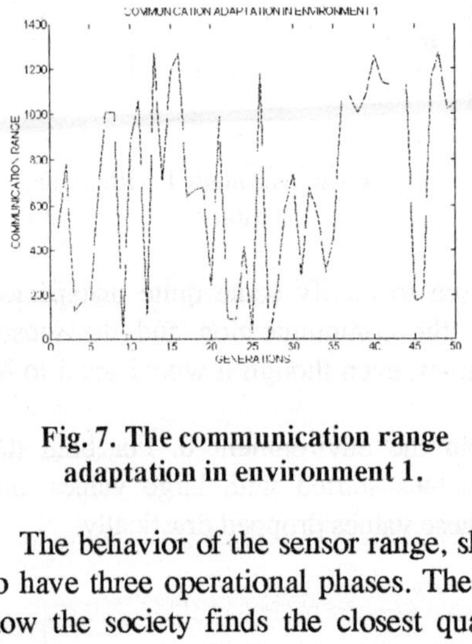

Fig. 7. The communication range adaptation in environment 1.

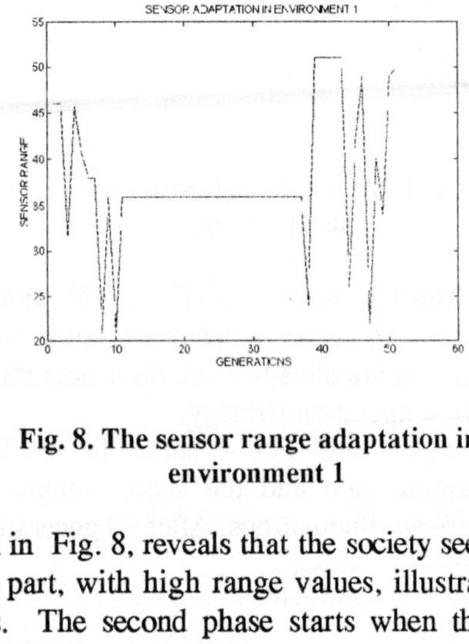

Fig. 8. The sensor range adaptation in environment 1

The behavior of the sensor range, shown in Fig. 8, reveals that the society seems to have three operational phases. The first part, with high range values, illustrates how the society finds the closest quarries. The second phase starts when these quarries are practically wiped out. In this phase the society is concentrating mostly on the sparsely located stones in the middle. In the sensor range, this can be seen as a period with a relatively small fixed value. The phase three contains the discovery of the farmost quarries and the sensor value will rise again.

Considering these experiments it seems that the society members, which are using GAs, are able to adjust their operation parameters, like sensor and communication ranges, to meet the challenges of the dynamic environment.

4.2 Society Level Adaptation

The society's overall performance was tested with and without GAs in two environments. The first one (environment 1) was the same as in the member level adaptation case (Fig. 4) and the other one (environment 2) is illustrated in Fig. 9.

The goal was to show that by using Genetic Algorithms, the performance of the society would gradually improve.

With GA: When GAs are used for optimization the convergence to the global minimum will usually take lots of generations. This is especially true, when the population size is small as it is in this case. The goal here was not to find the absolute optimum, but rather to show that the use of GAs will indeed improve the performance of the robot society in its foraging task. The chosen approach has very little flexibility concerning the environment because the evolution of the chromosome is totally attachted to a certain environment. Another problem arises from the fact, that during one generation the environment is changing all the time, and a fixed chromosome will just give an average value for the whole run. The advantage of this solution is its simplicity, the fitness function and the evolutionary operations are executed only once per run and still they will eventually "do the trick" and improve the performance of the society. These ideas were tested in two environments (env. 1, env.2). The result from environment 1 is shown in Fig. 10.

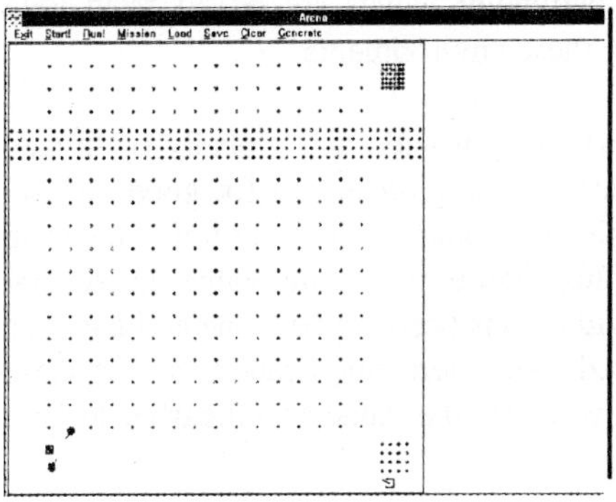

Fig. 9. The environment 2

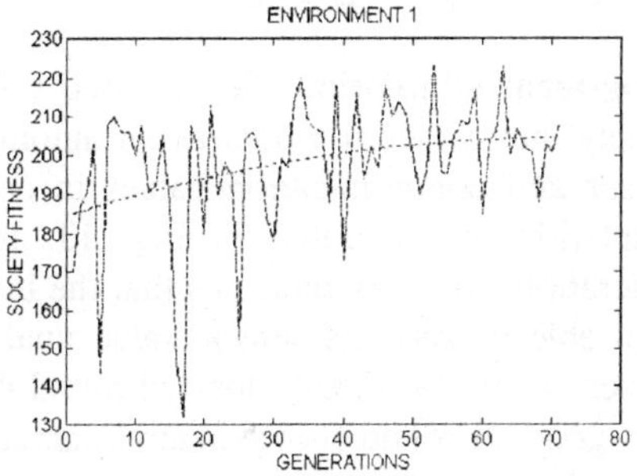

Fig. 10. The result from the environment 1 with GAs. The dashed line is a result from second-order curve fitting.

From Fig.10 it can be seen that there actually happens the society level adaptation. The downward spikes are caused by multiple death cases among the society members. One has to keep in mind that the increase in the society's fitness value is totally due to the improvements of the three best members (the rest seven are always chosen randomly). Fig. 11 illustrates that the GAs are not succeeding very well in environment 2, where the number of stones is very large. This is due to the fact, that it doesn't matter what kinds of values members have, they will anyway do relatively fine. The two smallest values are from generations, where the society was able to break through the barrigade. After the breakthrough some of the members got stuck or just couldn't navigate back to the base. The concept of having a stone barrigade in the environment is very interesting and deserves to be studied more closely. So far it seems that the structure of this barrigade has a very significant effect to the operation of the society (i.e. it is strongly related to the death of members) and thus to the usefulness of GAs.

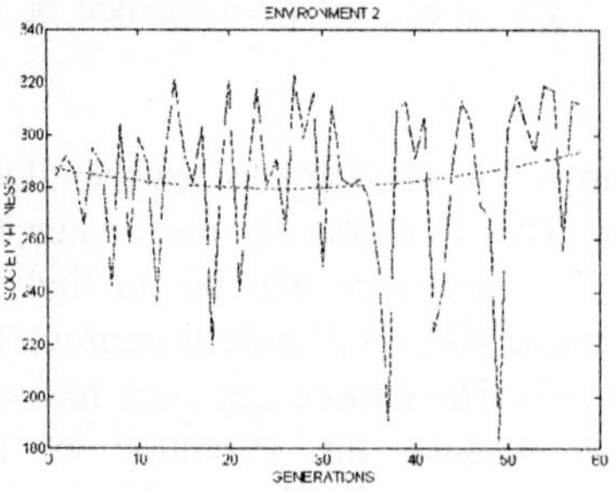

Fig. 11. The result from environment 2 with GAs.The dashed line is again result from second-order curve fitting

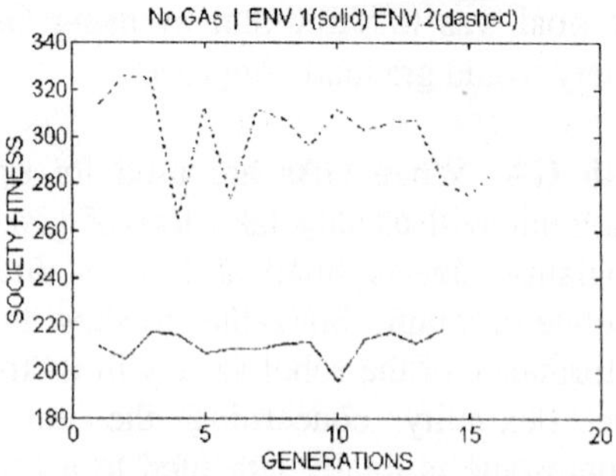

Fig. 12. The results from testing the society in both environments without GAs with fixed sensor and communication ranges

No GAs: Society's fitness curves, shown above in Fig. 12, illustrate that in both environment, a society with fixed parameters was able to perform quite well. The reason for this is of course, that the used parameter values are results from long testing and are thus good general choices for these environments.

Long-term adaptation: So the used GA will not provide any globally optimized society, but rather it works as an automatic searching procedure for good general sensor and communication values in certain environment. This useful long term adaptation is illustrated in Fig. 13. In this figure the same value in several generations in a row indicates that the individual has been the best one and has thus been able to keep its sensor value unaltered. An interesting feature is that these sensor values are mainly located round 40, which is the value found earlier to be a very good all-round choice in environment 1.

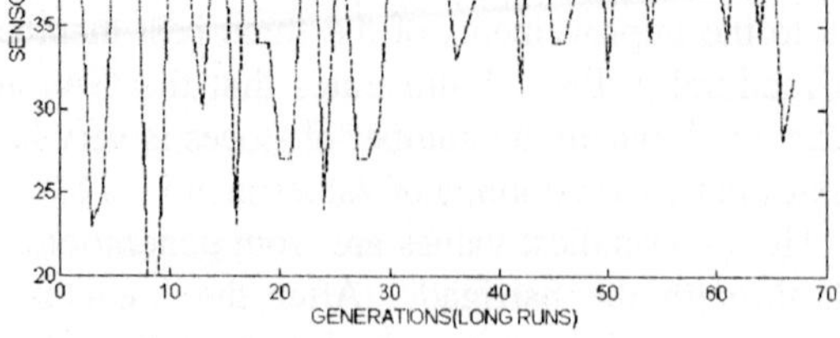

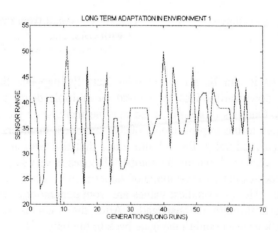

Figure 13: Long-term sensor adaptation in environment 1.

5 Conclusion and Future Work

A method for using Genetic Algorithms in optimizing the performance of a robot society has been developed. The model robot society was defined to be formed by multiple cooperative autonomous mobile robots with flexible communication and control structures. The concept was tested with several simulation runs, which confirmed that the detected performance improvements were due to the adaptation both at the member and at the society level. The member level adaptation demonstrated clearly, that a society member operating with a Genetic Algorithm was able to improve its operation in dynamic environment by adjusting its operational parameters (i.e. sensor and communication ranges). Some higher level (i.e. society level) adaptation was also verified to exist even though the used algorithm was not designed to produce an "optimally working" society. The value of the existing society level adaptation emerges from the long-term operational parameter adaptation, which enables an automatic parameter search.

Future work concentrates on the modification of the concept so that it could actually generate a society, which would be optimized for a given type of environment. This will be done by implementing the optimization with distributed GAs, where separate sub-societies are operating parallely, through local area network, with some form of "gene" migration between these populations. Method's suitability for various environments will be tested and an environment mapping will be included to the scenario. In the near future the results from this Genetic Algorithms based optimization will be tested with real robots foraging in an initially unknown restricted area. This will be the final testing, which should reveal how valid the concept actually is in real world's dynamic circumstances.

References

1. Seelay T: Honeybee Ecology A Study of Adaption in Social Life, Princeton University Press, New Jersey, USA, (1985).
2. Sudd J. H., Franks N. R.: The Behavioural Ecology of Ants, Chapman and Hall, New York, USA, (1987).
3. Brooks R. A.: A Robust Layered Control System For a Mobile Robot, IEEE Journal of Robotics and Automation, Vol. RA-2, No.1, (1986), 14-23.
4. Goss S., Deneubourg J.-L., Beckers R., Henrotte J.-L.: Recipes for Collective Movement, Proceedings of the Second European Conf. on Artificial Life, Preprints, (1993), 400-410.
5. Deneubourg J.-L., Theraulaz G., Beckers R.: Swarm-Made Architectures, Toward a Practice of Autonomous Systems, Proceedings of the First European Conf. on Artificial Life ed. by Varela F. J., Bourgine P., MIT Press/Bradford Books, (1992), 123-133.
6. Numaoka C., Takeuchi A: Collective Choice of Strategic Type, From Animals to Animats 2, Proceedings of the Second Intl. Conf. on Simulation of Adaptive Behavior, ed. by Meyer J-A, Roitblat H. L., Wilson S. W, MIT Press/Bradford Books, (1993), 469-477.
7. Premvuti S., Wang J.: A Medium Access Protocol (CSMA/CD-W) Supporting Wireless Inter-Robot Communication in Distributed Robotic Systems, Distributed Autonomous

Robotic Systems, ed. by Asama H., Fukuda T., Arai T, Endo I., Springer-Verlag, (1994), 165-175.

8. Hara F., Ichikawa S.: Effects of Population Size in Multi-Robots Cooperation Behaviors, Proceedings of the Intl. Symposium on Distributed Autonomous Robotic Systems, Wako, Saitama, Japan, (1992), 3-9.
9. Norelis F. R.: Toward a Robot Architecture Integrating Cooperation between Mobile Robots: Application to Indoor Environment, The Intl. J. of Robotics Research, Vol. 12, No1, (1993), 79-98.
10. Balch T., Arkin R. C.: Communication in Reactive Multiagent Robotic Systems, Autonomous Robots, Vol.1, No.1, (1994), 27-52.
11. Mataric M., Marjanovic M. J.: Synthesizing Complex Behaviours by Composing Simple Primitives, Proceedings of the Second Intl. Conf. on Artificial Life, Preprints, (1993), 698-707.
12. Mataric M. J.: Designing Emergent Behaviors: From Local Interactions to Collective Intelligence, From Animals to Animats 2, Proceedings of the Second Intl. Conf. on Simulation of Adaptive Behavior, ed. by Meyer J.-A., Roitblat H.L., Wilson S. W, MIT, Press/Bradford Books, (1993), 432-441.
13. Mataric M. J.: Interaction and Intelligent Behavior, Doctoral dissertation, M.I.T, (1994).
14. Kube C. R., Zhang H.: Collective Robotic Intelligence, From Animals to Animats 2, Proceedings of the Second Intl. Conf. on Simulation of Adaptive Behavior, ed. by Meyer J.-A., Roitblat H. L., Wilson S. W, MIT Press/Bradford Books, (1993), 460-468.
15. Fukuda T., Iritani G.: Intention Model and Coordination for Collective Behavior in Robotic System, Distributed Autonomous Robotic Systems, ed. by Asama H., Fukuda T., Arai T, Endo I., Springer-Verlag, (1994), 255-266.
16. Hackwood S., Beni G.: Self-organization of Sensors for Swarm Intelligence, Proceedings of the IEEE Intl. Conf. on Robotics and Automation, Nice, France, (1992), 819-829.
17. Kawauchi Y., Inaba M., Fukuda T.: Evolutional Self-organization of Distributed Autonomous Systems, Distributed Autonomous Robotic Systems, ed. by Asama H., Fukuda T., Arai T, Endo I., Springer-Verlag, (1994), 243-254.
18. Halme A., Jakubik P., Schönberg T., Vainio M.: The Concept of Robot Society and Its Utilization, Proceedings of the IEEE Intl.Workshop on Advanced Robotics, Tsukuba, Japan, (1993), 29-35.
19. Holland J. H.: Adaptation in natural and artificial systems, Ann Arbor, MI, The University of Michigan Press, (1975).
20. Goldberg D. E.: Genetic Algorithms in Search, Optimization, and Machine Learning, Addison-Wesley, Massachusetts, (1989).
21. Floreano D., Mondada F.: Active Perception, Navigation, Homing, and Grasping: An Autonomous Perspective, Proceedings of the From Perception to Action (PerAc´94), Lausanne, Switzerland, (1994), 122-133.
22. Nolfi S., Miglino O., Parisi D.: Phenotypic Plasticity in Evolving Neural Networks, Proceedings of the From Perception to Action (PerAc´94), Lausanne, Switzerland, (1994), 146-157.
23. Husbands P., Harvey I., Cliff D., Miller G.: The Use of Genetic Algorithms for the Development of Sensorimotor Control Systems, Proceedings of the From Perception to Action (PerAc´94), Lausanne, Switzerland, (1994), 110-121.
24. Schönberg T.: Simulator Studies of the Robot Society Concept, Licentiate Thesis, Helsinki University of Technology, (1993).
25. Vainio M.: Design of an autonomous small-scale mobile robot, HUTMAN, for robot society studies, Master's thesis, Helsinki University of Technology, (1993).
26. Koza J. R.: Genetic Programming: On the programming of computers by means of natural selection, MIT Press, (1992).

6. Societies and Collective Behavior

Spatial Games and Evolution of Cooperation

Robert M May, Sebastian Bohoeffer and Martin A Nowak

Department of Zoology, University of Oxford, Oxford OX1 3PS, UK

Abstract. The Prisoner's Dilemma is a widely employed metaphor for problems associated with the evolution and maintenance of cooperative behaviour. Here we outline a new approach, in which players – who are either pure cooperators or pure defectors – interact with neighbours in some spatial array. Such "spatial games" avoid problems associated with earlier work based on individual strategies, such as "tit-for-tat", and may provide a robust explanation for how various kinds of collective social behaviour is maintained.

1 Introduction

From Darwin's time to ours, one of the central problems of evolution has been the origin and maintenance of cooperative or altruistic behaviour.

Much attention has been given to the Prisoner's Dilemma (henceforth PD), as a metaphor for the problems surrounding the evolution of cooperative behaviour [1,2]. In its standard form, the PD is a game played by two players, each of whom may choose either to cooperate, C, or defect, D, in any one encounter. If both players choose C, both get a pay-off of magnitude R; if one defects while the other cooperates, D gets the game's biggest pay-off, T, while C gets the smallest, S; if both defect, both get P. With $T > R > P > S$, the paradox is evident. In any one round, the strategy D is unbeatable (being better than C whether the opponent chooses C or D). But by playing D in a sequence of encounters, both players end up scoring less than they would by cooperating (because $R > P$). Following Axelrod and Hamilton's pioneering work [2], many authors have sought to understand which strategies (such as "tit-for-tat") do best when the game is played many times between players who remember past encounters. These theoretical analyses, computer tournaments, and laboratory experiments continue, with the answers depending on the extent to which future pay-offs are discounted, on the ensemble of strategies present in the group of players, on the degree to which strategies are deterministic or error-prone (e.g. imperfect memories of opponents or of past events), and so on [1,3-5]. Fascinating though this work is, it is clear that simpler biological entities (self-replicating molecules, bacteria, and arguably most non-human animals) which exhibit cooperative interactions, cannot obey the restrictions – recognising past players, remembering their past actions, and anticipating future encounters whose pay-offs are not significantly discounted – which necessarily underpin such strategic analyses.

We have recently given a new twist to this discussion, by considering what happens when the game is played with close neighbours in some 2-dimensional spatial array: "spatial dilemmas". What follows is an outline of the main features of this work. Details are presented elsewhere [6-9].

We consider only 2 kinds of players: those who always cooperate, C, and those who always defect, D. No attention is given to past or likely future encounters, so no memory is required and no complicated strategies arise. These memoryless "players" — who may be individuals or organised groups — are placed on a 2-dimensional, nxn square lattice of "patches"; each lattice-site is thus occupied either by a C or a D. In each round of our game (or at each time step, or each generation), each patch-owner plays the game with its immediate neighbours. The score for each player is the sum of the pay-offs in these encounters with neighbours. At the start of the next generation, each lattice-site is occupied by the player with the highest score among the previous owner and the immediate neighbours. The rules of this simple game among n^2 players on an nxn lattice are thus completely deterministic. Specifically (but preserving the essentials of the PD), we choose the pay-offs of the PD matrix to have the values $R = 1$, $T = b$ ($b > 1$), $S = P = 0$. That is, mutual cooperators each score 1, mutual defectors 0, and D scores b (which exceeds unity) against C (who scores 0 in such an encounter). The parameter b, which characterises the advantage of defectors against cooperators, is thus the only parameter in the model; none of the findings are qualitatively altered if we instead set $P = \varepsilon$, with ε positive but significantly below unity (so that $T > R > P > S$ is strictly satisfied). In the figures that follow, we assume the boundaries of the nxn matrix are fixed, so that players at the boundaries simply have fewer neighbours; the qualitative character of our results is unchanged if we instead choose periodic boundary conditions (so that the lattice is really a torus). The figures are for the case when the game is played with the 8 neighbouring sites (the cells corresponding to the chess king's move), and with one's own site (which is reasonable if the players are thought of as organised groups occupying territory). As amplified below, the essential conclusions remain true if players interact only with the 4 orthogonal neighbours in square lattices, or with 6 neighbours in hexagonal lattices. The results also hold whether or not self-interactions are included [7]. Herz has, indeed, demonstrated formal equivalences between spatial games with and without self-interaction; these, of course, involve appropriate changes in the parameters T,R,P,S [10].

2 Results for the Basic Model

Using an efficient computer program in which each lattice-site is represented as a pixel of the screen, we have explored the asymptotic behaviour of the above-described system for various values of b, and with various initial proportions of C and D arranged randomly or regularly on an nxn lattice ($n = 20$ and up). The dynamical behaviour of the system depends on the parameter b; the discrete nature of the possible pay-off totals means that there will be a series of discrete transition-values of b that lead from one dynamical regime to another. These transition-values and the corresponding patterns are described in detail elsewhere [7]. The essentials, of the possible pay-off totals means that there will be a series of discrete transition-values of b that lead from one dynamical regime to another. These transition-values and the corresponding patterns are described in detail elsewhere [7]. The essentials, however, can be summarised in broad terms. If $b > 1.8$, a 2x2 or larger cluster of D will continue to grow at the corners (although not necessarily along the edges, for large squares); for $b < 1.8$, big D-clusters shrink. Conversely, if $b < 2$, a 2x2 or larger cluster of C will continue to grow; for $b > 2$, C-clusters do not grow. A particularly interesting regime is therefore $2 > b > 1.8$, where C-clusters can grow in regions of D and also D-clusters can grow in regions of C. As intuition might suggest, in this interesting regime we find chaotically varying spatial arrays, in which C and D both persist in shifting patterns. Although the detailed patterns change from generation to generation — as both C-clusters and D-clusters expand, collide, and fragment — the asymptotic overall fraction of sites occupied by C, f_C, fluctuates around 0.318 for almost all starting proportions and configurations.

Figure 1 illustrates a typical asymptotic pattern in this regime $2 > b > 1.8$, and shows the typical patterns of dynamic chaos found for almost all starting conditions in this regime. Figure 3a adds a temporal dimension to Fig 1, showing the proportion of sites occupied by C in successive time-steps (starting with 40% D). The asymptotic fraction, f_C, shown in Figure 3a is found for essentially all starting proportions and configurations, for these b-values.

Figure 2 is perhaps more in the realm of aesthetics than biology. Again $2 > b > 1.8$, but now we begin ($t = 0$) with a single D at the centre of a 99x99 lattice of Cs. The figure shows the consequent symmetrical pattern 200 time-steps later. Such patterns, each of which can be characterised in fractal terms, continue to change from step to step, unfolding a remarkable sequence — dynamic fractals. The patterns show every lace doily, rose window, or Persian carpet you can imagine.

As Fig 3b shows, the asymptotic fraction of C is as for the chaotic pattern typified by Figs 1 and 3a. Many of the dynamic features of the symmetric patterns typified by Fig 2 can be understood analytically. In particular, we can make a crude estimate of the asymptotic C-fraction, f_C, for very large such symmetrical patterns, by referring to the geometry of the D-structure. The D-structures are closed-boundary squares in generations that are powers of 2, $t = 2^n$; hence $f_C(t)$ has minima at generations that are powers of 2. These squares now expand at the corners and erode along the sides, returning to square shape after another doubling of total generations. On this basis, a crude approximation suggests that, i time steps en route from t to 2t, there will be roughly $4(2i)(2t+1-2i)$ C-sites within the D-structure of size $(2t+1+2i)^2$; hence the asymptotic C-fraction, f_C, for very large such symmetrical patterns is

$$f_C \approx 4 \int_0^1 s(1 - s)(1 + s)^{-2} ds = 12\ln 2 - 8 . \qquad (1)$$

This approximation, $f_C \approx 0.318$ .. , is indicated by the dashed line in Figs 3a and 3b. It agrees with the numerical results for the symmetric case, Fig 3b, significantly better than we would have expected. Why this approximation also works well for the irregular, spatially chaotic patterns, Fig 3a, we do not know [7].

Fig.1. The spatial Prisoner's Dilemma can generate a large variety of qualitatively different patterns, depending on the magnitude of the parameter, b, which represents the advantage for defectors. Both these figures are for the interesting region when $2 > b > 1.8$. The pictures are coded as follows: black squares represent a cooperator (C) that was already a C in the preceding generation; dark grey (the commonest square) represents a defector (D) following a D; light grey is a C following a D; and the lightest shading shows a D following a C. This simulation is for a 99x99 lattice with fixed boundary conditions, and starting with a random configuration with 10% D and 90% C. The Figure shows the asymptotic pattern after 200 generations: spatial chaos.

Fig. 2. Beautiful "fractal kaleidoscopes" ensue if the initial pattern is symmetric (the rules preserve such symmetry). Here the simulation is started with a single D at the centre of a 99x99 field of C, with fixed boundary conditions. The figure shows that pattern 200 generations later. The coding is as for Figure 1 (obviously, this and other figures are more beautiful if the coding is in colour [6-9]).

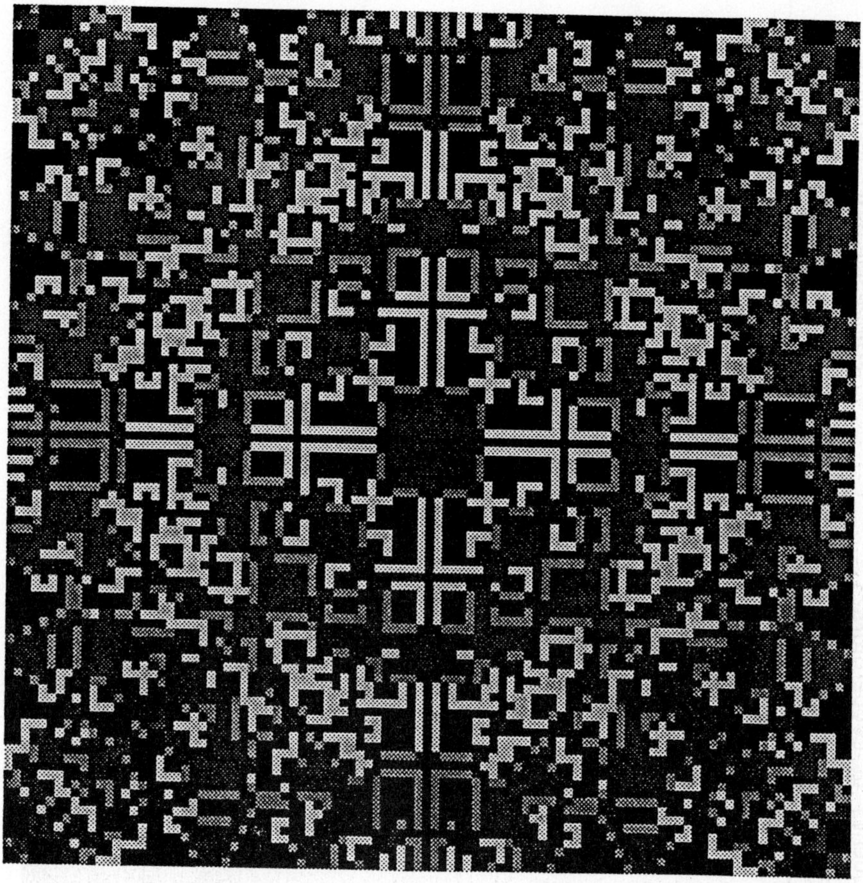

Fig. 3. (a) The frequency of cooperators, $f_C(t)$, for 300 generations, starting with a random initial configuration of $f_C(0) = 0.6$. The simulation is performed on a 400x400 square lattice with fixed boundary conditions, and each player interacts with 9 neighbours (including self). The dashed line represents $f_C = 12\ln 2 - 8 \approx 0.318$ (see text). (b) The frequency of C within the dynamic fractal generated by a single D invading an infinite array of C. At generation t, the width for the growing D-structure is 2t + 1, and Figure 4b shows the frequency of C, $f_C(t)$, within the square of size $(2t + 1)^2$ centred on the initial D-site, as a function of t. Again, the dashed line represents the approximation discussed in the text.

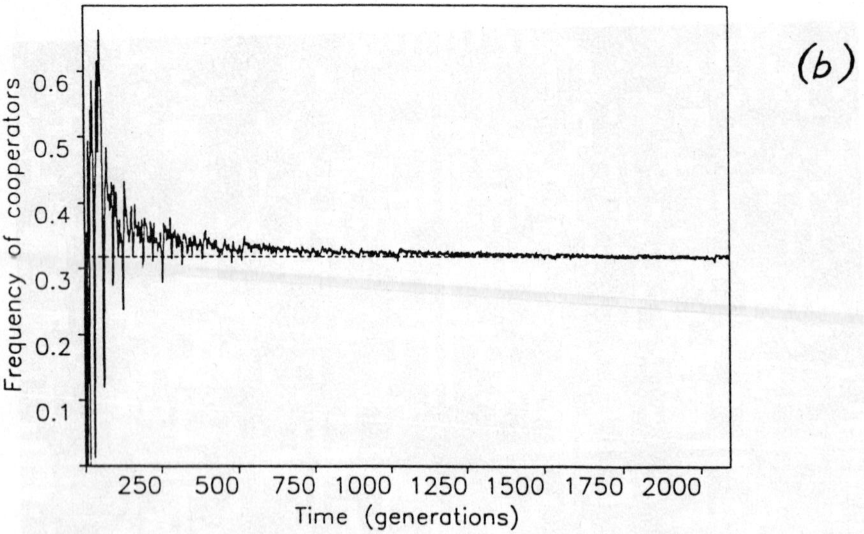

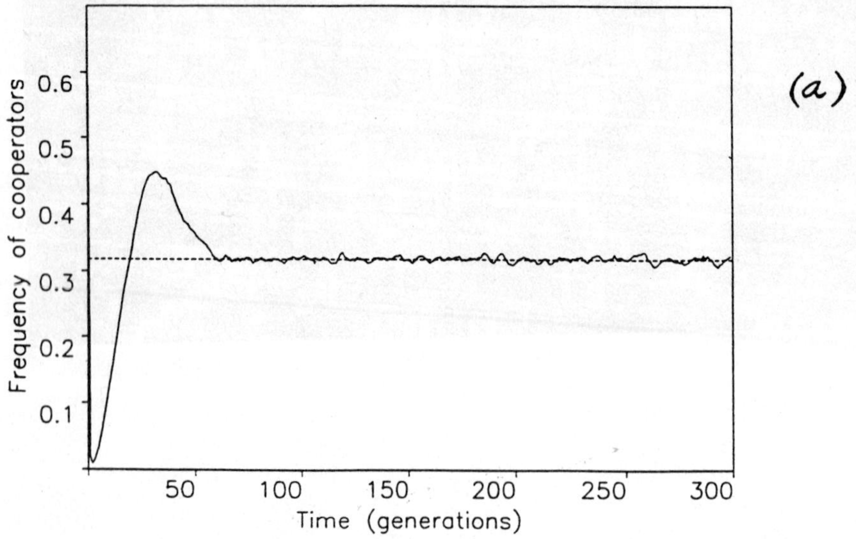

As explained more fully elsewhere [7], results similar to the above are found if we exclude self-interaction, and consider interactions only with the 8 nearest neighbours; here the "interesting" region is $5/3 > b > 8/5$. The symmetrical patterns analogous to Figure 2 are similarly kaleidoscopic, though different. The asymptotic C-fraction, f_C, is now around 0.299 for both symmetric and random starting conditions. For interactions only with the 4 orthogonal neighbours, again the same qualitative regimes are found (here the interesting regime is $2 > b > 5/3$ if self-interaction is included, and $3/2 > b > 4/3$ if not). Numerical studies suggest the asymptotic C-fraction, f_C, is around 0.374. Hexagonal arrays give complex patterns, but show less of the lacy, fractal character seen above, unless we weight the pay-offs from self-interactions somewhat more heavily than from neighbours (which is biologically plausible). In short, the above results seem robust.

3 Extensions of the Basic Model

The above-described work on spatial versions of the PD is based on several simplifying assumptions. We how indicate how the results can be generalised in several ways, using more realistic assumptions (with the original work representing a limiting case). First, we generalise the deterministic assumption that a given site is "won" by the neighbouring player with the largest total score, to allow for "probabilistic winning": the current site-holder or any relevant neighbour may win the site, with probabilities that depend to a specified extent on the relative score. Second, we go beyond the earlier analyses based on sites arranged as regular lattices, to spatially random distributions of sites (the game now being played with neighbours within some specified distance). We think these two generalisations are important, because real situations are likely to involve probabilistic winning (rather than the largest score always triumphing) and irregular arrays (rather than strictly symmetric lattices). A third extension is to continuous time (with individual sites "playing the game" with neighbours and being updated one by one) in contrast to discrete time (with the entire array simultaneously being updated each round). Huberman and Glance [11] have analysed this contrast between continuous and discrete time, but only for a single value of the "cheating-advantage" parameter; as spelled out below, we believe the conclusion drawn from this restricted analysis is misleading as to general properties of the system.

We now outline the results of each of these three generalisations of the basic model, and also sketch some other extensions.

4 Probabilistic Winning

We now introduce a degree of stochasticity into the contest for ownership of sites or cells [8,9]. At any given time, define $s_i = 0$ if site i is occupied by D, and $s_i = 1$ if C. Let A_i denote the pay-off to the occupier of site i, from playing the PD game with itself and with the ν_i neighbours with which it is defined to interact. Then we define P_j, the probability that site j is occupied by a C in the next round, to be

$$P_j = \sum_{i=1}^{\nu_j} A_i^m S_i \Big/ \sum_{i=1}^{\nu_j} A_i^m \qquad (2)$$

The parameter m then characterises the degree of stochasticity in the contest for sites. In the limit $m \to \infty$, we recover the deterministic limit studied earlier: site j will be C in the next round if the largest score among the sites $\{\nu_j\}$ is from a C-owned site, and D otherwise. In the opposite limit of $m \to 0$, we have random drift: the probability that site j will be C or D in the next round depends on the proportions of C and D in the current set of neighbours $\{\nu_j\}$. For m =1, the probability for site j to be C or D is linearly weighted according to the scores of the relevant contestants ("proportional winning").

The results are summarised (with the aid of colour figures) and extensively discussed elsewhere [8,9]. For all values of the parameter m, from deterministic winning to random drift, we find a clear qualitative pattern: for values of b close to unity, the system becomes all C; for relatively large b (approaching 2), the system becomes all D; but for a wide range of intermediate b-values, there are persisting polymorphisms of C and D. In these polymorphic cases, the proportions of C versus D tend to depend on the starting proportions for relatively small b, but for larger b the proportions are essentially independent of the initial configuration. Of course, the details do depend on the degree to which winning is probabilistic (as measured by m). As m decreases from very large values (essentially deterministic winning), D fares somewhat better in that b-values which for larger m gave polymorphisms now give all D. But as m decreases below unity, moving toward random drift, the band of b-values producing polymorphisms again widens.

5 Spatial Irregularities

We indicated earlier that the basic results were independent of whether players interact with 8 neighbours or with the 4 orthogonal neighbours in square lattices, or with 6 neighbours in hexagonal lattices. But spatial arrays in nature will rarely, if ever, have strict symmetry. We have therefore made extensive computer simulations of our spatial PD when the individual sites or players are distributed randomly on a plane [8]. Players interact with those neighbours which lie within some defined radius of interaction, r; this means, *inter alia*, that different sites can interact with different numbers of neighbours. Specifically, we generated the random array by starting with a 200×200 square lattice, and then letting some proportion (say, 5%) of the cells, chosen at random, be occupied by these players; these "active" cells thenceforth defined the random array. The "interaction radius", r, varied from 2 to 11 (measured in units of the original lattice) in different simulations.

As discussed in detail in [8], we explored our spatial PD for these irregular arrays, for various values of r and b, in the limit of deterministic winning ($m \to \infty$). As for the symmetric lattices, we found persistent polymorphisms of C and D for a range of intermediate b-values, provided r was not too big (conversely, if players interact with too many neighbours, the system became all D).The specific limit to r consistent with maintaining polymorphism, r_c, depends on b (eg, for b = 1.6, $r_c \sim 9$), with r_c

decreasing as b increases. In all cases, the patterns settled to become relatively static, and the proportion of C and D tended to depend on the initial configurations to some extent, especially for relatively small b.

6 Continuous Versus Discrete Time

Our original studies of spatial PDs were for discrete time, in the sense that the total pay-offs to each site were evaluated, and then all sites were updated simultaneously [6,7]. This corresponds to the common biological situation where an interaction phase is followed by a reproductive phase; although the game is usually played with individual neighbours in continuous time, at the end of each round of game playing the chips are cashed in, and the cashier pays in fitness coinage. Similar things happen in many other contexts (host parasitoid interactions, or prey-predator models, where dispersal and territory acquisition is followed by raising the young), resulting in biological situations in which individual events — like challenges for territories, eating and being eaten, and so on — occur in continuous time, yet the appropriate simple model is one with discrete time [12,13]. There are, however, some situations where it may be more appropriate to work in continuous time, choosing individual sites at random, evaluating all the relevant scores, and updating immediately. Huberman and Glance [11], indeed, suggest that "if a computer simulation is to mimic a real world system ... it should contain procedures that ensure that the updating of the interacting entities is continuous and asynchronous". We strongly disagree with this extreme view, believing that discrete time is appropriate for many biological situations, and continuous time for others.

Be this as it may, we have repeated out numerical studies of the PD played on symmetric lattices, with both deterministic and probabilistic winning, but now using continuous time (sequential updating of individual sites, chosen independently randomly) rather than the discrete time (synchronous updating) of the earlier work summarised above. As discussed more fully elsewhere, again with the aid of colour illustrations [8,9], the patterns which emerge are broadly similar to those found for discrete updating: for values of b which are neither too large or too small, C and D persist together in varying proportions (depending upon the specific values of the parameters b and m).

There are some differences of detail between the results found for continuous time and those for discrete time. For a small range of relatively large b values, continuous time leads to all D in circumstances where discrete time gives polymorphisms of C and D. In particular, for deterministic winning and b between 1.8 and 2, the result is polymorphism for discrete time and all D for continuous time. Huberman and Glance [11] consider only this single case, and from it they draw sweeping conclusions about the differences between continuous versus discrete time models. But when a broad range of values of b and m are explored, the similarities show plainly that these conclusions are mistaken [8,9]

7 Other Extensions.

Elsewhere, we have considered several other extensions [8]. One is to 3-dimensional arrays, either symmetric or irregular and with deterministic or probabilistic winning; the results are similar to the 2-dimensional ones. If some sites may become unoccupied ("death"), and remain so if surrounded by sites with low pay-offs, then C is easier to maintain. Suppose pay-offs to self-interactions are weighted by a parameter a, relative to pay-offs from interactions with neighbours (we have been considering $a = 1$, but we could have $a > 1$ or $a < 1$): polymorphisms can be maintained in the absence of self-interaction ($a = 0$), but only for relatively large values of m and for smaller values of b; for m = 1 ("proportional winning"), C cannot persist in the absence of self-interaction. A simpler "mean field" analysis is possible for models in which sites disperse propagules globally, in proportion to their total pay-off; polymorphisms of C and D can now be maintained for all b > 1, with the fraction of C being linearly proportional to the self-interaction parameter a. These extensions and refinements are discussed more fully in [8].

8 Conclusions

More generally, spatial effects can confound intuition about evolutionary games. Thus, for example, it can be seen that equilibria among strategies are no longer necessarily characterised by their having equal average pay-offs; a strategy with a higher average pay-off can converge toward extinction; and strategies can become extinct even though their basic reproductive rate (at very low frequency) is larger than unity. This is because the asymptotic equilibrium properties of spatial games are determined by "local relative pay-offs" in self-organised spatial structures, and not by global averages. Although this paper has focused on the PD, our overall conclusion is that interactions with local neighbours in 2- or 3-dimensional spatial arrays can promote the coexistence of strategies, in situations where one strategy would exclude all others if the interactions occurred randomly and homogeneously.

In summary, the PD is an interesting metaphor for the fundamental biological problem of how cooperative behaviour may evolve and be maintained. Essentially all previous studies of the PD are confined to individuals or organised groups who can remember past encounters, who have high probabilities of future encounters (with little discounting of future pay-offs), and who use these facts to underpin more-or-less elaborate strategies of cooperation or defection. The range of real organisms obeying these constraints is very limited. In contrast, the spatially-embedded models involve no memory and no strategies: the players are pure C or pure D. Deterministic interactions with immediate neighbours in 2-dimensional spatial arrays, with success (site, territory) going in each generation to the local winner, is sufficient to generate astonishingly complex and spatially chaotic patterns in which cooperation and defection persist indefinitely. The details of the patterns depend on the magnitude of the advantage accruing to defectors (the value of b), but a range of values leads to polymorphic patterns, whose nature is almost always independent of the initial proportions of C and D.

Theses studies suggest that deterministically-generated spatial structure within populations may often be crucial for the evolution of cooperation, whether it be among molecules, cells, or organisms. Other evolutionary games, such as Hawk-Dove [14], which recognise such chaotic or patterned spatial structure may be more robust and widely applicable than those that do not. More generally, such self-generated and complex spatial structures may be relevant to the dynamics of a wide variety of spatially extended systems: Turing models, 2-state Ising models, and various models for prebiotic evolution (where it seems increasingly likely that chemical reactions took place on surfaces, rather than in solutions).

References

1. Axelrod, R: The Evolution of Cooperation. Basic Books, (1984) New York
2. Axelrod, R. & Hamilton, W.D.: The evolution of cooperation. Science, **211** (1981) 1390-1396
3. Nowak, M.A. & Sigmund, K.: Tit for tat in heterogeneous populations. Nature **355** (1992) 250-253
4. Nowak, M.A. & Sigmund, K.: Chaos and the evolution of cooperation. PNAS, **90** (1993) 5091-5094
5. May, R.M.: More evolution of cooperation. Nature, **327** (1987) 15-17
6. Nowak, M.A. & May, R.M.: Evolutionary games and spatial chaos, Nature, **359** (1992) 826-829
7. Nowak, M.A. & May, R.M.: The spatial dilemmas of evolution. Int. J. Bifurc. Chaos **3** (1993) 35-78
8. Nowak, M.A., Bonhoeffer S. & May, R.M.: More spatial games. Int. J. Bifurc. Chaos **4** (1994) 33-56
9. Nowak, M.A., Bonhoeffer, S. & May, R.M.: Spatial games and the maintenance of cooperation. PNAS, **91** (1994) 4877-4881
10. Herz, A.V.M.: Collective phenomena in spatially extended evolutionary games. J. theor. Biol, **169** (1994) 65-87
11. Hubermann, B.A. & Glance, N.S.:Evolutionary games and computer simulations. PNAS, **90** (1993) 7712-7715
12. Godfray, H.C.J. & Hassell, M.P.: Discrete and Continuous insect populations in tropical environments. J. Anim. Ecol. **58** (1989) 153-174.
13. Maynard Smith, J. Models in Ecology. Cambridge University Press (1974) Cambridge
14. Maynard Smith, J. Evolution and the Theory of Games. Cambridge University Press,(1982) Cambridge

Aggressive Signaling Meets Adaptive Receiving: Further Experiments in Synthetic Behavioural Ecology

Peter de Bourcier and Michael Wheeler

School of Cognitive and Computing Sciences,
University of Sussex, Brighton BN1 9QH, U.K.
Telephone:+44 1273 678524
Fax:+44 1273 671320
E-Mail:peterdb@cogs.susx.ac.uk, michaelw@cogs.susx.ac.uk

Abstract. This paper describes our most recent investigations into aggressive communication. We perform experiments in a simple synthetic ecology, in which simulated animals (animats) are in competition over food. In the first experiment, each animat has an evolved signaling strategy — the degree to which that animat 'bluffs' about its aggression level. The form of artificial evolution used features no explicit fitness function. By varying the cost of signaling, we show that the general logic of the handicap principle (according to which high costs enforce reliability) can apply in the sort of ecological context not easily studied using formal models. However, because an animal's behavioural response to an incoming signal will be determined not only by the signal itself, but also by the degree of importance that that animal gives to the signal, we go on to introduce the concurrent evolution of signaling *and* receiving strategies. We discuss how, in this more complex scenario, the cost of signaling affects the reliability of the signaling system.

Keywords: aggression, animal signaling, behavioural ecology, communication.

1 Introduction

When animals of the same species come into conflict, the incidence of unrestrained battles is relatively low. Confrontations tend to revolve around ritualized signaling displays which, more often than not, allow those concerned to conclude matters without the need for actual physical combat. In this paper, we describe our latest work on aggressive signaling (see also [2, 12]).

Our study takes place within a theoretical framework that we call *Synthetic Behavioural Ecology* ('SBE'). Behavioural ecology is the discipline that aims to explain *why*, rather than *how*, animals behave as they do (see [6]). That is, behavioural ecologists try to identify the functional role that particular behaviours play in contributing to the survival and reproductive prospects (the Darwinian fitness) of an animal. The SBE-methodology (in common with much of A-Life) is to construct simple (although not trivial) synthetic ecologies. We then carry out experiments in these synthetic ecologies, with the specific goal of making a contribution to the scientific understanding of how ecological context influences the adaptive consequences of behaviour.

The primary aim of SBE is to contribute to ongoing work in the biological sciences, by providing a distinctive theoretical platform for testing hypotheses about the functional aspects of animal behaviour (cf. [11]). SBE may be able to play such a role, because it is pitched at a level in between abstract mathematical models and naturally occurring ecological contexts. However, we must stress that there is no suggestion that SBE provides any easy answers to the difficult problems faced by biologists in this area.

Two SBE-experiments are described. The first is a consolidation of our earlier work [2] on Zahavi's handicap principle [13]. The handicap principle states that the reliability of animal signals can be enforced if those signals cost the signaling creature something in fitness to make (see section 2). The significance of this first experiment to our overall project is that we have made several fundamental improvements to our experimental model. For example, the model of aggression has been revised, in order to bring it closer to current opinion in behavioural biology (see sections 2 and 3). And we have introduced an evolutionary scenario in which given (i) a large number of possible signaling strategies, and (ii) an initially random distribution of signaling strategies across a population of individuals, the subsequent distribution of signaling strategies evolves in response to ecological context (see section 3).

The second experiment introduces a significant complication into the model. In both experiments, the activity of each animat is, in part, determined by the values of the signals that that animat receives. In the first experiment, an incoming signal is not, in any way, 'interpreted' by the receiver. But it seems likely that, as well as signaling strategies, animals will evolve to have *receiving strategies;* that is, a receiver will give some weight (or degree of importance) to incoming signals. To investigate the effect of receiver tactics on aggressive communication systems, we introduce the concurrent evolution of individual signaling *and* receiving strategies.

2 Biological Background: Aggressive Signaling

Lorenz [7] characterized aggression as something akin to a 'spontaneous appetite,' such that, in the absence of the performance of aggressive acts, the tendency to behave aggressively increases with time. However, there is little empirical evidence in favour of the Lorenzian story. Moreover, if aggressive acts did occur purely as a consequence of endogenous changes in the animal, without the correlated presence of any external threat or adaptive goal, then it is hard to see how to make evolutionary-functional sense of aggression. The message is that aggression needs to be treated not as an end in itself, but as an adaptive phenomenon, with an adaptive purpose, such as to win or to defend a resource, (For a discussion of all these issues, see [1].)

The phenotypic traits that determine an animal's ability to win a fight (e.g., size) are called its resource holding potential ('RHP'). Signals which are biologically correlated with RHP (e.g., the deep croaks of large toads) cannot be faked. However, other signals — such as signals of *aggressive intentions* — are not necessarily reliable. Cheats who consistently signaled high levels of aggressive intent, whatever their actual intentions, could well prosper when confronted by 'trusting' opponents. Zahavi [13] has argued that the reliability of intention-signals could be increased if the animal concerned had to invest, in some way, in those signals. This idea — known as *the handicap principle* — is illustrated by the fact that a signal which is, for example, wasteful of energy is, as a consequence of that wastefulness, reliably predictive of the possession of energy; hence honesty is enforced [4]. To be reliable, signals of aggressive intent must be more costly in fitness terms than they strictly need be merely to communicate unambiguously the information at issue. Moreover, the costs involved must be differential. A specific signal indicating a particular level of intended escalation must be proportionally more costly to a weak individual than to a strong individual.

In confrontations involving aggressive signaling (as in many other scenarios), what counts as an adaptively fit individual strategy will be determined by the frequencies with which the various available strategies are adopted by the other members of the population; i.e., individual fitness is frequency-dependent. To investigate such situations, theoretical biologists employing formal models have come to use the concept of an *evolutionarily stable strategy* or 'ESS' [8]. An ESS is a strategy which, when adopted by most members of a population, means that that population cannot be invaded by a rare alternative strategy.

Zahavi's handicap principle has received support from ESS models (e.g., [3, 4]). But while ESS theory is, without doubt, a powerful framework for modeling signaling systems, existing ESS models, at least, are limited in their application to natural systems. For example, they do not allow for two-way information flow, or for situations in which there are many receivers of one signal [5]. We shall be concerned with just such a multi-agent context; i.e., a context in which there is two-way information flow and in which a signal is, in general, picked up by many receivers. Hence the experiments presented here are *early steps towards* a situation in which SBE may be of service to biology, by providing a methodology for bridging the explanatory gap between abstract mathematical models and natural environments.[1]

[1] For reasons of space, this summary of the relevant biological literature has been highly selective. A more comprehensive review, together with a more complete list of references, appears in [12]. That paper also includes a more detailed description of our basic experimental model.

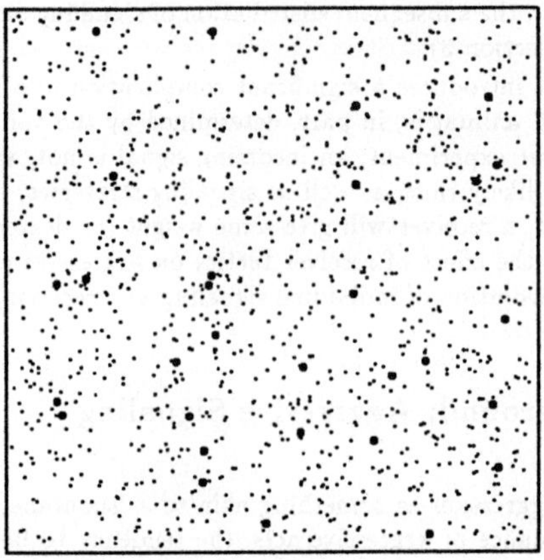

Fig. 1. *The world at the start of a typical run. The particles of food are shown as dots, and the animats are shown as filled-in circles. Both food particles and animats are placed randomly.*

3 Experimental Model

The synthetic world is two-dimensional and non-toroidal; i.e., it is a flat plane, the edges of which are barriers to movement. Its dimensions are 1000 by 1000 units (each animat being round and 12 units in diameter). Space is continuous.

The animats have two *highly* idealized sensory modalities. The 'visual' system is based on a 36-pixel eye providing information in a full 360 degree radius around the animat, with an arbitrarily imposed maximum range of 165 units. Each pixel returns a value corresponding to the proportion of that pixel's receptive field containing other animats. The 'olfactory' system employs principles similar to those used for vision, the only differences being that olfactory range is only 35 units, and that food particles are treated as point sources.

To begin a run, a number of animats (all with equal energy levels) and a number of food particles (all with equal energy values) are distributed randomly throughout the world (see figure 1). When an animat lands on a food particle, the energy value of the particle is transferred to the animat, thereby incrementing that animat's existing energy level. So, in order to maintain the resource, new food particles are added (with a random distribution) at each time-step. The resource is 'capped' so that food is never more plentiful than at the beginning of the run.

Animats receive energy from food, but they also pay a series of energy costs (a small existence-cost deducted at each time-step, and costs for fighting, moving, signaling, and reproducing — see below). If an animat's energy level sinks to 0, then it is removed from the world. Thus food-finding is an essential task, and, to encourage foraging, each animat has a hunger level (a disposition to move towards food) which changes in a way inversely proportional to an animat's energy level. Because the food supply is limited, animats are in competition for the available resources.

As well as a disposition to find food, each animat has a disposition to make aggressive movements (its 'aggression level'). An aggressive movement is defined as one in which one animat moves directly towards a second animat. If animats touch, a fight is deemed to have taken place, and the combatants suffer large energy reductions. When an animat makes an aggressive movement, its level of aggression is increased by an amount proportional to the previous aggression level. Conversely, a non-aggressive movement results in a decrease in an animat's level of aggression, by an amount proportional to the previous aggression level. Increases in aggression levels take place much more swiftly than decreases (see footnote 3 for the relevant parameter-values).

So the disposition to respond aggressively towards other animats decreases with time in the absence of aggressive encounters, and increases as soon as a confrontation develops (thereby making a further aggressive act more likely in the immediate future). This feature means that we are already working with a non-Lorenzian model of aggression. Moreover, in the ecological situation as a whole, aggression serves an adaptive purpose. Foraging behaviour is essential for survival. So it benefits an individual animat to inhabit an area which is not being foraged by other animats. Aggressive behaviour helps to 'defend' such an exclusive area. Of course, aggressive behaviour is reactively triggered by the presence of other animats within visual range; so animats do not, in any sense, 'plan' their aggression with respect to maintaining a foraging area. The adaptive significance of the behaviour is an emergent property of the ecological situation which we, as external observers, can identify, but which is not explicitly programmed into the animats themselves.[2]

Animats produce aggressive signals whenever at least one other animat is within visual range, and receive the aggressive signals of any other animats within that range. These signals are indicators, to receiving animats, of the *apparent* aggression of signalers (see below).

At each time-step, the direction in which each animat will move (one of the 36 directions) is calculated using the following probabilistic equation (each movement results in a small reduction in available energy):

$$p(d) = \frac{h.s(d) + a.v(d) + t(o).v(o) + c}{\sum_{i=1}^{n}(h.s(d_i) + a.v(d_i) + t(o_i).v(o_i) + c)}$$

where $p(d)$ is the probability that the particular animat will move in the direction d; n is the number of possible directions of movement; h is the animat's hunger level; $s(d)$ is the value returned by the olfactory system in direction d; a is the animat's aggression level; $v(d)$ is the value returned by the visual system for direction d; o is the direction 180 degrees off d — i.e., in the opposite direction to d; $t(o)$ is the threat (aggressive signal) that the animat perceives from other animats from the opposite direction to d; $v(o)$ is the value returned by the visual system in the opposite direction to d; and c is a small constant which prevents zero probabilities.

Roughly speaking, the effects of the movement equation, for an animat, I, can be summarized as follows: (i) the probability of I moving in the direction of food is proportional to I's degree of hunger; (ii) the probability of I moving in the direction of another animat is proportional to I's aggression level; (iii) the probability of I moving away from another animat is proportional to the threat which I perceives from that second animat; and (iv) if there are no other animats in I's visual field, and no food within I's olfactory range, then I will make a random movement.

Each member of the population has a *signaling strategy* which is under evolutionary control. Signals are produced in accordance with the calculation $S = A + ((C/100).A)$ where S is the value of the signal made, A is that individual's current aggression, and C is an individual-specific constant, the value of which is in the range 0-100. For each individual, C is specified by a bit string genotype. At the beginning of a run, a random population of genotypes is created, producing a random distribution of signaling strategies. However, when an animat achieves a pre-defined (high) energy level, it will asexually reproduce. The result is an only child placed randomly in the world. The offspring is given the same initial energy level as each member of the population had at the start of the run, and the corresponding amount of energy is deducted from the parent. The parent's genotype is copied over to the offspring, but there is a small probability that a genetic mutation will take place. So it is possible that the child will adopt a different signaling strategy to its parent. (Throughout the first experiment, an 8 bit genotype was used, and the mutation rate was set to be a 0.05 chance that a bit-flip mutation will occur *as each bit is copied.*) Notice that this form of artificial evolution features *no explicit fitness function.* Rather, the selection pressures imposed by the ecological context mean that different strategies will have different fitness consequences. Only those individuals adopting adaptively fit strategies will have a high probability of becoming strong enough, in energy terms, to reproduce.

Aggressive signals are displays for which a signaling animat has to pay via a deduction in energy. This cost increases as the level of aggression signaled increases, so that it costs more in energy to

[2] Although we shall not pursue the issue here, we consider the multi-agent dynamics observed in our model to amount to a minimal form of territorial behaviour (see [12]).

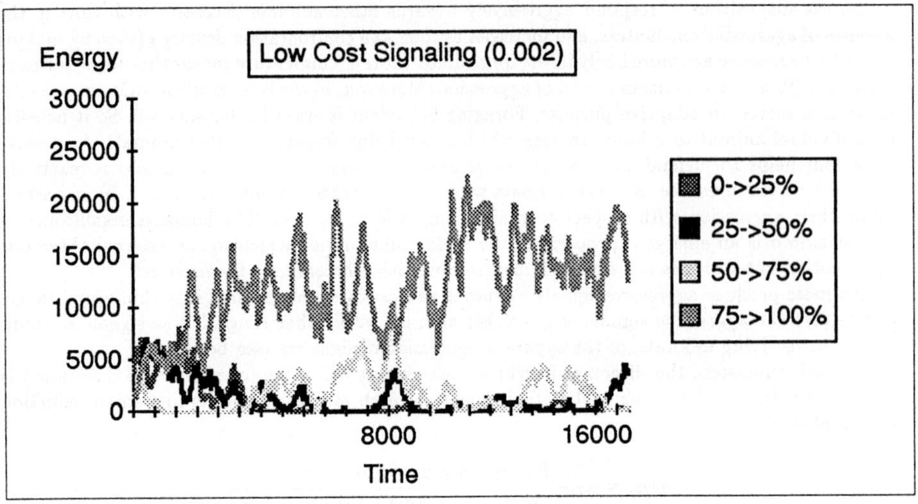

Fig. 2. *Low Cost Signaling — the total energy levels of the four sub-populations are plotted against time.*

make a more aggressive signal. Thus the costs involved are differential, in the sense required by the handicap principle, because, given a specific signal made by a high-energy individual, it will cost a low-energy individual proportionally more to produce the same signal.[3]

4 Results: The Effect of Cost on Signal Reliability

To uncover the trends in signaling behaviour, we partitioned the total population into four sub-populations: The first group included all those individuals producing signals indicating levels of aggression between 0% and 25% in excess of actual aggression, the second 25% to 50% in excess of actual aggression, the third 50% to 75%, and the fourth 75% to 100%. So the 0-25% group produce the most reliable signals, *in the sense that* the signals given out by those individuals are (relative to the signals produced by those in other groups) more accurate reflections of those individuals' dispositions to make aggressive movements. The fourth group are the most extreme bluffers. Each sub-population, at any one time, included all individuals adopting a strategy from the appropriate band, including any offspring. The simulation was run many times, with various values for the cost of signaling, and we shall now discuss two examples of the results obtained. (Throughout this paper, we have chosen to reproduce the results from individual runs, because this enables us to highlight certain interactions between the competing strategies — e.g., periods where dominant

[3] The values of the various parameters were set (largely as a result of trial and error) as follows: initial supply of food = 1200 particles; initial size of population = 30; initial energy level = 300; energy level at which reproduction takes place = 1000; initial level of aggression = 100; energy value of 1 particle of food = 45; rate of food replenishment = a maximum of 16 particles per time step; maximum supply of food at any one time = 1200; existence-cost = 1; movement-cost = 2; cost of fighting = 100 units of energy per time step of fight; increment to aggression level after making an aggressive movement = 1/10 of previous aggression level; reduction in aggression level after a non-aggressive movement = 1/100 of previous aggression level; constant preventing zero probabilities = 1.

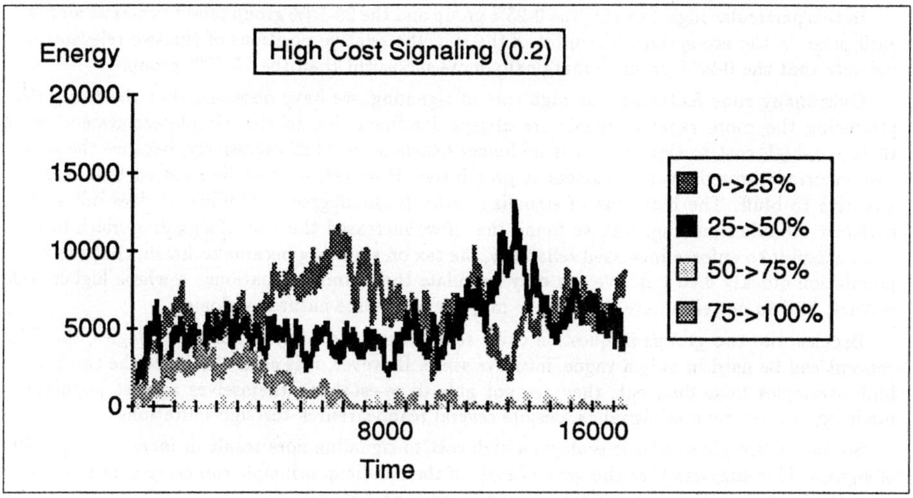

Fig. 3. *High Cost Signaling — the total energy levels of the four sub-populations are plotted against time.*

strategies are temporarily displaced. Such events would tend to be obscured by averages over large numbers of runs. These effects are particularly significant in the second set of experiments, after the introduction of receiving strategies. Each of the specific runs that we discuss is a typical example from the large number of runs that we have observed.)

At first, the cost of signaling was set to be low — only 0.002 units of energy deducted per unit of aggression signaled. Figure 2 shows plots of the total energy present in each of the 4 sub-populations, over the first 16000 time steps. We take it that 'total energy' is an intuitively satisfactory guide to general adaptive success. The decision that 16000 time steps was sufficient for the population-behaviour to stabilize was based on many observations of long runs.[4]

After an initial settling down period, the 75-100% group tend to dominate. Both the 0-25% group and the 25-50% group struggle to survive in this eco-system, and the 50-75% group mounts only one unsuccessful challenge (see the period around time = 8000). In every one of many runs at this cost, the most extreme bluffers have been by far the most successful sub-population. So, in this simple eco-system, when there is a low cost to signaling, the most adaptive strategy is to produce signals that indicate levels of aggression well in excess of actual aggression. And, on the basis of empirical observation, extreme bluffing appears to have the general character of an ESS. Following the initial decline of the low-bluffing strategies, these groups fail to re-establish themselves in the population, despite several reappearances due to fortuitous mutations. The results also suggest that reliable signaling would not be an ESS, as a population of reliable signalers could be invaded by high-bluffing mutants.

The cost of signaling was then increased to 0.2 units of energy deducted per unit of aggression signaled. Figure 3 shows the results of one representative run. Again the graph shows the total energy present in each of the 4 sub-populations, during the first 16000 time steps. (Due to the higher cost of signaling, the total energy present in the various sub-populations is generally less than in the low cost case.)

[4] In our second set of experiments (sections 5 and 6), a much longer time-span (64000 time steps) was required.

In this particular high cost run, the 0-25% group and the 25-50% group tend to co-exist alongside each other in the eco-system. For most of the run, the relative positions of the two relevant plots indicate that the 0-25% group is marginally more successful than the 25-50% group.

Over many runs featuring this high cost of signaling, we have observed that the two groups producing the more reliable signals are always dominant. So, in this simple eco-system, when there is a high cost to signaling, it is no longer beneficial to bluff excessively, because the energy cost incurred through such behaviour is prohibitive. However, it must be stressed that it is still adaptive to bluff. The high cost of signaling *restricts* the degree of bluffing, it does not enforce perfectly reliable signaling. But we found that if we increased the cost of signaling much further, in an attempt to enforce increased reliability, the tax on signaling became so harsh, that the whole population quickly died out. We can only speculate that similar situations — where higher costs restrict, but do not eliminate cheating — may occur in some natural ecologies.

Because the two groups in question cover such a wide spread of signaling strategies, the ESS-concept can be used in only a vague, intuitive sense. However, it is clear that once the two higher-bluff strategies have died out, they are not able to re-establish themselves against populations made up of more reliable signalers, despite several reappearances through mutations.

So, in this simple synthetic ecology, a high cost to signaling does result in *increased* reliability of signals. This suggests that the *general* logic of the handicap principle can carry over to the sort of multi-agent signaling systems not covered by existing ESS models.

5 The Introduction of Receiving Strategies

The conclusions from our first experiment constitute the beginning, rather than the end, of a story. To model more realistic signaling systems, we need to consider not only the behaviour of signalers, but also the behaviour of *receivers*. 'Receiver psychology' has become an increasingly important issue in the biological literature (e.g., [9, 10]).

In the experiment described above, the threat values for the movement equation were simply the values of the incoming aggressive signals. However, we now extend the experimental model, so that each animat has not only a genetically specified signaling strategy, but a similarly specified receiving strategy. Each animat's receiving strategy is determined by an individual-specific constant, K that 'weights' incoming signals, in order to produce the threat value for the movement equation. So $T = R.(K/100)$, where T is the threat value in the movement equation, R is the incoming signal, and K is an individual-specific constant, in the range 0-200. A K of 0 would result in that individual ignoring incoming signals; a K of 100 means that the value of the incoming signal itself is used as the threat value (i.e., equivalent to the situation in the first experiment); and a K of 200 results in incoming signals being doubled, and that resulting value being used as the threat value. Henceforth we shall speak of this weighting as the *degree of importance* assigned to incoming signals (the higher the value of K, the higher the degree of importance). So when an individual signals, the exact effect that that signal has on the behaviour patterns of animats within visual range is now also a function of those animats' receiving strategies.

To maintain a balance between the range of possible receiving strategies, and the range of possible signaling strategies, we now allow the existence of individuals who produce signals indicating levels of aggression *lower* than actual aggression. Individuals produce signals according to the calculation $S = A.(C/100)$, where S is the value of the signal made, A is that individual's current aggression, and C is an individual-specific constant, in the range 0-200. A C of 0 is equivalent to not making any signal, a C of 100 is equivalent to producing indicators of actual aggression, and a C of 200 is equivalent to producing signals indicating twice actual aggression.

The length of each individual's genotype is doubled, and encodes for both a signaling and a receiving strategy. We employ the same process of artificial evolution as was used in the first experiment (with the same system/rate of mutation).

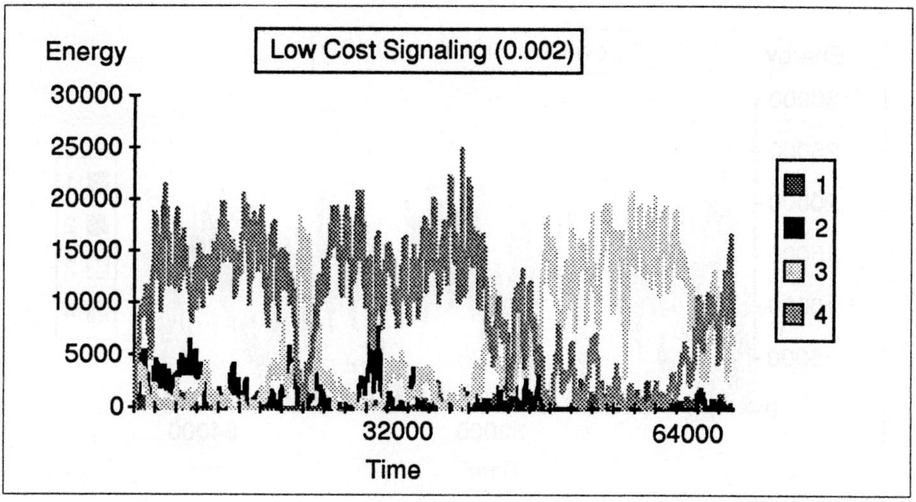

Fig. 4. *Low Cost Signaling — the total energy levels of the four* **RECEIVING** *sub-populations are plotted against time. See figure 5 for plots of the concurrently evolved signaling strategies.*

6 Results: Interacting Strategies

Once again, the population was divided into four sub-populations on the basis of signaling strategy. The groups were identified by ranges in the value of the individual-specific signaling-constant, C (Group 1: 0-50, Group 2: 50-100, Group 3: 100-150, Group 4: 150-200). The total energy present in each sub-population was then recorded against time. However, for this experiment, we also divided the population into four sub-populations on the basis of receiving strategy, and recorded the total energy present in each of those groups as well. The receiving groups were identified by ranges in the value of the individual-specific receiving-constant, K (Group 1: 0-50, Group 2: 50-100, Group 3: 100-150, Group 4: 150-200).

Our work on the interaction between signaling and receiving strategies is still in its early stages, and we do not pretend to have anything approaching a full understanding of the relatively complex population dynamics that unfold at different costs of signaling. Much more analysis needs to be done. However, there are some clear trends in the observed behaviour.

Once again, we began by setting the cost of signaling to be 0.002 units of energy deducted per unit of aggression signaled. Figure 4 shows plots of the total energy present in each of the 4 receiving sub-populations, over the first 64000 times steps of a typical run at this low cost of signaling. Figure 5 shows the equivalent plots for each of the 4 signaling sub-populations. (In the following discussion, we shall often speak of 'signalers' and 'receivers'; but it should be remembered that each individual is both a signaler and a receiver.)

The dominant signaling strategy for most of the period shown (and for low cost signaling cases in general) is signaling group 4 — the high bluffers (see figure 5). This was expected, given the low cost of signaling. However, it might have been thought that the failure of signals to reflect actual aggression would mean that it would pay individuals to ignore those signals, so that receiving group 1 would be the most successful. Then, with signals having little (if any) effect on receivers, the communication system would be in danger of breaking down (cf. [6]). But, in fact, the dominant

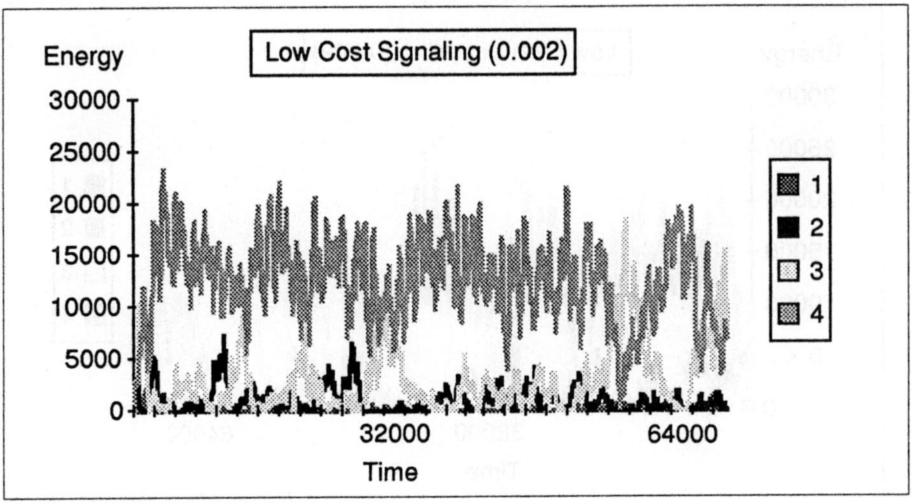

Fig. 5. *Low Cost Signaling — the total energy levels of the four* SIGNALING *sub-populations are plotted against time. See figure 4 for plots of the concurrently evolved receiving strategies.*

receiving strategy for most of the run is receiving group 4 — the group that gives the *highest* degrees of importance to incoming signals (see figure 4). The explanation for this seems to be that there is, in this simple ecology, just one source of information about the quality of potential adversaries, namely aggressive signals. So individuals do not have the option of switching to another, more reliable source of information, such as RHP. Under these circumstances, the strategy of giving a high degree of importance to incoming signals is less likely to lead a receiver into potentially costly fights. (The extent to which an individual ignores incoming signals is the extent to which that individual's behaviour towards other animats will tend to be driven directly by its own aggression level.) In that sense, there is selective pressure on receivers to give a high degree of importance to incoming signals. However, there is another side to the adaptive story.

In low cost runs, it is quite common to find the dominant signaling and receiving strategies being challenged by signaling group 3 and receiving group 3. For example, in the run chosen for discussion, the period following t = 46000 saw receiving group 3 temporarily displace group 4 as the dominant receiving strategy. We believe (from watching the distribution of strategies change whilst observing the aggressive interactions during such periods) that the explanation for this is as follows: Individuals who evolve to give less weight to incoming signals make a larger number of aggressive movements (because incoming signals produce lower threat values). This, in turn, increases the aggression levels of the animats adopting that policy (because when an animat makes an aggressive movement, its level of aggression is increased); so those animats produce bigger signals. Individuals giving high degrees of importance to the incoming signals then tend to retreat more often, giving the advantage (in terms of foraging areas) to those individuals adopting receiving strategies in group 3. There is then an improvement in the fortunes of signaling group 3. Why should this happen? On this point, we can, at present, offer no more than a conjecture. But it is plausible that it occurs as a direct result of the increase in the general aggression levels of individuals in receiving group 3. Higher aggression levels mean bigger signals; so a lower value of C (the individual-specific signaling-constant) will now produce signals equivalent to those previously

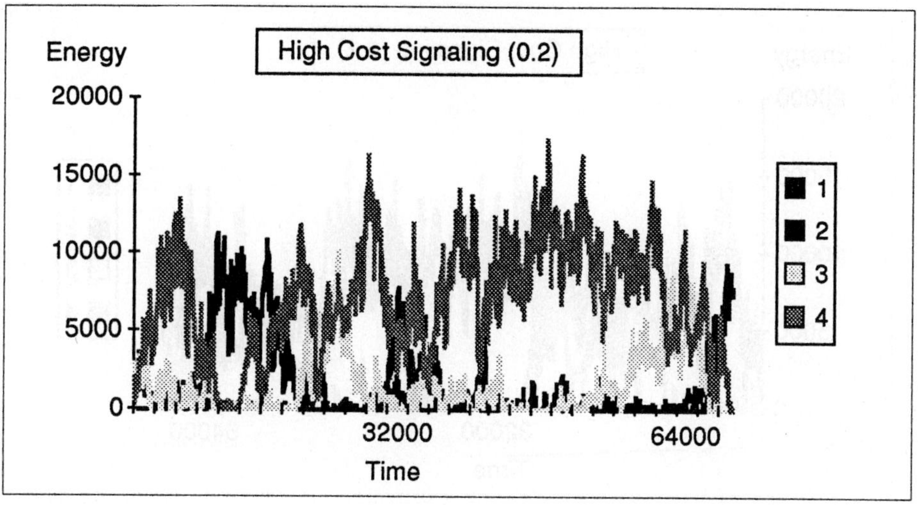

Fig. 6. *High Cost Signaling — the total energy levels of the four* RECEIVING *sub-populations are plotted against time. See figure 7 for plots of the concurrently evolved signaling strategies.*

produced with a higher value of C. Individuals with a lower C obtain a small saving in energy (and, thereby, a small fitness advantage) by signaling at a lower level. However, once the majority of the population have adopted the newly dominant signaling and receiving strategies, the adaptive advantage is lost, and the new arrangement is unstable against invasion by mutants who adopt the strategies which are generally dominant in low cost ecologies (i.e., signaling group 4/receiving group 4).[5]

As in the first experiment, the cost of signaling was then increased to 0.2 units of energy deducted per unit of aggression signaled. Figure 6 shows plots of the total energy present in each of the 4 receiving sub-populations, over the first 64000 times steps of the run. Figure 7 shows the equivalent plots for each of the 4 signaling sub-populations.

With a high cost to signaling, there is a powerful selection pressure for individuals to minimize the amount of energy that they expend on signaling. So it might seem relatively unsurprising that the dominant signaling group for most of the run is group 1 — the group producing the lowest-value signals (see figure 7). But now notice that these individuals are producing signals indicating levels of aggression *much less than* actual aggression. From this one might be led to conclude that the introduction of more complicated strategic possibilities has upset the operation of the handicap principle. Signals seem to be unreliable. Moreover, because signals are indicating apparent levels of aggression which are, in fact, less than actual aggression, there should be a higher probability that individuals will be drawn into costly fights. But, in fact, the scenario does *not* lead to an escalation in the number of fights; and, once we take account of the interaction between the signaling and receiving strategies, it becomes clear that something very like the handicap principle *does* operate.

[5] It may be that a suitable analysis, based on data from the simulation, could identify the two strategic arrangements — (i) signaling group 4/receiving group 4 and (ii) signaling group 3/receiving group 3 — as attractors in the dynamics of the self-organizing ecological system, with the first the much stronger of the two. We aim to investigate this possibility when we submit the results to further analysis.

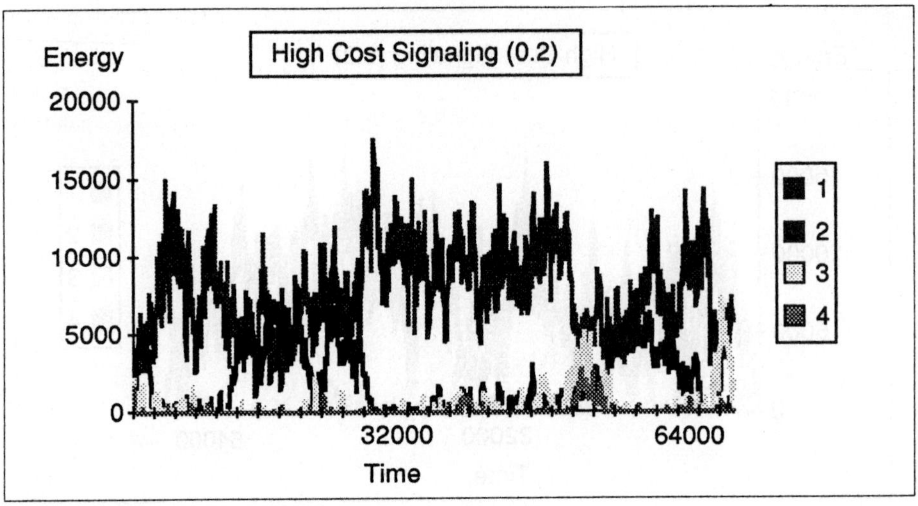

Fig. 7. *High Cost Signaling — the total energy levels of the four* SIGNALING *sub-populations are plotted against time. See figure 6 for plots of the concurrently evolved receiving strategies.*

The dominant receiving group for most of the run is group 4 — the group giving the *highest* degrees of importance to incoming signals (see figure 6). This is significant, because there is a theoretical equivalence, as far as the reactions of receivers are concerned, between situations where signals are small compared to actual aggression, but receivers give high degrees of importance to those signals, and situations where signals are higher, but receivers give those signals less importance. As an over-simplified example, consider that a one-way dyadic interaction in which the signaler's $C = 50$ and the receiver's $K = 200$ is equivalent to a similar interaction in which the signaler's $C = 100$ and the receiver's $K = 100$. Both result in a threat value for the receiving individual equal to the actual aggression of the signaling individual. But the key difference is that the signaling animat in the latter case would expend much more energy through signaling. Of course, there are ranges of different signaling and receiving strategies operating in the population at any one time; and communication events are two-way and, in general, multi-agent. Hence the situation is never that straightforward. But the point is that the ecology tends to evolve towards a situation which, in terms of receiver-behaviour, is *equivalent to* reliable signaling. And the adaptive advantage of that evolutionary solution is that the costs incurred by signaling animats are much lower than they would be if signals were reflections of actual aggression. (The limited adaptive success of the higher-signaling groups can be explained by the fact that, for most of the run, receivers are giving a high degree of importance to incoming signals. So, even where costs are high, there may well be some occasional advantage to producing bigger signals, which will have a high pay-off in terms of the subsequent behaviour of receivers.)

7 Conclusions

Although we hesitate to draw conclusions about the natural world from results obtained using our simple synthetic ecology, the evidence lends support to the following hypotheses: Where the fitness costs of signaling are low, signalers will tend to produce signals indicating levels of aggression well in excess of actual aggression. *If there is no alternative source of relevant information,* receivers may still pay heed to those signals. Where the fitness costs of signaling are high, the pressure on signalers to reduce the level of signaling may still lead to communication systems in which signals are not direct reflections of quality, in that signalers may produce signals indicating levels of aggression *lower* than actual aggression. However, if receivers evolve to give a high degree of importance to those signals, the effect would be to *compensate* for the actual values of the signals. As receivers, individuals would still behave *just as if* signals were direct reflections of aggression (so they would benefit from not being drawn into costly conflicts); as signalers, individuals would benefit from the low level of signaling.

Our work on the interaction between signaling and receiving strategies has only just begun. Even in this simple synthetic ecology, the dynamics of the communication system are made relatively complex by the need for animats to forage, the probabilistic nature of the behaviour, the constantly changing aggression levels, and the fact that many different signaling and receiving strategies can coexist in the population at any one time. We believe that these features of the experimental model are crucial in ensuring that the results of the simulation are non-trivial. But these same features complicate the process of explanation. Thus, whilst we believe that our proposed explanations of the observed behaviour are fundamentally correct, our conclusions would be strengthened by further analysis. That is our next task.

Acknowledgements

Peter de Bourcier is supported by the States of Jersey Education Committee. Michael Wheeler is supported by British Academy award no.92/1693. Thanks to our anonymous reviewers for helpful comments, and to Seth Bullock for many invaluable discussions.

References

1. J. Archer. *The Behavioural Biology of Aggression.* Cambridge University Press, Cambridge, 1988.
2. P. de Bourcier and M. Wheeler. Signalling and territorial aggression: An investigation by means of synthetic behavioural ecology. In D. Cliff, P. Husbands, J.-A. Meyer, and S. W. Wilson, editors, *From Animals to Animats 3: Proceedings of the Third International Conference on Simulation of Adaptive Behavior*, pages 463–72, Cambridge, Massachusetts, 1994. M.I.T. Press / Bradford Books.
3. M. Enquist. Communication during aggressive interactions with particular reference to variation in choice of behaviour. *Animal Behaviour*, 33:1152–1161, 1985.
4. A. Grafen. Biological signals as handicaps. *Journal of Theoretical Biology*, 144:517–546, 1990.
5. A. Grafen and R.A. Johnstone. Why we need ESS signalling theory. *Philosophical Transactions of the Royal Society: Biological Sciences*, 340:245–250, 1993.
6. J. R. Krebs and N. B. Davies. *An Introduction to Behavioural Ecology.* Blackwell Scientific, Oxford, 2nd edition, 1987.
7. K. Lorenz. *On Aggression.* Methuen, London, 1966.
8. J. Maynard Smith. *Evolution and the Theory of Games.* Cambridge University Press, Cambridge, 1982.
9. P.K. McGregor. Signalling in territorial systems: a context for individual identification, ranging and eavesdropping. *Philosophical Transactions of the Royal Society: Biological Sciences*, 340:237–244, 1993.
10. M. Stamp Dawkins and T. Guilford. The corruption of honest signalling. *Animal Behaviour*, 41(5):865–73, 1991.
11. I.J.A. te Boekhorst and P. Hogeweg. Effects of tree size on travelband formation in orang-utans: Data analysis suggested by a model study. In R. Brooks and P. Maes, editors, *Proceedings of Artificial Life IV*, pages 119–129, Cambridge, Massachusetts, 1994. M.I.T. Press.
12. M. Wheeler and P. de Bourcier. How not to murder your neighbor: Using synthetic behavioral ecology to study aggressive signaling. Technical Report 357, School of Cognitive and Computing Sciences, University of Sussex, 1994. Submitted to the journal *Adaptive Behavior.*
13. A. Zahavi. Mate selection — a selection for a handicap. *Journal of Theoretical Biology*, 53:205–214, 1975.

Modelling Foraging Behaviour of Ant Colonies

R. P. Fletcher * C. Cannings P. G. Blackwell

School of Mathematics and Statistics
University of Sheffield
Sheffield, U.K.

Email: c.cannings@sheffield.ac.uk *or* p.blackwell@sheffield.ac.uk
Telephone: +44 114 2824294 / +44 114 2824297
Fax: +44 114 2824292

Abstract

A major difficulty in understanding how an insect society functions is to be able to deduce collective activity from individual behaviour. Collective behaviour is not just the sum of each individual's behaviour, as patterns can emerge at the level of the society resulting from interactions between individuals or between individuals and the environment. A particular question concerns the organisations of recruitment and foraging in ant societies. When foraging ants discover food sources, ants are recruited to follow a scout trail according to communication with other individuals near the colony. Such communications might simply be of the source last visited, or of the ease of obtaining food, and may elicit either a fixed response or a random one. A program was written to simulate various recruitment strategies for ants operating in simple environments. The results obtained have been related to published experimental evidence, and in the simple case of identical food sources placed equidistant from an ants' nest, a mathematical model is presented to explain the curious behaviour of the ants that arises. Simulations showed that the most efficient strategy of recruitment for ants changes depending on the environment in which the ants operate. The program written to simulate ant recruitment could easily be improved and modified to incorporate more complicated aspects of ant recruitment and foraging.

Keywords: Ants, self-organization, symmetry-breaking, collective behaviour, optimal foraging.

1 Introduction

1.1 Self Organisation in Insect Societies

There are many areas of interest concerning social insect behaviour and the attempt to formulate laws describing why insect societies behave as they do. One of the major difficulties in understanding how an insect society functions is to be able to deduce collective activity from individual behaviour. It is not simply the case that collective behaviour is the sum of each individual's behaviour. How is it that individual ants appear so inefficient and disorganised, for example in their building activity, while at the same time highly elaborate nest structures are built? This paper deals with the organisations of foraging and recruitment in ants.

1.2 The Experimental Evidence

Deneubourg, Pasteels and co-workers (Deneubourg et al., 1987a, Pasteel et al., 1987) conducted a series of experiments and observed that ants in an apparently symmetric situation behaved, collectively, in an asymmetric way. When offered two identical food sources (which were constantly replenished) equidistant from an ants' nest, the ants at first exploited them evenly, until a sudden symmetry breaking occurred, after which one source was systematically more exploited than the other. Furthermore, from time to time the ants switched their attention to the source that had previously been neglected (Kirman, 1993). These experiments were repeated with three different

*Current address: Department of Applied Statistics, University of Reading

species of ant with similar results, and to ensure that no subtle changes in the food sources were causing the preference, the experiment was tried with one food source and two symmetric bridges leading to it. The same result was observed, but this time with asymmetric use of the two bridges. This symmetry breaking, or bifurcation, is a well known property of self organising systems, and cannot be inferred from analysis of one individual in isolation: the interaction between identical individuals must be analysed.

1.3 Foraging and Recruitment

There exist roughly 12,000 different ant species, exhibiting a variety of foraging and recruitment methods. Different ant species have different types of recruitment behaviour, depending in part on the distribution of food in the environment in which they operate. However, several rules are common to almost all ant species and also extend to many other insects. Typically, when an ant finds a source of food, be it a bird dropping, a small insect or even a bird, it will recruit other ants to that food source. The main exception to this rule is when the food source can be exhausted by a single ant and then obviously no recruitment need take place. Having found a source of food, an ant will return to its nest to deposit the food and invite other ants to return to the food supply. This is usually done through physical contact or chemical secretion. As recruitment takes place the trail is reinforced by successful recruits, and the richer the source, the greater the reinforcement. Many ant species leave a trail of pheromones which attract ants to the food supply, some even lay the chemical trail with varying intensity depending on the size of the available food supply. Ants may recruit just one ant to the source (tandem recruiting) or recruit a group of ants. Many ants will not be recruited but continue to forage for further food sources. It may be the case that the gain from letting an ant search is greater than the gain from recruitment by a successful forager. Of course this trade-off will differ between different ant species and food distributions. However, many ants which are non-collecting during a recruitment may simply be recruits that by chance lost the trail. This theory is supported by experiments (Pasteels et al., 1987), which found that the probability of losing a trail was smaller the higher the trail pheromone concentration.

2 Kirman's Model

Kirman (1993) suggested the following simple stochastic process, dealing only with tandem recruitment. Consider an environment in which there are only two food sources, source A and source B, which are identical. There are a total of n ants feeding at one or the other of them. (Clearly this model doesn't allow for the discovery of further sources or foraging.) We define the state of the system as the number of ants feeding at source A, and write X_t for the state of the system after t steps. At each step, two ants meet at random, i.e. two ants are randomly selected without replacement from the n. The first is recruited to the second's food source with probability γ. We also introduce a small probability ϵ that an ant changes its food source independently, without interaction with another ant. It is this small probability of 'mutation' which prevents the process from being absorbed at either extreme $X_t = 0$ or $X_t = n$.

This simple process $\{X_t\}$ constitutes a Markov chain, and is fact is well-known from other contexts (see e.g. Crow and Kimura, 1970). The main point of interest is the proportion of the time the process spends in each state, i.e. the equilibrium distribution of the Markov chain. The equilibrium distribution depends on the values of γ and ϵ, and is uniform when $\epsilon = \gamma/(n-1)$.

If $\epsilon < \gamma/(n-1)$ the equilibrium distribution has a 'U' shape, which corresponds to the situation arising in the experiments (the state of the system spends most of its time in the extremes). Note that, to obtain such a distributional form, γ could take any value provided it is less than one, provided the probability ϵ of self conversion is sufficiently small. This implies that it doesn't really matter how persuasive the individual ants are. The probability that a majority at one of the two food sources will decrease, decreases with the size of the majority, indicating that large majorities will be stable for a length of time.

If $\epsilon > \gamma/(n-1)$, i.e. the probability of self conversion is relatively high, then the equilibrium distribution will have a single mode: in this case the state of the system remains mostly around $n/2$.

Since large values of n are of interest, it is worth considering the asymptotic behaviour of this process. As the number of individuals n becomes larger and ϵ gets smaller, the equilibrium distribution of X_t/n approaches a symmetric beta distribution (see e.g. Kirman 1993).

Kirman's model is based on the view that ants are identical units, largely random and behaving independently of any past experience except the most recent. The model oversimplifies many aspects of ant behaviour and the environment, but nevertheless it manages to simulate the collective behaviour observed in the experiments.

3 A Detailed Model

Our current work involves a more detailed study in which the movements of ants to and from the nest and two (or more) identical food sources are explicitly modelled. An ant travels to a patch and feeds, and then returns from the patch to the nest. Thus, in the case of two patches, there are eight possible states for each ant at any given time: travelling to source A; feeding at source A; returning from source A; at the nest, having fed last at A; and similarly for source B. When the ant next sets out, it may go to the same patch, or it may change either completely at random (with some fixed probability ϵ), or because of recruitment by an ant feeding at another patch. The ant used as the potential recruiter is the ant which has arrived at the nest most recently—the 'last ant'—when the potential recruit—the 'current ant'—leaves the nest. The current ant will switch with probability γ to the patch used by the last ant.

In Section 4, we will describe some results from simulations of this model. In addition, some other refinements of the model are considered.

3.1 Feeding Times

The model above was first studied with constant feeding times. This is a little unrealistic: in practice, it seems likely that feeding times of ants will vary depending on the number of other ants exploiting the same food source. So as a simple alternative, the feeding time of an individual ant was made proportional to the total number of ants feeding at that source. This is reasonable as the more ants feeding, the more difficult it is to obtain resources, and the longer the feeding time. (A more sophisticated model of feeding time will be described shortly.)

3.2 Growth Rate

An unrealistic feature of the simple models, albeit consistent with the somewhat artificial experimental set-up of Deneubourg and co-workers, is the interpretation of identical patches. The two patches are identical in that they are both constantly replenished to keep them at the same size. The more ants feed at a source, the more food has to be supplied to that source to maintain the equality of the sources. Therefore, the more ants feeding at a source, the greater the flow of food into and out of that source. This situation seems highly unlikely to occur in the ant world: more realistic would be food sources with constant or variable growth rate.

Constant Growth Rate. The simplest form of growth for the food sources is constant growth. Each source grows at the same constant rate r.

Logistic Growth. A more realistic model for the growth of many sorts of resources is that in which

$$\frac{ds(t)}{dt} = rs(t)(k - s(t)),$$

giving the logistic equation

$$s(t) = \frac{k}{1 + (k/s(0) - 1)\exp\{-rt\}},$$

where $s(t)$ is the amount of resources at time t, given amount $s(0)$ at time 0 and assuming no removal by foraging, r is a parameter controlling growth rate, and k is the 'carrying capacity' i.e. the level the source would approach (asymptotically) if not exploited.

Feeding Time in Growing Patches. Given food sources which vary in size over time, according to their growth rate and the amount of resources removed by ants, a new function for the feeding time of an ant is required. This feeding time must now depend on the size of the food source as well as the number of ants exploiting it. The new function must satisfy certain obvious conditions. Firstly, the larger the source, the more easily the food can be removed and hence the smaller the feeding time, assuming the number of ants feeding remains constant. Secondly, for a source of a given size, food can be removed more quickly the smaller the number of ants exploiting it. The simple function implemented in the simulation will be

$$f(m,s) = m/s$$

where $f(\cdot,\cdot)$ is the feeding time, $m(\geq 1)$ is the number feeding at source, $s(> 0)$ is the size of the source. This feeding time will only be calculated if $s > m$ and m must be greater than 0, as there will always be at least the new arrival feeding at the source. If $s < m$, which means (since by definition each ant takes 1 unit of resources) that there is not enough food to satisfy the new ant, then that ant returns to the nest without feeding.

3.3 Communicating the Ease of Obtaining Food

In the simplest case, the only communication involved in recruitment is that one ant can communicate which patch it is currently using to another ant, which then switches its attention to that patch with some fixed probability. It may be the case, however, that ants relay information to each other concerning the ease of obtaining food, such as their last feeding time. Taking the probability of conversion to the last ant's source as

$$\gamma(T_C, T_L) = \frac{T_C}{T_L + T_C}$$

where T_L is the (most recent) feeding time of the last ant, and T_C is the (most recent) feeding time of the current ant, relays information about the ease of obtaining food.

4 Preliminary Results

4.1 Reproduction of Deneubourg's Experiment

First the scenario of Deneubourg's experiment was set up. Two identical food sources equidistant from the ants' nest were offered, each constantly replenished. This was simulated by taking the travelling times to and from the food sources as equal constants, and the feeding times at the two sources also as equal constants. Comparison with Kirman's model is achieved by counting the number of ants related to source A, i.e. in the first four states states described above, and comparing this with Kirman's X_t. The resulting state of the system (number of ants related to source A) is plotted against time, for every fiftieth transition of the ants (a transition corresponds to one ant changing state in the system). Figure 1 shows the state of the system when the probability of self conversion (mutation) is relatively high (ϵ=0.3) and the probability of being converted by another ant, is relatively low (γ=0.7). As expected, the state of the system fluctuates around $n/2$ (50 in this case, as n=100).

Figure 2 displays the corresponding graph when the probability of mutation is very low (ϵ=0.002) and the probability of being converted by another ant is high (γ=0.99). Now the system spends little time around state $n/2$ and a great deal of time in the extremes (that is with all 100 ants related to one of the sources). Notice how the ants switch their attention from one source to the other from time to time. It is this situation that was observed in the experiments by Deneubourg.

When the state of the system travels from one extreme to $n/2$ there is an equal probability that the state will switch or return back to its last extreme, due to the symmetry of the model, and this can be seen in the graphs, i.e. there is no bias in the system.

4.2 Variable Feeding Times

When feeding time is dependant on the number of ants present at a source, as described in section 3.1, the behaviour of the system is largely unchanged. There are some differences from the earlier model, but these differences are mainly attributable to changes in the time scales of the simulations. The stochastic nature of the process doesn't change and switching still occurs when the mutation probability is relatively small.

4.3 Constant Growth Rate

When patches grow at a constant rate, the results are again similar to the above cases. For the case when the mutation probability is relatively large, as expected the state fluctuates around $n/2$. Both sources remain roughly constant and equal in size, close to zero. This occurs because the ants are exploiting both sources faster than they are growing, and many ants will be returning from the sources without feeding as there are insufficient resources available for them to feed on. Had we chosen the growth rate to be very large, then the sources would have continued to grow indefinitely.

Now, keeping the same growth rate, but choosing the conversion and mutation probabilities to be high and relatively low respectively, again the ants switch their attention between the two sources. What is happening is that while most of the ants are concentrated at one source, the other isn't exploited much and so it grows. The source which has the ants' attention, cannot support them all feeding, and so its size is reduced to close to zero. After some random length of time, the ants switch their attention to the other source, and its size decreases whilst the first begins to grow again.

4.4 Logistic Growth

If the sizes of the food sources follow a logistic model, as described in Section 3.2, qualitatively similar behaviour is again obtained. As this is the most realistic of the models we have considered, the details of the results, and the question of the optimality of the behaviour observed, are discussed in depth in the following section.

5 Is Switching Optimal?

It is interesting to know whether the efficiency of an ant society increases when this asymmetric exploitation occurs, or whether the colony would benefit from more balanced exploitation of the food sources.

5.1 Evaluating the Efficiency of Recruiting Strategies

We need to decide upon some currency which we can use to compare alternative strategies. It seems reasonable to think of total energy input as the currency. An ant colony will want to maximise its total energy input, i.e. maximise the total amount of food returned to the nest. In nature, any food source will be exhausted after a while so it may be sensible to ensure that the flow of energy input doesn't fall below some minimum over a time interval. Furthermore, a real colony will have to consider how many workers should be allocated to foraging, rather than other tasks. We shall ignore these aspects, and just consider maximisation of the total energy input.

As we vary parameters in the simulations, the overall length of time it takes ants to complete the allotted transitions will vary. The currency used to compare alternative strategies will be

$$\bar{R} = \frac{\text{total amount of food taken}}{\text{total length of time taken}}.$$

It is also of interest to see how the rate of food removed varies over time, and so the rate of food removed is calculated over every interval of fifty transitions (denoted by R_t). The next four sections describe the effect on R_t of different mutation probabilities. The conversion probability and the parameters for the logistic growth remain fixed (r is chosen to be small so that switching will be observed in simulations of 100,000 transitions).

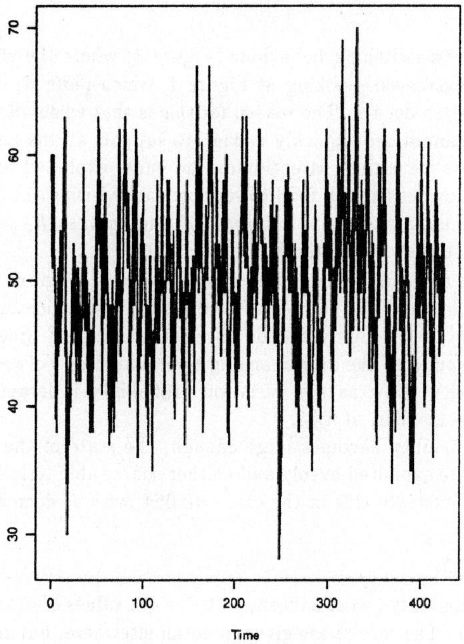

Figure 1: High self-conversion: small,rapid fluctuations

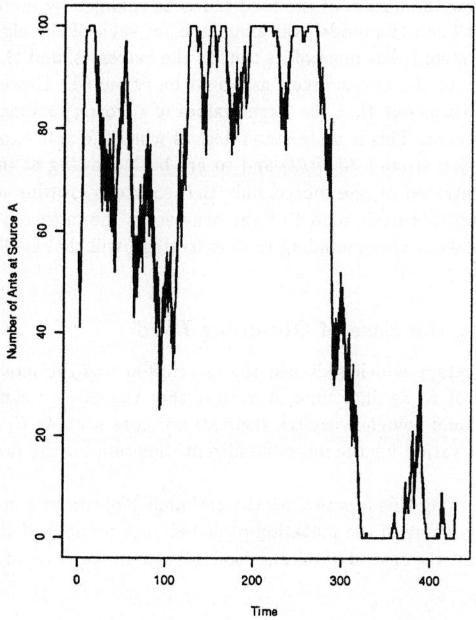

Figure 2: High tandem recruiting: switching between sources

5.2 Main results

The value ϵ=0.002 leads to switching behaviour (Figure 3) where the state of the system spends most of its time in the extremes. Looking at Figure 4, which plots R_t over time, it is noticeable that R_t peaks when a switch occurs. The reason for this is that when all the ants are concentrated at a source, this source cannot grow quickly enough to support all the ants feeding (R_t small). As the switch occurs the ants focus their attention on the under-exploited food source, which is at its carrying capacity, and provides enough food to support each visiting ant (R_t large). However, after a while this source diminishes and cannot support all the ants, so R_t decreases. The value of R_t (over a longer run than that shown in the figures) is 3.50.

With ϵ=0.008, again switching occurs, but more frequently (Figures 5 and 6). The peaks in R_t still correspond to switches and the number of peaks increases, which increases $\bar{R}$ to 6.50.

With ϵ=0.016, the system roughly spends an equal amount of time in each state (Figure 7). Peaks in Figure 8 correspond to the ants changing their attention between the two sources quickly (not necessarily a switch). Now as the mutation probability is larger, the state of the system fluctuates more and $\bar{R}$ is smaller, at 4.55.

If the mutation probability becomes large enough, the state of the system fluctuates around $n/2$, with both sources are exploited evenly and neither source able to fully support the ants feeding at it. Figures 9 and 10 illustrate this in the case ϵ=0.064, with $\bar{R}$ decreased to 2.18.

5.3 Varying ϵ and γ

With the conversion probability (γ) at 0.99 and at 0.7, eight values of ϵ (varying from 0.002 to 0.256) were used for simulation. The results are given in detail elsewhere, but confirm the above patterns, and also show that there is an interaction between the mutation and conversion probabilities, as expected.

5.4 Changes to the Food Source

In Section 5.2 it was clear that ants benefited (higher $\bar{R}$) if they switched their attention between the two food sources. Now the nature of the food source is changed by increasing the parameter r to 2. Keeping γ fixed at 0.99, the model was simulated for each of the eight values of ϵ. For the smaller values of ϵ the system spent most of its time in the extremes, and the larger values resulted in the equal exploitation of the two sources, as would be expected. However, if we consider the values of $\bar{R}$ obtained, it turns out that the larger values of ϵ, corresponding to equal exploitation, are more efficient in this case. This is understandable, as when the ants exploit the sources evenly, both sources remain in size around 30 ($k/2$) and so are both growing at maximum rate, whereas when the ants are concentrated at one source, only that source is growing at maximum rate.

This shows that using this basic model of ant behaviour, the most efficient strategy, and the observed qualitative behaviour corresponding to that strategy, will change depending on the nature of the food source.

5.5 Communicating the Ease of Obtaining Food

From Section 5.2 the strategy which switched the most often was the most efficient for the environment. In the graphs of R_t against time, it is clear that the closer together the peaks are, the larger $\bar{R}$ is. If the ants could somehow switch their attention as soon as R_t dropped then $\bar{R}$ would increase. This is the motivation for the more 'intelligent' function for the probability of conversion, given in Section 3.3.

However simulations using this function for the probability of conversion give equal exploitation of the two sources for any value of the mutation probability. A number of similar simple functions, relaying information about the ease of obtaining food, all caused the state of the system to fluctuate around $n/2$.

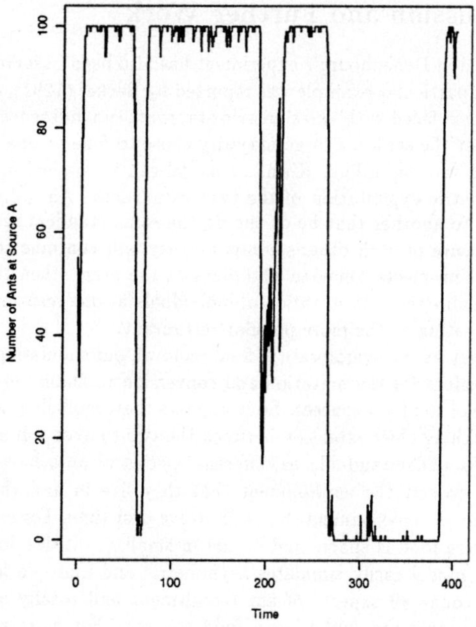

Figure 3: Switching between sources; $\epsilon = 0.002$

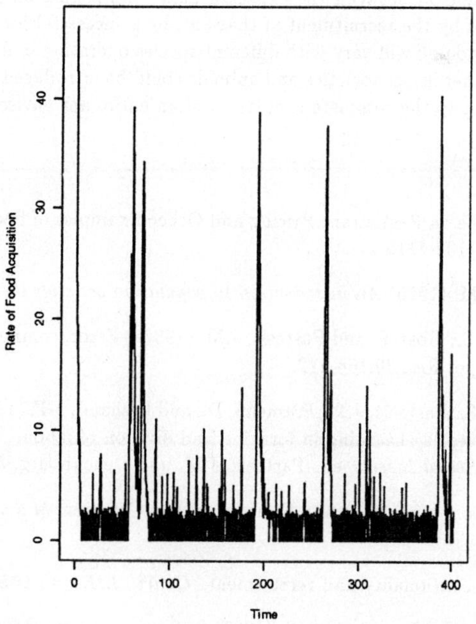

Figure 4: Variation in rate of food acquisition, R_t, when $\epsilon = 0.002$

6 General Discussion and Further Work

The phenomenon observed in Deneubourg's experiment has also been observed in markets and is of interest to economists. A particular example was reported by Becker (1991), concerning individuals dining at restaurants. When faced with two similar restaurants (no difference in the quality of food, and the prices were almost the same), a large majority chose to dine at one rather than the other, even though this involved waiting in line. Kirman's model can be applied directly and as with the ants explains the asymmetric exploitation of the two restaurants. An individual going to eat at one restaurant, suggests to another that he or she do the same (tandem recruitment). Assuming that both enjoy the pleasure of each other's company they will continue to eat out at the same restaurant until one of them meets someone who dines at the other, then this corresponds to the ants process. Becker explains the concentration of individuals at one restaurant due to the greater pleasure experienced by eating at the more popular restaurant.

Measuring the efficiency as the overall rate of food removal, our simulations have shown that the most efficient strategy (values for the mutation and conversion probabilities) for the ants changed depending on the nature of the food sources. So it appears that, modelling ant recruitment in this simple way, the ants switching their attention between the two sources will not always be the most efficient strategy. This is understandable as different species of ants have different recruitment and foraging techniques to suit the environment that they live in and the ants will not adapt immediately to changes in the environment, but will evolve over time. For example, ants operating in arid environments where food is sparse and occurs in small quantities do not recruit, but hunt singly. The recruitment model easily simulates asymmetric and multiple food sources. However it does not take into account all aspects of ant recruitment and totally ignores exploration for new food sources (all the ants are exploiting a food source). For most ant species, when ants are following a recruitment trail, some ants by chance lose the trail. These lost ants exploring the vicinity of a food source are not necessarily wasting their time, as they may be defending the source against competitors or foraging for other sources, close in time and space to the first one. So the program as it stands is very basic and there is scope for improvement and modification. The discovery of other food sources by foraging ants could be introduced and simulations carried out to investigate the trade-off between the expected gain from letting an ant forage for further sources and that obtained by the recruitment of that ant, by a successful forager, to a known food source. Obviously this trade-off will vary with different species operating in different environments. Disturbance caused by other insect societies and animals could be introduced and ant reproduction modelled, possibly leading to the complete simulation of an entire ant environment carried out by a computer.

References

Becker, G.S. (1991) A note on Restaurant Pricing and Other Examples of Social Influence on Price *J. Polit. Econ.*, **XCIX**, 1109-1116.

Crow, J.F. and Kimura, M. (1970) *An introduction to population genetics theory.* Harper and Row.

Deneubourg, J-L., Aron, S., Goss, S. and Pasteels, J.M. (1987a) Error, communication and learning in ant societies. Eur.J.Oper.Res.,30,168-172.

Deneubourg, J-L., Goss, S., Pasteels, J.M.,Fresneau, D., and Lechaud, J-P. (1987b) Self-organisation mechanisms in ant societies (2) Learning in foraging and division of labour. In *From Individual to Collective Behaviour in Social Insects*, ed. Pasteels, J.M. and Deneubourg, J-L. Pages 177-196.

Fletcher, R.P., (1994). Communication and Strategy in Ant Colonies. *M.Sc. dissertation, University of Sheffield.*

Kirman, A. (1993). Ants, rationality and recruitment. *Quart. J. Econ.*, **108**, 137-156.

Pasteels, J.M., Deneubourg, J-L. and Goss, S. (1987) Self-organisation mechanisms in ant societies (1) Trail recruitment to newly discovered food sources. In *From Individual to Collective Behaviour in Social Insects*, ed. Pasteels, J.M. and Deneubourg, J-L. Pages 155-176.

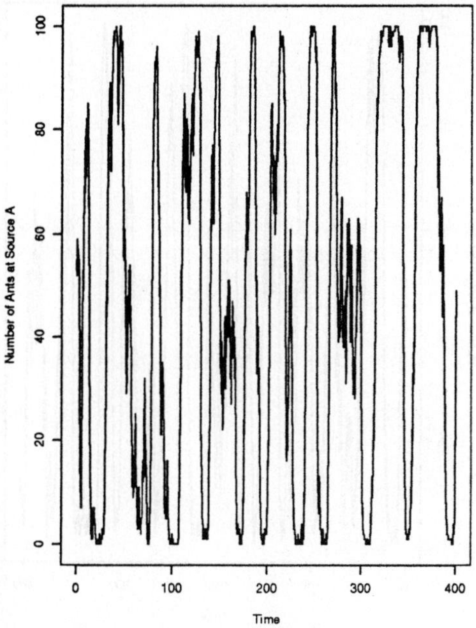

Figure 5: Switching between sources; $\epsilon = 0.008$

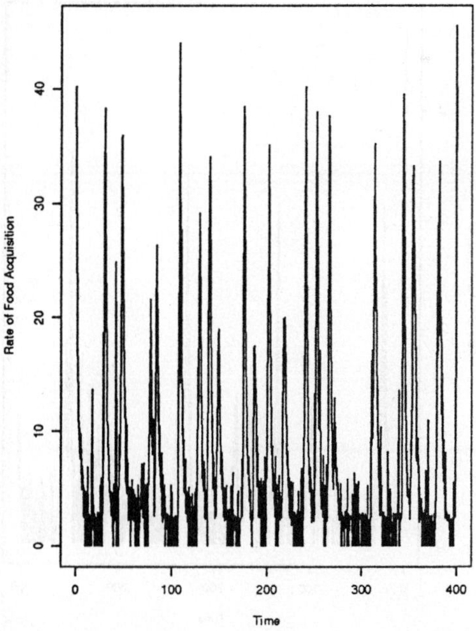

Figure 6: Variation in rate of food acquisition, R_t, when $\epsilon = 0.008$

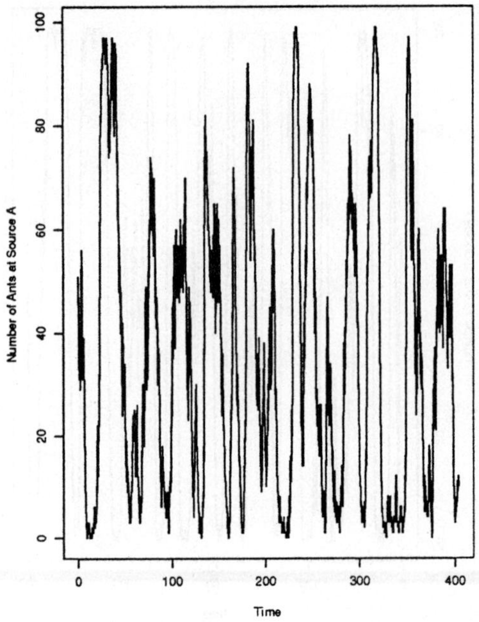

Figure 7: Switching between sources; $\epsilon = 0.016$

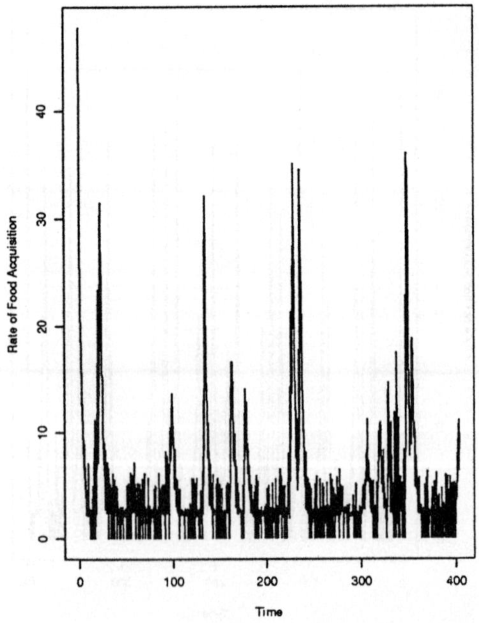

Figure 8: Variation in rate of food acquisition, R_t, when $\epsilon = 0.016$

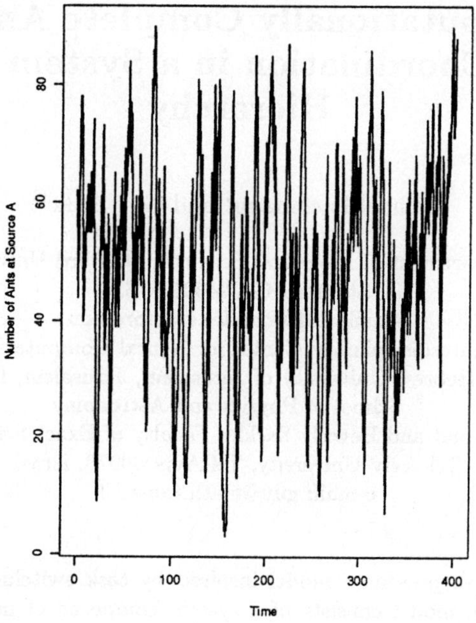

Figure 9: Switching between sources; $\epsilon = 0.064$

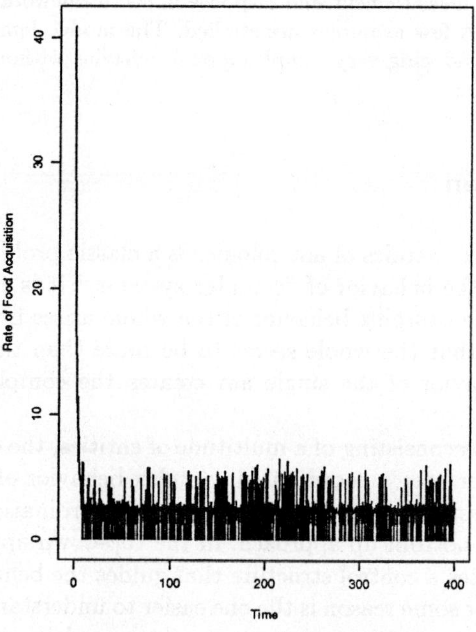

Figure 10: Variation in rate of food acquisition, R_t, when $\epsilon = 0.064$

The Computationally Complete Ant Colony: Global Coordination in a System with no Hierarchy

Michael Lachmann[1] and Guy Sella[23]

[1] Department of Biological Sciences, Stanford University
Stanford CA 94305, USA
e-mail: dirk@charles.stanford.edu
[2] Interdisciplinary Center for Neural Computation
The Hebrew University of Jerusalem, Jerusalem, Israel
[3] School of Physics and Astronomy
Raymond and Beverly Sackler Faculty of Exact Sciences
Tel-Aviv University, Tel-Aviv 69978, Israel
e-mail: guy@math.tau.ac.il

Abstract. We present a model inspired by task switching of ants in a colony. The model consists of a system composed of many identical ants. Each ant has a finite number of internal states. An ant's internal state changes either by interaction with the environment or by interaction with another ant. Analyzing the model's dynamics, we prove it to be computationally complete. This gives us a new perspective on the sophistication a colony can display in responding to the environment. A formalism for measurement and response in the framework of the model is presented. A few examples are studied. The model demonstrates the possibility of inducing very complex global behavior without any hierarchical structure.

1 Introduction

Understanding the dynamics of ant colonies is a classic problem in sciences that try to understand the behavior of "complex systems." It is a clear example of a system in which the complex behavior of the whole arises from the interactions of many parts, so that the whole seems to be more than the sum of its parts. What kind of behavior of the single ant creates the complex behavior of the whole colony ?

In many systems consisting of a multitude of entities, the composite behavior of the individuals creates a coordinated complex behavior of the whole system. There are two possible ways to achieve such a coordination – the top-down approach, and the bottom-up approach. In the top-down approach, a hierarchy in the entities creates a control structure that guides the behavior of the system. This approach is for some reason is the one easier to understand, perhaps because of the similarity between the apparent structure, and its manifestation in the causal structure of events. A manifestation of this approach is found in the way an army is coordinated, through a hierarchical command structure. In the bottom-up approach, the rules followed by the individual entities bring a self-organization of the system. Most systems in nature have elements of both approaches in them. The behavior of a modern free-market economy is composed both of some hierarchy imposed by corporations and the government, and of self-organization resulting from the action of independent agents in the free market.

In this paper we present a model that is inspired by the behavior of ants in a colony. The aim is twofold. The first is to show what level of coordination might be expected from an ant colony, even under very simplistic assumptions about the behavior of the ants. Thus, it is not claimed that an ant colony is fully described by the model, rather, as the ingredients of the model are contained in an ant colony, the colony will surely be able to achieve at least a response which is as sophisticated as the one described by our model. The second aim is to show one way in which a system can perform certain tasks or computations using the bottom-up approach. This insight will be useful in studying systems other than ant colonies.

Understanding the behavior of an ant colony requires understanding of global coordination without hierarchy. Even though the reproducing ant in the colony is called the queen, empirical work shows that in many cases she does not seem to be at the top of a hierarchy which rules the colony. The behavior of the colony seems to simply arise from the behavior of the individual workers. There are many other cases in nature where we face the same phenomenon. These include the behavior of colonies of other social animals such as bees or termites, but also the understanding of the action of the immune system or the interactions of the proteins and other molecules which make up the cell.

There are several approaches to studying the behavior of a colony of social insects. One is through understanding how the behavior of the colony might be affected by pheromone trails laid by ants (see Deneubourg and Gross, 1989). Several models show how this might lead to optimal, or observed, foraging patterns of the colonies (Deneubourg et al. 1989, 1990; Gross et al. 1990; Millonas 1994). Another approach is to understand the allocation of individuals to tasks in the colony. Individuals in a colony often engage in different tasks – foraging, nest-maintenace, brood care, etc. It has been shown that social insects may react to the environment by changing the proportion of individuals allocated to the various tasks in the colony (Calabi 1987; Gordon 1989; Robinson 1992). An example of this is recruiting ants for nest maintenance when there is a need to remove obstacles from the neighborhood of the nest. In this paper we follow the trail laid by the second approach, trying to understand mechanisms underlying the task allocation. It will, of course, be worthwhile to synthesize this work with the first approach in further research.

How does an ant colony react to its environment? Let us examine a reaction of a multi-cellular organism, such as our body, to the environment. Light signals are absorbed by photo-receptors in the retina cells. These signals are then transmitted through the central nervous system (CNS) to the brain, where they are assessed using information about the environment currently stored in the brain. A signal for a reaction is then transmitted through the CNS to regulating neurons controlling a muscle in the hand, for example. The ant colony might face a similar task. Some ants – patrolers – might gain information about a food source, and other ants will then need to be recruited to forage at that food source, potentially bringing into account some information about the current state of the colony, the hunger level in the brood for instance. This has to be achieved without a CNS connecting the various ants, and without a brain to store and assess the information.

Gordon et al. (1992) showed how certain behaviors of individuals in the colony may enable the colony to process information like a Hopfield net. The model assumed, however, that individual ants are able to measure global states of the colony, such as the proportion of ants allocated to certain tasks. Our model is based on the model presented by Pacala et al. (1994). In this model ants can engage in a task or be inactive. Ants doing a task can also be either "successful" or "unsuccessful" and can switch between these two according to how well the task is performed. Unsuccessful ants also have a certain chance to switch to be inactive, and successful ants had a certain chance to recruit inactive ants to their task. This is an example of how certain interactions of ants can give rise to global behavior in the colony.

In the model presented in this paper, the notion of a "successful" or "unsuccessful" ant engaging in a certain task is expanded to a general notion of a state that the ant is in. An ant in the colony can be in one of a finite number of states. Ants doing different tasks are always in different states, but ants doing the same task could also be in different states. Such a state might correspond, for example, to an ant foraging while hungry and successful. It is assumed that there are many more ants in the colony than states, so that it makes sense to talk about the fraction of ants in a certain state. The ways in which an ant can change its state are through interaction with the environment and through meeting another ant. The model is aspatial, and thus does not regard the place of ants in the colony or the pheromones left by the ants in certain places in the environment.

In Sect. 2 we present the master equation for this model, assuming an infinite number of ants in a colony. Then in Sect. 3, we show that this model can lead to a very complex behavior, including amplification, cycling, and potentially the performance of any computation that can be done by a finite Boolean network or by a finite Turing machine. We also present a different formulation of the model, in which the interaction with the environment is separated to "measurement". In Sect. 4 we present two examples. One is a test of how well a finite colony fits the predictions of the model. The other presents a model developed by Seeley et al. (1991) in order to show a possible solution to choosing between two alternate foraging sites in bees. It is shown how this solution can be stated within the framework of our model.

2 The Infinite-size Model

Our infinite-size model consists of the following assumptions:

- At any time, each ant can be in any of M states. Let $p_i(t)$, where $i = 1, \ldots, M$, be the fraction of ants in the colony that are in state i. Different tasks correspond to different states, but different states can correspond to the same task.
- An ant can switch tasks either spontaneously, or through social interaction. The probability of spontaneous switching is dependent on the environment. The probability per unit time for an ant in state i to switch to state j in an environment s, is denoted $A_{ij}(s)$. In a social interaction in which an ant in state i encounters an ant in state j, this ant has a probability B_{ijk} of switching to state k.
- The number of ants is large enough, and the number of states small enough, that $p_i(t)$ can be treated as a real number.

Under these assumptions, the master equation is[1]:

$$\frac{dp_i}{dt} = -\sum_{j=1}^{M} A_{ij}(s)p_i + \sum_{j=1}^{M} A_{ji}(s)p_j$$
$$+\,\alpha(N)\left(-\sum_{j,k=1}^{M} B_{ijk}e_{ij}(s)p_ip_j + \sum_{j,k=1}^{M} B_{jki}e_{jk}(s)p_jp_k\right), \qquad (1)$$

where $i = 1, \ldots, M$, and $e_{ij}(s)$ is the probability per time unit for an ant in state i to encounter an ant in state j. This probability can be a function of s, the environment. We will not deal with the determination of e_{ij}, and for the simplicity of the analysis it will be subsumed by B_{ijk}. $\alpha(N)$ is a factor to scale the speed at which ants meet relatively to the speed at which they switch states spontaneously, as the density changes as a result of a change in the number of ants in the colony, N.

Because p_i is the fraction of ants in state i, it is subject to the constraints $\sum_{i=1}^{M} p_i = 1$, $p_i \geq 0$. In Appendix A, it is shown that, assuming $A_{ij} \geq 0$ and $B_{ijk} \geq 0$, these constraints are maintained by this master equation.

In this model we omit the effects of space and finite size. These are briefly addressed in the example in Sect. 4. Furthermore, in the initial study of the model we have not included the changes to the environment resulting from its interactions with the ants, such as foraging, nest maintenance, and pheromone

[1] In general all variables in this equation could depend on the environment and on the density of the ants, but we believe the formulation given in the equation covers a wide range of cases. Thus B_{ijk} could depend on the environment, but we separated out cases in which this change could simply be called a change of the ant's state, or a change caused by changing the chance of the ants to meet (which is included in $e_{ij}(s)$) , or a change caused by the effect of the number of ants in the colony on the density of the ants (which is included in $\alpha(N)$).

marking. In the following sections we show the dynamics of this model to be very rich, even without these feedback mechanisms. Nevertheless, we believe understanding these mechanisms is important for understanding the behavior of an ant colony.

3 Interpreting the Master Equation

As we show in Sect.3.3, a colony in the model be *computationally complete* just as a computer or a neural network. This means that the colony can respond to the environmental input in a complex way. It is also shown that in this model the colony may exhibit stable fixed points, or periodic cycles.

The systems studied in the following sections are defined by the specific choices of A's and B's. Using the notation illustrated in Fig. 1 will be helpful in studying specific examples. An arrow (see Fig. 1a) shows that an ant in state 1 has a certain non-zero chance of going over to state 2; in other words, $A_{12} \neq 0$. The number under the arrow indicates the value of A_{12}, in this case 0.64. An arrow with a dashed line connected to it (see Fig. 1b) means that if an ant in state 1 meets an ant in state 2, the former has a non-zero chance of changing to state 3; $B_{123} \neq 0$. Again, the number under the arrow shows the value of B_{123}, 0.7.

3.1 Amplifier

In this section we consider a colony that exhibits a behavior like an amplifier. Two states (state 1 and 2) are affected by the environment, so that under certain conditions ants in state 1 tend to go to state 2, and under other conditions ants in state 2 tend to go to state 1. These will be called input states. This colony will behave like an amplifier, because the ratio of the ants in the input states will be amplified to states 3 and 4; thus, if $p_1/p_2 > 1$, then $p_3/p_4 > p_1/p_2$, and if $p_1/p_2 < 1$, then $p_3/p_4 < p_1/p_2$. The relationship of the states is illustrated in Fig. 1c. The equations for this system are

$$\frac{dp_1}{dt} = \frac{dp_2}{dt} = 0 \ , \tag{2}$$

$$\frac{dp_3}{dt} = -bp_3p_2 + ap_5p_1 \ , \tag{3}$$

$$\frac{dp_4}{dt} = -bp_4p_1 + ap_5p_2 \ , \tag{4}$$

$$\frac{dp_5}{dt} = -ap_5p_1 - ap_5p_2 + bp_3p_2 + bp_4p_1 \ . \tag{5}$$

They can be derived from the figure by adding, for each arrow with a dashed line that goes from i to j and has a dashed line at k, and a weight B_{ikj} – the

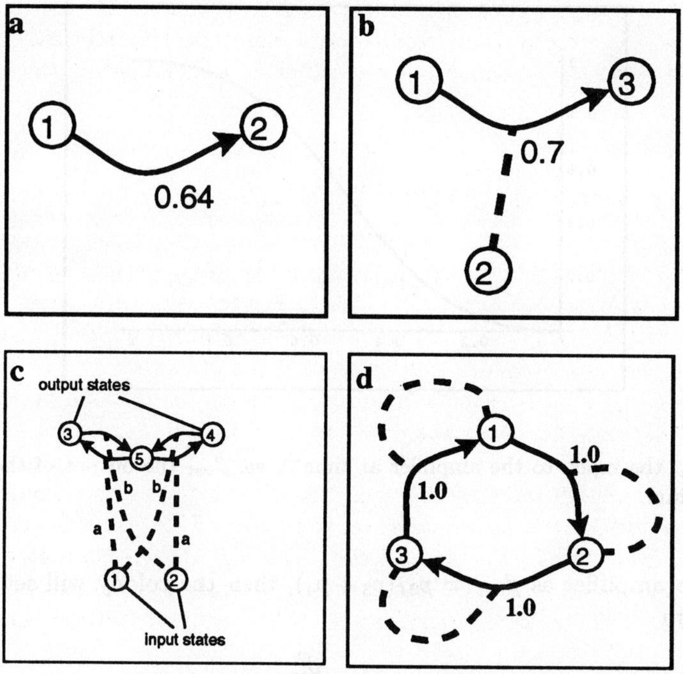

Fig. 1. a) Arrow representing spontaneous state transition. b) Transition through an encounter. c) Transition diagram of the amplifier. d) Transition diagram of the oscillator.

term $-B_{ikj}p_ip_k$ to dp_i/dt and the term $B_{ikj}p_ip_k$ to dp_j/dt. Given the initial conditions $p_i(0), i = 1, \ldots, 5$, the system will converge to the solution

$$p_1 = p_1(0) \ , \tag{6}$$

$$p_2 = p_2(0) \ , \tag{7}$$

$$p_3 = s\frac{a}{b}\left(\frac{1}{\alpha^2 + \alpha + 1}\right) \ , \tag{8}$$

$$p_4 = s\frac{a}{b}\left(\frac{1}{\alpha^{-2} + \alpha^{-1} + 1}\right) \ , \tag{9}$$

$$p_5 = s - p_3 - p_4 \ , \tag{10}$$

where $\alpha = p_1(0)/p_2(0)$, and $s = p_3(0) + p_4(0) + p_5(0)$. For this point one gets

$$\frac{p_3}{p_4} = \left(\frac{p_1}{p_2}\right)^2 \ . \tag{11}$$

Thus, the ratio of p_1 to p_2 was amplified. If we define the input to the amplifier as $\beta_{in} = p_1/(p_1 + p_2)$, which will be a number between 0 and 1, and the out-

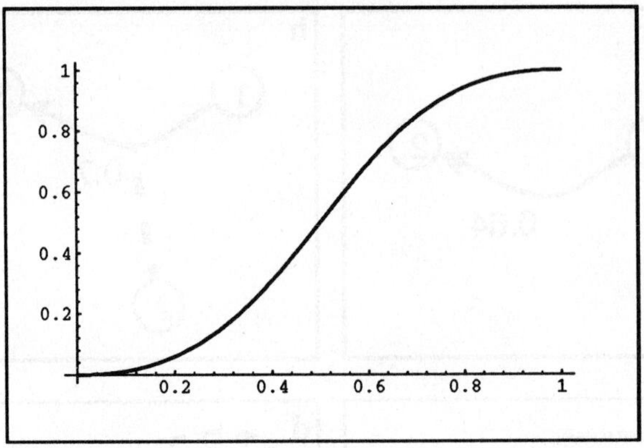

Fig. 2. β_{in}, the input to the amplifier at time 0, vs. β_{out} the output of the amplifier at equilibrium.

put of the amplifier as $\beta_{out} = p_3/(p_3 + p_4)$, then the colony will settle on the equilibrium

$$\beta_{out} = \frac{\beta_{in}^2}{2\beta_{in}^2 - 2\beta_{in} + 1} . \tag{12}$$

The graph of β_{out} vs. β_{in} is shown in Fig. 2.

It can be seen that this function is similar to the sigma function used in neural nets. Thus, the relation between the number of ants in states p_1 and p_2, which is represented by β_{in}, was amplified in a sigmoid-like manner to the relation between the number of ants in states p_3 and p_4, represented by β_{out}. By connecting several amplifiers "sequentially", by having the output states of one amplifier act as the input states of the other, one can approximate a step function, building a unit that acts much like a neuron.

3.2 Oscillator

The following colony has a periodic trajectory as a solution. The dynamics are depicted in Fig. 1d. The equations for this system are

$$\frac{dp_1}{dt} = -p_1p_2 + p_1p_3 \ , \tag{13}$$

$$\frac{dp_2}{dt} = -p_2p_3 + p_2p_1 \ , \tag{14}$$

$$\frac{dp_3}{dt} = -p_3p_1 + p_3p_2 \ . \tag{15}$$

This is an equation in the two-dimensional simplex $p_1 + p_2 + p_3 = 1$. It has four fixed points: $(1,0,0)$, $(0,1,0)$, $(0,0,1)$, $(\frac{1}{3}, \frac{1}{3}, \frac{1}{3})$. For solutions of these equations,

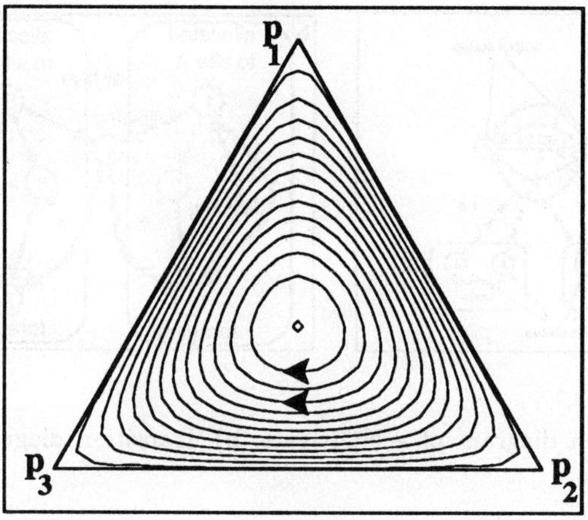

Fig. 3. Trajectories of solutions to equations 13–15 on the simplex.

$d(p_1p_2p_3)/dt = 0$, and therefore the trajectories lie on lines with constant $p_1p_2p_3$. From this it follows that all solutions, except for the four fixed points and the boundary of the simplex, are periodic. Figure 3 shows the trajectories on the simplex.

3.3 NOR-gate

The last example for possible behaviors of an ant colony under the assumptions of the model can be interpreted as a NOR-gate. A NOR-gate is a Boolean gate and has the property that with it any Boolean network can be built. Appendix B describes this gate. The relations of the states in the gate are depicted in Fig. 4. The full gate needs to feed the output through three amplifiers. The equations for this system are

$$\frac{dp_1}{dt} = \frac{dp_2}{dt} = \frac{dp_3}{dt} = \frac{dp_4}{dt} = 0 \ , \tag{16}$$

$$\frac{dp_5}{dt} = -p_1p_5 - p_3p_5 + p_4p_6 + p_2p_6 + \frac{1}{4}p_6 \ , \tag{17}$$

$$\frac{dp_6}{dt} = -p_4p_6 - p_2p_6 + p_1p_5 + p_3p_5 - \frac{1}{4}p_6 \ . \tag{18}$$

The inputs are defined as follows: $p_1 + p_2 = p_3 + p_4 = \frac{1}{3}$; if $p_1/p_2 > 3$, then the left input is 0 (see Fig. 4); if $p_1/p_2 < \frac{1}{3}$, then the left input is 1. If the ratio is between these two values, the input is undefined. The right input is defined similarly. The output is defined in the same way: if $p_5/p_6 > 3$, the output is 0; and if $p_5/p_6 < \frac{1}{3}$, the output is 1. In Appendix B it is shown that in all cases of a well-defined input (that is $p_1/p_2, p_3/p_4 < \frac{1}{3}$ or $p_1/p_2, p_3/p_4 > 3$), the output behaves as it should for a NOR-gate.

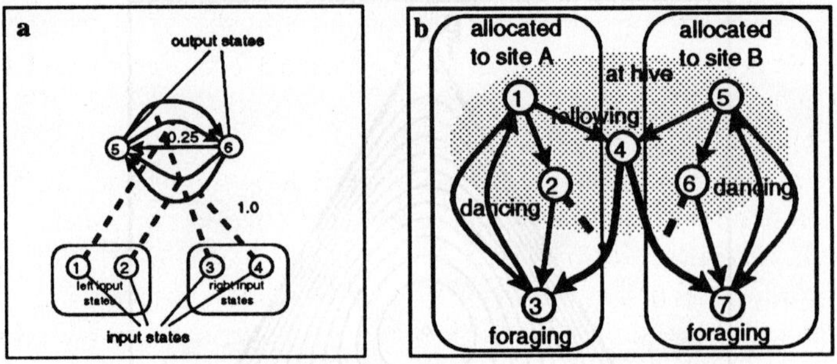

Fig. 4. a) Transition diagram of a NOR-gate. b) Transition diagram for choosing a foraging site by bees.

3.4 Computation Completeness and other Conclusions

The NOR-gate of Sect. 3.3 concludes the tools we need to prove that the system is computationally complete. Using this NOR-gate, one can build any Boolean circuit. Using a Boolean network of polynomial size (in the input size), with a clock of the sort built in Sect. 3.2, one can compute anything that a finite computer can (see Borodin 1977; Parbery 1994). Thus, such a system can do computation as much as silicon computers can.

A potential problem might arise because as the size of the memory of the system scales up, the number of potential states of each ant goes up, and with them the size of the transition matrices A and B; one might say that the individuals are not simple anymore. If we build the system with NOR-gates, then for an ant in a certain state, the number of other states it needs to interact with and that it can potentially reach, is small and does not scale up as the system scales up. Thus, most of the entries in these matrices are 0, and the ant just has to be supplied with the rule "do not interact with an ant in a state you do not know"; this might restore the "simplicity" of the ants.

By showing computational completeness of the model,we do not mean to say ant colonies perform universal computation. Computational completeness is a property characterizing the complexity of the dynamics, like being chaotic (discussion on the dynamical implications of systems with computational power could be found in C. Moore 1990). A use for the computational abilities described is briefly addressed in the discussion. Another implication of the computational completeness is that it can give us a new perspective on the level of sophistication an ant colony can display in its interaction with the environment. In Sect. 4.2 we describe a model presented by Seeley et al. for explaining the behavior of bees, in deciding between alternative foraging sites. The computational completeness tells us that even a more elaborate response, taking into account more inputs from the environment should not be surprising.

3.5 A Mechanism for Measurement and Response

Further understanding of the model can be reached by showing that for certain choices of A's and B's, one can interpret the dynamics as consisting of two processes: one in which the colony measures the environment, and another in which it uses the information from the measurement for a computation. Such a computation can result in a response to the measured environment.

A simple formalism is to denote each state, using two indices: $i = (m, n)$ where $(m = 1, \ldots, M$ and $n = 1, \ldots, N.)$ The first index, m, will be related to the measurement, and the second, n, to the computational state. We further assume that the A elements mediate the measurement: $A_{(m,n)(m',n')} = A_{m',m}\delta_{n,n'}$ (δ is the Kronecker delta function, $\delta_{ij} = 0 when i \neq j, \delta_{ii} = 1$), i.e the A elements do not change the computational index of the state, and

$$B_{(m,n)(m',n')(m'',n'')} = B_{n(m',n')n''}\delta_{m,m''} \ . \tag{19}$$

That is, the B elements do not change the measurement state. This assumption is made in order to simplify the following equations. Defining $P_m^{\text{mes}} = \sum_n p_{(m,n)}$ and $P_n^{\text{com}} = \sum_m p_{(m,n)}$, leads to the following equations for the change in P_m^{mes} and P_n^{com}:

$$\frac{dP_m^{\text{mes}}}{dt} = \sum_{m'} A_{m',m}P_{m'} - A_{m,m'}P_m \ , \tag{20}$$

$$\frac{dP_n^{\text{com}}}{dt} = \sum_{n',m'',n''} B_{n'(m'',n'')n}P_{n'}^{\text{com}}p_{(m'',n'')} - B_{n(m'',n'')n'}P_n^{\text{com}}p_{(m'',n'')} \tag{21}$$

The first set of equations are like standard master equations in statistical mechanics for P_m^{mes}s. One can interpret it as the colony measuring the environment with the result represented by the vector $(P_1^{\text{mes}}, \ldots, P_M^{\text{mes}})$. The second set of equations describes change in the computational states according to both the computational states and the measurement states.

It may be claimed that the term measurement is not simply a mathematical definition, that when one speaks of measurement one assumes the intervention of some intelligent entity that performs it. We do not claim that an ant colony does a measurement in this respect. Nevertheless, following a line of argument set by J. von Neumann (1955), we demonstrate the existence of a process having the physical manifestation of a measurement: One feature of measurement is the correlation induced between the state of the measuring apparatus and the state of the system measured. Furthermore, when a human performs a measurement, the perception of its result induces a correlation between the state of the brain and the state of the system. Another feature related to the process of measurement is the possibility of its result affecting the actions of whoever performed it. The possibility of these two features in the framework of the model was demonstrated.

As for the existence of "real" measurement, one may conceive an ant colony as an entity with purpose, and as such it might "qualify" to perform "real" measurements. Whether this is the case, is an interesting question, but beyond the scope of this work.

4 Examples

In this section we present two examples of systems exhibiting a behavior that can be explained by the described model. Only the first is an ant colony. The second is a bee colony, from a model done by Seeley et al. (1991).

4.1 Discrete Spatial Model

A first step in testing the applicability of this model to an ant colony is to change the model to a finite-size population. Another step is to take into account each individual ant's spatial position.

A_{ij} of (1) is interpreted now as the probability that any single ant in state i will change to state j per time unit. B_{ijk} is the chance for an ant in state i to go over to state k during an encounter with an ant in state j. A simulation was done using the following additional assumptions.

- N ants were put on a rectangular grid with cyclic boundary conditions.
- Each ant did a random walk with equal probabilities of walking in the eight possible directions (the eight nearest cells on the grid) and staying in its place. An ant never walked to a position already occupied by two other ants.
- The ants could be in M possible states. In each time unit, ants had a chance of changing their states according to A_{ij}. If two ants occupied the same grid point, they changed states according to B_{ijk}.

The simulation was run for the amplifier colony. Figure 5 shows β_{in} vs. β_{out} for colony sizes of 10, 20, 50, and 100 ants on a square grid of size 20. It can be seen that in this case the colony acts like an amplifier even when colony size is small.

4.2 Allocation of Bees to Alternate Foraging Sites

The model in this section is adapted from a model by Seeley et al. (1991), which explains how a bee colony can consistently allocate more foragers to the richer site, as they showed empirically. When the colonies were presented with two food sources at equal distances, more foraging effort was directed to the site with the richer food source. In the model, each foraging bee can be in one of seven states. A bee associated with a particular site can be in three states – at the hive, at the hive dancing, or foraging at that site. A bee can also not be associated with any site, in which case it is called "following" (see Fig. 4). The only difference between the models is that in the original model the bees spend a certain time at the site, whereas we use a Poisson process for the transition between foraging and being at the hive; at each moment a bee has the same chance of deciding to return to the hive. The results for this model are then, of course, exactly like those of Seeley et al. This model also demonstrates how a model with spatial components can be fit into the scheme of the model described here.

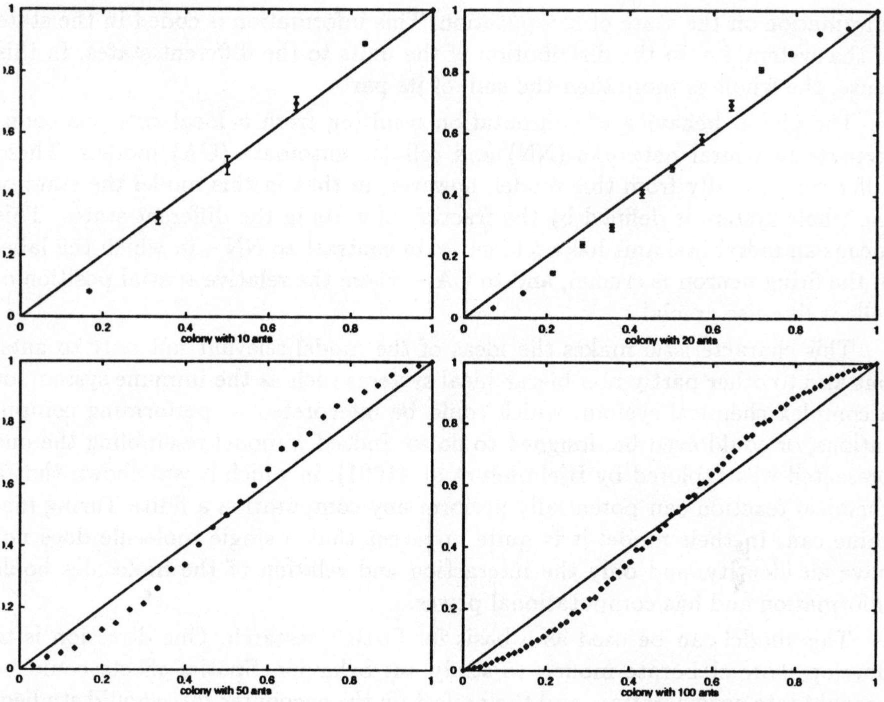

Fig. 5. β_{in} vs. β_{out} for a finite-size colony of size 10, 20, 50, and 100. Each point represents the mean of 10 simulation runs. The simulation was run on a square grid of size 20 by 20.

5 Discussion

Motivated by the manner in which an ant colony displays a highly coordinated global behavior in the absence of a hierarchical control structure, we studied a model system consisting of many identical units with a finite number of internal states. The state of the whole system is then determined by the fraction of units in each internal state. Following the bottom-up coordination scheme, the internal state of each unit was updated following a local rule. In this framework we demonstrated how such local rules can induce globally coordinated behavior. Furthermore, we addressed the question of how complex the global dynamics of such a system can be in this bottom-up approach. It is important to notice that even though the coordination is induced bottom-up, the collective complex behavior induced cannot be reduced to the level of the individual unit. Stated in computational terms, even though the system as a whole could perform a complex computation, no computation is performed by the single unit. This is necessary in the model (and maybe in an ant colony), since a single unit has no information on the state of computation. This information is coded in the state of the system, i.e. in the distribution of the units to the different states. In this sense, the whole is more then the sum of its parts.

The global behavior of computation resulting from a local rule has counterparts in neural networks (NN) and cellular automata (CA) models. These differ conceptually from this model, however, in that in this model the state of the whole system is defined by the fraction of units in the different states. This means an individual unit has no identity, in contrast to NN – in which the label of the firing neuron is crucial, and to CA – where the relative spatial position of cells is likewise crucial.

This characteristic makes the ideas of the model relevant not only to ants, but also to other partly non-hierarchical systems such as the immune system, or a complex chemical system, which could be interpreted as performing computations, or could even be designed to do so. Indeed a model resembling the one presented was explored by Hjelmfelt et al. (1991), in which it was shown that a chemical reaction can potentially perform any computation a finite Turing machine can. In their model it is quite apparent that a single molecule does not have an identity, and only the interaction and relation of the molecules holds information and has computational power.

This model can be used as a basis for further research. One direction is to develop more elaborate models to study ant behavior. Spatial effects could be brought into consideration, and their effect on the encounter rate should studied. An interesting question would be to study the complexity of response possible with respect to the number of ants in the colony (see also Gordon 1989, Pacala et al. 1994). A prediction which can be inferred from the model is that the statistics of task switching should show that when switching between tasks the ants must be going go through some minimal number of intermediate states, which might or might not be measurable otherwise.

Three related challenging questions are : Where are the local rules instructions (the analogs of the A's and B's), kept? How can these rules change (i.e. changing the A's and B's) so the colony can learn to react correctly to its environment? How did such a coordination evolve in the first place?

For the version of the model studied here it was assumed that ants are identical units. It seems to be the case that for many social insects, the genetic composition of the colony is not uniform. If the transition rates are genetically governed, this could potentially pose a problem to selection on these genes. One might think that selection would have minimized the genetic variation in the colony, but this is not the case, and some colonies even have multiple queens. From this we expect that ants have solved this problem in a different manner.

A principle we believe should be incorporated in the study of these questions, is closing reaction loops to enable feedback (in our case feedback to the colony through the environment). A schematic description of such a loop is: 1. The colony measures the environment (as described in Sect. 3.5). 2. The measurements outcome induces a coordinated response (as described in Sect. 3.5). 3. The colony responses by acting on the environment. 4. The action causes changes in the environment in a way correlated to the state of the colony, thus closing a loop. (Which is, in a sense, a loop in which the order in the whole system grows.) In this respect, incorporating pheromone trails laid by the ants seems natural. The time-wise separation drawn above between the different stages is, of course, artificial.

Considering the colony as a selection unit, the evolution of A's and B's could be studied by applying a selective pressure on a population of colonies. We believe the study of such an evolution, should also include a closed loop of reaction. Thus, not only having the selective pressure act on the population of colonies, but having the population of colonies change the selective pressure. How this should be done is indeed a challenging question.

6 Acknowledgments

ML wishes to Deborah Gordon for many discussions and insights, M. W. Feldman for many comments on drafts of the paper, F. Gruau and V. Schechter for aid during the infancy of the model, and J. Pritchard, J. Kumm, J. Doble, C. Bergstrom, D. Pollock for helping to make the paper readable. GS wishes to thank Itamar Pitowski for his guidance and enthusiasm in the pursue of this idea and others. GS wishes to thank Eshel Ben-Jacob and Ofer Sochat for their criticism which helped him understand the questions this model addresses, but more so the questions it does not. The work on the model began during the 7th complex systems summer-school in Santa-Fe. This work was supported by NIH grant GM 28016, and NASA grant NCC 2-778 to M.W. Feldman.

References

1. Borodin, A.: On relating time and space to size and depth. SIAM Journal on Computing **6**(4) (1977) 733–744
2. Deneubourg, J.L., S. Gross: Collective patterns and decision-making. Ethology Ecology & evolution **1** (1989) 295–311
3. Derrida B., and R. Meir 1988: Physical Review A **38** (1988) 3116–3119
4. Gordon, D. M.: Dynamics of task switching in harvester ant colonies. Behavioral Ecology and Sociobiology **31** (1989) 417–427
5. Gordon, D. M., B. Goodwin, and L. E. H. Trainor: A parallel distributed model of ant colony behavior. Journal of Theoretical Biology **156** (1992) 293–307
6. Jeanne, R. L.: The organization of work in Polybia occidentalis: The cost and benefits of specialization in a social wasp. Behavioral Ecology and Sociobiology **19** (1986) 333–341
7. Hjelmfelt, A., E. D. Weinberger, and J. Ross: Chemical implementation of neural networks and Turing machines. Proc. Natl. Acad. Sci. USA **88** (1991) 10983–10987
8. Marijuán, P. C.: Enzymes and theoretical biology: sketch of an informational perspective of the cell. BioSystems **25** (1991) 259–273
9. Millonas M. M.: Swarms, Phase Transitions, and Collective Intelligence. Artificial Life III, Ed. Christopher G. Langdon, SFI Studies in the Sciences of Complexity, Proc Vol. XVII Addison-Wesley (1994)

10. Moore C.: Generalized shifts: unpredectability and undecidability in dynamical systems. Non Linearity **4** (1991) 199-230
11. Pacala, S. W., D. M. Gordon, and H. C. J. Godfray: Effects of social group size on information transfer and task allocation. Evolutionary Ecology (in press)
12. Parberry, I: Circuit Complexity and Neural Networks. MIT Press. (1994) 31–43
13. Robinson, G. E.: Regulation of division of labor in insect societies. Annual Review of Entomology **37** (1992) 637–702
14. Seeley, T. D., S. Camazine, and J. Sneyd: Collective decision making in honey bees: How colonies choose among nectar sources. Behavioral Ecology and Sociobiology **28** (1991) 277-290
15. von Neuman, J.: Mathematical foundations of quantum mechanics , Princton University Press (1955)

Appendix A

In this appendix it is shown that for the model described in Sect. 2, if the constraints $\sum_{i=1}^{M} p_i = 1$ and $p_i \geq 0$ hold for $t = 0$, then they will hold always.

$$\frac{d}{dt}\sum_{i=1}^{M} p_i = \sum_{i=1}^{M} \frac{dp_i}{dt} \tag{22}$$

$$= \sum_{i=1}^{M}\left(-\sum_{j=1}^{M} A_{ij}(s)p_i + \sum_{j=1}^{M} A_{ji}(s)p_j + \alpha(N)\left(-\sum_{j,k=1}^{M} B_{ijk}p_ip_j + \sum_{j,k=1}^{M} B_{jki}p_jp_k\right)\right) \tag{23}$$

$$= -\sum_{i,j=1}^{M} A_{ij}(s)p_i + \sum_{i,j=1}^{M} A_{ji}(s)p_j + \alpha(N)\left(-\sum_{i,j,k=1}^{M} B_{ijk}p_ip_j + \sum_{i,j,k=1}^{M} B_{jki}p_jp_k\right) \tag{24}$$

$$= -\sum_{i,j=1}^{M} A_{ij}(s)p_i + \sum_{j,i=1}^{M} A_{ji}(s)p_i + \alpha(N)\left(-\sum_{i,j,k=1}^{M} B_{ijk}p_ip_j + \sum_{k,i,j=1}^{M} B_{ijk}p_ip_j\right) \tag{25}$$

$$= 0\ . \tag{26}$$

Hence $\sum_{i=1}^{M} p_i = 1$ always.

Next it will be shown that if $p_i = 0$, then $\frac{dp_i}{dt} \geq 0$, so that if initially $p_i \geq 0$, this will hold for later time also.

$$\frac{dp_i}{dt} = -\sum_{j=1}^{M} A_{ij}(s)p_i + \sum_{j=1}^{M} A_{ji}(s)p_j$$

$$+\alpha(N)\left(-\sum_{j,k=1}^{M} B_{ijk}p_i p_j + \sum_{j,k=1}^{M} B_{jki}p_j p_k\right) \tag{27}$$

$$= \sum_{j=1}^{M} A_{ji}(s)p_j + \alpha(N)\left(\sum_{j,k=1}^{M} B_{jki}p_j p_k\right) \tag{28}$$

$$\geq 0 \; . \tag{29}$$

This is because all quantities in the last expression are non-negative.

Appendix B

In this part it will be shown that the colony described in Sect. 3.3 behaves like a NOR-gate. For this, it has to have the following relations between the input and the output:

in_1	in_2	out
1	1	0
0	1	0
1	0	0
0	0	1

The input and output were defined such that, for example if $\frac{p_1}{p_2} > 3$ the first input is 0, and if $\frac{p_1}{p_2} < \frac{1}{3}$ the first input is 1. We also have $p_1 + p_2 = p_3 + p_4 = p_5 + p_6 = \frac{1}{3}$.This translates the same table to the following inequalities, which must hold at the steady state of the system.

$p_1 < \frac{1}{12}$ $p_3 < \frac{1}{12}$ output $= 0$
$p_1 > \frac{3}{12}$ $p_3 < \frac{1}{12}$ output $= 0$
$p_1 < \frac{1}{12}$ $p_3 > \frac{3}{12}$ output $= 0$
$p_1 > \frac{3}{12}$ $p_3 > \frac{3}{12}$ output $= 1$

At a steady state (16)-(18) give the following:

$$p_5(p_1 + p_3) = p_6(p_4 + p_2 + \frac{1}{4}) \; , \tag{30}$$

or

$$\frac{p_5}{p_6} = \frac{p_4 + p_2 + \frac{1}{4}}{p_1 + p_3} \; . \tag{31}$$

If we apply the first line of table 6 we get

$$\frac{p_5}{p_6} > \frac{\frac{3}{12} + \frac{3}{12} + \frac{1}{4}}{\frac{1}{12} + \frac{1}{12}} = \frac{9}{2} \; . \tag{32}$$

If this were amplified through three amplifiers, the relation between the output states will be $(((\frac{9}{2})^2)^2)^2$ which is greater than 3, so the output is 0, as required. The second line of the table gives

$$\frac{p_5}{p_6} > \frac{0 + \frac{3}{12} + \frac{1}{4}}{\frac{1}{3} + \frac{1}{12}} = \frac{6}{5} \; . \tag{33}$$

Amplifying this three times will give $(((\frac{6}{5})^2)^2)^2$ which is also greater than 3, so the output is again 0. The last line of the table gives

$$\frac{p_5}{p_6} < \frac{\frac{1}{12} + \frac{1}{12} + \frac{1}{4}}{\frac{3}{12} + \frac{3}{12}} = \frac{5}{6} \ . \qquad (34)$$

and amplifying this three times gives $(((\frac{5}{6})^2)^2)^2$ which is smaller than $\frac{1}{3}$, so the output is 1.

Mimicry and Coevolution of Hedonic Agents

Paul Bourgine[1] and Dominique Snyers[2]

[1] CREA, Ecole Polytechnique, 1 Rue Descartes, Paris, France
[2] LIASC, Télécom Bretagne, B.P. 832, F-29285 Brest Cédex, France
email: Dominique.Snyers@enst-bretagne.fr

Abstract. A hierarchy of cognitive agents is presented here in order to insist on the importance of interactions within a society of self teaching agents. In particular, this paper focuses on mimicry as the interaction mechanism of choice for hedonic agents. The dynamics of this mimetic type of interaction is studied both theoretically and by simulation. It leads to the coevolutionary dynamics best explained under the metaphor of an agent race on a landscape of sinking icebergs. A Darwinian selection mechanism is also included in the simulation model and the agents adapt their strategies by learning during one generation without being able to pass their findings directly to the next generation.

1 Introduction

A society of agents, either natural or artificial, is composed of autonomous situated agents whose primary goal is to maintain their viability in various and moving environments, i.e. to stay "alive". These agents interact not only with the environment itself but also with the other agents; their evolution is really a coevolutionary process. A well adapted agent can, for example, become handicapped after a modification of the environment or the composition of the society itself. It will then self-adapt to this new situation or be eliminated by the coevolutionary process. In fact, the dynamics of this society of agents is similar to the dynamics of an "immune system" which discriminates between the *self* and the *non-self*, the agents belonging to the society (those with a positive growth rate) and those having to be rejected (with a negative growth rate).

By considering the agents as interacting embodied minds whose goal is to survive and act in a changing environment despite their bounded cognitive capacity, we propose here a new framework to think about an agent society and we hope that this will lead to a new way of building artificial self-teaching agents. At the same time, we would like to push Artificial Life a little further toward the cognitive sciences. In section 2, we describe the hierarchy of agents with increasing cognitive capacity. Starting from reactive agents with pre-wired perception-action strategies, we successively describe the hedonic agents with their aptitude to choose between different strategies by anticipating rewards, and the eductive agents who have developed an additional semiotic capacity (capacity to manipulate symbols) to mentally explore new virtual strategies before deciding which one to follow. Only the first two levels will be covered in this paper and the eductive agent level is presented elsewhere (see [2] and [15]). In section 3, we will describe mimicry as the best way for hedonic agents to get access to new strategies. We will insist on the importance of mimicry in social sciences, economics and in Artificial Life. Section 4 will give a simple example of such a society of coevolving hedonic agents based on the imitation game from Kaneko[11] and whose mimetic interactions lead the whole society toward the edge of chaos as described in section 5.

2 A Cognitive Hierarchy of Agents

When studying self-organizing agents with emerging capacities, it is quite crucial to better understand how agents with more and more sophisticated cognitive capacities spontaneously emerge by evolution. An hierarchy of those cognitive levels is here proposed in an attempt to better grasp their mechanism of emergence.

2.1 Reactive Agents

At the bottom of this hierarchy lie agents with wired perception-action strategies corresponding to a stimuli-reflex type of behaviors. An agent receives some information about its environment through its sensory organs and reacts accordingly in a predefined way. Their sensory organs are often adapted to perform the categorization associated with the adequate action. Some bees, for example, can detect U.V light and this allows them to easily discriminate between the pistil and the flower since when observing a field of flower with a U.V filter, everything is uniformly white except the pistils. Insects indeed, are typical examples of reactive agents, and notwithstanding their limited sensorimotor behaviors, they build very complex social organizations. Some mathematical models of such behaviors have been successfully derived from the observation of ants, bees and termites [6] for example, and engineering applications of such principles have started to appear, e.g. in robotics [10] and optimization [5],[12] (see [1] for an overview of these techniques). But reactive agents with their quasi-invariant sensorimotor strategies can only survive in slowly changing environments.

2.2 Hedonic Agents

To allow for more flexibility and increase the agent robustness relatively to changing environments, one needs to add the capacity of learning new strategies. But how will this learning mechanism work? A pleasure/displeasure principle (the *hedonic* principle) is here postulated which leads the hedonic agent to revise its strategy according to some kind of anticipations about the pleasure associated with the strategy action. What is learnt therefore from the agent past experience, are these hedonic anticipations. Such postulate can be grounded in neurosciences, with the central nervous system of vertebrates synthesizing a huge number of chemical substances whose "rewarding" or "punishing" properties provide a general learning mechanism by selfreinforcement. Reinforcement learning techniques such as the *Q-learning* [17] , the *TD(λ) Learning* [16], the *Partition Q-Learning* [13] etc. provide incremental rules for the estimation of these hedonic anticipations during sensorimotor cycles. For continuous sensorimotor movements, a simple metadynamics to obtain hedonic strategies follows the *Stochastic Hedonistic Gradient Rule*: it consists in pushing the action dynamics in the direction of the hedonic anticipation gradient and in adding noise to ensure that all the possible strategies are experimented in non deterministic way ([3]). As in the case of reactive agents, perceptive invariant and intermediate categorizations of the perceptions are the key points to strategy building. But unlike reactive agents, the hedonic agents can now perform evolving categorizations of the perceptions in a moving environment. Through the history of interactions with its environment, the hedonic agent enacts a world of meanings which are pertinent for its pragmatic anticipations.

To summarize, the hedonic agents teaches itself the hedonic anticipations of a near future through an emergent categorization of the perceptions from its recent past and uses these anticipations for deciding which strategy to activate. But all these autoteaching processes suppose an active experimentation of the strategies. The question arises then of how the agent should react in face of completely new situations? Of course a random decision could be taken, but a better solution is often to observe the other agents and mimic their behavior. Mimicry is an excellent way indeed to deal with the bounded experience of agents by allowing them to share strategies. In order to be able to do so however, the hedonic agents must interact. The next sections of this paper will focus on this coevolutionary dynamics of mimicry that will make hedonic and socially admissible strategies to emerge.

3 Mimicry and Coevolution Dynamics

3.1 Mimicry

We have seen that due to their bounded experience the hedonic agents cannot have learnt a good strategy for every situation. By interacting with other hedonic agents however, they can increase their personal experience with the experience of others. This presupposes a certain level of consciousness called *primary consciousness* (see [7]): they can recognize other agents as similar, and therefore, they can imitate their behavior when facing a totally new situation. Mimicry is quite a general process that has not only been selected by natural evolution as a good mechanism for resisting predators (e.g. batesian mimicry), but is also present in the more elaborated economic agents we sometimes are. The study of mimetic interaction dynamics should therefore provide new insights to many mechanisms at work in economics. Mimetic interactions and their associated conflict resolution process has also been postulated by René Girard as one of the most fundamental mechanism in the structuration of many, if not all, human societies (see e.g [9]). This paper will focus on the mimetic interactions between self teaching-hedonic agents.

3.2 Hedonic Mimetic Agents

The hedonic agents are characterized by a strategy and a two level adaptation mechanism: a slow adaptation process of learning the anticipations to perception/action by a succession of hedonic selfreinforcements and a faster process of abandoning their current strategy by mimicking the successful strategy used by another hedonic agent in their vicinity.

A strategy is described by qualitative features, i.e the kind of perception to action binding it is able to perform, and by quantitative features of their perceptions and actions like their spatial extend, their speed etc. Such strategies therefore, can be seen as phenotypes in a phenotype space $\mathcal{P}$ defined as the sum of continuous spaces corresponding to the different types of such qualitative and quantitative features. Strategies are then evaluated in terms of their efficiency to maintain the agents adopting them alive in the moving environment. This strategy evaluation defines the *fitness or adaptation landscape* as an application from the strategy or phenotype space $\mathcal{P}$ to $\Re$, the set of real numbers. The fitness landscape itself is subjected to some intense variations along with the variations of the environment and the composition of the agent society.

3.3 Coevolutionary Dynamics

The evolution of a society of hedonic agents under mimetic interactions can be seen as a coevolutionary dynamics on such a moving fitness landscape.

The imitating agent can instantly switch from one strategy to another but small imitation errors occur that result into a slow phenotypic diffusion. In addition, the Darwinian evolution of the agent society also introduces a fast mechanism of selection in which the *selection rate* (or *growth rate* in terms of population dynamics) depends on the interactions between individuals and a slow mutation mechanism that amounts to a phenotypic diffusion.

These two dynamical systems are modelled by the same equation defined in terms of strategy density distributions. Let $x(q,t)$ be the density at time t of the strategy associated with a phenotype $q \in \mathcal{P}$. The temporal evolution of such a society of hedonic agents corresponds to the two level adaptation mechanism of the agent strategies as follows:

$$\frac{dx(q,t)}{dt} = w(q;\{x\})x(q,t) + D\,\nabla_q^2 x(q,t).$$

where $w(q;\{x\})$ is the growth rate depending on the phenotype q and on the population distribution $\{x\} = \{x(q^1,t), x(q^2,t), ...\}$, and D is the diffusion parameter (D is here supposed constant in all directions but this supposition can be released if necessary). With a positive $w(q;\{x\})$ growth rate the strategy q will be selected more often by the agents at time t. In the contrary with a negative value, it will start to be abandoned. *The sign of $w(q;\{x\})$ performs the "self-nonself" discrimination of hedonic agents society mentioned earlier and it differentiates between the agents belonging to the coevolutionary system and those having to be rejected.* An example of such a growth rate landscape, i.e. the application from the phenotype space $\mathcal{P}$ to $\Re$, will be given on Figure 3.

3.4 Coevolution as Hill Climbing on a Sinking Iceberg

When the growth rate $w(q;\{x\})$ can be associated with a Lotka-Volterra model, as it is the case in this paper, the solution of this differential equation is composed of a discrete set of eigen modes corresponding to viable strategies. We show indeed in a forthcoming paper that since the Lotka-Volterra model is associated with a compact operator it has a discrete spectrum. The associated growth rate landscape therefore can be compared to an ocean full of icebergs where the icebergs correspond to these eigen modes and where the ocean is the zero growth rate. The strategies with a positive growth rate correspond to the *tolerant* or visible part of the icebergs on which the agents have to climb on to avoid "falling into the water", i.e to become a strategy with a negative growth rate corresponding to the ***suppression area*** of the self-nonself discrimination. And indeed, the agents ***climb on these icebergs*** by adapting their strategies: this is the ***Fisher law*** detailed in Appendix A. But under the hypothesis that the agents evolve at the zero level of the growth rate, it can also be shown that the more individuals climb on an iceberg, the more ***it will sink***: this is the ***"red queen" effect*** (see [8]). Coevolution can really be seen as a hill climbing race between individuals on a sinking landscape.

The icebergs can also split into several smaller ones leading to new viable "species" of strategies at saddle points. They can also join together and make some species to die out and this without any mutation. Small mutations however often induce a tunneling effect. The coevolutionary dynamics indeed, usually sees phases of relative stability to be followed by phases of rapid changes where many agents change from one strategy to another, from one peak to another. This tunneling effect is induced by the small mutations.

4 Imitation Game

4.1 The Agents

To illustrate this mimetic coevolutionary dynamics we present a variation of Kaneko's imitation game [11] in which the agents are associated with a time series (a "song" generator):

$$s_{n+1} = 1 - q.{s_n}^2 = f(s_n)$$

where s_i takes values in $[0,1]$ and where q is a fixed parameter characteristic of the agent strategy (i.e its phenotype) taking a real value in $[0,2]$. The agents interact through a two player game from which the agent with the best capacity of approximating the other agent song (its time series) after a learning phase emerges as the winner. Every time an agent wins a game it increases the probability of its strategy to reproduce at the next generation. The density distribution of these agent strategies coevolves thus according to the differential equation of section 3. The growth rate originates from the interaction between agents and the diffusion term corresponds to some kind of mutation or errors during the strategy exchange between the agents.

In this simulation, we have considered a Darwinian type of coevolution mechanism. The agents adapt their strategies during their interactions with the other agents from the same generation but this developmental adaptation by learning is not transferred to the next generation. The fitness associated with an agent however depends on this developmental adaptation and a strategy is therefore selected for the new generation not only for its performance but also for its adaptability.

More specifically, our simulation runs as follows:

1. A society of agents is initialized with random strategies (i.e with characteristic parameter values randomly taken in $[0,2]$).
2. Each agent interacts with a given number of randomly chosen opponents. The evaluation of the agent growth rate is performed by iteratively splitting the population into a set of interacting random pairs. During these interactions the agents takes their turn in adapting their strategy by learning, but only the winner agent will carry this newly adapted strategy to the next interaction.

3. After this developmental phase in which the hedonic agents adapt their strategy-phenotype by interacting with others, the society is updated by simulated evolution in which the agents inherit the initial phenotype (i.e. the phenotype before the development of agents) selected with a probability proportional to their associated growth rate. This inheritance process however is not perfect here and undergoes some kind of mutation. We model these mutations by adding some noise from a normal random distribution centered around 0 and with a standard deviation μ. This leads to a slow diffusion in the strategy space.
4. The simulation then loops back to step two.

4.2 A Two Level Game

The agents interacts in pairs during a two level game composed of a learning phase followed by a contest, i.e the evaluation of the difference between time series. They take their turn in adapting their time series before the contest, Kaneko used the learning phase of the game just to skip the transients and to synchronize the time series of the players. We introduce here another learning mechanism in which the agents really adapt their own strategy during the interaction with the other agents by changing the value of their time series parameter by a classical gradient descent on the contest criteria:

$$\frac{dq}{dt} = -\epsilon \frac{\partial}{\partial q} \left[\sum_{i=0}^{m} \left(f_q(s_i(q)) - f_{q'}(s_i(q')) \right)^2 \right]$$

where f_q and $f_{q'}$ are the time series respectively associated with the learning agent and the non-learning agent, and where $s_t(q)$ is the value at time t of the time series associated with q. This learning phase starts with $s_0(q) = s_0(q')$ and lasts m steps. Thereafter, the Euclidian distance between the two agent time series is computed for a period $T - m$:

$$D\left(q,q'\right) = \sum_{t=m}^{T} \left(s_t(q) - s_t(q') \right)^2$$

with the initial $s_q(m)$ taken equal to $s_{q'}(m)$.

$D(q',q)$ is then computed next by making the time series q' the learner. The strategy associated with q is declared the winner when $D(q,q') < D(q',q)$.

4.3 Competition and Growth Rate

The growth rate of the strategy densities is related to the number of times the strategy has won. Under the hypothesis of *annealed disorder*, i.e when the agents have the same probability to meet any other agents, this can be modelled by a continuous Lotka-Volterra equation on the strategy density distribution (see Appendix B for more details). The growth rate is given by:

$$w(q;\{x\}) = a(q) + \int_Q b(q,q')x(q',t)dq'$$

where $a(q)$ is the internal dissipation parameter (e.g. dead minus birth rate), $b(q,q')$ describes the mimetic relationship between strategies q and q'. Figure 1 shows an example of such interaction landscape in the case of the imitation game described in the next section.

5 Simulation Results: Coevolution to the Edge of Chaos

The results presented here come from a simulation of the previous model with a population of 500 strategies evaluated during 100 games against randomly chosen opponent strategies from the same population. The parameter μ was set to 0.0001 and the adaptation parameter ϵ to

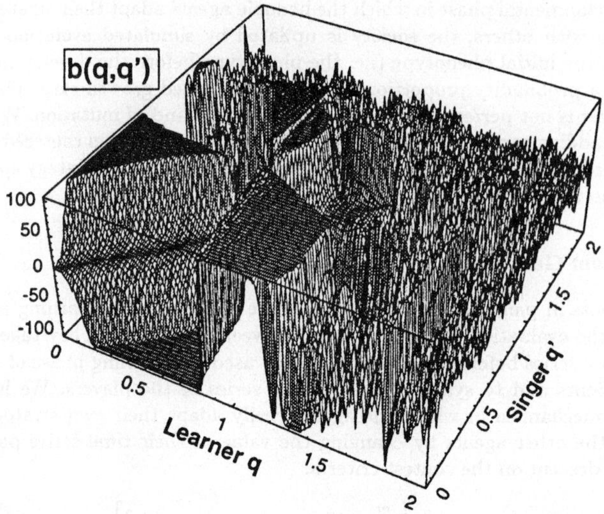

Fig. 1. Mean gain $b(q, q')$ over 100 games between strategies q and q' taken randomly on intervals sampled by 0.01 between 0 and 2. The game starts with random initial values $s_0(q)$ and $s_0(q')$. The x (resp. y) axis gives the strategy parameters q (resp. q') associated with the learner (resp. singer). The ridge around $q = 1.75$ corresponds to a robust learner strategy winning against a large range of singer strategies.

0.00005. The agent can thus change its strategy faster by learning than by mutation. The learning and the contest phases lasted 50 time series iterations.

Figure 2 shows the population evolution of the strategy densities during the first twenty generations. The quantification (or speciation) of the strategies clearly appears with a succession of emerging peaks starting at generation $t = 7$ around $q = 0.7$ associated with successive shifts toward $q = 0.75$ where a maximum is reached at $t = 15$. After that, the strategy densities in that vicinity start to decrease until they finally vanish after $t = 20$. On the other side, around $q = 1.75$, the same quantification appears but later. Starting from $t = 10$, a succession of peaks follows until the strategy around $q = 1.75$ invades most of the population after $t = 20$. (Let us note that the density distribution plot has been clipped here to 0.6 for scaling reasons). This tunnelling effect with the densities jumping from one peak to another is in full accord with the theory of section 3 (see also [8] and [4] for more details).

The selective pressure leading to this tunnelling effect of the strategies jumping from one peak to another can also be understood by the "sinking iceberg" metaphor that this simulation somewhat materializes. Figure 3 indeed shows the growth rate evolution during the first twenty generations. The sinking nature of the growth rate "icebergs" is clearly depicted with a first peak above the zero level around $q = 0.75$ attracting therefore many strategies. At $t = 12$ his peak starts to sink and vanishes at $t = 17$ with as consequence the dynamics observed on Figure 2 on the strategy densities. The other group of peaks appears less clearly because of the increasing uniformity of the population, but it starts to build up later around $q = 1.75$ with the same sinking behavior associated with the increase in the number of population member adopting such strategies.

Figure 4 shows the strategy density distribution for $t = 2$, $t = 10$ and $t = 14$. The negative curbs below the horizontal line correspond to the initial strategy before any learning and the positive curbs above correspond to the strategy distribution after the 100 adaptation steps. This clearly depicts the fact that the adaptation actually spreads the strategies around their original peaks. The curb at the bottom of this figure represents the Lyapounov exponent, the average

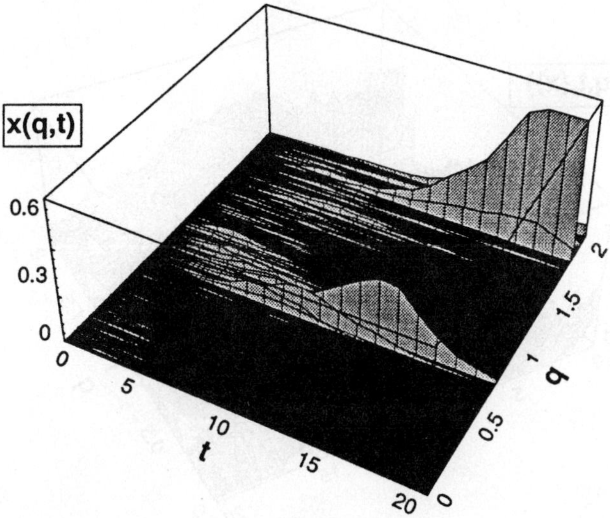

Fig. 2. The strategy density in the population at the beginning of the first twenty generations, i.e the density of the strategy before any adaptation from the learning phases. Starting from uniformly distributed strategies in the [0, 2] interval for q, it shows a fast quantification of the strategies around first the 0.75 parameter value and then around $q = 1.76$.

exponential divergence between closely adjacent points during one iteration of the time series ([14]). This a measure classically used in physics to identify chaotic behaviors in dynamical systems. Chaotic dynamics indeed are defined by a positive Lyapounov exponent (this correspond to the well known "butterfly effect" or sensibility to initial conditions). The big arrows on Figure 4 marks some points in the [0, 2] interval associated with a null Lyapounov exponent, i.e point at the edge of chaos. It can be observed that the emerging strategies corresponds to a time series with a characteristic dynamics at the edge of chaos as was already mentioned by Kaneko [11].

Figure 5 shows the individual strategy adaptation of the 500 agents during the developmental phase of evaluation (i.e 100 games) respectively at $t = 2$, $t = 10$, $t = 14$ and $t = 17$. This figure explains the spreading observed in Figure 4. Big arrows once again mark some points at the edge of chaos in terms of the zero of the Lyapounov function. Let us note that the time series $s_{n+1} = 1 - q.s_n{}^2$ is fractal, and that the Lyapounov exponent curb will therefore depend on the scale at which it was computed. (Here we use a 0.002 interval scaling factor.)

6 Conclusion

In describing a cognitive hierarchy of autonomous agents we dwelt on the importance of interactions between agents. For hedonic agent to successfully face new situations, for example, mimicry is a crucial mechanism to augment their personal skills with the experience of the other. We have seen how mimicry induces a coevolutionary dynamics that can be understood by the metaphor of the race between agents on a sinking iceberg landscape. We have given results of a simulation exhibiting, on a particular example, such kind of dynamics. Our model also included a Darwinian selection mechanism by which the agents developmental adaptation cannot directly be passed to the next generation. The fitness associated with an agent however depends on this adaptation and a strategy is therefore selected for the new generation not only for its performance but also for its adaptability during its development. The simulation dynamics was also shown to lead the system to the edge of chaos, in term of strategies associated with

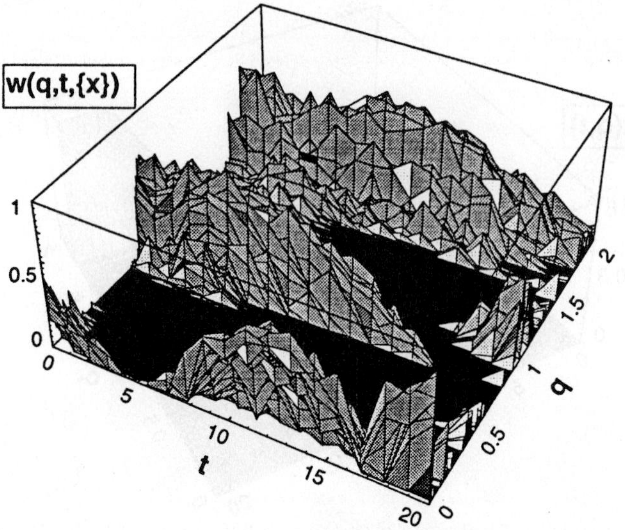

Fig. 3. The growth rate $w(q; \{x\})$ is computed here by sampling the $[0, 2]$ range by 0.002 intervals and by evaluating the Lotka-Volterra equation. The annealing order is simulated by evaluating the growth rate by picking up 100 strategies in the sampled intervals and by performing the game against an opponent randomly chosen within the current agent population. The negative values of the growth rate have been clipped for clarity reason. The "sinking iceberg" metaphoric dynamics can be observed with a fading selective pressure first around $q = 0.75$ first and then around $q = 1.75$. After $t = 20$, the population becomes too uniform and the growth rate landscape appears to be less characteristic.

a null Lyapounov exponent. We do not pretend to elevate this "life at the edge of chaos" type of behavior at the level of a general principle, we just observe it on this particular imitation game. This paper also tried to insist on the importance of connecting artificial life simulations with a theoretical model with both side shedding light on the other. We are currently pushing further this study of the mimetic coevolutionary dynamics and new research has been started to investigate the higher levels of the cognitive hierarchy.

References

1. E. Bonnabeau and G. Théraulaz. *Intelligence Collective.* Hermès, 1994.
2. P. Bourgine. Models of co-evolution in a society of autoteaching agents: from the society of hedonic mimetic agents to the society of eductive specular agents. In *Entretiens Jacques Cartier*, Lyon, 1993.
3. P. Bourgine. Viability and pleasure satisfaction principle of autonomous systems. In *Imagina-93*, 1993.
4. P. Bourgine and D. Snyers. Lotka-volterra coevolution at the egde of chaos. In *Proceeding of the Conference: Evolution Artificielle 94*, Toulouse, 1994.
5. A. Colorni, M. Dorigo, and V.Maniezzo. An investigation of some properties of an ant algorithm. In R. Männer and B. Manderick, editors, *Proc. of Parallel Problem Solving from Nature.* Elsevier Science Pub., 1992.
6. J.-L. Deneubourg and S. Goss. Collective patterns and decision-making. *Ethology Ecology and Evolution*, 1:295–311, 1989.
7. G. Edelman. *Bright Air, Brillant Fire: On the Matter of Mind.* Basic Books, 1992.
8. R. Feistel and W. Ebeling. *Evolution of Complex Systems: Selforganization, Entropy and Development.* Kluwer Academic,Dordrecht, 1989.
9. R. Girard. *Theater of Envy: William Shakespeare.* Oxford University Press, 1991.

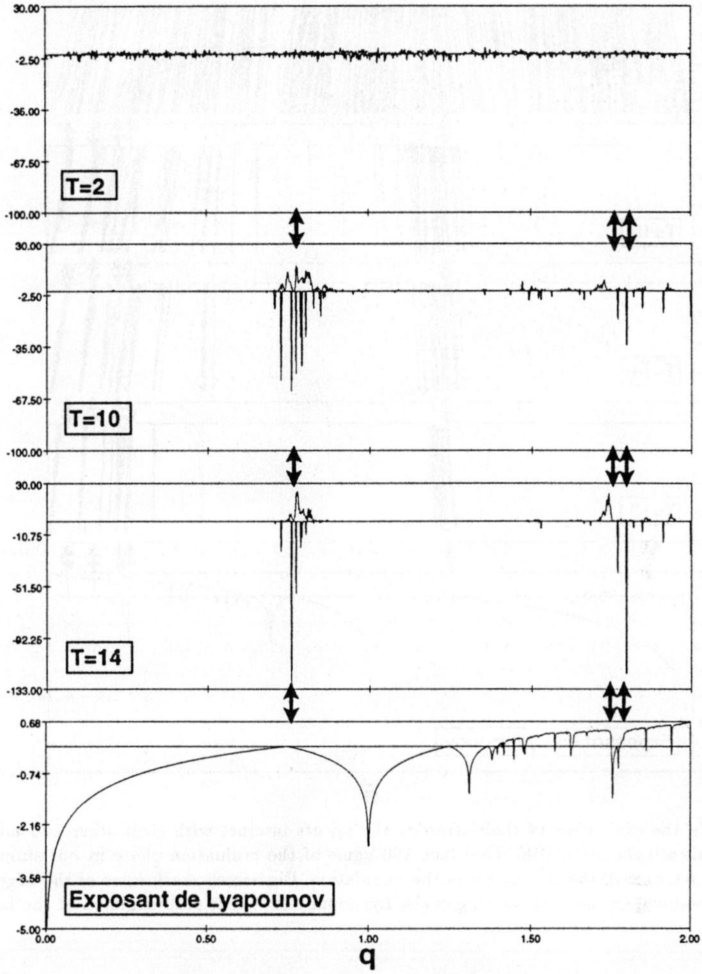

Fig. 4. Two strategy density distributions are depicted here at $t = 2$, $t = 10$ and $t = 14$: the density of strategies before any adaptation (negative values) and the density of strategies after the 100 games, i.e after the 100 adaptation steps (positive values above the horizontal lines). The bottom curb represent the Lyapounov exponent allowing us to mark by big arrows some of the q value associated with the edge of chaos, i.e with a zero Lyapounov exponent.

10. S. Goss and J.L. Deneubourg. Harvesting by a group of robots. In F. Varela and P. Bourgine, editors, *Toward a Practice of Autonomous Systems: Proc. of the first European Conference on Artificial Life*, Cambridge, MA, 1992. The MIT Press.
11. K. Kaneko and J. Suzuki. Evolution to the edge of chaos in an imitation game. In C. Langton, editor, *Artificial life III*, Redwood City, CA, 1993. Addison Wesley.
12. P. Kuntz and D. Snyers. Emergent colonization and graph partitioning. In *Proceeding of the Third International Conference on Simulation of Adaptive Behavior*, pages 494–500, Brighton, 1994. MIT Press.
13. Remi Munos and Joslyn Patinel. Partition q-learning. In J.L. Deneubourg, G. Nicolis, and H. Bersini, editors, *Self Organization & Life, from simple rules to global complexity: Proc. of the second European Conference on Artificial Life*, Brussels, Belgium, 1993.
14. H. G. Schuster. *Deterministic Chaos: An Introduction.* VCH Verlagsgesellschaft, 1989.

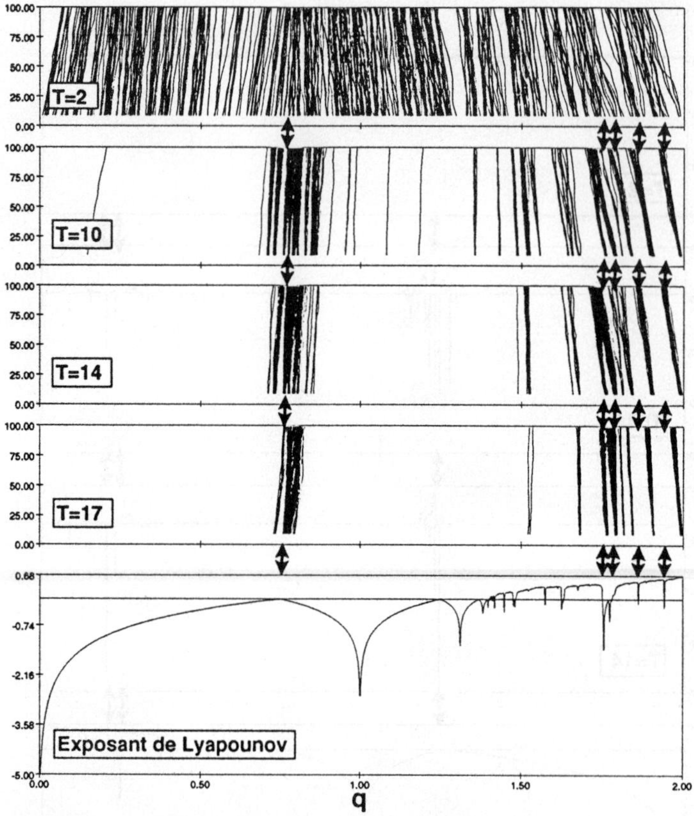

Fig. 5. During the evaluation of their strategy the agents interact with each other and adapt their strategies. This adaptation in 100 steps (the 100 game of the evaluation phase in our simulation) is depicted here for each of the 500 agents in the population. Big arrows mark some of the edge of chaos points corresponding to the zero crossing of the Lyapounov exponent curb depicted at the bottom.

15. D. Snyers, P. Kuntz, and P. Bourgine. Emergence de modèles de représentations sensori-motrices et coviabilité : l'exemple d'un jeu d'imitation. In *Proc. of Journées de Rochebrune*, Paris, 1995. Télécom Paris.
16. R. Sutton. Learning to predict by the methods of temporal difference. *Machine Learning*, pages 9–44, 1991.
17. R. Sutton. Reinforcement learning architectures for animats. In *Proc. of first International Conference on Simulation of Adaptive Behavior: From Animals to Animats*, Cambridge, MA, 1991. The MIT Press.

A The Fisher Law

Let us prove the Fisher law that says that small clusters of phenotypes climb on the growth rate landscape. The cluster barycenter $\bar{q}$ is defined by $\bar{q} = \frac{\int_\Gamma q\, x(q,t)dq}{\int_\Gamma x(q,t)dq}$ where Γ is the cluster support. The temporal evolution of the barycenters is given by the following equation for replicator systems (i.e. those systems with $\frac{\partial x(q,t)}{\partial t} = w(q;\{x\})x(q,t)$):

$$\frac{d\bar{q}(t)}{dt} = \frac{\int_\Gamma q\, \frac{\partial x(q,t)}{\partial t} dq}{\int_\Gamma x(q,t)dq} - \frac{\int_\Gamma qx(q,t)dq \int_\Gamma \frac{\partial x(q,t)}{\partial t} dq}{\left(\int_\Gamma x(q,t)dq\right)^2}$$
$$= \frac{\int_\Gamma (q-\bar{q}(t))\, \frac{\partial x(q,t)}{\partial t} dq}{\int_\Gamma x(q,t)dq}$$
$$= \frac{\int_\Gamma (q-\bar{q}(t))\, w(q,\{x\})\, x(q,t)dq}{\int_\Gamma x(q,t)dq}$$

For small clusters, $w(q;\{x\})$ is approximated by $w(q;\{x\}) = w(\bar{q};\{x\}) + \nabla_q w(\bar{q};\{x\}).(q - \bar{q}(t))$ and this leads to :

$$\frac{d\bar{q}(t)}{dt} = w(\bar{q};\{x\})\frac{\int_\Gamma (q-\bar{q}(t))\, x(q,t)dq}{\int_\Gamma x(q,t)dq} + \frac{\int_\Gamma (q-\bar{q}(t))\, [\nabla_q w(\bar{q};\{x\}) \,.\, (q-\bar{q}(t))]\, x(q,t)\, dq}{\int_\Gamma x(q,t)dq}$$

The first term is null by definition of barycenters and this equation simplifies as follows:

$$\frac{d\bar{q}(t)}{dt} = \frac{\int_\Gamma (q-\bar{q}(t))\, [\nabla_q w(\bar{q};\{x\}) \,.\, (q-\bar{q}(t))]\, x(q,t)\, dq}{\int_\Gamma x(q,t)dq}.$$

This indeed tells us that the small clusters climb on the growth rate landscape since:

$$\nabla_q w(\bar{q};\{x\}) \,.\, \frac{d\bar{q}(t)}{dt} = \frac{\int_\Gamma [\nabla_q w(\bar{q};\{x\}) \,.\, (q-\bar{q}(t))]^2\, x(q,t)\, dq}{\int_\Gamma x(q,t)dq} \geq 0\,.$$

B Annealing Disorder and Lotka-Volterra Models

The reproduction rate $w(q;\{x\})$ is proportional to the number of imitation games the agent has won during its developmental phase, i.e during t_{dev}, the duration of one generation:

$$w(q;\{x\}) \propto \int_Q \int_{t-t_{dev}}^{t} victory(q,q',t)dt\, dq'$$

where $victory(q,q',t)$ is the probability that q wins against q' at time t. This can also be expressed in terms of the probability for strategy q to be better than strategy q' at time t and the probability of q meeting q':

$$victory(q,q',t) = better(q,q',t)p(q' \mid q,t).$$

Two hypotheses are required to derive the Lotka-Volterra model from this:

1. the probability $better(q,q',t)$ must be time independent : $better(q,q',t) = b(q,q')$;
2. the conditional probability must be independent of q': $p(q' \mid q,t) = p(q' \mid t) \approx x(q',t)$. This is the annealing disorder condition: every strategy has the same probability to meet any other one.

The growth rate can then be expressed as follows:

$$w(q;\{x\}) \propto \int_Q b(q,q')\,[\int_{t-t_{dev}}^{t} x(q',t)dt]dq'.$$

By calling $\bar{x}(q',t) = \frac{1}{t_{dev}} \int_{t-t_{dev}}^{t} x(q',t)dt$ the mean distribution on t_{dev}, we get the continuous Lotka-Volterra equation:

$$w(q,\{x\}) \propto \int_Q b(q,q')\,\bar{x}(q',t)\, dq'$$

with the Lotka-Volterra $a(q)$ term equaling zero since the population size is constant here.

Evolution of Symbolic Grammar Systems

Takashi Hashimoto * and Takashi Ikegami **

Institute of Physics, College of Arts and Sciences, University of Tokyo, Komaba 3-8-1,Meguro-ku, Tokyo 153, Japan

Abstract. Evolution of symbolic language and grammar is studied in a network model. Language is expressed by words, i.e. strings of symbols, which are generated by agents with their own symbolic grammar system. By deriving and accepting words, the agents communicate with each other. An agent which can derive less frequent and less acceptable words and accept words in less computational time will have higher scores. Grammars of agents can evolve by mutationally processes, where higher scored agents have more chances to breed their offsprings with improved grammar system. Complexity and diversity of words increase in time. It is found that the module type evolution and the emergence of loop structure enhance the evolution. Furthermore, ensemble structure (net-grammar) emerges from interaction among individual grammar systems. A net-grammar restricts structures of individual grammar and determines their evolutionary pathway.

1 Introduction

Linguistic expressions are quite complex but may not be random. It is commonly assumed that one has to have an internal knowledge (hereafter *individual grammar*) of one's language when one can derive and recognize appropriately structured expressions. On the other hand, linguistic expressions are determined and restricted by a community of agents. Language is used by many speakers, not just a single speaker, the language as a whole is produced through interaction among various individual grammars. In this respect, a network determining grammar may be more important for linguistic expressions than internal knowledge. An individual grammar does not have a static form but dynamically changes: it can undergo changes induced by interactions with physical and cultural environment or conversations with other people. We have to discuss how is the grammar of a language is constructed through interaction among individual grammars and how does diversity and complexity of individual grammar evolve?

In the present paper, we will study an evolution of grammar in network through an agent model, where each agent has its own grammar and it communicates with. In our model the individual grammar is expressed by a symbolic generative grammar. When each grammar changes, the set of words it permits

* e-mail: `toshiwo@sacral.c.u-tokyo.ac.jp`
** e-mail: `ikeg@sacral.c.u-tokyo.ac.jp`

can change. The evolution of diversity of spoken words and such generative grammars will be discussed. Adequate automaton can accept the set of words which is accepted by a given symbolic grammar [1]. Hence the diversity of spoken words of a symbolic grammar system can be studied in terms of computational ability of automata.

According to N. Chomsky, the corresponding computational ability of symbolic language system is categorized into four different classes with respect to its grammar structure as follows [2]:

type 0 phrase structure grammar
type 1 context sensitive grammar
type 2 context free grammar
type 3 regular grammar.

A grammar in an upper hierarchy class generates a larger set of words than ones lower in the hierarchy. For example, a word set $\{0^n1^n|n \geq 1\}$ can not be derived by regular grammar (type 3) but can be derived by context free grammar (type 2) or ones even higher in the hierarchy, where xy is a concatenation of symbol x and y and x^n is n times concatenation of symbol x.

For practical situations, we have to deal with finite length of words with finite deriving processes. If we deal only with the finite set of symbols, i.e. $\{0^n1^n|N \geq n \geq 1\}$, this hierarchy does not always work. In computation theory, computational time to derive words are not bounded and no ensemble effect is considered. In this paper, we study practical ability to speak and recognize words. We need to figure out what kind of grammar has the practical ability to derive and accept words in finite time. Namely, the computational ability of a symbolic grammar and hierarchy will be studied in an ensemble of communicating agents.

Relations between different levels can only be clarified in a network and evolutionary context. If an upper structure in a network evolves to constrain individual grammars, we call it a *net-grammar*. A net-grammar system emerges from interactions between individual grammar systems rather than from one grammar system. A relationship between a net-grammar system and individual grammar systems will be discussed in this paper.

By taking each individual grammar as genotype and the set of generated words as the phenotype, we can regard a symbolic grammar system as a genetic system. In this paper, individual grammars can evolve through mutationally processes as well as genetic evolution.

Furthermore, autonomy of language is our main concern. It has been assumed that complexity of language is a mere reflection of complexity in the world we live, just as complexity of living systems is said to be the reflection of their complex environments. MacLennan has studied the communication among agents with simple rules[3]. His agents live in a particular local environment and communicate with each other by emitting signals. Those signals correspond to their external objects. Werner and Dyer have discussed the evolution of communication among the spatially distributed agents [4]. However, we believe that even without locality in space/information, systems can evolve and diversify their phenotype and genotype by some internal mechanisms. Examples can be found in evolutionary game [5, 6], Tierra world [7, 8]. We study evolution and diversification of sentences and grammars without external environments. Only conversation among agents can evolve grammar structure.

2 Modeling

2.1 Communication between Symbolic Grammar Systems

Agent. We express a communicating agent with a *generative grammar* by an ordered four-tuple:

$$G_i = (V_N, V_T, F_i, S) \ . \tag{1}$$

All agent have the same sets of nonterminal and terminal symbols, that is $V_N = \{S, A, B\}$, $V_T = \{0, 1\}$ respectively. Each agent is identified by index i. A symbol F_i is a set of rewriting rules peculiar to each agent, which is a finite set of ordered pair (α, β). The elements (α, β) in F are called *rewriting rules* and will be written $\alpha \longrightarrow \beta$. Here, α is a symbol over V_N. And β is an arbitrary finite string of symbols over $V_N \cup V_T$ not including the same symbol as α. The type of grammar which an agent can have is a context free or regular grammar here.

Communication. Agents communicate with each other by speaking and recognizing *words*, each composed of a finite string of symbols.

All agents derive words using their own rewriting rules. To derive a word a leftmost symbol equal to the left-hand side of a rewriting rule α is rewritten by the right-hand of the rule β. Derivation always starts with an initial symbol S. If a agent has more than two fitting rules in its rule set, the agent adopts one rule randomly. When no more nonterminal symbols are left in the derived word, a derivation terminates. And the derived word is spoken to all agents. An agent fails to speak a word when (i) the derivation does not finish within 60 rewriting steps or (ii) there is no applicable rule its rule set. The length of a word w (denoted by $|w|$) is given by the number of symbols in it. The possible largest length of a given word is M. The words longer than M are truncated after M-th symbol and then are spoken. The possible number of words (N_{all}) is limited to $2^{M+1} - 2$, and a full set of words speakable by an agent G_i is denoted by $L_{\text{sp}}(G_i)$.

Agents try to recognize words by applying their own rules in the opposite direction. If an agent can rewrite a given word back to the symbol S within 500 rewriting steps, we say that the agent can recognize the word. The language recognized by an agent G_i is denoted by $L_{\text{rec}}(G_i)$. Note that the inclusion relationship (i.e. $L_{\text{sp}}(G_i) \supseteq L_{\text{rec}}(G_i)$) holds, because of the truncation and limitation of rewriting steps.

2.2 Communication Game and Evolutionary Dynamics

We set a communication game in a network consisting of P agents. Each agent speaks in turn and each word is given to all the agents. Then every agent including the speaker tries to recognize the word. One time step consists of R rounds. For each round, every agent has a opportunity to speak.

Score. Each agent is ranked by three scores; speaking, recognizing and being recognized. A word spoken by the l-th agent to the m-th agent at a round c is denoted by $w_{lm}(c)$. The scores at round c is computed as follows:

For the factor of speaking, a score is given by

$$p_l^{\mathrm{sp}}(c) = \begin{cases} |w_{lm}(c)|/(\mathit{trend}+1) & \text{, for speaking a word } w_{lm}(c) \\ -3 \text{ ,} & \text{for failing to speak any word ,} \end{cases} \tag{2}$$

where ***trend*** is defined as the frequency of the word spoken in the last 10 time steps. An agent gets a higher value of $p_l^{\mathrm{sp}}(c)$ when he speaks a longer word and/or a less frequent word.

For the factor of recognition, a score is given by

$$p_{kl}^{\mathrm{rec}}(c) = \begin{cases} |w_{kl}(c)|/s & \text{, for recognizing a word spoken by } k\text{-th agent} \\ & \quad \text{in } s \text{ rewriting steps} \\ -|w_{kl}(c)| & \text{, for not recognizing a word spoken by } k\text{-th agent .} \end{cases} \tag{3}$$

A quick recognition of a long word provides a higher value of $p_k^{\mathrm{rec}}l(c)$.

For the factor of being recognized, a score is given by

$$p_{lm}^{\mathrm{br}}(c) = \begin{cases} |w_{lm}(c)|/P \text{ ,} & \text{if the spoken word is recognized by an by } l\text{-th agent} \\ -|w_{lm}(c)|/P & \text{, if the spoken word isn't recognized by } l\text{-th agent .} \end{cases} \tag{4}$$

Mutually recognizing agents will have high p_{kl}^{br}.

The total score in a time step is an average of a weighted sum of the three scores over R rounds:

$$p_l^{\mathrm{tot}} = \frac{1}{R}\sum_{c=1}^{R}\left(r_{\mathrm{sp}}p_l^{\mathrm{sp}}(c) + r_{\mathrm{rec}}\sum_{m=1}^{P} p_{lm}^{\mathrm{rec}}(c) + r_{\mathrm{br}}\sum_{k=1}^{P} p_{kl}^{\mathrm{br}}(c)\right) \text{ ,} \tag{5}$$

where r_{sp}, r_{rec} and r_{br} are the respective weighting coefficients. For example, if r_{br} is given a positive value, those who can be recognized by more agents get higher scores. But if the value is set negative, being recognized is no more favorable.

Mutations. In each time step, new agents are produced. The rule set of new agent is inherited from its ancestor's and suffers one of the following three mutation processes: a) *adding mutation* – a new rule is added, which is a modified rule of randomly selected from an ancestor's rule set. b) *replacing mutation* – a randomly selected rule is replaced with its modified rule. c) *deleting mutation* – a randomly selected rule is deleted.

The modifications of the selected rule are caused by, (i) replacing a symbol of the left-hand with the other nonterminal symbol or (ii) replacing a symbol in the right-hand with the other nonterminal/terminal symbol or (iii) inserting a symbol in the right-hand side or (iv) deleting a symbol from the right-hand.

Adding mutation is applied to agents within the rate m_{add}, if their scores exceed the average score. Replacing and deleting mutations are applied to all the agents within the rate of m_{rep} and m_{del}, respectively.

3 Results of Simulation

In this paper, a network consists of 10 agents ($P = 10$) and each agent tries to speak 10 times in each time step ($R = 10$). The score of the communication game is computed with the fixed parameters: $r_{\text{sp}} = 3.0$, $r_{\text{rec}} = 1.0$ and $r_{\text{br}} = -2.0$. Note that agents which can speak less acceptable words are benefited for a negative value of r_{br}. It is expected that a variety of the words in a population will increase. All the mutation rates are set at equal value 0.04 ($m_{\text{add}} = m_{\text{rep}} = m_{\text{del}} = 0.04$). The maximum length of a word is limited to 6 ($M = 6$), therefore the number of possible words N_{all} is 126.

Initially, all agents assumed to have the simplest grammar, i.e. a single rule with one symbol in the both hand side. They are classified as type 3 grammars due to Chomsky's hierarchy. At least, either a rule $S \rightarrow 0$ or a rule $S \rightarrow 1$ should be included to derive words.

3.1 Algorithmic Evolution

We find that evolution of grammar system is accelerated by the characteristic factors, one is a module type evolution and the other one is a loop evolution. Computational ability of an agent is measured by the ratio of recognizable words to the total number of possible words, i.e.

$$\text{Computational ability} = \frac{N(L_{\text{rec}}(G_i))}{N_{\text{all}}}, \tag{6}$$

where $N(L_{\text{rec}}(G_i))$ is the number of words which the agent G_i can recognize. Fig. 1 represents the example of evolution of the computational ability from the initial network. The computational ability, as well as the number of the distinct words spoken in the network, we call a *variety* of words, evolves with time.

A tree that displays the derivation path of a given word is called a derivation tree of the word. We put all possible derivation tree of a grammar system in a directed, connected graph. A structure of the graph expresses the algorithm of the grammar. Algorithmic evolution can be seen in the topological changes of this graph.

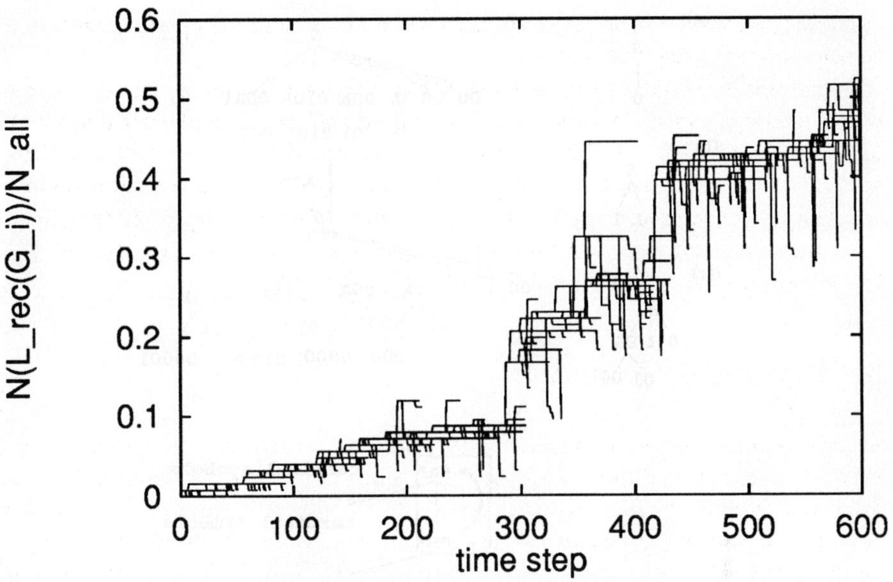

Fig. 1. Time step v.s. $N(L_{\rm rec}(G_i))/N_{\rm all}$. Each line connects one agent to oneself or its offsprings. It branches off by the mutations. A line terminates when the corresponding agent is removed. These lines show upward trend. In initial 200 time steps, computational ability gradually increases. After that, transitions to higher computational agent are frequently observed.

Evolution in Initial Period. It is shown in Fig. 1 that computational ability of agents slowly evolves during initial 200 time steps. In Fig. 2 (a)∼(c) the corresponding grammar systems are depicted in graph diagram. The initial agent has a weakest ability, having a direct derivation rule $S \to 0$ (Fig. 2(a)). The agent can increase the ability by the process of the adding mutation. Adding the rule $S \to 1$ to a production graph generates a branch structure (Fig. 2(b)). Further, the adding mutation evolves the multi branch structure (Fig. 2(c)).

Module Type Evolution. We find in Fig. 1 that an agent with the remarkably high ability (> 0.1) appears at time step 192. The change of grammar at this time step is sketched in Fig. 2(d). An acquired rule $A \to 00$ can double the acceptable size of the word of a grammar. Every intermediate word containing a symbol A can be rewritten by the rule $A \to 00$. In the sense that one common rule is used by many different words, we call the key rule a module rule. Evolution processes driven by module rules are called module type evolutions.

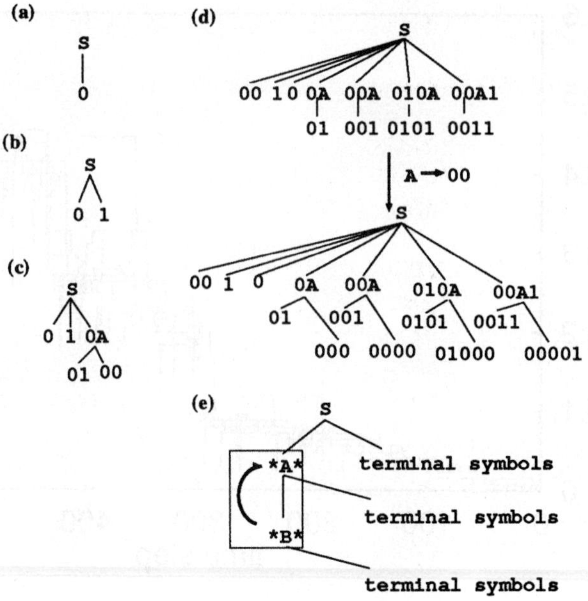

Fig. 2. The examples of grammar structure are shown by graph diagrams: (a) a sequential structure, (b) a branch structure and (c) a multi branch structure. In the (d) an example of module type evolution is shown. Acquiring a rule $A \rightarrow 00$, a grammar without bifurcation (upper tree) is evolved into one with bifurcated branches (lower tree). an example of grammar having a loop structure is schematized in (e). Asterisk stands for any symbols With this grammar, new agent can rewrite words $*A*$ into $*B*$ and vise versa. Words derived from such grammar can not represented in a tree form.

Emergence of Loop Structure. Grammar systems evolved by modules are not evolutionary stable in general. It is overpowered by more powerful grammar systems. Fig. 1 shows that a new agent with more powerful grammar appears in the population around time step 310. This agent has a loop structure in its grammar system (see Fig. 2(e)). A loop structure can derive a potentially infinite numbers of words recursively. A grammar with a loop structure is categorized as type 2 grammar or higher one in Chomsky's hierarchy.

3.2 Forming Ensemble with Common Word

An upper structure, which is named an *Ensemble with Common Word (ECW)*, emerges in the population of agents. The ensemble consisted of agents which can speak and recognize the common words. The other agents which can't speak or recognize the common words are less benefited than those in ECW.

When there exists ECW, even an agent of a high ability in a population will die out. At time step 403, an agent with the highest computational ability of the population dies out (see Fig. 1). Agents taking too much rewriting steps to recognize words decrease the fitness. We indicate this fact by Table 1. The rewriting steps to recognize the words at time step 400 are shown in this table.

Agents which cannot recognize frequent words in the population will be removed in order. An agent G_{306} (agent with ID 306), which has the second highest ability in the population, is removed first at time step 400. An agent G_{276} , which has the highest ability in the population, is removed at time step 403. It cannot recognize the word "00". To stay in the population, where a word "00" is the most commonly spoken, each agent should speak and recognize this word quickly. An ability to speak a certain frequent word quickly should be balanced with an ability to speak many words.

Numbers in bold font in Table. 1 represent the largest two rewriting steps to understand the words in the leftmost column. It is clear from this table that it takes much more time for agents G_{306} and G_{276} than the rest of agents. To take more steps to recognize commonly spoken words is disadvantageous for the agents G_{306} and G_{276}. If a group containing agents G_{306} and G_{276} constituted the majority and the words as "001011" or "010101" were the commonly spoken words, agents G_{306} and G_{276} will take an advantage.

Fig. 3 shows the phylogeny of agents at time step 400. It shows that the group consists of the agents G_{276} and G_{306} and that of the other agents forms the different lines. They form two different ECWs. The agents in the major ECW have lower computational ability than those in the minor ECW. Two ECWs conflict to survive in the network. Those in the major ECW behave cooperatively as the result by speaking and recognizing common words and get higher scores. At last all agent in the minor ECW is removed from the network. In this way the evolution toward the high computational ability is suppressed by forming ECW.

After removing agents G_{306} and G_{276} from a network, agents come to compete with each other within the same ECW. Proportional to the number of rewriting steps to recognize the commonly spoken words, the agents are removed from the network. In the ECW, a new agent with the high computational ability will emerge through algorithmic evolution.

3.3 Minimum All Mighty

We can make a *Minimal All Mighty* agent. It is an agent which can speak and recognize all possible words with the least number of rules. For example, a Minimal All Mighty agent has the rules such as:

$$S \rightarrow A, A \rightarrow SS, S \rightarrow 0, S \rightarrow 1 . \qquad (7)$$

This grammar is categorized as a type 2 grammar. It recognizes all the words very quickly and speaks all the words. However, it shows a low variety of words because of random adoption from plural fitting rules. A Minimal All Mighty agent cannot invade into the system composed of ECW because of lower variety of words. In the case of $r_{\rm rec} = 1.0, r_{\rm sp} = r_{\rm br} = 0.0$, an agent whose grammar

Table 1. **This table shows rewriting steps to recognize words (the left most column) spoken at time step 400. Simulation parameters are $r_{sp} = 3.0$, $r_{rec} = 1.0 and r_{br} = -2.0$. In the second column, the trend of each word (frequency of words in the last 10 time steps) is written. Numerals in the first row are ID of each agent at time step 400 in order that the earlier a agent is removed the lefter it is located. If the agent can't recognize the word, no numerals is put. Bold numerals represent the first two longest steps among agents.**

word	trend	**306**	302	307	**276**	305	301	299	294	290	284
001001	30			121	**185**			121	**145**	133	121
011001	20			157	**423**			166	**214**	190	157
11100	16			52	**245**			46	**58**	52	52
11	14		**7**		**12**	**7**	**7**				
110	12		13		**32**	**14**	**14**				
00110	26		53	**57**		**62**	48	43	54	49	53
110010	11		**174**	147				121	**202**	160	147
00	69	**3**	**3**	**3**		**3**	**3**	**3**	**3**	**3**	**3**
10	57	**7**		5	**7**	5	5		5	5	5
011010	24	**431**	210	164	**431**	120	180	166	251	209	164
001010	31	**233**	134	124	**236**	106	106	116	161	141	124
1110	24	**125**	45	27	**122**	60	60	24	29	27	27
01110	9	**122**	91	69	**192**	55	76	66	83	75	69
00101	25	**91**	65	58	**91**	52	49	57	66	63	58
111	30	**53**	21	13	**52**	24	24	13	13	13	13
0001	78	**20**	19		**20**	21	16	18	18	18	17
001011	14	**401**			**404**						
010101	10	**190**			**193**						

contains the rules same as rules in (7) has evolved. For short point such as low variety of speaking words has no effect on its fitness in this setting.

3.4 Punctuated Equilibrium

We have seen that our system shows rapid algorithmic evolution of the grammar systems in certain stages. On the other hand, algorithmic evolution is suppressed by forming ECW. Rapid algorithmic evolution follows quasi-equilibrium stages. Temporal evolution of amounts of handling information therefore shows punctuated equilibrium phenomena (Fig 4). Because handling information defined below is sensitive to the formation of ECW.

The handling information of the l-th agent is defined as the follows,

$$f_l = \frac{1}{RP^2} \sum_{c=1}^{R} \sum_{k=1}^{P} |w_{kl}^{\mathrm{rec}}(c-1)| \sum_{m=1}^{P} |w_{lm}^{\mathrm{rec}}(c)| \qquad (8)$$

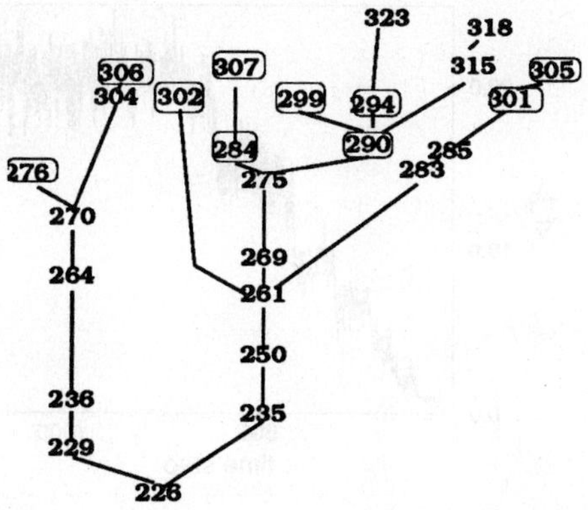

Fig. 3. This picture represents the phylogeny of agents at time step 400 (in oval boxes) from common ancestor (G_{226}). A number represents ID of each agent. A line is drawn from parent agent (lower) to its offsprings (upper). Two genetic series are bifurcated from the common root (G_{226}), the agents G_{306} and G_{276} and that of the other agents. They are forming different ECW. The agent G_{306} and G_{276} are both contained in the left series.

$$|w_{ij}^{\mathrm{rec}}(x)| = \begin{cases} 1 & \text{if } x = 0 \\ \text{the length of } w_{lm}(x) & \text{if } x > 0 \text{ and the word which spoken by the} \\ & i\text{-th agent is recognized by the } j\text{-th agent} \\ 0 & \text{otherwise .} \end{cases} \tag{9}$$

Information contents of a word is simply given by the length of the word as the first approximation. The initial amount of handling information, i.e. $|w_{ij}(0)|$, is defined as 1.

If an agent gets high f_l value, which suggests that the agent can recognize words spoken by the others and the agent's speaking words are recognized by the other agents. When some ECWs conflict with the other ECWs, the averaged handling information in a population, $\langle f_l \rangle = \sum_{l=0}^{P} f_l / P$, does not increase. After some ECWs occupy the whole network, long and new words will again be spoken to and recognized by the agents. Punctuated equilibrium phenomena in the amount of $\langle f_l \rangle$ is explained by the scenario.

4 Summary and Discussion

We have studied the evolution of symbolic grammar, by introducing a network model of communicating agents. Each agent has its own grammar system, being expressed as a set of rewriting rules. Via a combination of rules, each agent speaks words to the other agents and tries to recognize words spoken by the other agents. When mutationally dynamics of grammars are introduced, agents' grammar systems change in time. Generally, an agent can speak and recognize

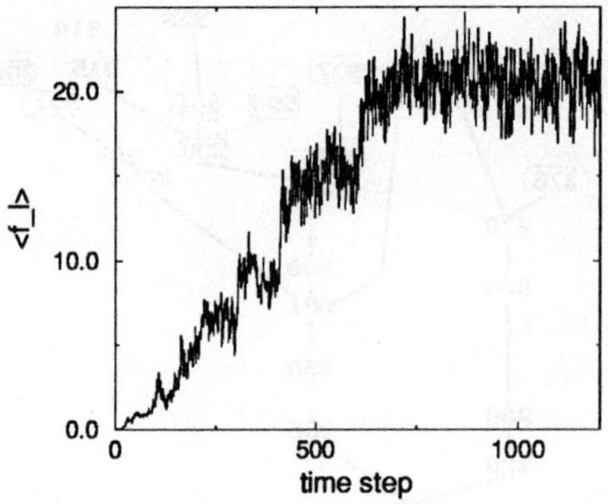

Fig. 4. time step v.s. the average handling information (see the definition in the text): In the first 700 time steps, evolution can be observed. The stepwise changes reflect alternate evolution of ECW and the algorithmic evolution.

more words if it has more rules. Hence agents with more rules can breed more. However the number of recognizable words is not a simple function of the number of rules.

In the present paper, we have shown that two processes are significant in evolution of grammar systems. One is a module type evolution. If a rule becomes a module, which means that it can be utilized by many words to generate nearly twice as many words as before. The number of recognizable words rapidly increases when this module emerges in a grammar system. The other one is a loop forming evolution. The grammar possessing a loop structure can derive recursively many words. It should be noted that a grammar with a loop structure cannot be represented in a tree shape. Namely, the grammar system climbs up Chomsky's hierarchy by one step from type 3 to type 2. Hence we regard such loop forming evolution as a single algorithmic evolution.

It is often believed that a grammar system in the higher hierarchy will perform better than grammars in the lower hierarchy. It may be argued that agents sharing the highest grammar system will dominate in population sooner or later. This argument would be valid if we evolved a grammar system without a network structure.

But this is not necessarily so for agents forming communication network. Synergetic behavior of agents generates a macro structure named ECW (Ensemble with Common Words). Within this ensemble, several words are spoken and recognized in common. In order to speak and recognize the common words more quickly, it is wiser to have the words as single rules (i.e. $S \rightarrow$ words). This becomes a restrictive condition for individual grammar systems. That is, any agent to survive in the ensemble has to evolve its rule set within this restriction.

This restrictive condition imposed by the ensemble structure disturbs the smooth evolution to the highest ability grammar systems. Therefore all mighty agent, i.e. it can speak and recognize all possible words, is difficult to emerge. The restriction and the algorithmic evolution occur in order, the average handling information shows a punctuated equilibrium evolution.

It is interesting to note that this restriction is not given as an individual rule, but spontaneously emerges from the evolution course of a network. The emergent restriction on each grammar system can be called a net grammar. If we define grammar as meta rule sets that give restrictions on a possible language set, this ensemble structure can be the example of such grammar. We have succeeded in showing the meta grammar structure and dynamics conducive to understand an actual language system.

Acknowledgments

The authors would like to thank N.Nishimura for critical reading of the manuscript. They thanks Eken S.Yoshikawa for stimulating discussions. One of the authors (T.H.) wish to express his gratitude to T.Yamamoto and N.Matsuo for helpful comments.

References

1. Hopcroft, John E. and Ullman, Jeffrey D.: *Introduction to Automata Theory, Languages and Computation* Addison-Wesley Publishing Co,, 1979.
2. Révész, György E.: *Introduction to Formal Languages* Dover Publications, Inc., 1991.
3. MacLennan, Bruce: Synthetic Ethology: An Approach to the Study of Communication in *Artificial Life II* Addison-Wesley Publishing Co,, 1992.
4. Werner, Gregory M. and Dyer, Michael G.: Evolution of Communication in Artificial Organisms in *Artificial Life II* Addison-Wesley Publishing Co,, 1992.
5. Lindgren, Kristian: Evolutionary Phenomena in Simple Dynamics in *Artificial Life II* Addison-Wesley Publishing Co,, 1992.
6. Ikegami, Takashi: From genetic evolution to emergence of game strategies Physica D **75** (1994) 310 – 327
7. Ray, Thomas S.: An Approach to the Synthesis of Life in *Artificial Life II* Addison-Wesley Publishing Co,, 1992.
8. Ray, Thomas S.: Evolution, complexity, entropy and and artificial reality Physica D **75** (1994) 239–263

Driven Cellular Automata, Adaptation and the Binding Problem

William Sulis
Department of Psychology, McMaster University
Hamilton, Ontario, Canada L8S 4K1

Abstract

The binding problem in neurobiology and the synchronization problem in distributed systems address a fundamental question, namely how can a collection of computational units interact so as to produce a stable response to environmental stimuli. It is proposed that the synchronization of responses of a complex system to a transient stimulus, so- called transient synchronization, provides a solution to this problem. Evidence is presented from studies involving inhomogeneous, asynchronous, adaptive cellular automata in support of this contention. In the transient synchronization approach, information is encoded as a distribution of response patterns in a global pattern space. The activation of such information requires a dynamic interaction between the system and its environment. Information is implicit rather than explicit. This permits a several orders of magnitude improvement in storage capacity over current models.

1 Introduction

The binding problem in neurobiology refers to understanding the mechanisms by which the activities of a disparate collection of feature detectors become bound together to form a unitary percept. Closely related to it is the synchronization problem in distributed parallel processing in which the activities of various processors must be integrated into a coherent global computation. Likewise in the study of swarms one seeks to understand the mechanisms by which the behaviours of the individual organisms become bound together so as to produce a coherent pattern of behaviour (such as the formation of a nest or a raiding party) which is recognizable at the collective level. A solution to the binding problem would thus have implications for many problems in the study of collective behaviour and collective intelligence.

Two different approaches to the binding problem have been proposed. The cardinal cell approach assumes that the output from disparate detectors converges upon cardinal cells which become active only when a specific configuration of features is present in a stimulus (for example the face of one's grandmother). Such cells appear to be rare based on neurophysiological studies. Combinatorial considerations show that there are not enough cells to encompass all of the patterns which can be learned.

The cell assembly approach replaces the cardinal cell with an assembly of cells. The presentation of a stimulus results in the synchronized firing of the cells in the assembly with zero phase delay. In recent years some evidence has been found in support of this approach both experimentally [1,8] and theoretically [4,21,5]. However, the need to avoid global synchronization requires that the network be segregated into weakly interacting clusters, significantly reducing the storage capacity [2]. In addition it is not clear that zero phase delay synchronization can propagate throughout the cortex, preserving the code.

Precise synchronization in the firing of spatially distributed neurons places severe constraints upon the nature of neuronal communication. There is abundant evidence now to support the contention that synaptic transmission is unreliable and neuronal firing patterns approximate a Poisson process [7,9,11]. Any solution to the binding problem must take these properties into account. One view is that the irregularity in firing patterns is due to deterministic chaos and reflects computation by the involved neurons. Theoretical work based upon the available evidence [9,10] suggests that neurons cannot operate as integrate and fire devices and must act as coincidence detectors. As a result, neurons would be capable of supporting a finely detailed temporal coding [11]. The other view is that neurons are fundamentally stochastic. As a result, neurons are capable only of supporting a rate code [7]. Neurons integrate and fire via a random walk mechanism.

These two views are presented as being mutually exclusive. However essentially the same experimental data is used to support both contentions so that the conclusions drawn depend crucially upon the specific features of the computational models used to interpret the data. Both views are fundamentally microscopic in orientation, presuming that the activity of individual neurons actually conveys meaningful information.

Virtually all attempts to solve the binding problem involve effectively static solutions. The cardinal cell and cell assembly models both associate a percept with either a fixed cell or a collection of cells. Precise zero phase synchronization among the cells of an assembly eliminates any dynamic structure within the assembly resulting in an essentially static view of information.

The approach presented below is fundamentally dynamic and macroscopic in orientation. It starts with the assumption that biological collectives are, first and foremost, complex adaptive dynamical systems. From this is derived a general language for the study of computational processes in dynamical systems and this is then applied to the binding problem. The binding problem is first formulated at the macroscopic level, leading naturally to a search for microscopic processes which could support it. The solution to the binding problem proposed here resolves the dichotomy entailed by the rate coding/temporal coding hypotheses. It is proposed that percepts or other behavioral responses may be stored as a distribution of spatiotemporal patterns in a general pattern space. Thus one has both temporal coding (in the individual patterns) and rate coding (in the pattern distribution). Stable representations can occur in the absence of synchronization between individual agents within the collective. Instead, binding occurs at the more general pattern level. A general cellular automaton model is used to illustrate this and it is shown that this solution is feasible within the context of inhomogeneity, asynchrony and adaptability expected with any biological collective.

An alternative dynamical system approach has recently appeared in the literature [20]. However it still represents memories according to the restricted framework of fixed point attractors and so remains fundamentally static in focus.

2 Computational Competence and Transient Languages

The approach taken here uses the transient language formulation of a dynamical system and the notion of computational competence [13,14,15,16]. Two related but distinct approaches have also appeared in the literature [3,22]. The standard formulation of a dynamical system does not readily yield information about its computational capability under natural ecological constraints. The transient language/dynamical automaton formulation of a dynamical system was introduced to rectify this problem. The trajectory of a dynamical system is represented as a sequence of finite duration transients with the dynamic of the system providing a natural automata action on this language of transients thus making the computational activity of the system readily accessible. More formally, a transient language (S, T, d) consists of a semigroup S, a positive abelian semigroup T and a mapping (duration) from $S \xrightarrow{d} T$ such that $d(ss') = d(s) + d(s')$. As a prototypical example, consider the set of all maps of the form $[0, a) \xrightarrow{f} \mathcal{R}$ where $a \in \mathcal{R}$, the real line. Given $[0, b) \xrightarrow{g} \mathcal{R}$, define fg as $[0, a + b) \xrightarrow{fg} \mathcal{R}$ where $fg(x) = f(x)$ if $x \in [0, a)$ and $fg(x) = g(x)$ if $x \in [a, a+b)$. Each such map represents a transient of a dynamical system, hence the name. A dynamical automaton $(\mathcal{S}, \mathcal{E}, T, \Delta)$ consists of two transient languages $\mathcal{S} = (S, T, d)$ and $\mathcal{E} = (E, T, d')$ and a dynamic Δ such that given $\psi \in S$ and $\eta \in E$, we have $\Delta(\psi, \eta) = \psi' \psi_\eta$ where $d(\psi) = d(\psi')$ and $d(\psi_\eta) = d'(\eta)$ and $\Delta(\psi, \eta\eta') = \Delta(\Delta(\psi, \eta), \eta')$.

Also define $P(\psi, \eta) = \psi_\eta$.

It should be obvious that every dynamical system gives rise to a corresponding dynamical automaton. It is assumed that a suitable metric ρ, such as the Hamming distance, is imposed upon the transient languages.

It should also be obvious that any experiment on a naturally occurring collective, be it swarm or brain is, of necessity, bounded in time and that the system under consideration actively interacts with its environment and so must be nonautonomous. The behavioral responses of a system and environmental stimuli both constitute dynamical transients and so the dynamical automaton formulation is the natural one in which to address computational issues in naturally occurring computational systems [17]. Moreover, any real experiment will of necessity involve variation in the stimuli and in the behavioral responses of the system under scrutiny.

In a typical experiment there will be an open set of stimuli Q reflecting the actions of the experimenter and the effects of the environment, an open set R of responses deemed representative of the appropriate response or correct computation, and some set Σ of possible initial histories. The system is successful or competent to carry out the required behaviour or computation if $P(\Sigma, Q) \subset R$. In the case of a perceptual stimulus, the percept becomes associated with some subset R of responses and so corresponds to a distribution of spatiotemporal patterns. The pairing (Q, R) will solve the binding problem if a) the pairing is unique, b) the pairing is robust against internal noise, and c) the pairing is robust over time, that is, repeat presentations of the stimulus produce the same association.

Transient synchronization provides a mechanism through which such binding can occur. Transient synchronization occurs when a transient stimulus applied to a dynamical system produces a set of responses which cluster closely in pattern space. In order to avoid an erroneous attribution of clustering to transient synchronization when actually due to mere statistical coincidence, we require that a stimulus η be said to synchronize the responses of a dynamical system if, given any two initial histories ψ, ψ', it follows that $\rho(\psi_\eta, \psi'_\eta) < 1/2\rho(|rand(\psi_\eta)|, |rand(\psi'_\eta)|)$ where $rand(\psi_\eta)$ and $rand(\psi'_\eta)$ are randomly generated patterns of same norm as ψ_η, ψ'_η.

Transient synchronization will clearly be robust against at least small degrees of noise. Evidence is presented below to demonstrate that transient synchronization exists in a general complex systems model and that it can provide a unique pairing of stimulus and response which is stable over time.

3 Previous Results

Transient synchronization was first detected in a study of tempered neural networks [12]. It was subsequently detected in coupled map lattices with input [15] and in driven cellular automata [16]. The latter study involved homogeneous, synchronous, 2 state, 3 neighbor cellular automata with input applied in discrimination mode (see below). It was demonstrated that the ability to induce transient synchronization was a function of input intensity and the symmetry class of the automaton to which it was applied. The symmetry class reflected the dominant symmetry present in the autonomous patterns produced by the automaton. The classes are uniform, linear, complex and chaotic [16] and were shown to be distinct from Wolfram's now classical classification scheme.

This study proved unsatisfactory for three major reasons. First, the input mode was unduly restrictive and the automata seemed unable to generalize their responses. Second, homogeneity and updating synchrony are implausible characteristics for any biological organism. Moreover there was the strong possibility that the transient synchronization observed was an artifact of the underlying temporal synchronization provided by the updating scheme. Finally, the automata were non-adaptive. This too is biologically implausible. The present study was thus undertaken to address these issues.

4 The Model

The cellular automaton chosen for the present study is the cocktail party automaton. This is an adaptive cellular automaton which can be controlled for the degree of adaptive response, inhomogeneity and asynchrony. Each cell is provided with both a state and a rule. Updating can be done either synchronously, or via a fixed asynchronous scheme, or via a stochastic asynchronous scheme. The state of the cell is first updated, then any input is applied to the cell according to the particular input mode. The rule of the cell is then updated. The rule can remain fixed or be updated according to the following adaptive scheme. Whenever the state of a cell is updated, a comparison is made between the response of the cell and that of all other cells possessing the same local neighborhood state configuration. The difference in the number of cells disagreeing and agreeing is calculated and the cell modifies its transition table entry to the opposite value if this difference exceeds a predetermined, individualized, fixed threshold. The cycle is then repeated.

Many of the results below hold if the rule is updated prior to the application of the input but this situation will not be discussed in this paper.

5 Input Modes

Each input to the host automaton was a complex spatiotemporal pattern derived from the output of a second, input automaton having identical lattice structure as the host. A fixed correspondence was established between the input and host cells. This provided a mapping between the output pattern of the input automaton and the cells of the host. The input pattern thus consisted of an array of state values, indexed by cell and by time. At time n for the host automaton, the row of the pattern corresponding to time n was sampled cell by cell and applied, according to the input mode, to the corresponding cell of the host automaton. The input automaton was chosen at random using varying combinations: homogeneous/inhomogeneous, linear/complex/chaotic rules, synchronous/asynchronous (random 20%), fixed/adaptive.

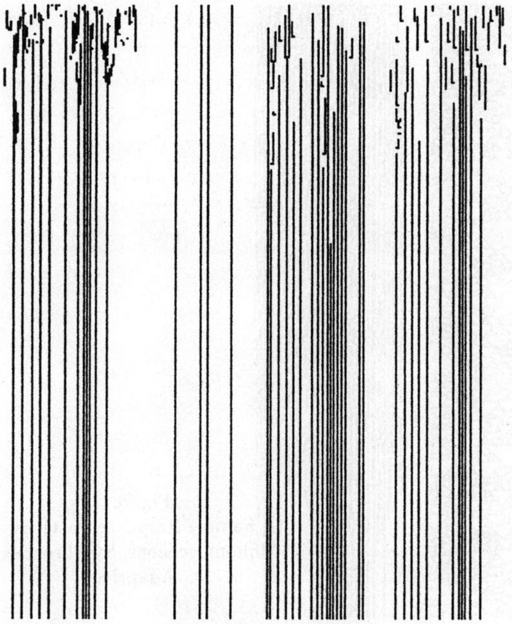

Layout of Figures
From Left To Right:
Input Pattern
Autonomous Run
Inferential Input Mode
Recognition Input Mode

Figure 1
Sample Response Patterns
Homogeneous, Synchronous,
Fixed

Figure 2
Sample Response Patterns
Inhomogeneous, Synchronous
Fixed

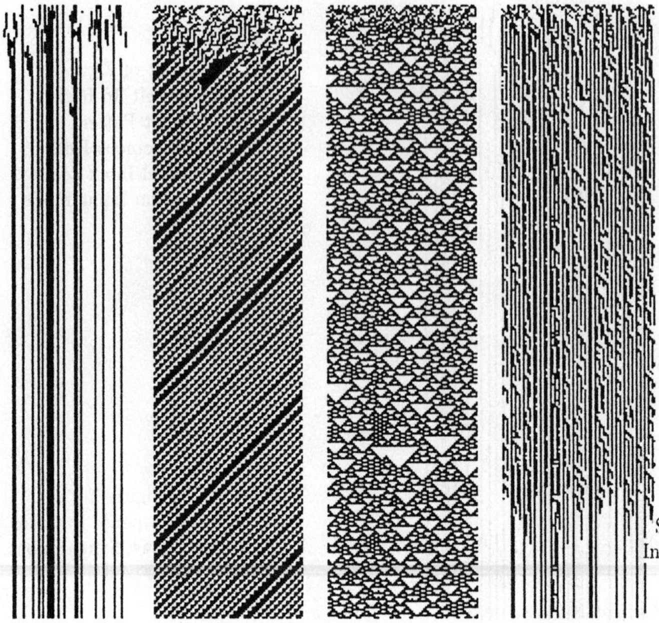

Figure 3
Sample Response Patterns
Inhomogeneous, Synchronous
Adaptive

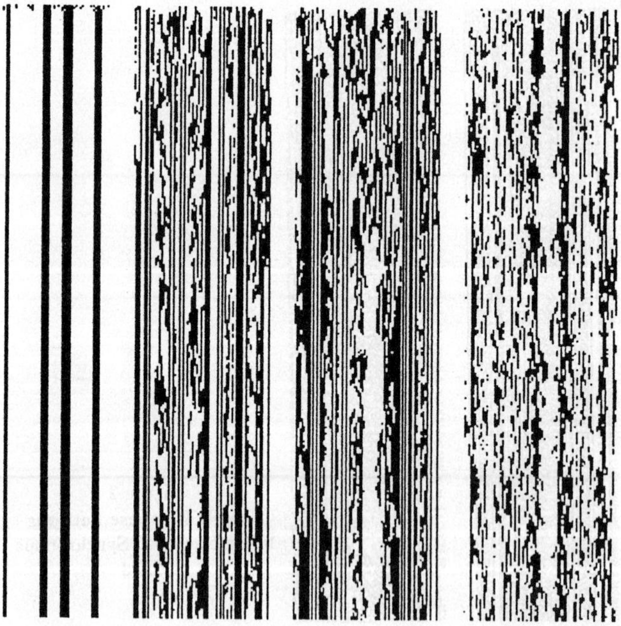

Figure 4
Sample Response Patterns
Inhomogeneous, Asynchronous
Fixed

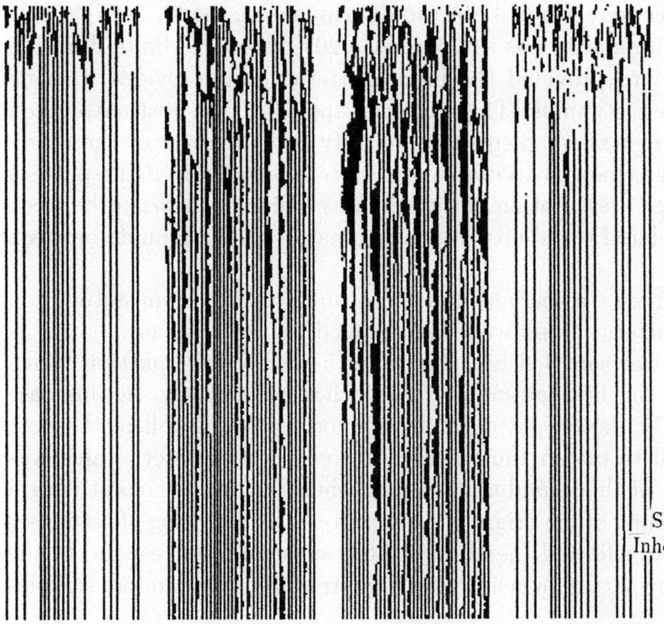

Figure 5
Sample Response Patterns
Inhomogeneous, Asynchronous
Adaptive

Figure 6
Sample Response Patterns
Inhomogeneous, Asynchronous
Adaptive

Three input modes were studied. In the discrimination mode, a single output from the input automaton was sampled at a 20% rate for a finite duration and this derived pattern provided the input pattern. This provided a single random, but correlated, stimulus. This was then applied to the host automaton as follows: the pattern was sampled in its entirety and the state of the corresponding host cell was changed to match the pattern value only if the pattern value was 1. Thus each presentation of the stimulus remained invariant between trials. This mode explored the ability of the automaton to discriminate between distinct samples.

In the inferential mode, a single complete output from the input automaton was stored as the pattern. This was then applied to the host automaton as follows: the pattern was sampled but this time the cells of the pattern which were to be input to the host automaton were chosen randomly, in this case at a 3% rate. Given a chosen pattern cell, the corresponding cell in the host had its state changed to match the pattern cell only if the pattern value was 1. Each presentation of the stimulus thus varied between trials, constituting a distinct random sampling of the original input pattern. A matching of responses under these conditions reflected the ability of the automaton to respond to the overall structure of the pattern which must be inferred from the random samples presented.

In the recognition mode, a single complete output from the input automaton was stored as the pattern. This was applied to the host as follows: the pattern was sampled as described above. The cells of the pattern which were to be applied to the host were again sampled randomly, this time at a 10% rate. This time, given a chosen pattern cell, the corresponding host cell had its state changed to match that of the pattern cell, regardless of value. Each presentation of the stimulus again varied between trials, constituting a random sampling of the original input. The automaton was studied for its ability to match its response to the pattern. Thus its capacity for pattern completion was studied. This provides one means of pattern recognition, hence the choice of the descriptor: recognition mode.

6 Results

For this study, a host automaton having 100 cells and toroidal boundary conditions was used. The automaton was simulated under input for 450 time steps. Only 2 state, 3 neighbor rules were considered.

Figures 1-6 provide representative samples of output from several different configurations of the cocktail party automaton in each of three input conditions: autonomous, inferential, recognition. Note the loss of pattern in the inferential mode and the eventual close approximation of patterns in recognition mode in Figures 1,3,5,6. The latter are clear examples of transient synchronization. Note that the presence or absence of transient synchronization or of pattern completion may depend upon the duration of the pattern stimulus, just as the observation of the presence or absence of a particular behaviour in a natural organism may be a function of observation time.

Below are displayed the results of four trials, each consisting of 100 presentations of a stimulus to a single inhomogeneous, asynchronous, adaptive cocktail party automaton using recognition mode. The stimulus varied between the trials. Displayed are the mean (variance) of the Hamming distances between responses and a fixed response template (S), between responses and input pattern (P), between each individual pair of responses (DP) and between each individual pair of randomly generated surrogate patterns having an identical distribution of pattern norms to that of the original sample (RP).

Ham	Synchronization			
Dist	Strong	Weak	Pseudo	Absent
S	4 (1.4)	31 (3.7)	18 (2.3)	48 (0.9)
P	3 (1.2)	28 (2.0)	21 (4.2)	44 (4.7)
DP	4 (2.1)	31 (5.5)	16 (3.9)	47 (5.1)
RP	39 (6.0)	49 (5.0)	17 (4.4)	49 (4.9)

One should immediately note that transient synchronization does exist. Moreover it is also clear that there exist stimuli which are not capable of inducing transient synchronization so that the concept is nontrivial. The case of pseudo synchronization illustrates need for fulfillment of the surrogate criteria.

Inspired by the work of Theiler [19], the determination of the distribution of interpair Hamming distances and that for a set of suitably constructed surrogates provides a possible means of detecting transient synchronization in vivo. Transient synchronization should show itself by a large peak in the low end of the Hamming distance scale. Multiple peaks might arise in the case of multiple synchronizing basins whereas the randomly generated surrogates would show only a single broad peak at the higher end of the scale.

To be truly useful as a solution to the binding problem, it is necessary to demonstrate that stability of the synchronization does indeed exist over time. Below are displayed the results of a simulation in which the automaton was presented with a stimulus followed by a random stimulus of equal duration followed by a second presentation of the stimulus, all via recognition mode. This was repeated for 100 different initial conditions. The table shows the mean (variance) of the Hamming distances between responses an initial response template and between the responses and the pattern. The same template was used for all three cases. The strong matching to the template shows that all of the responses to the fixed input stimulus lie within the same neighborhood about the template. Thus transient synchronization occurred on both presentations of the stimulus and the response set was the same in both cases. Thus the response

Input Pattern Dynamic	Automaton Dynamic Type								
	I/S/F			I/A/F			I/A/Ad		
	Mode			Mode			Mode		
	D	In	R	D	In	R	D	In	R
H/L/S/F	7(2.4)	36(4.3)	36(2.8)	24(2.1)	36(5.8)	34(5.4)	6(2.4)	5(1.9)	46(6.3)
H/L/S/Ad	1(0.4)	36(2.3)	38(3.9)	26(4.7)	37(3.0)	19(2.7)	29(4.5)	28(6.3)	47(2.0)
H/L/A/F	5(2.2)	37(6.1)	37(1.7)	34(6.0)	40(2.1)	24(3.5)	16(4.7)	15(3.9)	10(2.2)
H/L/A/Ad	2(1.0)	35(4.4)	40(2.3)	24(2.3)	35(2.8)	19(1.1)	26(6.5)	27(7.6)	11(3.3)
H/Co/S/F	4(2.5)	38(2.6)	30(5.6)	18(2.3)	38(6.0)	23(3.4)	29(5.6)	5(0.0)	2(1.2)
H/Co/S/Ad	5(1.3)	35(1.8)	39(5.1)	31(5.5)	37(1.3)	24(4.1)	22(4.5)	24(2.3)	21(4.6)
H/Co/A/F	3(1.0)	35(5.6)	39(3.2)	22(4.1)	33(5.1)	22(4.2)	11(3.4)	10(3.3)	2(0.8)
H/Co/A/Ad	4(1.8)	35(4.2)	38(6.1)	32(5.6)	35(4.0)	28(2.7)	6(2.5)	15(3.2)	11(2.4)
H/Ch/S/F	2(1.8)	40(5.3)	39(4.9)	22(2.9)	36(4.5)	30(3.6)	18(4.2)	14(0.8)	37(6.5)
H/Ch/S/Ad	3(1.9)	39(5.7)	38(3.4)	30(4.2)	40(4.3)	29(1.9)	11(2.1)	9(1.0)	18(2.2)
H/Ch/A/F	2(1.0)	38(0.0)	39(4.6)	30(4.3)	40(2.7)	30(4.6)	17(4.5)	16(2.9)	2(1.4)
H/Ch/A/Ad	1(1.0)	31(4.7)	37(2.0)	23(0.7)	39(5.6)	29(0.4)	6(2.2)	5(0.0)	6(2.3)

Table 1
Response-Response Template Hamming Distances

Input Pattern Dynamic	Automaton Dynamic Type								
	I/S/F			I/A/F			I/A/Ad		
	Mode			Mode			Mode		
	D	In	R	D	In	R	D	In	R
I/L/S/F	1(1.5)	37(3.6)	39(3.4)	18(4.1)	35(1.9)	19(4.2)	11(3.0)	8(2.7)	47(1.7)
I/L/S/Ad	5(2.9)	41(0.0)	34(4.2)	31(2.8)	36(2.5)	20(1.1)	13(2.4)	12(1.5)	47(3.5)
I/L/A/F	7(3.8)	36(5.8)	32(3.3)	25(2.7)	39(4.0)	20(1.5)	16(4.9)	18(4.0)	10(2.2)
I/L/A/Ad	2(1.2)	37(6.1)	35(2.6)	28(2.9)	42(2.3)	19(4.1)	19(4.6)	20(2.9)	10(2.9)
I/Co/S/F	2(1.2)	29(5.5)	38(5.9)	26(3.1)	37(5.2)	18(2.8)	9(3.9)	6(2.6)	3(0.7)
I/Co/S/Ad	2(1.0)	37(3.9)	37(1.9)	27(1.9)	35(3.7)	20(1.7)	13(3.8)	9(2.9)	3(1.3)
I/Co/A/F	1(1.2)	38(3.8)	36(5.5)	25(3.1)	37(5.4)	24(3.6)	15(2.2)	13(3.5)	3(1.6)
I/Co/A/Ad	4(1.5)	40(3.3)	38(3.3)	29(5.2)	39(4.4)	19(3.5)	37(5.6)	39(3.6)	3(1.4)
I/Ch/S/F	2(1.1)	34(0.0)	39(5.1)	26(4.1)	41(3.3)	30(2.5)	8(3.5)	7(2.6)	7(2.1)
I/Ch/S/Ad	2(1.3)	36(0.0)	38(6.1)	26(1.5)	41(6.0)	23(3.6)	16(4.4)	18(1.0)	22(3.6)
I/Ch/A/F	1(0.7)	38(2.2)	38(4.8)	29(1.5)	36(5.1)	29(5.1)	11(1.9)	10(3.0)	3(1.1)
I/Ch/A/Ad	3(0.8)	35(5.5)	41(3.8)	22(2.8)	38(4.1)	25(4.9)	13(3.8)	14(1.5)	4(2.7)

Table 2
Response-Response Template Hamming Distances

Legend for Tables
D-Discrimination, In-Inferential, R-Recognition
H-Homogeneous, I-Inhomogeneous, L-Linear, Co-Complex
Ch-Chaotic, S-Synchronous, A-Asynchronous
F-Fixed, Ad-Adaptive

Input Pattern Dynamic	Automaton Dynamic Type								
	I/S/F			I/A/F			I/A/Ad		
	Mode			Mode			Mode		
	D	In	R	D	In	R	D	In	R
H/L/S/F	50(6.7)	18(2.4)	30(5.4)	65(1.7)	18(2.8)	28(4.9)	90(8.9)	40(3.2)	43(6.1)
H/L/S/Ad	57(0.0)	17(3.1)	32(4.0)	52(3.2)	15(3.8)	12(2.5)	71(9.6)	43(9.2)	43(5.6)
H/L/A/F	52(4.2)	42(0.7)	30(5.3)	58(3.5)	41(5.1)	17(3.8)	85(8.5)	73(5.6)	8(1.0)
H/L/A/Ad	55(6.2)	25(4.2)	34(5.3)	54(4.6)	24(4.5)	12(2.1)	75(8.9)	47(8.8)	9(3.0)
H/Co/S/F	53(7.2)	31(5.1)	25(4.4)	65(2.2)	31(1.2)	17(1.8)	72(7.7)	13(3.5)	2(1.0)
H/Co/S/Ad	52(5.6)	27(3.9)	34(2.6)	54(1.8)	28(3.1)	17(4.2)	79(7.7)	81(5.4)	29(1.8)
H/Co/A/F	55(6.7)	18(2.7)	33(3.3)	62(5.3)	26(4.2)	15(2.8)	8.6(9.8)	36(3.4)	1(1.2)
H/Co/A/Ad	53(3.7)	27(1.2)	32(5.8)	52(6.6)	18(3.8)	21(4.4)	89(9.4)	85(2.5)	0(2.0)
H/Ch/S/F	55(3.4)	29(5.3)	34(3.7)	57(3.2)	27(5.2)	23(4.4)	80(6.6)	46(7.7)	38(5.9)
H/Ch/S/Ad	54(0.0)	26(0.0)	33(1.4)	61(4.7)	29(5.1)	22(4.1)	88(4.4)	64(6.9)	16(4.2)
H/Ch/A/F	56(0.0)	18(0.0)	34(2.6)	61(5.4)	26(4.6)	24(2.3)	81(9.5)	68(7.4)	1(1.4)
H/Ch/A/Ad	56(7.3)	38(4.3)	31(5.5)	65(4.3)	18(3.8)	22(3.5)	90(5.9)	35(5.0)	5(1.5)

Table 3
Response-Pattern Hamming Distances

Input Pattern Dynamic	Automaton Dynamic Type								
	I/S/F			I/A/F			I/A/Ad		
	Mode			Mode			Mode		
	D	In	R	D	In	R	D	In	R
I/L/S/F	55(3.6)	7(2.2)	33(4.4)	60(2.8)	35(5.2)	12(3.4)	86(8.5)	28(4.5)	44(3.5)
I/L/S/Ad	51(4.2)	29(5.4)	28(0.6)	55(5.3)	7(2.3)	13(1.1)	86(8.5)	77(8.2)	44(5.1)
I/L/A/F	48(6.9)	45(0.0)	25(2.4)	56(4.5)	29(0.9)	13(0.6)	81(5.8)	58(5.8)	8(2.7)
I/L/A/Ad	57(3.7)	14(1.8)	28(4.9)	55(5.4)	43(6.1)	13(1.9)	76(8.6)	63(3.4)	8(2.9)
I/Co/S/F	55(1.8)	34(0.0)	32(5.5)	63(1.5)	14(1.9)	11(3.1)	88(8.7)	48(0.7)	2(1.0)
I/Co/S/Ad	55(6.8)	31(0.0)	29(5.5)	60(5.5)	35(5.5)	13(2.1)	85(9.0)	50(4.1)	2(1.0)
I/Co/A/F	56(0.0)	18(2.8)	30(2.4)	59(7.3)	27(5.1)	17(2.1)	83(9.5)	48(5.5)	2(1.4)
I/Co/A/Ad	54(4.3)	31(5.1)	32(3.4)	51(5.5)	17(1.6)	12(3.3)	76(10.6)	68(6.7)	2(1.5)
I/Ch/S/F	56(0.8)	34(0.0)	34(2.9)	64(2.7)	32(4.7)	23(4.8)	89(4.2)	49(7.1)	6(2.0)
I/Ch/S/Ad	57(3.9)	24(1.0)	33(4.8)	52(3.7)	35(5.3)	16(4.0)	82(10.3)	56(4.6)	25(3.6)
I/Ch/A/F	56(0.0)	19(0.4)	33(2.6)	62(3.4)	23(4.7)	23(3.0)	88(9.5)	58(7.6)	2(1.3)
I/Ch/A/Ad	55(6.6)	30(5.2)	37(2.9)	60(5.7)	19(3.8)	20(2.0)	84(8.6)	40(0.9)	3(2.7)

Table 4
Response-Pattern Hamming Distances

set remains stable over time and bound to the stimulus and therefore can serve as a stable representation of the input stimulus.

Hamming Distance	Initial Stimulus	Random Stimulus	Repeat Stimulus
States	8 (2.5)	48 (1.2)	7 (3.2)
Pattern	6 (2.5)	48 (1.1)	5 (3.1)

Tables 1-4 present the results of a series of simulations. A host automaton was presented under 100 different initial conditions with an input stimulus. Each cell in the table shows either the mean (variance) Hamming distance between each response and an initial response template (Tables 1,2) or between each response and the input pattern (Tables 3,4). Simulations were carried out with three different host automaton configurations, 24 different input automaton configurations and three different input modes.

7 Discussion

7.1 Homogeneity

Given synchronous, nonadaptive updating, both the homogeneous and inhomogeneous automata demonstrate similar features under the presentation of a stimulus. Thus only results for the inhomogeneous case are presented in the tables. These automata demonstrate strong transient synchronization between individual response patterns when the stimulus is presented in discrimination mode. The variation between responses was less than 10% and frequently at little as 2-3% . There is no synchronization between the response and the pattern however. Thus in this mode, the automaton is able to represent and discriminate among a variety of random patterns. The representation bears little resemblance to the pattern yet remains dynamically bound to it. These automata demonstrated no significant synchronization in inferential or recognition mode.

It is interesting to examine the response patterns generated by these automata (Figures 1,2). In the absence of input, both types of automata generate patterns which are highly structured. The inhomogeneous patterns tend to consist of periodic domains, often with complex patterns being repeated within each domain, and with the individual domains separated by boundaries consisting of fixed cells. A previous study has shown that the number of such fixed cells varies as a function of the symmetry structure of the rules [18]. The introduction of an input tends to randomize the output, destroying the correlations which have long fascinated those interested in emergent computation. They too are an artifact of the simulation method and are not robust against noise.

7.2 Asynchrony

A major question was whether the transient synchronization observed in the synchronous case would also be present under asynchronous conditions. There are two general methods of introducing asynchrony into the updating scheme. In the first case, a fixed, nonsequential ordering of the cells is used to determine the sequence of updating. They can be updated one at a time or in small, synchronous, batches. In either case, several time steps are required in order to update each cell in the lattice. The second method involves randomly choosing cells, singly or in a batch, at each time step. In the first method the deterministic nature of the updating scheme still imposes a temporal relationship between the cells of the automaton. In the second method there is no temporal relationship between individual cells.

As might be surmised, the introduction of stochastic asynchrony into the host dynamic results in a loss of transient synchronization in discrimination and inferential modes. Nevertheless there are now, under recognition mode, several cases in which weak transient synchronization occurs. The response variation increases to 15-25 % , but in spite of this, reasonable matches between the responses and patterns can be seen (10-15% variation). Thus although the introduction of asynchrony inhibits transient synchronization, this can, in part, be compensated for through the nature of the input mode. The recognition mode provides more information, as it were, to the automaton, allowing it to produce a closer matching between its response and the pattern and thus permitting weakened transient synchronization.

The autonomous inhomogeneous asynchronous nonadaptive automata show a tendency towards domain formation as well but the behaviour within a given domain is more strongly aperiodic (Figure 3). There is again a tendency towards randomization of the response pattern, particularly in recognition mode.

7.3 Adaptation

The third concern to be addressed is that of adaptation. Most naturally occurring computational systems, such as a swarm or a nervous system, consist of a large number of individual agents, each of which is capable, to varying degrees, of modifying its pattern of response to its environment. This provides another level of variation with which a system must cope. If transient synchronization is to be a plausible solution to the binding or synchronization problem, then it will be essential that it be present in systems capable of making adaptive responses. Thus transient synchronization was sought in the fully adaptive, asynchronous, inhomogeneous automaton. This provided the most extreme situation available to this particular class of automata.

The results are quite encouraging. These automata were capable of transient synchronization in any of the three input modes. Moreover they produced quite different responses depending upon the nature of the input. Most intriguing is that they showed a very striking preference for inputs from automata which most closely resembled themselves, in this case inhomogeneous and complex, regardless of degree of synchrony and adaptation. This suggests that a network of such automata would show a high degree of transient synchronization in their individual responses, regardless of the specific details of each automaton. This might allow for the emergence of a higher level of pattern structure such as occurs in a swarm or a society.

The degree of pattern completion is remarkable. At times, a mere 2% sample of a pattern is sufficient to enable the automaton to match its response to the pattern with 92% accuracy. This occurs in the absence of a detailed representation of the pattern in the rules. Indeed the pattern space considered here is vast. Although one cannot give a precise estimate of the number of patterns which can induce transient synchronization, one should note that there are $(2^{100})^{450}$ total patterns (of 450 time step duration). There are some 30^{100} inhomogeneous, complex, synchronous, nonadaptive automata whose patterns the adaptive automata appear capable of recognizing. By way of contrast, a cell assembly version of the system could, at best, encode a number of patterns equal to the total number of subsets of cells which could constitute cell assemblies. Given the present example, this means a maximum of 2^{100} patterns. Adaptation and asynchrony appear to improve the dynamical response range of the automaton without increasing the amount of information which the automaton must carry within it in order to carry out its function. In effect, the information which the automaton needs is stored in its environment.

These asynchronous, adaptive automata are capable of producing highly varied patterns. For an example, see the input pattern of Figure 6.

7.4 Conclusion

Evidence has been presented to support the contention that transient synchronization provides a plausible dynamical mechanism capable of binding together the disparate activities of multiple semiautonomous, weakly coupled agents so as to produce a coherent, stable pattern of behaviour in response to an external stimulus. Transient synchronization has been shown to occur, nontrivially, in an inhomogeneous system of adaptive agents acting asynchronously. This addresses criticisms brought forward about prior research. The phenomenon of transient synchronization is compatible with current research and there is no a priori reason to believe that it will not be present in naturally occurring systems as well. However, it is expected that its detection will prove to be difficult since averaging and cross- correlation techniques would likely produce null results except in exceptional cases. Measurement of the distribution of interpair Hamming distances obtained from multiple trials, tested against suitably constructed random surrogates, is suggested as a possible method for its detection.

Unlike cell assembly theories which presume that information is somehow encoded within the geometrical structure of the system, the transient synchronization approach suggests that the information remains stored within the environment and that an appropriate response emerges out of the dynamical interaction between the system and its environment. Structural features determine in part the form of the response but the capacity to produce a response at all is implicit in the underlying dynamic. The system does not represent the environment as such in its structure. Instead it is at the dynamic level that a stable association becomes established between stimulus and response. As an aside, the storage of information within the environment may explain the catastrophic changes in cognitive function which are sometimes seen in dementing patients who are moved from home into the unfamiliar environment of the hospital or nursing home.

If one likes, one may think of the information as being encoded as a distribution of response patterns instead of as a collection of cells or as fixed connections between cells. The engram, if such exists, must occur at the level of the pattern space. Information in this model is thus latent, requiring a trigger for its elicitation. Such triggers may exist within the environment or may arise from interactions between subsystems within the larger system such as through internal pattern generators.

This approach to information storage permits, it is believed, a substantial increase in storage capacity, likely several orders of magnitude greater than the limits presently suggested for neural network or cell assembly models.

It is further conjectured that the role of the fast adaptations and modulations of synaptic function observed in vivo may provide the adaptive changes necessary to maintain transient synchronization in the face of nonstationarity, asynchrony and noise in the synaptic transmission process.

Work is underway to understand the basic mechanisms which underlie transient synchronization and to develop a general theory of collective behaviour based upon this phenomenon.

References

[1] Aertsen, Ad., Arndt, M. Curr. Opin. Neurobiol. 3 (1993) 586-594

[2] Cairns, D.E., Baddeley, R.J., Smith. L.S. Neural Comp. 5(2) (1993) 260-266

[3] Casdagli, M. In Nonlinear Modeling and Forecasting. Casdagli, M., Eubank, S. eds. (1992) 265-281 New York: Addison-Wesley

[4] Grannan, E.R., Kleinfeld, D., Sompolinsky, H. Neural Comp. 5(4) (1993) 550-569

[5] Koch, C., Schuster, H. Neural Comp. 4(2) (1992) 211-223

[6] Millonas, M. Santa Fe Institute Preprint. 93-06-039 (1993)

[7] Shadlen, M.N., Newsome, W.T. Curr. Opin. Neurobiol. 4 (1994) 569-579

[8] Singer, W. Ann. Rev. Physiol. 55 (1993) 349-374

[9] Softky, W.R., Koch, C. Neural Comp. 4(5) (1992) 643-646

[10] Softky, W., Koch, C. J. Neurosci. 13 (1993) 334-350

[11] Softky, W. Physics Today. 47(12) (1994) 11-13

[12] Sulis, W.: Tempered Neural Networks. In Proceedings of the International Joint Conference on Neural Networks '92. Vol. III (1992) 421-427 Baltimore: IEEE

[13] Sulis, W.: Emergent Computation in Tempered Neural Networks 1: Dynamical Automta. In Proceedings of the World Congress on Neural Networks '93. Vol. IV (1993) 448-451 New York: Lawrence Erlbaum

[14] Sulis, W.: Emergent Computation in Tempered Neural Networks 2: Computation Theory. In Proceedings of the World Congress on Neural Networks '93. Vol. IV (1993) 452-455 New York: Lawrence Erlbaum

[15] Sulis, W.: Computation in Complex Systems (unpublished manuscript)

[16] Sulis, W.: Driven Cellular Automata. In 1993 Lectures in Complex Systems, Santa Fe Institute. Stein, D., Nadel, L. eds. (1995) New York: Addison-Wesley (In Press)

[17] Sulis, W. World Futures. 39 (1994) 225-241

[18] Sulis, W. (Unpublished data)

[19] Theiler, J. et. al. Physica D. 58 (1992) 77-94

[20] Tsuda, I. Neural Networks. 5(2) (1992) 313-326

[21] Usher, M., Schuster, H.G., Neibur, E. Neural Comp. 5(4) (1993) 570-586

[22] Young, K., Crutchfield, J.P. Santa Fe Institute Preprint. 93-05-028 (1993)

7. Biocomputing

The Functional Composition of Living Machines as a Design Principle for Artificial Organisms

Christos Ouzounis[1], Alfonso Valencia[2], Javier Tamames[2], Peer Bork[3] and Chris Sander[4]
EMBL Heidelberg, Meyerhofstrasse 1, D-69012 Heidelberg, Germany. email: {author family name}@embl-heidelberg.de

[1] AI Center, SRI International, 333 Ravenswood Avenue, Menlo Park, CA 94025-3493, USA
[2] CNB, Universidad Autonoma de Madrid, Cantoblanco, Madrid 28049, Spain
[3] Max-Delbrück-Center for Molecular Medicine, Berlin-Buch, Germany
[4] European Bioinformatics Institute (EMBL-EBI), Hinxton, Cambridge CB10 1RQ, England

Abstract: How similar are the engineering principles of artificial and natural machines? One way to approach this question is to compare in detail the basic functional components of living cells and human-made machines. Here, we provide some basic material for such a comparison, based on the analysis of functions for a few thousand protein molecules, the most versatile functional components of living cells. The composition of the genomes of four best known model organisms is analyzed and three major classes of molecular functions are defined: energy-, information- and communication-related. It is interesting that at the expense of the other two categories, communication-related coding potential has increased in relative numbers during evolution, and the progression from prokaryotes to eukaryotes and from unicellular to multi-cellular organisms. Based on the currently available data, 42% of the four genomes codes for energy-related proteins, 37% for information-related proteins, and finally the rest 21% for communication-related proteins, on average. This subdivision, and future refinements thereof, can form a design principle for the construction of computational models of genomes and organisms and, ultimately, the design and fabrication of artificial organisms.

Introduction: Expanding the living machine metaphor

The machine metaphor for living systems came remarkably late in history, only long after vitalism was refuted. Less than two hundred years ago it was first realized that organisms are indeed composed of complex macro-molecular species, which are involved in intricate processes that manifest themselves as Life. Molecular components of living systems operate in coordinated manners that are thought to have emerged during evolution through chance and natural selection - but also other additional constraints (Kauffman, 1993). Thus, organisms, like machines, appear to function in a goal-oriented way. The philosopher Immanuel Kant was the first to propose a teleological behavior for organisms, and this metaphor was carried through scientific tradition into the realm of Artificial Life.

In recent years, we have witnessed the design and development of artificial computational systems which exhibit behaviours reminiscent of those in living systems. The machine metaphor can be said to be most successful in Artificial Life research in the following sense: as organisms can be conceived to operate like machines, in the same way, machines (or, their components) can be designed to operate as organisms. Despite strong and justifiable opposition to this thesis (Kampis, 1991) and the indisputable limitations of the metaphor, it is interesting that the construction of abstract models with some properties of living entities has been very successful. For example, neural networks are based on the principles on which neural systems are built (McCulloch & Pitts, 1943). Also, genetic algorithms behave approximately as genomes which are evolving by recombination, mutation and selection (Holland, 1975; Mitchell *et al.*, 1992). Finally, cellular automata have been proposed for simulating genome evolution and complexity (Ouzounis, 1988).

Yet, despite the considerable sophistication in these developments, it appears that computational modelling of living systems has been based on a few traditional principles and major ideas - such as self-reproducing cellular automata (Burks, 1970; Langton, 1984; Wolfram, 1984; Packard & Wolfram, 1985; Reggia *et al.*, 1993), emergent behavior (Sims, 1992; Kauffman, 1993), and back-propagating neural networks (Minsky & Papert, 1969); the essential components of these designs being cells (sites), pixels or graphs, and neurons, respectively. Living systems, on the other hand, use a vast repertoire of *qualitatively unique* molecular entities with an immense variety of structure, function and specificity. This is achieved with the use of protein molecules, the basic macro-molecular components of a living system which perform most of the actions for energy transformations, gene expression and evolution, and regulating communication between molecules and cells.

The dramatic progress in rapidly expanding fields such as cell and molecular biology, however, has not been adequately followed in Artificial Life, while modern biology has also contributed little directly to Artificial Life and interdisciplinary domains of science. In an effort to bridge the gap between Artificial Life and computational molecular biology, and more particularly protein sequence analysis, genome projects and molecular evolution, we are attempting here for the first time to quantify and compare the functional composition of the genomes of four model organisms from our current (albeit rather incomplete) knowledge, obtained through massive sequencing in genome projects (Bork *et al.*, 1994).

The main purpose of this study is to extend the concept of living machines further, through the analysis of their molecular components, gaining much insight towards the construction of more realistic computational models and - maybe in the distant future - even *actual* organisms (through design and engineering). The functional composition of living 'machines' is therefore investigated at the present rates of function prediction by homology, and some preliminary yet interesting conclusions can be drawn for the first time. In other words, we are drawing on the vast body of knowledge acquired through genome sequence analysis to further inspire and derive abstractions for ALife.

The problem can be stated clearly as follows: what is the genetic makeup of four of the best known model organisms in terms of their protein functions? How much genetic information is stored for basic survival (energy transformation and storage), reproduction (replication, repair, expression of genetic information) and communication with the cellular environment (intra- and extra-cellular communication, regulation and recognition/defence)? And, if an artificial genome was ever to be designed, built, or evolved, what might it be composed of, in accordance to living organisms?

Methods: Classification of protein function in four model genomes

Four sets from the genome data of model organisms were obtained: *Mycoplasma capricolum*, a small parasitic prokaryotic (bacterial) cell (Bork *et al.*, 1995); *Escherichia coli*, the model prokaryote for which most is known about its physiology, biochemistry and genetics than for any other organism (Riley, 1993); the yeast *Saccharomyces cerevisiae*, the model eukaryote whose genomic sequence is becoming known at a rapid rate (Ouzounis *et al.*, 1995); and finally sequences from the gene expression pattern from the brain of *Homo sapiens* (Adams *et al.*, 1993).

We have classified the proteins of known function in three broad classes: energy (meaning generation of energy in catabolic and anabolic pathways available to the cell), information (meaning storage, replication, evolution and transmission of genetic molecules) and finally communication (meaning intra- and extra-cellular molecular recognition, defence and translocation). The data represent some sizeable fraction of the total genomes, at least in the three micro-organisms studied (about 30% for *Mycoplasma*, more than 50% for *E. coli* and 5-10% for yeast) and a rather small fraction (<1%) of the expressed genes for human.

The analysis is limited by the fact that the data sets may not be strictly comparable, therefore special care has been taken to generate a consistent functional classification. From the two published reports for *E. coli* (Riley, 1993) and human (Adams *et al.*, 1993), the classifications have been converted to our simple (yet general) scheme. In addition, another class of abundantly expressed genes coding for structural proteins in human (Adams *et al.*, 1993), has been excluded for the purposes of this study, because there is no such comparable class in the other three model microorganisms. To confirm the arbitrary classification derived from these two reports, we have also classified all available *E. coli*, yeast and human proteins from Swiss-Prot (Bairoch & Boeckmann, 1991), a sequence database with excellent functional annotation. The automatic classification yielded results which were very similar to the manually generated ones. The automatic functional classification method will appear elsewhere (Tamames, Ouzounis, Sander & Valencia, in preparation).

The automatic system for function assignment allows a systematic treatment of the data and assesses the validity of the comparisons between different classification schemes. A brief description follows: (1) Creation of a training set of sequences and their functional classification. Expert human knowledge was used to classify all sequences from the three yeast chromosomes into the different classes. These lists of sequences were used as input training data; (2) Building of a dictionary with the individual keywords found in each of the functional classes. Keywords were extracted from the database annotation records (Bairoch & Boeckmann, 1991). Entries corresponding to the sequence itself or to the most similar sequences in the database were used; (3) Validation of the dictionaries. We have implemented three distinct strategies to check keywords used in the dictionaries, to improve specificity and sensitivity by: (a) cross-checking the class dictionaries to find unique identifiers, (b) cross-checking dictionaries against the database dictionary to improve consistency and (c) scoring each word in the dictionary by the number of occurrences in the training set; (4) Classification of sequences. Each new query sequence is classified by the number of keywords that its database entry shares with the different classes; and (5) Species classifications. A full set of sequences from a particular species, e.g. *E. coli*, were classified automatically. The distribution of the different classes was thus quantified.

The composition of the three published yeast chromosomes was merged; thus, chromosomes were treated as a unique entity (representing a random sample from the yeast genome). The three chromosomes included in the analysis were: chromosome III (Bork *et al.*, 1992b; Bork *et al.*, 1992a; Oliver *et al.*, 1992; Koonin *et al.*, 1994), chromosome XI (Dujon *et al.*, 1994; Dujon *et al.*, 1995) and chromosome VIII (Johnston *et al.*, 1994; Bork *et al.*, 1995). The detailed comparison of the functional classes in these three chromosomes and the recently sequenced chromosome II (Feldmann *et al.*, 1994) will appear elsewhere (Ouzounis *et al.*, 1995).

The functional assignment of these proteins has taken place either by direct experimental verification of the activity and action of protein products or, more frequently, by detection of homology to other proteins of known function. This methodology comprises the following steps: (i) availability of a large number of continuously updated sequence databases, (ii) sequence comparison searches with a protein of unknown function (e.g. newly sequenced chromosome proteins), (iii) sorting of significantly similar proteins from databases (usually but not always of known function) and (iv) functional deductions for the query protein on the basis of its similarity to the extracted (characterized) proteins (Bork *et al.*, 1994). Intelligent computational systems for the functional classification of protein sequences have been developed and contributed towards this goal (Scharf *et al.*, 1994). Currently, the rates of such prediction by homology have surpassed the 50% level (Bork *et al.*, 1994), that is, for any given data set, there is a probability that more than half of the proteins can be matched to another protein of known function. Some estimates of a finite protein universe, with a limited number of catalytic functions, are at present between 1,000 and 10,000 protein types (Orengo *et al.*, 1994). Therefore, we can conjecture that most organisms use about 5,000 proteins to perform most of their vital functions. In higher organisms some of these proteins have been duplicated and diverged significantly, therefore providing a much larger and more intricate spectrum of specialized functions, without, however, losing the signatures of common ancestry from a limited pool of primordial molecules.

Results: Functional composition of genomes in Real Life

Assuming that the four sets of protein functions are representative of the total genomic content, an assumption which is becoming ever stronger with some of the genome projects approaching completion, it is instructive to note the similarities and the differences between the four organisms, with respect to how much genomic information is devoted to energy transformations (or, more abstractly speaking, survival), replication and expression of genetic information (and more generally, evolution), and finally communication (which includes intra- and inter-cellular communication, regulation, defence and recognition) (Figure 1).

Interestingly, the two prokaryotic cells (*Mycoplasma* and *E. coli*) seem to devote as much as half of their total stored genetic information to proteins involved in energy transformations (Figure 1). It is remarkable that *E. coli* appears to do this to an extreme, while *Mycoplasma* (a parasitic organism feeding largely on nutrients provided by the host) follows the same strategy to a lesser degree. Even yeast appears to devote around 40% of its total genetic storage for the coding of proteins involved in energy transformations, while human does this to a remarkably reduced rate (Figure 1). However, it should be pointed out

Table 1

SPECIES CLASS (#%)	M. capricolum	E. coli	S. cerevisiae	H. sapiens
Energy	105 / 44.7%	915 / 59.7%	146 / 37.7%	167 / 26.2%
Information	110 / 46.8%	422 / 27.5%	151 / 39.0%	218 / 34.2%
Communication	20 / 8.5%	196 / 12.8%	90 / 23.3%	252 / 39.6%
Total*	235 / 100.0%	1533 / 100.0%	387 / 100.0%	637 / 100.0%

*percentage of the available data

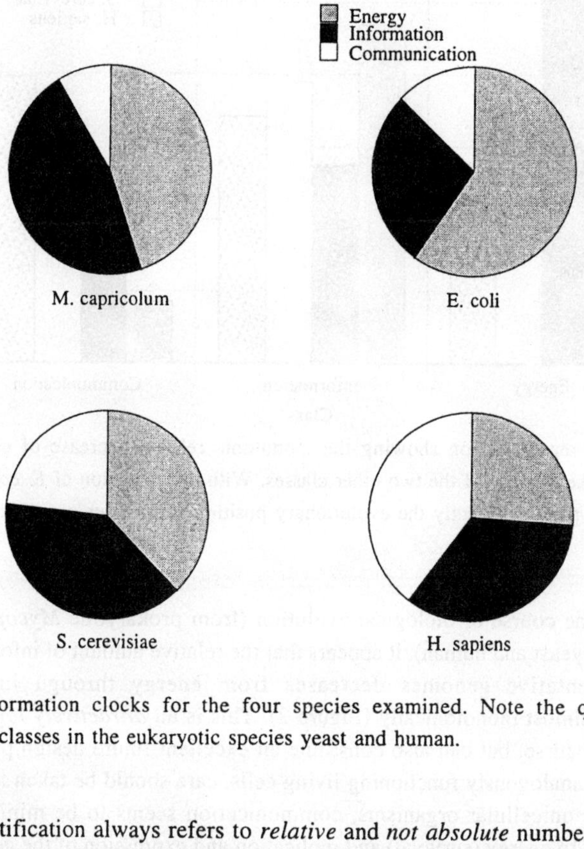

Figure 1: Information clocks for the four species examined. Note the quasi-homogeneous distribution of classes in the eukaryotic species yeast and human.

that, this quantification always refers to *relative* and *not absolute* numbers: that is, human sequences devoted to energy metabolism can be as many as 25,000 (from the total estimated number of 100,000 genes).

As far as information transmission is concerned, *Mycoplasma* dedicates almost half of its genomic content to store information for proteins involved in DNA replication and repair, RNA transcription and translation, and protein folding and transport (Bork *et al.*, 1995). Thus, it may be considered as the perfect replicator, spending most of its time for survival and reproduction. The content for communication-related proteins is minimal (Figure 1).

At the other extreme, the content of a human cell (and more specifically the expression pattern of a genome in a brain cell), has an even distribution for the three functional classes: using less storage for energy, it dedicates most of its storage for molecules involved in information transmission and communication within and between cells and the environment (Figure 1). It is remarkable indeed that a brain cell is not so special as far as storage of genetic information is concerned. This pattern can be predicted to persist when more data becomes available for human and other multicellular organisms.

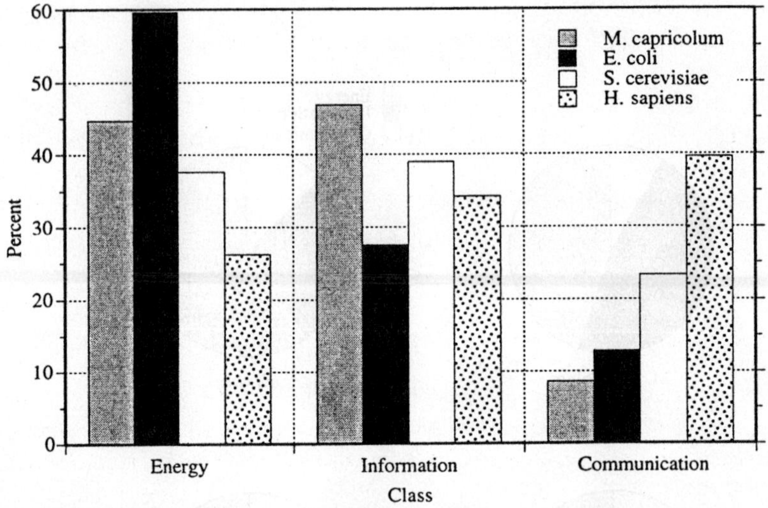

Figure 2: Another representation showing the monotonic relative increase of communication-related proteins at the expense of the two other classes. With the exception of *E. coli*, the decrease or increase follows almost perfectly the evolutionary position of the four organisms.

During the course of biological evolution (from prokaryotic *Mycoplasma* and E. coli to eukaryotic yeast and human), it appears that the relative amount of information stored in these representative genomes decreases from energy through information to communication, almost monotonically (Figure 2). This is an *attractively logical* aspect of the evolutionary course, but can also constitute an excellent future design principle: if we ever had to build analogously functioning living cells, care should be taken to follow these simple rules. For unicellular organisms, communication seems to be minimal and most storage is devoted to energy (survival) and replication and expression of the genetic material (evolution), while for higher multicellular organisms, communication with other cells and the environment is the main element of genomic information (Figure 2).

Discussion: Towards design principles for Artificial Life

Our results from this first glimpse into the functional composition of four model organisms support the idea that storage for communication (regulation and recognition) has increased during the phylogenetic course, and during the transformation of unicellular organisms to multicellular ones. This is an attractive hypothesis supported by the data, which can only be clarified when the entire genomes of (at least few) model organisms become known.

The prospect of storing functional information in self-replicating entities is a very interesting one. This is a prime example how progress in biological sciences can fertilize research on ALife, with the expectation that this research will further contribute towards a better understanding on the function and evolution of living systems.

As a design principle, the mean levels for the three functional classes in the four model organisms are interesting: energy-related proteins constitute 42% of the four genomes on average, information-related proteins are at an average of 37% and communication-related proteins the rest 21% of total. Thus, if we had to design an artificial cell (or, more realistically, an abstract model of it), these numbers may indeed form a useful guideline.

Another application lies in the modelling of genomes (Ouzounis, 1988), through genetic algorithms (Mitchell *et al.*, 1992), where *qualitatively different* information can be stored, instead of 'equal' sites with different values. In abstract terms, this would be the exploration of a hyper-surface (with as many dimensions as the qualitative different classes) instead of a rugged fitness landscape. The evolutionary course of such models can be recorded and the expansion or decline of these functional classes could be followed, according to the fitness functions and the selection process.

Our conclusions are two-fold: first, the analysis of biological 'machines' as networks of functional entities can now be done in almost quantitative detail (with protein functional classifications, but also other schemes, e.g. cellular localization patterns) and second, the derivation of abstractions for ALife can be anticipated, which will be useful both for biology and artificial intelligence. The modelling of cells as systems can be of great value for biological research while the inspiration of AI science to carry over analogies towards the simulation and design of living systems can only be beneficial.

Developments in both molecular biology and artificial life should be closely followed in the next years. With the prospect of knowing the genome composition of a few representative organisms, rising from about 1% in 1990 to completion by the year 2005, and the deeper understanding about function of living systems at the molecular level, we can envisage the emergence of a new science, which will focus on the study of organisms and machines alike, their structure and function, ecology and evolution, based on the principles of self-reproducing systems.

Note

Main concept, analysis, and exposition by CO; automatic cross-checking with the *E. coli* and human data sets by AV; automatic evaluation program by JT; *Mycoplasma* and yeast functional classification by CO and PB; additional ideas, co-ordination and editing by CS.

References

Adams MD, Kerlavage AR, Fields C & Venter JC (1993). 3,400 new expressed sequence tags identify diversity of transcripts in human brain. *Nature Genetics* **4**: 256-267.

Bairoch A & Boeckmann B (1991). The Swiss-Prot protein sequence databank. *Nucleic Acids Research* **19**: 2247-2249.

Bork P, Ouzounis C, Casari G, Schneider R, Sander C, Dolan M, Gilbert W & Gillevet PM (1995). Exploring the *Mycoplasma capricolum* genome: a small bacterium reveals its physiology. *Mol. Microbiol.* in press.

Bork P, Ouzounis C & Sander C (1994). From genome sequences to protein function. *Curr. Opin. Struct. Biol.* **4**: 393-403.

Bork P, Ouzounis C & Sander C (1995). Yeast chromosome VIII: identification of novel translation-associated protein families as inferred from bacterial sequences. Submitted.

Bork P, Ouzounis C, Sander C, Scharf M, Schneider R & Sonnhammer E (1992a). Comprehensive sequence analysis of the 182 ORFs of yeast chromosome III. *Protein Sci.* **1**: 1677-1690.

Bork P, Ouzounis C, Sander C, Scharf M, Schneider R & Sonnhammer E (1992b). What's in a genome? *Nature (London)* **358**: 287-287.

Burks AW (1970). Essays on cellular automata. University of Illinois Press, Urbana.

Dujon B, Alexandraki D & al. (1994). Complete DNA sequence of yeast chromosome XI. *Nature (London)* **369**: 371-378.

Dujon B, Bork P, Valencia A, Ouzounis C & al. (1995). Analysis of the ORFs of yeast chromosome XI. *Yeast* in press.

Feldmann H, Aigle M & al. (1994). Complete DNA sequence of yeast chromosome II. *EMBO J* **13**: 5795-5809.

Holland JH (1975). Adaptation in natural and artificial systems. University of Michigan Press, Ann Arbor.

Johnston M, Andrews S & al. (1994). Complete nucleotide sequence of *Saccharomyces cerevisiae* chromosome VIII. *Science* **265**: 2077-2082.

Kampis G (1991). Self-modifying systems in biology and cognitive science: a new framework for dynamics, information and complexity. Pergamon Press, Oxford.

Kauffman SA (1993). The origins of order: self-organization and selection in evolution. Oxford University Press, New York.

Koonin EV, Bork P & Sander C (1994). Yeast chromosome III: new gene functions. *EMBO Journal* **13**: 493-503.

Langton CG (1984). Self-reproduction in cellular automata. *Physica* **10D**: 135-144.

McCulloch W & Pitts W (1943). A logical calculus of the ideas immanent in nervous activity. *Bulletin of Mathematical Biophysics* **5**: 115-123.

Mitchell M, Forrest S & Holland JH (1992). The royal road for genetic algorithms: fitness landscapes and GA performance. In: *Toward a practice of autonomous systems - Proceedings of the 1st European Conference on Artificial Life* (Varela FJ & Bourgine P, eds.), pp. 245-254, Paris, MIT Press.

Oliver SG, van der Aart QJM & al. (1992). The complete DNA sequence of yeast chromosome III. *Nature (London)* **357**: 38-46.

Orengo CA, Jones DT & Thornton JM (1994). Protein superfamilies and domain superfolds. *Nature (London)* **372**: 631-634.

Ouzounis C, Valencia A, Bork P & Sander C (1995). Functional composition in three yeast chromosomes. Submitted.

Ouzounis CA (1988). The mathematics of nucleic acid sequences: stochastic analyses, pattern recognition, automata theory and fundamental problems in molecular evolution. M. Sc. Thesis, Department of Computer Science, University of York.

Packard NH & Wolfram S (1985). Two-dimensional cellular automata. *J. Stat. Phys.* **38**: 901-946.

Reggia JA, Armentrout SL, Chou H-H & Peng Y (1993). Simple systems that exhibit self-directed replication. *Science* **259**: 1282-1287.

Riley M (1993). Functions of the gene products of *Escherichia coli*. *Microbiol. Rev.* **57**: 862-952.

Scharf M, Schneider R, Casari G, Bork P, Valencia A, Ouzounis C & Sander C (1994). GeneQuiz: a workbench for sequence analysis. In: *Proceedings of the 2nd International Conference on Intelligent Systems for Molecular Biology (ISMB94)* (Altman R, Brutlag D, Karp P, Lathrop R & Searls D, eds.), pp. 348-353, Stanford, AAAI Press.

Sims K (1992). Interactive evolution of dynamical systems. In: *Toward a practice of autonomous systems - Proceedings of the 1st European Conference on Artificial Life* (Varela FJ & Bourgine P, eds.), pp. 171-178, Paris, MIT Press.

Wolfram S (1984). Cellular automata as models of complexity. *Nature (London)* **311**: 419-424.

Thermodynamics of RNA folding.
When is an RNA molecule in equilibrium?

Paul G. Higgs and Steven R. Morgan
University of Sheffield, Department of Physics,
Hounsfield Road, Sheffield S3 7RH, U.K.

Abstract: Most computational algorithms for RNA structure prediction are based on calculation of the minimum free energy secondary structure. For some sequences the structure predicted by phylogenetic comparison has a much higher free energy than the minimum free energy structure. It is therefore possible that some sequences do not fold to their minimum free energy state, but that the structure formed is governed by the kinetics of the folding process. Here we discuss the example of tRNA, where thermodynamic analysis suggests that the molecules are in equilibrium, and the example of SV-11, where the molecule folds to a metastable state determined by the folding kinetics. We have analysed the equilibrium thermodynamic properties of the complete set of transfer RNA sequences in the tRNA database. Random sequences were also generated having the same base composition and the same length distribution as the real sequences. Within the random sample, sequences with properties comparable to real tRNA molecules are rare. Secondary structure in real tRNA melts at higher temperatures than in random sequences, and over a narrower temperature range, i.e. the melting process is more cooperative. These results suggest that evolution has selected for sequences with thermodynamic properties substantially different from those of typical random sequences. We have written a Monte-Carlo programme to simulate the kinetics of secondary structure formation. Times for reorganisation of secondary structure are extremely long, and therefore the system may remain trapped in a particular region of configuration space. The sequence SV-11 from the Qb replicase system is known experimentally to fold to a metastable structure with free energy considerably higher than the groundstate structure. Our simulations show that if folding occurs during the growth of the sequence the molecule folds to a metastable state in which it alternates between a number of closely related structures. The structure was not observed to reorganise to the groundstate even after a very long simulation. If folding occurs only after completion of the whole molecule then the groundstate structure is often formed.

Introduction

Computational algorithms for prediction of secondary structure in RNA molecules are now widely available (Zuker, 1989; Jaeger et al, 1989; Abrahams et al, 1990; Higgs, 1993; Hofacker et al, 1994). Most of the programmes work by finding the minimum free energy secondary structure for a given sequence. In cases where strong evidence for the naturally occurring secondary structure exists, the minimum free energy structure predicted by these algorithms usually contains a substantial fraction of the helices thought to occur in nature.

A lot of effort has been made to refine the energy parameters in order to produce a larger percentage of correctly predicted helices, and substantial improvement has been obtained (Jaeger et al, 1989). There is, however, a limit to what one can expect to achieve by these refinements, since the algorithms take no account of the three dimensional structure of the molecules. Nevertheless, by analysis of large numbers of sequences on a statistical basis one can derive much useful information, despite the limitations of the algorithm.

There is also the possibility that the folding of real RNA molecules is governed by kinetics and not equilibrium thermodynamics. Whether or not a molecule reaches equilibrium will depend on the relaxation times for reorganisation of secondary structure. A molecule may become trapped in a metastable state if the energy barriers for reorganisation of the structure are too large to be overcome by thermal fluctuations in a reasonable time. Energy barriers and relaxation times are likely to increase rapidly with the length of the chain, so that it is very likely that kinetic effects are important for long RNA molecules. If this is the case, methods based on finding the minimum free energy structure will not give the correct answers. In this article we begin with a case where we believe that the molecules are really in thermal equilibrium, and then we discuss a particular case where we expect kinetic effects to play an essential part.

Equilibrium thermodynamics of transfer RNA molecules.

A large data base of transfer RNA molecules exists (Steinberg et al, 1993) containing sequences from a wide variety of organisms. It is possible to arrange almost all of these into the well known clover-leaf secondary structure, and there is strong experimental evidence that the clover-leaf structure is the naturally occurring structure for tRNA molecules. We thus have an ensemble of molecules with varying base sequences, which all fold to very similar secondary structures. It is natural to study the properties of this ensemble of molecules on a statistical basis. Part one of the tRNA database contains DNA sequences of tRNA genes, whilst part two contains the RNA sequences themselves. This work is based on the 509 sequences in section 2 at the time of writing. Sequences are classified according to the type of organism or organelle from which they originate: Viruses; Archaebacteria; Eubacteria; Chloroplasts; Mitochondria (Animals, Plants, and Single-celled organisms); and Cytoplasm (Animals, Plants, and Single-celled organisms).

The minimum free energy routine in the Vienna RNA library (Hofacker et al, 1994) was used to find the minimum free energy structure for each of the tRNA sequences, at temperatures of 37°C and 25°C. The clover leaf structure has four main helices. The percentage of correctly predicted clover leaf helices when averaged over the whole data set was 85% at 37°C, and 87% at 25°C. The percentage of molecules for which all four of the clover leaf helices were present in the minimum free energy structure was 65% at 37°C and 69 % at 25°C. There was considerable variation in these figures between the different classes of tRNA when these were considered separately (details given in Higgs, 1995).

The figure of 85% correct helices is comparable with previous studies of tRNA (Marlière, 1983; Jaeger et al, 1989) and with our previous work (Higgs, 1993). Up to 90% correct helices were found in a sample of 141 tRNAs analysed by Jaeger et al, 1989, after the energy parameters had been optimised. Predictions for longer molecules than tRNA are generally considerably worse than this, and a figure of 85% correct helices should be considered as fairly good from this point of view. It is generally assumed that almost all tRNAs have the clover leaf structure. A full 3d crystallographic analysis has only been done on a small number of tRNAs, so we cannot rule out the possibility that a few of the sequences do not fold to the clover leaf. However, the most likely reason that we do not get 100% correct predictions is that tertiary structure information has not been accounted for. Some of the alternative secondary structures of a molecule will be able to fold much more conveniently into a 3d globule than others. It seems very likely that in some cases where an alternative structure to the clover leaf has the lowest free energy in the secondary structure model, the clover leaf would in fact have the lowest free energy if tertiary structure were properly accounted for.

Recently there has been a considerable amount of theoretical work on random copolymers (see Miller et al, 1992, Sali et al, 1994, Bryngelson, 1994, Derrida and Higgs, 1994, and references therein). It is expected that a random polymer sequence will have a rugged energy landscape in configuration space, similar to what is found in other systems with random disorder, such as spin glasses. In other words, there will be many alternative low energy configurations (valleys in the landscape) separated by high energy configurations (ridges or barriers). The low energy valley-bottoms may represent very different secondary structures, but may have energies which are very close. We have pointed out (Higgs, 1993) that real molecules are far from random, since they have evolved under the action of natural selection. We may therefore expect the thermodynamic properties of real molecules to differ from those of random sequences, and it is interesting to compare the two.

Figure 1 shows the mean value of the minimum free energy at 37°C for each of the classes of tRNA in the database plotted against the content of C + G bases. The base composition differs widely between the different classes, and there is a rapid decrease in free energy as the C+G content increases. This is to be expected, since stacking free energies are more favourable for CG pairs. The standard deviations of the minimum free energies in each of the sets are around 3 kcal/mol. This shows that the differences between the mean values of the different sets are large compared with the spread of values for molecules in a given set, and hence that the differences between the different classes of tRNA are significant.

For each set of tRNAs, a set of approximately 500 random base sequences was generated with the same percentage base composition, and the same distribution of lengths as the real sequences. The minimum free energies for the random sequences are also given in figure 1. These are much less negative than the values for the corresponding real sequences. The difference is at least as large as the standard deviation of the values in each of the sets. This indicates that the real sequences have a groundstate which is much more stable than that of the random sequences relative to the unfolded state. In order for biomolecular reactions to take place, recognition must occur between the molecules, and this will only happen if the molecules are folded into the correct three dimensional shape. We would therefore expect most biomolecules to have a stable groundstate structure, rather than to be continually changing between many alternating structures. The observed difference between the minimum free energies of real and random RNA chains is evidence that evolution has selected molecules with groundstates which are more stable than average.

Further analysis of tRNA structure was done using a set of 337 sequences which we call Set 1 (see Higgs, 1995, for more details). Figure 2 shows the mean number of alternative secondary structures per molecule for sequences in Set 1 as a function of their free energy above the groundstate. This is calculated using the density of states algorithm discussed by Higgs, 1993. In figure 2 only structures which are local minima in free energy are included in the distribution. By cutting out the states which are not local minima we remove states which differ from each other by only a trivial change, such as unzipping of a single base pair from the end of a helix. Local minima represent alternative secondary structures which are substantially different from each other. It can be seen that the real tRNA molecules have far fewer alternative secondary structures with free energies close to the groundstate free energy than do the random sequences. Only states within a few kT of the groundstate will have a significant equilibrium probability, and it follows that if the number of alternative states in the low free energy tail of the distribution is decreased, then the stability of the groundstate is increased. From the data in fig. 2, the average number of local minima states within 0.5 kcal/mol of the groundstate is 1.18 for tRNA (i.e. 0.18 other states in addition to the groundstate), whilst within 2 kcal/mol there are an average of 1.57 states, and within 5 kcal/mol there are an average of 3.97 states. (Note that the thermal energy kT is 0.6 kcal/mol at 37°C). The equivalent numbers for the random sequences are much higher : 1.82, 4.06 and 14.08. These numbers are very model-dependent, but we expect the clear difference between tRNA and random sequences to be independent of the details of the model.

RNA secondary structure tends to melt on heating. The melting process can be followed exactly by calculating the "melting curve", or specific heat $C = \frac{\partial H}{\partial T} = -T\frac{\partial^2 G}{\partial T^2}$ where H is the total enthalpy, and G is the total free energy of the molecule. C is obtained by measuring G as a function of T using the partition function folding algorithm, and using a numerical procedure to take the second derivative (McCaskill, 1990; Hofacker et al, 1994). Melting curves are shown for several tRNAs and several random sequences in Higgs, 1995. It is found that the real tRNA sequences tend to melt at a higher temperature than the random sequences, and over a narrower range of

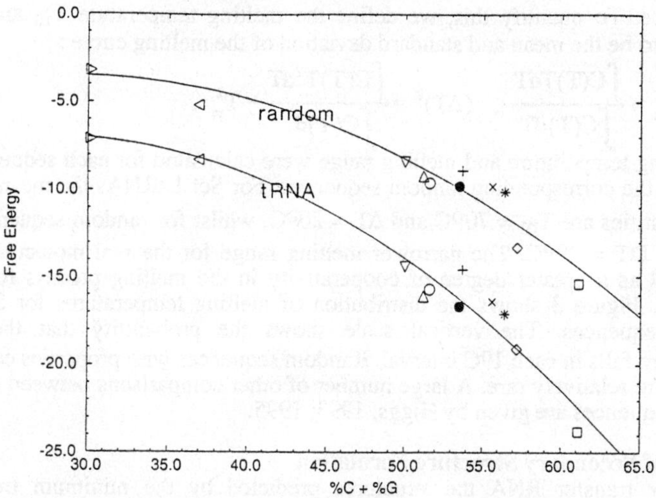

Fig. 1. Average value of the groundstate free energy (in kcal/mol) for tRNA and for random sequences shown as a function of proportion of C+G bases. Each symbol on the lower curve corresponds to a different class of tRNA from the database (details given in Higgs, 1995), and the matching point on the upper curve corresponds to random sequences with the same base composition and the same length distribution. Free energies are strongly dependent on base composition, but for each class of tRNA, the real sequences have much lower free energies than the random ones.

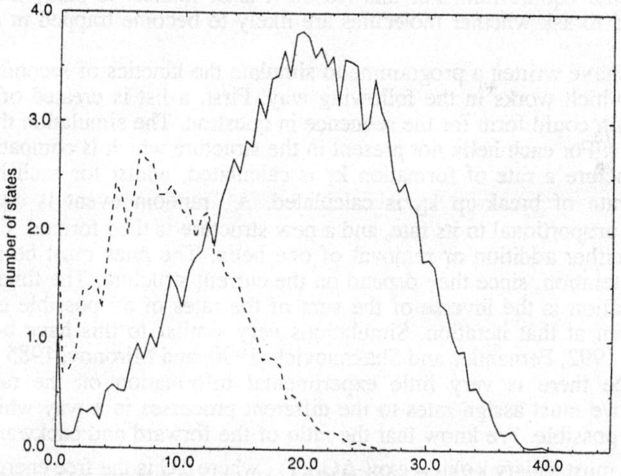

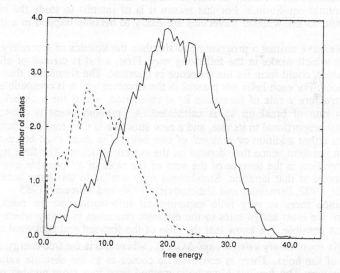

Fig. 2. Average distribution of local minima states relative to the groundstate for Set 1 tRNAs (solid line) and random sequences (broken line). The vertical scale gives the average number of states in each 0.5 kcal/mol interval. There are far fewer alternative states close to the groundstate in tRNA, and hence the groundstate structure is much more stable.

temperature. To quantify this, we define the melting temperature T_m and melting range DT to be the mean and standard deviation of the melting curve :

$$T_m = \frac{\int C(T)T dT}{\int C(T) dT}, \quad (\Delta T)^2 = \frac{\int C(T)T^2 dT}{\int C(T) dT} - T_m^2 .$$

The melting temperature and melting range were calculated for each sequence in Set 1, and for the corresponding random sequences. For Set 1 tRNAs the mean values of these quantities are $T_m = 70^oC$ and $\Delta T = 20^oC$, whilst for random sequences $T_m = 58^oC$ and $DT = 29^oC$. The narrower melting range for the real molecules may be interpreted as a greater degree of cooperativity in the melting process for the real molecules. Figure 3 shows the distribution of melting temperatures for Set 1, and random sequences. The vertical scale shows the probability that the melting temperature falls in each 1^oC interval. Random sequences with properties comparable to tRNA are relatively rare. A large number of other comparisons between tRNA and random sequences are given by Higgs, 1993, 1995.

Kinetics of Secondary Structure Formation

For transfer RNA the structures predicted by the minimum free energy algorithm are close to the clover leaf in most cases, and it seems reasonable to assume that the molecules fold to their lowest free energy states. However for longer molecules typically only 50% of the structure thought to exist from evidence of phylogenetic comparisons is predicted by the minimum free energy algorithms. In some cases the free energy of the phylogenetic structure is many times kT higher than the minimum free energy, so that if the molecule were in thermal equilibrium there would be a negligible probability of finding the molecule in the phylogenetic structure. We must therefore conclude that either the free energy parameters used in the programmes are seriously in error for larger molecules, or else the molecules are not in thermal equilibrium. For this reason it is of interest to study the kinetics of folding, and to ask whether molecules are likely to become trapped in a metastable structure.

We have written a programme to simulate the kinetics of secondary structure formation which works in the following way. First, a list is created of all possible helices which could form for the sequence in question. The simulation then proceeds by iterations. For each helix not present in the structure which is compatible with the current structure a rate of formation k_f is calculated, whilst for each helix already present a rate of break-up k_b is calculated. A random event is chosen with a probability proportional to its rate, and a new structure is then formed. Each iteration represents either addition or removal of one helix. The rates must be re-calculated after each iteration, since they depend on the current structure. The time represented by one iteration is the inverse of the sum of the rates of all possible events which could happen at that iteration. Simulations very similar to this have been done by Fernandez, 1992, Fernandez and Shakhnovich, 1990, and Mironov, 1985.

Since there is very little experimental information on the rates of helix formation, we must assign rates to the different processes in a way which seems as realistic as possible. We know that the ratio of the forward and backward rates for a given helix must satisfy $k_f/k_b = \exp(-\Delta G/kT)$, where ΔG is the free energy change on addition of the helix. There is considerable choice as to the absolute values of the rates, however. The familiar Metropolis method used in a large number of Monte-Carlo simulations assigns a rate 1 to all processes for which ΔG is negative, and a rate $\exp(-\Delta G/kT)$ to processes for which ΔG is positive. Another possibility , which we call Barrier Kinetics, is to suppose that the stacking free energy of the helix provides a barrier to helix removal, so that $k_b = \exp(-\Delta G_{stack}/kT)$. It therefore follows that the addition rate depends on the free energy change of the loops, $k_f = \exp(-\Delta G_{loops}/kT)$. (This is the choice made by Fernandez, 1992, and Mironov, 1985). A third possibility

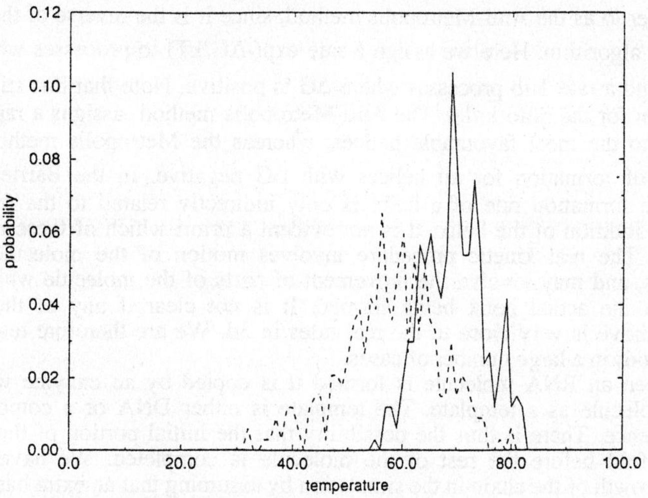

Fig. 3. Distribution of melting temperatures for tRNA (solid line) and random sequences (broken line). The vertical scale shows the probability that the melting temperature lies within each 1°C interval. Random sequences with melting temperatures as high as those of typical tRNAs are rare.

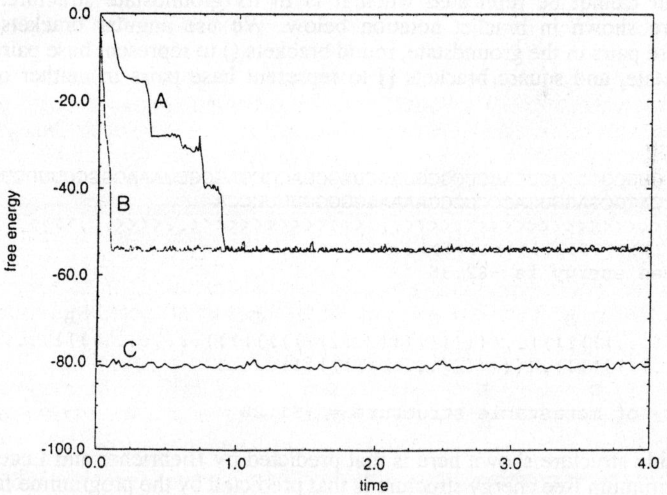

Fig. 4. Free energy as a function of time during the folding of SV-11. In the upper two curves folding occurs as the molecule is growing. (A) growth rate 100, (B) growth rate 1000. Both curves lead to the same metastable state. No further change in free energy is observed after the growth is complete. In curve C folding begins after the molecule is complete. The free energy drops rapidly to the groundstate value.

we will refer to as the Anti-Metropolis method, since it is the reverse of the standard Metropolis algorithm. Here we assign a rate exp($-\Delta G/kT$) to processes where ΔG is negative, and a rate 1 to processes where ΔG is positive. Note that this still satisfies the criterion for the ratio k_f/k_b. The Anti-Metropolis method assigns a rapid rate of formation to the most favourable helices, whereas the Metropolis method has the same rate of formation for all helices with ΔG negative. In the Barrier Kinetics method the formation rate of a helix is only indirectly related to the free energy change on addition of the helix. It is not evident a priori which of these methods is preferable. The real kinetic procedure involves motion of the molecule in three dimensions, and may involve the movement of parts of the molecule which are far away from the actual helix being formed. It is not clear if any of the methods proposed above is very close to the real rates in 3d. We are therefore testing all of these methods in a large number of cases.

When an RNA molecule is formed it is copied by an enzyme which uses another molecule as a template. The template is either DNA or a complementary RNA sequence. There is thus the possibility that the initial portion of the molecule begins to fold before the rest of the molecule is completed. We have therefore included growth of the chain in the simulation by assuming that an extra base is added onto the sequence at a rate k_g if the molecule has not yet reached its full length. In general we expect that the structure formed may depend substantially on the growth rate k_g and on the method used for assigning k_f and k_b. A full discussion of all these things will be given elsewhere. Here we confine ourselves to one particular molecule where the results are particularly clear and do not depend much on the details of the simulation.

The sequence SV-11 arises in experiments on in vitro replication of RNA by Qb replicase (Biebricher and Luce, 1992, and proceedings of ECAL 93 (Brussels)). The sequence is efficiently replicated by the enzyme when it is in a metastable structure, but cannot be replicated when it is in its groundstate structure. These structures are shown in bracket notation below. We use angular brackets <> to represent base pairs in the groundstate, round brackets () to represent base pairs in the metastable state, and square brackets [] to represent base pairs in neither of these structures.

```
SV-11 plus
GGGCACCCCCCUUCGGGGGGUCACCUCGCGUAGCUAGCUACGCGAGGGUUAAAGGGCCUUUCUCCCUCG
CGUAGCUAACCACGCGAGGUGACCCCCCGAAAAGGGGGUUUCCCA
<<<.<<<<<<<<<<<<<<<<<<<<<<<<<<<...<<<<<<<<<<<<<<..<<<<...>>>>..>>>>>>
>>>>>>>>...>>>>>>>>>>>>>>>>>>>..>>>>>>>>..>>>.
minimum free energy is -82.35

  A   B         B          C             C             A       D
(((((((((((...)))))))..((((((((((....))))))))))........)))).....(((((
(((.(....).)))))))).((((((((.....)))))))).....
           D       E          E
free energy of metastable structure = -51.20
```

The metastable structure shown here is that predicted by Biebricher and Luce, 1992, whilst the minimum free energy structure is that predicted by the programme from the Vienna library (Hofacker et al, 1994), which is very slightly different from the one shown by Biebricher and Luce. The molecule is almost palindromic, so that the groundstate structure is almost a perfect hairpin. The metastable structure has four large hairpins B,C,D, and E and one small helix A forming a multi-branched loop. The difference in free energies between these two structures is 31 kcal/mol, which is approximately 50 kT.

We have simulated the kinetics of folding of SV-11 using several different growth rates and several different methods for assigning the helix formation rates.

The following is a sequence of structures shown at intervals during the growth process in one typical run.

```
[[..(((((((...)))))))..]].........

[[..(((((((...)))))))..]]((((((((....))))))))..

....(((((((...)))))))..((((((((((....))))))))))....<<<<...>>>>......

.[[[(((((((...)))))))..((((((((((....))))))))))]]]..[[[[.......]]]]..
.....

....(((((((...)))))))..((((((((((....))))))))))......[[[........]]]((
(((........))))).....

[[..(((((((...)))))))..((((((((((....)))))))))).]]..[[[[.......]]]]((
(((........))))).................. 

....(((((((...)))))))..((((((((((....)))))))))).......[[[........(((((
(((........)))))))).((((((((.....))))))))]]]..
```

The final structure here is very similar to the predicted structure of the metastable state. Once growth is complete the molecule alternates between a number of similar structures. Some of these are shown below together with their free energies.

```
1.
....(((((((...)))))))..((((((((((....)))))))))).......[[[........(((((
(((........)))))))).((((((((.....))))))))]]]..
free energy of structure = -50.03
2.
....(((((((...)))))))..((((((((((....))))))))))....<<<<...>>>>...(((((
(((........)))))))).((((((((.....))))))))).....
free energy of structure = -54.58
3.
....(((((((...)))))))..((((((((((....))))))))))....[[[....]]]....(((((
(((........)))))))).((((((((.....)))))))).....
free energy of structure = -51.04
4.
....(((((((...)))))))..((((((((((....))))))))))......[[[[...]]]].(((((
(((........)))))))).((((((((.....)))))))).....
free energy of structure = -53.78
5.
....(((((((...)))))))..((((((((((....))))))))))).......[[[.....]]](((((
(((........)))))))).((((((((.....)))))))).....

6.
(((((((((((...)))))))..((((((((((....))))))))))......))))........(((((
(((........)))))))).((((((((.....)))))))).....
free energy of structure = -53.20
7.
....(((((((...)))))))..((((((((((....)))))))))).......[[.......]](((((
(((........)))))))).((((((((.....)))))))).....
free energy of structure = -51.05
8.
<<<.(((((((...)))))))..((((((((((....))))))))))..................(((((
(((........)))))))).((((((((.....)))))))).>>>.
free energy of structure = -46.32
9.
[[..(((((((...)))))))..((((((((((....)))))))))).]]...............(((((
(((........)))))))).((((((((.....)))))))).....
free energy of structure = -47.85
```

All these structures have similar free energies, and all contain the four helices B,C,D and E. Helix A is only present in structure 6. The growth rate of the molecule was varied by several orders of magnitude and the same metastable state was observed in each case. The metastable state can be thought of as a kind of restricted equilibrium in which many closely related structures in one valley of configuration space are found.

We also did simulations in which folding commenced with the molecule fully complete. The results of these simulations were more dependent on the details of the kinetics than were the simulations in which folding occurred during growth. When the Anti-Metropolis method was used the molecule folded very rapidly to the minimum free energy structure every time, whereas when the Metropolis method or the Barrier kinetics method was used the structure was much slower in relaxation towards a stable state. Sometimes the minimum free energy structure was formed, but in the folding pathway was much less reproducible than with the Anti-Metropolis method.

Figure 4 shows the free energy as a function of time for one run at growth rate 100 (arbitrary units), one run at growth rate 1000, and one run where folding began from the completed sequence. The first two curves show an approximately linear decrease of the free energy during the growth process, followed by no change once growth is complete. The molecules remain in the metastable state until the end of the simulation. The third curve shows a rapid drop to the groundstate, and then remains constant. The period of 4 time units shown here represents about 14000 Monte-Carlo iterations. One run of the simulation was continued for over 3.5×10^6 iterations after formation of the metastable state and no decay to the groundstate was observed. It is known experimentally that the molecules remain in the metastable state for at least a period of hours at room temperature before eventually converting to the groundstate. It may not be possible to run simulations for long enough to see this occur.

SV-11 is probably a very unusual sequence. It is only 115 bases long, which is not that much longer than tRNA (length 76). It seems unlikely that many sequences of this length will have such a well defined metastable state as SV-11. We are currently doing simulations on longer sequences. The longer the sequence, the longer the relaxation times for reorganisation of structure will be, and the more likely we are to find metastable states. If we wish to argue that the native structures of long RNA sequences such as 16S ribosomal RNA are not minimum free energy structures, then we need to show not only that the molecules become trapped in metastable states, but that there is an approximately reproducible kinetic pathway leading to the same metastable state each time. Things would be much easier if the molecules folded to their groundstates, since then the pathway would not matter. These results on SV-11 are significant since they show that there is a well defined kinetic pathway leading reproducibly to a particular metastable state in this case. It is therefore possible that this occurs for other molecules of biological significance. We note that evidence that certain molecules follow a folding pathway that does not lead to the minimum free energy state has been given by van Batenburg et al, 1994, using a genetic algorithm which mimics folding kinetics. Application of the genetic algorithm to the case of SV-11 gives results very similar to those obtained above using the Monte-Carlo simulation (Gultyaev et al. 1995. Also poster at this conference).

Acknowledgements :

I wish to thank Eke van Batenburg, Sasha Gultyaev, and Kees Pleij, for their hospitality and for very useful discussions during my recent stay in Leiden. I am grateful to the British Council and the Netherlands Organisation for Scientific Research for financial support for this visit.

References

Abrahams, J.P., van den Berg, M., van Batenburg, E. and Pleij, C. (1990) Prediction of RNA secondary structure, including pseudoknotting, by computer simulation. *Nucl. Acids Res.* **18**, 3035-44.

van Batenburg, F.H.D., Gultyaev, A.P. and Pleij, C.W.A. (1994) An APL-programmed genetic algorithm for the prediction of RNA secondary structure. *J.Theor. Biol.* (in press)

Biebricher, C.K. and Luce, R. (1992) In vitro recombination and terminal elongation of RNA by Qb replicase. *EMBO J.* **11**, 5129-35.

Bryngelson, J.D. (1994) When is a potential accurate enough for structure prediction? Theory and application to a random heteropolymer model of protein folding. *J. Chem. Phys.* **100**, 6038-45.

Derrida, B. and Higgs, P.G. (1994) Low temperature properties of directed walks with random self interactions. *J. Phys. A Math. Gen.* **27**, 5485-93.

Fernandez, A. and Shakhnovich, E.I. (1990) Activation energy landscape for metastable RNA folding. *Phys. Rev. A* **42**, 3657.

Fernandez, A. (1992) A parallel computation revealing the role of the in vivo environment in shaping the catalytic structure of a mitochondrial RNA transcript. *J. Theor. Biol.* **157**, 487-503.

Gultyaev, A.P. (1991) The computer simulation of RNA folding involving pseudoknot formation. *Nucl. Acids Res.* **19**, 2489-94.

Gultyaev, A.P., van Batenburg, F.H.D. and Pleij, C.W.A. (1995) The computer simulation of RNA folding pathways using a genetic algorithm. *J. Mol. Biol.* (in press).

Higgs, P.G. (1993) RNA secondary structure: a comparison of real and random sequences. *J. Phys. I France* **3**, 43-59.

Higgs, P.G. (1995) Thermodynamic properties of transfer RNA : a computational study. (in press).

Hofacker, I.L., Fontana, W., Stadler, P.F., Bonhoeffer, L.S., Tacker, M. and Schuster, P. (1994) Fast folding and comparison of RNA secondary structures (The Vienna RNA package). *Monatshefte für Chemie* **125**, 167-188.

Jaeger, J.A., Turner, D.H. and Zuker, M. (1989) Improved predictions of secondary structures for RNA. *Proc. Nat. Acad. Sci. USA* **86**, 7706-10.

Marlière, P. (1983) Computer building and folding of fictitious transfer RNA sequences. *Biochimie* **65**, 267-73.

McCaskill, J.S. (1990) The equilibrium partition function and base pair binding probabilities for RNA secondary structure. *Biopolymers* **29**, 1105-19.

Miller, R., Danko, C.A., Fasolka, M.J., Balazs, A.C., Chan, H.C. and Dill, K.A. (1992) Folding kinetics of proteins and copolymers. *J.Chem.Phys.* **96**, 768.

Mironov, A.A., Dyakonova, L.P. and Kister, A.E. (1985) A kinetic approach to the prediction of RNA secondary structures. *J. Biomol. Struct. Dyn.* **2**, 953-62.

Sali, A., Shakhnovich, E. and Karplus, M. (1994) Kinetics of protein folding. A lattice model study of the requirements for folding to the native state. *J. Mol. Biol.* **235**, 1614-36.

Steinberg, S., Misch, A. and Sprinzl, M. (1993) Compilation of tRNA sequences and sequences of tRNA genes. *Nucl. Acids Res.* **21**, 3011-15.

Zuker, M. (1989) Computer prediction of RNA structure. *Methods in Enzymology* **180**, 262-288.

An Artificial Life Model for Predicting the Tertiary Structure of Unknown Proteins that Emulates the Folding Process

Raffaele Calabretta[1,2], Stefano Nolfi[2] and Domenico Parisi[2]

[1] Centro di Studio per la Chimica del Farmaco, National Research Council
Department of Pharmaceutical Studies, University "La Sapienza"
Piazzale A. Moro 5, 00185 Rome, Italy

[2] Institute of Psychology, National Research Council
Viale Marx 15, 00137 Rome, Italy

e-mail: raffaele@caio.irmkant.rm.cnr.it
stefano@kant.irmkant.rm.cnr.it
domenico@gracco.irmkant.rm.cnr.it

Abstract. We present an "ab initio" method that tries to determine the tertiary structure of unknown proteins by modelling the folding process without using potentials extracted from known protein structures. We have been able to obtain appropriate matrices of folding potentials, i.e. 'forces' able to drive the folding process to produce correct tertiary structures, using a genetic algorithm. Some initial simulations that try to simulate the folding process of a fragment of the crambin that results in an alpha-helix, have yielded good results. We discuss some general implications of an Artificial Life approach to protein folding which makes an attempt at simulating the actual folding process rather than just trying to predict its final result.

1 Introduction

The prediction of the three-dimensional structure of proteins is a great challenge both for the difficulty of the task and for the importance of the problem. While computational approaches appear to be natural candidates to solve it, optimization techniques that try to predict the *result* of the folding process by ignoring the specificity of the *process* itself (Qian & Sejnowski, 1988; Fariselli *et al.*, 1993) have produced limited results. We claim that approaches in the spirit of Artificial Life (Alife) that try to reproduce, even if in extremely simplified ways, the natural processes as they actually occur could be more fruitful.

The protein folding problem presents many similarities with the kind of problems that have been investigated in the Alife literature in the last few years. Proteins, like the simple artificial creatures studied by several researchers in this field (Parisi *et al.*, 1990; Wilson, 1991; Taylor & Jefferson, 1994), are physical entities that have their own structure, which interact with an external environment (the solution), and which are made of sub-components which interact among themselves (the amino acids). In addition, proteins "behave" by folding into a stable structure and such "behaviour" depends on the interaction among the sub-components of the protein itself and between these sub-components and the external environment. Finally, as in most Alife models, to each individual protein corresponds a given fragment of DNA and the mapping between the genetic information and the final stable three-dimensional structure of the protein is very complex and non-linear (Langton, 1992). We will ask some interesting questions about the similarities/differences between a low-level mapping process such as protein folding and the overall developmental process of the organism.

2 The Protein Folding Problem

Many researchers have tried to predict the three-dimensional structure of proteins on the only basis of the amino acid sequence. The attempt has been defined as trying to decipher the second half of the genetic code (Gierasch & King, 1990). Success in this area would be the starting point for new research directions with promising results and possible applications in many fields (biology, genetics, drug-design, etc.).

Proteins chemically consist of the sequencing of structural units which are amino acids: each protein is constructed with the same twenty amino acids which are arranged according to a unique and well defined order. Each protein differs from any other in the number of amino acids linked together (generally between 50 and 3000) and the sequence in which the various amino acids occur. The amino acids are linked to each other by the peptide bond to form a typical linear polypeptide chain. The polypeptide backbone is a repetition of the basic unit common to all amino acids. What changes is the side-chain which is characteristic for each one of the twenty amino acids and is different in shape, bulk and chemical reactivity.

The protein structure can be discussed in terms of three levels of complexity. The primary structure refers simply to the linear amino acid sequence. The secondary structure describes the presence in the protein of regular local structure (alpha-helix and beta-sheets), built with segments of the protein chain. Finally, the tertiary structure represents the real three-dimensional structure of the entire protein. Thanks to the possibility of alternating the twenty amino acids, proteins differ in amino acid sequence (primary structure) and therefore in three-dimensional structure (tertiary structure). In other words, the primary structure of a protein, as it is codified exactly in DNA, contains all the information to determine the three-dimensional structure, on which the function of that protein finally depends. The proteins are necessary macromolecules for the normal deployment of almost all biological processes, but for this to happen it is necessary that the proteins, at the end of a folding process, assume their characteristic spatial structure, which varies from protein to protein. In fact, either during or after ribosomal biosynthesis of a protein as a linear chain of amino acids, the chain folds up rapidly until it assumes a stable and functional three-dimensional structure. A linear or randomly folded chain would not be biologically active.

During the folding process, amino acid chains can adopt an astronomical number of conformations: it would not be feasible for any protein to try out all of its conformations on a practical time scale. Nevertheless, proteins are observed to fold in 10^{-1}-10^{-3} seconds both in vivo and in vitro. The evident conclusion is that proteins do not fold by sampling all possible conformations randomly until the one with the lowest free energy is encountered. Instead, for the folding process to take place on a short time scale, it must be directed in some way which is yet unclear (Creighton, 1993).

On one hand, molecular biology methods have allowed us to identify the amino acid sequence of over 30,000 proteins (Swiss-Prot Data Bank; Bairoch & Boeckmann, 1992). On the other hand, by means of X-ray crystallography and nuclear magnetic resonance spectroscopy (NMR), we have been able to identify the high-resolution structure of only over 1,300 of them (Brookhaven Data Bank; Bernstein *et al.*, 1977). In the next few years the gap is expected to increase due to the great mass of data originated from the Human Genome Project.

3 Computational Approaches to Protein Folding

Currently there is an increasing interest in the field of computational approaches to protein folding. As Wodak and Rooman (1993) claim, this appears to be due to several factors:

(a) experimental mutagenesis studies have demonstrated that the overall fold of a protein is much more tolerant to sequence modification (Sondek & Shortle, 1990);

(b) analyses of known three-dimensional structures have revealed structural similarity for proteins with different functions (Farber & Petsko, 1990; Kabsch *et al.*, 1990);

(c) the number of known high-resolution protein structures has significantly increased allowing computational models to lie on more solid grounds;

(d) there is an widening gap between the increase in known protein sequences and the lack of information about the structure and function of most of them;

(e) finally, new computational approaches have been developed (Rumelhart & McClelland, 1986; Holland, 1975) that appear to be promising for the protein folding problem and computational power has increased significantly as well.

We will review some of the most significant attempts in this direction and then we will describe our own model.

3.1 Extracting Knowledge-based Potentials

Several researchers have used computational models to design pseudo-energy functions that represent a reduced description of detailed atomic force fields. These pseudo-energy functions or potentials are usually expressed as a sum of several terms and mostly ignore side-chain atomic details.

Examples of such potentials are:

(a) Residue-specific secondary structure propensities (i.e. the tendency of a given residue to fold in a helix, beta sheet or random coil structure; Rost *et al.* ,1994);

(b) Residue-residue potentials (i.e. the tendency of a given residue to end-up close to another one; Maiorov & Crippen, 1992);

(c) Hydrophobicity (tendency of a given residue to interact with water; Casari & Sippl, 1992);

(d) Phi-psi backbone angle probabilities (the probability that two subsequent amino acids can assume a certain relative position; Rooman *et al.*, 1991).

These pseudo-energy function potentials can be derived from known protein tertiary structures by using different computational methods (Statistics, Monte Carlo, Neural Networks, Genetic Algorithm).

The way in which statistics is used to extract potentials is straightforward: the probabilities of observing the parameter of interest are computed and then normalised to correct for sample bias and finally translated into scores (e.g. Bryant & Lawrence, 1993).

Neural networks, given their ability to classify noisy stimuli and generalize to new ones, have also been used to predict the secondary structure of proteins (e.g. Rost & Sander, 1994).

Powerful optimization methods can also be used. Maiorov and Crippen (1992), for example, used an optimization procedure to extract the residue-residue potentials. They derived the strengths of individual contacts starting with non-correct values and then changing such values so that the potential energy of any native structure in the training set would be lower than the potential of any alternative conformation generated from segments of known protein structures.

The extracted function potentials can in turn be used in order to build models which are able to predict the second or the tertiary structure of other sequences (see next paragraph). In other cases, potential extraction and prediction of tertiary structure of unknown sequences can be realized at the same time using a single model.

3.2 Application of Knowledge-based Potentials to Prediction of Folded Structures

The availability of knowledge-based potentials allows us to go beyond the classical approaches based on sequence alignment for predicting secondary and tertiary structure. The main idea is that the extracted potentials can be used to choose between alternative predicted structures by measuring which of them results in lower energy value (e.g. which of them best conforms to the known residue-structure propensity, residue-propensity, hydrophobicity, and virtually to any known potential). In other words, the knowledge based potentials that are extracted from known protein structures can be used to evaluate predicted protein structures.

There are two ways of using knowledge-based potential to predict the tertiary structure of sequences, a hybrid method that combines the classical alignment procedure with the use of potentials and a pure method that use the potential in order to derive the tertiary structure directly from the sequence.

The first approach involves scanning a library of sequences and corresponding known structure motifs in search of compatible sequence-structure combinations, i.e.

those which correspond to structures which best conform to the known potentials (see for example Sippl & Weitckus, 1992). In this case potentials are used to choose the best combination of chain folds present in the databases. The combinations of folds that best align with the given sequence and best conform to potentials are selected. This method produces good performance for proteins closely related to those present in the used database but, as the distance increases, performance progressively deteriorates and it becomes unreliable when the sequence identity is lower than 30% (Wodak & Rooman, 1993).

The second approach based only on potentials, by not restricting the space of possible tertiary structure to a known limited set, is much more demanding because it is necessary to assess the value of the potentials of a huge number of possible alternative configurations from which the correct fold needs to be singled out. In this approach an initial wrong structure configuration is chosen and then the structure is progressively modified for a given number of trials until the final configuration, which represents the predicted structure, is obtained. In each trial the actual structure is evaluated by using potentials in order to preserve good modifications (i.e. changes that result in a better configuration from the extracted potential point of view) and to reject bad modifications. The search in the conformation space of a given protein can be implemented by using different algorithms. In particular Monte Carlo (e.g. Godzik *et al.*, 1992) and genetic algorithms (e.g. Dandekar & Argos, 1994) have been used.

In the model of Dandekar and Argos (1994), an initial population of different hypothetical three-dimensional structures for a given sequence are generated. Each individual of the population consists of a vector of dihedral and rotation angles which in turn determines the folding of the main chain of the corresponding protein. Individuals are evaluated according to a set of extracted potentials (secondary structure propensities, presence of hydrogen bonds, hydrophobicity), and ad-hoc criteria (undesired overlapping of C atoms) by determining if and how much a given structure conforms to each potential or criterion. The sum of all these positive and negative contributions constitute the individual's fitness that determines which individuals are allowed to reproduce by generating copies of their vectors with the addition of mutations and combinations between two 'parent' vectors. By repeating this process for a certain number of generations, three-dimensional structures which have better and better fitness and closely resemble the actual folded structure may be obtained.

4 Emulating the Folding Process by Evolving Abstract Folding Potentials

We think that while using potentials extracted by folded sequences may be adequate in choosing between alternative final structures as is necessary in hybrid approaches that combine folded sub-parts of known protein structures, it may be less useful in "pure" or "ab initio" approaches in which the final folded state is progressively determined through successive modifications starting from the initial amino acid sequence. In fact, the type of conformations that a protein assumes during the folding process may differ from the final folded conformation (Creighton, 1978). In other words, it may happen that a structure, in order to reach its final stable state, is forced to pass through a state which, even if it does not resemble the final folded state, is crucial in order to reach the final state (cf. the controversy about this point between Creighton (1992) and Weissman & Kim (1992)). Dandekar and Argos (1994), for example, in order to limit search space, restricted the dihedral and rotation angles to a set of 7 standard conformations extracted from the topology of known folded proteins. However, it is not known whether *during* the folding process significant different conformations of angles occur.

In our own work, we used an "ab initio" method that does not use pre-extracted potentials and that tries to determine the tertiary structure of unknown proteins by modelling the folding process itself. In other words, we did not want only to predict the final tertiary structure of proteins but we also wished to model the temporal process of folding that results in such a structure. We are aware of the difficulty of the task and of the fact that our results are very preliminary. But we believe that the method can have some validity because a better understanding of the folding process itself, even in the limited case of very short sequences, can have useful results.

For the present time we, as many others (e.g. Lau & Dill, 1990; Unger & Moult, 1993; Šali *et al.*, 1994), have modelled the primary structure of proteins in an extremely simplified way. Amino acid side-chains are represented as spheres connected to the corresponding C_α of backbone with a link of fixed length (see Fig. 1); the backbone is represented as a chain of C_α atoms linked by pseudobonds between the C_α atoms of successive amino acid residues (for a similar approach, cf. Oldfield & Hubbard, 1994).

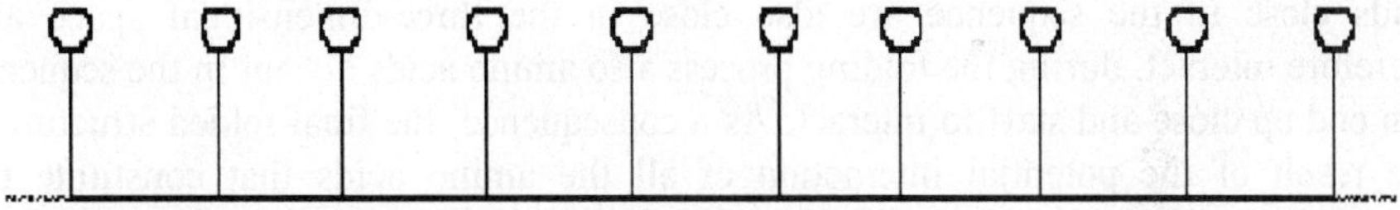

Fig. 1. Simulated protein at the beginning of the folding process.

The length of the link between the amino acid side-chain and the backbone is 25A and the pseudobond between two succeeding C_α is 15A (which approximates the average length in real proteins), but can slightly vary during the folding process. Different amino acid side chains all have the same dimensions but can differ in the way they interact with other amino acid side chains and possibly with other substances (e.g. water, but we have not explored this possibility yet). Side chains (spheres) by being attracted or repulsed by other side chains can move in the three-dimensional space. However, in doing so, because of the physical links, amino acid side chains can either (A) rotate around the backbone in the three dimensions modifying the angles between their link and the backbone and/or (B) bend the local portion of the backbone (see Fig. 2).

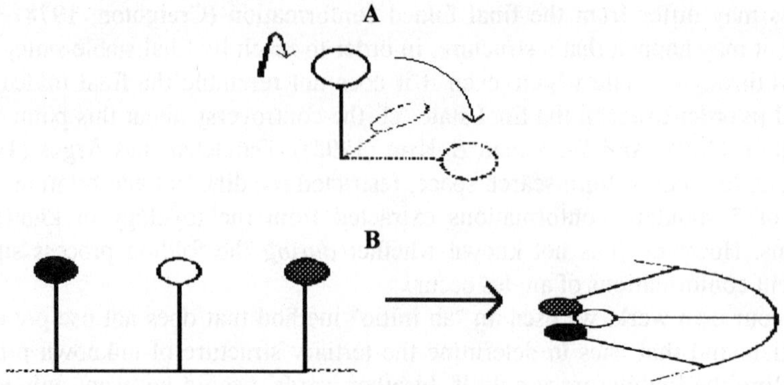

Fig. 2. A) A side-chain (sphere) rotating around the corresponding C_{α}. B) Side-chains that bend the backbone by reciprocal attraction.

A matrix of 20 x 20 values, which were initially randomly specified, determine for each amino acid how much it attracts or repulses other amino acids within a given distance (100A). The attraction or repulsion force is a function of both the value specified in the matrix and of the distance. The process starts with the backbone and the amino acids aligned (see Figure 1) then, depending on the types of amino acids and of the matrix of interaction forces, amino acids start to interact and as a consequence move and fold the backbone. Amino acids are let free to interact for 100 steps. During each cycle, all the interaction forces between neighbouring amino acids (spheres) are computed and then used to move and fold the structure. It is important to notice that while at the beginning of the folding process only amino acids close in the sequence are also close in the three-dimensional space and therefore interact, during the folding process also amino acids distant in the sequence can end up close and start to interact. As a consequence, the final folded structure is the result of the potential interaction of all the amino acids that constitute the sequence.

The problem now is how to determine the matrix of interaction forces in order to emulate the folding process. Once we have obtained a matrix able to fold primary structures into the right tertiary structures we can use such a matrix to predict the tertiary structure of unknown proteins by artificially folding them. For these reasons we can call this matrix of 'forces' *folding potentials*, i.e. potentials that do not extract regularities of known tertiary structures but instead represent 'forces' able to drive the folding process in order to produce correct tertiary structures.

In order to determine such folding potentials we used a genetic algorithm (Holland, 1975; Goldberg, 1989). We started with a population of 100 different matrices of folding potentials randomly generated that represent Generation 0. We then used these potentials to artificially fold proteins with known tertiary structures. In this way we obtained 100 different tertiary structures. The similarity of such tertiary structures with the known right tertiary structure was measured (see below) and used to determine which are the best individuals, i.e. the folding potentials that result in the best tertiary structures. The best 20 individuals were allowed to reproduce by generating 5 offspring each that are copies of the parent matrix of folding rules with the addition of mutations (i.e. random modifications of 10% of the folding potential values). These 20x5 individuals will constitute Generation 1. The process is then repeated for a certain number of generations. The folding potentials of each generation will tend to differ from the previous generation for 2 reasons: because they are the copies of the best individuals of the previous generation and because they receive mutations. Mutations may produce better or less good offspring with respect to the corresponding parents. However, selective reproduction will ensure that only individuals that received good mutations will be able to reproduce.

The evaluation of tertiary structures can be realized in different ways. For each residue segment one can measure the discrepancy of the alpha-carbon bend and torsion angles (see Oldfield & Hubbard, 1994) between the real known tertiary structure and the artificially folded structure produced by each individual matrix of folding potentials. The sum of these discrepancies represent a measure of the error produced by the corresponding folding potentials. Therefore the lower the error is, the higher the probability will be that the corresponding matrix of folding potentials will produce offspring. Alternatively, one can use the sum of discrepancies in distance measured between all combinations of C_{α} atoms in the three-dimensional space. In our first empirical attempts, the first method appeared to produce better results. In addition, one should decide whether to consider only the discrepancies at the end of the folding process (i.e. after 100 steps) or also as it is taking place.

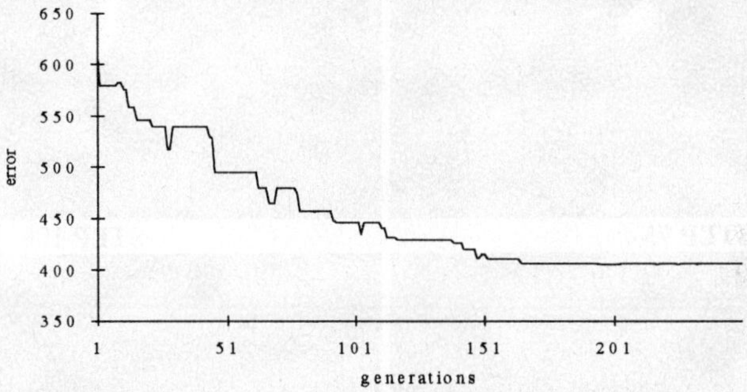

Fig. 3. Discrepancies between simulated folded proteins across generations in one of the most successful simulations. For each generation the error of the best individual of the population is shown.

We decided to pay attention to the discrepancies as the folding process is taking place but we weighted the discrepancies at the end of the folding process more in order to force the evolutionary process to select potentials that result in stable folded structures. Hence, the final evaluation of an individual is a weighted sum of the discrepancies throughout the folding process.

In a first attempt to test this model we have tried to simulate the folding process of a fragment of the crambin made of a sequence of 13 amino acids that result in an alpha-helix. We ran 10 simulations starting with different randomly generated folding potentials. As Fig. 3 shows, the error, i.e. the discrepancy between artificially folded structures and the real tertiary structure, progressively decreases across generations.

Fig. 4 shows six (not immediately) successive stages of the folding process generated by the evolved potentials.

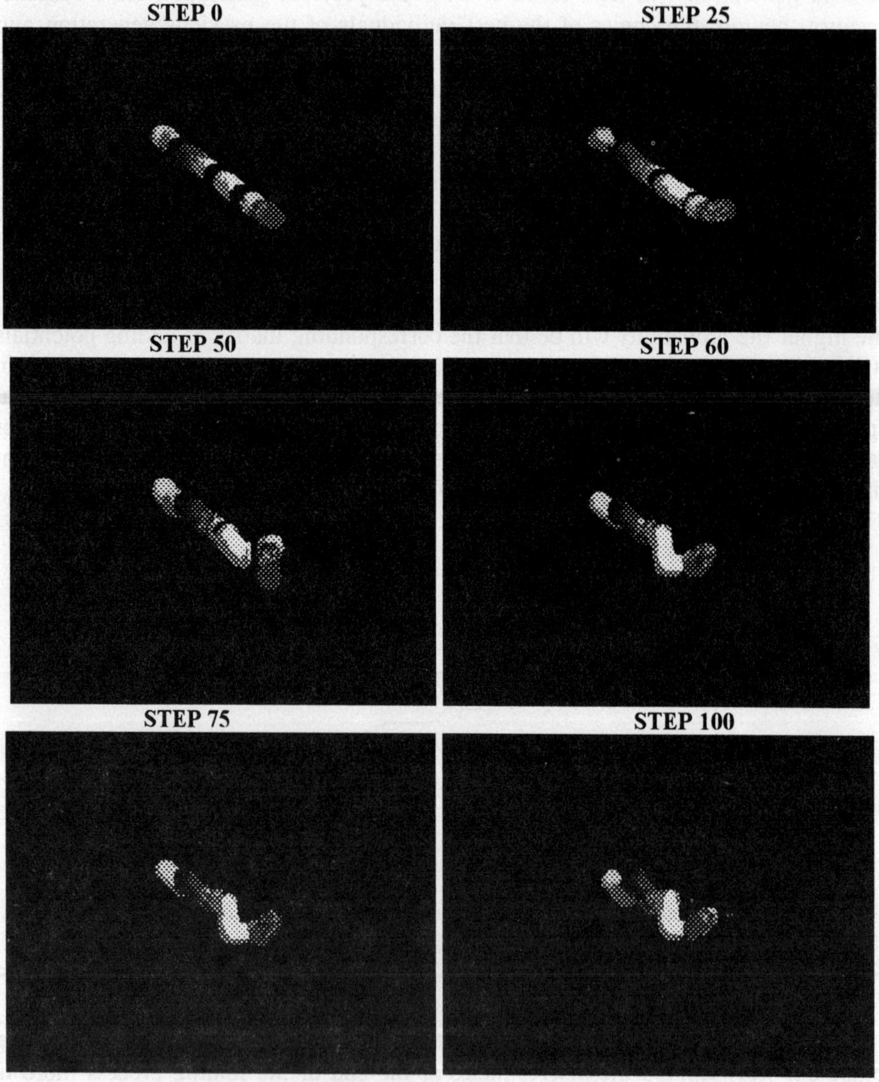

Fig. 4. Folding process of the best individual of the last generation of one of the most successful simulations. Only the backbone, represented as a chain of C_α atoms linked by pseudobonds, is displayed. For space reasons only 6 of the 100 time steps are shown.

A tertiary structure close to the expected one is obtained. In addition, it is interesting to note that in most of the simulations the tertiary structure stabilizes after a certain number of folding steps. How early the folding process reaches a stable state could be another component of the 'fitness formula' used to select folding potential matrices for reproduction.

5 Discussion

Artificial Life is an attempt at understanding all biological phenomena through their reproduction in artificial systems, e.g. computer simulations. More specifically, Artificial Life simulates life phenomena at various levels of biological entities (molecules, cells, organs, organisms, etc.) and tries to understand how phenomena at one level are related to phenomena at other levels. At the same time, Artificial Life is interested in determining similarities and differences in what happens at the various biological levels.

Computational approaches to the protein folding problem are often interpreted as alternative techniques for predicting the tertiary structure of proteins given their amino acid sequence. There is no implication that one is modelling or simulating the actual physico-chemical process that results in a given three-dimensional configuration starting from a linear sequence of amino acids. An Alife approach to protein folding suggests that one should try to model this process. The ability to predict the tertiary structure of unknown proteins should come as a by-product of these modelling efforts.

Assuming that one is modelling the amino acid sequence-to-tertiary structure mapping process, one can ask potentially useful questions about how this process is related to other biological processes and to their causes. For example, we have used a genetic algorithm to search for appropriate matrices of folding potentials that would give us the correct tertiary structure given an amino acid sequence. The genetic algorithm can be interpreted either as a search or optimization technique which is only 'inspired' by biological evolution or it can be taken to be a model of biological evolution. For example, when the genetic algorithm is applied to populations of neural networks, by using the genetic algorithm one may want to model the process of evolutionary change in a population of nervous systems, or organisms, or behaviors, etc., and to study such phenomena as the shape of evolutionary change (e.g. graduality or punctuated equilibria), evolutionary divergence, speciation, etc. Now, we can ask: When the genetic algorithm is applied to the protein folding problem, are we modelling some actual process of evolution which has taken place (and is taking place) at the molecular level and has shaped the mechanism that maps a linear amino acid sequence into a three-dimensional structure? Can the population of folding potential matrices be assimilated to a population of genotypes for neural networks?

The protein folding process can be viewed as a part of the larger process of mapping from the genotype to the phenotype of an organism which is called development. Can we find similarities between the process of protein folding which results in the 'adult' three-dimensional shape of the protein and the process of development which takes place during the developmental age of a multicellular organism and which result in the adult, mature form of the individual? For example, at the level of the organism all the successive phenotypical forms that are realized during development appear to be subject to an evaluation in terms of fitness. Is this the case also for the successive spatial conformations that a sequence of amino acid assumes before the final stable conformation? This problem is technically related to the choice of the 'fitness formula' when one is applying the genetic algorithm to the protein folding problem. We have adopted a fitness formula which takes into consideration all intermediate conformations but evaluates them only in function of their degree of approximation to the final conformation. Is this solution correct? We have advanced the hypothesis that during the folding process 'odd' conformations may appear that deviate from the final shape but are useful as stepping stones to arrive to the final shape. In this case a more sophisticated fitness formula would be more appropriate. (Notice that fitness formulae should not be necessarily decided by the researcher but can be viewed as evolvable - or co-evolvable - traits as any other trait; cf. Lund and Parisi, 1994). Although we cannot pretend that the intermediate states that are observed in our simulations actually reproduce the real intermediate states (Creighton, 1978), the novelty of our approach is that our model includes intermediate states that are usually ignored in computational attempts at predicting final three-dimensional protein structure.

In actual proteins the main force that drives the folding process appears to be hydrophobicity (i.e. the aversion for water of nonpolar residues) while Van der Waals interactions (i.e. interactions between dipoles), hydrogen bond interactions (i.e. sharing of an hydrogen atom between to two electronegative atoms), and electrostatics interactions in general appear to play a secondary role (Dill, 1990). However, as Dill states very clearly, driving forces are only half of the story. Another fundamental component that determines the folding process appears to be a opposing forces, e.g. the impossibility that two chain segments simultaneously occupy the same volume of space. Because the folding process involves the collapse of the chain from a large volume to a small one the role of this opposing force appear to be essential.

However, how the driving and the opposing forces produce the known three-dimensional structures is a controversial matter. Dill (1990) claims that any driving force, given the volume constraint (i.e. the fact that two elements cannot occupy the same space) would produce a structure with helices and sheets with hydrophobic interaction as the likely driving force. In fact, he claims that "there are very few possible ways to configure a compact chain, and most of them involve helices and sheet". In other words, he hypothesizes that the tertiary structure drives the secondary structures and not vice versa. Computer simulations appear to support this hypothesis because they show that the amount of secondary structure (helices and sheets) increases as a chain becomes increasingly compact (Dill, 1990). However this hypothesis is in contrast with NMR analysis which appear to suggest that "stable secondary structure first forms the framework necessary for the subsequent formation of the complete tertiary structure" (Udgaonkar & Baldwin, 1988). In addition, it remains to be determined what is the role of the other forces and why is the native structure unique given the fact that hydrophobicity cannot alone determine a unique native structure.

We think that models that reproduce the folding process like the one we have presented in this paper could shed some light on these issues. To pursue our objectives we certainly need to complicate our model by simulating the solution and by allowing the emergence of hydrophobic interactions between amino acids and the solution itself. This could be done by using an additional matrix that specifies for each amino acid the type and the strength of the hydrophobic interaction and by letting the genetic algorithm select the values contained in the matrices. It would then be interesting to observe which type of force will result the dominant one in the simulation, in particular if the hydrophobic forces will outnumber in strength the forces between amino acids.

We also claim that it might be misleading to try to predict the tertiary structures of unknown proteins by using minimization energy techniques based on potentials extracted by folded sequences. In fact, as we have observed, the type of conformations that proteins assume during the folding process may differ from final folded conformations. In addition, this approach requires that native conformations of proteins are at global energy minima (Anfinsen, 1973). But, as Baker & Agard (1994) have observed, "there are good reasons to think that the native states of proteins may not be at global energy minima.....there may be large regions of conformational space that are kinetically inaccessible in which a more stable state might exist". If this hypothesis is true, all computational efforts which try to find the global minimum of a specified potential function would be unable to predict the native state of proteins. We think that our approach which is not based on minimization of energy but tries to select a set of abstract forces which are able to induce the correct folding may avoid this problem.

References

Anfinsen, C. B. (1973). Principles that Govern the Folding of Protein Chains. *Science*, **181**, 223-230.

Bairoch, A. & Boeckmann, B. (1992). The SWISS-PROT Protein Sequence Data Bank. *Nucleic Acids Res.*, **20**, 2019-2022.

Baker, D. & Agard, D. A. (1994). Kinetics versus Thermodynamics in Protein Folding. *Biochemistry*, **33**, 7505-7509.

Bartel, D. P. & Szostak, J. W. (1993). Isolation of new ribozymes from a large pool of random sequences. *Science*, **261**, 1411-1418.

Bernstein, F. C., Koetzle, T. F., Williams, G. J. B., Meyer, E. F., Jr, Brice, M. D., Rodgers, J. R., Kennard, O., Shimanouchi, T., & Tasumi, M. (1977). The protein data bank: a computer-based archival file for macromolecular structures. *J. Mol. Biol.*, **112**, 535-542.

Bryant, S. H. & Lawrence, C. E. (1993). An Empirical Energy Function for Threading Protein Sequence through Folding Motif. *Proteins,* **16**, 92-112.

Casari, G. & Sippl, M. J. (1992). Structure-Derived Hydrophobic Potential. Hydrophobic Potential Derived from X-Ray Structures of Globular Proteins Is Able to Identify Native Folds. *J. Mol. Biol.*, **224**, 725-732.

Creighton, T. E. (1978). Experimental Studies of Protein Folding and Unfolding. *Prog. Biophys. Mol. Biol.*, **33**, 231-297.

Creighton, T. E. (1992). The Disulfide Folding Pathway of BPTI. *Science*, **256**, 111-112.

Creighton, T. E. (1993). *Proteins: Structures and Molecular Properties*. W. H. Freeman and Company, New York, NY.

Dandekar, T. & Argos, P. (1994). Folding the Main Chain of Small Proteins with the Genetic Algorithm. *J. Mol. Biol.*, **236**, 844-861.

Farber, G. K. & Petsko, G. A. (1990). The Evolution of α/β Barrel Enzymes. *Trends Biochem. Sci.*, **15**, 228-234.

Fariselli, P., Compiani, M. & Casadio, R. (1993). Predicting Secondary Structures of Membrane Proteins with Neural Networks. *Eur. Biophys. J.*, **22**, 41-51.

Gierasch, L. M. & King, J. (1990). *Protein Folding: Deciphering the Second Half of the Genetic Code*. American Association for the Advancement of Science, Washington, DC.

Goldberg, D. E. (1989). *Genetic Algorithms in Search, Optimization, and Machine Learning*. Addison-Wesley, Reading, MA.

Holland, J. J. (1975). *Adaptation in Natural and Artificial Systems*. University of Michigan Press, Ann Arbor, MI.

Kabsch, W., Mannherz, H. G., Suck, D., Pai, E. F. & Holmes, K. C. (1990). Atomic Structure of the Actin: DNase I Complex. *Nature*, **347**, 37-44.

Langton, C. G. (1992). Artificial Life. In *1991 Lectures in Complex Systems, SFI Studies in the Sciences of Complexity*, Lect. Vol. IV (Nadel, L. & Stein, D. eds.), Addison-Wesley, Reading, MA.

Lau, K. F. & Dill, A. (1990). Theory for protein mutability and biogenesis. *Proc. Natl. Acad. Sci. Usa*, **87**, 638-642.

Lehman, N. & Joyce, G. F. (1993). Evolution in vitro of an RNA enzyme with altered metal dependence. *Nature*, **361**, 182-185.

Lund, H. H. & Parisi, D. (1994). Simulations with an Evolvable Fitness Formula. *Technical Report* PCIA-1-94, C.N.R., Rome.

Lüthy, R., Bowie, J. U. & Eisenberg, D. (1992). Assessment of Protein Models with Three-Dimensional Profiles. *Nature*, **356**, 83-85.

Maiorov, V. N. & Crippen, G. M. (1992). A Contact Potential that Recognizes the Correct Folding of Globular Proteins. *J. Mol. Biol.*, **227**, 876-888.

Oldfield, T. J. & Hubbard, R. E. (1994). Analysis of C_α Geometry in Protein Structures. *Proteins*, **18**, 324-337.

Parisi, D., Cecconi, F. & Nolfi, S. (1990). Econets: Neural Networks that Learn in a Environment. *Network*, **1**, 149-168.

Qian, N. & Sejnowski, T. J. (1988). Predicting the Secondary Structure of Globular Proteins Using Neural Network Models. *J. Mol. Biol.*, **202**, 865-884.

Rooman, M. J., Kocher, J-P. A. & Wodak, S. J. (1991). Prediction of Protein Backbone Conformation Based on Seven Structure Assignments: Influence of Local Interactions. *J. Mol. Biol.*, **221**, 961-979.

Rost, B. & Sander, C. (1994). Combining Evolutionary Information and Neural Networks to Predict Protein Secondary Structure. *Proteins*, **19**, 55-72.

Rost, B., Sander, C. & Schneider, R. (1994). Redefining the Goals of Protein Secondary Structure Prediction. *J. Mol. Biol.*, **235**, 13-26.

Rumelhart, D. E. & McClelland, J. L. (1986). *Parallel Distributed Processing. Explorations In the Microstructure of Cognition*. MIT Press, Cambridge, MA.

Šali, A., Shakhnovich, E. & Karplus, M. (1994). Kinetics of Protein Folding. A Lattice Model Study of the Requirements for Folding to the Native State. *J. Mol. Biol.*, **235**, 1614-1636.

Sippl, M. J. & Weitckus, S. (1992). Detection of Native-Like Models for Amino Acid Sequences of Unknown Three-Dimensional Structure in a Data Base of Known Protein Conformations. *Proteins*, **13**, 258-271.

Sondek, J. & Shortle, D. (1990). Accommodation of Single Amino Acid Insertions by the Native State of Staphylococcal Nuclease. *Proteins*, **7**, 299-305.

Taylor, C. & Jefferson, D. (1994). Artificial Life as a Tool for Biological Inquiry. *Artificial Life*, **1**, 1-13.

Udgaonkar, J. B. & Baldwin, R. L. (1988). NMR evidence for an early framework intermediate on the folding pathway of ribonuclease A. *Nature*, **335**, 694-699.

Unger, R. & Moult, J. (1993). Genetic Algorithms for Protein Folding Simulations. *J. Mol. Biol.*, **231**, 75-81.

Weissman, J. S. & Kim, P. S. (1992). The Disulfide Folding Pathway of BPTI. *Science*, **256**, 112-114.

Wilson, S. W. (1991). The Animat Path to AI. In *From animals to animats: Proceedings of the First International Conference on Simulation of Adaptive Behavior* (Meyer, J.-A. and Wilson, S. W., eds), pp. 15-21, MIT Press, Cambridge, MA.

Wodak, S. J. & Rooman, M. J. (1993). Generating and testing protein folds. *Curr. Opin. Struct. Biol.*, **3**, 247-259.

Energy cost evaluation of computing capabilities in biomolecular and artificial matter

R. Lahoz-Beltra [1] and S.R. Hameroff [2]

[1] Laboratorio de Bioinformatica, Departamento de Matematica Aplicada, Facultad de Biologia, Universidad Complutense, Madrid-28040, Spain

[2] Advanced Biotechnology Laboratory, Department of Anesthesiology, University of Arizona, Health Sciences Center, Tucson, AZ 85724, USA

Abstract. Propierties which define living systems at the molecular level include self-organization, communication, adaptive behavior and computation. Logic functions may implement these propierties in biomolecules and therefore may be essential to living systems. Several logic systems from Boolean to Spencer-Brown algebra have been suggested to be applicable to molecular computation. Boolean equations are commonly implemented on silicon chips on computer, but may also exist in nature. For example, the *lac* and *arabinose* operons in *E. coli*, some genetic networks of the metazoan genome, the self-assembly of proteins (i.e. viruses like T4 bacteriophage, cellular cytoskeletal elements, etc) display Boolean logic and show the capability for symbolic logic manipulation in biological connectionist systems. In biological systems logic operations are carried out in nonlinear devices or automata (i.e protein, gen, neuron, etc) producing an output that represent a logical function. Combination of logical functions is the substrate for biocomputation: an emergent propierty of biological systems. In the present article we introduce a method to evaluate the *computational activity* of automata in the context of biomolecular and artificial (cellular automata) matter. The method is illustrated in two different situations, in the cellular automata realm as well as in a model of finite state protein based computation.

1 Introduction

Propierties which define living systems at the molecular level include self-organization, communication, adaptive behavior and computation. Logic functions may implement these propierties in biomolecules and therefore may be essential to living systems. Several logic systems, included Boolean algebra, have been suggested to be applicable to molecular computation. For example, in *The Laws of Form*,

Spencer-Brown [1] developed an algebraic logic (Table I) derived from two fundamental axioms based on quantum mechanical projection and annihilation/ creation operations. Siatkowski and Carter in [2] observed that various chemical systems can satisfy these two axioms and are candidates for logical operations at molecular level. Stern and Stern [3] developed a logical formalism capable of dealing with the ambivalence of atomic-molecular switching and Milch in [4] proposed a layered molecular approach for computation. Boolean algebra was introduced in 1847 by George Boole as a tool to logically manipulate postulates or statements by means of connective operators known as Boolean operators (OR, AND, NOT, etc). Such operators represent simple laws (Table I) which Boole considered as basic rules underlying the logic of nature. Operators establish a logical connection between related statements which determine their implication. Complex combinations of statements can be connected by these logic operators defining Boolean equations. These equations are commonly implemented on silicon chips on computer, but may also exist in nature. For example, the *lac* operon and the *arabinose* operon in *E. coli* [5], some genetic networks of the metazoan genome [6], the self-assembly of proteins (Figure 1) - for example, viruses like T4 bacteriophage - or cellular cytoskeletal elements like microtubules and microtubule associated proteins [7] display Boolean logic. These examples show the capability for symbolic logic manipulation in connectionist systems, demonstrating that a dichotomy need not exist between symbolic and connectionist systems [8].

Table I

Spencer-Brown	Boole	S.R.E. (1)
$\overline{a\rvert}$	not a	-
ab	a or b	$\dot{x} = \frac{(y+z)^2}{\theta+(y+z)^2} - kx$
$\overline{\overline{a\rvert}\ \overline{b\rvert}\rvert}$	a and b	$\dot{x} = \frac{(yz)^2}{\theta+(yz)^2} - kx$
$\overline{a\rvert}\,b$	a --> b	-

(1) S.R.E. = sigmoidal rate equations (see [6])

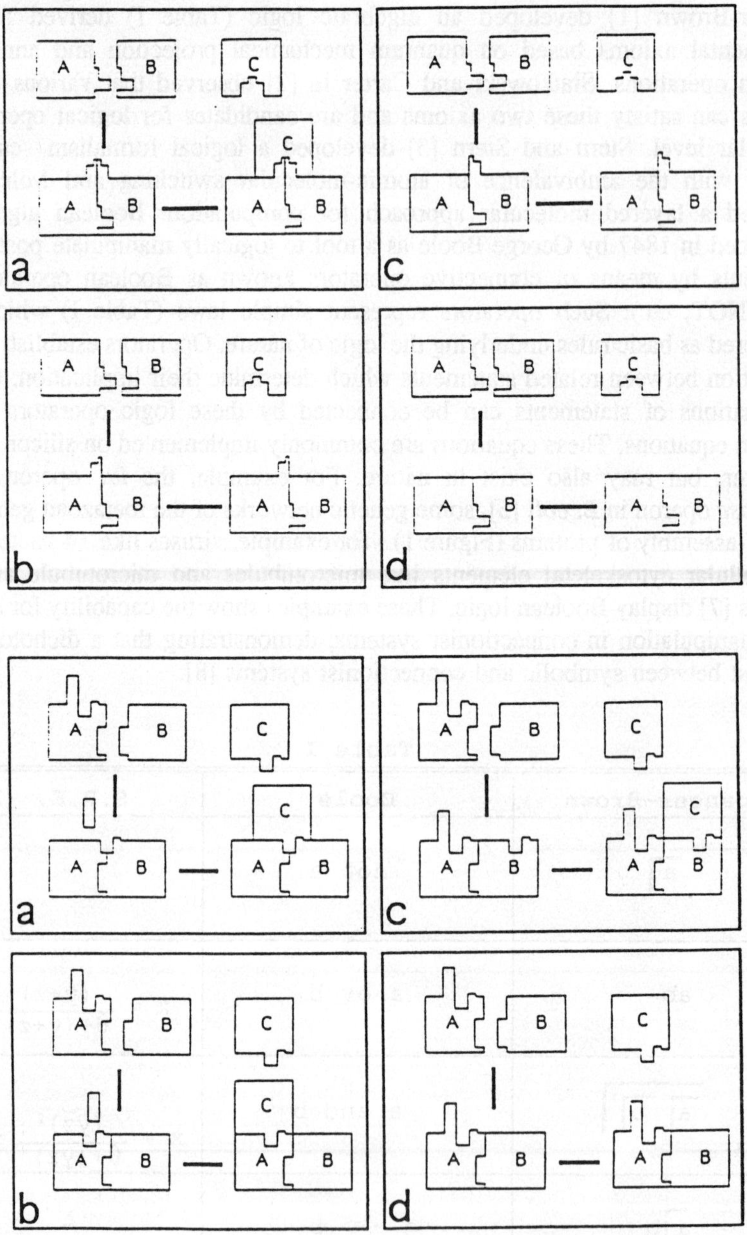

Figure 1.- Self-assembly in biomolecular systems can manifest behavior analogous to Boolean algebra. Boolean switching rules: sorting of subunits (A,B,C) with operators AND (top) and OR (bottom) (see [7]).

In 1986 Langton [9] introduced the concept of cellular automata as artificial matter which can be used to create cellular automata computers. In cellular automata a lattice subunits with discrete states interact only with nearest neighbors. Simple rules governing subunit neighbor interactions can lead to complex patterns in the lattice subunits capable of computation. Based on glider streams (patterns that move through the lattice unchanged) we could encode any binary number in the same way that digital pulses are in the bus of a computer. For instance, using this approach the two-state machine proposed by Banks in [10] is a good example of cellular automata computing machine. Poundstone in [11] has shown that is possible to build NOT and AND gates with the Conway's game of Life algorithm, and by using these logic gates a computer and a Turing machine could be build.

Thus, Boolean algebra is a system of basic logic which can exist both in nature (i.e. at the molecular level) and in artificial computation (i.e. cellular automata computer). In other words, computation is a general phenomena present in biomolecular and artificial matter. Information processing during computation either biomolecular or artificial matter exhibit the following essential features:

(a) At a given time, two or more streams of information interact [12].

(b) Information processing after or during information interaction is an irreversible phenomena (for a discussion about 3+2=5 operation and AND operator, see [13]).

(c) Organic and artificial matter may be in a state of nonequilibrium in order to be able of computation.

(d) The minimum of energy per single logical operation [14] is given by:

$$[1] \quad E = k\ T\ Ln(2)$$

where k is the Boltzmann constant and T is the temperature in Kelvin. At room temperature $E_{min} = 3 \times 10^{-14}$ erg.

Feature (a) is an obvious property in connectionist systems, for example the case of metabolic pathways or neural networks. However, features (b), (c) and (d) represent a set of restrictions to computing capabilities of matter. For instance, in biological systems the hydrolysis of ATP releases 30.5×10^{10} erg/mol. Assuming a theoretical cell where single logic operations involved in complex processes (i.e. genotype regulation, conformational changes in enzymes, etc) would need only 3×10^{-14} erg (E_{min}) then the total number of single logic operations per mol of ATP

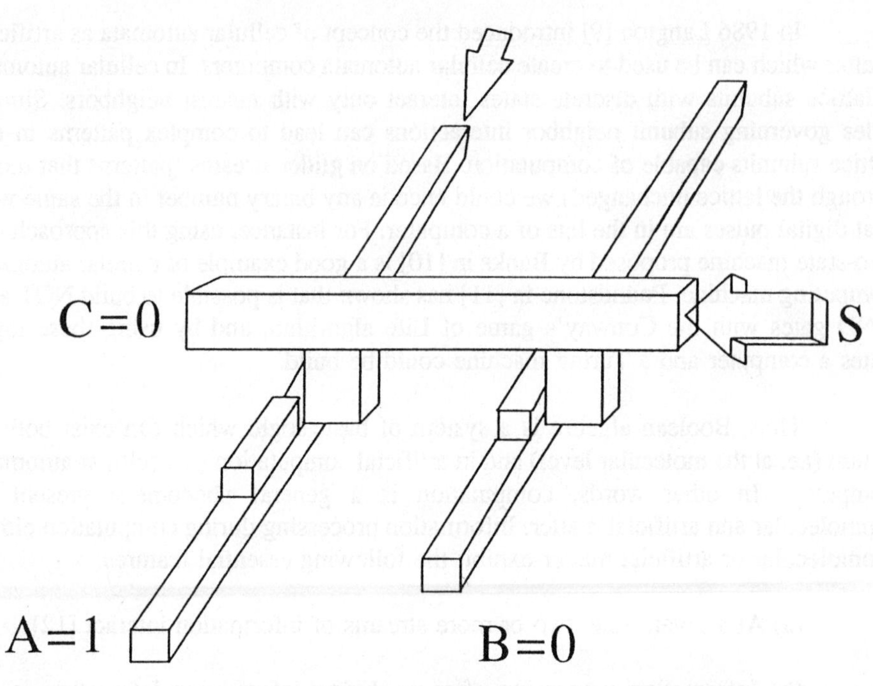

Figure 2.- Logic gate AND implemented with LOGO toy elements (modified from [18]). Bars A,B in resting position define an input equal to 0, otherwise (pushing the bar) input is defined equal to 1. If A and B bars are pushed then bar C moves (by the action of a spring) to the direction indicated by S arrow being the gate output equal to 1.

would be equal to 1.0166×10^{25}. This value seems very high however it is an unrealistic estimation since single logic operations in biological systems may dissipates a greater E value than single binary operations (eq.[1]). For example, let A be the statement 'substrate loses a proton', let B be the statement 'substrate bound to lysine' and let C be the implication 'reaction enzyme-substrate is successful'. If the statements A and B are connected by the operator AND then the implication will be determined by the logic AND function. Obviously, the type of AND function underlying in such enzyme-substrate reaction is closer to the mechanism depicted in Figure 2 than the AND function connecting statements A and B, resulting a value of energy per logical operation that may be above of E_{min}. A second example to illustrate these considerations is given by the spontaneous self-assembly of the tobacco mosaic virus (TMV) which could be translated to the pseudocode program showed in Figure 3. The TMV polymerization may dissipates a value of E above E_{min} being such polymerization closer to the mechanical description given by Pattee in [15] for a growing helical chain than the logical functions and operations showed in Figure 3.

```
{-----------------------------------}
{ Initialization                    }
{-----------------------------------}
BEGIN
  {
   Formation of a disc with 17 subunits
   of protein
                       }
  TEMPLATE(T)
END;

{-----------------------------------}
{ Polymerization_Subroutine         }
{-----------------------------------}

IF N = 1            { N, disc number }
  THEN
    BEGIN
      {
       RNA attaches to the TEMPLATE(T)
                                       }
      RNA_TEMPLATE(T);
      {
       New subunits form disc (T+1)
                                       }
      DISC(T+1);
      {
       Disc (T+1) polymerize onto the
       TEMPLATE(T)
                                       }
      START_POLYMER := RNA_TEMPLATE(T) + DISC(T+1)
    END;

IF N >= 2
  THEN
    BEGIN
      POLYMER := START_POLYMER;
      REPEAT
        BEGIN
          {
            RNA attaches to the disc (T+(N-1))
                                              }
          RNA_POLYMER(T+(N-1));
          {
            New subunits form disc (T+N)
                                              }
          DISC(T+N);
          {
            Disc (T+N) polymerize onto the
            disc (T+(N-1))
                                              }
          POLYMER := RNA_POLYMER(T+(N-1)) + DISC(T+N)
          {
            Increases N a unit
                                              }
          N := N + 1
        END;
      UNTIL (ALL RNA IS COILED INTO ITS GROOVE)
    END;
```

Figure 3.- Self-assembly of tobacco mosaic virus (TMV) translated to a pseudocode program.

In other words, in biological systems logic operations are carried out in devices (i.e protein, gen, neuron, etc) which could be described as automata. Such automata or their resulting structures (i.e network, lattice, assembly, etc) must be able to combine input signals in a nonlinear way producing an output that represent a logical function. Combination of logical functions is the substrate for biocomputation: an emergent propierty of biological systems.

2 Computational activity evaluation in automata

In the present article we introduce a definition to evaluate the *computational activity* of automata modeling biomolecular or artificial matter. Computation is understood in a broad sense as a logic operation (i.e transition from state S_1 to state S_2, $S_5 = S_3$ AND S_4, etc) and activity as the number of logical operations that takes place from time t-1 to t. The present definition is based on Hamming distance (number of different digit positions in two binary words of equal length) and in the minimum of energy required per logical operation given by von Neumann's expression [1]. The minimum of energy E per computational activity in an automata structure (i.e cellular automata lattice, genetic network, protein assembly, etc) that takes place from time t-1 to t, is given by the following expression:

$$[2]\quad E = k\,T\,\mathrm{Ln}(S) \sum_{i} S_i^{t} \underline{\vee}\, S_i^{t-1}$$

where k is the Boltzmann constant, T is the temperature in Kelvin, S is the number of automaton states and S_i the state of the automaton at i location. In the equation the sum term is the Hamming distance (defined with exclusive OR operator) between automata structures or patterns at time t and t-1. Since Hamming distance is related with the activity of the automata, and assuming an equal minimum of energy value per logical operation in each one of the automaton, then expression [2] evaluates the total computational activity achieved each time a cycle or iteration is updated.

2.1 Cellular automata

To illustrate the above definition Figures 4-7 show two different examples with cellular automata. Computer simulation experiments were performed with

Figure 4.- Cellular automata pattern displayed at iteration 30 for Conway's game of Life in two simulation experiments with initial density values D=0.10 (top) and D=0.25 (bottom).

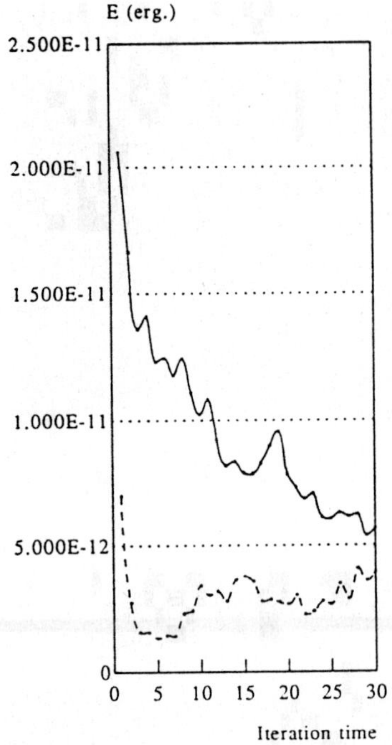

Figure 5.- Energy cost value per iteration for Conway's game of Life. The figure shows results obtained in two simulation experiments with initial density values D=0.10 (broken line) and D=0.25 (continuous line).

cellular automata defined on 25x80 lattice, with S=2 (two states, black or white) and at room temperature. Figures 4 and 5 show the obtained results for Conway's game of Life. Two simulation experiments were performed with different initial density values (D), equal to D=0.10 and D=0.25, in each one of the experiments. Figure 5 shows how in experiment with D=0.25, E decreases being the system close to equilibrium, whereas in the experiment with D=0.10 the E value oscillates being the system in a state of nonequilibrium. After thirty iterations the total energy in each one of the experiments was 8.754×10^{-11} and 2.851×10^{-10} for D=0.10 and D=0.25 densities, respectively. These results suggest that convergence of the system to equilibrium, which violates one of the essential propierties for computation, represent three times more of energy cost than a system in a dissipative state.

Figures 6 and 7 show similar experiments to described above but using cellular automata with directed aggregation. After thirty iterations the total energy in each one of the experiments was 5.737×10^{-10} and 2.949×10^{-10} for D=0.10 and D=0.25 densities, respectively. The obtained results indicate that for the cellular automata pattern with higher complexity (D=0.10) the cost of energy is two times the cost for the pattern with lower complexity (D=0.25).

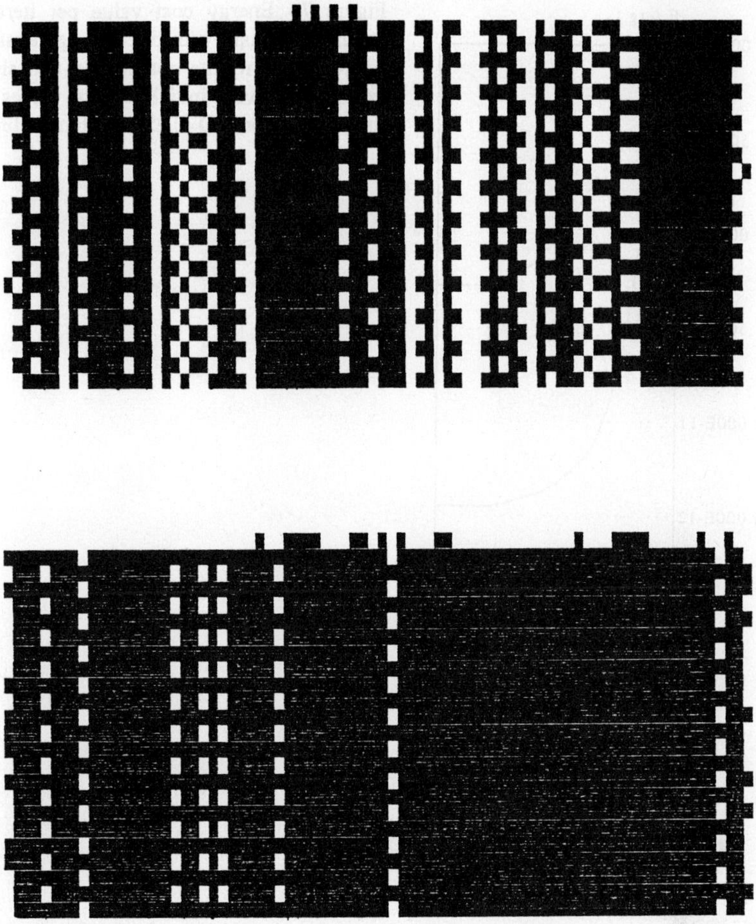

Figure 6.- Cellular automata pattern displayed at iteration 30 for cellular automata with directed aggregation in two simulation experiments with initial density values D=0.10 (top) and D=0.25 (bottom).

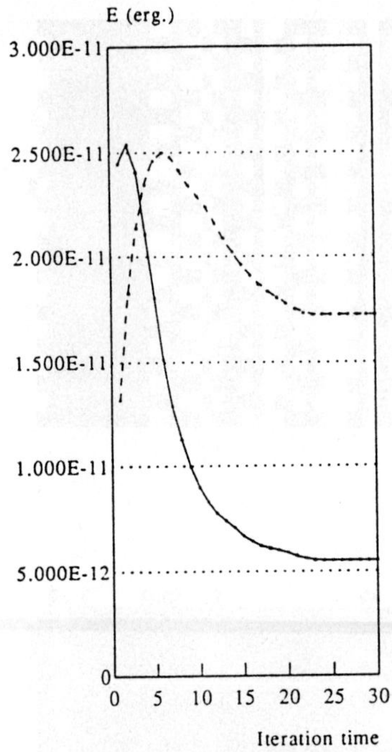

Figure 7.- Energy cost value per iteration for cellular automata with directed aggregation. The figure shows results obtained in two simulation experiments with initial density values D=0.10 (broken line) and D=0.25 (continuous line).

2.2 Molecular automata

In the field of molecular computing, computational activity in molecular automata (see [16] and [17]) could be evaluated based on E values. Computation based on protein conformational changes is one of the key mechanisms proposed to develop molecular computers. Consider a model where a protein is described as a Turing machine (Figure 8). In this model, changes in the protein environment (solvent, temperature, pH, ionic strength, neighborhood conformation, phosphorylation, ligand, etc) will lead to changes of conformation represented in Figure 8 by protein local movements modeling hydrogen bonds formation/breaking, hydrophobic interactions, electrostatic interactions, etc. Suppose that protein environment (E_i) is depicted as a *tape* moving left (L) or right (R) with squares holding environmental inputs coded as binary words (1111, 1110, 1101, etc), and the local movements of the protein (conformational change) are described by a discrete set of states (00, 01, 10, 11) representing their combinations (protein global movements) the *internal states* (IS_i) of the automaton. Having defined basic

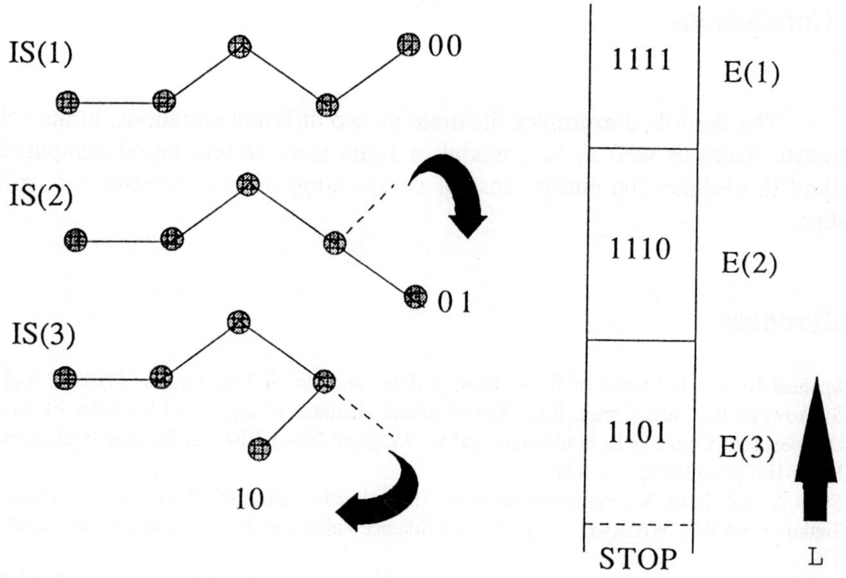

Figure 8.- Computation based on protein conformational changes is one of the key mechanisms proposed to develop molecular computers. The figure shows a model of protein described as Turing machine (for explanation see text).

components, in the example showed in Figure 8 (only one protein location) we have the following sequence of steps:

IS_1=00 E_1=1111 --> IS_2=01 E_1=1111 L
IS_2=01 E_2=1110 --> IS_3=10 E_2=1110 L
IS_3=10 E_3=1101 --> IS_3=10 E_3=1101 L.STOP

Including in the automaton (Figure 8) the three protein remaining locations, with 4 locations and 4 states (00, 01, 10, 11) per location, the computational activity of the protein can be evaluated by the following expression:

$$[3]\quad E = k\,T\,\mathrm{Ln}(4^4) \sum_{i=1}^{4} \sum_{j=1}^{2} S_{ij}^{t} \veebar S_{ij}^{t-1}$$

where S_{ij} is the bit value, i the protein location (i = 1...4) and j the bit location (j = 1,2).

3 Conclusions

The described examples illustrate in two different situations, in the cellular automata realm as well as in a model of finite state protein based computation a method to evaluate the energy cost of computation in biomolecular and artificial matter.

References

1. Spencer-Brown, G.: Laws of Form, George Allen and Unwin Ltd, London (1969) 1-141
2. Siatkowski, R.E. and Carter, F.L.: The chemical elements of logic, in Molecular Electronic Devices, F.L. Carter, R.E. Siatkowski and H. Wohltjen (eds.), Elsevier Science Publishers B.V., North-Holland (1988) 321-327
3. Stern,A. and Stern, V.: Fundamental logic, in Molecular Electronic Devices, F.L. Carter, R.E. Siatkowski and H. Wohltjen (eds.), Elsevier Science Publishers B.V., North-Holland (1988) 329-339
4. Milch, J.R.: Computers based on molecular implementations of cellular automata, in Molecular Electronic Devices, F.L. Carter, R.E. Siatkowski and H. Wohltjen (eds.), Elsevier Science Publishers B.V., North-Holland (1988) 303-317
5. Kauffman, S.A.: Assessing the possible regulatory structures and dynamics of the metazoan genome, in Kinetic Logic, Lecture Notes in Biomathematics, R. Thomas (ed.), Springer, Berlin **29** (1979) 30-36
6. Kauffman, S.A.: The self-organization of dynamical order in the evolution of metazoan gene regulatiuon, in The Living State, R.K. Mishra (ed.), John Wiley (1984) 291-307
7. Lahoz-Beltra, R., Hameroff, S.R., Dayhoff, J.E.: Cytoskeletal logic: a model for molecular computation via Boolean operations in microtubules and microtubule-associated proteins, BioSystems **29** (1993) 1-23
8. Fodor V. and Pylyshyn, Z.: Connectionism and cognitive architecture: a critical analysis in connections and symbols, in A Cognition Special Issue, S. Pinker and J. Mehler (eds.), MIT Press, Cambridge, MA (1988) 3-71
9. Langton, C.G.: Studying artificial life with cellular automata, Physica D **22** (1986) 120-149.
10. Banks, E.R.: Universality in cellular automata, I.E.E.E. 11th Ann. Sym. Switching and Automata Theory, Santa Clara, California (1970) 194-214
11. Poundstone, W.: The recursive universe: cosmic complexity and the limits of scientific knowledge. Contemporary Books, Chicago (1985) 1-252
12. Keyes, R.W.: Communication in computation, International Journal of Theoretical Physics **21**, Nos. 3-4 (1982) 263-273
13. Rietman, E.: Experiments in artificial neural networks, TAB Books, Advanced Technology Series (1988) 1-144

14. von Neumann, J.: Re-evaluation of self-reproducing automata, in The theory of self-reproducing automata, A.W. Burkes (ed.), University of Illinois Press, Urbana (1966)
15. Pattee, H.H.: Physical theories, automata, and the origin of life, in Natural Automata and Useful Simulations, H.H. Pattee, E.A. Edelsack, L. Fein, A.B. Callahan (eds.), Spartan Books, Washington (1966) 73-105
16. Conrad, M.: Molecular automata, in Lecture Notes in Biomathematics, Physics an mathematics of the Nervous System, M. Conrad, W. Guttinger and M. Dal Cin (eds.), Springer-Verlag, Heidelberg **4** (1974) 419-430
17. Hameroff, S.R., Dayhoff, J.E., Lahoz-Beltra, R., Samsonovich, A.V., Rasmussen, S.: Models for molecular computation: conformational automata in the cytoskeleton, IEEE Computer (special issue on Molecular Computing) **25**, No. 11 (1992) 30-39
18. Lampton, C.: Nanotechnology playhouse, The Waite Group (1993) 1-181

8. Applications and Common Tools

Contemporary Evolution Strategies

Hans-Paul Schwefel[1] and Günter Rudolph[2]

[1] University of Dortmund, Department of Computer Science, D-44221 Dortmund
[2] Informatik Centrum Dortmund at the University of Dortmund, D-44227 Dortmund

Abstract. After an outline of the history of evolutionary algorithms, a new $(\mu, \kappa, \lambda, \rho)$ variant of the evolution strategies is introduced formally. Though not comprising all degrees of freedom, it is richer in the number of features than the meanwhile old (μ, λ) and $(\mu+\lambda)$ versions. Finally, all important theoretically proven facts about evolution strategies are briefly summarized and some of many open questions concerning evolutionary algorithms in general are pointed out.

1 A Brief History of Evolutionary Computation

Since the spring of 1993, when the first issue of the international journal on Evolutionary Computation [1] appeared, more and more people have become aware of computer algorithms known to insiders since nearly thirty years as Genetic Algorithms (GAs), Evolution Strategies (ESs), and Evolutionary Programming (EP). As a common denominator the term Evolutionary Algorithms (EAs) has become rather widespread meanwhile.

Likewise common to all three approaches to computational problem solving, mostly in terms of iterative adaptation or stepwise optimization, is the use of biological paradigms gleaned from organic evolution. Exploring the search space by means of a population of search points that underlie variation, and exploiting the information gathered in order to drive the population to ever more promising regions, are viewed as mimicking mutation, recombination, and selection.

Apart from these common features, both the independent origins and the currently used incarnations of the GA, EP, and ES algorithms differ from each other considerably. Simulating natural phenomena and processes goes back to the time when electronic automata became available to academia. Artificial neural nets, homeostatic controllers, predictors and optimizers emerged, and if their environment could not easily be modeled in simple analytical terms, random number generators became attractive in order to bridge the knowledge gap towards deciding upon the next action within the circuits. Ashby's homeostat [2], Brooks' [3] and Rastrigin's [4] optimizers are witnesses of those days.

The first digital computers, however, were not that fast as often called in papers of that time. Thus artificial or simulated evolutionary processes to solve real-world problems [5, 6] had to give way to quicker problem solving methods relying upon simple computational models like linear or quadratic input-output relations. Another reason for the dominance of greedy algorithms has been their strong backing from theory. People like to know in advance whether the (exact) solution of a problem will be found with guarantee and how many cycles of an iterative scheme are necessary to reach that goal. That is why rather often problems have been fitted to the abilities of solution algorithms by rigorous simplification.

Besides Bremermann's simulated evolution [6], used for solving nonlinear systems of equations, for example, L. Fogel [7] devised a finite state automaton for prediction tasks by simulating evolutionary mechanisms. The current version of this approach, called Evolutionary Programming (EP), is due to D. Fogel [8] and is used for continuous parameter optimization. Discrete, even combinatorial search, adaptation, and optimization is the domain of Holland's [9] Genetic Algorithm (GA) which comprises many different versions nowadays. Independently of both these origins, Evolution Strategies (ESs) came to the fore as experimental optimization techniques [10, 11], e.g., to drive a flexible pipe bending or changeable nozzle contour successively into a form with minimal loss of energy. Similar to Evolutionary Operation (EVOP [12]) the variables were changed in discrete steps, but stochastically instead of deterministically. The earliest ES version operated on the basis of two individuals only, one parent and one descendant per generation.

Theory as well as practice, more and more turning to computer simulation instead of expensive real experiments, led to multimembered ESs operating on the basis of continuous decision variables [13–17] with $\mu > 1$ parents and $\lambda > \mu$ children per cycle.

The following section introduces an even more general ES version. In section 3 a short summary will be given to what theory can tell us already about reliability and efficiency. Finally, a couple of open questions will be pointed out, and EAs in general will be turned to again.

2 The $(\mu, \kappa, \lambda, \rho)$ Evolution Strategy

Though meanwhile many, sometimes specialized, ES versions exist, we will restrict ourselves to a rather canonical set of features here. Neither parallel nor multiobjective, neither discrete nor mixed-integer special forms are considered in the following – though they exist and have been used already in applications. A recent overview may be found in [17].

In the beginning, there existed two different forms of the multimembered evolution strategy, namely the $(\mu+\lambda)$ and the (μ, λ) ESs. The symbol μ denotes the number of parents appearing at a time in a population of imaginary individuals. The symbol λ stands for the number of all offspring created by these parents within one (synchronized) generation. The difference between both approaches consists in the way the parents of a new generation are selected.

In the $(\mu + \lambda)$ ES the λ offspring and their μ parents are united, before according to a given criterion, the μ fittest individuals are selected from this set of size $\mu+\lambda$. Both μ and λ can be as small as 1 in this case, in principle. Indeed, the first experiments were all performed on the basis of a $(1 + 1)$ ES. In the (μ, λ) ES, with $\lambda > \mu \geq 1$, the μ new parents are selected from the λ offspring only, no matter whether they surpass their parents or not. The latter version is in danger to diverge (especially in connection with self-adapting variances - see below) if the so far best position is not stored externally or even preserved within the generation cycle (so-called elitist strategy). We shall come back to that later on. So far, only empirical results have shown that the comma version has to be preferred when internal strategy parameters have to be learned on-line collectively. For that to work, $\mu > 1$ and intermediary recombination of the mutation variances seem to be essential preconditions. It is not true that ESs consider recombination as a subsidiary operator.

The (μ, λ) ES implies that each parent can have children only once (duration of life: one generation = one reproduction cycle), whereas in the plus version individuals may live eternally – if no child achieves a better or at least the same quality. The new $(\mu, \kappa, \lambda, \rho)$ ES introduces a maximal life span of $\kappa \geq 1$ reproduction cycles (iterations). Now, both original strategies are special cases of the more general strategy, with $\kappa = 1$ resembling the comma- and with $\kappa = \infty$ resembling the plus-strategy, respectively. Thus, the advantages and disadvantages of both extremal cases can be scaled arbitrarily. Other new options include:

- free number of parents involved in reproduction (not only 1, 2, or all)
- tournament selection as alternative to the standard (μ, λ) selection
- free probabilities of applying recombination and mutation
- further recombination types including crossover.

Though in the first ES experiments the object variables were restricted to discrete values, computerized ESs have mostly been formulated for the continuous case. An exception may be found in Rudolph [18]. The $(\mu, \kappa, \lambda, \rho)$ evolution strategy is defined here for continuous variables only by the following 19-tuple:

$$(\mu, \kappa, \lambda, \rho)ES := (P^{(0)}, \mu, \kappa, \lambda, \mathbf{rec}, p_r, \rho, \gamma, \omega, \mathbf{mut}, p_m, \tau, \tau_0, \delta, \beta, \mathbf{sel}, \zeta, t, \varepsilon) \quad (1)$$

with

$P^{(0)} := (\mathbf{a}_1, \ldots, \mathbf{a}_\mu)^{(0)} \in I^\mu$	$I := \mathbb{N}_0 \times \mathbb{R}^n \times$ $\times \mathbb{R}_+^{n_\sigma} \times [-\pi, \pi]^{n_\alpha}$	start population
$\mu \in \mathbb{N}$	$\mu \geq 1$	number of parents
$\kappa \in \mathbb{N}$	$\kappa \geq 1$	upper limit for life span
$\lambda \in \mathbb{N}$	$\lambda > \mu$ if $\kappa = 1$	number of offspring
$\mathbf{rec} : I^\mu \rightarrow I$		recombination operator
$p_r \in \mathbb{R}_+$	$0 \leq p_r \leq 1$	recombination probability
$\rho \in \mathbb{N}$	$1 \leq \rho \leq \mu$	number of ancestors for each descendant
$\gamma \in \mathbb{N}$	$1 \leq \gamma \leq n_x - 1$ $\gamma \geq \rho - 1$	number of crossover sites in a string of n_x elements
$\omega \in \{0, 1, 2, 3, \ldots\}$		type of recombination
$\mathbf{mut} : I \rightarrow I$		mutation operator
$p_m \in \mathbb{R}_+$	$0 < p_m \leq 1$	mutation probability

$\tau, \tau_0, \delta \in \mathbb{R}_+$	$0 \leq \delta \leq 1$	step length variabilities
$\beta \in \mathbb{R}_+$	$0 \leq \beta \leq \frac{\pi}{4}$	correlation variability
$\mathbf{sel} : I^{\mu+\lambda} \rightarrow I^{\mu}$		selection operator
$\zeta \in \mathbb{N}$	$2 \leq \zeta \leq \mu + \lambda$	tournament participators
$t : I^{2\mu} \rightarrow \{0, 1\}$		termination criterion
$\varepsilon \in \mathbb{R}_+^4$		accuracies required.

p_m, p_r, and ω may be different for object and strategy parameter variation.

The (n, f, m, G) optimization problem for continuous variables at hand may be defined as follows:

$$\text{Minimize } f(\mathbf{x}) \mid \mathbf{x} \in M \subseteq \mathbb{R}^n \tag{2}$$

with

$n \in \mathbb{N}$	dimension of the problem
$f : M \rightarrow \mathbb{R}$	objective function
$M = \{\mathbf{x} \in \mathbb{R}^n \vert g_j(\mathbf{x}) \geq 0 \ \forall\, j = 1, \ldots, m\}$	feasible region
$m \in \mathbb{N}_0$	number of constraints
$G = \{g_j : \mathbb{R}^n \rightarrow \mathbb{R} \ \forall\, j = 1, \ldots, m\}$	set of inequality restrictions.

$P^{(0)}$ denotes the initial population (iteration counter $T = 0$) of parents and consists of arbitrary vectors $\mathbf{a}_k^{(0)} \in I^{\mu}$. Each element of the population at reproduction cycle T is represented by a vector

$$\mathbf{a}_k^{(T)} = (\theta, \mathbf{x}, \sigma, \alpha)_k^{(T)} \in P^{(T)}, k \in \mathbb{N} \tag{3}$$

with

$\theta \in \mathbb{N}_0$	remaining life span in iterations (reproduction cycles), $\theta = \kappa$ at birth time
$\mathbf{x} \in \mathbb{R}^n$	vector of the object variables, the only part of $\mathbf{a}$ entering the objective function
$\sigma \in \mathbb{R}_+^{n_\sigma}$	so-called mean step sizes (standard deviations of the Gaussian distribution used for simulating mutations)
$\alpha \in [-\pi, \pi]^{n_\alpha}$	inclination angles, eventually defining (linearly) correlated mutations of the object variables $\mathbf{x}$.

The latter two vectors are called *strategy parameters* or the *internal model* of the individuals. They simply determine the variances and covariances of the n-dimensional Gaussian mutation probability density that is used for exploring the space of object variables $\mathbf{x}$.

One iteration of the strategy, that is a step from a population $P^{(T)}$ towards the next reproduction cycle with $P^{(T+1)}$, is modeled as follows:

$$P^{(T+1)} := opt_{ES}(P^{(T)}) \tag{4}$$

where $opt_{ES} : I^{\mu} \rightarrow I^{\mu}$ is defined by

$$opt_{ES} := \mathbf{sel} \circ (\mathbf{mut} \circ \mathbf{rec})^{\lambda}. \tag{5}$$

2.1 The recombination operator rec($p_r, \rho, \gamma, \omega$)

The recombination operator **rec** : $I^\mu \to I$ is defined as follows:

$$\mathbf{rec} := \mathbf{re} \circ \mathbf{co} \tag{6}$$

with

$$\begin{array}{lll} \mathbf{co} : I^\mu \longrightarrow I^\rho & \text{chooses } 1 \leq \rho \leq \mu \text{ parent vectors from } I^\mu & \\ & \text{with uniform probability} & \\ \mathbf{re} : I^\rho \longrightarrow I & \text{creates one offspring vector} & \\ & \text{by mixing characters from } \rho \text{ parents.} & \end{array} \tag{7}$$

Depending on ω, there are several ways to recombine parents in order to get an offspring:

$$\begin{array}{ll} \omega = 0 & \text{no recombination; this case always holds} \\ & \text{for } \mu = 1,\ p_r = 0, \text{ and/or } \rho = 1 \\ \omega = 1 & \text{global intermediary recombination} \\ \omega = 2 & \text{local intermediary recombination} \\ \omega = 3 & \text{uniform crossover} \\ \omega = 4 & \gamma \text{ point crossover.} \end{array}$$

Let $A \subseteq P^{(T)}$ of size $|A| = \rho$ be a subset of arbitrary parents chosen by the operator **co**, and let $\hat{\mathbf{a}} \in I$ be the offspring to be generated. If $A = \{\mathbf{a}_1, \mathbf{a}_2\}$, $\mathbf{a}_1$ and $\mathbf{a}_2$ being two out of μ parents, holds, recombination is called *bisexual.* If $A = \{\mathbf{a}_1, \ldots, \mathbf{a}_\rho\}$ and $\rho > 2$, recombination is called *multisexual.* Recombination in general is applied with probability p_r.

Recombination types may (often should) differ for the vectors $\mathbf{x}$ (object variables) of length n, σ (mean step sizes) of length n_σ, and α (correlation angles) of length n_α. In the following, $\mathbf{b}$ and $\hat{\mathbf{b}}$ of lengths n_x stand for the actual parts of $\mathbf{a}$ and $\hat{\mathbf{a}}$ at hand.

The components $\hat{b}_i \ \forall\, i = 1, \ldots, n_x$ of $\hat{\mathbf{b}}$ are defined as follows:

$$\hat{b}_i := \begin{cases} b_{k,i} & \text{no recombination, } k \in \{1, \ldots, \mu\} \text{ at random} \\ \frac{1}{\rho} \sum_{k=1}^{\rho} b_{k,i} & \text{global intermediary recombination} \\ u_i b_{k_1,i} + (1 - u_i) b_{k_2,i} & \text{local intermediary recombination,} \\ & k_1, k_2 \in \{1, \ldots, \rho\} \text{ for each offspring,} \\ & u_i \sim \mathbf{U}(0,1) \text{ or } u_i = 0.5 \\ b_{k_i,i} & \text{uniform crossover,} \\ & k_i \in \{1, \ldots, \rho\} \text{ randomly chosen for each } i \end{cases} \tag{8}$$

where $\mathbf{U}(v, w)$ denotes the uniform probability distribution with support $(v, w) \subset \mathbb{R}$. For other than uniform crossover forms, $1 \leq \gamma \leq n_x - 1$ crossover points are first chosen within the strings $\mathbf{b}_k$ at random, and then the offspring gets his $\gamma + 1$ parts of vector $\hat{\mathbf{b}}$ by turns from all of the $\rho \leq \gamma + 1$ parents involved.

2.2 The mutation operator $\mathbf{mut}(p_m, \tau, \tau_0, \delta, \beta)$

The mutation operator $\mathbf{mut} : I \rightarrow I$ is defined as follows:

$$\mathbf{mut} := \mathbf{mu_x} \circ (\mathbf{mu}_\sigma \times \mathbf{mu}_\alpha) \tag{9}$$

with $\mathbf{mu_x}$, $\mathbf{mu}_\sigma$, and $\mathbf{mu}_\alpha$ given below, separately.

Let $\hat{\mathbf{a}} = (\hat{\theta}, \hat{x}_1, \ldots, \hat{x}_n, \hat{\sigma}_1, \ldots, \hat{\sigma}_{n_\sigma}, \hat{\alpha}_1, \ldots, \hat{\alpha}_{n_\alpha})$ with $n, n_\sigma, n_\alpha \in \mathrm{I\!N}$ and $n_\alpha = (n - \frac{n_\sigma}{2})(n_\sigma - 1)$ be the result of the recombination step. The number n denotes the problem's dimension, and $1 \leq n_\sigma \leq n$ the number of different step sizes (standard deviations for mutating the object variables $\hat{\mathbf{x}}$). It may be worthwhile to investigate the new additional degree of freedom by choosing $0 < p_m \leq 1$ for mutating the step sizes.

- $\mathbf{mu}_\sigma : I \rightarrow \mathrm{I\!R}^{n_\sigma}$ mutates the recombined $\hat{\sigma}$:

$$\mathbf{mu}_\sigma(\hat{\mathbf{a}}) := \left(\hat{\sigma}_1 e^{z_1 + z_0}, \ldots, \hat{\sigma}_{n_\sigma} e^{z_{n_\sigma} + z_0}\right) =: \tilde{\sigma} \tag{10}$$

with

$$z_0 \in \mathbf{N}(0, \tau_0^2), \quad z_i \in \mathbf{N}(0, \tau^2) \ \forall\, i = 1, \ldots, n_\sigma. \tag{11}$$

$\mathbf{N}(\xi, \psi^2)$ denotes the normal distribution with mean ξ and variance ψ^2 (standard deviation ψ). For maximal rates of convergence in case of the so-called sphere model (see section 3), τ and τ_0 may be chosen according to the relationships:

$$\tau_0 = \frac{K}{\sqrt{p_m}} \frac{\delta}{\sqrt{2n}}, \quad \tau = \frac{K}{\sqrt{p_m}} \frac{1-\delta}{\sqrt{\frac{2n}{\sqrt{n_\sigma}}}} \tag{12}$$

where the constant K should reflect the convergence velocity of the ES (see section 3). So far, only $\delta = 0.5$ and $p_m = 1$ have been used, but other values may be worthwhile to be considered as well.

- $\mathbf{mu}_\alpha : I \rightarrow \mathrm{I\!R}^{n_\alpha}$ mutates the recombined $\hat{\alpha}$:

$$\mathbf{mu}_\alpha(\hat{\mathbf{a}}) := (\hat{\alpha}_1 + z_1, \ldots, \hat{\alpha}_{n_\alpha} + z_{n_\alpha}) =: \tilde{\alpha} \tag{13}$$

with

$$z_i \in \mathbf{N}(0, \beta^2) \ \forall\, i = 1, \ldots, n_\alpha. \tag{14}$$

Good results have been obtained with $\beta \approx 0.0873$ [$\approx 5°$], but the question whether β should be different for each z_i is still open.

- $\mathbf{mu_x}(\tilde{\sigma}, \tilde{\alpha}) : I \rightarrow \mathrm{I\!R}^n$ mutates the recombined object variables $\hat{\mathbf{x}}$, using the recombined and already mutated $\tilde{\sigma}, \tilde{\alpha}$ (for efficiency reasons only, otherwise the sequence of all variation steps and even of the cyclical selection and variation processes does not matter):

$$\mathbf{mu_x}(\tilde{\sigma},\tilde{\alpha})(\hat{\mathbf{a}}) := (\hat{x}_1 + cor_1(\tilde{\sigma},\tilde{\alpha}),\ldots,\hat{x}_n + cor_n(\tilde{\sigma},\tilde{\alpha})) =: \tilde{\mathbf{x}} \tag{15}$$

where $\mathbf{cor} := (cor_1,\ldots,cor_n)$ is a random vector with normally distributed, eventually correlated components, using $\tilde{\sigma}$ and $\tilde{\alpha}$. The components of **cor** can be calculated as follows [19]:

$$\mathbf{cor} = \mathbf{T}\,\mathbf{z} \tag{16}$$

where $\mathbf{z} = (z_1,\ldots,z_{n_\sigma})$ with $z_i \in \mathbf{N}(0,\tilde{\sigma}_i^2)\ \forall\, i = 1,\ldots,n_\sigma$ and

$$\mathbf{T} = \prod_{p=1}^{n_\sigma-1}\ \prod_{q=p+1}^{n_\sigma} \mathbf{T}_{pq}(\tilde{\alpha}_j) \tag{17}$$

with $j = \frac{1}{2}(2n_\sigma - p)(p+1) - 2n_\sigma + q$ and

$$\mathbf{T}_{pq}(\tilde{\alpha}_j) := \begin{pmatrix} 1 & 0 & & & & \cdots & & & & 0 \\ 0 & 1 & & & & & & & & \\ & & \ddots & & & & & & & \\ & & & 1 & & & & & & \\ & & & & \cos\tilde{\alpha}_j & & & -\sin\tilde{\alpha}_j & & \\ & & & & & 1 & & & & \\ \vdots & & & & & & \ddots & & & \vdots \\ & & & & & & & 1 & & \\ & & & & \sin\tilde{\alpha}_j & & & \cos\tilde{\alpha}_j & & \\ & & & & & & & & 1 & \\ & & & & & & & & & \ddots \\ & & & & & & & & 1 & 0 \\ 0 & & & & & \cdots & & & 0 & 1 \end{pmatrix} \tag{18}$$

with the terms $\cos\tilde{\alpha}_j$ and $-\sin\tilde{\alpha}_j$ in columns p and q and lines p and q, respectively. An efficient way of calculating (16) is the multiplication from right to left. $\tilde{\theta}$ is set to $\tilde{\theta} = \kappa$ for all offspring when they are created by means of recombination and mutation. Finally we have:

$$\tilde{\mathbf{a}}_k = (\tilde{\theta},\tilde{\mathbf{x}},\tilde{\sigma},\tilde{\alpha})\ \forall\, k = 1,\ldots,\lambda. \tag{19}$$

For constrained optimization the processes of recombination and mutation must be repeated as often as necessary to create λ non-lethal offspring such that

$$g_j(\tilde{\mathbf{x}}_k) \geq 0\ \forall\, j = 1,\ldots,m \text{ and } \forall\, k = 1,\ldots,\lambda. \tag{20}$$

This vitality check may already be part of the selection process, however.

2.3 The selection operator sel(ζ)

Natural selection is a term that tries to describe the final result of several different real world processes, i.e., from the test of new born individuals against natural laws (if not met, the trial is lethal) and other environmental conditions up to what is called mating selection. According to Darwin, selection mainly helps to avoid the Malthusian trap of food shortage due to overpopulation, the result of a normal surplus of births over deaths (this is neither reflected in GAs nor in EP). Others emphasize the reproduction success of stronger or more intelligent individuals, perhaps induced by Darwin's unlucky term 'struggle for life.'

Altogether, there are several ways of implementing selection mechanisms. Two typical selection operators will be presented here. They mainly differ in the selection pressure they exert on a population. Due to the strong impact of selection on the behavior of the evolutionary process, it is worthwhile to provide both schemes.

The *traditional deterministic ES selection operator* can be defined as:

$$\mathbf{sel} : I^{\mu+\lambda} \to I^{\mu}. \tag{21}$$

Let $P^{(T)}$ denote some parent population in reproduction cycle T, $\tilde{P}^{(T)}$ their offspring produced by recombination and mutation, and $Q^{(T)} = P^{(T)} \sqcup \tilde{P}^{(T)} \in I^{\mu+\lambda}$ where the operator $\sqcup$ denotes the union operation on multisets. Then

$$P^{(T+1)} := \mathbf{sel}(Q^{(T)}). \tag{22}$$

The next reproduction cycle contains the μ best individuals, i.e., the following relation is valid:

$$\forall\, a \in P^{(T+1)} : \nexists\, b \in Q^{(T)} \setminus P^{(T+1)} : b \overset{\kappa}{>} a \tag{23}$$

where the relation $\overset{\kappa}{>}$ (read: better than) introduces a maximum duration of life, κ, that defines an individual to be worse than an other one if its age is greater than the allowed maximum, κ, or if its fitness (measured by the objective function) is worse.

The definition of the $\overset{\kappa}{>}$ - relation is given by:

$$\mathbf{a}_k \overset{\kappa}{>} \tilde{\mathbf{a}}_\ell :\Leftrightarrow \theta_k > 0 \;\wedge\; f(\mathbf{x}_k) \leq f(\tilde{\mathbf{x}}_\ell). \tag{24}$$

In practical applications, where constraints can be evaluated quickly (e.g., in the case of simple bounds to the object variables), it may be advantageous to evaluate the constraints first. Thus, only if a search point lies within the feasible region (non-lethal individual), the time consuming objective function has to be evaluated. However, things may turn out to be just the other way round, i.e., the time consuming part of the evaluation lies in the check for feasibility (e.g., if a FEM is used to calculate the stresses and deformations of a mechanical structure, the result of which must be compared with given upper bounds). Then the selection process must be interwoven with the process of generating offspring by recombination and mutation.

At the end of the selection process, the remaining maximum life durations have to be decremented by one for each survivor:

$$\theta_k^{(T+1)} := \tilde{\theta}_k^{(T)} - 1 \;\; \forall\, k = 1, \ldots, \mu. \tag{25}$$

The *tournament selection* is well suited for parallelization of the selection process. This method selects μ times the best individual from a random subset B_k of size $|B_k| = \zeta$, $2 \leq \zeta \leq \mu + \lambda \;\; \forall\, k = 1, \ldots, \mu$ and transfers it to the next reproduction cycle (note that there may appear duplicates!). The best individual within each subset B_k is selected according to the $\overset{\kappa}{>}$ relation which was introduced in (24). A formal definition of the $(\mu, \kappa, \lambda, \zeta)$ tournament selection follows:
Let

$$B_k \subseteq Q^{(T)} \;\; \forall\, k = 1, \ldots, \mu \tag{26}$$

be random subsets of $Q^{(T)}$, each of size $|B_k| = \zeta$. For each $k \in \{1, \ldots, \mu\}$ choose $a_k \in B_k$ such that

$$\forall\, b \in B_k : a_k \overset{\kappa}{>} b\,. \tag{27}$$

Finally,

$$P^{(T+1)} := \bigsqcup_{k=1}^{\mu} \{\mathbf{a}_k^{(T+1)}\}. \tag{28}$$

2.4 The termination criterion $t(\varepsilon)$

The termination of the new evolution strategy should be handled in the same way as has been done within the older versions (see [17]) and will therefore not be described here.

2.5 The start conditions $P^{(o)}$

For reasons of comparability with GAs on the one hand and more classical optimization techniques on the other, there should be two distinct ways of setting up the initial population.

Case a: With given lower and upper bounds for all object variables (a prerequisite for all GAs, but not for ESs)

$$\underline{x}_i \leq x_i \leq \bar{x}_i \;\; \forall\, i = 1, \ldots, n \tag{29}$$

all μ parents at cycle $T = 0$ are arbitrarily distributed within the bounded region.

Case b: With a given start position $\mathbf{x}^{(0)}$ for the optimum seeking process that is assigned to one individual $\mathbf{x}_1$, the other $\mu - 1$ parents for the first iteration cycle are found by applying some kind of mutation process with enlarged step sizes $c\,\sigma^{(0)}, c > 1$, for example $c = 10$, by:

$$x_{k,i}^{(0)} = x_{1,i}^{(0)} + c\,\sigma_i^{(0)}\,z_i \quad \text{with } z_i \in \mathrm{N}(0,1)\ \forall\, i = 1,\ldots,n \text{ and } k = 2,\ldots,\mu. \tag{30}$$

One may increase c during this setup process if no constraints are violated and the objective function has been improved during the last step; otherwise c should be decreased.

2.6 The handling of constraints

During the optimum seeking process of ESs, inequality constraints so far have been handled as barriers, i.e., offspring that violate at least one of the restrictions are lethal mutations. Before the selection operator can be activated, exactly λ non-lethal offspring must have been generated.

In case of a non-feasible start position $\mathbf{x}^{(0)}$, a feasible solution must be found at first. This can be achieved by means of an auxiliary objective function

$$\tilde{f}(\mathbf{x}) = \sum_{j=1}^{m} g_j(\mathbf{x})\,\delta_j(\mathbf{x}) \tag{31}$$

with $\delta_j(\mathbf{x}) = -1$ if $g_j(\mathbf{x}) < 0$ and zero otherwise. Each decrease of the value of $\tilde{f}(\mathbf{x})$ represents an approach to the feasible region. As soon as $\tilde{f}(\mathbf{x}) = 0$ can be stated, then $\mathbf{x}$ satisfies all the constraints and can serve as a starting vector for the optimization proper.

3 Theoretical Results

All theoretical results heavily rely upon simplifications of the situation investigated, here with respect to the objective functions taken into consideration, as well as with respect to the optimum seeking algorithm. The gap between practical results in very difficult situations – so far these happen to be the sole justification for EAs – and theoretically proven results in rather simple situations – in which normally other solution techniques are preferable – remains huge. Nevertheless, an algorithm that is worth consideration for complex tasks should fulfill some minimal requirements in easy situations as well.

Global convergence in case of nearly arbitrary response surface landscapes (called effectivity or robustness) and high efficiency, i.e., a low number of function evaluations until achieving a specified approximation to the exact solution are maximal requirements that will remain in conflict with one another, forever. EAs have often been apostrophed as universal search methods since they do not try to gain advantage from higher order information like derivatives of the objective function and do not interpret intermediate results on the basis of a specific (e.g., linear or quadratic) internal model of the fitness landscape. As minimal requirements to them one must demand that they should never fail in simple cases and that they should provide some means to scale their behavior towards maintaining usefulness in more difficult situations, e.g., by choosing an appropriate population size and selection pressure. As long as no superior method is available, an EA arriving at one of the better local, though not global, optima is a useful search method.

Up until now, no all-embracing theory for the $(\mu, \kappa, \lambda, \rho)$ ES exists. However, some special cases like $\kappa = 1$ or $\kappa = \infty, \rho = 1$ or $\rho = \mu$, and $\mu = 1$ have been investigated thoroughly. The results will be summarized briefly in the following.

A very early global convergence proof for a $(1 + 1)$ ES with one parent and one descendant per generation, elitist selection, no recombination and normally distributed mutations without correlation has been given by Born [20]. No continuity, differentiability, or unimodality assumptions must be made. Except for singular solutions, global convergence with probability one in the limit of infinitely many mutations is guaranteed as long as the mutation variance is greater than zero in all directions. The same holds for the more general $(\mu + \lambda)$ ES [21]. More delicate is the non-elitist case of a $(1, \lambda)$ ES for which Rudolph [22] has developed sufficient conditions under which convergence is maintained. Like the canonical GA, the $(1, \lambda)$ ES with fixed mutation variances finally stagnates at a distance from the optimum that depends on σ (actually, it fluctuates around that position).

More interesting than all that is an answer to the question of the approximation velocity. Whereas this question is still open for GAs, except for a very special case (see [23]), the situation is somewhat better for ESs now. Linear convergence order, that is a constant increase of accurate figures of the object parameter or objective function values over the number of mutations or generations, has been proved by Rappl [24] for the $(1+1)$ ES when applied to a strongly convex fitness function like

$$f_1(\mathbf{x}) = c_0 + \sum_{i=1}^{n} c_i (x_i - x_i^*)^2 \tag{32}$$

which has been called spherical model if $c_i = 1 \;\; \forall i = 1, \ldots, n$. Most of the following results are valid not only for the spherical situation, but also for functions like $f_2(\mathbf{x}) = -e^{-f_1(\mathbf{x})}$ (a nightmare for quasi-Newton methods, which diverge everywhere in this case) and, approximately at least, also in case of higher even exponents than two in function f_1 [17].

One basic assumption for the proof has been the maintenance of the corresponding optimal mutation variances. For the $(1 + 1)$ ES this can be approximately achieved by applying the so-called 1/5 success rule. Following Rechenberg [13] the variance should be increased as soon as the observed success rate is greater than 1/5 and decreased if it turns out to be less than 1/5. This result was achieved for the spherical function above and for an endless inclined ridge model with rectangular cross-section of constant width perpendicular to the steepest descent direction, which is the main diagonal in case of

$$f_3(\mathbf{x}) = c_0 - \sum_{i=1}^{n} c_i x_i \tag{33}$$

if again $c_i = 1 \;\; \forall i = 1, \ldots, n$.

This is not the place to go into more details, but proportional control according to that rule has proven to lead to oscillatory behavior with some factor loss in convergence velocity against the optimal case. Linear convergence order, however, is maintained.

Rechenberg [16] claims linear convergence order for the (μ, λ) ES on the sphere model in case of optimal mutation variances. His law

$$\varphi = E\{\Delta r\} = C\sigma - \frac{n}{2r}\sigma^2 \tag{34}$$

for the expected difference $\Delta r = r^{(g-1)} - r^{(g)}$ between the Euclidean distances $r = r^{(g-1)}$ before and $r^{(g)}$ after one generation is an approximation to the very high-dimensional case ($n \gg 1$), as Beyer [25] has elaborated. In terms of dimensionless quantities $\breve{\varphi} = \frac{n}{r}\varphi$ and $\breve{\sigma} = \frac{n}{r}\sigma$ the law reads

$$\breve{\varphi} = C\breve{\sigma} - \frac{1}{2}\breve{\sigma}^2. \tag{35}$$

The maximal convergence rate $\varphi^* = \breve{\varphi}_{max} = \frac{1}{2}C^2$ corresponds to the optimal standard deviation $\sigma^* = \breve{\sigma}_{opt} = C$, a constant that of course still depends on μ, κ, λ, and ρ, at least.

An approximation with respect to λ, when μ, κ, and ρ equal one, has been found to be as simple as [21]:

$$C \approx \sqrt{2 \ln \lambda}. \tag{36}$$

Beyer [26] has investigated both uniform crossover and global intermediary multi-recombination with $\rho = \mu$. Though the laws (again approximations) look different

$$\breve{\varphi} = C\breve{\sigma} - \frac{1}{2\mu}\breve{\sigma}^2 \quad \text{(global intermediary)} \tag{37}$$

$$\breve{\varphi} = \sqrt{\mu}\, C\breve{\sigma} - \frac{1}{2}\breve{\sigma}^2 \quad \text{(uniform crossover)} \tag{38}$$

they yield the same maximum $\varphi^* = \frac{\mu}{2}\, C^2$ for $\sigma^* = \mu\, C$ in the former, $\sigma^* = \sqrt{\mu}\, C$ in the latter case. If one interprets φ as an approximation to the differential $\dot{r} = dr/dt$ (literally: the velocity of approaching the optimum $\mathbf{x}$ of the sphere model function [17]) one arrives at concluding $K = \varphi_*$ for the still open constant in relation (12). Based upon the observation that in the linear theory the convergence rate mainly depends on the ratio $\frac{\lambda}{\mu}$ if the population size is not too small, one might speculate about putting together what we know so far to the rather simple formula (for large n and λ, as well as not too small μ):

$$\varphi^* \sim \mu \ln \frac{\lambda}{\mu} \tag{39}$$

the first factor (μ) being due to the diversity of the population and exploited by recombination (so-called genetic repair [26]), the latter ($\frac{\lambda}{\mu}$) due to the selection pressure - two processes that compete with one another. A proof is still missing, however. The same holds for the influence of other strategy parameters like κ, p_r, and p_m, which were assumed to equal one in the considerations above.

Of great practical interest is the answer to the question how to achieve and maintain the optimal mutation variance σ^{*2}. The algorithm presented in section 2 tries to do it in a way called self-adaptation. It even allows for on-line learning of up to n different σ_i and $\frac{n}{2}(n-1)$ different α_j, thus presenting the ultimate degree of freedom for normally distributed mutations. No theory is available for that process, only a few experiments [27] have demonstrated that self-adaptation is possible under certain conditions. If these conditions are not observed the process fails and the resulting ES may not converge, even diverge in case of small values for κ.

The necessity of individually scaled σ_i can easily be derived from function f_1 in case of non-identical coefficients c_i. Nobody should be astonished that superfluous degrees of freedom, e.g., individual σ_i for the spherical model, where $\sigma_i = \sigma_0 \; \forall \; i = 1, \ldots, n$ would be the best choice, come at an extra cost. Introducing correlated mutations ($\alpha_j \neq 0$), however, does not slow down the progress further in this case. Correlations, properly learned, would help a lot in case of a hyperelliptical scene with main axes different from the coordinate axes like

$$f_4(x) = \sum_{i=1}^{n} \left(\sum_{j=1}^{i} x_j \right)^2 . \tag{40}$$

This still simple quadratic function poses heavy difficulties to all optimization algorithms that rely upon decomposability. Both the so-called coordinate (or Gauss-Seidel, or one-variable-at-a-time) search technique and many GA variants come into trouble with f_4. Major improvements require *simultaneous* changes of many if not all object variables. Correlated mutations help, but discrete recombination turns out to be disastrous in this case. Why? Without correlation and independent σ_i, the population of an ES tends to concentrate at one out of two positions where the curvature radii are smallest. With individual σ_i and correlation it spreads more or less to the whole temporal hypersurface $f(x^{(T)}) = const.$ Discrete recombination then often fails to produce descendants which keep to the vicinity of that hypersurface. Intermediary recombination of the object variables, even merely switching off recombination, heals that. That is why the recombination type and frequency should be incorporated into the set of internal strategy parameters, too. In nature there is no higher instance for controlling internal parameters in the way which has been proposed with the so-called nested or meta-evolution approach [16]. More natural seem to be simulations with several subpopulations, a concept that has been used for EA incarnations on parallel computers [28].

References

1. De Jong, K. (Ed.) (1993), Evolutionary computation (journal), MIT Press, Cambridge MA
2. Ashby, W.R. (1960), Design for a brain, 2nd ed., Wiley, New York
3. Brooks, S.H. (1958), A discussion of random methods for seeking maxima, Oper. Res. **6**, 244-251
4. Rastrigin, L.A. (1960), Extremal control by the method of random scanning, ARC **21**, 891-896
5. Favreau, R.F., R. Franks (1958), Random optimization by analogue techniques, Proceedings of the IInd Analogue Computation Meeting, Strasbourg, Sept. 1958, pp. 437-443
6. Bremermann, H.J. (1962), Optimization through evolution and recombination, in: Yovits, M.C., G.T. Jacobi, D.G. Goldstein (Eds.), Self-organizing systems, Spartan, Washington, DC, pp. 93-106
7. Fogel, L.J. (1962), Autonomous automata, Ind. Research **4**, 14-19
8. Fogel, D.B. (1995), Evolutionary Computation—toward a new philosophy of machine intelligence, IEEE Press, Piscataway NJ
9. Holland, J.H. (1975), Adaptation in natural and artificial systems, University of Michigan Press, Ann Arbor MI
10. Rechenberg, I. (1964), Cybernetic solution path of an experimental problem, Royal Aircraft Establishment, Library Translation 1122, Farnborough, Hants, Aug. 1965, English translation of the unpublished written summary of the lecture "Kybern tische Lösungsansteuerung einer experimentellen Forschungsaufgabe", delivered at the joint annual meeting of the WGLR and DGRR, Berlin, 1964
11. Klockgether, J., H.-P. Schwefel (1970), Two-phase nozzle and hollow core jet experiments, in: Elliott, D.G. (Ed.), Proceedings of the 11th Symposium on Engineering Aspects of Magnetohydrodynamics, Caltech, March 24-26, 1970, California Insti ute of Technology, Pasadena CA, pp. 141-148
12. Box, G.E.P. (1957), Evolutionary operation—a method for increasing industrial productivity, Appl. Stat. **6**, 81-101
13. Rechenberg, I. (1973), Evolutionsstrategie—Optimierung technischer Systeme nach Prinzipien der biologischen Evolution, Frommann-Holzboog, Stuttgart
14. Schwefel, H.-P. (1977), Numerische Optimierung von Computer-Modellen mittels der Evolutionsstrategie, Birkhäuser, Basle, Switzerland
15. Schwefel, H.-P. (1981), Numerical optimization of computer models, Wiley, Chichester
16. Rechenberg, I. (1994), Evolutionsstrategie '94, Frommann-Holzboog, Stuttgart
17. Schwefel, H.-P. (1995), Evolution and optimum seeking, Wiley, New York
18. Rudolph, G. (1994), An evolutionary algorithm for integer programming, in: Davidor, Y., H.-P. Schwefel, R. Männer (Eds.), Parallel problem solving from nature 3, Proceedings of the 3rd PPSN Conference, Jerusalem, Oct. 9-14, 1994, vol. 866 f Lecture Notes in Computer Science, Springer, Berlin, pp. 139-148

19. Rudolph, G. (1992), On correlated mutation in evolution strategies, in: Männer, R., B. Manderick (Eds.), Parallel problem solving from nature 2, Proceedings of the 2nd PPSN Conference, Brussels, Sept. 28-30, 1992, North-Holland, Amsterdam, pp. 105-114
20. Born, J. (1978), Evolutionsstrategien zur numerischen Lösung von Adaptationsaufgaben, Dr. rer. nat. Diss., Humboldt University at Berlin
21. Bäck, T., G. Rudolph, H.-P. Schwefel (1993), Evolutionary programming and evolution strategies—similarities and differences, in: Fogel, D.B., J.W. Atmar (Eds.), Proceedings of the 2nd Annual Conference on Evolutionary Programming, San Di go, Feb. 25-26, 1993, Evolutionary Programming Society, La Jolla CA, pp. 11-22
22. Rudolph, G. (1994), Convergence of non-elitist strategies, in: Michalewicz, Z., J.D. Schaffer, H.-P. Schwefel, and D.B. Fogel (Eds.), Proceedings of the 1st IEEE Conference on Evolutionary Computation, IEEE World Congress on Computational In elligence, Orlando FL, June 27-29, 1994, vol. 1, pp. 63–66
23. Bäck, T. (1994), Evolutionary algorithms in theory and practice, Dr. rer. nat. Diss., University of Dortmund, Department of Computer Science, Feb. 1994
24. Rappl, G. (1984), Konvergenzraten von Random-Search-Verfahren zur globalen Optimierung, Dr. rer. nat. Diss., Hochschule der Bundeswehr, Munich-Neubiberg, Department of Computer Science, Nov. 1984
25. Beyer, H.-G. (1995), Toward a theory of evolution strategies—the (μ, λ)-theory, submitted to Evolutionary Computation
26. Beyer, H.-G. (1994), Towards a theory of 'evolution strategies'—results from the N-dependent (μ, λ) and the multi-recombinant $(\mu/\mu, \lambda)$ theory, technical report SYS-5/94, Systems Analysis Research Group, University of Do tmund, Department of Computer Science, Oct. 1994
27. Schwefel, H.-P. (1987), Collective phenomena in evolutionary systems, in: Checkland, P., I. Kiss (Eds.), Problems of constancy and change—the complementarity of systems approaches to complexity, papers presented at the 31st Annual Meeting f the International Society for General System Research, Budapest, Hungary, June 1-5, International Society for General System Research, vol. 2, pp. 1025-1033
28. Rudolph, G. (1991), Global optimization by means of distributed evolution strategies, in: Schwefel, H.-P. and R. Männer (Eds.), Parallel problem solving from nature, Proceedings of the 1st PPSN Conference, Dortmund, Oct. 1-3, 1990, vol. 4 6 of Lecture Notes in Computer Science, Springer, Berlin, pp. 209-213

The Usefulness of Recombination

Wim Hordijk and Manderick

[1] Santa Fe Institute, 1399 Hyde Park, Santa Fe, NM 87501, email: wim@santafe.edu
[2] Computer Science Dept., Vrije Universiteit Brussel, Pleinlaan 2, B-1050 Brussel, Belgium, email: bernard@arti.vub.ac.be

Abstract. In this paper, we examine the usefulness of recombination from two points of view. First, the problem of crossover disruption is investigated. This is done by comparing two Genetic Algorithms with different crossover operators (one-point and uniform) to each other on NK-landscapes with different values of K relative to N, and with different epistatic interactions (random and nearest neighbor). Second, the usefulness of recombination in relation to the location of local optima in the fitness landscape is investigated.
There appears to be a clear relation between the type of fitness landscape and the type of recombination that is most useful on this landscape. Furthermore, there also is a clear relation between the location of local optima in the fitness landscape and the usefulness of recombination.

1 Introduction

The concept of a fitness landscape has proved to be very useful in thinking about evolutionary processes. Given a certain problem, a fitness landscape is the distribution of fitness values over the space of possible solutions for this problem, where the solutions have a neighborhood relation determined by some metric. For example, the genetic algorithm usually uses bit strings to encode solutions and the Hamming distance between such strings defines a metric. The fitness values are determined by some fitness function, which takes as input a genotype, the (genetic) coding of an individual, and returns a real value denoting this individual's fitness. The higher the fitness, the better a solution for the given problem this individual is.

By assigning a fitness value to every possible individual (or genotype), a more or less "mountainous" landscape arises, where the highest peaks denote the best solutions. An evolving population is now envisioned as adapting on such a landscape, in search for the highest peaks.

In Nature, *recombination* is used for constructing new genotypes during reproduction. This means that different parts of two parent genotypes are recombined to form two offspring genotypes. This recombination is copied in evolutionary algorithms that are used for solving difficult problems. It is implemented in the form of different crossover operators.

The role of crossover is a two-edged knife. On the one hand, crossover is used for exploration of the fitness landscape, because it can make large jumps across such a landscape (in terms of the metric by which the landscape is defined). On the other hand, there is the danger of crossover disruption, or the breaking apart of good solutions that were already found. In most of the studies on crossover, the emphasis is on finding a way to minimize this crossover disruption. This paper investigates both aspects of crossover: the power of exploration and the danger of disruption.

First, the problem of crossover disruption is dealt with. Next, the usefulness of recombination in relation to the location of optima in the fitness landscape is examined. With the results of these two investigations, the validity of the next statement made by Kauffman is assessed (see [5]):

> "*recombination is useless on uncorrelated landscapes but useful under two conditions: (1) when the high peaks are near one another and hence carry mutual information about their joint locations in the fitness landscape and (2) when parts of the evolving individuals are quasi-independent of one another and hence can be interchanged with modest chances that the recombined individual has the advantage of both parents*".

Different fitness landscapes are constructed using the NK-model (see [5]). The parameters of this model can be used to tune the landscape from smooth (small fitness differentials between neighboring points) to very rugged (large fitness differentials between neighboring points). The more rugged the landscape is, the less information the fitness of one point gives about the expected fitness of another point.

In Section 2, the NK-model is explained in more detail. Section 3 deals with the problem of crossover disruption by comparing two Genetic Algorithms, with different crossover operators, on different fitness landscapes generated by the NK-model. Section 4 then examines the relation between the usefulness of recombination and the location of optima. Finally, in Section 5 the validity of Kauffman's statement about the usefulness of recombination is assessed.

2 The NK-model

The *NK-model* was introduced by Kauffman to have a problem-independent model for constructing landscapes that can be tuned from smooth to rugged (see [5]). This tunability led [6] to use these landscapes to investigate the performance of genetic algorithms. The main parameters of the model are N, the number of parts, or genes, in the system, and K, the number of genes that epistatically influence a particular gene. Another parameter is A, the number of alleles every gene has. It appears that almost all properties of the NK-model are independent of A, so in this paper A will be set at 2, i.e., the genotypes are bit strings (strings of `0`'s and `1`'s). The NK-model is also explained in terms of bit strings (i.e., the two allele case).

Suppose every bit b_i $(i = 1, \ldots, N)$ in the bit string $\underline{b}$ is assigned a fitness w_i of its own. The fitness w_i of a bit b_i, however, does not only depend on the value (`0` or `1`) of this specific bit, but also on the value of K other bits b_j in the same bit string $(0 \le K \le N-1)$. These dependencies are called *epistatic interactions*.

So, the fitness contribution of one bit depends on the value of $K+1$ bits (itself and K others), giving rise to a total of 2^{K+1} possibilities. Since, in general, it is not known what the effects of these epistatic interactions are, they are *modelled* by assigning to each of the 2^{K+1} possibilities at random a fitness value drawn from the Uniform distribution between 0.0 and 1.0. Therefore, the fitness contribution w_i of bit b_i is specified by a list of random decimals between 0.0 and 1.0, with 2^{K+1} entries. This procedure is repeated for every bit b_i, $i = 1, \ldots, N$ in the bit string $\underline{b}$.

Having assigned the fitness contributions for every bit in the string, the fitness of the entire bit string, or genotype, is now defined as the average of the contributions of all the bits:

$$W = \frac{1}{N} \sum_{i=1}^{N} w_i$$

One further aspect of the NK-model characterizes how the K epistatic interactions for each bit are chosen. Generally, this is done in one of two ways.

The first way is by choosing them at random from among the other $N-1$ bits. This is called *random interactions.* It is important to note that no reciprocity in epistatic influence is assumed. This means that if the fitness of bit b_i depends on bit b_j, it is not necessary that the reverse also holds. So, the epistatic interactions for a bit are determined independent of the other bits.

The second way is by choosing the K neighboring bits as epistatic interactions. The $K/2$ bits on each side of a bit will influence the fitness of this bit. This is called *nearest neighbor interactions.* To make this possible, *periodic boundary conditions* are taken into account. This means that the bit string is considered as being circular, so the first and the last bit are each others neighbors. Note that for K=0 and $K = N - 1$, there is no difference between the two types of interactions. In the first case, the fitness of each bit depends only on its own value, and in the second case, the fitness of each bit depends on the value of all the bits in the string.

The fitness landscapes that result from this NK-model are called *NK-landscapes.* A low value of K gives rise to a rather smooth landscape (small fitness differentials between neighboring points), while increasing K, relative to N, results in an ever more rugged landscape (large fitness differentials). For K=0, the landscape is very smooth and contains only one peak, because for every bit either the value `0` or the value `1` has the highest fitness contribution, so there is one single bit string having the highest fitness contribution at all bit positions, and every other bit string can sequentially be changed to this optimal bit string. For K=$N - 1$, the landscape is completely random, because changing the value of one bit changes the fitness contribution of all bits. In this case there is no correlation at all between the fitness of neighboring points.

3 Crossover disruption

The notion of a *schema* is central to understand how a Genetic Algorithm works. A schema is a set of individuals in the search space, and the GA is thought to work by directing the search towards schemata containing highly fit regions of the search space. For example, the string `1**01*00*` is a schema, where a `*` means *don't care*, either value (`0` or `1`) is allowed. In schemata, `0` and `1` are called *defined bits*, the *order* is the number of defined bits, and the *defining length* is the distance between the first and the last defined bit. So, the schema in the example is of order 5 and has defining length 7.

According to the building block hypothesis [3, 2], a Genetic Algorithm works well when short, low-order, highly fit schemata (so-called building blocks) are recombined to form even more highly fit higher-order schemata. So, a GA works well when crossover is able to recombine building blocks to longer, higher-order schemata with a high fitness. On the other hand, it follows from the Schema Theorem [3, 2] that long, high-order schemata are more sensitive to crossover disruption than short, low-order ones. So, opposed to the usefulness of crossover in constructing longer, highly fit schemata, there is the danger of disrupting them again.

To investigate this construction-disruption duality, two types of crossover, one-point and uniform, are compared on NK-landscapes with different types of epistatic interactions, random and nearest neighbor. Uniform crossover is believed to be maximally disruptive, while one-point crossover is more conservative. But this depends highly on the type of epistatic interactions within a genotype.

With one-point crossover, a random crossover point between the first and last bit is chosen, and the parts of the two parents after this crossover point are swapped, thus creating two children.

With uniform crossover, for each bit position on the two children it is decided randomly which parent contributes its bit value to which child.

One-point crossover is more disruptive when the epistatic interactions in a bit string are randomly distributed than when they are the nearest neighbors: with random interactions almost every possible crossover point will affect the epistatic relations of almost all bits in a bit string, while with nearest neighbor interactions only the epistatic relations of the bits in the vicinity of the crossover point are affected.

With uniform crossover, however, there is a large chance that a good configuration of neighboring epistatically interacting bits will be disrupted. But when the epistatic interactions are random, uniform crossover can recombine good values for these interacting bits, while one-point crossover is unable to do this.

Next, the experimental setup for investigating the problem of crossover disruption is described, after which the results of the experiments are presented.

3.1 Experimental setup

Two Genetic Algorithms, with the following parameter values, are applied to different NK-landscapes: a population size of 50, deterministic tournament with tournament size s=3, a crossover rate $p_c = 0.75$, and a mutation rate $p_m = 0.005$

In deterministic tournament selection, s individuals are picked at random from the old population, and the fittest of them is put in the mating pool. This is repeated until the mating pool is as large as the old population. The tournament size s can be used to tune the selection pressure.

One GA uses one-point crossover (GA-ONEP), while the other GA uses uniform crossover (GA-UNIF). Both GA's are allowed to do a total of 10,000 function evaluations. Every 50 function evaluations, i.e. every generation, the maximum fitness in the population is recorded. NK-landscapes with the following values for N and K are taken: N=100, K=0, 2, 5, 25, 50, and 99. Both random and nearest neighbor interactions are considered. All results are averaged over 100 runs, each run on a different landscape, but with the same values for N and K.

3.2 Results

Figures 1 to 3 show the results of applying the two GA's to the different NK-landscapes. The abbreviation RND stands for random interactions, while NNI stands for nearest neighbor interactions. So, GA-ONEP (RND) means the Genetic Algorithm with one-point crossover applied to an NK-landscape with random epistatic interactions. Note that for K=0 and K=99 no distinction is made between random interactions and nearest neighbor interactions, because they are exactly the same for these values of K (see Section 2).

It appears that there are two phases in the search: the first phase consists of finding a good region in the fitness landscape, mainly by global search (crossover), and the second phase consists of trying to find the highest peaks within this region, mainly by local search (mutation). (See [4, 5] for a more detailed discussion about these phases in an adaptive search). Initially, the graphs are increasing rapidly (the first phase), but then they become gradually less steep (the second phase) until they are completely smooth, indicating that the highest peaks in a relatively good region are found. Only the completely random landscape (K=99, Figure 3) does not fit into this picture, because global search appears to be useless on this landscape (see again [4, 5]). Therefore, the case of K=99 is left out of the rest of the analysis.

Since the second phase in the search is dominated by local search (mutation), only the performance in the first phase of the search is evaluated here to examine the usefulness of recombination, which is a global search strategy. Differences in performance in the second phase are mainly a reflection of differences in performance in the first phase. Furthermore, the results are viewed in two ways: taking one type of GA and comparing random with nearest neighbor interactions, and taking one type of epistatic interactions and comparing GA-ONEP with GA-UNIF.

Table 1 presents the results from the first viewpoint. It shows for both types of GA's on which type of landscape (that is, with random interactions (RND) or with nearest neighbor interactions (NNI)) they are better able to find a good region in the landscape in the first phase of the search. "Better able" means either finding such a region faster, or finding a better region (that is, containing higher peaks), or both. An X means that there is no (significant) difference between the two types of epistatic interactions. The significance of the differences is checked with a statistical procedure called the *t-test* (see [1]).

	K=2	K=5	K=25	K=50
GA-ONEP	NNI	NNI	NNI/RND	NNI
GA-UNIF	X	RND	RND	X

Table 1. Comparison of random interactions (RND) with nearest neighbor interactions (NNI) for GA-ONEP and GA-UNIF in the first phase of the search. An entry RND means that the GA works better on a landscape with random interactions than on a landscape with nearest neighbor interactions. An X means that there is no difference.

The table shows that one-point crossover (GA-ONEP) works better on a landscape with nearest neighbor interactions than on a landscape with random interactions. So, one-point crossover is better able to combine configurations of nearby interacting bits (without disrupting them too much again), than configurations of random interactions. The entry NNI/RND for K=25 reflects the fact that the graph of GA-ONEP is initially increasing faster for landscapes with nearest neighbor interactions, but is overtaken by random interactions, for which it eventually finds a better region (that is, containing higher peaks), as can be seen in Figure 2.

The table shows furthermore that uniform crossover (GA-UNIF) works better with random interactions on NK-landscapes with intermediate epistasis (K=5 and K=25). Apparently, for very low and very high epistasis, uniform crossover is just as disruptive, no matter whether the epistatic interactions are randomly distributed or nearby.

Table 2 presents the results from the second viewpoint. It shows for both types of epistatic interactions which type of GA (GA-ONEP or GA-UNIF) is better able to find a good region in the landscape in the first phase of the search ("better able" in the same sense as in Table 1). Again, an X means no (significant) difference.

The table shows that for smooth and rugged landscapes (K=0, 2 and 5) uniform crossover (GA-UNIF) works better than one-point crossover (GA-ONEP) when the epistatic interactions are random. So, for low, random epistasis, uniform crossover is better able to combine building blocks, without disrupting them too much again, than one-point crossover. One-point crossover (GA-ONEP), however, works better than uniform crossover (GA-UNIF) for very rugged landscapes (K=25 and 50) when the epistatic interactions are random. So, for high random epistasis, uniform crossover becomes too disruptive.

	K=0	K=2	K=5	K=25	K=50
RND	GA-UNIF	GA-UNIF	GA-UNIF	GA-ONEP	GA-ONEP
NNI	GA-UNIF	X	GA-ONEP	GA-ONEP	GA-ONEP

Table 2. Comparison of GA-ONEP with GA-UNIF for random interactions (RND) and nearest neighbor interactions (NNI) in the first phase of the search. An entry GA-ONEP point means that one-point crossover works better on that particular landscape than uniform crossover. An X means that there is no difference.

For nearest neighbor interactions, it appears that one-point crossover (GA-ONEP) works better than uniform crossover (GA-UNIF) for K=5, 25 and 50 . As expected, uniform crossover is too disruptive, compared with one-point crossover, in these cases.

So, these results show that there is a clear relation between on the one hand the type of recombination that is used (one-point or uniform crossover) and the type, and also the amount, of epistatic interactions (random or nearest neighbor) on the fitness landscape, and on the other hand the usefulness of recombination.

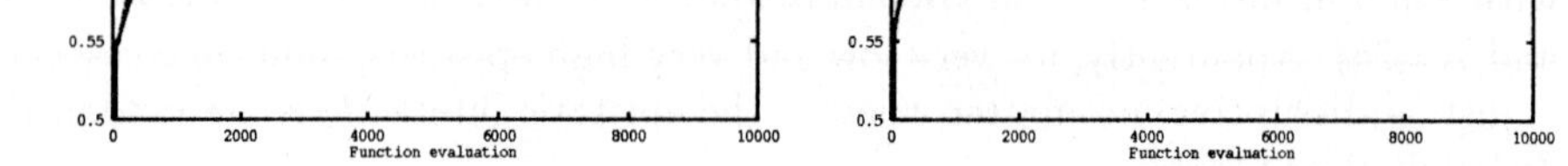

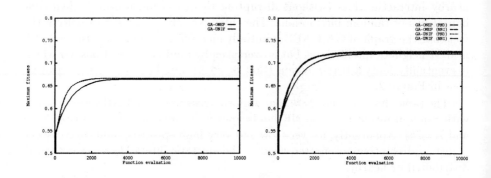

Fig. 1. Left: The maximum fitness of the two GA's on NK-landscapes with N=100 and K=0. Right: The maximum fitness of the two GA's on NK-landscapes with N=100 and K=2.

4 Recombination and the location of optima

In the previous section, we have investigated the role of crossover disruption. Here, we look at an alternative view on crossover which stresses its exploratory

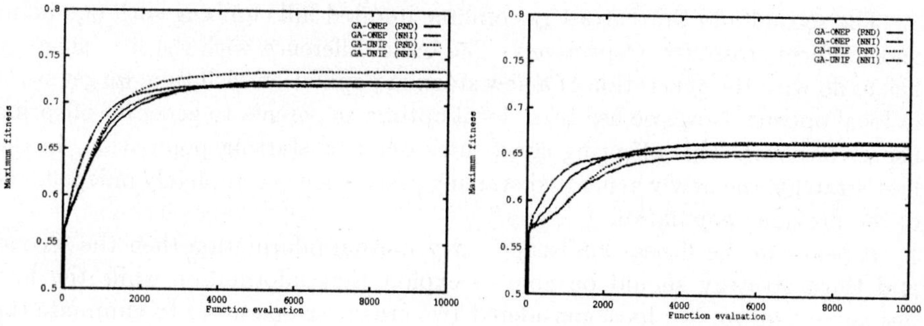

Fig. 2. Left: The maximum fitness of the two GA's on NK-landscapes with N=100 and K=5. Right: The maximum fitness of the two GA's on NK-landscapes with N=100 and K=25.

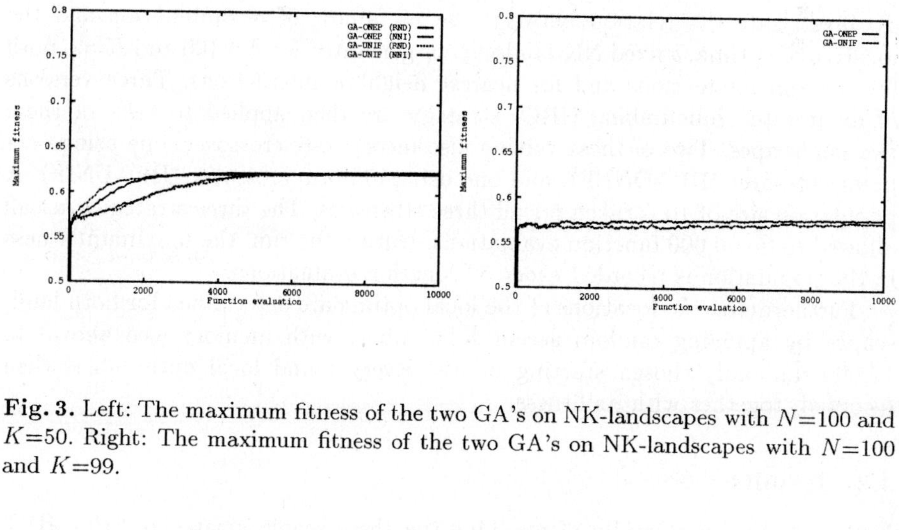

Fig. 3. Left: The maximum fitness of the two GA's on NK-landscapes with N=100 and K=50. Right: The maximum fitness of the two GA's on NK-landscapes with N=100 and K=99.

power. This view is exemplified by S. Kauffman. According to him, the first condition that has to be met, for recombination to be useful, is that the high peaks in the landscape are near one another and hence carry mutual information about their locations in the fitness landscape.

To examine to what extent this condition holds, three search strategies are compared. The first strategy is iterated hill climbing where we start with a population of bit strings and to each bit string we apply local hillclimbing until we get stuck in a local optimum. Then we randomly generate a new population and repeat local hillclimbing. This procedure is repeated until a predefined number of function evaluations is reached and during this procedure we keep track of the best fitness in each generation.

The second and third strategy combine iterated hill climbing with one-point and uniform crossover, respectively. The basic difference with the first strategy has to do with the generation of a new starting population each time we get stuck in local optima. Now, we use these local optima as parents to generate offspring using crossover. This offspring constitutes our new starting population. In the first strategy, the newly generated starting population is completely independent of the previous population.

If peaks in the fitness landscape carry mutual information then the second and third strategy should be able to exploit that information while the first one cannot do so. We have considered two crossover operators to eliminate the possible effect of crossover disruption.

The experiments described below are done on a fixed fitness landscape, where the relative positions of local optima are determined. Both random interactions and nearest neighbor interactions are considered.

4.1 Experimental setup

To investigate the relation between the usefulness of recombination and the location of optima, a fixed NK-landscape is generated for $N=100$ and $K=2$, both for random interactions and for nearest neighbor interactions. Three versions of an iterated hillclimbing (IHC) strategy are then applied to each of these two landscapes. Two of these version also incorporate crossover: one using one-point crossover (IHC-ONEP), and one using uniform crossover (IHC-UNIF). A population size of 10 is taken for all three strategies. The three strategies are all allowed to do 50,000 function evaluations. During the run, the maximum fitness in the population is recorded every 50 function evaluations.

Furthermore, the locations of the local optima are determined for both landscapes by applying random ascent hillclimbing with memory (see above) to 10,000 randomly chosen starting points. Every found local optimum is then recorded, together with its fitness.

4.2 Results

Figures 4 show the results of applying the three search strategies, IHC, IHC-ONEP and IHC-UNIF, to the fixed ($K=2$) fitness landscape for random and nearest neighbor interactions, respectively. It is clear that on the landscape with random interactions both crossover operators are useful. If crossover is applied to the population of local optima, the maximum fitness in the population stays relatively high, indicating that the locations of two optima give information about the locations of other optima. Also, the graphs appear to be gradually increasing. There is not much difference between IHC-ONEP and IHC-UNIF, and both outperform IHC.

For nearest neighbor interactions, however, the distinction is less clear. The IHC-ONEP and IHC-UNIF strategies appear to be just a little better than the IHC strategy during the search, but not much. Crossover contributes just a little in finding good regions in the landscape. Furthermore, the graphs are certainly not increasing, but instead appear to decrease a little after a while. Again, there is not much difference between IHC-ONEP and IHC-UNIF.

Figures 5 show the locations of local optima, relative to the best one found, for the landscapes with random interactions and nearest neighbor interactions, respectively. The fitness of the found optima is plotted against the (normalized) Hamming distance from the fittest local optimum that was found. For the landscape with random interactions 9,970 different local optima were found, while for the landscape with nearest neighbor interaction 10,000 different local optima were found.

There is a clear similarity between the two plots. The optima with a relatively higher fitness tend to be closer to the best optimum than optima with a relatively lower fitness. This shows a feature of the landscapes that Kauffman called a *massif central*: there is one place in the landscape where all the good optima are situated, surrounded by the less good optima (see [5]). This feature is the cause that crossover can help in finding a good region in the landscape: recombining the information of two optima gives a high chance of finding still better optima.

Besides the similarity, there is also one striking difference between the plots, though. For random interactions, the good optima are *much closer* to the best optimum (and thus to each other) than for nearest neighbor interactions. The number of optima with a (normalized) Hamming distance of 0.10 or less from the best optimum is 48 for random interactions, while it is only 2 for nearest neighbor interactions. This explains the difference in crossover performance (relative to no crossover) between the two landscapes. For nearest neighbor interactions, the good optima are just a little too far from the best optimum (and probably also from each other), to give enough information about the location of the highest peaks.

Kauffman already did this landscape analysis himself, and the plots shown here are very similar to his plots, which also show the similarity between random and nearest neighbor interactions (see [5]). But because he used different scales for both plots, the striking difference between the two is much harder to detect. At least, Kauffman does not say anything about it.

So, these results show that there also is a clear relation between the location of local optima in the fitness landscape, and the usefulness of recombination.

5 Conclusions

There appears to be a clear relation between on the one hand the type of recombination that is used and the type and amount of epistatic interactions on the fitness landscape, and on the other hand the usefulness of recombination.

In the first phase of a search, when a good region in the landscape is searched for by global search (i.e. crossover), one-point crossover works better when the epistatic interactions are nearby than when they are randomly distributed. In the latter case, one-point crossover is too disruptive. On NK-landscapes with intermediate values of K (K=5, 25) uniform crossover works better when the epistatic interactions are randomly distributed. For lower or higher values of

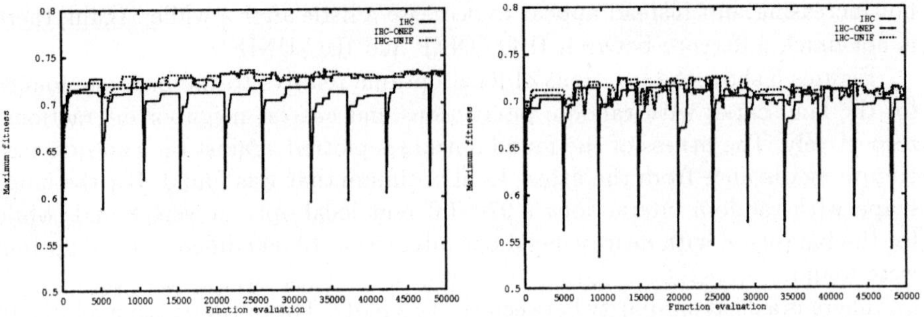

Fig. 4. Left: Comparison of 1-point (IHC-ONEP) and uniform (IHC-UNIF) crossover and no crossover (IHC) on an NK-landscape with N=100 and K=2, random interactions (RND). Right: Comparison of 1-point (IHC-ONEP) and uniform (IHC-UNIF) crossover and no crossover (IHC) on an NK-landscape with N=100 and K=2, nearest neighbor interactions (NNI).

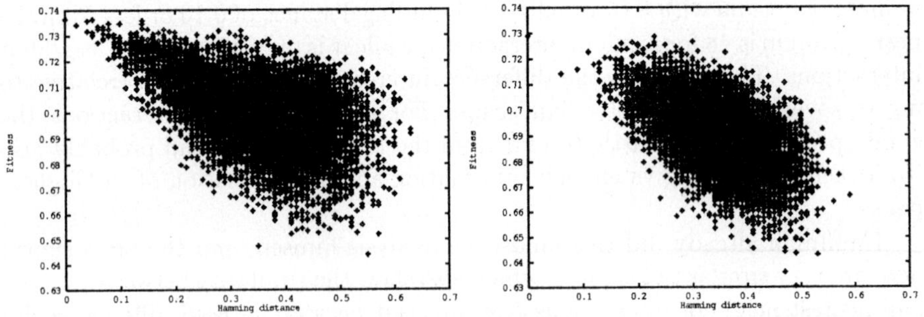

Fig. 5. Left: The correlation between the fitness of local optima and their (normalized) Hamming distance from the fittest local optimum found on an NK-landscape with N=100 and K=2, random interactions. Right: The correlation between the fitness of local optima and their (normalized) Hamming distance from the fittest local optimum found on an NK-landscape with N=100 and K=2, nearest neighbor interactions.

K, there is no difference between random and nearest neighbor interactions for uniform crossover.

On smooth and rugged landscapes (K=0, 2, 5) with random interactions, uniform crossover is faster than one-point crossover in finding good regions in the landscape. On very rugged landscapes (K=25, 50) with random interactions, however, one-point crossover is faster than uniform crossover. When the epistatic interactions are the nearest neighbors, one-point crossover is the better type in the first phase of the search on rugged to very rugged landscapes (K=5, 25, 50). As expected, uniform crossover is too disruptive in this case.

Furthermore, there is also a clear relation between the location of local optima in the fitness landscape, and the usefulness of recombination. Recombination is most useful when relatively high optima tend to be near each other. Recombining the information about the location of two optima gives a fair chance of finding even better optima. When the highest optima are not really close enough to each other, recombination becomes less useful.

With these conclusions, the validity of the next statement made by Kauffman can be assessed (see [5]:

> "*recombination is useless on uncorrelated landscapes but useful under two conditions: (1) when the high peaks are near one another and hence carry mutual information about their joint locations in the fitness landscape and (2) when parts of the evolving individuals are quasi-independent of one another and hence can be interchanged with modest chances that the recombined individual has the advantage of both parents*".

The first condition is validated, considering the conclusion above. The second condition, however, is not validated. As the above results show, the usefulness of recombination depends on the type of recombination that is used and the type and amount of epistatic interactions on the landscape. It is certainly not always necessary that the "parts of the evolving system are quasi-independent of one another". So, the results presented here do not fully support Kauffman's statement about the usefulness of recombination, but imply a more refined one.

References

1. L. J. Bain and M. Engelhardt. *Introduction to Probability and Mathematical Statistics.* Duxbury Press, 1987.
2. D. E. Goldberg. *Genetic Algorithms in Search, Optimization and Machine Learning.* Addison-Wesley, 1989.
3. J. H. Holland. *Adaptation in Natural and Artificial Systems.* MIT Press, 2nd edition, 1992.
4. W. Hordijk. Population Flow on Fitness Landscapes. Master's thesis, Erasmus University Rotterdam, 1994. Available via anonymous ftp at ftp.cs.few.eur.nl in pub/doc/masterstheses.
5. S. A. Kauffman. *Origins of Order: Self-Organization and Selection in Evolution.* Oxford University Press, 1993.
6. B. Manderick, M. de Weger, and P. Spiessens. The Genetic Algorithm and the Structure of the Fitness Landscape. In R. K. Belew and L. B. Booker, editors, *Proceedings of the Fourth International Conference on Genetic Algorithms*, pages 143–150. Morgan Kaufmann, 1991.

The Investigation of Lamarckian Inheritance with Classifier Systems in a Massively Parallel Simulation Environment

Eckhard Bartscht, Jens Engel and Christian Müller-Schloer

Inst. für Rechnerstrukturen und Betriebssysteme
Lange Laube 3
30159 Hannover
Phone: +49-511-762-4938
Fax: +49-511-762-4933
e-mail: bartscht@irb.uni-hannover.de

Abstract. In contrast to simulators for ecological processes as they are designed and implemented today, the *ParalLife* system which is introduced in this article can make use of problem inherent parallelism to speed up the simulation process. We describe how the simulated environment can be distributed over massively parallel computer architectures and what can be gained by doing so. We then describe Classifier Systems (CS) as one of the decision models available for Animats in *ParalLife*. A second topic of this paper is an experiment where Animats with CS improve their performance with Lamarckian Inheritance by recombining parental brain substructures.

1 Motivation

Over the last few year quite a large amount of simulators for ecological processes have been developed. Animats or adaptive autonomous agents with different properties scour two dimensional worlds for food, partners, enemies and obstacles.

Although today's computation power is at a point where such simulations can be done at a complexity promising interesting emergent behavior, experiments are still designed to fit into acceptable time frames. To follow the natural paragon where lots of activities happen in parallel it is desirable to find ways to parallelism in artificial environments, too. The objective is, that by doing so, experiments with a significantly higher number of simulated individuals are possible.

In this paper we present the *ParalLife* simulator which has been designed for execution on massively parallel architectures like nCube or workstation clusters running under PVM [Gei]. The paper starts with a short description of what the simulated worlds look like and the different kinds of decision models available in *ParalLife*. We then will have a look at the approach to parallelism chosen and on what was necessary to build a consistent simulation environment. The end of the first part of this paper is an evaluation of the speedup gained.

The last part of this paper descibes an experiments where Classifier Sytems as one of the available decision models in *ParalLife* have been used to solve a given problem. Used with lamarckian inheritance the Classifier Systems showed interesting problem partitioning behavior.

2 What makes the world

Simulated eco-systems normally consist of several different classes of objects like moving Animats, food, stones or obstacles. Instances of these classes are interacting with each other. All these objects share the same basic properties like size, age or coordinates. Animats differ from other objects in the sense that latter lack any kind of decision modules. The behavior of non-Animat objects is entirely determined by the physical laws programmed into the world.

Animats typically have to make decisions depending on the situation they are confronted with. They are living in a continuous sequence of perception and interaction, somehow have to get information about the environment, make their decisions and act in an appropriate manner. They furthermore should be able to learn from experience made in the past. An action initiated by an Animat always is restricted to the condition of the surrounding environment. If for example two different Animats saw the same piece of food the simulator has to make sure only one of them can get it. This means decisions made by an Animat only represent an intended action.

The plot of the world in such simulation environments can be considered a more or less complicated cellular automaton (CA) where each cell holds information of the object(s) situated on it. Physical restrictions like the one mentioned are handled by this CA array. Besides information about the objects it can be used to simulate all kinds of gradients like hights of sites or chemical gradients.

The basic functionality described above has been implemented in *ParalLife*. In the following section we will describe the structure of the Animats and other parts of the simulated environment. As interactions, sensors and effectors of Animats are not predetermined by the system, *ParalLife* offers a variety of possible experiments.

2.1 The structure of the objects

Animats in *ParalLife* mainly consist of four parts. First, a genetic memory which holds inherited information about the Animat itself. This information is not changed during the lifetime of the Animat. Second, it has a set of sensors and a set of effectors which are shapable according to the genetic information. Beside these structures each Animat has a brain model which may contain one or more decision models.

All Animats and the passive objects in *ParalLife* are instances of an arbitrary number of different classes. The constituent properties for each class may be specified by the user. As for the Animats, a class can be considered a species because recombination can only take place among individuals of the same class. Animats of different classes may not be recombined, because in general they are assigned to different decision models which might be of completely different structure, which would make a reasonable recombination impossible.

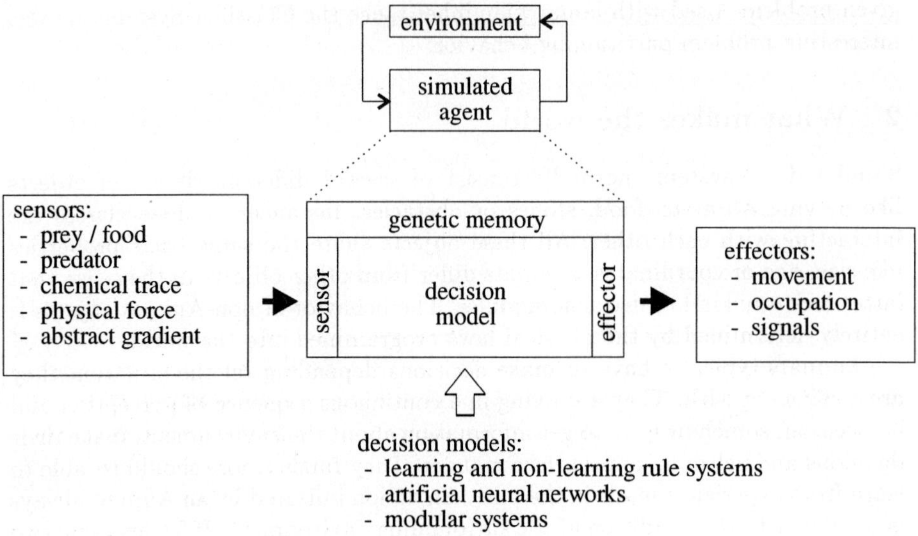

Fig. 1. The Animat and its environment

2.2 Decision models

ParalLife contains a library function interface for the decision models. These functions specify how to edit and visualize brain-related data, allowing the user to monitor learning and decision making. The most important function in this library is the one responsible for decision making. It gets the latest perception and external reward information as arguments and computes the appropriate next action based on former experience accumulated in the Animat's brain. Since the simulator provides both, a memory area for the brain's phenotype and one for the brain's genotype data, inheritance of brain information can either be done in a Darwinian or a Lamarckian manner.

It is not predetermined by the system what kind of decision models an Animat can have. So far several decision models have been developed or ported to *ParalLife* .

Class	*Decision model*
background	-
obstacle	-
herbivore	Artificial Neural Net
carnivore	Classifier System

Table 1. Example for the specification of classes

- simple non-learning rule systems
- Classifier Systems (a modified version of Riolo's CFS-C package)
- artificial neural networks

2.3 The "Environment"

Whenever one object tries to move onto another, an interaction takes place. With respect to the classes of the interacting object, the system looks up an interaction matrix which holds the names of user-defined functions handling the requested information. At the first glance a matrix like this seems to be inflexible. The objects in *ParalLife* though have the possibility to influence the course of the interaction in two different ways. On the one hand they can "decide" if the interaction shall be executed at all and on the other hand the selected action can be done according to the individual's phenotype, as each interaction function is invoked with a data structure containing the body data.

↓ *meets* →	background	obstacle	herbivore	carnivore
background	-	-	-	-
obstacle	-	-	-	-
herbivore	occupy	stumble	mate	get eaten
carnivore	occupy	stumble	eat	mate

Table 2. Example of an interaction matrix

3 Our approach to parallelism

A very simple approach to parallelism is to split the map of the world and scatter the parts over several processors. All objects contained in such part would then totally be calculated on the corresponding processor. This approach obviously has several drawbacks. If, for example, heaps of Animats are concentrating in a few plots the load between different processors would be extremely unbalanced. Furthermore, the limited amount of memory on MIMD machine nodes would drastically restrict the number of Animats situated in the plot, because brain models with learning capabilities require large data structures. To circumvent this, one would have to shrink the size of the world parts. This unfortunately implies an increase in inter-node communication when handling border transcending activities.

In contrast to this, our approach eliminates most of these problems. The idea is to seperate brain and body of the Animat. One group of the available processors performs the brain model's decision tasks while the other group is responsible for the simulated environment. By doing so, we are able to increase the number of Animats per plot. The major advantage lies in the fact that the number of decision models executed in parallel is only restricted by the number of available brain processors.

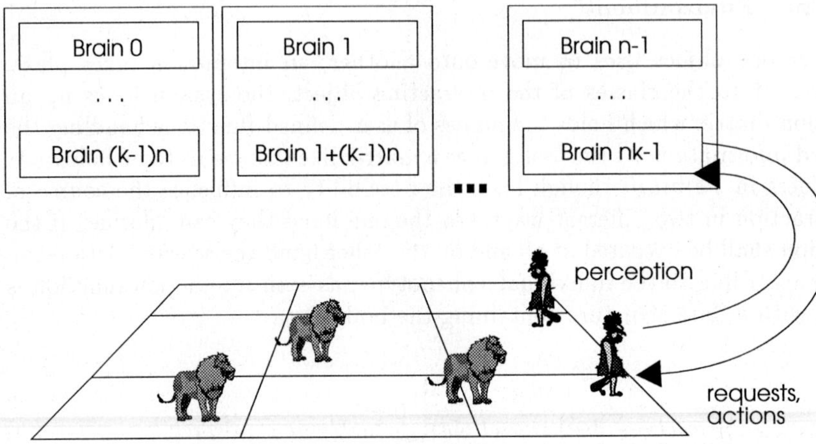

Fig. 2. Two levels of parallelism in *ParalLife*

Figure 2 shows the structure of a simulation environment with 9 world processors and n brain processors. The number k of "brains" calculated per processor depends on the number of animats in the simulation. In *ParalLife* it is made sure the decision models are equally distributed over the available brain processors.

4 The communication between body and brain

Due to the message passing hardware environment *ParalLife* has been implemented for some fairly complex communication protocols had to be designed. "Brains" for example have to communicate with the processors which currently hold the body of the Animats. They have to ask the world processors for environmental information and ask them for the execution of a selected action. No matter if the selected action is allowed or not the world processor has to notify the "brain" about the action really executed. Imagine an Animat intending running into a wall. Physical restrictions programmed within the world might prohibit this. The Animats brain though does not have the information about this physical limitations. It has to be informed if the selected action was possible to be executed or not when receiving an encoded version of the Animats new perception.

Communication between the different processing nodes can become quite complex when Animats are trying to recombine genetical information. Figure 3 shows an overview of what the simulator has to do when a mating interaction becomes necessary. It shows the communication between the world processor

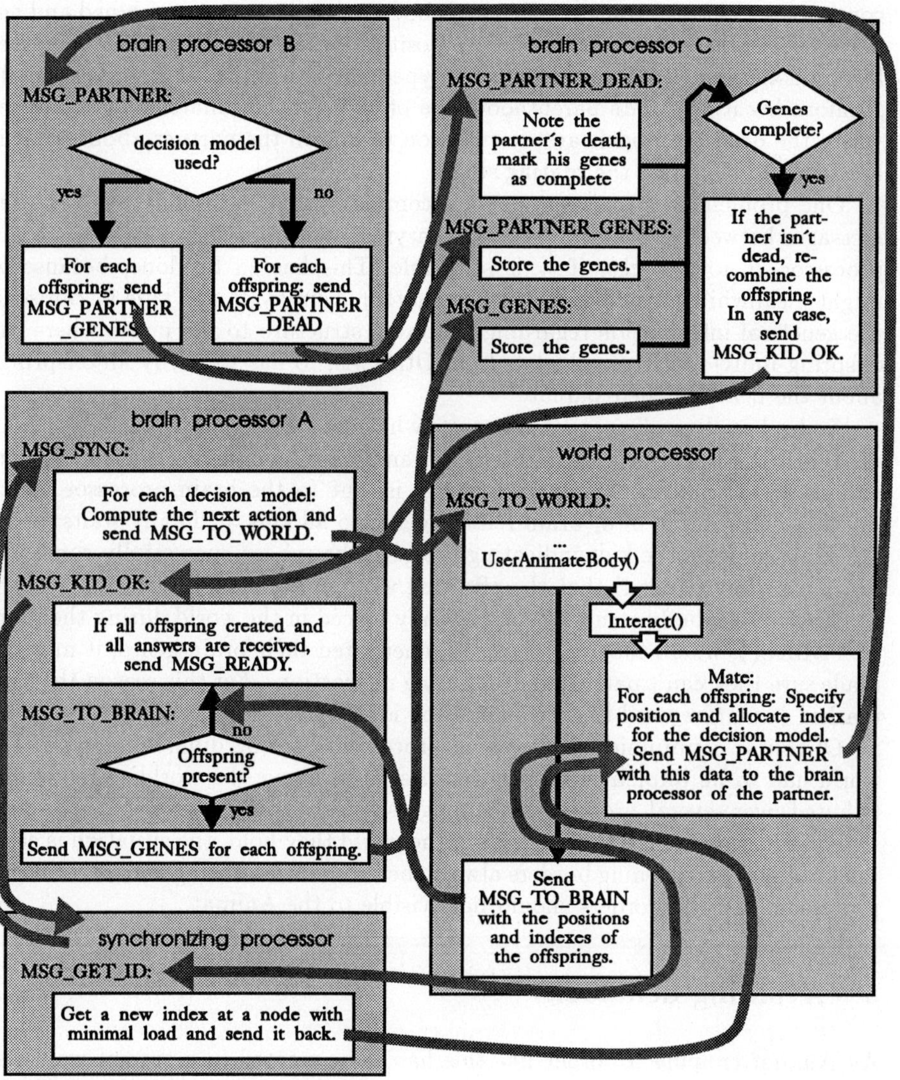

Fig. 3. Communication during inheritance. Short white arrows symbolize subroutine calls

holding the action initiating animat, the involved brain processors and a synchronization processor.

A mating interaction may be initiated by the individual named a_i which has its decision module data structures on processing node A. During the execution of the function **Mate()**, a_i requests the information, which processor the offspring's brains shall be calculated on. The information comes from a special syncronization node which holds all information that can't be distributed due to consistency reasons. Now the offspring's body structures are recombined and are placed somewhere around the current position of the mating partner b_j. After having done so a_i is sending a message typed **MSG_PARTNER** to b_j on node B informing it about its parenthood, the offspring's coordinates and the nodes where the decision models are executed on to inform the partner about what it has to do to complete the mating step.

One problem with message passing computer architectures is the fact that messages between processors can have varying times. For this reason b_j's brain processor B now checks, if b_j is still alife. This has to be done, because b_j might meanwhile have been killed bye another Animat. If still alife b_j can send the genetical information regarding the brain structure to the nodes where the offspring individuals are calculated on. Otherwise B has to notify all offsprings about the death of their parent.

Nodes handling offspring Animats now have to wait for the information from a_i. When the world processor currently handling a_i completed the world interactions of the mating Animat a message is sent to the brain processor which initiates the sending of a_i brain related genes to all kids. a_i finally waits for an OK message from the kids indicating the genes have been successfully combined or (in case b_j was dead) that the offspring will remove itself in the next step.

Because the body structures are already placed in the world during the function **Mate()**, where they not have yet inheritated their brain genes, it must be made sure they can't participate in mating interactions. For this reason they are marked *immature* until the recombination is complete.

Of course communication between processors holding different parts of the world had to be implemented, too. As described above the world's map is distributed over several processors. As long as world processors only have to deal with interactions regarding the part of the world they hold, the situation is easy. Looking over partitioning borders always means communicating with other world processors and the borders should be invisible to the Animat.

5 Avoiding deadlocks

An Animat crossing a border like this has to be treated in a special way. The world processor the Animat has been on so far isn't responsible for the body structure anymore and could free the place where the Animat was at. If though the place the Animat wants to go to on the new world processor is not free, the old coordinates have to be its fall back position. Therefore it has to be marked *transient* until received clearance to land on the target position. During this time, requests regarding the occupation of the Animats original position by others can't be answered by the world processor. The requesting Animat can only get the information to wait for a short while and ask again.

If several Animat's which are in a state of transition, try to change their position in a cyclic way, the simulator is in danger of creating a deadlock situation. The simplest version of a situation like this is when two Animats are trying to change their position over world processor borders. Both are in the state of transition and their corresponding world processor can't answer the request. To make this clear, the deadlock situation described here is one that could occur in the simulator itself, as a time step can't be completed until all requests by the Animats are completely handled. This kind of deadlock is fatal as it would stop the entire simulation process and has nothing in common with situations where Animats manoeuvre themselfs into situations where they can not move. The latter kind of deadlock situation is one occuring in the simulated world and *has* to be possible.

Whenever an Animat wants to move across a border, the handling world processor sends a message to the destination processor and waits for a receipt. If a request for movement, sent by another world processor arrives at that time, it is rejected by sending back a **MSG_TRY_AGAIN** message. This could result in an eternal sequence of requests and *try again*'s when the Animats requests are showing cyclic dependencies. To avoid this, the handling processor sends a "FLAG"-message to the destination processor. This message is passed along existing dependencies, until it reaches its sender again. If this is happening the deadlock is detected and will be eliminated by the withdrawal of the movement request, i.e. the Animat won't be moved.

6 Speedup evaluation

Several experiments have been done to test if our approach really offers an increase in execution speed. The parallel machine used for our experiments was an nCube II, a MIMD (multiple instruction, multiple data stream) system. It had 128 processing nodes which are connected as a 7 dimensional hypercube. The system can also be used in a way where several independent hypercubes of smaller dimensions can be accessed. Each processing node has 4MB of memory. Besides the CPU (operating at 20MHz) each node has a special network communications unit which is responsible for sending and routing messages.

In figure 4 the results from two series of experiments are shown. To measure execution time at various degrees of parallelism, we carried out experiments with 200 Animats randomly moving for 200 time steps in a world of size 160x160. The two series of experiments shown in figure 4 differ as follows: In series (a), the Animats used a simple non learning rule system, whereas series (b) dealt with individuals using a more complex decision model with learning capabilities (Classifier Systems with 100 classifer). We chose this simple experiments as the objective was to measure the speedup gained by using extra processing nodes.

The figure shows the execution times at various numbers of world and brain nodes. As one would expect, the reduction of execution time is more significant when using a more complex decision model. Having too few brain nodes causes the world nodes to wait for the end of the animats decision processes. In series (a) the execution of the decision models can not be speeded up very much as they are very simple.

One might wonder why there is no significant decrease in execution time when using two instead of one world nodes and why in some cases it is even increasing.

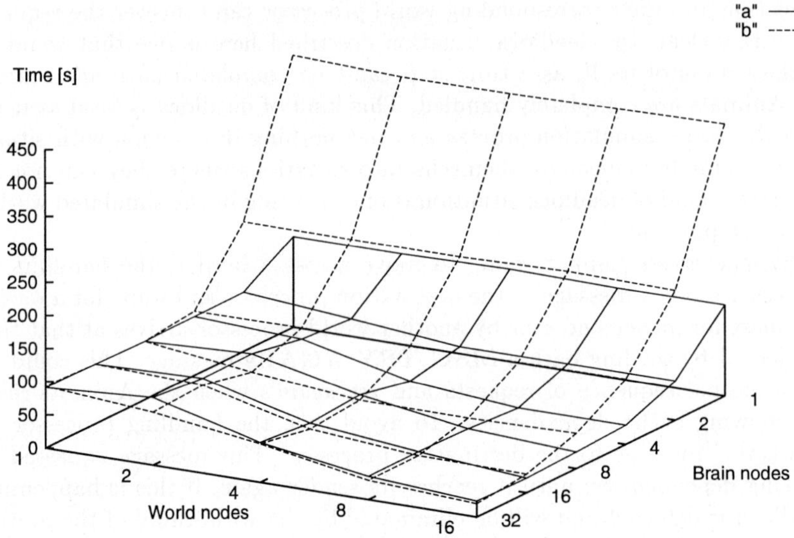

Fig. 4. Execution times for the experiments (a) and (b)

Communication between processors is not free. Obviously in this special case the decrease in execution time is fully used up by the extra time needed for communication. In case of only one world node though no communication is needed at all.

Another interesting result is the interdependency of world and brain node dimension. In all experiments we found that from a certain point on, additional brain nodes are only effective when using extra world nodes.

For experiment (b) an overall increase of processing power from 2 to 48 results in a speedup of 20.5 corresponding to an efficiency of 85%. In experiment (a) these values of course are not that high, but still interesting. The overall speedup from 2 to 48 processors is only 5.7, but good efficiency values can be seen when two or four brain processors are used instead of just one.

7 Classifier systems

The remaining part of this paper describes an experiment done with Classifer Systems as the Animat's decision model. Very shortly we now will describe the major properties of Classifier Systems. For a deeper introduction see [HR78] or [Gol89].

		world nodes									
		1	2	4	8	16	1	2	4	8	16
	1	137	137	140	137	138	410	404	405	408	411
brain	2	89	75	59	59	59	196	195	195	199	196
nodes	4	89	81	43	29	29	97	98	96	98	98
	8	88	95	59	34	25	88	67	51	50	53
	16	90	94	53	36	22	90	84	44	30	28
	32	91	102	60	35	24	93	80	58	29	20

Table 3. Execution times (in sec.) for system (a) on the left and (b) on the right.

7.1 General features

Classifier Systems (CS) are adaptive message passing rule systems based on Genetic Algorithms. Classifiers are rules, consisting of a condition part and a message part. If a Classifier finds messages in the system, matching its condition, it is activated and has the opportunity to send its message into the system. To do so the classifier typically must win a bidding competition for a space in a message list of limited size. Some of the messages in the list may have effects on the environment providing reward (or punishment). This now can be used to calculate the value of a certain classifier by giving strength to those classifiers which lead to the rewarding state. Genetic Algorithms are modifying and recombining the classifiers and thus change the behavior of the system.

Classifier Systems have features which make them interesting to use as decision models in Animats. They can be programmed to show a reasonable starting behavior which can continuously be improved by the GA. A second interesting property is that rules can be analyzed and found strategies can be understood. Besides these facts CS are even learning when the system is rarely rewarded.

7.2 CFS-C in *ParalLife*

CFS-C is a Classifier System implemented by Rick Riolo [Rio88]. It uses GA and several other operators for rule discovery as well. It assigns strength to the rules by using the Bucket Brigade Algorithm [Hol85]. The version of CFS-C ported to *ParalLife* differs from Riolo's original in some ways. First, it has been revised and some bugs in the original CFS-C, which appeared only on some operating systems, have been removed. Second an *implicit* Bucket Brigade (BB) algorithm has been used instead of CFS-C's original one. While Riolo's original BB allows payment between classifiers only when one classifier was explicitly activated by a message sent by another classifier in the preceding step, the implicit BB transfers strength from all message passing classifiers to the ones active in the following step. This makes sure credit is easily paid to classifiers which change the environmental state so that subsequent rules may become active and lead to external reward. Wilson used this BB version in [Wil85].

To implement CFS-C the basic working cycle had to be changed. While normaly a CFS-C working step ends with the payment of rewards, a time step in *ParalLife* ends with the generation of an action. In *ParalLife* the external rewards can't be given immediately. If for example an animat searches for food and "decides" to move on step ahead it has to make its move first (if allowed by the environment) and "see" if food can be found on that position. This means reward in *ParalLife* comes at the beginning of the following step.

8 Lamarckism and implicit problem partitioning

In this section we present an experiment with Lamarckian inheritance of classifier systems carried out in the *ParalLife* simulator as follows. Two Animats with CFS-C decision models perform a learning task and are recombined at various time steps to create offspring which go on performing the same task. The recombination is done in the following manner: The classifiers of the offspring decision models are copied from those of their parents using the roulette wheel parent selection method, i.e. a classifier C with strength S_C is drawn with probability

$$p(C) = S_C \;/\; \sum_{C'} S_{C'}$$

This means parents hand substructures of their acquired knowledge to their kids, thus one might call this kind of inheritance a Lamarckian one.

8.1 The Task

The Animats are confronted with an abstract environment called *Finite State World* (FSW), which consists of a number of states and transitions between these states. The Animat has the choice between several transition alternatives in each state, but has to select an appropriate one to reach a rewarding state. As the Animat is randomly placed on one of the possible start states it has to learn a path for each such state to achieve maximum performance.

Figure 5 shows the FSW used in this experiment. It has 24 states in total. Six are start states, six are final states including one reward state. For each episode i.e. a sequence from a start to a final state, a start state is randomly chosen.

8.2 Results

Figure 6 shows the performance of parents and offspring at various recombination times within the first 20,000 time steps. The curves show the percentage of correct episodes within the last 100 steps and show the mean of 12 runs with different random seeds. As one can see, the offspring show a better average performance than their parents. Already at step 3000, when the parent's performance still is bad, the offspring has about 80-90% of correct episodes.

So this form of inheritance causes the fact that the offspring do not need to start from scratch and immediately reach a performance not below their parent's

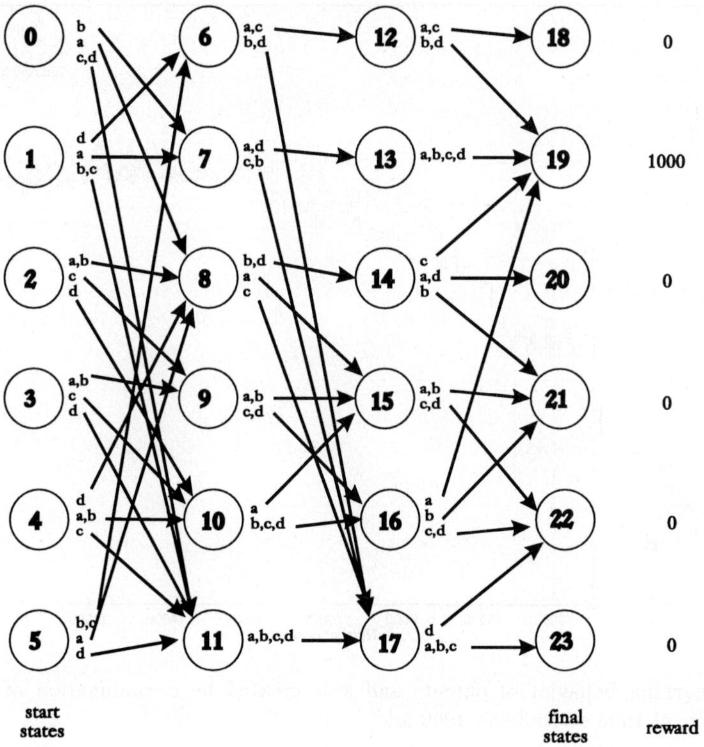

Fig. 5. The finite state world FSW

ones. As for the fact the offspring on average has an even better performance than their ancestors, we have got the following two possible explanations:

1. Let's assume that by recombining parental experiences the offspring *only* reaches performance values comparable to those of its best parent. The parental performance curves then of course would be worse, just because the less experienced parent has been taken into account when the mean was calculated. Even with this simple explanation there is the interesting point that it is possible to derive a global performance measure from low level learning structures such as classifiers, without explicitly evaluating the overall parental behavior.
2. If one assumes that, by being randomly placed in different start states, parents learned partially disjoint subsolutions for the problem (as long as the entire solution is not yet found), the offspring may benefit by having a solution for a bigger subproblem right from the beginning. This of course implies that parents found a somehow suitable partitioning for the overall problem.

No matter which of the explanations above we choose, in larger populations of Animats the offspring recombined in this manner has a greater amount of

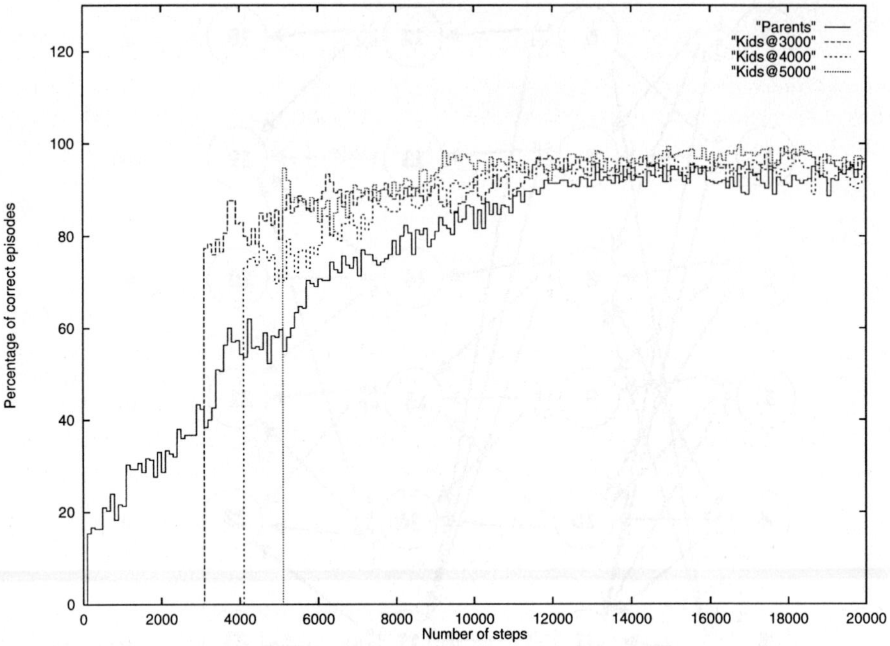

Fig. 6. Learning behavior of parents and kids created by recombination of parental information at time steps 3000, 4000 and 5000.

experience than their ancestors. It is possible to regard this kind of Lamarckian inheritance as a form of education, similar to the way Holland considers the process of adding *external* rules to classifier systems a sort of education [HHNT86]. Like others before (for example [AL92]) we found that obviously individuals using this form of "educational inheritance" have an advantage compared to those only adapting to the environment in a Darwinian fashion.

9 Summary and Outlook

In this paper we stated that a large amount of possible concurrency in today's simulators for ecological processes is still unused. It has been explained how a massively parallel computer architecture (nCube) can be used to gain increases in simulation speed. Some of the problems resulting from the distribution of the simulation over several processors have been mentioned and it has been shown that the speedup gained by this concept is worth the extra inter-processor communication.

It was shown that a variety of different decision models is available for *ParalLife* and we explained some of the features Classifier Systems have. In an experiment where a parental generation only had very short time to learn properties of the environment, the offspring generation with a new set of classifers, recombined from the parental one's showed a much better performance.

Future plans about *ParalLife* comprise a comparison of Classifier Systems and Artificial Neural Networks, which have been implemented for *ParalLife* lately. The concepts shown here can also be a starting point for new ways to an intelligent load balancing on massively parallel computer architectures.

References

[AL92] D. H. Ackley and M. L. Littman. A case for lamarckian evolution. In C. G. Langton, editor, *Artificial Life III*, 1992.

[Gei] Al Geist. *P(arallel) V(irtual) M(achine) user's guide and reference manual.*

[Gol89] David E. Goldberg. *Genetic Algorithms in Search, Optimization.* Addison-Wesley Publishing Comp.,Inc., 1989. IRB.

[HHNT86] John J. Holland, Keith J. Holyoak, Richard E. Nisbett, and Paul R. Thagard. *Induction, Process of Inference, Learning, and Discovery.* Cambridge, Massachusetts: The MIT Press, 1986.

[Hol85] John H. Holland. Properties of the Bucket Brigade Algorithm. In John J. Grefenstette, editor, *Proceedings of the First International Conference on Genetic Algorithms and their Applications*, 1985.

[HR78] J. H. Holland and J. S. Reitman. Cognitive Systems Based on Adaptive Algorithms. In D. A. Waterman and F. Hayes-Roth, editors, *Pattern-Directed Inference Systems.* New York: Academic Press, 1978.

[Mae94] Pattie Maes. Modeling adaptive autonomous agents. *Artificial Life*, 1:135–162, 1994. ISBN 1064-5462.

[Rio88] Rick L. Riolo. CFS-C: A Package of Domain Independent Subroutines for Implementing Classifier Systems in Arbitrary, User-Defined Environments. Technical report, Logic of Computers Group, Division of Computer Science and Engineering, University of Michigan, 1988.

[Wil85] Stewart W. Wilson. Knowledge Growth in an Artificial Animal. In John J. Grefenstette, editor, *Proceedings of the First International Conference on Genetic Algorithms and their Applications*, 1985.

Orgy in the Computer: Multi-Parent Reproduction in Genetic Algorithms

A.E. Eiben[1] C.H.M. van Kemenade[2] J.N. Kok[1]
gusz@cs.ruu.nl kemenade@cwi.nl joost@cs.ruu.nl

[1] Department of Computer Science
Utrecht University
P.O. Box 80089, 3508 TB Utrecht,
The Netherlands

[2] Department of Software Technology
CWI,
P.O. Box 94079, 1090 GB Amsterdam,
The Netherlands

Abstract. In this paper we investigate the phenomenon of multi-parent reproduction, i.e. we study recombination mechanisms where an arbitrary $n > 1$ number of parents participate in creating children. In particular, we discuss *scanning crossover* that generalizes the standard uniform crossover and *diagonal crossover* that generalizes 1-point crossover, and study the effects of different number of parents on the GA behavior. We conduct experiments on tough function optimization problems and observe that by multi-parent operators the performance of GAs can be enhanced significantly. We also give a theoretical foundation by showing how these operators work on distributions.

1 Introduction

Natural and artificial recombination mechanisms (applied in Evolutionary Computation) are rather different. However, they all agree on the number of parents: it is either 1 or 2. As for natural reproduction, the absence of multi-parent reproduction can be understood if we consider the practical difficulties. For instance, one could think of the problems of getting more individuals "in the mood" at the same time and the same place, or those of having sophisticated female mechanisms to keep sperm alive until the necessary number of mating acts and conceptions are performed. In artificial evolutionary systems the restriction on the number of parents is less obvious. In fact, we can expect advantageous changes in the behavior of the classical uniform crossover, as well as of the classical 1-point crossover if we generalize them to $n > 1$ parents. Explanation of such positive expectations is given after defining the n-ary operators in section 2. The main goals of this paper are:

- present two multi-parent recombination mechanisms: scanning crossover and diagonal crossover;
- study the effect of using more than 2 parents on the behavior of the GA for both mechanisms;
- compare scanning crossover to diagonal crossover.

The paper [2] just shows empirical results regarding scanning crossover combined with a generational genetic algorithm using fitness proportional selection. The structure of the paper is the following. After the Introduction we give a description of (different versions of) the scanning crossover mechanism and the diagonal crossover. In Section 3 our test suit consisting of difficult numerical optimization problems and the parameters of the applied GA are presented, furthermore the performance measures are discussed. Thereafter the results of the experiments are displayed in Section 4. In Section 5 we study the expected value and the variation of fitness of a chromosome from generation to generation by tracing expected values of Walsh products. Finally, in Section 6, we draw our conclusions and give an explanation of the results.

2 Multi-parent recombination mechanisms

2.1 Scanning crossover

The simplest form of the scanning crossover mechanism studied in [2] works by taking n parent strings and creating one child through investigating the j-th ($j = 1, ..., k$, where k is the chromosome length) gene of the parents and chosing one of them to be the j-th gene of the child. Notice that the way the choice is made about the gene to be inherited is not specified. This allows different versions of gene scanning, distinguished by different choice mechanisms. Possible problem independent choice mechanisms are for instance uniform random choice, voting or random choice biased by the fitness of the parents. Figure 1 illustrates occurrence based (voting) scanning for bit-pattern representation, where the allele with the highest number of occurences should be inserted in the child.

```
Parent 1: 1111010111000110
Parent 2: 1100010101000010
Parent 3: 0011101010101011
Parent 4: 0101010101100100
   Child : 1111010111000010
```

Fig. 1. Occurence based scanning crossover on bit patterns

The different choice mechanisms amount to a different level of bias in the genetic operator. In the meanwhile they all have in common that child construction is based on a larger (i.e. $n > 2$) sample of the search space than in classical

GAs, and that a promising gene is chosen for each position of the child. The definition of 'promising' can be problem independent (as in the above examples), but can also be based on on some problem specific heuristics. Therefore, scanning crossover is very well suited for being enriched with heuristics, moreover the presence and the influence of the incorporated heuristics is explicit. Let us note that the scanning crossover operator can also be adjusted for order based representation very easily as it is illustrated in [2].

In this paper we restrict ourselves to uniform scanning crossover, where the allele that is inserted in the child is chosen randomly by giving an equal chance to each parent to deliver its allele. Recall that the classical uniform crossover is a very disruptive operator. Applying an n-ary version can reduce the level of disruptivity by using a bigger sample of the search space and by creating only one child.

2.2 Diagonal crossover

Traditional crossover creates two children from two parents by splicing the parents along the single crossover point and exchanging the 'tails'. The basic idea behind diagonal crossover is to generalize this mechanism to an n-ary $(n-1)$-point crossover. Diagonal crossover selects $(n-1)$ crossover points resulting in n chromosome segments in each of the n parents and composes n children by taking the pieces from the parents 'along the diagonals'. For instance, the first child is composed by taking $substring_1$ from $parent_1$, $substring_2$ from $parent_2$, etc., while the second child would have $substring_1$ from $parent_2$, $substring_2$ from $parent_3$, etc. Figure 2 illuminates this idea.

1*a*	1*b*	1*c*	*parent – 1*
2*a*	2*b*	2*c*	*parent – 2*
3*a*	3*b*	3*c*	*parent – 3*

1*a*	2*b*	3*c*	*child – 1*
2*a*	3*b*	1*c*	*child – 2*
3*a*	1*b*	2*c*	*child – 3*

Fig. 2. Diagonal crossover with three parents

Notice that for $n = 2$ diagonal crossover coincides with the traditional 1-point crossover. The reason to expect that the use of more parents in diagonal crossover leads to improved GA performance is basically that the search becomes more more explorative, without hindering exploitation. The more explorative character is the result of having more crossover points and thus a higher level of disruptiveness, and the fact that using more parents there is more 'consensus' needed to focus the search to a certain region.

When considering the higher number of crossover-points in diagonal crossover the question obviously arises whether the same results could be achieved with the classical n-point crossover, which uses two parents. Therefore we decided to test this operator too, in order to see if the higher number of parents contributes to a better performance.

Genetic algorithms typically just use one gender, so do the multi-parent operators.

3 Test functions and setup of the experiments

We have decided to test multi-parent crossovers on tough optimization problems. We have chosen a test suit consisting of the second de DeJong function F2, the Ackley, the Griewangk, the Michalewicz, the Rastrigin and the Schwefel functions. All these functions, except for the F2 function, have a large number of local optima, which make those functions difficult to optimize. The defining formulas of these functions can be found in [4], [6], [9] and [11]. For each function we applied binary representation; the most important properties of the test functions and their representation are summarized in Table 1. Let us remark that the Michalewicz function is to be maximized, while all the others are to be minimized.

	F2	ACKL	GRIE	MICH	RAST	SCHE
dimension	2	30	10	2	20	10
chrom. length	22	600	200	33	400	210
global opt.	0	0	0	38.8503	0	0

Table 1. Properties of the test functions

In all of the experiments we used the same GA-setup, which is exhibited in Table 2.

When monitoring the performance we maintained different measures, namely efficiency (speed), and success rate (percentage of cases when an optimum was found). We measured speed by the total number of function evaluations (averaged over all runs). Let us note that the results on uniform scanning crossover deviate from those presented in [2] because there a generational GA with fitness proportional selection was used. Further tests, however, indicated that using a steady state GA with ranked selection yields better results.

4 Experimental results

We will review the results grouped around the two performance measures discused in the previous section: success rate and efficiency. Since 6 test functions

nr. of parents	2-10 or 2-15
GA type	steady state
selection mechanism	ranked bias 1.2
xover rate	0.7
mut. rate	1/chrom. length
pool size	200
max. nr. of func. eval.	70.000
termination cond.	optimum hit or population converged
averages over	100 runs

Table 2. GA parameters

and 2 performance measures would result in 12 Figures we only display some of them here. The complete date files can be obtained from the authors on request.

4.1 Success rates

Let us first consider the rate of success of the different operators, which is the most important measure from a strict optimization point of view. Table 3 shows the optimal versions of the genetic operators and the corresponding success rate for each test function; within brackets we displayed the success rate of the 2-parent versions.

test function	Scanning Xover #par	succ.	Diagonal Xover #par	succ.	N-point Xover #Xover points	succ.
F2	7	.91 (.73)	11	.88 (.38)	11	.84
Mic	10	.72 (.57)	15	.76 (.34)	15	.6
Schw	2	.015 (.015)	15	.24 (.00)	10	.1
Grie	10	.48 (.22)	14	.32 (.04)	10	.15
Ras	5	.10 (.00)	13	.28 (.00)	15	.06
Ackl	8	.90 (.84)	15	.89 (.00)	10	.24

Table 3. Optimal nr. of parents and corresponding success rates. Within brackets the results for 2 parents. (For n-point crossover the number of parents is always 2.)

It appears immediately that the optimal number of parents is always higher than 2, with one exception. Also, the gains achieved by using more than two parents are substantial, especially for the diagonal crossover. The figures within brackets show an interesting phenomenon too. Namely, on all tests functions the standard uniform crossover performs much better than 1-point crossover (diagonal crossover for two parents). At first sight one would expect uniform crossover to be too disruptive, as a single individual represents a vector of integers, so there is often a strong correlation between successive bits. Uniform crossover seems to disrespect these boundaries and turns out to be less sensitive for premature convergence. Looking at the results of diagonal crossover and n-point (2 parent) crossover we can see that the better performance of diagonal crossover is not only the consequence of applying more crossover points, but the higher number of parents contributes considerably.

As for the effect of different number of parents on success rates we observed that diagonal crossover yielded better results when the number of parents increased on all of the test functions. To illustrate this effect we show the success rates on the Griewangk and the Schwefel functions in Figure 3.

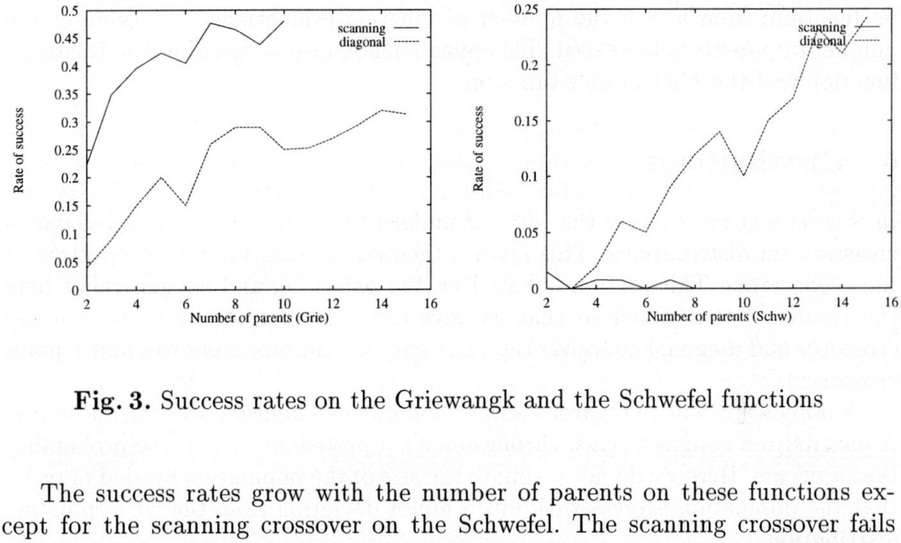

Fig. 3. Success rates on the Griewangk and the Schwefel functions

The success rates grow with the number of parents on these functions except for the scanning crossover on the Schwefel. The scanning crossover fails completely on the Schwefel function. This is a result of the large distance between the best and the second best optimum in this case. On the two-dimensional Michalewicz function and the F2 function the success rates of scanning crossover do not show that more or less monotonous growth that can be seen for diagonal crossover. Recall from Table 3, however, that the best performance was mostly obtained with more than two parents.

4.2 Efficiency

As for efficiency an increased performance means that it takes fewer function evaluations to reach a (sub)optimum if the number of parents increases. We observed this effect on the Ackley, the Griewangk and the Rastrigin functions for the diagonal crossover as well as for scanning crossover, see Figure 4.

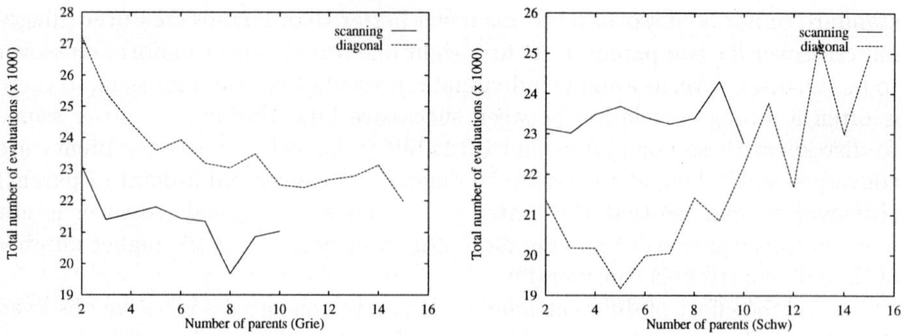

Fig. 4. Efficiency on the Griewangk and the Schwefel functions

Notice, however, the remarkable behavior of diagonal crossover on the Schwefel function: from $n = 5$ the number of function evaluations is growing as the number of parents is increased. This phenomenon also occured on the Rastrigin function and the Michalewicz function.

5 Distributions

In this section we consider the effect of uniform scanning crossover and diagonal crossover on distributions. This gives a theoretical basis for the comparison of these operators. This section is based on the paper [5], and we generalize here the results of that paper in that we give also formulae for uniform scanning crossover and diagonal crossover (and not only for uniform crossover and 1-point crossover).

A population will be regarded as a probability distribution over chromosomes. A distribution assigns to each chromosome x a probability $P(x)$: the probability that x occurs. Here we do not estimate the size of the population needed in order that the population evolves with but a minor deviation from the corresponding distribution.

Let f be the fitness function. We are interested in tracing the expected value of the fitness of an individual $E[f] = \sum_x f(x)P(x)$ from generation to generation and also its variance.

Instead of following the probabilities, we follow the expected values of Walsh products. The expected values of Walsh products are equivalent to probabilities in the sense that if we know the probabilities, then we can compute the expected values of the Walsh products and the other way around. Following the expected values of Walsh products is much more efficient than following probabilities in terms of computational effort, and the formulae are nicer and mathematically more tractable.

First we introduce some notation. We can interpret a chromosome x in three different ways:

1. as a bit string of length n, that is $x = x_1 \cdots x_n$, with $x_i \in \{0,1\}$ $i = 1, \ldots, n$,
2. as subset $\{i : x_i = 1\}$ of $\{1, \ldots, n\}$, or
3. as integer $\sum_{i=1}^{n} x_i \cdot 2^{i-1}$.

As an example, the bit string 10110 is equivalent to the set $\{1,3,4\}$ and to the integer 13. The main advantage of using different representations is that the formulae become simpler.

Given a bit string x and a set i, the Walsh product $R_{x,i}$ is either 1 or -1 and is computed as follows. Construct x' from x by replacing each 0 by -1. Then take the product of all those elements of x' whose index is in i. For example if $x = 10110$ and $i = \{1,2,5\}$, then $x' = 1\ -1\ 1\ 1\ -1$ and $R_{x,i} = 1 \cdot -1 \cdot -1 = 1$. Walsh products are used in the literature on genetic algorithms, for example to construct deceptive functions [3].

Formally, define the matrix of Walsh products R as follows ($x, i = 0, \ldots, 2^n - 1$)

$$R_{x,i} = \prod_{k \in i} (2x_k - 1).$$

These matrices can be constructed recursively as follows:

$$\mathcal{R}_0 = (1)$$

$$\mathcal{R}_{n+1} = \begin{pmatrix} \mathcal{R}_n & -\mathcal{R}_n \\ \mathcal{R}_n & \mathcal{R}_n \end{pmatrix}$$

Given a distribution, the expected values of Walsh products are (for $i = 0, \ldots, 2^n - 1$)

$$ER_i = \sum_x R_{x,i} P(x)$$

and, from the expected values of the Walsh products we can get the distribution back:

$$P(x) = \frac{1}{2^n} \sum_i R_{x,i} ER_i$$

The expected value of the fitness function $E[f]$ can be computed from the expected values of the Walsh products as follows. Define

$$r_i = \frac{1}{2^n} \sum_x R_{x,i} f(x)$$

Then

$$E[f] = \sum_i r_i ER_i$$

This is based on the fact that the fitness function can be written as a weighted sum over the Walsh products:

$$f(x) = \sum_i r_i R_{x,i}$$

We next give formulae how the expected values of Walsh products change under the genetic operators. In these formulae the primed values denote the values after applying the genetic operator.

Mutation (mutation rate p_m): Let $||i||$ denote the number of elements in the set i. Then

$$ER'_i = (1 - 2p_m)^{||i||} ER_i$$

Note, that if we applied only mutation to a distribution, then all ER_i would go to zero. This is in agreement with the intuition that we obtain a distribution in which every string has equal probability.

Uniform scanning crossover (k parents, crossover rate p_c): Let $\mathcal{S}_i = \{\langle i_1, \ldots, i_k \rangle : i_1 \cup \cdots \cup i_k = i \wedge \forall \alpha, \beta : \alpha \neq \beta \Rightarrow i_\alpha \cap i_\beta = \emptyset\}$. Then

$$ER'_i = (1 - p_c) ER_i + p_c \, (\frac{1}{k})^{||i||} \sum_{\langle i_1, \ldots, i_k \rangle \in \mathcal{S}_i} \prod_{j=1}^{k} ER_{i_j}$$

In order to find the expected value of Walsh product i after uniform scanning crossover, we have to take a sum over subsets of i. These subsets need to be disjoint, and their union should be i. It is of interest to note that it does not matter if we first do mutation and then uniform scanning crossover or the other way around.

Diagonal crossover (k parents, crossover rate p_c): Define the set $\mathcal{S}$ of crossover-points by $\mathcal{S} = \{\langle i_0, \ldots, i_k \rangle : i_0 = 1 < i_1 < \cdots < i_k = n + 1\}$. Then

$$ER'_i = (1 - p_c) ER_i + p_c \, \frac{1}{||\mathcal{S}||} \sum_{\langle i_0, \ldots, i_k \rangle \in \mathcal{S}} \prod_{j=1}^{k} ER_{i \cap \{i_{j-1}, \ldots, i_j - 1\}}$$

Note that the number of elements in the set $\mathcal{S}$ is significantly less than in the set $\mathcal{S}_i$ in the definition of the uniform scanning crossover. Hence, for diagonal crossover the expected values of the Walsh products are easier to compute than for the uniform scanning crossover.

Proportional selection: Let $xor(i, j)$ be the xor-operator on bit string. For example, $xor(1010, 1001) = 0011$. Then

$$ER'_i = \frac{\sum_j r_{xor(i,j)} ER_j}{\sum_j r_j ER_j}$$

Now we can use the expected values of Walsh products to trace the expected value of the fitness. Using the above transformation formulae, the value $E[f]$ can be traced from generation to generation. We took the length of the bit string $n = 10$, and the fitness function an 1-dimensional inverted rastrigin function with an optimal fitness of 100. We start from an initial distribution in which every string has equal probability. We took $p_c = 1$ and $p_m = 0$. Then in Figure 5 the expected values of the fitness of a chromosome for the uniform scanning

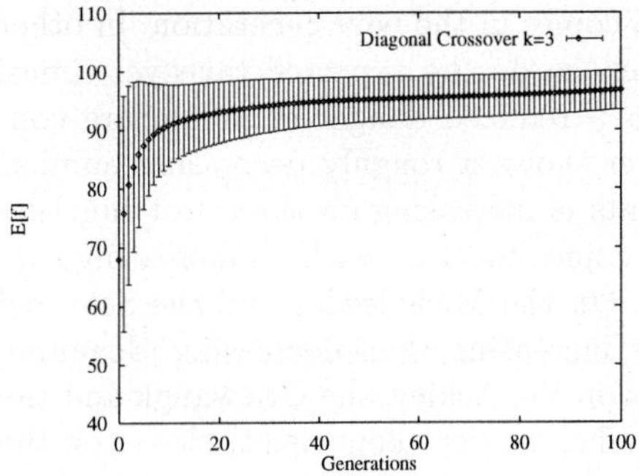

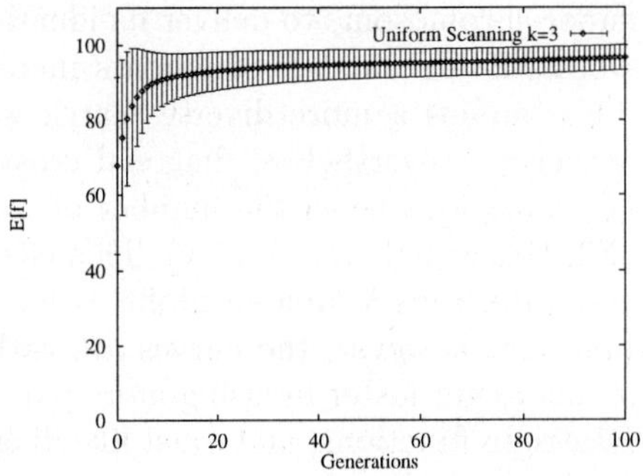

Fig. 5. Expected value of the fitness of a bit string (and its variance) from generation to generation. On the left is diagonal crossover and on the right is uniform scanning crossover.

crossover and the diagonal crossover with three parents are plotted. There is not much difference in the two graphs, and if we look at the underlying data we see that the difference (for each generation) of the expected values of the fitness is never bigger than one.

The second moment is given by $\sum_i ER_i \sum_j r_j r_{xor(i,j)}$, from which the variance can be computed in the standard way.

6 Conclusions and further work

The main conclusion that can be drawn from the experiments is that using more parents does increase GA performance. However, refinements of this statement are necessary due to the different multi-parent recombination mechanisms we tested and the different performance measures we considered.

As for the success rate (the percentage of cases when an optimum is found) we can conclude that both operators, i.e. scanning crossover and diagonal crossover, achieve their best performance when using more than 2 parents, see Table 3. The only exception is scanning crossover on the Schwefel function, which is probably a random effect, since scanning crossover has a very low success rate on this function. We observed a positive correlation between the number of parents and the success rate for diagonal crossover. For scanning crossover this is not always the case. An important conclusion from Table 3 is that the increased success rates of diagonal crossover are not simply the consequence of using more crossover points. The comparison between n-point crossover, which takes two parents, and diagonal crossover shows that the usage of more parents substantially contributes to the better results.

It seems that of all measures we considered more parents have the least effect on the efficiency, i.e. the total number of function evaluations. This is actually not surprising if we realize that using more parents makes it harder for a super-chromosome to deliver its identical copies in the next generation. In other words, if the number of parents increases, so do the expected takeover times. This implies a more diverse search with a reduced danger of premature convergence. Nevertheless, diagonal crossover shows a roughly decreasing number of evaluations when the number of parents is increasing on three test function (F2, Griewangk and Ackley). This effect comes together with an increasing success rate, thus a 'double profit' is made. On the Michalewicz and the Schwefel function, however, the curves are rather increasing than decreasing. Scanning seems to get faster by using more parents on the Ackley, the Griewangk and the Rastrigin functions, and - just like diagonal crossover - gets visibly slower on the Michalewicz function.

In explaining the above results two notions play an important role: disruptiveness and takeover times. Disruptiveness is a powerful means to prevent premature convergence. Nevertheless, a very disruptive operator might prevent the global optimum from being found. This stresses the importance of a 'right proportion' of disruptiveness. Increasing the number of parents results in an increase of the disruptiveness of an operator, as less schemata will be preserved. So, by tuning the number of parents we can tune the level of disruptiveness by little steps.

The comparison between diagonal crossover and 2-parent n-point crossover provides evidence that we are also dealing with another effect that is not related to disruptiveness. Take-over times might be an important parameter. When recombination takes part between n parents the chance that a complete copy of one of the parents occurs among the offspring becomes smaller as n increases, that is a larger fraction of the population has to be centered around an optimum before the complete population converges to this optimum. Thus, when using more than two parents, the time it takes for a single good individual to take over the complete population will increase.

When looking for the best operator for function optimization we cannot appoint a clear winner. As the figures in Table 3 show, scanning crossover wins on F2, the Griewangk and the Ackley functions, while diagonal crossover is better on the Michalewicz, the Schwefel and the Rastrigin functions. However, it should be noted that diagonal crossover is a cheaper operator. Namely, uniform scanning crossover requires the generation of many random numbers, which makes it computationally more expensive. An interesting fact is that the profit of more parents for diagonal crossover is the highest on the problems with long chromosomes cq. high dimensional problems.

Finally, let us make a note to put these results in a broader context. Evolutionary computation consists of three main branches: Genetic Algorithms, Evolution Strategies and Evolutionary Programming. One of the main differences between GAs and the other two branches is the arity of the typically applied operators. Using binary operators, i.e. sexual reproduction, is inherent to GAs, while this is not the case in ES and EP. There are researchers who question the usefulness of sex in Evolutionary Computation, and indeed, in some experimental comparisons ES and EP exhibit better performance then GAs, [1]. Some recent publications show that GAs can be enhanced by new features, out of the traditional GAs paradigm, such as Lamarckian/Baldwinian effects or by applying a problem decomposition, [9], [11]. In this paper we follow another approach. We do not leave the GA paradigm, but rather boost it, that is we raise the extent of sexuality by using 'orgies', i.e. multi-parent reproduction. The results show that there are very promising possibilities within the GA paradigm.

Currently we are doing further research to obtain a better understanding of the behavior of the multi-parent operators. One of our tools is the theoretical model described in section 5. Furthermore we are trying to identify some guidelines for selecting the operator and choosing the most appropriate number of parents.

Acknowledgements: For our tests we used LibGA, provided by A. Corcoran.

References

1. T. Bäck and H.-P. Schwefel, *An Overview of Evolutionary Algorithms for Parameter Optimization,* Journal of Evolutionary Computation 1(1), 1993, pp. 1-23.
2. A.E. Eiben, P.-E. Raué and Zs. Ruttkay, *Genetic Algorithms with Multi-parent Recombination,* in Proceedings of ICEC/PPSN 3, eds. Y. Davidor, H.-P. Schwefel and R. Männer, LNCS 866, Springer-Verlag, 1994, pp. 78-87.
3. D.E. Goldberg *Genetic Algorithms and Walsh functions: Part I, A gentle introduction Complex Systems* 3:129-152, 1989.
4. D.E. Goldberg, *Genetic Algorithms in Search, Optimization and Machine Learning,* Addison-Wesley, 1989.
5. J.N. Kok, P. Floréen and M. Rasch *Tracing Expected Fitness Values in Genetic Algorithms using Bit products and Walsh Products,* Technical Report, Utrecht University, 1994.
6. Z. Michalewicz, *Genetic Algorithms + Data Structures = Evolution Programs,* Springer-Verlag, 2nd edition, 1994.
7. H. Mühlenbein, *Parallel Genetic Algorithms, Population Genetics and Combinatorial Optimization,* in Proceedings of ICGA-89, eds. J. Schaffer, Morgan Kaufmann, 1989, pp. 416-421.
8. K.F. Pál, *Selection Schemes with Spatial Isolation for Genetic Optimization,* in Proceedings of ICEC/PPSN 3, eds. Y. Davidor, H.-P. Schwefel and R. Männer, LNCS 866, Springer-Verlag, 1994, pp. 170-179.
9. M.A. Potter and K.A. De Jong *A Cooperative Coevolutionary Approach to Function Optimization,* in Proceedings of ICEC/PPSN 3, eds. Y. Davidor, H.-P. Schwefel and R. Männer, LNCS 866, Springer-Verlag, 1994, pp. 249-257.
10. G. Seront and H. Bersini, *In Search of a Good Evolution-Optimization Crossover,* in Proceedings of PPSN 2, eds. R. Männer and B. Manderick, North-Holland, 1992, pp. 479-488.
11. D. Whitley, V. Scott Gordon and K. Mathias *Lamarckian Evolution, the Baldwin Effect and Function Optimization,* in Proceedings of ICEC/PPSN 3, eds. Y. Davidor, H.-P. Schwefel and R. Männer, LNCS 866, Springer-Verlag, 1994, pp. 6-15.

A Simplification of the Theory of Neural Groups Selection for Adaptive Control*

S. Lobo, A. J. García-Tejedor[1] , R. Rodríguez-Galán, Luis López and A.García-Crespo

Area de Inteligencia Artificial, Departamento de Ingeniería, Universidad Carlos III
28911 Leganes, Madrid (SPAIN). Phone: (34-1) 348 11 87. Fax: (34-1) 624 94 65.
Email: gtejedor@ia.uc3m.es

Abstract. Mathematical models have been extensively used to model living organisms behaviour. Nonetheless, those models do not take into account the role that individual history plays into the establishment of neural structures responsible for its interaction with the environment. TNGS has been postulated as a global, comprehensive solution that models individual behaviour from both biological and evolutionary aspects. Besides, it provides a non symbolic approach to learning processes that does not require extensive prior knowledge from system designer.

This paper presents a simplification of TNGS oriented towards its use in adaptive control processes for chemical reactors. An oculomotor system has been implemented based on Darwin III automaton. Simplifications are made on the state equation that describes the dynamic behavior of every processing element. They are driven by a chaotic study on the equation for the weight modification. Based on very simple assumptions ("seeing is better that not seeing"), the system learns to trace a randomly moving object within its vision field. Simulation present data obtained under different assumptions. The evolution of the distance between the center of the input image and the center of the Visual Retina is displayed.

Keywords: Adaptive control, TNGS, neural Darwinism, biological models, real-time applications, sensor-effector coordination

---

* This work has been partially supported by BRITE/EURAM project nº BE-7686 PSYCHO (Powerful SYstems for identification and Control of Highly nOn-linear processes using neural networks).

[1] To whom correspondence should be addressed.

1 Introduction

The use of intelligent systems to control mechanical devices have been extensively tested. However, the elaboration of such systems requires specific knowledge on the environment where the system is placed, a detailed description of the actions to be carried out and the specification of the relations between these actions and the events in the world. This has been the dominant view in Artificial Intelligence. It can be described as the physical symbol hypothesis: for a system to have knowledge processing capabilities, it is necessary to implement a symbolic world model. It implies that the system's designer has a priori knowledge about the problem and problem solving methods. The system will use this knowledge, implemented as a symbol-relation system, to reach its goals[1].

Nevertheless, Gerald Edelman proposes a theory of behavior, based on concrete hypothesis on brain organization, that can constitute an alternative to the symbolic approach. The Theory of Neural Groups Selection (TNGS) provides the basis to develop systems whose a priori knowledge of the environment (or even self-knowledge) is reduced to very basic functionalities[2]. The underlying idea tries to reflect the way in which a child brain is able to self-organize and adapt to changing conditions finishing, after a learning period, with the abilities needed to perform very complex (mechanical) tasks[3]. In Edelman's theory, the designer is also involved. He still must specify those very basic functionalities (genetically coded value schemes) that constraint brain evolution by selecting only those behaviors that represent an adaptive advantage for the organism. Nonetheless, this is a lower level of intervention that the elaboration of a body of knowledge to be implemented into the intelligent system.

Several computational implementations of TNGS have been carried out by Edelman himself. They are known as the Darwin series, a set of recognition automata. The results of those simulations are in good agreement with the results predicted by the theory, but the high cost in terms of training time prevent the use from those implementations in real-time world applications. Darwin automata are not designed to efficiently carried out control tasks. They are devised as generic mechanisms that learn how to interact with their environment without detailed specifications. In fact, Darwin III shows us how selective and adaptive systems are able to drive an oculomotor system while adapting to the requirements of the environment[3]. Nevertheless, it would be very interesting to use these principles to implement a control system that does not require extensive description of either the devices to be controlled or their physical environment.

Within the BRITE/EURAM project PSYCHO, we are developing a generic neural controller for chemical reactors (in general, for highly non-linear processes) partially based on Darwin III oculomotor system. The main goals of this implementation are the understanding of the system and its temporal dynamics as well as how environmental constraints involve effector organs, neural architecture and sensors. In order to make the implementation computationally efficient, several decision have been adopted. They are related to the basic equation that defines the behavior of the processing elements (neurons) as proposed by Edelman. We provide a simplified equation which is not so intensively resource consuming during the training phase and so, can be foreseen as a possible control system for industrial applications. The paper describes the simplifications adopted on the Oculomotor Darwin III model. These simplifications are driven by a chaotic study on the equation that determines the weight modification in Edelman's model. Some preliminary results of the automaton abilities to center an image in its visual field are also presented. Further studies will replace the current basic functionality ("seeing is better that not seeing") in order to provide actions on the mechanical elements that regulates basic parameters of chemical reactors.

2 A basic overview of TNGS

The problem Edelman tries to address is how can an organism categorize the world just by interacting with its environment, and without previous knowledge. The answer, as described through the Theory of Neural Groups Selection (TNGS), is nonsymbolic systems that can develop their own ontology by means of selective mechanisms.

One of the first questions that a living organism must face (or that a brain theory must answer) is to categorize an unlabeled world. It means that the organism must create classification criteria extracted out of its interaction with the world, instead of receiving information from it. From a biological point of view, the only application of computational formalisms to human behavior could only show partial and slanted visions of the nature of the brain and its capabilities for unsupervised categorization[2]. We must turn to a more comprehensive theory that also includes epigenetic processes for the mind's physical substrate: the brain.

Edelman's theory is based on the definition of local collections of thousand of neurons strongly interconnected, the so called Neural Groups, in a specific region of the brain. The resulting formations (neural populations) are known as Repertoires, in reference to their collective potential to respond to a wide range of inputs. They are the result of neural systems self-organization due to competition and selection mechanisms[2]. These ideas already appeared in previous works on non-equilibrium evolutionary systems or on other biological systems, such as prebiotic evolution or Edelman's own work on antibody selection. There are three key concepts on TNGS: selection at development stage, selection due to experience (adaptation) and reentry.

Anatomical macroscopical order is generated during embyogenesis in a very specific way, accurately defined for each species. Within this global order, patterns for dendritical and axonal connections of neurons show enormous variations, both inside and between individuals. This structural diversity is genetically coded, but it could also be the result of epigenetic processes acting during the development, producing movement, differentiation, growth, division and death of the cells[4]. The general theory of brain operation that proposes TNGS affirms that the nervous system develops by selection on the pre-existing neural populations that compose a repertoire. Neural Groups are formed dynamically and correspond (approximately, but not necessarily) to regions of anatomical strong connectivity. The groups of neurons that form a particular group could vary along the time. In sum, differential growth of neural structures, accordingly to TNGS, is produced by:

- influence of morphoregulatory molecules during cell development.
- the context
- local environmental influences.

This *developmental selection* leads to the formation of the Primary Repertoires, originated during the development. They show a large diversity, that provides the basis for adaptation by experiential selection.

During neonatal development, the anatomical structure of the brain reaches a stationary configuration, while synaptic connections play a prominent role as the principal agents of the plasticity and the adaptation[5]. The functional and dynamic circuits are subjects to other mechanisms, like the *experiential selection.* These mechanism strength some synaptic connections, while some others are weakened. The result is the formation of the Secondary Repertoires which compete among them to show a better adaptation to the environment.

At this level of development, the organism must be able of extracting categories out of its interaction with the world, categories that necessarily must remain flexible. This is achieved by coupling the repertoires with the outside world through yet another type of repertoires genetically determined that constraint system development. They are called Value Repertoires and can be defined as basic evolutionary adaptations, genetically coded, that define behavioral goals. Value Repertoires are activated when a certain property, relevant for the survival of the organism, is found in its interaction with the environment. They influence in other areas in a fuzzy and modulatory way: they do not predefine the exact form a behavioral answer is executed, but rather they impose restrictions to the synaptic modifications that depend on the result of previous interactions with the environment[6]. Value Repertoires ensure that neural structures present in any organism are adequate the diversity of precise behaviors for their survival.

Besides being capable of recognizing stimuli and relevant categories within the environment, the nervous system should be also capable of associating them accordingly to spatial and temporal correlation. Stimuli association requires comparison, and therefore memory. For the association and comparison to take place, it is necessary to correlate the reconnaissance of different events with different times and places. This conjunction is achieved by means of the *Reentry* between neural groups[2]. Reentry is a dynamic process that evolves over time and implement the continuous and recursive interchange of signs between several repertoires by means of anatomical parallel connections. It reflects the spatiotemporal continuity of events and signals in the outside world.

TNGS has been illustrated by several computational models that implements, with a continuous increase of complexity, the basic features of the theory[7]. These implementations are the Darwin series of automata. Darwin I, the simplest one, explores the role of selective mechanism in the formation and properties of neural groups. Darwin II studies the interactions of cortical maps and its role, as reentrant maps, in categorization processes. Darwin III is the most advanced computer simulation and describes the organization of the nervous system and its expression through the phenotype. It addresses, by using simulated sensor and effector organs, the whole control cycle: monitoring, decision making and acting. Finally, Darwin IV or Nomad (Neurally Organized Mobile Adaptive Device) implements in a robot provided of physical sensors and effectors the theory already expressed in Darwin III.

The dynamic behavior of all processing elements in all the repertoires defined in Edelman's theory is described by the following equation:

$$S_i(t) = \left[(A + G + M)\phi(I_S)\right]\phi(D) + N + W \qquad (1)$$

where:

A: Total input given by *specific connections*. They are built accordingly to a given rule, for example, a topographical map

M: Total input given by *modulatory connections*. They receive the mean activity corresponding to all neurons in a certain input layer.

G: Total input given by *geometrical connections*. They are built accordingly to a special distance/density relationship and they are used to provide lateral inhibition.

I_s: Sum of all specific and geometrical connections that are fired by the input signal.

D: Depression, calculated as $vd\, s_i\,(t\text{-}1) + wd\, D(t\text{-}1)$ where

vd: depression increase rate

wd: depression decay rate

N: Noise.

W: Activity decay, defined by $ws_i\,(t\text{-}1)$.

ϕ(x): Sigmoidal function, $\phi(x) = 1 - 2x^2 + x^4$.

A second equation (not shown) is used by Edelman's model to describe the modification of weights connecting the processing elements. This equation has not been modified by us.

3 Model Implementation

The study we present in this paper is carried out through an automaton that implements a model of an oculomotor system, based on Edelman's Darwin III model. The basic assumption in its value repertoire (Fovea repertoire) is that "*seeing is better that not seeing*". Therefore the system tries to maximize the visual field by minimizing the distance between the center of the input image and the center of the Visual Retina. The Fovea repertoire becomes activated when there is a high activation in the center of the visual field. When it happens, then it associates related states of the visual and oculomotor repertoires reinforcing the correct movements that center the image.

We start from a non predefined neural network system, that is, a neural architecture where the matrix of connections (weight matrix) between neurons that link the input (sensitive maps) with the output (ocular motion) is assigned at random without a prior knowledge of the task that the system has to carry out.

We have simplified several terms appearing in Edelman's state equation in order to speed up all computations involved in training. Since our goal is to check the appropriateness of TNGS for on-line control, it is necessary to ensure a response time for the system within the limits imposed by the physical system to be controlled. We must also take into account that Edelman's theory is aimed to fully mimic biological processes, and it does not deal with computational efficiency. The modifications are based on the study of the dynamic of the system. This study was carried out through a chaotic analysis of the weight modification equation (results not shown). The aim of this analysis was to find out if neuron states behaves as strange attractors. The control of weight modification, main responsible of repertories instability, provokes steady neuron states. As a consequence, several modifications to the state equation were implemented.. The resulting modified equation is:

$$S_i(t) = [(A + G + M)]\phi(D) + N + W \tag{2}$$

that, respecting Edelman's original equation shows the following changes:

A: Fully connected repertoires have been used. This term represent now connections to every neuron coming from any other repertoire with a distinctive weight.

M: Connections to circular groups in different repertoires, each one with a different weight.

G: Geometrical connection to a different neural group with equal weight.

N: Gaussian Noise has been specifically used.

$\phi(I_s)$: Term $\phi(I_s)$ representing the sum over all specific and geometrical connections that are fired by the input signal has been also removed. This term is specially relevant to reach steady repertoires. In our model, we normalize repertories states after each stimulation. This mechanism leads to stable models under certains circumstances, with a lower computational cost.

To validate the result of these modifications, we have developed an "ad hoc" implementation of Darwin III Oculomotor Model based on the following set of repertoires (see Fig. 1) [8]:

- **VR Repertoire (Visual Retina):** This is the first repertoire that receives input signals from outside world. As it can be seen in Fig. 2, the image is the input to

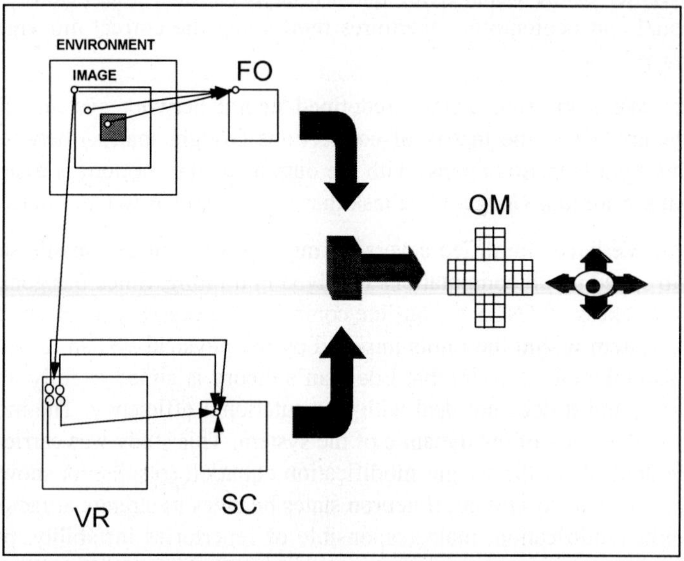

Fig. 1. Schematic diagram of the Oculomotor model implementation.

the oculomotor system. Figure 2.b represents a visual retina (VR), which is composed of two layers arranged in a matrix of 29x29 neurons: the first one is formed of excitatory neurons, and the second one is formed of inhibitory cells. It means that each layer is composed of 841 cells. The input image is represented by a real number matrix, where each cell represent an image pixel. Connections are organized in the following way: each neuron has a topographical connection (specific) to the corresponding position of the image (1:1 ratio) as depicted in figure 2. These connections have all the same weight. Local excitatory and lateral inhibitory connections yield better results in the output signal and increase image fixing in case an object enter the visual field.

The output of the VR is the neuron state matrix as defined by equation [2] for each element. These states influence the Superior Culliculus repertoire.

- **SC Repertoire (Culliculus Superior):** This repertoire has, as inputs, state values of the VR repertoire. Its output directly influences the OculoMotor repertoire (OM). SC is also a two layered repertoire, much in the same way that VR, with

256 neurons each layer. They are arranged in a 16x16 square matrix. The first layer is again an excitatory one. Local connections are formed in the same way as in VR. So this repertoire shows lateral inhibition to improve outputs.

Each SC neuron should be topologically connected to 1.86 VR processing elements. This ratio is a consequence of mapping a 29x29 surface (the VR) into a 16x16 layer (the size of the SC). Since it is not possible to reach this ratio, because we deal with integer number of neurons, we connect a SC cell with a 2x2 area in the VR repertoire. This implies that after a given number of neurons, two of them will share the same neuron, that its, there is an overlapping. It means that the A term in the state function is reinforced as it receives the input of four neurons. So, if a image is found within the visual field, the corresponding neuron in the SC will have an enhanced activation compared to a neuron whose visual field is empty.

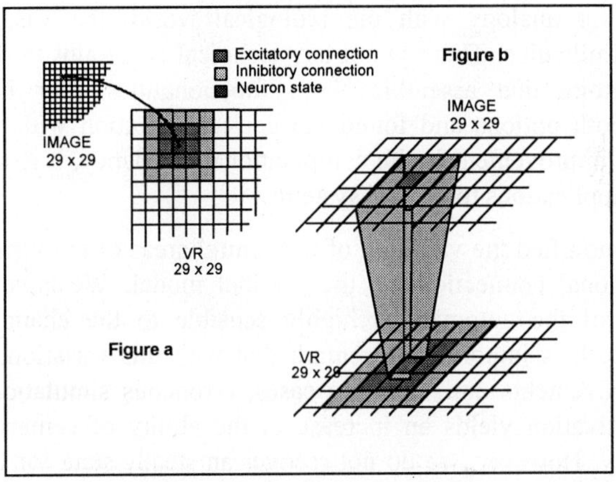

Fig. 2. Schematic diagram of a Visual Retina repertoire.

- **FO Repertoire (Fovea):** This repertoire gets as input the image found in the visual field. It is composed of 121 excitatory neurons, arranged in a 11x11 square matrix. Each excitatory neuron receives one topographically mapped connection from the entire visual field. besides, it receives too an additional connection from the central 15% and 3% of the visual field respectively. The output value influences directly the OculoMotor repertoire. It consists in the global average of the existing activity over the whole repertoire and it will be used as input value for heterosinaptic connections linking SC and OM. This value will allow the analysis of ocular movement in order to discover if this movement has centered the object in the visual field or, on the contrary, has increased the distance to the center of the image.

- **OM Repertoire (OculoMotor):** This repertoire gets the states of SC neurons and uses them as inputs. OM outputs directly provoke ocular movements which influence the visual field received as input by the Visual Retina. This repertoire is composed of 36 motor units divided in four groups with 9 units each, represented as 3x3 square matrix. This architecture is a noteworthy simplification of the ocular human anatomy. Each motor unit receives 256 excitatory connections from each of the SC neurons. It has also one inhibitory connection that prevent the concomitant activation of the effector pair that moves the eye in each of the Cartesian axis (up-down and right-left). These inhibitory connections cause sharp eye movements (lateral inhibition). Otherwise small differences in the state function of opposite neurons will not be translated into movements. Weights are randomly initialized following a normal distribution.

Besides this implementations, several others changes in the repertoire structure has been tested.

1. The SC repertoire has been eliminated and there is been directly mapping of the VR output the OM repertoire; outputs are very similar to those that are obtained through the complete model. Edelman employs two repertoires, VR and SC. They simulate, for analogy with the biological world, the Visual Retina and the Superior Culliculus. There is not any technical constraint in order to create only one repertoire that assembles both functionalities. Our implementation has checked both options and found the correct operation with a single repertoire. Given the improvement in the computational efficiency, it is expected to be used in future applications to control systems.
2. We have modified the variation of the central areas of the visual retina by adding two additional connections to the original model. We have observed that the evolution of the automata is highly sensible to the changes of the primary repertoires. It is important to remark that with this variation, we have obtained both improvements and, in some cases, erroneous simulations. An increase in weight activation yields an increase in the ability of centering the image in a fastest way. However, we do not recover an steady state for the repertoire when the system decreases the weights involved in the movement.
3. We have modified the OM repertoire; in this repertoire all the neurons receive connections from each one of the SC neurons. We observed that it is not necessary for all the neurons to have the same connections; even, it is possible to carry out random connections on the SC repertoire and still get satisfactory outputs. We have also checked that the net is fault tolerant: faced to the loss of part of its components, it still shows a quite stable behavior without retraining.

After these modifications, both in the different terms of the equations and in the architecture of the system, the outputs have been positives in an elevated number of situations. Compared to Edelman's system, computational time have been strongly reduced. The previous considerations will eventually permit the use of this model in industrial environments.

4 Simulation Results

Simulations show the dynamics of the simplified model while centering an object placed within the environment. The simulations start with an image

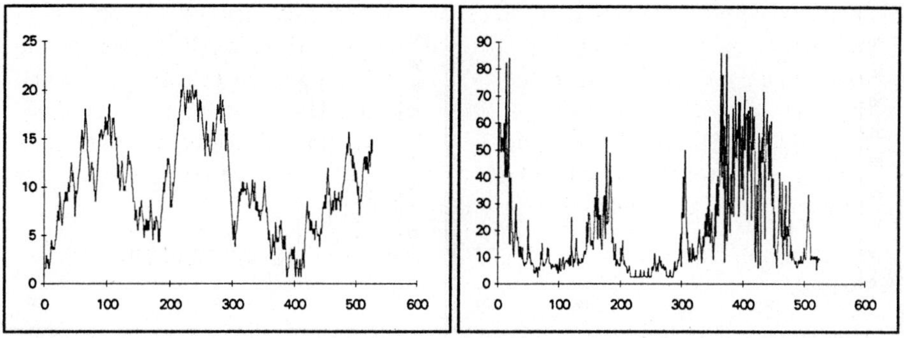

Fig. 3. A typical example of center distance (left) and FO activation (right) plots.

(29x29pixel size) randomly located in a 75x75 pixel field. The system must learn to focus an image driven by the OculoMotor repertoire, under supervision of the Fovea. The ideal process would finished when the system learns to trace a randomly moving object centering it in the visual field. Experiments were carried out on a 486 PC compatible under Linux operating system (45 minutes of CPU time) and on a HP 9000 Apollo workstation (10 minutes of CPU time with an average of 1 second per iteration).

A graphical output is displayed in the console. It shows the activation state of the different repertoires as well as retina movements over the visual area. Besides, two data sets are obtained. The first one represent the distance between the center of the object and the center of the Visual Retina that has to be minimized. A plot of this distance versus the number of training iterations will measure the learning evolution. This plot is represented in the left part of the following figures. The second data set represent the resulting activation of the FO repertoire, which is responsible of supervising the learning. The more centered the object be, the higher activation values will be obtained. This plot is depicted in the right side of the figures.

Both plots represent complementary information: whenever the system is able to center the object, lower distances will be obtained and higher activation values will be shown. On the contrary, if the Retina begins to lose the object, values on the left graph will increase more than the right graph decreases (see Fig. 3). This is due to the fact that the FO repertoire implements four refractary cycles after 8 continuous states of maximal activity. Thus, the decrease is not immediate. Other figures show experiments with different initial conditions (mostly randomly weight initialization) and noise influence.

Fig. 4 displays the results after an incorrect weight modification in the initial simulation time. High Foveation values (up to 60 iterations) are obtained (right plot). A random meeting with the image object is produced at iteration 100, but the system is unable of modifying the weights in order to keep track of its movements.

Figure 5 and 6 represent satisfactory simulations where the automaton is able of centering the visual object. The system is able to retrieve the image even after

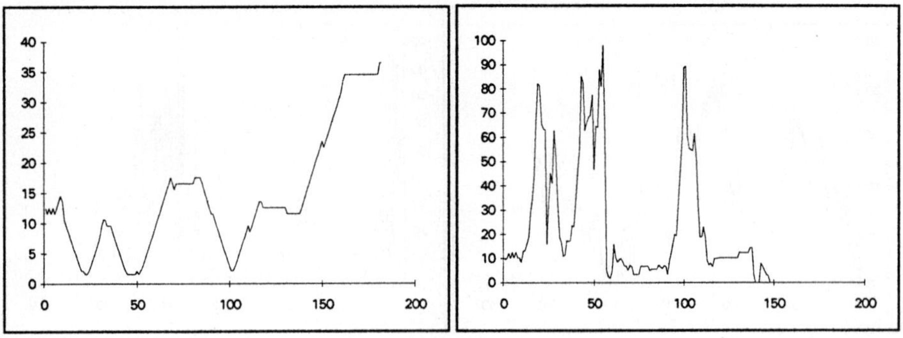

Fig. 4.

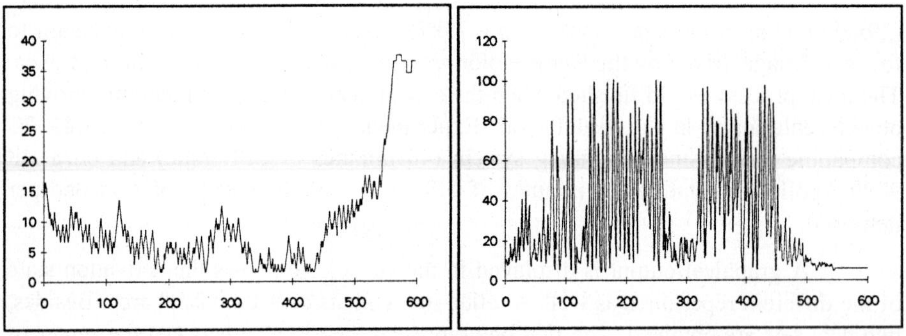

Fig. 5.

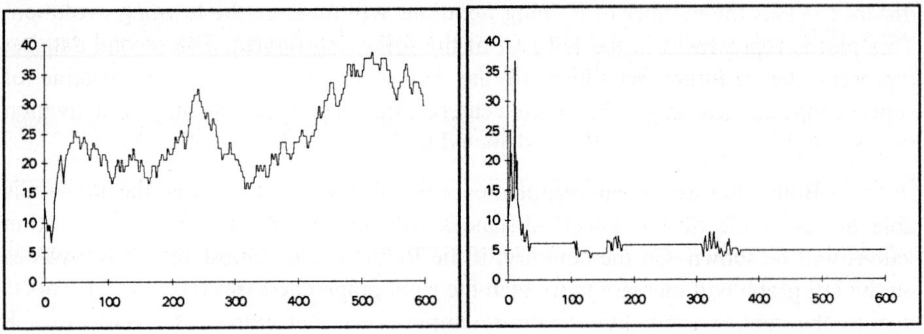

Fig. 6.

loosing it. At the end of the simulation, weights are saturated and so, the tracking capacity decreases.

Figure 7 shows a simulation where gaussian noise (used in the other cases) has been removed. Results shown the inability of the automaton to recover the object once it is lost due to absence of random ocular motion.

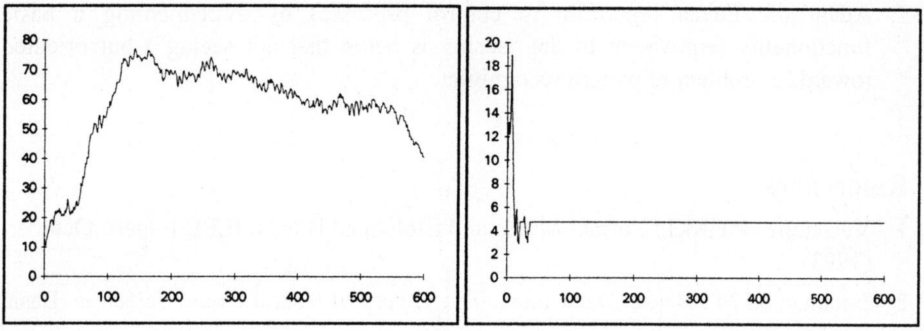

Fig. 7.

5 Conclusion and further research

In this paper we have presented a simplification of the Theory of Neural Groups Selection for adaptive control. A non-symbolic, generic model as the one proposed by Edelman would allow an easier modeling of control processes without intrusive intervention from the human designer. This can only be achieved if the resulting system is efficient in terms of computational time.

The basic assumption in our approach is to reuse the adequate functionalities of TNGS but removing those assumptions that, although relevant from a biological point of view, are not strictly necessary to implement the basic features of the theory. A system based on the oculomotor Darwin III automaton has been implement. The simplified state equation of the processing elements has been proven to be acceptable for the purposed pursued in this work.

There are several conclusions that can be brought out of the present study: TNGS is a neural theory that can be used in control at a reasonable cost in training time provided that the complexity of the problem does not exceed the computational capacity of the hardware. TNGS is probably not the best solution but it is a generic model that does not require specific task knowledge. Concerning temporal aspects, simplifications carried out on Edelman's model show that it is feasible to obtain efficient response time without a significant loss of functionalities. Finally, it has been proven that the Oculomotor implementation is fault tolerant: faced to the loss of part of its components, it still shows a quite stable behavior.

Within the PSYCHO project, we will keep on working on this modified Edelman model applied to adaptive control. We will center on the following points:

1. Implementation of categorization capabilities in the OculoMotor repertoire. It means to implement control capabilities by classifying the different states in the environment.

2. Adapt the Fovea repertoire to control processes by implementing a basic functionality (equivalent to the "seeing is better that not seeing") but oriented towards a problem of pattern recognition.

References

1. Verschure, P.F.M.J.: Formal Minds and Biological Brains. IEEE Expert. October. (1993).
2. Edelman, G.M.: Neural Darwinism: The Theory of Neural Group Selection. Basic Books. (1987).
3. Reeke, G.N., Finkel, L.H., Sporns, O., Edelman,G.M.: Synthetic Neural Modelling: A Multilevel Approach to the Analysis of Brain Complexity. The Neurosciences Institute & The Rockefeller University. (1988).
4. Kandel, E.R., Schwartz, J.H., Jessell, T.M.: Principles of neural science. Elsevier Science Publishing. (1991).
5. Zubay, G.: Biochemistry. Addison-Wesley. (1983).
6. Kuppers, B. O.: Molecular Theory of Evolution. Springer-Verlag. (1983).
7. Reeke, G.N., Sporns, O., Edelman G.M.: Synthetic Neural Modeling: The "Darwin" Series of Recognition Automata. Proceedings of the IEEE, vol. 78, n° 9, September. (1990).
8. Merzenich, M.M.: Progression of Change Following Median Nerve Section in the Cortical Representation of the hand in Areas 3b and 1 in Adult Owl and Squirrel Monkeys. Neuroscience. (1983).

Authors Index

Lecture Notes in Artificial Intelligence (LNAI)

Vol. 745: V. Roberto (Ed.), Intelligent Perceptual Systems. VIII, 378 pages. 1993.

Vol. 746: A. S. Tanguiane, Artificial Perception and Music Recognition. XV, 210 pages. 1993.

Vol. 754: H. D. Pfeiffer, T. E. Nagle (Eds.), Conceptual Structures: Theory and Implementation. Proceedings, 1992. IX, 327 pages. 1993.

Vol. 764: G. Wagner, Vivid Logic. XII, 148 pages. 1994.

Vol. 766: P. R. Van Loocke, The Dynamics of Concepts. XI, 340 pages. 1994.

Vol. 770: P. Haddawy, Representing Plans Under Uncertainty. X, 129 pages. 1994.

Vol. 784: F. Bergadano, L. De Raedt (Eds.), Machine Learning: ECML-94. Proceedings, 1994. XI, 439 pages. 1994.

Vol. 795: W. A. Hunt, Jr., FM8501: A Verified Microprocessor. XIII, 333 pages. 1994.

Vol. 798: R. Dyckhoff (Ed.), Extensions of Logic Programming. Proceedings, 1993. VIII, 360 pages. 1994.

Vol. 799: M. P. Singh, Multiagent Systems: Intentions, Know-How, and Communications. XXIII, 168 pages. 1994.

Vol. 804: D. Hernández, Qualitative Representation of Spatial Knowledge. IX, 202 pages. 1994.

Vol. 808: M. Masuch, L. Pólos (Eds.), Knowledge Representation and Reasoning Under Uncertainty. VII, 237 pages. 1994.

Vol. 810: G. Lakemeyer, B. Nebel (Eds.), Foundations of Knowledge Representation and Reasoning. VIII, 355 pages. 1994.

Vol. 814: A. Bundy (Ed.), Automated Deduction — CADE-12. Proceedings, 1994. XVI, 848 pages. 1994.

Vol. 822: F. Pfenning (Ed.), Logic Programming and Automated Reasoning. Proceedings, 1994. X, 345 pages. 1994.

Vol. 827: D. M. Gabbay, H. J. Ohlbach (Eds.), Temporal Logic. Proceedings, 1994. XI, 546 pages. 1994.

Vol. 830: C. Castelfranchi, E. Werner (Eds.), Artificial Social Systems. Proceedings, 1992. XVIII, 337 pages. 1994.

Vol. 833: D. Driankov, P. W. Eklund, A. Ralescu (Eds.), Fuzzy Logic and Fuzzy Control. Proceedings, 1991. XII, 157 pages. 1994.

Vol. 835: W. M. Tepfenhart, J. P. Dick, J. F. Sowa (Eds.), Conceptual Structures: Current Practices. Proceedings, 1994. VIII, 331 pages. 1994.

Vol. 837: S. Wess, K.-D. Althoff, M. M. Richter (Eds.), Topics in Case-Based Reasoning. Proceedings, 1993. IX, 471 pages. 1994.

Vol. 838: C. MacNish, D. Pearce, L. M. Pereira (Eds.), Logics in Artificial Intelligence. Proceedings, 1994. IX, 413 pages. 1994.

Vol. 847: A. Ralescu (Ed.) Fuzzy Logic in Artificial Intelligence. Proceedings, 1993. VII, 128 pages. 1994.

Vol: 861: B. Nebel, L. Dreschler-Fischer (Eds.), KI-94: Advances in Artificial Intelligence. Proceedings, 1994. IX, 401 pages. 1994.

Vol. 862: R. C. Carrasco, J. Oncina (Eds.), Grammatical Inference and Applications. Proceedings, 1994. VIII, 290 pages. 1994.

Vol 867: L. Steels, G. Schreiber, W. Van de Velde (Eds.), A Future for Knowledge Acquisition. Proceedings, 1994. XII, 414 pages. 1994.

Vol. 869: Z. W. Raś, M. Zemankova (Eds.), Methodologies for Intelligent Systems. Proceedings, 1994. X, 613 pages. 1994.

Vol. 872: S Arikawa, K. P. Jantke (Eds.), Algorithmic Learning Theory. Proceedings, 1994. XIV, 575 pages. 1994.

Vol. 878: T. Ishida, Parallel, Distributed and Multiagent Production Systems. XVII, 166 pages. 1994.

Vol. 886: M. M. Veloso, Planning and Learning by Analogical Reasoning. XIII, 181 pages. 1994.

Vol. 890: M. J. Wooldridge, N. R. Jennings (Eds.), Intelligent Agents. Proceedings, 1994. VIII, 407 pages. 1995.

Vol. 897: M. Fisher, R. Owens (Eds.), Executable Modal and Temporal Logics. Proceedings, 1993. VII, 180 pages. 1995.

Vol. 898: P. Steffens (Ed.), Machine Translation and the Lexicon. Proceedings, 1993. X, 251 pages. 1995.

Vol. 904: P. Vitányi (Ed.), Computational Learning Theory. EuroCOLT'95. Proceedings, 1995. XVII, 415 pages. 1995.

Vol. 912: N. Lavrač, S. Wrobel (Eds.), Machine Learning: ECML – 95. Proceedings, 1995. XI, 370 pages. 1995.

Vol. 918: P. Baumgartner, R. Hähnle, J. Posegga (Eds.), Theorem Proving with Analytic Tableaux and Related Methods. Proceedings, 1995. X, 352 pages. 1995.

Vol. 927: J. Dix, L. Moniz Pereira, T.C. Przymusinski (Eds.), Non-Monotonic Extensions of Logic Programming. Proceedings, 1994. IX, 229 pages. 1995.

Vol. 928: V.W. Marek, A. Nerode, M. Truszczynski (Eds.), Logic Programming and Nonmonotonic Reasoning. Proceedings, 1995. VIII, 417 pages. 1995.

Vol. 929: F. Morán, A. Moreno, J.J. Merelo, P.Chacón (Eds.), Advances in Artificial Life. Proceedings, 1995. XIII, 960 pages. 1995.

Lecture Notes in Computer Science

Vol. 893: G. Gottlob, M. Y. Vardi (Eds.), Database Theory – ICDT '95. Proceedings, 1995. XI, 454 pages. 1995.

Vol. 894: R. Tamassia, I. G. Tollis (Eds.), Graph Drawing. Proceedings, 1994. X, 471 pages. 1995.

Vol. 895: R. L. Ibrahim (Ed.), Software Engineering Education. Proceedings, 1995. XII, 449 pages. 1995.

Vol.896: R. N. Taylor, J. Coutaz (Eds.), Software Engineering and Human-Computer Interaction. Proceedings, 1994. X, 281 pages. 1995.

Vol. 897: M. Fisher, R. Owens (Eds.), Executable Modal and Temporal Logics. Proceedings, 1993. VII, 180 pages. 1995. (Subseries LNAI).

Vol. 898: P. Steffens (Ed.), Machine Translation and the Lexicon. Proceedings, 1993. X, 251 pages. 1995. (Subseries LNAI).

Vol. 899: W. Banzhaf, F. H. Eeckman (Eds.), Evolution and Biocomputation. VII, 277 pages. 1995.

Vol. 900: E. W. Mayr, C. Puech (Eds.), STACS 95. Proceedings, 1995. XIII, 654 pages. 1995.

Vol. 901: R. Kumar, T. Kropf (Eds.), Theorem Provers in Circuit Design. Proceedings, 1994. VIII, 303 pages. 1995.

Vol. 902: M. Dezani-Ciancaglini, G. Plotkin (Eds.), Typed Lambda Calculi and Applications. Proceedings, 1995. VIII, 443 pages. 1995.

Vol. 903: E. W. Mayr, G. Schmidt, G. Tinhofer (Eds.), Graph-Theoretic Concepts in Computer Science. Proceedings, 1994. IX, 414 pages. 1995.

Vol. 904: P. Vitányi (Ed.), Computational Learning Theory. EuroCOLT'95. Proceedings, 1995. XVII, 415 pages. 1995. (Subseries LNAI).

Vol. 905: N. Ayache (Ed.), Computer Vision, Virtual Reality and Robotics in Medicine. Proceedings, 1995. XIV, 567 pages. 1995.

Vol. 906: E. Astesiano, G. Reggio, A. Tarlecki (Eds.), Recent Trends in Data Type Specification. Proceedings, 1995. VIII, 523 pages. 1995.

Vol. 907: T. Ito, A. Yonezawa (Eds.), Theory and Practice of Parallel Programming. Proceedings, 1995. VIII, 485 pages. 1995.

Vol. 908: J. R. Rao Extensions of the UNITY Methodology: Compositionality, Fairness and Probability in Parallelism. XI, 178 pages. 1995.

Vol. 909: H. Comon, J.-P. Jouannaud (Eds.), Term Rewriting. Proceedings, 1993. VIII, 221 pages. 1995.

Vol. 910: A. Podelski (Ed.), Constraint Programming: Basics and Trends. Proceedings, 1995. XI, 315 pages. 1995.

Vol. 911: R. Baeza-Yates, E. Goles, P. V. Poblete (Eds.), LATIN '95: Theoretical Informatics. Proceedings, 1995. IX, 525 pages. 1995.

Vol. 912: N. Lavrač, S. Wrobel (Eds.), Machine Learning: ECML – 95. Proceedings, 1995. XI, 370 pages. 1995. (Subseries LNAI).

Vol. 913: W. Schäfer (Ed.), Software Process Technology. Proceedings, 1995. IX, 261 pages. 1995.

Vol. 914: J. Hsiang (Ed.), Rewriting Techniques and Applications. Proceedings, 1995. XII, 473 pages. 1995.

Vol. 915: P. D. Mosses, M. Nielsen, M. I. Schwartzbach (Eds.), TAPSOFT '95: Theory and Practice of Software Development. Proceedings, 1995. XV, 810 pages. 1995.

Vol. 916: N. R. Adam, B. K. Bhargava, Y. Yesha (Eds.), Digital Libraries. Proceedings, 1994. XIII, 321 pages. 1995.

Vol. 917: J. Pieprzyk, R. Safavi-Naini (Eds.), Advances in Cryptology - ASIACRYPT '94. Proceedings, 1994. XII,

Vol. 918: P. Baumgartner, R. Hähnle, J. Posegga (Eds.), Theorem Proving with Analytic Tableaux and Related Methods. Proceedings, 1995. X, 352 pages. 1995. (Subseries LNAI).

Vol. 919: B. Hertzberger, G. Serazzi (Eds.), High-Performance Computing and Networking. Proceedings, 1995. XXIV, 957 pages. 1995.

Vol. 920: E. Balas, J. Clausen (Eds.), Integer Programming and Combinatorial Optimization. Proceedings, 1995. IX, 436 pages. 1995.

Vol. 921: L. C. Guillou, J.-J. Quisquater (Eds.), Advances in Cryptology – EUROCRYPT '95. Proceedings, 1995. XIV, 417 pages. 1995.

Vol. 923: M. Meyer (Ed.), Constraint Processing. IV, 289 pages. 1995.

Vol. 924: P. Ciancarini, O. Nierstrasz, A. Yonezawa (Eds.), Object-Based Models and Languages for Concurrent Systems. Proceedings, 1994. VII, 193 pages. 1995.

Vol. 925: J. Jeuring, E. Meijer (Eds.), Advanced Functional Programming. Proceedings, 1995. VII, 331 pages. 1995.

Vol. 926: P. Nesi (Ed.), Objective Software Quality. Proceedings, 1995. VIII, 249 pages. 1995.

Vol. 927: J. Dix, L. Moniz Pereira, T. C. Przymusinski (Eds.), Non-Monotonic Extensions of Logic Programming. Proceedings, 1994. IX, 229 pages. 1995. (Subseries LNAI).

Vol. 928:V.W. Marek, A. Nerode, M. Truszczynski (Eds.), Logic Programming and Nonmonotonic Reasoning. Proceedings, 1995. VIII, 417 pages. (Subseries LNAI).

Vol. 929: F. Morán, A. Moreno, J.J. Merelo, P. Chacón (Eds.), Advances in Artificial Life. Proceedings, 1995. XIII, 960 pages. 1995. (Subseries LNAI).